Frontiers in Phytopathology

Frontiers in Phytopathology

N.G. Ravichandra

Professor & Head
AICRP (Nematodes)
Department of Plant Pathology
College of Agriculture
University of Agricultural Sciences, GKVK Campus
Bangalore, Karnataka

I.K. International Publishing House Pvt. Ltd.

NEW DELHI

Published by
I.K. International Publishing House Pvt. Ltd.
S-25, Green Park Extension
Uphaar Cinema Market
New Delhi–110 016 (India)
E-mail: info@ikinternational.com
Website: www.ikbooks.com

ISBN: 978-93-85909-06-1

Published by Krishan Makhijani for I.K. International Publishing House Pvt. Ltd., S-25, Green Park Extension, Uphaar Cinema Market, New Delhi–110 016 and Printed by Rekha Printers Pvt. Ltd., Okhla Industrial Area, Phase II, New Delhi–110 020.

To Anusha

Preface

Plant Pathology is a global and integrated discipline that develops and applies the fundamentals of plant-microbe interactions to advance plant health and productivity. The study of plant-microbe interactions plays a major role in building basic knowledge of biological systems. The indispensable role of plant pathology in the global society will become even more evident in the future due to increase in worldwide trade, travel and the inherent consequences of introduced invasive species and exotic diseases. Through an emphasis on plant health, plant pathologists develop broad concepts that are fundamental to many fields of science. These concepts will continue to have applications for the global advancement of human health.

The purpose of this book is to provide facts whereby the readers are introduced to the some new aspects pertaining to phytopathology and few established aspects with latest information. It presents comprehensive treatments of some fresh, less discussed topics like "Plant Pathology in the 'Omics' Era, Plant Disease Diagnostic Kits, Plant Quarantine and Pest Risk Analysis, Biosafety and Biosecurity, Intellectual Property Rights and World Trade Organization in relation to Phytopathology, Nanotechnology in Plant Disease Management and Innovations in Teaching, Research and Extension in Plant Pathology". Of late, there have been phenomenal developments in major issues of phytopathology like pathogenesis, defense mechanisms in plants and epidemiology. This book provides the latest updates on these topics. Each chapter is well supported by convincing tables, appropriate diagrams and illustrations at right places. A detailed bibliography provided at the end of the book helps for further reading.

Necessity of a book which gives a fairly comprehensive account of the new topics along with latest information on established aspects of phytopathology has been felt ever since the realization of significance of interdisciplinary approach. In this context, this book is designed to meet the requirement of students, teachers, research workers, extension personnel, policy makers and industries despite all those concerned and interested in phytopathology.

The author appreciates receiving suggestions and constructive criticism that would improve the quality of the book (e-mail: ravichandrang_3@yahoo.co.in).

N.G. Ravichandra

Acknowledgements

I am grateful to Dr. H. Shivanna, Hon'ble Vice Chancellor and Dr. T. Sheshadri, Director of Research, University of Agricultural Sciences, GKVK, Bangalore for their support and wishes.

I extend my special thanks to Dr. R.K. Walia, Project Coordinator, AICRP (Nematodes), Division of Nematology, Indian Agricultural Research Institute, New Delhi, India for the encouragement.

I thank Prof. B.M.R. Reddy, Professor & University Head, Dr. K.T. Rangaswamy, Professor & Head, Dr. T. Narendrappa, Dr. H.A. Prameela, Dr. N. Nagaraju, Dr. Y.M. Somashekhar, Dr. M.K. Prasanna Kumar, Dr. C.T. Manjula and Mr. B.R. Nagendra, Department of Plant Pathology, UAS, GKVK, Bangalore.

I would like to express my deep sense of gratitude to my beloved parents Sri. N. Gurushankaraiah and Smt. Parvathamma to whom I affectionately dedicate this book.

I wish to record my appreciation and affectionate thanks to my wife, Ms. Deepthi and daughter, Ms. Anusha for their affection, patience, keen interest and concern throughout the preparation of the manuscript. Words fail to acknowledge their concern and cooperation.

I am grateful to my In-laws Mr. M.S. Chandrashekariah and Smt. Kamalamma for their wishes and encouragement.

Mere words are insufficient to express my deep sense of gratitude to my brother-in-law Mr. Sudesh Kumar and sister Ms. Tara for their special care and moral support. I fondly acknowledge their continued motivation and affection being showered on me.

I appreciate the support rendered by N.G. Jayachandra, Dr. K.R. Nagaraj, Ms. Gayathri and Ms. Swapna.

N.G. Ravichandra

Contents

1 Plant Pathology in the 'Omics' Era

The revolution of 'omic' sciences has introduced integrated high-throughput approaches to address the understanding of the biochemical systems and their dynamic evolution. 'Omics' are depicting a comprehensive view which largely overcomes the merely descriptive approaches of the early proteomic and metabolomic era. The diagnosis of the causal agent of plant diseases is critical for its management; hence, misdiagnosis affects the management actions of the primary causal agent. At present, there are some plant diseases where the causal agent in unknown or the disease symptoms are associated with different pathogens or abiotic agents (F. Victoriano Lara, 2011). In this sense, metagenomics provides a new rationale and effective methodology to identifying the primary causative agents. The study of uncultured microbes may carry out by metagenomics, which gives a broad sight for the investigation of microorganism origins and function in the environment. Metagenomics is based on studies of ecological diversity of uncultured and cultural microorganisms using molecular biology. For the metagenomic analysis of microbial populations in phytopathology, the total content of nucleic acids involved with the disease is used.

We live in an 'omic' era. Remarkable advances in both analytical methodology and computer algorithms pave the way for the discovery of the molecular basis of biology. From the study of the genome to the proteome to the metabolome, our ability to correlate the function of genes as components of a complex interconnected network of organism and environment promises to unlock the secrets of organisms. That disease is a complex interaction of genetic makeup, gene expression, protein action and the environment mandates a comprehensive study of the complex organism at the level of the common language, chemistry. 'Omics' aim at the collective characterization and quantification of pools of biological molecules that translate into the structure, function, and dynamics of an organism or organisms. The term 'omics' refers to the comprehensive analysis of biological systems. Systems biology is 'biology' that focuses on complex systems of life.

The word genomics is said to appear in the 1980s and became widely used in 1990s. Modern usages of the term 'omics' derive from the term genome (hence genomics), a term invented by Hans Winkler in 1920, although the use of 'ome' is older, signifying the 'collectivity' of a set of things (Trishita Issar, 2009). The genome is the most fundamental part of many 'omics'. The suffix 'om' originated as a back-formation from 'genome,' a word formed in analogy with 'chromosome.' The word 'genomics' implies some hidden network among genetic elements. This network is regulated by several other 'omics' like proteomics, transcriptomics, metabolomics and physiomics.

Simply, high-throughput refers to just that, a technology in which a large or even exhaustive number of measurements that can be taken in a fairly short time period. 'Ome' and 'omics' are suffixes that are derived from genome (the whole collection of an organism's DNA, as a combination of 'gene' and 'chromosome') and genomics (the study of the genome). Scientists like to append to these to any large-scale system (or just about anything complex), such as the collection of proteins in a cell or tissue (the proteome), the collection of metabolites (the metabolome), and the collection of RNA that's been transcribed from genes (the transcriptome) (Reagan Kelly, 2008). A high-throughput analysis is essential considering data at the 'omic' level, that is to say considering all DNA sequences, gene expression levels, or proteins at once (or, to be slightly more precise, a significant subset of them). Without the ability to rapidly and accurately measure tens and hundreds of thousands of data points in a short period of time, there is no way to perform analyses at this level.

1.1 'OMIC' TECHNOLOGIES

'Omic' technologies are primarily aimed at the universal detection of genes (genomics), mRNA (transcriptomics), proteins (proteomics) and metabolites (metabolomics) in a specific biological sample. 'Omic' technologies have a broad range of applications. Genomic and transcriptomic research has progressed due to advances in microarray technology. Mass spectrometry is the most common method used for the detection of analytes in proteomic and metabolomic research. Data analysis is complex as a huge amount of data is generated and statistician and bioinformatician involvement in the process is essential (Horgan and Kenny, 2011).

The development of nontargeted analytical methods for everything from metabolites to genomes has revolutionized biological research and been a major driver for the adoption of systems-based approaches to biological research. The potential for 'omic'-based technologies to drive innovation in crop protection is powerful when applied to understanding plant-pathogen interactions and genetic variation among crop genotypes and populations of target pathogens. Faced with a decline in the discovery of new actives using conventional methods, 'omic' technologies linked to systems approaches are now being increasingly used across the life science area to identify new potential 'druggable' targets.

Some ideas on using 'omics' more extensively in 'conventional' crop protection strategies include, using molecular approaches to optimize the integrated use of agrochemicals with crop, varieties of differing genetic background (e.g., matching fungicide mode of action with host resistance), identification of new crop protection targets for intervention in pathogens, which may provide the basis for screening chemical and biological agents, understanding how beneficial endophytes and resistance elicitors enhance crop protection and thereby improve their efficacy, understanding and exploiting 'natural' plant protection strategies including pest and disease resistance, as well as embracing less studied interactions such as allelopathy, understanding the many resistance mechanisms reducing ability to deploy existing toxophore and how they may be counteracted, developing synergistic mixes and formulations of pesticides and directly identifying novel bioactive natural products through bioprospecting (Anonymous, 2013).

These 'omics' technologies enable a direct and unbiased monitoring of the factors affecting plant growth and development and provide the data that can be directly used to investigate the complex interplay between the plant, its metabolism, and also the stress caused by the environment or the biological threats [(insects, fungi, or other pathogens) (Fig. 1.1)]. Plant responses to stress are mediated via profound changes in gene expression which result in changes in the composition of plant transcriptome, proteome, and metabolome (Rosa M. Pérez-Clemente *et al.*, 2013).

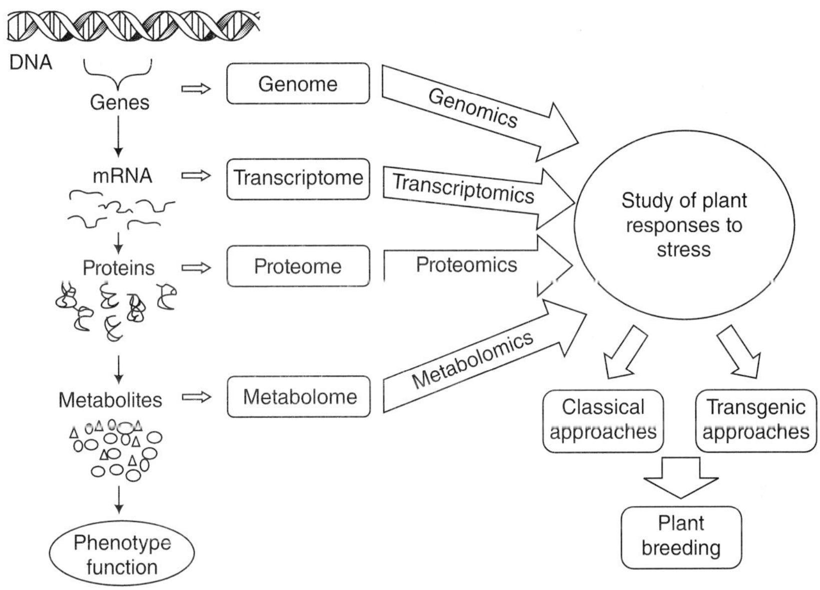

Fig. 1.1 Plant response to abiotic stress factors. Genomics, transcriptomics, proteomics, and metabolomics have enabled active analyses of regulatory networks that control abiotic stress responses. Such analyses increase our knowledge on plant responses and adaptation to stress conditions and allow improving plant breeding

1.2 TYPES OF 'OMICS'

Various types of 'omics,' in general, include antibodyome and antibodyomics, bacteriome, cardiogenomics, cellome and cellomics, diseaseome and diseaseomics, epigenome and epigenomics, foodome and foodomics, genome and genomics, glycome and glycomics, healthome and healthomics, herbome and herbomics, hygienome and hygienomics, immunolome and immunolomics, metabolome and metabolomics, neurogenome, nucleome, nutrigenomics, proteome and proteomics, sequenceome, transcriptomics, virusomics and many more (Trishta Issar, 2009).

'Omic' technologies adopt a holistic view of the molecules that make up a cell, tissue or organism. They are aimed primarily at the universal detection of genes (genomics), mRNA (transcriptomics), proteins (proteomics) and metabolites (metabolomics) in a specific biological sample in a nontargeted and nonbiased manner. This can also be referred to as high-dimensional biology; the integration of these techniques is called systems biology (Kell, 2007). The basic aspect of these approaches is that a complex system can be understood more thoroughly if considered as a whole. Systems biology and 'omics' experiments differ from traditional studies, which are largely hypothesis-driven or reductionist. By contrast, systems biology experiments are hypothesis-generating, using holistic approaches where no hypothesis is known or prescribed but all data are acquired and analyzed to define a hypothesis that can be further tested.

New advances resulting in 'omic' technologies have exploded the areas of genomics, transcriptomics, proteomics and metabolomics and revealed many fundamental processes driven by both pathogens and commensals. These advances in genome sequencing coupled with other 'omic' technologies such as proteomics and transcriptomics have revolutionized plant pathology. There are four major types of high-throughput measurements that are commonly performed, viz., genomic SNP analysis (i.e., the large-scale genotyping of single nucleotide polymorphisms), transcriptomic measurements (i.e., the measurement of all gene expression values in a cell or tissue type simultaneously), proteomic measurements (i.e., the identification of all proteins present in a cell or tissue type), and metabolomic measurements (i.e., the identification and quantification of all metabolites present in a cell or tissue type). Each of these four is distinct and offers a different perspective on the processes underlying disease initiation and progression as well as on ways of predicting, preventing, or treating disease.

1.2.1 Genomics

It is a new science that deals with the discovery and noting of all the sequences in the entire genome of a particular organism. The genome can be defined as the complete set of genes inside a cell. Genomics, is, therefore, the study of the genetic makeup of organisms (Anonymous, 2001). Genomics is an entry point for looking at the other 'omics' sciences. The information in the genes of an organism, its genotype, is largely responsible for the final physical makeup of the organism, referred to as the 'phenotype.' DNA in the genome is only one aspect of the complex mechanism that keeps an organism running. Genomics may be described as the comprehensive analysis of DNA structure and function (Trishita). Understanding biological diversity at the whole genome level will yield insight into the origins of individual traits and disease susceptibility. A gene by gene approach has been typically used to understand its function. The emergence of the novel 'omics' technologies is now allowing researchers to identify the genetics behind plant stress responses [(Table 1.1) (Rosa M. Pérez-Clemente *et al.*, 2013)].

Table 1.1 Genes Involved in Plant Responses to Stress

Gene	*Stress*
14.3.3 gene family (GF14b, GF14c)	Salinity, drought, fungal
MAPK	Abiotic and biotic stresses
MEKK1 and ANP1	Environmental stress
MPK3, MPK4, and MPK6	Abiotic stress (pathogens) and oxidative stress
CBF/DREB families (CBF1, CBF2, DREB2A)	Drought, cold, salinity
HVA1	Salinity and drought
Glycerol-3-phosphate acyltransferase gene	Cold
ICS	Pathogens, UV light
LOX	Wounding, drought, and pathogens
bZIPs family (e.g., ABF1, ABF2)	Drought, temperature, salinity
WRKY family (AtWRKY2, AtWRKY6, AtWRKY18)	Pathogens, wounding, salinity, temperature, drought, oxidative stress
ATAF	Wounding, drought, salinity, cold, pathogens

Numerous investigations show that plant defense response genes are transcriptionally activated by pathogens and also by different types of abiotic stress. It has been described that the induction of specific defense genes, in the response against certain pathogens, is dependent on specific environmental conditions, suggesting the existence of a complex signaling network that allows the plant to recognize and protect itself against pathogens and environmental stress. Similar induction patterns of members of the 14.3.3 gene family (GF14b and GF14c) by abiotic and biotic stresses such as salinity, drought, ABA, and fungal inoculation have been documented in rice (Chen *et al.*, 2006). The rice GF14 genes contain *cis*-elements in their promoter regions that are responsive to abiotic stress and pathogen attack. The 14-3-3s family genes are also subject to the regulation by certain transcript factors. Advances in plant genomics research have opened up new perspectives and opportunities for improving crop plants and their productivity (Rosa M. Pérez-Clemente *et al.*, 2013). The genomics technologies have been found useful in deciphering the multigenicity of biotic and abiotic plant stress responses through genome sequences, stress-specific cell and tissue transcript collections, protein and metabolite profiles and their dynamic changes, protein interactions, and mutant screens.

Genomics analyses of plant-associated microorganisms are as essential to understanding the development and suppression of plant diseases. Analyses of microbial genomes will complement those done on plant genomes (e.g., for *Arabidopsis*, rice, etc.) by providing new insights into the nature of plant-microbe interactions. The use of genomics has already been demonstrated through enhancing disease resistance by using viral components that can be incorporated into a host with current technology. Identifying critical genes in other pathogens will permit similar strategies to be developed for disease control. Similar strategies are receiving significant support for microbes of medical importance; as such efforts are put in identifying targets for drugs and suggesting new targets for drug development and diagnostics. The genetic potential and sensitivity of pathogens must be understood to effectively combat current and future disease threats (Anonymous, 2003).

Genomic SNP genotyping measures an organism's genotypes for several hundred thousand single nucleotide polymorphisms spread throughout the genome. Other assays exist to genotype ten thousand or so polymorphic sites that are near-known genes (under the assumption that these are more likely to have some effect on these genes). The genotyping technology is quite accurate, but the SNPs themselves offer only limited information. These SNPs tend to be quite common (with typically at least 5% of the population having at least one copy of the less frequent allele), and not strictly causal of the disease. Rather, SNPs can act in unison with other SNPs and with environmental variables to increase or decrease the risk of a disease (Kell, 2007). This makes identifying important SNPs difficult; the variation in a trait that can be accounted for by a single SNP is fairly small relative to the total variation in the trait. Even so, because genotypes remain constant (barring mutations to individual cells) throughout life, SNPs are, potentially, among the most useful measurements for predicting risk.

1.2.2 Applications of Genomics

Among several, major applications of genomics include disease diagnosis, confirmation of a diagnosis, confirmation of the existence of a disease in hosts, prediction of the risk of future disease in healthy hosts and genetic tests for carrier screening.

Since the pathogen profile of *Ralstonia solanacearum* was published 10 years ago, studies concerning this plant pathogen have taken a genomic and postgenomic direction. This was pioneered by the first sequenced and annotated genome for a major plant bacterial pathogen and followed by many more genomes in subsequent years. All molecular features studied now have a genomic flavor. In the future, this will help in connecting the classical field of pathology and diversity studies with the gene content of specific strains. In this review, we summarize the recent research on this bacterial pathogen, including strain classification, host range, pathogenicity determinants, regulation of virulence genes, type III effector repertoire, effector-triggered immunity, plant signaling in response to *R. solanacearum*, as well as a review of different new pathosystems (Peeters *et al.*, 2013).

1.3 TRANSCRIPTOMICS

Transcriptome is the set of all RNA molecules, including mRNA, rRNA, tRNA and other noncoding RNA produced in one or a population of cells. It can be applied to the total set of transcripts in a given organism or the specific subset of transcripts present in a particular cell type.

Transcriptomic measurements, often referred to as gene expression microarrays or 'gene chips,' are the oldest and most established of the high-throughput methodologies. The most common are commercially produced 'oligonucleotide arrays,' which have hundreds of thousands of small (25 bases) probes, between 11 and 20 per gene. RNA that has been extracted from cells is then hybridized to the chip, and the expression level of ~30,000 different mRNAs can be assessed simultaneously. More so than SNP genotypes, there is the potential for a significant amount of noise in transcriptomic measurements. The source of the RNA, the preparation and purification methods and variations in the hybridization and scanning process can lead to differences in expression levels; statistical methods to normalize, quantify, and analyze these measures has been one of the hottest areas of research (Kell, 2007). Gene expression levels influence traits more directly than SNPs and hence, significant associations are easier to detect. While transcriptomic measures are not as useful for predisease prediction (because the gene expression levels very far in advance of disease initiation are not likely to be informative because they have the potential to change so significantly), they are very well-suited for either early identification of a disease (i.e., finding the organisms, which have gene expression levels characteristic of a disease but which have not yet manifested other symptoms) or classifying hosts with a disease into subgroups (by identifying gene expression levels that are associated with either better or worse outcomes or with higher or lower values of some disease phenotype).

1.4 PROTEOMICS

Proteins are responsible for an endless number of tasks within the cell. The complete set of proteins in a cell can be referred to as its proteome and the study of protein structure and function and what every protein in the cell is doing is known as proteomics. The proteome is highly dynamic, and it changes from time to time in response to different environmental stimuli. The goal of proteomics is to understand how the structure and function of proteins allow them to do what they do, what they interact with, and how they contribute to life processes. An application of proteomics is known as protein 'expression profiling' where proteins are identified at a certain time in an organism as a result of the expression to a stimulus. Proteomics can be used to develop a protein-network map where interaction among proteins can be determined for a particular pathogen, to map protein modification to determine the difference between a wild-type and a genetically modified organism and also to study protein-protein interactions involved in plant defense reactions (*Example:* Identifying the proteins expressed in response to diseases like soybean cyst nematode).

Since proteins are directly involved in plant stress response, proteomics studies can significantly contribute to elucidate the possible relationships between protein abundance and plant stress acclimation. Several studies have already proven that the changes in gene expression at the transcript level do not often correspond with the changes at the protein level. The investigation of changes in plant proteome is highly important since proteins, unlike transcripts, are direct effectors of plant stress response. Proteins not only include enzymes catalyzing changes in metabolite levels but also include components of transcription and translation machinery [(Rosa M. Pérez-Clemente *et al.*, 2013) (Table 1.2)].

Table 1.2 Proteins and Enzymes Involved in Plant Responses to Stress

Proteins and enzymes	*Stress*
ERF family	Cold, drought, pathogen infection, wounding, ET, SA, and JA
bZIPs family (e.g., ABF1, ABF2)	Drought, temperature, salt
WRKY	Pathogens, wounding, salinity, temperature, drought, oxidative stress
MYB family (AtMYB15, AtMYB30, AtMYB33 AtMYB60, AtMYB96, AtMYB101 AtMYB15, and AtMYB108)	Biotic and abiotic stress (pathogens, drought, cold)
ABF	Drought
NAC	Drought, salinity, cold
MYC	Environmental stresses
LEA family (PMA 80, PMA 1959)	Salinity and drought
Heat shock proteins	Temperatures
LOX family (e.g., LOX1)	Wounding, drought, and pathogens
Glutathione peroxidase, superoxide dismutase, ascorbate peroxidases, and glutathione reductases	Oxidative stress

Studies of plant reaction upon stress conditions at protein level can significantly contribute to our understanding of physiological mechanisms underlying plant stress tolerance. In the last years, there has been an important progress in the knowledge of several families of plant transcription factors linked to plant stress responses, such as responses to pathogens. The most important ones are as follows.

- The ethylene-responsive-element-binding factors (ERFs). This protein family has been linked to a wide range of stresses; the RNA levels of specific ERF genes are regulated by cold, drought, pathogen infection, wounding or treatment with ethylene, SA or JA.
- Another important family of transcription factors is called 'basic-domain leucine-zipper (bZIP)' which are regulators of important plant processes including pathogen defense.
- WRKY proteins are a family of transcription factors that are unique to plants. Specific WRKY family members show enhanced expression and/or DNA-binding activity following induction by a range of pathogens, defense signals, and wounding.
- MYB proteins are key factors in regulatory networks controlling development, metabolism and responses to biotic and abiotic stresses. AtMYB30 encodes an activator of the hypersensitive cell death program in response to pathogen attack. AtMYB96 acts through the ABA signaling cascade to regulate water stress and disease resistance.

- MYC proteins are involved in the response of plants to unfavorable environmental conditions. This transcription factor family plays a role in the induction of apoptosis, important in the hypersensitive cell death program in response to pathogen attack. Another putative MYC target is the ornithine decarboxylase gene, involved in polyamines synthesis. On the other hand, MYC proteins activate the major ABA-dependent stress response.

Proteomics is similar in character to transcriptomics. The most significant difference is in regards to the measurements. Unlike transcriptomics, where the gene expression levels are assessed simultaneously, protein identification is done in a rapid serial fashion. After a sample has been prepared, the proteins are separated using chromatography, two-dimensional protein gels (which separate proteins based on charge and then size) or one-dimensional protein gels (which separate based on size alone) and digested, typically with trypsin (which cuts proteins after each arginine and lysine), and then run through mass spectroscopy. The mass spec identifies the size of each of the peptides, and the proteins can be identified by comparing the size of the peptides created with the theoretical digests of all know proteins in a database (Kell, 2007). This searching is the key to the technology, and a number of algorithms both commercial and open-source have been created for this. Unlike transcriptomic measures, the overall quantity of a protein cannot be assessed, just its presence or absence. Like transcriptomic measures, though, proteomic measures are excellent for early identification of disease. Proteomics focuses on gene products while genomics focuses on an organism's genetic makeup. Both are interrelated.

1.5 METABOLOMICS

Metabolomics is one of the newest 'omics' sciences. It refers to the complete set of low-molecular-weight compounds in a sample. These are the substrates and by-products of enzymatic reactions and have a direct effect on the phenotype of the cell. Metabolomics aims at determining a sample's profile of these compounds at a specified time under specific environmental conditions. The high-throughput measure of the metabolites is present in a cell or tissue. As with proteomics, the metabolites are measured in a very fast serial process. NMR is typically used to both identify and quantify metabolites (Kell, 2007). This technology is newer and less frequently used than the other technologies, but similar caveats apply. Measurements of metabolites are dynamic as are gene expression levels, and proteins and so are best suited for either early disease detection or disease subclass identification.

The possibility of monitoring a complete set of metabolites could largely improve the understanding of many physiological plant processes. This systematic study, defined as 'metabolomics,' is intended to provide an integrated view of the functional status of an organism [(Rosa M. Pérez-Clemente *et al.*, 2013) (Table 1.3)].

Table 1.3 Metabolites and Hormones Involved in Plant Responses to Stress

Metabolites and hormones	*Stress*
Abscisic acid, jasmonic acid, salicylic acid, polyamines, and others	Drought, salinity, cold
Proline, glycine-betaine, and other compatible osmolytes	Environmental stresses: drought, salinity, osmotic
Phytoalexins	Microbial pathogens
Terpenes	Toxins and pathogens
Phenolic compounds (coumarin, lignin, flavonoids, tannins, isoflavonoids)	Pathogens, oxidative stress, UV light
Alkaloids	Pathogens (predators)
Unsaturated fatty acids	Environmental stresses
ROS, malondialdehyde	Biotic and abiotic stresses
Phytochelatins and metallothioneins	Heavy metal intoxication

This technological tool, recently developed, includes targeted analysis, metabolic fingerprinting, and metabolite profiling.

The metabolome represents the downstream result of gene expression and is closer to phenotype than transcript expression or proteins. Extensive knowledge of metabolic flows could allow assessment of genotypic or phenotypic differences between plant species or among genotypes exhibiting different tolerance to some biotic or abiotic stresses. In addition, target metabolites have been analyzed as nutritional and/or agronomical biomarkers to classify different crop cultivars or to optimize growth conditions (Arbona *et al.*, 2010). In contrast to high-throughput methodology for the analysis of DNA, RNA, and proteins, current strategies for metabolite characterization still face significant obstacles. These challenges are largely caused by the high degree of chemical diversity among metabolite pools as well as the complexity of spatial and temporal distribution within living tissues. Plant metabolomics methodology and instrumentation are being developed at a rapid pace to address these analytical challenges. Like other functional genomics research, metabolomics generates large amounts of data. Handling, processing, and analyzing this data is a clear challenge for researchers and requires specialized mathematical, statistical, and bioinformatic tools. Further developments in this area require improvements in both analytical science and bioinformatics. Development of new analytical techniques is largely focused on increasing resolution and comprehensiveness, increasing speed and throughput of analytical assays and equipment miniaturization.

Matteo Lorito *et al.* (2010) presented an overview of the latest discoveries on the *Trichoderma* expressome and metabolome, of the complex and diverse biotic interactions established in nature by these microbes, and of their proven or potential importance to agriculture and industry. These fungi may affect the plant response by increasing its basic immunity or the microbe-associated molecular patterns (MAMPs)-triggered immunity (MTI), as well as reducing the effector-triggered susceptibility (ETS) and increasing the effector-triggered immunity (ETI). Different strategies have been followed to study the transcriptome of *Trichoderma*, viz., methods based on ESTs, search of specific gene groups (i.e., chitinases, peptidases, hydrophobins), subtractive hybridization (SSH and RaSH), and DNA arrays.

Proteomic analysis of biotechnologically important fungi has developed significantly only in the last decade, with relatively few cases studied compared with the numerous species whose genome has been sequenced. A biocontrol strain of *T. harzianum* has been the first filamentous fungus for which mass spectrometry [matrix assisted laser desorption/ionization-time of flight (MALDITOF) and liquid chromatography tandem mass (LC-MS-MS)] has been used to improve the method of protein identification after classical two-dimensional electrophoresis performed on a whole cell extract (Matteo Lorito *et al.*, 2010). The research focused on the proteome of biocontrol strains of *Trichoderma,* mainly *T. harzianum* and *T. atroviride,* permitted the identification of many key protein factors probably involved in the *Trichoderma*-fungal host or *Trichoderma*-plant interaction.

The *Trichoderma*-plant interaction proteome has received increasing attention in the last few years by different research groups. Marra *et al.* (2006) were the first to use proteomics to study the two- and three-way interaction between *Trichoderma* (*T. atroviride*), a plant (bean), and a pathogen (*Botrytis cinerea* or *Rhizoctonia solani*) by mapping and separately analyzing the intracellular proteomes of the three components, tested in all possible combinations. A large amount of data collected indicated that the set of differential proteins in the plant induced by *Trichoderma* was substantially different from those produced by the interaction with either one of the two pathogens.

The variety and the number of compounds found in the metabolome of different *Trichoderma* strains/species are astonishingly high and include lytic enzymes, metabolic intermediates, hormones and other signaling molecules, etc. but also many secondary metabolites (a few hundreds have been identified) with important biological functions. Secondary metabolites are natural compounds having different chemical structures and not directly involved in the primary metabolic fluxes of an organism, such as those related to normal growth, development, or reproduction. Instead, they support microbe survival and basic processes, such as competition, symbiosis, metal transport, differentiation, etc.

In the case of *Trichoderma* spp., some of the knowledge provided by functional genomics studies can be and are being directly implemented for improving the application of these beneficial agents. For instance, the selection of more effective or useful strains has been aided by the identification of the relevant genes and the knowledge of their expression pattern under different interaction conditions. Examples include the discovery of metabolites and molecular mechanisms that support the desirable *Trichoderma* activities/effects both outside (mycoparasitic, antimicrobial, degradation of toxins) and inside the plant (increased resistance to pathogens and abiotic stresses, enhanced photosynthetic efficiency, promotion of growth and development, etc.). Also, the molecular characterization of the plant physiological response to *Trichoderma* with the identification of plant cultivars and *Trichoderma* strain combinations to be recommended for use or not was given by Harman (2000). Metabolomics and proteomics data have been directly used to augment the effectiveness of these microbes and facilitate their implementation in crop management. Proteomics and metabolomics characterization of the extracellular fraction, associated with *in vivo* assays, has allowed the identification of useful compound combinations and of the fermentation conditions required to obtain them on an industrial scale. Together with advanced strain selection, this has led to the development and commercialization of new liquid formulations comprising spores, mycelia, and metabolites, highly effective and recommended both for soil and foliar treatments in diverse agricultural contexts. Matteo Lorito *et al.* (2013) conclude by quoting that the learning is on with the support of 'omics' research how to best utilize these natural tools (living microbes, metabolites, and genes) for meeting the next challenges of agriculture. A new green revolution is necessary, which requires alternative technologies to feed the fast-growing world population while reducing the input of chemical pesticides and fertilizers in our food chain and the environment.

1.6 IMPACT OF THE 'OMIC' TECHNOLOGIES

1.6.1 Detection of Plant Pathogens

The detection and identification of new pathogens presently can be accomplished through a number of traditional and modern techniques. Traditionally, it is done testing a susceptible host or through a set of tests for the characterization of the plant pathogen. This is also possible using molecular techniques such as polymerase chain reaction or nucleic acid hybridization. In this sense, metagenomics provides a new rationale and effective methodology for identifying the primary causative agents (Trewavas, 1999). The study of uncultured microbes may carry out by metagenomics, which gives a broad sight for the investigation of microorganism origins and function in the environment. Molecular techniques have revealed that more than one pathogen or pathogen isolation may be involved in a plant disease symptom; another contribution is the identification of some pathogens involved in a plant disease that were not identified until today. Plant pathogen genomics offers new knowledge about infection mechanisms, host-plant interactions, new molecular targets for fungicides, bactericides and so on.

Phytopathology in the 'omics'-era book has been designed as a textbook for a plant pathology bachelor or master degree course, as an intent to present in simple terms the most recent advances in the 'omic' sciences and their applications in plant pathology, viz., plant pathogen genomics, proteomics, transcriptomics, and metabolomics. This kind of scientific knowledge may contribute to a better understanding of plant pathogens, plant disease, and plant disease management. Genetic engineering has the potential to provide a cornucopia of beneficial plant traits, particularly an enhanced ability to withstand or resist attack by plant pathogens. New approaches to plant disease management are particularly important for pathogens that are difficult to control by existing methods.

1.6.2 Enhancing Resistance with Plant Genes

Scientists from all over the world are investigating the biochemical nature of, and the signals involved in, a plant's reactions to pathogen invasion and disease development. Plant resistance genes and the genes involved in resistance reactions are being identified and engineered into crop plants to protect them against plant diseases (Wambugu, 1999). This rapidly advancing field of investigation is described in this feature under enhancing a plant's resistance with genes from the plant kingdom.

1.6.3 Pathogen-derived Resistance

Plants can be protected from diseases with transgenes (genes that are engineered into plants) that are derived from the pathogens themselves, a concept referred to as pathogen-derived resistance (Hails 2000). For example, plant viral transgenes can protect plants from infection by the virus from which the transgene was derived. Genetic engineering of plants for viral resistance is a thriving area of research.

1.6.4 Antimicrobial Proteins

Another area of investigation involves peptides and proteins with antimicrobial properties that when produced by plants have the potential to strengthen plant resistance to fungal, bacterial and nematodes and other plant pathogens (Ferber, 1999). Pathogens like fungi contain genes encoding antimicrobial compounds, which can be utilized to improve plant resistance to pathogens.

1.6.5 Plantibodies

Although plants have mechanisms to protect themselves against pathogen attack, in contrast to animals, there is no 'immune system' *per se* in plants. With the advent of genetic engineering, plants can be engineered to express an antibody against a protein crucial for pathogenesis resulting in a level of immunity or resistance to the pathogen. Biotechnology is now a lightning rod for visceral debate, with opposing camps making strong claims of promise and peril. The debate involves not only scientific but also political, socio-economic, ethical, and philosophical issues (Sagar *et al.*, 2000). *Phytophthora infestans*, the cause of late blight, is the most devastating disease of potato worldwide. Recent years have seen a dramatic intensification in molecular biological studies of *P. infestans*, including the development of novel tools for transformation and gene silencing and the resources for genetical, transcriptional and physical mapping of the genome. Efforts are being intensified to use these resources to discover the genetic bases of pathogenicity, avirulence, and host-specificity.

1.6.6 High-Throughput Sequencing of the 16S rRNA Gene and Metagenomics

Sequencing of the 16S ribosomal RNA (16S rRNA) gene as a means to distinguish between species is a powerful tool when coupled with the development of high-throughput sequencing methods. This allows for the assessment of bacterial diversity in a wide range of communities and environmental niches within a short period of time. In fact, 'omic' technology, through high-throughput sequencing of 16S rRNA, has enabled the identification of new candidate beneficial microbes that, after further study and analysis, may be potential probiotic bacteria.

More recently, advances in nanotechnology have enabled sequencing of a single molecule without the need for PCR amplification. Benefits of this system include reduction of reagents required and the omission of possible sequence bias introduced by PCR amplification. Other major systems include the Helicos platform, which does not require amplification of the DNA template but sequences of single template molecules (Milos, 2008), the real-time DNA sequencing using single DNA polymerase molecules theoretically resulting in read lengths greater than 1,000 bp (Eid *et al.*, 2009) and DNA nanoarray sequencing (Drmanac *et al.*, 2010). Here are the limits to 16S rRNA sequencing. The 16S rRNA gene is highly conserved; therefore it can be difficult to distinguish between closely related species and strains within species using 16S rRNA sequencing. Metagenomics is one method that overcomes this by sequencing all the nucleic acids directly from a sample and is, therefore, not restricted to the 16S rRNA gene. In addition to sequencing individual strains, one of the main outcomes of high-throughput sequencing has been the emergence of metagenomics. Although this technology was originally applied to environmental biology, in the past five years new information has been established in regard to our microbiome and relationship of the human microbiome and diet.

Genome sequencing and subsequent comparative genomics of pathogenic bacteria led to the description of the pan-genome, which represents the total repertoire of genes for a bacterial species. The pan-genome also encompasses accessory or dispensable genes unique to single strains and genes present in more than one strain (Medini *et al.*, 2005). The pan-genome is dynamic with the dispensable genome, in particular, increasing as more genomes are sequenced. Therefore, the pan-genome can be orders of magnitude larger than the genome of a sequenced strain. The description of a pan-genome has only become available through advances in 'omic' technologies through sequencing of large numbers of genomes.

The integration of the 'omics' approaches is likely to enable researchers to reconstruct the whole cascade of cellular events leading to rapid responses and adaptation to the various abiotic stimuli. A well-focused approach combining molecular, physiological, and metabolic aspects of plant stress tolerance is required to increase knowledge on the effects of gene expression and to understand whole plant phenotype under stress. A better understanding of the underlying physiological processes in response to different abiotic stresses can drive the selection of the appropriate promoter or transcription factor to be used for transformation.

In addition, the use of genetic and genomic analysis to identify DNA molecular markers associated with stress resistance can facilitate breeding strategies for crop improvement. This approach is particularly useful when target characters are controlled by several genes, as in the case of abiotic stress tolerance. These 'omics' approaches could be combined with the potential to map different quantitative trait loci (QTLs) contributing to a given agronomical trait and to identify linked molecular markers. This will open the possibility to transfer simultaneously several QTLs and to pyramid QTLs for several agronomical traits in one improved cultivar.

1.7 EVIDENCES ON THE APPLICATION OF 'OMIC' TECHNOLOGIES IN PLANT PATHOLOGY

Major factors involved in prediction of disease effects on ecosystems include the effect of the pathogen on host survival, physiology, behavior and/or reproduction, life stages of a host vulnerable to a pathogen, proportion of individuals/biomass infected at a site, spatial extent and distribution of infection, rate of pathogen effects on hosts in relation to rate of response/recovery by hosts of individuals replacing hosts, frequency and duration of pathogen impact, and functional similarity of infected individuals versus replacements.

Major evidence on the application of 'omic' technologies in plant pathology has been briefed hereunder.

Molecular factors enabling microbial pathogens to cause plant diseases have been sought with increasing efficacy over three research eras that successively introduced the tools of disease physiology, single-gene molecular genetics, and genomics. Jeffrey A Kimbrel *et al.* (2010) developed a unified model of the interactions of biotrophic and hemibiotrophic pathogens, which posits that successful pathogens typically defeat two levels of plant defense by translocating cytoplasmic effectors that suppress the first defense while evading the second defense. As is predicted from this model and confirmed by sequence pattern-driven discovery of large repertoires of cytoplasmic effectors in the genomes of many pathogens, the coevolution of hemi-biotrophic pathogens and their hosts has generated pathosystems featuring extreme complexity and apparent robustness. These findings highlight the need for a fourth research era of systems biology in which virulence factors are studied as pathosystem components and pathosystems are studied for their emergent properties.

To date, there is no automated system to predict genome-wide plant-pathogen interactions and mechanisms to visualize these networks in a user-friendly way. Under an ongoing project on predicting Protein Interaction Networks (PINs) in the model plant host-pathogen system, *Arabidopsis-Pseudomonas syringae*, unsupervised learning (Bayesian-based) and supervised learning (machine learning) techniques have been used to develop novel algorithms and predict genome-scale PINs, including visualization of these networks in a Cytoscape environment (Rakesh Kaundal, 2015). By integrating diverse data types or properties such as Plant-Associated Microbe Gene Ontology (PAMGO) annotations, protein domain interactions such as intraspecies protein domain profiles, topology or proximity in intraspecies Protein-Protein Interaction (PPI) networks, protein sequence similarity (interologs and orthologs), correlated gene expression, phylogenetic relationship, and the available experimental evidence into a computational framework, our results to predict host-pathogen interactions show more than 95%

prediction accuracy. The models have been further implemented as a web-based resource, AP-*i*NET. The users could analyze their 'query' host/pathogen sequence data to predict the interactions using this tool. The positive interaction pairs could be visualized in AP-*i*NET implemented as a Cytoscape plug-in. Some novel interaction pairs are being validated experimentally *in vivo* using yeast two-hybrid and BiFC (Bimolecular Fluorescence Complementation) techniques, in collaboration with the Noble Foundation. The major interest was to apply this knowledge to develop such computational models in agriculturally relevant crop systems, and help the plant science community in guiding cost-effective experimental strategies to detect host-pathogen PPIs and drive research on how pathogens infect host cells.

Following bioinformatics software pertaining phytopathogens have been developed (Rakesh Kaundal, 2015).

(1) **AP-*i*NET:** It is a bioinformatics system for predicting and visualizing genome-wide Protein Interaction Networks (PINs) in the *Arabidopsis-Pseudomonas syringae* model interaction system.

(2) **DoBlast:** A parallelized BLAST server for genome-scale annotations. Large-scale sequence data analysis could be finished in minutes using automated parallel computing.

(3) **RB-Pred:** This is a first of its kind worldwide. This server forecasts rice leaf blast severity based on the weather parameters for general use to plant pathologists and farming community.

During the past decade, however, genomic technologies have revealed numerous additional bacterial genes that more subtly influence transformation. Members of the genus *Agrobacterium* cause the neoplastic diseases crown gall, hairy root, and cane gall on numerous plant species. Extensive genetic analyses conducted in the 1980s identified key bacterial genes involved in virulence. The results of these genomic analyses allowed scientists to develop a more integrated view of how *Agrobacterium* interacts with host plants. In a similar manner, genomic technologies have identified numerous plant genes important for *Agrobacterium*-mediated genetic transformation. Knowledge of these genes and their roles in transformation has revealed how *Agrobacterium* manipulates its hosts to increase the probability of a successful transformation outcome. The first wave of *Pseudomonas syringae* next-generation genomic studies has revealed insights into host-specific virulence and immunity, genome dynamics and evolution, and genetic and metabolic specialization (Heath E O' Brien *et al.*, 2011). These studies have further enhanced our understanding of type III effector diversity, identified an atypical type III secretion system (T3SS) in a new clade of nonpathogenic *P. syringae*, identified metabolic pathways common to pathogens of woody hosts and revealed extensive genomic diversity among strains that infect common hosts. In general, these discoveries have illustrated the utility of draft genome sequencing for quickly and economically identifying candidate loci for more refined genetic and functional analyses.

Xanthomonas is a large genus of Gram-negative bacteria that cause disease in hundreds of plant hosts, including many economically important crops. Pathogenic species and pathovars within species show a high degree of host plant specificity and many exhibit tissue specificity, invading either the vascular system or the mesophyll tissue of the host. Robert P Ryon *et al.* (2011) discussed the insights that functional and comparative genomic studies are providing into the adaptation of this group of bacteria to exploit the extraordinary diversity of plant hosts and different host tissues.

Cultivation-independent terminal restriction fragment length polymorphism analysis of 16S ribosomal DNA directly amplified from plant tissue DNA was compared with molecular characterization of isolates to examine the influence of plant stress, achieved by infection with *Erwinia carotovora* subsp. *atroseptica,* on the endophytic population in two different potato varieties (Birgit Reiter *et al.,* 2002). Community analysis clearly demonstrated increased bacterial diversity in infected plants compared to that in control plants. The pathogen stress had a greater impact on the bacteria population than the plant genotype had. Partial sequencing of the 16S rRNA genes of isolated endophytes revealed a broad phylogenetic spectrum of bacteria, including members of the alpha, beta, and gamma subgroups of the Proteobacteria, high- and low-G+C-content Gram-positive organisms, and microbes belonging to the Flexibacter-Cytophaga-Bacteroides group. Screening of the isolates for antagonistic activity against *E. carotovora* subsp. *atroseptica* revealed that 38% of the endophytes protected tissue culture plants from blackleg disease.

1.8 FUTURE NEEDS

Major thrusts for future investigations may include a better understanding of integrated 'omic' studies of host and pathogen responses, as well as communities of soil and plant-associated microbes, gene expression in plants and pathogens in response to climatic factors, multifactor studies of climate change effects, better models of adaptation rates, better data and models related to dispersal, current levels of intraspecific diversity, strength of selection under different climate change scenarios and heritability of traits, long-term large-scale records of pathogen and host distributions, models of regional processes that incorporate disease data and models describing dispersal of propagules and vectors, integrated multidisciplinary international networks for data collection and synthesis.

1.9 CONCLUSION

The recently born 'omic' sciences, intended as the systematic definition of subsets of specific biomolecules, are finding applications to plant pathogens. The dawn of a new era in plant pathology and plant protection is upon us. 'Omic' technology has rewritten the scope of scientific investigation, broadened the avenues to resistant plants and challenged us to take safe and careful steps. Plant disease has a major impact on agricultural and natural systems. Systems may change more rapidly than in the past, requiring more research and policy attention. Plant pathologists have a long history of bringing together different scales of biology and are noted for having a balanced focus on molecular biology while never losing sight of more complex systems. This need for a careful balance between a systems biology and reductionist approach is becoming increasingly important in the study of plant-microbial interactions. The costs for metagenomic studies are rapidly increasing. Molecular factors enabling microbial pathogens to cause plant diseases have been sought with increasing efficacy over three research eras that successively introduced the tools of disease physiology, single-gene molecular genetics, and genomics (David J. Schneider and Alan Collme, 2010). From this work emerged a unified model of the interactions of biotrophic and hemibiotrophic pathogens, which posits that successful pathogens typically defeat two levels of plant defense by translocating cytoplasmic effectors that suppress the first defense (surface arrayed against microbial signatures) while evading the second defense (internally arrayed against effectors). As is predicted from this model and confirmed by sequence pattern–driven discovery of large repertoires of cytoplasmic effectors in the genomes of many pathogens, the coevolution of (hemi) biotrophic pathogens and their hosts has generated pathosystems featuring extreme complexity and apparent robustness. These findings highlight the need for a fourth research era of systems biology in which virulence factors are studied as pathosystem components and pathosystems are studied for their emergent properties.

Transcriptomics, proteomics and, more recently, metabolomics are three of these techniques that have impacted on fungal plant pathology. Used individually, each of these techniques can generate a plethora of data that could occupy a laboratory for years. When used in combination, they have the potential to comprehensively dissect a system at the transcriptional and translational level. Transcriptomics, or quantitative gene expression profiling, is arguably the most familiar to researchers in the field of fungal plant pathology. Microarrays have been the primary technique for the last decade, but others are now emerging. Proteomics has also been exploited by the fungal phytopathogen community, but perhaps not to its potential. A lack of genome sequence information has frustrated proteomics researchers and has largely contributed to this technique not fulfilling its potential. The coming of the genome sequencing era has partially alleviated this problem. Metabolomics is the most recent of these techniques to emerge and is concerned with the nontargeted profiling of all metabolites in a given system (Kar-Chun Tan *et al.*, 2009). Metabolomics studies on fungal plant pathogens are only just beginning to appear, although its potential to dissect many facets of the pathogen and disease will see its popularity increase quickly.

Like any other new technology, much still needs to be done before the full potential of 'Omic' technology is realized. As more and more technology sources become available, studies to evaluate the risks associated with 'omic' technology must be intensified. Findings from such investigations must be easily accessible to all. 'Omic' technology's vast perspective is being confronted, and this astounding view has expanded the very foundation of our understanding of plant pathogens.

2 Pathogenesis

Pathogenesis refers to the origin, development and resultant effects of a disease, from the initial appearance of disease all the way to its end stages. The pathogenesis can also describe the origin and development of the disease and whether it is acute, chronic or recurrent. The study of pathogenesis is important to diagnose and manage diseases. The term 'pathogenesis' comes from the Greek words for 'disease' and 'beginning.' The origins of disease are the first step in pathogenesis. Studying pathogenesis can also provide insight into the ways in which diseases spread and potentially contribute to the development of a program which is designed to slow the spread of disease.

2.1 DISEASE DEVELOPMENT

Disease development in brief involves a number of distinct events including the dissemination of the pathogen, attraction of the pathogen to the host, prepenetration, penetration of the pathogen into the host, invasion and spread of the pathogen, reproduction of the pathogen and the survival of the pathogen.

2.1.1 Chain of Infection

For infection and disease to occur in an individual, a process involving six related components must occur. This process has been referred to as the 'chain of infection.' The six steps or 'links' in the chain include an etiologic agent, reservoir, portal of exit, mode of transmission, portal of entry and a susceptible host. This module helps in understanding infectious disease transmission. To stop the spread of disease, one or more of these links must be broken. The infection process extends from the germination or multiplication of an infective propagule within or on the surface of a potential host to the time the pathogen establishes some form of parasitic relationship with its host. The nature of the host-parasite interaction established is determined by the genotypes of the host and the pathogen and is influenced by the environment.

2.1.2 Etiologic Agents

Several categories of biological agents can cause infectious diseases. Each has its own particular characteristics. Major types of agents are fungi, bacteria, viruses, viroids, nematodes and phanerogamic plant parasites.

2.1.3 Phenomenon of Infection

The 'infection process' can be divided into three phases: preentry, entry, and colonization. It encompasses the germination or multiplication of an infective propagule in or on a potential host through to the establishment of a parasitic relationship between the pathogen and the host (Ravichandra, 2013). The process of infection is influenced by properties of the pathogen, the host, and the external environment. If any of these factors inhibits any of the stages of the infection process, the pathogen will not cause disease in the host. While some parasites colonize the outside of the plant (ectoparasites), pathogens may also enter the host plant by penetration, through a natural opening (like a stomatal pore) or via a wound. The symptoms of the diseases produced by these pathogens result from the disruption of respiration, photosynthesis, translocation of nutrients, transpiration and other aspects of growth and development.

2.1.4 Preentry Events

Depending upon the plant pathogen activity, the plant pathogens are classified into two categories, active invaders, and passive invaders.

Active Invaders	*Passive Invaders*
1. Pathogens that make an aggressive effort to gain entry into intact host cells	1. No aggressive effort
2. They do not require help from any external agency to gain entry into host cells	2. Require help from external agencies like insect vectors or wounds caused by agricultural implements
Example: Phytopathogenic fungi, phytonematodes phanerogamic parasites	*Example:* Plant viruses, phytopathogenic bacteria

2.2 INFECTION PROCESS IN FUNGI

The preentry or prepenetration phase of infection by fungi usually commences with the germination of a spore and includes the growth of the resulting germ tube on the surface of the plant. Some pathogens form specialized infection structures (e.g., appressoria and infection cushions) while others enter by direct hyphal growth through wounds or natural openings. Among major phytopathogens, motile ones, like few Oomycota, Chytridiomycota, and Plasmodiophoromycota, bacteria and nematodes must negotiate the external environment before entering the relative security of the host, usually through natural openings or wounds. However, viruses, viroids, and phytoplasmas often establish relationships with arthropod or fungal vectors and have minimal contact with the environment on the plant surface. On the other hand, phytonematodes enter plant tissues through the feeding organ; stylet or ectoparasites may just stay outside the tissue and absorb cell contents by the stylet.

Germination is essentially the change from low metabolic rate to a high metabolic rate and involves a change from near dormancy to intense activity; for this an energy source is needed such as a carbohydrate or fat reserve in the propagule. Fungal invasion is chiefly by germ tubes or structures derived from them. In some fungi like *Rhizoctonia solani* and *Armillariella mellea,* the hypha act in a concerted way to achieve the penetration. In *Rhizoctonia solani,* the fungus on coming in contact with root surface, first forms infection cushions and appressoria and from these multiple infections takes place by means of infection pegs. In *Armillariella mellea,* the hyphae form the rhizomorphs and only these can cause infection.

The survival of the propagules in the external environment is sensitive to interactions between the pathogen and host and the biological, physical and chemical components of the surrounding environment (David Guest and John Brown, 1997). To become infected, a plant must first be inoculated. Inoculation is the process whereby spores or other infective propagules of a parasite come into contact with potential hosts. Plant surfaces differ between species and cultivars in their chemical, physical and biological characteristics. Root surfaces are different from leaf surfaces. For example, the type of wax that forms the cuticle, the thickness of the cuticle, the surface topography, size, shape and number of leaf hairs and stomata are all species-specific characteristics of leaves. Leaf surfaces are subject to extremes of temperature, ultraviolet radiation, and moisture. Roots are buffered against these environmental extremes by the soil environment. However, root pathogens face intense biological competition for nutrients.

Preentry or prepenetration phase includes inoculation, spore germination, germ tube growth and formation of infection while entry or penetration phase involves the actual entry and colonization or postpenetration phase includes colonization, first symptoms of the disease and the production of spores and their dispersal. The initial contact between infective propagules of a parasite and a potential host plant is termed 'inoculation.' Pathogens use a variety of stimuli to identify a suitable entry point. Several fungi use topographical cues on the plant surface to guide them towards a likely stomatal site. Once the hypha reaches a stoma, volatile compounds escaping from the pore appear to provide a signal for the formation of a specialized penetration structure, the 'appressorium.' Sugars, amino acids, and minerals secreted by plants at the leaf surface can nonspecifically trigger spore germination or provide nutrition for the pathogen. Some pathogenic spores will not germinate in the absence of these substances. Pathogen development is influenced by temperature, moisture, light, aeration, nutrient availability and pH. The conditions necessary for survival and successful infection differ between pathogens.

Predation by soil fauna and hyperparasitism of fungal pathogens, such as that of *Phytophthora cinnamomi* by species of *Gliocladium* and *Trichoderma,* also affect prepenetration growth (David Guest and John Brown, 1997). These interactions have sometimes been exploited in biological control programs. Surface topography is used by some pathogens to locate suitable infection sites. Germ tubes of the wheat stem rust fungus (*Puccinia graminis tritici*) and the wheat powdery mildew pathogen (*Blumeria graminis*) grow at right angles to parallel ridges on leaf surfaces. Because of the staggered arrangement of stomata along wheat leaf veins, this increases the chances of germ tubes finding stomata. Once a germ tube reaches a stoma, yet unidentified volatile compounds escaping from the stomatal pore initiate the differentiation of a specialized penetration structure, the appressorium. Hyphae of *Rhizoctonia solani* grow along the valleys above epidermal cell junctions where they penetrate after forming infection cushions. Conidia of the banana anthracnose pathogen, *Colletotrichum musae* settle in these valleys and germinate to form appressoria.

Valleys in the cuticle surface connect to ectodesmata which are pits in the outer wall of underlying epidermal cells where the plasma membrane is exposed. Plant metabolites leak from ectodesmata and accumulate in valleys on the leaf surface, providing nutrients for pathogens and saprophytes. Physical stimuli like scratches on glass, aluminum, gold or polythene surfaces also initiate appressorium formation in some pathogens (e.g., cereal rusts), presumably by imitating the surface topography of host surfaces.

Very few microorganisms that come into contact with a plant surface develop a successful parasitic relationship. Either their nutrient requirements are not satisfied, or their growth and development is inhibited. Sugars, amino acids, and minerals secreted onto the leaf surface (the phylloplane) often stimulate the germination and growth of pathogens. Leaf and petal surfaces contain secretions that nonspecifically stimulate the germination and attract the directional movement or growth of leaf pathogens, including *Botrytis cinerea.* Motile cells of the vascular wilt pathogen *Ralstonia solanacearum* are attracted to stomata by chemical gradients and subsequently enter the host through stomatal apertures. Germinating seeds produce a similar mix of stimulatory plant exudates which influence infection by pathogens such as *Pythium ultimum.* The rhizosphere (the zone of soil adjacent to and influenced by roots) also contains exudates. The concentration of exudates is the highest adjacent to the root surface, and the gradient of plant diffusates away from the root directs the movement of zoospores, bacteria and some nematodes towards the root surface. The rhizosphere supports a higher population of saprophytic and sometimes antagonistic microorganisms than the rest of the soil. These microorganisms may affect pathogen development.

Numerous fungi, oomycetes and parasitic plants produce feeding structures, known as 'haustoria.' They are formed within living cells of the host organism, assumed to play a major role in the absorption of plant metabolites and thus infection. Haustoria of rust fungi can be purified biochemically. Several genes that are expressed in haustoria have been identified. An amino acid transporter protein of the rust fungus, *Uromyces fabae* on broad bean is specifically localized in haustoria, which confirms a role in nutrient assimilation. Signal transduction pathways allow the pathogen to recognize and respond to the plant surface/plant tissue environment. Proteins involved in these signal transduction pathways can then be predicted to be essential for pathogenicity. These signal transduction pathways are often conserved between pathogenic and nonpathogenic organisms.

Puccinia striiformis is an obligate biotrophic fungus, which during infection forms an intimate relationship with the host plant. This includes the formation of highly specialized structures, haustoria, which are formed between the cell wall and plasma membrane of host cells in leaves (Chris Khadgi Sørensen, 2012).

Haustoria play an important role in the acquisition of water and nutrient from the host and are further involved in signaling processes, including suppression and induction of plant defense. Due to its biotrophic nature P. *striiformis* cannot be grown in culture, and infection structures can thus only be studied in leaf tissue. In this study, a classical staining technique was used in combination with advanced 2-photon microscopy to study haustoria in fixed-whole leaves. This allowed 3D reconstruction of fungal structures and revealed several characteristics of haustoria not previously reported. The haustorial body gradually changed from small and spherical to highly irregular and apically branched, taking up significant space in the host cell. The haustorial neck area, which is thought to play an important role in the formation of haustoria, showed significant changes during haustoria development. These results indicate that haustorium formation is a highly dynamic process, which should be considered in future studies on the function and constitution of haustoria of *P. striiformis*.

Pathogen development is influenced by temperature, moisture, light, aeration, nutrient availability and pH. Each species of the pathogen and each strain of many pathogenic species has its own distinct requirements for parasitic development. The spores of most parasitic fungi germinate only in the presence of free moisture. However, conidia of some powdery mildew fungi, like many saprophytic molds, germinate at a relative humidity as low as 70%. Most plant pathogens develop rapidly at temperatures between 15 and 25°C, although some species prefer colder or warmer temperatures. These temperatures may not always coincide with the optimum temperature for disease severity, as this is a product of the effects on the host-pathogen interactions.

The movement of zoospores of *Phytophthora cinnamomi* in the rhizosphere of potential host roots illustrates the interacting environmental factors that determine pathogen development. Zoospores are aerobic and swim towards the air-water interface where oxygen concentrations are highest. They also swim towards higher sugar concentrations and respond to electrical charges on the root surface. Wound pathogens such as *Pythium ultimum* have a positively charged anterior flagellum that is attracted to negatively charged surfaces around root hairs and wound sites. In contrast zoospores of most species of *Phytophthora* have a negatively charged anterior flagellum and are attracted to the positively charged surface at the zone of root elongation. At each stage of the journey, the zoospores are susceptible to antagonistic and parasitic microbes or inhibitory plant exudates and also to temperature and moisture changes.

2.2.1 Entry

Pathogens exploit every possible pathway to enter their host, although individual species of pathogen tend to have a preferred method. Fungal pathogens often use direct penetration of the plant surface to enter the host. This requires adhesion to the plant surface, followed by the application of *pressure* and then enzymatic degradation of the cuticle and cell wall, to overcome the physical barriers presented by the plant's surface. During the degradation of the cuticle and wall, a succession of genes are switched on and off in the pathogen, so that cutinase, followed by cellulase, and then pectinase and protease are produced, attacking the cuticle, cell wall, and middle lamella in the order that they are encountered. The pressure needed for the hypha to penetrate the cell wall is achieved by first firmly attaching the appressorium to the plant surface with proteinaceous glue. The cell wall of the appressorium then becomes impregnated with melanin, making it watertight, and capable of containing the high turgor pressure that builds up within the appressorium. The point of the appressorium that is in contact with the cuticle is called the 'penetration pore' and the wall is thinnest at this point. The increasing turgor pressure causes the pore to herniate, forming a 'penetration peg,' which applies huge pressure to the host cuticle and cell wall.

Pathogens penetrate plant surfaces by direct penetration or indirectly through wounds or natural openings (Fig. 2.1). Bacteria enter plants mostly through wounds and less frequently through natural openings. Viruses, viroids, mollicutes, fastidious bacteria enter through wounds made by vectors. Fungi, nematodes, and parasitic higher plants enter through direct penetration and less frequently through natural openings and wounds.

2.2.1.1 Direct penetration: Breakdown of physical barriers and chemical barriers

2.2.1.1.1 Through noncutinized surfaces

(1) *Seedlings:* Grain smut of jowar (*Sphacelotheca sorghi*), Loose smut of jowar (*Sphacelotheca cruenta*), Downy mildew of jowar and bajra (*Sclerospora graminicola*), Wheat bunt disease (*Tilletia caries, Tilletia foetida*)
(2) *Root hairs:* Wilt causing fungi (*Fusarium* sp.), Clubroot of cabbage (*Plasmodiophora brassicae*), Root rot of cotton (*Phymatotrichum omnivorum*)
(3) *Buds:* Pea rust fungi (*Uromyces pisi*), Witches broom of cherries (*Taphrina cerasi*)
(4) *Flowers:* Loose smut of wheat (*Ustilago nuda tritici*), Long smut of jowar (*Tolyposporium ehrenbergi*), Bunt of rice (*Neovossia horrida*), Ergot of rye (*Claviceps purpurea*)
(5) *Leaves:* Basidiospores of white pine blister rust fungus (*Cronartium ribicola*) germinate and grow down into branches and leaves, where aecia are produced.
(6) *Nectaries:* Fire blight of apple (*Erwinia amylovora*)
(7) *Stalk ends:* *Penicillium italicum, Theilaviopsis paradoxa* (Postharvest disease fungi)

2.2.1.1.2 Through cutinized surfaces

Cuticle: Leaf spot of spinach (*Cercospora beticola*), early blight of solanaceous plants (*Alternaria solani*), Tikka disease of groundnut (*Cercospora personata*)

Pathogens entering through natural openings and pathogens that directly penetrate the plant surface require much more sophisticated mechanisms involving adhesion, the formation of appressoria or other infection structures and infection pegs.

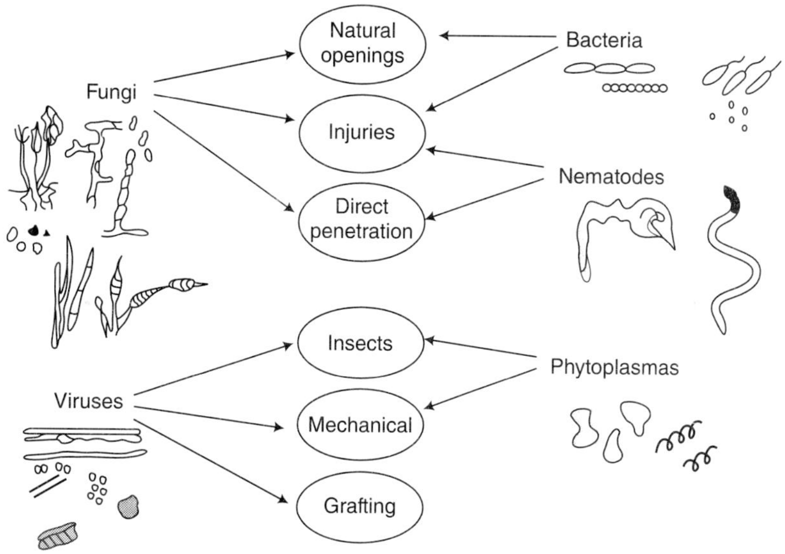

Fig. 2.1 Modes of entry of plant pathogens

2.2.1.2 *Indirect penetration*

It is through wounds or natural openings (Stomata - *Puccinia graminis tritici;* lenticels-*Sclerotinia fructicola, Streptomyces scabies, Phytophthora arecae,* and hydathodes - *Xanthomonas ocampestris* pv. *campestris).*

2.2.2 Penetration

Pathogens have evolved several mechanisms that include structural and/or enzymatic components to enter into their plant hosts (Agrios, 2005). Many fungi, such as *Alternaria alternata, Colletotrichum graminicola, M. oryzae, Pyrenophora teres,* and many oomycetes, such as *P. infestans* and *Phytophthora cinnamomi,* develop appressoria to directly penetrate plant cuticles (Shaowu Meng *et al.,* 2009). An appressorium is a highly specialized structure that differentiates from the end of a symbiont germ tube. It is a swollen, dome-shaped or cylindrical organ, from which a narrow penetration peg emerges to rupture the plant cuticle and cell wall. The penetration peg extends and forms a penetration hypha to penetrate through the epidermal cells and emerge into the underlying tissue. In some instances, penetration is driven by astoundingly high turgor pressures within the appressoria. Generation of turgor is due in part to a thick inner cell wall layer of melanin in mature appressoria. In other instances, cell wall degrading enzymes may play a primary role in or may facilitate the penetration process. Appressoria produced by some fungi, such as rust fungi, do not penetrate directly through the cuticle, but gain entry through stomata.

Plant diseases caused by fungi and oomycetes result in significant economic losses every year. Although phylogenetically distant, the infection processes by these organisms share many common features. These include dispersal of an infectious particle, host adhesion, recognition, penetration, invasive growth, and lesion development (Shaowu Meng *et al.*, 2009). Previously, many of these common processes did not have corresponding Gene Ontology (GO) terms. For example, no GO terms existed to describe processes related to the appressorium, an important structure for infection by many fungi and oomycetes. In this chapter, we have identified common features of the pathogenic processes of fungi and oomycetes and created a pathogenesis model using 256 newly developed and 38 extant GO terms, with an emphasis on the appressorium and signal transduction. This set of standardized GO terms provides a solid base to further compare and contrast the molecular underpinnings of fungal and oomycete pathogenesis.

Sixty-four new GO terms were developed to describe the biological process of penetration into the host, which were formed into two groups (Shaowu Meng *et al.*, 2009). The first group included 43 new GO terms related to infection structures established on the outside of the host tissue, such as appressoria, hyphopodia, infection cushions and haustorium mother cells. The second group had 21 new terms related to specialized structures that directly pierce the surface of the host, for example, penetration pegs, penetration hyphae and haustorium necks. Two hundred fifty-six new GO terms were developed to annotate genes or gene products involved in common pathogenic processes in fungi and oomycetes, including spore dispersal, host adhesion, recognition, penetration, and invasive growth. These new GO terms provide the opportunity to apply a standard set of terms to annotate gene products of fungi, oomycetes, and their associated hosts, as well as those of other plant-associated pathogens and their hosts. The ability to compare and contrast these annotations for widely different plant-associated microbes and their hosts, using a standardized vocabulary, will greatly facilitate the identification of unique and conserved features of pathogenesis across different kingdoms. In addition, such comparisons should provide insight into the evolution of pathogenic processes.

Successful pathogenesis requires a number of coordinated processes whose genetic bases remain to be fully characterized. Schreiber *et al.* (2012) utilized a high-throughput, liquid media-based assay to screen transposon disruptants of *Pseudomonas syringae* pv. *maculicola* ES4326 to identify genes required for virulence in *Arabidopsis*. Many genes identified through this screen were involved in processes such as type III secretion, periplasmic glucan biosynthesis, flagellar motility, and amino acid biosynthesis. A small set of genes did not fall into any of these functional groups, and their disruption resulted in context-specific effects on in planta bacterial growth.

2.2.3 Direct Penetration

Probably the most common mode of entry by fungal pathogens is by direct penetration of the plant surface (Fig. 2.2). Plant surfaces are complex and robust and resistant to penetration by the vast majority of microorganisms. During penetration, pathogen genes are expressed in a coordinated sequence that results in adhesion, followed by the application of physical pressure to the plant surface and the enzymic degradation of the cuticle and different layers of cell walls. Physical and chemical signals switch pathogen genes on and off precisely, so that cutinase, followed by cellulase, then pectinase and protease digest the host cuticle, cell wall, and middle lamella in the order they are confronted.

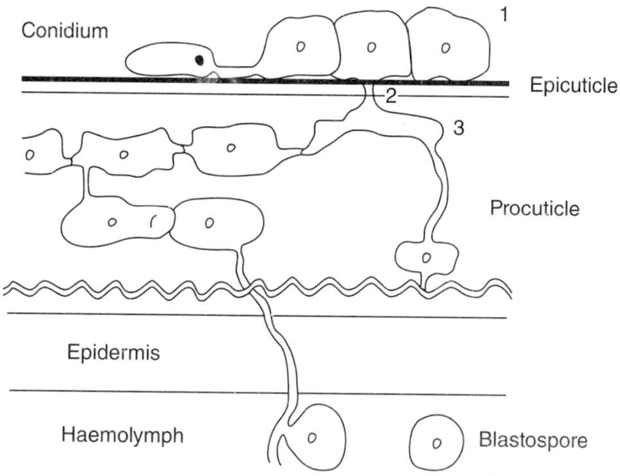

Fig. 2.2 Penetration of host cuticle by a Deuteromycetes entomopathogens 1 = appressorial complex, 2 = penetration peg, 3 = penetration plate

Enzymic degradation of the cuticle by fungal enzymes is necessary for adhesion and penetration by some pathogens. Inhibition of cutinase activity using enzyme inhibitors or cutinase-specific polyclonal antibodies reduces infection of unwounded pea stems by *Fusarium solani* and of unwounded pawpaw fruits by *Cotletotrichum gloeosporiotdes*. These inhibitors do not affect penetration if the cuticle of the inoculated pea stem or pawpaw fruit is first mechanically damaged (David Guest and John Brown, 1997).The pathogenicity of nonpathogenic cutinase-deficient mutants of *C. gtoeosporioides* is restored in the presence of added cutinase. *Mycosphaerella* sp. normally only pathogenic to wounded pawpaw fruit became able to enter unwounded fruit when purified cutinase from *C. gloeosporioides* was added to the inoculum. Pathogenicity was also restored when the fungus was transformed with a cutinase gene from *Nectria haematococca*. These findings provide solid evidence, although not conclusive proof, that cutinase activity is required for pathogenesis.

Root pathogens face a surface composed of suberized tissue, the periderm. Suberin is extremely resistant to enzymic degradation and to physical pressure. However, the take-all pathogen of cereals and the root-rot fungus of many trees, *Arrnitlaria mellea*, penetrate root surfaces using specialized runner hyphae or rhizomorphs, respectively, that apply intense pressure to the infection point. Secondary cell walls are degraded by only a few wood rotting fungi, forcing most pathogens to grow in between cells, or within their lumens, as happens with vascular pathogens.

2.2.3.1 *Breakdown of physical barriers*

Viruses have no physical force or enzyme system of their own to overcome structural or chemical barriers of the host and, therefore, come in contact with the host protoplasm only through wounds. Bacteria are mostly weak parasites and cannot employ force to effect penetration. Fungi and nematodes are the only groups of plant pathogens that employ force for direct penetration of the host. Fungi penetrate host plants directly through a fine hypha produced directly by the spore or mycelium or through a penetration peg produced by an appressorium. These structures exert pressure on the surface which results in stretching of the epidermis which becomes thin. Then the infection peg punctures it and affects its entry.

2.2.3.2 *Breakdown of chemical barriers*

The host is provided with defense mechanisms against invasion which include the presence of cuticular layer on the epidermis, lack of suitable nutrients for the pathogen in the host cells, presence of inhibitory or toxic substances in the host cells and exudation of substances toxic to pathogen or stimulatory to antagonists of the pathogen [*Example*: The glands in leaf hairs of Bengal gram contain maleic acid which is antifungal and provide resistance to infection by the rust fungus (*Uromyces ciceris arietini*)]. Similarly, protocatechuic acid and catechol in the red scales of onion provide resistance to onion smudge pathogen, *Colletotrichum circinans*. To overcome these physical and chemical barriers, the fungi produce various enzymes, toxins organic acids and growth regulators.

2.2.4 Indirect Penetration–Through Natural Openings

Plants have many natural openings that facilitate gas exchange between plant tissues and their environments. These openings are also used by pathogens as entry points. Major among them are stomata, hydathodes, lenticels, and nectaries.

2.2.4.1 *Stomata*

Stomata are microscopic pores formed by pairs of guard cells in the epidermis of terrestrial plants and are essential for gas exchange with the environment and controlling water loss. Accordingly, plants regulate stomatal aperture in response to environmental conditions, such as relative humidity, CO_2 concentration, and light intensity. Stomatal openings are also a major route of pathogen entry into the plant and plants have evolved mechanisms to regulate stomatal aperture as an immune response to bacterial invasion. Stomata represent a major route of pathogen invasion, and recent studies have begun to shed light on the signal transduction cascades underlying bacterial regulation of stomatal closure and opening. It was suggested that stomatal closure was a functional output of both PAMP-triggered and effector-triggered immunity. Because stomata respond to both abiotic and biotic signals, pathogens might exploit abiotic environmental conditions, such as high humidity and/or produce virulence factors to actively suppress stomatal closure as part of their infection strategy (Weiqing Zeng *et al.*, 2011). In the future, it would be important to learn about the detailed molecular and biochemical mechanisms by which PAMP, SA, and ABA signaling cascades intersect each other at the molecular level and to elucidate how different pathogen virulence factors inhibit stomatal closure. Knowledge derived from such studies could lead to novel methods of managing plant disease.

There is variation in the behavior of germ tube at the time of penetration through the stomata. In *Puccinia graminis tritici,* the uredospore germinates and forms a germ tube which on approaching stoma swells at the tip to form an appressorium in the stomatal aperture. From the appressorium, a blade like wedge grows through the stomatal slits and swells inside to form a substomatal vesicle from which the haustoria penetrating the cells are produced. In *Peronospora destructor* infecting onion leaves, the germ tube continues to grow after the formation of the first appressorium. In *Pseudoperonospora cubensis,* the hyphae penetrate the stomatal aperture and swell to form a substomatal vesicle from which in turn other hyphae grow to form haustoria in the adjacent cells of the leaves. *Mycosphaerella musicola* forms a small structure called stomatopodium over the pore of the stoma after growing for few days on the surface of the leaf. A hypha then arises from it which grows into the substomatal chamber and swells to form a vesicle, which in turn gives rise to hyphae which invade palisade tissues.

Plant pathogenic bacteria and nematodes often enter through open stomatal pores when the leaf surface is covered with a film of moisture. Fungi may penetrate stomata by simple undifferentiated hyphal or germ tube growth through the open pore, as has been observed with *Alternaria helianthi, Fulvia fulva,* and *Cochliobolus uictorine.* The wheat stem rust pathogen, *Puccinia graminis trittci,* forms a swollen appressorium over the stomatal aperture, and a fine penetration hypha passes between the guard cells into the airspace under the stomatal pore where it forms a substomatal vesicle. Infection hyphae emerge from this vesicle and form haustoria. Infection structures formed by *P. striiformis* (stripe rust of cereals) are inconspicuous while those formed by *Rhizoctonia solani* are complex infection cushions.

In *Peronospora destructor* infecting onion leaves, the germ tube continues to grow after the formation of the first appressorium. In *Pseudoperonospora cubensis,* the hyphae penetrate the stomatal aperture and swell to form a substomatal vesicle from which in turn other hyphae grow to form haustoria in the adjacent cells of the leaves. *Mycosphaerella musicola* forms a small structure called stomatopodium over the pore of the stoma after growing for few days on the surface of the leaf. A hypha then arises from it which grows into the substomatal chamber and swells to form a vesicle, which in turn gives rise to hyphae which invade palisade tissues.

The stomata of plants regulate gas exchange and water transpiration in response to changing environmental conditions. New work reveals that stomata also have an important role in host defense. Maeli Melotto *et al.* (2006) showed that stomata close upon detection of potential microbial pathogens to prevent the infection of the leaf interior. Moreover, pathogenic bacteria have evolved strategies to suppress the closure of stomata. They also observed the ability of *P. syringae* to move toward open stomata when inoculated on the leaf surface. Stomata are formed by a pair of specialized cells known as guard cells, whose adjustable aperture regulates gas exchange and controls water transpiration. Although it remains to be determined how *P. syringae* can discriminate between open and closed stomata, this may be driven by a chemotactic process that could involve sensing of nutrients derived from the leaf interior and subsequent directed bacterial movement that is propelled by the flagellin-based motility apparatus. Directed motility by chemotaxis is required for virulence and competitive fitness of *Ralstonia solanacearum*, which invades host plants via their roots. *R. solanacearum* is attracted by diverse amino acids and organic acids present in exudates from roots of their host plants; mutants lacking core proteins that regulate chemotaxis exhibit reduced virulence despite retaining normal motility.

Stomata-based immunity might also have relevance for other classes of pathogens, such as parasitic fungi. Epiphytic hyphae of the basidiomycete rust fungus move directionally toward stomata to enter the leaf interior. Also, chitosan, a β-1,4-linked glucosamine derived from fungal cell walls, induces stomatal closure. Unlike bacteria, however, many parasitic fungi including rusts must penetrate the plant cell wall to accommodate specialized feeding structures in plant cells for nutrient uptake. Recent evidence suggests that fungal entry into plant cells is restricted by two secretory pathways that likely deliver defensive compounds into the extracellular space upon PAMP recognition. Ascomycete powdery mildew fungi appear to engage a plasma membrane protein of the host to manipulate these secretory pathways for entry into plant cells.

Foliar diseases have been reported to decrease crop water-use efficiency (WUE) substantially, yet the effects of plant pathogens are seldom considered when methods to improve WUE are debated. Michael K. Grimmer *et al.* (2012) reviewed the effects of foliar pathogens on plant water relations and the consequences for WUE. The effects reported varied between host and pathogen species and between host genotypes. Some general patterns emerged, however. Higher fungi and oomycetes cause physical disruption to the cuticle and stomata and also cause impairment of stomatal closing in the dark. Higher fungi and viruses are associated with impairment of stomatal opening in the light. A number of toxins produced by bacteria and higher fungi have been identified that impair stomatal function. Deleterious effects are not limited to compatible plant–pathogen interactions. Resistant and nonhost interactions have been shown to result in stomatal impairment in light and dark conditions. Mitigation of these effects through a selection of favorable resistance responses could be an important breeding target in the future. The challenges for researchers are to understand how the effects reported from work under controlled conditions translate to crops in the field, and to elucidate underlying mechanisms.

The hypothesis that stomatal wax plugs in *Agathis robusta* (Araucariaceae) protect leaves against fungal invasion by preventing hyphae entering the stomatal pore was tested by Mansour A. Mohammadian *et al.* (2009). Leaves with intact wax plugs and leaves from which wax plugs had been experimentally removed were inoculated with either *Botrytis cinerea* or *Alternaria solani*. Interactions of fungal hyphae with stomata were subsequently analyzed by scanning electron microscopy. Wax plugs blocked the penetration of fungal hyphae into stomata of *A. robusta* in 100% of encounters. In contrast, hyphae readily penetrated stomata of leaves from which wax plugs had been removed or where the wax plugs were damaged. On rare occasions, hyphae were also observed to penetrate directly through the cuticle. Florin rings around the external surface of stomata also acted to deflect fungal hyphae from the stomatal opening. Formation and degradation of wax plugs were also examined to assess the period over which they offer protection from fungal invasion. Wax plugs formed in young leaves in spring and were complete by summer. There was subsequent degradation of wax plugs in the following winter, and they then regenerated again in spring. However, regeneration of wax plugs appeared to decline after the leaves reached two years of age. The data were discussed with reference to the adaptive significance of wax plugs in conifers and other plants. Other examples include *Xanthomonas campestris* pv. *malvacearum* (Black arm of cotton), *Xanthomonas phaseoli* (Bacterial leaf spot of green gram), *Phytophthora infestans* (Late blight of potato), *Albugo candida* (White rust of crucifers) and uredospores of *Puccinia graminis tritici* (Black stem rust of wheat).

2.2.4.2 Hydathodes

These are pores at the margins of leaves, continuous with the end of xylem vessels. The hydathode is the apparatus of guttation that consists of stomata-like pores in the epidermis and the epithem, which has a large chamber with masses of thin-wall parenchyma cells and a sheath layer surrounding its tissues. Under particularly humid conditions when evapotranspiration rates are low, root pressure forces guttation droplets, consisting of xylem fluid, onto the margin of the leaf. Bacteria such as the fireblight pathogen of pomes, *Erwinia amylovora* and the leaf spot pathogen of many plants, *Xanthomonas campestris* pv. *campestris*, become suspended in guttation droplets and enter the leaf when these droplets retreat into hydathodes as the relative humidity drops. Fungi can also penetrate open stomata without the formation of any specialized structures (David Guest and John Brown, 1997). Some fungi form a swollen appressorium over the stomatal aperture, and a fine penetration hypha enters the airspace inside the leaf, where it forms a substomatal vesicle, from which infection hyphae emerge and form 'haustoria' in surrounding cells. Also vulnerable to pathogen invasion are hydathodes, pores at the leaf margin that are continuous with the xylem. Under particularly humid conditions, droplets of xylem fluid (guttation droplets) can emerge at the surface of the leaf where they can be exposed to pathogenic bacteria, which then enter the plant when the droplet retreats into the hydathode as the humidity decreases.

Xanthomonas oryzae pv. *Oryzae* (Xoo) and *Xanthomonas oryzae* pv. *oryzicola* (Xoc) cause bacterial blight (BB) and bacterial leaf streak (BLS) of rice, respectively. Xoo enters the rice leaf typically through hydathodes at the leaf tip and leaf margin (David O. Niño-liu *et al.*, 2006). Cells on the leaf surface may become suspended in guttation fluid as it exudes at night and enter the plant by swimming or passively as the fluid is withdrawn into the leaf in the morning. Bacteria multiply in the intercellular spaces of the underlying epithem, then enter and spread into the plant through the xylem. Xoo may also gain access to the xylem through wounds or openings caused by emerging roots at the base of the leaf sheath. Within the xylem, Xoo interacts with xylem parenchyma cells. The pathogen moves vertically through the leaf through primary veins but also progresses laterally through commissural veins. Within a few days, bacterial cells and EPS fill the xylem vessels and ooze out from hydathodes, forming beads or strands of exudates on the leaf surface, a characteristic sign of the disease and a source of secondary inoculum. Xoc, by contrast, penetrates the leaf mainly through stomata multiplies in the substomatal cavity and then colonizes the intercellular spaces of the parenchyma. Like Xoo, Xoc may also gain access through wounds, but it remains restricted to the apoplast of the mesophyll tissue and does not invade the xylem. Xoc also exudes from natural openings in the leaf in chains or strands, or under moist conditions as small beads of ooze. Yellow exudate on the leaf surface is a typical sign of bacterial leaf streak, and as is the case for Xoo, it may fall into irrigation water or be dispersed by wind, rain, insects or other means, and contribute to the spread of the disease.

2.2.4.3 Lenticels

Lenticels are raised pores that facilitate gas exchange across the bark of woody plant tissues. Although they adequately exclude most microorganisms, pathogens such as *Spongospora subterranea* (powdery scab of potatoes), Penicillium *expansum* (blue mold of apples) and *Monilinia fructicola* (brown rot of stone fruits), enter through lenticels. A lenticel is an opening that allows gasses to be exchanged between air and the inner tissues of a plant. They can be commonly found in bark and on the outside of fruits. Some are more noticeable than others. Ectodesmata, styles and nectaries also provide potential entry points for specialized pathogens. Surprisingly, pathogens appear unable to follow the route taken by pollen tubes through the stigma and style to invade the ovaries. Bacteria can enter, and exit, from stomata along the sides of the style. The alternative pathway for pathogen entry is via a preexisting opening in the plant surface. This can be a natural opening or a wound. Pathogenic bacteria and nematodes often enter through stomatal pores when there is a film of moisture on the leaf surface.

Blackleg is caused by the bacterium *Erwinia carotovora* subsp. *atroseptica*. The blackleg organism may also infect lenticels, causing them to be slightly sunken and brownish to black in color. The infected lenticels can be up to 1/4 inch in diameter. The tissues under the infected lenticels are brownish and usually dry. This extends less than 1/8 inch into the tuber flesh. *Pseudomonas syringae* pv. *actinidiae* (Psa) is a bacterium whose virulent form (Psa-V) causes severe infections of kiwifruit, particularly *Actinidia chinensis* 'Hort16A.' Lenticels on kiwifruit canes function as pores, allowing gaseous exchange. The lenticellular structure penetrates through the periderm, potentially allowing bacterial entry and subsequent cortex infection (Larsen *et al.*, 2012). Bacteria have been observed inside and directly below lenticels from ca 3-year-old woody tissue from the field. To investigate this pathway of bacterial infection further, lenticels from three wood ages were inoculated with a strain of Psa-V at 109 cfu/ml. Brown staining was observed on lenticels three days after inoculation on the youngest wood (less than one-year-old). Lenticels were sectioned two and four weeks after inoculation, and isolations were conducted from sterilized tissue after four weeks. Psa was visible inside lenticels of the youngest wood two weeks postinoculation and was also isolated from the youngest tissue after four weeks. Bacteria were not observed in lenticels of older wood and Psa was not isolated.

Isolates of *Penicillium* spp. were collected regularly from 2001 to 2003 from the surfaces of apple fruits, pre- and postharvest, and from the atmosphere of orchards and storage rooms in France (Amiri and Bompeix, 2005.). *Penicillium* spp. were not detected from the atmosphere of conventional orchards, while their density did not exceed 50 spores m^{-3} in the atmosphere of organically managed orchards. *Penicillium* spp. were seldom detected on apple surfaces in the orchard. The density of *Penicillium* on apples increased from 10 to 50 spores cm^{-2} after one month in storage to 300–400 spores cm^{-2} after six months. *Penicillium expansum* (30–62%) and *P. solitum* (6–45%) were the most prevalent species on apple or in storage rooms. Other species of *Penicillium* isolated included *P. commune, P. verrucosum, P. chrysogenum, P. rugulosum* and *P. digitatum*. Apple fruits were also surveyed for wounds and the number of open lenticels using the sulfur dioxide test. The number of open lenticels per cm^2 of apple surface varied from 0·5 on cv. Boskoop to 4·4 on cv. Golden Delicious. An average of 13 and 2·1% of lenticels, respectively, were infected when they were inoculated by *P. expansum* and *P. verrucosum*.

Ectodesmata, styles and nectaries also provide potential entry points for specialized pathogens. Surprisingly, pathogens appear unable to follow the route taken by pollen tubes through the stigma and style to invade the ovaries. Bacteria can enter and exit, from stomata along the sides of the style.

2.2.4.4 Nectaries

These are glands that secrete sweet liquid to attract and feed pollinators. They are important entry points for bacteria. Morphology and histology of the floral nectary were studied between in 10 pear cultivars, susceptible to *Erwinia amylovora* to various degrees (Farkas and Zs. Orosz-Kovács, 2006). The nectary was receptaculo-ovarial in each pear cultivar studied; the gland was located on the adaxial side of the plate-like hypanthium and on the apical part of the ovary. The nectary was automorphic, protruding out of the hypanthial tissue on the apical and in most cultivars on the basal part, too. In most cultivars, the nectariferous tissue continued along the style, too, where the accumulating nectar could stay in the flower for a longer time, thus increasing the possibility of infection by *E. amylovora*. The size of the nectary was bigger in non/less susceptible cultivars than in highly susceptible ones. No relationship was found, however, between the thickness of the nectary, the thickness of the glandular tissue and the degree of susceptibility. The nectary stomata were sunken below the level of epidermal cells, on the basis of which they belong to the xeromorphic ecological type. Thus, nectar can accumulate in the small cavity above the guard cells, ensuring a favorable environment for the growth of bacteria, provided that both sugar composition and concentration of nectar is optimal. The size of nectar chambers, located under nectary stomata, was usually larger in tolerant cultivars than in susceptible ones.

2.2.4.5 Wounds

So-called wound pathogens gain entry without forming specialized infection structures. They are unable to penetrate intact surfaces, even if supplied with nutrients. Wounds caused by farm operations, hail storms, or insects will help in the entry of different plant pathogens into the host cells. Organisms which cause storage diseases and ripe rots will enter through the wounds caused by farm operations. (*Example: Rhizopus, Gloeosporium, Aspergillus, Penicilium, Colletotrichum, Diplodia,* etc.). Weak parasites enter through the wounds caused by hail storms and freezing (*Example: Macrophomina phaseolina*). Pathogen causing brown rot of fruits (*Sclerotinia fructicola)* enters through the wounds caused by insect punctures. Similarly, the causal organism of Dutch elm disease (*Ceratostomella ulmi)* enters through the wounds caused by elm bark beetle.

Penicillium digitatum (blue mold of citrus) only infects fruit wounded during harvest or postharvest handling. Postharvest handling damage allows the entry of many fruit and vegetable rot pathogens, including *Rhizopus nigricans* and *Sclerotium rolfsii.* Wounds also occur naturally. Wounds caused by farm operations, hail storms, or insect punctures, etc., will help in the entry of different plant pathogens into the host cells. Organisms that cause storage diseases and ripe rots will enter through the wounds caused by farm operations (*Example*: *Rhizopus, Gloeosporium, Aspergillus, Penicilium, Colletotrichum, Diplodia,* etc.). The points of lateral root emergence enable the fungus *Chalara elegans* to enter tobacco roots. Leaf scars provide similar infection sites for the apple canker pathogen, *Nectria gattigena.* Damage also results from wind, hail, frost or extreme heat or from the effects of insects, herbivores or humans. Most plant viruses enter through wounds, such as those created by their insect vectors.

The mechanical wounding of plant leaves, which is one of the first steps in pathogen infection and herbivore attack, activates signal transduction pathways and airborne signals to fend-off harmful organisms. The mechanisms by which these signals promote plant immunity remain elusive (Yuri L. Dorokhov *et al.*, 2012). It was demonstrated that plant-leaf wounding results in the synthesis of a cell wall enzyme, pectin methylesterase (PME), causing the plant to release methanol into the air. Gaseous methanol or vapors from wounded PME-transgenic plants induced resistance to the bacterial pathogen, *Ralstonia solanacearum* in the leaves of nonwounded neighboring 'receiver' plants. To investigate the mechanism underlying this phenomenon, the methanol inducible genes (MIGs) were identified in *Nicotiana benthamiana,* most of which fell into the category of defense genes. Genes were selected and isolated, viz., *noncell-autonomous pathway protein (*NCAPP*),* β*-1,3-glucanase (*BG*),* and the previously unidentified MIG-21. It was later, demonstrated that BG, MIG-21, and NCAPP could enhance cell-to-cell communication and *Tobacco mosaic virus (*TMV) RNA accumulation. Moreover, gaseous methanol or vapors from wounded plants increased TMV reproduction in 'receivers.' Thus, methanol emitted by a wounded plant enhances antibacterial resistance as well as cell-to-cell communication that facilitate virus spreading in neighboring plants.

2.2.5 Colonization

A successful infection requires the establishment of a parasitic relationship between the pathogen and the host, once the host has gained entry to the plant (David Guest and John Brown, 1997). There are two broad categories of pathogens are biotrophs (those that establish an infection in living tissue) and necrotrophs (those that kill cells before colonizing them, by secreting toxins that diffuse ahead of the advancing pathogen) (Table 2.1). These two kinds of pathogens are also sometimes known as 'sneaks' and 'thugs,' because of the tactics they use to acquire nutrients from their hosts. The toxins produced by necrotrophs can be specific to the host or nonspecific. Nonspecific toxins are involved in a broad range of plant-fungus or plant-bacterial interactions, and will therefore not usually determine the host range of the pathogen producing them. Necrotrophs often enter the plant through wounds and cause immediate and severe symptoms. An intermediate category of the parasite is the hemibiotrophs, which start off as biotrophs and eventually become necrotrophic, employing tactics from both classes of the pathogen.

Table 2.1 Major Characteristics of Idealized Necrotrophic and Biotrophic Plant Parasites

Necrotrophs	*Biotrophs*
1. Obtain nutrients only from dead host cells	Obtain nutrients from living host cells
2. Kill host cells in advance of invasion	Sustain host cell viability during parasitism
3. Do not form specialized infection structures and cause immediate and severe symptoms	Produce specialized infection structures (e.g., haustoria) and cause only limited host damage
4. Often enter through wounds	Enter through natural openings or by direct penetration
5. Inoculum contains few food reserves	Inoculum contains rich food reserves
6. Competitive saprophytes	Poor saprophytes
7. Weakened, unthrifty hosts most susceptible	Fast-growing plants most susceptible
8. Wide host range *Example: Pythium* spp., *Phytophthora cinrnamomi*, *P. citrophthora*	Narrow, specialized host range *Example:* Powdery mildews, Downy mildews, Rusts, *Phytophthora infestans*, *Venturia inaequalis*, *Agrobacterium* spp., plant viruses, phytonematodes

Pathogens that colonize the surface of plants, extracting nutrients through haustoria in epidermal or mesophyll cells are termed ectoparasites. The haustoria are the only structures that penetrate the host cells. Some parasites colonize the area between the cuticle and the outer wall of the epidermal cells, penetrating host epidermal and mesophyll cells with haustoria. These are called 'subcuticular infections.' Pathogens can also form colonies deeper in the plant tissues. These are 'mesophyll and parenchyma infections,' and can be necrotrophic, hemibiotrophic or biotrophic relationships. Necrotrophs do not produce specialized penetration structures. Instead, they kill host cells by secreting toxins, and then degrade the cell wall and middle lamella, allowing their hyphae to penetrate the plant cell walls and the cells themselves. In hemibiotrophic infections, intercellular hyphae can form haustoria in living mesophyll cells, but as the lesion expands under favorable conditions, those heavily parasitized cells at the inner, older part of the colony collapse and die. A similar sequence of events can take place in plants infected by burrowing nematodes. Viruses, mildews and rusts develop specialized biotrophic relationships with their hosts. Intercellular hyphae of downy mildew colonize host mesophyll cells and form haustoria. The mildew sporulates and the infected cells eventually die, although necrosis is delayed and contained, compared to that caused by necrotrophic pathogens. Rust fungi can also delay senescence in infected cells while they sporulate. Vascular infections usually cause wilting and discoloration as a result of the physical blockage of infected xylem vessels. True vascular wilt pathogens colonize the vascular tissue exclusively, although other pathogens can cause the same symptoms if they infect the vascular system as well as other tissues. There are a few pathogens that manage to achieve systemic infection of their host. For example, many viruses can spread to most parts of the plant, although not necessarily all tissues. Some downy mildews can also systemically infect their host by invading the vascular tissue and growing throughout the host, causing deformation, rather than necrosis. Finally, there are some pathogens that complete their entire life cycle within the cells of their host, and may spread from cell to cell during cytokinesis. These are endobiotic infections.

The characteristics of idealized biotrophic and necrotrophic relationships are summarized in Table 2.1 (David Guest and John Brown, 1997). In reality, this distinction is not always clear. Every host-pathogen combination is unique, and few archetypal necrotrophs or biotrophs exist. Symbiotic fungi, such as mycorrhizal fungi and endophytes and symbiotic bacteria such as *Rhizobium* spp., represent highly coevolved examples of biotrophic parasitism. Most interactions are hemibiotrophic, initially biotrophic, but at later stages of parasitism, necrotrophic. Hemibiotrophs utilize parasitic strategies of both biotrophs and necrotrophs. *Phytophthora palmivora, Fulvia fulva,* and *Colletotrichum lindemuthianum* at first colonize and parasitize living cells. Nevertheless, these cells eventually collapse and die, triggering necrotrophic growth that often coincides with sporulation.

Infection of plants by pathogenic bacteria can generally be considered in terms of three interrelated phases, viz., population build-up, competition and migration of bacteria at the plant surface, bacterial entry into plant tissue and the migration of bacteria within the plant to and from regions of multiplication.

2.2.5.1 Population build-up

The presence of epiphytic pathogens on host plants does not imply that disease will necessarily develop, and many cases have been reported where quite high levels of pathogenic bacteria were present on symptomless foliage. This has been noted, for example, for *Pseudomonas syringae* pathovars on red maple and snap beans and for *Erwinia amylovora* on apple and pear blossom.

In other situations, the presence of epiphytic bacteria does lead to disease development. The presence of *Pseudomonas syringae* pv. *morsprunorum* as an epiphyte on cherry foliage leads to canker formation. The relationship between epiphytic occurrence and disease development has subsequently been investigated for a wide range of bacterial pathogens by monitoring naturally occurring populations and carrying out experimental inoculations of plant surfaces. These studies have shown that plant infection and disease development depend on a number of factors, including the particular host–pathogen combination, critical environmental conditions, physiological stress of the host plant and the attainment of minimal threshold levels of the pathogen.

The entry of bacteria into the plant during the infection process leads to various types of interaction, observable at the level of the whole plant, constituent tissues or individual cells. These interactions have been investigated experimentally by artificial infiltration of intact plants or by the use of *in vitro* systems.

2.2.5.2 Inoculation of intact plants

During the determination of the effect of different bacteria in determining the nature of the plant response, three main types of results are noticed, viz., hypersensitive reaction, where there is typically a rapid death of the plant cells, with no spread of bacteria to surrounding tissues. This reaction was induced by a range of bacteria comprising various pathovars of *Pseudomonas syringae.* The disease reaction involving a delayed host cell response, with the spread of bacteria to other parts of the plant, was induced by *Pseudomonas syringae* pv. *tabaci* and resulted in wildfire disease and no observable reaction, after infiltration of the saprophytic bacterium *Pseudomonas fluorescens.*

Although compatible phytopathogenic bacteria share a common ability to spread and multiply within the host plant, the manner in which they do this and the effect they have on the host plant (disease) vary considerably. This chapter considers general aspects of disease induction, different types of disease that are caused by plant pathogenic bacteria and the range of bacterial characteristics that are important in disease development.

The ability of plant pathogenic bacteria to cause disease in a particular host plant depends on many features, including environmental aspects, plant physiology and development, and the expression of pathogenicity and virulence factors by the bacterial cells. Environmental factors are important in the development of plant disease for their direct effects on infection and for their indirect effects in determining the physiological status of the plant. The various aspects of the plant that affect disease development include nutritional status, photoperiodic conditioning, and stage of maturity and development. Levels of macronutrients have been shown to be important in plant susceptibility to *Erwinia stewartii,* where elevated levels of N and P increased susceptibility and high levels of Ca and K increase resistance.

The ability of plant pathogenic bacteria to survive and multiply outside and inside plants, and to cause disease, is determined to a large extent by their genetic constitution. The genetic analysis of plant pathogenic bacteria currently involves the application of molecular techniques for the identification and investigation of bacterial genes that are important in all of these aspects, and will be considered first. Following sections discuss the role of specific genes and gene systems in the activity of plant pathogenic bacteria in relation to the determination of compatibility and incompatibility, disease virulence, and nonpathogenic characteristics. The final part of this chapter deals with the occurrence and role of plasmids in these bacterial cells.

2.2.5.3 *Molecular genetics: identification and investigation of bacterial genes*

Bacterial genes, occurring on either chromosomal or plasmid DNA, are involved in the determination of a wide range of phenotypic characteristics. In recent years, new techniques of molecular biology have been particularly successful in the genetic analysis of plant pathogenic bacteria and have been described in detail in a number of recent texts. The major objectives of molecular genetics are identification and isolation (cloning) of specific genes with defined functions.

In natural environments, where a particular host species occurs within mixed vegetation, the development, and spread of disease is probably limited to some extent by the separation of individual plants within the area. This constraint does not apply in the crop situation, where localized infection and progression of disease within the homogeneous plant population can occur rapidly. In this artificial situation, where the natural balance between pathogen and host does not apply, special control measures often have to be adopted if the large-scale occurrence of disease and consequent major crop loss are to be avoided. These measures fall into four main categories: chemical control, biological control, breeding of resistant cultivars and sanitary procedures.

2.2.5.4 *Disease physiology*

While necrotrophs have little effect on plant physiology since they kill host cells before colonizing them, biotrophic pathogens become incorporated into and subtly modify various aspects of host physiology, such as respiration, photosynthesis, translocation, transpiration and growth and development. The respiration rate of plants invariably increases following infection by fungi, bacteria or viruses. The higher rate of glucose catabolism causes a measurable increase in the temperature of infected leaves. An early step in the plant's response to infection is an oxidative burst, which is manifested as a rapid increase in oxygen consumption, and the release of reactive oxygen species, such as hydrogen peroxide (H_2O_2) and the superoxide anion (O^{2-}). The oxidative burst is involved in a range of disease resistance and wound repair mechanisms.

2.2.5.5 *Link to rapid active defense*

In resistant plants, the increase in respiration and glucose catabolism is used to produce defense-related metabolites via the pentose phosphate pathway. In susceptible plants, the extra energy produced is used by the growing pathogen. Pathogens also affect photosynthesis, both directly and indirectly. Pathogens that cause defoliation rob the plant of photosynthetic tissue, while necrotrophs decrease the photosynthetic rate by damaging chloroplasts and killing cells. Biotrophs affect photosynthesis in varying degrees, depending on the severity of the infection. A biotrophic infection site becomes a strong metabolic sink, changing the pattern of nutrient translocation within the plant and causing a net influx of nutrients into infected leaves to satisfy the demands of the pathogen. The depletion, diversion and retention of photosynthetic products by the pathogen stunts plant growth and further reduced the plant's photosynthetic efficiency. In addition, pathogens affect water relations in the plants they infect. Biotrophs have little effect on transpiration rate until sporulation ruptures the cuticle, at which point the plant wilts rapidly. Pathogens that infect the roots directly affect the plant's ability to absorb water by killing the root system, thus producing secondary symptoms such as wilting and defoliation. Pathogens of the vascular system similarly affect water movement by blocking xylem vessels. Growth and development, in general, are affected by pathogen infection, as a result of the changes in source-sink patterns in the plant. Many pathogens disturb the hormone balance in plants by either releasing plant hormones themselves, or by triggering an increase or a decrease in synthesis or degradation of hormones in the plant. This can cause a variety of symptoms, such as the formation of adventitious roots, gall development, and epinasty (the down-turning of petioles).

2.2.6 Growth and Development

Because pathogens affect the source-sink patterns in plants, normal growth and development are affected. In addition to the direct damage to plant tissue and nutrient deprivation, many pathogens disturb the hormone balance in plants. Plant growth and development depends on subtle shifts in the balance between auxin, gibberellin, cytokinin, ethylene and abscisic acid. Pathogens may disturb the hormone balance by releasing plant hormones or by inducing increased or decreased synthesis or degradation of plant hormones. Gibberellic acid, the first plant hormone described, was discovered during a study of bakanae, or 'foolish seedling' disease of rice. Infected seedlings are unusually tall, and flowering is suppressed. Following the discovery in the 1930s that cell-free culture filtrates of the pathogen, *Gibberella fujikuroi,* cause the same symptoms, the active chemical was identified as gibberellic acid. Gibberellic acid was subsequently shown to be produced by plants, where it naturally functions as a plant hormone. Epinasty, the down turning of petioles, is a symptom typical of many systemic bacterial infections such as bacterial wilt caused by *Ralstonia solanacearum.* The fact that the addition of ethylene can also induce this symptom suggests that this plant stress hormone is involved in symptom development. Green islands formed around rust pustules result from sustained cytokinin levels in infected cells. Cytokinin generates a metabolic sink and maintains cell viability in parasitized cells while nonparasitized cells surrounding the pustule senesce and die prematurely.

Hormonal disturbances are also involved in the appearance of witches' broom and broomsticking symptoms caused by phytoplasmas (e.g., big bud diseases) and some fungi, such as *Crinipellis perniciosa* (witches' broom of cocoa), and *Oncobasidium theobromae* (vascular-streak dieback of cocoa). Vascular wilt infections upset auxin and ethylene balances in plants, inducing epinasty and the formation of adventitious roots on the lower stem. Galls and club-roots caused by fungi, nematodes, insects, and bacteria develop because of excessive cell growth (hypertrophy) and division (hyperplasia). Perhaps the best-studied example of symptoms associated with hormonal disturbance is crown gall, a disease of a wide range of plants caused by the bacterium *Agrobacterium tumefactens*. Genes from the pathogen, after incorporation into the plant genome, condition the overproduction of hormones that induce cell division and gall formation. Giant cells formed in roots infected by the root knot nematode, *Meloidogyne* species result from the breakdown of cell walls as well as excessive cell growth.

Members of the kingdom fungi comprise numerous plant pathogens, including the causal agents of many agriculturally relevant plant diseases such as rust, powdery mildew, rice blast and cereal head blight. Data from recent sequencing projects provide deep insight into the genomes of a range of fungi that infect different organs of monocotyledonous or dicotyledonous hosts and that have diverse pathogenic lifestyles. These studies have revealed that, similar to sequenced phytopathogenic oomycetes, these plant parasites possess very plastic and dynamic genomes, which typically encode several hundred candidate secreted effector proteins that can be highly divergent even among related species. A new insight is the presence of lineage-specific genes on mobile and partly dispensable chromosomes that are transferred intraspecifically and possibly interspecifically, thereby constituting pathogenicity and host range determinants. Convergent lifestyle-specific adaptations have shaped the parasite genomes to maximize pathogenic success according to the different infection strategies employed (Schmidt and Panstruga, 2011).

To exploit plants as living substrates, biotrophic fungi have evolved remarkable variations of their tubular cells, the hyphae. They form infection structures such as appressoria, penetration hyphae, and infection hyphae to invade the plant with minimal damage to host cells. To establish compatibility with the host controlled secretory activity and distinct interface layers appear to be essential. *Colletotrichum* species switch from initial biotrophic to necrotrophic growth and are amenable to mutant analysis and molecular studies. Obligate biotrophic rust fungi can form the most specialized hypha: the haustorium. Gene expression and immunocytological studies with rust fungi support the idea that the haustorium is a transfer apparatus for the long-term absorption of host nutrients (Mendgen and Hahn, 2002).

Pathogenesis is a complex progress involving the pathogen's inherent capabilities and multiple factors that govern penetration and infection of a host plant. The host plant has various defense mechanisms that must be inactivated or nullified before a disease can be developed (Robab Davari, 2012). This interaction between host and pathogen also depends on the environmental factors. The pathogen's invasion mechanisms include producing the cell wall and middle lamella dissolving enzymes, toxins, enzymes to degrade host tissue and defense substances, and rapidity in host plant infection. *S. sclerotiorum* facilitates penetration and colonization of host plant tissues by secreting multiple pectinolytic enzymes. In sunflower, *Sclerotinia sclerotiorum* produces the appressoria and the hyphal strands branched upon contact of the pathogen with the host surface. Moreover, a direct penetration of fungal hyphae was observed through the cuticle within 12 hours of inoculation. Microscopic observation of inoculated tissues after 24 hours revealed that fungal hyphae have developed both inter and intracellular layer. Moreover, the fungal hyphae growth was incremental among and inside the host cells. The host cells were completely colonized by fungal mycelium 48 hours after inoculation, leading to a tissue collapse. The hyphae invaded the dead host tissues and emerged from the host tissue after extensive colonization. The present study has detailed the infection processes and pathogen development both at plant surface and inside the host tissues.

2.2.7 Exit of Pathogen

After invasion and colonization of the host, the pathogens come out of the host to maintain the continuity of the infection chain or disease cycle and escape death due to overcrowding. Once the pathogens exit from the host, they survive and are disseminated to other hosts and continue the infection cycle. Viruses can exist only with the living protoplasm and hence disseminated through their animate vectors like insects, fungi, and nematodes. The bacteria ooze out in the form of slime on the host surface from where they can be disseminated through water and insects. However, the fungi have the most elaborate system of exit, grow out on the host surface and produce repeating spores (secondary inoculum), usually asexually, under favorable conditions. The spores thus formed are disseminated through the wind, water, soil, seed, and vegetative propagating material and agricultural implements. Phytonematodes, however, come out of dead tissues, either infest another healthy tissue/plant part or survive in the soil.

Dothideomycetes is the largest and most ecologically diverse class of fungi that includes many plant pathogens with high economic impact. Currently 18 genome sequences of *Dothideomycetes* are available, 14 of which are newly described allowing unprecedented resolution in comparative analyses (Ohm *et al.*, 2012). These 18 organisms have diverse lifestyles and strategies of plant pathogenesis. Three feed on dead organic matter only, six are necrotrophs (killing the host plant cells), one is a biotroph (forming an association with and thus feeding on the living cells of the host plant cells), and 8 are hemibiotrophs (having an initial biotrophic stage, and killing the host plant at a later stage). These various lifestyles are also reflected in the gene sets present in each group. For example, sets of genes involved in carbohydrate degradation and secondary metabolism are expanded in necrotrophs. Many genes involved in pathogenesis are located near repetitive sequences, which are believed to speed up their evolution. Blocks of genes with conserved gene order were identified. In addition to this we deduce that the mechanism for mesosynteny, a type of genome evolution particular to *Dothideomycetes*, is by intrachromosomal inversions.

The genome and transcriptome analyses of *Colletotrichum higginsianum* infecting *Arabidopsis thaliana* and *Colletotrichum graminicola* infecting maize was reported by Richaard J O'Connell *et al.* (2012). Comparative genomics showed that both fungi have large sets of pathogenicity-related genes, but families of genes encoding secreted effectors, pectin-degrading enzymes, secondary metabolism enzymes, transporters, and peptidases are expanded in *C. higginsianum*. Genome-wide expression profiling revealed that these genes are transcribed in successive waves that are linked to pathogenic transitions: effectors and secondary metabolism enzymes are induced before penetration and during biotrophy, whereas most hydrolases and transporters are upregulated later, at the switch to necrotrophy. These findings showed that preinvasion perception of plant-derived signals substantially reprograms fungal gene expression and indicate previously unknown functions for particular fungal cell types.

Bacteria have many export and secretion systems that translocate cargo into and across biological membranes. Seven secretion systems contribute to pathogenicity by translocating proteinaceous cargos that can be released into the extracellular milieu or directly into recipient cells. Chang *et al.* (2014) described these secretion systems and how their complexities and functions reflect differences in the destinations, states, functions, and sizes of the translocated cargos as well as the architecture of the bacterial cell envelope. They examined the secretion systems from the perspective of pathogenic bacteria that proliferate in plant tissues and highlight examples of translocated proteins that contribute to the infection and disease of plant hosts.

2.2.8 Major Steps in Pathogenesis

Fungi must find the host and the appropriate entry site, develop specialized infection structures, overcome preformed and induced host defenses, take up nutrients, grow and colonize host tissue.

2.2.8.1 Finding the host

It may be either passive vs. active and involves steps like wait in favorable site for the host; dispersal by animal vectors and use of dispersal stages like spores, zoospores, etc.

2.2.8.1.1 Chemotaxis of Phytophthora zoospores

Tactic response allows zoospores to reach infection sites in roots of host plants, positive chemotaxis towards the plant, compounds (*Example: P. sojae* to isoflavones), electrotactic responses also noted (weak electric fields generated by roots). *Phytophthora* and *Pythium* zoospores swim towards amino acids and sugars. *Phytophthora palmivora* marked is attracted to the root tip (electrotaxis) while *Pythium aphanidermatum* is attracted to wounds (Sophien Kamoun, 2010).

2.2.8.2 Developmental processes and infection

Morphological changes are essential to penetration and adaptation to plant issue. Many fungal/oomycete pathogens develop highly specialized infection structures in response to signals such as contact with the plant surface and plant tissue environment. Most important structures are appressoria and haustoria. Not all fungal plant pathogens produce specialized infections structures. Forcible entry is seen in rice blast fungus. A spore from *Magnaporthe grisea* has germinated on the surface of a rice leaf and formed a dome-shaped appressorium. The appressorium has to breach the thin but tough rice leaf cuticle to invade the leaf and cause disease (Talbot, 1999).

2.2.8.3 Penetration of plant epidermis

Infection process by appressoria can involve enzymatic action; external matrix around appressoria contains cutinase, cellulases, and other enzymes to help soften the cuticle, thereby aiding adhesion and penetration. However, some fungi can physically force their way through plant cuticles. Fungi such as *Colletotrichum* and *Magnaporthe* produce appressoria with tough melanin-pigmented cell walls. Appressoria generate pressures between 6-8 Mega Pa (30-40x pressure of a car tire), produce hydrostatic turgor by accumulating molar concentrations of glycerol, and can puncture artificial plastic membranes. Melanin mutants are nonpathogenic and cannot accumulate turgor pressure.

2.3 INFECTION BY BACTERIA

Bacteria can enter through cuts our other areas of damage that occur in the leaves or stems due to wind bending the plant, objects hitting the plant, aphids sucking the plant sap or animals grazing on it. The symptoms then depend on which internal tissues and structures where the infection takes a hold. Although thousands of species of bacteria can cause disease in animals and humans, a much smaller number, probably about one hundred, are able to infect plants and damage them. Bacterial plant diseases are generally due to bacilli, bacteria that are shaped by rods, some of which are Gram-negative bacteria and some of which are Gram-positive. The majority of plant diseases due to bacterial infection occur in parts of the world that have a tropical weather pattern.

Once a species of pathogenic bacteria gains entry to a plant it causes one of four main problems that leads to a less than a healthy plant. Some bacteria produce enzymes that break down the cell walls of plants anywhere in the plant. They do this to break open to the cell to gain access to the nutrients inside, but the plant cells affected die quickly, causing parts of the plant to start rotting. This is why plant diseases that develop due to cell wall degrading enzymes are generally known as a type of 'rot.' Some bacteria produce toxins that are damaging to plant tissues generally, usually causing early death of the plant. Others produce large amounts of polysaccharide sugars that have long chains and are very sticky. As these travel in the water carrying vessels, the xylem, they block the narrow channels, preventing water getting from the plant roots up to the shoots and leaves, again causing rapid death of the plant. Finally, some bacteria produce proteins that mimic plant hormones. These lead to overgrowth of plant tissue and tumors form. These grow rapidly, taking up valuable nutrients and energy resources that the plant would otherwise use for its own tissue growth, and it becomes weakened and susceptible to attack by other pathogens such as fungi.

Infection of plants by pathogenic bacteria can generally be considered in terms of three interrelated phases, viz., population build-up, competition and migration of bacteria at the plant surface; bacterial entry into plant tissue and migration of bacteria within the plant to and from regions of multiplication (David C Sigee, 2009).

2.3.1 Build-up and Activity of Epiphytic Populations

The presence of epiphytic pathogens on host plants does not imply that disease will necessarily develop, and many cases have been reported where quite high levels of pathogenic bacteria were present on symptomless foliage. This has been noted, for example, for *Pseudomonas syringae* pathovars on red maple and snap beans and for *Erwinia amylovora* on apple and pear blossom. In other situations, the presence of epiphytic bacteria does lead to disease development. The relationship between epiphytic occurrence and disease development has subsequently been investigated for a wide range of bacterial pathogens by monitoring naturally occurring populations and carrying out experimental inoculations of plant surfaces. These studies have shown that plant infection and disease development depend on various factors including the particular host—pathogen combination, critical environmental conditions, physiological stress of the host plant and the attainment of minimal threshold levels of the pathogen.

Pathogenic bacteria establishing compatible interactions with plants also obtain nutrients from the host upon colonization. This process, which often involves the participation of hydrolytic enzymes and toxins, provokes plant injury, disease or even death. In spite of these different outcomes, accumulating evidence suggests that rhizobia and plant pathogens use similar 'weapons' to invade the host (Maria J Soto, Juan Sanjuan and Jose Oliverares, 2006).

Aerial plant organs, particularly leaf tissues, provide an important niche for microbial colonization referred to as the phyllosphere. This niche supports the survival and, often, the proliferation of microbes such as fungi and bacteria with diverse lifestyles including epiphytes, saprophytes, and pathogens. Foliar pathogens of plants face unique challenges in interacting with and exploiting their hosts (William Underwood *et al.*, 2007). Upon introduction to leaf surfaces, potential pathogens must survive an initial epiphytic phase on the surface of the leaf, during which they are likely exposed to a harsh environment and multiple stresses including desiccation, UV irradiation and nutrient limitation (Beattie and Lindow, 1995; Lindow and Brandl, 2003). Epiphytes complete their life cycles on the surface of leaves, but foliar pathogens must enter the leaf tissue to initiate pathogenesis. Entry into host tissue is likely a critical first step in the establishment of infection. To gain access to the intercellular spaces and internal leaf tissues, microbial pathogens must cross the cuticle and epidermis. Many fungal plant pathogens have the ability to directly penetrate the epidermis using cuticle- and cell wall-degrading enzymes, mechanical force, or both. Bacteria, however, cannot directly penetrate the leaf epidermis and must enter leaf tissues through natural openings or wounds on the leaf surface.

Stomata dominate in number among natural openings in the aerial part of the plant. Other natural openings include hydathodes in leaf margins, nectarthodes in flowers, and lenticels in stems and roots (Huang, 1986). Entry of bacteria into leaf tissues through natural openings has been generally perceived as a passive event, where bacteria lack active mechanisms for gaining entry, and plants similarly lack active mechanisms for preventing entry. This perception, along with the widespread use of popular laboratory inoculation techniques that deliver bacteria directly into the leaf intercellular space between mesophyll cells, has contributed to our lack of knowledge regarding the transition from an epiphytic lifestyle on the leaf surface to a pathogenic lifestyle within the leaf intercellular space. However, recent evidence suggests that entry of bacteria into leaf tissue is more complex and dynamic than the simple act of swimming into the leaf through passive openings (Melotto *et al.*, 2006).

2.3.2 Inoculation of Intact Plants

The effect of different bacteria in determining the nature of the plant response was initially demonstrated by Klement *et al.* (1964), who artificially infiltrated leaves of tobacco with a range of bacterial species and observed three main types of results, viz., hypersensitive reaction, where there is typically a rapid death of the plant cells, with no spread of bacteria to surrounding tissues. This reaction was induced by a range of bacteria comprising various pathovars of *Pseudomonas syringae*. The disease reaction involving a delayed host cell response with the spread of bacteria to other parts of the plant was induced by *Pseudomonas syringae* pv. *tabaci* and resulted in wildfire disease and, finally, no reaction was observed after infiltration of the saprophytic bacterium *Pseudomonas fluorescens*.

Although compatible phytopathogenic bacteria share a common ability to spread and multiply within the host plant, the manner in which they do this and the effect they have on the host plant (disease) vary considerably. This chapter considers general aspects of disease induction, different types of disease that are caused by plant pathogenic bacteria and the range of bacterial characteristics that are important in disease development.

2.3.3 Induction of Bacterial Disease

The ability of plant pathogenic bacteria to cause disease in a particular host plant depends on many features, including environmental aspects, plant physiology and development, and the expression of pathogenicity and virulence factors by the bacterial cells.

2.3.4 Environmental and Physiological Factors Affecting Disease Development

Environmental factors are important in the development of plant disease for their direct effects on infection) and for their indirect effects in determining the physiological status of the plant. The various aspects of the plant that affect disease development have been well discussed and include nutritional status, photoperiodic conditioning, and stage of maturity and development. Levels of macronutrients have been shown to be important in plant susceptibility to *Erwinia stewartii*, where elevated levels of N and P increased susceptibility, and high levels of Ca and K increased resistance.

The ability of plant pathogenic bacteria to survive and multiply outside and inside plants, and to cause disease, is determined to a large extent by their genetic constitution. The genetic analysis of plant pathogenic bacteria currently involves the application of molecular techniques for the identification and investigation of bacterial genes that are important in all of these aspects, and will be considered first. Following sections discuss the role of specific genes and gene systems in the activity of plant pathogenic bacteria in relation to the determination of compatibility and incompatibility, disease virulence, and nonpathogenic characteristics. The final part of this chapter deals with the occurrence and role of plasmids in these bacterial cells.

2.3.5 Molecular Genetics – Identification and Investigation of Bacterial Genes

Bacterial genes, occurring on either chromosomal or plasmid DNA, are involved in the determination of a wide range of phenotypic characteristics. In recent years, new techniques of molecular biology have been particularly successful in the genetic analysis of plant pathogenic bacteria and have been described in detail in a number of recent texts. The major objectives of molecular genetics include identification and isolation (cloning) of specific genes with defined functions.

In natural environments, where a particular host species occurs within mixed vegetation, the development, and spread of disease is probably limited to some extent by the separation of individual plants within the area. This constraint does not apply in the crop situation, where localized infection and progression of disease within the homogeneous plant population can occur rapidly. In this artificial situation, where the natural balance between pathogen and host does not apply, special control measures often have to be adopted if the large-scale occurrence of disease and consequent major crop loss are to be avoided. These measures fall into four main categories: chemical control, biological control, breeding of resistant cultivars and sanitary procedures.

To exploit plant nutrients, phytopathogenic bacteria have evolved sophisticated infection strategies (Roberto Buonaurio, 2008). A number of biotrophic Gram-negative bacteria are able to cause diseases in plants through a type III secretion system, composed of a protruding surface appendage, known as the hrp pilus, through which the bacterium injects effector proteins into the host cells to manipulate plant cells, in particular, to suppress plant defenses. A unique infection strategy in plant-bacterial interactions is adopted by *Agrobacterium tumefaciens*, which genetically transforms its host by transferring T-DNA from its tumor-inducing plasmid to the chromosome of a plant cell. This transfer is mediated by the pilus T, belonging to the type IV secretion system. Other important virulence factors of phytopathogenic bacteria are phytotoxins, extracellular polysaccharides, phytohormones and plant-cell-wall-degrading enzymes. These latter factors are fundamental to the pathogenesis of necrotrophic bacteria. Plants defend themselves from bacterial attacks through a multilayered system of passive and active defense mechanisms, which can interfere with entry of bacteria into the plant tissue and restrict bacterial growth when the ingress has been gained. A number of hypotheses have been advanced to explain the restriction of the bacterial growth observed during the hypersensitive reaction and systemic acquired resistance development.

Except for a few rare cases, phytopathogenic bacteria provoke diseases in plants by penetrating into host tissues. Penetration occurs through natural openings, such as stomata, hydathodes, lenticels, nectarthodes, stigma, etc., or through wounds (Scurtleff and Averre, 1997). Bacteria colonize the apoplast that is the intercellular spaces or xylem vessels, causing parenchymatous and vascular or parenchymatous-vascular diseases, respectively. Besides the endophytic habitat, some bacterial species also have the capacity to survive as epiphytes on plant surfaces (phylloplane, rhizoplane, carpoplane, etc.). From an epidemiological point of view, this poses a particular danger because they are quick to infect plants in favorable conditions. Once inside plant tissues, bacteria may implement two main attack strategies to exploit the host plant nutrients: biotrophy, in which the plant cells are kept alive as long as possible, and bacteria extract nutrients from live cells, and necrotrophy, in which bacteria kill plant cells and extract nutrients from dead cells. Interactions of bacteria with plants can be either incompatible or compatible. Incompatible interactions occur when the bacterium encounters a nonhost plant (nonhost resistance) or a resistant host plant (cultivar-specific resistance) and they are frequently associated with a hypersensitive response, i.e., a rapid, programmed death of plant cells that occurs at the site of infection. Compatible interactions occur when the bacterium infects susceptible host plants causing disease symptoms.

A bacterium could be pathogenic yet have varying degrees of virulence. Pathogenicity and/or virulence of Gram-negative plant pathogenic bacteria are strictly dependent on the presence of secretion apparatuses in their cells, through which they secrete proteins or nucleoproteins involved in their virulence in the apoplast or inject into the host cell. To date, five secretion systems, numbered from I to V have been described in both animal and plant Gram-negative pathogenic bacteria. In addition, it is now emerging that the expression of virulence factors is under the control of quorum sensing, a communication mechanism by which bacteria regulate the expression of certain genes in response to their population density.

The pathogenicity of a number of biotrophic Gram-negative bacteria in the genera *Pseudomonas, Xanthomonas, Ralstonia, Erwinia,* and *Pantoea* is mainly due to their ability to produce a type III secretion system (T3SS), also called injectisome (Desvaux *et al.*, 2004), by which the bacterium injects proteins involved in its virulence into plant cells (T3SS effectors).

It is worth remembering that if the effector leads to the development of disease symptoms in the plant it is called a virulence protein, while if it triggers a defense reaction that leads to the HR, it is referred to as an avirulence protein. The T3SS is encoded by hrp (HR and pathogenicity) and hrc (HR and conserved) genes, whose mutations eliminate bacterial pathogenicity in susceptible host plants and the ability to elicit HR in nonhost or cultivar-specific resistant plants.

2.3.6 Infection Process *in Agrobacterium tumefaciens*

The particular infection modality of *Agrobacterium tumefaciens*, the causal agent of crown gall in many woody and herbaceous dicotyledonous plants, is worthy of separate discussion. This Gram-negative soil bacterium is able to genetically transform healthy host cells in tumoral cells by inserting a part of its DNA (transferred DNA; T-DNA), contained on the tumor-inducing (Ti) plasmid, into the plant genome (Sheng and Citovsky, 1996). T-DNA carries genes involved in the synthesis of plant growth hormones (auxins and cytokinins) and synthesis and secretion of amino compounds called opines, tumor-specific compounds that are exclusively assimilated by the pathogen as major carbon and nitrogen sources. The actively growing tumoral cells for the high plant growth hormone levels and the opine they produce render the tumor a favorable biological niche for *A. tumefaciens* growth.

At least three genetic components are required for tumorigenesis: (i) T-DNA; (ii) the virulence (*vir*) region located on the Ti plasmid, which genes respond to specific plant-released signals, to generate a copy of the T-DNA and mediate its transfer into the host cell; and (iii) a suite of chromosomal virulence (*chv*) genes, involved in bacterial chemotaxis toward-and attachment to the wounded plant cell wall (Sheng and Citovsky, 1996). In natural infections, *A. tumefaciens* cells present in the soil, reach wounds (the main infection sites) of roots and the subterranean part of the stem by the means of flagella, attracted by the compounds (e.g., sugars) released from plant wounds. The importance of motility in the infection process is illustrated by the fact that mutations in genes encoding flagellin abolish motility and reduce tumorigenesis (Chesnokova *et al.*, 1997; Hawes and Smith, 1989).

Binding of agrobacteria to plant surfaces, an essential stage for establishing a long-term interaction with the host, is dependent on several plant factors and takes place in two steps (Matthysse and Kijne, 1998). The first, rather weak and reversible, may involve a variety of bacterial polysaccharides (e.g., cyclic glucans synthesized by chvA and chvB), which bind host polysaccharides. Mutations in the *chv* genes reduced the binding of the agrobacteria to cultured plant cells and abolished tumorigenesis (Brencic and Winans, 2005). The second binding step requires the synthesis of bacterial cellulose, which causes a tight, irreversible binding and formation of bacterial aggregates on the host surface (Matthysse, 1983.)

Once inside the host-cell cytoplasm, the T-DNA is thought to exist as a mature T-complex (T-complex), in which the entire length of the T-strand molecule is coated with numerous VirE2 molecules, which confer to the T-DNA the structure and protection needed for its journey to the host-cell nucleus (Tzfira and Citovsky, 2006). There is evidence that the T-complex, like many DNA viruses, uses the plant cytoskeleton as a track for its subcellular movement toward the nucleus. Once inside the nucleus, the T-complex needs to travel to its integration point and be uncovered from its escorting proteins before integration into the host genome. VIP1 interacts with the H2A histone, a plant chromatin protein essential for T-DNA integration, suggesting it is involved in targeting the T-complex to the site of integration. Furthermore, there is evidence that the stripping of the T-complex is due to proteasomal degradation. The molecular mechanism underlying T-DNA integration is still under debate. One hypothesis is that host DNA repair machinery, by converting the T-strand molecule to double-stranded (ds) T-DNA, may recognize these molecules as broken DNA fragments and incorporate them into the host genome (Tzfira and Citovsky, 2006).

2.3.7 Other Bacterial Virulence Factors

The severity of disease symptoms caused by bacteria in plants is determined by the production of a number of virulence factors, including phytotoxins, plant-cell-wall-degrading enzymes, extracellular polysaccharides, and phytohormones. In general, phytopathogenic bacterial strains mutated in any virulence factor more or less reduce their virulence, while their pathogenicity remains unchanged.

2.4 INFECTION PROCESS IN PLANT VIRUSES

Plant viruses are biotrophic pathogens that need a living tissue for their multiplication and thus, in the infection–defense equilibrium, they do not normally cause plant death. In some instances, virus infection may have no apparent pathological effect or may even provide a selective advantage to the host, but in many cases it causes the symptomatic phenotypes of disease (Viente Pallas and Juan Antonio Garcia, 2011). These pathological phenotypes are the result of interference and/or competition for a substantial amount of host resources, which can disrupt host physiology to cause disease. This interference/competition affects a number of genes, which seems to be greater the more severe the symptoms that they cause. Induced or repressed genes belong to a broad range of cellular processes, such as hormonal regulation, cell cycle control and endogenous transport of macromolecules, among others. In addition, recent evidence indicates the existence of an interplay between plant development and antiviral defense processes, and that interference among the common points of their signaling pathways can trigger pathological manifestations.

More than one thousand viruses are currently known to be potentially capable of infecting plants. Despite a large number of possible combinations, the development of the disease is an exception rather than a common outcome and thus, in most cases, plants are capable of counteracting the harmful effects of viruses. This resistance is owed to the absence of essential host susceptibility factors (passive resistance) or to the existence of several defense layers that the virus has to overcome. First, the virus needs to overcome a series of preexisting physical and chemical barriers in plants. If a pathogenic virus succeeds in overcoming this first line of defense, it would have to face the nonspecific defensive reactions with which the plant responds to some molecular patterns that are common to different pathogens (Jones and Dangl, 2006). If a virus has evolved to acquire virulence factors to counteract this basal defense, it is in a position to be able to trigger infection. In many cases, however, plants are able to recognize these virulence factors and create a new, more specific resistance layer that is only induced when faced with viruses expressing this virulence factor. A virus can cause productive infection only in those plants that have not developed specific defensive responses to its virulence factors.

A virus not only needs to escape the defenses that plants erect but must also tackle different processes to complete its productive cycle (Maule *et al.*, 2002). The initiation of this cycle depends on the nature of the genetic material of the virus. Positive-polarity RNA viruses are the most abundant in the plant kingdom. For these viruses, genomic RNA must be uncoated and translated after viral particles have entered the plant cell, and both processes are highly coordinated. There also seems to be some kind of coupling between the synthesis of viral proteins and the assembly of some of these proteins with genomic RNA and host factors to form replication complexes. The next stage of the virus cycle entails its movement to neighboring cells and its dissemination throughout the plant. Interactions of viral and cellular factors may not only contribute to facilitate these viral infection steps and help to establish optimum infection susceptibility conditions but may also indirectly affect host physiological processes. Although many viral infections progress efficiently without symptom development, induction of plant defense mechanisms, their suppression by counteracting viral strategies and the cooption of host factors required for virus replication, and movement can confer a pathological character upon the viral infection. There is experimental evidence of the individual contribution of these elements in different viral infections; nonetheless, a model that includes them in the specific development of a particular pathology is lacking (Culver and Padmanabhan, 2007). This review does not intend to explain how different viral plant diseases develop but to describe some specific examples of viral and plant factors that contribute to viral pathogenesis.

Since the possible interference of viruses with cellular transport processes is a potentially effective form of altering plant physiology, it is intuitive to believe that the translocation of viruses among cells throughout the plant body strongly influences the pathogenesis process. Virus multiplication and movement are necessary for the symptoms of the disease to develop. Thus, the rate and extent to which these processes occur can be primary determinants of symptom development. The infective viral cycle in a susceptible host mostly begins through epidermal cells, or through roots, as a result of either mechanical damage or being assisted by biological vectors (e.g., insects, nematodes, fungi, etc.). Once the first viral genome replication cycles have been completed, the progeny viruses must be capable of translocating from one cell to another until they reach the vascular system, through which the viruses could invade the distal plant parts. The first of these phases is known as local or cell-to-cell movement, and the second is called systemic or vascular virus movement (Waigmann *et al.*, 2004). In both types of movement, but especially in the first one, the involvement of the virus-encoded movement proteins (MP) is essential (Fernadez-Calvino *et al.*, 2011). MPs can act by forming ribonucleoprotein complexes with the viral genome or tubular structures that hold virions to allow them to cross plasmodesmata (Lucas, 2006). One of the most common ways to restrict the invasion of a given virus is to block its cell-to-cell and long-distance movements. Thus, alterations in the viral movement function have a direct effect on the symptomatology. Most of the data that correlate this function with pathogenesis originate from studies conducted with natural or artificial mutants of the corresponding MPs, or with pseudorecombinant viruses.

Viruses need living tissue for their multiplication and thus do not normally cause the death of the host, although there are exceptions. A large body of evidence has recently shown that to accomplish their life cycle; plant viruses need to confront plant defense mechanisms and to hijack the functions of different host factors. As a consequence, viral components must interact and/or interfere with host components that, in turn, in some instances would cause an alteration in the plant physiology resulting in the development of symptoms. Indeed, recent discoveries have evidenced that plant development is affected by plant–virus interactions, which interfere with a broad range of cellular processes, such as hormonal regulation, cell cycle control and endogenous transport of macromolecules, among others. One important landmark along this line of experimentation has been the demonstration of an interplay between plant development and antiviral defense processes, and that the interference among the common points of their signaling pathways can trigger pathological manifestations (Chapman *et al.*, 2004; Gomez *et al.*, 2009; Kassachau *et al.*, 2003).

2.4.1 Infection Cycle

2.4.1.1 Entry

The entry process includes attachment of virions to a cell surface and a subsequent step for the virus to enter the cell or for viruses to be directly deposited into the cytoplasms in the case of plant viruses. Various ways viruses can enter into host plant cells which are briefed below.

2.4.1.1.1 Insect feeding

Most viruses are transmitted by animals bites (most of the times, insect feedings). An insect can pick up viruses from an infected plant and transmitted to a healthy individual. Other organisms such as nematodes, fungi and occasionally parasitic plants can also act as vectors for plant viruses.

2.4.1.1.2 Mechanical injuries

Mechanical transmission refers to virus infection through wound openings that are mechanically introduced, such as farming machinery, rainstorm and hail storms with closely spaced plants. In the laboratory, such wounds can be induced by brushing the plant surface with an abrasive. A mechanical transmission is probably less important for most viruses in nature than for experimental studies of plant viruses. In laboratory and greenhouses, abrasive materials such as carborundum and diatomaceous earth (Celite) are used to make wounds on the leaf surface.

2.4.1.1.3 Seeds and other propagation materials

Some viruses can be passed on from one plant to its progenies through either seeds or pollens. Other propagation materials such as buds for grafting can also pass viruses along.

2.4.2 Gene Expression

2.4.2.1 Uncoating

Uncoating process strips away the protective shell of the viruses and exposes viral genetic information to host transcription and translation machinery. The process is not well understood. Presumably factors in host cells (lysosomes, local pH, and local ionic concentration) are responsible for the removal of viral capsid proteins. In some instances, capsid proteins of the tobacco mosaic virus were shown to be removed by host ribosomes. Initially, a few capsid protein subunits were removed to allow ribosomes to engage the viral RNA. As the ribosomes translate the viral RNA, they strip away the rest of the capsid protein and eventually disassemble the virus particles. This process is called 'cotranslational disassembly.'

2.4.2.2 Translation (protein synthesis) and transcription (mRNA synthesis)

Genes necessary for virus life cycle are expressed. Some of the genes are expressed before the viral replication (early genes), and others are expressed later (late genes). Depending on the genome organization of various viruses, translation or transcription may occur before the other:

2.4.2.2.1 Translation before transcription

Genomes of all the (+)-strand RNA viruses are in fact mRNA themselves. They can be translated directly by the host ribosomes to produce viral proteins. Viral genomes are extremely small, and yet they have to pack enough genetic materials to complete the infection cycle of viruses. Thus, several genes may be encoded on a single viral RNA. However, eukaryotic mRNA is monocistronic in nature, i.e., only the first open reading frame is translated. To circumvent this restriction, viruses utilize various strategies to package and express their genomes. These include genome segmentation: Viral genome is broken into several pieces. Each piece of the genome contains one or a few open reading frames (ORFs) and effectively becomes a monocistronic mRNA; terminator read-through and ribosomal frameshifting that allows translation of two consecutive genes to a larger fusion protein and proteolytic processing (polyprotein cleavage). An ORF encoding several genes is expressed as a large polyprotein, which is then cleaved into smaller, functional proteins by an array of proteases encoded by viruses.

Subgenomic RNA (mRNA) synthesis: Some viral genes located internally in the viral RNA are not accessible to the host translation machinery. Thus, they need to be transcribed from the genomic RNA into subgenomic RNAs (mRNA) for gene expression.

2.4.2.2.2 Transcription before translation

For dsRNA viruses, (–)-strand RNA viruses, and DNA viruses, the first step in the gene expression is transcription. mRNAs encoding virus-specific proteins are first made and then serve as templates for translation. Dictated by the nature of viral genomes, these viral genomes cannot serve as mRNA directly. Therefore, an intermediate process (transcription) is taken before viral genes can be expressed. Transcription of most DNA viruses is accomplished entirely by host DNA-dependent RNA polymerase. However, all ds- or (-) RNA viruses must carry the necessary enzymes within the virus particles. These enzymes transcribe mRNA from the viral ds- or (–) RNA.

2.4.2.3 Replication

Replication of viruses is as diverse as viruses themselves. Viral replication varies depending on the type of viral genomes.

2.4.2.4 dsDNA viruses that replicate through dsDNA

None of plant viruses are true dsDNA viruses in this sense. However, this group contains numerous viruses that infect various organisms. Replication of the viruses is in all cases by the semiconservative method favored by cellular genomes. Among the viruses of Eukaryotes, replication mainly occurs in the nucleus, using cellular enzymes such as polymerases, methylases, etc.

2.4.2.5 ssDNA viruses that replicate through ds DNA intermediates:

Two plant virus groups, geminiviruses and banana bunchy top virus contain circular, single-stranded (ss) DNA. Replication of all of the viruses requires formation of a 'replicative form' (RF) double-stranded DNA intermediate: this is formed soon after infection, almost certainly by the host cell DNA polymerases engaging in 'repair' of the ssDNA. In the case of circular genomes, these get converted into covalently closed circular double-stranded (cccds) (plasmid-like)-DNA in the nucleus and become associated with nuclear proteins and complexes such as nucleosomes. Transcripts are then made from this cccds DNA for gene expression. Synthesis of progeny viral DNA is carried out by a rolling circle mechanism using the complementary strand as a template. Newly synthesized DNA are nicked and ligated to form a unit length, circular progeny DNA. Occasionally, longer than unit-length of viral DNA can be detected in the infected nucleus.

2.4.2.6 dsRNA viruses

Double-stranded RNA viruses include the animal-, plant-, and insect-infecting Reoviridae. Reoviruses have 10-12 segments of dsRNA per genome, all encapsidated in a single particle. When a sufficient amount of viral capsid proteins has accumulated, the (+) strand RNA transcripts are packaged within an opened subviral particle (inner core). Once the core is formed with the complete set of viral RNA segments, the (-) viral strands are synthesized to complete the formation of dsRNA.

2.4.2.7 ssRNA viruses with (+) and (–)-sense genomes

This group of viruses contains majority of plant RNA viruses. The genome of these viruses can serve directly as messenger RNA for protein translation. Upon infection of cells, the genomic RNA of positive-strand RNA viruses is translated into the necessary proteins for viral replication. These proteins with one or more host proteins form a replicase complex which recognizes and attaches to the 3′ of the positive strand RNA (promoters). The genomic RNA is then copied to a complementary RNA of negative polarity (minus strand) by the polymerase activity of the replicase. The same enzyme is also responsible for transcribing the genome into messenger RNA (subgenomic RNA). Subsequently, this newly synthesized minus RNA serves as a template to direct the synthesis of new plus-strand viral RNAs. Replication of (–)-strand RNA viruses is similar, except that the viral RNA has to be transcribed into mRNA for protein translation upon entry into the viral cell.

dsDNA viruses which replicate via greater-than-genome-length ssRNA: Upon the entrance of the virion into plant cells, the virus disassembles. The released viral DNA is transported into the nucleus and is repaired to form a supercoiled dsDNA minichromosome. This supercoiled DNA is capable of serving as a template for transcription. A longer than unit-length RNA with terminal repeats of about 180 nt (35S RNA) is transcribed by RNA polymerase II. This transcript is then exported into the cytoplasm and serves as a template for reverse transcription as well as translation. The reverse transcriptase in association with capsid proteins (possibly other proteins) forms a replication complex in the viroplasm. The reverse transcription is likely to take place in a virus-like structure. Reverse transcription of the 35S RNA is primed by tRNA met around nucleotide 500 to 600 of the 35S RNA. DNA synthesis continues until it reaches the 5 termini of the 35S RNA.

Subsequently, RNase H activity within the reverse transcriptase digests the RNA in the DNA: RNA hybrid. The ssDNA will base pair with the terminal portion of 35S RNA containing the repeat. The strong stop DNA is extended by the reverse transcriptase until it joins the 5′ terminus of the DNA. The process completes the formation of a second strand and creates a gap. There are one or two missing nucleotides at the gap. Again the RNase H activity of the reverse transcriptase digests away RNA in the DNA: RNA hybrid. However, two G-rich polypurine RNA fragments are undigested around nucleotide 1600 and 4200. These two RNA fragments serve as a primer for second strand DNA synthesis of the first and second strands. During the synthesis of strand, there is template switch event since the second strand overlaps with the gap in the strand.

2.4.2.8 Assembly

When both viral capsid proteins and viral genomic nucleic acids are produced in large quantities, the proteins encapsidate viral genomes within a protective shell (virions) by a self-polymerization process. The virion assembly process can be reproduced in test tubes given purified viral capsid proteins and viral nucleic acids.

2.4.3 Cell-to-cell Movement

To infect the entire host plants, viruses will have to move from initially infected cells to neighboring cells (cell-to-cell movement) and to spread from inoculated leaves to other leaves through plant vascular systems (long distance translocation). These two processes involve different mechanisms. Viral proteins are actively involved.

Plant cells are surrounded by thick layers of cell walls. Each cell, however, is interconnected by plasmodesmata. Plasmodesmata are membrane-lined cylindrical pores across the cell wall. The pore size of the plasmodesmata is usually very small, in the range of less than one nanometer (nm), much bigger than virus particles. To move out of the initially infected cells, viruses need to overcome the plasmodesmata barrier. Recent experiments have shown all viruses encode a special protein call movement protein that is able to increase the pore size of the plasmadosmata and target the viral RNA-protein complex into the plasmodesmata.

2.4.3.1 Long distance translocation

Long distance translocation is the spread of viruses from an inoculated leaf to other parts of the plant through the vascular system. The movement of the virus appears through the phloem of the vascular system. As a result, viruses flow with the carbohydrate materials in the plant. Systemic symptoms of a viral infection often occur in young growing areas such as young leaves at a plant tip or young roots. In some viruses, long distance translocation requires functional capsid proteins. But this is not the case in some other viruses. Capsid proteins may be required for virion formation. Viral nucleic acids in virions may be better protected from the adverse environment in the phloem tissues.

2.5 INFECTION PROCESS IN PHYTONEMATODES

Nematode parasitism of plants is a complex and dynamic interaction and may involve sensory abilities, host-finding behavior, recognition phenomenon, cytopathology, hatching stimuli, attraction to the host, penetration, feeding site formation and host response

(Ravichandra, 2008). The highly specialized sedentary endoparasites that have very complex and intimate relationships with their hosts. Many species of plant parasitic nematodes have evolved complex relationships with their hosts, inducing profound alterations in the morphology of cells from which they feed. The availability of modern molecular techniques has recently enabled several breakthroughs in understanding plant-nematode interactions. The complex nature of plant-nematode interactions, the life cycle and obligate biotrophy of the parasitic nematodes and a limited amount of affected material, even in a compatible interaction have hindered progress on elucidating mechanisms.

The plant root environment is composed of the rhizoplane and associated complexes of microorganisms both around and in the root itself. In addition to products of metabolism such as CO_2 and heat, anionic flux consisting mainly Potash, Sodium, Chloride, Hydrogen and Hydroxide at some times, is evident. Various carbohydrates and organic acids may also occur. Root exudates stimulate several activities of nematodes.

2.5.1 Egg Hatching

Plant nematode interaction starts with egg hatching process itself. Some plant parasitic nematodes like Golden cyst nematode, *Globodera rostochiensis,* and Soybean cyst nematode, *Heterodera glycines* need hatching stimuli for their hatching. Root exudates serve as hatching factor for *Globodera rostochiensis*. Juveniles of *Meloidogyne* are attracted by reducing substances and migration to the zone of cell elongation of the root may be due to the lower oxidation-reduction potential in this region or to root exudates. Moisture influences *Heterodera* larval emergence rate from cysts and larval migration toward roots. Juvenile orientation is greatly influenced by root exudates. The stimulatory effect of soybean root exudates on *H. glycines* egg hatch is caused by glycinoclepin- A, a dibasic acid.

Hatching factors required for hatching includes 4-6 moderately strong organic acids for *G. rostochiensis* and Glycinoeclopin A, B, and C for *Heterodera glycines.* The nematodes may orient toward a root from some distance (*Example*: 75 cm or farther perhaps using CO_2 and temperature as collimating stimuli). When close enough to a plant root, klinotaxic responses may predominate. The nematode's chance of successfully infecting the root would be further enhanced if it remained infective over a period of several days. These results may help explain why root infection sometimes occurs in the field when nematodes are undetectable.

2.5.2 Host Recognition

The recognition event is defined as the initial event in cell-cell communication that elicits a defined biochemical, physiological or morphological response. The nematode may respond to a wide array of stimuli to a particular stimulus or cue active at a particular stage of their development and interaction with the host. The chemical and physical stimuli important in host finding and cell recognition may be different for a nematode in search of a feeding site within root or plant tissue (Ravichandra, 2008). It is known that particular nematode species aggregate around particular zones of the root (*Example*: *Meloidogyne* and *Paratrichodorus* congregate and penetrate about the meristematic region behind the root cap). Juveniles also enter where secondary roots emerge and at surface cracks and lesions of enlarged roots. The nematodes may be responding to kairomones or chemical cues that induce a favorable behavioral and physiological response in the receiver. Perhaps after being directed to within several centimeters of the root by heat or CO_2 gradients, nematodes respond to another cue, the kairomone, which triggers their next behavioral response. This stage would initiate the recognition of the host by the nematode.

The intrigues and complexities involved in the host recognition by the plant parasitic nematodes in most cases remain unresolved. However, two theories were proposed for this process, viz., the orientation of nematodes to plants - the attraction of plant parasitic nematodes by the plant root exudates - and random movement of nematodes, trapped in free water in root vicinity.

2.5.3 Feeding Site Establishment

The host nematode interaction is a complex development system with genes from both sources interacting and failure of the association may result at any stage, from the initial parasitism of the first cell to the maturation of the adult nematode. The failure of the compatible response could result from the synthesis of metabolic inhibitors, the death of feeding site or the failure of host tissue to keep pace with nutritive demands of the nematode and all of these are, in one way or another, resistant responses. The multimechanismic nature of compatibility-incompatibility is evident in the differential development of *Heterodera glycines* on soybean, kidney bean, and other plants. The rates of development of juvenile development remained same on these hosts, but young females on kidney bean were often small and poorly developed and produced fewer eggs than on other hosts. Responses to compatible sedentary endoparasites include the development of the coenocytic cells and syncytia. The susceptible response may begin by recognition of the parasite by the plant, followed by a rapid accommodation of the parasite through genome-modifying sequences utilizing energy or the susceptible response may begin with a passive failure of the host to recognize the pathogen and respond to limit its development. Various models provide scenarios for recognition of self, facilitating the development of the pathogen or nonself, leading to restriction of the pathogen.

2.5.4 Host Penetration

Initial disturbance to the plant is caused when the nematode comes to the root surface. Most plant parasitic nematodes are known to produce enzymes like cellulase and B-1,4-endoglucanases to soften the cell wall before penetrating the plant cell. Pectinases, known to degrade pectin in middle lamellae, are also produced by *Pratylenchus zeae, Heterodera trifoli, Meloidogyne* spp., *Radopholus similis* and *Ditylenchus dipsaci.* Before penetration, nematode presses its lips to the plant root surface and thrusts backward and forward by using the stylet. By doing this process, if gets compatible response nematode penetrate the plant root. It can be concluded that penetration is by both enzymatic softening and mechanical piercing. Root tip, a region of cell differentiation and elongation and root hairs are the main regions of the plant roots to penetrate by the plant parasitic nematodes.

2.5.5 Feeding

Secretions flow from the dorsal gland through the lumen of the spear and into the host cell following penetration of the spear. Ingestion of cellular contents begins with rhythmic pulsations of the cardia and adjacent portions of the intestine. This may include pulsation of the posterior pharynx assisted by occasional pumping of the median bulb. The introduction of secretions into the host cell during salivation probably contributes to extraoral digestion prior to ingestion. The duration of salivation may be the longest phase of the feeding process, as in *Trichodorus similis,* in which a relatively short time is required for ingestion. In *Hemicycloiphora,* the pumping of the oesophageal bulb caused some fluids to flow back and forth between the host cell and the esophagus. Saliva was produced continuously and was presumably mixed with the cell contents. It is evident that there is an additional influence of saliva, secretions or enzyme in the process of disturbance by mechanical damage during penetration and feeding.

2.5.6 Migration Through Tissues

Genera of nematode differ in their mode of migration through host tissues. For example, in *Xiphenema and Longidorus,* only the stylet will be pierced and cause damage; in *R. similis formation of tunnels* and cavities in both intra- and extracellular ways; in root-knot and cyst forms, they migrate through the cortex-stelar region. Nematodes may disturb host cells in many ways. Some may disturb only a few cells during feeding or penetration; some move through the tissues feeding on cells as they move by leaving tunnels and cavities behind them, and some may fall in between these two categories. Host plant hence suffers to varying degrees depending on such feeding behavior. *Xiphinema and Longidorus,* which are ectoparasites, damage the plant by penetrating and feeding on cells with the stylet. Larvae of *Nacobbus* feed either on the surface of the root or enter the root to feed. *Hoplolaimus coronatus* may feed endo- or ectoparasitically on cotton roots. *Trichodorus christiei* sometimes may penetrate to half a body length as they force apart the loosely adhering cells at the root cap. *Helicotylenchus* may migrate for distances up to 75 cm from the pseudostem of bananas while *R. similis* forms tunnels and cavities by lysis of cells as they progress through the root cortex. Such feeding may result in a significantly more damage to the plant. On the other hand, a large number of *Partylenchus* can occur in roots without causing any appreciable damage to the growth of the plant, which is mainly because nematodes feed on cortical cells and leave the vascular tissues undamaged.

2.6 NEMATODE SECRETIONS AND EXCRETIONS

Phytonematodes have been reported to produce either one or other kind of enzymes for particular tasks like cell wall degradation and extraoral digestion. The host may not be just a passive substrate on which the nematodes' secretions act, instead, nematodes may have developed systems for directing the host's metabolism or its initiating physiological processes that achieve the same objective without the direct injection of an enzyme. *Ditylenchus dipsaci* is normally associated with host cell separation, which produces a large amount of pectinase. Different symptoms produced in different host plants by populations of this nematode may be caused by different pectolytic enzymes associated with these populations. The granular exudation, which adult female of *Meloidogyne javanica* injects from the dorsal esophageal gland into host cells, may contain material capable of synthesizing enzymes only when placed in the cytoplasm of the host. Plant parasitic tylenchid nematodes are well adapted for parasitism. In addition to a protrusible stylet, these nematodes have three large and complex esophageal secretary glands, one dorsal and two-subventral (Ravichandra, 2008). These secretions are spherical membrane bound secretary granules (dense core vesicles) formed in both dorsal and subventral esophageal glands.

2.6.1 Nature of Secretions

Elucidating the chemical nature of nematode secretions and characterizing their biological activity remains one of the most intriguing and challenging areas of future research in Nematology. Analysis of secretion composition showed the presence of proteins, glycoproteins, amino acids, like lysine, histidine and arginine, carbohydrates and cytokinins like compounds and enzymes like cellulose, proteinase, and B-1,4-endoglucanases. These secretions play an important role in penetration, induction, and maintenance of feeding sites and digestion process of plant-nematode interaction.

Therefore, types of tissues damaged by the nematodes are important and in general, disturbance of cells in the stele or the region of the root tip are likely to have the most influence on the plant nematodes injury to plants may be a mechanical, i.e., penetration and movement through plant tissues. It may include the following.

(a) Cellular changes: Death of cells (necrosis), changes in growth of cells.
(b) Physiological changes in host: Interruption in uptake and flow of water and nutrients from roots, interruption in the flow of food from leaves to roots.
(c) Create openings for entry of other microorganisms.
(d) Interaction with other disease-producing agents.
(e) Transmission of other disease-producing agents.
(f) Increase susceptibility to environmental stress.

2.7 ROLE OF ENZYMES, TOXINS, GROWTH REGULATORS AND POLYSACCHARIDES IN PATHOGENESIS

The term pathogenesis means step-by-step development of a disease and the chain of events leading to that disease due to a series of changes in the structure and/or function of a cell/tissue/organ being caused by a microbial, chemical or physical agent. The pathogenesis of a disease is the mechanism by which an etiological factor causes the disease. The term can also be used to describe the development of the disease, such as acute, chronic and recurrent. The word comes from the Greek pathos, 'disease' and genesis, 'creation.'

There are several chemical weapons secreted by pathogens that are utilized as they carry out their activities. These weapons include enzymes, toxins, growth regulators and polysaccharides.

2.7.1 Role of Enzymes

Enzymes are large protein molecules that occur within living cells and act as a catalyst in chemical reactions. Because enzymes are present in nearly all chemical reactions that occur within plant cells, and because there are a huge number of such reactions in plants of different species, genera, and families, there are countless types of plant enzymes. Pectin or pectic substances are major chemical constituents of wall layers and entire middle lamella, whereas, in other layers, cellulose is found in good amounts. Besides these two major components, other components such as hemicelluloses, lignin and some amount of protein are also present. Main components of the cell wall are pectic substances, cellulose, hemicelluloses, lignin and a small quantity of protein. The epidermis of plants is covered by a cuticle, whose major chemical substance is cutin in addition to a cuticular wax.

2.7.1.1 Cuticular wax

Plant waxes are found as granular or rod-like projections or as a continuous layer outside/within the cuticle. The wax formation is a continuous process, and it is not a terminal phase in the development of leaf. Cuticular waxes are made up of long chain molecules of paraffin, hydrocarbons, alcohols, ketones and acids. Most of the fungi and parasitic higher plants penetrate wax layers by means of mechanical force alone.

2.7.1.2 Cutin

It is an insoluble polyester of unbranched derivatives of C16 and C18 hydroxy fatty acids. Cutin is admixed with waxes on the upper side and with pectin and cellulose on the lower side. Cutinases break cutin molecules and release monomers as well as oligomers from insoluble cutin polymer. The cutinase reaches its highest concentration at the penetrating point of the germ tube and at the infection peg of the appressorium forming fungi (*Example*: *Colletotrichum gloeosporioides, Sphaerotheca pannosa, Venturia inaequalis, Helminthosporium victoriae,* etc.)

2.7.1.3 Pectic substances

These are major components of middle lamella (intercellular cement that holds in place the cells of plant tissues). They also make up a large portion of the primary cell wall in which they form an amorphous gel filling the spaces between cellulose microfibrils. The enzymes that degrade pectic substances are known as pectinases or pectolytic enzymes.

Pectinases and pectolytic enzymes are pectin methyl esterases (PMEs), polygalacturonases (PGs) and pectin lyases (PLs).

2.7.1.3.1 Pectin methyl esterases

Pectin methylesterases catalyze the demethylesterification of cell wall polygalacturonans. In dicot plants, these ubiquitous cell wall enzymes are involved in important developmental processes including cellular adhesion and stem elongation. It breaks ester bonds and removes methyl groups from pectin leading to the formation of pectic acid and methanol (CH3OH). Cell-wall pectin methyl esterification can influence plant resistance because highly methyl-esterified pectin can be less susceptible to the hydrolysis by pectic enzymes such as fungal endopolygalacturonases (PG). Pectin is secreted into the cell wall in a highly methylesterified form and, here, is de-methyl esterified by pectin methyl esterase (PME). The activity of PME is controlled by specific protein inhibitors called PMEI; consequently, an increased inhibition of PME by PMEI might modify the pectin methyl esterification. To test the possibility of improving wheat resistance by modifying the methyl esterification of pectin cell wall, durum wheat transgenic lines were produced expressing the PMEI from *Actinidia chinensis* (AcPMEI). The expression of AcPMEI endows wheat with a reduced endogenous PME activity, and transgenic lines expressing a high level of the inhibitor showed a significant increase in the degree of methyl esterification. These lines showed a significant reduction of disease symptoms caused by the fungal pathogens *Bipolaris sorokiniana* or *Fusarium graminearum*. This increased resistance was related to the impaired ability of these fungal pathogens to grow on methyl-esterified pectin and to a reduced activity of the fungal PG to hydrolyze methyl-esterified pectin. In addition to their importance for wheat improvement, these results highlight the primary role of pectin despite its low content in the wheat cell wall.

2.7.1.3.2 Polygalacturonases

Split pectin chain by adding a molecule of water and breaks the linkage between two galacturonan units. These enzymes catalyze reactions that break α-1,4-glycosidic bonds. Polygalacturonases (PGs) are produced by fungal pathogens during early plant infection and are believed to be important pathogenicity factors (D'ovido *et al.*, 2004). Polygalacturonase-inhibiting proteins (PGIPs) are plant defense proteins which reduce the hydrolytic activity of endoPGs and favor the accumulation of long-chain oligogalacturonides (OGs) which are elicitors of a variety of defense responses. PGIPs belong to the superfamily of leucine reach repeat (LRR) proteins which also include the products of several plant resistance genes. A number of evidence demonstrate that PGIPs efficiently inhibit fungal invasion.

Polygalacturonases otherwise referred to as pectic hydrolyases are pectic enzymes that hydrolyse pectic substances into their monomeric units. They are known to be responsible for the hydrolytic cleavage of the B-1,4-glycosidic of the galacturonan, a moiety of pectic substances (Arotupin Daniel Juwon *et al.*, 2012). However, extracellular polygalacturonases catalyze the hydrolytic cleavage of terminal X-1,4-glycosidic bonds (pectic acid) releasing galacturonic acid as the main product. Polygalacturonases appeared to be the most frequently encountered pectic enzymes. They are formed in the majority of plant tissues particularly in ripening fruits. Also many plant-pathogenic and saprophytic microorganisms produced polygalacturonases. Most fungi that exhibited pectolytic activity produce polygalacturonases either as sole pectic enzymes as observed in *Sacchromyces fragilis* (Phaff, 1966) or in association with either or both pectinmethylesterase and pectin lyase. The critical role of these enzymes in the degradation of the host middle lamella and cell walls, leading to plant tissue maceration and cellular death had been documented. The infection of cotton seedlings by *Rhizoctonia solani*, soft rot yam and sweet potato, brown rot of apple by *Monilinia* species and deterioration of tomato by *Botryoplodia theobromae* Pat have been reported to produce polygalacturonases and other cell wall degrading enzymes. The culture filtrates of *Collectotrichum lagnarum* exhibited polygalacturonase activity.

2.7.1.3.3 Pectin lyases

A pectin lyase also known as pectolyase is a naturally occurring pectinase, a type of enzyme that degrades pectin. Pectin lyases are the only known pectinases capable of degrading highly esterified pectins (like those found in fruits) into small molecules via β-elimination mechanism without producing methanol, which is toxic, in contrast to the combination of PG and PE, which are normally found in commercial products. In addition, the presence of undesirable enzymatic activity in commercial pectinases may be detrimental to aroma because they are responsible for producing unpleasant volatile off flavor.

Pectin lyase acts on the pectic substances that occur as structural polysaccharides in the middle lamella and primary cell walls of higher plants. This enzyme has potential applications in food, paper and textile industries. Since new applications of this enzyme are emerging, there is a scientific need to explore the important aspects of the enzyme specifically the catalytic efficiency and possible sources. Split pectin chain by removing a molecule of water from the linkage, thereby breaking it and releasing products with unsaturated double bonds. These pectin enzymes can be exopectinases (break only terminal linkage) or endopectinases (break pectin chain to random sites). Pectin degradation results in liquefaction of the pectic substances and weakening of cell walls, leading to tissue maceration (*Example*: Soft rot bacterium, *Erwinia caratovora* subsp. caratovora and other fungi like *Botrytis cinerea, Sclerotium rolfsii*, etc.).

Pectinases have multiple biological functions. Pectate lyases can act as extracellular virulence agents and their role in the release of cell wall oligogalacturonides is important for activation of plant defense mechanisms (Jun Cao, 2012). Furthermore, pectate lyases may be important for fruit ripening and softening, as well as plant growth and development. Similarly, polygalacturonases contribute to pectin disassembly during many stages of plant development, such as those that require cell separation. Polygalacturonase genes also have roles in pollen maturation and pollen tube growth, as well as intine and exine formation. In addition, like pectate lyase and polygalacturonase, pectin lyase can enhance the reconstituted expansin-induced extension of the apical (elongating) segments of cucumber hypocotyls. Most pectin lyases are produced by microorganisms (such as *Aspergillus, Penicillium*, and *Fusarium*). In these microorganisms, expression of pectin lyase gene is generally induced by medium pH, carbon sources, and pectin, and is generally repressed by glucose.

The early steps in the interaction between a pathogen and a host are confined to the phase when the pathogen gains entry into the host tissues, into the cells or between the cells by dissolving the cell walls. At this stage, the ability of the pathogen to degrade the cell wall constituents such as cellulose, pectin, hemicelluloses and protein by secreting the appropriate hydrolytic enzymes seems important (Subramanian, 1969). Although evidence for such enzymatic breakdown of cell walls has been obtained in some cases such as soft rot, damping-off of seedlings, foot-rot of cereals, the role of these extracellular enzymes of the pathogen in other diseases is less clear.

The other facet of this problem, namely, the reaction of host tissues to the action of these pathogen factors is also important in the development of the disease. Several enzymes are either stimulated or inhibited in and around the infection loci. For example, oxidases such as polyphenolase and peroxidase show increased activity in several host-pathogen interactions studied. Increased activity of these enzymes could have been owing to their activation consequent to the release from 'inactive' combinations with other cellular components, or it could mean a synthesis of these enzymes *de novo* in response to the activity of the pathogen. Be that as it may, the excessive activity of these enzymes would bring about the death of cells and tissues if the stimulation is violent enough as encountered in 'hypersensitive' reaction. Perhaps, plants can afford to lose the diseased portions in this battle without any serious repercussion on the rest of the portions.

2.7.1.3.4 Pectic enzymes

Plant cell walls are primarily polysaccharide in composition. A simple but major pathogenic mechanism in plants involves degradation of the cell wall by a battery of polysaccharidases secreted by pathogens. Most of the degradative enzymes are glycoside hydrolases, which degrade the cellulose and pectate matrices by the addition of water to break the glycosidic bonds. The pectate network is also degraded by polysaccharide lyases, which cleave the glycosidic bonds via a β-elimination mechanism (Steven R. Herron *et al.*, 2000). To better understand the latter virulence mechanism, research has been carried out on pectate lyase C, a pectolytic enzyme secreted by the pathogenic bacterium *Erwinia chrysanthemi*. The story of pectate lyase C illustrates how structural techniques have contributed to a detailed understanding of polysaccharide recognition and the lyase cleavage mechanism. In the process, a novel protein structural fold and a unique catalytic role for an arginine have been discovered. The structural results have also provided the first atomic description of a pectate fragment, which differs considerably from the popular view in conformation as well as the mode of interactions with Ca^{2+} ions. Finally, the growing structural database of pectolytic enzymes is enabling researchers to elucidate subtle structural differences that are responsible for the specific recognition of a unique oligosaccharide sequence from a heterogeneous mix in the plant cell wall. Such knowledge will ultimately lead to a better understanding of the characteristics that render the host susceptible to attack by a particular pathogen.

2.7.1.3.5 Degradative enzymes of plant cell wall components

The plant cell wall is composed of two types of polysaccharide matrices: the pectate network and the cellulose network. The pectate network consists of the smooth region composed of homogalacturonans and the hairy region composed of highly branched rhamnogalacturonans. The cellulose network consists of microfibrils composed of 1-4-linked β-D-glucan units as well as xyloglucan and arabinoxylan, the two hemicelluloses that coat the cellulose microfibrils to prevent excessive aggregation. Not only does the basic saccharide unit differ in the two types of network, but the saccharide units in the pectate network also have a uronic acid or an esterified uronate moiety at the C-6 position of the saccharide units. Enzymes, which degrade the pectate network, belong to two classifications: glycoside hydrolases and polysaccharide lyases (Steven R. Herron *et al.*, 2000). Glycoside hydrolases incorporate a water molecule via a general acid catalysis during the cleavage of the glycosidic bond between the two saccharide units. In contrast, polysaccharide lyases cleave the glycosidic bond via a β-elimination reaction that removes a proton. The final product contains an unsaturated bond between C-4 and C-5 of the saccharide unit at the nonreducing end. Enzymes, which degrade the cellulose network, all function as glycoside hydrolases. Generally, hydrolases have acidic pH optima, using aspartic and glutamic acid groups during catalysis, whereas lyases have basic pH optima, using catalytic amino acids that are still under active investigation.

The degradation of plant cell walls by secreted enzymes from pathogens has been a fertile area of research for many years. Although it is well known that certain pathogens are effective against specific hosts, the mechanisms of host–pathogen interactions remain elusive. In certain organisms, such as *Erwinia chrysanthemi*, the genetic organization and regulation of many secretory saccharidases have been elucidated (Reverchon *et al.*, 1991). One surprising finding is that many pathogenic organisms secrete multiple isozymes of the same enzyme, but the transcription of the genes is often independently regulated. Since 1990, research has been directed at the protein level, to understand the structure and enzymatic mechanism of the degradative cell wall enzymes. Additional efforts are being made to elucidate the molecular basis for protein recognition of the composition and conformation of the individual saccharide units in the substrate. Another major question, yet to be resolved, is whether there is isozyme-specific recognition of a unique polysaccharide sequence and if this specificity plays an important role in pathogen–host cell interactions. The work will be illustrated by the research on *Erwinia chrysanthemi* pectate lyase C (p*elC*).

2.7.1.3.6 Three-dimensional structures of plant cell wall degradative enzymes

PelC is one of the several isozymes secreted by *E. chrysanthemi* that cleave the pectate component of the plant cell wall, causing soft rot disease in a variety of crops (Barras *et al.*, 1994). The *pelC* gene has been cloned into *Escherichia coli*, with the latter constructs exhibiting maceration activity comparable to *E. chrysanthemi*. By using the recombinant form from *E. chrysanthemi*, the three-dimensional structure of *pelC* has been determined to fold into a novel topology, not observed before 1993 or predicted during the first 30 years of crystallography (Yoder *et al.*, 1993). *PelC* folds into a large right-handed coil termed the parallel β helix. There are eight coils in the helix, and each is comprised of three β strands, connected by three turns with unique features (Fig. 2.3). When the coils are stacked, the structure has the appearance of three parallel β sheets, stabilized by an extensive network of interstrand hydrogen bonds. Another notable feature of the *PelC* fold is the internal organization of the amino acids that form the core of the parallel β helix. All of the amino acids, which are oriented toward the interior, are regularly aligned with amino acids from neighboring coils, giving rise to long ladders of hydrophobic, aromatic, or polar amino acids. In contrast, the exterior amino acids are randomly oriented and comprise loops, of varying length and composition, which protrude from the central core. From the perspective of an effective plant virulence factor, the most important feature of the parallel β helix fold is the stability that it confers upon an enzyme that must function in the hostile extracellular environment.

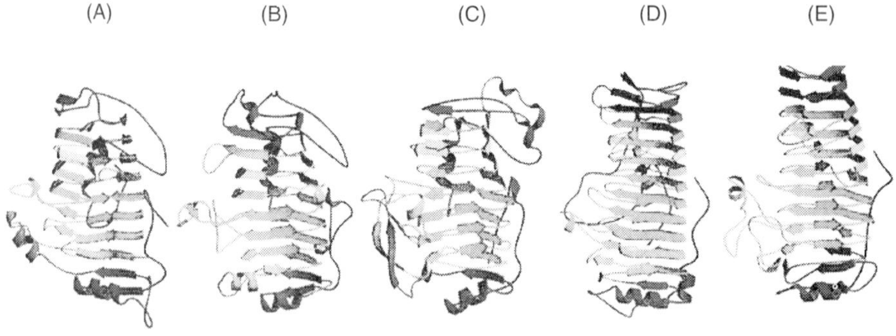

Fig. 2.3 Five examples of plant cell wall degradative enzymes that fold into a parallel β helix motif. The predominant secondary structural features of the proteins are illustrated as β strands, and the coils represent β helices. (A) *E. chrysanthemi* pectate lyase C; (B) *E. chrysanthemi* pectate lyase E; (C) *A. niger* pectin lyase B; (D) *E. carotovora* polygalacturonase; and (E) *A. aculeatus* rhamnogalacturonase A

2.7.1.3.7 Significance of multiple isozymes for pathogenesis

One enigma, yet to be resolved in the pectate lyase field, is the function of multiple, independently regulated pectate lyase isozymes and the role, if any, in pathogenesis (Steven R. Herron *et al.*, 2000). The composite results of the structural studies suggest that the answer may lie in the heterogeneous nature of the substrate. The pectate component is composed of repeating units of negatively charged galacturonic acid. The uronic acid group on C-6 is frequently esterified with a methyl group. Methylation neutralizes the negative charge of an individual Gal*p*A unit and, consequently, alters the surface charge of the pectate polymer. The percentage and the positional sequence of methylated Gal*p*A units (mGal*p*A) vary during the life cycle of the plant and from one type of plant to another. Thus, a pathogen that secretes a battery of pectic enzymes, each of which uses a similar catalytic mechanism, but recognizes a different sequence of methylated and nonmethylated oligogalacturonate units, would be expected to have a broader host range.

The structural studies provide tentative evidence that the pectate lyase isozymes cleave the same *in vitro* substrate but differ in the preferred *in vivo* substrate. In the majority of enzyme families, the most conserved region, in terms of primary and tertiary structure, is the active site and the surrounding region. The pectate lyase family appears to be somewhat different. Comparisons of the pectate lyase structures have revealed that, although the immediate region around the catalytic site is highly conserved, the substrate groove extending on either side is structurally the most diverse (Lietzke *et al.*, 1994).

The diversity arises from the variation in the sequence and length of the loops protruding from the parallel β barrel. One set of loops, termed the T3 loops, form the binding pockets required for recognition of multiple saccharide units in the substrate. The surface charges in *pelC* are localized to an elongated groove that complements the shape and length of an oligosaccharide composed of repeating Gal*p*A units (Yoder *et al.*, 1993). In contrast, the analogous region in *pelE* is highly charged only around the primary Ca^{2+} binding site, with a predominance of neutral groups lining the remaining portion of the groove (Lietzke *et al.*, 1994). Analogous differences are found in three-dimensional models constructed from the sequences of other pectate lyase isozymes. Thus, polygalacturonic acid may serve as a suitable *in vitro* substrate for all isozymes because the saccharide binding pockets near the active site recognize only galacturonic acid moieties. The structural differences between *pelC, pelE,* and other isozymes at more distant subsites suggest that the optimal *in vivo* substrate may differ in length, charge, and possibly type of saccharide unit from one isozyme to another. Differences among isozyme substrates need not be as great as a different type of saccharide unit. Smaller differences, such as the degree of methylation of the pectate component, may be sufficient to explain the physiological significance of multiple pectate lyase isozymes.

Steven R. Herron *et al.* (2000) concludes that the story of *pelC,* although incomplete, illustrates how structural information can contribute to an understanding of the multifaceted steps of pathogenesis. In the process, a novel protein structural fold and a unique catalytic role for an arginine have been discovered. The structural results have also provided the first atomic description of a pectate fragment, finding the conformation as well as the mode of interactions with Ca^{2+} ions to be considerably different from the popular view. The growing structural databases of pectolytic enzymes should allow researchers to elucidate the structural motifs that recognize specific types of saccharides. The initial attempt to elucidate the structural principles that permit an enzyme to distinguish between pectate and pectin is the first step in understanding how saccharidases recognize a unique oligosaccharide sequence from a heterogeneous mix in the plant cell wall. Such knowledge should ultimately lead to a better understanding of the unique properties of the host that render it more vulnerable to attack by a given pathogen.

The cell wall-degrading enzymes polygalacturonase and pectate lyase have been suggested to be crucial for penetration and colonization of plant tissues by some fungal pathogens. Ralph A. Dean and William E. Timberlake, (1989) observed that *Aspergillus nidulans* (-*Emericella nidulans*), a saprophytic Ascomycete, produced levels of these enzymes equal to those produced by soft-rotting *Erwinia* species. Induction of polygalacturonase and pectate lyase in *A. nidulans* required substrate and was completely repressed by glucose. Surprisingly, inoculation of excised plant tissues with *A. nidulans* conidia lead to the formation of necrotic, water-soaked lesions within which the organism sporulates. Thus, *A. nidulans* has phytopathogenic potential. The release of glucose and other sugars from wounded tissues may repress pectolytic enzyme production and limit disease development. They tested creA204, a mutation that relieves glucose repression of some *A. nidulans* carbon utilization enzymes, for its effect on the production of pectolytic enzymes. The creA204 failed to relieve catabolite repression of polygalacturonase or pectate lyase and had no effect on disease severity.

Cutinases, cellulases, pectinases and lignases are often secreted by the pathogenic organism. Fungi, nematodes, and bacteria are all known to produce one or more of the above enzymes in specific pathogen-host combinations. Viruses and viroids are generally not considered to secrete enzymes, although some viruses may encapsidate an enzyme in their particle. Pathogenic organisms either continually secrete enzymes or upon contact with the host plant.

In fungal plant pathogenesis, enzymes play a crucial role, and they are involved in the external and internal interactions (Ales Lebeda *et al.*, 1999). To restrict the development of fungal pathogens, the plants formed many defense mechanisms. They built mechanical barriers from lignin, suberin, and callose and produced a lot of antimicrobial compounds with low molecular weight like phenols, chinons, alkaloids and others. Enhanced production of some enzymes and their activity increasing is one of the most important processes in plant defense. These enzymes frequently occur in many isoforms and are involved in the synthesis of defense substances or have a direct antimicrobial activity.

The first surface an organism comes into contact with is cuticle and the cell wall of the plant. The cuticle is comprised of a complex wax, cutin, which impregnates the cellulose wall. The cell wall is comprised of cellulose, which makes up the structural framework of the wall, along with the matrix molecules hemicellulose, glycoproteins, pectin, and lignin. Thus, penetration into living parenchymatous tissues and degradation of the middle lamella is due to the action of one or more enzymes which degrade these chemical substances. Cutinases degrade the cutin on the cuticle layer presoftening the tissue for mechanical penetration or as a first step in tissue degradation. Studies have shown that several fungi and at least one bacterial species produce cutinases. Further, evidence indicates that cutinases are continually produced, albeit in low concentrations, with degradation products often inducing even higher levels of cutinase secretion.

Studies have shown that in some organisms, cutinase production may be linked to virulence. Pectic substances comprise the middle lamella and also form an amorphous gel between the cellulose microfibrils in the primary cell wall. Pectin degrading substances often termed pectinases or pectolytic enzymes include pectin methylesterases (PME), Polygalacturonases (PG) and pectin lyases or transeliminases. Pectin methylesterases remove small groups such as methyl groups (CH3) often altering solubility and thus affecting the rate of chain splitting by polygalacturonase and pectin lyase. Polygalacturonases split chains by adding a molecule of water while pectin lyases split chains by removing a water molecule from the linkage. Pectin degrading enzymes are involved in a wide range of plant diseases particularly in the soft rot diseases (*Example*: *Erwinia carotovora, Sclerotinia sclerotiorum* and *Rhizoctonia solani*). Organisms such as these, as well as, elaborate pectic enzymes lead to tissue maceration. In fact, these enzymes are sometimes referred to as macerating enzymes.

2.7.1.4 Cellulases

Cellulose is the major framework molecule of the plant cell wall existing as microfibrils with matrix molecules (glycoproteins, hemicelluloses, pectins, lignins) filling the spaces between the microfibrils and cellulose chains. Cellulases have been shown to be produced by many pathogenic fungi, bacteria, and nematodes. Cellulolytic enzymes play a role in softening and disintegration of cell walls. No doubt cellulolytic enzymes are involved in the invasion and spread of the pathogen, but also are instrumental in the collapse of cells and tissues. Indirectly cellulolytic enzymes participate indirectly in disease development by releasing soluble sugars that may be used as nutrients by pathogens and also may be involved in the release of materials in the vascular system interfering with transport or translocation of water. Cellulose is degraded by cellulases. Cellulase one (C1) attacks native cellulose by cleaving cross-linkages between chains. A second cellulase (C2) also attacks native cellulose and breaks into shorter chains. These shorter chains are then attacked by Cx enzyme, which degrade them into disaccharide, cellobiose. Finally, cellobiose is degraded by the enzyme, β-glucosidase into glucose. Cellulase degrading enzymes play a role in softening and degradation of cell wall material and facilitate easy penetration and spread of the pathogen in the host. (*Example*: Basidiomycetes fungi).

The ubiquitous oomycete *Pythium oligandrum* is a potential biocontrol agent for use against a wide range of pathogenic fungi and an inducer of plant disease resistance. The ability of *P. oligandrum* to compete with root pathogens for saprophytic colonization of substrates may be critical for pathogen increase in the soil, but other mechanisms, including antibiosis and enzyme production, also may play a role in the antagonistic process. Karine Picard *et al.* (2000) used transmission electron microscopy and gold cytochemistry to analyze the intercellular interaction between *P. oligandrum* and *Phytophthora parasitica*. The growth of *P. oligandrum* towards *Phytophthora* cells correlated with changes in the host, including retraction of the plasma membrane and cytoplasmic disorganization. These changes were associated with the deposition onto the inner host cell surface of a cellulose-enriched material. *P. oligandrum* hyphae could penetrate the thickened host cell wall and the cellulose-enriched material, suggesting that large amounts of cellulolytic enzymes were produced. Labeling of cellulose with gold-complexed exoglucanase showed that the integrity of the cellulose was greatly affected both along the channel of fungal penetration and also at a distance from it. They measured the cellulolytic activity of *P. oligandrum* in substrate-free liquid medium. The enzymes present were almost as effective as those from *Trichoderma viride* in degrading both carboxymethyl cellulose and *Phytophthora* wall-bound cellulose. *P. oligandrum* and its cellulolytic enzymes may be useful for biological control of oomycete pathogens, including *Phytophthora* and *Pythium* spp., which are frequently encountered in field and greenhouse production.

Uredospores of the obligatory biotrophic broad bean rust fungus *Uromyces viciae-fabae* form infection structures on artificial membranes providing a thigmotropic signal (Heiler *et al.*, 1993). In nature, these are essential for invasion of the host plant through the stomata. This experimental system was used to analyze the production of cellulolytic enzymes during the differentiation of rust infection structures. Low cellulase activity was detected in dormant spores and germ tubes. Enzyme activity increased during appressorium development and reached a maximum when infection hyphae and haustorial mother cells were formed. Seven cellulolytic enzymes were separated by chromatofocusing on DEAE = Si500 and PBE 94 exchangers. At least two cellulases, characterized by isoelectric points of 7.1 and 7.3, were identified as endocellulases. The neutral enzyme forms increased from 3.3% of the total activity at appressorium formation to 36.6% when infection hyphae were formed (18 h p.i.), and to 45.4% in haustorial mother cells (24 h p.i.). Cellulase activity of the 'rust' fungus is neither substrate-inducible nor catabolite repressible. The regulation of these enzymes appears to be strictly controlled by the differentiation of infection structures and is, therefore, distinct from cellulases of necrotrophs and saprophytes investigated thus far.

Cellulases are produced by rust fungi. The expression of cellulase activity by *U. viciae-fabae* is controlled by mechanisms active during differentiation of infection structures (Deising & Mendgen, 1992). Substrate induction and catabolite repression are not involved in the regulation of cellulase activity. These findings are in contrast to results obtained from other fungi and support the concept that polymer-degrading enzymes in obligate biotrophs and necrotrophs or saprophytic fungi are regulated by essentially different mechanisms.

In a study on cellulolytic and pectolytic activity of *Cylindrocarpon destructans*, Dahm and Strzelczyk (1987), found that only one of the twenty isolates studied exhibited cellulolytic activity. The total activity of this isolate was similar in media with CMC and powdered cellulose. The specific activity was, however, two times higher with powdered cellulose. All isolates identified as pathogenic to fir and to pine produced pectolytic enzymes. Not all of them, however, exhibited exo- and endo-PMG activity. In general, an increase of total activity of exo-PMG was accompanied by an increase in the specific, activity. Of the nonpathogenic isolates, only one did not show pectolytic activity. There exist no significant differences in pectolytic activity between the isolates pathogenic and nonpathogenic to fir and pine. Also, the isolates belonging to both groups were not cellulolytic except one nonpathogenic.

2.7.1.5 Hemicellulases

Hemicelluloses are complex polysaccharide polymers that link the ends of pectic compounds to cellulose microfibrils. Since hemicelluloses are such a diverse group of polymers such as xyloglucans, glucomannans, glactomannans, arabinoglucans, etc., several hemicellulases have been identified in many plant pathogenic fungi. The mechanism by which they participate in cell wall breakdown is not clear, nor is it known how they contribute to pathogenesis. These are the major constituents of the primary cell wall and also seen in middle lamella and secondary cell wall. The hemicellulose polymers include primarily xyloglucan but also glucomannans, galactomannans, arabinogalactans, etc. Hemicelluloses link the ends of pectic polysaccharides and various points of the cellulose microfibrils. Hemicellulases degrade hemicelluloses and depending on the monomer released from the polymer on which they act; they are termed xylanase, galactanase, glucanase, arabinase, mannose, and so on (*Example: Sclerotinia sclerotiorum, Sclerotinia fructigena*).

2.7.1.6 Cell wall proteins

Cell wall proteins are similar to other proteins, except that they are rich in amino acid, hydroxyproline. Five classes of structural proteins are found in cell walls: extensins, proline-rich proteins (PRPs), glycine-rich proteins (GRPs), Solanaceous lectins and arabinogalactan proteins (AGPs). Proteins are degraded by means of enzymes, proteases or proteinases or peptidases. Cell wall proteins are essential constituents of plant cell walls involved in modifications of cell wall components, wall structure, signaling, and interactions with plasma membrane proteins at the cell surface. The nature of cell wall proteins is as varied as the many functions of plant cell walls. With the exception of glycine-rich proteins, all are glycosylated and contain hydroxyproline (Hyp) (Cassab, 1998). Again excepting glycine-rich proteins, they also contain highly repetitive sequences that can be shared between them. The majority of cell wall proteins is cross-linked into the wall and probably has structural functions, although they may also participate in morphogenesis. On the other hand, arabinogalactan proteins are readily soluble and possibly play a major role in cell-cell interactions during development. The interactions of these proteins between themselves and with other wall components are still unknown, as is how wall components are assembled. The possible functions of cell wall proteins are suggested based on repetitive sequence, localization in the plant body, and the general morphogenetic pattern in plants.

The discovery and development of novel plant cell wall degrading enzymes is a key step towards more efficient depolymerization of polysaccharides to fermentable sugars for the production of liquid transportation biofuels and other bioproducts (Brian C. King *et al.*, 2014). The industrial fungus *Trichoderma reesei* is known to be highly cellulolytic and is a major industrial microbial source for commercial cellulases, xylanases and other cell wall degrading enzymes. However, enzyme-prospecting research continues to identify opportunities to enhance the activity of *T. reesei* enzyme preparations by supplementing with enzymatic diversity from other microbes. Genomic analysis of lignocellulose-degrading fungi shows that a single species can have the genetic capacity to produce many different enzymes with similar functional designations (cellulase, xylanase, and others). For example, the genome of the phytopathogen *Magnaporthe grisea* is predicted to encode at least 30 enzymes in six GH families for the degradation of cellulose and 44 enzymes in 11 families for the degradation of hemicelluloses.

Most plant-associated microbes (both pathogenic and saprophytic) that break down plant cell walls have the genetic capacity to produce enzymes for the degradation of the major structural polysaccharides found in the cell wall, namely cellulose, xylan and pectin (Brian C. King *et al.*, 2014). In particular, plant pathogens have intimate relationships with their hosts, requiring penetration of the cell wall and colonization of living host tissue. Many pathogenic fungi actively kill and degrade plant tissue and utilize liberated carbohydrates for growth and reproduction. Plants produce proteins to inhibit microbial CWDE as one mechanism of disease resistance, and this interaction may drive the evolution of unique enzymes in phytopathogens.

2.7.1.7 Ligninases

Lignin is a phenylpropanoid which is found in the middle lamella and secondary cell wall of plants. More than anything else, lignin confers the tough, woody nature to woody tissues. Only about 500 species of fungi are capable of decomposing wood. Most lignin degradation is by basidiomycetes known as white-rot fungi. These fungi produce ligninases that enable the fungi to utilize lignin. Lignin is found in the middle lamella, as well as in the secondary cell wall of xylem vessels and the fibres that strengthen plants. It is an amorphous, three-dimensional polymer made up of the basic structural unit, phenylpropanoid. Lignin forms by oxidative condensation (C-C and C-O bond formation) between phenylpropanoid units or substituted cinnamyl alcohols (p-coumaryl alcohol, coniferyl alcohol, and sinapyl alcohol). White rot fungi (Basidiomecetes) secrete one or more ligninases which enable them to utilize lignin (*Example: Xylaria, Chaetomium, Alternaria, Cephalosporium,* etc.).

Ligninase is the original term encompassing many different types of oxidative, extracellular fungal enzymes which catalyze the breakdown of lignin which is commonly found in the cell wall of plants. Instead of the term ligninase, the term lignin-modifying enzymes (LMEs) should be used, since these enzymes are not hydrolytic but oxidative (electron withdrawing) by their enzymatic mechanisms. LMEs include peroxidases, such as lignin peroxidase, manganese peroxidase and versatile peroxidase and several phenol-oxidases of laccase type. Peroxidases are haem-containing enzymes that use hydrogen peroxide as the electron acceptor to catalyze a number of oxidative reactions. Peroxidases are found in bacteria, fungi, plants and animals. Fungal ligninases are extracellular heme enzymes involved in the degradation of lignin. They include lignin peroxidases (LiPs), manganese-dependent peroxidases (MnPs) and versatile peroxidases, which combine the substrate-specificity characteristics of the other two. These substructures are efficiently decomposed by wood-rotting fungi employing various enzyme systems. The mechanisms of these systems are extremely complex and highly interactive. It is now known in considerable detail how the three main groups of wood-rotting fungi, that is, white-rot, brown-rot, and soft-rot fungi attack and degrade lignocellulosic materials. The enzyme mechanisms involved in the degradation of cellulose and the hemicelluloses are investigated and known in depth, and also the complex mechanisms of lignin degradation are known in some detail (Eriksson and Bermek, 2009).

2.7.1.8 Lipids

Various types of lipids occur in all plant cells. The most important ones are phospholipids and glycolipids. These lipids contain fatty acids, which may be saturated or unsaturated.

Lipolytic enzymes, called lipases (phospholipases, glycolipases) hydrolyze lipids and release fatty acids.

2.7.1.9 Starch

Starch is the main reserve polysaccharide found in plant cells. It is a glucose polymer and exists in two forms: amylose, a linear molecule, and amylopectin, a highly branched molecule. Starch is degraded by the enzyme amylases.

2.7.1.10 Proteinaceous enzymes

In the course of evolution, plants have elaborated protective mechanisms that allow them to successfully resist different kinds of unfavorable conditions including insects and phytopathogenic microorganisms. The most important components of all protective mechanisms are proteinaceous compounds. These include enzymes such as β-1,3-glucanases and chitinases, inhibitors of proteases and α-amylases, lectins, and also other proteins and peptides which have antimicrobial activity (Valueva and Mosolov, 2004).

Many phytopathogenic microorganisms produce active extracellular proteinases that along with other enzymes play an important role in pathogenesis, e.g., polygalacturonases, pectolyases, and xylanases. It was demonstrated that the *Colletotrichum lindemithianum*, when grown on plant cell walls or on artificial nutrient medium, secretes an active protease of 25 kD molecular weight and pH optimum at 8.6. This was the first extracellular proteinase of a plant pathogen obtained in its pure form. In recent years, many extracellular proteases produced by phytopathogenic microorganisms have been isolated and characterized to some extent. Among these serine proteinases prevail, but there are enzymes belonging to other mechanistic classes. All known serine proteinases of phytopathogens can be divided into trypsin-like and subtilisin-like enzymes. The first group contains proteinases that are produced by *Cochliobolus carbonum*, *Verticillium dahliae*, *Stagonospora* (*Septoria*) *nodorum* and *Phytophtora infestans*. Subtilisin-like enzymes are secreted by *C. carbonum*, *P. infestans*, *Acremonium typhium*, *Magnaporthe poae*, *Trichoderma harzianum* and *Fusarium oxysporum*.

Among extracellular proteinases of phytopathogens, the aspartic proteinases are fairly widespread. These include the enzymes produced by *Botrytis cinerea* (Valueva and Mosolov, 2004) *Cryphonectria parasitica* (endothiapepsin) and *Glomerella cingulata*. Cysteine proteinase is secreted by the fungus *Pyrenopeziza brassicae*. Metalloproteinases include a family of Zn-dependent bacterial enzymes belonging to the genus *Erwinia*. One of these proteinases extracted from *Erwinia carotovora* subsp. carotovora is similar in its properties to thermolysin from *Bacillus thermoproteolyticus*. Extracellular proteinases apparently play an active role in the process of pathogenesis. For instance, it has been revealed that in *P. brassicae* nonpathogenic mutants are unable to produce extracellular cysteine proteinase. Recovery of pathogeneses in these mutants was accompanied by the recovery of their ability to produce the proteinase. An important role in disease progression is also played by aspartic proteinase of the fungus *B. cinerea*, which is a wide profile pathogen.

Proteinases found in pathogens can also play an active role in the degradation of other proteins involved in plant protection, for instance, such enzymes as chitinase and β-1,3-glucanase (Alexander *et al.*, 1994). Proteinases of phytopathogenic microorganisms can also perform other specific functions. For instance, in the bacterium *E. chrysanthemi* extracellular metalloproteinase catalyzed the transformation of pectate lyase into the mature form of this enzyme, which is crucial for plant tissue maceration. It has been assumed that certain peptides released by the action of extracellular proteinases of phytopathogenic microorganisms can act as elicitors, activating plant protection reactions. Proteinase inhibitors in plants are able to suppress the enzymatic activity of phytopathogenic microorganisms.

2.8 ROLE OF TOXINS IN PATHOGENESIS

Toxins are compounds that are produced by the pathogens and cause part or all of the symptoms of a disease. Genetic and biochemical studies revealed that at least in part of the plant-pathogen interactions toxins are the determinants of specificity. In such cases, resistance or susceptibility to the fungus correlates with insensitivity or sensitivity to the toxin (Slavov, 2005). Microbial toxins have been the objects of extensive studies as possible pathogenicity or virulence factors for the producer pathogens. Toxins are considered to be the special weapons of the plant pathogens to evade or overcome the inherent resistance strategies of host plants (Kimura *et al.*, 2001). Plants themselves have a broad spectrum of defense barriers to protect themselves from invading organisms. Pathosystems are very diverse, and there is neither a single model of plant-pathogen interactions nor a sample and common resistant mechanism. In some plant diseases, phytotoxins may play a critical role in the development of the disease symptoms.

Microbial toxins are metabolites produced by plant pathogens, which play a role in host-pathogen interactions and disease expression. They are low molecular weight substances produced by some pathogens which are capable of reproducing symptoms similar to that found in natural infections in plants (Bilgram and Dube, 1976). Several plant pathogens often damage their host (plants) tissues by producing toxic metabolites, which induced various symptoms such as necrosis, chlorosis, wilting, water soaking and eventually the death of plants. These toxic metabolites also known is one of the weapons used by pathogen inducing disease condition in susceptible host plants. Many pathogens are known to produce toxins both *in vitro* and *in vivo*, and these toxins have been implicated in the symptom development on the host tissues (Amusa, 2006).

Several phytotoxic metabolites have been found associated with pathogens including bacteria and fungi, which, causes symptoms similar to those caused by the pathogen. Such toxic metabolites include pinolidoxin from *Ascochyta pinodes*, deoxyradicin and maculosin from *Alternaria helianthi* and *Alternaria alternata* (Stierle *et al.*, 1988). Identified metabolites from other pathogens include piricularin from *Piricularia oryzea*, victorin from *Cochliobolus vitoriae*, phaseolotoxin from *Pseudomonas syringae* pv. *phaseolicola*, a toxin from *Periconia circinata*, saccharitoxin from *Helmithosporium sacchari*, cercosporin from *Cercospora* spp. Phytotoxic metabolites of most of these pathogens have been reported to play a significant role in pathogenesis (Amusa, 2006).

Several characteristics have been used for the classification of toxins that affect plants. Such features include their chemistry. Based on this, some phytotoxins are regarded as low molecular weight peptides, others have terpenoid structures, and still others contain carbohydrates (Amusa, 1991). However, few other structures are known for toxins that play an unquestionable role in plant disease. Another form of classification is based on the producing organism. This is, however, of no predictive value since more than one type of phytotoxins can be produced by one organism. Phytotoxin classification has also been based on biological activities such as enzyme inhibitors, antimetabolites, membrane-affecting compounds (Amusa, 2006). However, the widely accepted classification is that based on toxic selectivity to plant genotypes (host selective or nonhost selective) and on the general role in disease development.

Toxins have been implicated in plant disease as far back as deBary who advanced a theory of plant disease often termed the 'toxin theory.' A primary tenant of the toxin theory is that a toxin elaborated by a pathogen may produce all of the symptoms of the disease. As more information was developed the theory was largely discarded. As we will see a little later in this discussion, the discovery of the toxin victorin, a host-specific toxin, revived interest in the toxin theory of plant disease. Toxins may act directly on living host cells, damaging or even killing the host. Some toxins are active on a wide range of plant species (nonhost-specific) or in some cases, as with the toxin victorin (host-specific).

A toxin can be defined as a microbial metabolite excreted (exotoxin) or released by lysed cells (endotoxin) which in very low concentration is directly toxic to the cells of the suscept (host). The term toxin is used for a product of the pathogen, its host, or pathogen-host interaction which even at very low concentration directly acts on living host protoplasm to influence disease development or symptom expression. Toxins are different from enzymes in that they do not attack structural integrity of host tissues but affect the metabolism of the host because the toxins will act on protoplast of the cell.

2.8.1 Toxin Hypothesis

It states that a toxin should produce all symptoms characteristic of the disease; correlation of toxin sensitivity with susceptibility to pathogen and toxin production by the pathogen will be directly related to its ability to cause disease. Except, victorin, the toxic metabolite of *Cochliobolus victoriae*, the vast majority of toxins associated with plant diseases fail to exhibit all the above characters (Luke and Wheeler, 1955).

2.8.2 Classification of Toxins

According to the source of origin, toxins are divided into three broad classes, namely, pathotoxins, vivotoxins, and phytotoxins (Wheeler and Luke, 1963).

2.8.2.1 Pathotoxins

These are the toxins which play a major role in disease production and produce all or most of the symptoms characteristic of the disease in susceptible plants. Most of these toxins are produced by pathogens during pathogenesis. A pathotoxin is a chemical of biological origin, other than an enzyme and plays an important causal role in a plant disease. Most pathotoxins are produced by plant pathogenic fungi or bacteria, but some are produced by higher plants, and one has been reported to be the product of an interaction between a plant and a bacterial pathogen. Some pathogen-produced pathotoxins are highly selective in that they cause severe damage and typical disease symptoms only on plants susceptible to the pathogens that produce them. Others are nonselective and are equally toxic to plants susceptible or resistant to the pathogen involved. A few pathotoxins are species-selective and are damaging to many but not all plant species. In these instances, some plants resistant to the pathogen are sensitive to its toxic product [(*Example*: Victorin: *Cochliobolus victoriae (Helminthosporium victoriae)*)], the causal agent of Victoria blight of oats. This is a host-specific toxin. It fully meets the strict requirements of a pathotoxin.

2.8.2.1.1 Victorin

The fungus, *Cochliobolus victoriae (Helminthosporium victoriae)*, secretes a toxin known as victorin that is toxic to susceptible oat leaves at very low concentrations. Victorin is an unusual, halogen containing cyclic pentapeptide (Fig. 2.4). Toxin synthesis is inherited at a single locus TOX3, and it would seem reasonable to hypothesize that it is synthesized in a manner analogous to other cyclic peptide toxins, that is, by a nonribosomal (cyclic) peptide synthetase (Jonathan E. Markham and Jacques Hille, 2001). Enzymes must also be present to synthesize the novel amino acids required by such a peptide synthetase, but unfortunately to date no reports on the identification of any genes necessary for victorin synthesis are known.

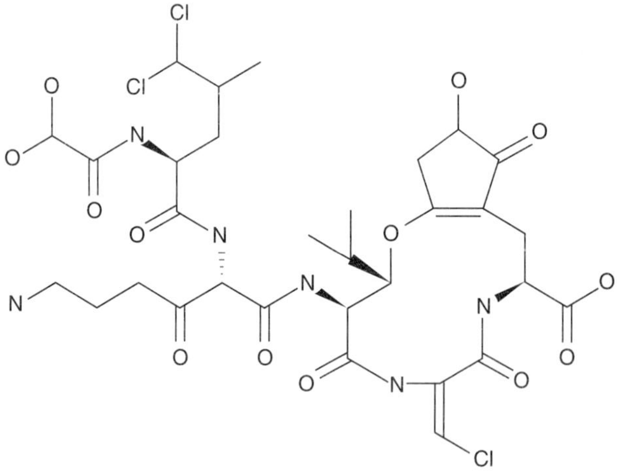

Fig. 2.4 Victorin

Victorin is required for *C. victoriae* to infect its host successfully. Fungal strains that do not produce victorin are not pathogenic. Victorin also reproduces the disease symptoms in the absence of the fungus. Oat sensitivity to victorin is dominant and is determined by a single plant gene, designated *Vb*. Therefore, the study of susceptibility and also host range is simplified to the study of the interaction of a single fungal metabolite, victorin, with the product of a single dominant oat gene *Vb*. The structure of the most prevalent form of victorin has been identified as a cyclized pentapeptide of 814 D. A biologically active, I^{125}-victorin derivative was produced and used to search for oat victorin binding proteins. Several binding proteins were found, one of which binds victorin in leaf slices from susceptible but not resistant cultivars. The gene encoding this protein was cloned and identified as the P protein component of the glycine decarboxylase complex (GDC), an important enzyme complex in the photorespiratory cycle. Picomolar victorin concentrations inhibit GDC in leaf slices, and micromolar concentrations inhibit GDC activity *in vitro*.

The host-selective toxin victorin is produced by *Cochliobolus victoriae*, the causal agent of victoria blight of oats. Victorin has been shown to bind to the P-protein of the glycine decarboxylase complex (GDC) in mitochondria and induces defense-related responses such as phytoalexin synthesis, extracellular alkalization, and programmed cell death. However, evidence demonstrating that the GDC plays a critical role in the onset of cell death is still lacking, and the role of defense-like responses in the pathogenicity has yet to be elucidated. Cytofluorimetric analyses, using the fluorescein (VicFluor) or bovine serum albumin–fluorescein derivative of victorin (VicBSA), demonstrated that victorin-induced cell death occurs before these conjugates traverse the plasma membrane (Yasuomi Tada *et al.*, 2005). As with native victorin, VicBSA clearly elicited apoptosis-like cell death, production of phytoalexin, extracellular alkalization, and generation of nitric oxide and reactive oxygen intermediates. These results suggested that the initial recognition of victorin takes place on the cell surface, not in mitochondria and leads to the activation of a battery of victorin-induced responses. Pharmacological studies showed that extracellular alkalization is the essential regulator for both victorin- and VicBSA-induced cellular responses. A model was proposed where victorin might kill the host cell by activating an HR-like response, independent of the binding to the GDC, through ion fluxes across the plasma membrane.

Previously, victorin was shown to be bound specifically to two proteins of the mitochondrial glycine decarboxylase complex, at least one of which binds victorin only in toxin-sensitive genotypes *in vivo* (Duroy A. Navarre and Thomas J. Wolpert, 1999). This enzyme complex was involved in the photorespiratory cycle and was inhibited by victorin, with an effective concentration for 50% inhibition of 81 pM. The photorespiratory cycle began with ribulose-1,5-bisphosphate carboxylase/oxygenase (Rubisco), and victorin was found to induce a specific proteolytic cleavage of the Rubisco large subunit (LSU). Leaf slices incubated with victorin for 4 hours in the dark accumulated a form of the LSU that is cleaved after the 14th amino acid. This proteolytic cleavage was prevented by the protease inhibitors E-64 and calpeptin. Another primary symptom of victorin treatment is chlorophyll loss, which along with the specific LSU cleavage was suggestive of a victorin-induced, senescence-like response. DNA from victorin-treated leaf slices showed a pronounced laddering effect, which is typical of apoptosis. Calcium appeared to play a role in mediating the plant response to victorin because $LaCl_3$ gave near-complete protection against victorin, preventing both leaf symptoms and LSU cleavage. The ethylene inhibitors amino-oxyacetic acid and silver thiosulfate also gave significant protection against victorin-induced leaf symptoms and prevented LSU cleavage. The symptoms resulting from victorin treatment suggested that victorin causes premature senescence of leaves.

2.8.2.1.2 Other examples

(1) *Host selective/specific:* T-toxin (*Helminthosporium maydis* race T); HC-toxin (*Helminthosporium carbonum);* HS-toxin (*Helminthosporium sacchari*); Phyto-alternarin (*Alternaria kikuchiana);* PC-toxin (*Periconia circinata*).

(2) *Nonselective/Nonspecific:* Tentoxin (*Alternaria tenuis);* Tabtoxin or wildfire toxin (*Pseudomonas tabaci); Phaseolotoxin* (*Pseudomonas syringae* pv. *Phaseolicola).* Table 2.2 gives the major differences between host-specific and nonhost-specific toxins.

Table 2.2 Differences between Host-Specific and Nonhost-Specific Toxins

Host specific	*Nonhost specific*
1. Selectively toxic only to susceptible host of the pathogen	1. No host specificity and can also affect the physiology of those plants that are normally not infected by the pathogen
2. Primary determinants of disease	2. Secondary determinants of disease
3. Produce all the essential symptoms of the disease	3. Produce few or none of the symptoms of the disease
4. *Example:* Victorin, T-toxin	4. *Example:* Tentoxin, Tabtoxin

(3) *Produced by plant or plant X pathogen interaction:* Amylovorin: *Erwinia amylovora* (Fire blight of apple and pears)

2.8.2.2 Phytotoxins

These are the substances produced in the host plant due to host-pathogen interactions for which a causal role in disease is merely suspected rather than established. These are the products of parasites which induce few or none of the symptoms caused by the living pathogen. They are nonspecific, and there is no relationship between toxin production and pathogenicity of the disease-causing agent (*Example*: Alternaric acid–*Alternaria solani*).

The ability of a pathogen to infect and invade a compatible host may be facilitated by the production of toxins that induce cell death in the proximity of the invading organism (Dangl and Jones, 2001). These toxins were also reported to play important roles in inhibiting the physiological processes in cells surrounding the point of infection, enabling the spread of the disease (Staskawicz *et al.*, 2001). Gaumann (1950) has earlier suggested that some pathogens would be unsuccessful if the toxin did not kill the cells in advance of the fungus and permitted it to establish itself continually on dead or dying cells and produce more toxins. While Baker *et al.* (1997) reported that the virulence of an organism is sometimes enhanced by its ability to produce phytotoxins that kill cells in the tissue surrounding the point of infection. In some plant diseases, especially with yam anthracnose, toxins often produced a more rapid and extensive invasion by the pathogen than would be in the case in the absence of toxins (Amusa, 1991). Amusa *et al.* (1993) reported the extraction of phytotoxic metabolites from *Colletotrichum gleoesporioides* infected yam leaves. The extracted phytotoxic substance-induced necrotic lesion was similar to the symptoms induced by the pathogens on healthy yam leaves. Phytotoxins often act as the initiation factor for successful pathogenesis. Spores of some fungal pathogen have been associated with phytotoxin production, which probably kill cells of susceptible host paving the way for the penetration of the germ tube.

Most of the phytotoxic metabolites acts by modifying the metabolism of the host plants while some are toxic to the plant tissues once accumulated and poison the plant tissues. A phytotoxin secreted by *Psuedomonas syringae* pv. *tabaci*, the pathogen inducing wildfire disease of tobacco, drastically modifies the amino acid metabolism of the plant with the eventual accumulation of ammonia in tobacco leaves, which causes extensive blighting. Interestingly, the pathogens that synthesize the phytotoxin remain unaffected by the toxin (Balasubramanian, 2003). The development of an effective phytotoxin for use in plant disease control will require a comprehensive understanding of the pathogen(s) involved including its virulence and the biology of the target host plant.

Phytotoxins often act as the initiation factor for successful pathogenesis. Spores of some fungal pathogen have been associated with phytotoxin production, which probably kill cells of susceptible host paving the way for the penetration of the germ tube (Amusa, 2006). All known host specific toxins can be detected from the spore germinating fluids of each virulent pathogen but not from those of the avirulent ones. Thus, specificity found to be characteristically associated with host-specific toxin suggests the early participation of toxin at the site of initial contact with inoculated and host surface. Several phytotoxins are now known, beyond a reasonable doubt, to be the determinant factor in pathogenesis, and some can even act as reliable surrogates for the pathogen that produce them. The partially purified metabolites of *Colletotrichum* spp. induced necrotic lesion of varying sizes on leaves and stems of susceptible hosts. While the phytotoxic metabolites of *Colletotrichum graminicola, C. truncatum,* and *C. lindemutianum* inhibited seed germination in respective host crops.

The applications of phytotoxins and toxic containing culture filtrates of plant pathogens combined with tissue culture techniques successfully resulted in obtaining resistant lines in some crops as banana, carnation, wheat, tobacco, and grapevine. This approach is an important complement to classical breeding methods. Although many resistant breeding lines were obtained by *in vitro* selection, there are not many cultivars in agricultural production whose resistance is based on these techniques. The interest of utilizing the *in vitro* methods for improving resistance to plant pathogens remains. They could be considered as methods to contribute to classical selection methods and in combination with them to accelerate obtaining of disease resistant plants since *in vitro* breeding doesn't meet the antipathy of public. Very precise preliminary bioassays and inoculation experiments must be conducted before choosing the successful combination plant pathogen – selective agent – host plant selection level (Slavov, 2005).

2.8.2.3 *Vivotoxins*

These are the substances produced in the infected host by the pathogen and/or its host which functions in the production of the disease, but is not itself the initial inciting agent of the disease. (*Example*: Fusaric acid: Wilt causing *Fusarium* sp., Lycomarasmin: *Fusarium oxysporum* f. sp. *lycopersici*, Piricularin – *Pyricularia oryzae*).

2.8.2.4 *Other Alternaria alternata pathotype toxins*

A. alternata f. sp. *fragariae*, which infects strawberry, and *A. alternate* f. sp. *citri*, pathotype tangerine, produce toxins with epoxydecatrienoic ester backbones, designated AF- and ACT-toxin, respectively (Otani *et al.*, 1995). This structure is also present in AK-toxin, indeed certain forms of AF- and ACT-toxin are also toxic to AK-toxin susceptible cultivars of Japanese pear. Cross pathotype hybridization studies and PCR cloning have revealed that the strawberry and tangerine pathotypes contain homologs of the genes implicated in AK-toxin biosynthesis (Masunaka *et al.*, 2000). The arrangement of the loci and structure of other genes involved in toxin biosynthesis in these fungi awaits further research.

2.8.2.5 *Photoactivated Perylenequinone toxins in plant pathogenesis*

Phytotoxic compounds play multiple roles in plant disease. These range from toxins that facilitate infection by suppressing normal plant defense pathways to ones that alter normal metabolic processes and symptom expression to toxins that directly kill the cells of the host, allowing for colonization of dead tissue (Margaret E. Daub and Kuang-Ren Chung, 2009). Among the most intriguing of the well-studied plant pathogen toxins are the photoactivated perylenequinones. These substances, the product of the polyketide pathway in ascomycete fungi, are colored compounds that are converted to their toxic state through photoactivation. Although studied most for their involvement in plant pathogenesis, photoactivated perylenequinones have also been recovered from saprophytic species, suggesting that they may have broad functions in fungi.

All of the perylenequinone toxins that have been characterized are synthesized by fungi in the phylum Ascomycota. Many of the producing fungi are found within the Dothideomycetes, but other classes are also represented [(Table 2.3) (Margaret E. Daub and Kuang-Ren Chung, 2009)]. Many of the identified perylenequinone producers are plant pathogens, but these compounds have also been isolated from saprophytic species and a lichen. Many compounds and analogs have been identified, but all share the same 3,10-dihydroxy- 4,9-perylenequinone chromophore. Interestingly, the fungal compounds show similarities to compounds isolated from plants and protozoans. These include the photoactive compound hypericin (Giese, 1980) that is produced by St. John's wort, a popular medicinal herb used for the treatment of depression. In nature, animals that feed on *Hypericum* species develop a disease known as hypericism that is characterized by skin lesions, changes in behavior, and sometimes convulsions and death. Moderate to severe photosensitivity has been reported in humans taking St. John's wort for the treatment of depression (Rodriguez-Landa and Contreras, 2003). In addition, some species of protozoa produce photoactivated extended quinines that are involved in photomovement responses (Giese, 1981).

Table 2.3 Photoactivated Perylenequinones Produced by Fungi

Perylenequinone references	*Fungal species*
Alteichin Alterlosins Altertoxins	*Alternaria alternata* *Alternaria eichorniae* *Stemphylium botryosum*
Cercosporin Isocercosporin Acetylisocercosporin	*Cercospora* spp. *Scolecotrichum graminis* *Stagnospora convolvuli*
Calphostin C Cladochrome Ent-isophleichrome Phleichrome	*Cladosporium cucumerinum* *Cladosporium cladosporioides* *Cladosporium herbarum* *Cladosporium phlei*
Elsinochromes	*Elsinoe* spp. *Stagnospora convolvuli*
Hypocrellins Isohypocrellin	*Hypocrella bambusae* *Shiraia bambusicola* *Graphis hematites*
Hypomycin A	*Hypomyces* spp.
Shiraiachromes	*Shiraia bambusicola*
Stemphyltoxins	*Stemphylium botryosum*
Other perylenequinones	*Bulgaria inquinans*

2.8.2.5.1 Cercosporin

Cercosporin is one of the most widely studied toxins. Several photoactivated perylenequinones are produced by a number of major fungal plant pathogens, including species of *Alternaria, Cercospora, Cladosporium, Elsinoe,* and *Hypocrella,* among others. The toxins produced by these pathogens are red in color and share a similar structure. Of the perylenequinone toxins, the best studied are the hypocrellins and cercosporin. The hypocrellins have been studied, not for their role in plant disease, but rather due to their pharmaceutical potential as anticancer agents through photodynamic tumor therapy. Cercosporin, by contrast, has been shown to play an important role in *Cercospora* diseases of diverse hosts. Cercosporin is a red pigment, and toxin-deficient mutants can be visually identified on agar plates.

Cercosporin and the other perylenequinone toxins are photoactivated and lack toxicity in the dark. In the light, these compounds absorb light energy and are converted to an energetically activated triplet state. The triplet molecule then reacts with oxygen and results in the generation of toxic, activated oxygen species such as singlet oxygen ($^1O_2$)

and superoxide (O_2^-) (Fig. 2.5). This property classifies cercosporin as a photosensitizer, a term that dates to the early 20th century and was used to describe compounds that

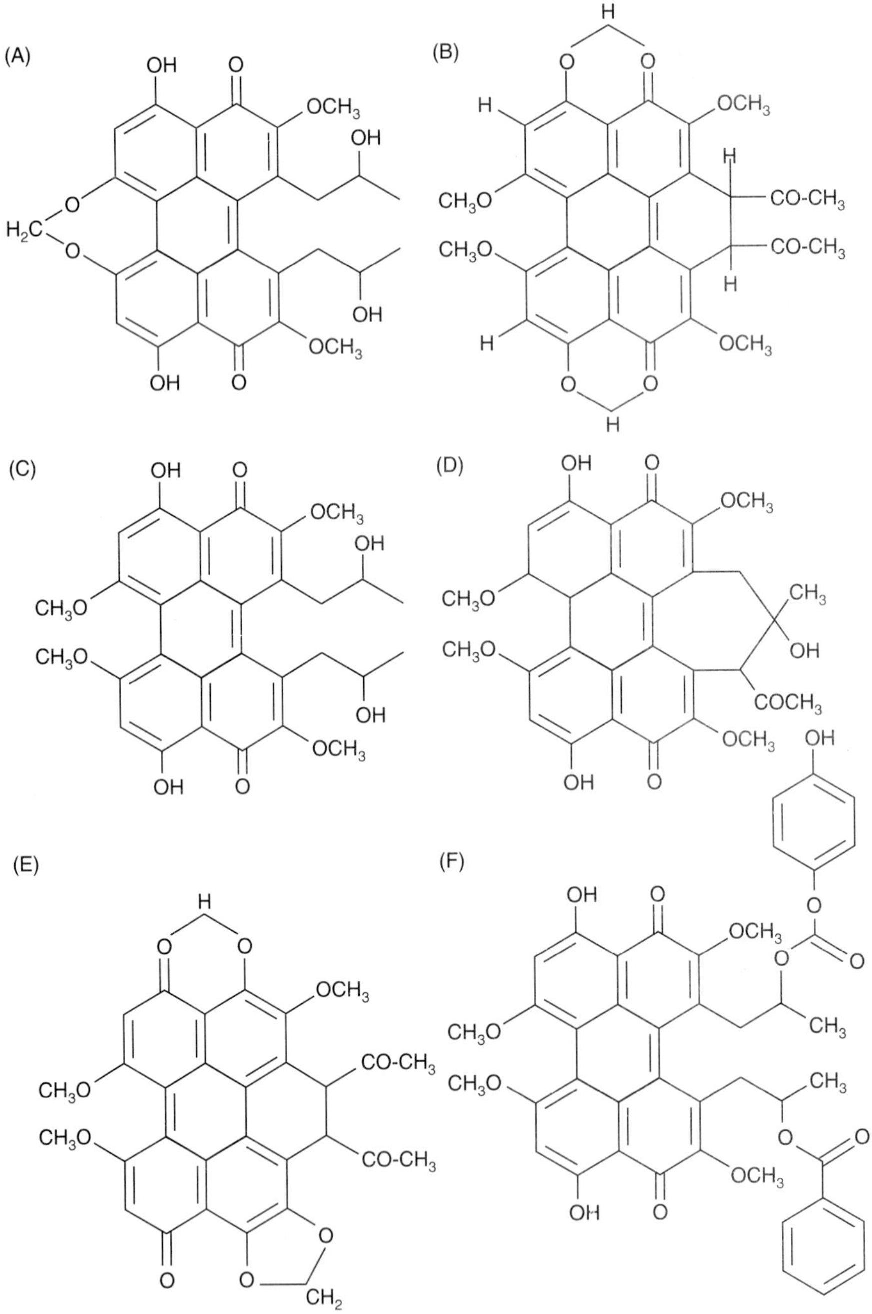

Contd...

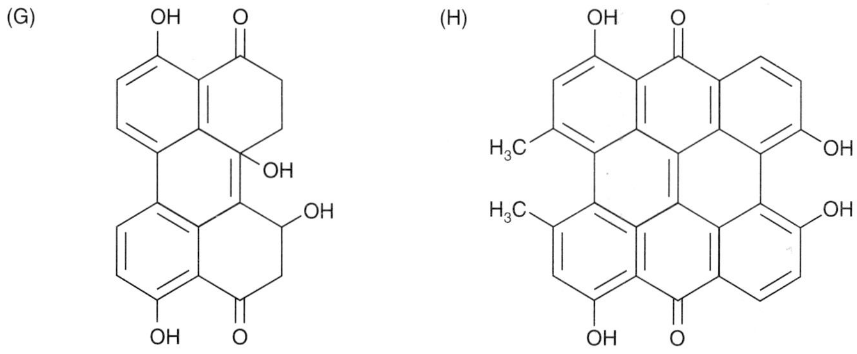

Fig. 2.5 Structure of photoactivated perylenequinones produced by fungi (A-G) and H (Plants)- A(Cercosporin), B (Elsinochrome), C (Phleichrome), D (Hypocrellin), E (Hypomycin), F (Calphostin), G (Altertoxin), H (Hypericin from a herb, *Hypericum perforatum*)

'sensitized' cells to visible wavelengths of light. Photosensitizers are structurally diverse and include common dyes such as methylene blue and acridine orange as well as natural products such as porphyrins, flavins, chlorophyll, and extended quinones. Toxicity of photosensitizers is due to oxidative damage to lipids, proteins, and nucleic acids, with the cellular target dependent on where the photosensitizer molecule localizes in cells (Valenzeno and Pooler, 1987). Cercosporin is a membrane sensitizer and a potent producer of singlet oxygen. Exposure of plant cells and tissues to cercosporin results in peroxidation of the membrane lipids, leading to membrane breakdown and death of the cells. It is hypothesized that membrane damage allows for leakage of nutrients into the leaf intercellular spaces, allowing for fungal growth and sporulation (Daub and Chung, 2007).

Production of cercosporin is mediated by multiple signal transduction pathways. Studies using pharmacological inhibitors implicated the involvement of calcium and calmodulin signaling in cercosporin production (Chung, 2003.). A study also revealed that cercosporin production was associated with a gene encoding an MAP kinase kinase kinase in *C. zeae-maydis* (Shim and Dunkle, 2003). In addition, cercosporin biosynthesis is regulated by another transcription factor, *CRG1*. The *CRG1* gene was originally identified as a gene involved in cercosporin resistance, but disruption mutants also showed a reduction in production. Cercosporin plays an important role in *Cercospora* pathogenicity and lesion formation in several crop species. Thus, any means of reducing cercosporin toxicity promises to have an important application for managing Cercospora leaf spot diseases in the field. In addition, cercosporin is very toxic due to the production of reactive oxygen species. Breeding for Cercospora disease resistance has been difficult since cercosporin has general toxicity to cells. Through our studies, we found that biosynthesis and regulation of cercosporin is a complicated process (Daub. and Chung, 2007). Also, the mechanisms by which the *Cercospora* fungi operate to protect themselves against cercosporin toxicity likely involve multiple factors.

2.8.2.5.2 Mode of action of perylenequinone toxins

The fungal perylenequinone toxins fit the classification of a group of compounds known as 'photosensitizers,' which signifies the ability of these compounds to 'sensitize' living cells to visible wavelengths of light. Beyond the perylenequinones and synthetic dyes, there are many classes of compounds that have photosensitizing activity. These include numerous compounds isolated from plants such as chlorophylls, coumarins, thiophenes, and acetylenes, compounds important to human health such as riboflavin and porphyrins, and other commonly used dyes such as methylene blue, rose bengal, and acridine orange (Spikes 1989). Due to their toxicity, photosensitizers are being investigated for possible utility as herbicides and insecticides, as antiviral agents, and in photodynamic tumor therapy (Hudson and Towers 1991).

Photosensitizers are colored and absorb visible wavelengths of light. The absorbed light energy causes the photosensitizer to be converted to the energetically activated and long-lived 'triplet state.' Triplet state photosensitizers may react through radical or energy transfer reactions. In the so-called 'type I reaction' the triplet sensitizer reacts through a reducing substrate, leading to the formation of a reduced photosensitizer (Margaret E. Daub and Kuang-Ren Chung, 2009). The reduced photosensitizer in turn may react with cellular macromolecules such as lipids, proteins, or nucleic acids, causing oxidative damage and leading to the production of lipid peroxides and other free radical compounds. The reduced photosensitizer may also react with oxygen, leading to the production of reduced and free radical forms of oxygen including superoxide (O_2^-), hydrogen peroxide (H_2O_2), or the hydroxyl radical (OH).

Photosensitizers may also react directly with oxygen via energy transfer, a reaction known as a 'type II reaction.' These reactions result in the production of the highly toxic singlet oxygen ($^1O_2$). In cells, the type of damage resulting from photosensitization is due to where the photosensitizer molecule localizes (Ito 1981). Lipid-soluble photosensitizers, for example, generally cause membrane damage whereas those that localize in nuclei damage DNA Photoactivated perylenequinone toxins play an important role in diseases caused by *Cercospora* species and likely also by a diversity of other fungal pathogens that synthesize these interesting compounds. Significant advances have been made in the understanding of the mode of action of these toxins, their biosynthesis and regulation, and possible mechanisms involved in resistance to their toxic effects. Although efforts thus far to engineer resistance in crop plants to cercosporin and *Cercospora* pathogens have not yet been successful, the strategy of targeting cercosporin in genetic engineering efforts remains a promising strategy for developing crop species resistant to these damaging pathogens (Margaret E. Daub and Kuang-Ren Chung, 2009).

2.8.2.6 Toxin classification based on their specificity

The idea that toxins play causal roles in plant diseases is attractive and dates back to the time of deBary. When one reduces the toxin theory to its most elementary form, it can be stated that all of the symptoms of a given disease result from the direct action of a toxic product of the pathogen in that disease. Tenents of this theory can be tested by the criteria like the toxin, applied at concentrations which could be reasonably expected in or around the diseased plant. It produces in a susceptible host all the symptoms characteristic of the disease. The pathogen and the toxin exhibit similar suscept (host) specificity; while, the ability of the pathogen to produce the toxin varies directly with its ability to cause disease, and a single toxin is involved.

The first toxin isolated that fulfilled the above criteria was that produced by *Cochliobolus (Helminthosporium) victoriae. Helminthosporium* leaf blight of oats appeared in 1945 with the introduction and widespread use of the oat variety 'Victoria' and its derivatives which contained the *Vb* gene for resistance to the crown rust disease. In the Victoria lines, the fungus infects the basal portions of plants, and as it infects, the fungus elaborates a toxin that is carried to the leaves, causing a severe leaf blight, and death of the plant. All other oat varieties and other plant species are either immune or sensitivity to the toxin was proportional to their susceptibility to the fungus. In addition to the toxin producing external symptoms identical to that induced by the pathogen, internal biochemical and histochemical induced by the toxin are identical to that induced by the pathogen. Chemically, the toxin is a complex chlorinated cyclic pentapeptide. Ultrastructural studies demonstrate the primary target of the toxin is the plasma membrane. Here, it apparently binds to proteins and in some yet to be determined mechanism alters the metabolism of susceptible cells.

2.8.2.7 Host-specific/Host-selective toxins

These are the metabolic products of the pathogens which are selectively toxic only to the susceptible host of the pathogen [*Example*: Victorin, T-toxin, Phytoalternarin, etc. (Table 2.4)]. Host-selective toxins are known determinants of compatibility in plant–fungus interactions and provide a powerful model for understanding the specificity of these associations (Jonathan E. Markham and Jacques Hille, 2001). The identification of genes required for toxin biosynthesis has shown that the genes are unique to the toxin-producing species and cluster in complex loci. These loci may have been acquired by horizontal gene transfer. Many, if not all, host-selective toxins act by disrupting biochemical processes and in several cases the resulting cell death has the characteristics of programmed cell death. This ability to make dead tissue from living has enabled these facultative saprophytic fungi to become plant pathogens.

Host-selective toxins are known determinants of compatibility in plant–fungus interactions and provide a powerful model for understanding the specificity of these associations. The identification of genes required for toxin biosynthesis has shown that the genes are unique to the toxin-producing species and cluster in complex loci. These loci may have been acquired by horizontal gene transfer. Many, if not all, host-selective toxins act by disrupting biochemical processes and in several cases the resulting cell death has the characteristics of programmed cell death. This ability to make dead tissue from living has enabled these facultative saprophytic fungi to become plant pathogens.

Table 2.4 Fungi Producing Host-Selective Toxins

Organism	*Toxin*	*Structure*
Cochliobolus carbonum	HC-toxin	Cyclic tetrapeptide
C. heterostrophus	T-toxin	Polyketide
C. victoriae	Victorin	Cyclic pentapeptide
Alternaria alternata f. sp. *lycopersici*	AAL-toxin	Aminopentol ester
A.alternata f. sp. *kikuchiana*	AK-toxin	Epoxy-decatrienoic acid backbone
A. alternata f. sp. *fragariae*	AF-toxin	Epoxy-decatrienoic acid backbone
A.alternata f. sp. *citri tangerine*	ACT-toxin	Epoxy-decatrienoic acid backbone
A. alternata f. sp. *mali*	AM-toxin	Four members cyclic depsipeptide

Fungi can fail to be pathogenic because they induce defenses. Induced resistance involves detection of the fungus and the mounting of an increased localized and systemic defense against fungal invasion. This involves the biosynthesis of pathogenesis-related proteins and phytoalexins. The hypersensitive response implicates cell death as a mechanism of resistance, but some necrophytic or facultative saprophytic fungi utilize cell death to their advantage. By secreting toxins which are able to kill cells of susceptible hosts, these fungi are able to infect, colonize and feed off their host and in the meantime complete their own life cycle. In this case, cell death is associated with susceptibility, and resistance is mediated by insensitivity to the effects of the toxin. In most cases, application of purified toxin is sufficient to cause cell death in susceptible plants. These toxins are an absolute requirement for successful infection of susceptible hosts and are therefore compatibility factors (Briggs and Johal, 1994).

Host-selective toxins are known determinants of compatibility in plant–fungus interactions and provide a powerful model for understanding the specificity of these associations. The identification of genes required for toxin biosynthesis has shown that the genes are unique to the toxin-producing species and cluster in complex loci. These loci may have been acquired by horizontal gene transfer (Jonathan E. Markham and Jacques Hille, 2001). Many, if not all, host-selective toxins act by disrupting biochemical processes and in several cases the resulting cell death has the characteristics of programmed cell death. This ability to make dead tissue from living has enabled these facultative saprophytic fungi to become plant pathogens.

By killing the plant cells with a secreted, soluble toxin, the fungi are able to circumvent the innate defenses of living plant cells (Van Loon, L.C., 1997). In all cases examined so far, a single plant gene has conferred a stable form of resistance (or susceptibility) to this type of infection strategy, but several genes are associated with toxin production in the fungus. Thus the gene-for-gene interaction model does not apply for this type of plant–fungus interaction. The toxins and host sensitivity to them are a major determinant of the fungus' host range, and so they are commonly referred to as host-selective toxins (Pringle and Scheffer, 1964). The genera *Alternaria* and Cochliobolus contain species that produce a variety of host-selective toxins and, as a result, have different host ranges. These species provide a unique model for understanding host-selectivity and the strategies deployed by pathogens to allow infection. Both genera also contain saprophytic species or usually isolates of the same species, which have no known specific plant host.

Cochliobolus and *Alternaria* contain several pathogens which show enhanced pathogenicity towards their hosts. The enhanced pathogenicity in these species is linked to toxin production because nontoxin producing strains can be just as pathogenic as toxin producing strains if they are inoculated on toxin-sensitive host varieties in the presence of the host-selective toxin (Comstock and Scheffer, 1973). Toxins are usually a variety of acyl or ester derivatives of common structures with varying degrees of toxicity. As it is not always clear if a single molecular species is being used and because there is no consensus in the literature on the use of plural vs. singular names, we refer to all toxins in this review in the singular. While toxins are required for pathogenicity; they are not necessarily the sole determinants of pathogenicity. Extracellular proteins, enzymes, and nonhost selective toxins may all play a role in overall pathogenicity (Tonukari *et al.*, 2000). It is vital for our understanding of these diseases to know what genetic changes have created the enhanced pathogenicity. The *Cochliobolus* species are common pathogens of monocotyledons, which include important crop species such as rice, maize and oat (Agrios, 1997). *Cochliobolus* (anamorph *Bipolaris*) species are true fungi of the *Ascomycota*. Most isolates of *Cochliobolus* are non- to only weakly-pathogenic, but strains with increased pathogenicity due to the production of host-selective toxins have emerged.

The genera *Alternaria* and *Cochliobolus contain* species that produce a variety of host-selective toxins and, as a result, have different host ranges [(Table 2.4) (Jonathan E. Markham and Jacques Hille, 2001)]. These species provide a unique model for understanding host selectivity and the strategies deployed by pathogens to allow infection.

2.8.2.8 *Toxins promoting pathogenicity of Cochliobolus and Alternaria species*

These genera contain several pathogens which show enhanced pathogenicity towards their hosts. The enhanced pathogenicity in these species is linked to toxin production because non-toxin-producing strains can be just as pathogenic as toxin producing strains if they are inoculated on toxin-sensitive host varieties in the presence of the host-selective toxin (Comstock and Scheffer, 1973). Toxins are usually a variety of acyl or ester derivatives of common structures with varying degrees of toxicity. While toxins are required for pathogenicity; they are not necessarily the sole determinants of pathogenicity. Extracellular proteins, enzymes, and nonhost selective toxins may all play a role in overall pathogenicity (Annis and Goodwin, 1997). Recent progress in the isolation of genes involved in toxin biosynthesis has shown that the arrangement of toxin biosynthesis genes follows a similar pattern in this diverse range of fungi and suggests a common mechanism by which toxin producing strains may have arisen.

2.8.2.8.1 HC-toxin

HC-toxin is a cyclic tetrapeptide of structure cyclo (D-Pro-L-Ala-D-Ala-L-Aeo), where Aeo stands for 2-amino-9,10-epoxi-8-oxodecanoic acid (Fig. 2.6). It is a determinant of specificity and virulence in the interaction between the producing fungus, *Cochliobolus carbonum*, and its host, maize. HC-toxin qualifies as one of the few microbial secondary metabolites whose ecological function in nature is understood. Reaction to *C. carbonum* and HC-toxin is controlled in maize by the Hm1 and Hm2 loci. These loci encode HC-toxin reductase, which detoxifies HC-toxin by reducing the 8-carbonyl group of Aeo. HC-toxin is an inhibitor of histone deacetylases (HDACs) in many organisms, including plants, insects, and mammals (Walton, 2006), HC-toxin appears to induce the degradation of defense gene transcripts in maize. Another leaf spot disease of maize is caused by the pathogen *C. carbonum*. Races 2 and 3 are weakly virulent, producing small chlorotic or necrotic flecks on leaves, but race 1 is extremely pathogenic on susceptible varieties causing large necrotic lesions on leaves (Pitkin *et al.*, 2000). The increased pathogenicity of race 1 is due to the production of HC-toxin, a cyclic tetrapeptide containing D-amino acids (Kawai *et al.*, 1983). It was the first host-selective toxin to be identified and consequently provides a model for studies concerning other host-selective toxins.

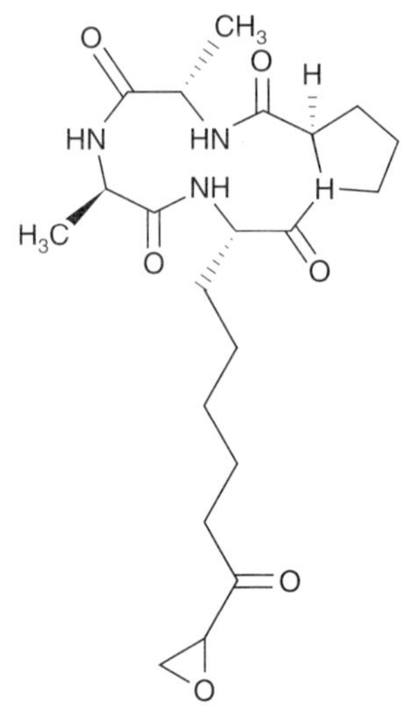

Fig. 2.6 HC-toxin

The genes for HC-toxin biosynthesis have been mapped to a single locus, *TOX2*, which, depending upon the isolate is located on either a 3.5 Mbp or 2.2 Mbp chromosomes. The locus covers at least 540 kbp and is thought to contain all the genes required for toxin biosynthesis. The central gene for toxin biosynthesis encodes the enzyme HC-toxin synthetase (HTS1), a 570 kDa nonribosomal peptide synthase. Other genes, *TOXCTOXD, TOXF*, and *TOXG*, encode enzymes which may synthesize the substrates for HC-toxin synthetase (Ahn and Walton, 1997). *TOXA* is closely linked to *HTS1* and encodes a putative HC-toxin efflux carrier (Pitkin *et al.*, 1996). All these genes, except *HTS1*, are under the transcriptional influence of *TOXE*.

Some effects of HC-toxin on host plant tissues could be construed as beneficial. Many of these effects are directly opposed to those induced by other host selective toxins.

Instead of inducing a general leakage of electrolytes, HC-toxin increases uptake of organic and inorganic solutes by susceptible maize roots and stimulates uptake, accumulation, and reduction of nitrate. In contrast to the effects of most *HSTs* on cell membranes, HC-toxin induces membrane hyperpolarization of susceptible maize cells rather than depolarization. Further, instead of killing protoplasts, HC-toxin increases the long-term viability of leaf protoplasts a few days beyond that of protoplasts incubated in the absence of HC-toxin. Most of the effects of HC-toxin also are observed with tissues of the resistant genotype but only at toxin concentrations 100- to 200-fold higher than those that affect the susceptible genotype (Thomas J. Wolpert *et al.*, 2002).

2.8.2.8.2 T-toxin

T-cytoplasmic male sterile (T-cms) maize is sensitive to T-toxin, a polyketide synthesized by the fungus *Cochliobolus heterostrophus*, which is responsible for Southern Corn Leaf Blight disease (Wise *et al.*, 1999). T-Toxin is a trichothecene that, unlike the other trichothecenes that inhibit protein synthesis, also causes plant plasma membrane leakage of electrolytes at low concentrations.

The genes for T-toxin biosynthesis are located at the locus TOX1, which has become separated by a chromosome translocation into two loci, TOX1A found on chromosome 12;6 and TOX1B on chromosome 6:12. The loci are inseparably linked to a four-armed linkage group which causes the TOX1 locus to inherit as a single locus. Karyotype comparisons between race T and non-T-toxin producing race show 1.2 Mbp more DNA in race T, which maps to the TOX1A and TOX1B loci (Jonathan E. Markham and Jacques Hille, 2001). Two genes have been identified which are necessary for T-toxin biosynthesis, a polyketide synthase (*PKS1*) that is a multifunctional enzyme similar to fatty acid synthases and *DEC1* which is thought to act as a decarboxylase to remove one carbon as CO_2 from the even-numbered carbon chain to produce the odd numbered final T-toxin (Yoder, 1998). T-toxin, the HST produced by *C. heterostrophus* (*Helminthosporium maydis* = *Bipolaris maydis*) race T and *Mycosphaerella zeae-maydis* (*Phyllosticta maydis*), is certainly the best-characterized HST in regard to its site and mode of action (Lindgren, 1997). Both organisms produce a group of structurally related polyketides that are selectively toxic to maize. Toxin-producing isolates are considerably more virulent than nonproducing isolates, and mutations in structural proteins required for toxin production clearly reduce virulence (Thomas J. Wolpert *et al.*, 2002). T-2 is a trichothecene mycotoxin (Fig. 2.7). It is a naturally occurring mold by-product of *Fusarium* spp. the fungus which is toxic to humans and animals.

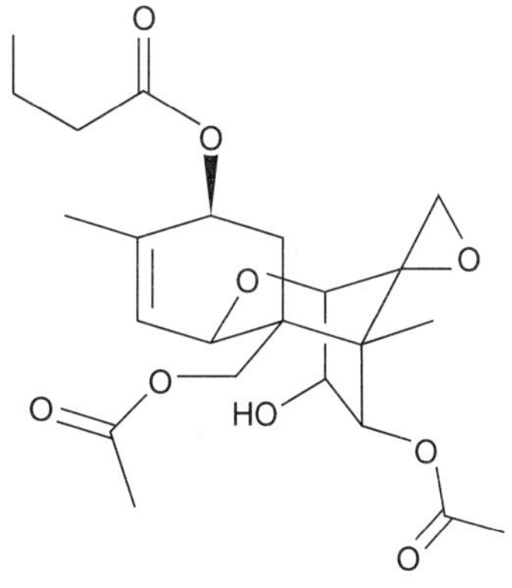

Fig. 2.7 T-2 mycotoxin

The toxin appears to have two modes of action in T mitochondria, viz., as an uncoupler and as an inhibitor of electron transport at Complex I. For example, NADH- or succinate-dependent respiration

(O_2 uptake) in isolated T mitochondria is stimulated by HmT toxin and not coupled to ATP production. The toxin may act at multiple sites on the inner mitochondrial membrane or that the toxin increased membrane permeability in T mitochondria (Marcia J. Holden and Heven Sze, 1984). T-toxin increases membrane permeability specifically in T-mitochondria. Ca^{2+} uptake into T mitochondria driven by either malate succinate, or ATP generated a Ca^{2+} gradient that was dissipated by either the Ca^{2+} ionophore (A23 187) or HmT toxin.

2.8.2.8.3 Victorin

The fungus *Cochliobolus victoriae* causes Victoria blight disease in oats. The fungus secretes a toxin known as victorin that is toxic to susceptible oat leaves at very low concentrations (Wolpert *et al.*, 1995). Victorin is an unusual halogen-containing cyclic pentapeptide. Victorin-C is a heterodetic cyclic peptide (Fig. 2.8). Toxin synthesis is inherited at a single locus TOX_3, and it would seem reasonable to hypothesize that it is synthesized in a manner analogous to other cyclic peptide toxins, that is, by a nonribosomal (cyclic) peptide synthetase. Enzymes must also be present to synthesize the novel amino acids required by such a peptide synthetase, but unfortunately to date no reports on the identification of any genes necessary for victorin synthesis are known. Victorin appears to induce many of the plant responses classically associated with avirulence elicitors such as callose deposition, the respiratory burst, lipid peroxidation, ethylene evolution, extracellular alkalinization phytoalexin synthesis and KC efflux. Thus, genetic and physiological studies support the perception that victorin functions as an elicitor to induce components of a resistance response similar to those induced by avirulence factors. Victorin causes premature senescence of leaves.

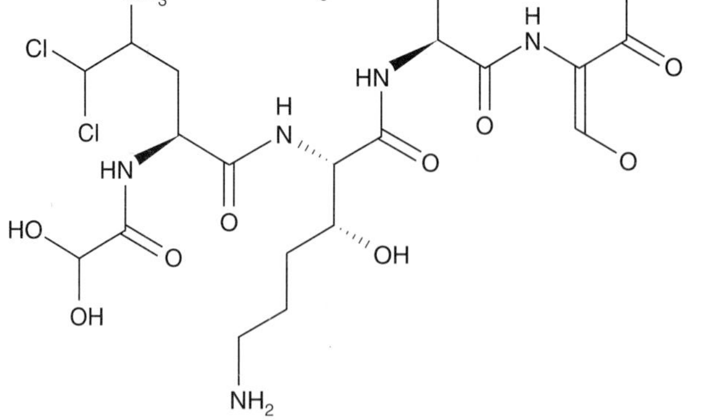

Fig. 2.8 Victorin-C

Evidence is accumulating that the hypersensitive cell death response often associated with induced resistance in gene-for-gene interactions is a type of programmed cell death (PCD) (Thomas J. Wolpert *et al.*, 2002). PCD is a genetically controlled, organized form of cellular suicide that normally functions to eliminate unnecessary or aged cells during morphogenesis (i.e., by cell selection) or cellular maturation. It is a 'normal' form of cell death essential to cellular homeostasis in multicellular organisms. Thus, PCD is distinct from necrotic cell death, which typically arises as a result of severe cellular damage. Necrosis does not entail the participation of the cell in its own destruction, as does PCD, but rather arises from a loss of function. The distinction between PCD and necrosis may be envisioned as the difference between 'organized disassembly' and 'death from damage.' Victorin induces a form of PCD that shares many of the biochemical and morphological characteristics of apoptosis, including DNA laddering, heterochromatin condensation, cell shrinkage and protease activation. The signaling events leading to the host response to victorin also share characteristics with signaling events in resistance, including changes in calcium homeostasis and the participation of kinases and ethylene (Ishihara *et al.*, 1996). Further, victorin contributes to mitochondrial dysfunction (Curtis and Wolpert, 2002). This finding is noteworthy because, as mentioned above, mitochondrial dysfunction is commonly associated with the induction of PCD and because voctorin had previously been shown to bind to members of the mitochondrial enzyme complex, glycine decarboxylase (GDC) (Navarre and Wolpert, 1995).

2.8.2.8.4 Alternaria species

The fungal genus *Alternaria* contains at least ten host-selective toxin-producing plant pathogens which are all pathogens of dicotyledons (Jonathan E. Markham and Jacques Hille, 2001). The largest group of pathogens is proposed to be all pathotypic variants of the species *Alternaria alternata*, a very common, imperfect fungus (Nishimura and Kohmoto, 1983). Molecular phylogenetics by analysis of DNA–DNA reassociation kinetics, rDNA regions and internal transcribed spacer regions from different *Alternaria alternata* pathotypes support this proposal.

2.8.2.8.5 AAL-toxin

The fungus *Alternaria alternate* f. sp. *lycopersici* causes *Alternaria* stem canker of susceptible tomatoes. The fungus produces an aminopentol ester toxin which is very similar in structure to fumonisin, a toxin first identified in *Fusarium moniliforme*. The consumption of food contaminated by fumonisin is implicated in the development of several human and animal diseases. Identification of the genes involved in AAL-toxin (Figure 2.9) biosynthesis is not possible by traditional genetic methods because *A. alternata* has no known sexual cycle. By using REMI however, it has been possible to create AAL-toxin deficient mutants (Jonathan E. Markham and Jacques Hille, 2001). These mutants are not pathogenic, confirming the requirement of the toxin for pathogenesis. The number of mutants recovered is 1 per 100 screened, which suggests that there may be many genes involved, or that there is instability in the locus during the REMI transformation procedure. Isolation of the sequences adjacent to the insert has not been reported in *A. alternata* f. sp. *lycopersici*, but this technique has been successful for other *A. alternata* pathotypes.

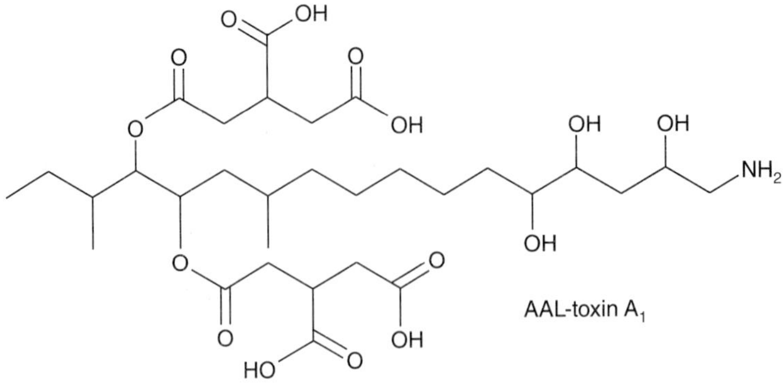

Fig. 2.9 AAL toxin

The AAL-toxins induce an apoptotic-like response in toxin-sensitive tomato, as demonstrated by TUNEL-positive cells, DNA laddering, and the formation of apoptotic-like bodies (Wang *et al.*, 1996). This PCD response, as in the HR, appears to involve calcium and ethylene and therefore, exhibits similarities to an induced resistance response. The AAl-toxins are part of a larger group of compounds structurally related to sphingosine and sphinganine and are referred to as sphinganine analog mycotoxins, or SAMs. Other SAMs, known as fumonisins are produced by the unrelated fungus *Fusarium verticillioides* (syn. *F. moniliforme*) and other *Fusarium* species Fumonisin B1 (and other fumonisins) is as selectively toxic to the AAL-toxin sensitive tomato genotypes as the AAL-toxins. Both fumonisin B1 and TA inhibit ceramide synthase in rat hepatocytes and microsomal preparations from green tomato fruit, with a concomitant alteration in the concentration of sphingoid bases (Thomas J. Wolpert *et al.*, 2002).

2.8.2.8.6 AM-toxin

Alternaria blotch is a worldwide disease of susceptible apple caused by *A. alternata* f. sp. *mali* which is known to produce AM-toxin, a four-member cyclic depsipeptide (Figure 2.10). Cloning of the genes involved in toxin biosynthesis is underway with the isolation of a gene for AM-toxin synthetase (AMT), a potential cyclic peptide synthetase (Johnson *et al.*, 2000). A new assay was developed to measure the necrotic activity of AM-toxin analogs using cultured leaves from meristem cells (Miyashita *et al.*, 2001). This method was not only more sensitive to AM-toxin I but also more reliable than the previous one that used tree leaves due to the homogeneous nature of cultured leaves and to the method of application of toxins. A structure-activity relationship of AM-toxin analogs synthesized was developed. Most residues and the macrocyclic ring structure were strictly recognized by AM-toxin putative receptor, whereas the L-Ala binding subsite of the receptor allowed for side chain structures with various stereoelectronic properties. One of the most homologous genes of AM-toxin synthetase is HC-toxin synthetase from *C. carbonum*. Similarly, as for HC-toxin synthetase, there is also evidence for more than one copy of AMT in the *A. alternata* f. sp. *mali* genome located together on one chromosome. Such an arrangement of genes is highly similar to the characterized TOX2 locus for HC-toxin biosynthesis. It will be interesting to see whether the small chromosome of *A. alternata*

f. sp. *mali contains* a locus with all the genes necessary for AM-toxin biosynthesis and is also unique to the apple pathotype.

AM-toxin I (4 µg/mL), a host-specific toxin of *Alternaria mali,* caused permeability changes in susceptible apple and Nijisseiki pear leaves within 5 min of treatment (Pyoyun Park *et al.*, 2011). The first effect of the toxin appeared on plasma membranes of susceptible apple cells. One hour after toxin treatment, slight invaginations were evident in plasma membranes near plasmodesmata. Six hours after treatment, the spaces between the cell wall and invaginated plasma membrane contained lomasome-like vesicles, membranous fragments, and desmotubules extending from plasmodesmata. The membranous materials appeared to originate from the plasma membrane. Thirty-one hours after treatment, necrotic cells of susceptible leaves showed the general disruptions of cellular membranes which might be caused by secondary effects of the toxin. The similar changes in plasma membrane and plasmodesmata also were found in Nijisseiki pear cells treated with AM-toxin I (4 µg/mL) for 1 h and 6 h. The invaginations of the plasma membrane at plasmodesmata were the first ultrastructural modifications in the toxin-treated Nijisseiki cells. AM-toxin I did not affect the ultrastructure of resistant apple and Chojuro pear cells. These results indicate that the initial sites for the toxin may be on the plasma membrane of the susceptible apple cells. Furthermore, it was suggested that the plasma membrane modifications are associated with permeability changes in the toxin-treated, susceptible plant cells except for original host cells treated with the toxin.

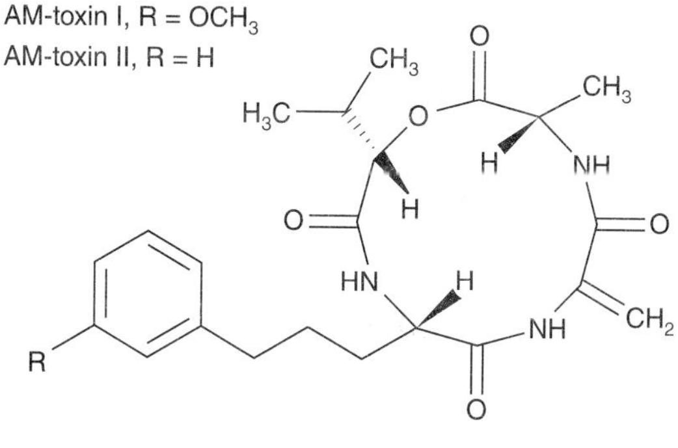

Fig. 2.10 AM-toxin

2.8.2.8.7 AK-toxin or phytoalternarin

Alternaria alternata f. sp. *kikuchiana* causes a black spot on susceptible varieties of Japanese pear. It produces AK-toxin which is an epoxy-decatrienoic ester (Fig. 2.11). The binding site of AK-toxin, a host-specific toxin against Japanese pear, was searched for in the membrane fractions of the pear leaves, using 3H-labeled AK-toxin I methyl ester (Okada *et al.*, 2000). Binding activity, which was displaceable by the unlabeled ligand, was observed for a microsomal fraction from a toxin-susceptible cultivar, Nijisseiki. However, the binding was also observed for those from toxin-resistant cultivars, Kosui and Hosui. Detection of the specific binding failed for the plasma membrane fraction which was prepared from microsomal fraction of the toxin-susceptible cultivar by aqueous two-phase separation.

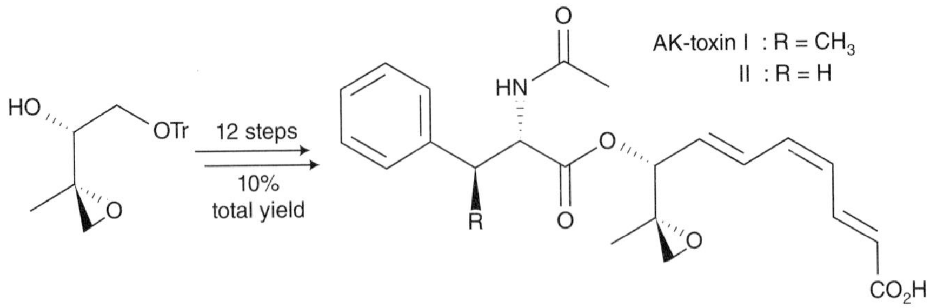

Fig. 2.11 AK-toxin

The filamentous fungus *Alternaria alternata* includes seven pathogenic variants (pathotypes) which produce different host-selective toxins and cause diseases on different plants (Ai Imazaki *et al.*, 2010). The Japanese pear pathotype produces the host-selective AK-toxin, an epoxy-decatrienoic acid ester, and causes a black spot of Japanese pear. It was shown that genes, *AKT1, AKT2, AKT3,* and *AKTR* were involved in AK-toxin biosynthesis. *AKT1, AKT2,* and *AKT3* encode enzyme proteins with peroxisomal targeting signal type 1 (PTS1)-like tripeptides, SKI, SKL, and PKL, respectively, at the C-terminal ends. The peroxisome localization of *Akt1, Akt2,* and *Akt3* was verified by using strains expressing the N-terminal green fluorescent protein (GFP)-tagged versions of the proteins. To assess the role of peroxisome function in AK-toxin production, Aa*PEX6 was isolated* which encodes a peroxin protein essential for peroxisome biogenesis, from the Japanese pear pathotype and made Aa*PEX6* disruption-containing transformants from a GFP-*Akt1*-expressing strain. The ΔAa*PEX6* mutant strains did not grow on fatty acid media because of a defect in fatty acid β oxidation. The import of GFP-Akt1 into peroxisomes was impaired in the ΔAa*PEX6* mutant strains. These strains completely lost AK-toxin production and pathogenicity on susceptible pear leaves. This proved that peroxisomes were essential for AK-toxin biosynthesis. The ΔAa*PEX6* mutant strains showed a marked reduction in the ability to cause lesions on leaves of a resistant pear cultivar with defense responses compromised by heat shock. It was suggested that peroxisome function was also required for plant invasion and tissue colonization in *A. alternata.* Mutation of Aa*PEX6* caused a marked reduction of conidiation.

Restriction enzyme-mediated integration (REMI) mutagenesis was used to tag genes required for toxin biosynthesis (Tanaka *et al.*, 1999). Protoplasts of a wild-type strain were treated with a linearized plasmid along with the restriction enzyme used to linearize the plasmid. Of 984 REMI transformants recovered, three produced no detectable AK-toxin and lost pathogenicity on pear leaves. Genomic DNA flanking the integrated plasmid was recovered from one of the mutants. With the recovered DNA used as a probe, a cosmid clone of the wild-type strain was isolated. Structural and functional analyses of an 8.0-kb region corresponding to the tagged site indicated the presence of two genes. One, designated *AKT1,* encodes a member of the class of carboxyl-activating enzymes. The other, *AKT2,* encodes a protein of unknown function. The essential roles of these two genes in both AK-toxin production and pathogenicity were confirmed by transformation-mediated gene disruption experiments. DNA gel blot analysis detected

AKT1 and *AKT2* homologs not only in the Japanese pear pathotype strains but also in strains from the tangerine and strawberry pathotypes. The host-specific toxins of these two pathotypes are similar in structure to AK-toxin. Homologs were not detected in other pathotypes or nonpathogenic strains of *A. alternata,* suggesting acquisition of *AKT1* and *AKT2* by horizontal transfer.

2.8.2.8.8 Other alternaria alternata pathotype toxins

A. alternata f. sp. *fragariae,* which infects strawberry and *A. alternate* f. sp. *citri,* pathotype tangerine, produce toxins with epoxydecatrienoic ester backbones, designated AF- and ACT-toxin, respectively (Otani *et al.,* 1995). This structure is also present in AK-toxin; indeed certain forms of AF- and ACT-toxin are also toxic to AK-toxin susceptible cultivars of Japanese pear. Cross pathotype hybridization studies and PCR cloning have revealed that the strawberry and tangerine pathotypes contain homologs of the genes implicated in AK-toxin biosynthesis (Masunaka *et al.,* 2000).

2.8.3 Mycotoxins in Plant Pathogenesis

Mycotoxins are defined as low molecular weight fungal metabolites that are toxic to vertebrates. Mycotoxins can have dramatic adverse effects on the health of farm animals and humans that eat contaminated agricultural products. Mycotoxicology has not been a traditional field of plant pathological research. Mycotoxin research has historically been performed by natural product chemists, mycologists, animal toxicologists, and human disease epidemiologists. The apparent lack of specificity of mycotoxins has hindered the acceptance of a role for mycotoxins in plant pathogenesis. In addition, mycotoxin contamination was perceived to be a postharvest problem of stored grain. But it is now well established that many mycotoxin-producing fungal species cause plant disease under field conditions. It thus becomes logical to ask whether mycotoxins themselves play a role in plant pathogenesis in addition to their role in animal diseases. A wide variety of fungal metabolites are both mycotoxic and phytotoxic (Table 2.5). The identification of toxins as probable virulence factors by fungal genetic analysis should be confirmed by plant genetic analysis. If the production of a toxin increases pathogen virulence then increased host resistance to the toxin should increase host resistance to the disease (Anne E. Desjardins and Thomas M. Hohn, 1997).

Table 2.5 Major Mycotoxins

Fungus	*Substrate*	*Mycotoxin*
Aspergillus flavus, A. parasiticus	Maize, groundnut, cotton	Aflatoxin
A. ochraceus, Penicillium viridictum	Wheat, barley, maize	Ochratoxin
Fusarium moniliforme, F. culmorum, F. avenaceum, and *F. nivale*	Wheat, barley, maize	Fumonisins
Fusarium moniliforme, F. equiseti, F. culmorum, F. solani, F. avenaceum, F. roseum, and *F. nivale*	Wheat, barley, maize, oats, rye.	Trichothecene

Contd...

Contd...

Fusarium culmorum, F. graminearum, F. oxysporum, F. roseum, F. moniliforme,	Wheat, barley, maize, sorghum	Zearalenone
Aspergillus carneus, A. terreus, Penicillium citrinum, P. hirsutum,	Rice, wheat, barley, maize	Citrinin
Aspergillus clavatus, Penicillium expansum, Botrytis, P. roquefortii, P. claviforme, P. griseofulvum	Apple and fig	Patulin
Claviceps fusiformis	Rye	Ergot alkaloids

Some lesser known mycotoxins include alternariol, cyclopiazonic acid, diacoumarol, ergotamine, fumitremorgen, fusaric acid, fusariocin, islanditoxin, luteoskyrin, neosolaniol, oosporein, penicillic acid, penitrem, phomopsin, rubratoxin, sporidesmin, tremorgens, etc.

2.8.3.1 *Ergot alkaloids*

The most notorious mycotoxicosis in human history is ergotism, which is caused by consumption of grain, usually rye, contaminated with sclerotia of *Claviceps purpurea.* Ergotism has been known for more than 2,000 years. In 1918, the alkaloid ergotamine was isolated from sclerotia of *C. purpurea* and proven to be a potent vasoconstrictor. Sclerotia can contain a complex mixture of biologically active alkaloids, which are the principal causative agents of ergot poisoning (Beardall and Miller 1994) (Table 2.6). Ergot alkaloids represent a large family of mycotoxins that are derived from both amino acid and isoprenoid precursors and include clavines, simple derivatives of lysergic acid, and structurally complex ergopeptines such as ergotamine. The core structural feature of ergot alkaloids is the ergoline nucleus, the branch point step in ergot alkaloid biosynthesis, is catalyzed by the prenyltransferase, 4-(dimethylallyl) tryptophan (DMAT) synthase, from dimethylallyl diphosphate and tryptophan. The gene encoding DMAT was recently isolated from *C. purpurea* (Tsai *et al.* 1995), but no information is available concerning the possibility that additional ergot alkaloid pathway genes may be closely linked, as occurs in other fungal toxin pathways.

Ergot is the name given to the sclerotium of the fungus *Claviceps purpurea* and some other *Claviceps* species that infect many wild kinds of grass and cereals. These hard black tuber-like bodies consist of a compact mass of hyphae and are the resting stage of the fungus. These ergots produce a range of up to 40 different alkaloids that can lead to the formation of lysergic acid if the ergot ferments. Lysergic acid causes hallucinations, agitation, and other symptoms. Ergot in rye consumed by the population was the main cause of Holy Fire, '*ignis sacer,*' or St. Anthony's fire in the 8th to 16th centuries in Europe, and the effects included gangrenous ergotism, burning sensations and hallucinations. Occasional outbreaks have continued to occur, usually associated with humid summers. Instances occurring in the last 30 years include outbreaks in Ethiopia and India. Ergot was first recognized as a fungus in the 18th century and was known as a source of drugs ('a treasure house' of pharmacologically active compounds) by the beginning of the 19th century. The main ergot alkaloids are ergotamine, ergocornine, ergocristine, ergocryptine, ergometrine and agroclavine.

Table 2.6 Physical and Chemical Properties of Selected Ergot Alkaloids

	Molecular formula	*Molecular weight*	*Melting point (°C)*
Ergotamine	$C_{33}H_{35}O_5N_5$	581.6	212–214 (dec)
Ergocornine	$C_{31}H_{39}O_5N_5$	561.7	182–184 (dec)
Ergocristine	$C_{35}H_{39}O_5N_5$	609.7	165–170 (dec)
Ergocryptine	$C_{32}H_{41}O_5N_5$	575.7	173 (dec)
Agroclavine	$C_{16}H_{18}N_2$	238.2	210–212 (dec)
Ergometrine	$C_{19}H_{23}O_2N_3$	325.4	175–180 (dec)

*dec = decomposition

The ergot sclerotium contains high concentrations (up to 2% of dry mass) of the alkaloid ergotamine (Fig. 2.12), a complex molecule consisting of a tripeptide-derived cyclol-lactam ring connected via an amide linkage to a lysergic acid (ergoline) moiety, and other alkaloids of the ergoline group that are biosynthesized by the fungus. Ergot alkaloids have a wide range of biological activities including effects on circulation and neurotransmission Ergot alkaloids can be classified into two classes, viz., derivatives of 6,8-dimethylergoline and lysergic acid derivatives (Eadie, 2003).

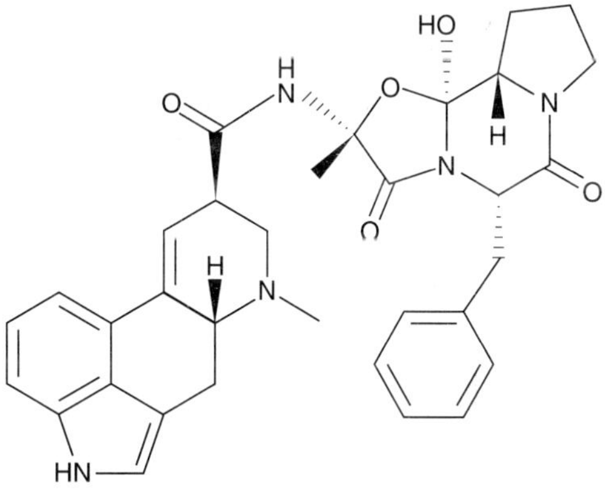

Fig. 2.12 Ergotamine

Biosynthesis of the ergopeptines is catalyzed by a nonribosomal peptide synthetase that employs D-lysergic acid, D-proline, and two additional amino acids as substrates. For ergotamine, these unspecified amino acids are L-alanine and L-phenylalanine. Genes encoding specific peptide synthetases involved in ergopeptine biosynthesis have not been identified in *C. purpurea*. However, portions of genes for putative peptide synthetases have been amplified by polymerase chain reaction from *C. purpurea* and the closely related species *Acremonium coenophialum* (Panaccione 1996). Outbreaks of ergotism occur in animals that eat grain contaminated with *C. purpurea* and other *Claviceps* spp. Similar mycotoxicoses occur in livestock that graze on pastures of certain fescue and ryegrass species that are infected with various *Acremonium* endophytes. These endophytic fungi appear to enhance growth, disease resistance, and drought tolerance of their grass hosts, but also contaminate them with ergot alkaloids that produce gangrene, convulsions, and other neurological disorders in animals that graze infected pastures.

2.8.3.2 *Aflatoxins*

With Turkey X disease and the discovery of aflatoxins, the modern era of mycotoxicology began in England in 1960. Toxicity of animal feeds containing contaminated peanut meal led to the deaths of more than 100,000 turkeys, and of other farm animals, by acute liver necrosis. Scientists in England quickly identified the toxin-producing organism as *Aspergillus flavus* and the toxic agents as a group of related bisfuranocoumarins that were named aflatoxins B1, B2, G1, G2, etc. [(Fig. 2.13) (Reddy and Waliyar, 2012)]. Subsequent studies have shown that aflatoxins are potent liver toxins and liver carcinogens in a wide variety of animals, causing hepatocellular carcinomas in some species at dietary levels below 1.0 µg per kg of feed. Human exposure to aflatoxins can result from consumption of contaminated peanuts, corn, and other agricultural commodities, but also from consumption of meat, milk, and eggs from animals that have consumed contaminated feeds. The occurrence of aflatoxins in milk is of particular concern worldwide (Marasas and Nelson, 1987).

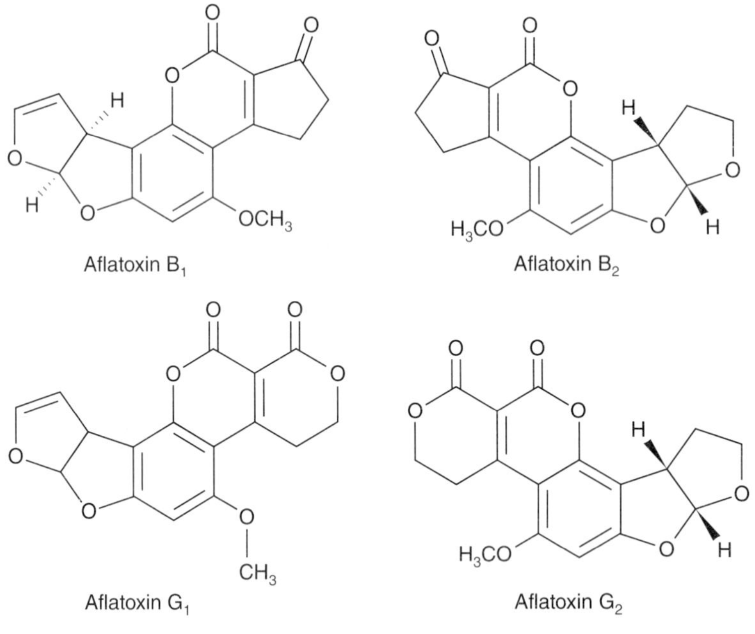

Fig. 2.13 Structures of the major aflatoxins

The various food products contaminated with aflatoxins include cereals like maize, sorghum, pearl millet, rice and wheat, oilseeds such as groundnut, soybean, sunflower and cotton, spices like chillies, black pepper, coriander, turmeric and ginger; tree nuts such as almonds, pistachio, walnuts and coconut, and milk and milk products.

Groundnuts are often contaminated with aflatoxins B_1 and B_2, less often with aflatoxins B_1, B_2, G_1, and G_2, so it is important to have analytical values that represent the total aflatoxin content.

Aspergillus fumigatus is known to produce various immunosuppressive mycotoxins including gliotoxin (Kamel and Watanabe, 2005). However, none of these mycotoxins has been confirmed as being directly related to the pathogenesis of aspergilli. Recent studies have made substantial progress in the determination of mycotoxins as virulence factors. Gliotoxin was found to be produced much faster than previously believed under certain culture conditions, such as high oxygen content, which is close to the environment in the host. Gliotoxin was also found to be detectable in the sera of aspergillosis mice and aspergillosis patients. Based on these findings, it is evident that gliotoxin is produced in the infected organs of patients with aspergillosis at a significant level. In addition to these known mycotoxins, *A. fumigatus* produces many mycotoxins apparently different from known toxins. From the aspect of gene analysis, the deletion of laeA was found to block the expression of metabolic gene clusters such as sterigmatocystin, and the gene is also expected to be related to the production of gliotoxin.

Aflatoxins are produced by *A. flavus* and *A. parasiticus,* whereas a wide range of *Aspergillus* spp. produce the aflatoxin precursor sterigmatocystin, which also is an animal toxin and carcinogen (Table 2.7). The aflatoxin/sterigmatocystin pathways of *Aspergillus* spp. are perhaps the most thoroughly studied fungal polyketide pathways. The first step in the biosynthesis of sterigmatocystin/aflatoxin is catalyzed by a type I polyketide synthase (Chang *et al.,* 1995). In contrast to most polyketide synthases, which utilize acetate as a starter unit, the starter unit for the aflatoxin/sterigmatocystin enzyme is hexanoate (Brobst and Townsend, 1994). The synthase reaction product and first stable intermediate in the pathway is norsolorinic acid, which undergoes a complex series of modifications to yield sterigmatocystin and finally, aflatoxin. Studies of the sterigomatocystin pathway in *A. nidulans* have shown that the gene encoding the polyketide synthase (*pksST*) is part of a gene cluster containing at least 25 pathway-related genes (Brown *et al.,* 1996). This gene cluster occupies a 60-kb region and contains genes for regulatory factors in addition to all of the required pathway enzymes. The genes for the aflatoxin pathways in *A. flavus* and *A. parasiticus* are similarly organized and in most cases appear to contain closely related homologs of the sterigmatocystin pathway genes (Anne E. Desjardins and Thomas M. Hohn, 1997).

Both *A. flavus* and *A. parasiticus* are pathogenic on a variety of plant species, although *A. flavus* predominates on most hosts except peanuts. Aflatoxin production is widespread in both species; field strains of *A. parasiticus,* in particular, rarely lose the ability to produce aflatoxins (Payne, 1983). Aflatoxins, their metabolites, the aflatoxin-8,9-epoxide and the generated ROS causes deleterious effects on the various body organs and body systems including the development of cancers especially the liver cancer mainly due to AFB_1 exposure. Aflatoxins are also responsible for the suppression of both the humoral and cell-mediated immunity and thus making individuals susceptible to infectious diseases. Aflatoxins also responsible for the malabsorption of various nutrients thus leading to nutritional deficiencies, impaired immune function, malnutrition and stunted growth and hence the development of kwashiorkor and marasmus in infants. Aflatoxins also can affect almost all the different body systems and hence the health of the affected individuals especially in poor developing nations of South East Asia and sub-Saharan Africa where there is poor food harvesting, processing and storage thus allowing the growth of mold on them.

Table 2.7 Summary of the Major Aflatoxins Produced by the *Aspergillus* Species of Molds

Difuranocoumarins	*Type of aflatoxin*	*Aspergillus species*
Difurocoumarocyclopentenone series	Aflatoxin B_1 (AFB_1)	*A. flavus*, *A. arachidicola*, *A. bombycis*, *A. minisclerotigenes*, *A. nomius*, *A. ochraceoroseus*, *A. parasiticus*, *A. pseudotamarii*, *A. rambellii*, *Emericella venezuelensis*
	Aflatoxin B_2 (AFB_2)	*A. arachidicola*, *A. flavus*, *A. minisclerotigenes*, *A. nomius*, *A. parasiticus*
	Aflatoxin B_{2a} (AFB_{2a})	*A. flavus*
	Aflatoxin M_1 (AFM_1)	*A. flavus*, *A. parasiticus*; metabolite of aflatoxin B_1 in humans and animals and comes from a mother's milk
	Aflatoxin M_2 (AFM_2)	Metabolite of aflatoxin B_2 in milk of cattle fed on contaminated foods
	Aflatoxin M_{2A} (AFM_{2A})	Metabolite of AFM_2
Aflatoxicol (AFL)	A. flavus, metabolite of AFB_1 Aflatoxicol M_1	Metabolite of AFM_1
Difurocoumarolactone series	Aflatoxin G_1 (AFG_1)	*A. arachidicola*, *A. flavus*, *A. minisclerotigenes*, *A. nomius*, *A. parasiticus*
	Aflatoxin G_2 (AFG_2)	*A. arachidicola*, *A. flavus*, *A. minisclerotigenes*, *A. nomius*, *A. parasiticus*
	Aflatoxin G_{2A} (AFG_{2A})	Metabolite of AFG_2
	Aflatoxin GM_1 (AFG_1)	*A. flavus*
	Aflatoxin GM_2 ($AFGM_2$)	Metabolite of AFG_2
	$AFGM_{2A}$	Metabolite of $AFGM_2$
Aflatoxin B_3 (AFB_3)	Aspergillus species not defined	
	Parasiticol (P)	*A. flavus*
	Aflatrem	*A. flavus*, *A. minisclerotigenes*
	Aspertoxin	*A. flavus*
	Aflatoxin Q_1 (AFQ_1)	Major metabolite of AFB_1 in *in vitro* liver preparations of other higher vertebrates

2.8.3.3 *Trichothecenes*

For more than 100 years, both acute and chronic mycotoxicoses in farm animals and in humans have been associated with consumption of wheat, rye, barley, oats, rice, and maize contaminated with *Fusarium* spp. that produce trichothecene toxins. A number of *Fusarium* spp. produce trichothecenes, viz., *F. acuminatum, F. crookwellense, F. culmorum, F. equiseti, F. graminearum* (*G. zeae*), *F. lateritium, F. poae, F. sambucinum* (*G. pulicaris*), *F. solani*, and *F. sporotrichioides* (Marasas *et al.*, 1984; El-Banna *et al.*, 1984; and Clark *et al.*, 1995). Trichothecene-producing *Fusarium* spp. are destructive pathogens and attack a wide range of plant species. The acute phytotoxicity of trichothecenes and their occurrence in plant tissues also suggest that these mycotoxins play a role in the pathogenesis of *Fusarium* on plants.

The trichothecenes (Fig. 2.14) constitute a family of more than sixty sesquiterpenoid metabolites produced by a number of fungal genera, including *Fusarium, Myrothecium, Phomopsis, Stachybotrys, Trichoderma, Trichothecium*, and others. The term trichothecene is derived from trichothecin, which was the one of the first members of the family identified. All trichothecenes contain a common 12,13-epoxytrichothene skeleton and an olefinic bond with various side chain substitutions.

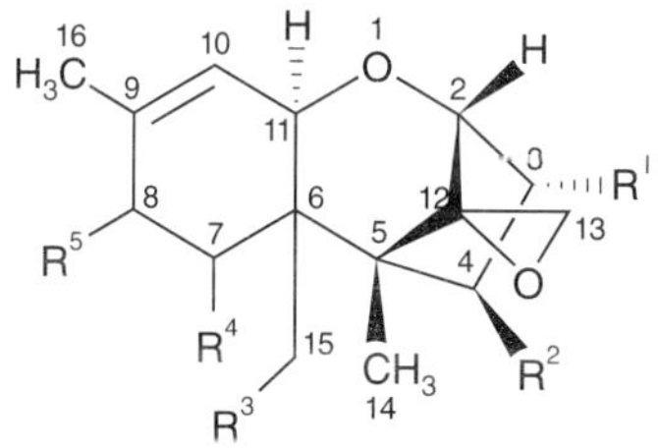

Fig. 2.14 Trichothecenes

Experiments with chemically pure trichothecenes at low dosage levels have reproduced many of the features observed in moldy-grain toxicoses in animals, including anemia, immunosuppression, hemorrhage, emesis, and feed refusal. Historical and epidemiological data from human populations indicate an association between certain disease epidemics and consumption of grain infected with *Fusarium* spp. that produce trichothecenes. In particular, outbreaks of a fatal disease known as alimentary toxic aleukia, which has occurred in Russia since the nineteenth century, have been associated with consumption of overwintered grains contaminated with *Fusarium* spp. that produce the trichothecene T-2 toxin (Anne E. Desjardins and Thomas M. Hohn, 1997). In Japan, outbreaks of a similar disease called akakabibyo or red mold disease have been associated with grain infected with *Fusarium* spp. that produce the trichothecene deoxynivalenol and related compounds.

There is more direct evidence that trichothecenes were responsible for recent human disease outbreaks in India and Japan, where trichothecenes were detected in the toxic grain samples themselves (Beardall and Miller, 1994). In addition, symptoms produced by the trichothecene diacetoxyscirpenol in clinical trials conducted with terminally ill cancer patients were similar to reported symptoms of alimentary toxic aleukia and akakabibyo (Anonymous, 1983). Trichothecenes constitute a large family of sesquiterpene epoxides that inhibit eukaryotic protein synthesis. The biosynthesis of trichothecenes by *Fusarium* spp. proceeds from the hydrocarbon trichodiene through a complex series of steps to trichothecenes such as diacetoxyscirpenol, deoxynivalenol, and T-2 toxin.

2.8.3.4 *Fumonisins*

Fumonisins are amino polyalcohols and are structurally similar to the long-chain base backbones of sphingolipids. Fumonisins inhibit the activity of sphingosine N-acetyltransferase, which leads to the accumulation of toxic sphingoid bases (Fig. 2.15). The toxicity of maize contaminated by *F. moniliforme* has been well documented for more than 100 years. A disease of farm animals known as moldy corn poisoning or blind staggers was first described in the USA in 1850. The most dramatic manifestation of moldy corn disease is leucoencephalomalacia, a fatal brain disease of horses, donkeys, mules and rabbits. In 1988, a South African research group reported the isolation of fumonisin B1 from cultures of *F. moniliforme* (sexual stage, *G. fujikuroi* mating population A) (Anne E. Desjardins and Thomas M. Hohn, 1997). The structural similarity of fumonisins to the long chain sphingolipid bases suggests that fumonisin biosynthesis may be similar to sphingolipid biosynthesis. The latter begins with the condensation, catalyzed by serine palmitoyltransferase, of an amino acid with a fatty acyl-CoA. If fumonisin B1 is synthesized in a similar manner, then alanine would replace serine and an 18-carbon fatty acyl-CoA would replace palmitoyl-CoA. Isotope-feeding studies determined that alanine is a biosynthetic precursor of fumonisin B1 and that the polyalcohol moiety is derived from acetate (Blackwell *et al.*, 1996).

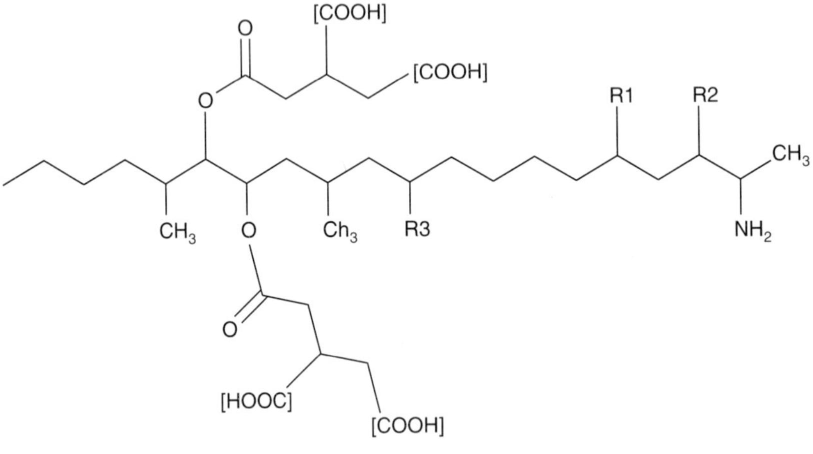

Fig. 2.15 Structural formula of fumonisin B_1-B_4: Fumonisin B1: R1= OH; R2= OH; R3= OH; Fumonisin B2: R1= H; R2= OH; R3= OH; Fumonisin B3: R1= OH; R2= OH; R3= H; Fumonisin B4: R1= H; R2= OH; R3= H

Fumonisins are produced by several members of the *G. fujikuroi* species complex, including serious pathogens of maize, sorghum, millet, and rice, but the most consistent and important producer of fumonisins is *G. fujikuroi* mating population A [*Example*: *Fusarium verticillioides* (formerly *Fusarium moniliforme* = *Gibberella fujikuroi*), *Fusarium proliferatum*, and *Fusarium nygamai*, as well as *Alternaria alternata* f. sp. *Lycopersici]*.

The high frequency (>95%) of fumonisin production among strains of *G. fujikuroi* mating population A from maize and the high frequency of fumonisin contamination in maize raise the possibility that fumonisins play a role in the pathogenesis of maize. Further indirect evidence is provided by the structural similarity of fumonisins to AAL-toxin, which plays a role in the pathogenesis of *Alternaria alternata* f. sp. *lycopersici* on certain genotypes of tomato (Winter *et al.*, 1996). In addition, pure fumonisins at low concentrations have been shown to cause necrosis and other symptoms in maize seedlings, tomato seedlings, and other plants (Lamprecht *et al.*, 1994). Fumonisins are acutely toxic to the liver and kidney of a wide range of experimental animals. Consumption of feed contaminated with fumonisins or an intravenous injection of pure fumonisin B1 can produce a fatal lung edema in pigs. Although the role of fumonisins in some moldy corn diseases of livestock has now been well established, their role in human diseases and, most particularly, their carcinogenic potential in humans are much more difficult to determine.

2.8.4 Other Mycotoxins

A number of other mycotoxins warrant closer scrutiny with respect to their role in plant pathogenesis. These include a diverse array of metabolites produced by *Fusarium*, *Aspergillus*, and *Penicillium* spp. (Anne E. Desjardins and Thomas M. Hohn, 1997).

2.8.4.1 Fusarium mycotoxins

Mycotoxins produced by *Fusarium* spp. are of two general types, the nonestrogenic trichothecenes, including DON, nivalenol, T-2 toxin, and diacetoxyscripenol, and the mycoestrogens, including Zearalenone (ZEN) and zearalenol.

Zearalenone and zearalenol are both estrogenic resorcylic acid lactone compounds produced by the fungi *Fusarium* spp. (Diekman and Green, 1992). Zearalenone {(6-[10-hydroxy-6-oxo-*trans*1-undecenyl]-B-resorcyclic acid lactone) (Fig. 2.16)}, a secondary metabolite from *Fusarium graminearum* (teleomorph *Gibberella zeae*) was given the trivial name zearalenone as a combination of *G. zeae*, resorcylic acid lactone, -ene (for the presence of the C-1′ to C-2 double bond), and -one, for the C-6′ ketone. Zearalenone is better classified as a nonsteroidal estrogen or mycoestrogen. Sometimes it is called a phytoestrogen. Zearalenones are the strongly estrogenic polyketides produced by *F. graminearum* and related species, viz., *F. culmorum*, *F. equiseti* and *F. crookwellense*.

Consumption of feeds contaminated with zearalenones causes severe reproductive and fertility problems in animals (Marasas *et al.*, 1984). The enniatins and beauvericins constitute a family of cyclic depsipeptides that are produced by many *Fusarium* spp. and demonstrate toxicity to both vertebrates and plants. Despite their structural dissimilarity to the steriodal estrogens, ZEN and several of its derivatives possess estrogenic activity. ZEN undergoes a folding such that hydroxyl or potential hydroxyl groups become appropriately orientated to facilitate binding to tissue receptors that normally bind estrogens. Similar binding affinities for ZEN have been determined for the estrogen receptor in sheep and calf uterus (Diekman and Green, 1992).

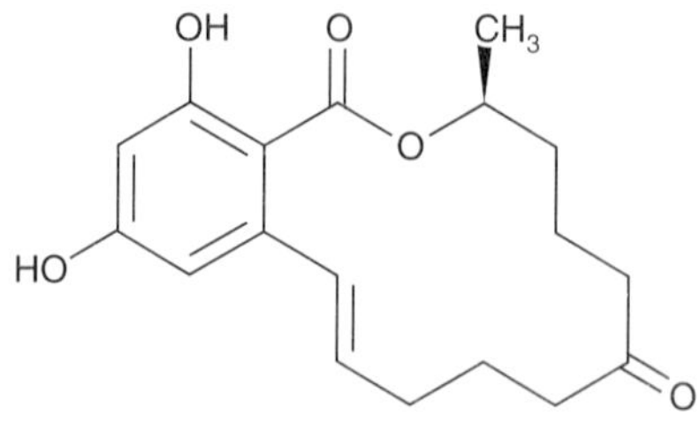

Fig. 2.16 Zearalenone

2.8.4.2 *Moniliformin*

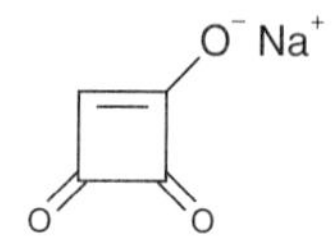

Fig. 2.17 Moniliformin

This is an unusual cyclobutane derivative with phytotoxicity and mycotoxicity, especially to avian species and has been reported to inhibit mitochondrial oxidative enzymes (Leslie *et al.*, 1996). Strains of *G. fujikuroi* associated with the Bakanae disease of rice produce particularly high levels of moniliformin [(Marasas *et al.*, 1986) (Fig. 2.17)]. Moniliformin is formed in many cereal types by a number of *Fusarium* species that include *F. moniliforme, F. avenaceum, F. subglutinans, F. proliferatum, F. fujikuroi* and others. Four new moniliformin-producing species of *Fusarium* were reported by Rabie *et al.* (1982), viz., *F. acuminatum, F. concolor, F. equiseti,* and *F. semitectum*. Isolates of *F. acuminatum and F. concolor* produced large amounts of moniliformin (3.4 and 9.5 g/kg, respectively), whereas isolates of the other three species yielded less than 30 mg/kg. Moniliformin is mainly cardiotoxic and causes ventricular hypertrophy. Moniliformin actually causes competitive inhibition of the activity of pyruvate dehydrogenase enzyme complex of respiratory reaction, which prevents pyruvic acid, a product of glycolysis, to convert to acetyl CoA. Moniliformin has been found in nature in cereal grains together with other toxins such as deoxynivalenol (DON), zearalenone (ZEA) and fusarin C. Moniliformin is produced by approximately 50% of the isolates tested, and its presence might mask unknown toxins. This study indicates that moniliformin found in the culture at a concentration of 1,000 ppm (1,000 μg/g) or higher and when incorporated into a rat diet, most likely accounts for the toxicity observed (Hamed K. Abbas *et al.*, 1990).

2.8.4.3 *Patulin, Ochratoxins, and Citrinin*

These mycotoxins are produced by *Aspergillus, Penicillium,* and *Byssochlamys*. Patulin is a secondary metabolite synthesized by a number of anamorphic fungal genera. Patulin is a cyclic tetraketide with phytotoxic activity [(McKinley and Carlton 1991) (Fig. 2.18)]. It was originally considered an antibiotic but now recognized as a mycotoxin displaying diverse toxic effects: immunosuppression, embryo and maternal toxicity in mice, and genotoxicity. Patulin can be produced by a wide range of fungi, including numerous *Aspergillus, Penicillium,* and *Byssochlamys* species, growing on fruits, viz., apples, pears, grapes and other fruit. In whole fruits, a visual inspection will usually identify poor-quality items. The principal risk arises when the unsound fruit is used for the production of juices and other processed products. It has also been reported in vegetables, cereal grains and silage. *Penicillium expansum* appears to be the mold usually responsible for patulin in apple juice. The major source of patulin in the food supply is juice from apples infected with *P. expansum*. The gene encoding 6-methylsalicylic acid synthase, a polyketide synthase involved in patulin biosynthesis, has been isolated from *P. patulum* and *P. urticae* (Wang *et al.*, 1991). It is most commonly found in rotting apples. In general, the amount of patulin in apple products is viewed as a measure of the quality of the apples used in production. In addition, patulin has been found in other foods such as grains, fruits and vegetables also. Isoepoxydon dehydrogenase is an important enzyme in the multistep biosynthesis of patulin, and its gene is present in other fungi that may potentially produce the toxin.

Ochratoxins are chlorinated cyclic pentaketides with potent liver and kidney toxicity and carcinogenicity [(Prelusky *et al.*, 1994) (Fig. 2.19). Ochratoxins are produced by several *Aspergillus* and *Penicillium* spp. and can occur in a wide variety of grains and other plant products including the meat of animals consuming contaminated feeds. Ochratoxin-A is the most prevalent and relevant fungal toxin of this group, while ochratoxins B and C are of lesser importance Ochratoxins have been reported to be carcinogens and protein synthesis inhibitors. Ochratoxin-A is known to occur in commodities such as cereals, coffee, dried fruit, and red wine. It is possibly a human carcinogen and is of special interest as it can be accumulated in the meat of animals. Thus, meat and meat products can be contaminated with this toxin. Exposure to ochratoxins through diet can cause acute toxicity in mammalian kidneys. Human exposure has been confirmed by measurement of ochratoxin and its metabolites in serum (Fink-Gremmels *et al.*, 1995).

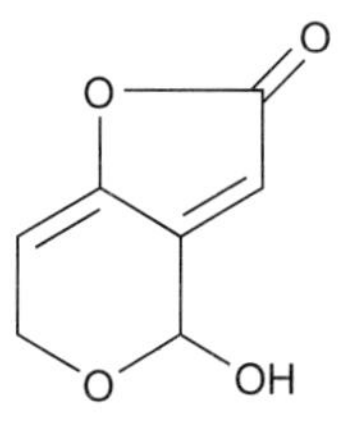

Fig. 2.18 Patulin

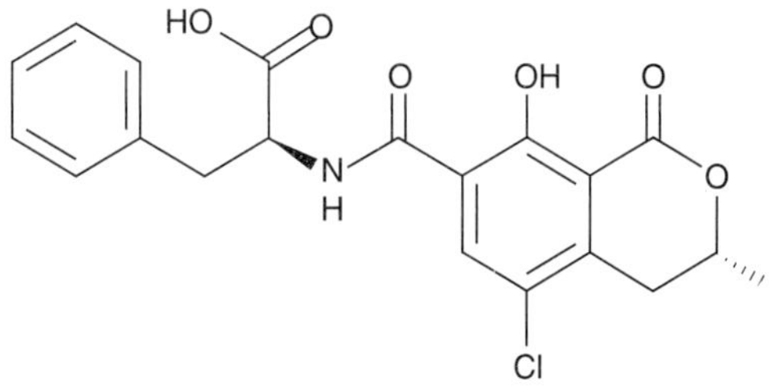

Fig. 2.19 Ochratoxin-A

2.8.4.3.1 Citrinin

It is a toxin that was first isolated from *Penicillium citrinum*, but has been identified in over a dozen species of *Penicillium* and several species of *Aspergillus*. Some of these species are used to produce human foodstuffs such as cheese (*Penicillium camemberti*), sake, miso, and soy sauce (*Aspergillus oryzae*). Citrinin is associated with yellow rice disease in Japan and acts as a nephrotoxin in all animal species tested. Although it is associated with many human foods (wheat, rice, corn, barley, oats, rye, and food colored with Monascus pigment), its full significance for human health is unknown. Citrinin can also act synergistically with Ochratoxin-A to depress RNA synthesis in murine kidneys.

Citrinin (Fig. 2.20) was first isolated from *Penicillium citrinum* prior to World War II; subsequently, it was identified in over a dozen species of *Penicillium* and several species of *Aspergillus* (e.g., *Aspergillus terreus* and *Aspergillus niveus*), including certain strains of *Penicillium camemberti* (used to produce cheese) and *Aspergillus oryzae* (used to produce sake, miso, and soy sauce). More recently, citrinin has also been isolated from *Monascus ruber* and *Monascus purpureus,* industrial species used to produce red pigments. Citrinin has been associated with yellow rice disease in Japan. It has also been implicated as a contributor to porcine nephropathy. Citrinin acts as a nephrotoxin in all animal species tested, but its acute toxicity varies in different species. The 50% lethal dose for ducks is 57 mg/kg; for chickens it is 95 mg/kg, and for rabbits it is 134 mg/kg. Citrinin can act synergistically with ochratoxin A to depress RNA synthesis in murine kidneys. Wheat, oats, rye, corn, barley, and rice have all been reported to contain citrinin. With immunoassays, citrinin was detected in certain vegetarian foods colored with *Monascus* pigments. Citrinin has also been found in naturally fermented sausages from Italy.

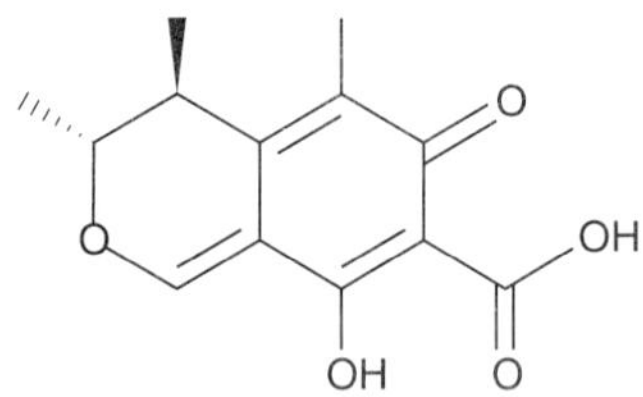

Fig. 2.20 Citrinin

Paenibacillus polymyxa, a Gram-positive low-G+C spore-forming soil bacterium, belongs to the plant growth-promoting rhizobacteria. The swarming motility of *P. polymyxa* strain E681 was greatly induced by a secondary metabolite, citrinin, produced by *Penicillium citrinum* KCTC6549 in a dose-dependent manner at concentrations of 2.5–15.0 μg mL^{-1} on tryptic soy agar plates containing 1.0% (w/v) agar (Soo-Young Park *et al.*, 2008). Flagellum staining showed that citrinin activated the production of flagella by *P. polymyxa*. This result was supported by reverse transcriptase-PCR analysis of gene expression, which showed increased transcriptional levels of *sigD* and *hag* homologs of *P. polymyxa* E681 in the presence of citrinin. The results presented here show that a mycotoxin, citrinin, has a newly identified function of inducing bacterial motility by transcriptional activation of related genes. This finding contributes to our understanding of the interactions between bacteria and fungal strains in nature.

2.8.4.4 Nonspecific/Nonselective toxins

These are the metabolic products of the pathogen, but do not have host specificity and affect the protoplasm of many unrelated plant species that are normally not infected by the pathogen. [*Example*: Ten-toxin, Tab-toxin, Fusaric acid, Piricularin, Lycomarasmin and Alternaric acid].

2.8.5 Effect of Toxins on Host Tissues

2.8.5.1 Changes in cell permeability

Toxins kill plant cells by altering the permeability of plasma membrane, thus permitting loss of water and electrolytes and also the unrestricted entry of substances including toxins. The cellular transport system, especially, H^+/K^+ exchange at the cell membrane is affected.

2.8.5.2 Disruption of normal metabolic processes

The major processes include increase in respiration due to disturbed salt balance, malfunctioning of enzyme system (Example: Pyricularin inhibits polyphenol oxidase) and uncoupling of oxidative phosphorylation).

2.8.5.3 Interfere with the growth regulatory system of host plant

Example: Restricted development of roots induced by *Fusarium moniliforme.*

2.8.5.3.1 Tentoxin

Tentoxin is a natural cyclic tetrapeptide (Fig. 2.21) produced by phytopathogenic *Alternaria alternata*. It selectively induces chlorosis in several germinating seedling plants. Therefore, tentoxin may be used as a potential natural herbicide. Tentoxin was first isolated from *Alternaria alternata* (syn. *tenuis*) and characterized by George Templeton *et al.* in 1967. Tentoxin has also been used in recent research to eliminate the polyphenol oxidase activity from seedlings of higher plants (Duke and Vaughn, 1982.). At a concentration of 2 ppm, the toxin is effective. Acid hydrolysis of the toxin produces leucine, glycine, and N-methylalanine. It causes a reduction in chlorophyll content, inhibits cyclic photophosphorylation and stomatal closure in sensitive plants.

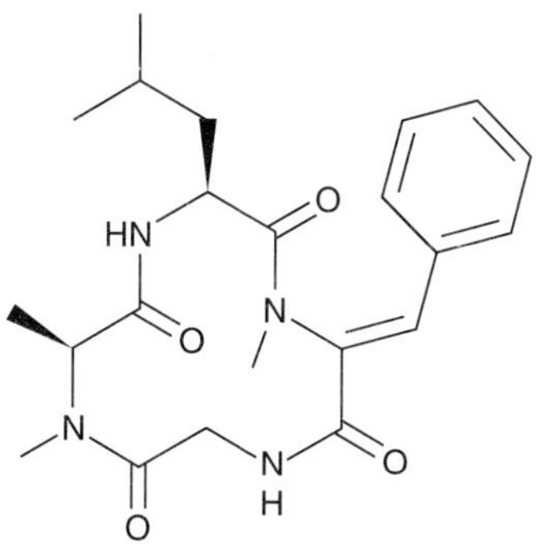

Fig. 2.21 Tentoxin

The disease induced by this pathogen is primarily a seedling disease in a wide range of plant species. Seedling death results when greater than one-third of the leaf area become chlorotic, and reduce vigor with less than that amount of leaf chlorosis. The toxin is a cyclic tripeptide that binds to and inactivates a chloroplast-coupling factor protein involved in energy transfer and also the inhibition of light-dependent phosphorylation of ADP to form ATP.

Tentoxin inactivated photophosphorylation and coupling factor 1 (CF_1) ATPase in lettuce, a sensitive species (John A. Steele *et al.*, 1976). This effect was due to binding of tentoxin with CF_1 at a single site (affinity constant 1.3 to 20 × 10^7 M^{-1}). Neither AMP nor adenyl-5'-yl imidodiphosphate appeared to bind to this site. In radish, an insensitive species, 20 times more tentoxin was required for 50% inhibition of photophosphorylation. In this species CF_1ATPase was unaffected by tentoxin, and its CF_1 bound tentoxin only weakly (affinity constant less than 1 × 10^4 M^{-1}). The sensitivity of photophosphorylation to tentoxin was correlated with chlorosis sensitivity in six other species examined.

Tentoxin-treated mung bean plants were shown to lack chloroplast polyphenol oxidase (PPO) by enzymatic, electrophoretic and cytochemical analysis. Incorporation of PPO (a protein coded by nuclear DNA into the plastid may occur via concentration of the protein into inner envelope-derived vesicles (Kevin C. Vaughn and Stephen O. Duke., 1981). PPO integration into the plastid is apparently blocked by a tentoxin treatment although fraction I protein (and hence the proteins for chloroplast ribosome production) is not affected by this fungal toxin. Both apical and etiolated plastids from teotoxin-treated plants lack PPO. Thus, it is unlikely that the primary effect of tentoxin is due to the binding of the chloroplast coupling factor, as previously supposed.

Holland *et al.* (1997) clarified the mechanism of tentoxin-induced chlorosis in *Nicotiana* spp. seedlings, who observed that chlorosis did not correlate with the inhibition of chloroplast ATP synthesis *in vivo,* since it occurs at tentoxin concentrations far higher than that required for the inhibition of photophosphorylation measured in the same seedlings. However, tentoxin-induced chlorosis did correlate with *in vivo* overenergization of thylakoids. They demonstrated that tentoxin induced overenergization in intact plants and isolated thylakoids, probably via multiple interactions with ATP synthase. Furthermore, gramicidin D, a protonophore that relieves overenergization, also relieved chlorosis. Two lines of evidence suggested that reactive oxygen species may be involved in the process of chlorosis: ascorbate, a quencher of oxygen radicals, significantly protects against chlorosis, whereas transgenic *Nicotiana* spp. mutants overexpressing chloroplast superoxide dismutase were partially resistant to tentoxin-induced chlorosis. It was proposed that chlorosis in developing seedlings resulted from overenergization of thylakoids, which lead to the generation of oxygen radicals.

Tentoxin and to a lesser extent, dihydrotentoxin (both at 10 mmol m^{-3}) reduced stomatal opening in epidermal strips of *Commelina communis* in the light but not in darkness (Dahse *et al.*, 1990). This effect was significantly greater in the normal air than in CO_2-free-air. Fusicoccin overcame the tentoxin effect. However, tentoxin did not inhibit stomatal opening in the light in epidermal strips of *Paphiopedilum harrisianum,* a species which lacks guard cell chloroplasts. It is concluded that tentoxin exerts its action on stomata not by an ionophorous effect in the plasmalemma of guard cells but by the inhibition of photophosphorylation in their chloroplasts. The effects of DCMU and tentoxin on guard cells are discussed in terms of their effects on chloroplasts and the extent to which energy is supplied from this organelle during the stomatal opening in the light. The results indicate that neither photophosphorylation nor noncyclic electron transport in guard cell chloroplasts is essential for stomatal opening.

2.8.5.3.2 Tabtoxin

Tabtoxin is strongly antimicrobial and functions by inhibiting glutamine synthetase and ornithine carbamoyltransferase, respectively. Genetic analysis has revealed the mechanisms responsible for toxin biosynthesis. Tabtoxin is derived from the lysine biosynthetic pathway, whereas syringomycin, syringopeptin, and phaseolotoxin biosynthesis require peptide synthetases. Activation of phytotoxin synthesis is controlled by diverse environmental factors including plant signal molecules and temperature.

Tabtoxin is a monocyclic β-lactam produced by *P. syringae* pv. *tabaci*, *coronafaciens*, and *garcae* [(Mitchell, 1991) (Fig. 2.22)]. The dipeptide toxin contains tabtoxinine-β-lactam (TβL) linked by a peptide bond to threonine. Although tabtoxin is the primary intracellular metabolite produced, the chlorosis-inducing activity occurs only after hydrolysis of the peptide bond by aminopeptidases of plant or bacterial origin. Cleavage of the peptide bond in tabtoxin releases TβL, the toxic moiety (Levi and Durbin 1986).

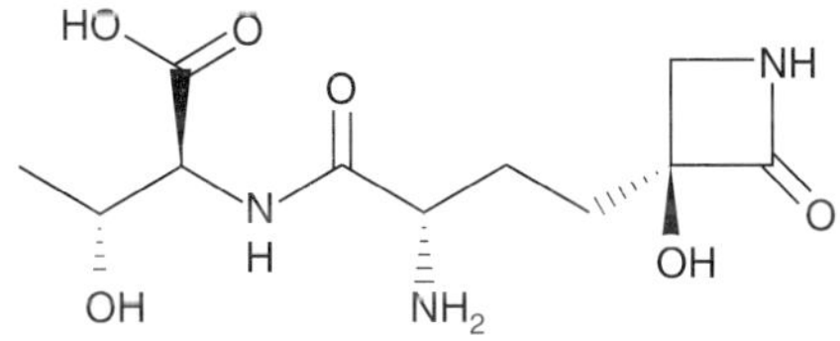

Fig. 2.22 Tabtoxin

TβL irreversibly inhibits glutamine synthetase. The inhibition of glutamine synthetase has at least two major effects; first, the enzyme is presumably the major route for glutamine synthesis in plants, and second, the enzyme is the only way to efficiently detoxify ammonia. A variety of harmful effects has been attributed to ammonia in plants, including disruption of the thylakoid membrane of the chloroplast and uncoupling of photophosphorylation. Protection of the bacteria from the toxin has been associated with the adenylation of glutamine synthetase, which renders the target enzyme less susceptible to inactivation by TβL (Knight *et al.*, 1986). A second potential detoxification mechanism involves the production of β-lactamases which hydrolyze the β-lactam ring of TβL to liberate the nontoxic metabolite, tabtoxinine.

Pseudomonas syringae pv. *coronafaciens*, a pathogen of oats, was mutagenized with Tn5 to generate mutants defective in tabtoxin production (Barta *et al.*, 1992). From a screen of 3,400 kanamycin-resistant transconjugants, seven independent mutants that do not produce tabtoxin (Tox-) were isolated. Although the Tn5 insertions within these seven mutants were linked, they were not located in the previously described tabtoxin biosynthetic region of *P. syringae*. Instead, all of the insertions were within the *P. syringae* pv. *coronafaciens lemA* gene. The *lemA* gene is required by strains of *P. syringae* pv. *syringae* for pathogenicity on bean plants (*Phaseolus vulgaris*). In contrast to the phenotype of a *P. syringae* pv. *syringae lemA* mutant, the Tox-mutants of *P. syringae* pv. *coronafaciens* were still able to produce necrotic lesions on oat plants (*Avena sativa*), although without the chlorosis associated with tabtoxin production. Northern (RNA) hybridization experiments indicated that a functional *lemA* gene was required for the detection of a transcript produced from the *tblA* locus located in the tabtoxin biosynthetic region. Marker exchange mutagenesis of the *tblA* locus resulted in the loss of tabtoxin production. Therefore, both the *tblA* and *lemA* genes are required for tabtoxin biosynthesis, and the regulation of tabtoxin production by *lemA* probably occurs at the transcriptional level.

Eighteen spontaneous, nontoxigenic (Tox-) mutants of *Pseudomonas syringae* pv. *tabaci* caused symptomless infections in tobacco leaves (Turner and Taha, 1984). One of these mutants was selected for detailed study. Infected tissue displayed none of the physiological alterations (loss of chlorophyll, accumulation of ammonia, inactivation of glutamine synthetase) that characterized the effect of tabtoxin in infections caused by the wild-type. However, early growth in, and invasion of inoculated leaves by the Tox-mutant was not detectably different from that of the wild-type. Also, infectious cells of both cultures were transmitted with equal efficiency from inoculated to healthy seedlings in water splash. Therefore, the Tox-mutant was a nonpathogenic parasite. One important difference between the parasitic abilities of these cultures was revealed in the steady decline in numbers of the Tox-mutant in tobacco leaves from three days after inoculation whereas numbers of the wild-type continued to increase, or remained constant, after three days. The population of the wild-type in inoculated tissue grew to an apparent upper limit of 1·6 × 107 cells cm^{-2} leaf from inocula containing between 5 × 103 and 1·0 × 107 cells. This upper limit may have been determined by the amount of free nutrient in the apoplast. The Tox- and wild-type differed not only in toxigenicity but also in that growth of the former was inhibited by 30 μmol mL^{-1} tabtoxin whereas growth of the latter was unaffected or even stimulated by as much as 3·5 μmol mL^{-1}.

2.8.5.3.3 Fusaric acid

Fusaric acid is a picolinic acid derivative (Fig. 2.23). It is typically isolated from various *Fusarium* species and has been proposed for various therapeutic applications. However, it is primarily used as a research tool. Its mechanism of action is not well understood. It likely inhibits dopamine beta-hydroxylase (the enzyme that converts dopamine to norepinephrine). It may also have other actions, such as the inhibition of cell proliferation and DNA synthesis.

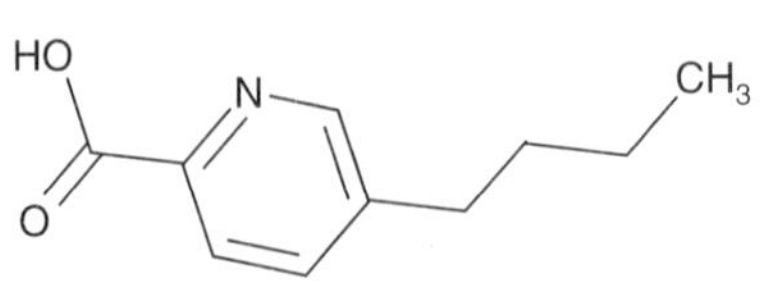

Fig. 2.23 Fusaric acid

Fusaric acid is a mycotoxin with low to moderate toxicity, which is of concern since it might be synergistic with other co-occurring mycotoxins. Fusaric acid is widespread on corn and corn-based food and feeds and is frequently found in grain, where *Fusarium* spp. are also isolated. Fusaric acid (5-butylpicolinic acid) was first discovered during the laboratory culture of *Fusarium heterosporum* Nees by Yabuta *et al.* (Yan *et al.*, 1937). This compound was one of the first fungal metabolites implicated in the pathogenesis of tomato wilt symptoms caused by *F. oxysporum* f. sp. *lycopersici* Schlecht. Emend. Snyd & Hans (Bacon *et al.*, 1996). In addition to the suggested role in plant pathogenesis, fusaric acid is potentially toxic to animals. Fusaric acid is mildly toxic to mice, and it has several important. *Fusarium moniliforme* Sheld (sexual stage *Gibberella fujikuroi* (Sawada) Ito in Ito & K. Kimura) of the section *Liseola* is associated with corn as a nonobligate and usually symptomless endophyte. The endophytic association of corn with this fungus is a cause for concern since several mycotoxins may be produced during its symptomless endophytic colonization. *F. moniliforme* produces fumonisins, fusarins, moniliformin, and beauvericin.

Fusaric acid is known to enhance the toxicity of other mycotoxins in terms of both animal and plant toxicities. This result could be due to the presence of numerous *Fusarium* spp. and strains within the contaminated feed samples or to the fact that with multiple mycotoxins being produced any analysis that focuses on only a single toxin is unlikely to show a strong correlation between toxicity and the amount of any single mycotoxin. Reports of multiple mycotoxins being present in some toxic samples are not uncommon, and work of this type should be encouraged, especially in light of the widespread ability of strains of numerous *Fusarium* species to synthesize fusaric acid (Bacon *et al.*, 1996).

Fusaric acid (5-butylpicolinic acid) was first discovered during the laboratory culture of *Fusarium heterosporum* and was one of the first fungal metabolites implicated in the pathogenesis of wilt symptoms of plants especially under adverse conditions (Bacon Charles *et al.*, 2006). In addition to a primary role in plant pathogenesis, fusaric acid is mildly toxic to mice and has other pharmacological properties. During tests for control of the fungus *F. verticillioides* and reduction of the mycotoxin fumonisin B1, it was determined that fusaric acid was produced in planta and appears to control the growth of the biocontrol endophytic bacterium, *Bacillus mojavensis,* without any apparent symptoms of a disease. Since fusaric acid is considered a wilt toxin, we examined it in planta production and role in the wilt of field maize. Using plants infected with fusaric acid producing and nonproducing strains of *F. verticillioides,* they isolated, identified, and measured fusaric acid in roots of seedlings grown with and without drought stress. It was determined that fusaric acid was produced in planta at the same concentrations regardless of drought stress, and there were no symptoms of wilt disease in the one field maize cultivar tested. Perhaps its major importance was like an antibiotic against *B. mojavensis,* other biocontrol species and endophytic competing bacteria that co-occur with *Fusarium* species and in the soil, suggesting that fusaric acid does not function solely as a wilt toxin in maize.

Suppression of soil-borne diseases by biocontrol agents involves complex interactions among biocontrol agents and the pathogen and between these microorganisms and the plant (Landa *et al.*, 2002). They studied the diversity among strains of fluorescent

Pseudomonas spp., *Bacillus* spp., and *Paenibacillus* sp. for their sensitivity to fusaric acid (FAc) and phytoanticipins from different host plants, the diversity of pathogenic and nonpathogenic *Fusarium oxysporum* isolates for their sensitivity to phytoanticipins, and the influence of FAc on the production of pyoverdine by fluorescent *Pseudomonas* spp. tolerant to this compound. There was a great diversity in the response of the bacterial strains to FAc; however, as a group, *Bacillus* spp. and Paenibacillus macerans were much more sensitive to FAc than *Pseudomonas* spp. FAc also affected the production of pyoverdine by FAc-tolerant *Pseudomonas* spp. strains. Phytoanticipins differed in their effects on microbial growth, and sensitivity to a phytoanticipin varied among bacterial and fungal strains. Biochanin A did not affect the growth of bacteria, but coumarin inhibited the growth of *Pseudomonas* spp. strains and had no effect on *Bacillus circulans* and *P. macerans*. Conversely, tomatine inhibited the growth of *B. circulans* and *P. macerans*. Biochanin A and tomatine inhibited the growth of three pathogenic isolates of *F. oxysporum* but increased the growth of three nonpathogenic *F. oxysporum* isolates. Coumarin inhibited the growth of all pathogenic and nonpathogenic *F. oxysporum* isolates. These results indicated that complex interactions can occur among plants, pathogens and biological control agents in the rhizosphere and on the root surface. Also, these results helped to explain the low efficacy of some combinations of biocontrol agents, as well as the inconsistency in achieving disease suppression under field conditions.

2.8.5.3.4 Piricularin

Piricularin, one of the toxic substances from *Pyricularia oryzae*, produces brown necrotic spots and inhibits seedling growth (Tamari and Kaji, 1959). It exists in two froms, pyricularin and α-picolinic acid (Fig. 2.24). At low concentrations, it inhibits the germination of the conidia of *P. oryzae*, which produces a piricularin binding protein, a copper oxidase that binds [iricularin and destroys its fungi-toxicity but not its phytotoxicity. This toxin induces an increase in the concentrations of polyphenols and oxidases and its toxicity is counteracted by Chlorogenic acid, one of the principal polyphenols of the rice plant (Mehrothra, 1981). A novel phytotoxin, tenuazonic acid, has also been reported in rice plants. Piricularin increases respiration and growth at low doses but inhibits at higher doses. Aspargine was the best nitrogen source for the *in vitro* production of toxins by the pathogen followed by urea, ammonium sulphate and sodium nitrate.

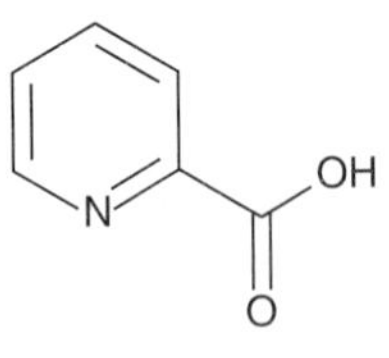

Fig. 2.24 α-picolinic acid

Piricularin plays an important role in the blast-fungal infection together with picolinic acid suppressing the resistant reaction, i.e., the hypersensitive reaction, of the host tissue to the infection (Kinjiro TAMARI *et al.*, 1966). To determine whether or not piricularin takes part in the blast fungal infection, they investigated the effect of a piricularin-detoxifying substance on the resistance of rice tissue to this infection. As a pircularin-detoxifying substance ferulic acid (4-hydroxy-3-methoxycinnamic acid) was chosen, which is a minor ingredient of polyphenols of the rice plant since it is easy to obtain by synthesis and it has not an anti-blast-fungal activity. It was observed that the application of ferulic acid to the rice plant caused an increase in the tissue resistance to the blast-fungal infection.

Several facts supported the view that the increase in the resistance to the infection of rice plants treated with ferulic acid depends upon the piricularin-detoxification, viz., (a) Neither ferulic acid itself nor its oxidized product by oxidase exhibits any anti-blast-fungal activity at 1/5000 dilution with the pH level of rice juice. (b) Not only ferulic acid but its oxidized product by oxidase also had a piricularin-detoxifying ability. (c) The juice pressed from the rice plant applied with ferulic acid also possessed a piricularin-detoxifying ability, exhibiting no inhibitory effect on the germination and growth of blast fungal spores. (d) The application of ferulic acid caused a respiration-rise of the rice plant, but this respiration-rise did not couple with oxidative phosphorylation which was related to the resistant reaction of the host tissue against the infection. The toxic cation of piricularin is mostly due to its inhibitory effects on the activity of various enzymes like peroxidases, catalase, cytochrome oxidase and ascorbic acid oxidase, at low concentrations of 1/100,000-1/200,000. The toxin stimulates the respiration and growth of plants, at still low doses of 1/1,600,000. Because of this stimulatory effect, an increase in all the fractions of organic phosphorus, particularly nucleic acid, phosphorus and protein phosphorus is noticed, apart from an increase in the ribosomal RNA (Mehrothra, 1985).

2.8.5.3.5 Lycomarasmin

Lycomarasmin is a small molecule of empirical formula C9 H15 O7 N3~, and that it yielded on hydrolysis two amino acids, glycine, and aspartic acid. Since these constituents would leave only three carbon atoms and one nitrogen atom unaccounted for and since mild hydrolysis led to the formation of one molecule of ammonia, a labile grouping was indicated in the molecule and indeed, the failure of the toxin to give a blue color with ninhydrin or N2 in the Van Slyke reaction with HN02, suggested that the unidentified three carbon and one nitrogen did not belong to a conventional amino acid such as alanine. The instability of lycomarasmin was well illustrated by the finding that heating it in water, or merely allowing the free acid to stand for long periods in aqueous solution, led to the formation of a biologically inactive compound, C9H1207N2, which differed from the toxin by the elements of NH3. This latter compound was designated Substance I. Lycomarasmin can be extracted from the culture filtrates of *Fusarium lycopersici.* Other acids like pyruvic acid, fusaric acid, and glycine some or all wilt toxins are also produced by *Fuisarium oxysporum.* The biosynthesis of this pyridine alkaloid has been established (Popplestone and Unrau, 1973). It is a polypeptide with molecular weight 277.3.

Topical application of fusaric acid and lycomarasmin results in rapid wilting and necrosis of leaves. Whether the wilting symptoms exhibited by infected plants particularly during high transpiration stress are directly due to the systemic presence of these alkaloids or due to tylosis are not entirely clear. It is known that high concentrations of these wilt toxins are only produced in still cultures. The minimum dose of this toxin needed to induce injury to plants like tomato shoot is 150 mg/kg fresh weight. The structure of lycomarasmin (Figure 2.25) indicates that at least some parts of the molecule could well be derived from simple amino acids or their deaminated keto analogs. A subdivision of the molecule into its apparent structural units gives rise to a series of well-known metabolic intermediates all of which could conceivably serve as immediate biogenetic precursors. Thus, acid hydrolysis of lycomarasmin gives rise to aspartic acid, pyruvic acid, and glycine (I), some or all of which might serve as building units.

This toxin yields glycine and aspartic acid on hydrolysis and has been shown to be a derivative of asparagine. Thus, the third N atom, which had previously not been assigned position, was located. Because of the reactions which lycomarasmin underwent, a structural formula for it was proposed which contained the new amino acid, alpha-hydroxyalanine. This was attached to a common N atom (i.e., the amino group) to the amino group of glycylasparagine. A synthetic product was isolated from the reaction of ethyl-alpha-acetoxy-alpha-bromopropionate and the methyl ester of glycylasparagine, which had properties in common with lycomarasmin, including quantitatively the same biological activity. Some reasons were advanced for regarding lycomarasmin as a much-simplified model of proteins.

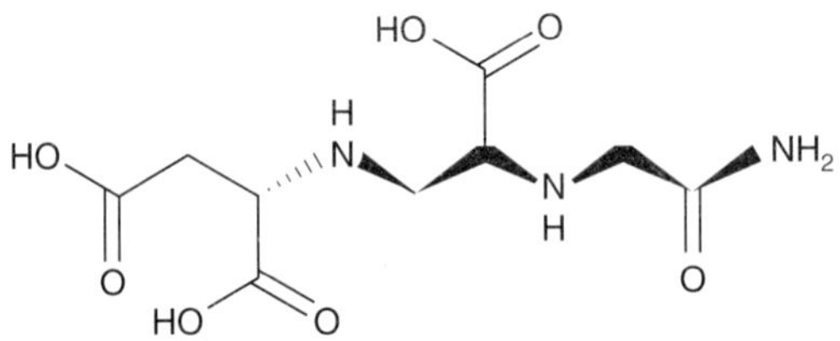

Fig. 2.25 Lycomarasmin

A number of new derivatives of aspartic acid and asparagine have been prepared to examine the effects of various structural changes on the activity of lycomarasmin, which is a peptide derived from asparagine, glycine, and probably a-hydroxyalanine (Woolley, 1948). It has been shown that quantitative change in the constitution of Richard's medium causes a corresponding change in the quantity of toxins produced by the various strains of *F. lycopersici*. Thus, the virulent strain, producing lesser amounts of toxins (used in the widest sense of the word) in Richard's medium, produces more than the avirulent strain when the quantities of carbon and nitrogen are varied in the medium. With glycine as the exclusive carbon source in the medium, *F. lycopersici* produces all the three known toxins (vasinfuscarin, fusaric acid and lycomarasmin).

Thus, glycylserylaspartic acid, glycyl-P-aspartylserine, acetyllactylglycylasparagine, cY, a-diacetaminopro-pionylglycylasparagine, the corresponding aspartic acid compound, 01-hydroxy-cu-acetaminopropionylglycylaspartic acid, and pyruvylglycylaspartic acid were synthesized. These, along with known compounds such as glycylaspartic acid, glycylasparagine, aspartic acid, asparagine, glycine, serine, pyruvylglycine, serylglycylaspartic acid, serylglycylglutamic acid, and glutathione, were compared with lycomarasmin for the ability to cause excised tomato leaves to wilt and curl. Several of the peptides were somewhat less active in this respect than was lycomarasmin, but one of them, ar-hydroxy-cr-acetaminopropionylglycylaspartic acid, was about equal to the natural toxin in activity. Some produced no detectable effect. A comparison was made of the relative effect on the activity of changing the position of an amino acid in a peptide with that of altering the nature of the amino acid.

2.8.5.3.6 Alternaric acid

Alternaric acid, a metabolic product of some strains of *Alternaria solani*, is highly phytotoxic. Alternaric acid is an antifungal and natural product, which bears a substituted glycolic acid in the functionally and stereochemically dense core of the molecule (Fig. 2.26). It is optically inactive, unsaturated dibasic acid and it inhibits the growth of the fungal germ tubes without actually preventing germination.

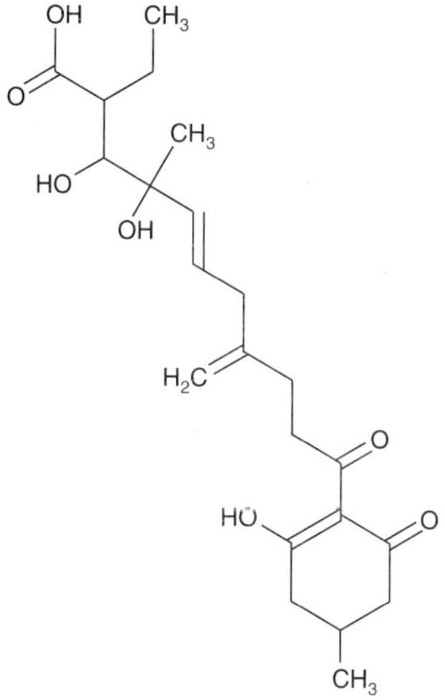

Fig. 2.26 Alternaric acid

In 1960, the planar structure of alternaric acid (Bartels-Keith, 1960), a phytotoxin isolated from *Alternaria solani*, causal fungus in the early blight disease of potato and tomato, was determined by using classical chemical methods. However, the total stereostructure of 1 remained to be elucidated. *A. solani* also produces several secondary metabolites, and some of them were isolated in our laboratory. After that, alternaric acid was shown to contribute to the disease development of the host by *A. solani* in a manner similar to the mode of action of the group of compounds classified as host-specific toxins, although all of the requirements as a primary disease determinant were not fulfilled.

If introduced into the vascular system of cut shoots of tomato or potato, this toxin travels upwards and causes necrotic lesions of the stem, petioles, and leaf blades, very similar in appearance to the lesions appearing in some phases of natural attack by *A. solani*. Alternaric acid also produces similar lesions on plants outside the host range of *A. solani*. Associated with the production of visible lesions is a disturbance of the water-balance of shoots. Transpiration is increased and, in spite of a simultaneous but smaller increase in water uptake; this eventually leads to complete desiccation of the shoots (Brian *et al.*, 2008). A strain of *A. solani* known to produce alternaric acid in synthetic culture media produces a substance with similar biological properties when inoculated into tomato fruits. Since other workers have shown that some of the symptoms of *A. solani* attack are toxigenically induced it is tempting to suggest that alternaric acid is the toxin concerned. On the other hand, there is no correlation between virulence of strains of *A. solani* and their ability to produce alternaric acid in synthetic culture media or tomato fruits. Some highly virulent strains produce little or no alternaric acid under any conditions tested. There is some evidence that other toxins besides alternaric acid are produced by *A. solani*, not necessarily in the same proportions by all strains.

2.8.5.3.7 Phaseolotoxin

This is the toxin involved in one of the bacterial bean blights called 'halo blight.' Symptoms of the disease incited by the bacterium can be produced by the toxin alone. Within cells, the toxin is enzymatically cleaved releasing phosphosulfinylornithine which is the toxic moiety. Cellular effects are a result of the inactivation of the enzyme ornithine carbamoyltransferase.

Phaseolotoxin is a modified tripeptide [Nδ-(N'-sulfodiaminophosphinyl)-ornithyl-alanyl-homoarginine] (Fig. 2.27) produced by certain strains of *Pseudomonas syringae* pv. *phaseolicola*, *Pseudomonas syringae* pv. *actinidiae* and strain *Pseudomonas syringae* pv. *syringae* CFBP 3388 (Bender *et al.*, 1999) Phaseolotoxin is a reversible inhibitor of the enzyme ornithine carbamoyltransferase, which catalyzes the formation of citrulline from ornithine and carbamoylphosphate in the arginine biosynthetic pathway. Phaseolotoxin is an effective inhibitor of OCTase activity from plant, mammalian, and bacterial sources and causes a phenotypic requirement for arginine. Additionally, phaseolotoxin inhibits the enzyme ornithine decarboxylase (EC 4.1.1.17), which is involved in the biosynthesis of

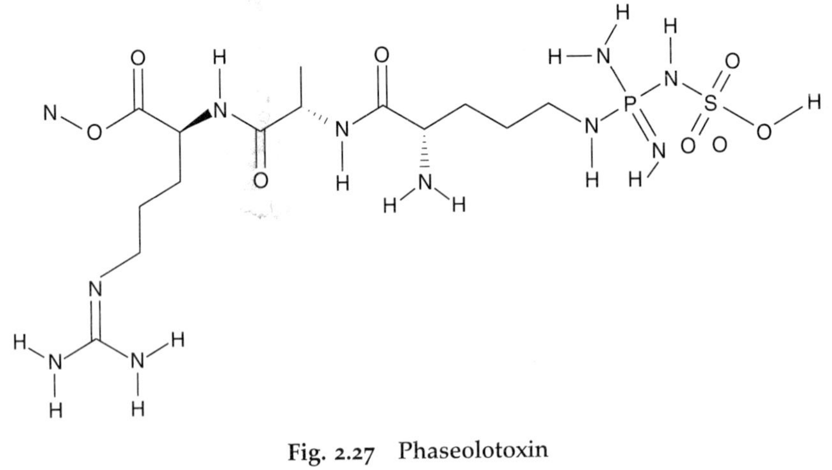

Fig. 2.27 Phaseolotoxin

polyamines. Phaseolotoxin is an effective inhibitor of OCTase activity from plant, mammalian, and bacterial sources and causes a phenotypic requirement for arginine.

Tourte and Manceau (1995) isolated a bacterial strain, CFBP 3388 from Vetch (*Vicia sativa* L.) which was identified as *P. s.* pv. *syringae* on the basis of nutritional and biochemical patterns which were obtained with classical tests and the Biolog™ system. It caused necrotic symptoms typical of *P. s.* pv. *syringae* on bean leaves and pods after artificial inoculation. However, the isolate caused a citrulline-reversible inhibition of *E. coli* in phaseolotoxin bioassay. Furthermore, with CFBP 3388 DNA as template a 1900 bp DNA fragment, specific for the phaseolotoxin DNA cluster of *P. s.* pv. *phaseolicola* was amplified by PCR. This is the first demonstration that an isolate of *P. syringae* that is not pv. *phaseolicola* can produce phaseolotoxin.

2.8.6 Overall Role of Pathogen Toxins in Plant Pathogenesis

The interaction between plant pathogens and their hosts is extremely complex. There are many factors affecting the plant disease development, plant, and pathogen physiology, metabolism of the host plant, physical environment and the way it impacts the plant growth and development, biochemistry and genetics of the pathogen-plant interactions (Slavov, 2005). An area of special interest in the specificity of these interactions is the role of pathogen toxins in the disease development. Toxins are compounds that are produced by the pathogens and cause part or all of the symptoms of a disease. They are of various chemical types and include peptides, glycoproteins, polysaccharides, organic acids, fatty acids and derivatives, polyketides and terpenoids (Turner, 1984). The best-known mycotoxins are coming from the fungi of genera *Aspergillus, Penicillium, Fusarium, Alternaria, Claviceps, Stachybotrys, Myrothecium, Phoma,* and *Diplodia*. There are also toxin-producers among bacteria, like those from genera *Xanthomonas, Pseudomonas,* and *Clavibacter* (Švàbova and Lebeda, 2005). There are large differences between the toxin-producing pathogens and mode of plant-pathogen interactions. The role of a toxin as a disease determinant is proved by the occurrence of the toxin in an infected plant and the ability of the toxin alone to elicit at least part of the symptoms of the disease (Hamid and Strange, 2000).

The major visible symptoms caused by the toxins are chlorosis, necrosis, and wilting. Some of the toxins have general phytotoxic properties and are active toward a broad range of plant species. Those are nonhost-specific toxins. They contribute to the virulence or symptom development in the disease in which they occur, but are not primary determinants of host range. Many non-host-specific toxins as brefeldin A, tenuasonic acid, tentoxin, and zinniol are produced from several *Alternaria* species. They exert their phytotoxic activities through different modes (Thomma, 2003). Other toxins are host-specific ones and affect only certain plant varieties or genotypes (Osbourn, 2001). The host-specific toxins (HSTs) are playing a role in determining the host range of specificity of plant pathogens and can act as an agent of virulence of those pathogens. HSTs appear to be specific for individual plant species, and in some cases their effects are mediated by gene-for-gene interactions. Most of the HSTs exist as families of compounds, each number of which are produced in different amount and have a different potency. They are acting in very low concentrations ranging from 10 μM to 1 μM. HSTs can be very diverse by chemical structure even when they are produced by pathogens belonging to one genus, as it is in species *Alternaria alternata*.

Different pathotypes of this plant pathogen produce different toxins known as AK-, AF-, ACR-, ACT-, ACTG-, AM-, and AAL-toxins, which differ in their structure and host range (Otani *et al.*, 1995). Despite the differences in chemical structure of those toxins, *Alternaria* pathotypes cause similar infection patterns. Species of genus *Cochiobolus* produce victorin, T-toxin, HC-toxin or HS toxin, which differ in the chemical structure. Some fungi that are not closely related can produce structurally very similar toxins as it is in the case of T-toxin of *Cochiobolus heterostrophus* race T and PM-toxin from *Phyllosticta maydis*. One pathogen, tangerine type of *Alternaria alternata* is known to produce more than one class of HSTs - ACT-, and ACTG-toxins (Kohmoto *et al.*, 1993). The role of toxins in infection and colonization of the host plant by the pathogen is of critical importance. It was proved in experiments with a nontoxic mutant of the pathogens as it was in the case of *Pseudomonas syringae* pv. *tabaci* (Turner and Taha, 1984.). The toxigenicity is partially or fully determinant of pathogenicity, for example of victorin and *Helmintosporium victoriae* (Scheffer, 1976.). This toxin determines the host range and parasitic ability of the pathogen.

2.8.6.1 *Toxins produced by plant or plant X pathogen interaction–Amylovorin*

Amylovorin is a high-molecular-weight polysaccharide, characterized as a host-specific toxin associated with *Erwinia amvlovora*-infected tissues. The common characteristic of infection of rosaceous plants by *Erwinia amylovora*, the fire blight pathogen, is the appearance of wilt, necrosis, and the production of copious amount of exudates, or 'ooze.' In the 1930s, experiments established that the ooze was bacterial in origin and responsible for the induction of wilt symptoms in pear shoot cuttings (Dwayne *et al.*, 2013). A majority of the field samples of *E. amylovora* examined by Billing were encapsulated by the ooze, with less than 1% having no capsule. The ooze was named polysaccharide amylovorin, to reflect its toxic effect on plant tissues. Amylovorin was subsequently renamed amylovoran to be consistent with polysaccharide nomenclature. In due course, the linkages among *E. amylovora*'s capsule, slime, polysaccharides, ooze, and pathogenicity were made. The growth of the bacterium on sucrose-, glucose-, or sorbitol-enriched media results in the production of excess quantities of two exopolysaccharides (EPSs), one acidic and one neutral. The acidic EPS, amylovoran, may form a capsule, slime, and/or float free in the liquid medium. Amylovoran is a heterogeneous polymer consisting of repeating units of one glucuronic acid and four galactose residues. The *ams* region of the genome controls the production of the EPS, with *rcsA* and *rcsB* genes being required for synthesis. Mutation in *rcsA* or *rcsB* results in reduced amylovoran production. The neutral EPS, levan, was synthesized in the presence of sucrose by the enzyme levansucrase, which cleaves the sugar and polymerizes fructose into a polyfructan (β-2,6-D-fructofuranan). Levan synthesis took place extracellularly through the action of the enzyme encoded by the *lsc* gene.

The strongly hydrophobic bacteriocin amylovorin L471 from *Lactobacillus amylovorus* DCE 471 was isolated and purified to homogeneity from complex culture broth by a novel, rapid and simple three-step protocol including (i) ammonium sulfate precipitation, (ii) chloroform/methanol extraction/precipitation and (iii) reversed-phase HPLC, the only chromatographic step involved (Callewaert *et al.*, 1999). The molecular mass of the peptide was determined to be 4876.9 Da by electrospray mass spectrometric analysis. N-terminal amino acid sequencing identified 35 amino acid residues as being identical to the N-terminal sequence of lactobin A, a bacteriocin from another *L. amylovorus* strain. These nonidentical strains produce bacteriocins that display small differences in molecular mass and inhibitory spectrum. The amino acid sequence of amylovorin L471 shared significant homology with lactacin X, one of the two bactericidal peptides produced by *Lactobacillus johnsonii* VPI11088. A purified amylovorin L471 preparation permitted confirmation of the inhibitory spectrum previously established with a crude extract. It displayed a bactericidal mode of action on lactobacilli after an extremely rapid adsorption to the target cells. Two *Listeria* spp. were only weakly sensitive. Amylovorin L471 appeared to be produced constitutively. Ethanol not only stimulated specific bacteriocin production but also prevented adsorption of the bacteriocin molecules to the producer cells upon prolonged fermentation.

The effects of both oxygen and carbon dioxide on growth of and product formation by *Lactobacillus amylovorus* DCE 471, a promising new sourdough starter culture, were assessed through controlled, *in vitro* fermentation experiments, using a temperature of 37 °C and a constant pH of 5.4 (Neysens and De, 2011). It was seen that aeration affected both cell growth and amylovorin L production. At aeration rates of 11 min^{-1} and more, the bacterial population was subjected to oxidative stress as reflected by biphasic growth patterns. During the first growth phase, the maximum specific growth rate increased with increasing aeration rates stabilizing at the highest oxygen concentrations. The maximum obtainable cell yields decreased. During the second growth phase, the amylovorin L production was stimulated at the highest aeration. However, amylovorin titers were never higher in the presence of oxygen compared with the anaerobic fermentations. Carbon dioxide did not influence cell growth of *L. amylovorus* DCE 471. The maximum specific growth rate and the biomass concentrations were merely affected. On the other hand, the maximum soluble bacteriocin titers coincided with the highest carbon dioxide flow rates. These results indicated that mild aeration of type II sourdoughs might enhance both cell yield and amylovorin L production by *L. amylovorus* DCE 471, thereby contributing to the competitiveness of the strain. Growth in an ecosystem together with yeasts producing carbon dioxide might exert a positive effect on the production of amylovorin L.

2.9 ROLE OF GROWTH REGULATORS IN PLANT PATHOGENESIS

Plant growth and response to environmental cues are largely governed by phytohormones. The plant hormones ethylene, jasmonic acid, and salicylic acid (SA) play a central role in the regulation of plant immune responses. In addition, other plant hormones, such as auxins, abscisic acid, cytokinins, gibberellins, and brassinosteroids, that have been thoroughly described to regulate plant development and growth, have recently emerged as key regulators of plant immunity (Nicolas Denance *et al.*, 2013). Plant hormones interact in complex networks to balance the response to developmental and environmental cues and thus limiting defense-associated fitness costs. The molecular mechanisms that govern these hormonal networks are largely unknown. Moreover, hormone signaling pathways are targeted by pathogens to disturb and evade plant defense responses.

The resistance response is regulated by phytohormones, which are small molecules which synergistically and/or antagonistically work in a complex network to regulate many aspects of plant growth, development, reproduction, and response to environmental cues (Jaillais and Chory, 2010.). Recent progress has been made in understanding the complex hormone network that governs plant immunity, giving rise to a database containing information on the hormone-regulated genes (e.g., in *Arabidopsis thaliana*) and the phenotypic description of hormone-related mutants. In parallel, it has been found that pathogens have developed sophisticated molecular mechanisms to deregulate the biosynthesis of hormones and/or to interfere with hormonal signaling pathways, thus, facilitating the overcoming of plant defense mechanisms. The essential roles of salicylic acid and ethylene/jasmonic acid mediated signaling pathways in resistance to pathogens are well described.

Hormones such as auxins and abscisic acid (ABA), originally described by their function in the regulation of plant growth processes and the response to abiotic stresses, have recently emerged as crucial players in plant–pathogen interactions. All the phytohormone pathways are linked to each other in a huge, complex and still obscure network. For example, ET, ABA, auxin, gibberellins, and cytokinins pathways are considered as hormone modulators of the SA–JA signaling backbone (Jaillais and Chory, 2010.).

Growth regulators are of two types, growth promoting substances and growth inhibiting substances. Auxins, gibberellins, and cytokinins are growth promoting substances, whereas, ethylene and abscisic acid are major growth inhibiting substances. The imbalance in growth promoting and growth inhibiting substances causes hypertrophy (excessive increase in cell size) and atrophy (decrease in cell size). Symptoms may appear as tumors, galls, knots, witches broom, stunting, excessive root branching, defoliation and suppression of bud growth. There is an increasing evidence that both the pathogen and the host have the capacity to synthesize various PGRs. Alterations in the levels of PGRs and disease susceptibility or resistance reaction are associated with plant-pathogen interaction (Singh *et al.*, 1997). Some investigations indicated that naphthalene acetic acid is a potential antifungal agent (Michniewicz and Rozej, 1988). Auxin strongly inhibited mycelial growth, sporulation, and spore germination of *Fusarium culmorum in vitro* NAA, indole acetic acid (IAA), 2,4-diphenol acetic acid (2,4,D) and abscisic acid (ABA) were exogenously applied to control *Alternaria solani* caused early blight of potato. Auxins such as IAA, naphthalene acetic acid ethyl ester and N-metatotyl phthalamic acid reduced *Botrytis* blight of cut rose flowers (Elad, 1995).

2.9.1 Growth Promoting Substances

2.9.1.1 Auxins

Auxins are a group of molecules including IAA (indole-3-acetic acid) that regulate many aspects of plant development, such as apical dominance, root gravitropism, root hair, lateral root, leaf, and flower formation, and plant vasculature development (Jaillais and Chory, 2010). It is continuously produced in young meristematic tissue and moves rapidly to older tissues. If auxin concentration is more, its concentration is reduced by the enzyme, IAA oxidase. Auxins are the major growth regulator in plants and are defined as growth regulators whose major mode of action is cell elongation; they resemble indole-3-acetic acid in their activity. The naturally occurring auxin is indoleacetic acid [(IAA) (Fig. 2.28)].

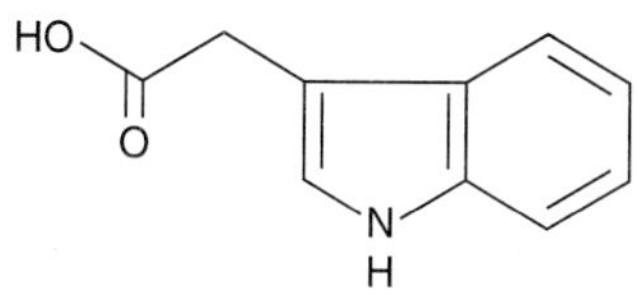

Fig. 2.28 Auxin (IAA)

Increased levels of auxin occur in plants infected with fungi, bacteria, viruses, mollicutes and nematodes. Increased IAA resulted in hypertrophy and decreased IAA results in atrophy. Increased IAA may be due to inhibition of IAA oxidase (*Example: Ralstonia solanacearum*, the causal agent of wilt of Solanaceous plants, induces a 100 fold increase in IAA level in diseased plants). Increased plasticity of cell walls as a result of high IAA levels renders the pectin, cellulose and protein components of the cell wall more accessible to pathogen degradation. The increase in IAA levels may also inhibit lignifications of tissues. Increased IAA levels have been reported in plants infected with the following pathogens: *Phytophthora infestans* (late blight of potato), *Ustilago maydis* (maize smut), *Plasmodiophora brassicae* (clubroot of crucifers), *Sclerospora graminicola* (downy mildew of sorghum), *Agrobacterium tumefaciens* (crown gall of apple), and *Meloidogyne* (root knot nematode).

Both direct and indirect effects of auxins on the regulation of pathogen resistance responses in plants have been described. Indirect effects may be caused by auxins regulation of development-associated processes, such as cell wall architecture, root morphology, and stomata pattern. For example, treatment of rice with IAA impaired the resistance to *Xanthomonas oryzae* pv. *Oryzae* probably as a consequence of the activation of the biosynthesis of cell wall-associated expansins that lead to cell wall loosening, which facilitates pathogen growth.

Many pathogenic microbes and plant growth promoting rhizobacteria have evolved complete pathways for auxin biosynthesis with tryptophan as the main precursor. Auxin-producing phytopathogenic bacteria are mostly, but not exclusively, gall-inducing microbes. They include, for instance, *Agrobacterium tumefaciens, Agrobacterium rhizogenes, Erwinia chrysanthemi, Erwinia herbicola, Pseudomonas fluorescens, P. putida, Pseudomonas savastanoi, P. syringae, R. solanacearum* and *Rhodococcus fascians*. In *R. solanacearum*, auxin biosynthesis is governed by HrpG, a major regulator of bacterial virulence and response to metabolic signals. In *Agrobacterium tumefaciens*, two genes required for conversion of tryptophan to auxin are localized in the T-DNA region of the Ti plasmid injected into plant cells. Auxin biosynthesis is necessary for tumor gall formation and pathogenicity of *Agrobacterium.* Auxins negatively regulate the expression of genes necessary for the transfer of *Agrobacterium* T-DNA in plants and also inhibit the growth of several bacterial species *in vitro* (Nicolas Denance *et al.*, 2013). Auxin biosynthesis in fungal pathogens seems to be limited to a few species. In *Ustilago maydis*, *U. esculenta*, and *U. scitaminea* auxin is produced. Additionally, other fungi have enzymatic tools to produce auxins, such as *Colletotrichum gloeosporioides* f. sp. *aeschynomene*, *Colletotrichu macutatum*, and *F. proliferatum*.

Phytonematodes have developed the capacity to sense and respond to chemical signals of host origin, and the ability to orientate towards plant roots enhances the nematode's chance of survival. Root exudates contain a range of compounds which mediate belowground interactions with pathogenic and beneficial soil organisms. Chemical components of root exudates may deter one organism while attracting another and these compounds alter nematode behavior and can either attract nematodes to the roots or result in repellence, motility inhibition or even death (Curtis, 2008). *In vitro* plant signals present in root exudates, triggered a rapid alteration of the surface cuticle of *Meloidogyne incognita* and the same changes were also induced by indole-acetic acid (IAA). IAA bound to the chemosensory organs of *M. incognita*, and it was opined that IAA might act as a signal that orientates the nematode on the root surface in the rhizosphere and/ or inside the root tissue and thereby promoted nematode infection.

Auxins can negatively impact plant defense by interfering with other hormone signaling pathways or with PTI. The bacterial PAMP flg22, a peptide from flagellin protein (Pel and Pieterse, 2013) induces an *Arabidopsis* microRNA (miR393), which negatively regulates the mRNA levels of auxins receptors TIR1 (transport inhibitor response 1), AFB2 (auxin signaling F-box 2), and AFB3. Thus, the flg22-triggered suppression of auxin signaling leads to increased resistance to the bacterium *Pseudomonas syringae* pv. *tomato* DC3000 (*Pst*DC3000) and also to the oomycete *Hyaloperonospora arabidopsidis.* The flg22-induced resistance to these biotrophic pathogens was explained by the observed induction of the SA signaling pathway. Supporting this hypothesis, it was found independently that treatment of *Arabidopsis* leaves with flg22 induces salicylic acid accumulation (Jaillais and Chory, 2010).

The conjugated auxin–aspartic acid (IAA–Asp) has been recently reported to play a key role in regulating resistance to the necrotrophic fungus *Botrytis cinerea* and *Pst*DC3000. In *Arabidopsis*, tomato, and *Nicotiana benthamiana* infected with these pathogens there is an enhanced expression of *GH3.2* and *GH3.4* genes, which encode two enzymes required for conjugation of auxins with Asp. Thus, upon pathogen infection, accumulation of IAA–Asp takes place, promoting the development of disease symptoms in infected plants (González-Lamothe *et al.*, 2012). The negative effects of auxins on the activation of plant resistance are further supported by the observed enhanced susceptibility of auxin-treated rice to *X. oryzae* and of auxin-treated *Arabidopsis* to *Pst*DC3000 and *Fusarium culmorum*. Disruption of auxin signaling in *Arabidopsis* mutants, such as *axr1*, *axr2* and *axr3*, leads to enhanced resistance to *F. oxysporum* (Jaillais and Chory, 2010). Nevertheless, auxins have also been shown to positively regulate *Arabidopsis* immunity as *axr2-1* and *axr1-1* mutants were more susceptible than wild-type plants to the necrotrophic fungi *B. cinerea* and *Plectosphaerella cucumerina*.

One of the biosynthetic pathways of auxins is partially shared with those required for the biosynthesis of tryptophan-derived antimicrobials, such as indole glucosinolates and camalexin. This might lead to competition for the biosynthetic precursor of auxin and antimicrobials (Grubb. and Abel, 2006). The recently characterized *Arabidopsis wat1* (*walls are thin1*) mutant exhibits specific enhanced resistance to vascular pathogens such as *Ralstonia solanacearum*. This response was associated to a misregulation of tryptophan derivatives (i.e., lower levels of auxin and indole glucosinolates) specifically in roots, resulting in enhanced levels of SA, which is, like tryptophan, a chorismate-derivative. Collectively, these data demonstrate that auxins play a central role in balancing plant resistance responses.

2.9.1.2 Gibberellins

First isolated from *Gibberella fujikuroi* (Conidial stage: *Fusarium moniliforme*), the causal agent of bakanae or foolish seedling disease of rice. Infected seedlings show abnormal elongation due to excessive elongation of internodes. Best known gibberellin is Gibberellic acid. A century ago, a rice farmer in Asia noticed some exceptionally tall seedlings growing in their paddies. Before these rice seedlings could mature and flower, they grew so tall that they toppled over. In Japan, this aberration in growth pattern became known as bakanae ('foolish seedling disease') disease of rice. In 1926, Kurosawa, a Japanese scientist discovered that the disease was caused by a fungal pathogen, *Gibberella fujikuroi*. By the 1930s, Japanese scientists had determined that fungus produced hyperelongation of rice stems by secreting a chemical, which was given the name gibberellin. Gibberellins are normal constituents of green plants and also produced by several other microorganisms. The best-known gibberellin is gibberellic acid. In the past years, scientists have identified more than 80 different gibberellins, many of them occurring naturally in plants. Spraying of diseased plants with gibberellin overcomes some of the symptoms (stunting) caused by several virus- or mollicute-pathogens indicating that gibberellin involves in disease development. Its main functions include cell elongation, stem and root elongation, promote flowering and growth of fruits. It also induces IAA synthesis. IAA and GA act synergistically (*Example*: *Sclerospora sacchari*, the causal agent of downy mildew of sugarcane induces GA production).

Upon pathogen attack, the plant defense response is mediated by a set of connected signal transduction pathways. Experiments were conducted to observe the role of the plant hormone, in particular, gibberellin (GA), in the response of rice to infection by the sedentary endoparasitic nematode, *Meloidogyne graminicola* and the migratory endoparasitic nematode *Hirschmanniella oryzae* (Mohammod Hossain, 2010). Exogenous application of GA was done several times on Nipponbare and GA might have a negative effect on defense in Nipponbare against *M. graminicola*. On the other hand, an inhibitory effect to *M. graminicola* in Taichung 65 was found when GA was applied exogenously. There was a significant effect of GA in reducing infection in Nipponbare against *H. oryzae* but the experiment was not repeated to confirm this result. A significantly higher number of *H. oryzae* per plant was found in GA-deficient mutant waitoC compared to WT and a lower number in GA-insensitive mutant gid1-3 plants. Experiments with *M. graminicola* showed that galling was significantly and reproducibly lower in mutants, whether it was GA-insensitive or GA-deficient than in wild type (Taichung 65). There were even significantly lower galls in these GA-deficient mutants after GA treatment than in nontreated mutant plants. There was no difference in the expression of GA signaling gene *Osgid1-3* after *M. graminicola* infection in Taichung 65, neither in root tip nor leaf tissue. The expression of GA signaling gene *Osgid1-7* was significantly higher only in the whole root at 2 dpi but not in leaves or root tips. Even this gene was not differentially expressed in root and leaf tissues at a later stage (6 dpi) of infection. The nematode inhibited biosynthesis of *OsGA20ox* in leaves at 2 dpi and in both of leaves and roots at 6 dpi. The GA response gene *AK073385* was also less expressed in root tip at 6 dpi. These might help nematodes to continue their infection in Taichung 65.

Gibberellins play a vital role in plant growth and development. However only 12 fungi, associated with plants and/or soil have been reported as GA producers (Muhammad Hamayun *et al.*, 2010). Endophytic strains of the *Cladosporium sphaerospermum* and *Penicillium citrinum* have been reported as GA producers. Genus *Cladosporium* includes several plant-associated saprophytic and some plant pathogenic species. Some species are widely distributed dematiaceous molds, and some are predominant in tropical and subtropical regions. Some *Cladosporium* species such as *C. fulvum* are pathogenic to plants, causing scab diseases as well as leaf spots and blights. *Cladosporium* sp. MH-6 is a novel GA-producing fungus that secretes comparable amounts of GA to *F. fujikuroi*. Therefore, this isolate might have the potential for use as a biofertilizer with minimal environmental risks. There are two classes of gibberellins, the 19-carbon gibberellins, and the 20-carbon gibberellins. The 19-carbon gibberellins, formed from 20-carbon gibberellins, are the biologically active forms. Gibberellins also vary according to the position and number of hydroxyl groups linked to the carbon atoms of the *ent*-gibberellane skeleton (Fig. 2.29). Hydroxylation has a profound influence on biological activity.

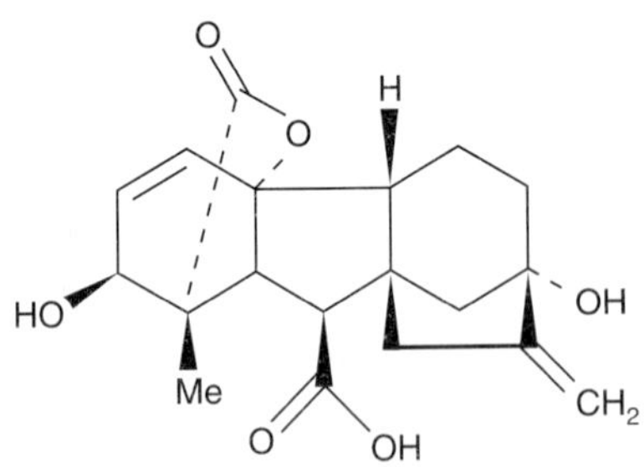

Fig. 2.29 Gibberellic acid

The causative agent *Gibberella fujikuroi* contains several different types of gibberellin (abbreviated to GA, after gibberellic acid, the first form discovered), some of which also exist in plants. Plants possess many other unique gibberellins, and collectively there are well over 100 identified compounds. This intimidating complexity can be reduced to a comprehensible level by realizing that each species contains only about 25 of these gibberellins and that most gibberellins are biosynthetic intermediates or inactive end-products, and not active in their own right. There are many steps and enzymes involved in building up the 19- and 20-carbon gibberellin molecules from five-carbon mevalonic acid. Several parallel pathways exist differing only in the number of hydroxyl (–OH) groups. Hydroxyl groups are the key to gibberellin functions: some positions (3b) are generally essential for activity, whereas others (2b) completely abolish it. Inactivation by conjugation to glucose also occurs. Gibberellin synthesis takes place mainly in developing leaves and stems, in developing seeds and during germination. Gibberellins function in dormancy release and germination, as well as in growth promotion (e.g., stem elongation, fruit tissue expansion).

2.9.1.3 Cytokinins

Cytokinins are plant-specific chemical messengers (hormones) that play a central role in the regulation of the plant cell cycle and numerous developmental processes (Schmülling, 2004). The first cytokinin discovered was an adenine (aminopurine) derivative named kinetin. Cytokinins are present in all plant tissues. They are abundant in the root tip, the shoot apex, and immature seeds. Their endogenous concentration is in the low nM range. Cytokinins may also act on the cell that produced them (autocrine signaling). Cytokinins are necessary for cell growth and differentiation. It inhibits the breakdown of proteins and amino acids and thereby inhibits senescence, and they have the capacity to direct the flow of amino acids and other nutrients towards high cytokinin concentration. Cytokinin activity increases in club root, in crown galls and in rust infected bean leaves (*Example*: Green islands are formed around infection in bean (*Phaseolus vulgaris*) leaves infected by *Uromyces phaseoli*).

In many ways, cytokinins are opposites of auxins, being synthesized in roots but with most dramatic effects on shoot development. However, shoot tissues can also produce cytokinins as can developing seeds. A classic example of the latter is coconut milk, the copious liquid endosperm from coconut seed, which is still a popular cytokinin source in plant tissue culture media. Cytokinins were originally named for their ability to promote cell division, but they also function in the initiation of new shoot structures, dormancy release and retardation of senescence. Cytokinins are derivatives of adenine, one of the purine bases found in all DNA and RNA. Indeed, cytokinins were originally thought to be products of transfer RNA (tRNA) breakdown. Cytokinins, such as zeatin and isopentenyl adenosine (IPA) have been isolated from plants.

Cytokinins have emerged as a major factor in plant-microbe interactions during nodule organogenesis and pathogenesis. Microbe-originated cytokinins confer abnormal hypersensitivity of cytokinins to plants, augmenting the sink activity of infected regions. However, recent findings have shed light on a distinct role of cytokinins in plant immune responses. Plant-borne cytokinins systemically induce resistance against pathogen infection. This resistance is orchestrated by endogenous cytokinin and salicylic acid signaling. Cytokinins are also produced by cyanobacteria, some plant pathogenic bacteria (e.g., *Agrobacterium tumefaciens, Pseudomonas savastanoi, Rhodococcus fascians*) and the slime mold *Dictyostelium discoideum*. Naturally occurring cytokinins are adenine derivatives with a side chain at the N6-position (Fig. 2.30). The structure and conformation of the N6-attached side chain can markedly influence the biological activity of the cytokinin (Schmülling, 2004). The cytokinin zeatin was named after the genus of corn, *Zea,* in which it was discovered.

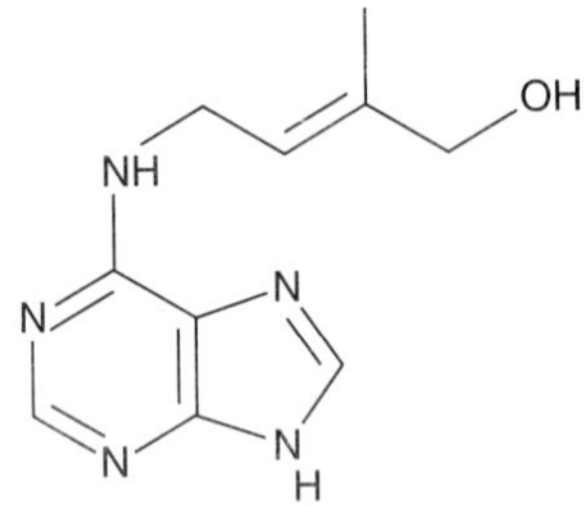

Fig. 2.30 Cytokinin zeatin

Depending on the structure of the N6-substituent, cytokinins are classified as isoprenoid or aromatic cytokinins. The biological activities of both classes are qualitatively similar, but they may differ quantitatively in different processes. Isoprenoid cytokinins are the most abundant class.

Cytokinins are produced by several plant pathogenic bacteria and play a role in pathogenicity. One such pathogen is *Agrobacterium tumefaciens,* the causative agent of the crown gall disease. During the infection process, *A. tumefaciens* transfers a small stretch of DNA, the T-DNA, to the host plant, where it becomes integrated into the nuclear genome. The T-DNA harbors an *IPT* gene, which is expressed in the host cell and causes cytokinin overproduction. This leads, together with enhanced auxin content, to tumorous cell proliferation. Other cytokinin-synthesizing pathogens are *Pseudomonas syringae,* which induces gall formation and *Rhodococcus fascians,* which causes fasciation and a growth abnormality called witches´ broom disease. The root-nodule forming and nitrogen-fixing plant symbiont, *Rhizobium* species is also known to produce cytokinin.

It is known that plants utilize hormone signal molecules to regulate plant development and respond to the environment. Historically, direct evaluation of the role of hormones in various plant signaling pathways was hampered by extremely low levels of hormones present in any tissue. New technologies based on the discovery that certain plant proteins can be used as markers for the presence of the group of plant hormones known as cytokinins are allowing the investigation of the role of cytokinins in plant development and responses to the environment at the molecular and plant physiology levels (Prasad *et al.,* 2004). An attempt was made to identify and correlate the presence of endogenous cytokinins in different regions of developing roots and roots exposed to symbiotic nodulating bacteria or pathogenic root knot nematodes. As assayed by protein expression for cytokinin responsiveness, the hormones appeared to regulate spatially and temporally in the formation of lateral roots. A role for cytokinins was implicated in the formation of nitrogen-fixing nodules, and it was found that a cytokinin response was present subsequent to nematode root penetration and migration. In addition, experiments with roots engineered to destroy the cytokinin signal revealed that those roots had increased amounts of lateral roots, but fewer nodules and nematode galls compared to control roots. Taken together, the results demonstrated an intricate balance of cytokinin signals in different cells of root tissues and a role for cytokinins in the formation of nitrogen-fixing nodules and nematode-induced galls.

Cytokinin which was analyzed by immunoaffinity chromatography (IAC), high-performance liquid chromatography (HPLC) and radioimmunoassay (RIA) or enzyme-linked immunosorption assay (ELISA) has been used to study two separate topics: the role of tRNA in bacterial cytokinin biosynthesis and the changes in cytokinin concentration which occur during cereal grain development (Morris *et al.*, 1993). Transfer RNA isopentenylation in the gall-forming plant pathogen *Agrobacterium tumefaciens* is encoded by the chromosomal miaA locus. Mutation of miaA reduces tRNA isopentenylation significantly, and the preliminary data suggested that turnover of isopentenylated tRNA is responsible for low-level secretion of free N6-isopentenyladenine (iP) by the bacteria. However, the major route of cytokinin biosynthesis by gall-forming plant pathogenic bacteria is not via tRNA turnover but by direct biosynthesis mediated by dimethylallylpyrophosphate: 5′-AMP transferase (DMAPP: AMP transferase) encoded by such genes as ipt, tzs (from *A. tumefaciens*) or ptz (from *Pseudomonas savastanoi*). Analysis of cytokinin levels in developing wheat and rice grains in the period immediately following pollination showed large transient increases in zeatin (Z) and zeatin riboside (ZR) which coincided with the period of maximum endosperm cell division reported by others. Detailed analyses of maize kernels, where development can be staged readily, showed that Z and ZR concentrations peaked nine days after pollination (DAP). During the period, 8-10 DAP; cytokinin oxidase underwent a significant increase in specific activity, indicating that cytokinin catabolism was enhanced as endosperm cell division ended.

2.9.2 Growth Inhibiting Substances

2.9.2.1 Ethylene

Ethylene ($CH_2 = CH_2$) is one of the simplest hydrocarbons. It comes in a gaseous form, and similar to many other hydrocarbons, it is colorless and flammable. It consists of two double-bonded carbon atoms with hydrogen atoms occupying the other bond positions. This gives ethylene a chemical formula of C_2H_4. The structural shape of the ethylene molecule is linear (straight line) due to the presence of the double bond in the center of the molecule (Fig. 2.31). Ethylene does, however, have a distinctive property that makes it relative well recognized. Ethylene has a sweet, musky smell which makes it identifiable in the air. This is typically seen in pure ethylene gas, and can fade as ethylene mixes with other chemicals and gasses.

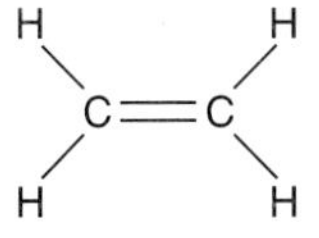

Fig. 2.31 Ethylene molecule

Ethylene exerts a variety of effects on plants, viz., chlorosis, leaf abscission, epinasty, stimulation of adventitious roots, fruit ripening and increased permeability of cell membranes. Ethylene production in infected tissues can be dramatically induced. This induction is largely dependent on activation of the ethylene biosynthetic pathway in plant tissues. Genes encoding several key enzymes involved in the ethylene biosynthesis are highly activated at the transcriptional level. It has not been shown that ethylene is produced directly by plant pathogenic fungi and bacteria.

The role of ethylene in the hormonal regulation of plant development has been well established (Leendert C. van Loon *et al.*, 2006). In addition, it has been implicated in biotic stress, both as a virulence factor of fungal and bacterial pathogens and as a signaling compound in disease resistance. This apparent discrepancy has stimulated research on the effects of various types of pathogens on mutant and transgenic plants that are impaired in ethylene production or perception. It has become clear that ethylene differentially affects resistance against pathogens with different lifestyles and plays an important role in mediating different types of induced resistance. It is a gaseous plant hormone, which plays a role in plant development, defense, and climacteric fruit ripening. Both genetic and biochemical evidence suggest that the response of plants to ethylene is mediated by a specific ethylene receptor.

Ethylene has been regarded as a signal in the plant for wounding and senesces responses. Recent studies show that ethylene together with another signal component jasmonic acid may play an essential role in plant defense responses of several pathosystems and induces ethylene production in infected tissues. Ethylene induces expression of certain types of *PR*-genes and some signal components involved in defense signaling. Ethylene insensitive mutants block several *PR*-gene-expression, and ethylene-insensitive mutants can be beneficial against some pathogens but deleterious to resistance against other pathogens in a specific a gene-for-gene manner. Other chemical weapons include polysaccharides, plant defense suppressors, transporters, etc. Plant viruses and viroids are not known to produce any substances themselves, but they adapt, induce and manipulate the host metabolism to replicate themselves.

Example: Ethylene is involved in the premature ripening of fingers in banana infected by *Pseudomonas solanacearum*, the causal agent of moko disease of banana. Ethylene was also detected in leaf epinasty symptom of the vascular wilt syndrome [*Example: Fusarium oxysporum* f. sp. *lycopersici* (wilt in tomato)].

Enhanced ethylene production is an early, active response of plants to the perception of pathogen attack and is associated with the induction of defense reactions (Tzeng and DeVay, 1985). It is generally assumed that ethylene production during stress contributes to stress alleviation, but several plant pathogenic fungi and bacteria are capable of producing ethylene as a virulence factor, which improves their ability to colonize plant tissues. For instance, the ability of the bacterial leaf pathogen *Pseudomonas syringae* pv. *glycinea* to proliferate in the leaves of its host plant soybean is impaired in mutants that lack the capacity to produce ethylene, which indicate that ethylene produced during infection promotes disease rather than alleviates it. Indeed, ethylene is responsible for the epinasty and defoliation caused by the soil-borne fungus *Verticillium dahliae* in cotton and for the stunting and chlorosis of cucumber infected by cucumber mosaic virus (Marco and Levy, 1979), yet, similar to the defense-regulating compounds salicylic acid and jasmonic acid, plant-derived ethylene is generally considered to be involved in resistance.

Tomato plants infected with the *Meloidogyne javanica* produce ethylene at a rate several-fold higher than the uninfected plants (Itamar Glazer *et al.*, 1984). This increase coincided with an increase in 1-aminocyclopropane-l-carboxylic acid (ACC) level in the plant's roots and leaves. When excised roots were treated with amino-etoxyvinylglicine (AVG) or amino-oxyacetic acid, the nematode-induced ethylene production was inhibited, indicating that the nematode infection induced ethylene production by accelerating the rate of ACC formation from S-adenosyl-methionine. Altering the rate of ethylene production in the infected roots did not affect nematode development up to the 3rd stage of juveniles, nor did it affect nematode penetration and initiation of gall formation. However, development of adult nematodes was markedly inhibited by both ethylene stimulators and inhibitors, whereas the rate of gall growth was accelerated by stimulators of ethylene production and suppressed when the production or action of the hormone was inhibited. Examination of fractured galls with SEM revealed that the hypertrophied cortical parenchyma of the galls treated with AVG was less developed than the untreated control. On the other hand, the parenchyma of galls treated with a supraoptimal concentration of auxin was larger in diameter than that of the untreated galls, due to the induction of ethylene production by the auxin. In all treatments, the vascular cylinder was not affected in size and structure, nor were the coencytes induced by the nematode.

Several studies indicate that when ethylene is applied before inoculation with a pathogen, it reduces or has no effect on disease development, whereas disease development is accelerated when plants are treated with ethylene after infection. Thus, it seems that the timing of the exposure of plants to ethylene can determine whether resistance is stimulated or reduced. Ethylene perception is also required to limit growth and systemic spread of pathogens in the nonhost resistance of tobacco plants. Ethylene can induce certain types of pathogenesis-related proteins or phytoalexins, and, through stimulation of the phenylpropanoid pathway, can rigidify cell walls in various plant species (Abeles *et al.*, 1992). In general, ethylene appears to stimulate and enhance defense responses. Unlike salicylic acid and jasmonic acid, which have a relatively restricted range of effects on plants, ethylene affects almost all stages of plant development but does not normally induce resistance or activate defense-related genes. As a gas, ethylene can diffuse rapidly in plant tissues but also escapes readily into the atmosphere. High levels of ethylene can be produced locally at infection sites and gradients could be sensed by surrounding cells, which then activate appropriate defense programs in response to the local concentrations of signaling compounds. This suggests that ethylene-regulated resistance responses depend on the spatial interactions of multiple signals interacting in a network-type fashion. Evidence for the functioning of such a network is becoming available through transcriptome analyses (Maleck *et al.*, 2000) which indicate that in response to environmental cues, multiple genes are regulated in complex ways by the interplay of several regulatory factors. The spatial analysis could benefit from the application of laser-assisted microdissection (Day *et al.*, 2005) of tissues at varying distances from the infection front. Such temporal and spatial analyses should aid in further delineating the involvement of ethylene in plant defense responses and the elucidation of the underlying mechanisms.

The plant hormones salicylic acid (SA), jasmonic acid (JA), and ethylene (ET) play crucial roles in the signaling network that regulates induced defense responses to biotic stresses (Antonio Leon-Reyes *et al.*, 2009). Antagonism between SA and JA operates as a mechanism to fine-tune defenses that are activated in response to multiple attackers. In *Arabidopsis thaliana*, 'Nonexpressor of Pathogenesis-Related Genes1' (*NPR1*) was demonstrated to be required for SA-mediated suppression of JA-dependent defenses. Because ET is known to enhance *SA/NPR1*-dependent defense responses, the role of ET was investigated in the SA-JA signal interaction. Pharmacological experiments with gaseous ET and the ET precursor 1-aminocyclopropane-1-carboxylic acid showed that ET potentiated *SA/NPR1*-dependent *Pathogenesis-Related1* transcription, while it rendered the antagonistic effect of SA on methyl jasmonate-induced *PDF1.2* and *VSP2* expression *NPR1* independent. This overriding effect of ET on *NPR1* function in SA-JA crosstalk was absent in the *npr1-1/ein2-1* double mutant, demonstrating that it is mediated via ET signaling. Abiotic and biotic induction of the ET response similarly abolished the *NPR1* dependency of the SA-JA signal interaction. Furthermore, JA-dependent resistance against biotic attackers was antagonized by SA in an *NPR1*-dependent fashion only when the plant-attacker combination did not result in the production of high levels of endogenous ET. Hence, the interaction between ET and *NPR1* plays an important modulating role in the fine tuning of the defense signaling network that is activated by pathogen and insect attack. A model was suggested in which ET modulates the *NPR1* dependency of SA-JA antagonism, possibly to compensate for enhanced allocation of *NPR1* to function in SA-dependent activation of *PR* genes.

Ethylene evolution occurs concomitantly with the progression of disease symptoms in response to many virulent pathogen infections in plants. A tomato mutant impaired in ethylene perception,'never ripe', exhibited a significant reduction in disease symptoms in comparison to the wild type after inoculations of both genotypes with virulent bacterial (*Xanthomonas campestris* pv. *vesicatoria* and *Pseudomonas syringae* pv. *tomato*) and fungal (*Fusarium oxysporum, F.* sp*lycopersici*) pathogens (Steven T. Lunda *et al.*, 1998). Bacterial spot disease symptoms were also reduced in tomato genotypes impaired in ethylene synthesis (1-aminocyclopropane-1-carboxylic acid deaminase) and perception (14893), thereby corroborating a reducing effect for ethylene insensitivity on foliar disease development. The reduction in foliar disease symptoms in '*never ripe*' plants was a specific effect of ethylene insensitivity and was not due to reductions in bacterial populations or decreased ethylene synthesis. *PR-1B1* mRNA accumulation in response to *X. c. vesicatoria* infection was not affected by ethylene insensitivity, indicating that ethylene is not required for defense gene induction. Broad tolerance of diverse vegetative diseases may be achieved via engineering of ethylene insensitivity in tomato.

2.9.2.2 Abscisic acid

It exerts dormancy in seeds, closure of stomata, inhibition of seed germination and growth and stimulated germination of fungal spores. It is one of the factors involved in stunting of plants. Abscisic acid is a single compound unlike the auxins, gibberellins, and cytokinins. It was called 'abscisin II' originally because it was thought to play a major role in abscission of fruits. At about the same time, another group was calling it 'dormin' because they thought it had a major role in bud dormancy. The name abscisic acid (ABA) was coined by a compromise between the two groups. Though ABA generally is thought to play mostly inhibitory roles, it has many promoting functions as well.

Abscisic acid (ABA) is an isoprenoid compound that regulates developmental processes, such as seed development, desiccation, and dormancy (Nicolas Denance *et al.*, 2013). In addition, the function of ABA as a regulator of abiotic stress has been thoroughly described. ABA has also emerged as a complex modulator of plant defense responses (Sánchez-Vallet *et al.*, 2012). ABA can function as a positive or a negative regulator of plant defense depending on the plant–pathogen interaction analyzed. ABA-impaired (biosynthesis or signaling) mutants in tomato (*sitiens*) and *Arabidopsis* (*abi1-1, abi2-1, aba1-6, aba2-12, aao3-2,* and *pyr1pyl1pyl2pyl4*) were shown to overexpress defensive-signaling pathways, leading to enhanced resistance to different pathogens such as *B. cinerea, P. syringae, F. oxysporum, Plectosphaerella cucumerina,* and *Hyaloperonospora parasitica.* Negative interactions of ABA with the major hormones involved in plant defense response (SA, JA, and ET) have been described by means of exogenous hormone treatments. ABA plays a direct role in regulating R (resistance) protein activity. ABA and exposition of plants to high temperature both reduce the nuclear accumulation of SNC1 (suppressor of *npr1-1*, constitutive1) and RPS4 (resistant to *Pseudomonas syringae* 4) compromising disease resistance to *P. syringae* (Mang *et al.*, 2012). Abscisic acid (Fig. 2.32) can also positively regulate the resistance to some pathogens, such as *Alternaria brassicicola, R. solanacearum,* and *Pythium irregulare,* as ABA-deficient and -insensitive mutants (*abi1-1, abi2-1, abi4-1, aba1-6, aba2-12, aao3-2,* and *npq2-1*) were found to be more susceptible than wild-type plants to these pathogens.

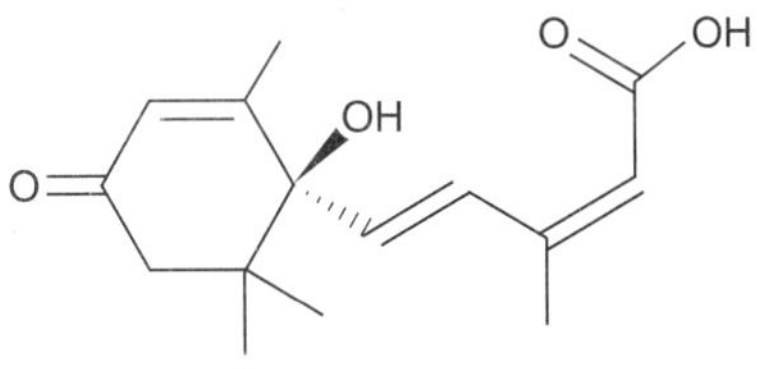

Fig. 2.32 Abscisic acid

Several fungal species produce ABA, including *B. cinerea, Rhizoctonia solani, Ceratocystis fimbriata,* and *Rhizopus nigricans.* Unlike plants, fungi such as *B. cinerea* and *Cercospora* sp. use the mevalonate pathway to produce ABA. It is a key hormone in *Arabidopsis* response to *R. solanacearum* infection, as 40% of the genes up-regulated during the development of wilting symptoms were related to ABA, including those encoding proteins for ABA biosynthesis [i.e., 9-*cis*-epoxycarotenoid dioxygenase3 (NCED3)] or signaling [i.e., ABA-insensitive1 (ABI1) and ABI5. More recently, it has been shown that preinoculation of *Arabidopsis* with an avirulent strain of *R. solanacearum* activates plant resistance to virulent isolates of this bacterium, and this resistance was correlated with the enhanced expression of ABA-related genes that resulted in a hostile environment for the infection development (Nicolas Denance *et al.*, 2013). These results suggest that ABA may be used in biological control of bacterial wilt caused by *R. solanacearum*.

The plant hormone abscisic acid (ABA) plays important roles in many aspects of plant development, in the regulation of stomatal aperture, and in the initiation of adaptive responses to various environmental conditions (Brigitte Mauch-Mani and Felix Mauch, 2005.). Adaptation to drought, low temperature and salinity is regulated by the combinatorial activity of interconnected ABA-dependent and ABA-independent signaling pathways (Shinozaki *et al.*, 2003). The effect of the abiotic stress hormone, abscisic acid, on plant disease resistance, is a neglected field of research. With few exceptions, abscisic acid has been considered a negative regulator of disease resistance. This negative effect appears to be due to the interference of abscisic acid with biotic stress signaling that is regulated by salicylic acid, jasmonic acid, and ethylene, and to an additional effect of ABA on shared components of stress signaling. However, recent research shows that abscisic acid can also be implicated in increasing the resistance of plants towards pathogens via its positive effect on callose deposition.

In addition, ABA also has a role in biotic stress: In barley host-pathogen interactions with powdery mildew, ABA has been shown to reduce penetration resistance of barley leaves after floating leaves in ABA solution by an unknown mechanism. Defense against abiotic and biotic stresses is crucial for the fitness and survival of plants under adverse or suboptimal growth conditions. The phytohormone abscisic acid (ABA) is not only important for mediating abiotic stress responses but also plays a multifaceted and pivotal role in plant immunity. ABA may also be widely involved in plant responses to biotic stresses caused by a broad range of plant pathogens. Many early studies relied on the application of exogenous ABA to increase plant ABA levels, although exogenous ABA may not exert the same physiological effects as endogenous ABA and this may potentially complicate the interpretations of the roles of ABA in plant-pathogen interactions. Studies on the *cds2-1D* mutant, which was identified from a forward genetic screen, showed that activation of NCED5, a key enzyme regulating *de novo* ABA biosynthesis, led to elevated endogenous ABA levels and to enhanced susceptibility to infection by *P. syringae* (Jun Fan *et al.*, 2009). The observations that overexpression of NCEDs in transgenic Arabidopsis recapitulated *cds2-1D* disease phenotype and that genetic lesions in the ABA biosynthesis pathway abolished the enhanced bacterial growth in the *cds2-1D* mutant pinpoints the role of ABA in promoting susceptibility to this hemibiotrophic bacterial pathogen. Susceptibility was increased by application of exogenous ABA at physiological levels or by endogenous accumulation in response to mild water stress, thereby demonstrating the physiological relevance of ABA-mediated suppression of resistance to this hemibiotrophic bacterial pathogen. A potentially important link between abiotic stress and the level of host susceptibility to disease was established.

In addition to the important role of abscisic acid (ABA) in abiotic stress signaling, basal, and high ABA levels appear to have a negative effect on disease resistance (Bob Asselbergh *et al.*, 2008). Using the ABA-deficient sitiens tomato mutant and different application methods of exogenous ABA, the influence of this plant hormone on disease progression of *Erwinia chrysanthemi* was demonstrated. This necrotrophic plant pathogenic bacterium is responsible for soft rot disease in many plant species, causing maceration symptoms mainly due to the production and secretion of pectinolytic enzymes. On wild-type (WT) tomato cv Moneymaker, *E. chrysanthemi* leaf inoculation resulted in maceration both within and beyond the infiltrated zone of the leaf, but sitiens showed a very low occurrence of tissue maceration, which never extended the infiltrated zone. A single ABA treatment prior to infection eliminated the effect of pathogen restriction in sitiens while repeated ABA spraying during plant development rendered both WT and sitiens very susceptible. Quantification of *E. chrysanthemi* populations inside the leaf did not reveal differences in bacterial growth between sitiens and WT. Sitiens was not more resistant to pectinolytic cell-wall degradation, but upon infection it showed a faster and stronger activation of defense responses than WT, such as hydrogen peroxide accumulation, peroxidase activation, and cell-wall fortifications. Moreover, the rapid activation of sitiens peroxidases was also observed after application of bacteria-free culture filtrate containing *E. chrysanthemi* cell-wall-degrading enzymes and was absent during infection with an out *E. chrysanthemi* mutant impaired in the secretion of these extracellular enzymes.

Studies involving plant-nematode interactions provide an opportunity to unravel plant defense signaling in root tissues. Nahar *et al.* (2012) characterized the roles of salicylate (SA), jasmonate (JA), ethylene (ET) and abscisic acid (ABA) in plant defense against the rice root nematode, *Hirschmanniella oryzae* in the monocot model plant rice (*Oryza sativa*). Experiments with exogenous hormone applications, biosynthesis inhibition, and mutant/transgenic lines were executed to test the effect on *H. oryzae* parasitism in rice roots. It was observed that an intact ET, JA, and SA biosynthesis pathway was a prerequisite for defense against *H. oryzae*. By contrast, exogenous ABA treatment drastically compromised the rice defense towards this nematode. Gene expression analyses using quantitative reverse transcription polymerase chain reaction (qRT-PCR) demonstrated that the disease-inducing effect of ABA was likely to be the result of an antagonistic interaction between this hormone and the SA/JA/ET-dependent basal defense system. Collectively, in rice defense against *H. oryzae*, at least three pathways, namely SA, JA and ET, are important, while ABA plays a negative role in the defense. It was further suggested that the balance of ABA and SA/JA/ET signaling was an important determinant of the outcome of the rice-*H. oryzae* interaction.

The *in vitro* effects of abscisic acid (ABA) and nitric oxide (NO) on the nematode-trapping fungus *Drechslerella stenobrocha* AS6.1 were examined (Ling-Ling XU *et al.*, 2010). The average number of traps (constricting rings) per colony and the percentage of nematodes (*Caenorhabditis elegans*) trapped were greatly increased by the addition of ABA but greatly suppressed by addition of sodium nitroprusside (SNP, an NO donor) to corn meal agar. The suppressive effect of SNP was not negated by the addition of an NO synthase competitive inhibitor (l-naphthylacetic acid, L-NNA) or an NO-specific scavenger [2-(4-carboxyphenyl)-4,4, 5,5- tetramethylimidazoline-1-oxyl-3-oxide, cPTIO]. When added without SNP however, L-NNA and cPTIO caused moderate increases in trap number and trapping. The results indicated that the trap formation and nematode-trapping ability of *D. stenobrocha* were enhanced by ABA but decreased by exogenous NO.

2.9.3 Other Plant Growth Regulators

Various other plant growth regulators include Jasmonic acid, Salicylic acid, Polyamines, Brassinolides, Nitrobenzene, Seaweed products, etc., among which, few are very important.

2.9.3.1 Jasmonic acid

Jasmonic acid (JA) is a fatty-acid-derived plant hormone that is similar in overall structure, to physiologically active small molecules from animals called prostaglandins. In plants, jasmonic acid is firmly associated with pathogen defense pathways. The physical stimuli of certain insects can trigger the synthesis of JA, which then functions to increase expression of genes involved in defending the plant. Microbial and viral pathogens can also trigger JA synthesis. The study of JA-mediated events in the plant cell is of interest to plant pathologists who wish to engineer transgenic plants that are disease-resistant.

Jasmonic acid and salicylic acid are essential compounds in the pathogen- and wound-signaling pathways accompanying induced expression of acidic and basic pathogenesis-related (PR) protein genes, respectively (Tomoya Niki *et al.*, 1998). However, on the effect of exogenously supplied SA and JA in the induction of *PR* gene expression, conflicting results have been obtained using various plant materials at different developmental stages. There is no clear evidence on these effects in the presence of both signals at the same time. The effect of SA on the wound- and JA-induced basic *PR* gene expression was analyzed and that of JA on SA-induced acidic Pi gene expression in mature tobacco leaves. Wound-induced accumulation of transcripts for all four basic *PR* genes tested was enhanced in the presence of MeJA, and inhibited in the presence of SA. On the other hand, expression of all three acidic *PR* genes tested was induced by SA and was inhibited by MeJA. Using antibodies raised against acidic PR-1 and PR-2 proteins, these effects were confirmed at the protein level. These results indicated that JA works as an inducer of basic *PR* genes, and also as an inhibitor of acidic *PR* genes, while SA does the opposite.

The jasmonates are key signal compounds in the elicitation process leading to *de novo* transcription and translation and, ultimately, to the biosynthesis of secondary metabolites in plant cell cultures. To deter pathogenic microorganisms and herbivores, plants have developed an inducible chemical defense system. It is known that the induced synthesis of low molecular weight compounds can be provoked by exposing cultured cells to fungal cell wall fragments. Heidrun Gundlach *et al.* (1992) demonstrated that endogenous jasmonic acid and its methyl ester accumulate rapidly and transiently after treatment of plant cell suspension cultures of *Rauvolfia canescens* and *Eschscholtia caisfornica* with a yeast elicitor. Thirty-six plant species tested in cell suspension culture could be elicited with respect to the accumulation of secondary metabolites by exogenously supplied methyl jasmonate. The addition of methyl jasmonate initiates *de novo* transcription of genes, such as phenylalanine ammonia lyase, that are known to be involved in the chemical defense mechanisms of plants. These data demonstrated the integral role of jasmonic acid (Fig. 2.33) and its derivatives in the intracellular signal cascade that begins with the interaction of an elicitor molecule with the plant cell surface and results, ultimately, in the accumulation of secondary compounds.

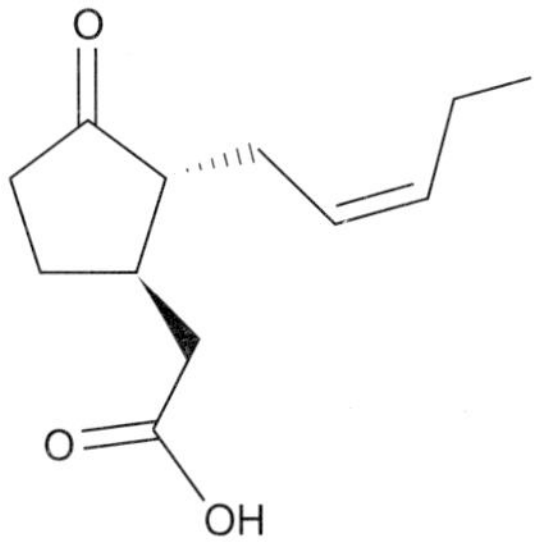

Fig. 2.33 Jasmonic acid

Jasmonic acid is a plant hormone that plays an important role in a wide variety of plant physiological processes. The plant pathogenic fungus, *Lasiodiplodia theobromae* also produces JA. However, its biosynthesis in this fungus has yet to be explored (Tsukada *et al.*, 2010). Administration of NaOAc into *L. theobromae* established that JA in this fungus originates from a fatty acid synthetic pathway. The methyl ester of 12-oxo-phytodienoic acid (OPDA) was detected in the culture extracts of *L. theobromae* by GC-MS analysis. This finding indicated the presence of OPDA, a known intermediate of JA biosynthesis in plants, in *L. theobromae*. (2)H NMR spectroscopic data of JA produced by *L. theobromae* with the incorporation of linolenic acid showed that five deuterium atoms remained intact. In plants, this is speculated to arise from JA being produced by the octadecanoid pathway. However, the observed stereoselectivity of the cyclopentenone olefin reduction in *L. theobromae* was opposite to that observed in plants. These data suggested that JA biosynthesis in *L. theobromae* was similar to that in plants but differed in the facial selectivity of the enone reduction.

Responses of resistant (*Mi-1/Mi-1*) and susceptible (*mi-1/mi-1*) tomato to root-knot nematodes infection were monitored using cDNA microarrays and the roles of salicylic acid (SA) and jasmonic acid (JA) defense signaling were evaluated in these interactions (Kishor K. Bhattarai *et al.*, 2008). Array analysis was used to compare transcript profiles in incompatible and compatible interactions of tomato roots 24 hr after nematode infestation. The *jai1* and *def1* tomato mutant, altered in JA signaling, and transgenic tomato line *NahG*, altered in SA signaling, in the presence or absence of the nematode resistance gene *Mi-1*, were evaluated. The array analysis identified 1,497 and 750 genes differentially regulated in the incompatible and compatible interactions, respectively. Of the differentially regulated genes, 37% were specific to the incompatible interactions. *NahG* affected neither *Mi-1* resistance nor basal defenses to nematodes. However, *jai1* reduced tomato susceptibility to nematode while not affecting *Mi-1*resistance. In contrast, the *def1* mutant did not affect nematode susceptibility. These results indicated that JA-dependent signaling did not play a role in the *Mi-1*-mediated defense. However, an intact JA signaling pathway was required for tomato susceptibility to nematodes. In addition, low levels of SA might be sufficient for basal and *Mi-1*resistance to nematodes.

2.9.3.2 *Salicylic acid*

Salicylic acid (SA) is a phenolic phytohormone and is found in plants with roles in plant growth and development, photosynthesis, transpiration, ion uptake and transport. SA also induces specific changes in leaf anatomy and chloroplast structure. SA is involved in endogenous signaling, mediating in plant defense against pathogens. It plays a role in the resistance to pathogens by inducing the production of pathogenesis-related proteins. It is involved in the systemic acquired resistance in which a pathogenic attack on one part of the plant induces resistance in other parts. The signal can also move to nearby plants by salicylic acid being converted to the volatile ester, methyl salicylate.

The function of SA in activating resistance against pathogens has been thoroughly described. In *Arabidopsis*, SA is synthesized from chorismate (a precursor of tryptophan and, consequently, of auxins) *via* two pathways, either through phenylalanine or isochorismate. This second pathway, in which SID2/ICS1 (salicylic acid induction-deficient 2/isochorismate synthase 1 is involved, is activated upon pathogen infection, such as *Erysiphe* or *P. syringae*, and after plant recognition of pathogen effectors or PAMPs. Deficiency of SA biosynthesis in *sid2-1* mutant leads to reduced resistance response in *Arabidopsis* plants. SA (Fig. 2.34) is a regulator of plant resistance to biotrophic and hemibiotrophic pathogens, such as *Hyaloperonospora arabidopsidis* and *P. syringae*, and it also regulates systemic acquired resistance, a well-studied type of induced resistance. In addition, SA is a central regulator of immunity. It interacts with other signaling pathways (e.g., ET and JA pathways), as a strategy to induce the proper resistance responses and to reduce the associated fitness costs (Nicolas Denance *et al.*, 2013).

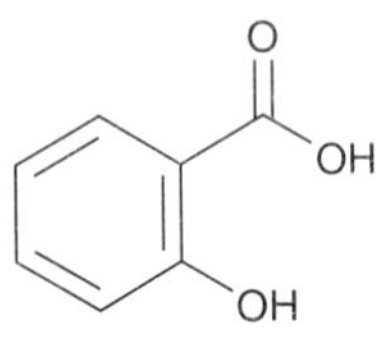

Fig. 2.34 Salicylic acid

Significant insight has been gained in the past year into the roles of salicylic acid in plant-pathogen interactions. The ability to accumulate SA has been shown to be essential for systemic acquired resistance in tobacco plants (Vernooji *et al.*, 1994). It has been shown that SA is apparently not a systemic, vascular-mobile signal, but rather is required for signal transduction at the local level. Its mode of action may include inhibition of catalase activity, leading to increased levels of hydrogen peroxide. After a hypersensitive response to invading pathogens, plants show an elevated accumulation of SA-induced expression of plant defense genes and systemic acquired resistance (SAR) to further infection by a broad range of pathogens (Marianne *et al.*, 2000). There is compelling evidence that SA plays a crucial role in triggering SAR. Tobacco was transformed with two bacterial genes coding for enzymes that convert chorismate into SA by a two-step process. When the two enzymes were targeted to the chloroplasts, the transgenic (CSA, constitutive SA biosynthesis) plants showed a 500- to 1,000-fold increased accumulation of SA and SA glucoside compared to control plants. Defense genes, particularly those encoding acidic pathogenesis-related proteins, were constitutively expressed in CSA plants. This expression did not affect the plant phenotype, but the CSA plants showed resistance to viral and fungal infection resembling SAR in nontransgenic plants.

Systemic induction of pathogenesis-related proteins in tobacco, which occurs during the hypersensitive response to tobacco mosaic virus (TMV), may be caused by a minimum 10-fold systemic increase in endogenous levels of SA (Yalpani *et al.*, 1991). This rise in SA parallels PR-1 protein induction occurs in TMV-resistant Xanthi-nc tobacco carrying the *N* gene, but not in TMV-susceptible (nn) tobacco. By feeding SA to excised leaves of Xanthi-nc (NN) tobacco, it was shown that the observed increase in endogenous SA levels was sufficient for the systemic induction of PR-1 proteins. TMV infection became systemic and Xanthi-nc plants failed to accumulate PR-1 proteins at 32°C. This loss of hypersensitive response at high temperature was associated with an inability to accumulate SA. However, spraying leaves with SA-induced PR-1 proteins at both 24°C and 32°C. SA was most likely exported from the primary site of infection to the uninfected tissues. A computer model predicted that SA should move rapidly in the phloem. When leaves of Xanthi-nc tobacco were excised 24 hr after TMV inoculation and exudates from the cut petioles were collected, the increase in endogenous SA in TMV-inoculated leaves paralleled SA levels in exudates. Exudation and leaf accumulation of SA were proportional to TMV concentration and were higher in light than in darkness. Different components of TMV were compared for their ability to induce SA accumulation and exudation: three different aggregation states of coat protein failed to induce SA, but unencapsidated viral RNA elicited SA accumulation in leaves and phloem. These results supported the hypothesis that SA acts as an endogenous signal that triggers local and systemic induction of PR-1 proteins and, possibly, some components of systemic acquired resistance in NN tobacco.

Compatible plant-nematode interactions involve the formation of an elaborate feeding site within the host root that requires the evasion of plant defense mechanisms by the parasite. Little is known regarding plant defense signaling pathways that limit nematode parasitism during a compatible interaction. Therefore, *Arabidopsis thaliana* mutants perturbed in salicylic acid biosynthesis or signal transduction were utilized to investigate the role of SA in inhibiting parasitism by the beet cyst nematode, *Heterodera schachtii*. It was determined that SA-deficient mutants (sid2-1, pad4-1, and NahG) exhibited increased susceptibility to *H. schachtii*. In contrast, SA-treated wild-type plants showed decreased *H. schachtii* susceptibility. The npr1-2 and npr1-3 mutants, which are impaired in SA signaling, also showed increased susceptibility to *H. schachtii*, whereas the npr1-suppressor mutation sni1 showed decreased susceptibility. Constitutive pathogenesis-related gene-expressing mutants (cpr1 and cpr6) did not show altered susceptibility to *H. schachtii*; however, constitutive *PR* gene expression was restricted to cpr1 shoots with wild-type levels of PR-1 transcript present in cpr1 roots. It was determined that *H. schachtii* infection elicited SA-independent PR-2 and PR-5 induction in wild-type roots while PR-1 transcript and total SA levels remained unaltered. This was in contrast to shoots of infected plants where PR-1 transcript abundance and total SA levels were elevated. It was concluded that SA acted via NPR1 to inhibit nematode parasitism which, in turn, was negatively regulated by SNI1. An inverse correlation between root basal PR-1 expression and plant susceptibility to *H. schachtii* was noticed, and it was suggested that successful cyst nematode parasitism might involve a local suppression of SA signaling in roots.

2.9.3.3 Polyamines

Polyamines are small polycationic molecules found ubiquitously in all organisms and function in a wide variety of biological processes. In the past decade, molecular and genetic studies using mutants and transgenic plants with an altered activity of enzymes involved in polyamine biosynthesis have contributed much to a better understanding of the biological functions of polyamines in plants. A polyamine is an organic compound having two or more primary amino groups-NH2. This class of compounds includes several synthetic substances that are important feedstocks for the chemical industry. It also includes many substances that play important roles in both eukaryotic and prokaryotic cells, such as putrescine, cadaverine, spermidine and spermine (Figures 2.35–2.38).

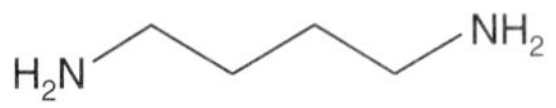

Fig. 2.35 Putrescine

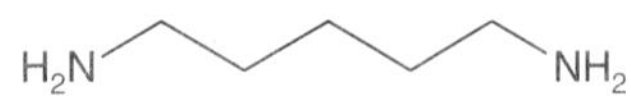

Fig. 2.36 Cadaverine

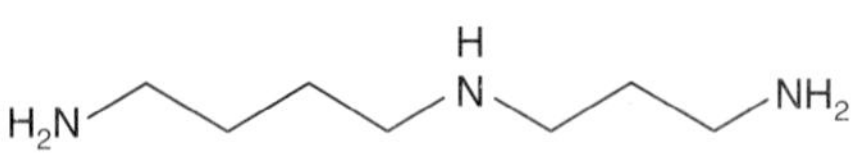

Fig. 2.37 Spermidine

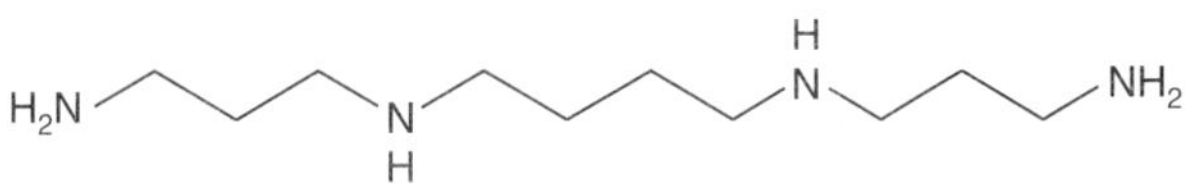

Fig. 2.38 Spermine

Ubiquitous in nature, polyamines are a group of aliphatic amines, cationic at neutral pH, that are essential for cell growth and viability. Because of their positive charge, polyamines are able to bind by electrostatic linkages to many cellular macromolecules, including DNA, RNA, and proteins. Spermidine is essential for the survival of *Arabidopsis*e *mbryos*. One of the reasons may lie in the fact that spermidine serves as a substrate for the lysine → hypusine posttranslational modification of the eukaryotic translation initiation factor 5A, which is essential in all eukaryotic cells (Taku Takahashi and JunIchi Kakehi, 2010). Spermine is not essential but plays a role in stress responses, probably through the modulation of cation channel activities, and as a source of hydrogen peroxide during pathogen infection. Thermospermine, an isomer of spermine, is involved in stem elongation, possibly by acting on the regulation of upstream open reading frame-mediated translation.

The diamine putrescine, the triamine spermidine, and the tetraamine spermine are ubiquitous in plant cells while other polyamines are of more limited occurrence. Their chemistry and pathways of biosynthesis and metabolism are well characterized (Galston and Sawhney, 1990). They occur in the free form as cations, but are often conjugated to small molecules like phenolic acids and also to various macromolecules. Their titer varies from approximately micromolar to more than millimolar and depends greatly on environmental conditions, especially stress. In cereals, the activity of one of the major polyamine biosynthetic enzymes, arginine decarboxylase, is rapidly and dramatically increased by almost every studied external stress, leading to 50-fold or greater increases in putrescine titer within a few hours. The physiological significance of this increase is not yet clear, although most recent work suggests an adaptive, protective role. Polyamines produced through the action of ornithine decarboxylase, by contrast, seem essential for DNA replication and cell division. The application of exogenous polyamines produces effects on patterns of senescence and morphogenesis, suggesting but not proving a regulatory role for polyamines in these processes. The evidence for such a regulatory role is growing.

Polyamines are unique as they are effective (and are applied) in relatively high concentrations. Typical concentrations range from 5 to 500 mg/L. Polyamines such as putrescine, spermidine or spermine influence flowering and promote plant regeneration. Transgenic tomato (*Solanum lycopersicum*) lines overexpressing yeast spermidine synthase (*ySpdSyn*), an enzyme involved in polyamine (PA) biosynthesis, were developed (Savithri Nambeesan *et al.*, 2012). These transgenic lines accumulated higher levels of spermidine (Spd) than the wild-type plants and were examined for responses to the fungal necrotrophs *Botrytis cinerea* and *Alternaria solani*, bacterial pathogen *Pseudomonas syringae* pv *tomato* DC3000, and larvae of the chewing insect tobacco hornworm (*Manduca sexta*). The Spd-accumulating transgenic tomato lines were more susceptible to *B. cinerea* than the wild-type plants; however, responses to *A. solani*, *P. syringae*, or *M. sexta* were similar to the wild-type plants. Exogenous application of ethylene precursors, *S*-adenosyl-Met, and 1-aminocyclopropane-1-carboxylic acid or PA biosynthesis inhibitors reversed the response of the transgenic plants to *B. cinerea*. The increased susceptibility of the *ySpdSyn* transgenic tomato to *B. cinerea* was associated with down-regulation of gene transcripts involved in ethylene biosynthesis and signaling. These data suggested that PA-mediated susceptibility to *B. cinerea* was linked to interference with the functions of ethylene in plant defense.

Polyamine (PA) biosynthesis inhibitors, difluoromethylornithine (DFMO), difluoromethylarginine (DFMA), methylglyoxal bis-(guanylhydrazone) (MGBG) and bis-(cyclohexylammonium) sulphate (BCHA) have been tested for their effects on colony diameters at different intervals after inoculation of four plant pathogenic fungi (*Helminthosporium oryzae, Curvularia lunata, Pythium aphanidermatum and Colletotrichum capsici*) (Rajam and Rajam, 1996). All these inhibitors, except DFMA, had strongly retarded the growth of four fungi in a dose- and species-dependent fashion, and *H. oryzae* and *C. lunata* were found to be most sensitive to the effects of PA inhibitors. *P. aphanidermatum* and *C. capsici* were relatively insensitive and required rather high concentrations of inhibitors to get greater inhibition of mycelial growth, except DFMA, which had a stimulatory effect on the growth of these two fungi. However, DFMA had greatly suppressed the growth of *H. oryzae* and *C. lunata*. The effect was generally more pronounced with MGBG than with DFMO and BCHA, and 1 mM Put completely prevented the inhibitory effects of 1 and 5 mM DFMO. Analysis of free and conjugated PAs in two sensitive fungi (*H. oryzae* and *C. lunata*) revealed that Put was present in highest concentrations followed by Spd and Spm, and their levels were greatly reduced by DFMO application, and such inhibitions were totally reversed by exogenously supplied Put. In fact, PA titers were considerably increased by 1 mM Put alone and in combination with 1 mM DFMO. These results suggest that PA inhibitors, particularly DFMO and MGBG may be useful as target-specific fungicides in plants.

The levels of polyamines in leaves of *Gynura aurantiaca* DC and tomato, *Lycoperskon esculentum* cv Rutgers, infected with citrus exocortis viroid (CEVd) or treated with silver nitrate or ethephon (2-chloroethylphosphonic acid) were measured by HPLC in relation to the development of symptoms (Jose M. Belles *et al.*, 1991). Previously it had been demonstrated that treatment of *G. aurantiaca* plants with silver nitrate or ethephon closely mimicked the effects of viroid infection in the plants. In the studies reported here, a marked decrease in putrescine level was observed in plants infected by CEVd or treated with silver ions or ethephon. There was no significant change in either spermidine or spermine levels. Treatment of *G. aurantiaca* plants with specific inhibitors of ethylene biosynthesis (aminoethoxyvinylglycine, Co^{2+}) or ethylene action (norbomadiene) prevented the decrease of putrescine associated with silver nitrate treatment and had no effect on spermidine or spermine levels. The development of viroid-like symptoms, the production of associated pathogenesis-related proteins, and the rise in protease activity induced by silver nitrate, were all suppressed by exogenous application of putrescine.

Cyst nematodes are sedentary endo-phytoparasites that cause dramatic cellular changes in the plant root to form feeding cells, so-called syncytia. 10A06 is a cyst nematode secretory protein that is most likely secreted as an effector into the developing syncytia during early plant parasitism (Tarek Hewezi *et al.*, 2010). A homolog of the uncharacterized soybean cyst nematode (*Heterodera glycines*), the *10A06* gene, was cloned from the sugar beet cyst nematode (*H. schachtii*), which is able to infect *Arabidopsis* (*Arabidopsis thaliana*). Constitutive expression of *10A06* in *Arabidopsis* affected plant morphology and increased susceptibility to *H. schachtii* as well as to other plant pathogens. Using yeast two-hybrid assays, Spermidine Synthase2 (SPDS2), a key enzyme involved in polyamine biosynthesis, was identified, as a specific 10A06 interactor. In support of this protein-protein interaction, transgenic plants expressing *10A06* exhibited elevated SPDS2 mRNA abundance, significantly higher spermidine content, and increased polyamine oxidase (PAO) activity. Furthermore, the SPDS2 promoter was strongly activated in the nematode-induced syncytia, and transgenic plants overexpressing *SPDS2* showed enhanced plant susceptibility to *H. schachtii*. In addition, in planta expression of *10A06* or SPDS2 increased mRNA abundance of a set of antioxidant genes upon nematode infection. These data lend strong support to a model in which the cyst nematode effector 10A06 exerts its function through the interaction with SPDS2, thereby increasing spermidine content and subsequently PAO activity. Increasing PAO activity results in stimulating the induction of the cellular antioxidant machinery in syncytia. An apparent disruption of salicylic acid defense signaling as a function of 10A06 was observed. Increased antioxidant protection and interruption of salicylic acid signaling were regarded as key aspects of 10A06 function in addition to other physiological and morphological changes caused by altered polyamines, which are potent plant signaling molecules.

2.10 ROLE OF POLYSACCHARIDES IN PATHOGENESIS

Polysaccharides are polymeric carbohydrate molecules composed of long chains of monosaccharide units bound together by glycosidic bonds. They range in structure from linear to highly branched (Fig. 2.39). Examples include storage polysaccharides such as starch and glycogen and structural polysaccharides such as cellulose and chitin. Nutrition polysaccharides are common sources of energy. Many organisms can easily break down starches into glucose; however, most organisms cannot metabolize cellulose or other polysaccharides like chitin and arabinoxylans. These carbohydrate types can be metabolized by some bacteria and protists. Polysaccharides have a general formula of $C_x(H_2O)_y$ where x is usually a large number between 200 and 2500. Considering that the repeating units in the polymer backbone are often six-carbon monosaccharides, the general formula can also be represented as $(C_6H_{10}O_5)_n$ where $40 \leq n \leq 3000$.

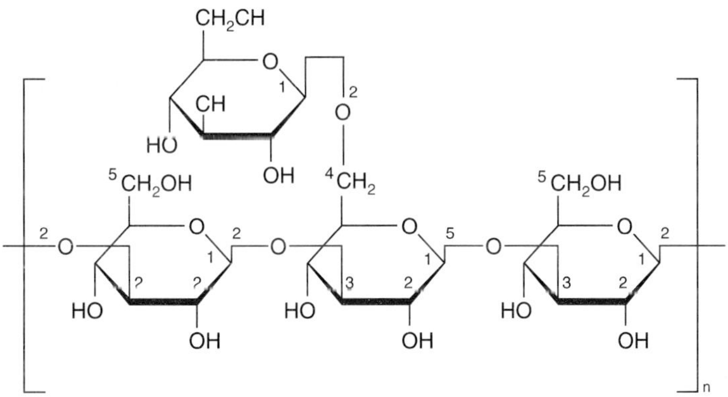

Fig. 2.39 Polysaccharides

Plant biomass is the most abundant organic matter on earth and the major substrate for the majority of fungal species. It consists mainly of polysaccharides, but also contains proteins and the aromatic polymer lignin. Plant polysaccharides can be divided into plant cell wall polysaccharides (cellulose, hemicelluloses, pectin) and storage polysaccharides (e.g., starch, insulin, gums). They consist of many different monomeric components that are attached to each other by a variety of linkages. Fungi, bacteria and nematodes release varying amounts of mucilaginous substances that coat their bodies and provide an interface between the outer surface of the microorganism and its environment. The role of slimy polysaccharides is of utmost importance in wilt diseases. In the vascular wilts, large polysaccharide molecules released by the pathogen in the xylem causes mechanical blockage of vascular bundles and initiate wilting as in *Ralstonia solanacearum*. The high degree of structural complexity of plant cell wall polysaccharides has led to suggestions that some components might function as latent signal molecules that are released during pathogen infections and elicit defensive responses by the plant. However, there has been a paucity of genetic evidence supporting the idea that variation in cell wall composition plays a role in the outcome of host-pathogen interactions. Recently, several genetic studies have provided new lines of evidence implicating cell wall polysaccharides as factors in host-pathogen interactions. The polysaccharides-rich cell walls of plants perform major functions like maintenance of tensile strength and allowing plant growth.

Fungi, bacteria, nematodes and possibly other pathogens constantly release varying amounts of mucilaginous substances that coat their bodies and provide the interface between the outer surface of the microorganism and its environment (Agrios, 2005). Exopolysaccharides appear to be necessary for several pathogens to cause normal disease symptoms either by being directly responsible for inducing symptoms or by indirectly facilitating pathogenesis by promoting colonization or by enhancing survival of the pathogen. The role of slimy polysaccharides in plant disease appears to be particularly important in wilt diseases caused by pathogens that invade the vascular system of the plant. In vascular wilts, large polysaccharide molecules released by the pathogen in the xylem may be sufficient to cause a mechanical blockage of vascular bundles and thus initiate wilting. Although such an effect by the polysaccharides alone may rarely occur in nature, when it is considered together with the effect caused by the macromolecular substances released into the vessels through the breakdown of host substances by pathogen enzymes, the possibility of polysaccharide involvement in the blockage of vessels during vascular wilts becomes obvious.

Filamentous fungi obtain their nutrients by degrading dead or living plant material. Plant material consists of different cell wall and storage polysaccharides. Due to the complex structure and the variety of plant polysaccharides, filamentous fungi secrete a wide range of plant polysaccharide degrading enzymes into the environment to degrade these polysaccharides into monosaccharides that can be taken up into the fungal cell and be used as carbon source. A study was conducted that focused in particular on pentose-related regulators in different fungi and the enzymatic spectrum fungi produce to degrade polysaccharides (Battaglia, 2011). A novel transcriptional regulator involved in L-arabinose release and catabolism was identified in the industrially important fungus *Aspergillus niger*. Functional analysis of AraR and XlnR confirmed the previously suggested antagonistic relation between the two regulators involved in D-xylose and L-arabinose catabolism. The presence of the AraR/XlnR-regulator system was shown to be restricted to *A. niger* and other species belonging to the order of Eurotiales. Comparative analysis of the function of the AraR/XlnR-regulator system in *A. niger* and the model fungus *Aspergillus nidulans* revealed that homologous regulatory systems could differ significantly in related fungi. The regulatory role of AraR in pentose catabolism was different between these two Aspergilli while this was not the case for XlnR. A comparison of the function of the XlnR ortholog (Xlr1) from the plant pathogenic fungus *M. oryzae* to *A. niger* XlnR showed that Xlr1 regulated D-xylose catabolism but had no major role in L-arabinose catabolism. Since no ortholog for AraR is present in the *M. oryzae*, it suggests that another regulator must be involved in the regulation of L-arabinose release and catabolism. In addition, it was shown that Xlr1 has a role in plant pathogenicity. Finally, this Thesis describes the first analysis of CAZy (Carbohydrate Active enzymes) families of a zygomycete fungus (*Rhizopus orzyae*). Comparative analysis of the CAZy repertoires and growth preferences of *R. orzyae* and a selection of ascomycete and basidiomycete species demonstrated highly specialized plant and fungal cell wall degrading abilities for *R. oryzae*, different from ascomycetes and basidiomycetes.

Degradation of plant cell wall polysaccharides is of major importance in the food and feed, beverage, textile, and paper and pulp industries, as well as in several other industrial production processes. Enzymatic degradation of these polymers has received attention for many years and is becoming a more and more attractive alternative to chemical and mechanical processes. Over the past years, much progress has been made in elucidating the structural characteristics of these polysaccharides and in characterizing the enzymes involved in their degradation and the genes of biotechnologically relevant microorganisms encoding these enzymes (Ronald P.de Vries and Jaap Visser, 2001). The members of the fungal genus *Aspergillus* are commonly used for the production of polysaccharide-degrading enzymes. This genus produces a wide spectrum of cell wall-degrading enzymes, allowing not only complete degradation of the polysaccharides but also tailored modifications by using specific enzymes purified from these fungi.

Ralstonia solanacearum, the causal agent of bacterial wilt of diverse plants, produces copious extracellular polysaccharide (EPS), a major virulence factor. The function of EPS in wilt disease is uncertain (Annett Milling *et al.*, 2011). Bacterial wilt-resistant plants can specifically recognize EPS from *R. solanacearum*. Leading hypotheses are that EPS physically obstructs plant water transport or that EPS cloaks the bacterium from host plant recognition and subsequent defense. Tomato plants infected with *R. solanacearum* race 3 biovar 2 strain UW551 and tropical strain GMI1000 upregulated genes in both the ethylene (ET) and salicylic acid (SA) defense signal transduction pathways. The horizontally wilt-resistant tomato line Hawaii7996 activated expression of these defense genes faster and to a greater degree in response to *R. solanacearum* infection than did susceptible cultivar Bonny Best. However, EPS played different roles in resistant and susceptible host responses to *R. solanacearum*. In susceptible plants, the wild-type and *eps*$^-$ mutant strains induced generally similar defense responses. But in resistant Hawaii 7996 tomato plants, the wild-type pathogens induced significantly greater defense responses than the *eps*$^-$ mutants, suggesting that the resistant host recognizes *R. solanacearum* EPS. Consistent with this idea, purified EPS triggered significant SA pathway defense gene expression in resistant, but not in susceptible, tomato plants. In addition, the *eps*$^-$ mutant triggered noticeably less production of defense-associated reactive oxygen species in resistant tomato stems and leaves, despite attaining similar cell densities *in planta*.

Recent studies have shown that living and heat-killed cells of the rhizobacterium, *Rhizobium etli* strain G12 induce in potato roots systemic resistance to infection by the potato cyst nematode, *Globodera pallida (*Reitz *et al.*, 2000*)*. To better understand the mechanisms of induced resistance, focusing was done on identifying the inducing agent. Since heat-stable bacterial surface carbohydrates such as exopolysaccharides (EPS) and lipopolysaccharides (LPS) are essential for recognition in the symbiotic interaction between *Rhizobium* and legumes, their role in the *R. etli*-potato interaction was studied.

EPS and LPS were extracted from bacterial cultures, applied to potato roots, and tested for activity as an inducer of plant resistance to the plant-parasitic nematode. Whereas EPS did not affect *G. pallida infection,* LPS reduced nematode infection significantly in concentrations as low as 1 and 0.1 mg mL^{-1}. Split-root experiments, guaranteeing a spatial separation of inducing agent and challenging pathogen, showed that soil treatments of one-half of the root system with LPS resulted in a highly significant (up to 37%) systemic induced reduction of *G. pallida* infection of potato roots in the other half. The results clearly showed that LPS of *R. etli* G12 act as the inducing agent of systemic resistance in potato roots.

3 Defense Mechanism in Plants

Disease is the exception rather than the rule in natural plant communities. Several pathogens are unable to attack most plants; they have a restricted host range. Assuming environmental conditions favor pathogen development, the resistance or susceptibility of a plant to a particular pathogen depends on interrelated factors, the substrate requirements of the pathogen and the response of the plant to the pathogen. During evolution, plants have developed mechanisms to cope with and adapt to different types of stress, including microbial infection. Once the stress is sensed, signaling pathways are activated, leading to the induced expression of genes with different roles in defense (Inés Ponce de León and Marcos Montesano, 2013).

Two broad groups of pathogens, necrotrophs, and biotrophs were distinguished by their different substrate requirements. Necrotrophs are 'thugs' in the sense that they kill plant cells before parasitizing them. Host and parasite cells cannot coexist harmoniously. Thus, an incompatible cellular relationship between the parasite and host is essential for disease development. If the toxins used to kill host cells are not released at the right time, place or concentration, or if a particular host genotype is insensitive to the toxin, host cells will not die. The necrotroph will be unable to colonize or reproduce, and the plant will be resistant. Two types of necrotrophic pathogens exist, i.e., those with a wide host range involving many plant species and those with a host range restricted to a few plant species or even to cultivars within a species. The key difference between these two types of necrotroph is the specificity of the toxin(s) produced. Necrotrophs with a broad host range secrete toxins that act on metabolic targets common to many plants. In contrast, the pathogenic ability of necrotrophs that release host-specific toxins are conditioned by the gene that encodes the ability to produce the toxin and by a gene in susceptible cultivars of the host that encodes sensitivity to that toxin. Host specific necrotrophs usually form a pathogenic race or pathotype structure where some races can attack some cultivars within a species but not others. If the gene that conditions sensitivity to a particular host-specific toxin is absent from a cultivar, that cultivar will be resistant to the disease caused by that pathogen.

Biotrophs, on the other hand, are obligate parasites that obtain nutrients from living cells. Consequently, they must establish a compatible cellular relationship with their hosts. Biotrophs act as 'sneaks.' They typically infect through natural openings or by directly penetrating their host's surface. They mostly then grow between the cells of their host and only penetrate host cell walls (but not host cell membranes) to form food-absorbing haustoria. The pathogen develops without eliciting the host's defense responses or by spreading in advance of the plant's ability to activate its defense responses. The level of specialization required to establish this type of relationship usually means that biotrophs have a restricted host range and a well-defined pathogenic race structure. If host cells die in advance of invasion by a biotrophic pathogen, the plant will be resistant because the pathogen is unable to establish a parasitic relationship.

A second factor that influences whether a parasitic relationship will become established is the way that the plant under challenge responds. Some interactions between individual pathogen propagules and plant cells may lead to successful pathogen establishment while others may not. In this chapter, it will become evident that resistance or susceptibility of a whole plant and plant communities is the sum of many individual cellular interactions. Plants that are resistant restrict or retard the development and reproduction of an overwhelming majority of individual pathogen propagules that attack it. In this sense resistance is quantitative-resistant hosts prevent or slow the development and reproduction of a higher proportion of pathogen propagules than susceptible hosts. For plant breeding, the response of a plant to pathogen inoculation is often categorized as either 'resistant' or 'susceptible,' although from a cellular perspective this distinction is not always so clear.

Resistance and susceptibility are more accurately portrayed as the extremes of a continuum upon which most host-parasite interactions sit. Resistance may be expressed in many ways, from the inhibition of propagule germination and penetration, the killing of pathogens before establishment, to the restriction or retardation of colony development and reproduction once the pathogen has established. For example, different genes for stem rust resistance in wheat act at different stages of the host-parasite interaction. Some cause the rapid death of the pathogen following attempted penetration, others allow initial infection, but prevent haustorial development and starve the pathogen, while the 'slow-rusting' genes allow parasitism and pathogen reproduction, but at a much slower rate than in susceptible cultivars. Each type of interaction provides useful resistance for plant breeders because they all delay the onset of epidemics and reduce yield losses.

The early steps involved in the establishment of a host-pathogen relationship are delicate and sensitive to environmental factors, including the presence of other microorganisms. The host-parasite-environment interaction is mediated by a complex interchange of signals. Plants respond to pathogen attack by erecting a highly coordinated series of molecular, cellular and tissue-based defense barriers. All plants have the capacity to activate these defenses. However, if they are activated too little, too late, or in the wrong place, they will fail to restrict the pathogen, and the plant will be susceptible. Pathogens respond by escaping or suppressing plant defense responses or by rendering these responses impotent, for example by detoxifying plant antibiotics.

The interaction of pathogen nutrient requirements and host responses leads to five possible outcomes if environmental conditions favor infection.

(i) No relationship is established when the plant and the pathogen ignore each other. For example, a spore of a fungus may germinate, but because the host does not provide essential requirements for pathogen development, the resulting hyphen fails to penetrate or establish a parasitic relationship. The fungus dies when its energy reserves are exhausted. The plant does not react in any way and is resistant by default. It is a nonhost.

(ii) The plant is antagonistic to the pathogen when it secretes inhibitory compounds into its environment that prevents pathogen development. For example, the stubble of some basics releases 'biofumigants' into the soil that prevent the hatching of nematode eggs and inhibit the growth of some root-infecting fungi. Asparagus and marigolds (*Tagetes* spp.) secrete substances into the rhizosphere that are toxic to nematodes and provide useful protection against nematodes when interplanted with nematode susceptible plants like tomato. Many plants secrete phenolic compounds onto their leaf surfaces that not only discourage herbivore feeding, but also inhibit many microorganisms, including potential pathogens. In this relationship, the pathogen fails to develop and has no observable effect on the metabolism of the host plant. In some cases, such as in the quiescent infection of ripening avocado fruit with *Colletotricltum gloeosportoides*, plant antagonists only temporarily inhibit pathogen development. Spores germinate to form appressoria, but their development is arrested by fungistatic substances in the peel. After harvest, these substances are enzymically degraded, and the appressorium germinates to form infection hyphae. Eventually, anthracnose lesions develop. This type of interaction involving a quiescent stage in pathogen development is common among the stem end rot pathogens of avocado and mango (e.g., *Dothiorella domtnicana, Phomopsis* spp. and *Colletotrichum gloeosporioides*).

(iii) The pathogen is antagonistic to the plant when it secretes compounds that damage the plant. For example, *Periconia circinata* infects the roots of sorghum, but only those strains of the fungus that produce the host-specific toxin, periconin, induce symptoms of milo disease, but only in cultivars that are sensitive to this toxin. Similarly, some strains of *Alternaria alternata* release host-specific toxins that kill cells of susceptible host species and cultivars. For example, a strain of the fungus that is pathogenic on tomato produces AAl-toxin, to which tomato is uniquely sensitive. Strains producing AAM-toxin attack apples, AAK-toxin producing strains affect Japanese pears, AAC toxin-producing strains affect citrus and so on. The tomato, apple, and Japanese pear strains are not pathogenic to citrus because citrus is only sensitive to the AAC-toxin. *Cochliobolus victoriae* produces the toxin victorin that causes a severe seedling blight on susceptible cultivars of oats but has little effect on resistant cultivars or other plant species. Resistance is the result of insensitivity to the toxin produced by the pathogen. If this insensitivity is common to all cultivars within a plant species, that species is said to be a nonhost.

(iv) Mutual antagonism between plant and pathogen results in the inhibition or death of both the host tissue and pathogen. For example, incompatible interaction between the stem rust pathogen, *Puccinia gramtnis* f. sp. *tritici* and resistant cultivars of wheat cause the death of both host and pathogen cells.

(v) Mutual adjustment leads to a compatible cellular relationship between the host and pathogen. Symbiotic relationships between mycorrhizal fungi and plant roots and between nitrogen-fixing prokaryotes and plant roots are examples of mutually beneficial interactions. Endophytic fungi and bacteria colonize the intercellular spaces of plant tissue, apparently without damaging their host cells. Many stem end rot pathogens have an endophytic phase in leaves and twigs before they infect fruits. Biotrophic pathogens, like the mildews and rusts, grow and reproduce on living host tissue. However, the diversion of nutrients to the invading pathogen adversely affects the growth of the host, even though host cells are not killed. Broadly speaking, passive defense mechanisms are those that are present before contact with the pathogen, while active defense mechanisms are activated only after pathogen recognition. In reality, this distinction is not always clear, as many preexisting defenses are modified after infection.

Most biotrophic and hemibiotrophic pathogens can only cause disease in a relatively small group of host plants because of the slightly different set of specialized genes and molecular mechanisms required for each host-pathogen interaction (Freeman and Beattie, 2008). The host range refers to the plant species on which a pathogen is capable of causing disease. For example, brome mosaic virus infects grasses such as barley but not legumes. A plant species that does not show disease when infected with a pathogen is referred to as a nonhost plant species for that pathogen. Organisms that do not cause disease in any plant species, such as the saprophytic bacterial species *Pseudomonas putida*, are referred to as nonpathogens. When a pathogen is capable of causing disease in a particular host species, two outcomes are possible, a compatible response is an interaction that results in disease while an incompatible response is an interaction that results in little or no disease at all. Although a particular plant species may be a susceptible host for a particular pathogen, some individuals may harbor genes that help recognize the presence of the pathogen and activate defenses. For example, some tomato cultivars show disease when infected with the bacterial pathogen *Pseudomonas syringae* (a compatible response), but others (cultivar Rio Grande) are capable of recognizing the bacteria and limiting disease via resistance (an incompatible response). Disease resistance exists as a continuum of responses ranging from immunity (the complete lack of any disease symptoms) to highly resistant (some disease symptoms) to highly susceptible (significant disease symptoms).

Defense mechanisms in plants against pathogens can be categorized in several ways. Most common one is structural or chemical defenses.

(i) **Preexisting/preinfectional structural:** Waxes, hairs, epidermis, cuticle, natural openings, etc.

(ii) **Preexisting biochemical:** Antifungal, antimicrobial compounds released by the host plant and its environment, viz., amino acids, nucleotides, sugars, lipids, organic acids, enzymes, alkaloids, flavonones, phenols, saponins, phenolic glycosides, a wide range of inorganic, toxic materials, etc.

(iii) **Postinfectional structural:** Lignified cells, cork layers, abscission layers, tyloses, gum deposition, etc.

(iv) **Postinfectional biochemical:** This invovles toxic material production in response to infection, phenolic compounds, phytoalexins, proteins and enzymes, formation of substrates resistant to pathogenic enzymes (viz., pectinolytic enzymes like pectin methyl esterase, pectin glycosidase, polygalacturonase and polymethyl galacturonase) by inhibiting pathogen enzymes (production of phenolics, terpenes, lignin, tannins, proteins as enzyme inhibitors), defense through detoxification of pathogen toxin, by altered biosynthetic pathways, defense triggered by previous invaders, hypersensitivity reaction, etc.

Major defense mechanisms have been briefly discussed hereunder by correlating with the latest published information.

3.1 STRUCTURAL DEFENSES

3.1.1 Cuticle

The cuticle plays an important role in plant interactions with pathogens and with their surroundings. The cuticle acts as both a physical barrier against physical stresses and pathogens and a chemical deterrent and activator of plant defense response. The cuticle is a protective layer covering nearly all aerial parts of the plant. The cuticle not only acts as a physical barrier to biting insects, UV radiation, environmental stresses, penetrating fungi and other pathogens, but its chemical composition greatly contributes to the plant's defense against different types of pathogens. The cuticle is produced and secreted by epidermal cells and can be found as the outer layer of the epidermis and in extracellular spaces. The cuticle is largely composed of cutin, a polyester polymer which is embedded in and overlaid with various types of waxes (Nawrath, 2006).

The plant cuticle is a hydrophobic layer that covers the aerial surface of plants and forms the first line of contact with the environment. In the epidermis of young leaves, the cuticle is present as a highly repellent wax layer, termed procuticle, which matures with the leaf. The mature cuticle is composed of cutin and cuticular wax (Ye Xia *et al.*, 2009). They reported an active role for the plant cuticle in SAR-related molecular signaling. It was observed that an acyl carrier protein, ACP4, was required to perceive the mobile SAR signal in distal tissues of Arabidopsis. Although ACP4 plants generated the mobile signal, they failed to induce the systemic immunity response. Defective SAR in ACP4 plants was not due to impairment of salicylic acid (SA)-, methyl SA-, or jasmonic acid-mediated plant hormone signaling pathways but was associated with the impaired cuticle of ACP4 leaves. Other cuticle-impairing genetic mutations or physical removal of the cuticle also compromised SAR. This cuticular requirement was relevant only during mobile signal generation and its translocation to distal tissues.

Recent evidence suggests that the composition of the cuticle affects stomata and trichome development and, in addition, plays a role in organ fusions (Bird and Gray, 2003). For example, epidermal fusion phenotype of the fdh mutant has been suggested to be due to a change in the permeability of the cuticle (Lolle *et al.*, 1997). Similarly, diffusion of an unknown inhibitor of guard cells through the stomata has been proposed to control the development of stomata (Bird and Gray, 2003). The lipids present in the outermost layer of pollen grains and the cuticle on the epidermal surface are also suggested to play a direct role in the regulation of water transfer between the pollen grain and stigma surface (Lolle and Pruitt, 1999). More recently, cutin has been associated with basal resistance to necrotrophic pathogens. Both lacs2 and gpat4 gpat8 plants are defective in cutin content, but while lacs2 plants show enhanced resistance to *Botrytis* (Bessire *et al.*, 2007), gpat4 gpat8 plants are susceptible to *Alternaria* (Li *et al.*, 2007). These observations, together with our current findings implicate the cuticle in a more active signaling role in plant defense.

Systemic immunity in plants not only requires the proper generation and transmittance of a signal at the site of primary infection but also depends upon its reception and decoding in the distal tissues. The cuticle or a derived component in the distal tissues is required for this perception of the mobile SAR signal. A mutation in ACP4, a critical component of FA biosynthesis, impairs SAR because it affects cuticle formation in the leaf (Ye Xia *et al.*, 2009).

The cuticle is generally regarded as a mere passive physical barrier against pathogen invasion, but it has also been recognized as a potential source of signaling and elicitor molecules (Reina-Pinto and Yephremov, 2009). Plant cutin monomers trigger cutinase secretion in pathogenic fungi, and cutin and wax components initiate appressorium formation and penetration in appressorium-forming pathogens (Katrien Curvers *et al.*, 2010). It has been postulated that a defective or thin cuticle encourages the plants to constitutively express defense-related mechanisms and to secrete antifungal compounds to the plant surface, thereby inhibiting *B. cinerea* growth. In addition, cuticle metabolic pathways might directly modulate plant-pathogen interactions by interacting with hormonally regulated defense pathways or with complex lipid signaling pathways leading to hypersensitive cell death (Katrien Curvers *et al.*, 2010).

Once plant pathogens have penetrated the cuticle, they secrete hydrolases that target the plant cell wall (Jakob *et al.*, 2007) that is mainly composed of cellulose, hemicellulose, and pectin (35% of total dry weight). Pectin consists mainly of the polysaccharides homogalacturonan and rhamnogalacturonan I and II. Homogalacturonans are linear chains of α-(1–4)-linked D-GalA residues that can be methylesterified at C-6. Rhamnogalacturonan I and II are more complex, branched polysaccharides. *B. cinerea* is typically regarded as a pectinolytic pathogen because it possesses efficient pectinolytic machinery, including a variety of polygalacturonases and pectin methylesterases (PMEs), some of which are important virulence factors. It was suggested that plant defense elicitors released by *Botrytis cinerea* or hydrolytic enzymes producing cell wall-derived elicitors may diffuse faster through the more permeable *sitiens* cuticle. Alternatively, monomers released from a defective cuticle might induce signaling events leading to the activation of plant defense mechanisms. In favor of this hypothesis, pathogenesis-related gene expression is higher in *sitiens* plants under noninfected conditions than that of the wild type.

In contrast to the link between cuticle permeability and *B. cinerea* resistance in *Arabidopsis* and tomato leaves, in tomato fruits a less permeable cuticle leads to *B. cinerea* resistance. Fruits of the tomato mutant '*delayed fruit deterioration*' have a denser cutin matrix and higher wax and cutin contents than the wild-type cv Ailsa Craig and are highly resistant to *B. cinerea,* unless the cuticle is damaged (Saladié *et al.,* 2007). This indicates that there is a different defense mechanism against *B. cinerea* operating in tomato leaves as compared with fruits, which have a 10-fold thicker cuticle. Components of cuticle originate from fatty acid precursors, and their metabolism is orchestrated by a complex pathway which involves a large number of condensing enzymes reductases, putative decarbonylase enzymes, and others. This complex metabolic pathway is regulated by various transcription factors; the first of these to be identified and characterized for its role in cuticle metabolism was SHN1/WIN1, a member of the *Arabidopsis thaliana* APETALA2 (AP2)-domain transcription factor (Kannangara *et al.,* 2007).

There is growing interest in understanding the roles played by the cuticle and cuticle components in plant cell structure and plant defense. Various studies have generated a significant amount of data on cuticle-related genes and their products (i.e., cuticle-related enzymes) in tomato and other fruit-bearing species (Wang *et al.,* 2011). Among the various roles played by the cuticle during the plant's life cycle, protecting plant cells from pathogens is perhaps one of its most important. The cuticle functions as both a physical and biochemical barrier. The cuticle's mechanical strength is largely provided by the cutin matrix, a polymer formed by three-dimensional crosslinking of covalent bonds (Kolattukudy, 1980). It has been suggested that one of the cuticle's primary functions is to protect the plant surface from possible external mechanical damage caused by biting insects or growing fungal hyphae (Muller and Riederer, 2005).

In addition to the role of cuticle as a physical barrier, evidence suggests that its components, which operate as pathogenesis signals for the invading pathogens, may also function as triggers for the plant's defense responses (Bajar *et al.,* 1991). Xia *et al.* (2009) demonstrated that intact cuticle is required for induction of systemic acquired resistance-related molecular signaling. An impaired SAR response in plants mutated in the acyl carrier protein ACP4 (required for SAR signaling) is due to the defective cuticle and not to impaired salicylic acid- or jasmonic acid-mediated signal perception and transduction pathways. Several other studies have suggested that cuticle permeability may lead to increased plant resistance. Thus, for example, the *Arabidopsis* CUTE (overexpressing fungal cutinase) transgenic line and the bdg and lacs2 mutants have all been shown to be defective in the integrity of their cuticular layer. Although these plants exhibit higher cuticular permeability than their wild-type counterparts, they are all fully immune to some necrotrophic pathogens, but not others (Chassot *et al.,* 2007). The Arabidopsis mutant lacerate (lcr), which exhibited intermediate permeability as compared to lacs2, also displayed intermediate resistance to *Botrytis cinerea* while the hth mutant, with lower cuticle permeability, was as susceptible to *B. cinerea* as wild-type plants (Bessire *et al.,* 2007). A strong link between higher cuticle permeability, reactive oxygen species (ROS) accumulation and resistance to *B. cinerea* has also been demonstrated (L'Haridon *et al.,* 2011). Yet, in contrast to these reports, the rst1 mutant, which exhibits elevated levels of cuticular lipids but normal cuticular permeability, displays enhanced resistance to *B. cinerea* and enhanced susceptibility to the biotrophic fungus *Erysiphe cichoracearum* (Mang *et al.,* 2009). Thus, the functional aspects of cuticle structure and composition during plant interactions with fungal pathogens may depend on as yet unidentified factors since not all *Arabidopsis* mutants that exhibit permeable or altered cuticles demonstrate the same response to infection by various fungal pathogens.

Cuticle production in tomato plants is regulated by several transcription factors, including SlSHINE3, an ortholog of the *Arabidopsis* WIN SHN3. A SlSHINE3-overexpressing (SlSHN3-OE) and silenced (Slshn3-RNAi) lines and a mutant was used in SlCYP86A69 (Slcyp86A69), a direct target of SlSHN3 to analyze the roles of the leaf cuticle and cutin content and composition in the tomato plant's defense response to the necrotrophic foliar pathogen *Botrytis cinerea* and the biotrophic bacterial pathogen *Xanthomonas campestris* pv. *vesicatoria* (Kobi Buxdorf *et al.*, 2014). It was shown that SlSHN3, which is predominantly expressed in tomato fruit epidermis, also affected tomato leaf cuticle, as morphological alterations in the SlSHN3-OE leaf tissue, resulted in shiny, stunted and permeable foliage. SlSHN3-OE leaves accumulated 38% more cutin monomers than wild-type leaves, while Slshn3-RNAi and Slcyp86A69 plants showed a 40 and 70% decrease in leaf cutin monomers, respectively. Overexpression of SlSHN3 resulted in resistance to *B. cinerea* infection and *X. campestris* pv. *vesicatoria* correlated with cuticle permeability and elevated expression of pathogenesis-related genes *PR1a* and *AOS*. Further analysis revealed that *B. cinerea*-infected Slshn3-RNAi plants were more sensitive to *B. cinerea* and produced more hydrogen peroxide than wild-type plants. Cutin monomer content and composition differed between SlSHN3-OE, Slcyp86A69, Slshn3-RNAi and wild-type plants and cutin monomer extracted from SlSHN3-OE plants altered the expression of pathogenesis-related genes in wild-type plants.

All plant tissues contain preformed structural barriers that help limit pathogen attachment, invasion, and infection. The cell wall is a major line of defense against fungal and bacterial pathogens. It provides an excellent structural barrier that also incorporates a wide variety of chemical defenses that can be rapidly activated when the cell detects the presence of potential pathogens. All plant cells have a primary cell wall, which provides structural support and is essential for turgor pressure, and many also form a secondary cell wall that develops inside of the primary cell wall after the cell stops growing. The primary cell wall consists mostly of cellulose, a complex polysaccharide consisting of thousands of glucose monomers linked together to form long polymer chains. These chains are bundled into fibers called microfibrils, which give strength and flexibility to the wall. The cell wall may also contain two groups of branched polysaccharides: cross-linking glycans and pectins. Cross-linking glycans include hemicellulose fibers that give the wall strength via cross-linkages with cellulose. Pectins form hydrated gels that help 'cement' neighboring cells together and regulate the water content of the wall. Soft-rot pathogens often target pectins for digestion using specialized enzymes that cause cells to break apart: these organisms are extremely common, and anyone who has seen fruits or vegetables become brown and 'mushy' have seen these pathogens in action.

Many cell walls also contain lignin, a heterogeneous polymer composed of phenolic compounds that give the cell rigidity. Lignin is the primary component of wood, and cell walls that become 'lignified' are highly impermeable to pathogens and difficult for small insects to chew. Cutin, suberin, and waxes are fatty substances that may be deposited in either primary or secondary cell walls (or both) and outer protective tissues of the plant body, including bark.

3.1.2 Lignin

Lignin is the generic term for a large group of aromatic polymers resulting from the oxidative combinatorial coupling of 4-hydroxyphenylpropanoids (Fig. 3.1). These polymers are deposited predominantly in the walls of secondarily thickened cells, making them rigid and impervious. In addition to developmentally programmed deposition of lignin, its biosynthesis can also be induced by various biotic and abiotic stress conditions, such as wounding, pathogen infection, metabolic stress and perturbations in cell wall structure. Because lignin protects cell wall polysaccharides from microbial degradation, thus imparting decay resistance, it is also one of the most important limiting factors in the conversion of plant biomass to pulp or biofuels. The removal of lignin from plant biomass is a costly process; hence, research efforts are now aimed at designing plants that either deposit less lignin or produce lignins that are more amenable to chemical degradation. The phenylpropanoid-derived polymer lignin cross-links plant secondary cell walls to provide mechanical strength and hydrophobicity to the vascular system as well as contributes to defense against biotic stress.

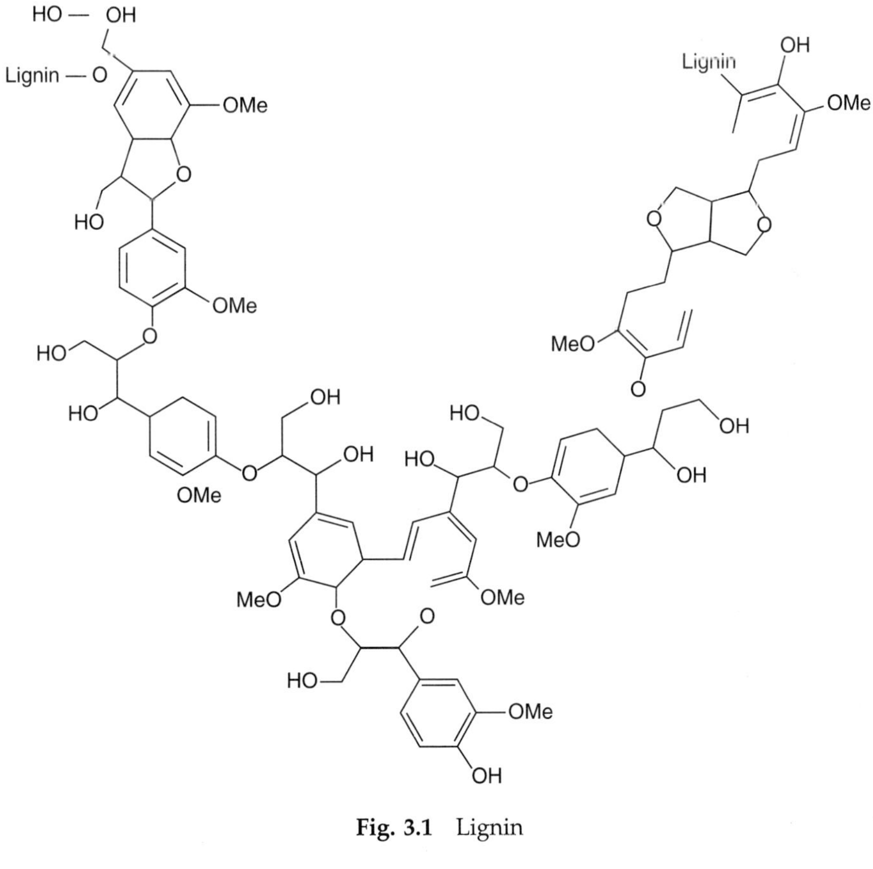

Fig. 3.1 Lignin

Lignin is a strong type of molecule that provides plants with a defensive structure similar to that of fibrous proteins. It acts as a barrier and can be found in wood and is characteristically found in plants that have recently endured pathogen attack. The insolubility and complexity of the lignin polymer make it resistant to degradation by most pathogens. Variation in lignin content, composition and location is likely to affect these essential processes.

Lignin is a ubiquitous polymer present in cell walls of all vascular plants, where it rigidifies and strengthens the cell wall structure through covalent cross-linkages to cell wall polysaccharides. The presence of lignin makes the cell wall recalcitrant to conversion into fermentable sugars for bioenergy uses. Therefore, reducing lignin content and modifying its linkages have become major targets for bioenergy feedstock development through either biotechnology or traditional plant breeding (Sattler *et al.*, 2013). In addition, lignin synthesis has long been implicated as an important plant defense mechanism against pathogens, because lignin synthesis is often induced at the site of pathogen attack. Lignin modifications influence the susceptibility of a range of plant species to their associated pathogens, and there are several implications for the development of feedstocks for the second-generation biofuels industry. Surprisingly, there are some instances where plants modified in lignin synthesis may display increased resistance to associated pathogens.

On the basis of global gene regulation of phenylpropanoid metabolism-related genes, phenylpropanoid metabolism was deduced to be involved in the cotton defense response (Li *et al.*, 2011). A closer look at the expression of these genes, enzyme activity, and lignin levels revealed differences between resistant and susceptible cotton plants. Both types of plants showed an increased level of expression of lignin synthesis-related genes and increased phenylalanine ammonia lyase (PAL) and peroxidase (POD) enzyme activity after inoculation with *V. dahliae*, but the increase was greater and faster in the resistant line. Histochemical analysis of lignin revealed that the resistant cotton not only retains its vascular structure but also accumulated high levels of lignin. Furthermore, the quantitative analysis demonstrated increased lignification and cross-linking of lignin in resistant cotton stems. Overall, a critical role for lignin was believed to contribute to the resistance of cotton to disease.

Lignin content and enzymes involved in the cation were measured in leaf discs of reed canary grass (*Phalaris arundinacea* L.) inoculated with *Helminthosporium avenae* and floated on water or solutions of cycloheximide (25 Ag/mL) (Carroll and Robert, 1976). Fungal germ tubes did not penetrate localized lignified swellings, which formed beneath penetration sites, in the outer epidermal wall of discs floated on water. Within 18 hours, inoculated discs on water had a higher Iignin content and higher activity of the enzymes phenylalanine ammonia lyase, tyrosine ammonia lyase, hydroxyl cinnamate-CoA ligase and peroxidase than noninoculated discs on the water. When inoculated tissues were floated on cycloheximide solutions, increases in lignin content and enzyme activities associated with lignin biosynthesis were inhibited, and the tissue was susceptible to fungal penetration. Lignin biosynthesis at the site of attempted fungal penetration may play an important role in the resistant response of reed canary grass to leaf-infecting fungi.

The role of *Ganoderma* on oil palm as the cause of a white rot disease has been neglected. This type of attack involves the degradation of lignin as the defining mode. Much more knowledge is required on the biochemical structural of the macromolecules in oil palm, and how they vary within and between species. In addition, the ligninolytic enzymes in the *Ganoderma* involved in oil palm disease require being determined (Robertobert *et al.*, 2009). They felt that an obvious approach for control was to develop oil palm with altered lignin so that it is resistant to attack. The technology existed in other plants, and other forms of transformation were undertaken with oil palm. Oil palm lags somewhat behind the other crops. However, an oil palm could be produced with resistant lignin to *Ganoderma* disease.

3.1.3 Cutin

It is a macromolecule, a polymer consisting of many long-chain fatty acids that are attached to one another by ester linkages, creating a rigid three-dimensional network (Fig. 3.2). Cutin is formed from 16:0 and 18:1 fatty acids with hydroxyl or epoxide groups

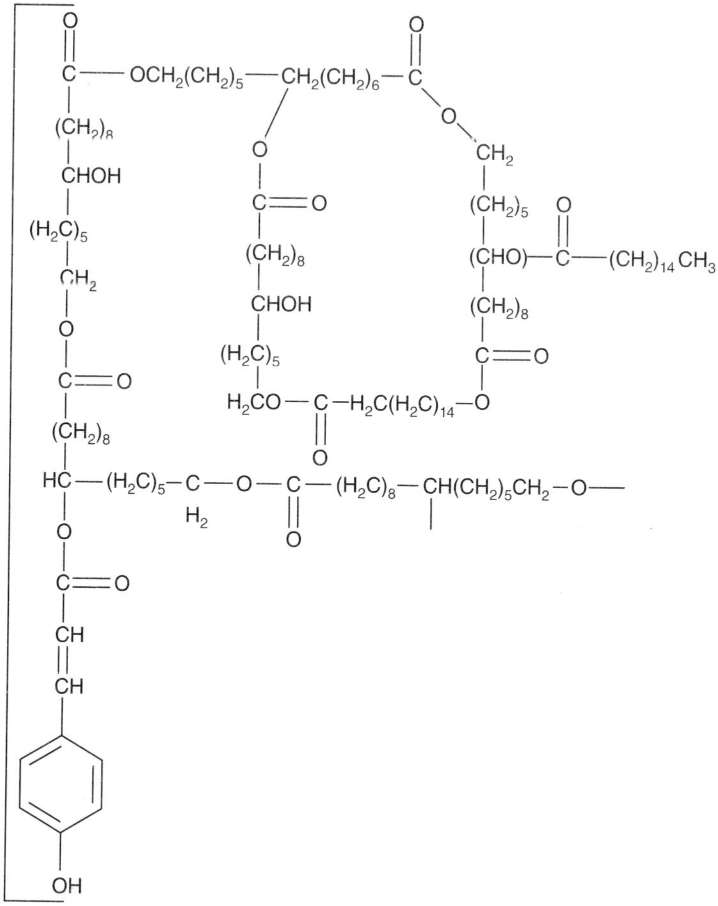

Fig. 3.2 Model cutin monomeric unit

situated either in the middle of the chain or at the end opposite the carboxylic acid function. Cutin is a principal constituent of the cuticle, a multilayered secreted structure that coats the outer cell walls of the epidermis on the aerial parts of all herbaceous plants. The cuticle is composed of a top coating of wax, a thick middle layer containing cutin embedded in wax (the cuticle proper), and a lower layer formed of cutin and wax blended with the cell wall substances pectin, cellulose, and other carbohydrates (the cuticular layer). It has been suggested that in addition to cutin, the cuticle may contain a second lipid-polymer made up of long-chain hydrocarbons, which has been named cutan (Jeffree, 1996).

Cutin is a major structural component of the cuticle of plants. Analysis of the function of cutin *in vivo* has been limited because no mutants with specific defects in cutin have been characterized. Therefore, transgenic Arabidopsis plants were generated that express and secrete a cutinase from *Fusarium solani* f. sp. *pisi*. *Arabidopsis* plants expressing the cutinase in the extracellular space showed an altered ultrastructure of the cuticle and an enhanced permeability of the cuticle to solutes (Patrick Siebera *et al.*, 2000). In addition, pollen could germinate on fully differentiated leaves of cutinase-expressing plants but not on control leaves. These differences coincided with strong postgenital organ fusions. The junctions of the fusions contained pectic polysaccharides. As the fused organs grew apart from each other, organ deformations and protrusions of epidermal cells developed at positions with high mechanical stress. These results demonstrated that an intact cutin layer not only is important for plant–environment interactions but also prevents fusions between different plant organs and is, therefore, necessary for normal epidermal differentiation and organ formation.

Hypocotyls from etiolated cucumber (*Cucumis sativus* L.) seedlings were gently abraded at their epidermal surface and cut segments were conditioned to develop competence for H_2O_2 elicitation (Markus Fauth *et al.*, 1998). Alkaline hydrolysates of cutin from cucumber, tomato, and apple elicited H_2O_2 in such conditioned segments. The most active constituent of cucumber cutin was identified as dodecan-1-ol, a novel cutin monomer capable of forming hydrophobic terminal chains. Additionally, the cutin hydrolysates enhanced the activity of a fungal H_2O_2 elicitor, similar to cucumber surface wax, which contained newly identified alkan-1,3-diols. The specificity of elicitor and enhancement activity was further elaborated using some pure model compounds. Certain saturated hydroxyl fatty acids were potent H_2O_2 elicitors as well as enhancers. Some unsaturated epoxy and hydroxy fatty acids were also excellent H_2O_2 elicitors but inhibited the fungal elicitor activity. Short-chain alkanols exhibited good elicitor and enhancer activity, whereas longer-chain alkan-1-ols were barely active. The enhancement effect was also observed for H_2O_2 elicitation by ergosterol and chitosan. The physiological significance of these observations might be that once the cuticle is degraded by fungal cutinase, the cutin monomers may act as H_2O_2 elicitors. Corrosion of cutin may also bring surface wax constituents in contact with protoplasts and enhance elicitation.

Cutin monomers generated by the action of fungal cutinase were shown to act as signals for fungal gene activation and also to contribute to induction of fungal appressoria. Plant surface wax constituents can also trigger several steps in fungal development.

Taken together, these results provide convincing evidence that constituents of the plant cuticle represent important potential signals for the recognition of plant surfaces by fungal pathogens. In contrast, a signaling role of cutin monomers in the plant's defense strategy has only recently been recognized. Root pretreatment of rice plants with monoepoxy linoleic and linolenic acids was shown to induce resistance of the leaves against *Pyricularia oryzae*, the imperfect form of *Magnaporthe grisea* (Namai *et al.*, 1993). It was also shown that a topical spray of cutin monomers on leaves of barley and rice enhanced resistance against *Erysiphe graminis* f. sp. *hordei* and *M. grisea*, respectively. Since the cutin monomers used in these experiments exhibited no apparent fungicidal effect, the observed protection was taken as evidence that the cutin monomers induced acquired resistance in the plants. That plant cells can perceive free cutin monomers was shown in a model system consisting of suspension-cultured potato cells, which exhibited a transient alkalinization of the culture medium, similar to known elicitors of defense responses.

3.1.4 Waxes

Waxes are not macromolecules, but rather complex mixtures of long-chain acyl lipids that are extremely hydrophobic (Fig. 3.3). Waxes are the waterproofing component of the plant cuticle and are therefore essential for life in an aerial environment. They are embedded in cutin or suberin polymers and continue as an amorphous layer on the outer surface of the plant. In many plants, epicuticular wax crystalline structures overlay this layer, which gives the plant surface a glaucous or gray appearance. Plants deficient or altered in surface waxes appear shiny, glossy, or bloomless. This distinct and easily observed phenotype has been widely used in the isolation of mutants defective in wax production. The most common components of waxes are straight-chain alkanes and alcohols of 25 to 35 carbon atoms (Table 3.1). Long-chain aldehydes, ketones, esters and free fatty acids are also found in waxes. The waxes of the cuticle are synthesized by epidermal cells. They leave those cells as droplets that pass through pores in the cell wall by an unknown mechanism. The wax that forms the outer coating of the cuticle often crystallizes in an intricate pattern of rods, tubes, or plates. Certain patterns of these microstructures enhance water repellency by increasing the roughness of the wax surface. This roughness prevents water from forming large contact areas with, and thus from adhering to the already hydrophobic surface. Thus, water droplets form instantly on contact and carry away contaminating particles, cleaning the plant's surface (Neinhuis *et al.*, 1992). This phenomenon was first described for the leaves of the leguminous lotus *Lotus japonicus* and for that reason it is sometimes referred to as the 'lotus effect.'

Two different methods were used to proof if tulip wax layer had an influence on *Botrytis* resistance (Antonio Leon Reyes *et al.*, 2005). Wet inoculation method (standard method) with the removal of the wax layer and dry inoculation method with no removal of the wax layer were applied in cultivars Christmas Marvel, Leen van der Mark and Ile de France. The wet inoculation showed that Christmas Marvel was susceptible, Leen van der Mark was partial resistant, and Ile de France was resistant. However, when the dry method was applied to the same cultivars the level of resistance changed, leading to an increase in the level of resistance in Christmas Marvel, a decrease in Ile de France and equal levels in Leen van der Mark compared to the wet method. These results suggest that wet and dry inoculation altered the level of resistance, meaning that differences in the epicuticular wax layer provide an extra defense strategy against the pathogen in Christmas Marvel, enhanced infection in Ile de France and it had no effect in Leen van der Mark. Moreover, differences in the amount of wax layer in several species of the genus *Tulipa* are present. In general *T. fosteriana* was more susceptible to *B. tulipae* than *T. gesneriana*. The results showed that *T. fosteriana* had less wax layer than *T. gesneriana*. However, there was no complete correlation between the amount of wax layer and resistance in all tested cultivars.

Table 3.1 Components of Wax

Compound	*General structure*
n-Alkanes	$H_3C\ (CH_2)n\ CH_3$
Ketones	R^1COR^2
Secondary alcohols	$R^1\ CH(OH)R^2$
B-diketones	$R^1COCH_2COR^2$
Monoesters	R^1COOR^2
Primary alcohols	RCH_2OH
Aldehydes	RCHO
Alkanoic acids	RCOOH
Dicarboxylic acids	$HOOC(CH_2)nCOOH$
w-Hydroxy acids	$HOCH_2(CH_2)n\ COOH$

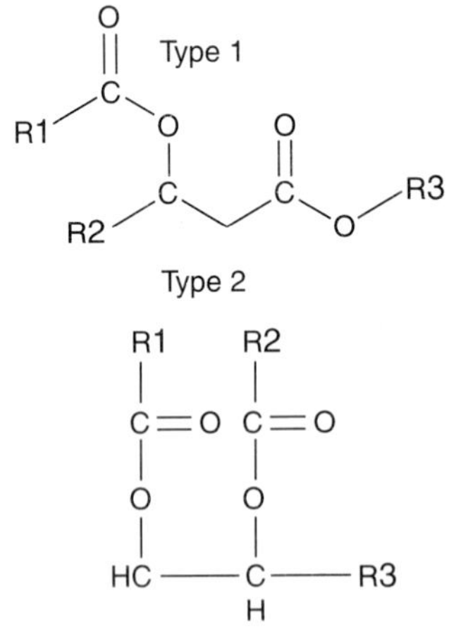

Fig. 3.3 Waxes of Type 1 and Type 2

Curvularia eragrostidis, a causal agent of head blight on the weed, *Digitaria sanguinalis,* did not cause disease on the turfgrass, *Festuca arundinacea.* Different extracellular esterase isoenzymes were detected in saprophytic and parasitic phases during the fungal germination (Fei Wang *et al.*, 2006). The epicuticular waxes of *D. sanguinalis* were more efficient to induce the secretion of esterases from the fungus than that of *F. arundinacea* but were more rapidly degraded by the fungal enzymes. Component analysis indicated that the epicuticular waxes from *D. sanguinalis* were mostly composed of alcohols, with 54.3% being 9,12-Octadecadien-1-ol. The main component of *F. arundinacea* waxes was alkyl compounds, with 49.8% being an olefin, 9-Tricosence. More long-chained esters were found in *D. sanguinalis* waxes, which were easier to be digested than those in *F. arundinacea* waxes by extracellular esterases of the fungus. Epicuticular waxes play a role in varying pathogenicity of *C. eragrostidis* on *D. sanguinalis* and *F. arundinacea.*

The presence of proteins in plant cuticular waxes has been known for a long time. A study was undertaken to investigate the presence of defense-related proteins in the wax of the 'Carnauba' (*Copernicia cerifera*) tree, an economically important tree of the palm family in the northeastern region of Brazil (Marco Antonio *et al.*, 2002). The presence of protein was shown in different fractions isolated from this wax. A DEAE-Sepharose anion exchange column was employed for further discrimination of proteins from different fractions. The samples were visualized by SDS-Tricine-gel electrophoresis. The acidic and basic fractions were recovered by freeze-drying and fractions with antifungal activity identified by assaying for inhibition of fungi growth, β-1,3-glucanase and chitinase activities. The study revealed the presence of enzymatic activities in different wax fractions. Isolated proteins were submitted to automated N-terminal amino acid sequencing and obtained sequences showed the presence of proteins with a high degree of homology to class III chitinases and to β-1,3-glucanase. Different physiological factors related to the intensity of floral malformation in twelve mango cultivars differing in their degree of resistance to malformation were investigated (Saboki Ebrahim *et al.*, 2013). The results revealed that thickness of leaf blade had no significant correlation with the intensity of floral malformation, but mango resistant cultivars Bhadauran and Elaichi showed thicker leaf blade when compared with highly susceptible cultivar like Eldon. Leaf wax content showed highly significant negative correlation with the intensity of mango floral malformation ($r = -0.88$).Results revealed that highly susceptible cultivars have less wax thickness (315–375 g/cm^2) when compared to resistant cultivars (484–522 g/cm^2).

To investigate the role of epicuticular wax in resistance of cocoa to *Phytophthora* pod rot, twelve cocoa genotypes with and without epicuticular wax on pod were inoculated with *P. palmivora* and *P. megakarya* (Nyadanu *et al.*, 2012). The wax layers were removed by washing the surfaces of leaf and pod in chloroform for 30 seconds. The level of resistance of cocoa genotypes was higher in leaves and pods with wax layer than in chloroform washed leaves and pods (wax removed). Cocoa genotypes with a higher amount of wax were more resistant than cocoa genotypes with a lesser amount of wax. These results suggested that epicuticular wax layer provides extra defense against *Phytophthora* species. There was a significant difference in lesion number and lesion size of cocoa genotypes after removal of their waxes suggesting that other factors are also involved in the resistance of cocoa to *Phytophthora* species. Cocoa genotypes with a higher amount of cuticular waxes on the surfaces of their leaves and pods retained the smaller amount of water and took a shorter time for moisture to evaporate from their surfaces than cocoa genotypes with a lesser amount of wax. Water retention and time to drying of water varied significantly among the cocoa genotypes with intact wax layers. However, there were no significant differences among cocoa genotypes for moisture retention and time to drying of water from leaf and pod surfaces after washing leaves and pods in chloroform. This indicated hydrophobicity of epicuticular waxes on leaf and pod surfaces of cocoa.

3.1.5 Suberin

Suberin is a polymer whose structure is poorly understood. Like cutin, suberin is formed from hydroxy or epoxy fatty acids joined by ester linkages. However, suberin differs from cutin in that it has dicarboxylic acids (Fig. 3.4), more long-chain components, and a significant proportion of phenolic compounds as part of its structure. Suberin is a cell wall constituent found in many locations throughout the plant. In the Casparian strip of the root endodermis, it forms a barrier between the apoplast of the cortex and the stele. Suberin is a principal component of the outer cell walls of all underground organs and is associated with the cork cells of the periderm, the tissue that forms the outer bark of stems and roots during secondary growth of woody plants. Suberin also forms at sites of leaf abscission and in areas damaged by disease or wounding fatty substances.

The unique chemical composition of suberin, its density of deposition, and location determine its physiological roles, one of which is to act as a barrier to penetration by pathogens (Lulai and Corsini, 1998). Suberin can serve as a nearly impermeable solute barrier and a physical barrier to fungal penetration. There are also indications that soluble compounds associated with the suberin polymer, such as phenolics or wax components, may themselves act as antifungal agents. The epidermis, which contains a nonlamellar or diffuse suberin, is in direct contact with the environment and is often the site of initial penetration by soil-borne pathogens; as such, it offers the first line of defense against pathogenic attack. The suberized endodermis, on the other hand, serves as the last line of defense before pathogens invade the vascular cylinder and spread throughout the plant. It is not only preexisting suberin that acts as a barrier; there are several reports of fungal or viral attack eliciting the deposition of suberin in the walls of cells in and around the site of penetration, serving to limit the spread of infection (Kolattukudy and Espelie, 1989). It can be argued that suberin and suberization play a major role in plant defense against pathogens, and could be a useful target for the development of increased resistance in specific plant-pathogen interactions.

Pathogen strains with novel virulence characteristics, arising from gene-for-gene selective pressure, reinitiate the cycle of resistant variety development. While several studies have indicated that suberization might be part of the general response of plants to controlling disease infection, the extent to which this occurs in the soybean root system is unknown. Even more fundamentally, little is known about the pattern(s) of suberin deposition in soybean root tissues and whether there is any correlation between the extent or nature of preformed suberin in roots of different cultivars and their degree of resistance to *P. sojae* (Raymond Thomas *et al.*, 2012).

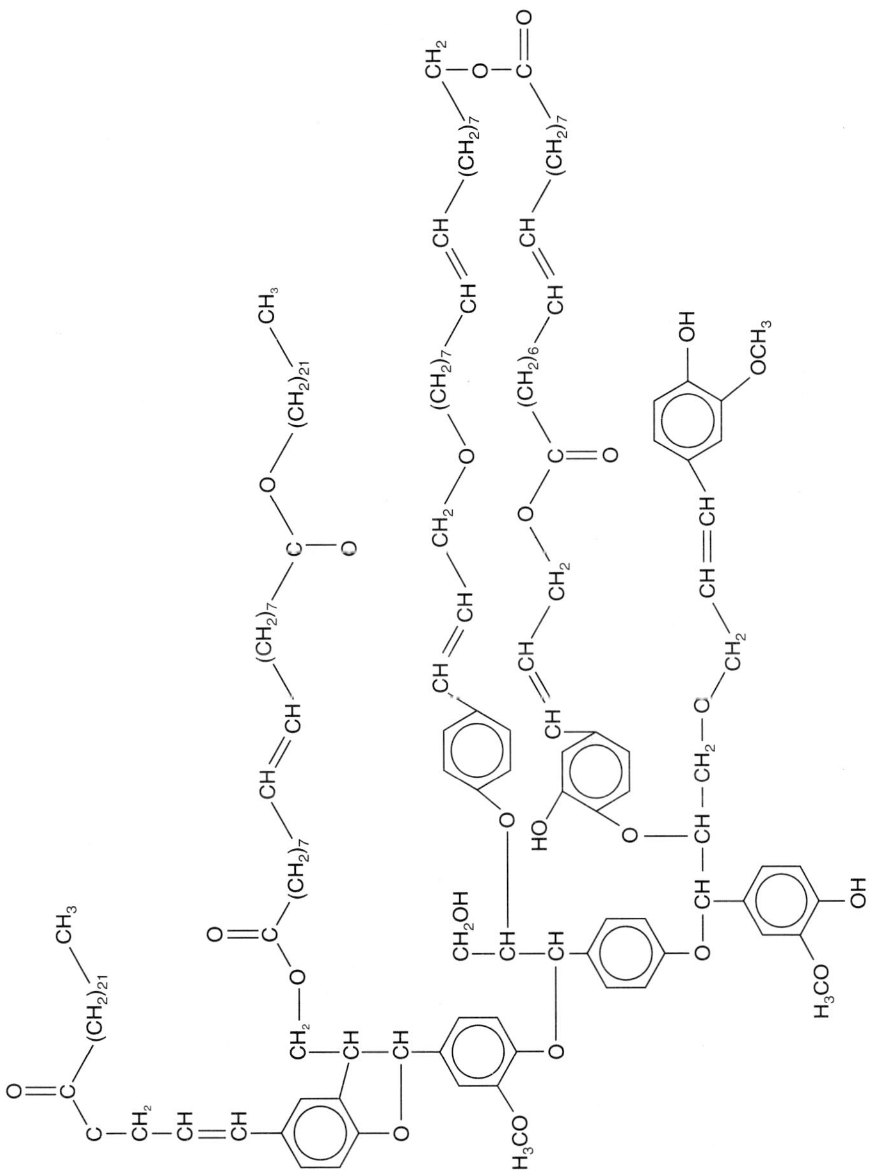

Fig. 3.4 Suberin polymer

The anatomical distribution and chemical composition of soybean root suberin as well as its relationship to partial resistance to *Phytophthora sojae,* 'a root rot-inducing pathogen,' were described (Raymond Thomas *et al.*, 2012). Soybean roots contain a state I endodermis (Casparian bands only) within the first 80 mm of the root tip and a state II endodermis (Casparian bands and some cells with suberin lamellae) in more proximal regions. A state III endodermis (with thick, cellulosic, tertiary walls) was not present within the 200 mm long roots examined. An exodermis was also absent, but some walls of the epidermal and neighboring cortical cells were suberized. Chemically, soybean root suberin resembles a typical suberin and consists of waxes, fatty acids, ω-hydroxyl acids, *a*,ω-diacids, primary alcohols, and guaiacyl- and syringyl-substituted phenolics. Total suberin analysis of isolated soybean epidermis/outer cortex and endodermis tissues demonstrated significantly higher amounts in the endodermis compared to the epidermis/outer cortex; increased amounts in the endodermis as the root matured from state I to state II; increased amounts in the epidermis/outer cortex along the axis of the root; and significantly higher amounts in tissues isolated from a cultivar (Conrad) with a high degree of partial resistance to *P. sojae* compared with a susceptible line (OX760-6). This latter correlation was extended by an analysis of nine independent and 32 recombinant inbred lines (derived from a 'Conrad' × OX760-6 cross) ranging in partial resistance to *P. sojae.* Strong negative correlations (−0.89 and −0.72, respectively) were observed between the amount of the aliphatic component of root suberin and plant mortality in *P. sojae*-infested fields.

When plants are wounded, they form a protective layer next to the exposed surface to prevent dehydration and potential penetration by opportunistic pathogens. This physical barrier termed suberin comprises a specific cell wall modification characterized by both a poly(phenolic) domain and wax-embedded poly(aliphatic) domain (Fawzi and Mark, 2003). Suberin is also formed developmentally and is found in the dermal cells of underground tissues, the Casparian band and the cork cells of bark tissue. The oxidative coupling of the poly(phenolic) component of suberin is thought to follow a peroxidase/H_2O_2 free radical process analogous to that of lignifications. It was recently demonstrated in the laboratory that H_2O_2 is essential for suberization in potatoes, but the origin of the H_2O_2 has remained an open question. In response to wounding, potato tubers generate reactive oxygen species (ROS) in association with suberization. Immediately following wounding, an initial burst of ROS occurs, reaching a maximum within 30 to 60 min. In addition to this initial oxidative burst, at least three other massive bursts occur at 42, 63 and 100 h post-wounding. These latter bursts are associated with wound healing and are probably involved in the oxidative cross-linking of suberin poly phenolics. The source of ROS is likely to be a plasma membrane NADPH-dependent oxidase immunorelated to the human phagocyte plasma membrane oxidase. The initial oxidative burst does not appear to be dependent on new protein synthesis, but the subsequent bursts are associated with an increase in oxidase protein components. Oxidase activity is enhanced *in vitro* by hydroxycinnamic acids and conjugates associated with the wound healing response in potato.

The lipophilic biopolyester suberin forms important boundaries to protect the plant from its surrounding environment or to separate different tissues within the plant (Rene

Hofer *et al.*, 2008). In roots, suberin can be found in the cell walls of the endodermis and the hypodermis or periderm. Apoplastic barriers composed of suberin accomplish the challenge to restrict water and nutrient loss and prevent the invasion of pathogens. Despite the physiological importance of suberin and the knowledge of the suberin composition of many plants, very little is known about its biosynthesis and the genes involved. A detailed analysis of the *Arabidopsis* aliphatic suberin in roots at different developmental stages was conducted. This study demonstrated some variability in suberin amount and composition along the root axis and indicated the importance of ω-hydroxylation for suberin biosynthesis. Using reverse genetics, the cytochrome P450 fatty acid ω-hydroxylase CYP86A1 (At5g58860) was identified as a key enzyme for aliphatic root suberin biosynthesis in *Arabidopsis*. The corresponding *'horst'* mutants showed a substantial reduction in ω-hydroxyacids with a chain length $<C_{20}$, demonstrating that CYP86A1 functions as a hydroxylase of root suberized tissue. Detailed expression studies revealed a strong root specificity and a localized expression in the root endodermis. Transgenic expression of CYP86A1 fused to GFP distributed CYP86A1 to the endoplasmic reticulum, indicating that suberin monomer biosynthesis takes place in this subcellular compartment before intermediates are exported in the apoplast.

Plants form suberin as a natural, durable barrier to infection. The suberization response was determined by assessing the accumulation of two major biopolymers that comprise suberin; polyaliphatics and polyphenolics (Lulai, 2005). A similar suberization response was found in potato tubers that had been naturally infected by wilt pathogen *Verticillium dahliae* in the field. This simple model system provided a versatile tool to investigate the physiological responses of potato tuber to the colonization of vessel elements and resist the spread of the disease. This process of suberization occurred inside the tuber without a wound stimulus. Postharvest wounding was known to cause potato tuber cells to form a suberin layer as part of the healing process and to protect against infection. However, in these experiments, the signal to initiate suberin formation was derived from the plant-pathogen interaction.

Cell walls contain proteins and enzymes that actively work to reshape the wall during cell growth yet thicken and strengthen the wall during induced defense. When a plant cell detects the presence of a potential pathogen, enzymes catalyze an oxidative burst that produces highly reactive oxygen molecules capable of damaging the cells of invading organisms. Reactive oxygen molecules also help strengthen the cell wall by catalyzing cross-linkages between cell wall polymers, and they serve as a signal to neighboring cells that an attack is underway. Plant cells also respond to microbial attack by rapidly synthesizing and depositing callose between the cell wall and cell membrane adjacent to the invading pathogen.

3.1.6 Callose

Callose deposits, known as 'papillae,' are polysaccharide polymers that impede cellular penetration at the site of infection and these are often produced as part of the induced basal defense response. Callose is a (1→3)-β-D-glucan that is widely distributed in higher plants and is readily recognizable in tissue sections through the formation of an intense yellow, UV light-induced fluorescence with the aniline blue fluorochrome (Stone *et al.*, 1985) (Fig. 3.5). During normal plant growth and development, callose is found as a transitory component of the cell plate in dividing cells, is a major component of pollen mother cell walls and pollen tubes, and is found as a structural component of plasmodesmatal canals. Callose also has been observed in abscission zones and on sieve plates in the dormant phloem. In addition to its role in normal growth and development, callose is deposited between the plasma membrane and the cell wall after exposure of plants to a range of abiotic and biotic stresses, including wounding, desiccation, metal toxicity, and microbial attack.

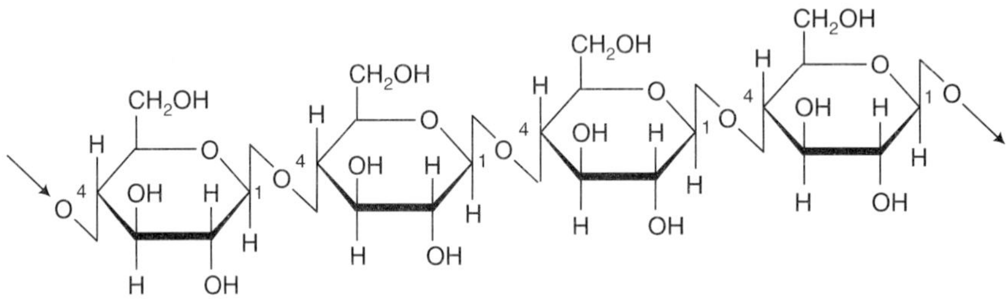

Fig. 3.5 Callose

Callose is a multifaceted defense response that is controlled by distinct signaling pathways, depending on the environmental conditions and the challenging pathogen-associated molecular pattern (Luna *et al.*, 2011). Callose deposition in *Arabidopsis* has emerged as a popular model system to quantify the activity of plant immunity. However, there has been a noticeable rise in contradicting reports about the regulation of pathogen-induced callose. The robustness of callose deposition under different growth conditions was examined and in response to two different pathogen-associated molecular patterns, the flagellin epitope Flg22, and the polysaccharide chitosan. Based on a commonly used hydroponic culture system. It was observed that variations in growth conditions had a major impact on the plant's overall capacity to deposit callose. This environmental variability was correlated with levels of hydrogen peroxide (H_2O_2) production. Depending on the growth conditions, pretreatment with abscisic acid stimulated or repressed callose deposition. Despite a similar effect of growth conditions on Flg22- and chitosan-induced callose, both responses showed differences in timing, tissue responsiveness and colocalization with H_2O_2. Further, the mutant analysis revealed that Flg22- and chitosan-induced callose differ in the requirement for the NADPH oxidase RBOHD, the glucosinolate regulatory enzymes VTC1 and PEN2, and the callose synthase PMR4.

The role of callose, in the form of papillae, in plant defenses during interactions with pathogens has been controversial. In incompatible interactions, results have generally indicated that callose is important for penetration resistance. The susceptibility of *Arabidopsis* to nonadapted fungal pathogens increased because of the inhibition or absence of callose deposition (Jacobs *et al.*, 2003). Hence, it was anticipated that an improved resistance exists to the avirulent powdery mildew, Bgh, in *Arabidopsis* lines showing increased pathogen-induced callose deposition. The complete penetration resistance of the generated 35S:PMR4-GFP lines, which indeed showed elevated early callose deposition after Bgh inoculation, confirmed our assumption. In compatible interactions, the role of callose in plant defense seemed to be dependent on the infection strategy of the pathogen. Whereas higher callose accumulation contributes to resistance against nectrophic fungal pathogens, such as *Alternaria brassicicola* and *Plectosphaerella cucumerina*, recent results have suggested that callose deposition is irrelevant for plant resistance to adapted biotrophic powdery mildews (Consonni *et al.*, 2010). However, these results proved that this assumption cannot be generalized (Christian Voigt, 2013). In 35S:PMR4-GFP lines, elevated amounts of callose were deposited at sites of attempted Gc penetration at early time points in the infection process. The increased callose formation provided complete resistance to the virulent, biotrophic powdery mildew Gc. This clearly indicates that callose synthesis and deposition can play an active role in plant resistance to adapted powdery mildews.

Regarding the enzymatic function of callose synthases, previous studies have proposed that callose synthase activity is not regulated by gene expression but rather by subcellular control of preexisting enzymes (Jacobs *et al.*, 2003). Plasma membrane proteins, which are involved in the synthesis of a structural component, including callose, the synthesis of haustorial encasements and the synthesis of pathogen-induced papillae, are transported via multivesicular bodies (MVBs) and exosomes. The treatment of *Arabidopsis* leaves with brefeldin A, which inhibits vesicle trafficking, prior to powdery mildew infection, blocked the deposition of callose at sites of fungal penetration (Nielsen *et al.*, 2012). This indicates the involvement of MVBs in the transport of components for callose synthesis to the penetration site.

Quantification of callose deposits is a useful measure for the activities of plant immunity and pathogen growth by fluorescence imaging. For robust scoring of differences, this normally requires many technical and biological replicates and manual or automated quantification of the callose deposits. However, previously available software tools for quantifying callose deposits from bioimages were limited, making batch processing of callose image data problematic (Ji Zhou *et al.*, 2012). In particular, it is challenging to perform large-scale analysis on images with high background noise and fused callose deposition signals. CalloseMeasurer, an easy-to-use application was developed, that quantified callose deposition, a plant immune response triggered by potentially pathogenic microbes. Additionally, by tracking identified callose deposits between multiple images, the software could recognize patterns of how a given filamentous pathogen grows in plant leaves. The software was evaluated with typical noisy experimental images and could be automatically executed without the need for user intervention. The automated analysis was achieved by using standard image analysis functions such as image enhancement, adaptive thresholding, and object segmentation, supplemented by several novel methods which filter background noise, split fused signals, perform edge-based detection and construct networks and skeletons for extracting pathogen growth patterns. To efficiently batch process callose images, they implemented the algorithm in C/C++ within the Acapella™ framework. Using the tool they could robustly score significant differences between different plant genotypes when activating the immune response. Examples were provided for measuring the *in planta* hyphal growth of filamentous pathogens.

Xanthan is the major exopolysaccharide secreted by *Xanthomonas* spp. (Maximina *et al.*, 2006). Despite its diverse roles in the bacterial pathogenesis of plants, little is known about the real implication of this molecule in *Xanthomonas* pathogenesis. In this study, it was shown that in contrast to *Xanthomonas campestris* pv. *campestris* strain 8004 (wild type), the xanthan minus mutant (strain 8397) and the mutant strain 8396, which was producing truncated xanthan, failed to cause disease in both *Nicotiana benthamiana* and Arabidopsis (*Arabidopsis thaliana*) plants. In contrast to wild type, 8397 and 8396 strains induced callose deposition in *N. benthamiana* and *Arabidopsis* plants. Interestingly, treatment with xanthan but not truncated xanthan suppressed the accumulation of callose and enhanced the susceptibility of both *N. benthamiana* and *Arabidopsis* plants to 8397 and 8396 mutant strains. Finally, in concordance, it was demonstrated that treatment with an inhibitor of callose deposition previous to infection induced susceptibility to 8397 and 8396 strains. Thus, xanthan suppression effect on callose deposition is important for *Xanthomonas* infectivity.

Some plant cells are highly specialized for plant defense. Idioblasts ('crazy cells') help protect plants against herbivory because they contain toxic chemicals or sharp crystals that tear the mouthparts of insects and mammals as they feed.

3.1.7 Idioblast

An idioblast is an isolated plant cell that differs from neighboring tissues. They have various functions such as storage of reserves, excretory materials, pigments, and minerals. They could contain oil, latex, gum, resin, tannin or pigments, etc. (Fig. 3.6). Among the most common idioblasts are those that synthesize crystals of calcium oxalate. Some can contain mineral crystals such as acrid tasting and poisonous calcium oxalate or carbonate or silica. Any of the tissue or tissue systems of plants can contain idioblasts. Idioblasts are divided into three main categories: excretory, tracheoid, and sclerenchymatous. Idioblasts can contain biforine cells that form crystals. The chemicals are excreted by the plant and stored in liquid or crystalline form. In bundles, they are known as 'druse' and as crystals they can be of the raphide (needle) form (Gary, 2009). When the end of an idioblast is broken, the crystals or another substance is ejected by internal water pressure. Idioblasts of calcium oxalate appear to have no function other than a poisonous deterrent to herbivores.

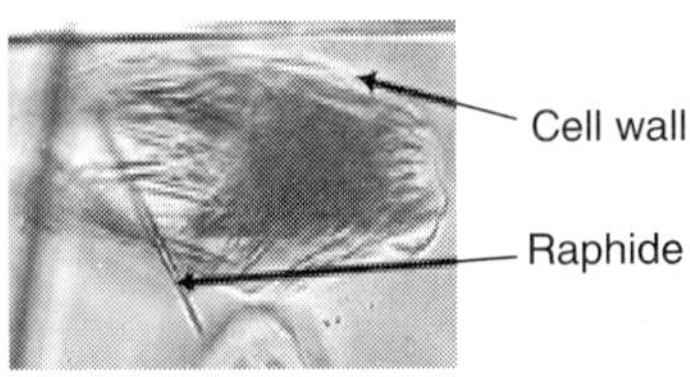

Fig. 3.6 Idioblast cell from ***Sanseveria***

Several biochemical and localization studies have shown that the acidic isoforms of the tobacco pathogenesis-related proteins, PR-1a, -1b and -1c are secreted into the extracellular spaces of leaves in response to pathogen infection or chemical treatment. The differential accumulation of these proteins in the vacuoles of specialized cells known as crystal idioblasts was reported (Dixon *et al.*, 1991). *In situ* hybridization analysis indicated that crystal idioblasts expressed the PR-1 genes at the mRNA level and suggested that PR-1 proteins were synthesized by these cells. Transgenic plants which constitutively express a chimeric gene encoding an acidic PR-1b isoform also accumulated PR-1 protein in the extracellular spaces and within crystal idioblast vacuoles. Analysis of mRNA derived from these transgenic plants indicated that expression of the introduced PR-1b gene was responsible for the accumulation of PR-1 protein in these two distinct locations. The synthesis and accumulation within crystal idioblasts of PR-1 proteins, which are secreted by other cell types, indicated that idioblasts sort these proteins in a unique manner. This suggested that protein sorting in higher plants might be modulated in a cell-specific manner.

There are many classes of idioblasts including pigmented cells, sclereids, crystalliferous cells, and silica cells. Pigmented cells often contain bitter-tasting tannins that make plant parts undesirable as a food source. Young red wines often contain high levels of tannins that give the wine a sharp, biting taste. Sclereids are irregularly-shaped cells with thick secondary walls that are difficult to chew: the rough texture of pear fruit (*Pyrus* spp.) is caused by thousands of sclereid stone cells that can abrasively wear down the teeth of feeding animals. Stinging nettles (*Urtica dioica*) produce stinging cells shaped like hypodermic needles that break off when disturbed and inject highly irritating toxins into herbivore tissues. Some stinging cells contain prostaglandins, hormones that amplify pain receptors in vertebrate animals and increase the sensation of pain. Crystalliferous cells contain crystals of calcium oxalate that may tear herbivore mouthparts when chewed and can be toxic if ingested. Members of the genera *Philodendron* and *Dieffenbachia* are very common tropical house plants that contain large amounts of these cells. Humans and pets who chew the leaves of these plants may experience a burning sensation in the mouth and throat that is often accompanied by swelling, choking, and an inability to speak. For these reasons, species of *Dieffenbachia* are commonly called dumb cane. Grasses and sedges contain rows of silica cells in their epidermal layers which give strength and rigidity to the growing leaf blades and deter feeding by chewing insects.

3.2 PLANT TISSUES AND SPECIALIZED APPENDAGES

3.2.1 Epidermis

The epidermis constitutes the outermost protective tissue system of leaves, floral parts, fruits, seeds, stems, and roots of plants until they undergo considerable secondary growth. It is the first line of defense against invading pathogens and consists of both specialized and unspecialized cells. Epidermis is usually closely packed, without intercellular spaces or chloroplasts. The outer walls, which are exposed to the atmosphere and usually thickened, and may be covered by a waxy, waterproof cuticle which are made up of cutin. Apart from the normal epidermal cells there are also stomata in the epidermis of leaves and stem. A stoma is an opening (pore) which is bounded by two bean-shaped cells called guard cells. The guard cells differ from normal epidermal cells in that they have chloroplasts, and the cell walls are thickening unevenly; the outer wall is thin, and the inner wall (nearest the opening) is thick. The thin-walled epidermal cells of roots give rise to root hairs. Hair-like outgrowths may also be found in the epidermis of leaves and stems. The epidermal cells of aerial plant parts are often covered by a waxy cuticle that not only prevents water loss from the plant but also prevents microbial pathogens from coming into direct contact with epidermal cells and thereby limits infection. The cuticle can be relatively thin (aquatic plants) or extremely thick (cacti). The hydrophobic nature of the cuticle also prevents water from collecting on the leaf surface, an important defense against many fungal pathogens that require standing water on the leaf surface for spore germination. However, some fungal pathogens including *Fusarium solani* produce cutinases that degrade the cuticle and allow the fungi to penetrate the epidermis.

Tissue-specific or regulated expression of transgenes is desirable to prevent pleiotropic side effects of putatively harmful transgene products as well as loss of energy resources due to the unnecessary accumulation of transgene products (Altpeter *et al.*, 2005). The epidermis-specific expression would be useful for many defense-related genes directed against attack by fungal pathogens that enter the plant body by direct penetration through the epidermis. In an approach to enhance the resistance of wheat to the powdery mildew pathogen *Blumeria graminis* f. sp. *tritici,* a novel epidermis-specific promoter was developed and used for expression of two defense-related genes. A 2.3 kb fragment of the wheat GstA1 promoter in combination with an intron-containing part of the wheat WIR1a gene was found to drive strong and constitutive transient expression in the wheat epidermis. This promoter-intron combination was used for overexpression of oxalate oxidase 9f-2.8 and TaPERO peroxidase, two defense-related wheat genes expressed in inner leaf tissues. Expression studies of several transgenic lines by *in situ* oxalate-oxidase staining, RNA, and protein blot analyses, as well as real-time PCR, demonstrated strong and constitutive transgene expression in the shoot epidermis. Transient, as well as stable overexpression of the *TaPERO* peroxidase gene in wheat epidermis under the control of the GstA1i promoter, resulted in enhanced resistance to *Blumeria graminis* f. sp. *tritici,* whereas oxalate-oxidase overexpression had no effect on either system. The data suggested that the wheat GstA1 promoter in combination with the WIR1a intron was useful for transgenic approaches to fungal disease resistance in cereals.

Interspersed among the many unspecialized cells of the epidermis are guard cells which regulate gas exchange through small openings called stomata. These pores allow carbon dioxide to enter the leaf for use in photosynthesis while restricting excessive water loss from the plant. Stomatal pore size is highly regulated by plants, and guard cells can participate in defense by closing in response to the presence of MAMPs. The first line of protection from an initial pathogen infection is achieved through passive defense mechanisms such as physical and chemical barriers. The plant epidermis is covered by a waxy cuticle, and each cell is surrounded by a complex cell wall containing highly cross-linked polysaccharides, proteins and phenolic compounds that bacterial pathogens must overcome to access cell nutrients. To breach these barriers, many plant pathogenic bacteria produce a range of extracellular virulence factors such as cutin-degrading enzymes and cell wall-degrading enzymes. These enzymes are typically secreted via a type II secretion system and include cell wall-degrading enzymes such as cellulases, pectinases, and endoglucanases. Such enzymes are particularly important in causing soft-rot diseases induced by bacteria in the *Erwinia* genus.

The host-endophyte interaction between roots of aspen (*Populus tremuloides*) and *Cryptosporiopsis radicicola* was examined primarily by transmission electron microscopy (Akihiko and Wei, 2009). Hyphae growing on the exterior of the inoculated roots had a thick, electron-dense, adhesive sheath. At hyphal contact and penetration, host epidermal cells exhibited a series of defense responses like the formation of papillae and partition walls, general wall thickening and walling-off of internal hyphae. In papilla formation, loop-shaped, rough endoplasmic reticula (rER) gave rise to globose secretory vesicles that accumulated around and then fused to the developing papilla. Unlike papillae, general wall thickening was associated with the Golgi apparatus (GA) that produced cell wall materials; 1–3 layers of Golgi cisternae were in contact with or in the immediate proximity (mostly within 0–0.5 μm) of and lying parallel to the host cell wall, where they budded out numerous subglobose vesicles that fused directly to the host cell wall and made it thicker. Partition wall formation and walling-off of internal hyphae also were common; the former was associated with an extended single cisterna, which was indistinguishable from rER or individual cisternae of GA, and in the latter phenomenon internal hyphae were encased by an electron-dense material containing numerous ribosomes and membranous elements that were derived apparently from proliferated rER. These pronounced defense responses protected the stele and contributed to making *C. radicicola* endophytic rather than pathogenic.

In a study, segments from dark-grown cucumber (*Cucumis safivus*) hypocotyls were used to understand defense reactions occurring upon fungal infection and induced by elicitors in the same tissue (Jiirgen Siegrist *et al.*, 1994). The segments were rendered resistant to infection by *Colletotrichum lagenarium* either by growing the seedlings in the presence of dichloroisonicotinic acid (DCIA) or by preincubation of the cut segments with DCIA, salicylic acid (SA), or 5-chlorosalicylic acid (5CSA). This resistance appeared to be due mainly to inhibition of fungal penetration into epidermal cells. In the resistant hypocotyl segments, the fungus-induced, at the time of attempted penetration, an increased deposition of phenolics, which were visualized by autofluorescence. These phenolics were located mainly in the epidermal cell wall around and the emerging papillae below appressoria and were quantified either as lignin-like polymers by the thioglycolic acid method or as 4-OH-benzaldehyde, 4-OH-benzoic or 4-coumaric acid liberated upon treatment with alkali at room temperature. Pretreatment with DCIA, SA and 5CSA induced little chitinase activity, but this activity greatly increased in resistant tissues upon subsequent infection. These observations indicated that resistance was associated with an improved perception of the pathogen stimulus resulting in the enhanced induction of diverse defense reactions. When the cut segments were pretreated with DCIA, SA, or 5CSA and then split and incubated with chitosan fragments, the deposition of cell wall phenolics was also enhanced. These pretreated and split segments also exhibited an increase in the rapid production of activated oxygen species induced by an elicitor preparation from *Phytophthora megasperma* f. sp. *glya*. Pretreatment of the segments with methyl jasmonate neither induced resistance nor enhanced induction of cell wall phenolics upon fungal infection, although it was observed in the corresponding split segments some increase in chitosan-induced cell wall phenolics and elicitor-induced rapid production of activated oxygen species.

3.2.2 Trichomes

Trichomes (leaf hairs) are specialized epidermal cells found on aerial plant parts that may provide both physical and chemical protection against pathogens. The velvety appearance of dusty miller (*Senecio cineraria*) is caused by thousands of tiny trichomes covering the plant's surface. It is hypothesized by biologists that trichomes evolved as a defense mechanism of the cannabis plant against a range of potential enemies. Trichomes, from the Greek meaning 'growth of hair,' act as an evolutionary shield, protecting the plant and its seeds from the dangers of its environment, allowing it to reproduce. These adhesive sprouts form a protective layer against offensive enemies.

Trichomes can run the gamut in structure, appearance, and texture. Some trichomes are frail, some are coarse, some are branched like tree limbs, others star-shaped, some are long and straight, others are short and curly (Fig. 3.7). Just as mammal hair serves various protective purposes, including insulation and camouflage, so do trichomes. Trichomes can be insulating by keeping frost away from leaf cells. They can help reduce evaporation by protecting the plant from wind and heat. In many cases, trichomes protect plants from herbivorous insects that may want to feed on them. And in some cases, if the trichomes are especially stiff or irritating, they may protect a plant from larger herbivores. Glandular trichomes are specialized hairs found on the surface of about 30% of all vascular plants and are responsible for a significant portion of a plant's secondary chemistry (Joris *et al.*, 2012). Glandular trichomes are an important source of essential oils, i.e., natural fragrances or products that can be used by the pharmaceutical industry, although many of these substances have evolved to provide the plant with protection against herbivores and pathogens. The storage compartment of glandular trichomes usually is located on the tip of the hair and is part of the glandular cell, or cells, which are metabolically active. Trichomes and their exudates can be harvested relatively easily, and this has permitted a detailed study of their metabolites, as well as the genes and proteins responsible for them. This knowledge now assists classical breeding programs, as well as targeted genetic engineering, aimed to optimize trichome density and physiology to facilitate customization of essential oil production or to tune biocide activity to enhance crop protection. The metabolic diversity found within plant glandular trichomes, with the emphasis on those of the Solanaceae and of the tools available to manipulate their activities for enhancing the plant's resistance to pests was provided.

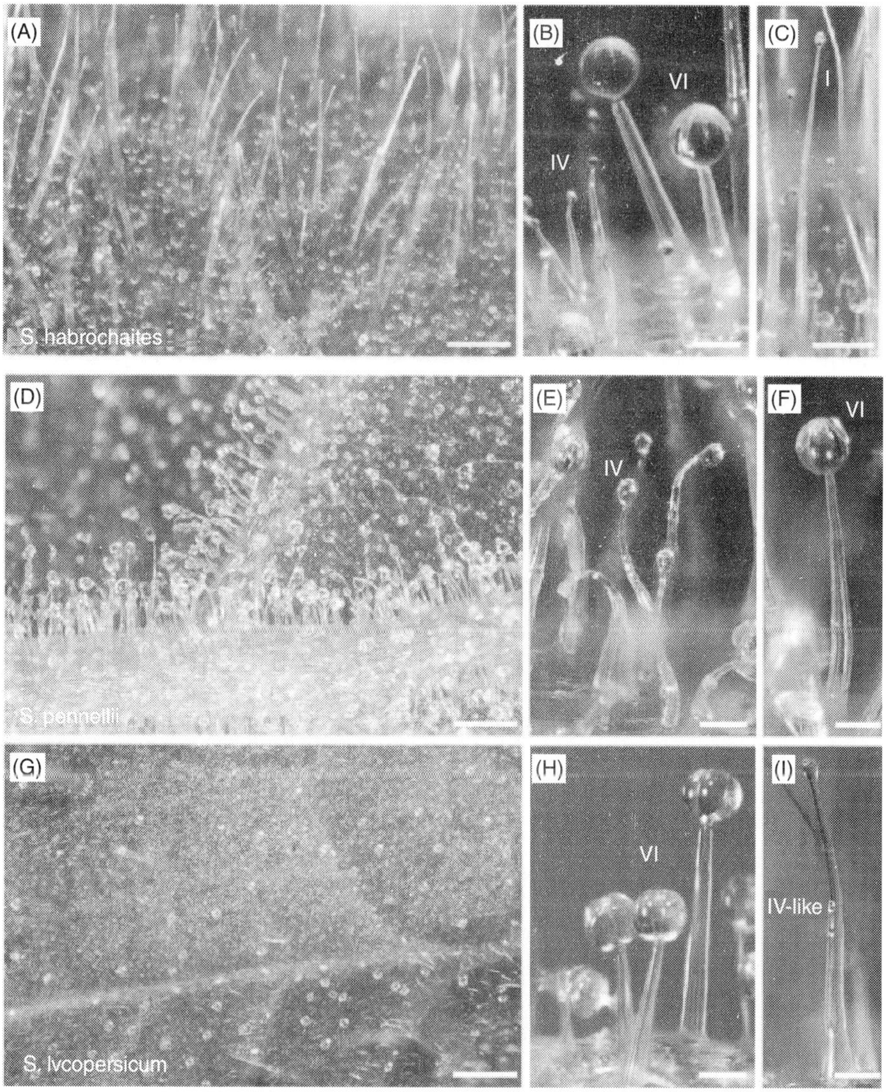

Fig. 3.7 Glandular trichomes in section ***Lycopersicon***. Wild accessions have high densities of glandular trichomes that confer resistance to several pests. Panel (A) shows the leaflet surface of *Solanum habrochaites* acc. LA 1777 with high densities of glandular trichome types IV and VI (B), and type I (C). The surface of *Solanum pennellii* acc. LA 716 is also covered by type IV trichomes (D, E) producing and secreting acyl sugars. This accession also has type VI trichomes, but in low density (F). Panel (G) shows the surface of *Solanum lycopersicum* cv. Moneymaker. Cultivated tomato has a low density of type VI trichomes (H) and type I trichomes. Sometimes, type IV-like trichomes (I) are observed on stems, veins, and on the leaflet edges. White bars represent 500 μm in panel A, C, D, and G. In panels B, E, F, H, and I, bars represent 50 μm.

The potential role of host trichomes in resistance to parasitism by *Cuscuta pentagona* was examined (Justin *et al.*, 2010). Plant trichomes have been long appreciated as the first line of defense against insect herbivores and more recently pathogens. It was hypothesized that trichomes could also defend against parasitic plants based on our observations that tomato trichomes become denser with age, notably on hypocotyls which are the first area contacted by *Cuscuta* seedlings, and these trichomes could act as a physical barrier to *C. pentagona* seedlings. To test this hypothesis, seedlings of *C. pentagona* were allowed to attach to 25-day-old tomato plants (*Solanum lycopersicum* 'Halley 3155') in a climate controlled growth chamber. Of 20 trials conducted, in six (30%) the parasite seedling was completely blocked by trichomes and was unable to reach the host stem, the parasite perished in each of these. Type I glandular trichomes, which are several millimeters long with a glandular tip, were primarily responsible for the blocking effect. Thus, trichomes could defend against parasitic plants in a manner analogous to herbivores by physically obstructing their movement. Interestingly, the effectiveness of trichomes was also dependent on the age of the host since those on younger tomato plants (<20 days old) are too sparse to impede *Cuscuta* seedlings.

Interest in trichome-produced compounds, combined with technical breakthroughs in analytical equipment and the possibility of applying genomic approaches, has greatly increased the understanding of the biochemical pathways that operate in trichomes, as well as the products they generate. Sequencing of Expressed Sequence Tag (EST) libraries generated from mRNA from isolated trichomes has resulted in large databases, which, in combination with metabolite profile analysis of glandular trichomes and proteomics, has led to a much more detailed general insight into the biosynthesis of these specialized metabolites than obtained previously via chemical-analytical methods (Wagner *et al.*, 2004). Based on analyses of EST databases, it has been suggested that trichomes operate mostly as a self-supporting system and have highly active biochemical pathways for both primary and secondary metabolism. In glandular trichomes of tomato and tobacco, genes encoding enzymes and proteins related to photosynthesis and carbon fixation are significantly expressed (Cui *et al.*, 2001), indicating that at least some of the carbon necessary for secondary metabolism can be fixed within trichome cells. Earlier studies in cultivated tobacco had also indicated that glandular trichomes possibly can fix carbon and produce sugar and diterpenoids (i.e., duvatrienediol) independent from the rest of the plant, although a role for additional carbon imported from the tissues below the trichome cannot be excluded.

It has been suggested that the total amount of secondary metabolites produced by glandular cells could be related to their capacity to fix carbon since exudates from photosynthetically active glandular cells can constitute up to 20% of the leaf dry weight biomass in wild Solanaceae (Slocombe *et al.*, 2008) while exudates from species with leucoplasts instead of chloroplasts in their trichomes, like the Lamiaceae and Fabaceae, contribute less than 2% to the leaf dry weight. Leucoplasts are nonpigmented plastid-type organelles specialized for *de novo* biosynthesis of (precursors for) the metabolites that often will be secreted. However, photosynthetically active glandular trichomes are probably supplemented with carbon substrates (e.g., CO_2, sucrose, glucose) as well, because their own primary metabolism is most likely incapable of meeting the huge carbon demands. Besides carbon, other compounds, like nitrogen, phosphate, and micronutrients, are required for metabolism as well. Import of amino acids into trichomes is minimal, begging the question of how trichomes acquire their nitrogen necessary for a large amount of proteins synthesized in trichomes. Possibly, nitrogen can be recycled by aminotransferases in glandular trichomes, but it cannot be excluded that nitrogen and other essential substances are imported via the trichome stalk, as well. EST analyses have played an important role in identifying enzymes of trichome secondary metabolism, for instance, in the synthesis of geraniol and eugenol in basil; the synthesis of methyl ketones monoterpenes and sesquiterpenes in *Solanum* species; and, xanthohumol synthesis in hops (Joris *et al.*, 2012).

Tremendous progress in the availability of genomic data has allowed for the discovery of genes in various biosynthetic pathways involved in trichome-produced compounds. However, the full potential of trichomes has not been exploited even remotely since plant secondary metabolism is complex and multilayered while our knowledge of the precise actions of the different members of large gene families and of the rate-limiting steps in pathways is still too incomplete to make the outcome of such manipulations easily predictable (Joris *et al.*, 2012). Thus, these minute glandular trichomes may soon prove to be the ideal vehicles for targeted modification of the versatile secondary metabolism of many plant species to customize essential oil production and enhance biocide-based protection of crops.

3.2.3 Lignified Cells

Lignin is a natural, branched plant biopolymer generated by radical coupling of hydroxycinnamyl subunits called monolignols, mainly coniferyl (CA), sinapyl (SA), and p-coumaryl alcohols (p-CA) [Figs. 3.8 (a)-(c)] and creates, together with hemicelluloses a glueing matrix for cellulose microfibrils in tracheary elements (TEs) and fibres of higher plants (Kaisa Marjamaa *et al.*, 2009). Lignification in TEs and fibres occurs along the secondary cell wall formation which is initiated in tree xylem from cell corners and middle lamellae. Eventually, these cells die through programmed cell death. Lignin

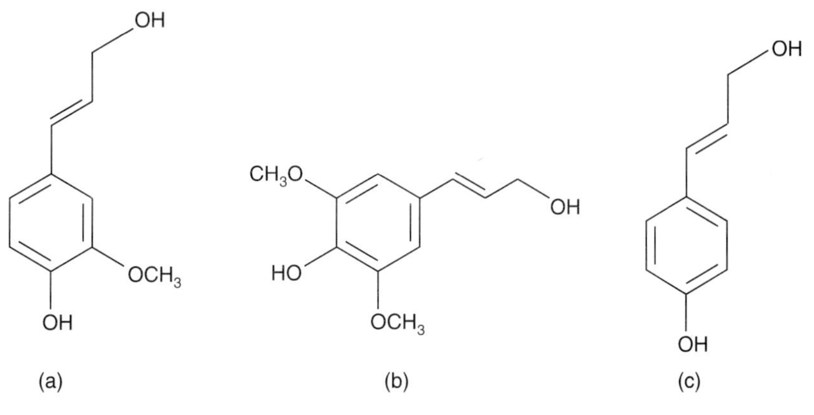

Fig. 3.8 (a) Coniferyl alcohol, (b) sinapyl alcohol and (c) *p*-coumaryl alcohol

provides mechanical strength and resistance to pathogens and makes the cell walls impermeable to water, thus enabling the transport of solutes via tracheids in the xylem tissue. In addition to lignification during growth, lignification also occurs as a stress response.

Lignification is a cell wall fortifying process which occurs in xylem tissue in a scheduled manner during tissue differentiation. Enzymes and the genes responsible for lignin biosynthesis were studied with an emphasis on lignin polymerizing class III secretable plant peroxidases (Kaisa Marjamaa *et al.*, 2009). The aim was to understand the cell and molecular biology of the polymerization of lignin especially in tracheids and vessels of woody species, but much of the experimental evidence comes from herbaceous plants. Class III peroxidases pose many problems for empirical work as their encoding genes are variable, their substrate specificities are wide, and the half-life of many of the isozymes is very long. However, there is some evidence for the role of specific peroxidases in lignin polymerization through antisense mutants in tobacco and poplar and from tissue and cell culture lines of *Picea abies* and *Zinnia elegans.* Peroxidase enzyme action has been shown by substrate specificity studies and, for example, RT-PCR results have pointed out that many peroxidases have tissue-specific expression patterns. Tissue-level location of gene expression of some peroxidases has been studied by *in situ* hybridization and their cellular localization with antibodies and using EGFP-fusion genes. From these, it was concluded that, although many of the xylem class III peroxidases have the potential for functioning in the synthesis of the lignin polymer, the combined information of catalytic properties, expression, and localization could reveal differences in the significance of different peroxidases in the lignification process.

Lignification renders the cell wall more resistant to mechanical pressure applied during penetration by fungal appressoria and water resistant and thus less accessible to cell wall-degrading enzymes. Lignification is necessary for the structural integrity of plant cell walls, and that is crucial for plant development (Boerjan *et al.*, 2003), but the monomeric composition of lignin can vary depending on the developmental process: thus, defense lignin accumulated by an elicitor treatment was shown to be significantly different from lignin in vascular tissues.

Lignification may have several roles in defense. Five ways have been suggested on hindering fungal growth by lignification through plant tissue (Ride, 1978). Lignin may make walls more resistant to mechanical penetration. It is generally held that lignin increases the resistance of walls to compressive forces such as would be expected at the growing point of a penetration peg; Lignification of the wall at the point of attack may render it resistant to dissolution by fungal enzymes. Induced lignification of wheat leaves makes them resistant to enzyme degradation. Lignification of walls may restrict diffusion of enzymes and toxins from the fungus to host, and water and nutrients from the host to fungus, in essence starving the fungus; Low molecular weight phenolic precursors of lignin and free radicals produced during polymerization may inactivate fungal membranes, enzymes, toxins, and elicitors. Many phenolic compounds have antifungal activity. The hyphal tip may become lignified and lose plasticity necessary for growth. Each of these roles can be tested and, therefore, used to evaluate lignification as a resistance mechanism. Thus, if lignification is to play a primary role in restricting pathogen development, it must occur early in the host-pathogen interaction, and it must be localized near the invading pathogen. Endogenous enzymes chitinase, β-1, 3-glucanases and lignin content in plant leaves can be used as biochemical markers for identifying plant varieties resistant to fungal infection or other biotic and abiotic factors. Also by transferring pathogenesis-related proteins such as chitinases, β-1,3-glucanases genes can induce resistance in plants to various photogenes (Kaisa Marjamaa *et al.*, 2009).

The cell wall determines the shape of plant cells and is also the primary interface for pathogen interactions. The structure of the cell wall can be modified in response to developmental and environmental cues, for example, to strengthen the wall and to create barriers to pathogen ingress. The ectopic lignin 1-1 and 1-2 (eli1-1 and eli1-2) mutations lead to an aberrant deposition of lignin, a complex phenylpropanoid polymer (Cano-Delgado *et al.*, 2003). It was shown that the eli1 mutants occurred in the cellulose synthase gene *CESA3* in *Arabidopsis thaliana* and caused reduced cellulose synthesis, providing further evidence for the function of multiple CESA subunits in cellulose synthesis. The reduced levels of cellulose synthesis caused by mutations in cellulose synthase genes and in genes affected cell expansion activate lignin synthesis and defense responses through jasmonate and ethylene and other signaling pathways. These observations suggested that mechanisms monitoring cell wall integrity can activate lignification and defense response.

For a long time, it has been believed that lignification has an important role in host defense against pathogen invasion. By using an RNAi gene-silencing assay (Nazmul *et al.*, 2009) showed that monolignol biosynthesis plays a critical role in cell wall apposition (CWA)-mediated defense against powdery mildew fungus penetration into diploid wheat. Silencing monolignol genes led to supersusceptibility of wheat leaf tissues to an appropriate pathogen, *Blumeria graminis* f. sp. *tritici* (Bgt), and compromised penetration resistance to a nonappropriate pathogen, *B. graminis* f. sp. *hordei.* Autofluorescence of CWA regions was reduced significantly at the fungal penetration sites in silenced cells. An important role for monolignol biosynthetic genes in effective CWA formation against pathogen penetration was indicated. It was shown that silencing of monolignol genes also compromised penetration resistant to Bgt in a resistant wheat line.

3.2.4 Cork Layers

Cork (phellem) is a multilayered dead tissue protecting plant mature stems and roots and plant healing tissues from water loss and injuries. Cork cells are made impervious by the deposition of suberin onto cell walls. Although suberin deposition and cork formation are essential for the survival of land plants, molecular studies have rarely been conducted on this tissue. Cork, or phellem, which is the technical term for cork, is formed by the phellogen (cork cambium) (Fig. 3.9). Cork formation involves proliferation and commitment of the phellogen derivatives, cell expansion and extensive deposition of suberin and waxes, and an irreversible program of senescence ending in cell death. Various pathogens like fungi, bacteria, viruses, nematodes, etc., are known to frequently induce the formation of cork layers beyond the point of infection, as a result of stimulation of host cells by the substances secreted by the pathogen. Cork layers are impenetrable by pathogen due to thickness and strength of suberized cell walls of cork; these block the spread of toxic substances secreted by the pathogen and prevent damage to the underlying tissues; and these stop the flow of nutrients and water from the healthy to the infected zone and thus ensure isolation of the pathogen by depriving it of nutritional source.

Sometimes cork layers are lignified, and this imparts additional toughness to their cell walls. Cork layers not only prevent the spread of the pathogens but also carry out functions similar to that of lignin. As a postinfectional mechanism of defense, quick reaction of the host cells to infection and ability of the plant to produce material for cock formation are necessary prerequisites for the cork layer formation. It is important to note that all plants cannot form cork layers. Formation of cork layers is common in common scab of potato, necrotic lesions of tobacco mosaic, *Rhizopus* rot of sweet potato, etc.

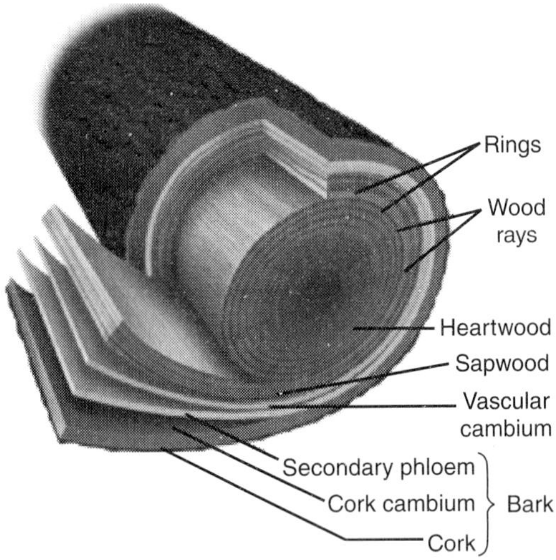

Fig. 3.9 Cork layer

Sometimes cork layers are lignified, and this imparts additional toughness to their cell walls. Cork layers not only prevent the spread of the pathogens but also carry out functions similar to that of lignin. As a postinfectional mechanism of defense, quick reaction of the host cells to infection and ability of the plant to produce material for cock formation are necessary prerequisites for the cork layer formation. It is important to note that all plants cannot form cork layers. Formation of cork layers is common in common scab of potato, necrotic lesions of tobacco mosaic, *Rhizopus* rot of sweet potato, etc.

3.2.5 Abscission Layers

Abscission is the process in which plants intentionally shed their leaves, flowers, buds, seeds, fruit, branches or roots. This may occur for a variety of reasons, for example, if the organ is senescing or stressed. If a leaf is attacked by a pathogen or becomes diseased, a plant may quickly shed the organ rather than going through the usual (longer) process of recycling nutrients. These layers are noncellular and denote the empty space between infected and healthy tissues (Singh, 1984) (Fig. 3.10). This space is the result of the disintegration of middle lamella of parenchymatous tissues and thinning of the cell walls. Hence, the infected portion gets separated and drops down. Shot hole symptoms are mostly due to these layers. Abscission layer is also known as 'amputative resistance.'

Abscission is a highly coordinated developmental process that results in shedding of organs at predetermined sites called abscission zones (Shimon Meir *et al.*, 2011). They suggested an abscission model with the formation of a postabscission transdifferentiation of a protective layer as the last step in the process. The present report expands the repertoire of genes activated in the tomato flower abscission zone (AZ), which are likely to be involved in defense responses. They identified four different defense-related genes, including *Cysteine-type endopeptidase*, *α-Dioxygenase 1 (α-DOX1)*, HopW1-1-Interacting protein2 *(WIN2)* and Stomatal-derived factor-2 *(SDF2)*, that are newly associated with the late stage of the abscission process. The late expression of these genes, induced at 8–14 h after flower removal when pedicel abscission was already in progress, was AZ-specific, and was inhibited by treatments that prevented pedicel abscission, including 1-methylcyclopropene pretreatment or IAA application. This information supported the activation of different defense responses and strategies at the late abscission stages, which might enable efficient protection of the exposed tissue toward different environmental stresses.

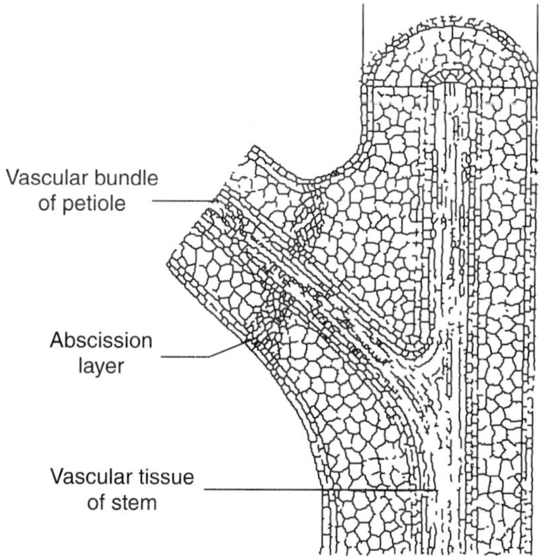

Fig. 3.10 Abscission layer

Stem-end rot (SER) caused by *Diplodia natalensis* is a major postharvest disease of fresh Florida citrus degreened with ethylene to enhance fruit appearance. A significant increase in disease incidence by the use of ethylene at concentrations above those needed for degreening may be related to the activity of the abscission enzymes, polygalacturonase

(PG) and cellulase (CX). Quiescent mycelia of *D. natalensis* in necrotic tissue on the surface of the calyx and disk (button) at the stem-end of the fruit grow into the fruit upon separation of the button from the fruit during abscission (Eldon Brown and Jacqueline K. Burns, 1998). Activity of the abscission enzymes in oranges was enhanced by high ethylene (0.055 mL/L), and a larger number of cells were degraded within the abscission layer by their activity than at lower (0.002 mL/L) more typical degreening concentrations. Commercial enzymes or partially purified preparations of the abscission enzymes added to abscission areas of debuttoned oranges before inoculation with mycelia of *D. natalensis* caused a significant increase in SER. Fruit dipped in 2,4-dichlorophenoxyacetic acid (2,4-D) or silver thiosulfate (STS) before degreening were more resistant to SER when inoculated following degreening. The addition of these metabolic inhibitors or cycloheximide (CHI) to the fruit abscission zone after removal of the button during early stages of degreening (18 h) with high ethylene also caused a significant reduction in SER. The metabolic inhibitors had no effect on SER if applied at 60 h after terminating the ethylene treatment. Reductions in the activity of PG and CX were obtained by applying 2,4-D or STS to fruit before degreening, and STS or CHI to fruit abscission zones at early stages of degreening (18 h).

3.2.6 Tyloses

Tyloses are outgrowths of xylem parenchyma cells into the lumens of adjoined vessels via vessel-parenchyma pits. Tyloses occur widely among plant species and are induced by environmental stimuli such as wounding and pathogen infection. Tyloses impair xylem function by blocking vessels, but they are also a component in wound healing and may inhibit the intrusion and spread of pathogens.

Tyloses are suberized structures that develop from ray parenchyma cells and project through pits to occlude the lumina of tracheids and vessels (Figs. 3.11 and 3.12). A microscopic study of tyloses showed that they originate in the parenchyma cells associated with the xylem and from there invade the xylem lumen. As tyloses are formed, there were changes in the cell wall, associated with the accumulation of crystalline structures. The cells surrounding the blocked xylem contained more polyphenolic compounds than cells of intact xylem (José A. Del Rio *et al.*, 2003). Changes in the level and composition of polyphenolic compounds were related to the severity of the infection. Tannin concentrations also increased with increasing numbers of xylem vessels containing tyloses. Root extracts added to fungal culture media inhibited mycelial growth of *Phaeomoniella chlamydospora, Phaeoacremonium aleophilum,* and *P. inflatipes*. Inhibition was greater with extracts from roots with tyloses and was also correlated with polyphenol content.

Vascular occlusion and a subsequent limitation of the upward movement of *Ophiostoma ulmi sensu lato* may result from embolism (i.e., entry of air into the vessels), accumulation of fungal hyphae, and the formation of gels (gums) and tyloses in the xylem vessels and the tracheids (Sinclair and Campana, 1978). Gels (gums) are vascular plugs with a fibrillose structure formed by material secreted from adjacent xylem parenchyma through membranes of half-bordered pit pairs. Although the precise chemical composition is not yet known, pectin, hemicellulose, and phenolics appear to be basic components of gels. Tyloses are outgrowths of the parenchyma contact cells surrounding the xylem vessels that protrude through the pit cavities into the lumen of the vessel.

The nucleus of the parenchyma cell migrates into the tylose. After tylose formation, the reacting parenchyma cell remains in continuous contact with its protrusion. Tyloses (and gels) are sporadically found in fibers. Besides in response to pathogen infection, tyloses can be formed after injury or in aging elm vessels that become nonfunctional. The partial or complete embolism usually detected during these processes is thought to give rise to a massive increase in xylem water potential (Van Bel and Van der Schoot, 1989). Subsequently, the contact cells adjoining the vessels absorb water until the elasticity of their cell walls prevents any further swelling. Contact cells form a protective layer containing pectic polysaccharides after completion of the secondary cell wall. This layer is situated either unilaterally against the paratracheal cell wall or along the whole cell wall with a thickened deposit on the vessel side. Although pit membranes of contact cells lack a secondary cell wall, they are covered by the protective layer. Once the pressure built up by the excessive water uptake has exceeded a certain limit, the protective layer is no longer able to contain the osmotic pressure of the contact parenchyma cells. As a consequence, these cells bulge through the pit cavities into the xylem vessels to form tyloses.

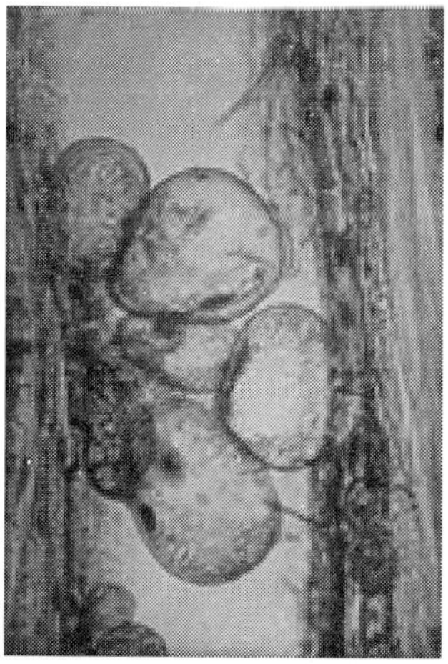

Fig. 3.11 Tyloses in a xylem vessel of an elm infected with *O. ulmi*

Fig. 3.12 Tyloses in an elm xylem vessel

Embolism inducing a change in water potential may be the main cause triggering the formation of vessel occlusions. In the latter case, the xylem becomes nonfunctional before any other form of plugging becomes evident. Tylosis formation may preserve the tissue pressure by turning the embolized-vessel gas space into the incompressible parenchyma. There appears to be an inverse relationship between water potential and xylem embolism;

the lower the water potential, the higher the xylem embolism. The primary wall of the tylose was shown to contain cellulose and a small amount of pectin. The flexibility of this wall permits expansion of the tylose. Increased esterification of the primary wall after tylosis maturation might improve resistance to attack by fungal pectinases. Secondary wall formation was rarely detected in tyloses of the Americal elm.

However, a suberized layer was formed against the primary wall layer (or internal to the secondary layer, when present) in tyloses that had completed their expansion. Suberin is a protective coating, consisting of long chain fatty acids, hydroxylated fatty acids, long-chain alcohols, and phenolics (primarily ferulic acid). The phenolics are thought to bind the lipid fraction of the suberin (up to half of the total suberin) to the cell wall suberized tyloses are randomly formed in *U. Americana.* Occasionally, the internal suberized layer appears to be covered with an additional layer containing β(1(4)-D-glucans.

In continuation with the protective layer, the primary cell wall of expanding tyloses secretes pectin microfibrils into the vessel elements. The material appears to be pushed towards areas that cannot be clogged completely by the tylosis primary wall, such as chambers of bordered and half-bordered pits, void spaces left between the tylose and the vessel wall, the vicinity of the rim of vessel perforation plates, and in vessel areas where the curving angle of the secondary wall is very pronounced. In places where the tylosis primary wall exhibits close contact with the xylem vessel wall, the pectic layer is very thin and hardly discernible. Besides a complete blockage of the vessel elements, the secreted pectin may promote cohesion of the tylosis wall to other structures.

José A. Del Rio *et al.* (2001) analyzed the formation of tyloses in grape infected with *Phaeomoniella chlamydospora, Phaeoacremonium aleophilum,* and *P. inflatipes.* A microscopic study showed that tyloses originated in parenchyma cells associated with the xylem and from there invaded the xylem lumen. As tyloses are formed, there were changes in the cell wall, associated with the accumulation of crystalline structures. The cells surrounding the blocked xylem contained more polyphenolic compounds than the cells of intact xylem. Changes in the level and composition of polyphenolic compounds occurred in relation to the severity of the infection. Tannin concentrations also increased with increasing numbers of xylem vessels containing tyloses. Root extracts added to fungal culture media inhibited mycelial growth of *Phaeomoniella chlamydospora, Phaeoacremonium aleophilum,* and *P. inflatipes.* Inhibition was greater with extracts from roots with tyloses and was also correlated with polyphenol content. Extracts of the leaves, stems, and berries of vines treated with Brotomax, which increases the biosynthesis of phenolic compounds, inhibited mycelial growth of the fungi compared to untreated control plants.

3.2.7 Gum Deposition

Deposition of gum in cells along the border of infected tissues is a protective mechanical barrier. Infection by vascular wilt fungi often leads to the coating of cell walls with vascular gels that fill the lumina of infected vessels. These gels are normally formed within 48 hours but are often dissolved by fungal enzymes in most susceptible cultivars.

These gels might arise from the perforation plates, end walls and pit membranes by a process of distention of pectinaceous material, hemicelluloses and other carbohydrates in the primary wall and middle lamella (Van de Molen *et al.*, 1977) or due to the synthesis of new carbohydrates and other material that are secreted into the vessels to form gels.

Xanthan (Fig. 3.13) is the major exopolysaccharide secreted by *Xanthomonas* spp. Despite its diverse roles in the bacterial pathogenesis of plants, little is known about the real implication of this molecule in *Xanthomonas* pathogenesis. Xanthan gum is a natural anionic polysaccharide and an important industrial biopolymer derived from the bacterial coat of *Xanthomonas campestris*). It is made up of pentasaccharide subunits, forming a cellulose backbone with trisaccharide side chains composed of mannose (β1,4) glucuronic-acid(β1,2)mannose attached to alternate glucose residues in the backbone by α1,3 linkages. This biopolymer is of great commercial significance because of its superior pseudoplasticity, thixotropy, viscosity, and biosafety.

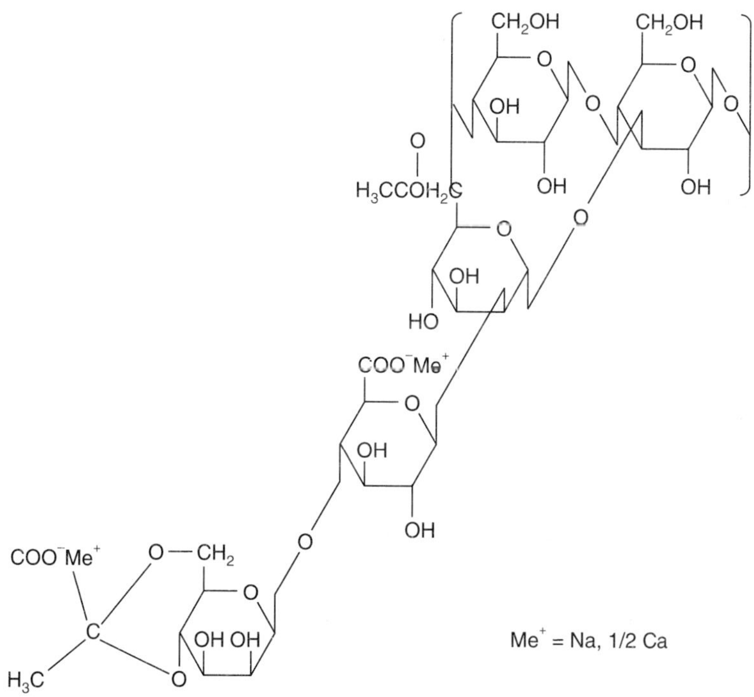

Fig. 3.13 Xanthan

It was shown that in contrast to *Xanthomonas campestris* pv. *campestris* strain 8004 (wild type), the xanthan minus mutant (strain 8397) and the mutant strain 8396, which produces truncated xanthan, failed to cause disease in both *Nicotiana benthamiana* and Arabidopsis (*Arabidopsis thaliana*) plants (Maximina H. Yun *et al.*, 2006). In contrast to wild type, 8397 and 8396 strains induced callose deposition in *N. benthamiana* and rabidopsis plants. Interestingly, treatment with xanthan but not truncated xanthan suppressed the accumulation of callose and enhanced the susceptibility of both *N.*

benthamiana and *Arabidopsis* plants to 8397 and 8396 mutant strains. Finally, in concordance, it was observed that treatment with an inhibitor of callose deposition previous to infection induced susceptibility to 8397 and 8396 strains. Thus, xanthan suppression effect on callose deposition seems to be important for *Xanthomonas* infectivity.

3.3 CHEMICAL DEFENSES

Defense mechanism requires an energy investment from plant's behalf. The same energy could be used to maximize individual's inclusive fitness, but natural selection favored plants with defenses, so many plants used a chemical which allowed them to regulate the synthesis of defense compound, therefore investing energy only when needed. Chemical defenses allowed plants to better defend themselves against a wide range of pathogens. But plants organs that are frequently attacked accumulate high constitutive levels of toxins for defense present all the time (Wittstock and Gershenzon, 2002). Plant defenses have a negative effect on the growth, development or survival of the pathogens. There is a huge diversity of the methods used by plants to defend themselves. Although many plant defenses are similar, over evolutionary time each plant has modified the chemicals to better fit their environment and defend themselves against specialized pathogens in their environment.

On the other hand, pathogens have developed a variety of mechanisms to circumvent plant toxins. Recent studies have improved the understanding of both metabolic detoxification and behavioral mechanisms for avoiding toxins. Examples of the detoxification of plant toxins by pathogens include plant-pathogenic fungi that are able to metabolize the saponins of their hosts (Osbourn, 1996). Recent work has elucidated the detoxification enzymes of pathogenic isolates of the fungus *Stagonospora avenae* growing on *Avena sativa*. The antifungal 26-desglucoavenacosides released from the steroidal saponins avenacoside A and B upon pathogen attack of *Avena sativa* are sequentially hydrolyzed by three fungal enzymes, one α-rhamnosidase and two β-glucosidases, resulting in a strong reduction in antifungal activity. The evolution of plant defenses has allowed them to persist on land and interact effectively with all the organisms. The defenses are really similar in their basic structure and defense, they all could have originated from the same source but over time they evolved to better fit their environment. Even though the plant defenses protect them from a wide range of pathogens, there are still many organism that have evolved to deal with plant defenses. The chemical warfare has been going on between pathogens and plants for millions of years, it is responsible for the huge diversity and different in chemical defenses found between plants but since all the chemical defenses evolved from a common ancestor, there are many similarities in the basic structure of defense compound as well.

Glucosidases have been suggested to be involved in fungal detoxification of α-tomatine, a steroidal glycoalkaloid of tomato fruits, by the cleavage of sugar residues attached to the basic alkaloid skeleton. Recent investigations using a tomatinase-deficient mutant of the fungus *Septoria lycopersici* revealed that although the targeted replacement of the tomatinase gene made the fungus more sensitive to α-tomatine the mutant fungus was still able to grow on high α-tomatine concentrations. These findings indicate that other detoxification mechanisms must be involved in the α-tomatine resistance of this pathogen (Martin-Hernandez *et al.*, 2000).

As the discovery of plant defense genes seems set to continue at its present rapid pace, the studies discussed above likely represent just the beginning of a long series of investigations on plant defense using genetically modified plants or plant enemies (Ute Wittstock and Jonathan Gershenzon, 2002). Given the chemical complexity of plants, the ability of the experimenter to manipulate their phenotypes or the phenotypes of their enemies precisely is crucial to demonstrating the actual role that postulated defense adaptations or counter-adaptations play in plant–pathogen interactions.

Plant chemicals can be divided into two major categories: primary metabolites and secondary metabolites. Primary metabolites are substances produced by all plant cells that are directly involved in growth, development, or reproduction. Examples include sugars, proteins, amino acids, and nucleic acids. Secondary metabolites are not directly involved in growth or reproduction, but they are often involved with plant defense. These compounds usually belong to one of three large chemical classes: terpenoids, phenolics, nitrogen compounds and alkaloids.

3.3.1 Terpenoids (Terpenes)

The terpenoids, sometimes called isoprenoids/isoprenes, are a large and diverse class of naturally occurring organic chemicals similar to terpenes, derived from five-carbon isoprene units assembled and modified in thousands of ways. Most are multicyclic structures that differ from one another not only in functional groups also in their basic carbon skeletons. These lipids can be found in all classes of living things, and are the largest group of natural products and related compounds formally derived from isoprene units. They contain oxygen in various functional groups. This class is subdivided according to the number of carbon atoms in the same manner as are terpenes. The skeleton of terpenoids may differ from strict additivity of isoprene units by the loss or shift of a fragment, generally a methyl group. The terpenes, or terpenoids, constitute the largest class of secondary metabolites. Most of the diverse substances of this class are insoluble in water, and terpenes are toxins and feeding deterrents to many pathogens, thus play important defensive roles in the plant kingdom.

Terpenoids are organic chemicals derived from five-carbon isoprene units; over 10,000 types of terpenoids are known Terpenoids are similar to each other in the organic chemical they are derived but greatly differ from one another in their basic carbon skeleton and functional groups. Monoterpenes and diterpenes, which are the main components of essential oils, act as allelopathic agents, attractants in plant-plant or plant-pathogen/herbivore interactions or repellants (Grassman, 2005). This type of defenses is mostly associated with volatile essential oils, deactivation and enzyme inhibition and poison in leaves. Terpenoids also defend the plants against pathogens by evaporating at the wound region causing the resin acid to undergo oxidative polymerization and crystallization thus prevents pathogens from entering the plants.

So terpenoids are really important chemical in plant defense by helping it defend itself against pathogens, the chemical is similar in its function but has a different effect on different organisms. They occur in all plants and represent the largest class of secondary metabolites with over 22,000 compounds described. The simplest terpenoid is the hydrocarbon isoprene (C_5H_8), a volatile gas emitted during photosynthesis in large quantities by leaves that may protect cell membranes from damage caused by high temperature or light (Fig. 3.14). Terpenoids are classified by the number of isoprene units used to construct them. For example, monoterpenoids consist of two isoprene units, sesquiterpenoids (three units), diterpenoids (four units), and triterpenoids (six units).

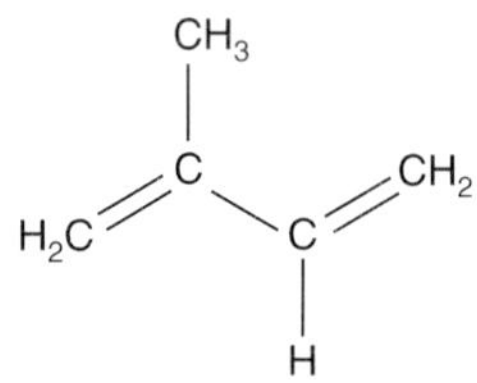

Fig. 3.14 Isoprene

Monoterpenoids and sesquiterpenoids are the primary components of essential oils, which are highly volatile compounds that contribute to the fragrance (essence) of plants that produce them. Essential oils often function as insect toxins, and many protect against fungal or bacterial attack. Mint plants (*Mentha* spp.) produce large quantities of the monoterpenoids menthol and menthone which are produced and stored in glandular trichomes on the epidermis. Pyrethrins are monoterpenoid esters produced by chrysanthemum plants that act as insect neurotoxins. Many commercially available insecticides are actually synthetic analogs of pyrethrins, called pyrethroids, including the insecticides permethrin and cypermethrin. Pine tree resin contains large quantities of the monoterpenoids α- and β-pinene, which are potent insect repellents; these compounds give the organic solvent turpentine its characteristic sharp odor.

3.3.2 Monoterpenoids

Monoterpenoids are not just used as insecticides. Many spices, seasonings, condiments and perfumes are made using essential oils that function as insect toxins in plants but are relatively harmless to humans. Examples include peppermint and spearmint (*Mentha* spp.), basil (*Ocimum* spp.), oregano (*Origanum* spp.), rosemary (*Rosmarinus* spp.), sage (*Salvia* spp.), savory (*Satureja* spp.), thyme (*Thymus* spp.), black pepper (*Piper* spp.), cinnamon (*Cinnamomum* spp.), and bay leaf (*Laurus* spp.).

The antifungal activity of twelve monoterpenes (Fig. 3.15), camphene, (R)-camphor, (R)-carvone, 1,8-cineole,cuminaldehyde, (S)-fenchone, geraniol, (S)-limonene, (R)-linalool, (1R,2S,5R)-menthol, myrcene and thymol was evaluated against four plant pathogenic fungi *Rhizoctonia solani, Fusarium oxysporum, Penecillium digitatum* and *Asperigallus niger* by using mycelial growth inhibitory technique (Gehan I. Kh. Marei *et al.*, 2012). (S)-limonene and thymol were examined for their inhibitory effects on pectin methyl esterase (PME), cellulose and polyphenol oxidase (PPO) of tested fungi. Thymol was the most potent antifungal compound against the four test fungi with EC50 values of 33.50, 50.35, 20.14 and 23.80 mg/L on *R. solani, F. oxysporum, P. digitatum* and *A. niger*, respectively. The antifungal activity of thymol was comparable to a reference fungicide, carbendazim (S)-limonene and 1,8-cineole exhibited pronounced antifungal activity against the four tested fungi. The most effective antifungal compounds thymol and (S)-limonene showed a strong inhibitory effect on the activity of PME and cellulase but revealed no inhibitory effect on PPO. The results showed that PME was more sensitive than cellulase to thymol and (S)-limonene. This is the first report of the inhibitory effects of monoterpenes thymol and (S)-limonene on PME, cellulase, and PPO. The results indicated that monoterpenes may cause their antifungal activity by inhibiting PME and cellulase. The strong antifungal activity of thymol, (S)-limonene and 1,8-cineole reported in this study indicated that these compounds have a potential to be used as fungicides.

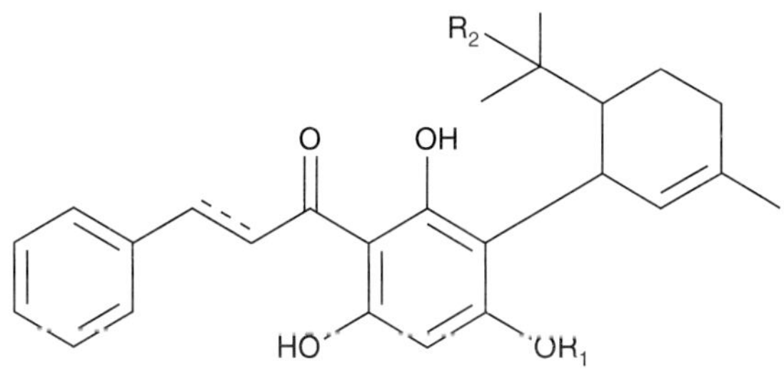

R1 = H, Me
R2 = H, OH

Fig. 3.15 Monoterpene

Twenty-two naturally occurring monoterpenoid compounds were selected for antifungal evaluation against two of the most common postharvest pathogens on temperate fruits, *Botrytis cinerea* and *Monilinia fructicola* (Rong and Ting, 2000). Experiments were carried out to examine the separate effects of these compounds on spore germination and mycelial growth. The growth inhibition test was conducted by incorporating the compounds into the medium, or by exposing the fungi to the vapor of the monoterpenoids. The results showed that carvacrol and thymol, the only two phenolic monoterpenoids, were the most potent inhibitors to the two pathogens in the germination and growth inhibition tests. Both of the compounds completely prevented the spore germination and mycelial growth of *B. cinerea* and *M. fructicola* at 100 μg/mL. Even at 10 μg/mL medium (0.25 mg/Petri dish) in the volatility test, carvacrol and thymol had 85% and 82% inhibition at 48 h against *B. cinerea,* and 38% and 57% against *M. fructicola*, respectively. Other monoterpenoids such as eugenol, citronellol, geraniol, citral, (−)-perillaldehyde, citronellal, (−)-perillyl alcohol and (−)-menthol were good growth inhibitors although their activities toward the two pathogens were slightly different, with higher inhibition found against the *M. fructicola.*

Detoxification of an antifungal monoterpene terpinolene by the phytopathogenic fungus *Botrytis cinerea* afforded hydroxylated metabolites 2,3-dihydro-3,6-dihydroxy-terpinolene (39%) and 2,3-dihydro-1α,3α-dihydroxy-terpinolene (20%), respectively (Afgan Farooq *et al.*, 2002). Terpinolene showed good levels of antifungal activity while both the metabolites were inactive against another plant pathogenic fungus *Cladosporium herbarum*.

3.3.3 Diterpenoids

They include gossypol, a terpenoid produced by cotton (*Gossypium hirsutum*) that has strong antifungal and antibacterial properties. Diterpenes constitute some of the commonly encountered chemical classes of phytoalexins (Fig. 3.16). In cotton (*Gossypium hirsutum*), pathogens and insect herbivores trigger the synthesis of the sesquiterpene δ-cadinene and numerous subsequent oxidation products, which, collectively, afford protection against these attackers (Eric A. Schmelza *et al.*, 2011). Similarly, in rice (*Oryza sativa*), pathogen infection induces the levels of transcripts encoding key biosynthetic enzymes, namely, an *ent*-copalyl diphosphate synthase (*ent*-CPS; OsCPS2) and a syn-CPS (OsCPS4), which lead to diterpenoid phytoalexin production. These inducible diterpenoids have established roles in protecting against the ascomycete fungus *Magnaporthe grisea* the causal agent of rice blast disease, which is responsible for 10–30% of losses in production.

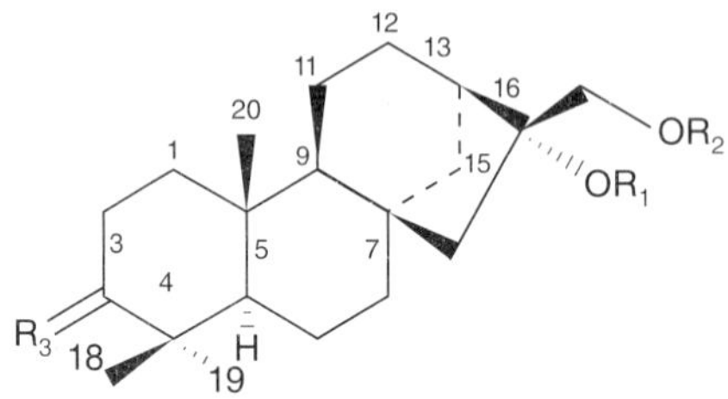

Fig. 3.16 Diterpene

Gibberellins (GAs) are a class of diterpenoid phytohormones essential for many processes of plant growth and development, such as seed germination, stem elongation, leaf expansion and flower development. The growth-promoting activity of GA in plants was initially recognized by observing the 'bakanae disease' in rice (foolish rice), in which exogenous bioactive GA produced by *Gibberella fujikuroi* causes hyperelongation of shoots. The roles of endogenous GAs in plant development have been elucidated using a series of mutants that are defective in an enzyme in the GA biosynthesis pathway. For example, severe GA-deficient mutants of *Arabidopsis*, such *as*ga1-3, ga2-1 *and* ga3-1, are nongerminating, male-sterile and extremely dwarfed (Koornneef and van der Veen, 1980).

The recent completion of the genome sequence analysis of *Neurospora crassa* (Galagan, 2003) has revealed that this fungus possesses a large number of putative genes for terpenoid secondary metabolism, as do plants. Part of the diversity of fungal terpenoids was realized earlier because of the production of several classes of terpenoids, including such biologically active compounds as fusicoccin from *Fusicoccum amygdali* and aphidicolin from *Phoma betae*. Therefore, functional expression of fungal enzymes in plants may provide new possibilities for useful terpenoid productions through metabolic engineering. Unlike fungal terpenoids, which are produced through the cytosolic mevalonate pathway, the majority of plant diterpenes including GAs are synthesized through the methylerythritol phosphate (MEP) pathway in plastids (Okada, K. Saito *et al.*, 2000). Thus, a possible challenge in functional expression of fungal diterpene cyclases in plants would be the necessity of the enzyme to be targeted to appropriate subcellular locations.

3.3.4 Triterpenoids

These are similar in molecular structure to plant and animal sterols and steroid hormones. Triterpenes are a type of terpene containing 30 carbon atoms (Fig. 3.17). Triterpenes are assembled from a five-carbon isoprene unit through the cytosolic mevalonate pathway to make a thirty-carbon compound. Some triterpenes are steroidal in nature. Cholesterol, phytosterols, and phytoecdysteroids are triterpenes. Triterpenoid saponins are a diverse group of natural products in plants and are considered defensive compounds against pathogenic microbes and herbivores.

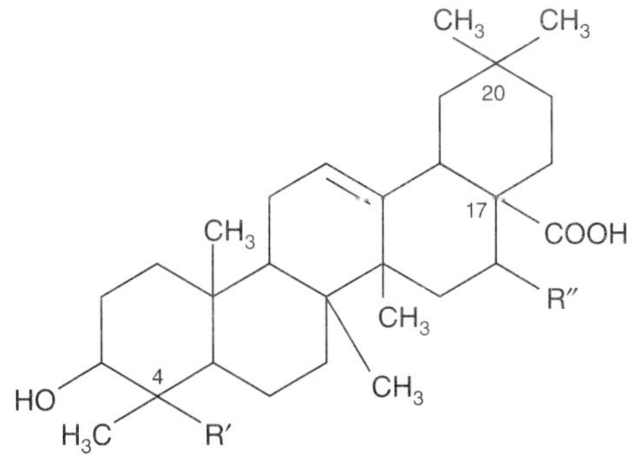

Fig. 3.17 Triterpene

Triterpenoids are a very diverse group of natural products with wide distribution and particularly high chemical diversity in plants. They include compounds such as betulinic acid, the avenacins, and glycyrrhizin, which have important biological functions and medicinal properties (Zhonghua Wang *et al.*, 2011). The biosynthetic pathway toward triterpenoids proceeds by joining six isoprene units together to form the branched long-chain hydrocarbon squalene. In prokaryotes, squalene is directly cyclized into hopanoid triterpenes, whereas, in eukaryotes, it is first activated into 2,3-epoxysqualene and then cyclized. The cyclizations are highly region specific and stereospecific, establishing the final carbon structure of the triterpenoid products.

3.3.5 Phytoanticipins

Most plants synthesize antimicrobial compounds as part of normal plant development, called, phytoanticipins or synthesize such compounds *de novo* when challenged by microorganisms, known as phytoalexins. The presumed role of these plant antibiotics is to protect the plant from disease. However, many phytopathogenic fungi have enzymes that can detoxify the phytoanticipins or phytoalexins produced by their host (Hans D. VanEtten *et al.*, 1995). This may be a means that these pathogens have evolved to circumvent resistance mechanisms based on the production of plant antibiotics. Many of the phytoanticipin- and phytoalexin-detoxifying enzymes produced by phytopathogenic fungi have biochemical and regulatory properties that would indicate the phytoanticipins and phytoalexins produced by their host are their normal substrates. In addition, their activity, enzymatic products, or transcripts can be detected in infected plant tissue suggesting that they are functioning *in planta* during pathogenesis. Specific mutations have been made by transformation-mediated gene-disruption procedures that eliminate the ability of *Gaeumannomyces graminis* var. *avenae*, *Gloeocercospora sorghi*, and *Nectria haematococca* to detoxify the phytoanticipins or phytoalexins produced by their hosts. The effect of these mutations on pathogenicity indicated a requirement for detoxifying enzymes in *G. graminis* var. *avenae* but not in *G. sorghi* or *N. haematococca*. Tomatine (lycopersicin), saponin, coumarin (1,2-benzopyrone), and the constitutive isoflavone biochanin A (5,7-dihydroxy-4'-methoxyisoflavone) (Figs. 3.18—3.20) are some of the phytoanticipins.

They may be excreted into the external environment (e.g., rhizosphere or phylloplane), accumulate in dead cells, or they may be sequestered in vacuoles in an inactive form. The dead cells of brown onion skins contain the quinones catechol and protocatechuic acid, which inhibit germination of spores of the smudge pathogen, *Colletotrichum circinans*, and the neck rot pathogen, *Botrytis cinerea.* White onions do not produce these compounds and are susceptible to smudge. *Aspergillus niger* is insensitive to these inhibitors and attacks both white and brown onions. Avocado rootstocks resistant to root rot caused by *Phytophthora cinnamom* secrete borbinol, an antimicrobial phenolic compound, into the rhizosphere. The secretion of nematode-inhibiting substances into the rhizosphere surrounding asparagus and marigold roots has already been mentioned. Symptoms of anthracnose of avocado, caused by *Colletotrichum gloeosporiordes*, only develop on ripe fruit. The peel of unripe avocado fruit contains antifungal lipids called dienes that prevent appressorial germination. As these dienes are gradually metabolized during fruit ripening to less toxic compounds, quiescent appressoria germinate and susceptibility to anthracnose increases. In anthracnose-resistant cultivars, diene breakdown is blocked following infection, so that antifungal levels are sustained for longer periods. The resistance of immature apples and pears to scab caused by *Venturia inaequalis* and *V. pirina* respectively correlates with the presence of the phenolic compounds chlorogenic acid, phloridzin, arbutin and iso-chlorogenic acid in the outer layers of the fruit. These compounds also contribute to the bitter taste of unripe apples and pears and, as the fruit ripens and sweetens, it also becomes more susceptible to scab.

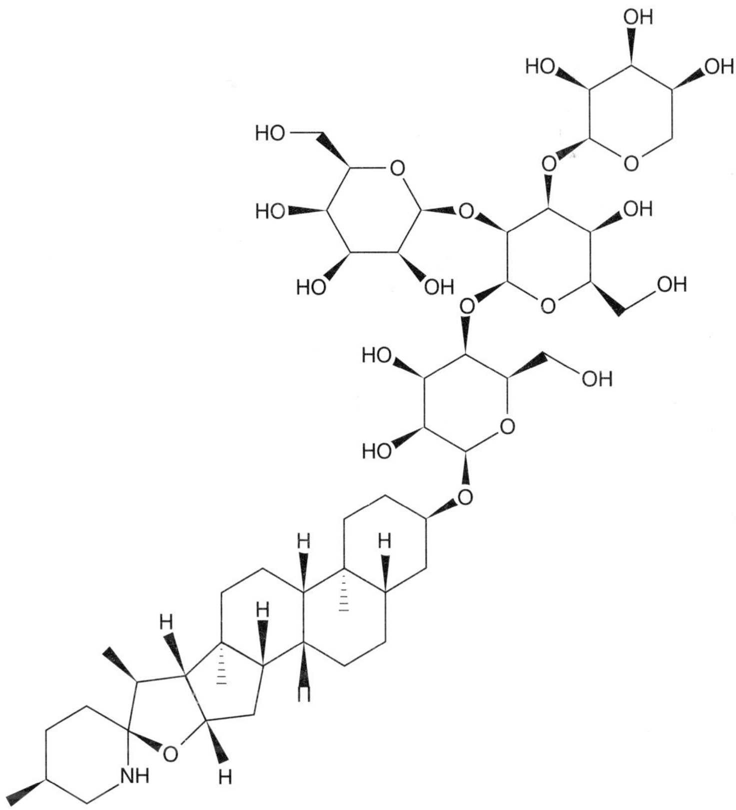

Fig. 3.18 Tomatine

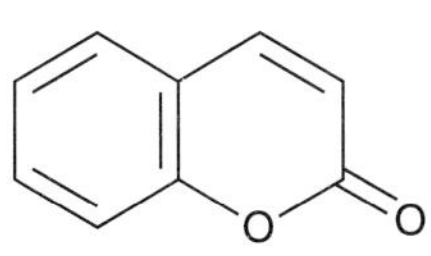

Fig. 3.19 Coumarin

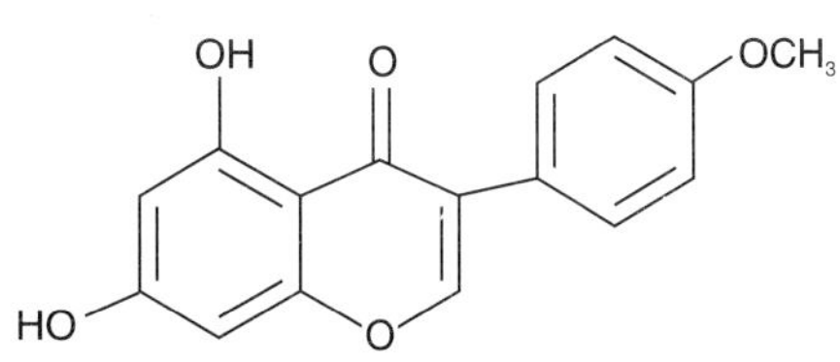

Fig. 3.20 Biochanin A

3.3.5.1 Saponins

Saponins are glycosylated phytoanticipins that are found in a wide range of plant species and can be divided into three major groups, triterpenoid, steroid or steroidal glycoalkaloid, depending on the structure of their aglycones (Rocío González-Lamothe *et al.*, 2009) (Fig. 3.21)]. These substances have detergent (soap-like) properties and disrupt the cell membranes of invading fungal pathogens. Saponins consist of triterpenoid, steroid or steroidal glycoalkaloid molecules bearing one or more sugar chains ~-3. They are characterized by their surfactant properties and give stable foam when shaken with water. The detergent action of saponins was recognized hundreds of years ago when extracts of plants such as *Saponaria officinalis* were used to make soap. The wheat pathogen *Gaeumannomyces graminis* is unable to infect oats that contain avenacins, a class of triterpenoid saponins. However, some fungal pathogens have developed countermeasures to these plant defenses: *Botrytis cinerea, Fusarium oxysporum,* and *Septoria lycopersici* are all capable of degrading saponins and causing disease in susceptible saponin-producing plants. Because they have potent antimicrobial activities, it is proposed that the natural role of these molecules in plants is to confer protection against potential pathogens. The saponins studied in deepest detail in relation to their potential role in defense are avenacin and α-tomatine. Avenacins are oat root saponins. The antifugal activity of avenacin is associated with its ability to form complexes with sterols present in fungal membrane leading to pore formation and loss of membrane integrity. The localization of the major avenacin, avenacin A-1, in the epidermal cell layer of oat root tips and the emerging lateral root initials, suggests a role as a chemical barrier. Moreover, the capacity of *Gaeumannomyces graminis* var *avenae* to detoxify avenacin A-1 has been shown to be essential for its interaction with oat. Fungal mutants lacking the saponin-detoxifying enzyme avenacinase showed increased sensitivity to avenacin A-1 and were no longer able to infect. Saponin-deficient mutants also showed compromised resistance to several pathogens, indicating that avenacins provide a preformed chemical defense to pathogen attack. Interestingly, accumulation of avenacin biosynthesis pathway intermediates in oat roots results in callose accumulation, a well known defense mechanism which suggests that phytoanticipin accumulation may also trigger other defense responses. This implies that the antimicrobials may work in defense-related signaling processes and strengthen the relevance of these compounds as biotechnological weapons against pathogen infection

Saponins occur in a great many plant species and have been implicated as preformed determinants of resistance to fungal attack. A number of fungi that succeed in breaching these antimicrobial plant defenses produce saponin-detoxifying enzymes (Anne Osbourn, 1996). The importance of one of these (avenacinase, produced by *Gaeumannomyces graminis*) in determining host range has been demonstrated. Recently, avenacinase has been shown to be closely related to another saponin-detoxifying enzyme, tomatinase, which is produced by the tomato pathogen *Septoria lycopersici,* suggesting that common mechanisms for saponin detoxification may be widespread in phytopathogenic fungi.

Saponins such as avenacin A-1 from oat roots (a triterpenoid) and e-tomatine from tomato (a steroidal glycoalkaloid) each has a single sugar chain attached to carbon C-3 via oxygen, and so are referred to as monodesmosidic saponins. For some bisdesmosidic saponins, specific plant glycosyl hydrolases that carry out this conversion have been identified (Lttning and SchlSsser, 1975). The best characterized of these enzymes is the oat enzyme avenacosidase, which activates avenacosides A and B, the bisdesmosidic steroidal saponins found in oat leaves. In healthy oat leaf tissue, avenacosidase is confined to the plastids 6, while avenacosides A and B are located in the vacuole 7. Wounding of plant tissue resulting from mechanical damage or pathogen attack causes a breakdown of compartmentalization, allowing the enzyme to come into contact with its saponin

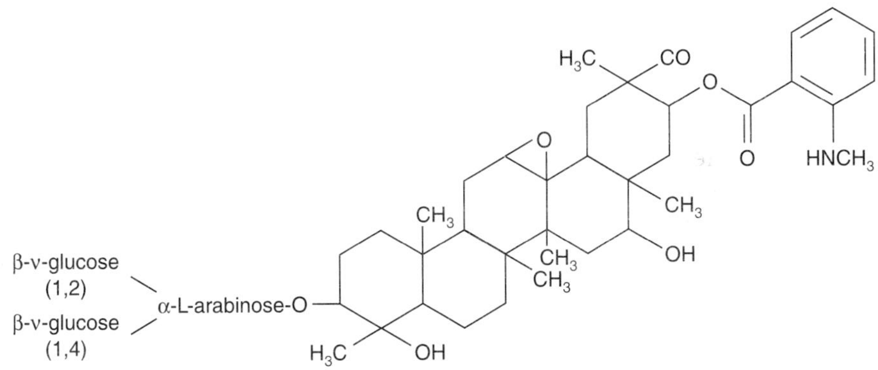

Fig. 3.21 Avenacin A-1

substrates. Although bisdesmosidic saponins such as the avenacosides lack antimicrobial activity, they are still regarded as preformed inhibitors because their conversion to the monodesmosidic form does not require induction of gene expression in response to pathogen attack.

The primary mode of action of saponins towards fungi is similar to that of polyene antibiotics, and involves the formation of complexes with membrane sterols. This results in pore formation and loss of membrane integrity. The precise way in which saponins become incorporated into membranes is unclear, although various models have been proposed (Anne Osbourn, 1996). The nature of the aglycone and the oligosaccharide moiety of the saponin molecule are both likely to contribute to the membraneolytic properties of individual saponins. In the case of the steroidal glycoalkaloids, antifungal activity has been shown to be pH-dependent. The susceptibility of membranes to saponin action may be influenced by sterol composition since detailed sterol-binding studies involving the steroidal glycoalkaloid α-tomatine indicate a requirement for membrane sterols with free 3-hydroxyl groups. The possibility that saponins may also have more subtle effects on fungi in addition to their membraneolytic action cannot be excluded, although little is known about this area.

An alternative mechanism of action of saponins has been proposed in which the aglycone, rather than the saponin itself, is believed to be the membraneolytic agent. Here the sugars are regarded as a solubilizing group aiding delivery of the saponin to the membrane, where membrane-bound glycosyl hydrolases are then presumed to activate the saponin by converting it to the aglycone. This view has received little support because it is incompatible with evidence showing that the free aglycone oftomatine does not bind cholesterol (Arneson and Durbin,1968) and that saponins affect the permeability of artificial membranes (which lack hydrolytic enzymes).

3.3.5.1.1 Fungal strategies for dealing with saponins

Pathogens such as *Cladosporium fulvum* may achieve this simply by circumventing the host plant saponin altogether. This biotrophic fungus restricts its growth to the intercellular spaces of tomato leaves, thus avoiding the release of tomatine. Other tomato-infecting fungi may lower the pH at the infection site to levels at which tomatine is ineffective as a membranolytic agent (SchSnbeck and Schbsser, 1976).

In general, however, there appear to be two main mechanisms of resistance to fungi to saponins. The first of these involves intrinsic resistance by virtue of membrane composition. Oomycetes such as *Pythium* and *Phytophthora* lack membrane sterols and are highly resistant to saponins *in vitro.* These fungi can incorporate exogenous sterols into their membranes, and show an increase in sensitivity to saponins when sterols are added to the growth medium. Conversely, fungi that normally contain sterols in their membranes become more resistant to saponins when they are grown in the presence of inhibitors of sterol biosynthesis (Olsen, 1973). The importance of membrane composition in determining the ability of fungi to infect saponin-containing plants has been illustrated by experiments involving a tomato-attacking isolate of *Fusarium solani.* The wild-type isolate was pathogenic only to ripe tomato fruits (which contain relatively low levels of tomatine) and was unable to infect green fruits (which are rich in tomatine). However, mutants of this isolate defective in sterol biosynthesis showed increased resistance to α-tomatine and were able to infect green fruit. The second major mechanism of resistance of fungi to saponins involves the production of saponin detoxifying enzymes.

3.3.5.1.2 Saponin-detoxifying enzymes

A number of phytopathogenic fungi produce specific glycosyl hydrolases that remove sugar molecules from the glycosyl chains at the C-3 carbon position, to give products which are less toxic to fungal growth. Some of these enzymes are constitutively expressed while others are induced by their saponin substrates. Despite the widespread distribution of saponins in the plant kingdom, detailed studies of saponin detoxification by fungi have been restricted primarily to pathogens of oat and tomato. This is probably because the structures and antifungal properties of oat and tomato saponins are well documented, and the saponin profiles of these plants relatively simple (in contrast to alfalfa, for example, where over 20 different saponins have been identified). The importance of saponin-detoxifying enzymes in determining fungal pathogenicity to plants has recently been demonstrated for oat-attacking isolates of the soil fungus *Gaeumannomyces graminis,* which detoxify the oat root saponin avenacin A-1 (Bowyer, 1995).

If saponin-'saponinase' combinations do have widespread significance for determining the host range of phytopathogenic fungi, then this phenomenon may provide the basis for novel crop protection strategies which give control of a wide range of pathogens (Anne Osbourn, 1996). Avenacinase and tomatinase are both extracellular enzymes and may present attractive, accessible targets for inhibitor-based approaches. The manipulation of plant saponin biosynthetic pathways may also offer potential for disease control, but first it is important to establish the contribution of saponins to disease resistance. The isolation of plant mutants defective in saponin biosynthesis will allow this to be tested directly.

Saponins are secondary plant metabolites that occur in a wide range of plant species (Figen Mert-Turks, 1995.). They are stored in plant cells as inactive precursors but are readily converted into biologically active antibiotics by plant enzymes in response to pathogen attack. These compounds can also be regarded as 'preformed,' since the plant enzymes that activate them are already present in healthy plant tissues (Osbourn, 1996). The natural role of saponins in plants is thought to be protection against attack by pathogens and pets. The production of low-molecular-weight antimicrobial molecules within plants is one component of defense against pathogens. Among them, preformed antimicrobial compounds (phytoanticipins) are the first biochemical barriers against pathogens. Saponins (a group of phytoanticipins) are present constitutively in plants and play important roles in plant defense. A variety of biological roles has been postulated for different saponins, involving allelopathic activity, defense against insects and pathogens. The mechanism of antifungal action of saponins is not well understood, but it is believed that they complex with sterols in the cell membrane, leading to pore formation and consequent loss of membrane integrity. Saponins that have been studied in detail in relation to their potential role in plant defense against attack by phytopathogenic fungi are triterpenoid avenacins in oat roots and the steroidal glycoalkaloids α-tomatine in the leaves of tomato.

Saponins are glycosylated compounds that are widely distributed in the plant kingdom and can be divided into three major groups; a triterpenoid, a steroid, or a steroidal glycoalkaloid. Triterpenoid saponins are found primarily in dicotyledonous plants but also in some monocots, whereas steroid saponins occur mainly in monocots, such as the Liliaceae, Droscoraceae, and Agavaceae and certain dicots, such as foxglove.

3.3.5.2 Avenacins

The avenacins are a family of four structurally related molecules, avenacins A-1, B-1, A-2, and B-2 (Fig. 3.22). Avenacin A-1 is the most abundant of the four in extracts from young oat roots, comprising around 70% of the total avenacin content (Figen Mert-Turks, 1995). Avenacins, like many other saponins, have potent antifungal activity. The antifungal properties of saponins come from the ability of these molecules to complex with sterols in fungal membranes, so causing pore formation and loss of membrane integrity. Evidence of affinity for membrane sterols comes from electron microscopy studies of saponin-treated membranes, which were found to contain permanent lesions. These lesions are thought to be a micelle-like aggregation of saponins and cholesterol in the plane of the membrane, possibly with the saponin molecules arranged in a ring with their hydrophobic moieties combined with cholesterol around the outer perimeter. The carbohydrate residues of avenacin play an important role in neutralizing fungal pathogens. These initial findings were confirmed by a number of studies using other saponins and related compounds called glycoalkaloids (Armah *et al.*, 1999).

Avenacins that are synthesized in the roots and that provide protection against soil-borne pathogens are acylated molecules that are conjugated with either *N*-methyl-anthranilic acid or benzoic acid (Fig. 3.22). The major avenacin found in oat roots is avenacin A-1. Acylation of avenacins is carried out by the serine carboxypeptidase-like acyltransferase SAD7, which requires an acyl glucose donor.

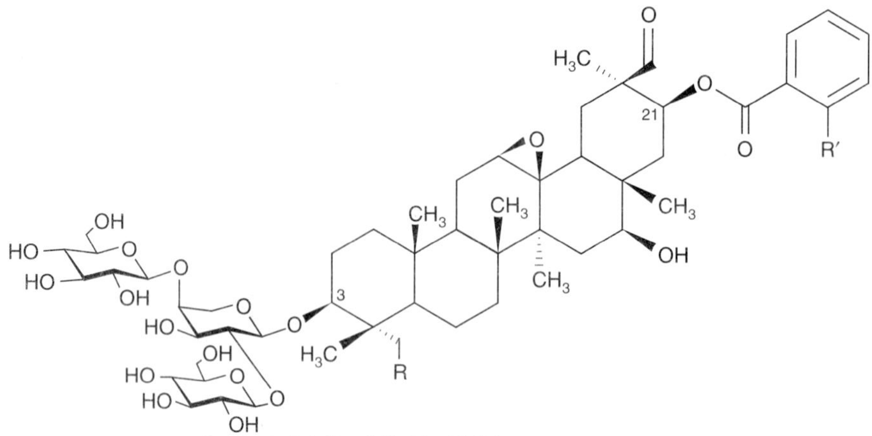

Avenacin A-1 R = OH, R′ = NHMe
Avenacin B-1 R = H, R′ = NHMe
Avenacin A-2 R = OH, R′ = H
Avenacin B-2 R = H, R′ = HMe

Fig. 3.22 Avenacins

3.3.5.3 α-Tomatine

The major saponin in tomato is the steroidal glycoalkaloid, α-tomatine (Fig. 3.23). α-Tomatine is, like the avenacins, present in healthy plants in its biologically active form. The levels of this saponin are particularly high in the leaves, flowers and green fruits of tomato (Roddick, 1974). It is assumed that α-tomatine is present in tomato leaves in the concentration around 1 mM which is sufficient to inhibit the growth of many nonpathogens of tomato. Therefore, it would be expected that this molecule could protect the tomato leaves from fungal pathogens. However, a number of tomato pathogens, including *Septoria lycopersici, Botrytis cinerea, Fusarium oxysporum* f. sp. *lycopersici, Verticillium albo-atrum* and *Alternaria solani,* are able to produce an enzyme that detoxifies α-tomatine. The ability of hydrolyzing sugar from α-tomatine is found to be common in *S. lycopersici, B. cinerea, F. oxysporum* f. sp. *lycopersici*. The *in vitro* experiments indicated that the fungal pathogens of tomato are considerably more tolerant to the compound than the nonpathogens pointing out they had evolved together. Despite the considerable variation in α-tomatine levels in the genus, Lycopersicon, relationship between saponin content and disease resistance are not well documented. It has, however, been suggested that α-tomatine may play a secondary role in the variety specific resistance of tomato to incompatible races of the biotroph, *Cladosporium fulvum,* and that release of the saponin from leaf cells as a consequence of an incompatible interaction may act to kill or contain the pathogen (Dow and Callow, 1978).

Several plants use secondary metabolites against insects and pathogens attack. Among such metabolites known glycoalkaloids are the most studied. Plants from Solanaceae family are the most abundant on those alkaloids (Gomes *et al.*, 2014). Despite alfa-tomatine is a known glycoalkaloid, its specific action against *Moniliophthora perniciosa,* fungi

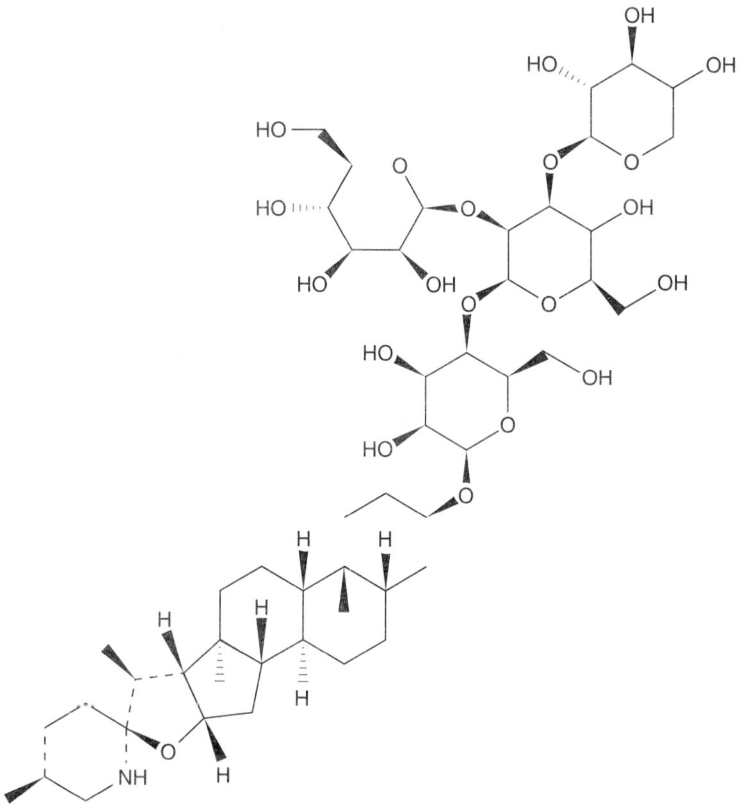

Fig. 3.23 α-tomatine

responsible for witches broom disease in cocoa plantations used; α-tomatine was infusion-extracted from *Solanum lycopersicum* leaves, and its action against the fungi growth was shown on the bioassays. In these tests, *M. perniciosa* growth was totally inhibited but other fungus tested, the inhibition was not totally or did not present effect on growth, showing in this way, the specificity of α-tomatine to *M. perniciosa*. The advantages of work with a natural molecule for the fungi control is the stability of the molecule, its resistance to high temperature and pressure and for its use, the simple method of extraction used. The confirmation of the α-tomatine molecule was done by High-Performance Liquid Chromatography and Mass Spectrometry in comparison to the Sigma standard.

Antifungal saponins occur in many plant species and may provide a preformed chemical barrier to attack by phytopathogenic fungi. Some fungal pathogens can enzymatically detoxify host plant saponin, which suggests that saponin detoxification may determine the host range of these fungi (Bowyer *et al.*, 1995). A gene encoding a saponin detoxifying enzyme was cloned from the cereal-infecting fungus *Gaeumannomyces graminis*. Fungal mutants generated by targeted gene disruption were no longer able to infect the saponin-containing host oats but retained full pathogenicity to wheat (which does not contain saponins). Thus, the ability of a phytopathogenic fungus to detoxify a plant saponin can determine its host range. *Sapindus mukorossi* saponin, which is effective against *Rhizoctonia bataticola* and *Sclerotium rolfsii*, can be exploited for the development of a natural fungicide. A sugar moiety is a prerequisite for the antifungal activity of triterpenic saponin (Supradip Saha *et al.*, 2010)

α-Tomatine is a steroidal glycoalkaloid present in Solanaceous plants, principally in a number of *Lycopersicon* and *Solanum* species. High concentrations are found in leaves, stems, roots and green fruit of tomato plants, suggesting a possible role in resistance to pathogens. The toxic effects of α-tomatine are attributed to its ability to complex with membrane sterols, causing pore formation and leakage of cell contents (Manuel Ruiz-Rubio *et al.*, 2014). Pathogenic fungi to tomato are usually less sensitive to tomatine than pathogens of other plant species and saprophytes. Fungi have evolved two principal mechanisms to avoid the toxic effect of tomatine, either changing the composition of the cell membrane or producing specific tomatine-detoxifying enzymes, known as tomatinases. Tomatinases have a distinct mechanism of action and molecular and catalytic properties in different pathogen species. Up to now, the knowledge of α-tomatine metabolism by phytopathogenic fungi is almost exclusively limited to the removal of sugar residues from the molecule, although recent studies in *Gibberella pulicaris* demonstrated that further hydroxylations of the aglycone moiety tomatidine occur, after the action of tomatinase. Tomatinase-encoding genes have been cloned from *Septoria lycopersici* and *Fusarium oxysporum*. Both genes do not share sequence homology, suggesting that different tomato pathogens have independently developed the ability to cleave tomatine by convergent evolution. Expression of *F. oxysporum* tomatinase was fully repressed in the presence of glucose, indicating a possible role in nutrition besides detoxification of tomatine, although in *Botrytis cinerea* glucose repression of tomatinase was not observed.

Saponin detoxification enzymes from pathogenic fungi are involved in the infection process of their host plants. *Fusarium oxysporum* f. sp *lycopersici*, a tomato wilt pathogen, produces the tomatinase enzyme Tom1, which degrades α-tomatine to less toxic derivates. To study the role of the Tom1 gene in the virulence of *F. oxysporum*, targeted disruption, and overexpression of the gene was performed (Yolanda Pareja-Jaime *et al.*, 2008). The infection process of plants inoculated with transformants constitutively producing Tom1 resulted in an increase of symptom development. By contrast, tomato plants infected with the knockout mutants showed a delay in the disease process, indicating that Tom1, although not essential for pathogenicity, is required for the full virulence of *F. oxysporum*. Total tomatinase activity in the disrupted strains was reduced only 25%, leading to β_2-tomatine as the main hydrolysis product of the saponin *in vitro*. In silico analysis of the *F. oxysporum* genome revealed the existence of four additional putative tomatinase genes with identities to tomatinases from family 3 of glycosyl hydrolases. These might be responsible for the remaining tomatinase activity in the ΔTom1 mutants. These results indicated that detoxification of α-tomatine in *F. oxysporum* was carried out by several tomatinase activities, suggesting the importance of these enzymes during the infection process.

3.3.6 Phenolics

Phenolics are another large class of secondary metabolites (Fig. 3.24) produced by plants to defend themselves against pathogens. They are produced primarily via the shikimic acid and malonic acid pathways in plants and include a wide variety of defense-related compounds including flavonoids, anthocyanins, phytoalexins, tannins, lignin, and furanocoumarins.

Phenolics have been in the center of many discoveries on plant defenses to pathogens, including bacteria, fungi, and viruses. More recently, this fascinating group of metabolites was also pointed out as a target of invading microbes to defeat such defenses (Daayf *et al.*, 2012). Findings were summarized in several host-pathogen interactions involving the oomycete, *Phytophthora infestans* and soil-borne pathogens such as *Verticillium dahliae*, depicting the role of phenolics in plant defenses against them, and how they target such response to settle their infection. The defenses described included the upregulation of genes in the phenylpropanoid pathway, and the accumulation of phenolics such as scopoletin (a coumarin), *p*-coumaric acid methyl ester (a hydroxycinnamic acid derivative), or rutin (a flavonoid) in response to several pathogens. Counterdefense mechanisms developed by these pathogens, such as downregulation of defense genes and degradation of phenolics, were described.

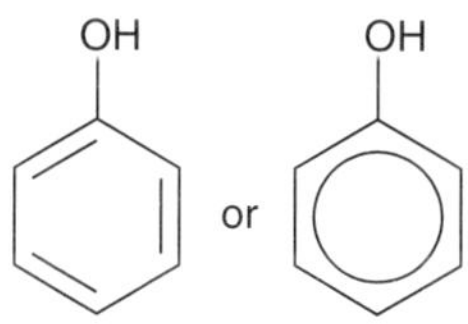

Fig. 3.24 Phenol

Phenolics consist of 6-carbon ring bonded to a hydroxyl group; over 6,000 types of phenolics have been discovered so far. They are similar to each other in that they all have a similar basic structure; they are found all over the plant, sometimes giving the plant different colorations, and they all cause a chemical reaction making digestion of the plants hard for herbivores, this way preventing them from using the plant as a food source. But like other defenses, they all have similarities in the basic structure and the type of basic defense they provide for the plant, but the chemicals also differ in the way that they have specialized in dealing with different predators over evolutionary time (Vincenzo Lattanzio *et al.*, 2006). Tannins and favonoids are two important phenolics. Phenolics are important plant defenders against pathogens. They have antiseptic properties that allow them to kill pathogens very effectively. Salicylic acid is a well known Phenolic that is very important in defense signaling against pathogens (Nicholson and Hammerschmidt, 1992). Salicylic acid causes an oxidative burst when the plant is attacked by a pathogen altering the host chemistry; the alternation can allow the plant to kill the affected cells before the pathogens can spread to other areas of the plant (Bostock *et al.*, 2000). Phenolics function very similarly in the defense they provide against pathogens, they act as signaling and antiseptic agents, but the responses can different based on the pathogen.

Plants need phenolic compounds for pigmentation, growth, reproduction, resistance to pathogens and many other functions. Therefore, they represent adaptive characters that have been subjected to natural selection during evolution. Plants synthesize a greater array of secondary compounds than animals because they cannot rely on physical mobility to escape pathogens and have therefore evolved a chemical defense against such pathogens. Phenol itself is a natural product, but most phenolics have two or more hydroxyl groups. Unless they are completely esterified, etherified or glycosylated, plant phenolics are normally soluble in polar organic solvents. Most phenolic glycosides are water-soluble, but the corresponding aglycones are usually less so. With a few exceptions, water solubility increases with the number of hydroxyl groups present. Some phenolics are solubilized in sodium hydroxide and sodium carbonate but in alkaline medium their oxidation is enhanced and, therefore, treatment with alkaline solvents should either be performed under N2 or preferably avoided. Phenolics with only a few hydroxyl groups are soluble in ether, chloroform, ethyl acetate, methanol, and ethanol (Vincenzo Lattanzio *et al.*, 2006).

In contrast with a basic metabolism that refers to the anabolic and catabolic processes required for cell maintenance and proliferation, secondary metabolism refers to compounds present in specialized cells that are not directly essential for basic photosynthetic or respiratory metabolism but are thought to be required for plants survival in the environment. Secondary metabolites apparently act as a defense against pathogens and signal compounds (to attract pollinating or seed dispersing animals), as well as protecting the plant from ultraviolet radiation and oxidants. Generally, the role of phenolic compounds in defense is related to their antibiotic, antinutritional or unpalatable properties. Besides their involvement in plant-animal and/or plant-microorganism relationships, plant phenolics also have key roles as the major red, blue and purple pigments, as antioxidants and metal chelators, as signaling agents both above and below ground between plant and other organisms, and as UV light screens. Phenolics are often stored at strategically important sites where they play a signaling role, and often a direct role, in defense. Several studies have indicated a high degree of compartmentation of phenolic compounds and the enzymes involved in their biosynthesis. Phenolics usually accumulate in the central vacuoles of guard cells and epidermal cells as well as subepidermal cells of leaves and shoots. Furthermore, some phenolics are found covalently linked to plant cell wall; others occur in waxes or on the external surfaces of plant organs. It has been suggested that a flavonoid-DNA complex provides a mutual protection against oxidative damage (Onyilagha and Grotewold, 2004.). Plant phenolics may be divided into two classes: (i) preformed phenolics that are synthesized during the normal development of plant tissues and (ii) induced phenolics that are synthesized by plants in response to physical injury, infection or when stressed by suitable elicitors such as heavy metal salts, UV-irradiation, temperature, etc. (phytoalexins). Induced phenolics may also be constitutively synthesized but, additionally, their synthesis is often enhanced under biotic or abiotic stress (Hammerschmidt, 2003).

Preformed antibiotic compounds that occur constitutively in healthy plants are likely to represent inbuilt chemical barriers to herbivorous and fungal enemies and may protect plants against attack by a wide range of potential pests and pathogens. In contrast, induced defense compounds are synthesized in response to biotic stress as part of the plant defense response and are restricted to the damaged tissue (Vincenzo Lattanzio *et al.*, 2006). Both tolerance and resistance traits require the reallocation of host resources. Therefore, defensive chemicals are considered to be costly for plants because of the resources consumed in their biosynthesis or the ecological consequences of their accumulation. One way for a plant to reduce these costs is to synthesize defense compounds only after an initial damage by a pathogen or insect: this strategy could be risky because the initial attack may by too rapid or too severe for an effective defense response. Therefore, plants that are likely to suffer frequent and/or serious damage may be better off investing mainly in constitutive defenses whereas plants that are rarely attacked may rely predominantly on induced defenses (Koricheva *et al.*, 2004).

There is considerable evidence that preformed defenses are a major component of nonhost resistance, particularly in nondomesticated plants. Plants contain preformed peptides, proteins and secondary metabolites (phenolics, sulfur compounds, saponins, cyanogenic glycosides, and glucosinolates). The multicomponent defense response induced after the pathogen attack requires a substantial commitment of cellular resources, including extensive genetic reprogramming, because the induced expression of a large number of defense-related genes is essential for plants to counter pathogen attack. Many defense-related genes encode proteins possessing antifungal and antimicrobial activities or enzymes that catalyze defense metabolites, known as phytoalexins. Thus, defense responses are kept under tight genetic control and are activated only when the plant detects a potential invader (Vincenzo Lattanzio *et al.*, 2006). Other responses of invaded cells result from allosteric enzyme activation initiating tissue reinforcement by oxidative crosslinking, apposition of callose and lignins. The initial signal perception also leads to the production of an endogenous systemically translocated signal that has the function to activate defense mechanisms in parts of the plant remotely located from the initial site of interaction, providing durable protection against challenge infection with a broad range of pathogens. Other defense genes encode these regulatory proteins, important for defense signal transduction, and this form of induced defense response is referred to as systemic acquired resistance [(SAR) (Mysore and Ryu, 2004)].

3.3.6.1 Preformed antifungal phenolics

Preformed antibiotic compounds such as phenolic and polyphenolic compounds are ubiquitous in plants and play an important role in nonhost resistance to filamentous fungi (Vincenzo Lattanzio *et al.*, 2006). The term 'phytoanticipin' has been proposed to distinguish these preformed antifungal compounds from phytoalexins, which are synthesized from remote precursors in response to pathogen attack. Some antibiotic phenolics are stored in plant cells as inactive bound forms but are readily converted into biologically active antibiotics by plant hydrolyzing enzymes (glycosidases) in response to pathogen attack. These compounds can also be regarded as preformed antibiotics since the plant enzymes that activate them are already present but are separated from their substrates by compartmentalization, enabling rapid activation without a requirement for the transcription of new gene products (Osbourn, 1996). In such cases, free phenolics are likely to be much more toxic to the invading organism than the bound forms. In addition, even if preformed antifungal phenolics are present in healthy plants at levels that are anticipated to be antimicrobial, their levels may increase further in response to a challenge by pathogens.

The distribution of preformed antifungal phenolics within plants is often tissue specific, and there is a tendency for many lipophilic compounds (e.g., flavone and flavonols methyl ethers) to be located on the plant surface (e.g., in leaf wax and bud exudates) or in the cytoplasmic fraction within the epidermal cells, suggesting that they may indeed act as deterrents to pathogens. In general, however, preformed antifungal phenolics are commonly sequestered in conjugated form, usually with glycosidic attachments, in vacuoles or organelles in healthy plants (Katagiri *et al.*, 2002). Biotrophs may avoid the release of preformed antibiotics by minimizing the damage to the host whereas necrotrophs are likely to cause a substantial release of these compounds. The first demonstrated example of phenolics providing disease resistance was the case of onion scales accumulating sufficient quantities of catechol (I) and protocatechuic acid (II) to prevent onion smudge disease, *Colletotrichum circinans.* The colored outer onion scales of resistant onion varieties contain enough of these two phenols to reduce spore germination of *C. circinans* to below 2% while susceptible varieties lack these compounds, and the germination rate is over 90%. Adequate levels of chlorogenic acid (III) account for the resistance of potato tubers against *Streptomyces scabies, Verticillium alboatrum,* and *Phytophthora infestans* while at low concentrations it stimulates the growth of *P. infestans* and *Fusarium solani* var. *coeruleum.* Small concentrations of benzaldehyde (IV) totally inhibited spore germination of *Botrytis cinerea* and germination of *Monilia fructicola* (Wilson and Wisniewski, 1989).

Preexisting antifungal phenolics are simple phenols, phenolic acids, flavonols and dihydrochalcones (Figs. 3.25–3.27). In addition, many flavones and flavanones have been shown to be active against fungal pathogens commonly found during the storage of fruits and vegetables, i.e., *Aspergillus* sp., *B. cinerea* and *F. oxysporum* (Weidenbörner *et al.*, 1990). Lipophilicity and/or the presence of at least one acidic hydroxyl group are,

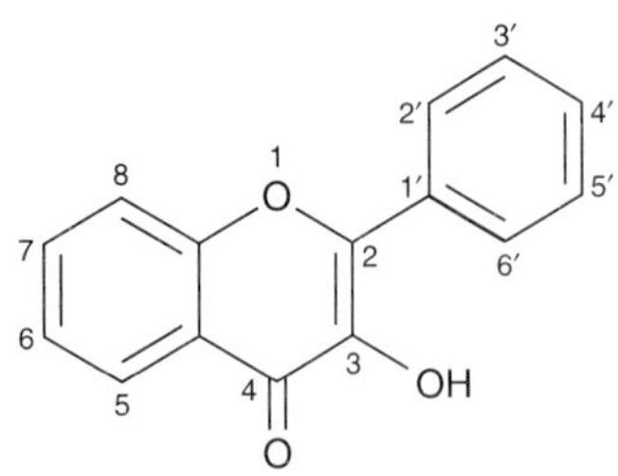

Fig. 3.25 Flavonols

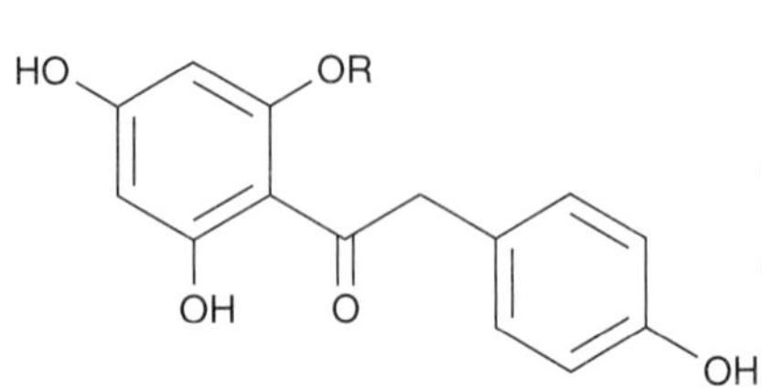

Fig. 3.26 Dihydrochalcone

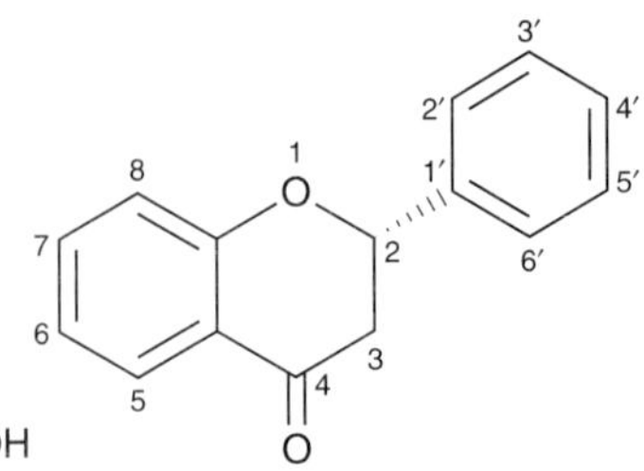

Fig. 3.27 Flavanone

therefore, considered to be a structural feature essential for a good antifungal activity. Lipophilicity allows active phenols to penetrate biological membranes while hydroxyl groups may act by uncoupling oxidative phosphorylation.

Toxicity of tannins, hydrolysable tannins, and proanthocyanidins, usually estimated by the measurement of the reduction of the *in vitro* growth of mycelium, is well documented for several filamentous fungi, for example, *B. cinerea, Aspergillus niger, Colletotrichum graminicola, Gloeophyllum trabeum, Trichoderma viride,* and *Penicillium* spp. Tannins (Fig. 3.28) are quite potent antibiotics. In temperate trees, tannins and related phenolic compounds preserve heartwood from fungal decay and inhibit extracellular hydrolases from invading pathogens, thus preventing their rapid development in the plant. It is possible that inhibition of extracellular fungal enzymes (cellulase, pectinase, laccase, xylanase, etc.), nutrient deprivation of substrates (metal complexation, protein insolubilization) and action on fungal membranes (inhibition of oxidative phosphorylation) are effectively involved in tannin toxicity (Vincenzo Lattanzio *et al.*, 2006). Finally, lignans, a phenolic class of dimeric phenylpropanoid units linked by the central carbons of their side chains, also play a role in plant-fungus interactions. Some of the fungistatic activity of lignans is attributable to their inhibition of the extracellular fungal enzymes, cellulase, polygalacturonase, glucosidase and laccase.

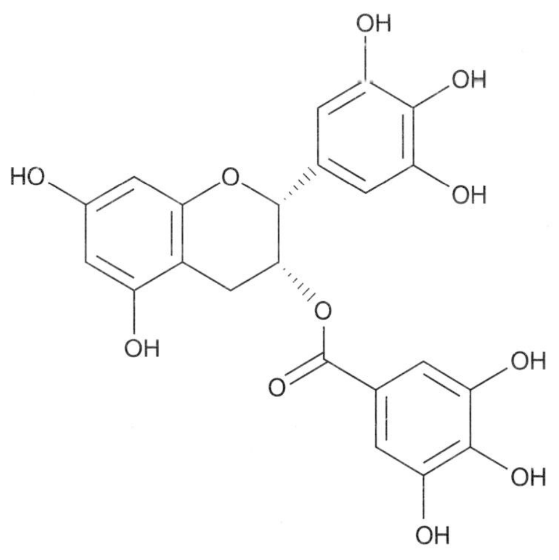

Fig. 3.28 Tannin

3.3.6.2 *Induced disease resistance*

When a pathogen manages to overcome constitutive defense barriers it may become subject to recognition at the plasma membrane of the plant cell. Activation of inducible plant defense responses is probably brought about by the recognition of invariant pathogen-associated molecular patterns (PAMP) that are characteristic of whole classes of microbial organisms: PAMP perception systems trigger signaling cascades whose recognition is very likely to activate defense responses in natural plant-pathogen encounters (Nürnberger and Lipka, 2005). Plants respond to pathogens by activating broad-spectrum innate immune responses that can be expressed locally at the site of pathogen invasion as well as systemically in the uninfected tissue. Locally induced defense responses, which restrict pathogen infections to the site of attempted ingress are characterized by a hypersensitive response, a complex, early defense response that causes necrosis and cell death to restrict the growth of a pathogen (Vincenzo Lattanzio *et al.*, 2006).

If preexisting antifungal phenolics are not sufficient to stop the development of the infectious process, plant cells usually respond by increasing the level of preexisting antifungal phenols at the infection site, after an elicited increased activity of the key enzymes (PAL and chalcone synthase) of the biosynthetic pathway (Yedidia *et al.*, 2003). The increased level of phenolics provides an adequate substrate to oxidative reactions catalyzed by PPO and/or POD that, consuming oxygen and producing fungi-toxic quinones, make the medium unfavorable to the further development of pathogens. Therefore, although many phenolic compounds present in plant tissues may not have any antimicrobial activity per se, there are several cases recorded where the oxidation products of preexisting phenolics might have antimicrobial activity, which is often associated with an inhibition of the cell wall degradation by extracellular enzymes produced by pathogens. In addition, many simple low molecular phenolic compounds present in plants may be readily polymerized by oxidation to yield brown tannin-like substances (melanins) containing quinonoid groups. These can also precipitate protein and cross-link to other polymers. They are often formed in necrotic cells after the invasion by a pathogen, as shown by the browning, which takes place in and around the area. Thus, oxidized phenolics in resistant varieties of apple play an important role in the restricted lesion formation associated with the brown rot disease of fruits caused by *Sclerotinia fructigena* or with rotting of stored 'Golden Delicious' apples caused by *Phlyctaena vagabunda* (Lattanzio *et al.*, 2003).

Postinfectional accumulation of preexisting phenolics, especially phloridzin and chlorogenic acid (Figs. 3.29 and 3.30) which are the best substrates of apple PPO (KM = 1.351 and 1.139 mM, respectively), provides an adequate substrate for the increased PPO activity (Vincenzo Lattanzio *et al.*, 2006). Polyphenol oxidase (Fig. 3.31), consuming oxygen and producing fungi-toxic quinones, makes the medium unfavorable to further development of pathogens. In fact, *in planta* bioassays showed that when a crude extract of apple PPO was added to a spore suspension of *P. vagabunda* that contains 10-3 M of each apple phenolics, an inhibition of fungal spore germination was observed.

These bioassays also showed a potential synergistic effect of phloridzin and chlorogenic acid, thus demonstrating that the simultaneous presence of chlorogenic acid and catechin (Fig. 3.32) in a model system increases the oxidation rate of phloridzin in the presence of polyphenol oxidase. This synergistic effect should probably be considered in the overall defensive strategy of apple against fungal attack. Finally, when in long stored apples these primary responses (increased phenolic level and activation of apple PPO in infected tissues), which counteract the further development of the pathogen, fall during storage, a simultaneous increase in apple rot was observed.

Antifungal activity of oxidized phenolics may also be related to the necrotic reaction, e.g., the oxidative polymerization involving phenolic compounds, amino acids, and

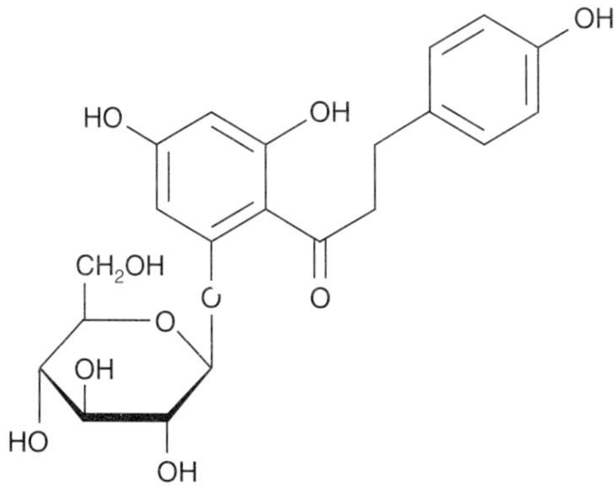

Fig. 3.29 Phloridzin

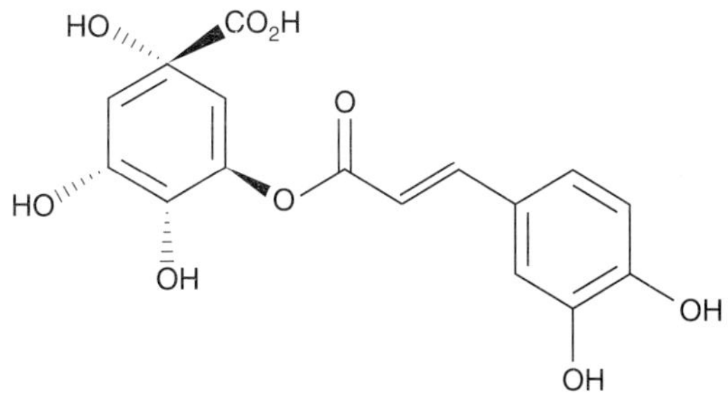

Fig. 3.30 Chlorogenic acid

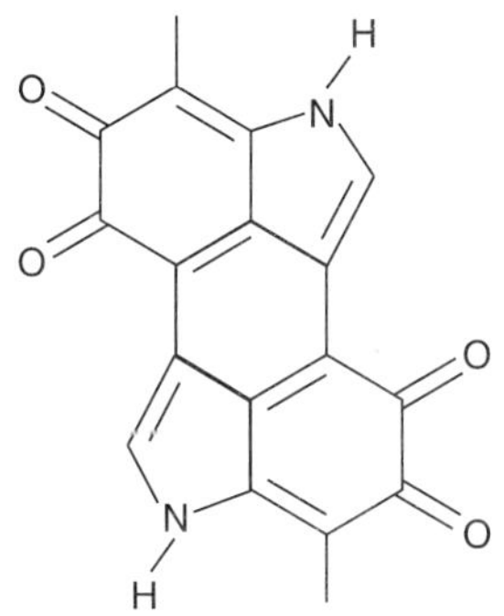

Fig. 3.31 Polyphenol oxidase

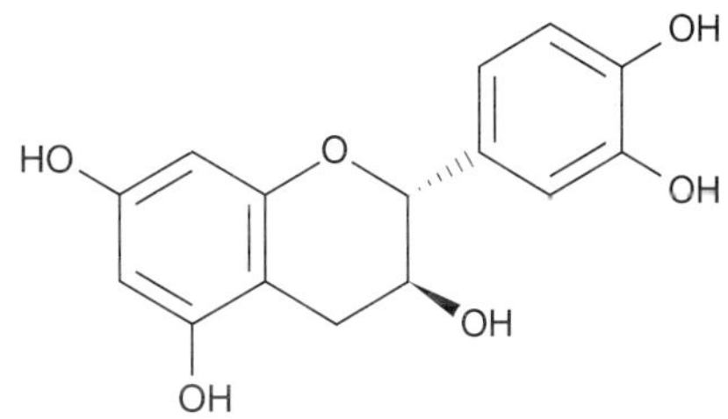

Fig. 3.32 Catechin

proteins that yield brown melanins. This reaction results in the formation of an impermeable barrier to pathogenesis by plant parasites and a decrease of nutrients essential to the fungal development (Vincenzo Lattanzio *et al.*, 2006). Secondary responses also include the release of toxic phenols that are normally stored as less toxic glycosides in the vacuoles of the plant cells, the formation of lignin – a biopolymer resistant to the degradation by most microorganisms, the accumulation of cell wall appositions such as papillae, and, finally, the synthesis of specific antibiotics such as phytoalexins (Nicholson and Hammerschmidt, 1992).

3.3.6.3 *Lignification*

Lignin is a three-dimensional phenolic structure resulting from the free radical polymerization of *p*-coumaryl, coniferyl, and sinapyl alcohols, within the plant cell wall, even if there are now many examples showing that other phenolics can be incorporated into lignins. As lignin polymerizes, it forms covalent crosslinks with carbohydrate and protein and renders cell walls highly resistant to mechanical and enzymatic disruption thus acting as a major line of defense against pathogens, insects, and other herbivores (Boerjan *et al.*, 2003). Deposition of lignin has been hypothesized to interfere with the enzymatic hydrolysis and mechanical penetration of plant tissue by fungal pathogens and may also impair the movement of water and diffusible molecules between the plant and fungus and then help to starve a pathogen. Lignin precursors themselves might exert a toxic effect on pathogens or, by binding to fungal cell walls, make them more rigid and impermeable, thus hindering further growth or uptake of water and nutrients. Thus, *in vitro* coniferyl alcohol is toxic to fungi at low concentration. Moreover; it has been proposed that the polymerization of lignin precursors by free radicals in the intercellular space might also lead to a lignification of pathogen structures: each of these possible events provides a testable hypothesis, and some of these mechanisms have been demonstrated. Pectic fragments have been identified as elicitors for the accumulation of histochemically detected lignin-like materials in cucumber hypocotyls and suspension cultures of castor bean (Vincenzo Lattanzio *et al.*, 2006). Other elicitors of lignin or lignin-like substances include chitosan (Fig. 3.33) extracts from fungal cell walls and fungal lipids.

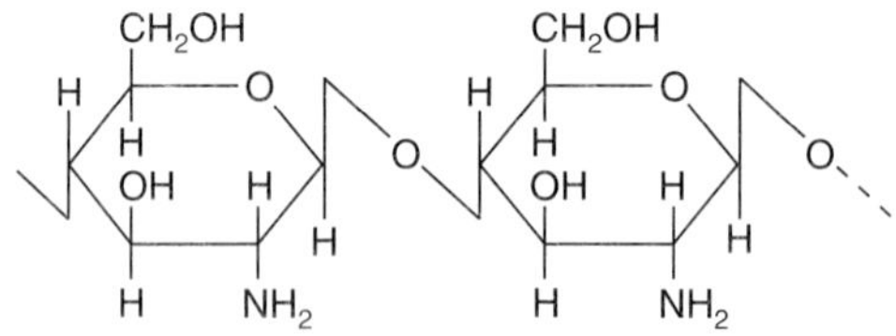

Fig. 3.33 Chitosan

3.3.6.4 Phytoalexins

Phytoalexins are antimicrobial, low-molecular-weight secondary metabolites that are both synthesized by and accumulated in plant cells as a result of the interaction between the metabolic systems of the host and a pathogen and that require *de novo* expression of the enzymes involved in their biosynthetic pathway (Paxton, 1980). They are broad spectrum inhibitors and are chemically diverse with different types characteristic of particular plant species. Phytoalexins tend to fall into several classes including terpenoids, glycsteroids, and alkaloids; however, researchers often find it convenient to extend the definition to include all phytochemicals that are part of the plant's defensive arsenal. Phytoalexins produced in plants act as toxins to the attacking organism. They may puncture the cell wall, delay maturation, disrupt metabolism or prevent the reproduction of the pathogen in question. Their importance in plant defense is indicated by an increase in susceptibility of plant tissue to infection when phytoalexin biosynthesis is inhibited. Mutants incapable of phytoalexin production exhibit more extensive pathogen colonization as compared to wild type. As such, host-specific pathogens capable of degrading phytoalexins are more virulent than those unable to do so (Glazebrook *et al.*, 1994).

The concept of phytoalexins that is antimicrobial substances of low-molecular-weight produced by plants in response to infection or stress was proposed as early as 1940 (Müller and Börger, 1940) in the study of the interaction between potato and *Phytophthora infestans*. Briefly, the inhibition of the development of a race of *P. infestans* which normally develops profusely on potato tubers following infection of the potato tubers with another race of *P. infestans* able to initiate the hypersensitive reaction was caused by a 'principle' produced by the hypersensitively-reacting plant cells. The principle was named a phytoalexin. The term phytoalexin was derived from Greek to mean warding-off agents in plants' and the proposal of this elegant concept was made after deliberating two important phenomena in plant pathology. The first one is the active response of the plant cell to attempted infection; the second one is the acquisition of resistance by plants after exposure to microorganisms (Philippe Jeandet *et al.*, 2013). Phytoalexins form part of the defense mechanisms of plants against phytopathogenic microorganisms. These are antibacterial, antifungal and antiviral compounds showing phytotoxic activities as well. It is now clear that phytoalexins exhibit toxicity across much of the biological spectrum, prokaryotic and eukaryotic, and that their activities are by no means confined to phytopathogenic fungi though considerable work has been done on plant/fungus interactions (Smith, 1982).

Phytoalexins are chemically diverse. Examples include simple phenylpropanoid derivatives, flavonoid- and isoflavonoid-derived phytoalexins, sesquiterpenes, and polyketides. Figure 3.34 includes few representative structures. Phytoalexins may be biosynthetically derived from one or several primary biosynthetic pathways. Polyketides such as 6-methoxymellein and sesquiterpenes such as capsidiol are derived from the acetate-malonate and acetate-mevalonate pathways, respectively (Ray Hammerschmidth, 1999).

Although the studies on phytoalexins diversity do not directly shed new light on the role of phytoalexins in defense, the research has provided us with a wealth of information on new chemistry as well as information on the distribution of phytoalexin types and how the spectrum of phytoalexin structures can change as plants are domesticated.

Phytoalexins have long been recognized to as being important in the defense mechanisms of plants against phytopathogenic microorganisms. Their potential biological properties have stimulated a ferment of activity concerning the biosynthesis and the metabolism of these compounds in the plant. Research has gone forward because these compounds are thought to help agriculturally and economically important crop plants withstand colonization by pathogens. In many instances, a close correlation has been found between phytoalexin production and resistance to diseases. Use of modern molecular biology tools for elucidating the control mechanisms of phytoalexin synthesis and for engineering disease-resistant plants is based on the expression of stress- or disease-related genes. Genetic engineering of phytoalexins for disease resistance requires manipulation of a single or a few genes directly involved in their biosynthetic pathways or involved in their signaling/regulatory pathways.

The phytoalexins produced by a range of crop plants from Brassicaceae, Fabaceae, Solanaceae, Vitaceae and Poaceae, which includes the very recently identified kauralexins and zealexins produced by maize and the biosynthesis and regulation of phytoalexins produced by rice (Ahuja *et al.*, 2012). The major phytoalexin in alfalfa is the pterocarpan medicarpin, synthesized from phenylalanine via the isoflavonoids pathway. The elicitation of medicarpin in alfalfa cell suspension cultures exposed to crude elicitor from cell walls of *Colletotrichum lindemuthianum* is preceded by increases in the activity of all

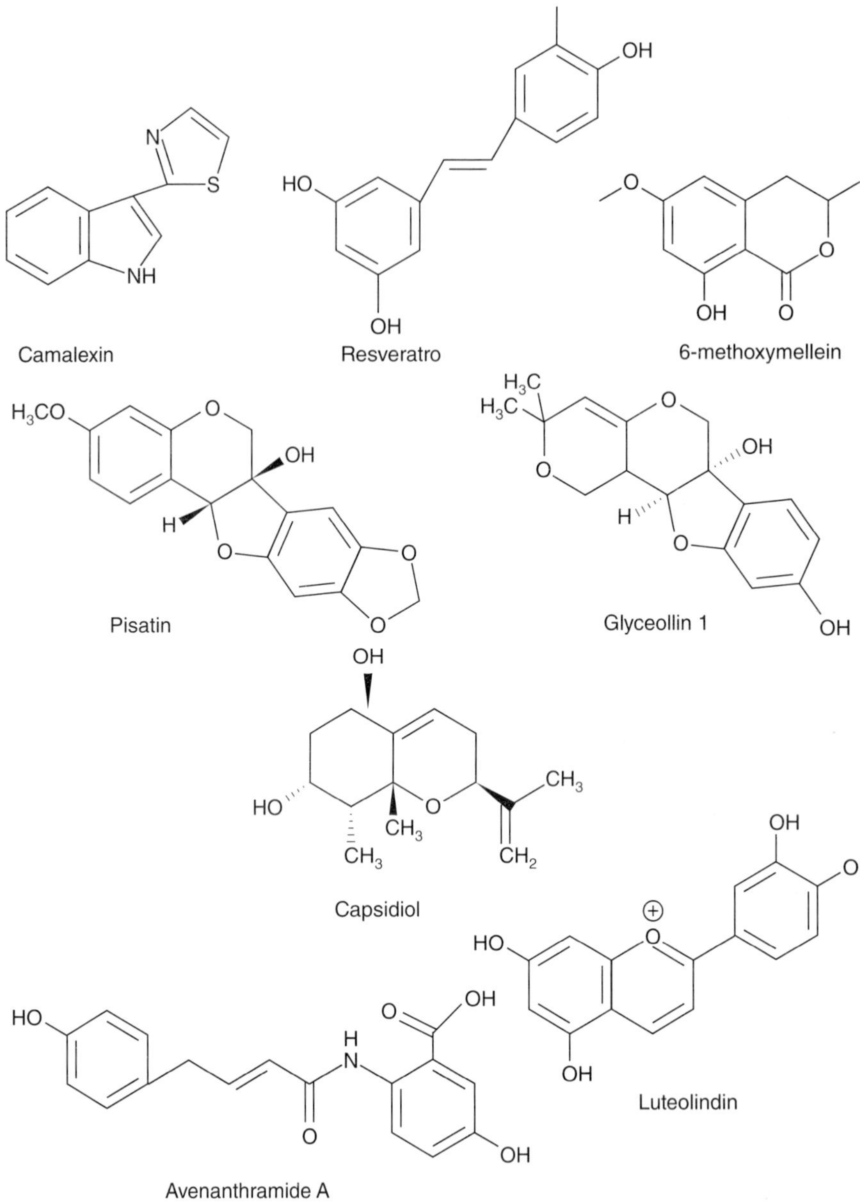

Fig. 3.34 Representative phytoalexin structures

11 enzymes required for its biosynthesis and their transcriptional activation. The genes encoding L-phenylalanine ammonia-lyase (PAL), chalcone synthase (CHS) and chalcone reductase (CHR) were the most rapidly activated, within 10-20 min after the elicitation of the isoflavonoids biosynthetic pathway (Salles *et al.*, 2002).

Phytoalexins are substances not detectable before infection and are considered to inhibit the further development of most attacking pathogens of the species under consideration. The molecules that signal plants to begin the process of phytoalexin synthesis are called elicitors. Of late, the term 'elicitor' refers to both biotic elicitors (such as complex carbohydrates from fungal and plant cell walls, and microbial enzymes) and abiotic elicitors (such as heavy metals, autoclaved ribonuclease, chloroform, detergents, freezing, heating, exposure to UV light). There is some evidence that cell wall fragments, active as elicitors, are released by a plant enzyme that is activated by damaging plant cells.

The main features that relate phytoalexin synthesis and metabolism include rapid accumulaiton of phytoalexins in response to infection to an antimicrobial level in resistant plants while there is either lesser or slower accumulation in susceptible plants. When the accumulation of phytoalexins is either increased or decreased by manipulation of the experimental conditions, the plant becomes either more resistant or more susceptible. Some pathogens can metabolize phytoalexins to less toxic products. In addition, some pathogens have developed mechanisms to suppress the defense response of the host plant, making the plant unable to perceive the pathogen entering it. This leads to a failure in the hypersensitive response that is often considered a prerequisite for induced disease resistance (Vincenzo Lattanzio *et al.*, 2006). However, the role of phytoalexins in stopping pathogen development in host tissues is far from clear for many host-pathogen interactions. Some very good information has been obtained from those systems where phytoalexin concentration at the site of infection has been quantified. Knowing when, where, and to what extent these antimicrobial compounds accumulate in response to infection is a critical step in determining if a phytoalexin is casually related to defense. In sorghum, taking advantage of the fact that the deoxyanthocyanidin phytoalexins produced in response to fungal infection are visible pigments, by utilizing adequate analytical techniques it has been observed that the timing of phytoalexin accumulation and the quantity of phytoalexins at the infection site is more than sufficient to stop the development of fungal infection in the invaded tissues (Nicholson and Wood, 2001).

The phytoalexins of different plants are generally unique structures. However, closely related plants often have phytoalexins more closely chemically related than plants phylogenetically more distant. In addition, it must be emphasized that plants not only synthesize phytoalexins, but plants also catabolize phytoalexins: this is advantageous because phytoalexins are toxic to the plants that synthesize them. The maximum levels of phytoalexins accumulated in a plant upon challenge by a microbe or elicitor are determined by the rate of catabolism as well as the rate of synthesis of the phytoalexins (Hammerschmidt, 1999). Clearly, not all plants produce phytoalexins. However, phytoalexins have been found in both Gymnospermae and Angiospermae, and within the latter group in both Monocotyledoneae and Dicotyledoneae. The structures of phytoalexins are often unique at the family level: the majority of phytoalexins produced by the members of the family Leguminosae are isoflavonoids (90 out of 113 identified structures), and the commonest isoflavonoid subclass are the pterocarpans such as medicarpin and glyceollin II (XII), whereas phytoalexins from Vitaceae seem to constitute a rather restricted group of molecules belonging to the stilbene family, the skeleton of which is based on the *trans*-resveratrol structure [(3,5,4'-trihydroxystilbene) (Jeandet *et al.*, 2008)]. Dicotyledonous species represent the majority of plants from which such compounds have been identified. Some monocotyledonous species producing phenolic phytoalexins are rice (sakuranetin (XIII)), oat (avenalumin I (XIV)), sugarcane (piceatannol (XV)), and sorghum (3-deoxyanthocyanidins) (Nicholson and Wood, 2001).

Sorghum (*Sorghum bicolor*) synthesizes a complex mixture of 3-deoxyanthocyanidin phytoalexins in response to inoculation with the nonpathogenic fungus *Bipolaris maydis*. The sorghum phytoalexins have been identified as 3-deoxyanthocyanidins: luteolinidin, 5-methoxyluteolinidin, apigeninidin, caffeic acid ester of arabinosyl 5-*O*-apigeninidin (XVI) and 7-methoxyapigeninidin. These compounds accumulate in inclusion bodies in epidermal cells that migrate towards the site of fungal penetration. The inclusions eventually release their contents, killing both the fungus and the cell that synthesized them (Wharton and Nicholson, 2000). Research on the silicon-mediated accumulation of flavonoid phytoalexins in cucumber revealed that silicon was involved in the increased resistance of cucumber to powdery mildew by enhancing the antifungal activity in infected leaves (Fawe *et al.*, 1998). This antifungal activity was attributable to the presence of low-molecular-weight metabolites. One of these metabolites, described here as a phytoalexin, was identified as a flavonol aglycone rhamnetin (3,5,3,'4'-tetrahydroxy-7-*O*-methoxyflavone). This is the first report of a phytoalexin for this chemical group in the plant kingdom and of a flavonol phytoalexin in cucumber, a chemical defense long believed to be nonexistent in the family Cucurbitaceae.

Some evidence that phytoalexins are effective in providing plants with disease resistance has been obtained by transferring foreign phytoalexins expression from one plant to another, and such approach has been undertaken in the case of stilbenes and isoflavonoids (Dixon and Steele, 1999). Stilbene phytoalexins, which include resveratrol, require only a single enzyme, stilbenes synthase, to link the two universally available precursors, malonyl-CoA and *p*-coumaroyl-CoA, for their synthesis. Pathogen resistance has thus been engineered into alfalfa by the transfer of a peanut cDNA encoding resveratrol synthase. Such plants constitutively accumulate a glucoside of the novel phytoalexins resveratrol (*trans*-resveratrol-3-*O*-β-D-glucopyranoside) and exhibit vastly reduced lesions following infection by the leaf spot pathogen *Phoma medicaginis*. Similarly, the constitutive overexpression of isoflavone O-methyltransferase in transgenic alfalfa resulted in a more rapid and increased production of the phytoalexins formononetin (XVII) and medicarpin after infection by *P. medicaginis*, with a consequent amelioration of the symptoms (He and Dixon, 2000). However, it is not so easy to do the same for many other phytoalexins. For instance, for nonleguminous plants many new enzymes would be required to produce isoflavonoid phytoalexins so that many genes would have to be transferred. Therefore, an alternative way of using the knowledge of phytoalexin production in plants is to stimulate the plants, even susceptible genotypes, to produce their own phytoalexins rapidly and in large quantity by inoculation with nonpathogenic microorganisms or with an incompatible strain of pathogen. For example, systemic resistance to *Pyricularia oryzae* in rice has been induced by inoculation with *Pseudomonas syringae* pv. *syringae*, which causes a hypersensitive response in rice (Grayer and Kokubun, 2001).

Likewise, phytoalexin (glyceollin I, II and III, and coumestrol) induction and accumulation in soybean cotyledon tissue was observed using four species of *Aspergillus*:

A. sojae, A. oryzae, A. niger and *A. flavus*. All the four *Aspergillus* species tested elicited phytoalexin accumulation in living soybean cotyledons. Results from a time course study indicated that maximum concentrations of the phytoalexin glyceollin, 955 μg/g fresh weight (fw), occurred at day 3 in soybean cotyledon tissue inoculated with *A. sojae*. Other *Aspergillus* species caused an accumulation of glyceollin at significantly lower levels. A maximum concentration of coumestrol of 27.2 μg/g fw was obtained from soybean cotyledons inoculated with *A. niger* (Boue´ *et al*., 2000).

Philippe Jeandet *et al.* (2013) reviewed the efforts that have been done for engineering phytoalexin production in plants as well as provide information on how phytoalexin accumulation can be modulated through engineering of plant hormones, defense-related markers or elicitors. They stress upon the exploitation of new techniques for metabolic engineering in the next years to come. Methodologies for generating high-quality libraries of enzyme variants and novel high-throughput screening (HTS) technologies will open the way for the engineering of enzymes for the biosynthesis of various phytoalexins with potent biological activities. Specifically, HTS technologies can rapidly lead to the identification of genes which modulate a particular biosynthesis pathway. Gathering all genes encoding for a biomolecular pathway will allow the assembly of genetic constructs for the synthesis of a given phytoalexin.

3.3.6.5 *Flavonoids*

These are one of the largest classes of phenolics. Flavonoids are a major class of oxygen-containing heterocyclic natural products that are widespread in green plants. Generally, they are found as plant pigments in a broad range of fruits and vegetables. These are C15 compounds composed of two aromatic rings linked by a three-carbon bridge with a carbonyl functional group located at one end of the bridge (Bohm, 1998). The basic structural feature of flavonoid compounds is the 2-phenyl-benzopyrane or flavane nucleus, which consists of two benzene rings (A and B), linked through a heterocyclic pyrane ring (C) (Fig. 3.35).

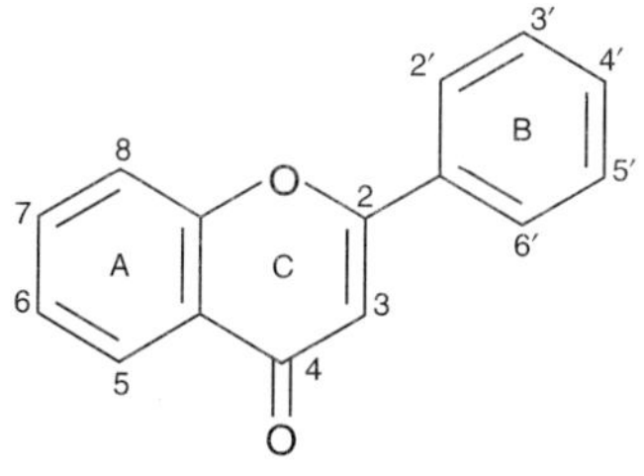

Fig. 3.35 The skeleton structure of the flavones

Flavonoids have been recognized as having a protective effect in plants against microbial invasion by plant pathogens. Flavonoid-rich plant extracts have been used for centuries to treat human disease. Isolated flavonoids have been shown to possess a host of important biological activities, including antifungal and antibacterial activities. The potential of naturally occurring flavonoids as antiinfective agents has been recognized (Qudsia Kanwal *et al*., 2009).

Flavonoids are ubiquitous in photosynthesizing cells and are commonly found in fruit, vegetables, nuts, seeds, stems, flowers, tea, wine, propolis, and honey. Increasingly, this class of natural products is becoming the subject of antiinfective research, and many groups have isolated and identified the structures of flavonoids possessing antifungal, antiviral and antibacterial activity (Tim Cushnie and Andrew, 2005). Moreover, several groups have demonstrated synergy between active flavonoids as well as between flavonoids and existing chemotherapeutics. Reports of activity in the field of antibacterial flavonoid research are widely conflicting, probably owing to inter- and intra-assay variation in susceptibility testing. However, several high-quality investigations have examined the relationship between flavonoid structure and antibacterial activity, and these are in close agreement. In addition, numerous research groups have sought to elucidate the antibacterial mechanisms of action of selected flavonoids. The activity of quercetin, for example, has been at least partially attributed to inhibition of DNA gyrase. It has also been proposed that sophoraflavone G and (−)-epigallocatechin gallate (Figs. 3.36 and 3.37) inhibit cytoplasmic membrane function and that licochalcones A and C inhibit energy metabolism.

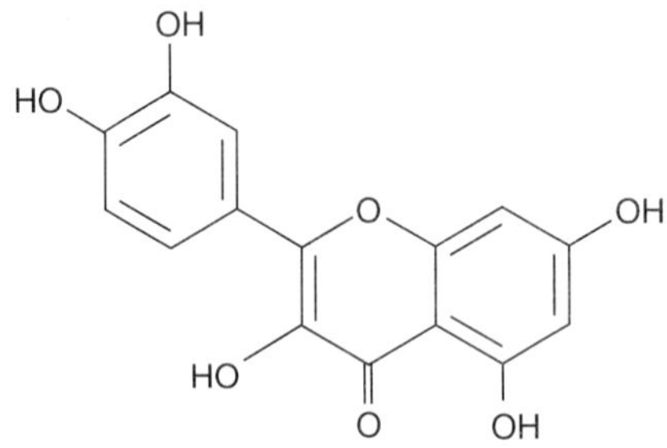

Fig. 3.36 Quercetin

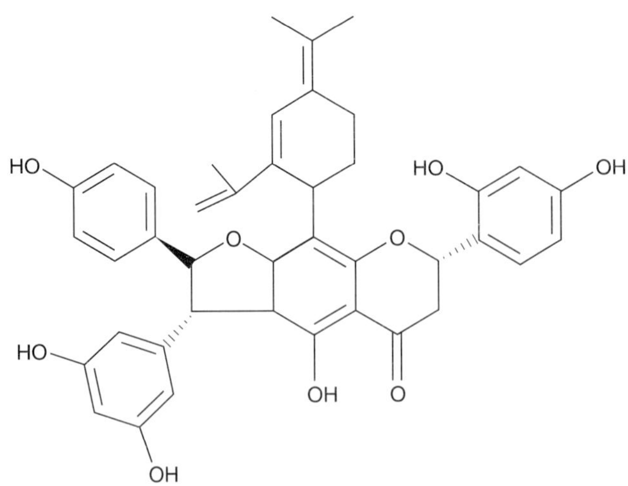

Fig. 3.37 Sophoraflavone G

Other flavonoids whose mechanisms of action have been investigated include robinetin (Figs. 3.38 and 3.39), myricetin, apigenin, rutin, galangin, 2,4,2-trihydroxy-5-methylchalcone and lonchocarpol A. These compounds represent novel leads, and future studies may allow the development of a pharmacologically acceptable antimicrobial agent or class of agents. Flavonoids may inhibit *nucleic* acid synthesis, cytoplasmic membrane function, and energy metabolism.

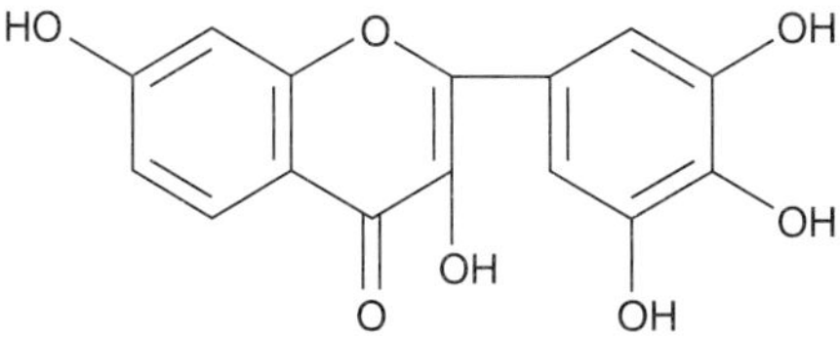

Fig. 3.38 Myricetin

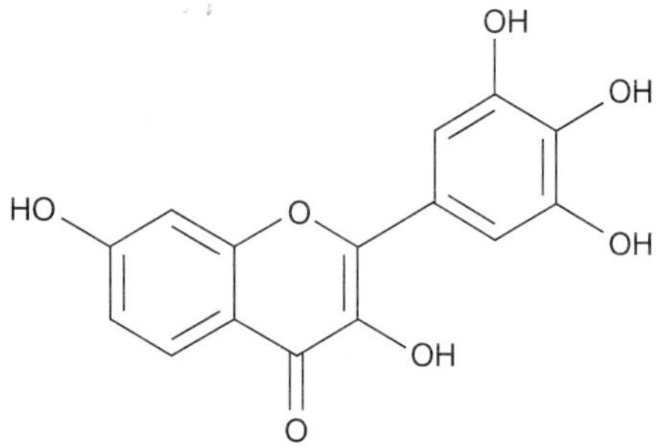

Fig. 3.39 Robinetin

3.3.6.6 Major antimicrobial flavonoids

3.3.6.6.1 Flavones and their glycosides

Acacetin, Apigenin, Baicalin, Baicalein Chrysin, Gardenin A, (demethylated), Genkwanin, Luteolin, Luteolin 7-glucoside, 7,8-Dihydroxyflavone, 5,5-Dihydroxy-8,2,4-trimethoxyflavone, 5-Hydroxy-7,4-dimethoxyflavone, 5,7,4-Trihydroxy-3,5-dimethoxyflavone, 6,7,4-Trihydroxy-3,5-dimethoxyflavone, Galangin, Kaempherol, 3-*O*-methylquercetin, Morin, Myricetin, Quercetagetin, Quercetagetin-7-arabinosyl-galactoside, Quercetin, Quercetin-3-*O*-(2-galloyl)-larabinopyranoside, Quercetrin, Robinetin, Rutin, 3,2-Dihydroxyflavone and 3,6,7,3,4-Pentahydroxyflavone

3.3.6.6.2 Flavan-3-ols

Catechin, Epicatechin gallate, Epigallocatechin, Epigallocatechin gallate, 3-*O*-octanoyl-(+)-catechin and 3-*O*-octanoyl-(−)-epicatechin

3.3.6.6.3 Flavanon3-ols

Dihydrofisetin and Dihydroquercetin

3.3.6.6.4 Isoflavones

6,8-Diprenylgenistein and Sophoraisoflavone A

3.3.6.6.5 Flavanones and their glycosides

Lonchocarpol, Naringenin, Naringin, Pinocembrin, Ponciretin, Sophoraflavanone, 3-Methyleneflavanone and 5,7,4-Trihydroxy-8-methyl-6-(3-methyl-[2-butenyl])-(2*S*)-flavanone

3.3.6.6.6 Chalcones

Licochalcone, Licochalcone, 2,4,2-Trihydroxychalcone and 2,4,2-Trihydroxy-5-methylchalcone

3.3.6.6.7 Flavan-3,4-diols and anthocyanidins

Leucocyanidin and Pelargonidin chloride

3.3.6.6.8 Flavans

6,4-Dichloroflavan and 7-Hydroxy-3,4-(methylenedioxy)flavan

Existing structure–activity data suggest that it might be possible, for example, to prepare a potent antibacterial flavanone by synthesizing a compound with halogenation of the B ring as well as lavandulyl or geranyl substitution of the A ring. Also, it is worth noting that the rapid progress which is being made toward the elucidation of flavonoid biosynthetic pathways (Dixon *et al.*, 1998) may soon allow the production of structural analogs of active flavonoids through genetic manipulation. Screening of these analogs might lead to the identification of compounds that are sufficiently potent to be useful as antifungal, antiviral or antibacterial chemotherapeutics. In addition to the structural alteration of weak and moderately active antimicrobial flavonoids, an investigation into the mechanisms of action of these compounds is likely to be a productive area of research. Such information may assist in the optimization of a lead compound's activity, provide a focus for toxicological attention and aid in the anticipation of resistance. Also, characterization of the interaction between antimicrobial flavonoids and their target sites could potentially allow the design of second generation inhibitors.

Five flavonoids, viz.,(–)-epicatechin-3-*O*-β-glucopyranoside (1), 5-hydroxy-3-(4-hydroxylphenyl) pyrano[3,2-*g*]chromene-4(8*H*)-one (2), 6-(*p*-hydroxybenzyl) taxifolin-7-*O*-β-D-glucoside (tricuspid) (3), quercetin-3-*O*-α-glucopyranosyl-(1→2)-β-glucopyranoside (4) and (–)-epicatechin (2-(3,4-dihydroxyphenyl)-3,4-dihydro-2*H*-chromene-3,5,7-triol) (5), were isolated from the leaves of mango [(*Mangifera indica* L.) (Qudsia Kanwal *et al.*, 2009)]. The antibacterial activity of different concentrations of these flavonoids (100, 300, 500, 700, 900 and 1000 ppm) was evaluated against four bacterial species, namely, *Lactobacillus* sp., *Escherichia coli*, *Azospirillium lipoferum* and *Bacillus* sp. All the tested concentrations of the five flavonoids significantly reduced the growth of all the five tested bacterial species. However, differences in the antibacterial activity of the flavonoids were evident. Compound 1 exhibited the lowest antibacterial activity, resulting in a 7–75% reduction in the growth of the different bacterial species. Compound 5 showed the greatest antibacterial activity and the different concentrations reduced the bacterial growth by 45–99.9%. *A. lipoferum* and *Bacillus* sp. showed the highest susceptibility to this compound. Compounds 2–4 also depicted pronounced antibacterial activity. Different concentrations of these compounds decreased bacterial growth by 52–96%. From the present study, it can be concluded that Compound 5 is the most effective of the tested flavonoids against *A. lipoferum* and *Bacillus* sp.

3.3.6.7 *Anthocyanins*

These are colorful water-soluble flavonoids pigments produced by plants to protect foliage from the damaging effects of ultraviolet radiation. Anthocyanins are responsible for the showy colors of many plants and are present in high concentrations in flowers, fruits, and the leaves of deciduous plants in fall. Anthocyanins are plant secondary metabolites synthesized by the flavonoid pathway, and play important roles in flower pigmentation, allelopathy, or UV protection (Himeno Misako *et al.*, 2013). They are also important as antioxidant molecules to protect plant cells against damage by reactive oxygen species. Indeed, the production of anthocyanins in autumn foliage reduces the risk of photo-oxidative damage and delays leaf senescence. In the plants, anthocyanidins can be classified into six categories according to the number and position of hydroxyl and methoxyl groups on the flavan nucleus, including pelargonidin, cyanidin, delphinidin, peonidin (Figs. 3.40—3.43), petunidin, and malvidin (Tahara, 2007).

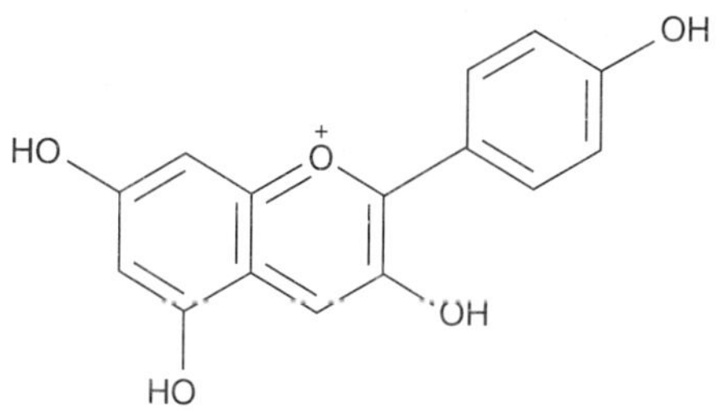

Fig. 3.40 Pelargonidin

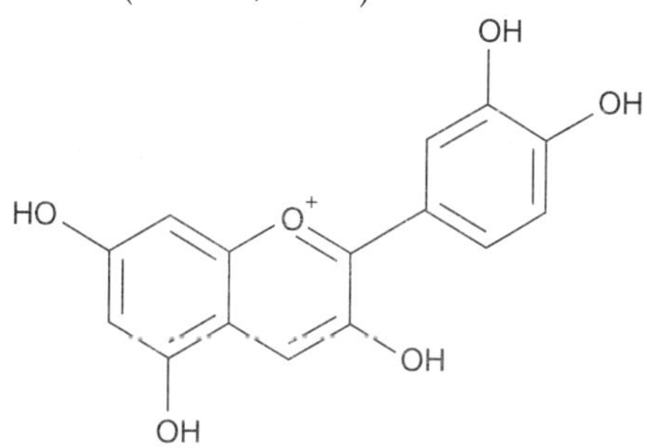

Fig. 3.41 Cyanidin

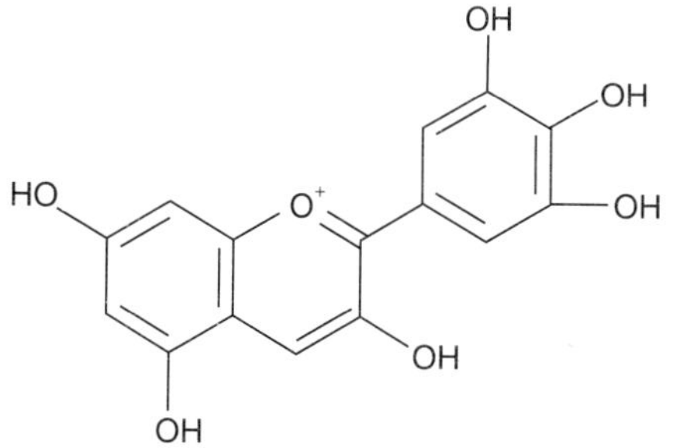

Fig. 3.42 Delphinidin

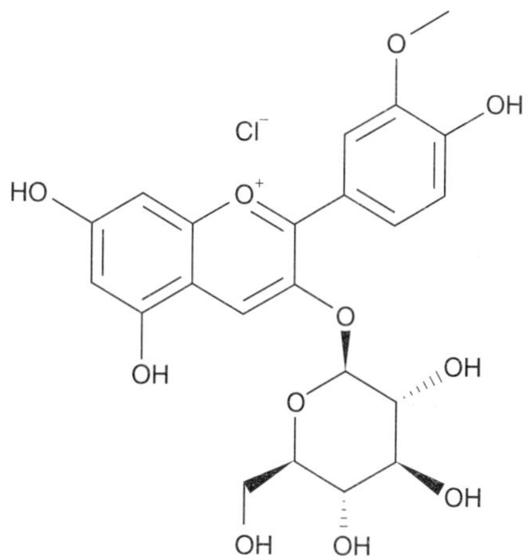

Fig. 3.43 Peonidin

Anthocyanin biosynthesis genes have been identified in several plant species including *Arabidopsis thaliana* and *Petunia hybrida*. The biosynthetic pathway for anthocyanin production can be divided into upstream and downstream sections. In general, most plant symptoms are recognized as harmful effects on host plants, and little is known about positive aspects of symptoms for infected plants. Himeno Misako *et al.* (2013) reported the beneficial role of purple top symptoms, which were characteristic of phytoplasma-infected plants. First, by using plant mutants defective in anthocyanin biosynthesis, they demonstrated that anthocyanin accumulation was directly responsible for the purple top symptoms and was associated with a reduction of leaf cell death caused by phytoplasma infection. Furthermore, it was revealed that phytoplasma infection led to significant activation of the anthocyanin biosynthetic pathway and a dramatic accumulation of sucrose by about 1000-fold, which could activate the anthocyanin biosynthetic pathway.

As the most visible role of anthocyanins is to impart colors, the adaptive significance of anthocyanins in plant reproductive organs is invariably attributed to the attraction of seed dispersers and pollinators. However, few studies suggested that anthocyanins in vegetative tissue might also function in plant defense although the evidence for such a function is not particularly strong (Martin Schaefera *et al.*, 2008). It was investigated whether anthocyanins contributed to fruit defense against fruit-rot fungi and was observed that the risk of fruit-rot in grape varieties infected with *Botrytis cinerea* decreased with increasing anthocyanin contents. Moreover, anthocyanin contents directly inhibited growth rates of nine fruit-rot fungi on agar plates. Anthocyanins reduced fungal growth by 50% in the concentrations that typically characterize unripe blackberries and by 95% in the concentrations that typify ripe blackberries. It was concluded that anthocyanins in fruits not the only function to attract animal vectors but that they also contribute to fruit defense. The antifungal activity of anthocyanins might more widely explain their occurrence in various plant organs such as vegetative tissue and roots.

Hipskind *et al.* (1996) showed that a final stage of resistance expression involves the accumulation of a pigment in uninfected, healthy epidermal cells that surround restricted lesions on leaves of the resistant cultivar. The pigment was characterized by plasma desorption mass spectrometry and found to have a structure consistent with that of cyanidin 3-dimalonyl glucoside. The function of this zwitterionic anthocyanin was proposed to involve the protection of uninfected, healthy plant tissue from the toxic, oxidative metabolites that accumulate during the expression of resistance. Microscopic examination of leaf tissue showed that the pathogen ceased to grow in the resistant cultivar at 18 h postinoculation, a time considerably prior to pigment accumulation. Thus, the pigment accumulates in cells that are affected but not infected by the pathogen.

3.3.6.8 Tannins

Tannins are naturally occurring complex plant polyphenols. Their main characteristic is that they bind and precipitate proteins (Fig. 3.44). They can have a large influence on the nutritive value of many foods eaten by humans and feedstuff eaten by animals. Tannins are common in fruits (grapes, persimmon, blueberry, etc.), in tea, in chocolate, in legume forages (trefoil, etc.), in legume trees (*Acacia* spp., *Sesbania* spp., etc.), in grasses (sorghum, corn, etc.). Tannins are widely distributed in the plant kingdom. They are common both in gymnosperms and angiosperms. Within angiosperms, tannins are more common in dicotyledons than in monocotyledons. Tannins are located mainly in the vacuoles or surface wax of the plants. In these sites, they do not interfere with plant metabolism. Only after cell breakdown and death can they act and have metabolic effects. Tannins can be obtained from almost any kind of green plants, plants low level and high level it with the content and quality varies. The effectiveness of antibacterial compounds found in plant tannins, for example, guava leaves is influenced by the concentration of tannins. The higher levels of tannin antibacterial activity will increase.

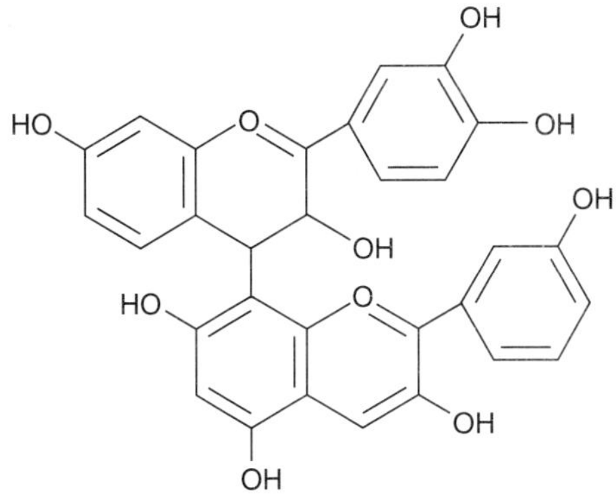

Fig. 3.44 Tannin

Preliminary studies on the antibacterial activity of guava leaf has been implemented by Faharani (2008), who observed that the water extract of guava leaves have antibacterial activity against *E. coli* at a concentration of 40%, whereas in *S. aereus* extract showed no inhibitory activity, and the active compound is believed to have antibacterial activity of flavonoids, tannins, and saponins. Guava leaves contain tannin by 9%, which can be used as an antibacterial. Tannins can be used as an antibacterial because it has a phenol group so that the tannins have properties like alcohol and is antiseptic that can be used as an antimicrobial component.

Meigy Nelce Mailoa *et al.* (2014) conducted a study to determine the antimicrobial activities of tannin extracts from guava leaves against pathogens microbial. The method used for qualitative analysis of the tannins was formed by the intensity of the color, blackish green FeCl3 compounds. Quantitative analysis of tannins used variations of organic solvent (ethanol with a concentration of 30%, 50%, and 70%). Levels of tannins in the sample solution were calculated with Tannates Acid Equivalent (TAE). Test of antimicrobial activities using diffusion agar method. The results showed levels of tannins in leaves of guava with ethanol 30%, which is 2.351 mg/g, ethanol 50% which is 1.728 mg/g, ethanol 70% is 1.835 mg/g. The results showed that the tannin extracts can inhibit the growth of *Escherichia coli, Pseudomonas aureginosa, Staphilococcus aureus, Aspergillus niger and Candida albicans.*

Three strains of pathogenic bacteria were treated with condensed tannins (CT) purified from eight different woody plant species to investigate their inhibition effect on the growth of these bacteria *in vitro* (Min *et al.*, 2008). *Escherichia coli, Klebsiella pneumoniae,* and *Staphylococcus aureus* were tested against low (0,2, 4 and 8 mg CT/ml) and high dose levels (0, 50, and 100 mg CT/ml) of CT extracted from different plant species. When exposed to purified tannin extracts at 4 mg extract/mL dosage, growth inhibition of *S. aureus* was dose dependent manner and observed in the following order: Shinnery oak

> Post oak >Locust > Blackjack oak _ Skunk bush > Sericea lespedeza > commercial Quebracho _ Sumac > Plum. The extracts from Shinnery and Post oaks were particularly inhibitory against *S. aureus*, having growth inhibition zones exceeding 23 mm at 8 mg tannin extract/mL. *S. aureus* and *E. coli* exhibited dose dependency and susceptibility ($P < 0.01$) when exposed to 4 mg/mL of the following tannin monomers which exhibited differential inhibitory activity: catechin > ellagetannin - tannic acid - epi-catechin -gallotannin. In the presence of high dose levels at 0, 50, and 100 mg tannin extract/mL, inhibition zones of growth were varied among plant species. The findings indicated that source and concentration are important factors that influence the antimicrobial activity of tannins. Because some of the plant tannin extracts are highly inhibitory to selected pathogens, they may provide alternatives and supplements to conventional antimicrobial feed additives.

In a survey of nonhost plant secondary metabolites regulating motility and viability of zoospores of the *Aphanomyces cochlioides*, Md. Tofazzal Islam *et al.* (2002) found that stem bark extracts of *Lannea coromandelica* remarkably inhibited the motility of zoospores followed by lysis. Bioassay-guided fractionation and chemical characterization of *Lannea* extracts by MALDI-TOF-MS revealed that the active constituents were angular type polyflavonoid tannins. Commercial polyflavonoid tannins, Quebracho and Mimosa, also showed identical zoosporicidal activity. Against zoospores, the motility-inhibiting and lytic activities were more pronounced in *Lannea* extracts (MIC 0.1 micro gm /mL) than in Quebracho (MIC 0.5 micro gm/mL) and Mimosa (MIC 0.5 micro gm/mL). Scanning electron microscopic observation visualized that both *Lannea* and commercial tannins caused lysis of cell membrane followed by fragmentation of cellular materials. Naturally occurring polyflavonoid tannin merits further study as potential zoospore regulating agent or as a lead compound. They further concluded that polyflavonoid tannins from *L. coromandelica* and commercial sources have motility-inhibitory and unique lytic effects *in planta* against the zoospores of a phytopathogenic oomycete, *A. cochlioides*. Plant polyflavonoid tannins could be useful for managing the infestation of crops by the notorious oomycete phytopathogens.

3.3.7 Furanocoumarins

The furanocoumarins, or furocoumarins, are a class of organic chemical compounds produced by a variety of plants in response to pathogen or herbivore attack. They are activated by ultraviolet light and can be highly toxic to certain vertebrate and invertebrate herbivores due to their integration into DNA, which contributes to rapid cell death. In fact, grapefruit juice contains small quantities of furanocoumarins, which greatly increase the absorption of certain drugs into the bloodstream from the intestines. Some medicines carry warning labels cautioning people to avoid drinking grapefruit juice while taking the drugs to avoid an accidental overdose. They are biosynthesized partly through the phenylpropanoid pathway and the mevalonate pathway, which is biosynthesized by a coupling of dimethyl pyrophosphate (DMAPP) and 7-hydroxycumarin (umbelliferone). The chemical structure of furanocoumarins consists of a furan ring fused with coumarin. The furan may be fused in different ways producing several isomers. The compounds that form the core structure of the two most common isomers are psoralen and angelicin.

Derivatives of these two core structures are referred to respectively as linear and angular furanocoumarins. Many furanocoumarins are toxic and are produced by plants as a defense mechanism against various types of pathogens.

Sixty-two strains of *Gibberella pulicaris* (anamorph: *Fusarium sambucinum)* from diseased plants and soil were tested for tolerance of the furanocoumarin xanthotoxin *in vitro* (Anne *et al.*, 1989). Twenty-one (88%) of the plant-derived strains and two (5%) of the soil-derived strains were highly tolerant of xanthotoxin. Sixteen selected strains were tested further against 16 furanocoumarins or furanocoumarin precursors. All plant-derived strains tested were highly tolerant of and, in most cases, able to completely metabolize all 16 compounds. Most soil-derived strains tested were tolerant of furanocoumarin precursors but sensitive to certain furanocoumarins. Linear compounds methoxylated at C-8 appeared more toxic than both those unsubstituted and those with longer-chain ethers. Tolerance of psoralen, angelicin, xanthotoxin, pimpinellin and isopimpinellin (Figs. 3.45 and 3.46) correlated in large part with their metabolism. All strains that were highly virulent on *Pastinaca sativa* root were tolerant of xanthotoxin, which is corroboration that xanthotoxin is a phytoalexin in *P. sativa.*

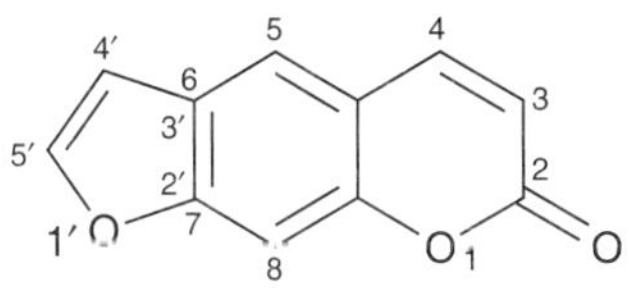

Fig. 3.45 Psoralen

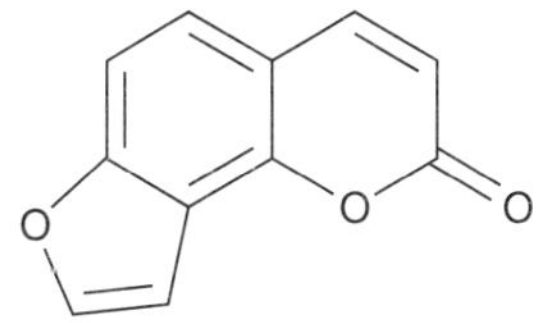

Fig. 3.46 Angelicin

Furanocoumarins constitute a subfamily of coumarin compounds with important defense properties against pathogens and insects, as well as allelopathic functions in plants. Furanocoumarins are divided into two subgroups according to the alignment of the furan ring with the lactone structure: linear psoralen and angular angelicin derivatives. Determination of furanocoumarin type is based on the prenylation position of the common precursor of all furanocoumarins, umbelliferone, at C6 or C8, which gives rise to the psoralen or angelicin derivatives, respectively. Karamat *et al.* (2014) identified a membrane-bound prenyltransferase PcPT from parsley (*Petroselinum crispum*) and characterized the properties of the gene product. PcPT expression in various parsley tissues is increased by UV irradiation, with a concomitant increase in furanocoumarin production. This enzyme has strict substrate specificity towards umbelliferone and dimethylallyl diphosphate, and a strong preference for the C6 position of the prenylated product (demethylsuberosin), leading to linear furanocoumarins. The C8-prenylated derivative (osthenol) was also formed, but to a much lesser extent. The PcPT protein was targeted to the plastids *in planta*. The introduction of this PcPT into the coumarin-producing plant *Ruta graveolens* showed increased consumption of endogenous umbelliferone. Expression of PcPT and a 4-coumaroyl CoA 2′-hydroxylase gene in *Nicotiana benthamiana,* which does not produce furanocoumarins, resulted in the formation of demethylsuberosin, indicating that furanocoumarin production may be reconstructed by a metabolic engineering approach. The results demonstrated that a single prenyltransferase, such as PcPT, opened the pathway to linear furanocoumarins in parsley, but might also catalyze the synthesis of osthenol, the first intermediate committed to the angular furanocoumarin pathway, in other plants.

3.3.8 Nitrogen Compounds

A large variety of plant secondary metabolites have nitrogen as part of their structure. Most nitrogenous secondary metabolites are synthesized from common amino acids. Nitrogen compounds are very important defense mechanisms in plants. They are mostly divided into alkaloids, cyanogenic glocosides, and glucosinolates.

3.3.8.1 Alkaloids

These are a large class of bitter-tasting nitrogenous compounds that are found in many vascular plants and include caffeine, cocaine, morphine, and nicotine. They are derived from the amino acids aspartate, lysine, tyrosine, and tryptophan, and many of these substances have powerful effects on animal physiology. They are a group of naturally occurring chemical compounds that mostly contain basic nitrogen atoms. This group also includes some related compounds with neutral and even weakly acidic properties. Some synthetic compounds of similar structure are also attributed to alkaloids. In addition to carbon, hydrogen and nitrogen, alkaloids may also contain oxygen, sulfur and more rarely other elements such as chlorine, bromine, and phosphorus. True alkaloids contain nitrogen in the heterocycle and originate from amino acids (Table 3.2). Their characteristic examples are atropine, caffeine, nicotine, theobromine, morphine and capsaicin (Figs. 3.47—3.51).

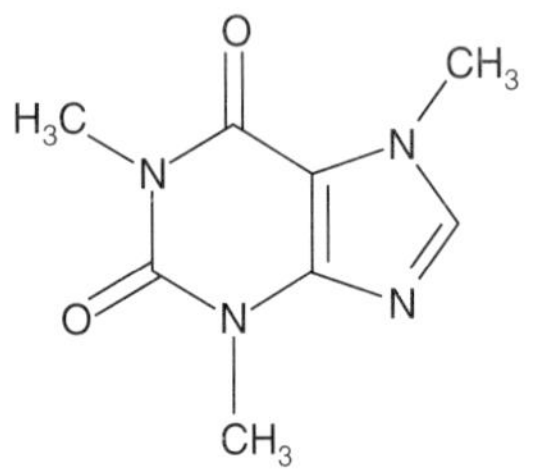

Fig. 3.47 Caffeine

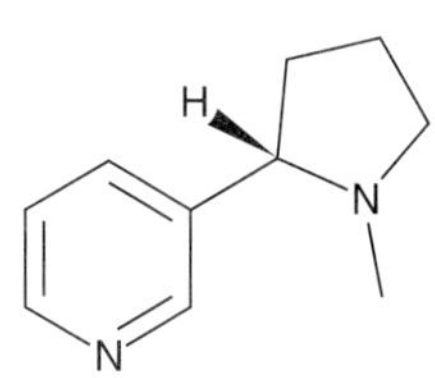

Fig. 3.48 Nicotine

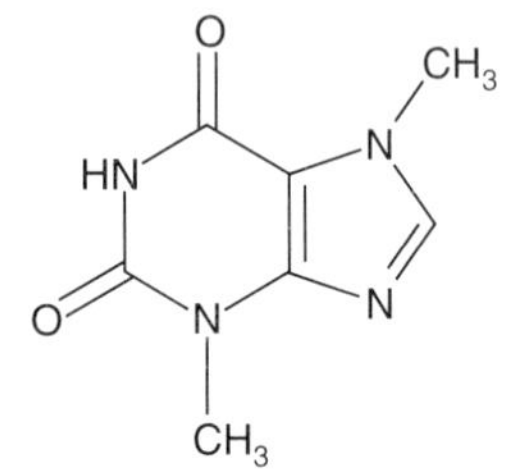

Fig. 3.49 Theobromine

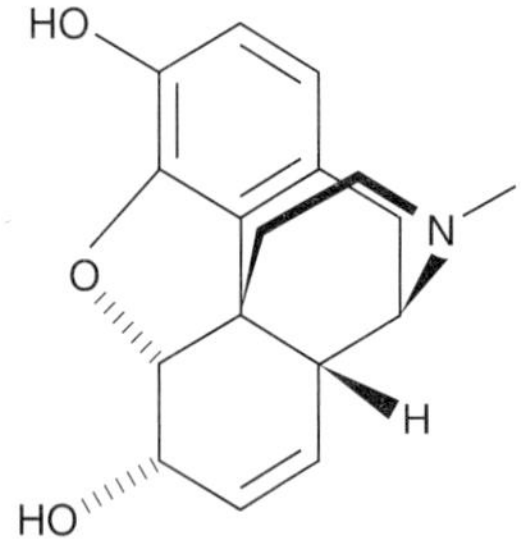

Fig. 3.50 Morphine

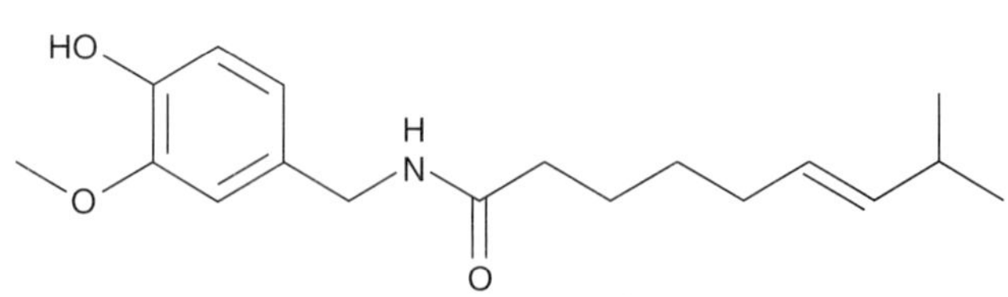

Fig. 3.51 Capsaicin

Members of the nightshade family (*Solanaceae*) produce many important alkaloid compounds. Caffeine is an alkaloid found in plants such as coffee (*Coffea arabica*), tea (*Camellia sinensis*) and cocoa (*Theobroma cacao*). It is toxic to both insects and fungi. In fact, high levels of caffeine produced by coffee seedlings can even inhibit the germination of other seeds in the vicinity of the growing plants, a phenomenon called allelopathy. Allelopathy allows one plant species to 'defend' itself against other plants that may compete for growing space and nutrient resources. Nicotine is an alkaloid that is produced in the roots of tobacco plants (*Nicotiana tabacum*) and transported to leaves where it is stored in vacuoles. It is released when herbivores graze on the leaves and break open the vacuoles.

Table 3.2 Major Types of Alkaloids, Their Amino Acid Precursors with Common Examples

Alkaloid class	*Structure*	*Biosynthetic precursor*	*Examples*
Pyrrolidine	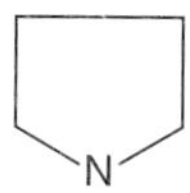	Ornithine(aspartate	Nicotine
Tropane	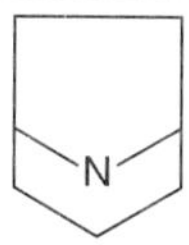	Ornithine	Atropine Cocaine
Piporidine	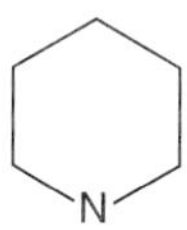	Lysine (or acetate)	Coniine
Pyrrolizidine	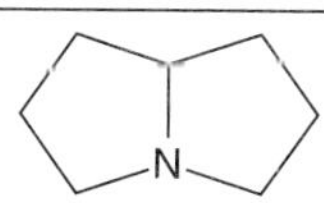	Ornithine	Retrorsine
Quinolizidine	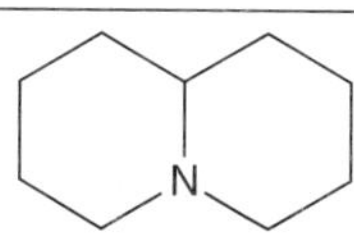	Lysine	Lupinine
Isoquinoline	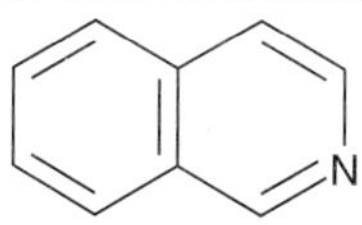	Tyrosine	Codeine Morphine
Indole	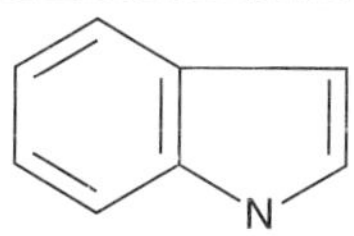	Tryptophan	Psilocybin Reserpine Strychnine

Atropine is a neurotoxin and cardiac stimulant produced by the deadly nightshade plant (*Atropa belladonna*). Although it is toxic in large quantities, it has been used medicinally by humans in small amounts as a pupil dilator and antidote for some nerve gas poisonings. Theobromine, formerly known as xantheose, is a bitter alkaloid of the cacao plant and is found in chocolate, as well as in a number of other foods, including the leaves of the tea plant, and the kola nut. Capsaicin and related capsaicinoids produced by members of the genus *Capsicum* are the active components of chili peppers and produce their characteristic burning sensation in hot, spicy foods.

Species of several families within the angiosperms are known for their production of nitrogen-based secondary metabolites called pyrrolizidine alkaloids (PAs). PAs have toxic, deterrent and/or repellent effects on a wide range of generalist herbivores to reduce or prevent damage. Our understanding of the selection factors that drive chemical diversification of PAs in plants is still incomplete. It has been suggested that the diversity of PAs in plant species has evolved under selective pressures by generalist herbivores. However, long before herbivores existed, plants already had to cope with the attack of microorganisms. Therefore, we suggest that also microbial pathogens played an important role in the evolution of the secondary metabolite diversity. If so there should be differences in the effectiveness of the individual PA types against microorganisms as well as differences in adaptation of microorganisms to the different PAs (Lotte and Johannes A. van Veen, 2011). They further stated that the existing evidence on the role PAs play in plant defense against microorganisms was scarce especially in comparison to ecological studies on insects.

PA mixtures affect microorganisms *in vivo* and *in vitro* and variation existed in antimicrobial effects of different PAs (Hol *et al.,* 2003). Several PAs tested showed structurally related PAs differed in their effects on microorganisms, but some PAs were more effective than others. High levels of jacobine were associated with lower fungal diversity in the rhizosphere of the plant of its precursor, but the primary PA senecionine did not have this effect. There were indications that adaptations of microorganisms to PAs occur, in particular, microorganisms that were isolated from PA-containing plants showed a higher tolerance and more rapid adaptation. The relationship between PAs and microorganisms needs further investigation before the strength of microorganisms as selection factor can fully be assessed. More is known about the effect of PAs on fungi. Also, here most studies were carried out *in planta*. Of the 145 bioassay-tests with different fungal species, 61% showed significant growth inhibition caused by the different PAs. Antifungal activity against *Fusarium monilifermel* was observed by the PA europine produced by *Heliotropium bovei*. PAs produced by *Heliotropium subulatum* showed to be active against *Aspergillus fumigatus, Aspergillus niger, Rhizoctonia phaseoli* and *Penicillium chrysogenum* (Lotte and Johannes A. van Veen, 2011).

Adaptation is known in microbiology as a development of resistance. Many resistance mechanisms are known, such as making it more difficult for the toxin to enter the microorganism and chemical modification of the toxic compound which results in inactivation of the toxin. Adaptation by fungi to secondary metabolites is known for a variety of compounds such as steroidal glycoalkaloids (Morrissey and Osbourn, 1999) and phytoalexins. Adaptation by inactivating the toxin could be followed by the ability to utilize the compounds as a substrate for energy and biomass production or storage for their own defense against attackers. Endophytic symbionts, for instance, need to cope with the secondary metabolites of their host plant, like *Rhizobium* specimens. They form symbioses with members of the leguminoseae, such as lupines, which produce large amounts of alkaloids, in particular, quinolizidine alkaloids (Harborne, 2003). Werner *et al.* (1997) showed that endophytic fungi were able to metabolize the polyamine alkaloid aphelandrine. Several fungi were capable of using part of the polyamine alkaloid as a nitrogen source. Nearly all fast-growing endophytes of the family of *Nectriaceae* were able to degrade and to grow on this alkaloid while slow-growing endophytes could only partially degrade aphelandrine.

Bioassay-guided fractionation of the crude extract of the whole plant of *Macleaya cordata* R. Br. led to the isolation of four alkaloids, which were identified as sanguinarine (1), chelerythrine (2), protopine (3) and α-allocryptopine (4) on the basis of their physicochemical and spectrometric data (Liu *et al.*, 2009). Compounds 1 and 2 demonstrated a significant antifungal activity against the six test fungi with median inhibitory concentrations (IC50) ranging from 0.47 to 6.13 μg/mL. Compound 1 was the most effective with an IC50 of 0.47 μg/mL on *Rhizoctonia solani*. Furthermore, compounds 1 and 2 also demonstrated strong antibacterial activity, with IC50 values ranging from 5.01 to 11.3 μg/mL, and minimum inhibitory concentrations (MIC) were ranging from 8.0 to 32.0 μg/mL. This is the first report on the activity of the alkaloids from *M. cordata* against economically important phytopathogenic fungi and bacteria. The presence of defense compounds such as PAs plays a role in the selection processes that shape the soil-borne microbial community present in the rhizosphere. This selection might favor those microorganisms that are tolerant or resistant to these defense compounds or in some cases even can degrade or utilize them. The highest PA concentrations in the roots were found in root cortex and the lateral roots instead of the vascular cylinder (Lotte and Johannes, A. van Veen, 2011).

The tissue in which plant defense compounds are stored is often crucial for their effectiveness. Hol *et al.* (2003) studied the distribution of PAs over different root parts. They observed that the concentration in the main root cortex was five times higher than the concentrations in the vascular cylinder. This suggested that a first line of defense against microbial attackers from the outside world might be created by tissue containing the highest concentration of PAs. The mature root parts were probably well protected because of its importance for regrowth after damage. Therefore, from an evolutionary point of view it is reasonable that they contained the highest PA concentrations.

3.3.8.2 *Cyanogenic glycosides*

They are a particularly toxic class of nitrogenous compounds that break down to produce hydrogen cyanide (HCN), a lethal chemical that halts cellular respiration in aerobic organisms. Plants that produce cyanogenic glycosides (cyanogens) also produce enzymes that convert these compounds into hydrogen cyanide, including glycosidases and hydroxynitrile lyases, but they are stored in separate compartments or tissues within the plant. When herbivores/pathogens feed on these tissues, the enzymes and substrates mix and produce lethal hydrogen cyanide. Major examples of cyanogens include amygdalin, prunasin, dhurrin and linamarin (Figs. 3.52–3.55).

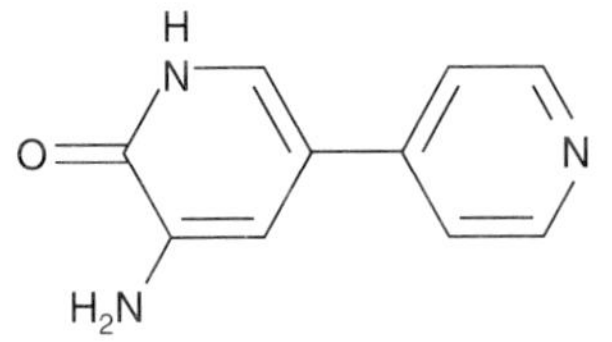

Fig. 3.52 Amygdalin

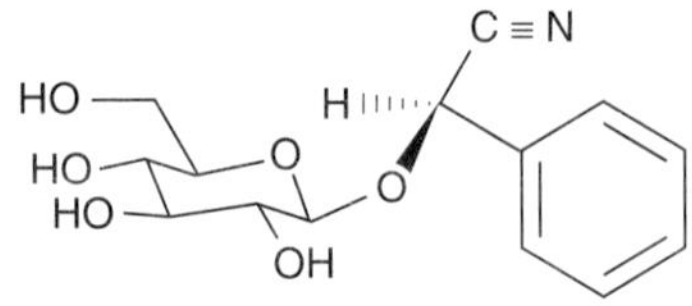

Fig. 3.53 Prunasin

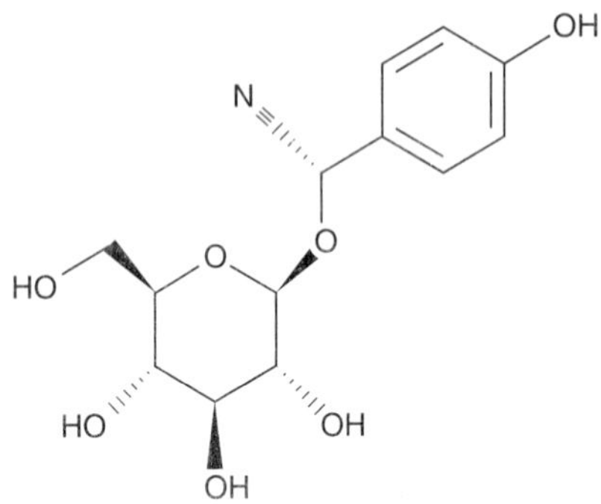

Fig. 3.54 Dhurrin

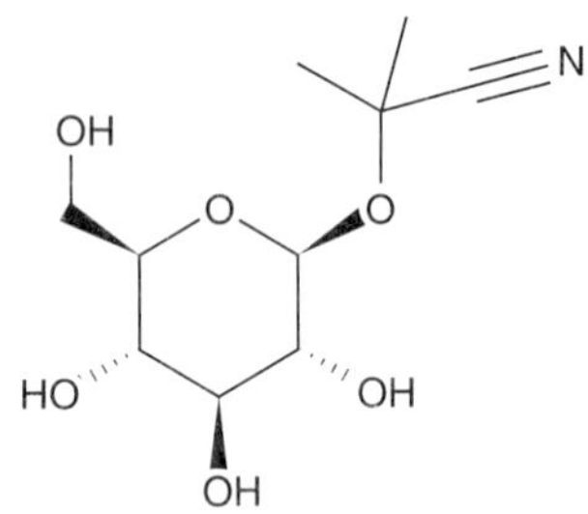

Fig. 3.55 Linamarin

Cyanogenic glycosides (CNglcs) are bioactive plant products derived from amino acids. Structurally, these specialized plant compounds are characterized as α-hydroxynitriles (cyanohydrins) that are stabilized by glucosylation. In recent years, improved tools within analytical chemistry have greatly increased the number of known CNglcs by enabling the discovery of less abundant CNglcs formed by additional hydroxylation, glycosylation, and acylation reactions (Gleadow and Moller, 2014). Cyanogenesis-the release of toxic hydrogen cyanide from endogenous CNglcs-is an effective defense against generalist herbivores but less effective against fungal pathogens. In the course of evolution, CNglcs have acquired additional roles to improve plant plasticity, i.e., establishment, robustness, and viability in response to environmental challenges. CNglc concentration is usually higher in young plants when nitrogen is in ready supply, or when growth is constrained by nonoptimal growth conditions. They further state that efforts are under way to engineer CNglcs into some crops as a pest control measure, whereas in other crops efforts are directed toward their removal to improve food safety. Given that many food crops are cyanogenic, it is important to understand the molecular mechanisms regulating cyanogenesis so that the impact of future environmental challenges can be anticipated.

The toxicity of HCN to aerobic organisms implies a positive correlation between cyanide potential and plant defense. Thus, the presence of cyanogenic glucosides would be expected to combat fungal attack if the incipient attack HCN releases or if the intact cyanogenic glucoside would inhibit fungal growth. Fungal pathogens of cyanogenic plants are generally tolerant to HCN as mediated by their ability to convert HCN into formamide in a reaction catalyzed by formamide hydrolyase. A recent study in lima bean *Phaseolus lunatus* (Ballhorn *et al.*, 2010) provides further results documenting that plant defenses do not always act synergistically but may be trade-offs against each other due to negative interactions at the biochemical level. The fungal pathogen, *Colletotrichum gloeosporioides* benefited from the presence of the cyanogenic glucosides because of its ability to effectively detoxify HCN and because the concentration of gaseous HCN liberated from the damaged plant cells was high enough to inhibit the plant polyphenoloxidase based defense system of the plant. Bacteria also have various options to cope with the presence of cyanogenic glucosides and their toxic degradation products. β-cyanoalanine synthase catalyzes the reaction between free HCN and cysteine to form β-cyanoalanine (Gleadow and Moller, 2014).

Two iridoid glycosides, aucubin and catalpol, present in ribwort plantain (*Plantago lanceolata*), were tested to confer *in vivo* resistance to the biotrophic fungal pathogen *Diaporthe adunca* using plants from *P. lanceolata* lines that had been selected for high- and low-leaf iridoid glycoside concentrations for four generations (Biere *et al.*, 2004). The lines differed approximately threefold in the levels of these compounds. Plants from the high-selection line showed enhanced resistance to *D. adunca*, as evidenced by a smaller lesion size and a lower fungal growth rate and spore production. It was suggested that the hydrolysis products formed after tissue damage following fungal infection mediate pathogen resistance. It was concluded that iridoid glycosides in *P. lanceolata* could serve as broad-spectrum defenses and that selection for pathogen resistance could potentially result in increased resistance to generalist insect herbivores and vice versa, resulting in diffuse rather than pairwise coevolution.

3.3.8.3 Glucosinolates

The glucosinolates are natural components of many pungent plants such as mustard, cabbage, and horseradish. The pungency of those plants is due to mustard oils produced from glucosinolates when the plant material is chewed, cut or otherwise damaged. These natural chemicals most likely contribute to plant defense against pests and diseases, They also known as mustard oil glycosides and are sulfur-containing compounds synthesized by members of the mustard family (*Brassicaceae*) and produce cyanide gas when broken down by enzymes, thioglucosidases. Glucosinolates constitute a natural class of organic compounds that contain sulfur and nitrogen and are derived from glucose and an amino acid. They are water-soluble anions and belong to the glucosides. Some examples include glucotropaeolin, gluconasturtiin, and glucoraphanin (Figs. 3.56–3.58).

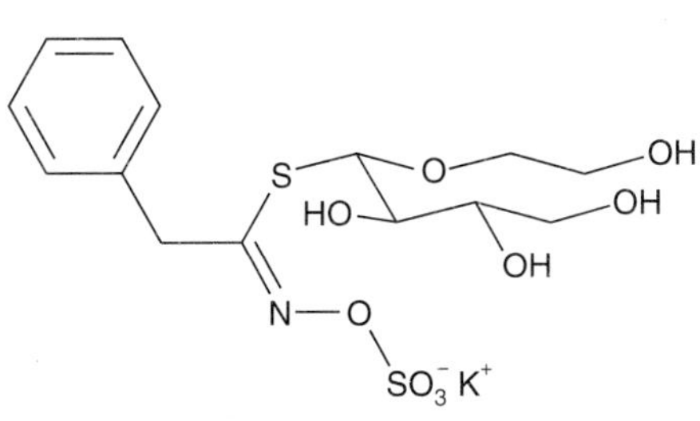

Fig. 3.56 Glucotropaeolin

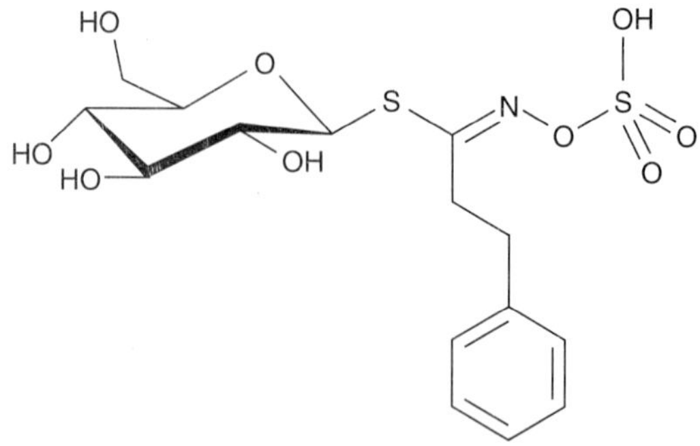

Fig. 3.57 Gluconasturtiin

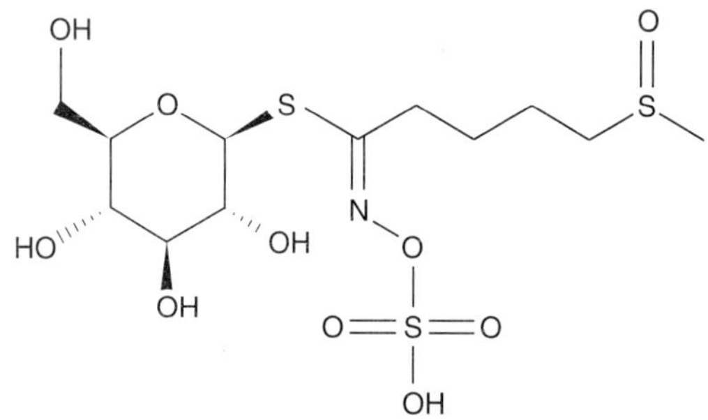

Fig. 3.58 Glucoraphanin

Aires *et al.* (2009) evaluated the *in vitro* antibacterial effects of glucosinolate hydrolysis products (GHP) against plant pathogenic microorganisms, namely, *Agrobacterium tumefaciens, Erwinia chrysanthemi, Pseudomonas cichorii, Pseudomonas tomato, Xanthomonas campestris* and *Xanthomonas juglandis*. Using a disc diffusion assay, seven different doses of 10 GHP were tested against each bacterium. The results showed that the isothiocyanates were potent antibacterials, whilst the other GHP were much less efficient. Moreover, the antibacterial effects were dose-dependent, increasing with the dose applied; 2-phenylethylisothiocyanate and sulforaphane showed the strongest inhibitory effects. The overall results show a great potential for using the isothiocyanates as an alternative tool to control undesired bacterial growth in plants. Glucosinolate hydrolysis products and more specifically the isothiocyanates: benzylisothiocyanate, 2-phenylethyliso-thiocyanate, the isothiocyanate mix and sulforaphane, were effective phytochemicals against the *in vitro* growth of the phytopathogenic bacteria. The antibacterial activity exhibited by these phytochemicals reinforced their potential as alternatives to the traditional chemical control of phytopathogenic bacteria.

Glucosinolates are β-thioglucoside N-hydroxysulfates containing a side chain and a β-D-glucopyranosyl moiety. Upon cellular disruption, glucosinolates are hydrolyzed to various bioactive breakdown products by the endogenous enzyme myrosinase. Isothiocyanates and indole glucosinolate metabolites (in particular indol-3-carbinol) are two major groups of autolytic breakdown products of glucosinolates. It is believed that glucosinolates can confer resistance to *Brassica* crops against pests and diseases. The quality of *Brassica rapa* crop is highly affected by the disease caused by the bacteria *Xanthomonas campestris* pv. *campestris* (Xcc). Glucosinolates and phenolic compounds can confer resistance to *Brassica* crops against pests and diseases, but few works have been done to evaluate their role in Xcc resistance (Pablo Velasco *et al.*, 2013). Gluconapin and its ITC varieties had an antibacterial effect on the development of *Xanthomonas*, and this effect was strongly dependent on the concentration applied. Methanolic extracts from *B. rapa*, containing glucosinolates and phenolic compounds, inhibited the growth of these bacteria. The concentration of gluconapin is higher in resistant plants than in the susceptible ones, and there is an induction of gluconapin, some flavonoids and sinapic acid 48 to 72 h after inoculation. Gluconapin played a role in the constitutive resistance to Xcc, while gluconapin, some flavonoids, and hydroxycinnamic acids were induced by a Xcc infection, but it was not clear if this induction confers resistance to this disease.

3.3.9 Proteins and Enzymes

The hypersensitive reaction to a pathogen is one of the most efficient defense mechanisms in nature and leads to the induction of numerous plant genes encoding defense proteins. These proteins include structural proteins that are incorporated into the extracellular matrix and participate in the confinement of the pathogen; enzymes of secondary metabolism, for instance those of the biosynthesis of plant antibiotics, pathogenesis-related (PR) proteins which represent major quantitative changes in soluble protein during the defense response (Stintz *et al.*, 1993). The PRs have typical physicochemical properties that enable them to resist to acidic pH and proteolytic cleavage and thus survive in the harsh environments where they occur: vacuolar compartment or cell wall or intercellular spaces. Since the discovery of the first PRs in tobacco, many other similar proteins have been isolated from tobacco but also from other plant species, including dicots and monocots, the widest range being characterized from hypersensitively reacting tobacco. Based first on serological properties and later on sequence data, the tobacco PRs have been classified into five major groups. Group PR-1 contains the first discovered PRs of 15-17 kDa molecular mass, whose biological activity is still unknown, but some members have been shown recently to have antifungal activity. Group PR-2 contains three structurally distinct classes of 1,3-β-glucanases, with acidic and basic counterparts, with dramatically different specific activity towards linear 1,3-β-glucans and with different substrate specificity. Group PR-3 consists of various chitinases-lysozymes that belong to three distinct classes, are vacuolar or extracellular, and exhibit differential chitinase and lysozyme activities. Some of them, either alone or in combination with 1,3-β-glucanases, have been shown to be antifungal *in planta* and *in vivo* (transgenic plants), probably by hydrolyzing their substrates as structural components in the fungal cell wall. Group PR-4 is the less studied, and in tobacco contains four members of 13-14.5 kDa of unknown activity and function. Group PR-5 contains acidic-neutral and very basic members with extracellular and vacuolar localization, respectively, and all members show sequence similarity to the sweet-tasting protein thaumatin. Several members of the PR-5 group from tobacco and other plant species were shown to display significantly *in vitro* activity of inhibiting hyphal growth or spore germination of various fungi probably by a membrane permeabilizing mechanism.

Depending on their function during the defense response, proteins can be grouped into three classes. The structural proteins in the first class participate in strengthening and repairing of the cell wall or modification of the properties of the extracellular matrix. The second class of proteins exhibit direct antimicrobial activities or catalyze the synthesis of antimicrobial compounds. The third class comprises of proteins, which function in plant defense is not well known (Ales Lebeda *et al.*, 1999).

Extensins are structural proteins rich in hydroxyproline (HPRGs). They have the central role in the organization of the primary cell wall of plants. All extensins are basic, highly glycosidated proteins. Extensins polypeptides are synthesized inside cells, secreted as soluble monomers to the apoplast and incorporated into the cell wall. Another group of structural proteins are GRPs, rich in glycine. The expression of genes for HPRGs and GRPs is regulated by signals produced during physiological or environmental stress including wounding and pathogen attack. Currently, the research is also aimed at leucine-rich repeat proteins (LRR proteins), which have a significant role in plant defenses. They may be involved either as resistance proteins or as proteins required for resistance proteins to function. However, LRR proteins may also function as PR proteins.

3.3.9.1 Pathogenesis-related/PR-proteins and their classification

Pathogenesis-related (PR) proteins are proteins produced in plants in the event of a pathogen attack. They are induced as part of systemic acquired resistance. Infections activate genes that produce PR proteins. Some of these proteins are antimicrobial, attacking molecules in the cell wall of a bacterium or fungus. Others may function as signals that spread 'news' of the infection to nearby cells. Infections also stimulate the cross-linking of molecules in the cell wall and the deposition of lignin, responses that set up a local barricade that slows spread of the pathogen to other parts of the plant. By changing their physiological conditions, higher plants protect themselves from various stresses including an attack by pathogens, wounds, chemical applications, and air pollutants like ozone, UV rays, and other harsh growing conditions. These protective reactions are termed 'defense responses' of higher plants and the proteins actively synthesized in accordance with this reaction are termed 'defense-related proteins.'

Plants display an innate pathogen-specific resistance by producing responses like an oxidative burst of the cell, change of cell wall composition that prevent infection and *de novo* synthesis of compounds like phytoalexin and pathogenesis-related proteins. All these responses can be triggered by exposing the plant to virulent, avirulent, and nonpathogenic microbes, or artificially with low molecular weight and sometimes volatile molecules like such as salicylic acid, jasmonate (Borad and Sriram, 2008), 2,6-dichloro-isonicotinic acid or benzo(1,2,3) thiadiazole-7-carbothioic acid S-methylester (BTH). These types of resistance are called Systemic Acquired Resistance (SAR) or Induced Systemic Resistance (IAR). Among all induced responses, production of 'Pathogenesis-Related (PR) proteins' is most important because they can lead to the increased resistance of the whole plant against a pathogenic attack.

3.3.9.1.1 Classification of PRs

Originally, five main groups of PRs (PR-1 to PR-5) were characterized by both molecular and molecular genetic techniques in tobacco, numbered in order of decreasing electrophoretic mobility. Each group consists of several members with similar properties (Table 3.3). Group PR-1 is the most abundant, reaching up to 1-2% of total leaf proteins. PRs of group 5 share significant amino acid sequence homology with the sweet tasting protein in the fruits of the tropical plant *Thaumatococcus daniellii,* and have been named thaumatin-like (TL) proteins. Osmotins of the same group display similarity to TL proteins (Singh *et al.,* 1987). Thereupon, in 1994 a unifying nomenclature for PRs was proposed based on their grouping into families sharing amino acid sequences, serological relationships, and enzymatic or biological activity. By then eleven families (PR-1 to PR-11) were recognized and classified for tobacco and tomato, with the families PR-8 and PR-10 being also present in cucumber and parsley, respectively. Later three novel families

Table 3.3 Recommended Classification of Pathogenesis-Related Proteins (PRs)

Family	*Type member*	*Properties*
PR-1	tobacco PR-1a	Antifungal ?, 14-17kD
PR-2	tobacco PR-2	Class I, II, and III endo-β-1,3-glucanases, 25-35 kD
PR-3	tobacco P, Q	Class I, II, IV, V, VI, and VII endochitinases, about 30 kD
PR-4	tobacco R	Antifungal, *win*-like proteins, endochitinase activity, similar to prhevein C-terminal domain, 13-19 kD
PR-5	tobacco S	Antifungal, Thaumatin-like Proteins, Osmotins, Zeamatins, Permeatins, Similar to α-amylase/trypsin inhibitors
PR-6	tomato inhibitor I	Protease inhibitors, 6-13 kD
PR-7	tomato P69	Endoproteases
PR-8	cucumber chitinase	Class III chitinases, chitinase/lysozyme
PR-9	lignin-forming peroxidase	Peroxidases, peroxidase-like proteins
PR-10	parsley PR-1	Ribonucleases, B-v-1-related proteins
PR-11	tobacco class V chitinase	Endochitinase activity
PR-12	radish Ps-AFP3	Plant defensins
PR-13	*Arabidopsis* THI2.1	Thionins
PR-14	barley LTP4	Non-specific lipid transfer proteins (ns-LTPs)
PR-15	barley OxOa (germin)	Oxalate oxidase
PR-16	barley OxOLP	Oxalate-oxidase-like proteins
PR-17	tobacco PRp27	Unknown

(PR-12, PR-13, and PR-14) were recognized in radish, *Arabidopsis*, and barley, respectively. Germins and germin-like proteins (GLPs) have been classified as PR-15 and PR-16; PR-16 has been isolated from hot pepper during the resistance response to bacterial and viral infection.

Large numbers of small, basic, cysteine-rich antimicrobial proteins are produced by many organisms throughout all kingdoms. They display a great variety in their primary structure, in species specificity, and in the mechanism of action (Leiter *et al.*, 2005). There are more than 13 different pathogenesis-related proteins are known to us. Antifungal PR proteins are of great biotechnological interest because of their potential use as food and seed preservative agents and for engineering plants for resistance to phytopathogenic fungi (Borad and Sriram, 2008). Various studies have revealed that transgenic plants overexpressing genes of the PR-1, PR-2, PR-3, and PR-5 families mediate host plant resistance to phytopathogenic fungi. Coexpression of multiple antifungal protein genes in transgenic plants seems to be more effective than an expression of single genes.

Enzymes indispensable for biosynthesizing low-molecular-weight, antibiotics (phytoalexin), isoflavone reductase, for example, can also be considered to be defense-related proteins. On the other hand, many of the reserve proteins accumulated in seeds and fruits are considered to have a constitutive defense function against microbial pathogens and invertebrate pests in addition to their storage function. These inducible or constitutive defense mechanisms of higher plants are relatively conserved during the course of evolution. Accordingly, most plants produce or accumulate structurally and functionally similar protective proteins under certain situations, irrespective of their morphological differences. For instance, PR proteins, which have been found in many plant species to date, are classified. Recently, protective (proteins of higher plants have drawn much attention from plant breeders (Fritig *et al.*, 1998). Because defense-related proteins usually provide a plant with resistance to stresses, varieties that are apt to intensively induce such proteins can be agriculturally valuable. Less toxic substances that cause crops to express defensive proteins are also being investigated as a new type of agrochemical. Moreover, some T-defense-related proteins will be constantly expressed in genetically modifies plants. However, several of these proteins have been proven to be latex allergens and cross-reactive allergens in fruits, vegetables, and pollen. As such, it is probable that the crop content of allergenic proteins responsible for 'oral allergy syndrome' (OAS) or similar syndromes has increased in response to traditional plant-breeding or the creation of stress-resistant varieties through the use of gene recombination techniques. When a genetically modified new crop is brought to market as food, its safety, including its possible allergenicity, must be strictly examined and reviewed in advance. The sequence of similar itierologic or immunologic relationships and enzymatic properties are the bases of this classification (van Loon *et al.*, 1999).

Other major plant proteins with defensive activities against pathogens include 2S storage albumins, patatin, ribosome-inactivating proteins (RIPs), polygalacturonase inhibitor proteins (PGIPs), nonenzymatic chitin-binding proteins (hevin, lectins, AMPs, AFPs, etc.), cystatins, prolamin storage proteins, vicilin (7S) storage proteins, etc.

Rhizoctonia solani is one of the most destructive pathogens causing foot rot disease on tomato. In a study by (Parissa and Saeed, 2012), the molecular and cellular changes of a partially resistant (Sunny 6066) and a susceptible (Rio Grande) tomato cultivar after infection with necrotrophic soil-borne fungus *R. solani* were compared. The expression of defense-related genes such as chitinase (LOC544149) and peroxidase (CEVI-1) in infected tomato cultivars was investigated using semiquantitative reverse transcription-polymerase chain reaction (RT-PCR). This method revealed elevated levels of expression for both genes in the partially resistant cultivar compared to the susceptible cultivar. One of the most prominent facets of basal plant defense responses is the formation of physical barriers at sites of attempted fungal penetration. These structures are produced around the sites of potential pathogen ingress to prevent pathogen progress in plant tissues. Formation of lignin was investigated, as one of the most important structural barriers affecting plant resistance, using thioglycolic acid assay. A correlation was found between lignification and a higher level of resistance in Sunny 6066 compared to Rio Grande cultivar. These findings suggested the involvement of chitinase, peroxidase, and lignin formation in defense responses of tomato plants against *R. solani* as a destructive pathogen.

3.3.9.2 Relevance of PRs to disease resistance

Presently, it is still difficult to assign a causative role of PRs in plant resistance to pathogens. The reason for this is that the numerous data on PRs as disease resistance factor are mostly of correlative character. Following four lines of supporting evidence can be outlined (Aglika Edreva, 2005).

(i) Stronger accumulation of PRs in inoculated resistant as compared to susceptible plants. Besides previous data, substantiating this statement, differential responses of resistant/susceptible plants were recently reported in tomato plants, inoculated with *Cladosporium fulvum*, *Phytophthora infestans*-infected potato, *Venturia inaequalis*-inoculated apple, *Pseudomonas syringae*-infected grapevine, *Xanthomonas campestris* pv. *vesicatoria* and TMV-Po-infected hot pepper etc. Additional resistance gene(s) against *Cladosporium fulvum* present on the Cf-9 introgression segment have been shown to be associated with strong PR-protein accumulation (Laugé *et al.*, 1998). In some pathosystems mRNAs for certain PRs members accumulate to similar levels in compatible and incompatible interactions, but the maximum level of expression is reached much faster in the latter (Aglika Edreva, 2005).

(ii) Important constitutive expression of PRs in plants with a high level of natural disease resistance. This correlation was observed in several pathosystems, such as apple – *Venturia inaequalis*, tomato – *Alternaria solan*, and potato – *Phytophthora infestans*.

(iii) Significant constitutive expression of PRs in transgenic plants overexpressing PR genes accompanied by increased resistance to pathogens. Thus, increased tolerance to *Peronospora tabacina* and *Phytophthora parasitica* was demonstrated in tobacco overexpressing PR1a gene. Transgenic rice and orange plants overexpressing thaumatin-like PR-5 possessed increased tolerance to *Rhizoctonia solani* and

Phytophthora citrophthora, respectively while transgenic potato overexpressing PR-2 and PR-3 had improved resistance to Phytophthora infestans. PR-2 and PR-3 genes, coding for β-1,3-glucanase and chitinase, respectively, confer resistance of carrot to several fungal pathogens. The simultaneous expression of tobacco β-1,3-glucanase and chitinase genes in tomato plants results in increased resistance to fungi. On the contrary, silencing of PR-1b gene in barley facilitates the penetration of the fungal pathogen *Blumeria graminis* f. sp. *hordei* in the leaves.

(iv) Accumulation of PRs in plants in which resistance is locally or systemically induced. It can be stated that PRs are recognized as markers of the systemic acquired resistance (SAR), and PR genes are involved in the list of SAR-genes. It has largely been demonstrated that SAR and the accompanying set of PRs are induced by different pathogens, as well as by a range of chemicals predominantly in a salicylic acid-dependent pathway; SAR is active against a broad spectrum of pathogens. Some SAR-inducing chemicals, such as benzothiadiazole (BTH), β-aminobutyric acid (BABA) or 2,6-dichloroisonicotinic acid (DCINA) are harmless commercially supplied compounds and have promising practical application as novel tools in plant protection (Aglika Edreva, 2005.) PRs members induced in resistant or SAR-expressing plants, as well as PRs from transgenic resistant plants, exhibit high antimicrobial activity.

3.3.9.3 *Pathogenesis-related (PR) protein 1*

The first PR-1 protein was discovered in 1970. Since then, a number of PR-1 proteins have been identified in *Arabidopsis, Hordeum vulgare* (barley), *Nicotiana tabacum* (tobacco), *Oryza sativa* (rice), *Piper longum* (pepper), *Solanum lycopersicum* (tomato), *Triticum* sp. (wheat) and *Zea mays* (maize) (Liu and Xue, 2006). These PR-1 having 14 to 17 kD molecular weight and mostly of basic nature. Nonexpressors of Pathogenesis-Related Genes1 (NPR1) regulate systemic acquired resistance via regulation pathogenesis related 1 (PR-1) in *Arabidopsis thaliana*. The interaction of nucleus-localized NPR1 with TGA transcription factors, after reduction of cysteine residues of NPR 1 by salicylic acid (SA) results in the activation of defense genes of PR-1. PR-1 proteins have antifungal activity at the micromolar level against a number of plant pathogenic fungi, including *Uromyces fabae, Phytophthora infestans,* and *Erysiphe graminis*. The exact mode of action of the antifungal activities of these proteins are yet to be identified, but a PR-1-like protein, helothermine, from the Mexican banded lizard have been found to be interacting with the membrane-channel proteins of target cells, inhibiting the release of Ca^{2+} (Borad and Sriram, 2008).

3.3.9.3.1 Glucanase (PR2)

Glucanases are enzymes that break down a glucan, a polysaccharide made of several glucose subunits. As they perform hydrolysis of the glucosidic bond, they are hydrolases. β-1,3-glucanase, an enzyme that breaks down β-1,3-glucans such as callose or curdlan. β-1,3-Glucanases (βGlu) exist as multiple structural isoforms that differ in size, isoelectric point, primary structure, cellular localization, and pattern of regulation. β (1,3)-glucanases are a class of hormonally and developmentally regulated plant hydrolytic enzymes, which are also induced upon pathogen infection, suggesting a role of these hydrolases in the defense response to pathogens. Several β(1,3)-glucanases were purified which were present in tobacco leaves from control plants and plants treated with salicylic acid or infected with *Pseudomonas syringae* and studied in detail the subcellular localization of the different isoforms (Van Den Bulcke *et al.*, 1989). Partial protein sequence analysis demonstrated that each of the different isoforms had a unique amino acid sequence and was therefore encoded by a different gene. We have also demonstrated that two of these isoforms, similar to the cytokinin/auxin-regulated isoforms previously isolated from tobacco cell suspensions, are located in the central vacuole. Upon salicylic acid treatment or *P. syringae* infection, three secreted isoforms were induced, belonging to the so-called pathogenesis-related proteins. These pathogenesis-related β(1,3)-glucanases were all distinct from each other and also different from the vacuolar isoforms. It was demonstrated that the vacuolar isoforms were not secreted into the extracellular spaces of the plant following pathogen infection, suggesting that any function they play in the plant defense response is restricted to an intracellularly coordinated defense process.

Antifungal class I [β]-1,3-glucanases are believed to be part of the constitutive and induced defenses of plants against fungal infection. Unexpectedly, mutants deficient in these enzymes generated by antisense transformation showed markedly reduced lesion size, lesion number, and virus yield in the local-lesion response of Havana 425 tobacco to tobacco mosaic virus (TMV) and of *Nicotiana sylvestris* to tobacco necrosis virus (Beffa *et al.*, 1996). These mutants also showed decreased severity of mosaic disease symptoms, delayed spread of symptoms, and reduced yield of virus in the susceptible response of *N. sylvestris* to TMV. The symptoms of disease in the responses of both plant species were positively correlated with [β]-1,3-glucanase content in a series of independent transformants. Taken together, these results provide direct evidence that [β]-1,3-glucanases function in viral pathogenesis. Callose, a substrate for [β]-1,3-glucanase, acts as a physical barrier to the spread of the virus. Callose deposition in and surrounding TMV-induced lesions was increased in the [β]-1,3-glucanase-deficient, local-lesion Havana 425 host, suggesting as a working hypothesis that decreased susceptibility to virus resulted from increased deposition of callose in response to infection. These results suggested novel means for protecting plants against viral infection, based on antisense transformation with host genes.

They also defend plants against fungal pathogens either alone or in association with chitinases and other antifungal proteins. They are grouped in the PR-2 family of pathogenesis-related (PR) proteins. Use of β-1,3-glucanase genes as transgenes in combination with other antifungal genes is a plausible strategy to develop durable resistance in crop plants against fungal pathogens. These genes sourced from alfalfa, barley, soybean, tobacco, and wheat have been coexpressed along with other antifungal proteins, such as chitinases, peroxidases, thaumatin-like proteins and α-1-purothionin, in various crop plants (Balasubramanian *et al.*, 2012).

Trichoderma harzianum secretes α-1,3-glucanases when it is grown on polysaccharides, fungal cell walls, or autoclaved mycelium as a carbon source (simulated antagonistic conditions). One of these enzymes, named AGN13.1 was purified and characterized

(Hassane Ait-Lahesan *et al.,* 2001). The enzyme was monomeric and slightly basic. AGN13.1 was an exo-type α-1,3-glucanase and showed lytic and antifungal activity against fungal plant pathogens. Northern and Western analyses indicated that AGN13.1 is induced by conditions that simulated antagonism. It was proposed that AGN13.1 contributes to the antagonistic response of *T. harzianum.*

Only class I vacuolar isoforms of tobacco βGlu and Chn were effective in promoting the lysis of hyphal tips and inhibiting the growth of *Fusarium solani.* These effects were greatly enhanced by using combinations of βGlu I and class I Chn. In contrast, the class II βGlu PR-2a, PR-2b, and PR-2c did not exhibit antifungal activity either alone or in any combinations tested. Similar studies with tomato βGlu and Chn have shown that the vacuolar class I isoforms, but not the secreted class II isoforms inhibit the growth of *Alternaria solani, Trichoderma viride,* and *Phytophthora infestans* and that none of the combinations of βGlu and Chn tested inhibited growth of the tomato pathogen *Cladosporium fulvum.*

The accumulation of the pathogenesis-related (PR) proteins β-1,3-glucanase and chitinase and structural defense responses were studied in leaves of wheat either resistant or susceptible to the hemibiotrophic pathogen *Septoria tritici* (Nandini P. Shetty *et al.,* 2009). Resistance was associated with an early accumulation of β-1,3-glucanase and chitinase transcripts followed by a subsequent reduction in level. Resistance was also associated with high activity of β-1,3-glucanase, especially in the apoplastic fluid, in accordance with the biotrophic/endophytic lifestyle of the pathogen in the apoplastic spaces, thus showing the highly localized accumulation of defense proteins in the vicinity of the pathogen. Isoform analysis of β-1,3-glucanase from the apoplastic fluid revealed that resistance was associated with the accumulation of an endo-β-1,3-glucanase, previously implicated in defense against pathogens, and a protein with identity to ADPG pyrophosphatase (92%) and germin-like proteins (93%), which may be involved in cell wall reinforcement. In accordance with this, glycoproteins like extensin were released into the apoplast and callose accumulated to a greater extent in cell walls, whereas lignin and polyphenolics were not found to correlate with defense. Treatment of a susceptible wheat cultivar with purified β-1,3-glucan fragments from cell walls of *S. tritici* gave complete protection against disease, and this was accompanied by increased gene expression of β-1,3-glucanase and the deposition of callose. Collectively, these data indicate that resistance is dependent on a fast, initial recognition of the pathogen, probably due to β-1,3-glucan in the fungal cell walls, and these results in the accumulation of β-1,3-glucanase and structural defense responses, which may directly inhibit the pathogen and protect the host against fungal enzymes and toxins. The different resistance of cotton (*Gossypium hirsutum* L.) cultivars to the crude toxin of *Verticillium dahliae* (VD) was correlated with the activities of chitinase and β-1,3-glucanase in callus cells (LI Ying-Zhang *et al.,* 2003). The activities of chitinase and β-1,3-glucanase in the callus cells treated with the VD-toxin were increased to the higher level at an earlier time point in resistant cultivars than these in the susceptible cultivars. Exogenous salicylic acid (SA) induced the accumulation of chitinase and β-1,3-glucanase, which resulted in the resistance of callus cells to the VD toxin. Western blot using a polyclonal antibody against β-1,3-glucanase identified 28 kD protein that was induced by VD-toxin, SA, or VD-toxin plus SA. β-1,3-Gs have a molecular mass in the range from 33 to 44 kDa (Hong and Meng, 2004; and Saikia *et al.*, 2005). Wounding, hormonal signals like methyl jasmonate and ethylene (Wu and Bradford, 2003), pathogen attack like fungus *Colletotrichum lagenarium* and some fungal elicitors releases from pathogen cell wall can also induce α-1,3-Gs in the various parts of the plant. The enzyme α-1,3-Gs was found to be strongly induced by ultraviolet (UV-B; 280–320 nm) radiation in primary leaves of French bean (*Phaseolus vulgaris*) so that UV-induced DNA damage is a primary step for the induction of α-1,3-Gs (Borad and Sriram, 2008). These enzymes are found in wide variety of plants like *Arachis hypogaea* (peanut), *Cicer arietinum* (chickpea), *Nicotiana tabacum* (tobacco), etc. and those having resistivity against various fungi like *Aspergillus parasiticus, A. flavs, Blumeria graminis, Colletotrichum lagenarium, Fusarium culmorum, Fusarium oxysporum, fusarium udum, Macrophomina phaseolina* and *Treptomyces sioyaensis.*

3.3.9.3.2 Chitinases (PR3)

These are hydrolytic enzymes that break down glycosidic bonds in chitin. As chitin is a component of the cell walls of fungi and exoskeletal elements of some animals (including worms and arthropods), chitinases are generally found in organisms that either need to reshape their own chitin or dissolve and digest the chitin of fungi or animals. Chitinases are digestive enzymes that break down glycosidic bonds in chitin. Chitinases (E.C 3.2.2.14) are glycosyl hydrolases with the sizes ranging from 20 kDa to about 90 kDa (Fig. 3.59).

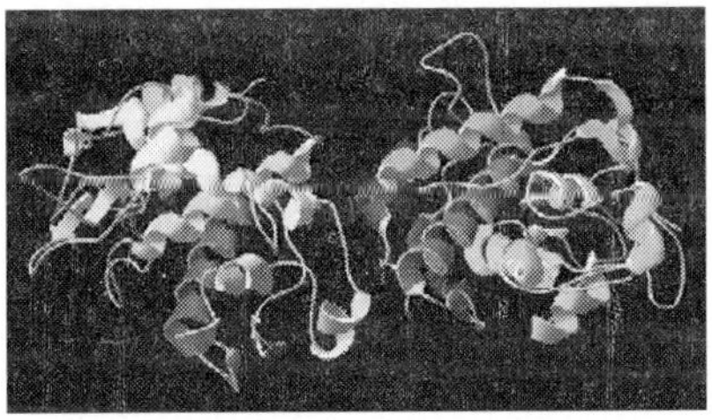

Fig. 3.59 Chitinase structure

Antifungal activity of chitinase, ChiCW and synergistic interactions between ChiCW with fungicides were investigated (Huang and Chen, 2008). Conidial germinations of phytopathogenic fungi, *Alternaria brassicicola, Botrytis elliptica,* and *Colletotrichum gloeoporioides,* were inhibited by ChiCW, but *A. longipes* was not. In addition, ChiCW showed synergistic effect with fungicides Switch (cyprodinil + fludioxonil) and tebuconazole to inhibit fungal conidial germinations. The level of synergism of ChiCW with tebuconazole was higher than that with Switch. The results indicated that ChiCW may exhibit a higher level of synergism with fungicides that have a primary effect upon membranes. The *Autographa californica* nucleopolyhedrovirus chitinase A (AcMNPV ChiA) is a chitinolytic enzyme with fungicidal and insecticidal properties. Its expression in transgenic plants enhances resistance against pests and fungal pathogens.

Antimo Di Maro *et al.* (2010) exploited tobacco for the production of a biologically active recombinant AcMNPV ChiA (rChiA), as such species is an alternative to traditional biological systems for large-scale enzyme production. The protein was purified from leaves using ammonium sulfate precipitation followed by anion exchange and gel filtration chromatography. Transgenic plants produced an estimated 14 mg kg^{-1} fresh leaf weight, which represents 0.2% of total soluble proteins. The yield of the purification was about 14% (2 mg kg^{-1} fresh leaf weight). The comparison between the biochemical and kinetic properties of the rChiA with those of a commercial *Serratia marcescens* chitinase A indicated that the rChiA was thermostable and more resistant at basic pH – two positive features for agricultural and industrial applications. It was shown that the purified rChiA inhibited spore germination and growth of the phytopathogenic fungus *Alternaria alternata*. The data indicated that tobacco represents a suitable platform for the production of rChiA, an enzyme with interesting features for future applications as 'eco-friendly' control agent in agriculture.

The exploitation of chitinase with respect to plant defense can be done by a variety of ways. The enzyme can also be used in a free or immobilized form to kill fungi and insects in affected areas (Sharma *et al.*, 2011). The microorganisms producing chitinase can also be used in the soil as rhizobacterial population or alternatively the gene encoding chitinase can be inserted in the native microflora of soil. Chitinase of *Serratia marcescens* was inserted into *Pseudomonas fluorescens* which is normally present as normal flora of soil and obtained resistance in radish plants against *Fusarium oxysporum* infection. Chernin *et al.* (1997) were also able to decrease the onset of *Rhizoctonia solani* infection in cotton plants by using recombinant *E. coli* in rhizosphere expressing chitinase gene of *Enterobacter agglomerans*. Abroad-spectrum antifungal activity in *E. coli* was demonstrated expressing chitinase gene of barley against *Botrytis cinerea* (blight of tobacco), *Pestalotia theae* (leaf spot of tea), *Bipolaris oryzae* (brown spot of rice), *Alternaria* sp. (grain discoloration of rice), *Curvularia lunata* (Leaf Spot of Clover) and *Rhizoctonia solani* (sheath blight of rice). Besides these, the phenomena of synergism (the interaction of organisms or proteins or elements that when combined, produce a total effect that is greater than the sum of the individual element contributions) can be meaningful to fasten the killing of pathogenic organisms.

Alternatively, the production of transgenic plants overexpressing chitinase gene had been also demonstrated to get resistance against pathogens. It was achieved by manipulating the activity of extracellular enzymes through the construction of overproducing mutants, enzyme-negative mutants or even transgenic plants expressing the enzyme. Broque *et al.* (1991) showed an increased ability to survive in tobacco plants in *Rhizoctonia solani* infected soil and delayed development of disease symptoms in tobacco seedlings by expressing chitinase. The reduction in the occurrence of *Rhizoctonia solani* infection in transgenic tobacco plants in which two bacterial chitinase gene were overexpressed at high levels was demonstrated. Besides the immunity against fungal pathogen, overexpression of chitinase was also found be effective to raise resistance plants against bacterial pathogens, salinity stress, and heavy metals stress.

It is not only the microbial chitinase but the plant chitinase had been also used to improvement plant health by various workers. In oilseed rape (*Brassica napus* var. *oleifera*), the importance of chitinase was also shown by various researchers. The tolerance could be increased in these plants against *Cylindrosporium concentricum, Phoma lingam, Sclerotinia sclerotiorum* infection by expressing chitinase gene. Transgenic wheat plants expressing chitinase of plants were also raised. Enhanced resistance to powdery mildew infection in wheat was observed upon expression of chitinase of barley (Oldach *et al.*, 2001). In another experiment the transgenic wheat lines carrying a combination of a wheat β-1, 3-glucanase and chitinase exhibited delayed symptoms of *Fusarium* Head Blight.

Because chitin is a structural component of the cell wall of many phytopathogenic fungi, extensive research has been conducted to determine whether plant chitinases have a role in defense against fungal diseases. Plant chitinases have different degrees of antifungal activity of several fungi *in vitro. In vivo,* although rapid accumulation and high levels of chitinases (together with numerous other pathogenesis-related proteins) occur in resistant tissues expressing a hypersensitive reaction, high levels also can occur in susceptible tissues (Zamir K. Punja and ye-Yan Zhang, 1993). Expression of cloned chitinase genes in transgenic plants has provided further evidence for their role in plant defense. The level of protection observed in these plants is variable and may be influenced by the specific activity of the enzyme, its localization and concentration within the cell, the characteristics of the fungal pathogen, and the nature of the host-pathogen interaction. The expression of chitinase in combination with one or several different antifungal proteins should have a greater effect on reducing disease development, given the complexities of fungal-plant cell interactions and resistance responses in plants.

A part of the cell walls of various fungi including plant pathogens consists of chitin and β-1,3-,1-6 glucan. Many chitinases (EC 3.2.1.14) and β-1,3-glucanases (EC 3.2.1.39) produced by plants can inhibit fungal growth probably by dissolving the tips of germ tubes and hyphae. It is known that a combination of chitinase and β-1,3-glucanase shows stronger antifungal activity to a wider range of fungi that when each of them acts separately (Mauch *et al.*, 1988). Moreover, it has been reported recently that one race-specific elicitor AVR4 of the tomato pathogenic fungus, *Cladosporium fulvum,* which is a chitin-binding protein, can protect fungi against plant chitinases. On the other hand, tomato evolved to recognize the AVR4 through the R gene product Cf-4 induces race-specific/variety-specific resistance, which completely shuts out of the fungal spread. Correspondingly, *C. fulvum* evaded the resistance further by mutating into AVR4, which is no longer recognized by CF-4 but still keeps the chitin-binding activity. The interaction between tomato and *C. fulvum* is a very interesting example that plants have established more sophisticated systems, variety-specific resistance or gene-for–gene resistance by the coevolution of host (resistance) and pathogen (pathogenicity) from the defense mechanism mediated by carbohydrate recognition, which seems to be obtained prior to the evolution into plant and animal since it is conserved in both kingdoms. Plant chitinase genes have evolved more rapidly especially around the catalytic site than other genes, suggesting the hypothesis that plant chitinases may coevolve with fungi in response to variation in pathogen defense against chitinolytic activity (Bishop *et al.*, 2000).

Chitinase proteins are widely distributed across diverse biological systems. Chitinases hydrolyze chitin, chitosan, lipochitooligosaccharides, peptidoglycan, arabinogalactan and glycoproteins containing N-acetylglucosamine. Analyses of genome-wide sequence and microarray expression profiling show that chitinase genes are represented by large families, and the individual member genes are expressed in diverse conditions. Chitinase proteins are members of the group of the pathogenesis-related proteins that are strongly induced when host plant cells are challenged by pathogen stress, and thus chitinases constitute an important arsenal of plants against fungal pathogens. Transgenic plants have been produced that overexpress chitinases alone or in conjunction with other defense-related proteins. The phenotype analyses of such plants have shown enhanced disease resistance in a large number of cases. Apart from defense against pathogen stress, chitinases are implicated in relationships between plant cells and fungi (e.g., mycorrhizae associations) and bacteria (e.g., legume/*Rhizobium* associations).

Most of Chitinase have a molecular mass in the range of 15 kDa and 43 kDa. Chitinase can be isolated from *Cicer arietinum* (chickpea), *Cucumis sativus* (cucumber), *Hordeum vulgare* (barley) (Kirubakaran and Sakthivel, 2006), *Nicotiana tabacum* (tobacco) *Phaseolus vulgaris* (black turtle bean), *Solanum lycopersicum* (tomato) and *Vitis vinifera* (grapes). Chitinases can be divided into two categories: Exochitinases, demonstrating activity only for the nonreducing end of the chitin chain; and Endochitinases, which hydrolyze internal α-1,4-glycoside bonds. Chitinase and β-1,3-Glucanase are differentially regulated by Wounding, Methyl Jasmonate, Ethylene, and Gibberellin. These Chitinases have significant antifungal activities against plant pathogenic fungi like *Alternaria* sp. For grain discoloration of rice, *Bipolaris oryzae* for brown spot of rice, *Botrytis cinerea* for blight of Tobacco, *Curvularia lunata* for leaf spot of clover, *Fusarium oxysporum, F. udum, Mycosphaerella arachidicola, Pestalotia theae* for leaf spot of tea and *Rhizoctonia solani* for sheath blight of rice. The main substrate of Chitinases is chitin - a natural homopolymer of α-1,4-linked N-acetylglucosamine residues (Borad and Sriram, 2008). The mode of action of PR-3 proteins is relatively simple, i.e., chitinases cleave the cell wall chitin polymers *in situ*, resulting in a weakened cell wall and rendering fungal cells osmotically sensitive.

3.3.9.3.3 Chitin binding protein (CBP, PR4)

All chitin binding proteins do not possess antifungal activities. CBP can be isolate from plant *Beta vulgaris* (sugar beet), *Hydrangea macrophylla* (hortensia), *Nicotiana tabacum* (tobacco), *Piper longum* (pepper), *Solanum lycopersicum* (tomato) and *Solanum tuberosum* (potato) and bacteria like *Streptomyces tendae.* Expression of the CACBP1 chitin-binding protein isolated from a cDNA library of pepper [(*Capsicum annuum* L.) (CACBP1)] gene was rapidly induced in the incompatible interactions upon pathogen infection, ethephon, methyl jasmonate or wounding [(experimental model plant pepper) (Borad and Sriram, 2008)]. CBP shows a strong inhibitory effect against fungi *Aspergillus* species, *Cercospora beticola, Xanthomonas campestris* and many more and several crop fungal pathogen (Yang and Gong, 2002).

3.3.9.3.4 Thaumatin-like protein (TLP, PR5)

Thaumatin-like proteins comprise polypeptides classes that share homology with thaumatin, sweet protein from *Thaumatococcus danielli* (Bennett) Benth (Cornelissen *et al.*, 1986). Thaumatin-like proteins can be isolated from *Hordeum vulgare* (barley), *Actinidia deliciosa* (kiwifruit), *Zea mays* (maize), *Pseudotsuga menziesii* (Douglas firs), *Nicotiana tabacum* (tobacco), *Solanum lycopersicum* (tomato) and *Triticum* sp. (wheat). Constitutive levels of thaumatin-like protein are typically absent in healthy plants, with the proteins being induced exclusively in response to wounding or pathogen attack like *Uncinula necator* and *Phomopsis viticola*. Although the specific function of many PR5 in plants is unknown, they are involved in the systemic acquired resistance and in response to biotic stress, causing the inhibition of hyphal growth and reduction of spore germination, probably by a membrane permeabilization mechanism and/or by interaction with pathogen receptors (Thompson *et al.*, 2007).

Linusitin is a 25 kDa thaumatin-like protein isolated from flax seeds. Linusitin shows antifungal activity against *Alternaria alternata* by the mechanism of membrane permeabilization. The concentration of protein and lipid and composition of the cell wall of fungi play a major role in these mechanisms (Borad and Sriram, 2008).

3.3.9.3.5 Ribosome inactivating protein (RIP, PR10)

RIP has an inherent antifungal activity. It has been isolated from *Arachis hypogaea* (peanut), *Mirabilis expansa* (mauka), *Nicotiana tabacum* (tobacco), *Pisum sativum* (pea), *Solanum surattense* (nightshade) and *Volvariella volvacea* (mushroom) having a molecular mass of around 30 kDa. Numerous RIPs have been identified but among them some have antifungal activity. RIP isolated from tobacco, known as TRIP, releases adenine residues from the ribosomal and nonribosomal substrata. This is the probable mode of action of inhibition of translation in many fungi like *Cytospora cankar*, *Fusarium oxysporum*, *Pestalotia* sp. and *Trichoderma reesei* and bacteria like *Ervinia amylovora*, *Pseudomonas solancearum*, *Rhizobium leguminosarum*, *Salmonella typhimurium*, and *Shigella asonei* (Kim *et al.*, 2001).

3.3.9.3.6 PR 11-17

Tobacco 'class V' chitinase (PR 11), Radish Rs-AFP3 (PR 12), *Arabidopsis* THI2.1 Barley LTP4 (PR 13), Barley LTP4 (PR 14), Barley OxOa (germin) (PR 15), Barley OxOLP (PR 16) and Tobacco PRp27 (PR 17) are the additional PR groups.

3.3.9.3.7 Other proteins

Theis *et al.* (2003) investigated the inhibitory effects of the antifungal protein (AFP) from *Aspergillus giganteus*. AFP is a highly basic (pI 8.8) polypeptide of 51 amino acids with a high content of cysteine, tyrosine, and lysine residues. MICs of AFP were determined and ranged from 0.1 μg/mL against *Fusarium oxysporum* to 200 μg/mL against *Aspergillus nidulans*. They also showed that the growth inhibitory effect of the AFP is caused by permeabilization of the fungal membranes by using an assay based on the uptake of the fluorescent dye SYTOX Green and ethidium bromide displacement assays. It was found that AFP can inhibit the chitin synthesis by the *in situ* chitin synthase activity assays. These three results indicate that AFP causes cell wall stress and disturbs cell integrity by inactivating chitin synthase that results in membrane permeability (Borad and Sriram, 2008).

Inducible defense-related proteins have been described in many plant species upon infection with oomycetes, fungi, bacteria, or viruses, or insect attack. Several types of proteins are common and have been classified into 17 families of pathogenesis-related proteins (PRs). Others have so far been found to occur more specifically in some plant species. Most PRs and related proteins are induced through the action of the signaling compounds salicylic acid, jasmonic acid, or ethylene, and possess antimicrobial activities *in planta* through hydrolytic activities on cell walls, contact toxicity, and perhaps an involvement in defense signaling (van Loon *et al.*, 2006). However, when expressed in transgenic plants, they reduce only a limited number of diseases, depending on the nature of the protein, plant species, and pathogen involved. As exemplified by the PR-1 proteins in *Arabidopsis* and rice, many homologous proteins belonging to the same family are regulated developmentally and may serve different functions in specific organs or tissues. Several defense-related proteins are induced during senescence, wounding or cold stress, and some possess antifreeze activity. Many defense-related proteins are present constitutively in floral tissues, and a substantial number of PR-like proteins in pollen, fruits, and vegetables can provoke allergy in humans. The evolutionary conservation of similar defense-related proteins in monocots and dicots, but also their divergent occurrence in other conditions, suggest that these proteins serve essential functions in plant life, whether in defense or not. Proteins participating in defense mechanisms after pathogen attack are generally called pathogenesis-related proteins (PR-proteins). They are divided into five classes according to their solubility, biochemical activity, and other chemical properties.

Plant immunity involves several layers of defense that enable plants to recognize potential threats and mount the appropriate defense responses. Plants recognize the presence of potential pathogens by detecting common features present on pathogen surfaces and molecules resulting in microbe-associated molecular patterns or pathogen-associated molecular patterns (MAMPs/PAMPs), such as flagellin, lipopolysaccharide (LPS) and elongation factor Tu (EF-Tu). MAMP/PAMP recognition triggers a first layer of basal resistance resulting in PAMP-triggered immunity (PTI). Some pathogens are able to deploy effector proteins to counteract this first layer of defense and, consequently, trigger a second layer of plant defense resulting in effector-triggered immunity (ETI), also known as gene-for-gene resistance, which occurs upon recognition of the pathogen effectors by the plant resistance proteins. Both PTI and ETI induce stomatal closure and, in most cases, ETI triggers a rapid cell death called the hypersensitive response (HR); both responses limit access and multiplication of the pathogen]. Elicitation of the HR cell death is regulated by reactive oxygen species (ROS) (Lee *et al.*, 2013).

Heterotrimeric G-proteins have been proposed to be involved in many aspects of plant disease resistance, but their precise role in mediating nonhost disease resistance is not well understood (Lee *et al.*, 2013). Heterotrimeric guanine nucleotide-binding proteins (G-proteins hereafter) consisting of three distinct subunits, Gα, Gβ, and Gγ, are conserved in all eukaryotes, and regulate a multitude of physiological processes (Zhang *et al.*, 2012). In the inactive state, the Gα subunit binds guanosine diphosphate (GDP) and remains associated with the Gβγ dimer. External stimuli sensed by the cell surface-localized G-protein-coupled receptors (GCPR) trigger the activation of G-proteins by facilitating an exchange of guanosine triphosphate (GTP) for GDP, resulting in GTP-bound Gα and freed Gβγ dimer. Both these entities can interact with downstream targets of specific signal transduction pathways.

The roles of specific subunits of heterotrimeric G-proteins were evaluated using knock-out mutants of Arabidopsis Gα, Gβ, and Gγ subunits in response to host and nonhost *Pseudomonas* pathogens. Plants lacking functional Gα, Gβ, and Gγ1Gγ2 proteins displayed enhanced bacterial growth and disease susceptibility in response to host and nonhost pathogens. Mutations of single Gγ subunits Gγ1, Gγ2, and Gγ3 did not alter bacterial disease resistance. Some specificity of subunit usage was observed when comparing host-pathogen versus nonhost pathogen. Overexpression of both Gα and Gβ led to reduced bacterial multiplication of nonhost pathogen *P. syringae* pv. *tabaci* whereas overexpression of Gβ, but not of Gα, resulted in reduced bacterial growth of host-pathogen *P. syringae* pv. *maculicola*, compared to wild-type Col-0. Moreover, the regulation of stomatal aperture by bacterial pathogens was altered in Gα and Gβ mutants but not in any of the single or double Gγ mutants (Zhang *et al.*, 2012). Taken together, these data substantiated the critical role of heterotrimeric G-proteins in plant innate immunity and stomatal modulation in response to *P. syringae*. The genes encoding heterotrimeric G-protein are induced by host and nonhost pathogens, and the role of these genes in plant innate immunity is independent of the SA-mediated defense pathway.

PR-proteins with enzymatic activity are β-1.3 glucanases and chitinases. They occur in healthy plants mainly in the apoplast. β-1.3-glucanases participate in the decomposition of glucans like callose which occurs in plant tissues as one of the components of wall modifications involved in resistance responses. Together with chitinases they release components from cell walls of plant pathogenic fungi. Glucans and N-acetyl-glucosamines (parts of chitin) serve as elicitors for the synthesis of these defense substances. Enhanced production of them was noted in more cases of plant-pathogen interaction, in connection with both hypersensitive reaction and systemic acquired resistance. Deposition of callose in response to pathogen attack or mechanical injury is involved in structural defenses (Ales Lebeda *et al.*, 1999). The enzyme callose-synthetase catalyzes the formation of β-glucans which are components of extracellular matrix. In this case, it would be interesting to study the regulation of these processes-resolution and synthesis of compounds composed of β-glucans. Peroxidases are frequently associated with plant defense against pathogens. Peroxidases catalyze the oxidation of substrates like phenol and its derivates, by hydrogen peroxide. They are responsible for the radical dehydrogenation of sinapilalcohol and koniferylalcohol during the lignin synthesis.

3.3.9.4 Polygalacturonase inhibitor proteins

Plants have evolved finely tuned defense mechanisms to restrict cellular infection by microbes. Most of the host plant - pathogen battles take place at a formidable barrier, the plant cell wall. Cell wall degrading enzymes are necessary for pathogens that do not have specialized penetration structures. Necrotrophic pathogens also require cell wall degrading enzymes during the later stages of the invasion process to continue to supply fragments of cell wall polymers as nutrients. Polygalacturonases (PGs) are the most extensively studied cell wall degrading enzymes. They specifically cleave the linkages between 1,D-galacturonic acid residues in nonmethylated homogalacturonan, a major component of pectins present in both primary cell wall and the middle lamella that hold together neighboring plant cells. The interaction of polygalacturonase inhibitor proteins (PGIPs) with their fungal PG counterparts partly determines the effectiveness of the fungal penetration of the plant cell wall (Jason and Davis, 2005).

PGs are important to the pathogenicity of several fungal and bacterial pathogens and are often one of the first proteins that the pathogen secretes when penetrating the plant cell wall. Deletion of PG genes was performed in several fungi and the effect on virulence reduction was studied. There was an obvious reduction in the disease lesion development in cotton bolls with the deletion of the PG *pecA* gene in *Aspergillus flavus* Link: Fr.; whereas restoration of this gene in an *A. flavus* strain lacking *pecA* returned the lesions to their expected size a deletion of the endopolygalacturonase gene *Bcpg1* was demonstrated in *Botrytis cinerea* resulted in a reduced growth rate of lesions during secondary infection (Jason and Davis, 2005).

In *Alternaria citri*, which causes postharvest fruit disease in a range of citrus cultivars, disruption of endopolygalacturonase resulted in reduced disease symptoms on citrus fruit, whereas the disruption of the homologous gene in *Alternaria alternata* (Fr.:Fr.) caused no disease abnormalities, resulting instead in normal virulence on citrus leaves. Also, the disruption of *enpg-1*, a major endopolygalacturonase gene of the chestnut blight fungus *Cryphonectria parasitica* (Murrill) Barr had no effect on the pathogenicity of this fungus toward chestnut trees. While PGs are produced by fungal pathogens to degrade the homogalacturonan component of the plant cell wall, the levels of PGIP correlate, in some cases, with an increased resistance of plants to their fungal pathogen. Increasing susceptibility of ripening pear fruits to *Dothiorella gregaria* and *B. cinerea* correlated with a decline in the concentration of PGIP. In raspberry, the level of PGIP was maximal in immature green fruits, which were more resistant than mature fruits to fungal attack (Jason and Davis, 2005).

3.3.9.5 Plant oligogalacturonides as signaling molecules

Another indirect contribution to the plant defense response is the production of signaling molecules. The signaling molecules are the larger oligomeric breakdown products that result when the PGIPs inhibit the PG's cleavage of oligogalacturonides (OGs) in the plant cell wall. When a PGIP binds to and inhibits a target PG, the hydrolysis of the homogalacturonan substrate is restricted, resulting in larger breakdown products of the polymer. It has been proposed that when PGIP is not active, PG action is not inhibited and generates smaller cleavage products that have little elicitor activity. Although widely discussed in the literature, this hypothesis is yet to be tested *in planta*. OGs varying in chain length between 10 and 15 have been shown to be elicitors of further defense responses as determined by *in planta* assays (Jason and Davis, 2005). Studies confirm that fungal PGs can cause substantial damage in situ during the infection process by cleaving plant cell wall polymers, leading to cell dissociation and tissue collapse. *In planta* expression of endopolygalacturonase II of *Aspergillus niger* showed a 25% reduction in GalUA content with a concurrent dwarfing of transgenic plants (Capodicasa *et al.*, 2004). Work by Boudart *et al.* (2003) further supports elicitor activity with the expression of *CLPG1* from *Colletotrichum lindemuthianum* in tobacco leaves transformed by an *Agrobacterium* transient delivery system.

To begin to understand some OG induction mechanisms, recent studies in transgenic *Arabidopsis* seedlings expressing the calcium reporter aequorin have shown that OGs stimulate a substantial, transient increase in the cytosolic calcium concentration along with an initial rapid generation of H_2O_2, followed by a much larger oxidative burst of H_2O_2. Intracellular increases in these signaling molecules have increased the expression of defense-related genes such as chalcone synthase, glutathione S-transferase, phenylalanine ammonia-lyase, and pathogen-related protein 1. Application of calcium flux inhibitors inhibited both increases in calcium and H_2O_2 while inhibitors of NADPH oxidase blocked only the oxidative burst. *Arabidopsis* NADPH oxidase mutants were generated by disrupting the *AtrbohD* and *AtrbohF* gene functions (Jason and Davis, 2005; and De Lorenzo *et al.*, 2003) described the full genome analysis of *Arabidopsis* responses to oligogalacturonide elicitors and fungal infection. It was shown that the *AtPGIP1* gene expression was unaffected in *nahG* transgenic, *npr1, ein2, jar1,* and *coi1* mutants of *Arabidopsis*. The results of these experiments imply that OGs can activate gene expression during fungal infection independently of salicylic acid, ethylene, or jasmonate. *In planta* expression of fungal and plant PGs have offered some exciting information about the role of this enzyme in cell separation and elicitor release. Boudart *et al.* (2003) transformed a *Colletotrichum lindemuthianum* endopolygalacturonase gene, *CLPG1*, into *Nicotiana tabacum* L. 'Samsun NN,' which lacks detectable PGIP activity against *CLPG1*. One of the most noted effects was the release of OGs from the homogalacturonan domain of pectin and the progressive degradation of pectin in the junction zone and the middle lamella, ultimately leading to cell separation (Jason and Davis, 2005).

The induction of signaling molecules continues to be acknowledged as one possible method for the plant to recognize and respond to any direct attack on its cell wall. These events elicit a variety of responses in the cell that need careful identification to accurately document their involvement in the signaling mechanism. PG-inhibiting proteins defend the plant against specific PGs of pathogenic fungi. A PGIP molecule binds and inhibits a molecule of PG that is secreted by an invading fungus. Each PGIP–PG interaction occurs with a different affinity for each combination, resulting in various efficiencies of inhibition. PGs can exist as a gene family from a single pathogen species. Each isoform of PGIP has a set of fungal PGs that it inhibits while having little to no effect on others

(De Lorenzo and Ferrari, 2002). Pathogenic fungi are not the only producers of PG since both plants and bacteria are known to produce a variety of PGs. However, known PGIPs are likely to inhibit only fungal PGs, where no interaction between *Phaseolus vulgaris* PGIPs and either plant or bacterial PGs with the exception of the exopolygalacturonase from corn pollen. Although there is a variety of PGs produced by numerous fungi, each plant species regularly encounters only a few pathogens and has perhaps evolved a specific arsenal against them. In one study, the highly divergent PGIPs from bean, tomato, and pear all fully inhibited the growth of *B. cinerea* while varying in inhibition activity against the PGs of other fungal pathogens (Jason and Davis, 2005).

3.3.9.5.1 PGIP–PG binding interactions

PGIP functions by binding. To function effectively, it must closely interact with the structure of an invading PG such that it either blocks PG and pectin from interacting effectively or distorts the PG for the same effect. The exact mechanism of binding between PGIP and PG is not known, and it is possible to have multiple mechanisms, both between isoforms and between different PGs on the same isoform (Jason and Davis, 2005). A purified cotton PGIP was found to have different mechanisms of inhibition against PG from *Aspergillus niger* and *Verticillum dahliae* Kleb, a wilt-causing fungus that affects a variety of plant species. Against *Aspergillus niger* PG, the cotton PGIP showed noncompetitive inhibition while inhibition of *V. dahliae* PG showed allosteric interaction with its inhibitor (James and Drubery, 2001). Recent work by Kemp *et al.* (2004) supports the role of *Phaseolus vulgaris* PGIP-2 functioning as either an inhibitor or activator of *Aspergillus niger* polygalacturonase A and polygalacturonase C in a pH-dependent manner in the presence of homogalacturonan. They, therefore, suggest a reevaluation of the conventional description of PGIPs as PG-inhibiting proteins. Fish and Madihally (2004) mathematically modeled the inhibitor activity and relative binding affinities of the PG–PGIP system with two models based on both reversible and irreversible binding. Both models were validated using inhibition data from both pure and crude protein preparations isolated from developing cantaloupe fruit.

Polygalacturonase inhibitor proteins (PGIPs) act to thwart fungal penetration of the plant cell wall as an induced plant defense to early fungal attack. PGIPs are ubiquitous cell wall localized proteins that specifically inhibit the activity of fungal polygalacturonases (PGs), some of which are important fungal pathogenicity factors (Jason and Davis, 2005). They also indirectly contribute to the defense response by modulating the PG cleavage of pectin polymers, thereby forming longer-chain oligogalacturonides that are proposed to induce further defense responses within the plant. PGIPs and plant resistance genes are characterized by their leucine-rich repeat motif, which provides specific recognition of pathogen-derived molecules such as PGs. The high level of polymorphism between both PGIP and PG proteins provides clues toward understanding the functional evolution of plant leucine-rich repeat proteins and their ligands. This knowledge will be important for the development of directed disease control strategies for crop protection.

3.3.10 Microbe-Associated Molecular Patterns or Pathogen-Associated Molecular Patterns (MAMPs/PAMPs)

Entry of pathogens and the subsequent activation of inducible defense responses or disease symptom development is primarily prevented by performed barriers on the plant surface such as wax layers, rigid cell walls, cuticular lipids, antimicrobial enzymes or secondary metabolites. Pathogens which overcome these defensive layers are systematically encountered and defense response is elicited by two interconnected mechanisms, namely, microbial (or pathogen)-associated molecular patterns (MAMP/PAMP)-triggered immunity (MTI/PTI) and effector-triggered immunity (ETI) (Mehanathan Muthamilarasan and Manoj Prasad, 2013).

Pathogen-associated molecular patterns, or PAMPs, are molecules associated with groups of pathogens, that are recognized by cells of the innate immune system. These molecules can be referred to as small molecular motifs conserved within a class of microbes. They are recognized by toll-like receptors (TLRs) and other pattern recognition receptors (PRRs) in both plants and animals. They activate innate immune responses, protecting the host from infection, by identifying some conserved nonself molecules. Bacterial lipopolysaccharide (LPS), an endotoxin found on the bacterial cell membrane of a bacterium, is considered to be the prototypical PAMP. LPS is specifically recognized by TLR-4, a recognition receptor of the innate immune system. Other PAMPs include bacterial flagellin (recognized by TLR-5), lipoteichoic acid from Gram positive bacteria, peptidoglycan, and nucleic acid variants normally associated with viruses, such as double-stranded RNA (dsRNA), recognized by TLR-3 or unmethylated CpG motifs, recognized by TLR-9 (Jones and Dangl, 2006). Although the term 'PAMP' is relatively new, the concept that molecules derived from microbes must be detected by receptors from multicellular organisms has been held for many decades, and references to an 'endotoxin receptor' are found in much of the older literature.

The term 'PAMP' has been criticized on the grounds that most microbes, not only pathogens, express the molecules detected; the term microbe-associated molecular pattern, or MAMP, has therefore been proposed. A virulence signal capable of binding to a pathogen receptor, in combination with an MAMP, has been proposed as one way to constitute a (pathogen-specific) PAMP. Plant immunology frequently treats the terms 'PAMP' and 'MAMP' interchangeably, considering their recognition to be the first step in plant immunity, PTI (PAMP-triggered immunity), a relatively weak immune response that occurs when the host plant does not also recognize pathogenic effectors which damage it or modulate its immune response. The plant immune system is composed of surveillance systems that perceive several general microbe elicitors, which allow plants to switch from growth and development into a defense mode, rejecting most potentially harmful microbes. The elicitors are essential structures for pathogen survival and are conserved among pathogens. The conserved microbe-specific molecules referred to as microbe- or pathogen-associated molecular patterns (MAMPs or PAMPs), are recognized by the plant innate immune systems pattern recognition receptors (PRRs). General elicitors like flagellin, elongation factor Tu (EF-Tu), peptidoglycan (PGN), lipopolysaccharides (LPS), Ax21 (Activator of XA21-mediated immunity in rice), fungal chitin, and β-glucans from oomycetes are recognized by plant surface localized PRRs. Several of the MAMPs and their corresponding PRRs have, in recent years, been identified (Newman *et al.*, 2013).

MAMPs include oligogalacturonides, ergosterol, bacterial flagellin, Pep-13, xylanase, cold-shock protein and lipopolysaccharides (LPS).Recognition of particular MAMPs and subsequent elicitation of immune responses are confined to plant species belonging to a single family. For instance, EF-Tu recognition is reported only in the Brassicaceae family. Conversely, other MAMPs including chitin, LPS and flagellin trigger immune responses in various host species, even if there is some degree of specificity and perception efficacy for a plant family or species (Zipfel *et al.*, 2006). Many plant pathogens produce lytic enzymes to infringe the structural barriers of plant tissues. The products such as cell wall fragments, cutin monomers, and peptides, which are generated as a consequence, can function as endogenous elicitors called damage-associated molecular patterns (DAMPs). These DAMPs characteristically emerge in the apoplast and serve as danger signals to induce innate immunity similar to MAMPs (Henry *et al.*, 2012).

3.3.10.1 Pattern Recognition Receptors (PRR)

These are the plant receptors which perceive MAMPs/DAMPs and can be divided into surface and intracellular receptors. The surface receptors called PRRs are known to detect both MAMPs and effectors. The PRR family includes receptor-like kinases (RLK) and receptor-like proteins (RLP). RLK resides in the plasma membrane and comprises of a putative extracellular ligand-binding domain, a single transmembrane domain and an intracellular serine/threonine kinase domain (Mehanathan Muthamilarasan and Manoj Prasad, 2013). Although various MAMPs and their corresponding PRRs have been reported, only a few are well characterized (Table 3.4).

Table 3.4 MAMPs/DAMPs and their Corresponding PRRs

Source	*MAMPs/DAMPs*	*Epitope*	*PRR*	*PRR Type*
Bacteria	Flagellin	flg22	FLS2	LRR-RLK
	Cold-shock proteins	RNP-1 motif	unidentified	unidentified
	Xoo derived	unidentified	Xa26	LRR-RLK
	Ef-Tu	elf8	EFR	LRR-RLK
	Peptidoglycan hairpin	HrpZ	unidentified	unidentified
	Ax21	axYs22	Xa21	LRR-RLK
	lipopolysaccharides	unidentified	unidentified	unidentified
	Peptidoglycan	PGN	LYM1 & LYM3	LysM-RLP

Fungi	Ergosterol	unidentified	unidentified	unidentified
	Cerebrosides	sphingoid base	unidentified	unidentified
	Xylanase	TKLGE pentapeptide	LeEIX2	LRR-RLP
	Necrosis-inducing proteins (NLP)	unidentified	unidentified	unidentified
	Chitin	chitin oligosaccharides	CERK1	LysM-RLK
	β-glucan	β-heptaglucan	GEBP	unidentified
Oomycetes	β-glucan	hepta-β-glucoside	GnGBP	unidentified
	Lipid-transfer proteins (elicitins)	unidentified	unidentified	unidentified
	Transglutaminase	Pep13	unidentified	unidentified
	Cellulose-binding elicitor lectin (CBEL)	unidentified	unidentified	unidentified
DAMPs	Prosystemin	systemin	SR160	LRR-RLK
	PEPR1	Pep1	PEPR1	LRR-RLK
	Homoglacturonan	oligogalacturonides	WAK1	EGF-RLK

PAMPs represent structures that are essential for microbial life and that are typically harbored by invading pathogens. These include cell surface constituents but may also be secreted enzymes or proteins normally located in the cytoplasm. A broad array of structurally diverse PAMPs has been described originating from fungal, oomycete and bacterial pathogens. Most of these PAMPs are oligosaccharides, glycopeptides, and peptides. Some of these patterns such as Pep-13, xylanase, and cold-shock protein are only perceived by a narrow range of plant species belonging to only one plant family (Guillaume Henry *et al.*, 2012). A representative example is EF-Tu in the family Brassicaceae. By contrast, other PAMPs such as chitin, LPS and flagellin trigger defense responses in many host species even if there is some degree of specificity and perception efficacy for a plant family or species as in the case of flagellin (Zipfel, 2006).

PAMPs are perceived at the plant cell surface by high-affinity receptors typically consisting of an extracellular ligand-binding domain with leucine-rich repeats (LRR), a single transmembrane domain and an intracellular serine/threonine kinase signaling domain. They are referred to as receptor-like kinases (RLK). Receptor-like proteins (RLPs) are similarly structured but lack the cytoplasmic kinase domain. In *Arabidopsis*, 610 RLKs and 56 RLPs have been identified. A large number of genes encoding RLKs and RLPs are transcriptionally induced upon PAMP treatment, illustrating the large diversity of such perception systems and suggesting their potential role in the defense.

In a more indirect way, plants can also detect the presence of pathogens through the perception of endogenous compounds that have been released from structural barriers or other macromolecules by lytic enzymes produced by the invader or by the host itself.

Such DAMPs typically appear in the apoplast and may thus, like PAMPs, play the role of the signal for danger to induce innate immunity. For instance, oligogalacturonides are released by microbial enzymes and putatively recognized by the receptor WAK1 (D'Ovidio *et al.*, 2004). Emission of these endogenous signals allows disrupted or injured cells to communicate their damage to the tissue or systemically to all organs. Systemin is formed in damaged tomato leaves and is further perceived as a primary signal for systemic defense induction. Similarly, the 23-residue peptide AtPep1 is released from precursor proteins in response to wounding and triggers an innate immune response in *Arabidopsis via* recognition by the PEPR1 receptor.

Considering the large and still increasing number of molecular patterns and effectors harbored by pathogens, it is clear that plants have evolved to mount very efficient nonself recognition systems. As they also develop performing and specific receptors to detect endogenous DAMPs, the concept of 'stranger' recognition could be extended to 'danger' recognition. The perception of all these signals appears to trigger the same stereotypical defense program even if it varies in terms of kinetic and strength of the response between PTI and ETI (Boller *et al.*, 2009).

Successful pathogens, however, have evolved multiple virulence factors to suppress MAMP-triggered immunity. Specifically, diverse pathogenic bacteria have employed the type III secretion system to deliver a repertoire of virulence effector proteins to interfere with host immunity and promote pathogenesis. Plants challenged by pathogens have evolved the secondary plant innate immunity (Ping He *et al.*, 2006). In particular, some plants possess the specific intracellular disease resistance (R) proteins to effectively counteract virulence effectors of pathogens for effector-triggered immunity. This potent but cultivar-specific effector-triggered immunity occurs rapidly with localized programmed cell death/hypersensitive response to limit pathogen proliferation and disease development. Remarkably, bacteria have further acquired virulence effectors to block effector-triggered immunity. This review covers the latest findings in the dynamics of MAMP-triggered immunity and its interception by virulence factors of pathogenic bacteria.

The perception of different MAMPs by specific pattern-recognition receptors (PRRs) induces the defense responses that contribute plant immunity to both pathogenic and nonpathogenic microbes. In mammals, MAMPs are recognized by a limited number of PRRs including Toll-like receptors (TLRs) and CATERPILLER/NOD-like receptors (NLRs), which detect distinct microbial components and activate immune responses (Ting and Davis, 2005; Akira *et al.*, 2006). Successful pathogens have acquired multiple virulence factors to suppress host immunity for their survival. Many Gram-negative pathogenic bacteria inject a set of effector proteins into host cells through the type III secretion system (T3SS) to interfere with host innate immunity (Grant *et al.*, 2006). However, some of these specific virulence effector proteins are counteracted by the disease resistance (R) proteins in plants. Hundreds of plant genes encode putative R proteins that share conserved nucleotide-binding and leucine-rich repeat domains (NB-LRR) for pathogen sensing and defense signaling (Meyers *et al.*, 2003).

Plants perceive several general elicitors from both host and nonhost pathogens. These elicitors are essential structures for pathogen survival and are for that reason conserved among pathogens. These conserved microbe-specific molecules also referred to as Microbe or Pathogen Associated Molecular Patterns (MAMPs or PAMPs), are recognized by the plant innate immune systems Pattern Recognition Receptors (PRRs). General bacterial elicitors, like lipopolysaccharides (LPS), flagellin (Flg), elongation factor Tu (EF-Tu), cold shock protein (CSP), peptidoglycan (PGN) and the enzyme superoxide dismutase (SodM) are known to act as MAMPs and induce immune responses in plants or plant cells. The corresponding PRRs for some of these bacterial elicitors have, in recent years, been identified

Peroxidases participate in the synthesis of flavons, stilbens, and other phenolic secondary metabolites. They are represented by many of the isoenzymes. Peroxidase polymorphism could be also used as a biochemical marker related to the different levels of field resistance. Peroxidase participates in processes which occur in the extracellular matrix. Their association with the cell wall was confirmed (Ales Lebeda *et al.*, 1999).

Plants have evolved innate immune systems that recognize the presence of pathogens and initiate effective defense responses, whereas successful pathogens have evolved effector proteins that can suppress host immune responses. Furthermore, effectors can themselves act as elicitors and can be disabled by the host. The studies on plant–pathogen interactions had described several interesting information such as unique immune strategies adopted by plants against highly pathogenic strains, R-protein-mediated surveillance and memory of immunity in SAR. Considerable progress has been made in understanding the role and mechanism of RNA silencing, substantiating the fact that RNA silencing operates as an innate antiviral defense in higher plants. The development of powerful RNAi *in planta* assays, the isolation and further characterization of RNA silencing mutants, and the investigation of viral-encoded silencing suppressors provide an exciting and fascinating research ground for the future. Obviously, a better understanding of plant–pathogen interactions and the molecular details of how plants are able to heritably protect their offspring against potential enemies in their environment without making changes in their DNA sequence will pave the way for the improvement of disease resistance in economically important crops, thereby ensuring food security, an issue of global importance (Mehanathan Muthamilarasan and Manoj Prasad, 2013).

3.3.11 Enzymes

In plant pathogenesis, enzymes play a crucial role, and they are involved in the external and internal interactions. To restrict the development of pathogens like fungi, the plants formed many defense mechanisms. They built mechanical barriers from lignin, suberin, and callose and produced a lot of antimicrobial compounds with low molecular weight like phenols, chinons, alkaloids and others. Enhanced production of some enzymes and their activity increasing is one of the most important processes in plant defense. These enzymes frequently occur in many isoforms and are involved in the synthesis of defense substances or have a direct antimicrobial activity. The details on the role of enzymes in plant pathogenesis have been discussed in another chapter.

Cutinases, cellulases, pectinases and lignases are often secreted by the pathogenic organism. Fungi, nematodes, and bacteria are all known to produce one or more of the above enzymes in specific pathogen-host combinations. Viruses and viroids are generally not considered to secrete enzymes, although some viruses may encapsidate an enzyme in their particle. Pathogenic organisms either continually secrete enzymes or upon contact with the host Plant. The first surface an organism comes into contact with is cuticle and the cell wall of the plant. The cuticle is comprised of a complex wax, cutin, which impregnates the cellulose wall. The cell wall is comprised of cellulose, which makes up the structural framework of the wall, along with the matrix molecules hemicellulose, glycoproteins, pectin, and lignin. Thus, penetration into living parenchymatous tissues and degradation of the middle lamella is due to the action of one or more enzymes which degrade these chemical substances.

Cutinases degrade the cutin on the cuticle layer presoftening the tissue for mechanical penetration or as a first step in tissue degradation. Studies have shown that several fungi and at least one bacterial species produce cutinases. Further, evidence indicates that cutinases are continually produced, albeit in low concentrations, with degradation products often inducing even higher levels of cutinase secretion. Studies have shown that in some organisms, cutinase production may be linked to virulence. Pectic substances comprise the middle lamella and also form an amorphous gel between the cellulose microfibrils in the primary cell wall. Pectin degrading substances often termed pectinases or pectolytic enzymes include pectin methylesterases (PME), polygalacturonases (PG) and pectin lyases or transeliminases. Pectin methylesterases remove small groups such as methyl groups (CH_3) often altering solubility and thus affecting the rate of chain splitting by polygalacturonase and pectin lyase. Polygalacturonases split chains by adding a molecule of water while pectin lyases split chains by removing a water molecule from the linkage. Pectin degrading enzymes are involved in a wide range of plant diseases particularly in the soft rot diseases [Examples: bacterial soft rot due to *Erwinia carotovora*, leaf drop of lettuce and water soft rot of crucifers (*Sclerotinia sclerotiorum*) and damping off in various seedlings due to *Rhizoctonia solani*]. Organisms such as these, as well as, elaborate pectic enzymes lead to tissue maceration. In fact, these enzymes are sometimes referred to as macerating enzyme.

Cellulase is any of several enzymes produced chiefly by fungi, bacteria, and protozoans that catalyze cellulolysis, the decomposition of cellulose and some related polysaccharides; specifically, the hydrolysis of the 1,4-β-D-glycosidic linkages in cellulose, hemicelluloses, lichenin and cereal β-D-glucans. Cellulases break down the cellulose molecule into monosaccharides ('simple sugars') such as β-glucose or shorter polysaccharides and oligosaccharides. The name is also used for any naturally occurring mixture or complex of various such enzymes, that act serially or synergistically to decompose cellulosic material. Cellulose is the major framework molecule of the plant cell wall existing as microfibrils with matrix molecules (glycoproteins, hemicelluloses, pectins, lignins) filling the spaces between the microfibrils and cellulose chains. Cellulases have been shown to be produced by many pathogenic fungi, bacteria, and nematodes. Cellulolytic enzymes play a role in softening and disintegration of cell walls. No doubt cellulolytic enzymes are involved in the invasion and spread of the pathogen, but also are instrumental in the collapse of cells and tissues. Cellulolytic enzymes participate indirectly in disease development by releasing soluble sugars that may be used as nutrients by pathogens and also may be involved in the release of materials in the vascular system interfering with transport or translocation of water.

A hemicellulose, also known as polyose, is any of several heteropolymers (matrix polysaccharides), such as arabinoxylans present along with cellulose in almost all plant cell walls. While cellulose is crystalline, strong, and resistant to, hemicellulose has a random, amorphous structure with little strength. It is easily hydrolyzed by dilute acid or base as well as myriad hemicellulase enzymes. Hemicelluloses are complex polysaccharide polymers that link the ends of pectic compounds to cellulose microfibrils. Since hemicelluloses are such a diverse group of polymers such as xyloglucans, glucomannans, glactomannans, arabinoglucans, etc., several hemicellulases have been identified in many plant pathogenic fungi. The mechanism by which they participate in cell wall breakdown is not clear, nor is it known how they contribute to pathogenesis.

Lignin is a phenylpropanoid which is found in the middle lamella and secondary cell wall of plants. More than anything else, lignin confers the tough, woody nature to woody tissues. Most lignin degradation is by basidiomycetes known as white-rot fungi. These fungi produce ligninases that enable the fungi to utilize lignin. Hydrolytic enzymes are produced by some plants in response to pathogens and often accumulate in extracellular spaces where they degrade the cell walls of pathogenic fungi. Chitinases are enzymes that catalyze the degradation of chitin, a polymer with a backbone similar to cellulose that is present in the cell walls of true fungi. Glucanases are enzymes that catalyze the degradation of glycosidic linkages in glucans, a class of polymers similar to cellulose that is present in the cell walls of many oomycetes (water molds). *In planta* analysis has verified the antifungal properties of these compounds, and transgenic plants expressing high levels of these enzymes exhibit increased resistance to a wide range of both foliar and root pathogens. Lysozymes are hydrolytic enzymes that are capable of degrading bacterial cell walls

A study was conducted to scrutinize the response of banana (var. Grand Naine) plants when interacting with dead or live pathogen, *Fusarium oxysporum* f. sp. *cubense*, a causative agent of Panama disease (Janki *et al.*, 2013). The response of plants was evaluated in terms of induction of defense-related marker enzyme activity, namely, peroxidase (POX), polyphenol oxidase (PPO), β-1,3 glucanase, chitinase, and phenolics. Plant's interaction with live pathogen resulted in early induction of defense to restrain penetration as well as antimicrobial productions. However, pathogen overcame the defense of plant and caused disease. Interaction with dead pathogen resulted in escalating defense response in plants. Later on plants inoculated with dead pathogen showed resistance to even forced inoculation of the live pathogen. Results obtained in the present study suggested that dead pathogen was able to mount defense response in plants and provide resistance to Panama disease upon subsequent exposure. Therefore, preparation from dead pathogen could be a potential candidate as a biocontrol agent or plant vaccine to combat Panama disease.

Root and crown rot of cucumber caused by *Pythium aphanidermatum* can be suppressed by various rhizobacteria or PGPR (plant growth promoting rhizobacteria). When cucumber roots were treated with *Pseudomonas corrugata* 13 or *Pseudomonas aureofaciens* 63–28, phenylalanine ammonia-lyase (PAL) activity was stimulated in root tissues in 2 days and this activated accumulation lasted for 16 days after bacterization. Peroxidase (PO) and polyphenol oxidase (PPO) activities were increased in roots 2–5 days after bacterization with *P. corrugata* strain 13. After bacterized cucumber roots had been challenged with *P. aphanidermatum*, the enzyme activities of PAL, PO and PPO increased as the disease developed on the roots. These accumulations peaked 4–6 days after pathogen inoculation. A split root system demonstrated that the three enzymes could be systemically induced by the *Pseudomonas* strains 63–28 and 13, as well as *P. aphanidermatum*. Furthermore, the isoperoxidase native PAGE (polyacrylamide gel electrophoresis) analysis indicated that the peroxidase isomer forms in cucumber roots induced by rhizobacteria were different from that in roots infected with *P. aphanidermatum*. These results suggest that the plant defense enzymes could be stimulated in cucumber roots which have been colonized by nonpathogenic rhizobacteria or in a compatible interaction between cucumber and *P. aphanidermatum*. The mechanisms of PO activation by the rhizobacteria may be different from those of pathogen infection.

Most widely used *Trichoderma* strains against fungal diseases of plants are T. *harzianum, T. virens*, and *T. viride*. The process of parasitizing phytopathogenic fungi includes several stages, viz., the host-directed growth of *Trichoderma*; recognition of the host and attachment; synthesis and secretion of antibiotics and extracellular enzymes hydrolyzing cell wall components; penetration of hyphae and lysis of the host. Ultrastructural and histochemical studies demonstrate that enzymes of *Trichoderma* cause localized lysis of cell walls of phytopathogenic fungi at points of contact between the antagonist and the host (Saba Hasan *et al.*, 2014) (Table 3.5). The presence of chitin and/or fibrils of β-glucan, embedded in the protein matrix, in the cell walls of many phytopathogenic fungi, suggests that the lysis of their hyphae in the course of parasitism is likely affected by β-glucanases, chitinases, and proteases. Hence, to be able to degrade phytopathogen cell walls, the antagonistic *Trichoderma* induces the production of extracellular hydrolytic enzymes, responsible for the direct attack against the pathogen. These enzymes have been reported mainly in isolates of *T. harzianum*. Most of the work has been carried out on strains of *T. viride, T. reesei, and T. harzianum*. These strains have been studied for their ability to produce enzymes, namely, endoglucanases, exoglucanases, and cellobiase, which act synergistically in the conversion of cellulose to glucose (Saba Hasan *et al.*, 2014).

An enzymatic system that is involved in cell wall degradation by an antagonistic organism is β-glucan degrading enzymes. Two mechanisms of glucan degradation have been reported: exo- and endo-glucanases, both of which act synergistically in glucan degradation. β-glucan degrading enzymes are classified according to the type of β-glucosidic linkages: 1,4-β-glucanases (including cellulases), 1,3-β-glucanases, and 1,6-β-glucanases. Because 1,3-β-glucan is a structural component of fungal cell walls, the production of extracellular 1,3-β-glucanases has been reported as an important enzymatic activity in biocontrol microorganisms. In addition to chitin and glucans, filamentous fungi cell walls contain proteins. Thus, the production of proteases may play a role in antagonism. There are a number of chitinolytic enzymes produced by *Trichoderma harzianum* and that there are multiple forms of chitinases (Saba Hasan *et al.*, 2014).

Table 3.5 Hydrolytic Enzymes of Trichoderma Spp. Degrading the Cell Walls of Phytopathogenic Fungi in the Course of Mycoparasitism

Enzyme	*Substrate*	*Product*
Endochitinase	Chitin, chitooligosaccharides	Chitotetraose, chitotriose, and chitobiose
Chitin 1,4-β-chitobiosidase, or exochitinase		Chitobiose
N-acetyl-β-D-glucosaminidase	Chitooligosaccharides and their derivatives	N-acetyl-β-D-glucosamine
Endo-β-1,3-glucanase	Laminarin	Laminaritetraose, laminaribiose, and glucose
Exo-β-1,3-glucanase	Laminarin and β-1,3-oligosaccharides	Glucose

Lipids and lipid metabolites influence pathogenesis and resistance mechanisms associated with plant-microbe interactions. Some microorganisms sense their presence on a host by perceiving plant surface waxes, whereas others produce toxins that target plant lipid metabolism. In contrast, plants have evolved to recognize microbial lipopolysaccharides (LPSs), sphingolipids, and lipid-binding proteins as elicitors of defense response (Jyoti Shah, 2005). Recent studies have demonstrated that the plasma membrane provides a surface on which some plant resistance (R) proteins perceive pathogen-derived effectors and thus confer race-specific resistance. Plant cell membranes also serve as reservoirs from which biologically active lipids and precursors of oxidized lipids are released. Some of these oxylipins, for example, jasmonic acid (JA), are important signal molecules in plant defense. Arabidopsis thaliana is an excellent model plant to elucidate the biosynthesis and metabolism of lipids and lipid metabolites, and the characterization of signaling mechanisms involved in the modulation of plant defense responses by phytolipids. This review focuses on recent studies that highlight the involvement of lipids and lipid metabolites, and enzymes involved in lipid metabolism and modification in plant disease resistance.

Many plants and seeds contain proteins that specifically inhibit pathogen and pest enzymes by forming complexes that block active sites or alter enzyme conformations, ultimately reducing enzyme function. These proteins are generally small and rich in the amino acid cysteine. They include defensins, amylase inhibitors, lectins and proteinase inhibitors. Unlike simple chemicals such as terpenoids, phenolics, and alkaloids, proteins require a lot of plant resources and energy to produce. Consequently, many defensive proteins are only made in significant quantities after a pathogen or pest has attacked the plant. Once activated, however, defensive proteins and enzymes effectively inhibit fungi, bacteria, nematodes, and insect herbivores.

3.3.12 Defensins

Defensins are small cationic peptides of 45–54 amino acid residues with a conserved signature of cysteines, which can form three to four disulfide bridges. Plant defensins are cationic peptides that are ubiquitous within the plant kingdom and belong to a large superfamily of antimicrobial peptides found in several organisms collectively called defensins. The primary structure of these peptides includes 45 to 54 amino acid residues with considerable sequence variation. At the level of three-dimensional structure, they are small and globular, composed of three antiparallel β-sheets and one α-helix, which is highly conserved among these peptides. The three-dimensional structure is stabilized by four disulfide bridges formed by eight strictly conserved Cys residues. Two of these bridges compose the Cys-stabilized α-helix β-strand motif, which is found in other peptides with biological activities. Plant defensins present numerous biological activities, such as inhibiting protein synthesis, ion channel function and α-amylase and trypsin activity; impairing microbial, root hair and parasitic plant growth; mediating abiotic stress and Zn tolerance; altering ascorbic acid redox state; stimulating sweet taste sensation; serving as epigenetic factors; affecting self-incompatibility; and promoting male reproductive development. Some of these biological activities, such as microbial growth inhibition and sweet taste induction, coupled with a scaffold that provides these peptides with incredible physicochemical resistance to harsh environments and the potential for simple amino acid substitution, raise the opportunity to improve the function of defensins or introduce new activities, endowing these peptides with great biotechnological and medical significance (Carvalho and Gomes, 2011). Plant defensins exhibit a conserved tertiary structure that consists of a triple-stranded antiparallel β-sheet and one α-helix that are stabilized into a compact shape by the disulfide bridges. Variations in the amino acids are reflected by small conformational changes in the tertiary structure that contribute to the broad range of biological activities in these proteins. Only one amino acid substitution can change the spectrum of activity exhibited by these peptides.

They are termed plant defensins because they are structurally related to defensins found in other types of organism, including humans. In *Arabidopsis thaliana,* at least 13 putative plant defensin genes (PDF) are present, encoding 11 different plant defensins. Two additional genes appear to encode plant defensin fusions. Plant defensins inhibit the growth of a broad range of fungi but seem nontoxic to either mammalian or plant cells. Antifungal activity of defensins appears to require specific binding to membrane targets (Bart P.H.J. Thomma *et al.,* 2002). Plant defensins are small (45–54 amino acids) highly basic cysteine-rich peptides that are apparently ubiquitous throughout the plant kingdom. All plant defensins identified so far have eight cysteines that form four structure stabilizing disulfide bridges. Study of the three-dimensional structure of a number of plant defensins has shown that the structure comprises a triple-stranded β-sheet with an α-helix in parallel.

They are and function as host defense peptides. They are active against bacteria, fungi and many enveloped and nonenveloped viruses. They consist of 18-45 amino acids including six (in vertebrates) to eight conserved cysteine residues. Cells of the immune system contain these peptides to assist in killing phagocytosed bacteria, for example, in neutral granulocytes and almost all epithelial cells. Most defensins function by binding to the microbial membrane and, once embedded, forming pore-like membrane defects that allow efflux of essential ions and nutrients. Defensins are one of the largest and most studied families of antimicrobial peptides. More than 300 defensins have been identified so far [(Table 3.6) (Ariane F. Lacerda *et al.*, 2014)] and they are represented in a range of organisms including mammals, birds, invertebrates, plants and recently in the ebony-cup fungus. Plant defensins MsDef1 and MtDef4 (Figs. 3.60 and 3.61) are potent inhibitors of the growth of several filamentous fungi including *Fusarium graminearum.*

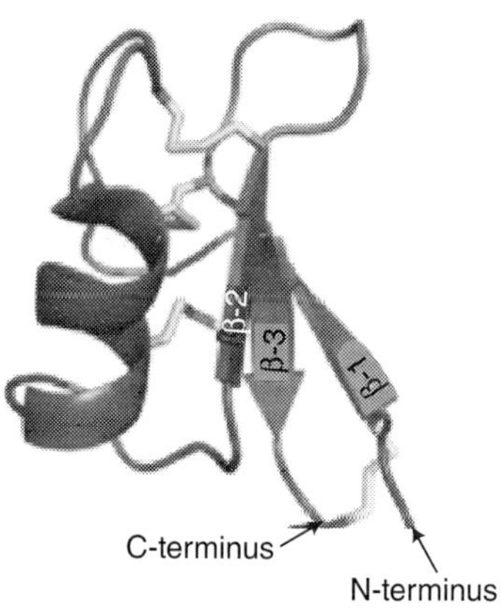

Fig. 3.60 MsDef1

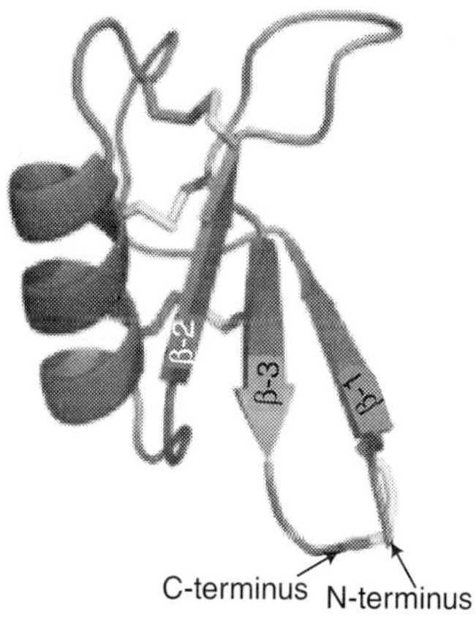

Fig. 3.61 MsDef4

Defensins are an integral part of the plant innate immune system. Most plant defensins already characterized show a constitutive pattern of expression with up-regulation in response to pathogen attack, injury and some abiotic stresses. Defensins were first isolated from the endosperm of barley (*Hordeum vulgare*) and wheat (*Triticum aestivum*). They are widely distributed and may be present in most plants. Defensins are best characterized in seeds, but can be found in virtually all types of plant tissues including leaves, pods, tubers, fruit, roots, bark, and floral tissues. They exhibit a wide range of biological activities that serve to inhibit the growth of many fungi and bacteria. Some defensins also inhibit digestive proteins in herbivores. The precise mechanisms employed by plant defensins to inhibit fungi and bacteria are still being characterized, but they appear to act upon molecular targets in the plasma membrane of pathogens. These defensins may inhibit preexisting ion channels or form new membrane pores that disrupt cellular ion balance. Digestive enzyme inhibitors are proteins that block the normal digestion and absorption of nutrients by vertebrate and invertebrate herbivores.

α-amylase inhibitors are proteins commonly found in legumes that bind to amylase enzymes and inhibit starch digestion.

Table 3.6 Some Plant Defensins and Their IC_{50} Concentrations Against Fungal Targets

Plant defensin	*Organism*	*Target organism*	*IC50(μM)*
PgD5	*Picea glauca*	*Verticillium dahliae*	0,4
Defensin-like peptide	*Phaseolus vulgaris*	*Mycosphaerella arachidicola*	3,9
NmDef02	*Nicotiana megalosiphon*	*Fusarium oxysporum*	1
Pdc1	*Zea mays*	*Fusarium graminearum*	0,75
Limyin	*Phaseolus limensis*	*Fusarium solani*	8,6
TvD1	*Tephrosia villosa*	*Pheaoisariopsis personata*	1,9
MtDef4	*Medicago truncatula*	*Fusarium graminearum*	0,75
MsDef1(alfAFP)	*Medicago sativa*	*Fusarium graminearum*	1,2
Psd1	*Pisum sativum*	*Neurospora crassa*	2
alfAFP	*Medicago sativa*	*Verticillium dahliae*	1
HsAFP1	*Heuchera sanguinea*	*Septoria tritici*	0,1
AhAMP1	*Aesculus hippocastanum*	*Leptosphaeria maculans*	0,1
RsAFP2	*Raphanus sativus*	*Pyricularia oryzae*	0,08

There are two major hypotheses that try to explain the mechanism of action of antimicrobial defensins, viz., the carpet model and the pore model. In both models, defenses are described to interact with the negatively charged molecules present in the cell membrane of pathogens, causing an increase of its permeabilization, leading to cell leakage and death by necrosis. While the carpet model emphasizes the pore formation of several peptides into the membrane, the pore model shows that peptides form oligomers that, then, form multiple pores into the cell membrane. However, there is an alternative hypothesis, where defensins do not damage the cell membrane, but interact with the phospholipids, leading to an increase of ion permeability, or even to the transportation of these peptides to the intracellular environment. Hence, they can also enhance reactive oxygen species (ROS) and activate programmed cell death (Ariane F. Lacerda *et al.*, 2014).

One of the first studies that attempted to highlight this class of plant antimicrobial defensins was carried out with two peptides isolated from Radish seeds, Rs-AFP1, and Rs-AFP2. Both peptides were assayed for 20 different plant pathogenic fungi and the lower protein concentration required for 50% inhibition of fungal growth (IC_{50}) was obtained by Rs-AFP2, when assayed against *Pyricularia oryzae* (Ariane F. Lacerda *et al.*, 2014). Its IC_{50} ranged from 0.08 to 5 μM. Osborn *et al.* (1995) assayed four AMP isolated from *Aesculus hippocastanum* (Ah-AMP1), *Clitoria ternatea* (Ct-AMP1), *Dahlia merckii* (Dm-AMP1), *Heuchera sanguinea* (Hs-AFP1) against eight different fungi in the presence, or absence, of inorganic ions. The lower IC_{50}, around 0.1 μM, was acquired when Ah-AMP1 was tested towards *Cladosporium sphaerospermum*, *Leptosphaeria maculans*, and

Septoria tritici. Hs-AFP1 presents the same antifungal activity when assayed against *Septoria tritici*. In all the studies, inorganic ions decreased IC_{50}. When visualized under a microscope, it was possible to see that such antifungal peptides caused distinct morphological changes during germ tube elongation and hyphae development, like multiple hyphae buds or the diminished of the rate of germ tube elongation.

Most plant defensins isolated to date exhibit antifungal activity against a broad range of fungi, including various plant pathogens. A common observation for most antifungal plant defensins, as well as for other types of defensin, is the reduction of antifungal activity when the cationic strength of the medium is increased. This antagonistic effect of cations is strongly dependent on the test fungus, indicating that electrostatic interactions probably change the conformation of the target site in the fungal membrane rather than the conformation of the defensin itself (Bart P.H.J. Thomma *et al.*, 2002).

3.3.12.1 *Mode of action*

Multiple antimicrobial molecules throughout the animal and plant kingdom, belonging to different families, display similar modes of action against a wide range of bacteria and fungi. These molecules, including defensins, share common properties including broad-spectrum antimicrobial activity and a cationic charge at physiological pH. There are currently two models describing the mode of antimicrobial activity of such cationic peptides. One model postulates the formation of multimeric pores within microbial membranes. After initial electrostatic binding of these positively charged peptides to negatively charged (phospho-) lipids on the target cell surface, they insert into the energized cell membrane and most likely form multimeric ion-permeable channels in a voltage-dependent manner. The second model postulates an electrostatic charge based mechanism of membrane permeabilization, as in the case of magainins, cecropins and HBD2, antimicrobial peptides isolated from frog skin, insect, hemolymph and human skin, respectively. They are thought to bind as monomers onto anionic lipid head groups of the membrane, thereby covering the membrane in a 'carpet like' manner. The subsequent neutralization of the anionic lipid head groups disrupts the integrity of the lipid bilayer, causing transient gaps and allowing ions and larger molecules to cross the membrane (Hoover *et al.* 2000). Recently, the theory that many cationic peptides exert their antimicrobial activity not only through permeabilization of the membrane but also through cytoplasmic targets has gained support. It was shown that apidaecins, short proline–arginine-rich insect peptides enter bacterial cells through stereospecific interactions with the outer membrane and, once into the interior of the cell affect protein synthesis (Castle *et al.* 1999). Apart from a working mechanism involving membrane permeabilization, some plant defensins have been reported to exert a different mode of action. It was demonstrated that gamma 1- and gamma 2-zeathionins, isolated from maize, inhibit sodium currents in a rat tumor cell line without changing the kinetics or voltage dependence of activation or inactivation of these currents. Similarly, some structurally related scorpion neurotoxins are known for blocking potassium channels. Since the overall structure of ion channels is conserved among eukaryotes, the native target of zeathionins may be an insect or fungal ion channels. It has been demonstrated that transgenic expression of plant defensins leads to protection of vegetative tissues against pathogen attack. Constitutive expression of a radish defensin clearly enhanced the resistance of tobacco plants to the pathogen, *Alternaria longipes.* It was shown that constitutive expression of an alfalfa defensin in potato provided a robust resistance against *Verticillium dahliae.*

At least two groups of antifungal plant defensins can be distinguished according to the morphogenic effects caused on treated fungal hyphae. The 'morphogenic' plant defensins cause reduced hyphal elongation with a concomitant increase in hyphal branching. The plant defensins from the Brassicaceae (including Rs-AFP1 and Rs-AFP2 from radish seeds) and Saxifragaceae (including Hs-AFPI from *Heuchera sanguinea* seeds) belong to this group (Osborn *et al.*, 1995). On the other hand, the plant defensins from the Asteraceae (including Dm-AMPl from dahlia seeds), the Fabaceae (including Ct-AMPl from *Clitoria tevnatea* seeds), and the Hippocastanaceae (including Ah-AMP1 from horse chestnut seeds) slow down hyphal extension but do not induce marked morphological distortions. These peptides are tentatively termed 'nonmorphogenic' plant defensins. Morphogenic and nonmorphogenic plant defensins also differ in their antifungal spectrum. A third subclass of plant defensins is formed by their homologs found in the Poaceae, including yl-P, yl-H, and Sial-3 from wheat, barley, and sorghum seeds, respectively. These plant defensins belong to the 'α-amylase inhibitor' type because they have been reported to inhibit α-amylases from insects and humans. At least yl-P, Sial, Sia2, and Sia3 were shown to be not or very weakly active against fungi (Osborn *et al.*, 1995). Conversely, antifungal plant defensins such as Pth-Sfl, Ah-AMP1, Ct-AMP1, Dm-AMPI, Hs-AFP1, and Rs-AFP2 (Willem *et al.*, 1995) do not exert a-amylase inhibitor activity. The a-amylase inhibitor-type plant defensins may possibly have evolved away from a role in protection against microorganisms toward protection against herbivores.

The three-dimensional structures of defensins are similar in that they consist of an α-helix and three antiparallel β-strands stabilized by four disulfide bonds. Plant defensins MsDef1 and MtDef4 are potent inhibitors of the growth of several filamentous fungi including *Fusarium graminearum.* However, they differ markedly in their antifungal properties as well as modes of antifungal action. MsDef1 induces prolific hyper branching of fungal hyphae, whereas MtDef4 does not. Both defensins contain a highly conserved γ-core motif ($GXCX_{3-9}C$), a hallmark signature present in the disulfide-stabilized antimicrobial peptides, composed of β2 and β3 strands and the interposed loop. The γ-core motifs of these two defensins differ significantly in their primary amino acid sequences and their net charge. In this study, we have found that the major determinants of the antifungal activity and morphogenicity of these defensins reside in their γ-core motifs. The MsDef1-γ4 variant in which the γ-core motif of MsDef1 was replaced by that of MtDef4 was almost as potent as MtDef4 and also failed to induce hyper branching of fungal hyphae. Importantly, the γ-core motif of MtDef4 alone was capable of inhibiting fungal growth, but that of MsDef1 was not. The analysis of synthetic γ-core variants of MtDef4 indicated that the cationic and hydrophobic amino acids were important for antifungal activity. Both MsDef1 and MtDef4 induced plasma membrane permeabilization; however, kinetic studies revealed that MtDef4 was more efficient in permeabilizing fungal plasma membrane than MsDef1. Furthermore, the *in vitro* antifungal activity of MsDef1, MsDef1-γ4, MtDef4 and peptides derived from the γ-core motif of each defensin was not solely dependent on their ability to permeabilize the fungal plasma membrane. The data indicated that the γ-core motif defines the unique antifungal properties of each defensin and may facilitate *de novo* design of more potent antifungal peptides (Uma Shankar Sagaram *et al.*, 2011).

Defensins and defensin-like peptides are functionally diverse and commonly presented as an immune reaction between plant and pathogen. *Trichoderma viride* and *Bacillus subtilis* as biological control agents were inoculated into the soil, to suppress the activity of the pathogenic fungi *Fusarium oxysporum* and *Rhizoctonia solani* on tomato (Hafez *et al.*, 2013). The up- and down-regulated genes were examined in both treated and nontreated plants, using differential display technique. In treated plants, many up-regulated genes with different molecular sizes, ranging from 50 to 7000 bp, were observed. Only four up-regulated genes were isolated from plants treated with *B. subtilis* + *R. solani*. The sequence analysis revealed that the identified genes were defensin genes; Amino Acid/Auxin Permease Family, Endopolygalacturonase PG1, Fructose-1,6-bisphosphatase, and glycosyl transferase. Moreover, chitinase gene, defensin genes (DF1 and DF2) were quantitatively determined using RT-PCR. The comparative expression level of the three induced genes was exponentially increased as a function of time, after the application of the biological control agents. However, while the expression levels of DF1 and DF2 were high in plants infected with either *F. oxysporum* or *R. solani* at the beginning of the experiment, the highest expression level of these genes was attained in the tomato plant treated with either *T. viride* or *B. subtilis*, after 24-hour postinoculation.

Jun Wu *et al.* (2009) cloned a plant defensin gene, named as Ovd, from *Orychophragmus violaceus* (L.) O. E. Schulz and subsequently introduced into *Brassica napus* genome by *Agrobacterium tumefaciens*-mediated transformation method in sense and antisense forms under the control of the CaMV 35S promoter. Genomic PCR analysis confirmed the integration of the sense and antisense Ovd into the host genome. Quantitative RT-PCR showed that the expression of Ovd in the sense plant line was stronger than nontransformed plant and antisense plant. The lesion size on detached leaves of the transgenic plants and nontransformed control caused by *Sclerotinia sclerotiorum* mycelia was examined. Lesion size was reduced in sense transgenic plants compared to nontransformed control (15–20% reduction area) while it was enlarged in antisense transgenic plants compared to nontransformed control (14–20% enlargement). The results showed that the overexpressing transgenic *B. napus* had higher resistance to *S. sclerotiorum*.

Antímicrobial peptídes (So-Dl-7) were isolated from a crude cell wall preparation from spinach leaves (*Spinacia olerácea* cv. Matador) and judged from their amino acid sequences, six of them (So-D2-7) represented a novel structural subfamily of plant defensins (group IV) (Ana Segura *et al.*, 1998). Group-IV defensins were also functíonally distinct from those of groups I—III. They were active at concentrations < 20 (J.M against Gram-positive (*Clavibacter michiganensis*) and Gram-negative (*Ralstonia solanacearum*) bacterial pathogens, as well as against fungi, such as *Fusarium culmorum*, *F. solari*, *Bipolaris maydis*, and *Colletotrichum lagenarium*. Fungal inhibition occurred without hyphal branching. Group-IV defensins were preferentially distributed in the epidermal cell layer of leaves and the subepidermal region of stems.

Antimicrobial cationic peptides are ubiquitous in nature and are thought to be a component of the first line of defense against infectious agents. It is widely believed that the killing mechanism of these peptides in bacteria involves an interaction with the cytoplasmic membrane. Cationic peptides from different structural classes were used in experiments with *Staphylococcus aureus* and other medically important Gram-positive bacteria to gain insight into the mechanism of action (Carol L. Friedrich *et al.*, 2000). The membrane potential-sensitive fluorophore dipropylthiacarbocyanine was used to assess the interactions of selected antimicrobial peptides with the cytoplasmic membrane of *S. aureus*. Study of the kinetics of killing and membrane depolarization showed that, at early time points, membrane depolarization was incomplete, even when 90% or more of the bacteria had been killed. CP26, a 26-amino-acid α-helical peptide with a high MIC against *S. aureus*, still had the ability to permeabilize the membrane. Cytoplasmic-membrane permeabilization was a widespread ability, and an action that may be necessary for reaching an intracellular target but in itself did not appear to be the killing mechanism. Transmission electron microscopy of *S. aureus* and *Staphylococcus epidermidis* treated with CP29 (a 26-amino-acid α-helical peptide), CP11CN (a 13-amino-acid, proline- and tryptophan-rich peptide), and Bac2A-NH_2 (a linearized version of the 12-amino-acid loop peptide bactenecin) showed variability in effects on the bacterial structure. Mesosome-like structures were seen to develop in *S. aureus*, whereas cell wall effects and mesosomes were seen with *S. epidermidis*. Nuclear condensation and abhorrent septation were occasionally seen in *S. epidermidis*. These experiments indicated that these peptides vary in their mechanisms of action and that the mechanism of action likely does not solely involve cytoplasmic-membrane permeabilization.

The beet cyst nematode, *Heterodera schachtii* induces a feeding site, called syncytium, in roots of host plants. In *Arabidopsis*, one of the genes whose expression is strongly induced in these structures is *Pdf2.1*, which codes for an antimicrobial plant defensin (Shahid Siddique *et al.*, 2011). *Arabidopsis* has 13 plant defensin genes. Besides *Pdf2.1*, the *Pdf2.2* and *Pdf2.3* genes were strongly expressed in syncytia and, therefore, the expression of all three *Pdf* genes was studied in detail. The promoter of the*Pdf2.1* gene turned out to be an interesting candidate to drive a syncytium-specific expression of foreign genes as RT-PCR showed that apart from the feeding site it was only expressed in siliques (seeds). The *Pdf2.2* and *Pdf2.3* genes were in addition expressed in seedlings, roots, leaves, stems, and flowers. These results were supported by the analysis of promoter: GUS lines. After infection with *H. schachtii*, all GUS lines showed a strong staining in syncytia at 5 and 15 dpi. This expression pattern was confirmed by *in situ* RT-PCR.

Drosomycin-type antifungal peptides (DTAFPs) are key innate immunity components of *Drosophila* and plants and confer resistance to fungal infection. Shunyi Zhu and Bin Gao (2014) reported the discovery of a multigene family of DTAFPs, comprising of 15 members (termed cremycin-1 to cremycin-15), in the fruit nematode *Caenorhabditis remanei*. Cremycins share highly similar amino-acid sequences and identical precursor organization to drosomycins. Of the 15 *cremycin* genes, 10 are found to be transcriptionally active, and 6 are upregulated after fungal challenge. Synthetic cremycin-5 is active on filamentous fungi and a series of clinical isolates of human pathogenic yeasts and exhibits low hemolysis and high serum stability. The specific distribution of DTAFPs in a clade of molting animals (Ecdysozoa), including Arthropoda, Nematoda and Tardigrada, together with the widespread presence in plants but the absence of fungi and protozoans, provides evidence for horizontal transfer of a disease resistance gene between plants and ecdysozoans.

3.3.13 Protease Inhibitors

Plant protease inhibitors (PIs) are small proteins, generally present at high concentration in storage tissues (up to 10% of protein content), but also detectable in leaves in response to the attack of insects and pathogenic microorganisms (De Leo *et al.*, 2002). PIs' contribution to plant defense mechanisms relies on inhibition of proteases present in insects' guts or produced by microorganisms, causing a reduction in the availability of amino acids necessary for their growth and development. As the role of inhibitors is simply achieved by the activation of single genes, several transgenic plants expressing PIs have been produced in the last 15 years and tested for enhanced defensive capacities, with particular efforts against pest insects. Among plant PIs, inhibitors active toward the four mechanistic classes of proteases have been described. PIs active against serine, cysteine and metallocarboxy-proteases are ubiquitous, while inhibitors active towards aspartic proteases have not been detected in seeds. The activity of PIs is due to their capacity to form stable complexes with target proteases, blocking, altering or preventing access to the enzyme active site.

The term 'protease' includes both 'endopeptidases' and 'exopeptidases' whereas, the term 'proteinase' is used to describe only 'endopeptidases.' Several nonhomologous families of proteinase inhibitors are recognized among the animal, microorganisms and plant kingdom. The majority of proteinase inhibitors studied in plant kingdom originates from three main families, namely, Leguminosae, Solanaceae and Gramineae (Table 3.7) (Paulraj K. Lawrence and Kripa Ram Koundal, 2001).

Proteases, though essentially indispensable to the maintenance and survival of their host organisms, can be potentially damaging when overexpressed or present in higher concentrations, and their activities need to be correctly regulated. An important means of regulation involves modulation of their activities through interaction with substances, mostly proteins, called 'protease inhibitors.' Some insects and many of the phytopathogenic microorganisms secrete extracellular enzymes and, in particular, enzymes causing proteolytic digestion of proteins, which play important roles in pathogenesis. Plants, however, have also developed mechanisms to fight these pathogenic organisms. One important line of defense that plants have to fight these pathogens is through various inhibitors that act on these proteolytic enzymes (Huma Habib and Khalid Majid Fazili, 2007). These inhibitors are thus active in endogenous as well as exogenous defense systems. Protease inhibitors active against different mechanistic classes of proteases have been classified into different families on the basis of significant sequence similarities and structural relationships. Specific protease inhibitors are currently being overexpressed in certain transgenic plants to protect them against invaders.

Table 3.7 Annotation of PIs in the Database is Organized According to the Well-known Classification of PI Families

Plant protease inhibitor family	*PLANT-PIs code*	*InterPro accession no.*
Bowman–Birk serine proteinase inhibitors	BBI	IPR000877
Cereal trypsin/α-amylase inhibitors	BRI	IPR001768
Cysteine proteinase inhibitors	CYS	IPR000010
Metallocarboxypeptidase inhibitors	MCI	Not available
Mustard trypsin inhibitors	MSI	Not available
Potato type I inhibitors	PI1	IPR000864
Potato type II proteinase inhibitors	PI2	IPR003465
Serpin	SPI	IPR000215
Soybean trypsin inhibitors (Kunitz)	KNI	IPR002160
Squash inhibitors	SQI	IPR000737

The most important components of all protective mechanisms are proteinaceous compounds. These include enzymes such as β-1,3-glucanases and chitinases, inhibitors of proteases and α-amylases, lectins, and also other proteins and peptides which have antimicrobial activity. For instance, damaging of tomato leaves (*Lycopersicon esculentum* by insects and microorganisms induced the synthesis of more than 20 different proteins including inhibitors of serine, cysteine, and aspartic proteinases and also a metallocarboxypeptidase (Figs. 3.62–3.64).

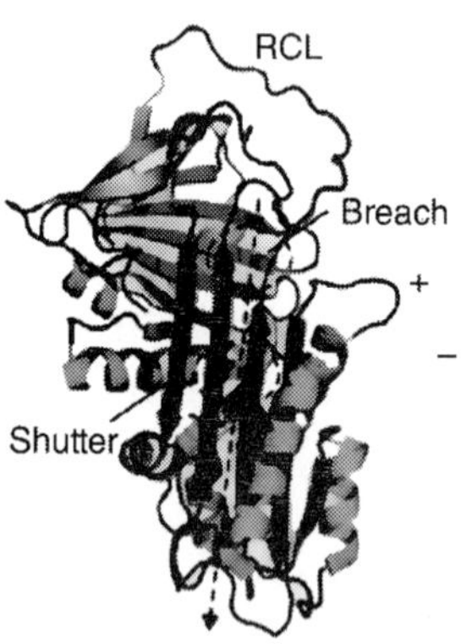

Fig. 3.62 Serpin

Fig. 3.63 Cysteine proteinase inhibitor

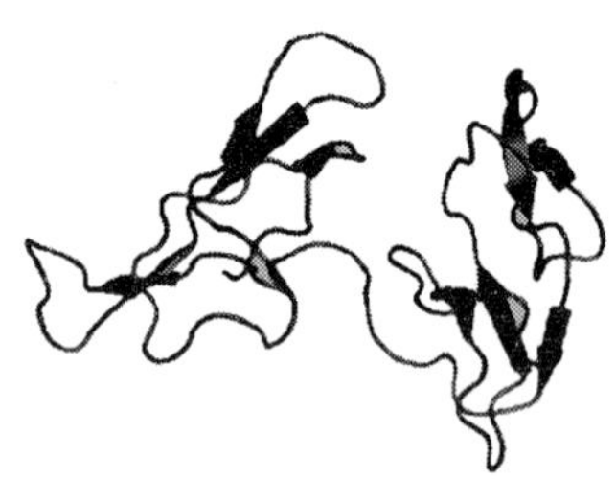

Fig. 3.64 Bowman-Birk Protease Inhibitor (BBI)

Many phytopathogenic microorganisms produce active extracellular proteinases that along with other enzymes play an important role in pathogenesis, e.g., polygalacturonases, pectolyases, and xylanases. Already in 1973, it was demonstrated that the phytopathogenic fungus *Colletotrichum lindemithianum,* when grown on plant cell walls or on artificial nutrient medium, secretes an active protease of 25 kD molecular weight and pH optimum at 8.6 (Valueva and Mosolov, 2004). This was the first extracellular proteinase of a plant pathogen obtained in its pure form. In recent years, many extracellular proteases produced by phytopathogenic microorganisms have been isolated and characterized to some extent. Among these serine proteinases prevail, but there are enzymes belonging to other mechanistic classes. All known serine proteinases of phytopathogens can be divided into trypsin-like and subtilisin-like enzymes. The first group contains proteinases that are produced by *Cochliobolus carbonum, Verticillium dahliae* (Valueva and Mosolov, 2004), *Stagonospora (Septoria) nodorum,* and *Phytophtora infestans,* microorganisms. Subtilisin-like enzymes are secreted by *C. carbonum, P. infestans, Acremonium typhium, Magnaporthe poae, Trichoderma harzianum* and *Fusarium oxysporum.* Among extracellular proteinases of phytopathogens, the aspartic proteinases are fairly widespread. These include the enzymes produced by *Botrytis cinerea, Cryphonectria parasitica* (*Endothia pepsin*) and *Glomerella cingulata.* Cysteine proteinase is secreted by the fungus *Pyrenopeziza brassicae.* Metalloproteinases include a family of Zn-dependent bacterial enzymes belonging to the genus *Erwinia.* One of these proteinases extracted from *Erwinia carotovora* subsp. *carotovora* is similar in its properties to thermolysin from *Bacillus thermoproteolyticus.*

Extracellular proteinases apparently play an active role in the process of pathogenesis. For instance, it has been revealed that in *P. brassicae,* a leaf pathogen damaging plants of Cruciferae, nonpathogenic mutants are unable to produce extracellular cysteine proteinase. An important role in disease progression is also played by aspartic proteinase of the fungus *B. cinerea,* which is a wide profile pathogen. Proteinase secretion by this microorganism was observed already at early stages of infection progression (before the formation of pectolytic enzymes begins) and was accompanied by the death of plant cells (Movahedy and Heale, 1990). Development of infection was significantly retarded by the initial treatment of *B. cinerea* spores with the aspartic proteinase inhibitor pepstatin. However, pepstatin did not affect spore germination. It has been recently revealed that a cell-free sample obtained from spore suspension and germination cysts of late blight causing agent (the oomycete *P. infestans*) causes plant tissue necrosis when injected into potato leaves. A correlation between the level of proteolytic activity in the sample and its necrotic action could be observed.

The plant apoplast forms a protease-rich environment in which proteases are integral components of the plant defense response. Plant pathogenic oomycetes, such as the potato (*Solanum tuberosum*) and tomato (*Lycopersicon esculentum*) pathogen *Phytophthora infestans* secrete a diverse family of serine protease inhibitors of the Kazal family. Among these, the two-domain EPI1 protein were shown to inhibit and interact with the pathogenesis-related protein P69B subtilase of tomato and was implicated in counterdefense (Miaoying Tian *et al.,* 2005). A second extracellular protease inhibitor, EPI10, from *P. infestans, was* described and functionally characterized. EPI10 contains three Kazal-like domains, one of which was predicted to be an efficient inhibitor of subtilisin A by an additivity-based sequence to reactivity algorithm (Laskowski algorithm). The *epi10* gene was up-regulated during infection of tomato, suggesting a potential role during pathogenesis. Recombinant EPI10 specifically inhibited subtilisin A among the major serine proteases, and inhibited and interacted with P69B subtilase of tomato. The finding that *P. infestans* evolved two distinct and structurally divergent protease inhibitors to target the same plant protease suggested that inhibition of P69B could be an important infection mechanism for this pathogen.

The oomycetes form one of the several lineages within the eukaryotes that independently evolved a parasitic lifestyle and consequently are thought to have developed alternative mechanisms of pathogenicity. The oomycete *Phytophthora infestans* causes late blight, a ravaging disease of potato and tomato. Little is known about processes associated with *P. infestans* pathogenesis, particularly the suppression of host defense responses. An extracellular protese inhibitor, EPI1, was described and functionally characterized from *P. infestans.* EPI1 contains two domains with significant similarity to the Kazal family of serine protease inhibitors. Database searches suggested that Kazal-like proteins are mainly restricted to animals and apicomplexan parasites but appear to be widespread and diverse in the oomycetes. Recombinant EPI1 specifically inhibited subtilisin A among major serine proteases and inhibited and interacted with the pathogenesis-related P69B subtilisin-like serine protease of tomato in intercellular fluids. The *epi1* and *P69B* genes were coordinately expressed and up-regulated during infection of tomato by *P. infestans.* Inhibition of tomato proteases by EPI1 could form a novel type of defense-counterdefense mechanism between plants and microbial pathogens. In addition, this study pointed to a common virulence strategy between *P. infestans* and several mammalian parasites, such as the apicomplexan *Toxoplasma gondii.*

3.3.13.1 Families of PIs

PIs have been grouped into families and subfamilies and different clans on the basis of sequence relationship and the relationship of protein folds of the inhibitory domains or units. An inhibitor domain is defined as the segment of the amino acid sequence containing a single reactive site after removal of any parts that are not directly involved in the inhibitor activity. On the basis of sequence homologies of their inhibitor domains, PIs have been classified into 48 families. Proteins containing a single inhibitor unit are termed simple inhibitors, and those that contain multiple inhibitor units are termed complex inhibitors. A total of 11 families belong to the latter category and contain between 2 and 15 inhibitory domains. Most of these are homotypic, containing inhibitor units from a single family, some are however heterotypic, and contain inhibitor units from different families (Huma Habib and Khalid Majid Fazili, 2007). On the basis of tertiary structure, 31 of the 48 families have been assigned to 26 clans, indicating that a large proportion of families shows no relationships in their three-dimensional structures. The families along with some representative examples are given in Table 3.8.

Table 3.8 Families of Plant Protease Inhibitors with Some Type Examples

Common name	*MEROPS Family/ subfamily*	*Type example*	*Source*	*Target Protease*
Kunitz (plant)	13A	Soybean Kunitz trypsin inhibitor	*Glycine max*	Trypsin, Chymotrypsin
Kunitz (plant)	I3B	Proteinase inhibitor A inhibitor unit	*Sagittaria sagittifolia*	Trypsin, Chymotrypsin, Kallikerin
		Kunitz subtilisin inhibitor	*Canavalia lineata*	Subtilisin-type microbial serine proteases
		Cathepsin D inhibitor	*Solanum tuberosum*	Cathepsin D, Trypsin
		Trypsin inhibitor	*Acacia confusa*	Trypsin and α-chymotrypsin
Cereal	I6	Ragi seed trypsin/_-amylase inhibitor	*Eleusine coracana*	α-amylase
		Barley trypsin/factor XIIa inhibitor	*Hordeum vulgare*	α-amylase, Trypsin
		Wheat trypsin/α-amylase inhibitor	*Triticum aestivum*	α-amylase, Trypsin
		Maize trypsin/factor XIIa inhibitor	*Zea mays*	Mammalian trypsin
Squash	I7	Trypsin inhibitor MCTI-1	*Momordica charantia*	Pancreatic elastase
		Trypsin inhibitor MCTI-II	*M. charantia*	Trypsin
		Macrocyclic squash trypsin inhibitor	*M. cochinchinensis*	Trypsin
		Trypsin inhibitor CSTI-IV	*Cucumis sativus*	Trypsin
Potato type I	I13	Chymotrypsin inhibitor I	*Solanum tuberosum*	Chymotrypsin, Trypsin
		Glutamyl peptidase II inhibitor	*Momordica charantia*	Glu *S. griseus* protease Subtilisin
		Subtilisin-chymotrypsin inhibitor CI-1A	*Hordeum vulgare*	Subtilisin, Chymotrypsin
		Wheat subtilisin/chymotrypsin inhibitor	*Triticum aestivum*	*B. lichenoformis* subtilisin, α-chymotrypsin

Contd...

Contd...

Mustard	I18	Mustard trypsin inhibitor	*Sinapis alba*	β-trypsin
		Mustard trypsin inhibitor-2	*Brassica hirta*	Bovine β-trypsin, α-chymotrypsin
		Rape trypsin inhibitor	*Brassica napus*	Trypsin, Chymotrypsin
Cystatin	I25B	Onchocystatin	*Onchocerca volvulus*	Cysteine proteinase
		Ovocystatin	*Gallus gallus*	Thiol proteases
		Oryzacystatin II	*Oryza sativa*	Cysteine proteinases
Kininogen	I25C	Metalloprotease inhibitor Sarcocystatin	*Bothrops jararaca*	Atrolysin C, Jararhagin
Bowman-Birk	I12	Bowman–Birk plant trypsin inhibitor unit 1	*Glycine max*	Trypsin, Chymotrypsin
		Bowman-Birk trypsin/chymotrypsin inhibitor	*Arachis hypogaea*	Trypsin, Chymotrypsin
		Sunflower cyclic trypsin inhibitor	*Helianthus annuus*	Trypsin, Cathepsin G, Elastase, Chymotrypsin, and thrombin
Potato type II	I20	Proteinase inhibitor II	*Solanum tuberosum*	Trypsin, Chymotrypsin
		Potato peptidase inhibitor II inhibitor unit 1	*Solanum tuberosum*	Trypsin, Chymotrypsin
		Tomato peptidase inhibitor II inhibitor unit 1	*Solanum lycopersicum*	Trypsin, Chymotrypsin
		Tomato peptidase inhibitor II inhibitor unit 2	*Solanum lycopersicum*	Trypsin, Chymotrypsin

Contrary to the earlier presented examples, in a number of cases dependency between extracellular proteinase activity and pathogenicity of the microorganism was not found. For instance, a decrease in pathogenicity of trypsin-deficient mutants of *C. carbonum* graminoid pathogen with deficiency of trypsin-like proteinases was not observed (Murphy and Walton, 1996). A recent study reveals that metalloproteinase of *Xanthomonas campestris* pv. *campestris* is able to cleave glycoproteins of extracellular matrix in turnip petals. Proteinases found in pathogens can also play an active role in the degradation of other proteins involved in plant protection, for instance, such enzymes as chitinase and β-1,3-glucanase. Proteinase inhibitors in plants are able to suppress the enzymatic activity of phytopathogenic microorganisms. Along with inhibitors of trypsin and chymotrypsin, many plants have proteins that act predominantly as inhibitors of microbial proteinases. A protein with molecular weight of 10 kD has been recently extracted from pumpkin fruit phloem exudation (*Cucurbita maxima* L.); it acted as an aspartic proteinase inhibitor. Besides pepsin, it also suppressed the activity of extracellular aspartic proteinase of the fungus *Glomerella cingulata*. The explanation is that cysteine proteinases play an active role in protein processing in many viruses. In this respect, cysteine proteinases become an attractive target for the development of efficient control means of plant viruses. It is important to note that proteinase inhibitors induced in response to infection can sufficiently differ from similar inhibitors present in a healthy plant.

PIs also exhibit a very broad spectrum of activity against pathogenic nematodes. Cowpea trypsin inhibitor (CpTi) inhibits the growth of nematodes, *Globodera tabacum, Globodera pallida* and *Meloidogyne incognita* (Williamson and Hussey, 1996). The buckwheat (*Fagopyrum sculentum*), trypsin/chymotrypsin inhibitor, interferes with spore germination and mycelium growth of the tobacco brown spot fungus, *Alternaria alternata*. Cysteine PIs from pearl millet (*Pennisetum glaucum*) inhibit the growth of many pathogenic fungi, including *Trichoderma reesei* (Huma Habib and Khalid Majid Fazili, 2007).

Vishnudasan *et al.* (2005) attempted engineering a plant serine proteinase inhibitor (pin2) gene into *Triticum durum* PDW215 by *Agrobacterium*-mediated transformation to combat cereal cyst nematode, *Heterodera avenae* infestation. Putative T(0) transformants were screened, and positive segregating lines analyzed further for the study of the stable integration, expression, and segregation of the genes. PCR, Southern analysis along with bar gene expression studies corroborated the stable integration pattern of the respective genes. The transformation efficiency was 3% while the frequency of escapes was 35.71%. chi(2) analysis revealed the stable integration and segregation of the genes in both the T(1) and T(2) progeny lines. The PIN2 systemic expression conferred satisfactory nematode resistance. The correlation analysis suggested that at $p<0.05$ level of significance the relative proteinase inhibitor (PI) values showed a direct positive correlation vis-à-vis plant height, plant seed weight and also the seed number.

Endogenous protease inhibitors (PIs) in the roots of Smooth Cayenne pineapple clones may affect the growth of the phytonematodes, *Rotylenchulus reniformis* (Rabovich *et al.*, 2009). In pineapple, reniform nematode population densities remain atypically near preplant levels for 6–9 months after pineapple planting. A potted plant experiment was conducted to determine if the PI present in pineapple roots affected nematode reproduction and possibly accounted for the observed nematode field population dynamics. Pineapple PI activity increased for the first six months after planting and was higher in nematode-inoculated plants. Nematode reproduction, however, was not correlated with PI activity. In a second experiment, PI activity was concentrated in the portion of the roots nearest the pineapple butt where nematode population densities were highest. The behavior of the PI in pineapple roots suggested a defensive role, such as systemic acquired resistance, against *R. reniformis*.

Defense system in plants may also be categorized as passive and active defenses, although the contents of these categories remain almost same as that of an earlier grouping of structural and chemical. Physical barriers include wax, cuticle, cell wall, stomata, lenticels and 'chemical' barriers comprise nutrient deprivation, pH, phytoanticipins and plant defensins. Active defenses include 'rapid' that comprises membrane function, oxidative burst, cell wall reinforcement, and hypersensitive cell death and phytoalexin accumulation. The other type, 'delayed' comprises pathogenesis-related proteins, systemic acquired resistance and pathogen containment.

Plant protease inhibitors (PPIs) are active against the insects and microbes which invade the plant. *Cassia* species possess anti-insecticidal and antimicrobial properties, and a study was taken up to investigate the antibacterial efficacy of a PPI present in the leaves of *Cassia fistula* (Arulpandi and Sangeetha, 2102). A PPI, fistulin, was isolated from the leaves of *C. fistula* and purified by gel filtration chromatography. The antibacterial activity of the purified fistulin was studied against five bacterial strains, namely, *Bacillus subtilis, Staphylococcus aureus, Klebsiella pneumoniae, Pseudomonas aeruginosa* and *Escherichia coli*. The PPI was found to be very active against *S. aureus, E. coli, B. subtilis*, and *K. pneumonia*, and its efficacy was comparable to the standard drug, streptomycin sulfate.

As the processing mechanism of all known potyviruses involves the activity of cysteine proteinases, Gutierrez-Campos *et al.* (1999) were interested to know whether constitutive expression of a rice cysteine proteinase inhibitor gene could induce resistance to two important potyviruses, tobacco etch virus (TEV) and potato virus Y (PVY), in transgenic tobacco plants. Tobacco lines expressing the foreign gene at varying levels were examined for resistance to TEV and PVY infection. There was a clear, direct correlation between the level of oryzacystatin message, inhibition of papain (a cysteine proteinase), and resistance to TEV and PVY in all lines tested. The inhibitor was ineffective against tobacco mosaic virus (TMV) infection because processing of this virus does not involve cysteine proteinases. These results showed that plant cystatins can be used against different potyviruses and potentially also against other viruses, whose replication involves cysteine proteinase activity.

3.3.14 Systemic Acquired Resistance (SAR)

The systemic acquired resistance (SAR) is a 'whole-plant' resistance response that occurs following an earlier localized exposure to a pathogen. SAR is analogous to the innate immune system found in animals, and there is evidence that SAR in plants and innate immunity in animals may be evolutionarily conserved. Plants use pattern-recognition receptors to recognize conserved microbial signatures. This recognition triggers an immune response. The first plant receptors of conserved microbial signatures were identified in rice [(XA21, 1995) (Song, 1995)] and in Arabidopsis (FLS2, 2000). Plants also carry immune receptors that recognize highly variable pathogen effectors. These include the NBS-LRR class of proteins. SAR is important for plants to resist disease, as well as to recover from disease once formed. SAR can be induced by a wide range of pathogens, especially (but not only) those that cause tissue necrosis, and the resistance observed following induction of SAR is effective against a wide range of pathogens, which is why SAR resistance is sometimes called 'broad spectrum.' SAR is associated with the induction of a wide range of genes, and the activation of SAR requires the accumulation of endogenous salicylic acid (SA). The pathogen-induced SA signal activates a molecular signal transduction pathway that is identified by a gene called *NIM1*, *NPR1* or *SAI1* (three names for the same gene) in the model genetic system *Arabidopsis thaliana*. SAR has been observed in a wide range of flowering plants, including dicotyledon and monocotyledon species. SAR can be activated in corn; however, widely adapted commercial like Benzothiadiazole may not be efficient against *P. sorghi* causing common rust.

In the SAR state, plants are primed/sensitized to more quickly and more effectively activate defense responses the second time they encounter pathogen attack. Since SAR depends on the ability to access past experience, acquired disease resistance is a paradigm for the existence of a form of 'plant memory.' Although the phenomenon has been known since the beginning of the 20th century, major progress in the understanding of SAR was made over the past sixteen years. This review covers the current knowledge of molecular, biochemical and physiological mechanisms that are associated with SAR. SAR is induced by most pathogens that cause tissue necrosis, either as a part of a hypersensitive response (HR) or as a symptom of the disease. One characteristic of SAR is the development of the enhanced resistance in the distal, uninoculated plant organs. Another hallmark of SAR is its activity against a broad and distinctive spectrum of pathogens which includes viruses, bacteria, oomycetes, nematodes and fungi. In addition, SAR confers a long-lasting protection that can last for weeks to months, and sometimes throughout an entire season. Thus, for SAR to be realized, a plant requires the ability to recall past experience.

3.3.14.1 Systemic SAR signaling

Early grafting experiments have shown that a primary infected leaf of a plant can produce a systemic signal that is graft transmissible from rootstock to the scion. The studies also revealed that the systemic signal induces SAR in the remote tissue in a species nonspecific manner (Uwe Conrath, 2006). The identity of the long-distance signal, however, remained unknown. Some experiments suggested that salicylic acid (SA) was the translocated signal, but they argued against such a role for SA. Recent work with the *Arabidopsis* defective in induced resistance1 (*dir1*) mutant indicated that wild-type DIR1 with sequence similarity to lipid transfer proteins might play a role in the generation and/or transmission of a mobile signal for SAR, possibly by interaction with a lipid-derived molecule. Also, in the *Arabidopsis* enhanced disease susceptibility 1 (*eds1*) and phytoalexin deficient 4 (*pad4*) mutants, which are both defective in putative lipases, SAR cannot be activated. Together, these findings suggest that lipid signaling might contribute to SAR.

Hydrogen peroxide has also been proposed to have a signaling role in SAR. Inoculation of lower leaves of *Arabidopsis* plants with avirulent *Pseudomonas syringae* pv. *tomato* (Pst) DC3000 (avrRpt2) elicited a prominent H_2O_2 burst in the infected leaves. The oxidative burst was followed by the appearance of lower levels of H_2O_2 in small groups of cells in uninoculated leaves. Gaseous methyl salicylate, a major volatile produced in tobacco leaves inoculated with tobacco mosaic virus, was recently shown to function as an airborne signal which induces disease resistance in both the infected and noninfected tissues of an infected plant, and also in neighboring plants. Similarly, experiments with an ethylene-insensitive, transgenic tobacco line implicated ethylene as a signal for SAR, at least in tobacco (Verberne *et al.*, 2003). Together, the above findings indicate a complex nature for SAR long-distance signaling that might involve various, different signals whose contribution to SAR might depend on the plant species.

3.3.14.1.1 Salicylic Acid: Endogenous Signal for SAR

Though the identity of the long-distance signal for SAR is unknown, it is appreciated that SA is needed to establish SAR in the remote tissue (Uwe Conrath, 2006). First compelling evidence for this came from studies with transgenic tobacco and *Arabidopsis* plants constitutively expressing a bacterial SA hydroxylase. These plants are unable to accumulate high levels of SA and do not acquire systemic resistance upon infection with necrotizing pathogens, presumably due to the destruction of the SA signal. More recent work with *Arabidopsis* mutants affected in either SA production or SA signaling confirmed the role of SA as an important signal in SAR. In addition, in transgenic tobacco and *Arabidopsis* plants, overproduction of SA enhanced the resistance to pathogens.

3.3.14.1.2 Chemical SAR activators

Since it has been appreciated that SA is an endogenous signal for the activation of SAR, there was increased characterization of synthetic chemicals able to mimic SA in SAR induction. 2,6-dichloroisonicotinic acid and its methyl ester (both are referred to as INA) were the first synthetic compounds shown to activate SAR. (Kessmann *et al.*, 1994). As INA is insufficiently tolerated by some crops, benzo (1,2,3)thiadiazole-7-carbothioic acid S-methyl ester (BTH) became an attractive synthetic SAR activator SA, INA, and BTH are assumed to activate SAR through a same signaling mechanism.

3.3.14.2 SAR genes

In tobacco and *Arabidopsis*, the establishment of SAR is associated with the expression of a set of so-called SAR genes, which include some of those encoding pathogenesis-related (PR) proteins. Some PR proteins have been identified as acidic β-1,3-glucanases (BGL2) and chitinases (PR-3), possibly able to hydrolyze microbial cell wall components. Therefore, the accumulation of PR proteins has often been proposed as the molecular basis for SAR (Uwe Conrath, 2006). However, over the past few years it became widely appreciated that the accumulation of PR proteins does not per se explain the SAR phenomenon. For instance, cloning of PR genes and plant transformation by now have not provided a single example in which an inducible acidic glucanase or chitinase, alone or in combination, enhances resistance to fungal pathogens. Thus, the contribution of PR proteins to SAR appears to be minor.

Over the past decade, a variety of mutants with compromised activation of SAR have been identified. The *Arabidopsis* mutant *npr1* is probably the most prominent of these mutants. *Npr1* accumulates wild-type SA levels in response to infection with avirulent pathogens but is unable to activate PR genes, or establish the primed state or develop biologically or chemically induced SAR. Thus, NPR1 is a likely key regulator of SAR and priming. This assumption has further been supported by two studies demonstrating that constitutive overexpression of NPR1 in transgenic plants did not lead to enhanced SA levels or constitutive expression of *PR* genes. Rather, these plants showed stronger *PR* gene expression after pathogen infection, and they also expressed greatly enhanced disease resistance. Interestingly, *npr1* shows enhanced susceptibility to some virulent pathogens and seems to be also involved in R gene-mediated disease resistance. In addition, NPR1 seems to play a key role in the SA-independent induced systemic resistance (ISR) response (Shah *et al.*, 1997) ISR is triggered by selected strains of saprophytic rhizobacteria and confers broad-spectrum disease resistance in the aerial parts of the plant.

It was shown that SAR required the signal molecule salicylic acid (SA) and was associated with accumulation of pathogenesis-related proteins, which are thought to contribute to resistance (Durrant and Dong, 2004). Using the model plant *Arabidopsis*, it was discovered that the isochorismate pathway is the major source of SA during SAR. In response to SA, the positive regulator protein NPR1 moves to the nucleus where it interacts with TGA transcription factors to induce defense gene expression, thus activating SAR. Exciting new data suggested that the mobile signal for SAR might be a lipid molecule. They discussed the molecular and genetic data that have contributed to the understanding of SAR and presented a model describing the sequence of events leading from initial infection to the induction of defense genes.

3.4 DEFENSE MECHANISM IN PHYTONEMATODES

Phytoematodes are protected by the cuticle and surface coat. The surface coat of the cuticle, consisting of lipid and protein molecules is shed as the nematode moves, shedding bacteria but also confusing the plants as to whereabouts of the nematode. Glutathione peroxidases on surface coats of nematodes reduce active oxygen plant defenses. Many phytonematodes produce glutathione S tranferases that detoxify endogenous toxic molecules. Nematodes produce superoxide dismutase that breaks down active oxygen plant defenses.

Chemical composition, origin, and biological role of the surface coat (SC) of plant-parasitic nematodes are described and compared with those of animal-parasitic and free-living nematodes. The SC of the plant-parasitic nematodes is 5-30 nm thick and is characterized by a net negative charge. It consists, at least in part, of glycoproteins and proteins with various molecular weights, depending upon the nematode species (Spiegel and Mcclure, 1995). The ability of its components and the binding of human red blood cells to the surface of many tylenchid plant-parasitic nematodes, as well as the binding of several neoglycoproteins to the root-knot nematode Meloidogyne, suggest the presence of carbohydrate-recognition-domains for host plants and parasitic or predatory soil microorganisms (*Pasteuria penetram* and *Dactylaria* spp., for example). These features may also assist in nematode adaptations to soil environments and to plant hosts with defense mechanisms that depend on reactions to nematode surfaces. Surface coat proteins can be species and race specific, a characteristic with a promising diagnostic potential.

Most plants are resistant to most pathogens; they have highly effective immune systems. Host defense mechanisms may be as extreme as programmed cell death, the hypersensitive response. All biotrophic pathogens must suppress host defenses (Huang *et al.*, 2005). The feeding site must be induced without host detection or induction of host defenses. Following the establishment of the feeding site by sedentary nematodes, it must be maintained for up to 5 or 6 weeks to allow the nematode to achieve its reproductive potential. That time scale is much greater than that required by many bacterial and fungal pathogens of plants. Failure to establish and maintain the feeding site may prevent reproduction and, therefore, is catastrophic to the nematode genotype. Consequently, there is strong selection pressure on nematodes to suppress host defenses. Plants characteristically deposit callose to strengthen cell walls at the point of invasion, including at the point of nematode stylet insertion. PAMP-triggered PTI, the first line of defense, may involve the production of salicylic acid (SA) as a signal to invoke defense mechanisms including callose thickening of cell walls and suppression of active oxygen defense responses (H_2O_2, superoxide) which may initiate localized programmed cell death, the hypersensitive response. Also, pathogen invasion triggers the jasmonic acid signaling pathway (JA) which stimulates the production of protease inhibitors and the release of toxins.

3.4.1 Effector Suppression of Plant Defenses and PTI

Invading bacteria and fungi, and probably nematodes, release effector molecules into plant cells to suppress PTI and render the plant susceptible to infection or invasion. PAMP-triggered PTI may result in the production of salicylic acid (SA) as a signal to invoke defense mechanisms. In that case, successful nematode infections would involve suppression of SA production, reduction of callose thickening of cell walls and suppression of active oxygen defense responses (H_2O_2, superoxide), and the hypersensitive response of localized programmed cell death. SA signaling is possibly disrupted by chorismate mutase produced in the esophageal glands. In the PTI signaling pathway, chorismate is converted to salicylic acid. Chorismate mutase from the nematode reduces chorismate and thus SA, so defense mechanisms are not triggered. Incidentally, like cellulases, chorismate mutase is an example of horizontal gene transfer from bacteria. Nematodes are the only metazoan with the enzyme. An alternative mechanism of PTI suppression by nematodes is the production of effectors which cause ubiquitin to attach to plant signal proteins and thus reduce their levels and effectiveness in triggering PTI responses.

3.4.2 The Evolutionary Response: Effector-triggered Immunity

The evolution of effector suppression of PTI has resulted in the evolution of immune receptors, with a nucleotide-binding domain and a leucine-rich domain (NB-LRR), in plants that recognize the effector molecules and activate effector-triggered immunity (ETI). However, successful pathogens have evolved next-generation effectors that suppress ETI. Plants have responded with more specific ETIs and the evolutionary treadmill continues. PTI responses to PAMPs and DAMPs are relatively general in their effect, but higher level ETIs are progressively more specific to individual pathogens. The cyclical evolutionary process of plant-nematode interactions with regard to plant immunity and susceptibility is depicted by the zig-zag-zig model (Jones and Dangl, 2006). Initially PAMPs trigger PTI, which reduces susceptibility. Then nematodes develop effectors that suppress PTI and plants evolve immunity responses to the effectors.

3.4.2.1 *Mechanisms of effector-triggered immunity*

In effect, the sources of specific ETIs as resistance genes. Thus, the Mi gene of tomato codes for receptors to the effector molecules introduced by root-knot nematodes to suppress plant defenses and, perhaps, facilitate the development of feeding sites. However, although they are known for some other pathogens, the nematode effector molecules that trigger ETIs have not yet been determined and are the focus of several active research programs. One possible candidate is the 30C02 effector protein of *Heterodera glycines* which may be involved in active suppression of host defenses (Hamamouch *et al.*, 2012). Although the nature of nematode effector products has still to be determined, the hypersensitive response of cells to the activated ETI effectively disrupts the feeding and development of sedentary endoparasitic nematode species.

3.4.3 Molecular Strategies for Nematode Resistance

In general, three basic strategies to create nematode resistance have been formulated: the use or reinforcement of natural plant defense mechanisms, the direct killing of nematodes, and interference with the plant-nematode interaction. The first strategy includes cloning of natural resistance genes and reintroduction of these genes into susceptible plants. This has been accomplished for the *Hs1^{pro-1}* gene of *Beta procumbens*, the wild relative of *B. vulgaris*, which is resistant to *Heterodera schachtii*. Transgenic plants had the same level of resistance as their wild-type counterparts (Elsen *et al.*, 2001).

In the second type of strategy, external and internal nematode structures are the target of lytic enzymes, such as chitinases and collagenases, and inhibitory or toxic proteins, such as protease inhibitors or *Bacillus thuringiensis* endotoxins. The aim of using chitinases is to increase the permeability of the nematode eggshell, which could be lethal to embryos. However, the chitin layer is not exposed directly to the environment but is protected by a vitelline layer. The same problem arises in the use of collagenases to degrade the cuticle of nematodes, as the cuticle is protected by the glycocalyx or surface layer. Protease inhibitors are part of the natural defense response of plants and have already proved useful to control insect pests. Their use as antinematode proteins is based on the assumption that nematodes digest food with proteases. Results so far, for sedentary nematodes only, are encouraging. The application of *B. thuringiensis* toxins is hampered by the fact that they are highly species-specific Interference in the plant-nematode interaction can be accomplished by the expression of a repelling protein, an inhibition of the induction or maintenance of feeding structures by sedentary nematodes, or by the expression of a protein that hampers host recognition. Antibodies, specifically targeted against nematode secretions responsible for feeding structure induction and maintenance, have not been successful in controlling *M. incognita* in transgenic tobacco. An alternative is to direct antibodies towards the structures and secretions of nematodes, responsible for chemoreception of host signals. However, clear indications of such an effect are not available.

Incompatibility (resistance) of plants to nematode parasites is conferred by repulsive chemicals in a few cases but rarely, if at all, by penetration barriers or nutritional deficiencies in the host plant. The major incompatibility mechanism(s) appear(s) to be the postinfectional occurrence of hypersensitive reactions. Although the role of preformed simple phenols in this type of incompatibility is obscure, available information suggests that inducibly formed structural and chemical (viz., phytoalexins) barriers to nematode development may be responsible (David T. Kaplan and Noel T. Keen, 1980). No firm conclusions may be made at present on the validity of possible mechanisms since much of the early work is difficult to interpret and contains procedural inadequacies. Thus, appraisal of incompatibility mechanisms to nematodes requires additional testing of the hypotheses, especially as to the existence of such mechanisms at the localized site where and when incompatibility is expressed. An understanding of incompatibility may be attained more easily by utilizing those experimental approaches that have recently been used to study incompatibility systems to microbial parasites.

Induction of defense enzymes phenol, peroxidase (PO), polyphenol oxidase (PPO), phenyl ammonia lyase (PAL), superoxide dismutase (SOD) and chitinase by *Pseudomonas fluorescens* isolate Pf1, against challenge inoculation of *Meloidogyne graminicola* in rice was studied *in planta* (Anita and Samiyappan, 2012). The activity of phenol, PO, PPO, PAL and chitinase was higher in the bacterized rice plants. The analysis of isoform profiles revealed unique PO, PPO and SOD isoforms induced in plants treated with *P. fluorescens*. The present study implies earlier and higher accumulation of phenols and defense enzymes, viz., PO, PPO, PAL and chitinase in rice root tissue resulting in significant reduction in nematode infection.

Compatible plant-nematode interactions involve the formation of an elaborate feeding site within the host root that requires the evasion of plant defense mechanisms by the parasite. Little is known regarding plant defense signaling pathways that limit nematode parasitism during a compatible interaction. Jin (2010) utilized *Arabidopsis thaliana* mutants perturbed in salicylic acid (SA) biosynthesis or signal transduction to investigate the role of SA in inhibiting parasitism by the beet cyst nematode *Heterodera schachtii*. It was determined that SA-deficient mutants (sid2-1, pad4-1, and NahG) exhibited increased susceptibility to *H. schachtii*. In contrast, SA-treated wild-type plants showed decreased *H. schachtii* susceptibility. The npr1-2 and npr1-3 mutants, which were impaired in SA signaling, also showed increased susceptibility to *H. schachtii*, whereas the npr1-suppressor mutation SNI1 showed decreased susceptibility. Constitutive pathogenesis-related (PR) gene-expressing mutants (cpr1 and cpr6) did not show altered susceptibility to *H. schachtii*; however, constitutive PR gene expression was restricted to cpr1 shoots with wild-type levels of PR-1 transcript present in cpr1 roots. *H. schachtii* infection elicited SA-independent PR-2 and PR-5 induction in wild-type roots while PR-1 transcript and total SA levels remained unaltered. This was in contrast to shoots of infected plants where PR-1 transcript abundance and total SA levels were elevated. It was concluded that SA acted via NPR1 to inhibit nematode parasitism which, in turn, was negatively regulated by SNI1. An inverse correlation existed between root basal PR-1 expression and plant susceptibility to *H. schachtii* and suggested that successful cyst nematode parasitism may involve a local suppression of SA signaling in roots.

3.4.3.1 *Defense enzymes*

Many defense enzymes are involved in defense reaction against plant pathogens. These include oxidative enzymes such as peroxidase (PO) and polyphenoloxidase (PPO), which catalyze the formation of lignin, and other oxidative phenols that contribute to the formation of defense barriers for reinforcing the cell structure. Enzymes such as phenylalanine ammonia-lyase (PAL) are involved in phytoalexin or phenolic compound biosynthesis. Such enzymes have been correlated with defense against pathogens in several plants (Binutu and Cordell, 2000).

3.5 REACTIVE OXYGEN SPECIES (ROS)

Reactive oxygen species play a central role in plant defense against various pathogens. Superoxide anion (O_2^-), hydrogen peroxide (H_2O_2), and hydroxyl radical are the three major forms of ROS. These molecules are highly reactive and toxic and can lead to the oxidative destruction of cells. The rapid accumulation of plant ROS at the pathogen attack site, a phenomenon called oxidative burst, is toxic to pathogens directly and can lead to a hypersensitive response (HR) that results in a zone of host cell death, which prevents further spread of biotrophic pathogens. In addition to the described direct effects, ROS can also serve as signals that lead to the activation of other defense mechanisms. Several mechanisms have been proposed for ROS generation in plants (Xuming Liu *et al.*, 2010). Of these mechanisms, the NADPH-dependent oxidase system has received the most attention because of its similarity to the mammalian oxidase system that initiates ROS production in phagocytes and B lymphocytes as a response to pathogen attack. In mammalian cells, NADPH-dependent oxidases catalyze the one-electron reduction of molecular oxygen to form O_2^-, which then undergoes dismutation to form H_2O_2 either spontaneously or catalyzed by superoxide dismutases. In plants, NADPH-dependent oxidases are also linked with O_2^- production in response to pathogen attack and wounding. Insertion mutations of two *Arabidopsis* (*Arabidopsis thaliana*) NADPH-dependent oxidase subunit genes, *AtrbohD* and *AtrbohF*, largely eliminate ROS production during disease resistance reactions to avirulent pathogens

Reactive oxygen species (ROS) are produced as a normal product of plant cellular metabolism. Various environmental stresses lead to excessive production of ROS causing progressive oxidative damage and ultimately cell death. Despite their destructive activity, they are well-described second messengers in a variety of cellular processes, including conferment of tolerance to various environmental stresses (Pallavi Sharma *et al.*, 2012). Whether ROS would serve as signaling molecules or could cause oxidative damage to the tissues depends on the delicate equilibrium between ROS production and their scavenging. Efficient scavenging of ROS produced during various environmental stresses requires the action of several nonenzymatic as well as enzymatic antioxidants present in the tissues.

All ROS are extremely harmful to organisms at high concentrations. When the level of ROS exceeds the defense mechanisms, a cell is said to be in a state of 'oxidative stress.' The enhanced production of ROS during environmental stresses can pose a threat to cells by causing peroxidation of lipids, oxidation of proteins, damage to nucleic acids, enzyme inhibition, activation of programmed cell death (PCD) pathway and ultimately leading to death of the cells (Maheshwari and Dubey, 2009)

Despite their destructive activity, ROS are well-described second messengers in a variety of cellular processes including tolerance to environmental stresses. Whether ROS will act as damaging or signaling molecule depends on the delicate equilibrium between ROS production and scavenging. Because of the multifunctional roles of ROS, it is necessary for the cells to control the level of ROS tightly to avoid any oxidative injury and not to eliminate them completely. Scavenging or detoxification of excess ROS is achieved by an efficient antioxidative system comprising of the nonenzymic as well as enzymic antioxidants (Noctor and C. H. Foyer, 1998). The enzymic antioxidants include superoxide dismutase (SOD), catalase (CAT), guaiacol peroxidase (GPX), enzymes of ascorbate-glutahione (AsA-GSH) cycle such as ascorbate peroxidase (APX), monodehydroascorbate reductase (MDHAR), dehydroascorbate reductase (DHAR), and glutathione reductase (GR). Ascorbate (AsA), glutathione (GSH), carotenoids, tocopherols, and phenolics serve as potent nonenzymic antioxidants within the cell. Various workers have reported increased activities of many enzymes of the antioxidant defense system in plants to combat oxidative stress induced by various environmental stresses. Maintenance of a high antioxidant capacity to scavenge the toxic ROS has been linked to increased tolerance of plants to these environmental stresses. Considerable progress has been made in improving stress-induced oxidative stress tolerance in crop plants by developing transgenic lines with altered levels of antioxidants.

Cell walls are also regarded as active sites for ROS production. The role of cell-wall-associated peroxidase in H_2O_2 generation has been shown. In horseradish, peroxidase associated with isolated cell walls catalyzes the formation of H_2O_2 in the presence of NADH. The reaction is stimulated by various monophenols, especially of coniferyl alcohol. Malate dehydrogenase was found to be the sole candidate for providing NADH. The generation of ROS by cell-wall-located peroxidases has been shown during hypersensitive response (HR) triggered in cotton by the bacterium *Xanthomonas campestris* pv. *malvacearum* and potassium (K) deficiency stress in *Arabidopsis* (Kim *et al.*, 2010). Diamine oxidases are also involved in the production of activated oxygen in the cell wall using diamine or polyamines (putrescine, spermidine, cadaverine, etc.) to reduce a quinone that autoxidizes to form peroxides.

3.5.1 Antioxidative Defense System in Plants

Plants possess complex antioxidative defense system comprising of nonenzymatic and enzymatic components to scavenge ROS. In plant cells, specific ROS producing and scavenging systems are found in different organelles such as chloroplasts, mitochondria, and peroxisomes. ROS-scavenging pathways from different cellular compartments are coordinated. Under normal conditions, potentially toxic oxygen metabolites are generated at a low level, and there is an appropriate balance between production and quenching of ROS. The balance between production and quenching of ROS may be perturbed by a number of adverse environmental factors, giving rise to rapid increases in intracellular ROS levels (Noctor *et al.*, 2002) which can induce oxidative damage to lipids, proteins, and nucleic acids. To avoid the oxidative damage, higher plants raise the level of endogenous antioxidant defense. Various components of antioxidative defense system involved in ROS scavenging have been manipulated, overexpressed or downregulated to add to the present knowledge and to understand the role of the antioxidant systems.

One of the earliest cellular responses following successful pathogen recognition is oxidative burst involving the production of ROS. Recognition of a variety of pathogens leads to the generation of, or its dismutation product H_2O_2 in the apoplast. Radwan and coworkers observed higher H_2O_2 and MDA concentrations in *Vicia faba* leaves infected with bean yellow mosaic virus than those of the corresponding controls. Several enzymes have been implicated in apoplastic ROS production following successful pathogen recognition. The use of inhibitors pointed to plasma membrane NADPH oxidases and cell wall peroxidases as the two most likely biochemical sources (Grant *et al.*, 2000). The expression of these enzymes is induced following recognition of bacterial and fungal pathogens. Although the primary oxidative burst following pathogen recognition occurs in the apoplast, ROS can be produced in other cellular compartments like mitochondria and chloroplast. The role of chloroplasts in the interaction between *Erwinia amylovora* and host plants was studied by using uracil as chloroplast ETC inhibitor. Uracil presence significantly reduced ROS generation during the pathogen-host interaction, and ROS generation corresponded with the appearance of necrosis in all cultivars (Sasaki *et al.*, 2004). Activation of the SIPK/Ntf4/WIPK cascade by pathogens actively promotes the generation of ROS in chloroplasts, which plays an important role in the signaling for and/or execution of HR cell death in plants. They concluded that chloroplast bursts occur earlier than NADPH oxidase burst, and mitochondria-generated ROS might be essential for accelerating the cell death process.

Differential regulation of antioxidant enzymes, in part mediated by SA, may contribute to increases in ROS and activation of defenses following infection. In tobacco, the reduction of CAT and APX activities resulted in plants hyperresponsive to pathogens (Mittler *et al.*, 1999). A significant increase in the activities of POD and CAT was observed in leaves of flax lines infected with powdery mildew. The increase in POD activity was much pronounced in tolerant lines than susceptible lines. Enhanced activities of POD, CAT, APX, and SOD were observed in *Vicia faba* leaves infected with bean yellow mosaic virus indicating that the ROS-scavenging systems can have an important role in managing ROS generated in response to pathogens.

3.5.2 Peroxidase

These are a large family of enzymes that typically catalyze a reaction of the form: ROOR′ + electron donor ($2e^-$) + $2H^+ \rightarrow$ ROH + R′OH. For many of these enzymes, the optimal substrate is hydrogen peroxide but others are more active with organic hydroperoxides such as lipid peroxides. Peroxidases can contain a hemi cofactor in their active sites or alternately redox-active cysteine or selenocysteine residues.

Peroxidases are frequently associated with plant defense against pathogens. Peroxidases catalyze the oxidation of substrates like phenol and its derivates, by hydrogen peroxide. They are responsible for the radical dehydrogenation of sinapilalcohol and koniferylalcohol during the lignin synthesis. Peroxidases participate in the synthesis of flavons, stilbens, and other phenolic secondary metabolites. They are represented by many of the isoenzymes. Peroxidase polymorphism could be also used as a biochemical marker related to the different levels of field resistance. Peroxidases participate in processes which occur in the extracellular matrix. Their association with the cell wall was confirmed. Peroxidases remove the toxic hydrogen peroxide from tissues; participate in the synthesis of phenolic compounds and the building of intermolecular bonds during the organization of the cell wall at the sites of infection by pathogens. The production of phenolic compounds includes synthesis of chinons, tanins, melanins and also the polymerization of lignin and suberin composing monomers. Tanins and melanins are dihydroxyphenol and chinon oligomers; these are toxic for pathogens. Lignin and suberin are involved to structural defenses. Peroxidase also participates in the synthesis of ethylene the concentration of which frequently increases in pathogenesis process. Generally, peroxidases enhance their activity after a pathogen attack because they participate in defensive lignification and synthesis of phenolic compounds effective against pathogens (Ales Lebeda *et al.*, 1999).

The term 'peroxidase' means an enzyme is catalyzing oxidoreduction between hydrogen peroxide and reductants ($H_2O_2 + AH_2 \rightarrow 2H_2O + A$). Peroxidases are found in animals, plants and microorganisms, and they are divided into three superfamilies based on their structural and catalytic properties (Table 3.9) (Susumu Hiraga *et al.*, 2001).

The first peroxidase superfamily consists of enzymes in animals such as glutathione peroxidase (EC 1.11.1.9), myeloperoxidase (EC 1.11.1.7), thyroid peroxidase (EC 1.11.1.7), eosinophil peroxidase (EC 1.11.1.7), lactoperoxidase (EC1.11.1.7) and a partial region of prostaglandin endoperoxide synthase (EC 1.14.99.1). Although glutathione peroxidase is categorized into the animal peroxidase superfamily, its activity is also detected in plants, and several cDNAs encoding its homologs have actually been isolated from plants. The second peroxidase superfamily consists of catalases (EC 1.11.1.6) in animals, plants, bacteria, fungi and yeast (Fig. 3.65). The third superfamily consists of peroxidases in plants, fungi, bacteria and yeast. Although the active sites of myeloperoxidase and cytochrome *c* peroxidase (a member of the animal and plant peroxidase superfamilies, respectively) have similar architectures, the overall primary sequences and 3-dimensional structures of these peroxidases are quite different, implying that these superfamily genes evolved from distinct ancestral genes.

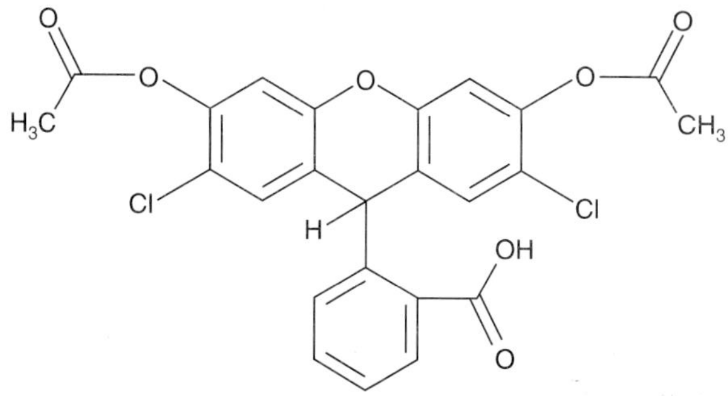

Fig. 3.65 Catalase

Peroxidases, a class of enzymes in animal, plant and microorganism tissues, catalyze oxidoreduction between H_2O_2 and various reductants. Class III plant peroxidase (POX) is a classical enzyme whose activity was described as early as 1855 and whose purification followed a few decades later (Theorell, 1951). Since POX is relatively stable at high temperature and its activity is easily measured using simple chromogenic reactions. It has been used as a model enzyme in the study of protein structure, enzyme reactions and enzyme functions, and it has been used for several practical applications. Studies have suggested that POXs play a role in lignification, suberization, cross-linking of cell wall structural proteins, auxin catabolism, self-defense against pathogens, salt tolerance and senescence (Abeles *et al.*, 1988). These functional assignments are based on the *in vitro* catalytic properties, expression profiles, localization and the characteristics of transgenic plants over or underexpressing a certain *pox* gene. However, it is difficult to define the specific functions of individual POXs because of their low substrate specificity *in vitro* and the presence of many isoenzymes.

Table 3.9 Classification of Peroxidases

Superfamily	*Class*	*Member (EC number)*	*Origin*	*Molecular weight (kDa)*
Animal peroxidase		Eosinophil peroxidase (EC 1.11.1.7)	Animal	50–75
		Lactoperoxidase (EC 1.11.1.7)	Animal	78–85
		Myeloperoxidase (EC 1.11.1.7)	Animal	79–150
		Thyroid peroxidase (EC 1.11.1.7)	Animal	90–110
		Glutathione peroxidase (EC 1.11.1.9)	Animal and plant	6–22 and 75–112
		Prostaglandin endoperoxide synthase (EC 1.14.99.1, partial)	Animal	115–140
Catalase		Catalase (EC 1.11.1.6)	Animal, plant, fungus, and yeast	140–530

Contd...

Contd...

Plant peroxidase	I	Cytochrome c peroxidase (EC 1.11.1.5)	Yeast and bacterium	32–63
		Catalase-peroxidase (EC 1.11.1.6)	Bacterium and fungus	150–240
		Ascorbate peroxidase (EC 1.11.1.11)	Plant	30–58
	II	Manganese-dependent peroxidase (EC 1.11.1.13)	Fungus	43–49
		Ligninase (EC 1.11.1.14)	Fungus	40–43
	III	Peroxidase (EC 1.11.17, POX)	Plant	28–60

3.6 DEFENSE AGAINST PATHOGEN INFECTION

Upon pathogen infection, plants newly synthesize a set of defense-related proteins such as PR proteins, which are induced in pathological or related situations. POXs have been reported to be induced by infection with fungi, bacteria, viruses, and viroids. Thus, specific pathogen-inducible POXs have been categorized into a family of PR proteins (PR-9). Many studies have indicated the importance of POXs in defense against pathogen infection. In tobacco, a positive correlation was found between POX activity and resistance to tobacco wildfire disease. A cationic POX isoenzyme was induced more rapidly and at a higher level when rice plants were infected with an incompatible rather than compatible strain of *Xanthomonas oryzae* pv. *oryzae*). The roles of POXs in defense are considered to be reinforcement of cell wall physical barriers comprising lignin, suberin, feruloylated polysaccharides and hydroxyproline-rich glycoproteins; enhanced production of reactive oxygen species as signal mediators and antimicrobial agents; and enhanced production of phytoalexin. Generally, multiple POXs are induced by pathogen infection, suggesting that each POX is involved in a specific defense process (Susumu Hiraga *et al.*, 2001).

Among the various POX isoenzymes in plants, several are induced by pathogen infection and wounding, suggesting the importance of POX in self-defense systems in plants. The importance of POX in pathogen defense was indicated earlier than the discovery of pathogenesis-related (PR) proteins as novel pathogen-inducible proteins, and POX was classified as PR-9. However, the signal transduction leading to the expression of pathogen- or wound-inducible genes has been more extensively studied for *PR* genes other than *pox* genes. Only recently has information been obtained on the regulated expression of pathogen- or wound-inducible *pox* genes in comparison with that of *PR* genes.

Increased peroxidase activity upon infection might be required for an additional deposition of lignin around the lesion court induced by the pathogen. Peroxidase is a key enzyme in the biosynthesis of lignin and other oxidized phenols. Among the proteins induced during the host plant defense, class III plant peroxidases (EC 1.11.1.7; hydrogen donor: H_2O_2 oxidoreductase, Prxs) are well known. They belong to a large multigene family, and participate in a broad range of physiological processes, such as lignin and suberin formation, cross-linking of cell wall components, and synthesis of phytoalexins, or participate in the metabolism of ROS and RNS, both switching on the hypersensitive response (HR), a form of programmed host cell death at the infection site associated with limited pathogen development (Almagro *et al.*, 2009).

Prxs are involved in a broad range of physiological processes throughout the plant life cycle (Fig. 3.66), probably due to the high number of enzymatic isoforms (isoenzymes) and to the versatility of their enzyme-catalyzed reactions (Passardi *et al.*, 2005). Thus, plant Prxs are involved in auxin metabolism, lignin and suberin formation, cross-linking of cell wall components, phytoalexin synthesis, and the metabolism of reactive oxygen species (ROS) and reactive nitrogen species (RNS).

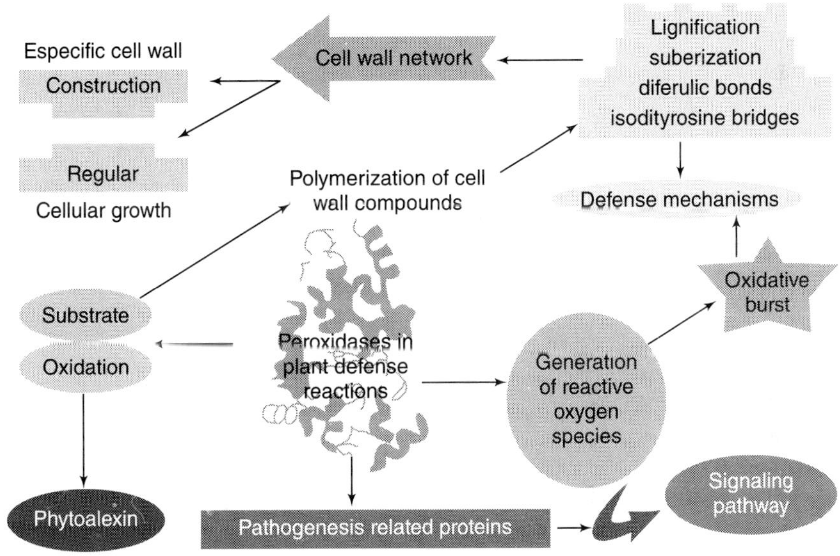

Fig. 3.66 Overview of the specific roles of plant Prxs in defense reactions

Plant Prxs are heme-containing enzymes which catalyze the single one-electron oxidation of several substrates at the expense of H_2O_2: $2RH + H_2O_2$ to $2R + 2H_2O$. Prxs are a well-known class of PR proteins, and so they are induced in host plant tissues by pathogen infection. They belong to the PR-protein 9 subfamily (van Loon *et al.*, 2006), and are expressed to limit cellular spreading of the infection through the establishment of structural barriers or the generation of highly toxic environments by massively producing ROS and RNS (Passardi *et al.*, 2005).

The corn smut, *Ustilago maydis* establishes a biotrophic interaction with its host plant maize. This interaction requires efficient suppression of plant immune responses, which is attributed to secreted effector proteins (Hemetsberger *et al.*, 2012). Previously they identified Pep1 (Protein essential during penetration-1) as a secreted effector with an essential role in *U. maydis* virulence. *Pep1* deletion mutants induced strong defense responses leading to an early block in the pathogenic development of the fungus. Using cytological and functional assays, it was shown that Pep1 functions as an inhibitor of plant peroxidases. At sites of Δ*pep1* mutant penetrations, H_2O_2 strongly accumulated in the cell walls, coinciding with a transcriptional induction of the secreted maize peroxidase POX12. Pep1 protein effectively inhibited the peroxidase driven oxidative burst and thereby suppresses the early immune responses of maize. Moreover, Pep1 directly inhibited peroxidases *in vitro* in a concentration-dependent manner. Using fluorescence complementation assays, a direct interaction of Pep1 and the maize peroxidase POX12 *in vivo* was noticed. Functional relevance of this interaction was demonstrated by partial complementation of the Δ*pep1* mutant defect by virus-induced gene silencing of maize POX12. It was concluded that Pep1 acts as a potent suppressor of early plant defenses by inhibition of peroxidase activity. Thus, it represents a novel strategy for establishing a biotrophic interaction.

Bacterial catalase-peroxidases are enzymes containing 0.5-1.0 heme per subunit. The identical subunits are generally 80 kDa in size, and the sequenced subunits of *E. coli, S. typhimurium,* and *B. stearothermophilus* contain 726-731 amino acid residues per subunit (Welinder, 1991). The heme-containing peroxidases of plants, fungi and yeast are monomeric, homologous and 290-350 residues in size. Analyses of the amino acid sequences indicate that the double length of the bacterial peroxidases can be ascribed to gene duplication. Each half is homologous to eukaryotic, monomeric peroxidase and can be modeled into the high-resolution crystal structure of yeast cytochrome *c* peroxidase. The comparisons and modeling have predicted that the C-terminal half does not bind heme, and bacterial peroxidases have one heme per subunit. The ten dominating helices observed in the yeast enzyme are highly conserved and connected by surface loops which are often longer in the bacterial peroxidases and yeast cytochrome *c* peroxidase has evolved more slowly than other known peroxidases. The study has revealed ten invariant residues and a number of highly conserved residues present in peroxidases of the plant peroxidase superfamily and provides a basis for rationally engineered peroxidases.

3.6.1 Polyphenol Oxidase (PPO)

Polyphenol oxidase (tyrosinase) is a bifunctional, copper-containing oxidase having both catecholase and cresolase activity (Fig. 3.67). Polyphenol oxidases (PPOs) are ubiquitous which use molecular oxygen to oxidize common ortho-diphenolic compounds such as caffeic acid and catechol to their respective quinine. PPO-generated quinones are highly reactive and may cross-link or alkylate proteins, leading to the commonly observed brown pigments in damaged plant tissues and plant extracts (Peter Constabel and Raymond Barbehenn, 2008).

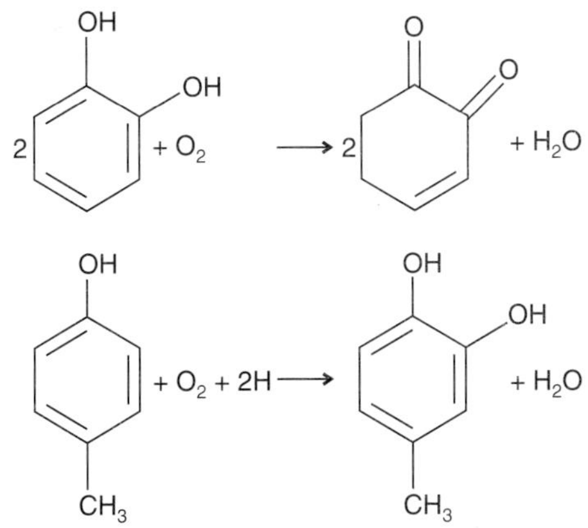

Fig. 3.67 Polyphenol oxidase

Direct evidence for a role of PPO in inhibiting pathogen ingress or growth comes from transgenic tomato plants with enhanced or suppressed PPO levels. Plant polyphenol oxidases (PPOs) are widely distributed, and well-studied oxidative enzymes and their effects on discoloration in damaged and diseased plant tissues have been known for many years. The cloning and characterization of PPO cDNAs from multiple plant species now allow for direct testing of defensive functions of PPO using transgenic plants. When challenged by the bacterial pathogen *Pseudomonas syringae* pv. *tomato,* PPO-overexpressing plants showed reduced bacterial growth, whereas PPO antisense-suppressed lines supported greater bacterial numbers (Thipyapong *et al.,* 2004). In poplar, infection with *Melampsora medusae* does not induce PPO but represses its expression together with many other herbivore defense genes (Miranda *et al.,* 2007).

The demonstration that PPO can inhibit bacterial plant pathogens suggests that, in addition to effects on some herbivores, PPO is important for inhibiting microbes introduced into damaged leaves via the mouthparts of feeding insects. These might be opportunistic bacteria or pathogens that are vectored by specific insects, although little is known about the microflora of caterpillar mouthparts and digestive systems. Thus, the distinction between pest and pathogen defenses may be artificial, and these may be seen as synergistic and complementary responses. This contrasts with the view that pest and pathogen defense responses are mutually exclusive. This hypothesis was based on early work on tomato showing that salicylic acid, a potent inducer of pathogen defense and systemic acquired resistance, inhibits the jasmonate-regulated herbivore defense response. Current models of defense signaling outline a complex and overlapping set of responses, regulated by jasmonate, ethylene, and salicylic acid-based signals, leading to several possible outcomes. Functionally, crosstalk is supported by the observation of herbivore induced resistance of *Arabidopsis* against microbial pathogens, including *P. syringae.* In tomato, the PPO-F gene is induced during both the wound response and infection with the pathogen *P. syringae* (Thipyapong *et al.,* 2004).

Several modes of actions were suggested including general toxicity of PPO-generated quinones to pathogens and plant cells, accelerating cell death; alkylation and reduced bioavailability of cellular proteins to the pathogen; cross-linking of quinones with protein or other phenolics, forming a physical barrier to pathogens in the cell wall; and finally, quinone redox cycling leading to H_2O_2 and other reactive oxygen species (Li and Steffens, 2002). While reactive oxygen species are known to be important factors in plant-pathogen interactions and defense signaling, and PPO is implicated in the formation of melanin-like polymers in potato black spot lesions. Total phenols with their oxidizing enzymes, peroxidase and polyphenol oxidase in diseased leaves of rose caused by *Alternaria tenuis* were measured at different progression periods of infection (Khatun *et al.*, 2009). There was an increase in the activity of peroxidase and polyphenol oxidase in host plants in response to infection, which was found to be coupled with an increased phenol content in disease leaf tissue as compared to the healthy tissues. This clearly revealed that the elicitation of the enzymes in response to infection is directly correlated with the resistance of the plant. Peroxidase and polyphenol oxidase mediate the oxidation of phenols, and oxidized phenols are highly toxic to the pathogen.

PO and PPO catalyze the oxidation of phenolic compounds through a PPO-PO-H_2O_2 system. A number of studies have found a correlation between PPO and the resistance response. PO itself was also found to inhibit the spore germination and mycelial growth of certain fungi (Joseph *et al.*, 1998). Peroxidase may be rapidly involved in the peroxidation of the substrate molecule, leading to the accumulation of highly toxic compounds (i.e., phenolic compounds), which may contribute to resistance via their antifungal potential (Ward, 1986).

Persistence of the soil seed bank requires both dormancy and resistance to seed decay organisms. However, there is little or no information evaluating biochemical responses of dormant weed seeds to pathogens (Patrick Fuerst *et al.*, 2011). Wild oat caryopses were incubated with four pathogenic fungal isolates to evaluate the response of the pathogen defense enzyme, polyphenol oxidase (PPO). Caryopsis PPO activity was induced by three *Fusarium* spp. isolates previously obtained from whole seeds incubated in the field whereas caryopsis PPO activity was decreased by a *Pythium* isolate. *Fusarium avenaceum* isolate F.*a*.1 caused the greatest PPO induction and was studied in more detail. When whole wild oat seeds were incubated on F.*a*.1, PPO activity was induced in seeds, hulls (lemma and palea), and caryopses. Incubation of whole seeds on F.*a*.1 gradually induced caryopsis PPO activity over an 8-d period, whereas incubation of caryopses on F.*a*.1 over a 4-d period caused a greater and more rapid induction of PPO activity. Very little PPO activity could be leached from untreated caryopses, but nearly all of the induced PPO activity in F.*a*.1-treated caryopses was readily leached when incubated in the buffer. In Western blots, both untreated and F.*a*.1-treated leachates contained a ~57-kilodalton (kD) protein, putatively the mature and relatively inactive form of PPO. However, lower molecular weight antigenic proteins between ~52 and ~25 kD were strongly induced in F.*a*.1-treated caryopses, with this induction being correlated with the increase in PPO activity. It was hypothesized that dormant weed seeds possess biochemical defenses against pathogens and more specifically, that proteolysis in the presence of fungal pathogens may release an activated form of PPO from the surface of wild oat caryopses and hulls.

Induction of defense enzymes phenol, peroxidase (PO), polyphenol oxidase (PPO), phenyl ammonia lyase (PAL), superoxide dismutase (SOD) and chitinase by *Pseudomonas fluorescens* isolate Pf1, against challenge inoculation of *Meloidogyne graminicola* in rice was studied *in vitro* (Anita and Samiyappan, 2012). The activity of phenol, PO, PPO, PAL and chitinase was higher in the bacterized rice plants. The analysis of isoform profiles revealed unique PO, PPO and SOD isoforms induced in plants treated with *P. fluorescens*. The present study implies earlier and higher accumulation of phenols and defense enzymes, viz., PO, PPO, PAL and chitinase in rice root tissue resulting in significant reduction in nematode infection.

3.6.2 Phenylalanine Ammonia-Lyase (PAL)

Phenylalanine ammonia-lyase is an enzyme that catalyzes a reaction converting L-phenylalanine to ammonia and transcinnamic acid [(Fig. 3.68) (Kilakkathi and Sindhu, 2010)]. Phenylalanine ammonia-lyase (PAL) is the first and committed step in the phenylpropanoid pathway and is therefore involved in the biosynthesis of the polyphenol compounds such as flavonoids, phenylpropanoids, and lignin in plants. Phenylalanine ammonia-lyase is found widely in plants, as well as some yeast and fungi, with isoenzymes existing within many different species. It has a molecular mass in the range of 270-330 kDa. The activity of PAL is induced dramatically in response to various stimuli such as tissue wounding, pathogenic attack, light, low temperatures, and hormones. Phenylalanine ammonia lyase is composed of four identical subunits composed mainly of α-helices, with pairs of monomers forming a single active site.

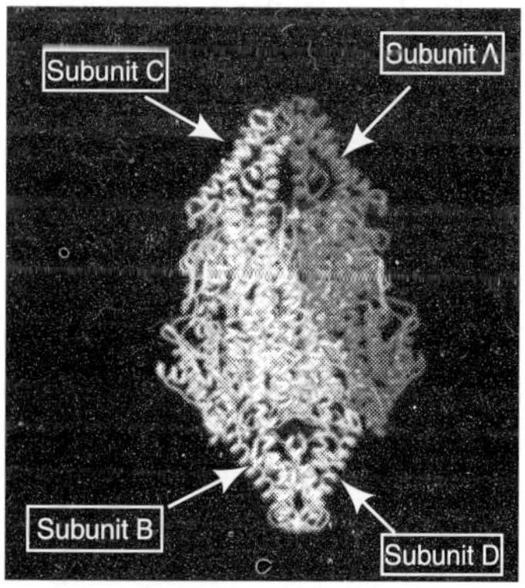

Fig. 3.68 Biological assembly of phenylalanine ammonia-lyase (PDB ID: 1T6J). This figure shows the four identical subunits that comprise the homotetrameric structure of PAL

Phenylalanine ammonia-lyase (PAL) catalyzes the first step in the biosynthesis of phenylpropanoids, which form a wide variety of secondary plant products. The transcription of PAL is regulated in response to various factors that induce the accumulation of flavonoids; lignin and compounds are thought to be involved in plant defense reactions. The 5′ upstream sequence of a PAL gene from *Phaseolus vulgaris* was fused to the coding region of the reporter gene encoding β-glucuronidase (GUS), and transformed into potato and tobacco plants. Histochemical analysis of GUS expression showed that the PAL promoter was active in specific cell types that accumulated phenylpropanoid derivatives in response to mechanical wounding, and also during normal development of the xylem and flower. In xylem that had undergone secondary thickening, GUS activity occurred in rays of cells thought to be the xylem parenchyma. It was postulated that PAL activity in these cells could provide intermediates for lignin synthesis in xylem vessels that had terminally differentiated.

In tuber tissues of potato (*Solanum tuberosum* L.) infected with an incompatible race of *Phytophthora infestans* (Mont.) de Bary, the activity of phenylalanine ammonia-lyase and the contents of free and bound salicylic acid (SA) considerably exceeded the corresponding indices in the tissues infected with a compatible race of the oomycete. The accumulation of the free form of SA apparently resulted from both enhanced SA biosynthesis and the liberation from the bound SA forms. SA accumulation in the incompatible host–pathogen combination presumes that SA participated in the local potato resistance to late blight (Panina *et al.*, 2005).

The activities of the enzymes chalcone synthase (CHS) and phenylalanine ammonia-lyase (PAL) were measured in leaf extracts obtained from four cultivars of the common bean [(AB 136, Rio Tibagi, Carioca, and Macanudo) (Bevan *et al.*, 1989)]. Two stages of plant development were examined: plantlets (V2) and the onset of blooming (R6). Initially, the plants were either treated with salicylic acid or inoculated with the delta race of *Colletotrichum lindemuthianum* (inductive fungus) and after three days they were evaluated for enzyme activity. Afterward, all plants were inoculated (challenged) with the virulent pathotype 33/95 of *C. lindemuthianum* except for the water control. Five days later, the activities of PAL and CHS were evaluated. There were significant changes in the activities of both enzymes three days after treatment with salicylic acid or inductive fungus when compared to the control. Five days after inoculation with the virulent pathotype 33/95 of *C. lindemuthianum* CHS activity in the Macanudo was similar to control plants that were not treated with salicylic acid or the inductive fungus but inoculated with 33/95 *C. lindemuthianum*. The increase in enzyme activity after challenge with 33/95 *C. lindemuthianum* was greatest for the salicylic acid treatment in the cultivar AB 136, followed by Rio Tibagi and Carioca.

Dae Sung Kim and Byung Kook Hwang (2014) identified the pepper (*Capsicum annuum*) *PAL* (*CaPAL1*) gene, which was induced in pepper leaves by avirulent *Xanthomonas campestris* pv. *vesicatoria* (*Xcv*) infection. *CaPAL1*-silenced pepper plants exhibited increased susceptibility to virulent and avirulent *Xcv* infection. Reactive oxygen species (ROS), hypersensitive cell death, expression of the salicylic acid (SA)-dependent marker gene *CaPR1*, SA accumulation, and induction of PAL activity were significantly compromised in the *CaPAL1*-silenced pepper plants during *Xcv* infection. Overexpression (OX) of *CaPAL1* in *Arabidopsis* conferred increased resistance to *Pseudomonas syringae* pv. *tomato* (*Pst*) and *Hyaloperonospora arabidopsidis* infection. *CaPAL1*-OX leaves exhibited restricted *Pst* growth, increased ROS burst and cell death and induction of *PR1* expression and SA accumulation. The increase in PAL activity in healthy and *Pst*-infected leaves was higher in *CaPAL1*-OX plants than in wild-type *Arabidopsis*. Taken together, these results suggested that CaPAL1 acts as a positive regulator of SA-dependent defense signaling to combat microbial pathogens via its enzymatic activity in the phenylpropanoid pathway.

To investigate the lignification process and its physiological significance under Karnal Bunt, the changes in enzymes responsible for lignifications likes, phenylalanine ammonia-lyase (PAL), were determined in resistant (HD-29) and susceptible genotype (WH-542) of wheat during different developmental stages (Shalini Purwar *et al.*, 2013). The PAL gene was cloned and sequenced. The expression of PAL gene was measured by means of semiquantitative RT-PCR. The enzyme was expressed constitutively in both the susceptible and resistant genotype. However, the activity was higher in all the developmental stages of resistant genotype, indicating that this genotype has a significantly higher basal level of these enzymes as compared to the susceptible line and could be used as marker(s) to define bunt resistance. The activity of PAL was significantly higher in WSv stage (Z=16). Structural comparisons based on alignments of all the protein sequences using the cluster W program and searches for conserved motifs using the MEME program have revealed broad conservation of main motifs characteristic of the plant PAL. MSA and phylogenetic analyses of different plants PAL demonstrated that all PAL cluster divided into two main clusters. The PAL also possessed a specific consensus sequences [GS]-[STG]-[LIVM]-[STG]-[SAC]-S-G-[DH]-L-x-[PN]-L-[SA]-x(2,3)-[SAGVTL]. The pathway might be associated with the enhancement of structural defense barrier due to lignifications of cell wall as evident from the enhanced synthesis of lignin in all the stages of resistant genotype. These results clearly indicated the possible role of enzymes of phenylpropanoid pathway metabolism provides genotype and stage dependent structural barrier resistance in wheat against bunt disease.

Cell-free culture filtrate of a biocontrol PGPR strain of *Bacillus subtilis* showed a concentration-dependent growth and conidiation inhibition of *Fusarium udum*, the incitant of pigeonpea wilt (Podile and Laxmi, 2008). *In vitro* interaction of cell free filtrate of AF 1 induced bulb-like structures in the hyphae of *F. udum*. Seed treatment with *B. subtilis* significantly reduced the incidence of pigeonpea wilt, but not with the cell free culture filtrate of AF 1. Seeds bacterized with *B. subtilis* AF 1 showed an increase in phenylalanine ammonia-lyase (PAL) from day 1, while the peroxidase activity increased from day 6.

3.7 RNA-MEDIATED INTERFERENCE/RNA SILENCING

It is a biological process in which RNA molecules inhibit gene expression, typically by causing the destruction of specific mRNA molecules. Historically, it was known by other names, including cosuppression, posttranscriptional gene silencing (PTGS), and *quelling*. Only after these apparently unrelated processes were fully understood did it become clear that they all described the RNAi phenomenon.

The RNA-mediated mechanism that is, the RNA silencing pathway has become a powerful tool for engineering resistant plants. RNA silencing referred to as gene-quelling in fungi and RNA interference (RNAi) in animals, is a conserved regulatory mechanism of gene expression that has been widely characterized in eukaryotic organisms. RNA silencing is a nucleotide sequence-specific process that induces mRNA degradation or translation inhibition at the posttranscriptional level (named PTGS in plants) or epigenetic modification at the transcriptional level, depended on RNA-directed DNA methylation (a process named RdDM in plants). The RNA silencing pathway is composed of a series of components: a dsRNA trigger; a processor called Dicer or a Dicer-like (DCL) protein; the processor product, small RNAs (siRNAs or miRNAs) of 21 to 24 nt (nucleotide) in length; an effector complex called RISC in which the Argonaute (AGO) protein is the key player. siRNAs-guided AGO-cleaved target RNA may be recognized by RNA-dependent RNA polymerase (RDR), which amplifies the dsRNA; and Suppressor of Gene Silencing (SGS), which stabilizes the dsRNA substrate for DCLs to produce secondary siRNAs and reinforce the RNA silencing process (Peragine *et al.*, 2004).

Besides the regulatory roles in plant development, the siRNA-mediated RNA silencing also functions as a natural antiviral defense mechanism, a process named virus-induced gene silencing (VIGS). Host RNA silencing machinery targets and processes the virus derived dsRNA, which is derived from pathogen replication or in a host RDR-dependent manner, into vsiRNAs (virus-derived siRNAs). The vsiRNAs are then recruited to host RISC complexes, which targets and inhibits gene expression and protein translation in the viral genome. To counteract the defense mechanism, many viruses encode a protein called viral suppressor of RNA silencing (VSR), which have been identified from diverse plant viruses. Recently, suppressor of RNA silencing was also identified in some bacteria (BSRs) VSRs, and BSRs may function in suppression of RNA silencing in different steps, either by binding siRNA duplex or by directly interacting with key components in RNA silencing. Some of them may combine various functions to realize a multilevel suppression (Duan *et al.*, 2012)

Double-stranded RNA (dsRNA) has been shown to play a key role as an inducer of different interference phenomena occurring in both the plant and animal kingdoms. It was shown that dsRNA derived from viral sequences could interfere with virus infection in a sequence-specific manner by directly delivering dsRNA to leaf cells either by mechanical inoculation or via an *Agrobacterium*-mediated transient expression assay (Tenllado and Díaz-Ruíz, 2001). Successful interference with the infection of plants by three viruses belonging to the tobamovirus, potyvirus, and alfamovirus groups was shown, demonstrating the reliability of the approach. It was suggested that the effect mediated by dsRNA in plant virus infection resembled the analogous phenomenon of RNA interference observed in animals. The interference observed was sequence specific, was dose dependent, and was triggered by dsRNA but not single-stranded RNA. Thes findings supported the view that a dsRNA intermediate in virus replication acts as an efficient initiator of posttranscriptional gene silencing (PTGS) in natural virus infections, triggering the initiation step of PTGS that targets viral RNA for degradation.

The injection into *C. elegans* of double-stranded RNA (dsRNA) corresponding to any one of its genes specifically depleted the mRNA molecules transcribed from that gene. In the absence of mRNA, the corresponding protein was not produced and, therefore, this ultimately resulted in the loss of function of the targeted gene. This phenomenon was known as RNA-mediated interference or, shortly, RNAi. The cells of organisms such as nematodes, plants, and humans do not produce dsRNA. However, many viruses produce dsRNA. Thus, the RNAi machinery seems to have evolved as a defense mechanism against viruses. Scientists have exploited RNAi as a powerful tool to disrupt the functions of specific genes to uncover their functions. Since its original discovery in *C. elegans,* RNAi has been observed in many other organisms as well. Now that the complete DNA sequence (genome) is available for many organisms, RNAi can be applied in all these organisms for revealing the functional biological information contained in their genomes. In fact, this has already revolutionized the field of functional genomics, an area of biology focused on the characterization of gene function in large scale (Timmons *et al.*, 2001).

An amazing aspect of RNAi in *C. elegans* is that it can be triggered when the worm feeds on bacteria that produce the dsRNA of the worm's gene. It was reasoned that a similar delivery of dsRNA of parasite's genes through host organisms was likely to induce RNAi in the parasites, and genes that are essential for the survival of the parasite are targeted, and then it may protect the host from infection. In addition, this approach could be a powerful tool to the functional genomics of parasites. Tobacco plant, a good plant model organism amenable to genetic transformation was engineered to produce dsRNA of two essential genes of the parasitic nematode *Meloidogyne incognita.* The transgenic tobacco plants very effectively resisted *M. incognita* infection. A closer observation of the nematodes in these plants revealed that their development was severely impaired. Further, these nematodes were specifically deficient in the mRNA of targeted genes, indicating that the dsRNA produced in plants did indeed trigger an RNAi response in the nematode.

An initial understanding of RNA silencing mechanisms and small RNAs was derived from model fungi. Now, new knowledge with practical implications for RNA silencing is beginning to emerge from the study of plant-fungus interactions (Nunes and Dean, 2012). Recent studies have shown that the expression of silencing constructs in plants designed on fungal genes can specifically silence their targets in invading pathogenic fungi, such as *Fusarium verticillioides, Blumeria graminis and Puccinia striiformis* f. sp. *tritici.* The important general aspects of RNA silencing mechanisms were highlighted and emphasized recent findings from plant pathogenic fungi. Strategies to employ RNA silencing to investigate the basis of fungal pathogenesis were discussed. Important aspects of the development of fungal-derived resistance through the expression of silencing constructs in host plants were also addressed as a powerful strategy to control fungal disease.

RNA silencing plays a critical role in plant resistance to viruses, with multiple silencing factors participating in antiviral defense. Both RNA and DNA viruses are targeted by the small RNA-directed RNA degradation pathway, with DNA viruses being also targeted by RNA-directed DNA methylation (Wang *et al.*, 2012). To evade RNA silencing, plant viruses have evolved a variety of counterdefense mechanisms such as expressing RNA-silencing suppressors or adopting silencing-resistant RNA structures. This constant defense-counter defense arms race is likely to have played a major role in defining viral host specificity and in shaping viral and possibly host genomes. Recent studies have provided evidence that RNA silencing also plays a direct role in viral disease induction in plants, with viral RNA-silencing suppressors and viral siRNAs as potentially the dominant players in viral pathogenicity. However, questions remain as to whether RNA silencing is the principal mediator of viral pathogenicity or if other RNA-silencing-independent mechanisms also account for viral disease induction. RNA silencing has been exploited as a powerful tool for engineering virus resistance in plants. Further understanding of the role of RNA silencing in plant-virus interactions and viral symptom induction is likely to result in novel antiviral strategies in both plants.

To counteract host defense, plant viruses encode viral suppressors of RNA silencing (VSRs) that target different effector molecules in the RNA silencing pathway. Evidence has shown that plants also encode endogenous suppressors of RNA silencing (ESRs) that function in the proper regulation of RNA silencing (Li *et al.*, 2014). The possibility that these cellular proteins can be subverted by viruses to thwart host defense is intriguing but has not been fully explored. It was reported that the *Nicotiana benthamiana* calmodulin-like protein *Nbrgs-CaM* is required for the functions of the VSR βC1, the sole protein encoded by the DNA satellite associated with the geminivirus *Tomato yellow leaf curl China virus* (TYLCCNV). *Nbrgs-CaM* expression is up-regulated by the βC1. Transgenic plants overexpressing *Nbrgs-CaM* displayed developmental abnormities reminiscent of βC1-associated morphological alterations. *Nbrgs-CaM* suppressed RNA silencing in an *Agrobacterium* infiltration assay and, when overexpressed, blocked TYLCCNV-induced gene silencing. Genetic evidence showed that *Nbrgs-CaM* mediated the βC1 functions in silencing suppression and symptom modulation and was required for efficient virus infection. Moreover, the tobacco and tomato orthologs of *Nbrgs-CaM* also possessed ESR activity and were induced by β-satellite to promote virus infection in these *Solanaceae* hosts. They further demonstrated that βC1-induced *Nbrgs-CaM* suppressed the production of secondary siRNAs, likely through repressing *RNA-Dependent RNA Polymerase 6*(*RDR6*) expression. *RDR6*-deficient *N. benthamiana* plants were defective in antiviral response and were hypersensitive to TYLCCNV infection. More significantly, TYLCCNV could overcome host range restrictions to infect *Arabidopsis thaliana* when the plants carried a*RDR6* mutation. These findings demonstrated a distinct mechanism of VSR for suppressing PTGS through usurpation of a host ESR and highlighted an essential role for RDR6 in RNA silencing defense response against geminivirus infection.

3.8 PROGRAMMED CELL DEATH (PCD) IN PLANT-PATHOGEN INTERACTIONS

Programmed cell death is the active process of cell death which occurs during development and in response to environmental cues, including pathogen attack. Often the interaction between a plant and a pathogen results in plant cell death. Plant cell death can occur when the pathogen unsuccessfully parasitizes the host as well as when the pathogen successfully causes disease (Alan M. Jones, 2001). The effects of host cell death on a pathogen may vary greatly depending on the lifestyle of the parasite. For example, some pathogens are obligate parasites that depend on living cells to grow and develop. Other pathogens may benefit from the release of nutrients from dead cells. Three strands of research came together to shape today's research agenda about plant PCD: terminal differentiation, senescence, and disease resistance. Cell death fulfills several essential functions in plant development: Senescence removes cells by recycling much of its carbon, nitrogen, and phosphorous; cell death is important in sculpting tissues such as the formation of lysigenous aerenchyma, flower primordia during floral abortion, and aleurone layers during germination; cells invaded by pathogens may be self-eliminated as part of a hypersensitive response against the pathogen; cell death also occurs during terminal differentiation and the classic example is the formation of vessel members and tracheids, collectively termed tracheary elements (TE); and cell death is programmed when the metabolism of cells is perturbed either by coping with abiotic stresses imposed upon it or by bioengineering.

The hypersensitive response (HR) is induced by certain plant pathogens and involves PCD to restrict the spread of pathogens from the infection site (del Pozo and Lam, 1998). Concurrent with the induction of cell death, the host activates a defense response. The cell death associated with the HR in several plant-pathogen systems has morphological similarities to animal apoptosis, which suggests that cell death mechanisms in plants and animals may share common components that lead to similar cellular events. Caspases are conserved cysteine proteases that regulate animal PCD; caspase activity or an involvement of caspases in cell death has yet to be reported in plants. In this work, we investigated the participation of caspases in HR cell death. Caspase-specific peptide inhibitors, Ac-YVAD-CMK and Ac-DEVD-CHO, could abolish bacteria-induced plant PCD but did not significantly affect the induction of other aspects of HR, such as the expression of defense genes. This result confirmed their previous model that cell death can be uncoupled from defense gene activation during HR. Caspase-like proteolytic activity was detected in tobacco tissues that were developing HR following infection with tobacco mosaic virus (TMV). These findings provided evidence for the presence of caspase-like plant protease(s) that participate in HR cell death.

Almost all plant cells have large vacuoles that contain both hydrolytic enzymes and a variety of defense proteins. Plants use vacuoles and vacuolar contents for programmed cell death (PCD) in two different ways: in a destructive way and in a nondestructive way (Hara Nishimura and Hatsugai, 2011). Destruction is caused by vacuolar membrane collapse, followed by the release of vacuolar hydrolytic enzymes into the cytosol, resulting in rapid and direct cell death. The destructive way is effective in the digestion of viruses proliferating in the cytosol, in susceptible cell death induced by fungal toxins, and in developmental cell death to generate integuments (seed coats) and tracheary elements. On the other hand, the nondestructive way involves fusion of the vacuolar and the plasma membrane, which allows vacuolar defense proteins to be discharged into the extracellular space where the bacteria proliferate. Membrane fusion, which is normally suppressed, was triggered in a proteasome-dependent manner. Intriguingly, both ways use enzymes with caspase-like activity; the membrane-fusion system uses proteasome subunit PBA1 with caspase-3-like activity, and the vacuolar-collapse system uses vacuolar processing enzyme (VPE) with caspase-1-like activity. The authors summarized two different ways of vacuole-mediated PCD and discussed how plants use them to attack pathogens that invade unexpectedly.

In plants, PCD is thought to be activated during the hypersensitive response to certain avirulent pathogens and in the course of several differentiation processes, a transgenic model system was described that mimicked the activation of programmed cell death in higher plants (Mittler *et al.*, 1995). In this system, expression of a bacterial proton pump in transgenic tobacco plants activated a cell death pathway that might be similar to that triggered by recognition of an incompatible pathogen. Thus, spontaneous lesions that resemble hypersensitive response lesions were formed, multiple defense mechanisms were apparently activated, and systemic resistance was induced in the absence of a pathogen. Interestingly, mutation of a single amino acid in the putative channel of this proton pump rendered it inactive with respect to lesion formation and induction of resistance to pathogen challenge. This transgenic model system may provide insights into the mechanisms involved in mediating cell death in higher plants. In addition, it may also be used as a general agronomic tool to enhance disease protection.

Programmed cell death (PCD) is essential for proper growth, development, and cellular homeostasis in all eukaryotes. The regulation of PCD is of central importance in plant-microbe interactions; notably, PCD and features associated with PCD are observed in many host resistance responses (Martin B. Dickman and Robert Fluhr, 2013). Conversely, pathogen induction of inappropriate cell death in the host results in a susceptible phenotype and disease. Thus, the party in control of PCD has a distinct advantage in these battles. PCD processes appear to be of ancient origin, as indicated by the fact that many features of cell death strategy are conserved between animals and plants; however, some of the details of death execution differ. Mammalian core PCD genes, such as caspases, are not present in plant genomes. Similarly, pro- and antiapoptotic mammalian regulatory elements are absent in plants, but, remarkably, when expressed in plants, successfully impact plant PCD. Thus, subtle structural similarities independent of sequence homology appear to sustain operational equivalence. The vacuole is emerging as a key organelle in the modulation of plant PCD. Under different signals for cell death, the vacuole either fuses with the plasmalemma membrane or disintegrates. Moreover, the vacuole appears to play a key role in autophagy; evidence suggests a prosurvival function for autophagy, but other studies propose a pro-death phenotype. It was suggested that it is not cell death as such but rather the processes leading to cell death that contribute to the outcome of a given plant-pathogen interaction.

The hypersensitive response (HR) is induced by certain plant pathogens and involves programmed cell death (PCD) to restrict the spread of pathogens from the infection site. Concurrent with the induction of cell death, the host activates a defense response. The cell death associated with the HR in several plant–pathogen systems has morphological similarities to animal apoptosis, which suggests that cell death mechanisms in plants and animals may share common components that lead to similar cellular events (Olga del Pozo and Eric Lam, 1998). Caspases are conserved cysteine proteases that regulate animal PCD; caspase activity or an involvement of caspases in cell death has yet to be reported in plants. It was investigated that caspases participate in HR cell death. Caspase-specific peptide inhibitors, Ac-YVAD-CMK and Ac-DEVD-CHO, could abolish bacteria-induced plant PCD but did not significantly affect the induction of other aspects of HR, such as the expression of defense genes. This result confirmed that cell death can be uncoupled from defense gene activation during HR. Caspase-like proteolytic activity was detected in tobacco tissues that were developing HR following infection with tobacco mosaic virus. These results provided evidence for the presence of caspase-like plant protease(s) that participate in HR cell death.

Programmed cell death (PCD) is a process aimed at the removal of redundant, misplaced, or damaged cells, and it is essential to the development and maintenance of multicellular organisms. In contrast to the relatively well-described cell death pathway in animals, often referred to as apoptosis, mechanisms and regulation of plant PCDare still ill-defined (Frank A. Hoeberichts and Ernst J. Woltering, 2002). Several morphological and biochemical similarities between apoptosis and plant PCD have been described, including DNA laddering, caspase-like proteolytic activity, and cytochrome *c* release from mitochondria. Reactive oxygen species (ROS) have emerged as important signals in the activation of plant PCD. In addition, several plant hormones may exert their respective effects on plant PCD through the regulation of ROS accumulation.

The interaction of plants with bacterial pathogens involves the manipulation of programmed cell death (PCD) pathways. During a resistance interaction, PCD is induced in a process termed the hypersensitive response (HR) which may function to limit pathogen spread. In susceptible plant-pathogen interactions, the pathogen both inhibits and/or induces host PCD depending on the infection stage and lifestyle of the pathogen (Timothy P Devarenne and Gregory B Martin, 2006). Genes/pathways regulating PCD in plants have been difficult to identify due to a lack of homologous sequences in plants for mammalian genes that control apoptosis and possibly due to functional redundancy. The plant PCD pathways and bacterial speck disease in tomato which is caused by *Pseudomonas syringae* pv. *tomato* (*Pst*) was studied and identified the tomato protein kinases Pdk1 and Adi3 as negative regulators of plant PCD. The plant Pdk1/Adi3 pathway appears to function similarly to the Pdk1/PKB (Akt) pathway in mammals which functions as a major apoptosis negative regulation pathway. They also discussed the regulation of Pdk1/Adi3 and targeting of this pathway during the tomato-*Pst* interaction for modulation of host PCD.

Recognition of pathogens in a gene-for-gene interaction is only one of several ways that programmed cell death is triggered. It is likely that many factors influence the control of cell death during pathogenesis as exemplified by the diversity of cell death patterns seen after pathogen attack and by the numerous cell death defense mutants in maize and Arabidopsis (Jean T. Greenberg, 1997). An emerging area of interest is how the metabolic state of the plant influences the cell death rate and patterns during plant-pathogen interactions. Indeed, alterations in sugar levels or modification of the ubiquity in pathway appear to trigger the HR spontaneously and alter cell death responses to pathogens.

3.9 PROTEOMICS

Proteomics, the high-throughput 'omic' technique has validated its role in precise identification and characterization of individual components of plant protein networks. A comprehensive understanding of plant response mechanism is essential to elucidate the key factors affecting plant performance under adverse conditions. Various proteomic approaches that are being exploited extensively for elucidating plant response largely include gel-based and mass spectrometry-based methods that involve both label-based and label-free protein quantification. Despite limitations in identifying low-abundance and hydrophobic proteins, exceedingly large or small proteins, as well as basic proteins, the gel-based method has been accepted as a global tool to unravel the underlying molecular mechanism of stress signal perception and transduction in plant defense responses. Moreover, recent advances in the mass spectrometry-based approaches provide a better opportunity to dissect stress signaling cascades. Nevertheless, complete protein extraction from different plant organs or subcellular organelles is the most challenging facet of plant proteome analysis. Plants perceive potential pathogens by recognizing pathogen-associated molecular patterns (PAMPs) through plasma membrane (PM) receptors. Recognition of flg22, a 22 amino acid PAMP derived from the bacterial flagellum, by the receptor-like kinase FLS2 induces defense signaling responses and contributes to innate immunity by restricting bacterial invasion (Carolina Escobar *et al.*, 2007). A quantitative phosphoproteomic analysis of *Arabidopsis* PM proteins revealed specific phosphorylation sites changing in response to treatment with flg22. By pursuing reverse genetic investigations of differentially phosphorylated protein candidates, we Scott C. Peck (2013) discovered that a specific syntaxin, SYP132, appears to be the cognate t-SNARE required for secretion of antimicrobial proteins and/or compounds, implicating protein secretion as a major determinant of resistance. Subsequent proteomic analyses of proteins secreted during infection by different genotypes of bacterial pathogens revealed a complex interplay between Type III secretion from the bacteria, host resistance responses, and alteration of the host secretome.

The term proteome refers to the complete set of proteins present in a cell, organ or organism at a given time (Wilkins *et al.*, 1995). Advances in proteomics have been made possible due to improvements in protein separation by two-dimensional gel electrophoresis (2-DE), multidimensional liquid chromatography (MudPIT), peptide sequencing by mass spectrometry (MS), and bioinformatics. However, progress in defining proteomes is expected to proceed at a slower pace than genome sequencing (Carolina Escobar *et al.*, 2007). The application of proteomics in plant pathology is being used to characterize pathogen virulence factors, as well as to identify changes in protein levels in plant hosts upon infection (Kav *et al.*, 2007). It is well known that nematode feeding site ontogeny is a reflection of extensive gene expression modification in infected root cells (Li *et al.*, 2009). In contrast, strategies aimed at studying the proteomic plant response to nematodes are still in their infancy. Technical advances in the quality and reproduction of 2-DE gels, software packages to process digitized images of gels, the development of nongel based high-throughput protein separation techniques, and analytical mass spectrometry should facilitate an increase in proteomic approaches to study plant-nematode interactions.

Root proteomics aimed at discovering plant defense-related proteins in roots have been studied in nematode-resistant cotton and coffee cultivars infected with *Meloidogyne paranaensis* and *M. incognita,* respectively. A 2-DE analysis comparing infected versus noninfected roots identified a class III chitinase of C. Arabica known to be involved in defense responses to pathogens (Jaubert *et al.,* 2002). Another differentially expressed protein in cotton was a quinone reductase 2 (QR2) that catalyzes the divalent reduction of quinones to hydroquinones to protect plant cells from oxidative damage. These findings emphasize the importance of root proteomics in the isolation of resistance and defense-related proteins against nematodes. Ultimately, transcriptomic, proteomic and metabolomic analyses will need to be integrated to elucidate the complex nature of the plant response to nematodes.

Phytobacteria cause diseases in important crops and seriously and negatively impact agricultural production. Therefore, an understanding of the mechanisms by which plants resist bacterial infection at the stage of the basal immune response or mount a successful specific R-dependent defense response is crucial since a better understanding of the biochemical and cellular mechanisms underlying these interactions will enable molecular and transgenic approaches to crops with increased biotic resistance (Tamara Zimaro *et al.,* 2011). In recent years, proteomics has been used to gain an in-depth understanding of many aspects of the host defense against pathogens and has allowed monitoring differences in abundance of proteins as well as posttranscriptional and posttranslational processes, protein activation/inactivation, and turnover. Proteomics also offers a window to study protein trafficking and routes of communication between organelles.

To gain an in-depth understanding of plant defense systems, there is a need to identify the diverse and complex signaling cascades and the multiple and interacting biochemical pathways activated by the pathogen. A number of studies addressing this complexity utilizing DNA microarrays or DNA chips have already been carried out with a view to investigate the global changes in the transcriptome caused by pathogens (Cernadas *et al.,* 2008). However, transcriptional changes are only a part of the response since they do not inform on posttranscriptional and posttranslational processes, protein activation, and turnover. This vital information can come from proteomics that allows monitoring differences in abundance of proteins present at the time of sampling and offers a window to study protein trafficking and routes of communication between organelles. Thus, complementary approaches such as proteome-based expression profiling are needed to obtain a more complete picture of the regulatory elements in plant-pathogen interactions.

Plants recognize PAMPs molecules on the surface of the pathogens by pattern recognition receptor proteins that are able to transduce signals that give rise to the innate immune response. A key goal of plant-pathogen interaction research is to decipher how the coordinated changes that cause PAMP-triggered immunity (PTI) occur. Proteomics studies have been performed in plants during PAMP-induced responses, and they have proven valuable in unraveling the mechanisms that mediate the early events leading to the basal immune response. Two of the most studied PAMPs are the Gram-negative bacterial lipopolysaccharide (LPS) and flagellin and its elicitor-active epitope, flg22.

Exhaustive proteomics analyses of *Nicotiana tabacum* cells elicited by LPS extracted from *Burkholderia cepacia* were done to identify cellular targets of the signal components induced by this molecule towards understanding the process of LPS perception and subsequent signal transduction (Gerber *et al.*, 2008). *B. cepacia* is not a tobacco pathogen, but the bacteria have a protective effect against subsequent inoculations with *Phytophthora nicotianae* and thus are considered a good candidate for studying elicitation of the immune response and, in particular, the proteins involved in LPS-induced signaling. Results obtained with cell suspension cultures were similar to those achieved with leaf discs and intact young leaves and hence provide an experimentally simpler and more reproducible system to do proteomics studies (Coventry and Dubery, 2001). Qualitative and quantitative changes in phosphoproteins were analyzed using two-dimensional electrophoresis in combination with a phosphoprotein-specific gel stain. A total of 27 phosphoproteins were identified from 32 excised gel spots. The identified phosphoproteins indicate that $LPS_{B.cep}$-induced signal perception/transduction involves G-protein-coupled receptor signaling, Ca^{2+}/calmodulin-dependent signaling pathways, H^+-ATPase regulation of intracellular pH, thioredoxin-mediated signaling, and phosphorylation of 14-3-3 regulatory proteins as well as phosphorylation of heat shock proteins, chaperones, and protein involved in biosynthesis and maintenance of nucleotide pools.

In an attempt to evaluate the differences between PTI, disease and ETI in the host-pathogen interaction model *A. thaliana* and *Pseudomonas syringae* pv. *tomato*, a proteomics analysis was undertaken. Leaves were either infected with wild-type DC3000 (susceptibility), the bacteria expressing the effector protein AvrRpm1 (DC3000 (avrRpm1)) that elicits an ETI mediated by the cognate R-gene RPM1, or an hrp mutant that elicits only basal responses. The analysis was restricted to total soluble protein, chloroplast-enriched, and mitochondria-enriched fractions (Jones *et al.*, 2007). Two antioxidant enzyme groups, the glutathione S-transferases (GSTs F2, F6, F7, and F8) and peroxiredoxins (PrxA, B, and IIE) showed differences in abundance after different times of bacteria infiltration implying that antioxidant enzymes play a critical role in modulating redox conditions of infected tissue with both pathogenic and nonpathogenic bacteria. Members of both enzyme groups showed multiple spots on gels that are consistent with specific posttranslational modifications of these enzymes. Regarding the GST group, GSTF2, GSTF6, and GSTF7 changed in abundance after bacterial inoculations independently of the strain used, while GSTF8 was the only GST to show specificity for the R-gene response. Regarding the peroxiredoxins proteins, differential changes were observed with PrxIIE and to a lesser extent PrxB after bacteria challenge. PrxA did not show significant changes. However, a lower molecular weight form of PrxA (PrxA-L) was reduced in response to wild type bacteria.

A more complete proteomics analysis using the same experimental system could separate components of the basal defense [leaves inoculated with a mutant in the hypersensitive response and pathogenesis (hrp) cluster] from disease and resistance responses following DC3000 and DC3000 (avrRpm1) inoculations. Moreover, since the proteomics of total soluble protein, chloroplast-enriched, and mitochondria-enriched fractions were analyzed, the experimental design also provided insights into the spatial dynamics of the response. In this study, a total of 73 differential spots representing 52 unique proteins were successfully identified (Tamara Zimaro *et al.*, 2011). The authors also correlated proteome changes with a transcriptomics analysis of the same response and observed that many of the changes in protein spot density occurred before significant transcriptional reprogramming between treatments was evident. Overall, the work showed that proteins that were subjected to significant changes after bacterial interactions were representative of two main functional groups, namely, defense-related antioxidants and metabolic enzymes. To date, experimental evidence suggests that PAMP-induced responses modify the host proteome and, in particular, proteins with a role in primary metabolism, redox modulation, molecular chaperones, cytoskeleton rearrangement, and signal transduction. In addition, the PAMP stimulus modifies mitochondrial and chloroplast proteome and reconfigures proteins in membrane rafts to enable efficient host signal transduction and downstream responses after the initial recognition.

The proteome of the model plant *Arabidopsis thaliana* infected with a necrotrophic fungal pathogen, *Alternaria brassicicola* was studied (Arup K. Mukherjee *et al.*, 2010). The *Arabidopsis–A. brassicicola* host–pathogen pair was developed as a model genetic system for incompatible plant–fungal interactions, in which the spread of disease was limited by plant defense responses. After confirming that a defense response was induced at the transcriptional level, they identified proteins whose abundance on 2-DE gels increased or decreased in infected leaves. At least 11 protein spots showed reproducible differences in abundance, increasing or decreasing during the progress of the infection. The pathogenesis-related protein PR4, a glycosyl hydrolase, and the antifungal protein osmotin were strongly up-regulated. Two members of the *Arabidopsis* glutathione S-transferase (GST) family increased in abundance in infected leaves. The spots in which these GST proteins were identified contained additional members of the GST family. Representation of GST family members in several protein spots migrating at similar molecular weight suggested posttranslational modifications. The signature of GST regulation might be specific for the type of plant–pathogen interaction. The proteomic view of the defense response to *A. brassicicola* could be compared with other types of plant–pathogen interactions, and to leaf senescence, identifying unique regulatory patterns.

Genetic and microarray analyses have provided useful information in the area of plant and pathogen interactions. *Pseudomonas syringae* pv. *tomato* DC3000 (Pst) causes bacterial speck disease in tomato. Previous studies have shown that changes in response to pathogen infection at transcript level are variable at different time points (Jennifer Parker *et al.*, 2011). This study provided not only proteomic changes between a resistant and a susceptible genotype but also information on changes between an early and a late time point. Using iTRAQ quantitative proteomics approach, 2,369 proteins were identified in tomato leaves, and 477 of them were determined to be responsive to Pst inoculation. Unique and differential proteins after each comparison were further analyzed to provide information about protein changes and the potential functions they play in the pathogen response. This information is applicable not only to tomato proteomics but also adds to the repertoire of now available for crop proteomic analysis and how it changes in response to pathogen infection.

3.10 TYPE III PROTEIN SECRETION SYSTEM

Many Gram-negative bacterial pathogens of plants and animals are dependent on a type III protein secretion system (TTSS). TTSSs translocate effector proteins into host cells and are capable of modifying signal transduction pathways (Avelina Espinosa and James R. Alfano, 2004). The innate immune system of eukaryotes detects the presence of pathogens using specific pathogen recognition receptors (PRRs). Plant PRRs include the FLS2 receptor kinase and resistance proteins. Animal PRRs include Toll-like receptors and nucleotide-binding oligomerization domain proteins. PRRs initiate signal transduction pathways that include mitogen-activated protein kinase (MAPK) cascades that activate defense-related transcription factors. This results in the induction of proinflammatory cytokines in animals, and hallmarks of defense in plants including the hypersensitive response, callose deposition and the production of pathogenesis-related proteins. Several type III effectors from animal and plant pathogens have evolved to counteract innate immunity. For example, the *Yersinia* Yop J/P cysteine protease and the *Pseudomonas syringae* HopPtoD2 protein tyrosine phosphatase inhibits defense-related MAPK kinase activity in animals and plants respectively. Thus, type III effectors can suppress signal transduction pathways activated by PRR surveillance systems. Understanding targets and activities of type III effectors will reveal much about bacterial pathogenicity and the innate immune system in plants and animals.

The immune system of plants seems to be far less complex. Because plants lack a circulatory system and mobile immune cells, they cannot use circulating immune receptors to detect nonself. Nonetheless, plants are capable of establishing immune responses that are highly specific, with restricted self-reactivity, and that often generate a lifelong 'memory' of the encountered pathogens. So, these features of vertebrate immunity can be achieved in plants using different immune strategies (Steven H. Spoel and Xinnian Dong, 2012). The initial obstacle that phytopathogens encounter is the plant cell wall, which can be reinforced by the deposition of callose (glucan polymers) following the activation of host defense pathways. The first active line of defense occurs at the plant cell surface when microorganism-associated molecular patterns (MAMPs) such as lipopolysaccharides, peptidoglycans, and bacterial flagellin are detected by pattern-recognition receptors (PRRs). Although there are some overall structural similarities between PRRs from plants and animals, such as the use of the leucine-rich repeat (LRR) domain for ligand binding, they are thought to have arisen through convergent evolution rather than divergent evolution. To circumvent pattern-triggered immunity, adapted pathogens can deliver effector molecules directly into the plant cell.

Plants have developed multiple layers of sophisticated surveillance mechanisms that recognize potentially dangerous pathogens and rapidly respond before those organisms have a chance to cause serious damage. These surveillance systems are linked to specific preprogrammed defense responses. Basal resistance, also called innate immunity, is the first line of preformed and inducible defenses that protect plants against entire groups of pathogens. Basal resistance can be triggered when plant cells recognize microbe-associated molecular patterns (MAMPs) including specific proteins, lipopolysaccharides, and cell wall components commonly found in microbes. The result is that living plant cells become fortified against attack. Nonpathogens as well as pathogens are capable of triggering basal resistance in plants due to the widespread presence of these molecular components in their cells.

Pathogens have developed countermeasures that are able to suppress basal resistance in certain plant species. If a pathogen is capable of suppressing basal defense, plants may respond with another line of defense: the hypersensitive response (HR). The HR is characterized by deliberate plant cell suicide at the site of infection. Although drastic compared to basal resistance, the HR may limit pathogen access to water and nutrients by sacrificing a few cells to save the rest of the plant. The HR is typically more pathogen-specific than basal resistance and is often triggered when gene products in the plant cell recognize the presence of specific disease-causing effector molecules introduced into the host by the pathogen. Bacteria, fungi, viruses, and nematodes are capable of inducing the HR in plants.

Once the hypersensitive response has been triggered, plant tissues may become highly resistant to a broad range of pathogens for an extended period of time. This phenomenon is called systemic acquired resistance (SAR) and represents a heightened state of readiness in which plant resources are mobilized in case of further attack. Researchers have learned to artificially trigger SAR by spraying plants with chemicals called plant activators. These substances are gaining favor in the agricultural community because they are much less toxic to humans and wildlife than fungicides or antibiotics, and their protective effects can last much longer. In addition to the hypersensitive response, plants can defend themselves against viruses by a variety of mechanisms including a sophisticated genetic defense system called RNA silencing. Many viruses produce double-stranded RNA or DNA during replication in a host cell. Plants can recognize these foreign molecules and respond by digesting the genetic strands into useless fragments and halting the infection. Plants that are infected with viruses will often exhibit chlorosis and mottling, but disease symptoms may eventually disappear if RNA silencing is successful, a process called recovery. In addition, the plant may retain a template of the digested genetic strand that can be used to quickly respond to a future attack by similar viruses, a process analogous to the memory of vertebrate immune systems.

4 Epidemiology

Plant disease epidemiology, in general, refers to the study of disease in plant populations. When a pathogen spreads to and affects many individuals within a population over a relatively large area and within a relatively short time, the phenomenon is known as an epidemic. An epidemic has been defined as any increase of disease in a population. A similar definition of an epidemic is 'the dynamics of change in plant disease in time and space'. The study of epidemics and factors influencing them is termed epidemiology. Epidemiology is concerned simultaneously with populations of pathogens and host plants as they occur in an evolving environment.

Plant disease epidemiologists strive for an understanding of the cause and effects of disease and develop strategies to intervene in situations where crop losses may occur. A typically successful intervention will lead to a low enough level of disease to be acceptable, depending on the value of the crop (Gr. *epi*, on; *demos*, people).

Plant disease epidemiology is often looked at from a multidisciplinary approach, requiring biological, statistical, agronomic and ecological perspectives. Biology is necessary for understanding the pathogen and its life cycle. It is also necessary for understanding the physiology of the crop and how the pathogen is adversely affecting it. Agronomic practices often influence disease incidence for better or for worse. Ecological influences are numerous. Native species of plants may serve as reservoirs for pathogens that cause disease in crops. Statistical models are often applied to summarize and describe the complexity of plant disease epidemiology so that disease processes can be more readily understood. For example, comparisons between patterns of disease progress for different diseases, cultivars, management strategies, or environmental settings can help in determining how plant diseases may best be managed. Policy can be influential in the occurrence of diseases, through actions such as restrictions on imports from sources where a disease occurs.

4.1 DEFINITION AND TYPES OF PREVALENCE

Epidemiology or epiphytology is the study of the outbreak of disease, its course, intensity, cause and effects and the various factors governing it. Based on the occurrence and geographical distribution they are classified as follows.

4.1.1 Endemic or Enphytotic

When a disease is more or less constantly occurring year after year in a moderate to severe form in a country or locality then it is called as endemic disease [*Example:* Wart disease of potato (*Synchytrium endobioticum*) is endemic in Darjeeling, Citrus canker (*Xanthomonas axonopodis* pv. *citri*) and sorghum rust (*Puccinia purpurea*)] in Asia.

4.1.2 Epidemic or Epiphytotic

It is a sudden outbreak of a disease periodically over a widespread area in a devastatingly severe form causing severe losses or complete destruction. This is constantly present in a locality, but it assumes severe form only on occasions. This is because of the occurrence of favorable environment responsible for the rapid development of disease [*Example:* Wheat stem rust (*Puccinia graminis tritici*) and powdery mildew (*Erysiphe graminis* var. *tritici*), late blight of potato (*Phytophthora infestans*), red rot of sugarcane (*Colletotrichum falcatum),* downy mildew of grapevine (*Plasmophora viticola*) and rice blast (*Pyricularia oryzae*)]. Certain diseases are endemic in one area and become an epidemic in another area [*Example:* Citrus canker is endemic in Asia but epidemic in the introduced place, Florida (USA)].The downy mildew of corn is an endemic disease in India but became an epidemic in the Philippines. An epidemic is the progress of the disease in time and space. Each epidemic has a structure whose temporal dynamics and spatial patterns are jointly determined by the pathosystem characteristics and environmental conditions. One of the important objectives in epidemiology is to understand such spatio-temporal dynamics via mathematical and statistical modeling.

4.1.3 Pandemic

When an epidemic disease spreads over continents or subcontinents and involves mass mortality, it is considered as a pandemic. The outbreak of black stem rust of wheat in India during 1947 is the best example of a pandemic disease.

4.1.4 Sporadic

Diseases which occur at irregular intervals over limited areas or locations are called sporadic. They occur relatively in few instances [*Example: Fusarium* wilt of cotton (*Fusarium oxysporum f* sp. *vasiinfectum*) grain smut of sorghum (*Sporisorium sorghi*) and loose smut of wheat (*Ustilago nuda*)].

The term 'epidemic' is used to denote the dynamics of disease, that is, the change in the amount of disease with time. By 'amount' of disease, we mean either incidence (the number of plants or plant parts affected), severity (the amount of tissue affected) or a combination of both. By this definition, there can be fast epidemics or slow epidemics, and there can even be 'negative' epidemics, in which there is a decrease in the amount of disease with time. For example, if we are using the proportion of leaves infected as a measure of disease incidence, and during a dry spell we get new, healthy leaves forming while the infected ones drop off, we could get a negative change in the amount of disease.

Associated with the change in the amount of plant disease with time is a change in the spatial distribution of lesions or of infected plants. A good working definition of an epidemic, therefore, is a change in plant disease in time and space. The term 'endemic' is sometimes used (erroneously) to mean the opposite of 'epidemic,' that is, to denote constant disease, because in areas where the pathogen and its host have coevolved over long periods of time, they often have reached equilibrium where there is little change in the amount of disease with time. In the strict sense 'endemic' means native to a particular area, and, therefore, we can have epidemics of endemic diseases.

4.2 ELEMENTS OF AN EPIDEMIC

Disease epidemics in plants can cause huge losses in yield of crops as well threatening to wipe out an entire species such as was the case with Dutch elm disease and could occur with sudden oak death. An epidemic of potato late blight, caused by *Phytophthora infestans*, led to the Great Irish Famine and the loss of many lives.

Commonly, the elements of an epidemic are referred to as the 'disease triangle': a susceptible host, pathogen, and conducive environment. For a disease to occur all three of these must be present. Below is an illustration of this point – where all three items meet there is a disease. The fourth element for an epidemic to occur is time. As long as all three of these elements are present disease can initiate, an epidemic will only ensue if all three continue to be present. Any one of the three might be removed from the equation, though. The host might out-grow susceptibility as with high-temperature adult-plant resistance, (Shultz and Line, 1992) the environment changes and is not conducive for the pathogen to cause disease, or the pathogen is controlled through a fungicide application for instance.

Sometimes the fourth factor of time is added as the time at which a particular infection occurs, and the length of time conditions remains viable for that infection, can also play an important role in epidemics. The age of the plant species can also play a role as certain species change in their levels of disease resistance as they mature – a process known as ontogenic resistance. If all of the criteria are not met, such as a susceptible host and pathogens are present but the environment is not conducive to the pathogen infecting and causing disease, the disease cannot occur. For example, corn is planted in a field with corn residue that has the fungus *Cercospora zea-maydis*, the causal agent of gray leaf spot of corn, but if the weather is too dry and there is no leaf wetness the spores of the fungus in the residue cannot germinate and initiate infection. Likewise, it stands to reason if the host is susceptible and the environment favors the development of disease but the pathogen is not present there is no disease. Taking the example above, the corn is planted into a ploughed field where there is no corn residue with *Cercospora zea-maydis*, the causal agent of gray leaf spot of corn, present but the weather means long periods of leaf wetness, there is no infection initiated. When a pathogen requires a vector to be spread then for an epidemic to occur the vector must be plentiful and active.

4.3 FACTORS GOVERNING EPIDEMIC OR ESSENTIAL CONDITIONS FOR AN EPIDEMIC

A disease is sometimes sporadic and assumes epidemic proportions under special circumstances. The essential conditions for an epiphytotic or the factors governing epidemics can be grouped under the three major heads, viz., nature of host, nature of the pathogen and environment. Of late, 'time' is also considered as an important factor.

An epidemic can only result from the cumulative effects of all the three factors mentioned above, acting simultaneously. Few pathogens are capable of assuming epiphytotic conditions while others are sporadic. The former group consists of late blight of potato, blast of rice, downy mildew diseases and rust diseases.

4.3.1 Host Factors

4.3.1.1 Susceptibility of the host

Plants have the ability to combat a disease which manifests itself as susceptibility or resistance. Plants are predisposed to the attack depending on their nature, environment, and stage of growth. The presence of susceptible varieties in an area may act as one of the causes of the epidemic. For example, late maturing varieties of groundnut are more susceptible to early leaf spot (*Cercospora arachidicola*) and late leaf spot (*Phaeoisariopsis*) than the early maturing varieties. Similarly, late maturing varieties of wheat are susceptible to loose smut (*Ustilago nuda tritici)* than the early maturing varieties. Early sown sugarcane varieties of sugarcane are more susceptible to leaf rust in Deccan canals in Bombay area than the late sown varieties. The wheat plant becomes susceptible to black rust (*Puccinia graminis tritici*) at the boot stage but is resistant when young. Susceptibility of rice plants to blast disease (*Pyricularia oryzae*) increases with the application of heavy doses of nitrogenous fertilizers. Cotton plants are susceptible to *Fusarium* wilt (*F. oxysporum* f. sp. *vasinfectum*) at soil temperatures of 26 to 28°C, brinjal to *Verticillium* wilt, *Verticillium dahliae* at 20°C. But crop plants are resistant to these soil-borne diseases at relatively lower or higher temperatures. Under the above conditions, the pathogen multiplies faster, cause infection and effectively uses its propagules for a quick, secondary spread causing an epidemic.

The susceptibility of the host plants, the virulence of the fungus or another pathogen, and the weather and other environmental conditions all affect the relative rate of increase (Van der Plant, 1965). They do so by affecting the time it takes a newly infected tissue to become infectious, the time tissue remains infectious, the infectiousness of the infectious tissue, and the susceptibility of healthy tissue to infection. These factors operate throughout the epidemic. Two other factors become increasingly important as the epidemic proceeds: the proportion of healthy, susceptible tissue remaining available for infection and the degree of uniformity of the population of host plants and their environment.

Umbellularia californica was one of the key infectious hosts of the exotic *Phytophthora ramorum*, which causes sudden oak death (SOD) in California and Oregon forests (Huberli *et al.*, 2012). This study provided a comprehensive analysis of the epidemiologically relevant parameters for SOD in California and southern Oregon, including potential differences between the two states. Experimental infection of *U. californica* leaves was optimal when leaves were wet for 6–12 h, the temperature was approximately 19°C and pathogen concentration was approximately 2·7 × 104 zoospores mL−1. Seasonal variation in host susceptibility and disease incidence was examined for two populations by inoculating detached leaves at 12 dates and by monitoring naturally infected leaves, respectively. Susceptibility of *U. californica* and disease incidence varied significantly in time, and the variation was highest for both in spring. Susceptibility of trees from 17 natural populations from California and southern Oregon was assessed in detached leaf inoculations. One California and three southern Oregon populations had significantly and repeatable lower average susceptibility in artificial inoculations, but differences among three selected California and Oregon populations were not significant in inoculations of seedlings grown from seed in a common garden. It was concluded that *U. californica* susceptibility had a large environmental component, yet still predicted potential disease severity in different sites especially where infestations were young or the pathogen had not yet arrived. The accuracy and utility of predictive risk models for *P. ramorum* could be enhanced by the inclusion of both the environmental and host susceptibility components.

Most mathematical models of plant disease epidemics ignore the growth and phenology of the host crop. Unfortunately, reports of disease development are often not accompanied by a simultaneous and commensurate evaluation of crop development. However, the time scale for increases in the leaf area of field crops is comparable to the time scale of epidemics (Ferrandino, 2008). This simultaneous development of host and pathogen has many ramifications on the resulting plant disease epidemic. First, there is a simple dilution effect resulting from the introduction of new healthy leaf area with time. Often, measurements of disease levels are made pro rata (per unit of host leaf area or total root length or mass). Thus, host growth will reduce the apparent infection rate. A second, related effect, has to do with the so-called 'correction factor,' which accounts for inoculum falling on already infected tissue. This factor accounts for multiple infections and is given by the fraction of the host tissue that is susceptible to disease. As an epidemic develops, less and less tissue is open to infection and the initial exponential growth slows. Crop growth delays the impact of this limiting effect and, therefore, tends to increase the rate of disease progress. A third and often neglected effect arises when an increase in the density of susceptible host tissue results in a corresponding increase in the basic reproduction ratio, R(0), defined as the ratio of the total number of daughter lesions produced to the number of original mother lesions. This occurs when the transport efficiency of inoculum from infected to susceptible host is strongly dependent on the spatial density of plant tissue. Thus, crop growth may have a major impact on the development of plant disease epidemics occurring during the vegetative phase of crop growth. The effects that these crop growth-related factors have on plant disease epidemics spread by airborne spores are evaluated using mathematical models, and their importance are discussed. In particular, plant disease epidemics initiated by the introduction of inoculum during this stage of development are shown to be relatively insensitive to the time at which inoculum is introduced.

4.3.1.2 *Aggregation and distribution of susceptible hosts*

Abundance of susceptible hosts in an area is one of the major causes of the spread of epidemics. Continuous cultivation of susceptible variety or varieties in an area that too in a large contiguous area help in the buildup of inoculum and improve the chances of epidemics. Under the above conditions, the pathogen increases the rate of multiplication of its propagules and repeats the disease cycles in a short span. Wind rice cultivation area in East Asian countries is exposed to a greater danger of epidemics by wheat black rust and rice blast respectively. The destructive epidemic of early and late leaf spots of groundnut in Bombay area (Gujarat and Maharashtra States) during 1912–1913 was mainly the result of the cultivation of local varieties in a larger area. Panama wilt (*Fusarium oxysporum* f. sp. *cubense*) susceptible variety, 'Son' in banana was responsible for the destructive epidemic in parts of Bombay area (Gujarat and Maharashtra) during 1936–1940. Countrywide cultivation of red rot (*Colletotrichum falcatum*) susceptible sugarcane varieties (local varieties like Pundya, Khajuria, etc.) practically made their cultivation impossible in Bombay area.

4.3.1.3 *Introduction of new host(s)*

Disease proneness in the host is induced by environment and other factors. The host is liable to vigorous attack and successful infection by the pathogen. A resistant or moderately resistant variety may become susceptible or highly susceptible. A susceptible variety may become highly susceptible when conditions favoring proneness exist and cause severe damage. Under the above conditions the pathogen multiplies faster, cause infection and produces more propagules for secondary spread. The introduction of an exotic cotton variety (C4 (Cambodia) caused an outbreak of bacterial blight (*Xanthomonas axonopodis* pv. *malvacearum*) and gray mildew (*Septoacylindrium gossypii*) in local variety, Deviraj, grown in Maharashtra area in India.

4.3.1.4 *Introduction of new collateral or alternate hosts*

Alternate hosts are those plants on which the heteroecious pathogens pass part of their life cycles. Similarly, collateral hosts are some wild plants in which the pathogen survives when the primary host is not available. Both alternate and collateral hosts are important in building up the primary inoculum to the next crop. They determine the course and intensity of an epidemic. Grass hosts (collateral hosts) of *Sclerospora sacchari, S. philippinensis* (downy mildews), *Pyricularia oryzae* (rice blast); *Ustilago scitaminea* (sugarcane smut) may produce abundant inoculum which aid in building up of epidemics. The outbreak of heteroecious blister rust of pine (*Cronartium ribicola*) in Europe and the USA happened due to import or introduction of *Pinus strobus* from the USA.

4.3.2 Pathogenic Factors

4.3.2.1 Introduction of a new pathogen

Some pathogens, epidemic in certain areas, may become quite aggressive and outbreak as an epidemic when introduced into the new area. For example, late *Phytophthora infestans* was epidemic in South America. This disease became epidemic when the infected tubers were introduced in Europe (in 1843–45). Fire blight (*Erwinia amylovora* in North America is endemic. Fire blight spread to Pacific coast fruit-growing areas of the USA in 1884 and subsequently it reached Canada. It reached New Zealand in 1919 and it appeared in England in 1957. The mode of introduction had been through fruit boxes. Coffee rust (*Hemileia vastatrix*) is indigenous to Ethiopia, where *Coffea arabica* is native. The disease spread to Sri Lanka in 1869, India in 1870, Sumatra in 1876, Java in 1878 and the Philippines in 1889. It also spread from Kenya to the Congo by 1918 and reached the Cameroons. From 1950 onwards, it spread to the reminder of West Africa. The mode of long distance transport of *H. vastatrix* is wind. Spores have been trapped at up to 1000 m above sea level up to 150 m from infected sites. Dutch elm disease (*Ceratocystis ulmi*) first reported in 1919 in Holland, spread throughout Europe and reached Great Britain in 1927. It was introduced to the eastern United States on elm logs imported from Europe.

4.3.2.2 Presence of aggressive strain of the pathogen

All the strains of a pathogen are not aggressive. Only the aggressive strains are capable of causing infectious diseases which spread like an epidemic. They are characterized by a rapid cycle of infection and cause successful infection in new hosts. Rapid cycle of infection is essential for successful infection, and it happens only by an aggressive strain of the pathogen [*Example: Puccinia graminis tritici* (wheat black rust) in India, stripe rust, bunt and loose smut of wheat in the USA and Europe]. The possibility of an outbreak of epidemics increases with the number of physiologic forms or pathogenic strains of the pathogen present in a locality.

Aggressiveness, the quantitative component of pathogenicity, and its role in the adaptation of plant pathogens were insufficiently investigated by Pariaud *et al.* (2009). Using examples of biotrophic and necrotrophic fungal pathogens of cereals and *Phytophthora infestans* on potato mainly, the empirical knowledge of the nature of aggressiveness components and their evolution in response to host and environment was reviewed. Means of measuring aggressiveness components are considered, as well as the sources of environmental variance in these traits. The adaptive potential of aggressiveness components was evaluated by reviewing evidence for their heritability, as well as for constraints on their evolution, including differential interactions between host and pathogen genotypes and trade-offs between components of pathogenicity.

Adaptations of pathogen aggressiveness components to host and environment revealed that selection for aggressiveness in pathogen populations could be mediated by climatic parameters, global population changes or remarkable population structures might be explained by variation in aggressiveness; and selection for quantitative traits could influence pathogen evolution in agricultural pathosystems and could result in differential adaptation to host cultivars, sometimes leading to erosion of quantitative resistance. Possible links with concepts in evolutionary ecology were suggested.

Isolates of the causal ascomycete of grapevine powdery mildew, *Erysiphe necator*, corresponded to two genetically differentiated groups (A and B) that coexist on the same host. This coexistence was analyzed by investigating temporal changes in the genetic and phenotypic structures of *E. necator* populations during three epidemics (Josselin Monetary *et al.*, 2008). Group A was present only at the start of the growing season, whereas group B was present throughout all three epidemics. Group A was less aggressive in terms of germination and infection efficiency but was more aggressive than group B in terms of the latency period, lesion diameter, and spore production. The results were consistent with a temporal differentiation of niches, preventing recombination, and suggest an association between the disease level and the frequencies of genetic groups.

4.3.2.3 *High birth rate of the pathogen*

Pathogen with high reproductive capacity and capable of rapid dissemination over wide areas mostly causes epidemics. The fungal members causing powdery mildews, downy mildews, rusts, blasts, blights, etc., produce an enormous amount of spores. These spores are easily dispersed by using water or insects and cause infections to new plants. The high degree of fecundity and the enormous amounts of inoculum produced by some common plant pathogens are given in Table 4.1.

Table 4.1 Fecundity Rates of Plant Pathogens

Sl. No.	*Pathogen*	*Extent of fecundity*
1.	Wheat stem rust (*Puccinia graminis tritici*)	Twenty five trillion uredospores in one hectare of wheat crops
2.	Wheat stem rust (*Puccinia graminis tritici*)	64,000 million aeciospores from aecial cups in a single barberry bush
3.	Cedar rust of apple	Two billion teliospores in a single gall
4.	A corn plant infected with downy mildew	225 million sporangia in one night
5.	A. grapevine infected by downy mildew	32,000 sporangia per sq. cm
6.	Bunt of wheat	6 to 12 million smut spores in a single kernel
7.	Smut of corn	125,000 billion smut spores in one hectare
8.	Chestnut blight	150,000,000 spores in a single spore horn
9.	*Fomes applanatus*	5,460 billion spores in a single fruiting body

4.3.2.4 Low death rate

Epiphytotics may also be caused by low death rate diseases. These diseases are caused by agents of systemic nature which are protected by plant tissues. As they are protected by plant tissues, the chances of high mortality are reduced to the minimum. In these diseases, the chief source of the accumulation of inoculum for epiphytotics is the diseased plant organ used for vegetative propagation (corms, setts, tubers, etc.). Here the buildup of epidemics is comparatively low compared to high birth rate diseases. When a particular area is planted and covered with diseased planting material the chances of occurrence of epiphytotics are very high [*Example:* virus and phytoplasma diseases in crops propagated through vegetative plant parts].

The transmission of insect-vectored pathogens is dependent on the population dynamics of the vector. Epidemiological models typically assume that birth and death rates of pathogen-free and inoculative vectors are equal, an assumption that is not true for all pathosystems. A series of simple and general epidemiological models were used to explore how assumptions about birth and death rates of vectors based on their infectivity status influence disease incidence (Mark S. Sisterson, 2009). With fixed death rate of pathogen-free vectors, increasing the death rate of inoculative vectors reduced vector density, the proportion of vectors that were inoculative, and the proportion of hosts infected. This effect was mediated by acquisition rate. Specifically, increasing the acquisition rate increased the proportion of vectors that were inoculative, thereby increasing the proportion of the vector population that experienced the increased death rate. With fixed birth rate of pathogen-free vectors, variation in the birth rate of inoculative vectors had little influence on disease incidence provided that the birth rate of pathogen-free vectors was much greater than their death rate. However, when the birth rate of pathogen-free vectors was only slightly greater than their death rate, large increases in the birth rate of inoculative vectors increased total vector density and disease incidence. The results indicated that assumptions about birth and death rates of vectors based on infectivity status could affect the vector population that in turn affects disease incidence.

4.3.2.5 Easy and rapid dispersal of the pathogen

The ability of the pathogen to cause epidemic depends both on the high birth rate and dispersal. The propagules of the pathogen produced should be dispersed for the development of an epidemic. It may happen by external agencies like wind, water, insects, mites and nematodes. Fungal spores/conidia are minute and light and resistant to adverse conditions. Fungal spores are mostly disseminated by wind. Bacteria are mostly disseminated by water or insects. Virus and phytoplasma diseases are mostly transmitted by insects, mites or nematodes. Epidemics are determined by the velocity of the wind, direction of the wind, moisture, relative humidity, temperature, presence and number of vectors and their rate of reproduction.

Disease spread has traditionally been described as a traveling wave of constant velocity. However, aerially dispersed pathogens capable of long-distance dispersal often have dispersal gradients with extended tails that could result in the acceleration of the epidemic front (Christopher C. Mundt *et al.*, 2010). They evaluated empirical data with a simple model of disease spread that incorporates logistic growth in time with an inverse power function of dispersal. The scale invariance of the power law dispersal function implies its applicability at any spatial scale; indeed, the model successfully described epidemics ranging over six orders of magnitude, from experimental field plots to continental-scale epidemics of both plant and animal diseases. The distance traveled by epidemic fronts approximately doubled per unit time; velocity increased linearly with distance and the exponent of the inverse power law was approximately 2. It was found that it also may be possible to scale epidemics to account for initial outbreak focus size and the frequency of susceptible hosts. These relationships improved understanding of the geographic spread of emerging diseases and facilitated the development of methods for predicting and preventing epidemics of plants, animals, and humans caused by pathogens that are capable of long-distance dispersal.

Metapopulation processes are important determinants of epidemiological and evolutionary dynamics of host-pathogen systems and are therefore central to explaining observed patterns of disease or genetic diversity (Papaïx *et al.*, 2014). In particular, the spatial scale of interactions between pathogens and their hosts is of primary importance because migration rates of one species can affect both spatial and temporal heterogeneity of selection on the other. In this study, we developed a stochastic and discrete time simulation model to specifically examine the joint effects of host and pathogen dispersal on the evolution of pathogen specialization in a spatially explicit metapopulation. They considered a plant-pathogen system in which the host metapopulation is composed of two plant genotypes. The pathogen was dispersed by airborne spores on the host metapopulation. The pathogen population was characterized by a single life history trait under selection, the infection efficacy. They found that restricted host dispersal could lead to high amount of pathogen diversity and that the extent of pathogen specialization varied according to the spatial scale of host-pathogen dispersal. The role of population asynchrony in determining the pathogen evolutionary outcomes was discussed. The spores of the disease need to be able to spread from one plant to another plant. Common means include water, wind, and human or animal activity. If conditions are not right, the disease can flourish on one plant but not be able to spread well. Human activity that spreads disease can include poor waste management practices and common practices such as mowing the lawn.

4.3.2.6 *Adaptability of the pathogen*

Pathogens have the capacity to adapt to adverse conditions. Fungi produce different types of spores like oospores, ascospores and smut spores (chlamydospores) which help in tiding over adverse conditions. Bacteria survive in diseased plant parts. Viruses and phytoplasmas live in collateral hosts or insect vectors in the absence of the suitable crop hosts, while various stages of phytonematodes can survive in soil and some form special dormant structures.

Within single populations, extensive long-term host demographic and disease monitoring has demonstrated the highly stochastic nature of the amplitude of disease incidence and prevalence from year-to-year, with high mortality associated with severe epidemic conditions (Burdon, 2012). Populations of *Linum marginale* are frequently composed of many different host resistance phenotypes distributed unevenly through the deme; sympatric pathogen populations are also typically diverse although not infrequently they may be dominated by just a few pathotypes. Year-to-year fluctuations in the incidence and frequency of different pathotypes provide strong circumstantial evidence for the effects of population bottlenecks and subsequent genetic drift. Analysis of multiyear data on disease prevalence has provided the first evidence that observed levels of asynchrony in disease dynamics may indeed be related to resistance structure in nature. At the metapopulation level, recent surveys of multiple hosts and pathogen populations have revealed significant spatial structure in the distribution of resistance such that host populations in close proximity were more closely related than more distantly placed ones. In contrast, reflecting the pathogen's greater mobility, the relatively little structure was detected at the same spatial scale in pathogen populations. Characterization of patterns of local adaptation in the Linum-Melampsora system illustrates that apparent levels of local adaptation of pathogens can vary unpredictably among populations depending on overall levels of host resistance, suggesting an important role for local history. It was shown how the coevolutionary interaction between hosts and pathogens can be further influenced by a local structure such that resistant hosts select for generally virulent pathogens while susceptible hosts select for more avirulent pathogens, strongly suggesting the possibility of an evolutionary trade-off between infectivity and virulence in this system. At larger geographic scales, comparisons between adjacent metapopulations have shown that differences in host mating system affect the way in which variation for resistance is partitioned within and among individuals and populations. These differences are likely to have a significant impact on the spatial patterning of resistance phenotypes within populations.

The potential for local adaptation between pathogens and their hosts has generated strong theoretical and empirical interest with evidence both for and against local adaptation reported for a range of systems (Thrall *et al.*, 2002). The used the *Linum marginale-Melampsora lini* plant-pathogen system and a hierarchical spatial structure to investigate patterns of local adaptation within a metapopulation characterized by epidemic dynamics and frequent extinction of pathogen populations. Based on large sample sizes and comprehensive cross-inoculation trials, their analyses demonstrated strong local adaptation by *Melampsora* to its host populations, with this effect being greatest at regional scales, as predicted from the broader spatial scales at which *M. lini* disperses relative to *L. marginale.* However, there was no consistent trend for more distant pathogen populations to perform more poorly. It was shown how the coevolutionary interaction between hosts and pathogens could be influenced by a local structure such that resistant hosts select for generally virulent pathogens while susceptible hosts select for more avirulent pathogens. Empirically, local adaptation was generally tested in two contrasting ways, pathogen performance on sympatric versus allopatric hosts and sympatric versus allopatric pathogens on a given host population. In situations where no host population was more resistant or susceptible than others when averaged across pathogen populations (and likewise, no pathogen population is more virulent or avirulent than others), results from these tests should generally be congruent. It was stated that was unlikely to be the case in the metapopulation situations that predominate in natural host-pathogen interactions, thus requiring tests that control simultaneously for variation in plant and pathogen populations.

4.3.3 Environmental Factors

The environmental conditions such as temperature, relative humidity, rainfall, duration and intensity of light, etc. play a very important role in causing epidemics. These are actually the deciding factors and influence almost all the stages of disease cycle. Favorable environmental conditions are needed for sporulation, liberation of spores, and dissemination of pathogen, germination, infection and establishment of the pathogen in the host. For example, persistent optimum temperature and moisture are needed for spore germination and entry of germ tube in the host. Similarly optimum temperature, moisture, light and specific nutrition is required for the development of the disease and sporulation of pathogen. Climate changes are in response to changes in the hydrosphere, biosphere and other atmospheric and interacting factors. Human activities driven by demographic, economic, technological and social changes have a major impact on climate change (Rayees A. Ahanger *et al.*, 2013). The climate influences the incidence as well as the temporal and spatial distribution of plant diseases. The main factors that control growth and development of diseases are temperature, light and water. The climate change affects the survival, vigor, rate of multiplication, sporulation, direction, and distance of dispersal of inoculums, the rate of spore germination and penetration of pathogens. Climate affects all life stages of the pathogen and host and clearly poses a challenge to many pathosystems. The environmental change, especially when combined with pathogen and host introductions, may result in unprecedented effects.

4.3.3.1 Rising temperatures and its effect on pathogens/diseases

Temperature has potential impacts on plant disease through both the host crop plant and the pathogen. Research has shown that host plants such as wheat and oats become more susceptible to rust diseases with increased temperature; but some forage species become more resistant to fungi with increased temperature (Coakley *et al.*, 1999). Many mathematical models that have been useful for forecasting plant disease epidemics are based on increases in pathogen growth and infection within specified temperature ranges. Generally, fungi that cause plant disease grow best in moderate temperature ranges. Temperate climate zones that include seasons with cold average temperatures are likely to experience longer periods of temperatures suitable for pathogen growth and reproduction if climates warm. For example, predictive models for potato and tomato late blight, caused by *Phytophthora infestans* show that the fungus infects and reproduces most successfully during periods of high moisture that occur when temperatures are between 45° F (7.2°C) and 80°F (26.8°C) (Wallin *et al.*, 1950). Earlier onset of warm temperatures could result in an earlier threat from late blight with the potential for more severe epidemics and increases in the number of fungicide applications needed for control.

4.3.3.2 *Moisture variations and effect on pathogens/diseases*

Moisture can impact both host plants and pathogen organisms in various ways. Some pathogens such as apple scab, late blight, and several vegetable root pathogens are more likely to infect plants with increased moisture – forecast models for these diseases are based on leaf wetness, relative humidity, and precipitation measurements. Other pathogens like the powdery mildew species tend to thrive in conditions with lower (but not low) moisture. More frequent and extreme precipitation events that are predicted by some climate change models could result in more and longer periods with favorable pathogen environments. Host crops with canopy size limited by lack of moisture might no longer be so limited and may produce canopies that hold moisture in the form of leaf wetness or high canopy relative humidity for longer periods, thus increasing the risk of pathogen infection (Coakley *et al.*, 1999). Some climate change models predict higher atmospheric water vapor concentrations with increased temperature – this also would favor pathogen and disease development.

4.3.3.3 *Rising CO_2 levels and effect on pathogens/disease*

Increased CO_2 levels can impact both the host and the pathogen in multiple ways. Some of the observed CO_2 effects on disease may counteract others. Researchers have shown that higher growth rates of leaves and stems observed for plants grown under high CO_2 concentrations may result in denser canopies with higher humidity that favor pathogens. Lower plant decomposition rates observed in high CO_2 situations could increase the crop residue on which disease organisms can overwinter, resulting in higher inoculum levels at the beginning of the growing season, and earlier and faster disease epidemics. Pathogen growth can be affected by higher CO_2 concentrations resulting in greater fungal spore production. However, increased CO_2 can result in physiological changes in the host plant that can increase host resistance to pathogens (Coakley *et al.*, 1999).

4.4 CLIMATE CHANGE AND ITS IMPACT ON PLANT DISEASE MANAGEMENT

While physiological changes in host plants may result in higher disease resistance under climate change scenarios, host resistance to disease may be overcome more quickly by more rapid disease cycles, resulting in a greater chance of pathogens evolving to overcome host plant resistance. Fungicide and bactericide efficacy may change with increased CO_2, moisture, and temperature. The more frequent rainfall events predicted by climate change models could result in farmers finding it difficult to keep residues of contact fungicides on plants, triggering more frequent applications. Systemic fungicides could be affected negatively by physiological changes that slow uptake rates, such as smaller stomatal opening or thicker epicuticular waxes in crop plants grown under higher temperatures. These same fungicides could be affected positively by increased plant metabolic rates that could increase fungicide uptake. It is not well understood how naturally-occurring biological control of pathogens by other microbial organisms could change as populations of microorganisms shift under changed temperature and moisture regimes, in some cases antagonistic organisms may out-compete pathogens while in others pathogens may be favored. Exclusion of pathogens and quarantines through regulatory means may become more difficult to authorities as unexpected pathogens might appear more frequently on imported crops.

The terms compound interest and simple interest are for explaining the rate of increase of pathogen. These terms were introduced by Van der Plank in 1963 in the book 'Plant Diseases-Epidemics and Control.' Based on the mode of multiplication of pathogen, the diseases are classified into two types, viz., monocyclic or 'simple interest' pathogens and polycyclic or 'compound interest' pathogens.

4.4.1 Monocyclic Pathogens

These pathogens complete only one infection cycle in a growing season of the host. The primary inoculum causes the disease, and there is no production of secondary inoculum to contribute to the development of the disease during the growing season. Consequently, the intensity of the disease depends on the quantity and quality of the primary inoculum present at the beginning of the growing season. Such pathogens can be compared with a simple interest monetary model in which 'interest' is not added to the 'capital investment' until the end of the season (or investment period). Monocyclic diseases lack secondary inoculum and secondary infections during the same year. Epidemics of simple interest pathogens develop relatively slowly but can be important, particularly in perennial crops where pathogen populations build-up over a number of years or in annual crops which are grown continuously in the same field. The disease progress curve of a typical simple interest disease is often a straight line that flattens out when host tissue available for infection becomes limiting. Monocyclic pathogens include the vascular wilt fungi (*Fusarium oxysporum* and *Verticilium dahlia)*, root rotting fungi such as *Gaeumannomyces graminis* (the cause of take-all in cereals) as well as the club root fungus of crucifers (*Plasmodiophora brasssicae*) and the flag smut fungus of wheat (*Urocystis agropyri*). Soil-borne nematodes such as the root knot and cyst nematodes and bacteria such as the vascular wilt pathogen *Ralstonta solanacearum* also exhibit simple interest disease epidemiology. Van der Plank considered that simple interest diseases involved pathogens with a 'low birth rate' (i.e., few propagules are produced) and a 'low death rate' (i.e., spores produced had a long survival period in soil).

4.4.2 Polycyclic Pathogens

They complete more than one infection cycle in a growing season. The time between inoculation and the appearance of symptoms, which may be as short as 3-4 days for some fungi such as *Alternaria helianthi* and the number of generation cycles that occur in a growing season is influenced by the interactions that occur between the host, the pathogen, and the environment. Many airborne pathogens such as the rusts, downy mildews, powdery mildews and the potato late blight fungus are polycyclic pathogens.

Polycyclic pathogens have a 'high birth rate' (many propagules produced) and a corresponding 'high death rate' (short-lived propagules and a high rate of wastage). The disease progress curves for compound interest diseases are typically sigmoidal or S-shaped, with the three phases of the curve being the initial lag phase due to the limited amount of inoculum present at the start of the epidemic (low levels of inoculum, abundant host tissue), the exponential or logarithmic phase when neither inoculum levels nor the amount of host tissue available for infection is limiting, and the decline or plateau phase due to the limited amount of host tissue available to become infected [(abundant inoculum, little host tissue remaining) (Fig. 4.1)]. Diseases caused by polycyclic pathogens can be described as compound interest diseases when using a monetary analogy. The 'interest' (the new inoculum produced) is continuously being added to the 'interest-bearing capital' (previous amount of inoculum).

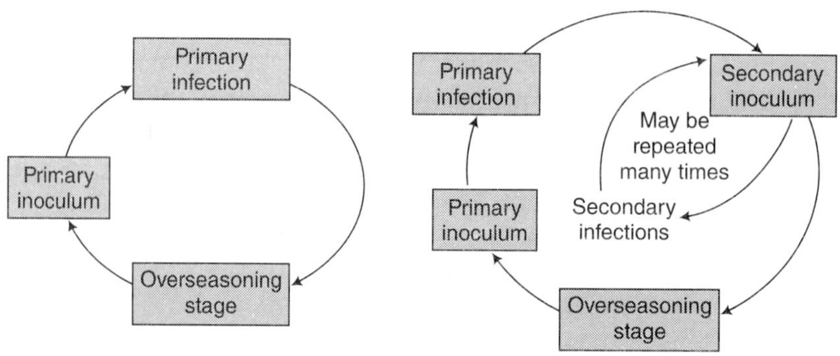

Fig. 4.1 Monocyclic (left) and polycyclic (right) plant diseases

4.4.3 Simple Interest Diseases

In simple interest diseases, the increase is mathematically analogous to simple interest in money. There is only one generation of the pathogen in the life of the crop. The primary inoculum is seed-borne or soil-borne. The secondary infection rarely occurs during the crop season. That is, the pathogens do not spread from plant to plant in one growing season. Simple interest diseases are caused by seed-or soil-borne smuts, like loose smut of wheat, covered smut of barley and soil borne fungi which attack roots, like wilt (*Fusarium oxysporum*) and root rot (*Rhizoctonia* spp.) diseases. Most of the smuts infect the seedlings, grow along with the growth of the plant and produce spores in the inflorescence on the maturity of the crop. There is no secondary spread from the smutted heads. These smut diseases are mostly systemic in nature. They do not produce propagules external to the host during the active season of the crop. Dispersal of propagules of these fungi is restricted by existing climatic and biotic conditions.

The concept of Simple Interest Diseases is most often applied to annual crops and some of their diseases when the crop is rotated with a crop that is not infected by the disease incitant. Vascular wilts caused by *Fusarium* sp. are characteristic. Plants become infected, and disease may ensue, but the propagules produced (asexual conidia, chlamydospores, and mycelia) remain in the infected plant. Vascular parasite while defined as simple interest disease may also be considered as polycyclic. They certainly produce large quantities on new infectious inoculum during the crop's growing season but are limited to being simple interest disease because the inoculum is retained inside the diseased plant. Inoculum may increase over time (seasons) if the susceptible host is replanted. Thus, the appropriate disease management strategy is crop rotation.

4.4.3.1 $Log_e[1/(1-x)]$

This term expresses the progress of a Simple Interest Disease. This term presumes that tissue that is diseased cannot be infected and uses the correction factor $(1 - x)$ to indicate that the amount of susceptible tissue has been reduced. The use of the Log_e term recognizes that biological entities, in this case, the pathodeme, increase in a geometric fashion not arithmetic.

4.4.3.2 Stalk rot of corn

It is exacerbated by the presence of *Fusarium moniliforme* and *Fusarium graminearum* as vascular parasites. Corn is planted in soil infested with *Fusarium* propagules. As the seed germinates its radicle and young roots come in contact with the germinating spores or a growing mycelium of the facultative parasite, *Fusarium*. Taking advantage of microwounds, the *Fusarium* infects the plant and ramifies intercellularly until it reaches the vascular tissue. Once it reaches the vascular tissue, it sporulates, and the spores are carried up the plant in the phloem stream. *Fusarium* is a 'good' parasite and seldom kills the corn plant except possibly after a severe environmental event (drought, heat, etc.). The mycelium and spores remain in the plants though out the season and are returned to the soil as the crop is harvested and the plant debris is incorporated back into the soil.

In the great plains of the USA, corn is grown to a large extent as a monoculture and without rotation. Consequently, overtime the inoculum has reached a biological equilibrium with the ecosystem and is both ubiquitous and stable. Therefore, it can be considered as a simple interest disease. Because of the large acreages planted and the lack of rotation it is unlikely that one can expect any reduction in this disease without a major effort to improve the genetic resistance of the host. With the advent of hybrid corn, great strides have been made to make corn more resistant to lodging (stalk collapse resulting in the plant's falling over) but little or no improvement in the genetic resistance of corn the *Fusarium* has been demonstrated.

4.4.3.3 Overwintering inoculum in simple interest diseases

The amount of potential new inoculum produced by a pathogen in a Simple Interest Disease can be quite substantial. It is, therefore, important to consider the ability of the pathogen, in its various forms, to successfully overwinter until the next season. Some pathogens do not survive well when exposed during the 'off-season' while others do not survive being incorporated into the soil. In either case it is important, from a disease control perspective to know the biology of both host and parasite to predict disease potential.

4.4.4 Compound Interest Diseases

In compound interest diseases, the rate of increase is mathematically analogous to compound interest in money. The pathogen produces an enormous amount of spores at a very rapid rate. These spores are disseminated rapidly by wind and infect the other plants. Both the inoculation and sporulation period are short so that the pathogen spreads from plant to plant during the same growing season. A new crop of spores is produced, disseminated, and the cycle is repeated fast. Thus, more generations of the pathogen are produced in the life of a crop [*Example:* late blight of potato, powdery mildews and rust diseases]. If we consider wheat stem rust caused by *Puccinia graminis tritici* as an example, the fungus produces uredospores in very large numbers (50,000 to 4,00,000 uredospores per uredosorus).

These spores are spread by wind and infect other plants. Each of the freshly infected wheat plants produces uredopustules within 5 to 7 days at 24°C. Thus within a week of the appearance of the first pustule in the crop, several thousand new pustules are formed which could repeat the process within a week. If the climatic conditions of about 24°C temperature and relative humidity remain for only a few weeks, the entire crop is severely affected by the disease. Reinfection occurs from inoculum generated within the diseased tissue. The expression of disease progress follows formulae common to calculating compounding interest in the financial markets. Because Compound Interest Diseases progress rapidly one must use a correction factor in the formulae that reflects the reduction in healthy, or 'infectible,' tissue. Compound Interest Disease progress curves tend to have a much steeper log phase of disease progress and, when very rapid, will actually reflect a 'tailing' due to healthy tissue limitations at the end of the season. The classic disease example is potato late blight caused by *Phytophthora infestans*. This disease progressed so rapidly across Ireland during the Potato Famine in the 1840's that the advancing margin of infected fields was documented to be moving at a rate of 1/4 mile per day.

4.4.5 Polyetic Diseases

Polyetic diseases are diseases, usually in perennial hosts that are compound interest diseases but with low r values. Dutch elm disease is caused by *Ceratocystis ulmi* is a good example of this type of disease. The pathogen may undergo many generations in a single year and reinfection of trees may occur, but the disease progress is sufficiently slow that death of the tree takes many years. The appearance of plants with Polyetic Diseases is that the disease causes severe debilitation over a period of years rather that rapid catastrophic collapse. In the case of Dutch Elm Disease, a collapse phase often follows drought stress.

4.5 EPIDEMIC CLASSIFICATION

Plant disease epidemics can be classified based on the 'source of the inoculum' that comes in contact with disease-free host plants.

4.5.1 First Type

Inoculum that can cause infections is produced during the current epidemic by the pathogen on or in plants that had been previously infected during the current epidemic. In case of Late blight, caused by *Phytophthora infestans,* spore (sporangium) lands (deposited) on a healthy leaf [contact] → infection → ultimately a lesion → production of sporangia → dispersal of spores → some land on healthy leaves (or healthy parts of same leaf) [contact] → infection → ... → ... .

Inoculum that can cause infections is produced during the current epidemic by the pathogen on or in plants (leaves, ...[individuals]) that had been previously infected during the current epidemic. New infections result in new inoculum that can result in new infections, and so on. Known as polycyclic diseases, the epidemics are known as polycyclic epidemics, that is, multiple disease cycles ('infection chains') – cycles of infection (with the new disease) – during the current epidemic. Sometimes they are aso known as compound-interest diseases. In many disciplines, 'spread from plant to plant' is the only scenario that would be considered an epidemic (as in 'disease increase').

4.5.2 Second Type

Inoculum that can cause infections is not produced during the current epidemic by the pathogen on or in plants (leaves, ...[individuals]) that had been previously infected during the current epidemic. The pathogen may still produce inoculum during the current epidemic, but it is not available to cause infections during the epidemic [*Example: Verticillium* wilt of potatoes (and early dying), caused by *V. dahlia.*]

Spores (microsclerotia) in soil → root contact with spores → infection of root, and systemic movement within plant → production of spores in vascular tissue → spore release only as plants decay in soil and on the surface (after harvesting).

Inoculum that can cause infections is not produced during the current epidemic by the pathogen on or in plants that had been previously infected during the current epidemic. Other scenarios also can be considered which preclude production of inoculum that can cause infections currently: insufficient time for the 'infection cycle' to repeat within the constraints of the growing season; crop matures and is harvested before the new inoculum is produced; long latent period (defined later); required conditions (host susceptibility, environmental conditions) only occur during narrow temporal windows – also resulting in insufficient time; *Fusarium* head blight of wheat; and infection only of flowers – at anthesis, or for a short time afterwards (a short time window).

A host species may only be susceptible to infection, but new inoculum is produced only on other susceptible host species. Virus vectors are only active (feeding on plants) once per growing season. Vectors may not be mobile (stay at the infected plant), or cannot move very far (fungal vectors, nematode vectors). The vector may be only able to acquire the virus from one host plant species and transmit to another species [*Example:* Tomato spotted wilt of tomato (thrips vector)]. Adults transmit the virus, but do not acquire the virus, and thrips do not (may not) reproduce on tomato. Larvae of thrips acquire the virus. Also, the inoculum of many pathogens simply cannot move very far (from plant to plant). In container media (pots in greenhouses), soil-borne inoculum cannot reach other plants.

Inoculum that can cause infections is not produced during the current epidemic by the pathogen on or in plants (leaves, …[individuals]) that had been previously infected during the current epidemic. They are known as monocyclic diseases and the epidemics known as monocylic epidemics. There is only one infection cycle per epidemic. It is also known as a simple interest disease: there is 'no spread from plant to plant'; disease intensity can be just as high with monocyclic; diseases as with polycyclic diseases – but disease progress curves look different; and modeling, analysis, and ultimately control strategies differ for the two classes of diseases.

4.5.3 Monocyclic vs. Polycyclic

In reality, sometimes the distinction disappears (or is diminished). It depends on host units of interest. For entire plants, then many soil-borne pathogens do cause monocyclic diseases (no plant-to-plant spread). However, there can be spread between roots (or along individual roots). Thus, inoculum produced in the current epidemic can result in new root infections in the same epidemic. Time units are important, as well. Diseases that are monocyclic during one growing season (a single epidemic) may be polycyclic over multiple seasons. If one considers the epidemic to be a long-term process, then the inoculum produced currently could ultimately cause new infections. For trees, the time scale is years (decades, century).

4.5.3.1 Polyetic

Monocyclic in one season, but polycyclic over several separate seasons

4.5.4 Primary vs. Secondary Infection

4.5.4.1 Primary infections

Result from contact between host plants and inoculum produced elsewhere, or in a different epidemic. Monocyclic epidemics consist only of primary infections. Primary infection occurs when the pathogen comes into contact with a susceptible host under favorable environmental conditions. Pathogens penetrate the surface of a plant directly or enter through wounds or natural openings. The bitter rot fungi, *Colletotrichum* spp., generally penetrate the apple fruit directly whereas the white rot fungus, *Botryosphaeria dothidea,* enters through cracks in the cuticle or lenticels on healthy twigs. Most fungal and bacterial pathogens require free water for spore germination; consequently, infection is favored by prolonged warm, wet periods with high relative humidity. Growth and development of a pathogen usually occur on or within infected plant tissue. Fungi grow and spread within their host by means of mycelium. Most fungi produce spore producing structures such as pycnidia or perithecia on or within the infected tissue. These structures give rise to secondary inoculum which eventually causes additional, or secondary, infections during the season. Bacteria spread by rapidly increasing in numbers. The spread of bacterial cells usually occurs when fissures or cracks develop on infected tissue exposing the cells (secondary inoculum) to the environment. Each pathogen has its own optimal conditions for growth and development which usually coincide with the optimal condition for infection. According to the general mode of primary infection plant diseases are recognized as soil borne, seed borne, including diseases carried with planting material, wind borne, insect borne, etc.

4.5.4.2 Secondary infections

Any infections that ultimately result from the primary infections in the current epidemic. Infections resulting from inoculum produced during the current epidemic. Secondary infections occur only in polycyclic epidemics. But there are also some primary infections in polycyclic epidemics. Primary infections initiate polycyclic epidemics (initial infections). Secondary infection results from spores or cells produced following primary infection or from other secondary infections. The secondary infection cycle can be repeated many times during the growing season. The number of cycles is dependent on the biology of the pathogen and its host and the duration of environmental conditions needed for infection. Apple powdery mildew, caused by the fungus *Podoshpaera leucotricha,* passes through many secondary cycles during the growing season until terminal buds set whereas cedar apple rust has no secondary cycle. *Erwinia amylovora* continues to cause infection as long as young succulent growth is available and warm, wet, and humid conditions prevail.

Overwintering or overseasoning is the ability of a pathogen to survive from one growing season to the next. Pathogens of apple survive the winter in a number of different ways. The fire blight bacterium overwinters in infected branch and trunk cankers. *Verturia inaequalis,* the cause of apple scab, *Mycosphaerella pomi,* the cause of Brooks spot, and *Alternaraia mali,* the cause of Alternaria blotch, survive from one season to the next in leaves on the orchard floor. *Botryoshpaeria dothidea,* the cause of white rot, overwinters in dead wood in the tree and on the ground and in mummied apples in the tree. *Gymnosporangium jumiperi-virginianae,* the cause of cedar apple rust, survives as a parasite of cedar trees, its alternate host. As it is well known, for a plant disease to occur, a susceptible host, a pathogen and favorable environmental conditions (and time, is some cases) must be present and interact with one another. If any one of these requirements is not met, a plant disease will not occur. At present, our ability to manipulate the environment is limited to only a few practices such as pruning to promote drying, bedding to improve soil drainage, and scheduling of irrigation. Severe disease outbreaks can be prevented by manipulating the host through the use of resistant cultivars and the pathogen through cultural practices and fungicidal or bactericidal sprays.

In the case of fungal diseases, the primary infections are usually caused by the resting or dormant structures of the fungal pathogen while the secondary spread occurs through the asexually produced, countless and short-lived spores. Although in the majority of plant diseases the above two phases occur sequentially, primary infections occurring only once in the season and secondary spread often repeating several times in the same season thus causing rapid. Transmission over wide areas, this type of cycle is not observed in many other diseases. In such cases, there is no secondary spread in the same season, i.e., the pathogen multiplies only once during the crop period and the spread of disease is observed only during the following crop season by way of increased primary infection. Mildews, leaf spots, blasts, blights, etc. are the common examples of former type while some smut disease of cereal or grain crops where black powder is formed in place of grains or inflorescence falls in the latter type. Control measures, therefore, in the former type need to be directed to avoid the primary infection and also to check the secondary spread.

An analysis by Jesús *et al.* (2013) revealed that each type of virus particle has a different probability of invading the host plant, at both the primary- and systemic-infection levels. Multipartite plant viruses were discovered because of discrepancies between the observed dose response and predictions of the independent-action hypothesis (IAH) model. The theory suggested that the number of genome segments predicts the shape of the dose-response curve, but a rigorous test of this hypothesis has not been reported. *Alfalfa mosaic virus* (AMV), a tripartite *Alfamovirus* and transgenic *Nicotianatabacum* plants expressing no (wild type), one (P2), or two (P12) viral genome segments were used to test whether the number of genome segments necessary for infection predicts the dose response. The dose-response curve of wild-type plants was steep and congruent with the predicted kinetics of a multipartite virus, confirming previous results. Moreover, for P12 plants, the data supported the IAH model, showing that the expression of virus genome segments by the host plant can modulate the infection kinetics of a tripartite virus to those of a monopartite virus. However, the different types of virus particles occurred at different frequencies, with a ratio of 116:45:1 (RNA1 to RNA2 to RNA3), which affected infection kinetics and required analysis with a more comprehensive infection model.

The interactions between hosts and pathogens produce the third component, disease. Each of these primary components of epidemics consists of subcomponents. The host may be an annual, a perennial or a tree; it goes through certain growth stages (seedling, tillering, blossoming); it is propagated by seed or vegetatively; it may be resistant or susceptible; it may react by producing lesions or a blight.

The subcomponents of the pathogen include pathogenicity (biotroph, necrotroph, toxins, mode of penetration); virulence (varietal specialization or race); sporulation (kind and amount of inoculum); dispersal (growth, by wind, water, and vectors); and survival (duration and form). The subcomponents of disease include infection (number of lesions, systemic); pathogenesis (presence and length of incubation period); lesion formation (size, rate, and toxins); infectiousness (time and amount of sporulation, amount of new inoculum); spread (infection gradient in host population); multiplication (length of infection cycle, duration and/or number of generations per season) and survival (longevity in months or years). As knowledge of the subcomponents of each epidemic is increased, the ability to predict the pattern of individual epidemics will also increase, which enables to interfere at the most appropriate stage epidemic with more efficient and more dependable methods of management.

4.5.4.3 *Course of epidemic*

The course of epidemic follows two distinct phases, viz., progressively destructive phase and the decline phase.

4.5.4.4 *Progressively destructive phase*

Some epidemics develop slowly (tardive) while others develop rapidly. Slow epidemics (or epiphytotics) usually occur among population caused by systemic pathogens. The pathogen multiplies slowly following the characters of simple interest disease. They belong to low death rate category and have less incubation period and sporulation period. However, the rapid epiphytotics are greatly influenced by environmental factors.

4.5.4.5 *Decline phase*

During the early stage, an epidemic spreads vigorously causing diseases in new hosts. After the development of the saturation stage, it shows a decline by itself. No epidemics may be due to nonavailability of suscepts, nonavailability of susceptible stages of the crop, unfavorable weather conditions, and reduction in aggressiveness of the pathogens. Generally, the hosts are prone to the disease at a specific developing stage. Once this stage is crossed in a plant, its proneness to infections is reduced or completely lost. Under the conditions the epidemic declines. The decline in the epidemic may also be due to unfavorable weather conditions for disease development. As a result, the future spread of the disease will be checked, and the epidemic will decline. Wheat crops in northern India usually get the attack of rusts in January to March.

Epidemics develop during these months. Although the plant remains prone to attack afterwards also, further development of the disease is checked because of the rise in temperature which is favorable for the pathogen. Due to the above mentioned and other causes, the aggressiveness of the pathogen may be reduced. When all susceptible individuals are destroyed by the pathogen, it may try to parasitize the remaining resistant individuals of the same species. In these adverse conditions, the pathogen may lose its power of successful infection, its reproduction may slow down, and the pathogen becomes less aggressive.

4.5.4.6 *Slow and rapid epiphytotics*

The form of an epidemic is decided by the nature of the pathogen, host, and the weather. Epidemic may develop slowly and is called 'tardive.' An epidemic which develops rapidly is called 'explosive.' In between these, intermediate forms of the epidemic may occur.

4.5.4.6.1 Slow epiphytotics

Slow epiphytotics occur among perennial (tree) populations. Infected host survives for several years before dying. Most of the characters of a simple interest disease are found in slow epiphytotics. The causal agent is mostly systemic. The pathogen multiplies slowly. Their movement from plant to plant is much slower. They are low death rate pathogen. In slow epiphytotics, crop sanitation is the best method, e.g., swollen shoot of cocoa. This disease spreads very slowly from tree to tree and still less from one garden to another garden. For instance, the incidence of 31% swollen shoot increased to 75% over a period of 2.5 years. As stated by Van der Plank (1959) the rate of multiplication of a systemic disease of trees is about tenfold a year whereas it is 10,000-fold in respect of herbaceous plants and it is of higher rates for local lesion pathogens, e.g., late blight of potato, wheat stem rust, etc.

4.5.4.6.2 Rapid epiphytotics

Rapid epiphytotics occur among annual crops. It is caused by nonsystemic pathogens with high birth rate. Several generations of the pathogen are produced within a short time. Rapid epiphytotics are largely governed by environmental factors compared to slow epiphytotics. Disease increase is rapid, and the disease rises to a peak in a short time and then show a sharp decline when the weather turns unfavorable or when the host becomes resistant due to maturity or due to restricted dispersal of propagules of the pathogen, e.g., apple scab. This type of epiphytotic is controlled by protective spraying or dusting with chemicals.

4.5.5 Types of Epidemics

Monocyclic epidemics are caused by pathogens with a low birth rate and death rate and meaning they only have one infection cycle per season. They are typical of soil born diseases such as *Fusarium* wilt of flax. Polycyclic epidemics are caused by pathogens capable of several infection cycles a season. These are most often caused by airborne diseases such as powdery mildew. Bimodal polycyclic epidemics can also occur. For example, in brown rot of stone fruits the blossoms and the fruits are infected at different times. For some diseases, it is important to consider the disease occurrence over several growing seasons, especially if growing the crops in monoculture year after year or growing perennial plants. Such conditions can mean that the inoculum produced in one season can be carried over to the next leading to a build of an inoculum over the years. In the tropics, there are no clear cut breaks between growing seasons as there are in temperate regions, and this can lead to accumulation of inoculum. Epidemics that occur under these conditions are referred to as polyetic epidemics and can be caused by both monocylcic and polycyclic pathogens. Apple powdery mildew is an example of a polyetic epidemic caused by a polycyclic pathogen and Dutch elm disease a polyetic epidemic caused by a monocyclic pathogen.

4.6 LANDSCAPE OF EPIDEMIOLOGY OF PLANT DISEASES

Many agricultural landscapes are characterized by a high degree of heterogeneity and fragmentation. Landscape ecology focuses on the influence of habitat heterogeneity in space and time on ecological processes. Landscape epidemiology aims at applying concepts and approaches originating from landscape ecology to the study of pathogen dynamics at the landscape scale. However, despite the strong influence that the landscape properties may have on the spread of plant diseases, landscape epidemiology has still received little attention from plant pathologists. Some recent methodological and technological progress provides new and powerful tools to describe and analyze the spatial patterns of host–pathogen interactions. Some important topics in plant pathology have been reviewed by Manuel Plantegenest *et al.* (2007) that may benefit from a landscape perspective. These included the influence of landscape composition on the global inoculum pressure, landscape heterogeneity on pathogen dynamics, landscape structure on pathogen dispersal and landscape properties on the emergence of pathogens and their evolution. Generally, the agricultural landscape consists of a changing mosaic of cultivated and uncultivated habitats. Landscape ecology that focuses on interactions between spatial patterns and ecological processes (Turner, 2005) appears relevant for providing concepts to build a new paradigm for agriculture with consequences on crop protection strategies.

Landscape epidemiology aims at identifying factors that influence the spatial spread of diseases between subpopulations of hosts. Recent developments in this field have resulted from progress in technologies for gathering and analyzing landscape structures combined with the realization of the importance of spatial structuring in epidemiology (Hess *et al.*, 2002). The number of studies devoted to the impact of landscape properties on human or animal diseases is now growing rapidly. Examples include studies of Lyme disease, alveolar echinococcosis and malaria in humans, and of sylvatic plague in prairie dogs and the hantavirus in deer mice. The same trends can be observed in studies devoted to agricultural pest dynamics and the evaluation of habitat manipulation for pest management. In surprising contrast, coupling landscape ecology and plant pathology have still received little attention, especially for economically important crop diseases. Nevertheless, some concepts originating from landscape ecology have been used in theoretical studies of plant disease dynamics. Metapopulation theory has been applied to theoretical studies of plant disease dynamics in a spatially structured host population. Percolation theory has also been used to study the spread of the soil-borne fungal pathogen *Rhizoctonia solani*. In both cases, the authors focused on the determination of thresholds for invasion and persistence of epidemics in metapopulations (Manuel Plantegenest *et al.*, 2007). Though the landscape approach is becoming more popular in forest pathology, empirical studies remain scarce.

4.6.1 Benefits of Landscape Epidemiology

From an applied point of view, landscape epidemiology studies could help identifying landscape characteristics impacting the *a priori* disease risk with two possible purposes, viz., designing crop protection strategies (like treatment advice based on decision-support systems) taking into account this *a priori* disease risk, and designing management strategies of landscape structures and agricultural systems with the aim to reduce the *a priori* disease risk. In both cases, a decrease in the need for curative interventions (like using chemicals) should be expected. Landscape impacts the composition on the global pathogen propagule pressure. Landscape composition refers to the relative amounts of each landscape element. A simple metrics of landscape composition is the relative frequencies of different types of patches. This composition determines the local abundance of potential reservoirs of inoculum that may obviously influence the global propagule pressure and hence the risk of infection of a plant. Those reservoirs may, in particular, consist of diseased individuals of the same host species or alternative, either cultivated or wild, hosts. The reservoirs may play different roles. They may simply influence the regional density of inoculum (Manuel Plantegenest *et al.*, 2007). They may also provide a refuge for pathogen during seasons or years unfavorable to disease spread. Finally, in the case of heteroecious pathogens, the presence of alternative hosts may be required to allow the pathogen to locally complete its life cycle.

Host density is recognized as a major factor driving disease epidemics, and theoretical studies have focused on determining a threshold below which a pathogen cannot invade a population of susceptible individuals. The dependence of plant disease incidence on the local density of the host is documented in forests or grasslands. Disease incidence could also depend on the abundance of phylogenetically related species in the nearest considering that the probability for two host species to share a pathogen increases with their phylogenetic proximity. Such a phylogenetic effect is likely to occur in cultivated plants in which closely related species, either cultivated or uncultivated, can be locally abundant. Despite being universally regarded as a chief driving factor, the effect of host density on disease dynamics has received little attention in crops even at the field scale and empirical studies remain scarce. A tendency towards increased disease severity at higher densities was observed, but this trend was not very consistent. For instance, a decrease in barley powdery mildew severity with increasing host density has been recorded (Manuel Plantegenest *et al.*, 2007).

In general, the effect of the presence of alternative hosts depends on the properties of the focal species relative to the entire community of alternative hosts. For example, if the focal host is a poor reservoir that does not multiply or transmit the pathogen effectively, then the presence of a more competent alternative host can largely increase the disease prevalence in the focal host. This effect was illustrated by showing that the presence of a highly susceptible host (*Avena fatua*) increased the prevalence of BYDV, a generalist virus, in several other species of annual wild grasses in experimental fields (Power and Mitchell, 2004). Such a source–sink interaction is also operating in a well-documented example: the invasion of bean dwarf mosaic virus (BDMV) in Argentina. In the 1970s, the soya bean acreage in Argentina increased sevenfold leading to the emergence of BDMV that threatened production of common bean. BDMV is vectored by the whitefly (*Bemisia tabaci*) and causes a severe disease on common bean (*Phaseolus vulgaris* L.) while being able to infect soya bean (*Glycine max* L.) without inducing symptoms (Wang *et al.*, 1999). The expansion of soya bean, a suitable reproductive host for *B. tabaci*, caused BDMV emergence in bean. Whereas common bean is not a preferred host of *B. tabaci*, it can feed and reproduce on this legume when populations are high or in the absence of its preferred breeding hosts. More generally, the indirect effects on the abundance of one species caused by the presence of a second one mediated by shared pathogens or predators (apparent competition) are recognized as an important factor influencing community dynamics and vegetation patterns (Morris *et al.*, 2004).

4.6.2 Role of Alternative Hosts

The role of alternative hosts on pathogen dynamics can be drastic when it governs pathogen survival at some critical point in its cycle. This effect is particularly obvious in the case of heteroecious pathogens that require two unrelated host plants to complete their life cycle as many rust species do. García-Guzmán and Wennström (2001) have shown that the occurrence of *Ochropsora ariae*, a heteroecious rust fungus, on *Anemone nemorosa* was strictly dependent on the presence in the nearest of the long-lived primary host *Sorbus aucuparia* (no diseased *A. nemorosa* was found farther than 90 cm away from *S. aucuparia*). They suggested that the short life duration of *A. nemorosa* infected by *O. aria* associated with the low dispersal capacity of the fungus resulted in the need for frequent reinfections from the alternative host to maintain disease in *A. nemorosa* populations. Wennström and Eriksson (1997) have shown that the heteroecious rust, *Gymnosporangium cornutum*, could not survive during winter in its secondary host *S. aucuparia* and that infection on this species relied on spore dispersal from its primary host *Juniperus communis* (in this case, spore dispersal is effective up to 50 m away from the primary host). The role of alternative hosts of cereal rusts (leaf rust caused by *Puccinia triticina*, stripe rust caused by *Puccinia striiformis* and stem rust caused by *Puccinia graminis*) on summer or winter survival of the pathogens has been considerably studied (Line, 2002).

Several new and powerful tools have been recently developed allowing acquisition and analysis of the spatial component of ecological data. They have been successfully used to identify landscape properties influencing the dynamics of human or animal diseases or the dynamics of crop pest populations. However, though the importance of considering the influence of landscape composition and configuration on disease is now recognized (Ostfeld *et al.*, 2005), empirical studies on plant disease dynamics remain scarce and mostly concern forest pathology. Consequently, whether or not landscape characteristics actually play a role in the epidemiology of most crop diseases remains largely unknown.

Inoculum density is considered by plant pathologists as a major factor determining the probability of occurrence and the severity of disease epidemics. The local abundance of source habitats and refuges has been proven to strongly influence the prevalence of crop disease. This suggests the importance of landscape composition on the dynamics of at least some diseases. Dispersal is also a key feature of disease epidemiology. Plant pathogens exhibit very diverse dispersal mechanisms, from soil-borne pathogens to aerially dispersed fungi and bacteria to vector-borne viruses. In all cases, dispersal scale defines the hierarchical structure of host–pathogen associations, by coupling the host and pathogen populations of connected demes and decoupling the dynamics of more distant ones (Gilbert, 2002). Landscape structure and heterogeneity are then likely to influence pathogen dispersal and hence disease dynamics. More specifically, some landscape elements may act as corridors facilitating the spread of plant disease or, conversely, as barriers limiting the dispersal of infective propagules. However, the sources distance and routes of dispersal of infective propagules are poorly known for most crop diseases. Though developing new tools to track the movement of pathogens at the landscape scale would surely be useful especially for air- or water-borne pathogens, the use of existing ones, such as molecular tools, would allow a great improvement of our knowledge in this field.

4.6.3 Landscape Ecology

Landscape properties may also influence evolutionary dynamics of pathogens. They may influence the evolutionary potential of pathogens by impacting population effective size and gene flow among local populations. Landscape composition is of special importance in heteroecious pathogens in which the reproductive mode and consequently the genetic diversity depends on the availability of the primary host. Finally, landscape characteristics may influence pathogen ecology and especially the range of hosts simultaneously or successively exploited and determine the evolution of important traits among which is virulence. Some of these questions have been addressed from a theoretical point of view; however, model predictions still remain to be validated on real data. Hence, coupling landscape ecology, landscape genetics and epidemiology is a promising approach. Many concepts and tools are yet available to handle efficiently the questions addressed by landscape epidemiology of plant diseases. It requires applying them to a large range of pathosystems to better understand the actual impact of landscape properties on crop disease emergence and dynamics.

From an applied point of view, safeguarding crop production is obviously of major importance for human beings. Agriculture faces several important challenges. It will have to rapidly increase crop production in response to increasing demand due to the growth of the human population. However, it will also at the same time have to adapt to global changes and to respond to the public demand for developing environmentally friendly practices. To achieve these goals, a better understanding of the interactions between agriculture and environment and in particular of plant disease ecology is required. The landscape appears as a key scale for this purpose. The success will require the translation of knowledge in landscape epidemiology into agricultural practices. Modeling is a useful approach to provide tactic decision-support tools allowing optimization of crop protection in real time and considering the particular landscape context or to help selecting strategies of ecological engineering for crop disease management based on agricultural practices and habitat manipulation. The conceptual framework provided by landscape ecology has yet been used to model crop disease dynamics. The few existing studies are still preliminary and mostly theoretical. A lot of work is still needed to develop tools that are truly useful for managers. Finally, a multidisciplinary approach is needed to simultaneously consider the main crops and their pests and diseases that coexist in a given landscape, to include technical and socio-economical considerations and to consider the multifunctional role of the landscape.

Plant disease epidemiology deals with diseases in plant populations. Over the years, it has become a vibrant field of science, achieving significant conceptual innovations with an important impact on the management of plant diseases. Plant disease epidemiologists have a strong scientific tradition in studying climate-pathogen-disease relationships and also in dealing with biodiversity. Understanding the factors that trigger the development of plant disease epidemics is helpful in creating and implementing effective strategies for disease management. Therefore, description and quantification of the disease cycle are the foundation of plant disease epidemiology and the key to developing effective disease management (Pushpinder Singh, 2011). Disease forecasts which assist growers in determining when to apply disease management techniques, is a logical and exciting application of plant disease epidemiology. It aids growers in making the best decisions among disease management alternatives. Since the development of plant disease epidemics depends upon favorable meteorological conditions, it seems reasonable to examine those conditions associated with a variation in disease severity from year to year. One goal of the analysis of meteorological and disease data is to develop simple models that use a minimum of variables so that the models are easy to use for the management of the disease. Also, the prediction models must be based on the meteorological variable that affects disease early enough in the growing season to allow an option of control. Nowadays, input data for the forecasts are usually derived from automatic weather stations (AWS) located in different regions. In India, late blight forecasting systems for the appearance of the disease have been developed both for hills and the plains and development of the disease support system (DSS) is underway. A computerized forecasting model 'JHULSACAST' has been developed for western UP, which has three important components—late blight prediction model, a computer programme for forecasting, and an interface to utilize weather data from an automatic weather station. The use of computer-based disease simulation models is quite common for the prediction of important diseases in the USA and other developed countries. Downy mildew caused by *Pseudoperonospora cubensis* is a limiting factor for successful cucurbits cultivation in Punjab conditions and the weather variables like maximum temperature and mean relative humidity during the period corresponding to SMW's 21 to 24 were identified as important components used in developing disease prediction model. An early appearance of disease preceding this period helped in the buildup of initial inoculum and rapid disease take-off if the subsequent conditions were favorable. A function of these two parameters termed as 'Humid Thermal Ratio' (HTR) was used to avoid the problem of multicolinearity in regression analysis. Overall the model based on HTR fit the data well as judged by its performance in the validation tests. The forecasts are formulated to be as objective as possible with the meteorological tools available. A degree of human experience and subjectivity will always be present in the forecast system until more disease/weather algorithms become available. The quality of disease site observations, confidence in the weather forecast data, forecaster experience, and model validation are all-important factors in determining how the model will function in its intended use and set the strategy for efficient disease control.

4.6.4 Population Genetics and Epidemiology

The potential for modern biology to identify new sources for genetical, chemical and biological control of plant disease is remarkably high. Successful implementation of these methods within globally and locally changing agricultural environments demands new approaches to durable control. This, in turn, requires fusion of population genetics and epidemiology at a range of scales from the field to the landscape and even to the continental deployment of control measures. It also requires an understanding of economic and social constraints that influence the deployment of control. An epidemiological framework was proposed by Christopher A. Gilligan (2008) to model invasion, persistence and variability of epidemics that encompasses a wide range of scales and topologies through which disease spreads. By considering how to map control methods onto epidemiological parameters and variables, some new approaches towards optimizing the efficiency of control at the landscape scale were introduced. Epidemiological strategies to minimize the risks of failure of chemical and genetic control were presented and some consequences of heterogeneous selection pressures in time and space on the persistence and evolutionary changes of the pathogen population were discussed. Some approaches towards embedding epidemiological models for the deployment of control in an economically viable framework were presented.

Knowledge of the incubation period of infectious diseases (time between host infection and expression of disease symptoms) is crucial to our epidemiological understanding and the design of appropriate prevention and control policies (Melen Leclerc *et al.*, 2014). Plant diseases cause substantial damage to agricultural and arboricultural systems, but there is still very little information about how the incubation period varies within host populations. The incubation period of soil-borne plant pathogens was focused, which are difficult to detect as they spread and infect the hosts underground and above-ground symptoms occur considerably later. Experiments were conducted on *Rhizoctonia solani* in sugar beet, as an example pathosystem, and used modeling approaches to estimate the incubation period distribution and demonstrate the impact of differing estimations on our epidemiological understanding of plant diseases. Measurements of the incubation period obtained in field conditions were presented, fit alternative probability models to the data, and showed that the incubation period distribution changes with host's age. By simulating spatially-explicit epidemiological models with different incubation period distributions, they studied the conditions for a significant time lag between epidemics of cryptic infection and the associated epidemics of symptomatic disease and examined the sensitivity of this lag to differing distributional assumptions about the incubation period (i.e., exponential *versus* Gamma). It was demonstrated that accurate information about the incubation period distribution of a pathosystem could be critical in assessing the true scale of pathogen invasion behind early disease symptoms in the field; likewise, it can be central to the model-based prediction of epidemic risk and evaluation of disease management strategies. It was concluded that reliance on observation of disease symptoms could cause a significant delay in detection of soil-borne pathogen epidemics and mislead practitioners and epidemiologists about the timing, extent, and viability of disease control measures for limiting economic loss.

Plant disease epidemics are cyclical phenomena, that is, they consist of repeated cycles of pathogen development in relation to the host and the environment. The inoculum, which might consist of fungal spores, bacterial cells, nematodes, viruses within an aphid vector, or some other propagules of a pathogen, gains entry into and establishment within the host tissues through the process of infection. The pathogen develops within the host and eventually begins to produce new inoculum, which, in turn, can be dispersed to new susceptible sites to initiate new infections. Pathogens that produce only one cycle of development (one infection cycle) per crop cycle are called 'monocyclic' while pathogens that produce more than one infection cycle per crop cycle are called 'polycyclic.' Generally in temperate climates there is only one crop cycle per year, so the terms 'monocyclic' and 'polycyclic' are based on the number of cycles per year. In tropical or subtropical climates, however, there can be more than one crop cycle per year, and it is important to remember that 'monocyclic' and 'polycyclic' are based on a single crop cycle. These same terms are used to describe the epidemics as well as the pathogens, so we often speak of a 'monocyclic epidemic' or a 'polycyclic epidemic'.

For some diseases, it is important to consider an epidemic over a period of many growing seasons. This is particularly true for perennial plants (forages, pastures, lawns, orchards, forests, etc.) or for annual crops that are grown in monoculture year after year. In these situations, the inoculum produced in one growing season carries over to the next, and there could actually be a buildup of inoculum over a period of years. In the tropics, there may not be clear-cut breaks between growing seasons such as we find in the temperate zones, and epidemics can be virtually continuous over periods of many years on such crops as bananas, coffee, and rubber trees. These epidemics are referred as 'polyetic' epidemics, regardless of whether the pathogen is monocyclic or polycyclic within each season.

Note 1: ***Basic reproductive number:*** The number of secondarily infected units (plants or plant parts) caused by the introduction of one infectious unit into a susceptible host population, curing that unit's period of infectiousness.

Note 2: ***Intraspecific diversity:*** Genetic variation occurring within a plant host and/ or pathogen population; such variation can determine the level of compatibility in a host–pathogen interaction as expressed in gene-for-gene recognition systems. It is used in disease control strategies based on multilines and varietal mixtures.

Note 3: ***Evolutionary epidemiology:*** An attempt to integrate population genetic and population dynamic approaches in epidemiology so that the evolutionary dynamics of plant pathogen populations can be studied and managed.

Note 4: ***Landscape epidemiology:*** A recognition that plant disease epidemics occur at the landscape scale, and are influenced by aspects of the wider environment including features such as topography, watersheds, human activity and the conjunction of cropland and semi-natural vegetation.

4.7 NEW TOOLS IN EPIDEMIOLOGY

Various tools are available in recent time, viz., molecular tools (Polymerase Chain Reaction (PCR), Enzyme Linked Immunosorbent Assay (ELISA), DNA Fingerprinting, etc. for rapid and accurate detection and identification of pathogens), data management (Geographic Information System, Global Positioning System, Remote Sensing etc., to assist in disease control strategies), geostatistics, disease modeling and forecasting (to predict the probability of outbreaks), and image analysis and information technology.

4.8 DISEASE PROGRESS CURVE

The progress of an epidemic measured in terms of the numbers of lesions/the amount of diseased tissue, or the numbers of diseased plants plotted over time is termed the disease progress curve. It is commonly used to quantify disease intensity over time

It is useful because it integrates the amount of disease over time, rather than at a particular time point. Major curve types include saturation type of curve (Three monocyclic diseases of different epidemic rates), sigmoidal curve (Polycyclic disease, such as late blight of potato) and bimodal curve (polycyclic disease, such as apple scab, in which the blossoms and the fruit are infected at different, separate times).

Disease on plants usually starts out at a low level, a small number of plants affected and a small amount of plant tissue affected, and it becomes of concern to us only when its incidence and severity increases with time. When we look at some examples of plant disease epidemics from the published literature, we not only notice that the incidence or severity starts near zero and then increases dramatically, but we also can discern some distinct patterns of development with time (Anonymous, 2104) (Figs. 4.2–4.9). For example, in *Phytophthora* blight of pepper seedlings (*Phytophthora capsici*) and *Fusarium* kernel rot (*Fusarium moniliforme,* currently *F. verticilloides*) of maize, disease progress is roughly linear (allowing for some minor deviations that we can consider random error).

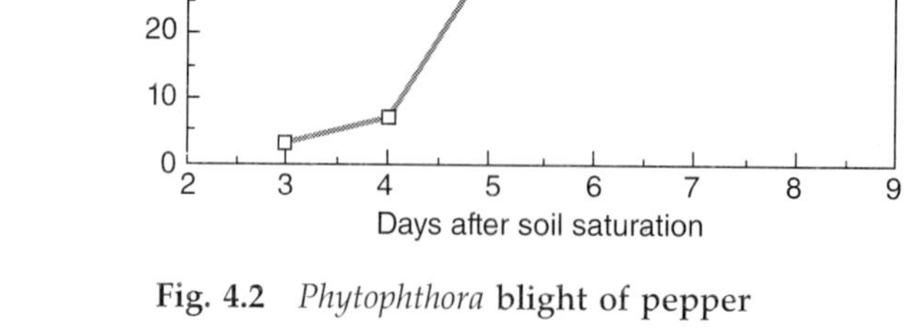

Fig. 4.2 *Phytophthora* blight of pepper

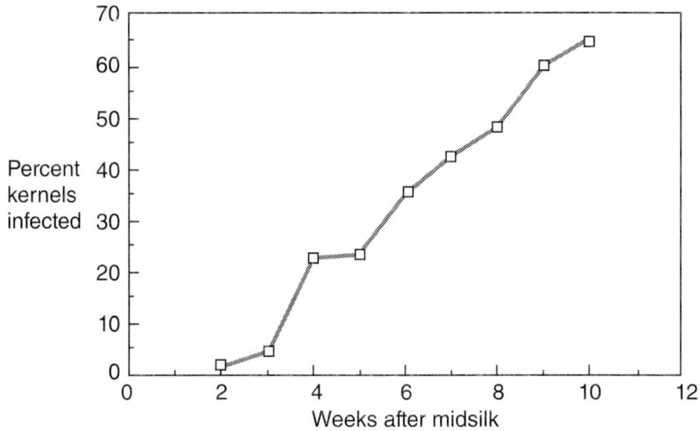

Fig. 4.3 *Fusarium* kernel rot of maize

On the other hand, in bean rust (*Uromyces phaseoli*) and gray leaf spot of corn (*Cercospora zeae-maydis*), there is a definite upward curve; that is, disease increases at an increasing rate, a curve we could call exponential.

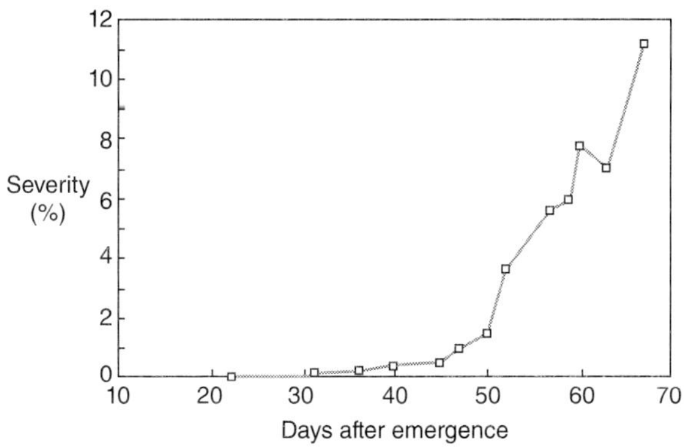

Fig. 4.4 Bean rust

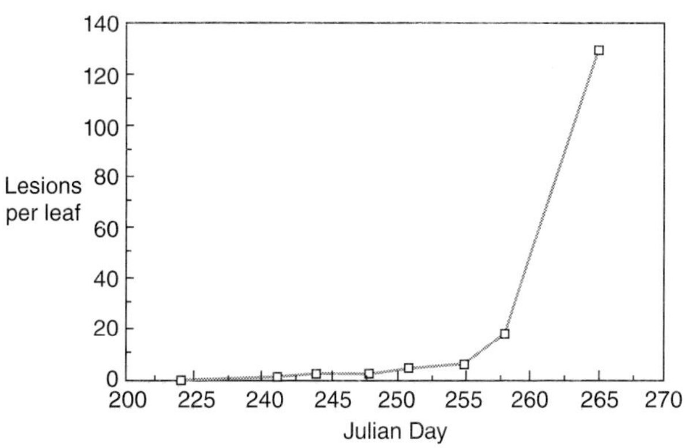

Fig. 4.5 Gray leaf spot of maize

Obviously, plant disease cannot continue to increase forever, and as the level of disease approaches 100%, the disease progress curve gradually flattens out. For example, in epidemics such as the infection of beans by *Sclerotium rolfsii* or the infection of tobacco by *Phytophthora nicotianae*, the disease progress starts out looking linear but slows down as it approaches a maximum.

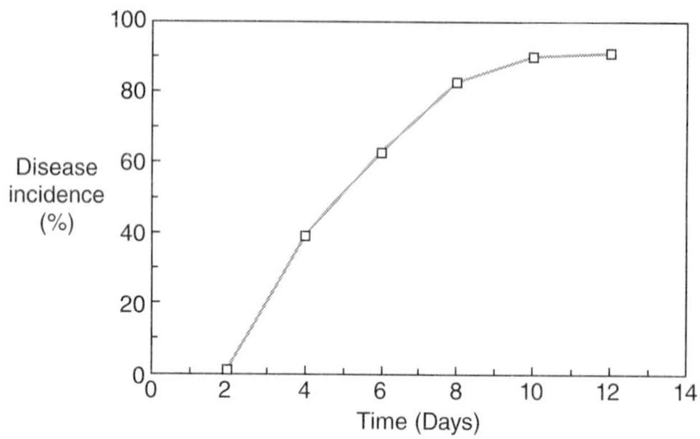

Fig. 4.6 *Sclerotium* **rolfsii** on beans

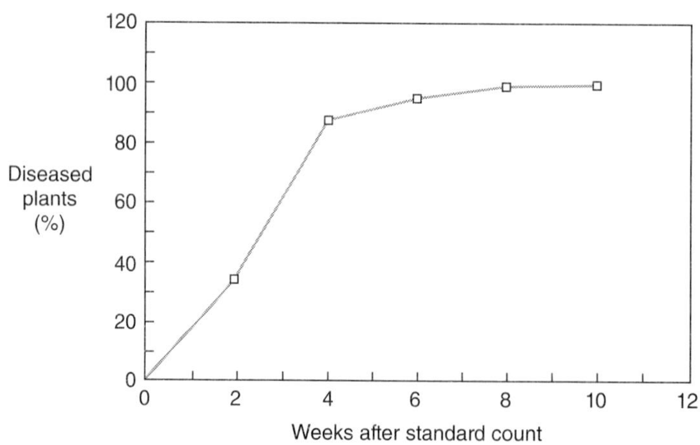

Fig. 4.7 Black shank on tobacco

Likewise, the disease progress curves of *Puccinia graminis* subsp. *graminicola* on ryegrass and *Pyrenophora teres* f. sp. *teres* on barley appear exponential at first, but as time goes on and the incidence and severity of disease approach 100%, the rate of disease progress gradually slows to zero, giving both curves a somewhat sigmoid shape ('S' shape).

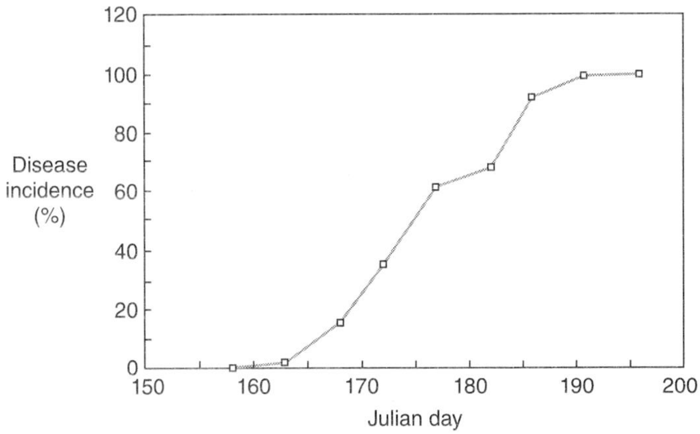

Fig. 4.8 Black stem rust on ryegrass

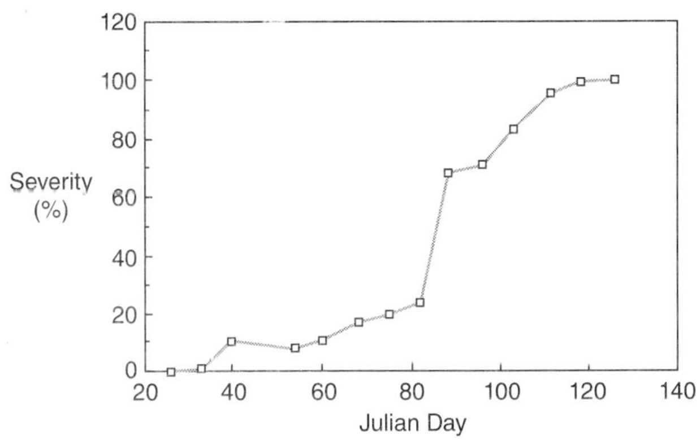

Fig. 4.9 Net blotch on barley

To be sure, not all examples of disease progress can be as neatly categorized as these, but, in general, plant disease epidemics tend to be either roughly linear or exponential in the early stages, and they tend to level off as they approach some limit. The impact of plant disease and the losses that it causes are a function of disease progress. To reduce this impact, there is a need not eliminate the disease, merely need to keep disease development below an acceptable level. That means that the progress of the disease and the factors that influence disease progress must be understood in quantitative terms. It is essential to know what kinds of diseases lead to linear disease progress and what factors affect the slope of the line (the rate of disease progress), and also what types of diseases tend to produce exponential disease progress curves and how we can reduce both the starting level of disease and the rate of epidemic development. Finally, it is important to know why epidemics sometimes level off and what imposes limits on their development.

Septoria tritici blotch (STB) caused by the ascomycete fungus *Mycosphaerella graminicola* (anamorph *S. tritici*) is currently one of the most serious foliar disease of wheat in Europe and other world regions characterized by temperate and wet environment during growing season. Twenty-five wheat genotypes with different resistance levels were evaluated in natural conditions for their reaction to *S. tritici* attack during the year 2009–2010 (Mirela Paraschivu *et al.*, 2013). The experimental design was a randomized block design with three replications. Disease rating was visually recorded by using the double-digit scale (00-99) at two different times and crop stages (Z53-1/4 head out and Z70-milk development). There were also calculated the Septoria Progress Coefficient (SPC) and Area under Disease Progress Curve (AUDPC) for each wheat genotype evaluated. SPC was low for the higher genotypes comparatively with the shortest ones leading to the conclusion that disease progress is higher as plant height is low ($r = -0{,}8584$). The same aspect was also emphasized by the negative correlation between plants height and disease progress height ($r = -0{,}7241$). The shortest genotypes showed higher necrosis percentage and AUDPC values. There was also a correspondence between genotype susceptibility and AUDPC showing that the most susceptible wheat cultivars recorded higher AUDPC values. Thus, Renan (295), Esential (231), Renesansa (234), Autan (186) recorded the highest AUDPC values while Aztec (45), Meunier (41) and Capo (20) had the best resistance reaction to *S. tritici* attack. Comparatively with the control Briana cultivar (AUDPC = 153), created to ARDS Simnic, the differences recorded by Renan, Renesansa Esential, Ciprian, Exotic, Cezanne, Aztec, Capo, and Meunier were statistically significant. There was also a negative correlation between AUDPC and yield ($r = -0{,}5140$). The Area under the Disease Progress Curve can be an efficient instrument to evaluate the epidemic development of foliar pathogen *S. tritici* considering each genotype susceptibility and specific architecture.

4.8.1 Disease Progress Over Time

Understanding how disease levels increase or decrease over time is one of the most basic elements of plant disease epidemiology and ecology. Statistical models are often applied to summarize and describe this complexity so that disease processes can be more readily understood (Madden *et al.*, 2007). For example, comparisons between patterns of disease progress for different diseases, cultivars, management strategies, or environmental settings can help in determining how plant diseases may best be managed. Understanding how populations of different types of plant pathogens increase in plant pathosystems is critical to developing appropriate management methods. Bacteria and viruses are capable of building up enormous populations in a very short time, only days or even hours under the right conditions. Fungi also can reproduce quickly, within days or weeks, which is important for fungal pathogens that cause polycyclic diseases, i.e., fungi that reproduce multiple times per season. Nematodes, meanwhile, often take a year or more to complete their reproductive cycle. These different characteristics will greatly influence disease progress over time.

The most useful measurement of disease progress depends on both the host and the pathogen of interest. Disease incidence or the number of diseased plants express as a proportion of the total number of plants assessed, is one common measurement. Disease incidence is often a useful measure for quantifying diseases for perennial crops, such as citrus, or for estimating crop losses due to a disease such as stalk rot in corn, where entire plant death from a single infection is possible. Disease severity, usually defined as the percentage (area) of diseased tissue present on an affected plant, is another common measure, particularly for evaluation of foliar diseases where the amount of disease present in the plant may be correlated to a yield loss estimate. Nematodes are typically quantified in terms of their population size, often the number of nematodes per volume of soil. Nematode damage is well correlated to population levels for several different crops. Despite differences in how a disease is measured, similar methods for describing and summarizing disease progress curves can be applied to a range of plant pathosystems. A measure of disease incidence or severity is typically plotted on the Y-axis as a function of time, which is plotted on the X-axis. Such disease progress curves provide a way of comparing disease across years or locations (Sparks *et al.*, 2008).

4.8.2 Calculating Area under Disease Progress Curve to Quantify Disease Progress

The area under the disease progress curve (AUDPC) is a useful quantitative summary of disease intensity over time, for comparison across years, locations, or management tactics. The most commonly used method for estimating the AUDPC, the trapezoidal method, is to discretize the time variable (hours, days, weeks, months, or years) and calculate the average disease intensity between each pair of adjacent time points (Madden *et al.*, 2007). We can consider the sample time points in a sequence $\{t_i\}$, where the time interval between two time points may be consistent or may vary, and we also have associated measures of the disease level $\{y_i\}$. We define $y(0) = y_0$ as the initial infection or the disease level at $t = 0$ (i.e., the first disease severity observation in our study). $A(t_k)$, the AUDPC at $t = t_k$, is the total accumulated disease until $t = t_k$, given by $A_k = \sum_{i=1}^{N_i-1} \frac{(y_i + y_{i+1})}{2}(t_{i+1} - t_i)$. To illustrate this concept, the exercise below shows the rectangular area S_i as the increase in accumulated disease during the time interval from t_{i-1} to t_i, $S_1 = A_1$ and $A_i = A_{i-1} + S_i$, $i \geq 2$.

Genetic effects on controlling stripe rust resistance were determined in two wheat crosses, Bakhtawar-92 × Frontana (cross 1) and Inqilab-91 × Fakhre Sarhad (cross 2) using Area Under Disease Progress Curve (AUDPC) as a measure of stripe rust resistance (Irfaq *et al.*, 2009). The resistant and susceptible genotypes for crosses were identified by initial assessment of 45 wheat accessions for stripe rust resistance. Mixed inheritance model was applied to the data analysis of six basic populations P1, F1, P2, B1, B2, and F2 in the crosses. The results indicated that AUDPC in cross 1 was controlled by two major genes with additive-dominance epistatic effect plus polygenes with additive-dominance epistatic effects (model E). Whereas in the case of cross 2, it was under the control of two major genes with additive-dominance epistatic effect plus additive-dominant polygenes (model E-1). An additive effect was predominant then all other types of genetic effects suggesting the delay in selection for resistance till maximum positive genes are accumulated in the individuals of subsequent generations. The occurrence of transgressive segregants for susceptibility and resistance indicated the presence of resistance as well as some negative genes for resistance to the parents. The major gene heritability was higher than the polygene heritability in B1, B2, and F2 for the crosses. The major gene, as well as the polygene heritability, was ranging from 48.99 to 87.12% and 2.26 and 36.80% for the two crosses respectively. The highest phenotypic variations in AUDPC (2504.10 to 5833.14) for segregating progenies (BC1, BC2, and F2) represent that the character was highly influenced by the environment.

Comparison of the linear coefficients from the regression analysis of 307 blast (*Pyricularia grisea*) disease progress curves of 42 rice genotypes, tested over a period of nine seasons, were made by fitting into the logistic and Gompertz models (Mohapatra *et al.*, 2008). The regression parameters Y-intercept (a), regression coefficient (b), its standard error (SEb), the coefficient of determination (R2) and the time required for the disease to reach a specific level of severity (T50) were estimated. The estimates of the linear coefficients following Gompertz model, revealed lower values of b, SEb and negative 'a'; higher values of T50 as well as R2 compared with those by logistic model. In general, the logistic apparent infection rates were higher than the Gompertz rates, the differences being wider for the slow-blasting genotypes compared with the fast blasting genotypes. The lower Gompertz infection rates accompanied with low negative 'a' values, resulted in higher T50 values, compared with the respective lower estimates through the logistic model. Both b and T50 values estimated by Gompertz model were more consistent among the slow-blasting genotypes over seasons. Out of the 307 disease progress curves, 91.2% fitted well into the logistic model and 90.8% into the Gompertz model, thus suggesting that both the models fit well into the rice blast pathosystem. The estimates of T50 were proportional to the corresponding apparent infection rates in both the models. The benefit of sanitation in terms of the epidemic to reach 0.5 levels was estimated to be 21 and 47 days following logistic and Gompertz model, respectively.

4.8.3 Monocyclic Epidemics

In general, there are three types of plant diseases that tend to produce only one infection cycle per host cycle, namely, postharvest diseases, diseases caused by soil-borne plant pathogens and rusts without a urediniospore stage.

4.8.3.1 Postharvest diseases

Not all postharvest diseases produce monocyclic epidemics, but in many cases the infections that result in storage rots have either already occurred before harvest or occur during the harvest, and postharvest handling before the product goes into storage. The rot progresses during storage, and new inoculum may be continuously produced, but unless the stored product is handled again to disperse the inoculum and create new infection sites by making small wounds, there are no new infection cycles. Note that the pathogen is not inherently monocyclic but is constrained by the environment to produce only a single cycle of infection. In other environments, these same pathogens might produce polycyclic epidemics. The Dutch elm disease is an example of a monocyclic pathogen giving rise to a polyetic epidemic. Note that while there is only one cycle of infection each year and disease progress within each year is roughly linear, the incidence of infected trees increases at an increasing rate from year to year.

4.8.3.2 *Relating disease progress to disease cycles*

The epidemics that progress roughly linearly or are linear at the start tend to be monocyclic epidemics. On the other hand, the diseases that increase at an increasing rate during the early part of the epidemic tend to be polycyclic epidemics.

Diseases caused by soil-borne plant pathogens: Many of the root rots, vascular wilts, and other diseases caused by soil-borne pathogens also produce only one infection cycle per crop cycle. The inoculum generally is some kind of survival structure resistant to desiccation or freezing, such as sclerotia, chlamydospores, or oospores in the soil or mycelium in crop residues. This inoculum is dispersed in the soil by plowing and fitting the land and turning under the crop residues. As the roots of the newly planted crop grow through the soil, they encounter the propagules of the pathogen embedded in the soil matrix and become infected. The epidemic progresses as new infections occur, but since any new inoculum that is produced is not dispersed until the soil is again cultivated, there is only one complete infection cycle per cropping cycle. To be sure, not all soil-borne plant pathogens produce monocyclic epidemics, and one must be very careful to understand the life cycle of each pathogen before drawing conclusions about its epidemiology.

4.8.3.3 *Demicyclic rusts*

Some rusts do not produce urediniospore stage (repeating stage) on a single host and the inoculum produced on one host species generally must infect a different host species. This alternation of hosts appears to have evolved in adaptation to the annual growth cycles of the hosts, and we see one infection cycle on each host per year. An example is the cedar-apple rust, where all the inoculum that infects apples comes from galls on red cedars (junipers), and all the inoculum that infects red cedars comes from apple leaves and fruit. The epidemic on apples occurs during a four- to six-week period of basidiospore production in the spring. A second monocyclic epidemic on red cedars occurs during a brief period of aeciospore production in late summer

4.8.4 Polycyclic Epidemics

For an epidemic to be considered polycyclic, there must be repeated complete infection cycles, that is, infection followed by pathogen development, new inoculum production, dispersal to new susceptible sites, and new infections, all within a single crop cycle. A good example is potato late blight, where a single cycle of infection, lesion development, sporulation, sporangium dispersal, and new infection can occur in as little as five days, and many overlapping cycles occur simultaneously during periods of favorable weather. Each cycle can produce more than a tenfold increase in numbers of sporangia landing on susceptible sites, and explosive epidemic results. Cereal rusts behave similarly; a single urediniospore can infect to produce a pustule from which hundreds of new urediniospores can be released to infect and produce hundreds of new pustules repeatedly throughout the season. Most plant diseases caused by bacteria are polycyclic, and many plant viruses, with the aid of their insect vectors, also can produce repeated cycles of infection in one season.

Combinations of monocyclic and polycyclic epidemics: Not every plant disease epidemic is clearly either monocyclic or polycyclic. Epidemics produced by fungi with two spore stages can have elements of both, sometimes in distinct phases, and sometimes occurring simultaneously. For example, *Venturia inaequalis* produces ascospores on the dead, infected leaves that have overwintered from the previous season. These ascospores are released over a period of six to eight weeks in the spring and infect the newly expanding apple leaves. Since no new ascospores are produced until the following spring, this component of the epidemic might be considered monocyclic. However, each leaf lesion within about ten days produces a second type of spores, conidia, that also can infect newly expanding leaves. Thus, for the early part of the season, a polycyclic epidemic is superimposed on a monocyclic epidemic. Since lesions produced by conidia cannot be distinguished from those produced by ascospores, the net effect appears to be a rapidly growing polycyclic epidemic.

An epidemic of tan spot of wheat, caused by *Pyrenophora tritici-repentis,* also comprises both polycyclic and monocyclic elements. The initial inoculum occurs on infected seed, and the epidemic starts as these seeds germinate to yield seedlings bearing leaf lesions. Conidia are disseminated by wind and splashing rain in repeated cycles of leaf infection throughout the period of crop development (the polycyclic phase). By the time flowering begins, the inoculum for the second phase of the epidemic has built up on the upper leaves of the plant, principally the flag leaf. The flowers and developing grains are susceptible to a relatively brief period of infection by the inoculum extant at the time of flowering, and the lesions on the glumes of the developing grains do not produce inoculum capable of infecting more grains. Seed infection, therefore, is a monocyclic phenomenon. It is these infected seeds that provide a major means of carryover of this pathogen from one crop to the next and the initial inoculum for the new crop.

4.9 MATHEMATICAL MODELS

Plant diseases represent one of the major threats to crops around the world because they carry healthy, economical, environmental and social problems (Contreras-Medina *et al.*, 2009). Considering this, it is necessary to have a description of the dynamics of plant disease to have sustainable strategies to prevent and diminish the impact of the diseases in crops. Mathematical tools have been employed to create models which give a description of epidemic dynamics; the commonly mathematical tools used are disease progress curves, Linked Differential Equation (LDE), Area Under Disease Progress Curve (AUDPC), and computer simulation. Nevertheless, there are other tools that have been employed in the epidemiology of plant diseases like statistical tools, visual evaluations, and pictorial assessment. Each tool has its own advantages and disadvantages. The nature of the problem and the epidemiologist necessities determine the mathematical tool to be used and the variables to be included in the model. To obtain sustainable practices for strategic and tactical management of diseases and also to decrease its environmental impact, it is necessary to understand the determining factors of epidemics (Jeger, 2004). Mathematical, statistical and other tools have been used to understand these factors by modeling the epidemic dynamics; the objective in modeling is to simplify the reality to summarize the process of the epidemic. Mathematical tools have been gaining popularity because they allow knowing a description of the epidemic dynamics and by consequence, to develop optimal forecasting and controlling mechanisms. The mathematical tools employed in plant disease epidemiology use several variable values as inputs; these variables are considered according to the nature of the problem and the questions to be answered (Van der Plank, 1982). Schoeny *et al.* (2007) proposed a predictive model of *Ascochyta* blight where an important variable in the model was the airborne inoculums. The variables used by the mathematical tools summarize the key characteristics of the epidemic dynamics. The most common mathematical tools used in plant disease epidemiology include disease progress curves, linked differential equations, the area under disease progress curve and computer simulations. Nevertheless, there are other tools to evaluate the disease progress.

4.9.1 Mathematical Tools Used in Plant Disease Epidemiology

Plant diseases epidemics are investigated according to variables of interest which are formulated as functions of external factors, for instance, temperature and rain (Van Maanen and Xu, 2003). The problem nature and epidemiologist specific questions determine the mathematical tool to be used for modeling plant disease epidemics (Contreras-Medina *et al.*, 2009). Next, the most common mathematical tools to describe epidemic dynamics are presented.

4.9.1.1 Disease progress curves

Disease progress curves show the epidemic dynamics over time (Agrios, 2005). This mathematical tool can be used to obtain information about the appearance and amount of inoculum, changes in host susceptibility during growing period, weather events and the effectiveness of cultural and control measures. Growth models provide a range of curves that are often similar to disease progress curves and represent one of the most common mathematical tools to describe temporal disease epidemics (Xu, 2006). The growth models commonly used are monomolecular, exponential, logistic and Gompertz (Fig. 4.10). A brief description of each growth model is presented below.

4.9.1.2 Monomolecular

This growth model is appropriate for modeling epidemics where there is no secondary spread within a growing season, meaning that the plant disease has a single cycle during growing season. This model is also called negative exponential model (Campbell and Madden, 1990).

4.9.1.3 Exponential

This model is also known as the logarithmic, geometric or Malthusian model. This growth model is appropriate when newly diseased (infected) individuals lead to more diseased (infected) individuals and has been used to model changes in disease prevalence on a geographic scale, it can be applied to describe the very early stages of most polycyclic epidemics.

4.9.1.4 Logistic

This was proposed firstly by Veshulst in 1838 to represent human population growth. A second type of logistic model was proposed by Van der Plank (1963), being more appropriate for most polycyclic diseases, meaning that there is a secondary spread within a growing season. This growth model is the most widely used for describing epidemics of plant disease (Contreras-Medina *et al.*, 2009).

4.9.1.5 Gompertz

This growth model is appropriate for polycyclic diseases as an alternative to logistic models. Gompertz model has an absolute rate curve that reaches a maximum more quickly and declines more gradually than the logistic models.

Figure 4.10 shows examples of disease progress curves represented by growth models, where it can be seen that Gompertz and logistic models have a characteristic sigmoid form and an inflection point meaning secondary inoculation or plant-to-plant spread within the crop in contrast to monomolecular model, which does not have an inflection point. The exponential model presents a very small value at the beginning comparing with the other models, and later it increases exponentially (Contreras-Medina *et al.*, 2009). In general, growth models that incorporate few variables to describe temporal disease dynamics have a good performance. However, this kind of model sometimes does not satisfy the acquiring process of the key characteristics because they frequently ignore relevant variables that affect the epidemic development [(Xu, 2006) (*Example:* host growth, fluctuating environmental condition, the length of the latent and infectious period, etc.)]. Van der Plank (1960) used exponential, monomolecular and logistic models to describe the development of epidemics. Xu (1999) used a logistic model to forecast and model the apple powdery mildew provoked by *Podosphaera leucotricha*.

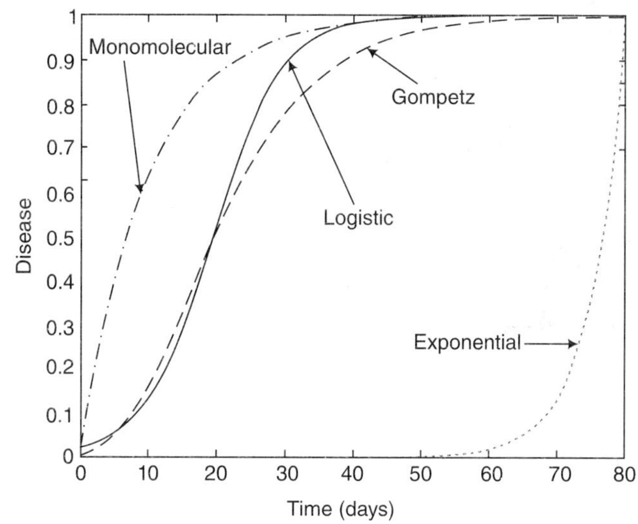

Fig. 4.10 Examples of disease progress curves represented by monomolecular, exponential, logistic and Gompertz models

4.9.1.6 Linked differential equations (LDE)

This mathematical tool achieves the description of plant disease epidemics by modeling each of the variables considered to be determinant in epidemic development and latter making the link between them. LDE allows the inclusion of new terms into the model as needed (Madden, 2006). When this technique cannot be easily integrated to obtain an analytical solution, it needs numerical methods to provide a numerical solution (Xu, 2006; Madden, 2006). Nowadays, the advances in computational technology have permitted that numerical solutions could be easily obtained through mathematically-oriented software like MATHCAD or MATHEMATICA (Madden, 2006). Usually, LDE is employed to investigate relationships of the plant disease dynamics in relation to the host, environment, and human intervention. A relatively simple linked differential equations for polycyclic disease involve variables like healthy tissue, latently infected, infectious and removed individuals (leaves, roots, plants) (Madden, 2006). Zhang *et al.* (2001) used four linked differential equations to illustrate the effect of synergism between plant viruses, where each differential equation represents one of the following four categories: healthy host plant, infected host plant with a virus kind A alone, infected with a virus kind B alone, and infected with both viruses A and B. The results obtained were used to understand an epidemic that has been ongoing in Uganda since the late 1980s, and the information about the development of Cassava Mosaic Disease (CMD) epidemic in Uganda and neighboring countries support the model results.

Plant disease simulation models are valuable tools in epidemiology. They organize available knowledge such that information gaps become clear. They also enable one to manipulate variables affecting disease to generate hypotheses. Finally, they can be used to illustrate epidemiological principles for educational purposes. The concept of model qualification, i.e., discovering the domain over which a validated model may be properly used, was illustrated with LATEBLIGHT, a mathematical model that simulates the effect of weather, host growth and resistance, and fungicide use on asexual development and growth of *Phytophthora infestans* on potato foliage (Andrade-Piedra *et al.*, 2005). Late blight epidemics from Ecuador, Mexico, Israel, and the United States involving 13 potato cultivars (32 epidemics in total) were compared with model predictions using graphical and statistical tests. Fungicides were not applied in any of the epidemics. For the simulations, a host resistance level was assigned to each cultivar based on general categories reported by local investigators. For eight cultivars, the model predictions fit the observed data. For four cultivars, the model predictions overestimated disease, likely due to inaccurate estimates of host resistance. Model predictions were inconsistent for one cultivar and one location. It was concluded that the domain of applicability of LATEBLIGHT can be extended from the range of conditions in Peru for which it has been previously validated to those observed in this study. A sensitivity analysis showed that, within the range of values observed empirically, LATEBLIGHT is more sensitive to changes in variables related to initial inoculum and weather than to changes in variables relating to host resistance.

4.9.1.7 *Mathematical tools used in epidemiology of plant disease: Advantages and disadvantages*

The primary objectives of a mathematical tool is to help to understand how the main factors affect the plant disease development to prevent disease appearing and the development of sustainable strategies for disease management. Models have been created using mathematical tools since the apparition of the first model (Van der Plank, 1960; Van der Plank, 1963). Next, some advantages and disadvantages of common mathematical tools are presented.

Disease progress curves that use growth models have the advantage of generally describing the disease progress in a good way by adding few variables (Xu, 2006), but sometimes it is not adequate to describe certain kind of pathologies because it needs to incorporate extra variables that are determinant in the pathosystem. The disadvantages of these models are the assumptions that the epidemiologist needs to make about the variables used by this mathematical tool. The resulting bias depends on the assumptions made, and if a violation of the assumptions is presented, it needs to be reinterpreted to obtain a correct result. An example of these disadvantages is presented by Rotem *et al.* (1988) when disease intensity was masked by the loss of infected leaves and the appearance of new healthy leaves.

Ignoring the growth-host effects could underestimate or even obtain negative rate values because, may be, the rate of host growth is faster than the disease development rate (Kushalappa and Ludwing, 1982). If disease increase is not as rapid as plant growth, the proportion of diseased tissue on the plant will appear to decrease. Analogously, if defoliation occurs as a result of disease, the severity of infected plant will appear to decrease. LDE is one of the most commonly used mathematical tools in modeling epidemics because it allows that new terms be easily included as needed to answer the question about the details of the pathosystem (Madden, 2006). In general, it is too difficult to obtain analytical solutions for this mathematical tool, but numerical solutions can be obtained. An advantage of this mathematical tool is the intrinsic relationship with other tools here mentioned like disease progress curves and computer simulations because important components of disease dynamics can be modeled by using growth models that latter could be linked and simulated using computer programs. A disadvantage of LDE is that the formulated equations using this tool are extremely troublesome for mathematical analysis, which makes difficult the exploration of different biological properties of host and pathogens or to know the results of different control strategies, on long-term disease development (Madden, 2006).

AUDPC is extensively used in analyzing disease progress curves to make comparisons between treatments, to know the relationship between yield losses and disease, to make a genetic analysis, to evaluate the effectiveness of chemical, biological and cultural controls over disease progress, and also for knowing the resistance of certain species to plant pathogens (Jeger, 2004). A disadvantage of AUDPC is that it depends on a good fitted disease progress curve to obtain correct results; this approach also can yield misleading results when it is summarized over a particular period instead of over a complete period.

The most common mathematical tools used to model plant disease epidemics are Disease progress curve, LDE, AUDPC and Computer simulations. Nonetheless, there are also other ways to give an evaluation of disease dynamics based on visual, pictorial assessment and statistical tools (Contreras–Medina *et al.*, 2009). The variables used by mathematical tools are chosen for relevance, pathosystem features, and epidemiologist necessities, which, according to the model objective, the epidemiologist can choose the mathematical tool solely or jointly with other different mathematical tools that give a better description of the reality in order to obtain a more accurate evaluation of the plant disease epidemics. The possible combination of the tools here presented can be made by the conjunction of several disciplines like image processing which could serve to obtain a better objective estimation of variables that can feed to the model allowing obtain a better fitness.

The main goal of plant disease management is to keep disease development below an acceptable level and therefore to understand disease progress in quantitative terms; it will be necessary to find some kind of mathematical model to describe the epidemic. A model should show how some variable, preferably something that we can count or measure in the real world, changes with time. From the management perspective, keeping track of new infections perhaps would be the most useful, but actually observing these microscopic events in most cases would be too difficult to be practical. Monitoring inoculum is also very useful for disease management, and for many diseases practical techniques have been developed for quantifying inoculum.

4.9.1.7.1 Monocyclic inoculum production

In monocyclic epidemics we are interested primarily in the inoculum present at the beginning of each season (initial inoculum). If we let Q_1 represent the quantity of initial inoculum at the beginning of the current season, it will be equal to the quantity of initial inoculum at the beginning of the previous season, Q_0, plus the increment that has resulted from the pathogen's growth and development during the season, $Q_1 = Q_0$ + increment. The increment will be a function of the quantity of last season's initial inoculum and a reasonable approximation is to make it a simple proportion of last season's initial inoculum, KQ_0, where K is a proportionality constant, $Q_1 = Q_0 + KQ_0$.

Embodied in K are all the factors affecting survival and growth of the fungus, propagule production, dispersal of inoculum, and death of the fungus. The value of K depends on a large number of factors, including environmental conditions, crop development, and cultural practices (Fig. 4.11). If there is a net increase in the inoculum from one season to the next, K will be positive. If, on the other hand, there is a net loss of inoculum, such as might occur during a rotation to a nonhost crop, K would be negative. To describe the changes in the initial inoculum from one season to the next in a polyetic epidemic, one can generalize the subscript that indicates the season, $Q_{T+1} = Q_T + KQ_T$. This equation can be solved repetitively, changing the subscript T, indicating time, with each successive season and making the current value of Q_{T+1} the value of Q_T in the subsequent season. To simplify the equation, we assume a constant K (an average over many seasons). The above equation gives the following graph.

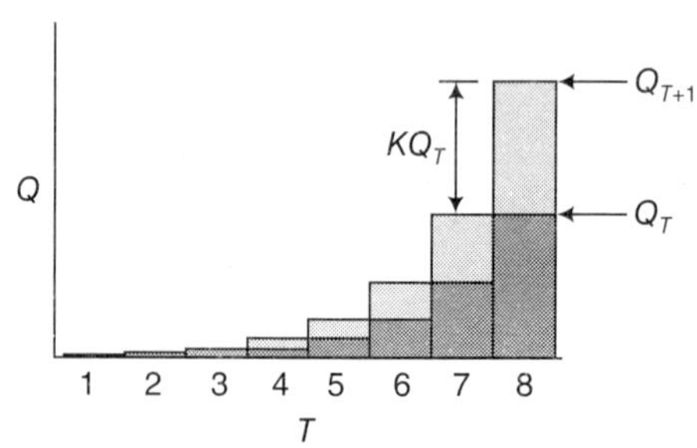

Fig. 4.11 Monocyclic inoculum production

Note that if K is positive, the increment (the light gray area on each bar) increases with the initial inoculum in each successive season, and the graph appears to curve upward.

4.9.1.7.2 Polycyclic inoculum production

For a polycyclic epidemic we can use essentially the same model as for a monocyclic pathogen viewed over several seasons, where instead of repeating the cycle season after season, we have many repeated cycles within the same season. The time step becomes days or weeks instead of years, and since the time steps are no longer necessarily one time unit (years), the time increment is given as ΔT – $q_{T+\Delta T} = q_T + q_T k \cdot \Delta T$ (Fig. 4.12). As a matter of notation, we will use a lower case q to represent the quantity of inoculum during the epidemic and a lower case k to represent the proportion by which inoculum increases in each time step. The units of k correspond to the units of T. For example, if the time is measured in days, the units of k would be proportion/day.

Inoculum production actually tends to occur irregularly in discontinuous, discrete infection periods of differing lengths, depending on the weather, and the value of k would likely be different for each infection period. However, in keeping with our objective of developing the simplest model possible to be useful as a management tool, we will simplify the above model by using uniform time steps and assuming a constant k. (Instead of having a k that varies according to the environmental conditions, one can use a value of k 'averaged' over the whole epidemic.) The above equation is rearranged to get: $q_{T+\Delta T} - q_T = q_T k \cdot \Delta T$. The change in the amount of inoculum in one time step, Δq, is simply the difference between the amount of inoculum at time T and the amount of inoculum at time $T + \Delta T$: $\Delta q = q_T k \cdot \Delta T$. Rearranging we get, $\Delta q / \Delta T = q_T k$. Now instead of advancing time in discrete steps, we will advance time continuously, making ΔT infinitesimally small: $dq/dt = qk$.

In this differential equation, dq is an infinitesimally small change in the quantity of inoculum, and dt is an infinitesimally small change in time. It tells us that the rate of change of the quantity of inoculum is proportional to the quantity of inoculum at any point in time. Using calculus, this equation can be integrated to $q = q_0 e^{kt}$

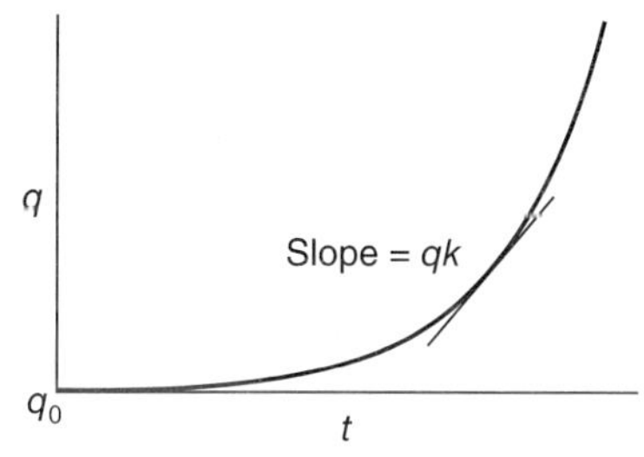

Fig. 4.12 Polycyclic inoculum production

This is the familiar exponential function, where q_0 is the initial inoculum, and e is the base of the natural (Napierian) logarithm. The instantaneous rate of change in q is dq/dt, the slope of the tangent to the curve at any point.

4.9.2 Modeling Disease Progress

While measurements of the quantity of inoculum provide a good assessment of the progress of the epidemic, direct observation of inoculum is not always feasible or even possible. With most plant diseases, however, we can see some kind of symptoms that allow us to count the numbers of plants infected or estimate the amount of tissue infected. To be sure, there is a time lag between infection and the appearance of visible symptoms, and the manifestation of symptoms can be affected by the environmental conditions, but, in general, the progress of symptoms parallels the progress of inoculum production. Therefore, the same models that we have used to describe the buildup of inoculum also can be used to describe disease progress. We will follow the notation convention of Van der Plank (1963) and use x to represent the proportion of infected plants or the proportion of infected plant tissue. Using a dimensionless proportion between zero and one to represent disease makes the mathematics far simpler than using specific units, such as the total population of infected plants or the total area of infected tissue, for example.

Monocyclic disease progress: If disease progress in monocyclic epidemics is linear, the slope of the disease progress curve is constant. Furthermore, if disease progress in a monocyclic epidemic is proportional to the amount of initial inoculum (which is itself constant during the epidemic), we can make the slope of the disease progress curve the product of initial inoculum and a proportionality constant. Therefore, we can describe a monocyclic epidemic with linear disease progress using the differential equation, $dx/dt = QR$, where, dx is an infinitesimally small increment in the proportion of disease, dt is an infinitesimally small time step, Q is the amount of initial inoculum, and R is a proportionality constant that represents the rate of disease progress per unit of inoculum. Since Q and R are both constant during the course of an epidemic, the slope, dx/dt, is constant, and disease progress is linear. Just as with the K constant in the model of monocyclic inoculum production, R has a value that represents the 'average' for the whole epidemic, a value that depends on many factors, such as aggressiveness of pathogen, the susceptibility of the host, the environmental conditions, etc. If the above differential equation is integrated, we get: $x = QRt$. Graphically a straight line which can be seen an intercept of zero and a slope of QR.

4.9.2.1 *Polycyclic disease progress*

In a model analogous to that of polycyclic inoculum production, the rate of change in disease is proportional to the amount of disease at any point in time. Therefore, in differential form, the equation to describe polycyclic epidemics is, $dx/dt = xr$.

As with the monocyclic model, x is a dimensionless proportion between zero and one and r is a constant that depends on the aggressiveness of pathogen, the susceptibility of the host, the environmental conditions, etc., averaged over the course of the epidemic. In this case, the slope, dx/dt, is proportional to x, and, therefore, disease progress increases with time at an increasing rate. In the integrated form, the model is $x = x_0e^{rt}$ where x_0 is the proportion of disease at the start of the epidemic and e is the base of the natural logarithm. Vanderplank (1963) called r the 'apparent infection rate' because it is based on the appearance of disease symptoms, which lag behind the actual infections. It is defined as the rate of disease increase per unit of disease and has the units of proportion per unit of time. The parameter x_0 is sometimes carelessly called initial inoculum, to which it is quantitatively related, but strictly speaking it is the initial disease (a proportion). Graphically the model has the familiar form of the exponential model.

4.9.2.2 *The upper limit to disease*

The above models assume unlimited growth of the disease, which, of course, is impossible; the proportion of diseased plants or diseased tissue cannot exceed one. We can adjust our models to address this issue by using a correction factor $(1 - x)$ to represent the proportion of healthy tissue remaining. A decrease in the remaining healthy tissue reduces the chance for new infections and, therefore, reduces the rate of disease progress.

As x approaches one, there is no healthy tissue left, and the rate of the epidemic slows to zero. The monocyclic model of disease progress, adjusted for the limit to disease is $dx/dt = QR(1 - x)$. Graphically an epidemic that starts out looking linear can be seen, but as x approaches 1, the slope decreases to zero.

In the polycyclic model a similar adjustment is made, $dx/dt = xr(1 - x)$. This model starts out approximately exponential, but its slope also decreases and approaches zero as x increases and approaches one. The result is a sigmoid-shaped curve. In reality, it is rare that disease incidence or severity exceeds 50%, and when it does, the disease progress curve is usually not quite sigmoid.

4.9.3 Estimating Model Parameters

4.9.3.1 Transformations

To apply our models to real epidemics, we would like to come up with numerical values for their parameters (R in the monocyclic model and x_0 and r in the polycyclic model). To estimate the parameters, we observe the epidemic, measure x at several times during the epidemic, and then plot x versus t. The difficulty comes in trying to fit nonlinear models to the data sets. It is far easier to transform the x's to get a linear model that can be fit with simple linear regression. The monocyclic model: In the case of the monocyclic model, if the observed x's are transformed to the natural logarithm of $1/(1 - x)$, and these transformed values are plotted against t, a straight line with a slope equal to QR is obtained. Then with an independent estimate of the initial inoculum, Q, R can be calculated.

4.9.3.2 Polycyclic model

If the observed x's in a polycyclic epidemic are transformed to the natural logarithm of $x/(1-x)$, and when the transformed values are plotted against t the result will be a straight line with a slope equal to r and an intercept equal to the natural log of $x_0/(1-x_0)$. In fitting models to observed data, it is important to select the model based on the known biology of the pathogen rather than simply on the shape of the curve. Likewise, one must not attempt to make inferences about the biology of the pathogen based on the shape of the curve and which model gives the 'best fit' of the data. There is random variability in each observation, and in the transformed models the data points on the ends of the line have undue weight in determining the fit to the model. It is quite possible to have a data set that fits both the monocyclic and polycyclic models equally well or to have a data set from a known monocyclic epidemic that gives a better fit to a polycyclic model and vice versa.

Plant pathogenic bacteria are heterotrophic organisms, which can develop on host plants as parasites They are not able to penetrate directly plant tissue. Infection usually occurs through natural openings like stomata, hydathodes, and nectaroides as well as injuries of various origin. The important places of infection appeared to be leaf scars (Piotr Sobiczewski, 2008).

After entering the plant and adapting to the new environment, bacteria start to multiply in intercellular spaces following dissemination into various plant tissues. Their further development is conditioned mainly by temperature and humidity. Optimum temperature and high relative humidity favor increase of bacteria population and appearance of disease symptoms as well as their intensity. Most of plant pathogenic bacteria can survive on plant surfaces also as epiphytes and some as saprotrophs. However, many bacteria infecting above ground organs are not able to survive on dead plant debris. When the infected tissue dies, the number of bacteria rapidly decreases, which leads to their total decline. In such situations, infection source can be drastically reduced, which is very important from the practical point of view. The bacteria infecting above-ground part of plants are disseminated by wind, rain, insects, and birds as well as infected plants. Pathogens of root system are spread mainly by soil solution and soil fauna. Man plays an important role through management practices as well. Some bacteria can be disseminated with nursery material, like seeds, scions, and seedlings. Certain bacterial pathogens like *Erwinia amylovora* can survive and multiply on living plant tissue only. On the other hand, pathogens like *Agrobacterium tumefaciens* or *Ralstonia solanacearum* are typical soil organisms. Although their main development occurs in host-plants, they can also survive for a long time in soil. The length of this period depends on abiotic factors like temperature, humidity, and availability of nutrients as well as on biotic factors, especially antagonistic bacteria and fungi. Necroses, blights, cankers, wilting, soft rots and tumors are the main symptoms of bacterioses. Thus, plant pathogenic bacteria can be divided into necrogens, macergens, and oncogens. The range of host plants of particular pathogens varies from very wide – consisting of several hundred plant species (like *Agrobacterium tumefaciens*) to very narrow, e.g., *Xanthomonas fragaria,* the pathogen of strawberry only.

Although seed production has been moved to semiarid regions to escape seed borne pathogens, seed-borne bacterial diseases continue to be problematic and cause significant economic losses worldwide (Gitaitis and Walcott, 2007). Infested seeds are responsible for the reemergence of diseases of the past, movement of pathogens across international borders, or the introduction of diseases into new areas. Considerable attention has been paid to improving the sensitivity and selectivity of seed health assays by using techniques such as flow cytometry and the polymerase chain reaction. There has also been progressing in understanding infection thresholds and how they influence seed sample size determination and ultimately the reliability of seed health testing. Disease development and dissemination of pathogens from contaminated seed lots can be predicted using formulas that take into account inoculum density and environmental pressures. In general, seeds infested with bacterial pathogens are distributed within a Poisson distribution. In a subset of contaminated seeds, bacteria are distributed in non-Gaussian distributions, for example, a lognormal distribution.

In five field experiments, carried out from 2002 to 2004, the infestation of potato plants and the contamination of soil with *Clavibacter michiganensis* subsp. *sepedonicus,* the causal agent of bacterial ring rot of potato, were investigated throughout the growing season (Kaemmerer *et al.,* 2007). Aiming at a detailed description of disease progress, for the first time potato tubers, stems and roots as well as soil were examined in parallel. The development of bacterial populations in plant organs and soil differed considerably within and between the experiments. Roots showed the highest and tubers the lowest incidence of infection of plant organs based on polymerase chain reaction (PCR) and immunofluorescence antibody staining (IFAS). Logistic regression analyses indicated a significant increase in the probability of infestation (in the case of soil: contamination) during the growing season for all investigated plant parts of highly infected plants and the respective soil samples. Results were discussed in terms of the suitability of particular plant organs and the determination of the best sampling time point for official testing. Furthermore, the risks are evaluated to contaminate machinery on the field and to spread bacterial ring rot through the soil.

Mathematical models of plant-virus disease epidemics were developed by Zhang and Holt (2001) where cross protection occurs between viruses or virus strains. Such cross protection can occur both naturally and through artificial intervention. Examples of diseases with continuous and discontinuous crop-host availability were considered, viz., citrus tristeza and barley yellow dwarf, respectively. Analyses showed that, in a single host population without artificial intervention, the two categories of host plants, infected with a protecting virus alone and infected with a challenging virus, could not coexist in the long term. For disease systems with continuous host availability, the virus (strain) with the higher basic reproductive number ($R(0)$) always excluded the other eventually; whereas, for discontinuous systems, $R(0)$ is undefined and the virus (strain) with the larger natural transmission rate was the one that persisted in the model formulation. With a proportion of hosts artificially inoculated with a protecting mild virus, the disease caused by a virulent virus could be depressed or eliminated, depending on the proportion. Artificial inoculation may be constant or adjusted in response to changes in disease incidence. The importance of maintaining a constant level of managed cross protection even when the disease incidence dropped was illustrated. Investigations of both pathosystem types showed the same qualitative result: that managed cross protection need not be 100% to eliminate the virulent virus (strain). In the process of replacement of one virus (strain) by another over time, the strongest competition occurred when the incidence of both viruses or virus strains was similar. Discontinuous crop-host availability provided a greater opportunity for viruses or virus strains to replace each other than did the more stable continuous cropping system. The process by which one Barley yellow dwarf virus replaced another in New York State was illustrated.

In Africa, only about five viruses (Rice stripe necrosis furovirus, Rice crinkle disease, Maize streak geminivirus strain A, African cereal streak virus, and Rice yellow mottle sobemovirus) have been known to infect rice in nature (Aboa and Ali Fadhilab, 2003). Most of the rice viruses are transmitted by either plant- or leafhoppers. A few are transmitted through mechanical sap inoculation, contact, soil, and seed while other viruses are transmitted by rice mealy bugs, beetles, aphids, and fungus (*Polymyxa graminis*). The causes of the virus epidemics and their pests' outbreaks are primarily based on the intensification of rice cultivation and the introduction of new, high-yielding varieties. It is also due to the increased use of irrigation, fertilizer, and pesticides, changing cultural practices, crop monoculture, and mechanical contamination. The management of rice virus diseases includes breeding and screening for varietal resistance, efficient cultural practices, and integrated pest and disease management strategies. Priority research gaps were also enumerated.

With regard to the epidemiology of diseases caused by the phytonematodes, the vast majority of plant parasitic nematodes possess a hollow stylet in their head that they use to puncture root cells and feed. The stylet can also be used to secrete enzymes into the plant that are important in the feeding process. Stylet structure can vary greatly in their size and shape between nematode genera and plays an important role in nematode taxonomy. Phytonematodes differ in how they feed on plant root. Some nematodes spend their entire lives in the soil feeding on roots from a single location or multiple locations as they move through the roots zone and are called ectoparasites. Other nematodes may spend most of their lives feeding inside the roots and are called endoparasites. Nematodes may compromise yield by competing with grain fill for photosynthates. In addition, the damage that they cause to roots may also impede the plant's ability to conduct water, even when adequate soil moisture is available.

The diversity of plant parasitic nematode genera can also dictate their distribution. For example, large genera such as *Longidorus* spp. and *Belanolaimus* spp. require sandy soil textures for movement. The damage caused by other smaller nematodes may also be amplified when the crop is grown in sandy soil, but many nematodes are not restricted by soil type in their distribution.

Most phytonematodes in soil do not usually migrate more than 1-2 meters from the root zone that they infect. In plantings, where root grafts may occur, as in the case of fruit trees, the nematodes may travel from plant to plant through roots. In most cases, the movement of lesion nematodes is defined as 'contagious,' small foci of infested areas gradually enlarge to encompass significant areas of disease. Areas of disease become more pronounced in adverse environmental conditions such as water and nutrient stress or if secondary pathogens simultaneously infect the roots. The spread of these nematodes within fields is usually accelerated by the cultural practices of the grower, such as soil cultivation. Nematodes can be introduced to noninfested sites by poorly sanitized farm equipment and contaminated planting stock (tubers, suckers, bulbs, rhizomes, seedlings), in the case of endoparasites.

4.10 EPIDEMIOLOGICAL MODELS

An epidemic is the progress of the disease in time and space. Each epidemic has a structure whose temporal dynamics and spatial patterns are jointly determined by the pathosystem characteristics and environmental conditions. One of the important objectives in epidemiology is to understand such spatiotemporal dynamics via mathematical and statistical modeling. A model is a simplification of reality and attempts to summarize the main processes, to put forward hypotheses and to verify their coherence and consequences. It also represents a trial to determine the minimal hypotheses which would allow minimal mathematical representation of real processes. In epidemiology, modeling aims to understand the main determinants of epidemic development to develop sustainable strategies for strategic and tactical management of diseases (van Maanen and X.-M. Xu, 2003).

Epidemiological models can be classified in several ways. For convenience, Kranz and Royle (1978) classified them into three types, descriptive, predictive and conceptual, according to their main objective. Descriptive models provide hypotheses or generalize experimental results, but they do not usually reveal the mechanisms underlying the processes. Predictive models, which are also descriptive, allow the prediction of the occurrence and the severity of epidemics. Both descriptive and predictive models use mathematical tools, such as simple or complex functions, regression, and differential equations, or simple decision models. The conceptual models, also known as explanatory or analytical models, allow the identification of problems by distinguishing cause from effect and quantify the effects of specific events on epidemic development. They are constructed as representations of underlying biological and ecological processes. These models may eventually lead to the development of complex simulation models. It should be pointed out that models can be disease-specific, but can also be very general. A descriptive model is often concerned with understanding and predicting the development of specific diseases, and thus is generally used for assisting growers in making tactical decisions in managing diseases. A conceptual model is often concerned with the theoretical understanding of generic features of epidemic development and thus is used more for making policy and strategic decisions. Recent advances in computer tools have made mathematical/statistical modeling more accessible and have led to the development of more complex models for many diseases.

Although early investigations of the epidemics caused by plant pathogens seldom included the demographics of the host population (Nik J. Cunniffe and Christopher A. Gilligan, 2010), replenishment of susceptible hosts is common in contemporary models. The simplest, and often the default, approach to modeling host growth is to introduce an exponential function for the net birth of susceptible hosts into an SIR epidemic model. A similar approach is widely used to model epidemics in animal populations. Unlimited growth of the host population is prevented by the density-dependent effect of the pathogen in killing hosts. Biological plausibility collapses; however, in the absence of the pathogen, as then the host population grows without bound.

Frequently, and especially when considering infection of individual leaves or roots, but also for invertebrate and some vertebrate and human populations; bounded growth functions are introduced to provide more realistic approximations of host–pathogen dynamics. Of particular interest in the case of viral pathogens of forest insect pests — which have been whether or not subjected to cyclic dynamics - are promoted by adding density dependence in host growth. In plant epidemiology, the focus has been on improving the realism of the approximation to host demography, by adding bounds to reflect the limited carrying capacity of the field and representing phenomena including disease-induced host responses. Epidemics of plant pathogens can be considered to be a practical platform and the context for theoretical work, as in addition to exemplifying systems for which host demography can markedly affect the outcome of epidemics, previous compartmental models of plant disease have been extensively and successfully confronted with data from both experimentally controlled systems and field trials. Two commonly used classes of functions in the context of host dynamics are typified by the monomolecular function and the logistic function. The monomolecular function introduces a simple upper bound on host growth while the logistic function introduces a nonlinear feedback in the net birth rate of the host population. It was shown that, although the parameters for host growth do not appear in the criteria for pathogen invasion, host demography does affect other pathogen dynamics, notably persistence and the approach to endemic levels. The responsiveness of the pathogen to control varies with host demography, even when the control is effected through changes in the epidemiological parameters and progressive increases in the intensity of disease control can lead to the counterintuitive result of an increase in endemic levels of the disease (Nik J. Cunniffe and Christopher A. Gilligan, 2010).

4.10.1 Problem Specification

Since models are most useful when directed at a clearly specified and well-defined problem, the crucial first step in modeling is to precisely define the question(s) to be addressed. The nature of the problem is likely to determine the most appropriate modeling techniques to use (Sutherst, 1993). In fungal epidemiology, possible objectives for modeling include (Norton *et al.*, 1993), predicting the timing of an event, such as when a disease infection is likely to occur; predicting the scale of an event, such as the severity of disease infection or crop loss, estimating the frequency or the probability of an event, such as monocyclic or polycyclic epidemics; and assessing and comparing the performance of different management strategies.

4.10.2 Mathematical Representation of Epidemic Development

Overall, modeling can be divided into three steps, viz., model development, model analysis, and hypothesis testing. While developing a model, biological characteristics of the pathosystem are expressed as mathematical relationships. In model analysis, epidemic dynamics is investigated in relation to the parameters of interest (or variables, which may be formulated as functions of external factors such as rain and temperature). Finally in hypothesis testing, the results from the model analysis are used to test or verify whether or under what conditions the hypothesis (i.e., the specified problem/question) is valid (van Maanen1 and X.-M. Xu, 2003). The mathematical expression of biological features of the system is critical, since it will, more or less, determine the techniques to be used in model analysis and hypothesis testing.

Alternatively, the mathematical relationships between the model parameters and treatment/environmental factors can be incorporated into the growth curve models and fitted to the observed data directly. More complex model fitting procedures are required for the latter approach. Most analyses of disease progress data rely on time as the independent factor. This may not be appropriate when data are collected in different years, seasons, locations, etc. A measure of heat-sum or degree-days provides an alternative method. This assumes that temperature is the most important factor driving growth rate of the host, the pathogen, and the disease. This method was used to model various epidemics, such as the cotton-*Verticillium* wilt system (Gutierrez *et al.*, 1983), powdery mildew on tomato, and more recently, take-all on wheat. Many other modifications to standard growth models are possible. For example, to take into account the temporal variability of host susceptibility to pathogens, Lalancette and Hickey (1986) modeled disease progress as a function of host growth, which was used to represent biological time. Van der Plank (1963) introduced a correction factor for the rate of disease increase based on the exponential change in mass of susceptible host tissue.

Plant disease mathematical models including continuous cultural control strategy and impulsive cultural control strategy were proposed and investigated (Meng and Li, 2010). This novel theoretical framework could result in an objective criterion on how to control plant disease transmission by replanting of healthy plants and removal of infected plants. Continuous replanting of healthy plants and removing of infected plants was taken. The existence and stability of disease-free equilibrium and positive equilibrium were studied, and continuous cultural control strategy was provided. Plant disease model with impulsive replanting of healthy plants and removing of infected plants was also considered. Using Floquet's theorem and small amplitude perturbation, the sufficient conditions under which the infected plant free periodic solution was locally stable were obtained. Permanence of the system was investigated. Under certain parameter spaces, it was shown that a nontrivial periodic solution emerges via a supercritical bifurcation. The findings were confirmed by means of numerical simulations. The modeling methods and analytical analysis can serve as an integrating measure to identify and design appropriate plant disease control strategies.

Mathematical models have been used to extend our understanding of plant disease epidemiology on a number of fronts. The models have allowed the formulation of hypotheses which can be tested in the field. These provide an opportunity for a more rational use of resources on expensive field trials and represent a step towards more sustainable and biologically sound control measures.

A simulation model of *Mycospaerella* on banana was developed. It was observed that an increase planting distance was much more likely to reduce disease caused by *M. musicola* than that caused by *M. fijiensis*. This might play a part, along with differential resistance, in the more severe damage caused by *M. fijiensis* to sparsely planted cooking and subsistence bananas. Another model was developed to examine the consequences of mixing systematic fungicides on the development of resistance. Results suggested that with fungicide mixtures, the dose of each of the components should be reduced to give a combined level of active ingredient similar to give a combined level of active components of the mixture when used alone. The model also indicated simple criteria by which fungicides suitable for use in mixtures may be identified. Two theoretical population models, developed at the University of Reading were studied in which seasonality had profound effects. The first model of competition among pathogen species predicted competitive exclusion of one species by another if interspecific competition is stronger than intraspecific. However, the introduction of an annual catastrophe (e.g., harvesting of the host), greatly relaxed the conditions for the coexistence of two or more species.

A second model of infection of a fungus by a mycovirus allowed only two solutions: either the virus dies out, or the fungus becomes totally infected. Introducing seasonality into this model could induce chaotic dynamics, with the maximum disease severity in any one year uncorrelated with that in the next. A stochastic, spatial model was developed based on the cocoa (*Theobroma cacao*)/witches' broom (*Crinipellis perniciosa*) pathosystem, where airborne basidiospores caused infections which developed into witches' brooms that subsequently became infectious themselves the following year. Simulated gradients of infection away from an area of diseased cocoa were examined under different conditions.

The modeling was useful in the interpretation of disease gradient data obtained in the International Witches' Broom Project (IWBP), where there was considerable speculation regarding the reasons for the wider variety of gradients obtained at different sites. The model showed that the dispersal parameter, reflecting differences in the scale of air movements, markedly affected the slope of the gradient, while the infection parameter determined the displacement of the front away from the source. Aging of the inoculum sources was found to reduce only slightly the advance of the front year by year. Another question raised by the IWBP data was the likely effect that the area of the sources might have on the gradient. With a small source, marginal increases in source size had a similar effect on raising the infection probability, i.e., the front was advanced, but the gradient did not change. As the source size grew, a point was reached where further increases had little or no effect. Studies on phytosanitation showed high efficiency of the removal of infected tissues was necessary to restrict disease spread to any significant degree. The effect of wind and splash on the passive dispersal of microbes was also reviewed. The review drew together a diverse area of multidisciplinary work and made an important contribution to the disease epidemiology literature.

Knowledge of the incubation period of infectious diseases (time between host infection and expression of disease symptoms) is crucial to the epidemiological understanding and the design of appropriate prevention and control policies. Plant diseases cause substantial damage to agricultural and arboricultural systems, but there is still very little information about how the incubation period varies within host populations. Leclerc *et al.* (2014) focused on the incubation period of soil-borne plant pathogens, which are difficult to detect as they spread and infect the hosts underground and above-ground symptoms occur considerably later. They conducted experiments on *Rhizoctonia solani* in sugar beet, as an example pathosystem and used modeling approaches to estimate the incubation period distribution and demonstrated the impact of differing estimations on the epidemiological understanding of plant diseases. They presented measurements of the incubation period obtained in field conditions, fit alternative probability models to the data, and show that the incubation period distribution changes with hosts' age. By simulating spatially-explicit epidemiological models with different incubation-period distributions, they studied the conditions for a significant time lag between epidemics of cryptic infection and the associated epidemics of symptomatic disease. The sensitivity of this lag to differing distributional assumptions about the incubation period (i.e., exponential vs gamma) was examined. It was demonstrated that accurate information about the incubation period distribution of a pathosystem could be critical in assessing the true scale of pathogen invasion behind early disease symptoms in the field; likewise, it could be central to the model-based prediction of epidemic risk and evaluation of disease management strategies. The results highlighted that reliance on observation of disease symptoms could cause a significant delay in detection of soil-borne pathogen epidemics and mislead practitioners and epidemiologists about the timing, extent, and viability of disease control measures for limiting economic loss.

Models which simulate the evolution of a plant disease during the season give important information to assess the seriousness of the situation. This activity precedes the choice of an appropriate action to be implemented for reducing the economic damage. Having effective models is a critical issue in modern agriculture, especially in Integrated Protection and Organic Farming, which rest on a set of plant disease management practices with low environmental impact. Considerable effort has gone into the study of models for the simulation of plant diseases evolution. Phenology models, population, and epidemiological models have been developed for several, diffused diseases (Susi *et al.*, 2004). Problems are still open, especially when the aim is that of including these models into decision support systems at the use of producers and agronomists. Phenology and population models have to be developed, choosing the most promising techniques. Moreover, requirements such as that of providing justification to the user of the results computed by a model or making the user aware of the accuracy of the model results become critical.

4.10.3 The Relative Infection Measure (RIM) Model

The RIM model is a mathematical method able to simulate the development phases of the *Venturia inæqualis*, an apple fungus present in many European Countries (Trapman, 1994). The model uses differential equation systems. The target of the model is to describe the infection risk during spring time. In particular, the model is based on two different submodels that are able to detect the development of two different stages for the fungus: the ascospore maturity (determining the quality and the quantity of the risk); the germination and the penetration of the fungus in the fruit (determining the effective damage for the orchards). Both stages are strictly related to the temperature, the rain, the photoperiod for a given latitude and the quantity of mature ascospores in the environment, a parameter not considered by previous models (Mills and La- Plante [1951]). The output of the entire process is the severity risk index for a given period of time including the model into an application. The mathematical model has been included in the RIMpro, a software system, experimented and used in different European Countries. The system is able to compute the risk degree related to an infection, given the meteorological values and the presence of ascospore ready to pass from the first stage (ascospore maturity) to the second and most dangerous stage (the germination and penetration). The tuning of the RIMpro respect to a specific geographical area can be synthesized in two main steps. Initially it uses a general model based on two sets of parameter: a first set of fixed parameters that are part of the differential equations structure, and a second set of modifiable parameters useful when the model has to be adapted to a particular area, describing environmental variables like temperature, rain, humidity, presence of ascospore in a given period of time. During the practical use, the set of possible values for the modifiable parameters are identified taking into account the experience and the data coming from the entomologists and the agronomists working in a given area. The use of the system in different environments produces information about the set of fixed parameters giving a validation of the mathematical model structure. Moreover, the feedback from the users contributes also to the software system improvement from the point of view of the model usability and integration with other software systems devoted to the automatic acquisition of the model input parameters values, for example, the acquisition and selection of the meteorological values.

4.10.4 Model for Viral Diseases

Rarely have plant virologists, vector entomologists, and crop specialists worked together in search of sustainable management practices for viral diseases. Historically, modeling approaches have been vector-based dealing with empirical forecasting systems or simulation of vector population dynamics. More recently, epidemiological models, such as those used in human/animal epidemiology, have been introduced in an attempt to characterize and analyze the population ecology of viral diseases. The theoretical bases for these models and their use in evaluating control strategies in terms of the interactions between host, virus and vector were considered (Jeger *et al.*, 2015). Vector activity and behavior, especially in relation to virus transmission, are important determinants of the rate and extent of epidemic development. The applicability and flexibility of these models were illustrated by reference to specific case studies, including the increasing importance of whitefly-transmitted viruses. Some outstanding research and methodological issues are considered.

Predicting epidemics of plant virus disease constitutes a challenging undertaking due to the complexity of the three-cornered pathosystems (virus, vector, and host) involved and their interactions with the environment. A complicated nomenclature is used to describe virus epidemiological models. A review by Roger *et al.* (2010) explains how the nomenclature evolved and provides a historical account of the development of such models. The process and steps involved in devising models that incorporate weather variables and data retrieval and are able to forecast plant virus epidemics effectively were explained. Their application to provide user-friendly, Internet-based decision support systems (DSSs) that determine when and where control measures are needed has been described. Finally, case studies of eight pathosystems were provided representing different scenarios in which modeling approaches have been used with varying degrees of effectiveness to forecast virus epidemics in parts of the world with temperate, Mediterranean, subtropical, and tropical climates.

In plant-virus disease epidemiology, dynamical models have invariably incorporated a bilinear inoculation rate that is directly proportional to both the abundance of healthy (susceptible) hosts and the abundance of infective vectors. Similarly, the acquisition rate is usually assumed to be directly proportional to the abundance of nonviruliferous vectors and that of infectious hosts (Zhang *et al.*, 2000). These bilinear assumptions have been questioned for certain human diseases and infection rates that incorporate power parameters of the variables have been proposed. Infection rates for plant-virus diseases that are of a more general form than the familiar bilinear terms are examined. For such diseases, the power parameter can be regarded as a measure of the spatial aggregation of the vectors or as a coefficient of interference between them, depending on the context. Field data of cassava mosaic virus disease (CMD) incidence were examined. When vector population density and disease incidence were high, disease progress curves over the first six months from planting could not be explained using models with bilinear infection rates. Incorporation of the new infection terms allowed the range of observed disease progress curve types to be described. New evidence of a mutually beneficial interaction between the viruses causing CMD and the whitefly vector, *Bemisia tabaci*, showed that spatial aggregation of the vectors was an inevitable consequence of infection, particularly with a severe virus strain or a sensitive host. Virus infection increased both vector fecundity and the density of vectors on diseased plants. It was postulated that this enhanced disease spread by causing an increased emigration rate of infective vectors to other crops. Paradoxically, within the infected crop, vector aggregation reduced the effective contact rate between vector and host and, therefore, the predicted disease incidence was less than when a bilinear contact rate was used.

4.10.5 Challenges in Modeling Plant Diseases

The underlying structure of epidemiological models and the questions that models can be used to address do not necessarily depend on the host organism in question. This means that certain preoccupations of plant disease modelers are similar to those of modelers of diseases in human, livestock and wild animal populations (Nik J. Cunniffe *et al.*, 2015). However, a number of aspects of plant epidemiology are very distinctive, and this leads to specific challenges in modeling plant diseases, which in turn sets a certain agenda for modelers. Nik J. Cunniffe *et al.* (2015) outlined a selection of 13 challenges that are specific to plant disease, as mentioned below.

(i) Linking epidemiological models to crop yield and ecosystem services
(ii) Temporal changes in host availability from plant organs to populations
(iii) Capturing host spatial structure even when data are limited beyond a single species
(iv) Multiple and alternate hosts
(v) Spillover and community ecology
(vi) Realistic dispersal models
(vii) Including meteorological and anthropomorphic drivers
(viii) Network models for human-mediated spread accounting for time-varying infectivity
(ix) Effects of vector preference on transmission beyond a single species' multiple strains

(x) Multiple pathogens and evolution
(xi) Using models to optimize detection
(xii) Optimizing dynamic controls in heterogeneous systems accounting for economics-moving optimal control theory to realistic landscapes
(xiii) Use of models by policy makers and stakeholders

Apart from these, others include models of biosecurity in the plant trade, models of disease for use in precision agriculture, methods to parameterize and test landscape scale models, making use of population geneticists' expanded molecular toolkit for assessing allelic and genotypic frequency variation in populations to link population models with the 'omics' revolution, using insights from soil imaging to improve models for soil-borne pathogens, modeling disease complexes and linking models of control to those from social science. This list is restricted to plant-oriented challenges. Nevertheless, there is a commonality of purpose and approach between models for analysis, prediction, and control of plant, human, livestock and wild animal epidemics. Much work during the past two decades has emphasized these similarities, but arguably there is still insufficient 'cross-talk' between modeling disciplines (Nik J. Cunniffe *et al.*, 2015). *Building better links between plant disease modelers and the wider epidemic modeling community is a significant challenge of its own.*

4.11 COMPUTER SIMULATION

Many computer simulation models have been developed in the past decades. Computer simulation is, in general, a natural extension of LDE modeling. In a computer simulation, model parameters in LDE are often assumed to be functions of external factors such as temperature and humidity. These functions can either be of simple linear type or complex nonlinear type. Computer simulation can be used to study both theoretical and applied problems. Using a stochastic simulation model, the relationships of spatiotemporal statistics with underlying biological, physical and biological factors have been successfully studied (Xu and Ridout, 1998; 2000; 2001). One of the first programs was the EPIDEM written by Wagonner and Horsfall (1969), and it was designed to simulate epidemics of early blight of tomato and potato caused by the fungus *Alternaria solani* and it resulted from modeling each stage of the life cycle of the pathogen as a function of the environment. Subsequently, several kinds of computer programs were created like MYCOS for *Mycospaherella* blight of chrysanthemums (McCoy, 1976), EPIMAY for southern corn leaf blight (Waggoner *et al.*, 1969), EPIVEN for apple scab caused by *Venturia inaequalis* (Kranz *et al.*, 1973) and a more general and flexible plant simulator called EPIDEMIC (Shrum, 1975) because it could be easily modified for another plant-pathogen system. Improvements of previously developed computer models continue being made and used for new applications (Jeger, 2004), for instance, epidemic model EPIMUL has been used to simulate the efficacy of various mixtures in relation to a range of different pathogen characteristics (Lannou *et al.*, 1994).

Nowadays, computer programs are used to support other techniques for predicting plant disease and to measure the impact of pathology on certain plant species based on particular variables like leaf wetness, duration of leaf wetness, primary inoculums, temperature, rainfall, and moisture. One example of this is PLASMO, which is used to measure the impact on the quality of *Plasmora viticola* due to leaf wetness duration (Marta *et al.*, 2005; Orlandini *et al.*, 2008). Spatio-temporal simulation models were proposed as EPIMUL (Kampmeijer and Zadoks, 1974), which laid the foundations for further developments (van Maanen1 and X.-M. Xu, 2003). Jeger (1986) discussed the advantages and disadvantages of the simulation approach in comparison to the analytical approach.

4.11.1 Visual Evaluation, Pictorial Assessment, and Statistical Tools

All the previously mentioned mathematical tools display descriptions of the disease dynamics and also give important characteristics about the epidemics as output (e.g., length of different infectious event, severity and infection intensity during growing season, etc.). Several works, when measuring plant resistance to certain kind of pathogens use scales to evaluate the severity of the plant or usually make the evaluation based on visual inspection. Anaya-Lopez *et al.* (2003) used a nine-level scale to measure qualitatively the severity of pepper disease provoked by geminiviruses, this scale has been previously used in other works (Contreras–Medina *et al.*, 2009). Another kind of evaluation, called pictorial disease assessment, has also been created based on standard diagrams that illustrate the development of the stages in the disease on small simple units (leaves and fruits) or on large composite units such as branches or whole plants, for instance, the modified Cobb scale, which is used to determine the percentage of possible rusted tissue. Holb *et al.* (2003) investigated the relationship between disease measurement, disease incidence, severity of apple scabs and their implications for the development of predictive models and thresholds levels; the disease assessment based on leave severity and fruit severity are made using a mathematical equation which classifies the fruit and leave severity into a six- and seven-level scale, respectively. Recently, certain works determine visually the severity of infections by evaluating the leaf area that presents oil spots, where the ability of the surveyors was standardized firstly by mean of cross-tests and also the severity was classified into a five-level scale. It is important to mention that there are other tools used in plant pathology. These are the statistical tools such as: including survival analysis, nonparametric analysis of disease association, multivariate statistical methods, neural networks, meta-analysis, Bayesian statistics, generalized linear models, linear mixed models, decision theory, etc.

These tools use normally distributed ordinal and discrete data, for instance, the linear mixed model, whereas the generalized linear mixed model works with discrete data. The statistical tools are employed depending on the necessity. For example, multivariate statistical tools are used to reduce the space of *n* variables into a low-dimensional space, to assess group differences and variable contribution and to describe and predict the relationship between two sets of variables. Survival analysis is a technique for analyzing data sets in which the time until an event occurs is the dependent variable containing censored observations (Contreras–Medina *et al.*, 2009).

4.11.2 Using Variables

A proposed model that uses mathematical tools should include the variables that represent the key features in the development of epidemics according to the objective of the investigation. For instance, temperature, moisture, susceptibility and resistance of host, initial inoculum, latent and infectious period and efficiency of conidia are examples of commonly included variables in used models to describe plant disease epidemics (Contreras-Medina *et al.*, 2009). The variables that the epidemiologists include depend on the pathosystems, the nature of the problem and the answers to be found (Xu, 2006). The central principle of plant pathology is the disease triangle, where the development of plant disease requires these three equally-relevant components, viz., a susceptible host, a virulent pathogen and a favorable environmental condition.

4.11.2.1 Variables included into different kinds of models

Some examples of variable included in various models are initial inoculum, host growth characteristics and temperature, rate of lesion increase, conversion rate of infectious into postinfectious tissue, initial proportion of infectious area, initial proportion of disease free area; latent infection, visible leaf area, infectious leaf area, no infectious leaf area, infection efficiency of conidia, incubation progress, latency progress, removal, colony growth, air temperature, rainfall, relative humidity, leaf wetness duration, initial inoculum, leaf area, spot area, sporulation area, viable spores and incubation, wind speed and direction, vapor pressure deficit, total duration rainfall, low growth rate, disease carrying capacity, infectious period, location and onset of primary infection, precipitation leaf wetness duration, and plant characteristics including stem density, plant geometry, mean distance between nodes and leaf area. Several kinds of variables have been included into certain models which are related to the development of plant diseases. Table 4.1 presents some examples of proposed models using mathematical tools here presented, and the variables included in the models.

As it can be seen, the variables are chosen according to the studied pathosystem and the objectives of the epidemiologist. It is generally assumed that the environment is the driven force for a disease. Several works use as driving variables those concerning weather like temperature, rainfall, moisture, leaf wetness, wind direction, radiation, among others. From this, pathogen variables that are involved directly with epidemic development can be formulated as a function of driving variables, e.g., latent and infectious period, initial inoculation, oil spoil survival (Calonnec *et al.*, 2008). Also, variables relating to host dynamics should be included to know how the pathogen dynamics are linked to the host dynamics and how pathogen affects the growth and reproduction of the host (Contreras-Medina *et al.*, 2009).

The computer simulation is a general natural extension of LDE because it depends on computations to obtain a numerical solution. In a computer simulation, some model variables in LDE are often assumed to be functions of external factors like temperature, humidity, radiation, wind direction, etc. This mathematical tool can be used to study theoretical and practical problems of certain kinds of pathologies (Agrios, 2005). One disadvantage of computer simulation is that generally each simulator just incorporates one disease to be simulated, restricting its functionality and also, a computer is required to achieve the task, having an intrinsic economical cost. Disease evaluation based on visual estimation has a clear disadvantage, because of the variation between different points of view are given by different evaluators. Several works have been developed where visual evaluation of host severity is made and classified into a multilevel scale according to the kind of damage that is presented in the host due to the disease evolution (Contreras-Medina *et al.*, 2009). This also provokes unsuitable strategies to forecast and prevent disease due to the subjectivity of the estimation. Similarly, pictorial assessment presents the disadvantage of subjectivity because it is a tool based on a visual evaluation that decreases the reliability of an evaluation.

Statistical tools are attractive because they allow to obtain a model where some drawbacks exist, such as when little is known about the structural form of complex relationships between response variables, when theoretical data presented by other investigation needs to be used or when the calculation probabilities for the parameter of interest needs to be calculated based on empirically derived prior probabilities in conjunction with the conditional probability of each possible outcome (Scherm *et al.*, 2006). The statistical tools have the disadvantage of supposing that the data are normally distributed, and also the theoretical models proposed by using this tool sometimes have nonexpected results in practice (Scherm *et al.*, 2006). Table 4.2 summarizes the advantages and disadvantages of the common tools used to describe the dynamics of plant disease.

Table 4.2 Advantages and Disadvantages of Common Tools Used to Describe Plant Disease Dynamics

Tool	*Advantages*	*Disadvantages*
Disease progress curves.	Often with two or three variables describe satisfactorily the disease.	Needs to make assumption to correct interpretation of results, sometimes ignore key components that affect disease development.
Linked Differential Equation.	New terms can be easily included into the model as needed.	Equation generated are extremely troublesome for mathematical analysis.
AUDPC (Area under disease progress curve).	It is a very useful alternative to fitting growth models, it can work with growth models.	Needs that the amount and duration of disease be proportional to damage, it may give misleading results when AUDPC is summarized over another period than just over a particular period.

Contd...

Contd...

Visual evaluation.	Practical evaluation, does not need a personal computer (PC), only evaluators are needed.	Subjectivity.
Pictorial assessment.	Practical evaluation, does not need a PC, only evaluators are needed.	Subjectivity
Statistical tools.	Probability calculation based on empirical knowledge, allows using theoretical data.	Supposes that data are normally distributed, the proposed models are mainly based on theory.

In a context of pesticide use reduction, alternatives to chemical-based crop protection strategies are needed to control diseases. Crop and plant architectures can be viewed as levers to control disease outbreaks by affecting microclimate within the canopy or pathogen transmission between plants. Modeling and simulation are a key approach to help analyze the behavior of such systems where direct observations are difficult and tedious. Modeling permits the joining of concepts from ecophysiology and epidemiology to define structures and functions generic enough to describe a wide range of epidemiological dynamics (Casadebaig *et al.*, 2012). Additionally, this conception should minimize computing time by both limiting the complexity and setting an efficient software implementation. In this paper, our aim was to present a model that suited these constraints so it could first be used as a research and teaching tool to promote discussions about epidemic management in cropping systems. The system was modeled as a combination of individual hosts (population of plants or organs) and infectious agents (pathogens) whose contacts are restricted through a network of connections. The system dynamics were described at an individual scale. Additional attention was given to the identification of generic properties of host-pathogen systems to widen the model's applicability domain. Two specific pathosystems with contrasted crop architectures were considered, ascochyta blight on pea (homogeneously layered canopy) and potato late blight (lattice of individualized plants). The model behavior was assessed by simulation and sensitivity analysis, and these results were discussed against the model ability to discriminate between the defined types of epidemics. Crop traits related to disease avoidance resulting in a low exposure, a slow dispersal or a de-synchronization of plant and pathogen cycles were shown to strongly impact the disease severity at the crop scale.

A computer simulation model (INHIBSIM) was developed to predict population dynamics of a bacterial biological control agent (*Pseudomonas fluorescens*) on the foliage of red pine (*Pinus resinosa*) seedlings, and subsequent effects on conidial germination of a fungal plant pathogen [(*Gremmeniella abietina*) (Gur R. Knudsen and George W. Hudler, 1987)]. The model was tested under controlled environment and field conditions. Recoverable populations of bacteria were extremely sensitive to leaf wetness but relatively insensitive to temperature. Germination of conidia was insensitive to small variations in bacterial concentration. Successful biological control of foliar plant pathogens using bacterial antagonists will require identification of agents that do not require dense concentrations to be effective and development of techniques (formulation, environmental manipulation) to enhance their survival. Predictive population models can identify and quantify these requirements, and reduce the costs of evaluating and developing potential biocontrol agents.

4.12 EXAMPLES OF FORECASTING MODELS

4.12.1 Development of Rust Epidemics

Rust development of epidemics (RustDEp) is a dynamic simulator of the daily progress of brown rust severity on wheat (Rossi *et al.*, 1997). It takes into account the proportion of spores required to be able to establish new infections influenced by temperature and leaf wetness (de Vallavieille-Pope *et al.*, 1995). The fact that the latent period depends on temperature and the fact that the infectious period depends on temperature and host growth stage (Tomerlin *et al.*, 1983). In the RustDEp model, the inputs of meteorological data are recorded by a weather station, allowing more accurate simulation of the disease progress (Rossi *et al.*, 1997).

4.12.1.1 Sclerotinia stem rot forecasting systems

Several approaches have been used to develop forecasting methods for *Sclerotinia* stem rot caused by *S. sclerotiorum*, including checklists (Thomas, 1984), risk point tables (Ahlers, 1989), serological tests (Jamaux and Spire, 1994; Lefol and Morrall, 1996) and petal infestation assessments (Turkinson and Morrall, 1993). Nevertheless, their practical use has not been adequately evaluated, and their accuracy was sometimes not satisfactory. Twengstrom *et al.* (1998) improved an existing risk point system using logistic regressions.

The factors that affect *Sclerotinia* infection were given points with regard to the risk of heavy infestations. The choice of risk factors and risk points in the improved model was made on the basis of statistical and biological considerations. New threshold values for spraying recommendations were determined. Mild *Sclerotinia* epidemics were predicted for <40 points and spraying was not recommended. If the risk point was ≥ 50, spraying was recommended. This system is a type of decision model, which has several advantages, including ease of use and distribution.

Epidemiological understanding of the sustainable deployment of disease control requires simple yet realistic models that capture not only the deterministic (average) behavior but also stochastic dynamics within and between epidemics. The models can be used to compare alternative strategies for control by mapping the effects of disease control onto the model parameters. Finding the appropriate balance between simplification and detail remains a serious challenge for epidemiological analyses of sustainable disease control (Chrisstopher A. Gilligan, 2008). Another challenge is how to deal with temporal and spatial heterogeneities. Temporal heterogeneities occur due to fluctuations in environmental variables within seasons that can appear to start and stop epidemics (Truscott and Gilligan, 2003). Fluctuations also impinge at the longer scale on the multiseasonal dynamics of many botanical epidemics with periods of survival interspersed between periods of parasitic activity on susceptible crops. This exposes the pathogen to quite different selection pressures within and between seasons and is particularly acute for pathogens in which reestablishment each season from resident inoculum has a marked effect on subsequent epidemics. Spatial heterogeneities arise from the mosaic of susceptible and nonhost crops encountered by a pathogen as it spreads at the landscape or regional scales (Gilligan, 2002). Surprisingly little attention has until recently been given to these important processes in understanding the dynamics of epidemics and the deployment of control.

SIX model with primary β_p and secondary β_s infection, the net birth of susceptibles and decay of inoculum, demographic stochastic version of SI model, SEIR model with secondary infection, latent (σ^{-1}) and infectious (μ^{-1}) periods for fixed host population size. Dispersal kernel (with parameters, θ) for the probability that susceptible (s) is infected by secondary infection from all infected hosts (i) within a neighborhood N_{hood} and by a constant source of primary infection (ϵ). Metapopulation with transmission between subpopulations in a neighborhood of interaction for a given strength of coupling (ϵ) in addition to transmission within subpopulations: model also shows separate density-dependent birth and death processes for susceptibles and criterion for invasion within a patch or subpopulation (R_p) and entire metapopulation (R_0). Effect of control agent on transmission rate and the infectious period for a single pathogen strain and for two pathogen strains typified by a fungicide-sensitive (I_s) and a fungicide-resistant (I_r) strain with fitness costs ρ_β, ρ_μ, in which the efficiency is a function of fungicide concentration (C). A criterion for the invasion of the resistant strain is given that links the ratio of reproductive numbers and the efficiency of control, for constant fungicide treatment. Note that all of the deterministic models can be readily adapted to stochastic treatments (Truscott and Gilligan, 2003).

Historically, there has been a tendency to focus on environmental variables, such as temperature and rainfall, often leading to multiple regression equations for disease progress. This reflected, in part, the ease with which abiotic variables could be measured relative to biotic variables such as infection and disease. Although these models may be useful as 'black-box' predictors of disease progress, they lack mechanistic interpretation of the epidemiological processes. Increasingly, therefore, mathematical models for botanical epidemics resemble those for animal and human diseases, centering upon the susceptible–infected–removed (SIR) framework, although the fine details may differ. The balance between primary and secondary infection is particularly important for many plant diseases (Gilligan and Kleczkowski, 1997). Primary infection is driven by ingress of inoculum from outside the system. For soil-borne pathogens, this often means infection from the reservoir of inoculum in soil; for aerial pathogens, it usually means allowance for inoculum coming from a different field or even a different region. Models for botanical epidemics also increasingly reflect seasonal dynamics accentuated by survival during intercrop periods. The emphasis on seasonal dynamics, together with the balance between primary and secondary infection, leads to SIX or SIR-X models for botanical epidemics in which X reflects the amount of 'free living' or simply surviving inoculum.

4.12.2 Role of Variability

Variability is an integral part of epidemics. It is manifest at a range of spatial and temporal scales. The timing, intensity and expansion of epidemics differ from one season to another: there are 'good' and 'bad' disease years for some pathogens while other pathogens generate epidemics in most years (Chrisstopher A. Gilligan, 2008). Even within the same season, the rates of spread of disease within and between adjacent fields sown to identical host varieties differ. Ideally, therefore, we would like to be able to predict not only the average trajectory of disease progress but also the variability. This centers on the estimation of the probability distribution for the probable occurrence of disease over time and possibly also over space in a population of susceptible sites. Individual sites can comprise entire farms, fields, orchards or glasshouses. By mapping the effects of alternative control strategies onto the probability distribution for disease progress, it is possible to predict not only the effect of treatment on the average amount of disease but also the risk of severe failure and in what proportion of fields or farms that failure might occur.

There are two broad types of variability. Demographic stochasticity relates to differences among individuals. For epidemics, it simply reflects the probabilistic nature of transmission of infection between an infected and a susceptible host under identical environmental conditions (Truscott and Gilligan, 2003). This leads to a series of stochastic events giving rise to different trajectories for different epidemics, even though the transmission and other parameters remain the same throughout the course of the epidemic. It is equivalent to saying that if epidemics were initiated in identical fields under the same initial conditions and exposed to identical environmental conditions, the resulting epidemics occurring in each field could still be markedly different, how different depends upon the inherent demographic stochasticity of the system.

In plant-virus disease epidemiology, dynamical models have invariably incorporated a bilinear inoculation rate that is directly proportional to both the abundance of healthy (susceptible) hosts and the abundance of infective vectors (Zhang *et al.*, 2000). Similarly, the acquisition rate is usually assumed to be directly proportional to the abundance of nonviruliferous vectors and that of infectious hosts. These bilinear assumptions have been questioned for certain human diseases and infection rates that incorporate power parameters of the variables have been proposed. Here, infection rates for plant virus diseases, which are of a more general form than the familiar bilinear terms are examined. For such diseases, the power parameter can be regarded as a measure of the spatial aggregation of the vectors or as a coefficient of interference between them, depending on the context. Field data of cassava mosaic virus disease (CMD) incidence were examined. When vector population density and disease incidence were high, disease progress curves over the first six months from planting could not be explained using models with bilinear infection rates.

Incorporation of the new infection terms allowed the range of observed disease progress curve types to be described. New evidence of a mutually beneficial interaction between the viruses causing CMD and the whitefly vector, *Bemisia tabaci*, showed that spatial aggregation of the vectors was an inevitable consequence of infection, particularly with a severe virus strain or a sensitive host. Virus infection increased both vector fecundity and the density of vectors on diseased plants. It was postulated that this enhanced the disease spread by causing an increased emigration rate of infective vectors to other crops. Paradoxically, within the infected crop, vector aggregation reduced the effective contact rate between vector and host and, therefore, the predicted disease incidence was less than when a bilinear contact rate was used.

Compared with many other plant pathogens, the ecology, and evolution of plant viruses is particularly complex because most are transmitted between plants by vectors. A three-way interaction exists between virus, host plant and vector, and transmission of viruses is controlled by two processes: acquisition of virus from infected plants; and inoculation of healthy host plants by vectors. Transmission of viruses by vectors is determined by many factors such as host susceptibility, the intensity of cropping, the number, mobility and life stage of the vectors, the spatial distribution and potency of virus-infected hosts, and various other environmental and climatic conditions. The movement and spatial distribution of vectors are important factors that affect the spread of plant viruses and are central to the mechanisms of disease progress. In plant-virus disease epidemiology, dynamical models have invariably incorporated a bilinear inoculation rate that is directly proportional to both the abundance of healthy (susceptible) hosts and the abundance of infective vectors. Similarly, the acquisition rate is usually assumed to be directly proportional to the abundance of nonviruliferous vectors and that of infectious hosts. These bilinear assumptions have been questioned for certain human diseases and infection rates that incorporate power parameters of the variables have been proposed. Infection rates for plant virus diseases that are of a more general form than the familiar bilinear terms were examined (Zhang *et al.*, 2000). For such diseases, the power parameter could be regarded as a measure of the spatial aggregation of the vectors or as a coefficient of interference between them, depending on the context. Field data of CMD incidence were examined. When vector population density and disease incidence were high, disease progress curves over the first six months from planting could not be explained using models with bilinear infection rates. Incorporation of the new infection terms allowed the range of observed disease progress curve types to be described. New evidence of a mutually beneficial interaction between the viruses causing CMD and the whitefly vector, *Bemisia tabaci*, has shown that spatial aggregation of the vectors was an inevitable consequence of infection, particularly with a severe virus strain or a sensitive host. Virus infection increased both vector fecundity and the density of vectors on diseased plants. It was postulated that this enhanced disease spread by causing an increased emigration rate of infective vectors to other crops. Paradoxically, within the infected crop, vector aggregation reduced the effective contact rate between vector and host and, therefore, the predicted disease incidence was less than when a bilinear contact rate is used.

Disease spread occurs at several spatial scales, e.g., from field to field, plant to plant, and leaf to leaf. So far, epidemiological models have largely overlooked the multiscale nature of epidemics. Gosme and Lucas (2009) proposed a model that simulates disease spread across multiple scales in a nested hierarchy. The model was based on the central ideas of hierarchy theory, i.e., the system is decomposed vertically into levels and horizontally into holons (elements at one level, which are complete systems when seen from the lower level), and higher levels are characterized by slower processes than lower levels. The model was individual-based, the individuals being the holons, which were either susceptible or infected. At each level, infections within one holon (i.e., infections between holons on the level below) occurred independently from the other holons, infections between holons happened at the higher level. The self-similarity of the model structure and processes across all levels allowed implementing the model with a simple recursive algorithm. The behavior of the model was studied using methods commonly applied to field data. Aggregation of the disease was characterized by the incidence-incidence relationship and the binomial power law, to study the effect of infectiousness at each level on disease aggregation. Sensitivity analyses showed that disease incidences at all levels were influenced by the infectiousness at any level, but infectiousness at higher levels had more effect than infectiousness at lower levels. It was also shown that increasing the probability of infection at a given level increased aggregation at higher level(s) and decreased aggregation at lower level(s). The results were consistent between incidence-incidence relationship and power law analysis, but the incidence-incidence relationship was more sensitive in detecting the differences in aggregation between treatments.

Models of epidemics in complex networks are improving the predictive understanding of infectious disease outbreaks. Nonetheless, applying network theory to plant pathology is still a challenge. Marco Pautassol and Mike (2014) summarized some key developments in network epidemiology that are likely to facilitate its application in the study and management of plant diseases. Recent surveys have provided much-needed datasets on contact patterns and human mobility in social networks, but plant trade networks are still understudied. Human (and plant) mobility levels across the planet are unprecedented—there is thus much potential in the use of network theory by plant health authorities and researchers. Given the directed and hierarchical nature of plant trade networks, there is a need for plant epidemiologists to further develop models based on undirected and homogeneous networks. More realistic plant health scenarios would also be obtained by developing epidemic models in dynamic, rather than static, networks. For plant diseases spread by the horticultural and ornamental trade, there is the challenge of developing spatiotemporal epidemic simulations integrating network data. The use of network theory in plant epidemiology is a promising avenue and could contribute to anticipating and preventing plant health emergencies such as European ash dieback.

Simulation models are undoubtedly necessary tools for understanding the dynamics of insect-borne virus diseases at a field level (Shun'ichi Miyai, 2000). Although great efforts have been made to develop simulation models for fungus diseases, considerably less research has been done on modeling virus disease epidemiology (Campbell and Madden 1990). Probably the disciplinary isolation of scientists studying various aspects of plant virus epidemiology is the main reason modeling has been used so seldom. Simulation models so far developed for insect-borne epidemiology are listed in Table 4.3. The vector species concerned are leafhoppers, planthoppers, and aphids. They model in various ways the effects of population dynamics of these vectors on disease spread.

The models of Berger and Ferriss (1989) and Miyai (1991) are partially stochastic, whereas the other models are wholly deterministic. In deterministic simulation models, computations are based on the expected values of the parameters and the model need be run only once with any particular set of parameter values. Simulation models that incorporate stochastic elements require a sequence of random numbers and the model need be run as many times as required by changing the sequence of random numbers.

Table 4.3 Simulation Models Developed for Some Insect-borne Plant Virus Diseases

Host	*Virus*	*Vector*	*Virus-vector relationship*
Rice	Rice dwarf virus	Leafhopper	Persistent
	Rice stripe virus	Planthopper	Persistent
Soybean	Soybean mosaic virus	Aphid	Nonpersistent
	Soybean dwarf virus	Aphid	Persistent
Subterranean clover	Subterranean clover stunt virus	Aphid	Persistent
Citrus	Citrus tristeza virus	Aphid	Semipersistent
General	—	Aphid	Nonpersistent

4.12.3 Model for Aphid-borne Nonpersistent Virus Disease

A discrete-time simulation model was constructed to study the spread of aphid-borne nonpersistent virus diseases within a field of crops. Although their original model is a rather general one which can handle multiple aphid species, in the following illustration, it is restricted for simplicity in such a way that the model deals with only one species (Marcus and Raccah, 1986). The spread of the virus was assumed to occur both by immigration of aphids into the field and by interplant movement of aphids within the field. In their model, therefore, aphids originating outside the field were discriminated from the ones within the field. Estimating roughly the parameters of the model from the laboratory and field experiments, they carried out some simulations to examine the effect of different control measures on consequent cucumber mosaic virus spread in a pepper field.

The concept of model qualification, i.e., discovering the domain over which a validated model may be properly used, was illustrated with LATE BLIGHT, a mathematical model that simulates the effect of weather, host growth and resistance, and fungicide use on asexual development and growth of *Phytophthora infestans* on potato foliage (Andrade-

Piedra *et al.*, 2005). Late blight epidemics from Ecuador, Mexico, Israel, and the United States involving 13 potato cultivars (32 epidemics in total) were compared with model predictions using graphical and statistical tests. Fungicides were not applied in any of the epidemics. For the simulations, a host resistance level was assigned to each cultivar based on general categories reported by local investigators. For eight cultivars, the model predictions fit the observed data. For four cultivars, the model predictions overestimated disease, likely due to inaccurate estimates of host resistance. Model predictions were inconsistent for one cultivar and one location. It was concluded that the domain of applicability of LATE BLIGHT can be extended from the range of conditions in Peru for which it has been previously validated to those observed in this study. A sensitivity analysis showed that, within the range of values observed empirically, LATE BLIGHT is more sensitive to changes in variables related to initial inoculum and weather than to changes in variables relating to host resistance.

Deterministic logistics models have been widely used to study pathogenicity mechanisms since the early 1950s and have focused on processes at higher biological resolution levels. In recent years, owing to the availability of high-throughput biological data and computational resources, interest in stochastic modeling of plant-pathogen interactions has grown. Stochastic models better reflect the behavior of biological systems. Most modern approaches to plant pathology modeling require molecular kinetics information. Unfortunately, this information is not available for many plant pathogens, including *P. infestans*. Boolean formalism has compensated for the lack of kinetics; this is especially the case where comparative genomics, protein-protein interactions, and differential gene expression are the most common data resources. The idea of the stochastic modeling of biological systems is not new, although traditionally, the mathematical frameworks used to represent and study these processes have been deterministic. This situation can be explained by taking into consideration the fact that quantitative and computational modeling usually requires the availability of important computational resources (Andrés Pinzón *et al.*, 2009).

These resources increase proportionally with the number of variables involved in the model; then apart from restrictions on the availability of biological information included in the model, there also exist restrictions on the availability of computational resources to perform a given simulation. In recent years, computational resources have become less restrictive. Moreover, stochastic processes are probabilistic in nature and thus require the use of more data as confidence in calculations depends on them. More data means more calculations and interconnections between variables. Thus, the availability of computational resources and biological data has restricted the use of stochastic approaches for decades, not only in plant pathology but in biological processes in general. Any top level biological observation implies an underlying molecular process, which is not isolated from its environment. This situation has always been evident to plant pathologists, as reflected in the conversion of environmental factors such water or humidity into model variables. The epidemiological research will find molecular plant-pathogen interaction models an important tool for describing disease spread and dynamics from its roots. In turn, molecular plant pathogen interactions cannot be modeled in isolation from environmental variables, largely analyzed by deterministic approaches since the early 1960s. Therefore, to reflect the real biological phenomena, it is crucial to take this information into account when a molecular interaction is considered.

Biological information that implies a networked structure can be represented by a Boolean formalism. This approach avoids the immediate necessity of chemical kinetics and the use of sets of equations (for example, differential equations) to run a simulation. Thus, Boolean networks are a viable and ideal strategy for the computational modeling of protein-protein interactions, metabolic networks, and differential expression data available today for organisms for which molecular kinetic information is not available. To date, differential gene expression data, protein-protein interaction and functional comparative analysis represent the only information available, not only for the majority of Oomycetes and their hosts but also for several other organisms. Here we argue that, due to the type of information available – although hidden Markov models, neural networks, and flux balance analysis have recently been used – a Boolean representation of plant-pathogen interactions between *P. infestans* and *S. tuberosum* is one of the most suitable approaches for computational modeling; an ongoing effort in our laboratory, based on microarray data for compatible interaction between these organisms. Once available, Boolean networks will allow kinetic information to be put back into the model and thus complement it with new information as it becomes available.

Quantitative representation and computational simulation of biological data are important tools for understanding complex biological networks and interactions. To date, the availability of efficient algorithms, biological information, and computational resources have opened the door to new insights into the analysis of such information. Bioinformatics, systems biology and its most representative tool, computational modeling, allow us to study complex plant pathogen interactions in a way unreachable to scientists two decades ago. Understanding of plant-pathogen interactions at the deterministic and stochastic, molecular and population levels requires a holistic approach, where any piece of available information is important. Today we are facing an integrative era of biological information, which approaches biological phenomena, not from their individual parts but their interactions (Andrés Pinzón *et al.*, 2009). Without a doubt, this approach reflects biological reality in a more convenient and realistic way, but it also brings new challenges as well as the necessity for new tools, which cover not only the biological sciences field but also engineering and mathematics.

Wan-Zhong Tan (2010) developed computer software for simulating the temporal development of plant disease epidemics using Richards, logistic, Gompertz, monomolecular, and exponential functions, respectively, and for predicting disease with a fitted model. The software was programmed using Visual Basic (VB6.0) and packaged with the Wise Installation System. The Fibonacci ('0.618') section strategy was used to find out the most appropriate value for the shape parameter (m) in Richards function simulation through looping procedures. The software program was repeatedly tested, debugged and edited until it was run through favorably and produced ideal outputs. It was named 'Epitimulator' based on the phrase 'epidemic time simulator' and has been registered by the National Copyright Department of China (Reg. no. 2007SR18489).

It can be installed and run on personal computers with all versions of Windows operational systems. Data of disease index and survey time are keyed in or imported from Access files. The output of fitted models and related data of parameters can be pasted into Microsoft Excel worksheet or Word document for editing as required and the simulated disease progress curves can be stored in separate graphic files. After being finally tested and completed, Epitimulator was applied to simulate the epidemic progress of corn northern leaf blight (*Exserohilum turcicum*) with recorded data from field surveys of corn crops, and the fitted models were output. Comparison of the simulation results showed that the disease progress was always best described by Richards function, which resulted in the most accurate simulation model. The result also showed that forecast of northern leaf blight development was highly accurate by using the computed progress model from Richards function.

New data on egg development and death rates and refinements of logic concerning the interaction of the nematode and host were incorporated into a simulation model of a *Meloidogyne arenaria* and grapevine system (Ferris, 1978). Simulations of field data improved, but other areas of weakness in the model were discovered. Two peaks in the egg population curve suggested that the nematode was able to complete two life cycles before host dormancy and declining temperatures limited physiological activity.

4.12.4 Nematode Egg Development and Death Rates

Experiments on *Meloidogyne arenaria* egg development relative to the temperature indicated the need for two regression models, one to describe the development and hatch rate of 74% of the egg population and the other to describe the slower hatch of the remainder. The simulator was modified to partition each cohort of newly deposited eggs into two developmental/hatch groups. Since temperature experiences of each daily cohort of eggs will vary, the deaths of each age group of eggs are determined individually. The database for the models was developed at constant temperatures (Ferris *et al.*, 1978) and the simulations involve fluctuating temperature. It could be argued, therefore, that the temperature-death rate models are somewhat misused in the simulator since the age of an egg is not a measure of time of exposure to constant temperature. However, temperature fluctuations in the soil are buffered, and changes are gradual. Consequently, use of these models is more reasonable than in an aerial system, where time-temperature interactions might be greater.

A computer-simulation model (MELSIM) of a *Meloidogyne*-grapevine system was developed by Ferris (1976) to attempt a holistic approach to the study of nematode population dynamics by using experimental data from controlled environmental conditions. A simulator with predictive ability was useful in considering pest management alternatives and in teaching. Rates of flow and interaction between the components of the system were governed by environmental conditions. Equations for these rates were determined by fitting curves to data from controlled environment studies. Development of the model and trial simulations revealed deficiencies in understanding of the system and identified areas where further research was necessary.

Various strains of naturally-occurring or recombinant bacteria offer promise for foliar application to crops for control of insect pests, plant pathogens, and weeds or to protect plants from frost injury (Guy and Louise—Marie, 1992). Research on microbial pest control agents for foliar application has progressed to where field tests are needed to evaluate their performance. Concerns about possible environmental effects of genetically engineered microorganisms applied to crop plants include uncertainty about their dispersal, survival, and interactions with indigenous organisms. Predictive models for bacterial dispersal and survival would significantly reduce costs and improve the efficacy of sampling efforts. A predictive model could prescribe optimum placement of sampling devices, probable detection thresholds for sampling and predict the effects of weather fluctuations and provide survival estimates to predict density-dependent environmental effects. Initial experiments were described to integrate a computer simulation model for dispersal of microbial aerosols with a model to predict bacterial survival and activity in phyllosphere habitats. The individual-based modeling approach was intended to accommodate environmental factors that are heterogeneous both in space and time.

Pseudomycin A is a cyclic lipodepsinonapeptide phytotoxin produced by a strain of the plant pathogenic bacterium *Pseudomonas syringae.* Like other members of this family of bacterial metabolites, it is characterized by a fatty acylated cyclic peptide with mixed chirality and lactonic closure. Several biological activities of Pseudomycin A are lower than those found for some of its congeners, a difference which might depend on the diverse number and distribution of charged residues in the peptide moiety (Coiro *et al.*, 1998). Hence, it was of interest to investigate its conformation in solution. After the complete interpretation of the two-dimensional NMR spectra, NOE data were obtained, and the structure was determined by computer simulations, applying distance geometry and molecular dynamics procedures. The conformation of the large ring of Pseudomycin A in solution included three rigid structural regions interrupted by three short flexible regions that act as hinges. The overall three-dimensional structure of the cyclic moiety was similar to that of previously studied bioactive lipodepsinonapeptides produced by other pseudomonads.

Over the past decade, the establishment of a phytoplasma classification scheme based on 16S rRNA restriction fragment length polymorphism (RFLP) patterns has enabled the accurate and reliable identification and classification of a wide range of phytoplasmas (Wei Wei *et al.*, 2007). Phytoplasmas are cell wall-less bacteria that cause numerous plant diseases. As no phytoplasma has been cultured in a cell-free medium, phytoplasmas cannot be differentiated and classified by the traditional methods which are applied to culturable prokaryotes. This classification scheme was expanded through the use of computer-simulated RFLP analysis, achieving rapid differentiation and classification of phytoplasmas. Over 800 publicly available phytoplasma, 16S rRNA gene sequences were aligned using the CLUSTAL_X program and the aligned 1.25 kb fragments were exported to pDRAW32 software for *in silico* restriction digestion and virtual gel plotting. Based on distinctive virtual RFLP patterns and calculated similarity coefficients, phytoplasma strains were classified into 28 groups. The results included the classification of hundreds of previously unclassified phytoplasmas and the delineation of ten new phytoplasma groups representing three recently described and seven novel putative 'Candidatus Phytoplasma' taxa.

4.13 GEOGRAPHIC INFORMATION SYSTEMS AND GEOSTATISTICS IN PLANT DISEASE EPIDEMIOLOGY

Plant disease management practices can be improved by putting epidemiological information in the same format as other farm information using a geographic information system (GIS). A GIS is a computer system capable of assembling, storing, manipulating and displaying data referenced by geographic coordinates. GIS can now be installed on any recent model desktop computer (a Pentium personal computer with at least 32 MB of RAM is adequate for most applications) and does not require an in-depth understanding of the statistical and mathematical basis of the technology (Merritt R. Nelson *et al.*, 1999). The large commercial push toward precision farming is based on combining GIS with sophisticated hardware for geographically referenced yield data and variable rate applications of fertilizers and other farm chemicals.

GIS can be adapted to any size operation, and data can be incorporated at any scale from a single field to an agricultural region. Many problems should be studied at more than one scale. Part of the art of GIS database development is to decide what scale to use and which kinds of information to include at each scale. The ability of government institutions, the nation's agricultural colleges and experiment stations, and private enterprise to provide spatially referenced information will open the way to farmers, pest control advisors, extension workers, and others to evaluate plant disease problems in a spatial context. New tools, including global positioning systems (GPS) and geostatistics, are available to use in connection with GIS.

4.13.1 Global Positioning System (GPS)

GPS receivers determine the location and are among the most important tools for spatially referencing agriculture data. GPS depends on a system of navigation satellites operated by the U.S. Department of Defense (the NAVSTAR system). These satellites broadcast signals containing time and position information. GPS receivers on the ground collect the satellite signals and determine the position in a spherical coordinate system such as latitude and longitude or a planar coordinate system such as the Universal Transverse Mercator system, common on terrestrial maps. An advantage of a planar coordinate system is that one can calculate distance and direction in a two-dimensional Euclidean system. Such calculations are required for geostatistical analyses. Elevation can be added if three-dimensional spatial considerations are important (Merritt R. Nelson *et al.*, 1999). GIS utilities are available to translate from one coordinate system to another at unannounced times. The Department of Defense degrades the freely available signal road cast by GPS satellites according to a practice known as 'selective availability' (SA), which is important to the GPS user because both the precision and the accuracy of GPS readings are reduced unless corrections are made. Accuracy is 'the closeness of a measured or computed value to its true value'; precision is 'the closeness of repeated measurements of the same quantity'. Without adapters, corrections cannot be made on the inexpensive handheld GPS units, so these devices cannot be depended on for horizontal accuracy better than 100 m and vertical accuracy better than 156 m. The accuracy specifications represent two standard deviations of distance from the actual antenna position to a set of position estimates made under a specified set of satellite conditions. In practice, accuracy can be in the range of 15 to 25 m, because the signal is not always degraded.

Differential GPS (DGPS) is used to improve accuracy to about 2 m. To achieve this accuracy, data from a receiver at a known position (base station) is combined with data from a receiver at the desired position (rover or remote station) to correct for SA and other sources of error. Precision farming equipment, such as yield monitors and variable rate applicators, depends on DGPS. DGPS can increase the cost and computer complication involved in obtaining point coordinates, so inexpensive handheld devices are preferable when DGPS accuracy is not required. The regional applications described in this article are done with inexpensive handheld GPS units because the interpretation of spatial patterns spanning more than 100 km does not require an accuracy of better than 100 m for point coordinates.

Even in smaller areas where greater accuracy is desired, a combination of tools can substitute for DGPS. Hand-drawn maps or USDA Farm Services Agency maps are frequently available. Handheld GPS units can be used to obtain coordinates of landmarks such as road intersections or irrigation wells, which can then be used as reference points to digitize the hand-drawn detailed map for display with other layers referenced to the same coordinate system. Similarly, greater precision can be achieved by staking out rectangular grids in a small area and then using a handheld GPS unit to relate the location of the grid as a whole to a wider area. The rapid growth in availability and affordability of handheld GPS units is due in part to the mass market for navigation applications by small boat owners. Most mass-produced GPS units are designed for nautical applications and the outdoorsman. Thus, most sources of handheld GPS units are marine or outdoor supply businesses.

4.13.2 Geographic Information System (GIS)

GIS relates the data collected by GPS to other sources of georeferenced information. GIS has the ability to integrate layers of spatial information and to uncover possible relationships that would not otherwise be obvious. The process of transforming one layer of spatial information to match the second layer is called registration (Merritt R. Nelson *et al.*, 1999). When a broad collection of data is registered in the same coordinate system the possibilities for sharing data and creatively analyzing the spatial relationships increase dramatically. GIS applications for agriculture are developing rapidly and include generic commercial packages such as ArcView (ESRI, Redlands, CA) or MapInfo Desktop (MapInfo Corp., Troy, NY) as well as more specific applications derived from these packages (e.g., Site-Specific Technology Development Group, Inc., Stillwater, OK

(ArcView), and Red Hen Systems, Fort Collins, CO (MapInfo). As with other types of computer software, compatibility between competing systems can be a problem, but data transfer protocols and generic data formats have been developed to work around this problem.

There are two main forms of GIS data, vector, and raster. Most GIS software can now handle both forms. In vector data sets, map features such as points, lines, and polygons are organized and manipulated in a database. In raster data sets, the data are organized as a matrix of numerical values and referenced spatially by row and column position. Header information or an associated computer file contains the spatial information that relates the rows and columns of the matrix to the GIS coordinate system. USGS digital elevation model data sets, satellite images, and some image files produced by scanners are examples of Raster data sets. GIS has been applied in agriculture for the spatial analysis of plant diseases (Orum *et al.*, 1997). Weltzien (1978.) introduced the term 'Geophytopathology' to describe studies of spatial patterns of plant diseases, the causal understanding of these patterns, and the geographic aspects of disease control. He gave a detailed description of the use of maps to illustrate the spread of plant diseases over regional or even continental scales and classified the types of maps that can be used for these purposes (Weltzien, 1978.). However, quantitative regional assessment requires analysis of large spatial databases. With the development of powerful computers and better software, GIS is the appropriate technology to manage such spatial data.

4.13.3 Geostatistics

Geostatistics focuses on the analysis of spatially distributed variables and the prediction or estimation of values at unsampled locations. Spatial analysis is divided into three broad types, geostatistical, lattice, and point (Merritt R. Nelson *et al.*, 1999). He describes geostatistics as a statistical model appropriate for estimates across continuous areas (characterized as processes with a continuous spatial index). Trend surface and inverse distant weighting methods of interpolation preceded geostatistics, but geostatistical considerations, implicitly or explicitly, underlie most attempts to create surface maps based on point samples or observations. A surface map is a map with an area shaded in a color or grayscale keyed to a variable. As with summary statistics (averages, medians, standard deviations) in nonspatial data analysis, when done well, surface maps allow the viewer to grasp the larger picture without being distracted by the noise and scatter of point data. Regional surface maps are appropriate when a variable (pathogen propagule density, disease incidence, insect vector abundance, etc.) exhibits positive spatial autocorrelation beyond the boundary of a single field. Spatial autocorrelation is the correlation of a variable with itself (auto) at various separation distances and directions (spatial). If adjacent fields are highly dissimilar with abrupt differences defined by field boundaries, surface maps that are interpolated without field boundary considerations can be misleading. Sometimes the absolute value of a variable is less important than whether it is above or below a biologically critical value. If a variable, such as pathogen propagule density, is uniformly above such a threshold, a surface map will not be helpful unless the uniformity is in contrast to patchiness at another place or time.

Geostatistics have been used in plant pathology to analyse the spatial distribution of plant disease epidemics, mainly at plot or field scales (Websterand Boag, 1992). They have also been used in a regional risk assessment program for the management of plant virus diseases in Mexico. The difficulties of spatiotemporal analyzes were explored in a geostatistical study of the development of cabbage downy mildew epidemics. Inoculum patterns at very small scales, such as along plant roots have also been quantified using geostatistics. A geostatistical analysis can be broken down into four phases, exploratory data analysis, modeling of the spatial autocorrelation (characteristics of the spatial continuity), estimating at unsampled locations across a surface (surface maps) and evaluation of the reliability of the estimates.

An easy-to-read explanation of geostatistics was provided by Isaaks and Srivastava (1989). In geostatistics, spatial autocorrelation is usually examined by variogram estimation and modeling. Procedures to develop experimental variograms and to fit model variograms are presented in several sources (Merritt R. Nelson *et al.*, 1999). The model variogram is incorporated into a procedure for surface interpolation known as 'kriging.' An advantage of kriging is that there are two outcomes, a surface map of the variable and a surface map of the kriging standard deviation, which provides a relative measure of confidence in the estimates. This combination is discussed later in the paper in an application involving a regional study of *Aspergillus flavus* strain composition. The most sophisticated geostatistical tool is a stochastic conditional simulation. This tool and its potential in crop protection have been explained in a regional study of corn rootworm in northwestern Iowa.

Maps show where a minimum loss strategy recommends the use of an insecticide for two different costs of the insecticide. The analysis can take into account not only the cost of acquiring and applying the insecticide but also an estimate of the environmental cost of applications (Rossi *et al.*, 1993). Because of the technical difficulties, few farmers, extension advisors, or regulators will apply stochastic conditional simulation in the near future. One might hope, however, that over time, the conditional simulation will be incorporated into software extending geostatistics to plant disease situations. Simple GIS maps of observed values overlaid on landmarks such as roads, rivers, towns, and district boundaries are sometimes sufficient to characterize a disease or pest problem. This can be done without geostatistics. Geostatistics adds the ability to model the range and directionality of the spatial autocorrelation, which can provide clues to underlying disease processes involving spread, climate, and alternate hosts for pathogens and their vectors. Surface maps remove noise from point data and can improve communication. Geostatistics adds an assessment of confidence. It also provides tools to create probability maps from descriptive, categorical variables using indicator kriging. A reasonable stepwise approach to learning this technology is to start with GPS and simple GIS applications and move into geostatistical analyses as experience with GIS is developed. A criticism of the utility of geostatistics in agriculture is that the mathematical framework in which it is usually presented is beyond many potential users.

4.13.4 Integrating GPS, GIS, and Geostatistics

GPS, GIS, and geostatistics can be used to advantage at different levels of management and expertise. For example, an extension agent with a handheld GPS unit could relay the coordinates of a serious plant disease situation to a specialist at a land grant college, who could use GIS to place the observation in a broader context with similar observations coming in from elsewhere. One of the most helpful applications of this technology is to store and query spatially referenced point data. The benefits of integration come from relating this simple format data to other data (roads, county boundaries, digital elevation models, hydrography, land ownership boundaries, and, most importantly, other agricultural data) that are increasingly available from public and private sources via the World Wide Web (Merritt R. Nelson *et al.*, 1999).

Where appropriate, point data can be interpolated to produce surface maps using geostatistics. GIS maps of plant diseases and risk of plant disease can be created in a number of ways. Experts, including farmers, can synthesize personal observations from memory and generate maps simply by sketching polygons over appropriately referenced backgrounds. For example, a risk map at the continental scale of gray leaf spot of corn based on observations of the NCR-25 technical committee on corn and sorghum diseases was presented as part of APSnet Feature, 1 through 31 May 1998. Spatially located field observations of incidence combined with weather data can be used effectively for certain aerially dispersed fungal diseases. Continental-scale risk maps for blue mold of tobacco have been published on the World Wide Web 3 days a week during the tobacco growing season since 1996 by the North American Blue Mold Forecast Center at North Carolina State University. GIS risk assessment maps based on kriging of field observations were part of a tomato virus management program at the regional scale in the Del Fuerte Valley, Sinaloa, Mexico (Nelson *et al.*, 1994).

4.13.5 Case Studies

4.13.5.1 Aspergillus flavus

A series of examples illustrate some of the ways in which GIS and geostatistics can be applied in an agricultural setting. Soil structure and chemistry information was linked to a point theme that is displayed as an overlay with field boundaries (polygon features) and roads (linear features) in the Texas Hill part of eastern Yuma County, Arizona. The points were color-coded according to the percentage of total sand in the soil. This was strictly a GIS illustration (no geostatistics). The important differences between GIS in this example and hand-drawn maps were the linkage of data to points and the use of a coordinate system. The x and y coordinates (determined from GPS readings in conjunction with digitized Farm Services Agency maps) were in the UTM coordinate system, which allows the points to be overlaid on similarly referenced polygon and linear features (Merritt R. Nelson *et al.*, 1999). Simple maps of observed values overlaid on spatial references such as roads and district boundaries were sometimes sufficient to characterize a disease or pest problem.

A. flavus frequently produces aflatoxins, carcinogens that can contaminate cottonseed and other important seed crops such as peanuts, corn, and nuts. The toxins are highly regulated to reduce their level in the human food supply. *A. flavus* isolates can be divided into two strains based on sclerotial size (S strain [small] and L strain [large]) (Cotty, 1989.). The S strain isolates are very toxigenic. As part of a management program to control aflatoxin using naturally occurring atoxigenic strains spatial patterns of the highly toxigenic S strain are being investigated. Understanding shifts in S strain incidence patterns in untreated areas will help in the interpretation of shifts observed in areas treated with atoxigenic L strain isolates.

4.13.5.2 *Tomato virus management*

A regional plan for tomato virus disease management was implemented with the support of the tomato processing industry in the Del jerte Valley in Sinaloa, Mexico. The plan involved a risk assessment procedure and mitigation of risk based on observations of virus hazards immediately surrounding fields at the time of transplant. Mitigation of risk involved either destroying alternate host plants near fields or changing the time and location of planting. Observations of virus disease symptoms in fields were made each year from 1991 through 1996. GIS displays show that there were recurring regional patterns. Regardless of the overall virus incidence level, the highest incidence occurred each year east of Los Mochis. Displays of the patterns were used to encourage cultural management efforts in the areas of highest risk and incidence. The encouragement of cooperation is a major value of GIS in a regional disease management program. Using GIS displays to avoid planting processing tomatoes in areas with a high risk of virus disease was part of an integrated pest management strategy that eliminated the use of pesticides for the control of insect vectors of tomato virus diseases by one producer (Bolkan and Reinert, 1994). To complement the virus data, a study of tomato pinworm adults was conducted using a grid of 110 pheromone traps placed throughout the region.

4.13.5.3 *Late blight of tomatoes and potatoes*

Geostatistics and GIS were also used to analyze the spatial patterns of genotypes of *Phytophthora infestans* (Mont.) de Bary and their relationship to late blight disease in the mixed tomato and potato production area, Del Fuerte Valley (Jaime- *et al.*, 1996). Population genetic studies of *P. infestans* have intensified worldwide following the appearance of the A2 mating type and metalaxyl insensitive isolates of the pathogen outside central Mexico in the 1970s (Fry *et al.*, 1992). The development of cellulose acetate techniques for rapid allozyme analyses has created the opportunity to obtain enough data to characterize the pathogen spatially and incorporate this information into management programs at the scale of an agricultural region such as the Del Fuerte Valley. Maps displaying the probabilities of occurrence of mating types and genotypes of *P. infestans* and disease severity on a regional scale were prepared. Probability maps of the GPI allozyme genotypes illustrated the results. The probability of occurrence of allozyme genotypes was estimated by indicator kriging. Indicator variables can be used for nonnumerical data, such as the presence or absence of a particular genotype, to generate probability maps by kriging. The maps of the probability of occurrence of *P. infestans* allozyme genotypes for the Del Fuerte Valley showed that the epidemics of late blight were largely dominated by only one genotype each year, which indicated strong clonal reproduction of the pathogen in the area.

Most genotype variables showed spatial autocorrelation in a range of 13,000 to 20,000 m in a valley that is 70,000 by 150,000 m. This suggests a restricted spread of some genotypes that can be visualized with the help of the probability maps generated by kriging. The idea of creating pathogen genotype probability maps is new, and its applicability in a plant disease management system has not been tested commercially. It might be useful where environmental, worker safety, and economic concerns lead farmers to weigh a variety of new factors, including pathogen genotype, in the selection of plant disease control measures.

Geographic information systems (GIS) provide valuable tools for monitoring, predicting, managing and fighting the spread of pests and diseases. The tools offer opportunities for cost-effective and efficient targeting of control interventions. In monitoring, GIS can be used to determine the spatial extent of a disease, to identify spatial patterns of the disease and to link the disease to auxiliary spatial data. GIS can also be used to predict the projected spread of diseases, to provide input for risk assessment models in pest control and in quantifying changing thresholds of pests and diseases due to climate change. To use GIS techniques at a larger scale, a protocol for data collection and management is essential (Bouwmeester *et al.*, 2010). They illustrated the use of GIS tools on data collected to identify critical intervention areas to combat the spread of Banana *Xanthomonas* wilt (BXW). In a survey covering the Great Lakes region, the on-farm incidence of the disease was monitored and precise GPS coordinates of each sampled field were recorded. This enabled accurate mapping of the disease and performing the various spatial analyses, permitting an understanding of the geographical distribution of BXW infection. Data on food security and dependency on banana to rural populations was linked to the BXW severity to target priority areas of interventions and maximize impact.

Geographic information systems have been used for geospatial data management and analysis, map production and spatial modeling. They also have the potential to incorporate climate and satellite data that could provide a spatial perspective on the risk of establishment of plant pathogens. The climatic suitability for the establishment of dwarf bunt, a disease that can cause market access restrictions, on wheat or other grass hosts in New Zealand was analyzed using a GIS approach (Kim and Beresford, 2009). Establishment risk for dwarf bunt in New Zealand was found to be very low. GIS was found to be a more versatile tool for modeling potential geographic distribution of organisms than conventional climate matching tools, such as CLIMEX.

The epidemiology of lettuce downy mildew was investigated extensively in coastal California. However, the spatial patterns of the disease and the distance that *Bremia lactucae* spores could be transported have not been determined. During 1995 to 1998, (Wu *et al.*, 2001) conducted several field- and valley-scale surveys to determine spatial patterns of this disease in the Salinas Valley. Geostatistical analyses of the survey data at both scales showed that the influence range of downy mildew incidence at one location on incidence at other locations was between 80 and 3,000 m. A linear relationship was detected between semivariance and lag distance at the field scale; although no single statistical model could fit the semivariograms at the valley scale. Spatial interpolation by the inverse distance weighting method with a power of 2 resulted in plausible estimates of incidence throughout the valley. Cluster analysis of geographic information systems on the interpolated disease incidence from different dates demonstrated that the Salinas Valley could be divided into two areas, north and south of Salinas City, with high and low disease pressure, respectively. Seasonal and spatial trends along the valley suggested that the distinction between the downy mildew conducive and nonconducive areas might be determined by environmental factors.

The global severity of potato late blight was estimated by linking two disease forecast models, Blitecast, and Simcast, to a climate database in a GIS (Hijmans *et al.*, 2000). The disease forecast models indirectly estimate late blight severity by determining how many sprays are needed during a growing season as a function of the weather. Global zonation of estimated late blight severity was similar for both forecast models, but Blitecast generally predicted a lower number of sprays. With both forecast models, there were strong differences between potato production zones. Zones of high late blight severity include the tropical highlands, Western Europe, the east coast of Canada and the northern USA, south-eastern Brazil and central-southern China. Major production zones with a low late blight severity include the western plains of India, where irrigated potato is produced in the cold, dry season, north-central China, and the north-western USA. Using a global GIS database of potato production, the average number of sprays was calculated by country. These averages were compared with estimates of current fungicide use. The results using Blitecast and Simcast were correlated, but only Blitecast estimates correlated with observed data for developed countries. The estimated number of sprays, whether from Blitecast or Simcast, did not correlate with the observed number of sprays in developing countries, and in a number of developing countries the predicted optimal number of sprays was much higher than the actual number observed. In these countries, increased access to host resistance and fungicides could have a strong economic impact.

Integrating remote sensing and geographic information systems (GIS) technologies offers tremendous opportunities for farmers to more cost effectively manage the causes of crop stress. Initial soybean cyst nematode population densities from 995 quadrants (2 m × 3 m each) were obtained from a soybean field near Ames, Iowa, in 2000 (Nutter, Jr. *et al.*, 2002). The percentage of sunlight reflected from each quadrant was measured weekly using a multispectral radiometer beginning in mid-May and continuing through mid-September. Aerial images were obtained at heights above the field ranging from 45 to 425 m on 12 dates during the soybean growing season. This was accomplished using color film and infrared film in conjunction with a filter to measure reflectance in the near-infrared region (810 nm). Satellite images (Landsat 7) were obtained for five dates during the 2000 growing season. Maps depicting initial nematode population densities, soybean yield, soy oil, and soy protein were generated using the GIS software program ArcView. Percentage reflectance (810 nm), aerial image intensity, and satellite image intensity data then were regressed against soybean yield, soy oil, and soy protein concentrations obtained from each geospatially referenced soybean quadrant. Percentage reflectance measurements explained up to 60% of the variation in initial nematode population densities within soybean quadrants and up to 91% of the variation in soybean yield. Aerial image and satellite image intensities explained up to 80% and 47% of the variation in soybean yield, respectively. Percentage reflectance data also explained 36% and 54% of the variation in oil and protein concentrations of the harvested soybeans, respectively. These results indicated that remote sensing coupled with GIS technologies may provide new tools to detect and quantify nematode population densities and their impacts on the quantity and quality of soybean yield.

4.14 REMOTE SENSING

Remote sensing is the science and art of acquiring information about material objects from measurements made at a distance without coming in physical contact with the objects. Reflectance measurements have been successfully used to distinguish between healthy and diseased plants in fields (Marvin E. Bauer, 1971). This is possible because healthy and diseased plants reflect energy differently. Some examples of diseases which have been found to be detectable with the aerial infrared film are wheat rust, potato late blight, and citrus foot rot.

Remote sensing in plant disease management is the practice of gathering information about a crops health without physically scouting the field. Typically, this occurs through images captured from aircraft or satellites, but there are also ground-based applications (Anonymous, 2005). The ability to view images of an entire field provides plant pathologists with greater precision and accuracy in disease assessment. By using remote imagery to differentiate between healthy and diseased plants, it is possible to determine how many acres are impacted by a particular disease. Remote sensing has the potential to save time previously spent randomly scouting fields for potential disease. Using images captured remotely enables to identify those areas that need closer surveillance. By targeting the area that is infected, farmers can spray the areas that are affected rather than the entire field. This reduces the amount of pesticide that is applied and the number of applications needed, reducing the cost of treatment and possible harm to the environment.

Remote sensing lends itself exceptionally well to the detection of anomalous locations within a field or orchard that have been differentially affected by weeds, diseases, or arthropod pests. In fact more than 35 years ago, ARS scientists were using aerial color-infrared photography for this purpose and relating their findings to laboratory spectra of pest damaged leaves (Paul J. Pinter *et al.*, 2003). CIR film was used to evaluate the effect of crop rotation and soil fumigation on a nematode (*Rotylenchulus reniformis*) occurring in Texas cotton fields. Cook *et al.* (1999) used multitemporal NIR videography to monitor the seasonal progression of the southern root knot nematode (*Meloidogyne incognita)* and its associated soil-borne fungi complex in kenaf (*Hibiscus cannabinus* L.). Of course, areas of reduced plant vigor could conceivably be caused by a number of factors unrelated to pests, so it is likely that additional spectral, spatial and temporal clues, provided within the context of a decision support system, will be required to uniquely identify the problem.

Examples in which ARS employed remote sensing technology for detecting crop disease and assessing its impact on productivity include using CIR photography to identify circular areas affected by cotton root rot, *Phymatotrichum omnivorum* and to estimate yield losses caused by black root disease in sugar beets. Cook *et al.* (1999) also demonstrated the potential for aerial video imagery to detect *P. omnivorum* in kenaf, a crop whose tall growth habit makes it almost impossible to survey from the ground. The TIR can provide early, sometimes previsual, detection of diseases that interfere with the flow of water from the soil through the plant to the atmosphere. As an example, Pinter *et al.* (1979) found that cotton plants whose roots were infected with the soil-borne fungus *P. omnivorum* and sugar beets infected with *Pythium apanidermatum* both displayed sunlit leaf temperatures that were 3 to 5°C warmer than adjacent healthy plants. The TIR was also useful for detecting root disease in red clover under irrigated conditions. Much more research is required when using remote sensing for identifying specific diseases or when separating them from other causes of plant stress. Hyperspectral techniques are likely to provide some assistance, but coupling existing techniques with weather driven computer models of disease development will probably provide the best approach.

Plant diseases influence the optical properties of plants in different ways. Depending on the host-pathogen system and disease-specific symptoms, different regions of the reflectance spectrum are affected, resulting in specific spectral signatures of diseased plants (Dehne, 2011). The aim of this study was to examine the potential of hyperspectral imaging and nonimaging sensor systems for the detection, differentiation, and quantification of plant diseases. Reflectance spectra of sugar beet leaves infected with the fungal pathogens *Cercospora beticola, Erysiphe betae,* and *Uromyces betae* causing Cercospora leaf spot, powdery mildew, and sugar beet rust, respectively, were recorded repeatedly during pathogenesis. Hyperspectral data were analyzed using various methods of data and image analysis and were compared to ground truth data. Several approaches with different sensors on the measuring scales leaf, canopy, and field were tested and compared. Much attention was paid to the effect of the spectral, spatial, and temporal resolution of hyperspectral sensors on disease recording. Another focus of this study was the description of spectral characteristics of disease-specific symptoms. Therefore, different data analysis methods were applied to gain a maximum of information from spectral signatures.

Spectral reflectance of sugar beet was affected by each disease in a characteristic way, resulting in disease-specific signatures. Reflectance differences, sensitivity, and best correlating spectral bands differed depending on the disease and the developmental stage of the diseases. Compared to nonimaging sensors, the hyperspectral imaging sensor gave extra information related to spatial resolution. The preciseness in detecting pixel-wise spatial and temporal differences was on a high level. Besides characterization of diseased leaves also the assessment of pure disease end members as well as of different regions of typical symptoms was realized. Spectral vegetation indices (SVIs) related to physiological parameters were calculated and correlated with the severity of diseases. The SVIs differed in their sensitivity to the different diseases. Combining the information from multiple SVIs in an automatic classification method with Support Vector Machines, high sensitivity and specificity for the detection and differentiation of diseased leaves was reached at an early stage. In addition to the detection and identification, the quantification of diseases was possible with high accuracy by SVIs and Spectral Angle Mapper classification, calculated from hyperspectral images. Knowledge from measurements under controlled condition was carried over to the field scale. Early detection and monitoring of Cercospora leaf spot and powdery mildew were facilitated. The results of this study contributed to a better understanding of plant optical properties during disease development. Methods will further be applicable to precision crop protection, to realize the detection, differentiation, and quantification of plant diseases in early stages.

Experiments were conducted by the Laboratory for Applications of Remote Sensing (LARS) at Purdue University to determine if southern corn leaf blight could be detected using remote sensing techniques. Six segments, 8 to 12 miles long and one mile wide, were selected from a flight line extending from Lake Michigan to the Ohio River in western Indiana (Marvin E. Bauer, 1971). The severity of blight infection in each corn field in these segments was rated by agronomists and plant pathologists. Several factors complicated the early detection from aircraft of a disease such as southern corn leaf blight. One was that since the disease started at the bottom of the plant, infection on the lower leaves may be hidden from the view of the camera or scanner by healthy upper leaves. Consequently, the infected leaves had little or no effect on the spectral measurements or the color infrared photography. Also, in fields with low populations or excessive weeds, the response from soil and weeds overshadowed that of the corn plants. The natural variability due to agronomic factors also contributed to the difficulty of detecting small amounts of infection. Some of these factors were differences in leaf angle, leaf color, and tassel color among different hybrids, and differences due to maturity, population, row direction and the vigor of things as available soil moisture and fertility.

The above Corn Blight Watch Experiment was the largest experiment conducted to date to determine the feasibility of crop disease detection with remote sensing over a large geographic area. The results indicated that early detection by remote sensing of diseases like southern corn leaf blight is unlikely at this time. It was demonstrated, however, that remote sensing can provide an accurate assessment of the situation when moderate-to-severe infection levels were present. This experiment provided experimenters a real-world situation in which to evaluate the capabilities of remote sensing. Much was learned about the mechanics required to operate a large information system utilizing remote sensing. Key problem areas were identified, and efforts are already underway to find solutions to these problems. Future research were aimed at finding new ways to obtain more information from remote sensing measurements and to make more efficient use of ground observations; for example, the use of calibrated data should extend the boundaries for which training fields are valid to include new sets of data from other times and/or places. Agricultural applications of remote sensing appeared promising for the near future. The projected applications of remote sensing to agriculture include the identification and area measurements of major crops, mapping soils characteristics, mapping disease and insect invasions, assessment of crop stresses and crop yield prediction.

A very large body of research spanning almost four decades has demonstrated that much of this required information is available remotely, via aircraft and satellite-based sensor systems. When combined with remarkable advances in Global Positioning System (GPS) receivers, microcomputers, geographic information systems (GIS), yield monitors and enhanced crop simulation models, remote sensing technology has the potential to transform the ways that growers manage their lands and implement precision farming techniques.

4.14.1 Remote Sensing and GIS

Integrating remote sensing and geographic information systems technologies offers tremendous opportunities for farmers to manage more cost effectively the causes of crop stress. Initial soybean cyst nematode (SCN) population densities from 995 2-x-3-m quadrants were obtained from a soybean field near Ames, Iowa, in 2000 (Nutter *et al.*, 2002). The percentage of sunlight reflected from each quadrant was measured weekly using a multispectral radiometer beginning in mid-May and continuing through mid-September. Aerial images were obtained at heights above the field ranging from 45 to 425 m on 12 dates during the soybean growing season. This was accomplished using color film and infrared film in conjunction with a filter to measure reflectance in the near-infrared region (810 nm). Satellite images (Landsat 7) were obtained for five dates during the 2000 growing season. Maps depicting initial SCN population densities, soybean yield, soy oil, and soy protein were generated using the GIS software program ArcView. Percentage reflectance (810 nm), aerial image intensity, and satellite image intensity data then were regressed against soybean yield, soy oil, and soy protein concentrations obtained from each geospatially referenced soybean quadrant. Percentage reflectance measurements explained up to 60% of the variation in initial SCN population densities within soybean quadrants and up to 91% of the variation in soybean yield. Aerial image and satellite image intensities explained up to 80% and 47% of the variation in soybean yield, respectively. Percentage reflectance data also explained 36% and 54% of the variation in oil and protein concentrations of the harvested soybeans, respectively. These results indicated that remote sensing coupled with GIS technologies may provide new tools to detect and quantify SCN population densities and their impacts on the quantity and quality of soybean yield.

Modern management of agricultural resources is a complex endeavor that is now benefiting from a convergence of technical advances in information sciences, geographic positioning capabilities, and remote sensing systems (Paul J. Pinter *et al.*, 2003). Much of the fundamental research relating spectral properties of soils and crops to agronomic and biophysical parameters has been accomplished by ARS researchers working collaboratively with NASA and university scientists in a variety of programs over the past four decades. Many aspects of crop management have already begun to benefit from applications of remote sensing technology. As growers gain more confidence in its use, additional opportunities will present themselves. The future brings tremendous prospects for integrating the spatially and temporally rich information provided through remotely sensed muiti- and hyperspectral imagery·with the capabilities of management-oriented crop simulation models.

There are two avenues in which remote sensing information either reflected or emitted radiation, can be incorporated into disease monitoring. The two approaches involve either direct or indirect methods of evaluating the disease occurrence and extent. Given the number of factors that cause variation in both the reflected and emitted radiation signals, it is unlikely that the direct monitoring method will be useful (Hatfield, 1990). Both the indirect and direct methods require *a priori* knowledge that a condition may exist. Given this knowledge then, one may use a direct monitoring program to measure the extent of a disease, *Phytophthora* spp. on soybeans or *Fusarium* spp. on beans, which cause a reduction in leaf area. The spatial sampling capability provides an assessment not possible with ground monitoring. The indirect method would involve interpretation of deviations from the expected case either in temporal or spatial patterns. For example, an increase in foliage temperature in a field with an adequate soil water supply would signal a potential problem that could invoke a monitoring effort. An unexplained change in leaf area or wilting resulting in a change in reflectance could signal a problem before complete infestation. The utilization of both indirect and direct methods will require imagination and dedication to the problem by a number of researchers.

4.14.2 New Methodologies

During recent years, new methodologies for remote analysis, such as remote and proximal sensing, have been added to the various standard techniques for plant disease monitoring. In particular, multispectral images, acquired at a high radiometric and geometric resolution, by means of digital airborne systems or by satellite, can be useful for monitoring some diseases of woody plants. (Fabi and Varvaro, 2009) reported the initial results of the application of one of these systems, called A.Sp.I.S. (Advanced SPectroscopic Imaging System) for monitoring the Dieback of hazelnut, in the province of Viterbo, as a case study, a bacterial disease which has affected hazelnut orchards in the province of Viterbo since the 1980's. The spatial data sources were aerial photographs taken during a decade of growing seasons, from 1996 to 2006. Moreover, satellite images (Quick Bird) were used to previously scan wide areas of the territory affected by the disease. The satellite images, once overlaid and manipulated by means of dedicated software (Image calculator – IDRISI™), were used as a basis for drawing new elaborated 'differential' digital images. The digital elaboration showed annually the plants which were suspected of dying the season before. The suspected plants were then automatically identified by means of E-cognition™, a segmentation GIS software which performs object-oriented image analysis. The results were then confirmed, comparing the A.Sp.I.S.

images and by proper ground activities performed *in loco*. The software was able to recognize the main part of dead or wilting plants when applied to the images. Finally, a protocol of investigation was set up, by means of vectorial GIS, to follow in time and space the local spread of the disease.

The applicability of remotely sensed images was explored by MInghua Zhang *et al.* (2005) to detect late blight of potato disease spectral anomalies for precision disease management. They used the indices approach and generated a 5-index image that we used to identify the disease in tomato fields based on information from field collected spectra and linear combinations of the spectral indices. Field results indicated that we were able to identify five clusters in the image space with small overlaps of a few clusters. Using the identified 5-cluster scheme to classify the tomato field images, we were able to successfully separate the diseased tomatoes from the healthy ones before economic damage was caused. Hence, the method based on a 5-index image may significantly enhance the capability of multispectral remote sensing for disease discrimination at the field level. The objective of the study was to develop a method for application of multispectral remote sensing to detect the diseased tomato plants. Several possible indices derived from red and NIR bands of the multispectral remote sensing data were evaluated for discriminating the diseased tomato plants. The method based on the effective indices was developed to explore the potential applications of the multispectral remote sensing to detect the diseased tomato plants. Therefore, analysis in the study mainly involved the three parts, viz., field spectral analysis to determine whether spectral features of diseased plants can be differentiated, numeric analysis to determine the best combination of red and NIR bands for effective separation of diseased plants, and validation analysis to test the developed image identification method for late blight on tomatoes with available ground truth information.

4.14.3 Major Examples

Remote sensing is a valuable tool for grape vineyard managers to assess several aspects of their viticultural operations, including pest management programs. Remote sensing refers to the use of aircraft or satellites to provide pictures or digital images of vineyards (Wildman *et al.*, 1992). Synoptic images taken from above vineyards often provide insight into patterns in vine growth that are more difficult to discern from the ground. Subtle differences that may go unnoticed in the field may be seen as significant in a remotely sensed image. Color infrared photography from aircraft flying above the vineyard is a form of remote sensing that has been in use for decades. Color infrared film records reflectance in the green and red wavelengths of visible light, and in the near infrared (NIR) range. It does not react to heat or temperature differences in plants. Green plants absorb most of the red light that reaches them, but they reflect both green and NIR wavelengths. Green plants are the only objects with this combination of absorption and reflectance. Thus they are the only objects that appear as a shade of red or magenta on the film. The shade depends on the cultivar, maturity, and health of the plant. The leaf's internal cell structure is responsible for the reflectance of NIR radiation, so conditions that alter cell structure, such as water stress or disease may lower the NIR reflectance. The future use of color infrared film photography will likely become less common with the decline in their availability with film manufacturers.

Remote sensing can identify and map areas of poor or variable grapevine growth and track their development over time, but it does not explain the causes of poor vine performance. For this, the vineyard manager must interpret the information provided by the imagery and follow up with on-the-ground investigations. Pierce's disease kills vines and often occurs in patterns depending upon the habitat of its insect vectors. In northern California, Pierce's disease is often associated with riparian areas, with the greatest vine damage occurring adjacent to the riparian zone (Fig. 4.13). This pattern is often most apparent in aerial images. *Armillaria* root rot kills vines, which is apparent in remote imagery (Fig. 4.14). It usually occurs in relatively small areas that might appear anywhere in a block as a result of infected roots or woody debris remaining in the soil. These weakened areas may never support healthy vines, but they rarely increase in size to any great degree (Wildman *et al.*, 1992). However, in smaller plantings, such areas may make up a significant proportion of the total number of vines.

Fig. 4.13 The pattern of Pierce's disease shown in this aerial photo is typical in grape-growing regions in which the blue-green sharpshooter is the principal vector of the bacterium that causes the disease. Vines closest to the riparian habitat of the sharpshooter leafhopper have died, while those farther from the stream are less affected

Vines weakened from phytonematode damage may have a similar appearance to drought-stressed vines in an NDVI image (Fig. 4.15). If a weak area increases in size in subsequent years, one might suspect a biological cause, such as phylloxera feeding on the roots, rather than a static difference in soil type. The differences between weak and healthy vines are very apparent in remote imagery, and patterns often emerge that are not apparent from ground level. The use of imagery can simplify diagnoses and help ensure proper treatment.

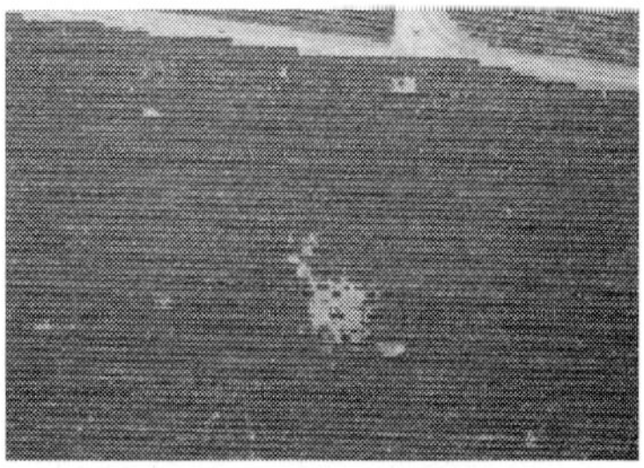

Fig. 4.14 *Armillaria* root rot has killed vines in an area where an oak tree was standing prior to the vineyard being planted. In contrast to phyloxera, the spread of this disease is relatively slow and progresses outward in concentric rings

Fig. 4.15 Nematode infestation. This NDVI image shows areas of depressed vine growth as yellow to brownish areas across the center of the image in all three main blocks

Wheat streak mosaic (WSM), caused by *Wheat streak mosaic virus* affects wheat (*Triticum aestivum* L.), other grains, and numerous grasses over large geographical areas around the world. To improve disease management and crop production, it is essential to have adequate methods for monitoring disease epidemics at various scales and multiple times. Remote sensing has become an essential tool for monitoring and quantifying crop stress due to biotic and abiotic factors (Mirik *et al.*, 2013), who explored the utility of Landsat 5 TM imagery for detecting, quantifying, and mapping the occurrence of WSM in irrigated commercial wheat fields. The infection and progression of WSM were biweekly assessed in the Texas Panhandle during the 2007–2008 crop years. Diseased-wheat was separated from uninfected wheat on the images using a subpixel classifier. The overall classification accuracies were >91% with kappa coefficient between 0.80 and 0.94 for disease detection were achieved. Omission errors varied between 2% and 14% while commission errors ranged from 1% to 21%. These results indicated that the TM image could be used to accurately detect and quantify disease for site-specific WSM management.

Remote detection of WSM using geospatial imagery may substantially improve monitoring, planning, and management practices by overcoming some of the shortcomings of the ground-based surveys such as observer bias and inaccessibility. Remote sensing techniques for accurate disease mapping offer a unique set of advantages including repeatability, large area coverage, and cost-effectiveness over the ground-based methods. Hence, remote detection is particularly and practically critical for repeated disease monitoring and mapping over time and space during the course of a growing season.

A study was undertaken to map plant stress caused by *Heterodera schachtii* in sugar beets using remote sensing data by (Schmitz *et al.*, 2010). Digital color infrared (CIR) airborne images and round truth data from a field experiment with increasing nematode densities and sugar beet varieties were used to perform a supervised classification with specific image processing software. The full image was divided into classes by nematode density and used to identify the spectral response patterns for each class. Data from the near infrared and red bands were used to calculate the vegetation index NDVI. The NDVI seems to be a good identifier for characterizing *H. schachtii* damage in sugar beets.

Remote sensing technology is being examined as a field-worthy means for predicting reniform nematode population densities in the cotton rhizosphere using hyperspectral reflectance data and self-organized maps. Gary W. Lawrence *et al.* (2003) determined the hyperspectral reflectance of cotton plants affected by different reniform nematode population levels in controlled microplot tests and in fields naturally infested with the reniform nematode, determined spatial distribution of the nematodes in the field and establish zones for variable rate nematicide applications, and also determined the cotton yield improvements utilizing variable rate nematicide applications. Hyperspectral reflectance data was collected from reniform infected cotton plants growing in controlled microplots and from naturally infested field locations. Hyperspectral signatures have been identified which correspond to reniform nematode colonization and level of infection of cotton plants. These signatures and reniform populations were classified using a supervised artificial neural network Self-Organized Map Architecture. Nematode contour maps were developed to demonstrate the spatial distribution of the nematode and to establish zones of differing nematicide rate. Cotton yield results were higher in the site-specific application treatments compared with the conventional single rate applications. The combination of hyperspectral reflectance and VRT applications of nematicides may someday provide our producers with a rapid nematode prediction method and reduce nematicide applications to specific in-field sites.

The methods and apparatus for pest management were studied by Gary Lawrence *et al.* (2012) using remote sensing technology. One aspect was related to a method for detecting plant-parasitic nematodes using hyperspectral reflectance data. Another aspect of the present invention related to a device for determining the population of reniform nematode in a target. The further aspect of the present invention was related to a method for managing nematode population with variable rate applications of nematicides.

The ability to use a time series of orthophotos (1953–2008) in combination with Landsat satellite imagery, including trend analysis, and GIS was investigated to identify the presence of vegetation impacted by *P. cinnamomi* at four sites in *Banksia* woodlands in Western Australia (Barbara A. Wilson *et al.*, 2012). Further, the historical extent and rate of spread of *P. cinnamomi* was assessed at one site between 1953 and 2008. The assessment identified that three of the four sites were affected by *P. cinnamomi*, results that were consistent with on-ground surveys. Investigation of disease progression at one site found a large increase in the area impacted between 1974 and 1988 and the rate of spread was highest between 1953 and 1963 (1.286 m $year^{-1}$) and lowest between 1997 and 2008 (0.526 m $year^{-1}$). The presented techniques provide a cost-effective tool to monitor broad-scale vegetation dynamics over time for management of this plant pathogen.

A systematic review of biosurveillance models that operate on select agents and can forecast the occurrence of a disease event was presented (Corley *et al.*, 2014). A disease event was defined as a biological event with a focus on the one health paradigm. These events were characterized by evidence of infection and or disease condition. They reviewed models that attempted to predict a disease event, not merely its transmission dynamics and considered models involving pathogens of concern as determined by the US National Select Agent Registry (as of June 2011). They searched commercial and government databases and harvested Google search results for eligible models, using terms and phrases provided by public health analysts relating to biosurveillance, remote sensing, risk assessments, spatial epidemiology, and ecological niche modeling. They identified 44 models, classified as one or more of the following: event prediction, spatial, ecological niche, diagnostic or clinical, spread or response and reviews. The model parameters like etiology, climatic, spatial, and cultural and data sources like remote sensing, nongovernmental organizations, expert opinion and epidemiological were recorded and reviewed. A component of this review included the identification of verification and validation (V&V) methods applied to each model if any V&V method was reported. All models were classified as either having undergone some verification or validation method, or no verification or validation. They also outlined an initial set of operational readiness level guidelines for disease prediction models based upon established 'Technology Readiness Level' definitions.

4.15 FORECASTING OF PLANT DISEASES

Forecasting of plant diseases refers to predicting the occurrence of plant disease in a specified area ahead of time so that suitable management measures can be planned/ undertaken in advance to avoid losses. Disease forecasts are predictions of probable outbreaks or increase in the intensity of the disease. It involves well-organized teamwork and expenditure of time, energy, and money. It is used as an aid to the timely application of chemicals. Among the first spray warning services to be established for growers, were the grapevine downy mildew forecasting schemes in France, Germany and Italy in the 1920s. Disease forecasting methods are available for the following plant diseases (Table 4.4).

Table 4.4 Disease Forecasting Methods for Some Diseases

Sl. No.	*Plant disease*	*Countries*
1.	Grapevine downy mildew	Australia, France, Germany, Greece, Italy, Romania, Spain, USSR, Yugoslavia
2.	Cucurbit downy mildew	USA
3.	Potato late blight	Australia, Brazil, Finland, France, Germany, Greece, Japan, the Netherlands, Norway, Peru, U.K, the U.S.S.R.

4.	Tobacco blue mold	Canada, USA
5.	Apple and pear scab	Australia, Canada, Netherlands, New Zealand, USA
6.	*Sugar beet* root rot (*Aphanomyces* sp.)	USA
7.	Wheat brown (Leaf) rust	USA
8.	Corn bacterial wilt (*Erwinia stewartii*)	USA
9.	*Sugar beet* curly top	USA

The principle behind plant disease forecasting systems is to determine the risk that a disease will occur, or that the intensity of the disease will increase (Campbell and Madden, 1990). Campbell and Madden (1990) outlined several attributes, which define a successful plant disease forecasting system, including reliability (use of sound biological and environmental data), simplicity (the simpler the system, the more likely it will be applied and used by producers), importance (the disease is of economic importance to the crop, but sporadic enough that the need for treatment is not a given), usefulness (the forecasting model should be applied when the disease and/or pathogen can be detected reliably), availability (necessary information about the components of the disease triangle should be available), multipurpose applicability (monitoring and decision-making tools for several diseases and pests should be available), and cost effectiveness (forecasting system should be cost affordable relative to available disease management tactics).

Plant disease forecasting systems often provide information about how a grower's management decisions can help to avoid initial inoculum or to slow down the rate of an epidemic. These two concepts are important because they often differentiate the risk for a monocyclic disease (having only one cycle of infection) versus polycyclic disease, where there are multiple infection cycles, and a forecasting system can be used to time appropriate management tactics, such as a foliar fungicide application (Madden *et al.*, 2007). It should be noted that some plant disease forecasts focus both on avoiding initial inoculum and also on reducing the rate of the epidemic during the season.

Many plant disease forecasting systems have emphasized forecasts based on major principles, with examples (Agrios, 2004; Campbell and Madden, 1990), viz., forecasts based on measures of initial inoculum or disease [*Example*: Stewart's disease of corn] ; forecasts based on favorable weather conditions for development of secondary inoculum [*Example:* Late blight of potato (description and simulation)] ; forecasts based on both initial and secondary inoculum [*Example:* Apple scab (description and simulation)] and an example of a multiple disease/pest forecasting system is the EPIdemiology, PREdiction, and PREvention (EPIPRE) system developed in the Netherlands for winter wheat that focused on multiple pathogens (Reinink, 1986).

Current examples of plant disease forecasting providing daily information on-line are available for two important plant diseases, *Fusarium* head blight of wheat and Asian soybean rust. Both systems provide background information on the disease, current management recommendations, as well as disease forecast information based on host, pathogen, and environmental factors important for making an accurate forecast.

The successful development of a plant disease forecasting system also requires the proper validation of a developed model. There is increased interest among plant disease modelers and researchers to improve producer profitability through validation based on quantifying the cost of a model making false predictions (positive and/or negative). This methodology is not necessarily a new concept, as historical systems, such as the Mills rules for apple scab, or those used for potato late blight and *Alternaria* leaf blight of carrots, were developed using prediction rules for plant disease management. An economic validation of a plant disease forecasting system requires the examination of two false predictions - false positive predictions in which a forecast was made for a disease when in fact no disease was found in a location, and false negative predictions in which a forecast was made for a disease not to occur when in fact the disease was found. These two types of false predictions may have different economic effects for producers (Madden, 2006).

Lastly, the range of disease forecasting models has expanded to include a Bayesian statistical approach. Throughout the rest of this document, an introduction to plant disease forecasting is presented through examples, many of which use the R programming environment (Garrett *et al.*, 2007). A brief introduction to some of the mathematical/ statistical approaches that have been used for developing plant disease forecasting systems is presented, followed by an introduction to how using rainfall and temperature may be applicable for developing a forecast model, and finally, four case studies are presented that highlight the aspects, viz., modeling the generic risk of infection due to a plant pathogen, use of differential equations to model the population dynamics of sugar beet cyst nematodes in order to understand long-term changes in nematode density, and to develop a model that increases the accuracy of an existing forecast model through new information on the percentage of mature pseudothecia.

Forecasting diseases is a part of applied epidemiology. Hence, knowledge of epidemiology (development of disease under the influence of factors associated with the host pathogen) is necessary for accurate forecasting. The factors of the epidemic and its components should be known in advance before forecasting is done. The information required for forecasting include host factors (prevalence of susceptible varieties in the given locality, the response of host at different stages of the growth to the activity of pathogen, e.g., some diseases are found during seedling stages while others attack grown up plants and density and distribution of the host in a given locality). Dense populations of susceptible variety invite quick spread of an epidemic. Growing susceptible varieties in scattered locations and that too in a limited area are less prone to epiphytotic. Pathogen factors include the amount of primary (initial) inoculum in the air, soil or planting material, dispersal of inoculum, spore germination, infection, incubation period, sporulation on the infected host, redispersal/dissemination of spores, perennating stages, and inoculum potential and density in the seed, soil and air. Environmental factors include temperature, humidity, light intensity and wind velocity.

4.15.1 Requirements for Disease Forecasting

There are five main requirements which must be satisfied before a useful and successful disease forecast is made.

(1) The disease must cause economically significant damage in terms of yield loss or quality: Damage assessment is essential to develop a strategy for controlling a disease, e.g., Annual estimation of yield loss caused by barley powdery mildew (*Erysiphe polygoni*) in England and Wales had ranged from 6 to 13%. Potato late blight can cause a yield loss of 28% if the disease reaches the 75% stage by mid-August. Diseases like apple scab and potato common scab reduce the quality of the produce lower the value of the harvested crop and cause considerable financial loss to the growers.

(2) Control measures must be available at an economically acceptable cost.

(3) The disease must vary each season in the timing of the first infections and its subsequent rate of progress. If it does not, there is no need for forecasting.

(4) The criteria or model used in making a prediction must be based on sound investigational work carried out in the laboratory and the field and tested for a number of years to establish its accuracy and applicability in all the locations where its use is envisaged.

(5) Growers must have sufficient manpower and equipment to apply control measures when disease warning is given. Long-term warnings or predictions are more useful than short-term warning or predictions.

4.15.2 Methods of Disease Forecasting

Disease forecasting requires field observations on the pathogen characters, collection of weather data, variety of the crop, and certain investigations and their correlations. Usually, the following methods are employed in disease forecasting.

4.15.2.1 Forecasting based on primary inoculum

Presence of primary inoculum, its density and viability are determined in the air, soil or planting material. The occurrence of viable spores or propagules in the air can be assessed by using different air trapping devices (spore traps). In the case of soil-borne diseases, the primary inoculum in the soil can be determined by monoculture method. The presence of loose smut of wheat, ergot of pearl millet and viral diseases of potato can be detected in the seed lots at random by different seed testing methods. Seed testing methods can be used to determine potential disease incidence and enable a decision to be made on the need for chemical seed treatment. The extent of many virus diseases is dependent on the severity of the preceding winter which affects the size of vector population in the growing season [*Example:* Sugar beet yellows virus].

4.15.2.2 *Forecasting based on weather conditions*

Weather conditions, viz., temperature, relative humidity, rainfall, light, wind velocity, etc., during the crop season and the intercrop season are measured. Weather conditions above the crop and at the soil surface are also recorded.

4.15.2.3 *Forecasting based on correlative information*

Weather data of several years are collected and correlated with the intensity of the diseases. The data are compared, and then the forecasting of the disease is done. Forecasting criteria developed from comparisons of disease observation with standard meteorological data have been provided for diseases like *Septoria* leaf blotch of wheat, fire blight of apple and barley powdery mildew.

4.15.2.4 *Use of computer for disease forecasting*

In some advanced countries, forecasting of disease is made by the use of computers. This system gives the results quickly. One such computer-based programme in the USA is known as 'Blitecast' for potato late blight.

4.15.3 Examples of Disease Forecasting

Few examples of well-developed forecasting systems are mentioned below.

4.15.3.1 *Early and late leaf spots of groundnut*

A technique has been developed for forecasting early and late leaf spots of groundnut in the USA. When the groundnut foliage remains wet for a period greater than or equal to 10 h and the minimum temperature is 21°C or higher for two consecutive days or nights, the disease development is forecasted. A computer programme has been developed in the USA. This is accurate and is widely used in the USA. The data on hours for the day with relative humidity (RH) of 95% and above and minimum temperature (T) during the RH observations for the period, for the previous five days are fed to the computer. Calculations are rounded to whole numbers. The T/RH index for each of the five days is calculated, e.g., when hours of the RH 95% equal 10 and the minimum temperature during the period equals 21.1°C the T/RH index is 2.0. The T/RH indices for days 4 and 5 are summed. If the total index exceeds 4, disease is forecasted. If the index is 3 or less, no disease is forecasted.

4.15.3.2 *Late blight of potato*

In the USA a forecasting programme has been developed for late blight of potato. The initial appearance of late blight is forecasted 7 to 14 days after the occurrence of 10 consecutive blight favorable days. A day is considered to be blight favorable when the five day average temperature is 25.5°C and the total rainfall for the last ten days is more than 3.0 cm. A computerized version (Blitecast) has also been developed in the USA for forecasting potato late blight. Blitecast is written in Fortran IV. When a farmer desires blight cast (Blitecast), he telephones the blight cast operator and reports the most recently recorded environmental data. The operator calls for the blight cast programmes in the computer, viz., typewriter terminal and feeds the new data into the computer. Within a fraction of second the computer analyzes the data and series of a forecast and spray recommendations to the operator who relays it to the farmer. The entire operation can be completed during standard three minutes telephone call. The system makes one of the four recommendations, viz., no spray, late blight warning, seven days spray schedule or five days spray schedule. The last five days spray schedule is issued only during severe blight weather. In West Germany, 'Phytoprog' is the programme used. It is based on measurements of temperature, relative humidity, and rainfall. Phytoprog provides a negative prognosis (an indication of when the usual routine spray application should be dispensed with).

4.15.3.3 *Blister blight of tea*

A system for predicting epidemics of blister blight of tea has been developed based on the number of spores in the air in the tea plantation and the duration of surface wetness on the leaves. The duration of sunshine is negatively correlated with the duration of surface wetness. The following prediction equation has been developed. $Y = 1.8324 + 0.8439\ X_1 + 0.9665\ X_2 - 0.1031\ X_3$ where, X_1 = log% infection $t_2 X_2$ = log% infection t_2 - log infection $t_1 Y$ = log of the number of spores in the air and $t_1 - t_2$ three weeks X_3 = mean daily sunshine for a 7-day period preceding t_2.

4.15.3.4 *Southern corn leaf blight*

'Epimay' is a system for forecasting Southern corn leaf blight (*Bipolaris maydis*) based on the conceptual model.

4.15.3.5 *Rice blast*

In India, forecasting rice blast is done by correlative information method. It is predicted on the basis of minimum night temperature 20 to 26°C in association with high relative humidity of 90% or above. Computer-based forecasting system has also been developed for rice blast in India.

4.15.3.6 *Wheat stem rust*

Forecasting wheat stem rust epidemic is done by analyzing the rain samples which give precise data for inoculum present in the air. Moreover, several wind trajectories are also prepared to survey the airborne primary inoculums and its deposition. It has been observed that primary inoculum comes from South India to the plains of Central and North India.

4.15.3.7 Brown stripe downy mildew of corn

The forecasting of brown stripe downy mildew of corn (*Sclerophthora raysise* var. *zeae*) which is restricted to India is done on the basis of average rainfall 100 to 200 cm or more accompanied by low temperature (25°C or less). Spore trapping techniques of acquisition of biological data for consecutive forecasting models are important. Spore traps have been widely used in to complete disease with weather conditions. Spore trapping is useful for understanding the epidemiology of a disease and behavior of the pathogens. This helps in developing models of dispersal of pathogens or epidemiology of the disease and to formulate methods of management. The methodology of spore trapping depends on the objectives of the worker like the biology of the pathogen, for infection forecasting, spore dispersal gradients, and management of the disease.

4.15.3.8 Cereal cyst nematode

Several simple agronomic practices such as crop rotation, the timing of sowing, no-till management, and resistant cultivars can reduce damage by the cereal cyst nematode (CCN), (*Heterodera avenae*), but the damage levels are unpredictable. David J. Bonfil *et al.* (2004) conducted a study to develop a fast assay that would forecast the potential damage to wheat by CCN in the fields. Forecast procedure was based on the SIRONEM bioassay with several modifications in the sampling and incubation of field soil samples and cultivation in them of wheat seedlings. For CCN disease assessment, a damage model that takes account of CCN infection and its damage to roots was developed. Field experiments were established to validate the resulting damage model and showed a correlation between the forecasted damage and field rating of CCN infection. This assay has advantages over other CCN estimation methods because it directly forecasts the potential crop damage in the field and it requires a relatively short period to obtain the data. According to the forecasted damage and weather conditions, the most appropriate crop management regime, especially of sowing, could be chosen for each field assayed.

4.15.4 Spore Traps

In epidemics of airborne plant diseases the number of spores of the pathogen landing on the plant which depends on the number of spores in the atmosphere above the crop is an important factor for the quantitative sampling of the atmosphere (number of spores per unit volume of air). For trapping and estimating these studies different types of traps are used.

The following spore traps are usually employed in the trapping of fungal spores:

4.15.4.1 Cylindrical rods or microscopic glass slides

It helps to gather data on the spore arrival in a locality. In this, the surface of the microscopic slide is smeared with grease and made sticky. In the method, quantitative estimation is not possible as the number of spores collected is very low.

4.15.4.2 *Hirst's volumetric spore trap*

In this instrument, air is sucked into at a controlled rate and impinged onto a glass slide moved by a clockwork mechanism past the orifice. It gives a continuous count of spores in 24 h. The number of spores per unit volume of air at any given time can thus be calculated.

4.15.4.3 *Rotorod sampler or rotorod spore trap*

It comprises of a 'U' shaped rod attached at its midpoint to the shaft of a small battery operate the electric motor. In this equipment, the surface of the rod is covered with a vaseline strip of transparent cellophane to catch spores which can be taken off and mounted on a glass slide. From the area of the strip and the speed of rotation, the volume of air samples can be calculated.

4.15.4.4 *Anderson cascade spore sampler*

It is a device where Petri plates with nutrient agar are used to collect the spores.

4.15.4.5 *Bourdillon slit sampler*

Air is sucked in a chamber by a vacuum pump which strikes the rotating Petri dish containing agar medium. The agar medium retains the spores sucked in the air. The concentration of viable spores is calculated after counting germinated spores in the medium.

4.15.4.6 *Burkard's day volumetric spore trap*

This device records spores in the air drawn by a pump on seven days basis on a cellophane strip wrapped on a drum rotating inside a chamber.

4.15.4.7 *Jet spore trap*

In the above sampling methods, the viability of the spores cannot be determined. To overcome this, living plants have been used as spore traps. In a jet spore trap in which spores are impacted in an air jet into a column of still air, through which they fall, to settle on leaf segments exposed at the base of the chamber. In this trap, suitable cultivars of host plants can be employed to determine the number of viable spores.

Plant diseases cost growers and consumers billions of dollars each year, by affecting both quality and quantity of food products. Traditional disease management techniques are often costly and may only be partially effective. Each type of disease-causing pathogen and the resulting epidemic it causes is characterized by its own unique biological system combined with specific spatial and temporal dynamics (Anonymous, 2001).

Fungal diseases, in particular, are greatly influenced by climatic conditions and weather patterns. Plant pathologists and meteorologists have often collaborated to develop disease forecasting or warning systems that attempt to help growers make economic decisions for managing diseases. These types of warning systems may consist of supporting a producer's decision-making process for determining cost and benefits for applying pesticides, selecting seed or propagation materials, or whether to plant a crop in a particular area. The primary principle of developing and utilizing disease forecasting systems is to determine the risk for a disease to occur and estimate whether the intensity of disease will increase. A successful system will allow both commercial growers and home gardeners to use the information to make timely decisions about practical management of diseases through fungicide applications or other management tactics.

Two forecasting systems for Nebraska crops have been developed and are currently being implemented for *Cercospora* leaf spot in sugar beets and scab of wheat. In addition to forecasting models, surveillance activities for airborne diseases which do not overwinter locally are extremely useful for timely management. The national sentinel monitoring program for soybean rust is serving this need for Nebraska producers.

Plant diseases and the pathogens that cause them are distributed spatially at different scales ranging from less than one meter to entire continents. The distribution of disease may result from local pathogen populations or the spread of disease within a geographic region (Molineros *et al.*, 2004). Locations or regions that have consistently high levels of the disease may have characteristics unlike those of the surrounding areas. For example, a particular section of a state may have a climate or cropping practices that are more conducive to pathogen survival, and/or disease development. These highly conducive areas can be called disease 'hotspots.' If patterns in these characteristics can be identified, it may provide valuable insights into the factors contributing to disease epidemics and could allow disease prediction models to account for the elevated risk levels in some regions of the country. Hotspot analysis represents a method of searching for consistent spatial patterns in the desired variable. The analysis evaluates three dimensions of patterns, viz., spatial, patterns within a landscape; temporal, persistence of high disease levels and spatial-temporal, addressing the migration of patches of disease or hotspots. FHB is a good candidate for hotspot analysis because of the availability of geo-referenced observations of disease intensity coupled with weather-driven models of disease biology, and information about crop/disease management practices (tillage, crop rotation). It was anticipated that hotspot analysis will allow us to identify areas with a consistently higher probability of epidemics, given consistent patterns in weather and cultural practices and whether these regions have expanded or changed over time. This analysis may also help assess factors that may contribute to the elevated levels of disease within these hotspots and suggest adjustments in crop management practices.

Frequent and serious biological disasters cause great damage to crop production. The plant disease data of multiform could be analyzed, and then a forecasting report was made (Ming-wang Shi *et al.*, 2009). The article, according to characteristics of the plant disease, introduced forecast system's design and key technologies realization based on data mining's plant disease system. The data mining function was divided into two kinds, description, and forecast. The data mining is a kind of deep level data analysis and can extract from the mass data.

Prediction of disease occurrence is a well-known historical theme and has begun to receive new interest due to internet-based prediction systems. The evaluation of these systems in a quantitative manner is an important step if they are to be used in modern agricultural production (Yuen *et al.*, 2003). Bayes' theorem is one way in which the performance of such predictors can be studied. In this way, the conditional probability of pest occurrence after a positive or negative prediction can be compared with the unconditional probability of pest occurrence. Both the specificity and the sensitivity of the predictive system are needed, along with the unconditional probability of pest occurrence, to make a Bayesian analysis. If there is little information on the prior probability of disease, most predictors will be useful, but for extremely common or extremely rare diseases, a Bayesian analysis indicates that a system predicting disease occurrence or nonoccurrence will have limited usefulness.

It is a known fact that a better knowledge of epidemiology and forecasting provides the basic information to design an efficient and workable plant disease management schedule (Naresh Mehta, 2014). Epidemiology and forecasting for the management of diseases of rapeseed and mustard were investigated. Weather variables including temperature, relative humidity (RH), rainfall, wind velocity and direction, leaf wetness duration and solar radiation influence various parameters of the infection process and disease development. Interaction among these parameters and disease development pave the way for the development of prediction models. It was observed that development of *Alternaria* blight was favored by the maximum temperature of 20–25°C, minimum temperature of 15°C, RH more than 90%, RH of more than 50%. For white rust average temperature of more than 15°C and RH more than 65% with intermittent rains was most effective for disease development. For downy mildew, a temperature range of 15–20°C with high RH was found the optimum. Leaf wetness duration of 4–6 h at 20°C and 6–8 h at 15°C was essential for the initiation of downy mildew. Powdery mildew development was favored by a temperature range of 16–28°C, mean RH less that 60% and dry weather, particularly during February-March. The white stem rot or *Sclerotinia* rot disease development was favored by high RH of more than 80%, the maximum temperature of 25°C and a minimum of 5–12°C. It was felt that the prediction models developed at one location may not fit at other locations, which indicates that data has to be generated for a longer period, and the model be tested at multilocation. The disease forecasting models must be developed by considering the crop variety, the prevalence of a specific pathotype and the microclimatic factors for higher efficiency.

Plant diseases and the pathogens that cause them are distributed spatially at different scales ranging from less than one meter to entire continents. The distribution of disease may result from local pathogen populations or the spread of disease within a geographic region. Locations or regions that have consistently high levels of the disease may have characteristics unlike those of the surrounding areas.

4.15.5 Mathematical Concepts for Disease Forecasting

Prediction of disease outbreaks enables the effective use of control measures, such as chemical or biological treatments, the prediction of crop yields and of the market potential for that crop. Disease forecasting involves the use of weather data and biological data to predict disease incidence. Usually, disease forecasting is only performed on economically important diseases, and as a method of cost reduction. If controlling a particular disease involves an expensive or time-consuming treatment, being able to predict outbreaks of the disease allows the treatment to be timed correctly, increasing its effectiveness, and reducing the cost compared to repeated treatments. Because environmental conditions vary from season to season, disease forecasting is necessary to predict the chance of disease in a certain set of conditions. The disease can be forecast using computer modeling and empirical correlations relating to weather conditions, levels of inoculum, test plots and site factors and the predictions can then be communicated to growers. Computer modeling of plant diseases uses systems analysis to accumulate all the factors that affect the development of a certain disease into a computer-based model and make predictions of the disease under different environmental conditions. A disease needs to be well understood to formulate an accurate model, and models based on diseases that we know little about are generally not very accurate. The more straightforward approach to developing empirical correlations between particular weather factors and disease has had considerable success. This does not attempt a complete modeling of all factors involved in a disease, but only those most important in affecting the disease. The accuracy of the model can then be measured statistically by comparing its predictions with what actually happens.

Monitoring the weather is the most important consideration in disease forecasting because of the overriding effect that weather has on disease development. While broad scale weather data has been used for disease forecasting, it is well known that microclimate within the crops has a more direct impact on disease. Devices have been developed to monitor microclimate factors like duration of leaf wetness and temperature and with time, they will be affordable and accurate enough for widespread use on individual farms. Synoptic weather forecasting charts can be used to predict 'critical periods,' the occurrence conditions favorable for disease development, so that farmers can spray their crop before it happens. There are now several self-calculating disease forecasting monitors available commercially that use environmental data and past season data to predict outbreaks of particular disease.

Some disease forecasting methods are based solely on monitoring inoculum levels, often as indicated by the amount of disease already present. This method can be successful when the disease is developing steadily under relatively uniform or predictable weather conditions, but not for diseases that can spread explosively in favorable conditions. Monitoring the amount of disease present can indicate whether the amount of disease is likely to exceed a certain threshold, at which point control measures become economical. There are numerous methods of directly monitoring the concentration of spores in the air as an indication of the chance of disease. Trapping vectors of diseases can also be useful in predicting the occurrence of viral diseases. In addition, estimation of populations of soil-borne pathogens by examining soil samples is necessary for predicting the outbreak of the diseases they cause. Monitoring systems can be combined with data specific to the site and the crop, such as soil type, topography, and irrigation levels, to increase the accuracy of predictions.

Test plots (or trap plots) of susceptible cultivars can be planted throughout a cropping area to give early warning of the arrival of inoculum or disease vectors. Alternatively, inoculation of the test plot with the pathogen can give an indication of favorable environmental conditions for disease development. Test plots are also useful for monitoring the occurrence of minor diseases on new cultivars.

The formation of a prediction is useful only if it can be 'communicated' to the growers who will be affected by it. General warnings for areas are broadcast over the radio or internet. Predictions based on monitoring by individual farmers or groups of farmers in an area remove the need for widespread communication systems. Computerized decision support systems based on local monitoring can educate and empower farmers when making decisions about their crops.

4.15.5.1 Application of maximum likelihood based methods for disease forecasting

With the development of more rapid computing technologies, it is possible to compute a large number of simulations to support new modeling methods for plant disease forecasting. The development of disease forecasting models is based on determining the value of unknown parameters for an appropriate model.

One method for estimation of parameters is maximum likelihood estimation (MLE), first developed by R.A. Fisher (Myung, 2003). The basic idea behind MLE is that one can calculate the probability of observing a particular set of data given the model of interest, and this is called the likelihood of a sample. The likelihood of observing the actual data for each of the potential parameter values is calculated to determine which set of parameter values maximizes the likelihood. Schmidt, (1993) studied the modeling of the population dynamics of the sugar beet cyst Nematode, *Heterodera schachtii* (Fig. 4.16).

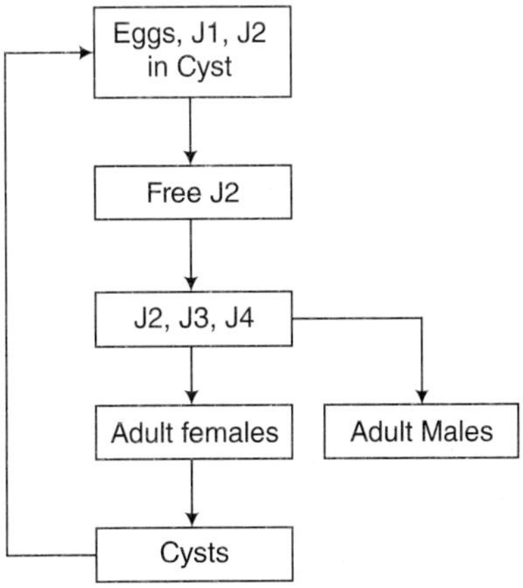

Fig. 4.16 Simple model of the sugar beet cyst nematode life cycle

Heterodera schachtii, the sugar beet cyst nematode, causes significant crop damage (Ravichandra, 2014). Female nematodes form cysts that, when mature, are able to overwinter in the soil. These cysts consist of the protective ectoderm of the female and several eggs. As the season progresses, the eggs hatch in the female to form J1 stage which then molt once and form J2.

When environmental conditions are conducive, J2 leave the cyst and find a plant root to feed on. The juveniles develop through J3 and J4 stages until adult males and females are formed. Adult males leave the roots and search for females. Adult females, when fertilized, form cysts. The number of nematodes in a field is highly dependent on two things temperature and the current and previous crop species. Relatively warm years may result in several generations of nematodes whereas cold years may only allow one generation of reproduction. Though the sugar beet nematode is somewhat generalist and can survive to some degree on many different hosts, it is greatly favored by planting sugar beets. Crop rotation may be an effective way to reduce the population of the sugar beet cyst nematode. Knowledge of previous crop rotation and temperature may be used to forecast the amount of disease in a given sugar beet field over a season (Schmidt *et al.*, 1993). In the model developed by Schmidt *et al.* (1993), probabilities are calculated for movement between each of the stages of a nematode life cycle.

Before the model is more thoroughly described, please note that the θ parameter is a crop-specific parameter that may take on a different value for each type of crop grown. In a more complete model set forth by the authors, θ is calculated separately for each component of the life cycle (the crop specific parameter is not necessarily equal for all components of the life cycle).

Factors affecting the ease or difficulty of forecasting and modeling cyst, root-ectoparasitic and soil surface nematodes were discussed (Jones and Kempton, 1980). To relate preplanting to postharvest populations of cyst nematodes, a version of the logistic curve was derived from first principles and modified to take account of the injury done to root systems as numbers increase, and of eggs carried over to the following year. Egg hatch, when potato or other crops are grown, appeared to be independent of population density. So also did root invasion over the population densities encountered in field plots. In the absence of enemies, population density was determined by competition for feeding sites induced by females in the root system, and by the size of the root system. Population oscillations about the equilibrium point were affected mainly by the proportion of the population that failed to hatch and the damage to the root system. The process of model building and the parameters required were outlined. The model included hypotheses relating to intraspecific (genetic) competition and interspecific competition. The limitations of models and the need to make them stochastic were discussed.

Multispectral remote sensing allows a wide range of application in precision farming. Remote sensing provides growers with yield assessments; shows yield variations across fields and give information about the growth rate at important development stages. This includes detection of stress due to drought and nutrient deficiency as well as a result of plant diseases or animal pests.

A study by Jacobi *et al.* (2011) was focused on the spatial distribution of fungal infestations in winter wheat and nematodes in sugar beet at different crop growth stages. Of particular interest was the detection of fungal diseases in early growth stages by the use of very high-resolution remote sensing (VHRRS) imagery. These images were taken from the QuickBird, the satellite with the highest available resolution up to 0, 6 m in the panchromatic mode and 2, 4 m in the multispectral mode. Due to the rare availability of the QuickBird images, alternatives to the satellite images were tested, which were aerial hyperspectral images (HyMap-sensor with 126 spectral channels), images taken by a NIR digital camera, which was carried by a model helicopter for aerial view or fixed to a stand to take images in close-up range. With image analysis, interpretation in relation to ground truth data and statistical analysis with GIS software the spatial spreading and infestation level of the fungal diseases in wheat and nematodes in sugar beet was mapped.

4.15.5.2 Disease severity and crop loss

Assessment or measurement of disease is the basis of epidemiology which is the study of disease at the level of populations of pathogens and hosts. It is also the basis of the study of the effects of disease on crop yield and of disease forecasting, which involves the prediction of the amount of disease that is likely to occur at some time in the future (John Brown and Philip Keane, 2012). It is usually not sufficient to determine whether a disease is present or absent. The critical information required is the amount of disease that is present. The disease often has to exceed a certain threshold before it reduces the yield of a crop. Small amounts have little effect on yield, and the disease may not be worth controlling. The amount of disease is measured as the proportion of the crop population (counted as individual plants or branches or leaves etc.) that is infected (disease incidence) or the proportion of the area of a plant or plant organ (e.g., leaf area) that is affected (disease severity). In some cases, the proportion of leaves or branches infected may provide a measure of disease severity. Although it is known that plant diseases reduce crop yields compared to yields that could be expected in the absence of disease, it is usually difficult to obtain accurate quantitative estimates of yield reductions caused by specific diseases.

Many diseases occur on senescing tissue which may not have been contributing much to yield. Plants are capable of compensating for the loss of a certain amount of leaf area, especially in crowded crop populations. The only diseases where the effect on yield is relatively easy to measure are those that kill entire trees in orchards or plantations (e.g., *Phytophthora* root or collar rot of citrus or apple), and those that destroy the actual harvested product, either just before harvest (e.g., the smuts, which destroy cereal grain, and fruit rots such as cocoa pod rot caused by *Phytophthora palmivora*) or after harvest (all postharvest rots of fruits and vegetables).

4.15.5.3 Methods for assessing disease incidence and severity

Disease assessment methods should provide objective measurements so that results obtained by different workers, from different locations, and different seasons are comparable, be simple and quick to use, be related to an identifiable growth stage of the crop and provide an adequate sample of the entire area of crop to which the assessment refers. Thus, disease assessment of a crop encompasses two main aspects-assessment of disease in samples from the crop and adequate sampling from the whole crop. The most common method of estimating disease severity is the use of a set of diagrams (disease assessment keys) of a crop (commonly leaves, but it could involve inflorescences, fruits or whole plants) showing different disease severities as blackened areas. Severity scales are adjusted to take into account the above concepts of visual perception. Samples of the crop are then compared with these diagrams to allow an assessment of severity. A disease index for a crop can be obtained by summing the individual disease assessments and then dividing this by the number of plants assessed. One standard example of descriptive and pictorial disease assessment keys is shown in Table 4.5 (British Mycological Society, 1948).

Table 4.5 Key for the Assessment of Late Blight of Potato Caused by *Phytophthora Infestans*

% Severity	*Symptoms*
0	Disease not seen in the field.
0.1	Only a few plants affected here and there. Up to one or two spots in 10.8 m.
1	Up to ten spots per plant or general light spotting.
5	About fifty spots per plant, or up to one leaflet in ten attacked.
25	Nearly every plant with lesions; plants still retaining normal form; fields may smell of blight but look green although every plant is affected.
50	Every plant affected and about half of the leaf area destroyed by blight; field looks green flecked with brown.
75	About three-quarters of the leaf area destroyed by blight; field looks neither predominantly brown or green. In some varieties, the youngest leaves escape infection so that the green is more conspicuous
95	Only a few leaves left green but stems green.
100	All leaves dead but stems dead or dying.

4.15.6 Assessment of Crop Losses

Having determined the amount of disease in relation to the growth pattern of a crop, the next step is to determine the effect of different levels of disease on the yield of the crop. The two main approaches used to relate disease intensity to crop losses are the experimental and the statistical (John Brown and Philip Keane, 2012).

4.15.6.1 The experimental approach

The experimental approach to crop loss assessment usually involves setting up experiments in which the level of disease in a crop is controlled using different levels of inoculation or treatments with biocides (fungicides, nematicides, bactericides). Often, an intensive spraying schedule is used to hold disease to a level close to zero, which allows the potential yield of the crop in the absence of disease to be determined. A reduced spray treatment may allow development of a moderate disease severity while withholding of treatment may allow development of a severe epidemic. This allows a comparison of epidemic progress and yields in the plots. From this a relationship between disease parameters and yield can be determined, allowing prediction of crop loss based on certain disease intensities at certain periods of crop growth. The value of this method depends on the assumption that the biocide treatment itself has no direct effect on yield. In some cases, this assumption is not valid. For example, sprays of copper-based fungicides used to control coffee rust often have a direct 'tonic' effect on yield of coffee, quite apart from any indirect effect they may have through control of rust. In these studies, it is important to determine the effect of a biocide on yield in the absence of disease.

4.15.6.1.1 Critical-point models

Critical point or single point models are based on a disease assessment that is made at a given time in the plant's life (the critical point) and then related to yield loss. Relationships between the amount of disease and percentage loss in yield are usually expressed in the form of linear regression equations where the independent variable is disease intensity (X) at the particular stage of crop development and the percentage loss in yield (Y) is the dependent variable. For example, the mean percentage reduction of grain yield attributable to powdery mildew in spring barley and oats in the United Kingdom can be expressed as $Y = 2.518$ where X is the percentage of mildew at the 10.5 stage of plant growth. Thus, if 25% of the plant is covered with mildew at the 10.5 stage of plant growth, one might expect a yield reduction of 12.5%. Similarly, loss in yield due to rice blast in Japan can be estimated by the equation $Y = 0.57X$ where X is the percentage of blasted necks (culm region immediately below the inflorescence) at 30 days after heading. In these examples, the disease severity was measured at a specific stage of plant growth (the critical point). A major criticism of critical point models is that they do not take into account the possibility that infection rates (r) may vary throughout an epidemic. They ignore the shape of the disease progress curve. The model cannot distinguish between a short, severe epidemic and a long, less severe epidemic if both happen to have the same amount of disease at the critical point. However, they may be very useful when disease progress is relatively uniform or when the amount of disease at a certain stage of crop development has a dominant effect on yield.

A model to estimate the damage caused by gray leaf spot of corn (*Cercospora zea-maydis*) was developed from experimental field data gathered during the summer seasons of 2000-01 and during the second crop season [January-seedtime] of 2001, in the southwest of Goiás state (Reis Erlei *et al.*, 2007). Three corn hybrids were grown over two seasons and on two sites, resulting in twelve experimental plots. A disease intensity gradient (lesions per leaf) was generated through the application, three times over the season, of five different doses of the fungicide propiconazole. From tasseling onward, disease intensity on the ear leaf (El), and El – 1, El – 2, El + 1, and El + 2, was evaluated weekly. A manual harvest at the physiological ripening stage was followed by grain drying and cleaning. Finally, grain yield in kg.ha^{-1} was estimated. Regression analysis performed between grain yield and all combinations of the number of lesions on each leaf type, generated thirty linear equations representing the damage function. To estimate losses caused by different disease intensities at different corn growth stages, these models should first be validated. Damage coefficients may be used in determining the economic damage threshold.

4.15.6.1.2 Multiple-point models

To develop multiple-point models, disease assessments are made at regular intervals during the development of the epidemic. An empirical equation obtained by multiple regression analysis can be used to estimate yield losses for disease progress curves showing different characteristics. The general form of the regression equation is $Y = a + b_1X_1 + b_2X_2 + ... + b_nX_n$, where Y is the loss in yield and X_n, = the amount of disease measured at the nth assessment time. The constants b can be determined from the crop loss experiments described above. Multiple-point models are especially useful in yield loss analyses where infection rates are variable, where the shape of the epidemic progress curve varies and for epidemics of different lengths. Because they take into account variations in epidemic progress, they are regarded as allowing more accurate yield predictions. In practice, the concepts of the critical point and multiple point models are often combined because the multiple point observations are often made at specially chosen critical stages in the development of the epidemic or the crop (i.e., stages at which the observations have greater predictive value for yield).

Sclerotinia stem rot (SSR) caused by *Sclerotinia sclerotiorum* is one of the most important diseases of canola (*Brassica napus*) in Golestan province, the leading canola producer in Iran. To assess the yield loss of canola caused by SSR, 80 fields were surveyed in four different regions of the province (Gorgan, Ali Abad, Kalaleh and Gonbad) during 2006–2007, and SSR intensity was recorded weekly in the fields (Mohammad Ali Aghajani *et al.*, 2013). Study of yield loss-SSR severity relationships by linear, nonlinear and multiple regression analyses with final intensity (Sf), time to initial symptoms (tis), Gompertz rate of disease progress (rG), and standardized area under disease progress curve (SAUDPC) as independent variables indicate that single point and integral models were significant ($P < 0.05$) only in three cases. Results of multiple point models which were performed using weekly recorded SSR intensities (S1, S2, ...), were significant in two cases, and a general model for 2007 survey was developed using S3 to S6. Eventually, response surface models were developed for each region by integrating this with SSR intensity variables (Sf or SAUDPC).

4.15.6.1.3 Area under disease progress curve (AUDPC)

The relationship between disease and yield loss can be determined by studying the area under the disease progress curve. This is based on similar logic to the multiple point model except that it involves the integration of the entire disease progress curve. The model assumes that yield loss is proportional to the amount of disease, and that yield reduction is proportional to the duration of the disease. Compared with the critical point model it has the advantage that it can distinguish between two epidemics with different areas under the curve but which have the same percentage severity at a critical time. The main disadvantage of the model is that it does not differentiate the severity of disease relative to time. James (1974) pointed out that when the model was applied to potato blight epidemics, it could not distinguish between epidemics which commenced at different stages in the plant's growth but which showed similar areas under the disease progress curve. He found that early light infections caused greater yield reductions than late; severe infections even though the areas under the disease progress curve were the same. This problem could readily be overcome by inclusion in the model of a factor which relates AUDPC to particular stages of crop development.

4.15.7 Statistical Approach

This approach to disease assessment involves statistical analysis of yields under different levels of disease that occur naturally in the field. The methods used include comparisons of yields obtained from crops in seasons with different disease intensities, comparisons of expected (based on, for example, experience under disease-free conditions on experimental stations) and actual yields, comparisons of the yields of a particular crop with the average yield for the region, and comparisons of epidemic progress and yield parameters between crops or portions of a crop showing different levels of infection or damage. These comparisons can be very useful because they are based on the real cropping situation in the field rather than on experimental plots, which often give artificially high yields. However, account has to be taken of the fact that in the field many other factors besides disease will affect the yield. Using statistical analysis of natural disease epidemics, critical point, multiple point and AUDPC models relating crop loss to epidemic development can also be determined.

4.15.7.1 Plant disease surveys

Another statistical approach to the assessment of crop loss involves the use of questionnaires to assess disease incidence or severity. Results of farmers' assessments of the disease can be related to the yields eventually obtained. Appraisals of disease and yield loss are often of limited value because of the difficulty of obtaining information concerning the prevalence and severity of diseases over large areas. For this reason, there is a special need for an increased use of plant disease surveys. The information obtained from plant disease surveys enables the relative importance of different diseases to be determined, which in turn enables rational allocation of resources for research as well as control and eradication programs. Disease surveys are also necessary to establish the geographic distribution of pests, pathogens, and their various pathotypes; the rate of spread of newly introduced pests and pathogens; the presence of other hosts of pathogens and pests and 'breakdowns' in disease resistance used in crop plants.

Knowledge about epidemiology and the impact of disease on yield is fundamental for establishing effective management strategies. Carisse *et al.* (2013) investigated the relationship between foliar strawberry mildew severity, *Podosphaera aphanis* airborne inoculum concentration, weather, and subsequent crop losses for day-neutral strawberry. The experiment was conducted at three, five, and four sites in 2006, 2007, and 2008, respectively, for a total of 12 epidemics. At each site, data were collected on 25 plants at 2-day intervals from the end of May to early October for a total of 60 to 62 samplings annually. First, seasonal crop losses were statistically described; then, a lagged regression model was developed to describe crop losses from the parameters that were significantly associated with losses. There was a strong positive linear relationship between seasonal crop losses and the area under the leaf disease progress curve (R(2) = 0.90) and daily mean airborne conidia concentration (R(2) = 0.86), and a negative linear relationship between crop losses and time to 5% loss (R(2) = 0.76) and time to 5% leaf area diseased (R(2) = 0.61). Among the 53 monitoring- and weather-based variables analyzed, percent leaf area diseased, log10-transformed airborne inoculum concentration and weather variables related to temperature were significantly associated with crop losses. However, polynomial distributed lag regression models built with weather variables were not accurate in predicting losses, with the exception of a model based on a combined temperature and humidity variable, which provided an accurate prediction of the data used to construct the model but not of independent data. Overall, the model based on log10-transformed airborne inoculum concentration did not provide accurate crop loss predictions. The model built using percent leaf area diseased with a time lag of 8 days ($n = 4$) and a polynomial degree of 2 provided a good description of the crop-loss data used to construct the model ($r = 0.99$ and 0.90) and of independent data ($r = 0.92$). For the 12 epidemics studied, 5% crop loss was reached when an average of 17% leaf area diseased was observed since the beginning of symptom development. These results indicated that information on foliar powdery mildew must be considered when making strawberry powdery mildew management decisions.

The amount of disease is generally referred to as disease intensity. On the other hand, disease severity is determined by a function of the degree of affection, colonization and damage of host tissue. The reduction in the amount of host development and growth is a function of disease severity, and yield realization is a function of host development. Thus, measurement of severity based on lesion number or lesion area may be less related to yield, but more to disease progress. Measurement of both actual and visible disease (Ghosh *et al.*, 2009) in terms of percent tissue affected and/or the green leaf area duration respectively gives a more precise conclusion on disease and yield loss in the presence of a pathogen. However, as translation of disease severity into crop biomass loss or yield loss is generally done by actual estimation and regression, or by including host physiological variables in yield or yield loss for more precise functions. The relationship between disease and yield is most often summarized as a simple empirical model that describes average crop performances is the presence of a pathogen. Such models may be robust and useful for surveys, but their use is usually constrained to the specific conditions under which the model was developed. Changes in production system usually invalidate the relationship. The alternative is to base the relationship on an epidemiological analysis of the pathogen population and a physiological concept of host growth and development. This review provides the knowledge and conceptual basis and discusses the limitations to progress in the development of such models. It is shown that a host-based assessment of disease is well suited to yield investigations and to multiple pest constraints and that disease is logically related to yield via radiation interceptions and radiation use efficiency (Gaunt, 1995).

This study has enabled indirect estimation and prediction of crop loss by developing significant (P = 0.05) linear prediction equations for severity in terms of intensity or occurrence, irrespective of the months of prevalence in leaf spots in general. In rose die-back and tar spot relation is linear between intensity and severity. Thus, severity can be predicted from the intensity and/or occurrence. When r values exceed 0.75, the function of severity may be considered bold enough to take up preventive and protective measures against diseases causing apprehension. Losses can easily be predicted as the relations are linear. These mathematical functions of severity on the intensity and even severity on the occurrence, if and where validated, can help in predicting a probable loss and forestalling further progress of disease through timely management measures as in supervisory management and appropriate IPM (Dasgupta, 1988). The methodology developed can be used as a simple technique to determine the diseases which are systemic or rapid and repetitive secondary infection (severity depends on occurrence), and local infection (severity depends on intensity). The mid values would indicate the role of primary and secondary inoculum.

Proportional loss models commonly used in disease surveys are based on the assumption that percent yield loss is the same in all crops, regardless of their yield potential (Helen O. Whelan, 1992). Estimates of regional crop loss may be inaccurate if the relationship between disease and yield loss is affected by crop yield potential. The importance of crop yield potential in disease: yield loss modeling was investigated, and models for more accurate regional crop loss estimates were developed, taking crop yield potential into account. Yield loss models were derived by regression analysis for each crop.

Single point, multiple point and area under curve models were derived from %DS and ~GLAI variables and proportional (%) and actual (gDM/m^2) grain yield. The effect of yield potential was determined by comparing regression equation coefficients for each crop with crop yield potential. An area under green leaf area index curve (AUGLAIC): actual yield model was best suited for determining the effect of yield potential on yield loss. This model was selected because AUGLAIC summarized the effect of disease on plant growth over the season and actual yield represented the crop yield potential in the absence of disease and the response of actual yield to disease. Crop yield potential did not affect actual yield loss caused by leaf rust.

Disease measured as AUGLAIC explained most of the variation in yield (R2 adj = 0.93) for all crops in both years. Assessment of GLAI is not suitable for estimation of regional crop loss because of the requirement for a rapid and low-cost method. The reflectance of NIR from the crop canopy was investigated as an alternative to GLAI measurements. Reflectance was correlated significantly ($P < 0.001$) with GLAI (r = 0.66 to 0.89) and green area index (r = 0.76 to 0.92). Reflectance measured at grain-filling (GS 85–87) explained most (R2 adj = 0.94) of the variation in yield for all crops in both years. The relationship between AUGLAIC and yield was validated with data from independent diseased and healthy barley crops. The AUGLAIC: yield model described the effects of disease on yield accurately but overestimated yield by 49 to 108% in the healthy crops.

Models based on accumulated PARij intercepted by green leaves explained the observed deviations in yield of these crops from the AUGLAIC: yield model. Accumulated PAR models accounted for differences in incident radiation, canopy structure, radiation interception by green leaves, radiation use efficiency and harvest index which are important in determining dry matter production and grain yield. Accumulated PAR models described the effects of disease on crop growth which were not represented by GLAI alone. Variation in crop yield potential at the regional scale is important in disease: yield loss modeling and can be accounted for by using either separate equations for each yield potential crop or crop category, robust models, the inclusion of a form function for yield potential or choice of disease and yield variables which integrate yield potential.

In the 1990s during wet seasons a new disease causing brown leaf spots on lettuce (*Lactuca sativa*) was found for the first time in many lettuce-growing areas of Austria and Germany. The causal agent, a new pathogenic species called *Septoria birgitae,* was thought to be responsible for total crop loss. To study how temperature, inoculum density and leaf wetness period influence disease incidence and severity of leaf spot on lettuce caused by *S. birgitae, in vivo* experiments were carried out in growth chambers and the field (Lohmeier *et al.,* 2013). Additionally, they evaluated the relevance of infected plant debris acting as a primary inoculum source in the soil for subsequent crops. *S. birgitae* produces spores over a wide temperature range between 5°C and 30°C, and can infect plants at temperatures between 10°C and 30°C, with an optimum between 20°C and 30°C. Spores of *S. birgitae* at a density of at least 10^3 conidia / mL were essential for disease outbreak on lettuce. Because leaf wetness is crucial for releasing conidia from pycnidia, they studied the impact of leaf wetness duration on disease development under various temperature conditions. For relevant leaf spot disease development on lettuce *in vivo,* a leaf wetness duration of at least 24 h and temperatures higher than 10°C were necessary. Leaf spot disease development in the field required several leaf wetness periods longer than 20 h at approximately 15°C at the beginning of crop cultivation. Incorporating *S. birgitae* infected plant debris in the soil as a primary inoculum was not relevant for leaf spot disease outbreak in the next year. However, in cases of continuous cropping of lettuce on the same field and the same season, *Septoria*-infected lettuce debris may become more relevant.

A diagrammatic scale to assess soybean rust severity, caused by *Phakopsora pachyrhizi,* was developed by (Cláudia V. Godoy *et al.,* 2006). Leaflets showing different severity levels were collected for determination of the minimum and maximum severity limits; intermediate levels were determined according to 'Weber-Fechner's stimulus-response law'. The proposed scale showed the levels of 0.6; 2; 7; 18; 42, and 78.5%. Scale validation was performed by eight raters (four inexperienced and four experienced), who estimated the severity of 44 soybean leaflets showing rust symptoms, with and without the use of the scale. Except for rater number eight, all showed a tendency to overestimate severity without the aid of the diagrammatic scale. With the scale, the raters obtained better accuracy and precision levels, although the tendency to overestimate was maintained. Experienced raters were more accurate and precise than inexperienced raters, and assessment improvements with the use of the scale were more significant for inexperienced raters.

Investigation of yield loss assessment in common bean due to anthracnose (*Colletotrichum lindemuthianum*) was conducted under field conditions using different cultivars having variable resistance (Baspa, resistant; Him-1, moderately resistant; Local, susceptible; Jawala, highly susceptible) (Sharma *et al.*, 2008). Field experiments revealed maximum disease in terms of incidence and severity on highly susceptible cultivar Jawala under both the situations of infection (seed borne infection and background contamination). The disease drastically affected the growth parameters and yield components in cultivar Jawala causing a significant reduction in yield of both the crops raised from infected seed (75%) and under background contamination (57.24%). Seed borne infection caused more yield losses than background contamination. The pod infection had a direct effect on seed yield as severity on pods ranging from 3–9 (0-(scale) caused 10.52–57.76 percent reduction in number of seeds per pod and 21.93–68.77 percent reduction in seed weight per pod. However, the determinate cultivars showed more reduction in yield as compared to indeterminate type, though both were susceptible. Common beans (*Phaseolus vulgaris* L.) commonly called as rajmash, which were generally cultivated during the *kharif* season and the humid environment predisposed the crop to attack by various fungal, bacterial and viral pathogens. Of these, bean anthracnose caused by *Colletotrichum lindemuthianum* (Sacc. and Magnus) Lams-Scrib is one of the most destructive diseases in the tropical and subtropical regions especially under cool and humid climates (Tu, 1988; Kumar *et al.*, 1999). The crop is vulnerable to the attack of the pathogen from seedling to maturity depending upon the prevalence of favorable environmental conditions.

Tomato pith necrosis (TPN) is caused mainly by soil-borne *Pseudomonas* spp., formerly known as *Pseudomonas corrugata*. Considerable phenotypic diversity and evidence of genomic variability were the basis of the characterization of two species within this group – *P. corrugata* and *P. mediterranea* sp. *nov*. Both have been isolated in several countries, including Portugal. There is no report on the relative aggressiveness of these two TPN-causing *Pseudomonas* species under commercial-like greenhouse conditions. In a study by Moura *et al.* (2005), severity and progress of pith necrosis caused by *P. corrugata* and *P. mediterranea* were compared under spring/summer greenhouse conditions in northern Portugal. In addition, a precise assessment of the yield of tomatoes (number, size, and weight of fruits) was carried out to determine the relationships among bacterial species, disease severity and yield loss. Tomato plants were stem-injected with either strain A344 of *P. corrugata* or strain A54 of *P. mediterranea*, 45 and 60 days after transplantation in the greenhouse. TPN symptoms were recorded throughout the growing period. The yield of tomato fruits was assessed weekly during the last seven weeks. Both strains showed a similar level of aggressiveness and caused the same type of symptoms. Both strains were responsible for significant yield reductions. However, no correlation ($p \geq 0.05$) was found between pith lesion length and fruit loss.

The yield losses due to common rust were found directly related to disease severity (Utpal Dey *et al.*, 2012). The avoidable yield losses due to disease were in the range of 11.75 to 60.53% in different spray schedule. Crop loss assessment at different fungicide spray schedule revealed that there was a significant reduction in yield when disease severity was more than 50%. Monthly climatic data (Minimum, maximum temperature, rainfall, humidity and number of rainy days, etc.) for the growing year 2010 were considered. The intensity of the disease was recorded by scoring all the randomly selected individual ten plants in each treatment using 1 to 5 scales. Further, the per cent disease index (PDI) was calculated with the scale above using the formula.

$$\% \text{ Disease index (PDI)} = \frac{\text{Sum of all disease ratings}}{\text{Total No. leaves observed} \times \text{Maximum grade}} \times 100$$

The percent avoidable loss over untreated control was calculated using the following formula:

$$\% \text{ Avoidable loss} = \frac{V_p - V_u}{V_p} \times 100$$

where, V_p = Values in protected plot; V_u = Values in unprotected plot

In a comprehensive survey, the frequencies of *Erwinia carotovora* subsp. *atroseptica* (Eca), *E. c.* subsp. *carotovora* (Ecc) and *Erwinia chrysanthemi* (Ech), based on the survey of 42 locations, were found to be 48%, 45%, and 7%, respectively (Hafiz Farhad Ali *et al.*, 2012). Regarding soil, diseased plant debris and seed potatoes as sources of inoculum for Eca, healthy plant material, potted unsterilized soil and healthy seed potatoes were inoculated with Eca, kept in screen-house and monitored for changes in bacterial population over a period of three months. It was found that seed potatoes were the most important source of primary inoculum as the bacterial population considerably increased in numbers throughout the monitoring period. The bacterial population in the inoculated plant debris did decrease, but the rate of decline was quite slow, and therefore, diseased plant debris could be considered as a good source of primary inoculum for the appearance of the blackleg disease in the next season.

Cephalosporium stripe caused by *Cephalosporium gramineum* can be a serious disease of winter wheat (*Triticum aestivum* L.) in the Pacific Northwest of the USA (Quincke *et al.*, 2012). Effects of *Cephalosporium* stripe on yield, test weight, protein, and kernel characteristics were examined using 12 winter wheat varieties in field plots inoculated and not inoculated with the pathogen. Averaged over varieties, inoculation decreased yield, test weight, kernel weight, and kernel diameter; grain protein and the standard deviations of kernel weight and kernel diameter were increased by inoculation. Grain yield of the susceptible check was reduced by as much as 41% with the addition of inoculum. The most resistant and the most susceptible varieties performed similarly for yield in the two environments while varieties with intermediate levels of resistance were sometimes inconsistent. There was a linear relationship between yield and % whiteheads (sterile heads caused by disease) in one environment and a curvilinear relation in the other.

Critical information in the assessment of disease is the amount of disease that is present. This can be measured as the proportion of a plant community that is diseased (disease incidence) or as the proportion of plant area that is affected (disease severity). Often, the disease has to exceed a certain threshold before it reduces the yield of a crop, but it is usually difficult to accurately estimate the yield reduction caused by a specific disease. For example, many diseases occur on senescing tissue that would not have contributed to the yield anyway. Easier diseases to assess are those that kill whole trees in orchards or plantations, and those that destroy the actually harvested product, such as fruit or grain. Disease and crop loss assessments are necessary for evaluating the economic impact of a disease and the benefit of particular control strategies. There is no point in implementing a control measure if it will cost more that the increased crop yield will return. The growth of the crop, its yield potential, the development of the disease and its impact on yield all have to be measured to predict the impact on yield of particular levels of disease. This information can be combined with predictions of likely disease levels to determine whether preventative treatments are worthwhile.

Assessment of the effect of disease on crop yield normally involves five steps, viz., developing a descriptive growth stage key for the particular crop species; developing methods to assess the incidence and severity of disease; developing statistically sound methods of sampling crop populations for assessment of the amount of disease; estimating the negative impact of particular levels of the disease on crop yield and quality; and evaluating the economic benefit from various methods available for reducing the amount of disease.

4.15.8 Assessment of Crop Growth and Development

The first step to quantify the effect of the disease is to develop a key that describes the growth and development of healthy plants during the growing season. It should describe the development, either from sowing to harvest, in the case of annual plants, or from season to season in perennial plants. Details drawings or photographs, showing characteristics of the various stages of development, including leaf formation, flowering, fruiting and senescence are needed. Standardized growth keys have been developed for a number of crop plants, enabling comparison between different countries and different conditions.

Disease assessment methods need to easily provide objective measurements for a specific growth stage of a crop so that data from different sources is comparable, and provide an adequate sample of the crop for assessment. Whether it is disease incidence, or disease severity, or both, that are measured, depends on the nature of the disease. An 'all or nothing' disease, for example, that inevitably kills any plant it infects, could be measured just by counting the number of plants that infected ('disease incidence'). However, in the case of a disease that causes varying degrees of damage to plants throughout the crop, a more complex measurement is needed, that assesses ('disease severity'). The disease incidence for biotrophic pathogens can be measured by counting the number of plants, leaves, flowers, etc. that are infected, but the disease severity is assessed by estimating the proportion of total photosynthetic area that is diseased. While this is generally less precise and less controllable than counting individual plants, it is usually a better predictor of crop loss. Because judging the proportion of diseased leaf by eye is unreliable, disease assessment keys, showing different disease severities as blackened areas, have been devised for various crops.

To produce a disease assessment key, the development of disease over the whole disease cycle and at different stages of plant growth must be studied to make standard prototype diagrams and/or descriptions. The accuracy of the key then needs to be tested, by assessing disease severity in the field using the key, and then assessing the same samples using accurate measurement techniques in the laboratory. There are also computer programs designed to train observers in disease severity assessment, by presenting images of diseased leaves, which the observer assesses, and comparing their result with the known level of disease in the diagram. This aims to reduce variation in results caused by different observers. These disease assessment methods will only be accurate if performed on a representative sample of the crop. Samples of crop units (plants, leaves, fruit, etc.) can be taken randomly from a crop, or standard quadrants can be placed in the crop, and all plants within the quadrant assessed. In a test plot, a part of the plot is usually assessed for disease. Taking samples only from the edge of a plantation will not necessarily be representative of the bulk of the crop. To determine how many samples need to be taken, it is possible to sample a number of times with progressively more samples and find the point at which the standard error is low. A disease that is uniformly spread throughout a crop will require fewer samples for an accurate assessment than a disease that has a patchy distribution throughout the crop.

4.16 CROP LOSS ASSESSMENT

Once the amount of disease has been determined, the next step is to assess, either experimentally or statistically, the effect of different levels of disease on the crop yield. Experimental assessment involves setting up test crops in which the level of disease is controlled. Monitoring crops with different levels of disease allow the comparison of epidemic progress and crop yield under different disease conditions. A relationship between disease parameters and yield can then be formulated, allowing prediction of crop loss for a certain level of disease at a particular point in the crop's growth. This method is dependent upon the assumption that the treatments used to keep the disease at a certain level have no effect on crop yield themselves. This might not always be the case. Similarly, assessments using susceptible and resistant cultivars rely on both cultivars having a similar yield to start with. Again, this might not be true, and must be determined fist in a disease-free environment. All aspects of experimental design need to be carefully considered to gain an accurate assessment of the effect of disease on crop yield. For example, harvesting methods under experimental conditions are often more efficient than under field conditions, giving the appearance of a higher crop yield. From crop loss assessment studies, models can be devised. Crop loss models are usually based on one of three types of disease assessment: disease at a critical point in development, disease at multiple points in development, and disease throughout crop development.

The statistical approach to assessing disease involves statistical analysis of crop yields under different levels of disease that occur naturally in the field. The levels of disease and the crop yields are monitored, but not manipulated and then yields from different seasons or areas with different levels of disease can be compared to determine the effect of disease on crop yield. The advantage of this method is the use of outcomes of real cropping situations in the field, not experimentally manipulated crops. The disadvantage of this approach is that in the field, the conditions are not controlled, and there are many factors that could affect yield besides disease. An alternative statistical approach to crop loss assessment is the use of questionnaires, filled in by the farmers that allow disease and pest incidence and severity to be related to the yields they produce.

Plant disease assessment or phytopathometry involves the measurement and quantification of plant disease and is therefore of fundamental importance in the study and analysis of plant disease epidemics (Cooke, 1998). The importance of accurate disease assessment methods was identified in early reviews on phytopathometry and crop loss assessment by Chester (1950) and Large (1966). Traditional methods of disease assessment, such as the use of pictorial keys derived from standard area diagrams to evaluate disease severity on a 0–100% scale, have now been joined by several new approaches made possible by rapid advances in computer technology. In addition, modern assays using immunological and molecular techniques for the identification, detection and quantification of plant pathogenic organisms are used. Other new approaches to phytopathometry have evolved in which remote sensing, image analysis and the detection of crop stress caused by disease (using changes in chlorophyll fluorescence and foliage temperature) are involved.

A study was conducted by Nzioki and Chirchir (2001) on sandy soil to assess yield loss caused by parasitic nematodes on sugarcane in Kibos. Three budded were planted in randomized split plot design with three replicates. Three varieties Co421, Co617, and Ken83-737 were selected. Two nematicides, Furadan and Nemacur were used at two rates: Furadan at 16.7 kg/ha (F1) and 8.3 kg/ha (F2) and Nemacur at 25 kg/ha (N1) and 12.5 kg/ha (N2) respectively. An untreated control where no nematicide was applied was included. The plots were treated with the nematicides before planting. The nematicides were applied in the furrows and mixed with soil. Nemacur was reapplied four months after planting at the 25 kg/ha and 12.5 kg/ha respectively. The recommended cultural practices were followed. Nematodes were extracted from the soil using Baeman's pan and filter method. Preliminary results showed that the nematicides had an effect on germination, tiller counts, plant height and nematode counts. The nematicides-applied treatments had increased germination, tiller counts, plant height and reduced nematode counts than the untreated control plots.

Alternaria blight caused by *Alternaria brassicae* (Berk.) Sacc. and *A. brassicicola* (Schw.) Wilts is responsible for poor seed quality, and yield losses in seed crop of Cauliflower (*Brassica oleracea* L. var. *botrytis* subvar. *cauliflora*). Yield loss due to the blight disease was studied in four cultivars, viz., Pant Gobhi (PG) 2, PG 3, PG 4 (early maturing cultivars) and Pant Subhra (a mid-maturing cultivar) (Lakshman Prasad and Karuna Vishunavat, 2006). Highest seed yield losses (55.93) were found in mid-maturing variety Pant Subhra, in spite of being a variety giving highest curd productivity. Variety Pant Gobhi 4 gave highest seed yield (46.13) amongst the early maturing cultivars. The curd area infection is found directly proportional to seed yield losses in all the cultivars. The maximum loss (49.97%) in seed test weight was in cultivar PG2 at all the degree of pod area infection while minimum loss (9.69%) in seed test weight was found in Pant Subhra.

4.16.1 Concepts and Definitions Related to Yield Levels, Production Situations, and Injuries

The concept of yield levels (potential, attainable, actual) and the factors which determine them provides a framework which has been, and still is, largely used to address the performances of agrosystems from the biophysical and socio-economical points of view (van Ittersum and Rabbinge, 1997). Figure 4.17 provides an overview of yield levels and the factors which determine them.

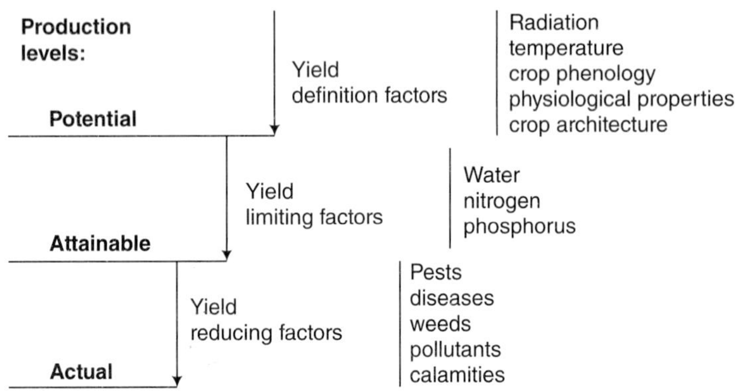

Fig. 4.17 Relationships among potential, attainable and actual yields and growth-defining, growth-limiting and growth-reducing factors (van Ittersum and Rabbinge, 1997)

The potential yield (*Yp*) of a crop is determined by *defining factors*: radiation, temperature, and morphological and physiological attributes determined by the genotype of the crop. The potential yield thus corresponds to the yield that would be produced by a crop grown under optimum conditions. The attainable yield (*Ya*) is determined by the defining factors in combination with the *limiting factors,* water and nutrients. The attainable yield corresponds also to the yield that would be produced by a crop when free of injuries. The combination of yield defining and yield limiting factors can be embedded in the concept of production situation (de Wit and Penning de Vries, 1982). 'The attainable yield of a given crop thus corresponds to the production situation under which this crop is grown.' The attainable yield can be reduced by the effect of reducing factors such as pest (disease, insects, and weeds) injuries. An injury is a visible, measurable symptom caused by a harmful organism (Zadoks, 1985). The resulting yield, obtained in a crop that has been injured by one or several pests, is defined by Rabbinge, (1993); the actual yield, therefore, is the crop yield actually harvested in a farmer's field.

Yield loss *(*YL*),* or damage (Zadoks, 1985), represents the difference between the attainable and the actual yield; that is, the yield lost from pests' injuries. Yield loss can be associated with individual pests as well as to multiple pests. The functional relationships between production situation, attainable and actual yields, yield losses and injuries are summarized in Fig. 4.18.

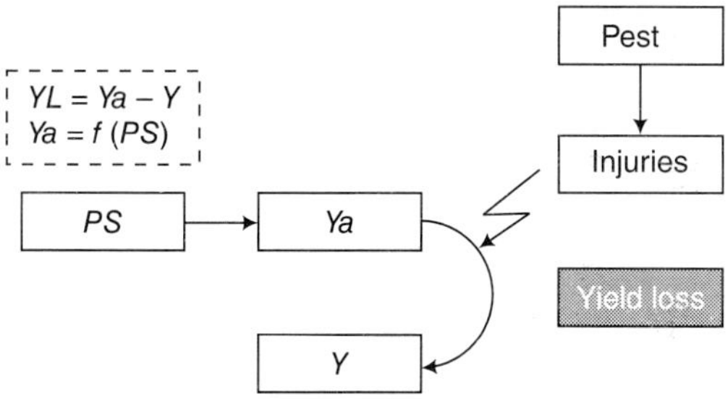

Fig. 4.18 Relationships among production situation (PS), attainable yield (Ya), and actual yield (Y); yield loss (YL) (Willocquet *et al.*, 1998)

Yield loss is frequently expressed as the fraction (percentage) of the attainable yield lost to pest injuries. It is then called *relative yield loss (RYL),* and is computed as $RYL = 100 \times [(Ya - Y)/Ya]$. The relationship between injury levels and the yield loss they cause is one important quantitative component analyzed when addressing yield losses. This relationship is called a 'damage function.' Production situations may correspond to varying levels of attainable and actual yields as illustrated in Fig. 4.18. For example, two different production situations (i.e., combinations of yield-defining and yield-limiting factors) may correspond to the same level of *Ya,* but to different levels of yield losses (i.e., different combinations of pest injuries), and therefore to different actual yields (PS1 and PS2); two production situations may correspond to different levels of *Ya*, but to the same level of actual yield (because yield losses are different: PS2 and PS3); two production situations may correspond to different levels of *Ya* and actual yield (PS3 and PS4), the ranking of yield levels between the two production situations being the same ($Ya_{PS3} > Ya_{PS4}$ and $Y_{PS3} > Y_{PS4}$); two production situations may correspond to different levels of *Ya* and actual yield (PS4 and PS5), the ranking of yield levels between the two production situations being opposite ($Ya_{PS4} > Ya_{PS5}$ and $Y_{PS4} < Y_{PS5}$).

This diversity of possibilities implies that the quantification of the relative role of the different factors determining the actual yield is the first step when aiming at improving agrosystems' performance.

4.16.1.1 *Stages to simulate yield losses, and possible outcomes*

Simulation of crop growth and yield affected by pest injuries can be made by incorporating into a crop growth model (such as GENECROP) the damage mechanisms corresponding to the injuries addressed. This new model is called GENEPEST. A three-stage approach allows the simulation of yield losses, i.e., simulation of noninjured growth, enabling one to model the attainable growth and attainable yield (*Ya*) of a crop under a given production situation. By definition, all injury levels are then set to zero; simulation of growth under specified levels of injuries to model the actual growth and actual yield (*Y*) and computation of yield losses, that is, the difference between simulated attainable and actual yields.

Note 1: Simulating growth and yield with levels of injuries corresponding to improved pest management (*Ym*) allows estimating yield that would be gained on the actual yield (*Y*) from this improvement in pest management (*Ym*–*Y*), thus providing a basis to guide strategic decisions such as research priorities in pest management.

Note 2: Yield losses can be simulated for a range of production situations, by setting the crop drivers (i.e., parameters and interpolation functions for crop growth) to values corresponding to each production situation, and proceeding to the three stages described above.

Note 3: Yield losses can be simulated for injuries considered individually and for combinations of injuries (i.e., grouped as predefined injury profiles), thus allowing ranking injuries according to their importance in terms of the yield losses they cause. Such results can help in ranking crop health problems and, again, help for guiding research priorities in pest management.

4.16.1.2 *Model drivers for pathogens injuries*

The damage mechanisms described above have been implemented into GENEPEST by considering several pathogens which provide a combination of the damage mechanisms described previously. The pathogens considered are described in Table 4.6.

Table 4.6 Examples of Pathogens Accounted for in GENEPEST[a]

Pest name	*Pest type*	*Driving function*	*Damage mechanism 1*	*Damage mechanism 2*
LD1	Foliar pathogen	Disease severity (fraction leaf area infected)	Light stealer	
LD2	Foliar pathogen		Light stealer	leaf senescence accelerator
LD3	Foliar pathogen		Light steaker	assunukate sapper
VIR	Virus	Disease incidence (fraction plants infected)	Photosynthetic rate reducer	
DEF	Defoliating insect	Daily fraction of leaf area defoliated	Tissue (leaf) consumer	
RDIS	Root-infecting pathogen	Disease severity (fraction root infected)	Photosynthetic rate reducer	
NBINS	Sucking insect	Nb of insects (per m^2)	Assimilate sapper	

4.17 SIMULATIONS

The STELLA model GENEPEST.STMX will allow to explore the model structure and equations, explore the model inputs, especially the driving functions of the different pests included, explore the model outputs, and run the model with varying levels of injury, which will allow to explore the effects of individual injuries on crop growth and yield apart from the effects of combined injuries on crop growth and yield.

Grain yield loss in rice (*Oryza sativa* L.) caused by blast disease, *Magnaporthe grisea* (Hebert) Barr, is a major concern of rice growers worldwide. Blast is considered as the most injurious disease of rice in Iran, resulting in severe loss especially to susceptible rice cultivars. To assess yield loss caused by blast pathogen and develop an appropriate model, different disease onsets and levels were simulated in the experimental field in a split-plot experimental design (Iran S. Mousanejad *et al.*, 2010). Independent variables including early diseased leaf area (X1), final diseased leaf area (X2), early neck blast index (N1), final neck blast index (N2), area under leaf blast disease progress curve (AUDPC1) and area under neck blast disease progress curve (AUDPC2) were taken as predictors and regressed to the loss in yield. Statistics as the coefficient of correlation (r), the coefficient of determination (R2), adjusted the coefficient of determination (aR2), standard error (SE), F and Durbin-Watson were considered in evaluating the resulting models. The most appropriate model was the one which predicts rice yield loss based on final diseased leaf area and final neck blast index. Modeling and predicting yield loss due to blast disease is necessary as a measure for practical disease management. A mathematical model, through which grain yield loss of rice, due to blast, can be determined, has not yet been developed in Iran. Hence, this study was carried out to develop a predictive crop loss model using various observed disease severities that lead to yield loss.

There are two approaches for predicting yield loss, viz., empirical damage functions and combined blast-crop simulation models. Empirical damage functions are the most common methods of estimating yield losses caused by blast disease. They are derived by applying regression analysis or nonlinear curve fitting techniques to the field data. The first equation is that of Kuribayashi and Ichikawa (1952). Goto (1965) reported several examples of functional disease-yield loss relationships for panicle blast. Katsube and Koshimizu (1970) determined the relationship as $Y = 0.57X$, where Y = percent yield loss and X = percent neck blast incidence 30 days after heading. In India, Padmanabhan (1965) reported that the apparent loss associated with neck infection at 1% incidence corresponded to 0.99% yield loss in a resistant variety and 1.22% yield loss in a susceptible one. In Brazil, Prabhu and De Faria (1982) described the relation between time of neck infection in days after milk stages (X) and loss in 1000-grain weight (Y) as $Y = 38.02X2 - 0.92$ ($r = 0.98$).

The representativeness of information on yield losses due to rice diseases in tropical Asia was studied (Serge Savary and Paul S. Teng, 1998). Published studies involving different groups of diseases (viral, bacterial, and fungal) and conducted in different rice production ecosystems were compared to help identify research gaps, the filling of which could improve current disease management in rice and help in developing strategies that fit the management needs of fast-evolving rice production systems in the future. Four criteria of representativeness of yield loss information were used: representativeness over time (the proportion of studies conducted over more than one crop cycle), representativeness over space (the proportion of studies conducted in more than one location), representativeness of scale (the proportion of studies conducted on the scale of plots or fields), and representativeness of injury (the standard deviation of the proportion of studies using inoculation, spontaneous infection, or chemical control). A strong imbalance in both the number and the representativeness of studies dealing with fungal, viral, and bacterial diseases was found. Most of the few studies of yield loss due to viral diseases (mainly rice tungro) were conducted on the scale of individual (potted) plants or were based on one-year data sets, often reflecting strong epidemics only. Studies of bacterial diseases were conducted in single locations only. There was an acute need to better document yield losses in rice ecosystems other than the irrigated ecosystem. While studies conducted in the upland, rainfed lowland, and deep-water rice ecosystems seem to have a high degree of representativeness, this could not compensate for their small number in view of the great diversity of these environments.

Different severities of northern leaf blight were induced by inoculating maize plants once at GS 4 and 5, three times at GS 4, 5 and 6, allowing natural disease development, and by applying fungicides to deter disease development on cultivars with different levels of resistance to *Exserohilum turcicum* (Adipala *et al.*, 1995). Percentage leaf area affected by northern leaf blight on the whole plant, ear, first, and second leaf above the ear leaf were assessed weekly for a total of six times at GS 8, 9.0, 9.1, 9.2, 9.3, and 9.4. Data obtained from the ear leaf, first leaf above the ear leaf, and second leaf above the ear leaf were used to develop a one-three-two weighted scale to account for leaf position and time of infection for the three leaves assessed were related to yield by regression analysis. The average disease severity estimated on the ear leaf was significantly ($P \leq 0.001$) correlated with severities estimated using other disease assessment methods and loss in grain yield. Overall, AUDPC yield-loss models using percentage leaf area affected on the ear leaf gave the best relationship (Y = 5835-135 AUDPC, R2 = 0.42, $P \leq 0.001$). Further analysis using leaves of various positions on the plant did not improve the yield-loss models. Critical point models, using percentage leaf area affected at GS 9.1 on A619 × A632 also gave good fit (R2 = 0.53).

To assess yield loss caused by blast pathogen and develop an appropriate model, different disease onsets and levels were simulated in the experimental field in a split-plot experimental design (Mousanejad *et al.*, 2010). Independent variables including early diseased leaf area (X1), final diseased leaf area (X2), early neck blast index (N1), final neck blast index (N2), area under leaf blast disease progress curve (AUDPC1) and area under neck blast disease progress curve (AUDPC2) were taken as predictors and regressed to the loss in yield. Statistics as the coefficient of correlation (r), the coefficient of determination (R2), adjusted the coefficient of determination (aR2), standard error (SE), F and Durbin-Watson were considered in evaluating the resulting models. The most appropriate model was the one which predicts rice yield loss based on final diseased leaf area and final neck blast index.

The soybean cyst nematode, *Heterodera glycines* (SCN) is of major economic importance and widely distributed throughout soybean production regions of the United States where different maturity groups with the same sources of SCN resistance are grown. Donald *et al.* (2006) assessed SCN-resistant and -susceptible soybean yield responses in infested soils across the north-central region. In 1994 and 1995, eight SCN-resistant and eight SCN-susceptible public soybean cultivars representing maturity groups (MG) I to IV were planted in 63 fields, either infested or noninfested, in 10 states in the north-central United States. Soil samples were taken to determine initial SCN population density and race, and soil classification. Data were grouped for analysis by adaptation based on MG zones. Soybean yields were 658 to 3,840 kg/ha across the sites. Soybean cyst nematode-resistant cultivars yielded better at SCN-infested sites but lost this superiority to susceptible soybean cultivars at noninfested sites. Interactions were observed among initial SCN population density, cultivar, and location. This study showed that no region-wide predictive equations could be developed for yield loss based on initial nematode populations in the soil and that yield loss due to SCN in our region was greatly confounded by other stress factors, which included temperature and moisture extremes.

Field studies were conducted in 2008 and 2009 to determine yield loss of cowpea (cv. ART98-12) due to natural infestation by *Meloidogyne incognita* using Carbofuran 3G at 2 kg a.i./ha and untreated as a check (Ayodele Adesina Adegbite, 2011). The yield of cowpea was found to be higher with the application of nematicide-carbofuran 3G at 2 kg a.i./ha. The percentage increase over control was 39.0 and 33.0% in the years 2008 and 2009, respectively. A significant reduction in the yield of cowpea in untreated plots was mainly attributed to direct damage to the root system by the feeding of root-knot nematode, *M. incognita*. The root knot nematode population in carbofuran treated plots was significantly lower than in check in the two years, also at harvest. In the check, the nematodes reproduced many folds during the cropping season. Higher nematode population in the untreated check decreased plant growth and consequently reduced the number of harvested seeds and other agronomic parameters.

The quantitative relationship between yields of spring oat cultivars Ogle and Noble and incidence of barley yellow dwarf (BYD) disease caused by barley yellow dwarf virus (BYDV-PAV-IL) was studied (Ellen M. Bauske *et al.*, 1997). Field plots were infested with *Rhopalosiphum padi* vectoring BYDV-PAV-IL. A linear critical point regression model was developed that related percentage of maximum yield to BYD incidence, assessed by estimating the percentage of plants with BYD symptoms at Feekes growth stage 10.5 in each plot. The model $y = 96.9 - 0.45x$ ($r_2 = 0.53$) best explained the relationship. Test weight was unaffected by BYD. Although Noble consistently sustained higher disease incidence than Ogle, covariate analysis indicated that the effect of BYD on yield was similar for both cultivars Studies of random versus focus inoculation of Noble oat, designed to determine the importance of compensation in reducing yield loss associated with BYD indicated that compensation did not play an important role.

The yield attributes were assessed on mungbean variety SML 134 by maintaining five grade of mungbean yellow mosaic virus (MYMV) infection (1–9) under natural epiphytotic conditions of Bathinda (Punjab) during 2004 and 2005 (Yadav and Brar, 2010). The result revealed that with the increased levels of MYMV severity, the corresponding yield attributes decreased significantly. Per plant reduction in seed yield varied from 22.3 to 61.7% under resistant (grade 1) to highly susceptible (grade 9) in comparison to healthy control. On an average 43.6% reduction in seed yield were assessed on a moderate level of susceptibility (grade 5) over healthy control. Out of 40 genotypes evaluated against MYMV, two genotypes, namely, NM 57 and NM 94 were found resistant and four genotypes (PBM 118, PBM 121, PBM 123 and A 3) were found moderately resistant. Resistant and moderately resistant genotypes possessing better phenotypic characters and yield attributes identified in these studies could be utilized in mungbean improvement programme.

A study was conducted to compare disease progress among the three cultivars throughout the season, to assess the effects of the disease on yield loss and crop canopy development as shown by sequential aerial infrared photography, and to establish any quantitative differences among the cultivars with respect to susceptibility to blight (Wallen and Galway, 1977). Percent ground cover has been utilized to assess growth, and to determine maturity, vigor, and yield potential. It was thought that this technique might detect differences in susceptibility not determined by other means. Disease progress curves were developed from the results of disease assessments for bacterial blight (*Xanthomonas phaseoli*) on Sanilac, Seafarer, and Kentwood field bean (*Phaseolus vulgaris*) cultivars in field plots over a 3-year period 1974–76. Although losses due to bacterial blight were substantial, mean 3-year losses were similar among the varieties, Seafarer 32%, Kentwood 32.2%, and Sanilac 33.1%. Crop canopy development was influenced by the onset and intensity of the disease, a hastening of maturity and senescence, and early defoliation of infected plants. In 1975, when yield losses were greatest, differences in canopy development between control plots and infected plots were also the greatest.

A collaborative study was carried out in four districts in Central Uganda to establish the factors contributing to the epidemic and to assess the impact of bacterial disease; Banana *Xanthomonas* wilt disease (BXW, caused by *Xanthomonas campestris* pv.) on banana yield (Karamura *et al.*, 2010). On average, 33% of the total banana mats were infected with BXW between 2001 and 2004. None of the banana cultivars were resistant to the disease. However, East African Highland cultivars (AAA-EAHB genome) were significantly less affected than cultivars with AAB and ABB genomes. The results showed that a participatory development communication campaign against the disease had a substantial impact on reducing disease incidence. Compared to preinfection levels, the total banana yield loss due to BXW infection was estimated at 30–52% between 2001 and 2004, hence a reduction in the amount of bananas harvested by farm households.

Field experiments were conducted to evaluate the relationship between severity of bacterial leaf blight of rice and yields. Epidemics of this disease with different disease progress curves were encouraged by manipulating the initial dates and frequencies of subsequent inoculations, the use of a bactericide, and selection of rice cultivars thought to differ in disease reaction to the causal organism, *Xanthomonas oryzae*. Late season epidemics (i.e., initiated after flowering) had no measurable effect on grain yield or yield components (Reddy *et al.*, 1979). Epidemics that began before panicle initiation significantly reduced grain yield, panicle fertility, and kernel weight. The discovery of a significant, linear relationship between blight severity at the soft dough stage of plant growth and grain yield allowed the construction of a critical point model to predict bacterial leaf blight-associated crop losses.

5 Plant Disease Diagnostic Kits

Diagnosis is the first step in implementing effective and economic disease management practices. Emerging, reemerging and endemic plant pathogens continue to challenge our ability to safeguard plant health worldwide. Further, globalization, climate change, increased human mobility, pathogen and vector evolution have combined to increase the spread of invasive plant pathogens. Early and accurate diagnoses and pathogen surveillance on local, regional and global scales are necessary to predict outbreaks and allow time for development and application of mitigation strategies. Plant disease diagnostic networks have developed worldwide to address the problems of efficient and effective disease diagnosis and pathogen detection, engendering cooperation of institutions and experts within countries and across national borders. Networking maximizes impact in the face of shrinking government investments in agriculture and diminishing human resource capacity in diagnostics and applied pathology (Sally A. Miller *et al.*, 2009). New technologies promise to improve the speed and accuracy of disease diagnostics and pathogen detection. Widespread adoption of standard operating procedures and diagnostic laboratory accreditation serve to build trust and confidence among institutions.

Plant disease diagnosis is the determination of the cause of a disease or syndrome in a plant or plant population, whereas detection refers to the identification of microorganisms or their products, e.g., toxins, in any number of substrates including plant tissues, soil, and water. Surveillance generally refers to the monitoring of plant systems for pathogens and/or diseases. Detection, surveillance, and diagnosis are all necessary components of programs designed to safeguard plant systems from invasive pathogens and identify causal agents of plant diseases as the first step toward disease management.

Since detection and accurate identification of plant pathogens is one of the most important strategies for controlling plant diseases to initiate preventive or curative measures, special interest should be taken in the early detection of pathogens in seeds, mother plants and propagative plant material to avoid the introduction and further spreading of new pathogens in a growing area where it is not present yet. For that reason, the availability of fast, sensitive and accurate methods for detection and identification of fungal pathogens is increasingly necessary to improve disease control decision making (Nieves Capote *et al.*, 2012). Traditionally, the most prevalent techniques used to identify plant pathogens relied upon culture-based morphological approaches.

These methods, however, are often time-consuming, laborious, and require extensive knowledge of classical taxonomy. Other limitations include the difficulty of some species to be cultured *in vitro*, and the inability to accurately quantify the pathogen (Goud and Termorshuizen, 2003). These limitations have led to the development of molecular approaches with improved accuracy and reliability. A high variety of molecular methods has been used to detect, identify and quantify a long list of plant pathogenic fungi. Molecular methods have also been applied to the study of the genetic variability of pathogen populations, and even for the description of new fungal species. In general, these methods are much faster, more specific, more sensitive, and more accurate, and can be performed and interpreted by personnel with no specialized taxonomical expertise. Additionally, these techniques allow the detection and identification of nonculturable microorganisms and due to its high degree of specificity, molecular techniques can distinguish closely related organisms at different taxonomic levels.

The advent of enzyme-linked immunosorbent assay (ELISA) and polymerase chain reaction (PCR) has caused a shift towards the use of molecular approaches to modern plant pathogen diagnostics. Nowadays, many techniques have been developed for the detection and identification of plant pathogens, each requiring its own protocol, equipment, and expertise. In addition, some of these techniques permit reliable quantification of the target pathogen as well and supply the information required to estimate potential risks regarding disease development, the spread of the inoculum and economic losses. In addition, since no culturing step is required, these techniques are equally suitable for the detection of culturable as well as nonculturable microorganisms. The major challenge at the moment is the develpoment of multiplex assays that allow accurate detection and quantification of multiple pathogens in a single assay. Apart from ELISA and PCR, some recent advances in molecular plant pathogen diagnostics include nucleic acid-based techniques, DNA microarray, membrane-based lateral flow assay, etc., with additional tests like analysis of fatty acids of organisms, carbohydrate utilization (i.e., BIOLOG test), and enzyme activity testing (i.e., pectinase, isozyme patterns).

This chapter emphasizes more on the development and use of 'diagnostic kits' against major plant diseases or pathogens, rather than elaborating 'disease diagnosis'.

Although capacity for traditional pathogen identification is generally insufficient to meet needs in both developed and developing countries, the creation of high-tech tools for plant pathogen diagnostics has expanded at a rapid rate. As mentioned above, field-ready serological tests such as lateral flow devices (LFDs) are commonly used as diagnostic tools to aid disease management decision making, to back up diagnoses based on symptoms, and as a triage tool to prescreen plants for specified target diseases (Smith *et al.*, 2008). LFDs and laboratory enzyme-linked immunosorbent assay (ELISA) kits are convenient and easy to use and have been adopted widely in developed countries. For example, an LFD for *Phytophthora* spp. detection has proven to be of significant value in the United Kingdom for prescreening woody plants for the absence of *P. ramorum/P. kernoviae* at the time of inspection. Samples testing positive are sent to a laboratory for follow-up testing with more specific laboratory tests such as PCR and pathogen culture.

Tools developed commercially for use in developed countries are generally too expensive for the same crops in poorer countries. Serodiagnostics are being used in limited situations for specific applications, such as brown rot (*Ralstonia solanacearum*) testing of seed potatoes. Adoption of ELISA kits, LFDs, and potentially other assays may expand when they are produced locally or regionally and can, therefore, be affordably priced. For example, ELISA kits are being developed by India's Ministry of Science and Technology for virus disease diagnosis in high-value crops including ornamentals, fruits, and spices.

DNA-based assays, particularly PCR and real-time/quantitative PCR (qPCR), are being adopted in diagnostic laboratories; ordinary PCR testing for many pathogens is now routine and affordable in the developed world. Quantitative PCR is more expensive than PCR as a result of high start-up costs and expensive materials, but it has been adopted for high-consequence pathogens, especially in regulatory circumstances, because of their high sensitivity and specificity. Both types of assay, however, are out of reach of all but elite biotechnology laboratories in the developing world. Tests utilizing isothermal amplification techniques may prove to be less costly and thus more appropriate than current DNA-based methods in these countries in the future. It is encouraging that within the field of human health greater strides have been made to recognize the need for a strategy for high-impact diagnostics in the developing world (Urdea *et al.*, 2006). Urdea *et al.* (2006) identified as crucial the need to develop diagnostic tests that can be performed at low-infrastructure sites that serve most of the global population, with support from more advanced laboratories, to identify where treatment is and is not required.

Electronic technology is also finding its way into disease diagnostic and pathogen surveillance systems. In addition to digital systems widely available for distance diagnosis, several free forms of software can be used to map and share the presence of diseases. One such example is Google Earth, a GIS system that can be employed by registered users to update pest distribution data. An application called RustMapper is being tested to map the distribution of wheat stem rust race UG99 across the world, coordinated by the International Maize and Wheat Improvement Center (CYMMIT) (commonly called by its Spanish acronym CIMMYT for *Centro Internacional de Mejoramiento de Maízy Trigo*)). Registered users of plant disease diagnostic network databases such as the University of Florida's Digital Diagnostic and Identification System (DDIS) also use Google Earth to map the distribution of diseases from samples received in network laboratories. In situations where internet access is lacking a novel alternative is to exploit the worldwide adoption of mobile phones to facilitate information exchange.

Visual examination, microscopy, culturing, a few simple biochemical tests, and ELISA are the mainstays for most routine diagnoses in many laboratories. When coupled with diagnostic references such as disease compendia, pathogen-specific manuals, and image databases, these techniques in the hands of trained diagnosticians or specialists are sufficient to provide answers in a reasonable amount of time at a manageable cost. One of the critical components of a modern, highly functioning diagnostic laboratories is access to high-speed internet. Web-enabled microscopes permit laboratories to network with each other and with experts worldwide in real time to identify pathogens and other microorganisms morphologically. Many diagnostic networks build and maintain expertise databases within their communications systems to facilitate knowledge sharing. For example, EPPO supports a searchable database on its website containing a list of laboratories by country and an expertise list by pathogen (Petter, 2008).

5.1 PLANT DISEASE DIAGNOSTIC NETWORKS

Although it is not possible to describe all of the networks of plant disease diagnostic laboratories and diagnosticians and other experts operating worldwide, several examples have been selected as case studies to demonstrate the benefits of networks in improving capacity, efficiency, and effectiveness in diagnostics. These networks share the goal of maximizing the ability of countries and multinational regions to improve their ability to prepare for and respond to threats to agricultural productivity and food security posed by plant pathogens, as well as to assist clientele in disease management decision making.

The National Plant Diagnostic Network (NPDN) was established in 2002 in response to the need to enhance agricultural security through protecting health and productivity of plants in agricultural and natural ecosystems in the USA. With support from the USDA-NIFA and through the collective efforts of many individuals representing Land Grant Universities, federal agencies, state departments of agriculture, and other stakeholders, the NPDN has grown into an internationally respected consortium of plant diagnostic laboratories. The specific purpose of the NPDN is to provide a cohesive, distributed system to quickly detect and identify pests and pathogens of concern. NPDN laboratories immediately report their findings to appropriate responders and decision makers. To accomplish this mission, the NPDN has invested in diagnostic laboratory infrastructure and training, developed an extensive network of first detectors through education and outreach, and enhanced communication among public agencies and stakeholders responsible for responding to and mitigating new outbreaks.

5.1.1 International Plant Diagnostic Network (IPDN)

It is funded by USAID as part of IPM CRSP program, beginning 2005, concentrates on plant disease and pest diagnostics (Miller, 2012). It has a network of laboratories in Africa, Asia and Central America. No regulatory or enforcement role, but communication with national plant health regulatory services is encouraged. Its major approach includes introducing distance diagnostic and data management web portal, to conduct training programs, to harmonize diagnostic protocols (SOPs) and to develop and test diagnostic assays.

Towards the ultimate goal of sustainably increasing capacity for plant disease and insect pest diagnostics in Africa and Central America, with emphasis on important staple and export crops such as banana, the International Plant Diagnostic Network (IPDN) was established in 2006. Among the objectives of the network described herein are to assess diagnostic capacity among African institutions, to develop and introduce a distance diagnostics and data management system, to enhance networking among diagnosticians and plant health management experts within East and West Africa and facilitate digital laboratory and field-based diagnostics, to harmonize laboratory diagnostic protocols within the region, and to develop capacity for the use of traditional and modern diagnostic tools and technologies through advanced training programs. One of the goals of the network is to interface with individuals and organizations, such as EAPIC, involved in information gathering to support pathogen and insect pest surveillance activities.

The IDPN was modeled after the United States National Plant Diagnostic Network (NPDN), which operates as a 'hub and spoke' system in five regions (Miller *et al.*, 2010). A hub and spoke system is an operational structure, in which a central laboratory (hub) coordinates activities amongst other laboratories (spokes) within the network.

5.1.2 The United States National Plant Diagnostic Network

The USDA Cooperative State Research, Education, and Extension Service (CSREES) responded to homeland security risk assessments in 2001 by developing in 2002 the United States National Plant Diagnostic Network [NPDN (http://www.npdn.org)], a consortium of land grant university (and some state departments of agriculture) laboratories in the United States and its territories. The network functions and goals include increasing capacity in all laboratories to a minimum standard, training diagnostic personnel, standardizing diagnostic protocols, increasing communication between partners such as the CSREES, APHIS, university and state department of agriculture personnel, and training field personnel as first detectors.

In addition, diagnosticians are populating an online diagnostic cookbook at http://wiki.bugwood.org/Diagnosticians cookbook; this resource is publicly accessible, but only approved personnel may write to the files, allowing for quality control and review. NPDN laboratories are involved in the development and testing of new diagnostic methods and work with researchers and the national reference laboratory personnel to improve protocols for regulatory pests. Diagnosticians are trained to process suspect select agents by the USDA APHIS Center for Plant Health Science Technology (CPHST) method development laboratory personnel, increasing NPDN laboratory capability and further standardizing select agent processing.

Additional training sessions held each year increase diagnostic personnel capability in techniques (PCR, virus detection, etc.) and specific pathogens and pests (*Fusarium* spp., stramenopiles, etc.). Technical training sessions occur in a traditional laboratory environment and through technological aids such as video-conferencing via Adobe Connect and streaming videos available on the NPDN websites. Further communication of protocols and diagnostic advice occurs between diagnosticians via email list servers, digital diagnostic services such as DDIS (www.ddis.ifas.ufl.edu) and PDIS (www.pdis.org), NPDN and American Phytopathological Society (APS) committee work, and networking during regional and national meetings. The NPDN has developed a system of continuous improvement through the promotion of training and standardization opportunities; this system aids in the professional development of both experienced and newly hired diagnosticians and increases the capacity of plant laboratories in the United States.

5.1.3 Mediterranean and European Plant Protection Organization

Since its inception in 1951, the Mediterranean and European Plant Protection Organization (EPPO) has expanded from 15 to 50 member countries. EPPO is an intergovernmental organization responsible for cooperation in plant protection in the European and Mediterranean region. The aim of EPPO is to develop a strategy against the introduction of pests that damage cultivated and wild plants through the harmonization and promotion of appropriate methods for pest control. The need for a standardized scheme for detection and diagnosis of organisms that were harmful to plants was recognized as a priority by EPPO to improve transparency and comparability of data.

5.1.4 National Reference Laboratories (NRLs) in the European Union

The role of NRLs will be to ensure compliance with regulations for plant phytosanitary issues and to fulfill diagnostic responsibilities through more rapid adoption and transfer of diagnostic methodologies. Development of a system of NRLs would also provide a sustainable basis for the maintenance of expertise and reference material required for diagnosis. Key responsibilities of each NRL would be to act as national contact points, provide guidance on collection and handling of samples, provide confirmatory diagnoses, develop and validate diagnostic protocols, and provide training through collaboration with other NRLs. The priority organisms for NRLs to diagnose have been identified as those that are quarantine pests with potentially devastating impacts on plant health and trade, requiring regular testing employing complex diagnostic tests.

5.1.5 Plant Pest Management Network

Taiwan has developed a comprehensive network of university and government diagnostic laboratories, research institutes, and field stations that serve the dual purpose of plant safeguarding and provision of diagnostic services for farmers. Pathogens are identified by laboratories in four universities and the Taiwan Agricultural Research Institute (TARI). Diagnostic services are provided at 30 stations associated with universities, district agricultural improvement stations, private organizations, TARI, and the Taiwan Agricultural Chemicals and Toxic Substances Research Institute (TACTRI). TACTRI also serves as an information management center for the network. Finally, local governments distribute plant disease information to farmers and work with local farmers and farmer organizations in the event of a potentially damaging disease outbreak. This system provides a means of passive disease monitoring and early notification of the presence of unusual or regulated pathogens through its close interactions with farmers.

5.1.6 Global Plant Clinic

The Global Plant Clinic (GPC) is a consortium of CABI Bioscience, Rothamsted Research, and Central Science Laboratory, United Kingdom. The GPC provides a cost-free diagnostic and advisory service for NPPOs in developing countries that provide diseased plant samples. This service has resulted in the publications of 40 NDRs over the last eight years. In 2001, the GPC initiated the establishment of mobile plant health clinics in several developing countries. Plant health clinics fulfill an advisory role in a cost-efficient and locally operated manner. They occur in public places such as markets on a regular basis where growers routinely arrive with diseased plant samples.

5.1.7 The International Plant Diagnostic Network

The International Plant Diagnostic Network [IPDN (http://www.intpdn.org)] was initiated in 2005 with the goal of fostering the development of local capacity for diagnostics through the establishment of communication and data sharing networks, training in classical and modern diagnostics and research into new diagnostic methods. The program is funded by the United States Agency for International Development (USAID) under the Integrated Pest Management Collaborative Research and Development Program (IPM CRSP). The IPDN is modeled after the United States NPDN, and as with that network, is comprised of a regional hub and local satellite diagnostic laboratories. Three regional programs have been established to date; in Central America (hub laboratory in Guatemala coordinated by a private company, Agro Expertos), East Africa (hub laboratory in Kenya, coordinated by Kenya Agricultural Research Institute), and West Africa (hub laboratory in Benin, coordinated by IITA). For each regional program satellite laboratories in three additional countries have been linked to hub laboratories. All of the laboratories have access to the Clinic Information Management System/Digital Distance Identification System (CIMS/DDIS) internet portal developed by the University of Florida (107). This portal allows data entry, storage and retrieval of samples received, sharing of digital images within the system and with outside experts, and storage and retrieval of management recommendations. Sample information remains confidential within a country, or when indicated, between the submitter and the diagnostician. An important objective of the IPDN is training diagnosticians in basic and advanced diagnostic methodologies. Working groups within the regional programs are developing SOPs that will be appropriate for local capacity and validated within the region. Reporting new diseases through international outlets is highly encouraged. For example, the West Africa regional program recently reported for the first time the occurrence of *Ralstonia solanaceraum*, the cause of a devastating wilt disease of tomato.

Diseases also involve a progression of symptoms that can vary significantly. The progression of symptoms is one of the most important characteristics associated with problems caused by biotic agents. Diseases can result in primary and secondary symptoms. For example, decayed roots on a tree may be a primary symptom while the toppling over of the tree or windthrow is a secondary symptom. At later stages of a disease, secondary invaders may also obscure the original disease symptoms so that symptoms observed at the later stages of the disease are not typical of the symptoms developed in response to the original pathogen.

5.1.8 Diagnostic Tests for Identification of Biotic Causal Agents

A major problem in identification of biotic causal agents is the inability of some infectious pathogens to grow on artificial media. Viruses, nematodes as well as some fungi (e.g., powdery and downy mildew causing agents) and some prokaryotes (e.g., phytoplasmas), require a living host to grow. In cases where the plant pathogen is difficult or impossible to grow on artificial media, other methods may be used for their detection, such as the use of serological tests for viruses. Viral identification is often accomplished utilizing ELISA (enzyme-linked immunosorbent assay) which is based on the binding of an antibody produced for a specific virus with the virus in the infected plant material. More tests are currently being developed using the polymerase chain reaction (PCR) for detection of specific organisms. These types of reactions take specialized equipment and reagents, and the tests are not commonly done outside diagnostic and research laboratories. Other techniques used for the identification of viruses include negative staining and electron microscopy to view the viral particles in plant tissue or suspensions.

5.1.9 Diagnostic Tests for Identification of Abiotic Plant Disease Causal Agents

It is extremely important to look for abiotic factors that may be important in observed symptoms. Soil and water tests may be necessary to determine pH, nutrient composition, salinity, and other factors such as pesticide residues that may induce various symptoms. It may also be important to get samples of plant tissue analyzed for nutrient content to determine if there is macro or micronutrient deficiencies or toxicities.

5.2 FINAL DIAGNOSIS

A diagnosis is a form of hypothesis testing, where the hypothesis is simply the identity of the disease, and a good diagnostician goes through multiple iterations of the scientific method (seeking evidence through testing that supports or refutes the hypothesis that s/he generates). These hypotheses are generated through observations of the plant, environment, and information from the grower. When all of the information is successfully collected, literature sources should be consulted to determine what is already known about diseases and disease-causing agents associated with the identified plant. Information can be obtained from published resources including plant disease compendia, plant disease indices, technical notes, commodity newsletters, online resources, and personal communication with plant disease experts. When no information is available on the specific plant, information on diseases and disease-causing agents of similar plants may be useful. There may also be rare cases where no information is available related to the disease. Then, extensive testing may be necessary to determine the identification of the plant pathogen. When this type of testing is required, it may take a long time to develop research-based control recommendations and control measures may have to be based on diseases of similar etiology. If these diseases have occurred in other areas of the world, control measures that have been previously developed in other areas may be useful.

Misdiagnosing a disease can be very costly if unnecessary or ineffective control measures are made. There are many real-life examples where proper identification of a disease was not made until after avoidable inputs were applied, management changed, or rumors started. It would be beneficial to growers, fieldmen and extension personnel to have rapid and simple disease identification test kits available to make accurate initial disease diagnoses in the field. These test kits can also be helpful in eliminating disease misdiagnoses and identifying whether the issue is physiological or chemical and not due to a pathogen. A research and extension program was developed to evaluate the use of commercially available pathogen test kits and determine the accuracy of each test kit. Using potatoes as a test model, multiple test kits for numerous diseases were used, assessed and compared to results from traditional diagnostic methods. Results indicated the kits tested were reliable, easy to use, and accurate. Some limitations were observed, but they were not restricting. Overall, the use of disease diagnostic test kits will provide an additional tool for greater accuracy, efficiency and sustainability (Nora Olsen, 2012).

Advances in molecular biology, plant pathology, and biotechnology have made the development of such kits possible. These kits are designed to detect plant diseases early, either by identifying the presence of the pathogen in the plant (by testing for the presence of pathogen DNA) or the molecules (proteins) produced by either the pathogen or the plant during infection. These techniques require minimal processing time and are more accurate in identifying pathogens. And while some require laboratory equipment and training, other procedures can be performed on-site by a person with no special training. So far, diagnostic kits have been designed to detect diseases in crops such as rice, potatoes, papaya, tomatoes, and banana. Similar kits are also increasingly important for identifying genetically modified organisms (GMOs) in shipments of conventional crops (Mahaletchumy Arujanan, 2009).

5.2.1 DNA-based Diagnostic Kits

DNA diagnostic kits are based on the ability of single-stranded nucleic acids to bind to other single stranded nucleic acids that are complementary in sequence (referred to as homologous). The tool used in DNA diagnostic kits is the Polymerase Chain Reaction (PCR). There are three steps involved in PCR. The DNA is first unwound, and its strands separated by high temperatures. As the temperature is lowered, short, single-stranded DNA sequences called primers are free to bind to the DNA strands at regions of homology, allowing the (Taq) polymerase enzyme to make a new copy of the molecule. This cycle of denaturation-annealing-elongation is repeated 30–40 times, yielding millions of identical copies of the segment. The primers in PCR diagnostic kits are very specific for the genes of a pathogen and amplification will occur only in diseased plants.

The primers in PCR diagnostic kits are very specific for the genes of a pathogen and DNA amplification will occur only in diseased plants. Several PCR-based methods have successfully been adapted for plant pathogen detection. Real-time PCR (RT-PCR) follows the general principle of polymerase chain reaction, its key feature is that the amplified DNA is quantified, using fluorescent dyes, as it accumulates in the reaction mixture after each cycle. It offers several advantages over normal PCR, including reduced risk of sample contamination, provision of data in real time and simultaneous testing for multiple pathogens. Real-time PCR protocols are among the most rapid species-specific detection techniques currently available.

DNA microarrays are also of great use for simultaneous pathogen detection. This is important, as plants are often infected with several pathogens, some of which may act together to cause a disease complex. Microarrays consist of pathogen-specific DNA sequences immobilized onto a solid surface. Sample DNA is amplified by PCR, labeled with fluorescent dyes, and then hybridized to the array (Alberts *et al.*, 1994).

PCR-based diagnostics is very sensitive compared to other techniques; detection of a small amount of DNA is possible. PCR can also help farmers detect the presence of pathogens that have long latent periods between infection and symptom development. Moreover, it can quantify pathogen biomass in host tissue and environmental samples, and at the same time detect fungicide resistance. PCR-based detection, however, is expensive compared to protein-based diagnostic methods, and also requires costly equipment. So far, PCR kits have been developed to detect a wide range of diseases including black Sigatoka disease in bananas, *Phytophthora* infestations in potatoes and *Fusarium* infection in cotton.

5.2.2 Protein-based Diagnostic Kits

The first step in a defense response reaction is the recognition of an invader by a host's immune system. This recognition is due to the ability of specific host proteins, called antibodies, to recognize and bind proteins that are unique to a pathogen (antigens) and to trigger an immune reaction (Alberts, *et al.*, 1994).

Protein-based diagnostic kits for plant diseases contain an antibody (the primary antibody) that can either recognize a protein from either the pathogen or the diseased plant. Because the antibody-antigen complex cannot be seen by the naked eye, diagnostic kits also contain a secondary antibody, which is joined to an enzyme. This enzyme will catalyze a chemical reaction that will result in a color change only when the primary antibody is bound to the antigen. Therefore, if a color change occurs in the kit's reaction mixture, then the plant pathogen is present.

The enzyme-linked immunosorbent assay (ELISA) method makes use of this detection system and forms the basis of some protein-based diagnostic kits. ELISA kits are very easy to use because test takes only a few minutes to perform, and does not require sophisticated laboratory equipment or training. There are already numerous ELISA test kits available on the market. Some of them detect diseases of root crops (e.g., cassava, beet, potato), ornamentals (e.g., lilies, orchids), fruits (e.g., banana, apple, grapes), grains (e.g., wheat, rice), and vegetables. ELISA techniques can detect ratoon stunting disease of sugarcane, tomato mosaic virus, papaya ringspot virus, banana bract mosaic virus, banana bunchy top virus, watermelon mosaic virus, and rice tungro virus.

One of the first ELISA kits developed to diagnose plant disease was by the International Potato Center (CIP). It can detect the presence of all races, biovars and serotypes of *Ralstonia solanacearum*, the pathogen that causes bacterial wilt or brown rot in potato.

They also developed a kit that samples for the presence of any of the following sweet potato viruses: SPFMV (sweet potato feathery mottle virus), SPCSV (sweet potato chlorotic stunt crinivirus), SPMSV (Sweet potato mild speckling virus), SPMMV (sweet potato mild mottle virus), SwPLV (sweet potato latent virus), SPCFV (sweet potato chlorotic fleck virus), SPCaLV (sweet potato caulimovirus), and C-6 (new flexuous rod virus). With even more advances in molecular biology and immunology, scientists and farmers alike will be able to improve plant disease diagnosis. Efforts are already underway to produce better diagnostic kits to detect pathogens in crops important to developing countries.

Diagnostic kits are an investment although, expensive, but the costs can be offset by gains, such as reduced crop losses and more environment-friendly crop-management practices. Their development should be made a priority by both the public and private sectors in developing countries.

5.2.3 On-site Plant Disease Diagnostics and Rapid Plant Disease Detection Products

On-site test kits have come a long way. Test kits are easy to use and are available for several diseases (viruses, fungi, and bacteria) important to floriculture crops. Growers should consider stocking up on a few test kits for the spring growing season. Test kits will confirm your disease diagnosis, allow you to detect diseases early, provide you the option to treat only when necessary and use the proper management of the disease. As a result, you will maintain good crop quality. The test kits are available in quantities of 25,100, and 1000, and contain everything needed except the samples. They can be used to diagnose a wide range of pathogens, viz., fungi, viruses, and bacteria.

Hydros Inc. was the first to develop a rapid diagnostic kit for on-site disease detection in plants, with no laboratory required. Other major kits include AgriCheck™; Hydros, Inc.'s AffiniTip™; ImmunoStrips (dip-stick tests) from AgDia to detect viruses such as Impatiens Necrotic Spot Virus (INSV), Calibrachoa Mottle Virus (CbMV), Tobacco Mosaic Virus (TMV), Tomato Spotted Wilt Virus (TSWV), and Cucumber Mosaic Virus CMV) to name a few. There is also a combination kit for INSV and TSWV. Immunostrips are also available for several important bacterial pathogens such as *Xanthomonas campestris* pv. *pelargonii, Ralstonia solanacearum* and *Clavibacter michaganensis* subsp. *michiganensis;* Alert kits from Neogen Corp. are available for *Pythium, Phytophthora* and *Rhizoctonia.*

Each company has kits that are used a little differently. However the same principle applies. Tests are based on antibodies that are designed to recognize antigens specific to each pathogen (i.e., INSV). Any antigen present in the plant sample is extracted by grinding in the extraction solution. The extracted antigen is then allowed to react with antibodies that capture specific antigens (when present). Early detection and success using test kits are only as accurate as the sample is taken. For example, if a plant has symptoms of INSV, a quick test will confirm your suspicions. Collect a sample from the part of the plant showing symptoms, including green or chlorotic tissues adjacent to the necrotic area. Even, when plants are systemically infected, INSV does not spread evenly throughout the plant, so careful selection of tissues is crucial to the success of the test.

5.3 SOME RAPID DIAGNOSTIC KITS

Several rapid diagnostic kits have been developed by various firms and institutes, most of which are popularly utilized to diagnose major diseases on different crops across the globe. Some major kits are briefed hereunder.

5.3.1 Pocket Diagnostic Kit

It is the market leader for on-site plant disease testing. They are easy to use, rapid test kits for diagnosing plant diseases caused by fungi, bacteria, and viruses, providing accurate results in 3–8 minutes. Pocket Diagnostic™ tests are designed for professional users working with protected and field crops. Pocket Diagnostic™ kits put the control in the hands of the growers for improved disease management. They give technical advisors greater certainty over advice; saving time and money for clients. They help plant health inspectors, and advisors make better decisions about disease management and reduce the need for unnecessary laboratory diagnoses. The 'Pocket Diagnostic™ Test Kit' is a Lateral Flow Device (LFD), also referred to as immunochromatographic strip tests, that applies a genus or species specific antisera/antibody used for detecting a range of plant pathogens (each test is relevant to one organism either by genus or species and is one time use only). The Pocket Diagnostic™ Test Kit is robust and simple to use and gives a result in approximately 3 – 10 minutes at a cost of between $10 and $20 per test (approximately 10–20% of the cost for current laboratory based testing) (John McDonald, 2010).

A novel isothermal DNA detection technology, RPA offers several advantages over other testing systems, including speed, sensitivity, specificity, and portability and is suitable for field-usable nucleic acid tests as well as high-throughput laboratory testing. The process employs enzymes capable of pairing oligonucleotide primers with homologous sequences in duplex DNA, directing DNA synthesis to defined points in sample DNA. A novel isothermal DNA detection technology, RPA offers several advantages over other testing systems, including speed, sensitivity, specificity, and portability and is suitable for field-usable nucleic acid tests as well as high-throughput laboratory testing. The process employs enzymes capable of pairing oligonucleotide primers with homologous sequences in duplex DNA, directing DNA synthesis to defined points in sample DNA.

Pocket Diagnostic Test Kits are available for the several pathogens, viz., *Phytophthora, Pythium*, Potato virus (five strains) *Rhizoctonia, Botrytis, Ralstonia solanacearum*, Cucumber mosaic virus, impatiens necrotic spot virus, plum pox virus, tomato mosaic virus, tomato spotted wilt virus, *Xanthomonas hortorum, Erwinia amylovora*, beet virus (two strains) and orchid virus screen.

The technology that sits behind the Pocket Diagnostic™ Test Kits and the LFD is based on a pathogen-specific antisera/antibody binding to an identified pathogen-specific protein (antigen). The pathogen-specific antibody is impregnated within the membrane at two locations, one at the point of sample input and one further along the membrane at the 'positive' (T-Line) site in the result window on the membrane cassette. The pathogen protein is attracted to the antibody and binds to this particle which is bound to blue latex beads. As the solution moves along the membrane carrying the antibody/ protein/latex combination it encounters a strip of antibody/latex impregnated membrane at the test result site (T line) and binds further to these particles, giving a blue line as a positive test result. The more of the pathogen protein in the sample, the 'stronger' the blue positive line (T line) appears.

If the sample does not contain the pathogen-specific protein, the blue latex beads do not move down the membrane, and no positive blue line appears resulting in a negative test response. To ensure confidence in the test an inert antibody moves within the solution to bind on a second test site (C Line) to show that the membrane is functional. If the control line (C Line) fails to turn blue, it can be assumed that the test has failed and needs to be undertaken again with a new test kit.

The Pocket Diagnostic Pepino mosaic virus test kit is used widely by growers of protected and field tomato crops for detection of the pathogen. Testing is usually carried out on symptomatic plants where the disease is suspected. Samples for testing should be taken from leaves with symptoms or skin from suspect tomato fruit. The tomato flesh should not be used as this inhibits the test. Sampling and testing can be carried out in a few minutes in the field or glasshouse. Pocket Diagnostic kits use a unique 'ball and bottle' extraction method which optimizes separation of the pathogen from the plant tissues in no more than 60 seconds. Results appear in the viewing window of the test device approximately 5 minutes after starting the test. The result is easy to read, allowing the user to make simple 'yes/no' interpretation. The Pepino mosaic virus test is designed to be used in the field or glasshouse, and can be used by operators with little or no training. Pocket Diagnostic test results form an important part of the information mix required for sound decision making in routine screening or disease control situations. PepMV tests are supplied in boxes of four. Each box contains four test devices, four extraction bottles, four pipettes and full instructions. Test devices are packed in individual foil pouches, meaning that there is no loss of quality once the outer box is opened. Pocket Diagnostic kits contain everything necessary to carry out testing.

5.3.2 AmplifyRP Kit

Agdia's AmplifyRP by Agdia testing platform uses a patented isothermal amplification technology, Recombinase Polymerase Amplification (RPA), which enables highly specific and sensitive DNA or RNA analysis of plant pathogens in any type of testing environment. AmplifyRP Discovery Kits provide all the reagents necessary to amplify DNA at a single operating temperature (39°C) lyophilized into individual reaction pellets. The end user can then design their own primers and probes for a specific target, mix them with our amplification reagents, and get results (generally in less than 30 minutes). RNA amplification is also possible with the addition of reverse transcriptase.

AmplifyRP technology achieves the sensitivity and specificity of PCR-based testing methods but has some clear advantages. Unlike PCR, AmplifyRP testing formats do not require thermal cycling. The entire process is completed at a single operating temperature (39°C) using a portable hot block or fluorometer. Additionally, AmplifyRP does not require a DNA or RNA purification step. Crude sample extracts can be prepared using a simple extraction buffer and can be tested directly. This makes the testing process very simple and saves the end user valuable time. AmplifyRP will be available in two formats, AmplifyRP Acceler8™ (Fig. 5.1) and AmplifyRP XRT. In the case of both formats, sample extraction and amplification is completed in 20 minutes or less. The Acceler8 format employs a special detection device that we call an amplicon detection chamber (Figure 5.2). Results are read visually on a lateral flow strip that is housed within the detection chamber. This chamber is also designed to prevent amplicons from escaping and contaminating future assays. The XRT format utilizes a portable fluorometer and provides a real-time fluorescent readout of results (much like real-time PCR). Agdia is currently working on several pathogens specific assays using AmplifyRP technology. We have demonstrated feasibility for all types of pathogens including viruses, bacteria, fungi, viroids, and phytoplasmas. In addition to pathogen-specific assays, Agdia offers Discovery™ platforms for both AmplifyRP formats.

The discovery platforms allow researchers to develop custom assays for plant pathogens using AmplifyRP technology. All amplification reagents are supplied to the end user in an easy-to-use format. The end-user can then design their own primers and probes for a specific target, mix them with our amplification reagents, and get results. If necessary, Agdia can also provide consultation to the end user for primer and probe development for a nominal charge.

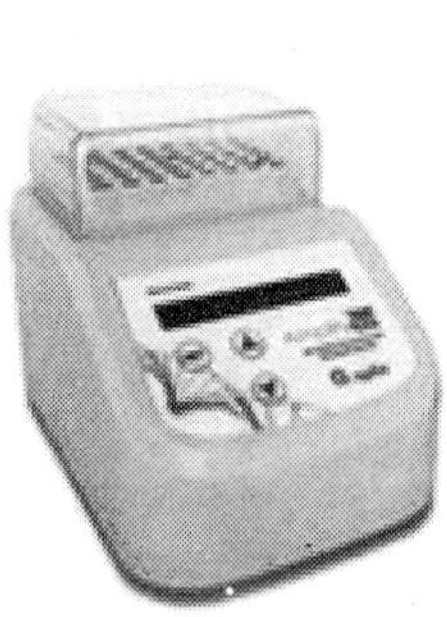

Fig. 5.1 AmplifyRP Acceler8 portable hot block

Fig. 5.2 AmplifyRP Acceler8 detection chamber

Quick, accurate diagnosis of plant pathogens allows growers to implement management practices at an earlier stage of the disease and choose appropriate management practices and crop protection products. However, disease identification can be difficult and requires knowledge, skill and experience.

If samples from a diseased plant need to be sent to a laboratory for identification or confirmation, the disease in the field may become more severe and widespread before results are received and appropriate action taken. Technology such as Enzyme-Linked ImmunoSorbent Assays (ELISA) has been developed and employed over the past 20–30 years for rapid pathogen identification and confirmation. There are several test kits now on the market for quick identification and confirmation of certain pathogens (Elaine Roddy, 2012).

5.3.3 Safeguard Biosystems *Phytophthora* Screening Test Kit

One of the easiest test kits to use is the lateral flow test kit such as the one developed by Safeguard Biosystems (Fig. 5.3). These 'easy-to-use' lateral flow ELISA tests have recently been developed for identifying or confirming specific disease causal agents in the field. Each kit contains disposable pipettes, vials of buffer extract solution with small ball bearings to macerate tissue; and lateral flow devices with a membrane partially coated with antibodies specific to a particular organism.

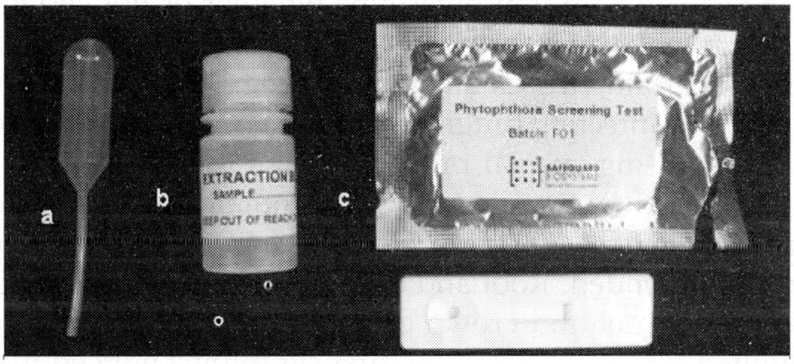

Fig. 5.3 Safeguard Biosystems *Phytophthora* screening test kit includes (a) pipette (b) vial with buffer extract solution and ball bearings and (c) a foil pouch containing the lateral flow device

To use the kits, a piece of infected plant tissue is placed into a vial containing buffer and small ball bearings, which is shaken for 90 seconds to macerate the infected tissue prior to dispensing 3-4 drops of the buffer solution containing the macerated diseased plant tissue onto the well of a lateral flow device. The lateral flow device is then allowed to incubate for 5 minutes before determining if the device detects the pathogen. Tests are considered valid if a blue line develops under the 'C' for control (right portion of the device window) on the lateral flow device regardless of whether or not a blue line developed under the test 'T' (Fig. 5.4). A test is determined to be valid and positive for the pathogen only when a blue line developed under both the 'T' and 'C' on the lateral flow device (Fig. 5.4 (b), (c),

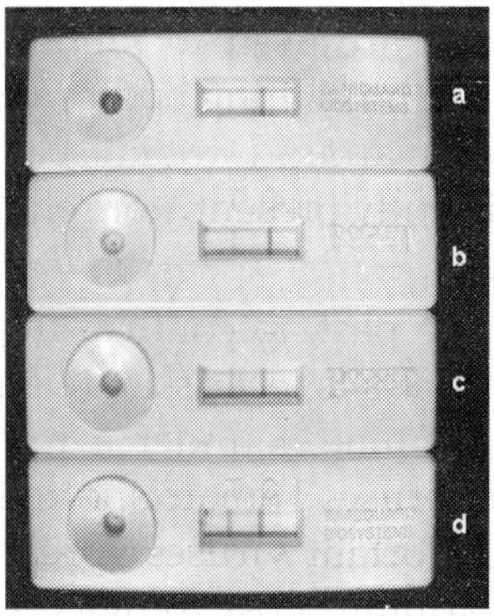

Fig. 5.4 Valid positive tests develop a blue line under the 't' and 'C' of the lateral flow device. The blue line under the 'T' develops (a) no reaction and, therefore, negative or (b) weak, (c) medium or (d) strong reaction and, therefore, positive

and (d). However, the line that develops under the 'T' can sometimes be weak, medium or strong depending upon the test.

The only test kit that appeared to work specifically, consistently and provided a clear, strong reaction was the *Phytophthora* Pocket Diagnostic Test Kit. It also performed well in diagnosing late blight caused by *P. infestans* on infected tomato leaf and fruit tissue in the field in both 2009 and 2010. Although woody tomato stem tissue with symptoms of late blight gave a weak reaction when placed in the vial containing buffer and the small ball bearings, the same test reacted better when the woody infected stem tissue was macerated more effectively with sandpaper or a mesh bag prior to placing into the vial. Since root rots of many woody ornamental plants and fruit trees can be caused by various *Phytophthora* spp., improved techniques to macerate the woody tissue would help improve this test. Regardless, this particular Pocket Diagnostic Test could be a valuable tool for potato, tomato and fruit growers who encounter diseases caused by *Phytophthora* from time to time.

Easy-to-use test kits can help you determine whether or not certain plant pathogens are present so you can make an informed decision (John N. Kabashima *et al.*, 1997). You can use kits to test for certain viruses infecting aboveground parts of virtually any plant. Some kits that detect root-infecting fungi are recommended for greenhouse and container-grown nursery plants. Using special procedures not discussed here, you can also use these kits to test for the same fungi in water and container media. The *Phytophthora* test kit has also been used in field soil assays and tests of field crops, but neither of these uses is currently recommended. Root and crown decay fungi *Phytophthora, Pythium* and *Rhizoctonia,* are common root and crown decay fungi that affect many different plants. Dull, yellow, or wilted leaves are often the first symptoms noticed in plants affected by these fungi. Infected plants become stunted and die when their roots and crowns decay. Similar symptoms are caused by other pathogens, insects, and noninfectious disorders so that specific diagnosis can be difficult. Fungal fruiting bodies such as spores and sclerotia are used to identify many pathogens, but *Phytophthora* and *Pythium* species produce no fruiting bodies visible to the naked eye. *Rhizoctonia* may form coarse web-like growths around rotted basal stems, but the tan-white fungal mycelia are neither consistently present, nor reliably indicative, and other fungal species also produce whitish mycelia.

In the past, to confirm the specific fungal species present one would have to send a sample of suspect tissue to a plant pathologist who would culture it on laboratory media and examine it under a microscope. The process would typically take 1 to 3 weeks. Tospoviruses Tomato spotted wilt virus (TSWV) and impatiens necrotic spot virus (INSV) are important viruses spread by thrips. These viruses under Tospoviruses have wide, overlapping host ranges and damage many agricultural and ornamental crops. Tospoviruses produce a broad range of symptoms, including stunting, yellow or white spotting, black or brown stem or leaf necrosis, ring spots, defoliation, vein necrosis, and dark or yellow line patterns, mottling, or mosaic.

For field diagnosis of viruses, growers commonly rely on recognition of damage symptoms, consultation of published lists of host plants, and knowledge of local disease history and how specific viruses develop and spread. Visual diagnosis of viruses, however, is often inaccurate. Symptoms vary by cultivar. Some plants serve as hosts for many different viruses. Multiple viruses can occur in a plant at the same time. Viral symptoms can easily be confused with the symptoms of nutritional disorders, herbicide damage, or diseases caused by bacteria or fungi. Some infected plants can be symptom-free for many weeks, showing symptoms only when the crop matures. Symptom-free plants can be a source of infection for nearby crops. Electron microscope analysis is the traditional method for expert diagnosis of viruses. However, electron microscopes are expensive, results take at least several days, similar viruses often cannot be microscopically distinguished, and few diagnosticians are available to perform this work. One can also use plant indicator series or bioassays to diagnose viruses, either by rubbing a solution from symptomatic plant tissue onto different indicator plants known to develop specific symptoms in reaction to certain viruses or by using their insect vectors for transmission. Indicator series are relatively slow, and necessitate the constant production of the indicator plants, and sometimes the insect vectors, that the bioassays require.

5.3.4 Pathogen Test/ELISA Kit

Now, instead of sending samples to a laboratory and waiting for answers, one can use quick and relatively simple, commercially available test kits to identify certain pathogens. Most of these field-useable test kits employ a serological technique called ELISA (enzyme-linked immunosorbent assay). This technology is routinely used in home pregnancy tests, blood tests for AIDS and hepatitis and sampling for pesticide residues. Fungal test kits, sold under the brand name Alert, can detect *Phytophthora, Pythium,* and *Rhizoctonia.* The QTA-Tospo brand test kits can diagnose impatiens necrotic spot virus and tomato spotted wilt virus. Other test kits are available that detect other fungi, certain bacteria (e.g., pathovars of *Xanthomonas campestris*), and many viruses, and new kits are constantly being introduced. Contact your local Cooperative Extension advisor or product suppliers for the latest information on pathogen test kits. A separate kit is available for each fungal genus and each tospovirus. You can run multiple tests at the same time. Each kit contains all necessary materials, including detector devices sensitized with antibodies that recognize and react with proteins from the target plant pathogen. Some test kits require refrigeration and must be used before an expiration date printed on the packaging.

Test kits for *Phytophthora, Pythium* and *Rhizoctonia,* can confirm pathogen presence in as little as 10 minutes. First collect tissue samples from the suspect plants, then macerate them by grinding between abrasive pads and treat them. After the sample treatment procedure is completed, a visible color change reaction in the detector device will indicate infection. A 'positive control' dot will darken to indicate that the test kit is working properly. If the target fungus is present in the sample, the 'pathogen detection' dot will also darken. If the pathogen detection dot does not darken, the target fungus is not present in the plant sample.

The virus test kits require more skill and patience than the root decay test kits. Virus kits must be used on a clean workbench area with running water. Samples must be set aside for hour-long incubation periods between certain steps.

5.3.5 *Phytophthora* and *Pythium* Test Kit

The Alert test kit for *Phytophthora* has been used to detect all common *Phytophthora* species. Pathogen detection can be accomplished when as little as 0.5 percent of a plant's roots are infected. The *Phytophthora* test does not react with other common soil-borne fungi, including species of *Fusarium, Mortierella, Trichoderma, Rhizoctonia, Rhizopus* and *Sclerotinia*. Exercise caution when interpreting test results. *Pestalotia* spp. may react weakly with the *Phytophthora* kit. These fungi mostly cause blight (spotting, dieback, and decay) of aboveground parts of trees, woody shrubs, and vines, so you may need to use characteristic damage symptoms, plant parts affected, and host lists to help you distinguish *Pestalotia* spp. from root decay fungi. *Pythium* and *Phytophthora* are closely related fungi, and at high concentrations they can cross-react (i.e., *Pythium* may cause a positive reaction on a *Phytophthora* kit and vice versa). In fact, a few *Pythium* species do not react at all with the *Pythium* kit and react only with the *Phytophthora* kit. We recommend running both tests on the same samples because the tests complement each other. These cross-reactions generally are not a problem, since cultural and chemical controls are the same for *Phytophthora* and *Pythium*. If some sort of regulatory decision is involved, however, you may require more exacting follow-up tests.

Plant health regulations to prevent the introduction and spread of *Phytophthora ramorum* and *P. kernoviae* require rapid, cost effective diagnostic methods for screening large numbers of plant samples at the time of inspection (Lane *et al.*, 2007). Current on-site techniques require expensive equipment, considerable expertise and are not suited for plant health inspectors. Therefore, an extensive evaluation of a commercially available lateral flow device (LFD) for *Phytophthora* species was performed involving four separate trials and 634 samples. The assay proved simple to use, provided results in a few minutes and on every occasion a control line reacted positively confirming the validity of the test. LFD results were compared to those from testing a parallel sample, using laboratory methods (isolation and real-time PCR). The diagnostic sensitivity of the LFD (87.6 percent) compared favorably to the standard laboratory methods although the diagnostic specificity was not as stringent (82.9 percent). There were a small number of false negatives, but for statutory purposes where all positive samples must be identified to species level by laboratory testing; overall efficiency was 95.6 percent as compared to visual assessment of symptoms of between 20–30 percent for *P. ramorum* and *P. kernoviae*. This work demonstrates the value of the lateral flow device for diagnosing *Phytophthora* species at the time of inspection and as a useful primary screen for selecting samples for laboratory testing to determine the species identification.

5.3.6 *Rhizoctonia* Test Kit

Rhizoctonia solani is an aggressive pathogen, but many other *Rhizoctonia* species are not. Some *Rhizoctonia* species are beneficial mycorrhizal fungi, and many are simply common soil fungi involved with the decomposition of organic matter. The Alert *Rhizoctonia* test does not currently distinguish between *R. solani* and other *Rhizoctonia* species. Since nonpathogenic types may colonize fine roots, complete reliance on this test could result in a mistaken diagnosis. Consider limiting the use of the *Rhizoctonia* test to bench-grown crops, including most greenhouse and shade house plants. Use kits on plants less than a year old that have not contacted unsterilized soil, since these plants are less likely to be colonized by nonpathogenic *Rhizoctonia* spp. Limit tissue sampling to lower stem and crown tissues to avoid the fine roots that can be colonized by nonpathogenic *Rhizoctonia* spp.

Besides diagnosing the cause of poor plant health, on-site test kits are useful for regular monitoring for viruses in symptom-free plants of susceptible crops. Root decay test kits can be used as part of routine monitoring. In particular, one can consider a regular testing program for 'mother blocks' and areas where cuttings or other new plants are received or shipped out for propagation. Routine testing can be especially useful when plants are young; most crops are more severely affected by pathogens if plants are infected during early growth. Be aware that negative results in a specific ELISA test do not rule out the possibility that another pathogen is causing disease. When testing for viruses, sample the youngest symptomatic leaves. When testing for decay fungi, samples should be collected from the leading edge of an infection, where apparently healthy tissue and infected tissue meet. Accurate diagnosis of certain problems may require other tests, including the sampling of water or media. Many tests can be performed only by an outside laboratory or are commonly performed off-site, such as indicator plant series bioassays for many viruses other than INSV and TSWV.

5.3.7 Tospovirus Test Kit

The regular sampling of symptom-free plants for the presence of the virus can be very beneficial when one is working with propagation material or crops that are highly susceptible, serve as symptom-free carriers of the virus or tend to develop delayed symptoms. On-site test kits allow growers to detect and confirm virus presence more quickly than is possible with outside laboratory tests. For propagators and growers of sensitive crops that will benefit from frequent testing, the on-site tests can be more economical than an outside laboratory.

Virus indicator plants are easy to use for detecting the presence of thrips that can transmit tospoviruses. Indicator plants may be used in combination with test kits. Indicator plants are used because they quickly develop blackish lesions when an infectious thrips feeds on their leaves, so they reveal when and where infectious thrips occur in growing areas. The petunias 'Burgundy Madness' and 'Blue Carpet' are good indicator plants for INSV and TSWV according to University of California research. Dark lesions develop on leaves of these petunias within about two to four days after feeding by infectious thrips. Indicator plants that detect virus presence in thrips are distinct from the test kits or the previously mentioned plant indicator series used to identify the virus species in crop plants. Indicator plants and test kits are complementary tools that can be used together in an overall virus monitoring program.

5.3.8 Root Decay Test Kit

Kits for detecting fungi were not developed and optimized for presymptomatic detection of disease. It is better to exercise caution if test kits are used for routine monitoring to decide whether or not root decay pathogens are developing, even though symptoms are not apparent. Since most *Phytophthora* species are plant pathogens, the regular sampling and testing of root tips in the bottom 1 or 2 inches of the root ball of container plants may work well for *Phytophthora*. Many *Pythium* and *Rhizoctonia* species are weak pathogens or are nonpathogenic, however. These nonpathogenic fungi commonly colonize very fine roots or dead roots killed by improper irrigation, solar heating of containers, or other causes.

Besides their use confirming the presence of infected plant tissue and possible routine monitoring use in crops, ELISA tests can detect root decay fungi in water and soil. Water tests can be especially important when you are using surface or recycled water for irrigation. Water or soil testing may require special sampling techniques; for more information, contact your Cooperative Extension advisor or agricultural product supplier.

5.3.8.1 *Establishing an overall disease monitoring program*

The test kits can be used in combination with other information to make good pest management decisions. Scout the crops regularly (at least weekly) for signs and symptoms of the disease. Each time that monitor for root disease select at least a few plants from different areas of each crop and remove the plants from their containers or gently scrape or wash away the soil. Examine the roots and crowns for browning, softness or other early indications of disease. Also, inspect aboveground parts for discoloration or wilting and fungal growths, which may indicate more advanced stages of the disease. Understand the conditions and practices that promote diseases and regularly scout for and remedy disease-promoting conditions and practices. Poor sanitation, inadequate drainage, and improper irrigation are primary conditions promoting root decay. Where virus is a concern, use yellow sticky traps (or possibly blue traps for thrips) throughout growing areas to provide early detection of insect vectors. Routinely scout growing areas and remove alternate hosts and reservoirs of disease and insect vectors, including weeds, crop residue, and old plants that will not be marketed.

5.3.9 The SAPS ELISA Kit for *Botrytis*

This kit has been developed as a low-cost kit to bring practical immunology into the classroom. The monoclonal antibody in the kit detects *Botrytis*, a fungal pathogen. *Botrytis* infects plant material (like strawberries, raspberries, tomatoes and flowers) and is of considerable importance in relation to crop losses in the commercial world of horticulture (Anonymous, 2015). The use of a plant pathogen in the kit avoids most of the restrictions associated with work with human pathogens. The principles explored give useful support in teaching students about monoclonal antibodies and topics in current syllabuses and new specifications relating to human health and disease. Using the kit for practical work will help students understand important uses of monoclonal antibodies for diagnosis of plant disease, medicine, and veterinary medicine. The SAPS ELISA kit contains all the specialized equipment (such as microwells, pastettes, and micropipettes) together with the necessary antibodies, the substrate and a Botrytis culture for reference. The materials provided are sufficient for five groups of students. The kit includes a Student Guide with full practical protocol and useful background material on ELISA systems, monoclonal antibodies, and *Botrytis* infection. The practical protocol is adapted to use raspberries (inoculated with *Botrytis*) as the basis for a practical investigation, and there are suggestions for other investigative work. A Technical Guide provides further support for teachers and technicians.

An 'In-the-Field' or 'Point-of-Use' diagnostic kit will detect the presence of specific fungi in plant crops. If a fungal pathogen is present in a plant, its DNA will cause the test kit to read positive. Current diagnostic kits give indirect evidence for the presence of fungal plant pathogens by detecting toxic byproducts. Since many of these by-products are a food safety concern, it is advantageous to directly detect the fungus at low levels before toxins accumulate.

5.3.10 Kit for Virus Detection

Using rapid diagnostic kits can be very beneficial for detecting viral diseases in greenhouse crops. These kits produce results in a short period of time and are reliable, sensitive, cost effective and above all, user-friendly. In fact, tests are usually as sensitive as their comparable ELISA test counterparts that are used in diagnostic laboratories. Results from rapid diagnostic kit tests are ready in about 10–30 minutes. A point-of-use diagnostic kit will allow detection of specific fungal pathogens before the visual disease is apparent. Diligent monitoring with point-of-use diagnostics will help to ensure a safe and high-quality food and feed supply. This kit Allows direct detection of a fungal plant pathogen before visible disease and toxic by-product accumulation. It allows for rapid on-site analysis so that disease monitoring is more efficient and allows for monitoring of multiple fungal plant diseases if the kit is formulated to detect multiple pathogens (Rex B. Dyer, 2015).

Before using these kits, it is important that to know the crops and what pathogens can infect them. One should also need to know if the crop is a local host for the virus, meaning the virus does not spread throughout the plant, or a systemic host, meaning the virus moves through the entire plant. Finally, the knowledge of symptoms the virus produces in the crop or if it produces any symptoms at all. Viruses often cause diseases that have visual manifestations in ornamental plants. However, there are viruses that do not produce symptoms at all and often appear to have no effect on the plant. These are called latent viruses, and the most common are ringspot viruses. But a word of caution, these viruses will often produce strong symptoms if the plant becomes stressed or is infected with another virus. Viral infection often reduces the number of cuttings that a stock plant can produce and can affect the flower color and quality of the plant. Virally infected plants are often unmarketable because of poor flower quality and the presence of symptoms. All of these problems combined lead to lost profits.

All of the current rapid diagnostic kits are designed to detect the pathogen in symptomatic plant tissue. You might obtain a positive result in asymptomatic tissue, but you cannot predict what tissue to sample to ensure a positive result. Also, a negative result does not assure that the tested plant does not contain the pathogen; it only a means that the plant tested negative at that point in time. This problem is due to latent infections. The four most common plant viruses that infect ornamental plants and other crops are cucumber mosaic virus (CMV), impatiens necrotic spot virus (INSV), tobacco mosaic virus (TMV) and tomato spotted wilt virus (TSWV) against which several diagnosis kits are available.

It is important to follow the specific instructions provided by the manufacturer for the kit. Even though instructions might be similar between kits, it has been observed that mixing the instructions or components from multiple kits could result in false results. All of the kits have a basic format for use. There is an extraction method followed by strip development and reading the results. There are two methods of extraction: The sample can be ground in a plastic mesh bag containing the extraction buffer or placed in a vial that contains the extraction buffer and some small beads and shaken.

There are also two formats for strip development. The ImmunoStrip is placed vertically within the mesh bag with the tip of the strip immersed in the extract. Development time is 30 minutes. The second method uses a cassette that remains in a horizontal position. The extract is pipetted into a well at one end of the cassette. Development time is 5–10 minutes. Neither test should be removed and read until the recommended time is over. The reason for this is that during the development of the sample test line the line might be difficult to see because the test area has a background color. The test area becomes a clear white zone after development is complete. This enhances the ability to observe faint positive results that might occur because of low titer.

All of the rapid diagnostic kits employ the same principal in their strip development. All rely on a strip control line that appears at the top of the test zone. The appearance of this line ensures that the strip is working properly. The failure of this line to develop indicates an invalid test, and the results should be ignored. If this happens, repeat the test and/or call the manufacturer to help troubleshoot the test.

A kit was developed by Akella Vani (2012) to detect CTV in citrus orchards. The kit is simple to operate can be used by the illiterate farmer in the field without any laboratory facilities as long as there is clean water. It involves crushing the leaf sample to get a few drops of sap which can be collected in a tube into which the stick coated with reagents is dipped followed by dipping in detection regent and detection substrate. A positive and a negative control are included for the reliability of the results. Washing of the dip stick in clean water are intermediary steps for which clean water can also be included in the kit. ELISA format is for the researcher and semi-quantitative and quantitative estimations can be done.

Planting disease-free material is important in managing this disease. Certification of nursery material is very important by government agencies. This kit is handy for use by semiskilled personnel for checking the presence of the virus. The ELISA format is rapid and cost-effective where up to 90 samples can be tested at one time. Secondly, early detection helps in removal and replanting of affected trees thereby preventing the spread of the disease. Orchards once affected continue to decline and die after about five years.

A diagnostic kit was used to test banana plant sap for the presence of banana bunchy top virus (Anonymous, 2006). A color change (dark wells) indicates the sap samples tested positive for BBTV. Each well in the plate represents a plant or sample.

5.3.11 RNeasy Plant Mini Kit

This kit is for purification of total RNA from plants and fungi. High-quality total RNA in 30 minutes, no phenol/chloroform extraction, no CsCl gradients, no LiCl or ethanol precipitation, excellent recovery of RNA and ready-to-use RNA for any downstream application, are some of its merits.

The RNeasy Plant Mini Kit includes QIAshredder spin columns for homogenizing and filtering viscous plant or fungal lysates, and RNeasy spin columns for purifying up to 100 μg of high-quality RNA using silica-membrane technology. Purification can be fully automated on the QIAcube. The kit can also be used in combination with a TissueRuptor or TissueLyser system, which provides efficient disruption and homogenization of plant samples.

George Monody (2007) developed an on-farm rapid diagnostic kit to detect MLND causative viruses. Other than laboratory diagnostic tests, there is no known on-farm rapid diagnostic kit to detect MLND causative viruses. Sampling and screening of farmer-retained seeds and certified seeds for MCMV and SCMV was carried out using Sandwich ELISA to prove whether the viruses proliferate in seeds. Sowing of infected seeds was taken up in a biosafety level 2 greenhouse to prove whether infected seeds could spread the virus to unaffected zones. Norgen's Plum Pox Virus RT-PCR Detection Kit was developed by Norgen Biotel Corp (Anonymous, 2006), which is a ready-to-use system for the isolation and detection of Plum Pox Virus from plant samples. First, the kit contains components for the rapid isolation of total RNA, including viral RNA, from plant samples using spin-column chromatography based on Norgen's proprietary resin (Nucleic Acid Isolation). Second, the kit contains Plum Pox Virus Master Mix and controls to allow for PCR amplification, as well as a Control Master Mix to allow for amplification of both an isolation control and a PCR control. The amplified PCR products are then detected using agarose gel electrophoresis. Alternatively, detection can be performed based on real-time PCR using melt curves (Detection).

The Plum Pox Virus Master Mix contains reagents and enzymes for the specific amplification of a 295 bp region of the viral genome. In addition, Norgen's Plum Pox Virus RT-PCR Detection Kit contains a second Mastermix, the Control 2X RT-PCR Master Mix, which can be used to identify possible PCR inhibition and/or inadequate isolation via a separate RT-PCR reaction with the use of the provided PCR control (PCRC) or Isolation Control (IsoC), respectively. This kit is designed to allow for the testing of 24 samples. The Plum Pox Virus RT-PCR Detection Kit is also available in a 96-well format for high-throughput analysis. The Plum Pox Virus RT-PCR Primer Set and Controls are also available separately for end point PCR detection.

5.3.11.1 *Features and benefits*

Major features and benefits include isolation of high quality, inhibitor-free RNA, isolation and detection of Plum Pox Virus in under three hours, the amplification bands are clearly distinct, and the melt curves do not overlap, high sensitivity and specificity, where the limit of detection is 100 genome copies and the Plum Pox Virus RT-PCR Primers, as well as the positive control and negative control can be purchased separately.

5.3.12 Kit to Detect Club Root Inoculum

A sensitive and reliable technique for the detection of club root inoculum is essential in preventing the spread of *Plasmodiophora brassicae*. The only commercial method available to Australian growers for testing soil or water for the presence of *P. brassicae* at present is a plant bioassay or a laboratory-based molecular assay. The detection of *P. brassicae* with a plant bioassay is time-consuming, labor-intensive and not sensitive enough to detect low concentrations of resting spores (<1000 spores). On the other hand, the molecular assay is extremely sensitive (<10 spores) but is also very expensive at approximately $ 450 per test. Neither is, therefore, useful as a supporting mechanism for routine soil testing to improve farm hygiene and prevent the spread of clubroot. Where the molecular test is most useful though is for the quantification of soil inoculum loads.

The test has also proven invaluable as a research tool enabling quantitative assessments of the impact of control measures on clubroot populations. In terms of commercial use, the assay has found a niche in seed testing where the high sensitivity has proven to be useful in tracing the source of new outbreaks (Robert Faggian *et al.*, 2007). Ideally, though, a diagnostic assay needs to be inexpensive, rapid and simple enough that it can be used on-farm by growers or other nonspecialist personnel. This would encourage more widespread use of diagnostics as a preplanting and predictive tool, to help prevent the spread of disease to clean areas, and to reduce chemical use by providing grower. A prototype on-farm kit has also been developed that allows the detection of *P. brassicae* in soil; the test is simple and rapid, although there are a number of issues to be resolved before it can be offered commercially.

5.3.13 ELISA-Kit for Detection of Viruses

A kit was developed by Pathibha Sharma (2014) to detect papaya ring spot virus and other related potyviruses. Polyclonal antibodies (PAbs) to recombinant nucleocapsid protein (rCP) of Papaya Ring Spot Virus (PRSV) were used in ELISA to detect potyviruses in the crude extract.

5.3.13.1 *Salient features*

The ELISA kit is sensitive to detect tospovirus infection in crude leaf extract. The sensitivity limit is about 10-25 ng of the virus. The kit has been validated by testing a large number of field samples. The conserved CP gene of PRSV was cloned, and an expression construct was developed in pET vector. rCP was expressed in *E. coli* and purified. PAbs were generated against rCP for broad spectrum detection of potyviruses. It is protein based, rapid, reliable, amenable to large-scale sampling, versatile, could detect natural infection of PRSV, Potato virus Y, Zucchini yellow mosaic virus, Bean common mosaic virus, Onion yellow dwarf virus and Chili veinal mottle virus affecting various crops.

5.13.3.2 Major benefits

Indigenous diagnostic antibodies to plant viruses are not commercially available in India. The major limitation in the production of antibodies is in the purification of adequate quality and quantity of antigen (native virus particle) from plant tissues. This limitation has been overcome by developing nucleocapsid protein expression construct, which can be used for the production of an unlimited quantity of antigen facilitating production of polyclonal antibodies, the key reagent in ELISA kit, at the commercial scale. Approximately, the antibodies can be raised by two months time. The technology will benefit entrepreneurs to produce the indigenous commercial kit. The diagnostic kit will help decision making on judicial application of insecticides for the control of PRSV and related viral diseases. The diagnostic kit will help in developing resistant cultivars. ELISA–Kits were also developed for detection of Cucumber mosaic virus, peanut mottle virus, and tospoviruses.

EnviroLogix, a leader in the development and manufacture of diagnostic test kits for GMOs and plant pathogens, released new diagnostic test kits for Tomato Apex Necrosis Virus (ToANV) and Tomato Spotted Wilt Virus (TSWV). EnviroLogix' QuickStix Kit for ToANV and QuickStix Kit for TSWV are immunoassay Lateral Flow Devices (LFD) for the detection of these plant pathogens in the field or laboratory. EnviroLogix also offers a high-throughput QualiPlate™ Kit for TSWV, a two-hour ELISA plate test for use in the laboratory. These test kits will help growers, breeders, seed producers and seed multipliers diagnose and initiate treatment for these severe tomato diseases (Anonymous, 2008).

ToANV (a torradovirus), confused for some time as a new strain of TSWV, was first identified in the state of Sinaloa, Mexico in 2007 and has since been reported in other areas. This virus has been described as causing 'significant mortality of plants and unmarketable fruit' resulting in a severe economic loss. Before the introduction of the QuickStix Kit for ToANV, there were no tools available to diagnose the disease in the field. Symptomatic samples had to be sent to diagnostic laboratories where complex tests often took days or weeks to develop results. Because the disease is so virulent and economically threatening, early detection and fast, accurate diagnosis are essential to the application of effective control measures.

There are two reasons for the development of a field test for ToANV. First, it has symptoms similar to TSWV (and other tomato diseases) and a test that can distinguish these diseases will help in making good treatment decisions. Second, ToANV is a new disease, and early and accurate detection in the field is critical to establishing the boundaries of the disease and to managing it. EnviroLogix' new QuickStix Kit for ToANV is a diagnostic test giving tomato growers and breeders the ability to detect the pathogen, initiate control measures and to protect the investment in their crops. In addition to field testing, the geographic and cyclical nature of TSWV and the need for large-scale testing requires a fast, high-throughput ELISA test. EnviroLogix developed the QualiPlate Kit for TSWV using removable strip plates that run in two hours. This test can be used to confirm field tests based on the QuickStix results, or it can be used independently to determine the presence of the pathogen in bulk samples.

The QuickStix Kits for ToANV and TSWV can be performed together using a common sample extraction process or may be used independently. They are quick, easy to use and accurate for on-site diagnostics. These kits, coupled with QualiPlate Kit for TSWV, will enable tomato growers, seed producers, seed multipliers, and independent testing laboratories to quickly and accurately determine the presence of either one or both of these diseases. Advances in the technologies to detect and diagnose plant pathogens have culminated in a variety of options for researchers in the laboratory and growers. After the application of ELISA to detect plant viruses was reported (Tamaki Uehara-Ichiki *et al.*, 2013). ELISA and its variations were utilized widely for diagnosing plant diseases. In the case of rice viruses, some reports showed that ELISA had higher sensitivity than classical methods, such as PHA and LAR, and could detect target virus in leaf extracts diluted from 10^{-3} to 10^{-5}.

Since PCR was devised in the late 1980s, the PCR, and its variations have contributed to more accurate detection and identification of plant pathogens. One of the advantages of these techniques is high sensitivity compared with serological or immunological methods. For example, the sensitivity of RT-PCR was 10^2-fold higher than dot-blot hybridization. The sensitivity of real-time RT-PCR was about 10^4-fold higher than ELISA and 10^3-fold higher than conventional RT-PCR. The sensitivity of RT-LAMP was about 10-fold higher than RT-PCR.

Current technologies, i.e., microarrays and next-generation sequencing also hold potential for use in diagnosing rice diseases. Microarrays are suitable techniques for high-throughput detection and identification since an almost limitless number of pathogen probes can be placed in a single array. The potential of array technology has not yet been realized because of various drawbacks that need to be resolved: sensitivity, cost, and lack of practical devices that can be used by nontechnical personnel. But, it does allow us to detect and identify many pathogens including those not proven at one time. Next-generation sequencing methods are being used to identify microbiomes within diseased plant tissue. This technique is largely dependent on software that can discriminate between the host plant and viral sequences, but it provides new opportunities in the areas of viral candidate pathogen discovery and viral ecology.

To select the most appropriate diagnostic methods, we need to focus on our objective. For large-scale analyses to evaluate incidence or for screening tests to monitor disease spread in fields, we should select methods using a user-friendly, evidence-based approach, an evaluation of cost per analysis and a calculation of posttest probability of disease. Therefore, conventional serological methods such as LAR are still used, and ELISA remains one of the most widely applicable methods to forecast virus diseases of rice in the field.

The RT-LAMP and real-time RT-PCR techniques are too expensive for routine analysis in large surveys, but they do enable the differentiation of closely related viruses such as SRBSDV and RBSDV and very early detection of disease before symptoms are visible. Such features can help farmers and plant health professionals to choose the best strategies to minimize potential damage. Particularly with real-time RT-PCR, applications such as screening for virus resistance and studying viral population dynamics, viral multiplication and virus–host interactions are feasible [(Tamaki Uehara-Ichiki *et al.*, 2013) (Fig. 5.5)]. Conventional detection methods have only provided information on the presence of target pathogens. But progress in the development of technologies to diagnose plant diseases may not only contribute to the control of rice viral diseases but also open opportunities to analyze potential pathogens/candidate pathogens and to develop a comprehensive understanding of the ecology of microorganisms in rice fields.

Figure 5.5 illustrates the outline of procedure, required time, cost comparison of reagent kits per diagnosis, and results of five techniques (Tamaki Uehara-Ichiki *et al.*, 2013). Index for the figure includes, (A) Latex agglutination reaction (LAR) to detect RSV in insect vectors. RT, room temperature. (a) Materials for LAR and DAS-ELISA. (1) Microtiter plate, (2) multisticks, (3) single stick. (b) Results of LAR. +, positive, –, negative. (B) DAS-ELISA to detect RSV in insect vectors. (a) The frozen insects in each well. (b) Homogenization of insects with mutisticks. (c) Visual assessment. (Yellow; positive). (C) Multiplex RT-PCR to detect and distinguish 10 rice viruses. (a) Agarose gel electrophoresis of RT-PCR products from healthy or infected rice plants using primer set I. Lane 1, infected with RDV (1106 bp); lane 2, infected with RSV (917 bp); lane 3, infected with RBSDV(734 bp); lane 4, infected with RTYV (631 bp); lane 5, infected with RTSV (504 bp); lane 6, healthy plant (339 bp). (b) Agarose gel electrophoresis of RT-PCR products from healthy or infected rice plants using primer ceII. Lane 1, infected with SRBSDV (1097 bp); lane 2, infected with RGDV (994 bp); lane 3, infected with RRSV (834 bp); lane 4, infected with RTBV (699 bp); lane 5, infected with RGSV (574 bp); lane 6, healthy plant (419 bp). (c) Agarose gel electrophoresis of RT-PCR products from rice plants infected with RDV and RSV. (D) RT-LAMP to detect and distinguish between SRBSDV and RBSDV. (a) Rice plants infected with SRBSDV. (b) Rice plants infected with RBSDV. (c) Healthy rice plants. Left, SRBSDV primer sets. Right, RBSDV primer sets. (E) Real-time RT-PCR to quantify SRBSDV in rice and insect vectors. (a) Samples with a higher density of virus yield an earlier rise in the fluorescence amplification curve, which corresponds to the density of RT-PCR products. The threshold cycle (Ct), calculated from the amplification curve, is used as an indicator of virus titer in the sample (higher virus titer results in lower Ct). (b) Virus titer is usually given as the relative density of virus titer to the expression of a housekeeping gene (e.g., actin and ubiquitin) from the host or vector.

To control rice diseases, developing accurate, quick methods to detect and diagnose the viruses in the host plants and any insect vectors of the viruses is very important. Based on an antigen–antibody reaction, serological methods such as latex agglutination reaction and enzyme-linked immunosorbent assay have advanced to detect viral particles or major proteins derived from viruses. They aid in forecasting disease and surveying disease spread and are widely used for virus detection at plant protection stations and research laboratories. From the early 2000s, based on sequence information for the target virus, several other methods such as reverse transcription-polymerase chain reaction

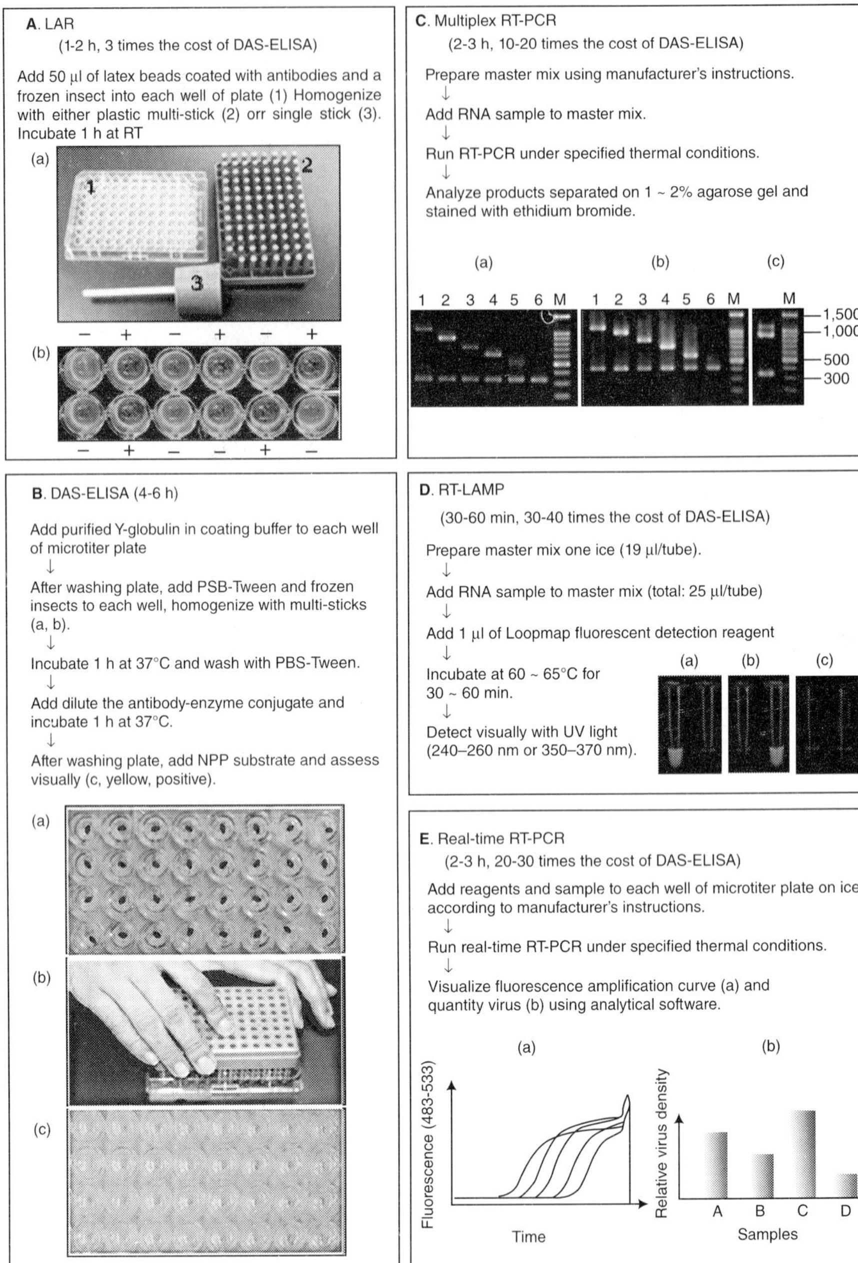

Fig. 5.5 Outline of procedure, required time, cost comparison of reagent kits per diagnosis and results of five techniques

(RT-PCR) and reverse transcription-loop-mediated isothermal amplification have been developed that are sensitive, rapid, and able to differentiate closely related viruses. Recent techniques such as real-time RT-PCR can be used to quantify the pathogen in target samples and monitor population dynamics of a virus, and metagenomic analyzes using next-generation sequencing and microarrays show potential for use in the diagnosis of rice diseases.

Several kits have been developed to detect Citrus Tristeza, Citrus Canker, Stubborn Citrus greening HLB, Citrus Gummosis, Citrus Variegated Chlorosis (Anonymous, 2008).

5.13.3.3 ELISA

It is the most popular diagnostic method, which allows running many samples during large-scale surveys for CTV, Xac, Xf and Sc control in nurseries and diagnostic laboratories.

5.3.14 Immunoprinting

It is an alternative diagnostic method for CTV. It allows easy sampling of citrus trees onto nitrocellulose membranes where the reactions take place. The positive result shows an image of the vascular tissue with dark blue spots in the vascular bundles, whereas the negative control remains uncolored. The use of membranes to capture and immobilize targets constitutes a good alternative to extract preparation. The development of direct tissue blot immunoassay, immunoprinting-ELISA or tissue print-ELISA in plant pathology and its application to CTV again revolutionized and simplified the detection of the virus. Immunoprinting-ELISA allows the sensitive analysis of thousands of samples in an easy way, without the need of extract preparation, by directly printing sections of plant material on nitrocellulose membranes.

Extract preparation is the most limiting factor for large-scale plant virus testing. Direct tissue print of fresh cross sections of tender shoots or leaf petioles on cellulose membranes, allows the collection of samples and testing of a large number of plants (1,250 plants per team of two workers per day). The printed membranes can be analyzed in the field, mailed or kept for several months before testing (Cambra *et al.*, 2000). The analysis is performed by a simple and fast (3 h) direct ELISA protocol using a mixture of citrus tristeza virus (CTV)-specific, alkaline phosphatase conjugated monoclonal antibodies 3DF1 and 3CA5 or using a mixture of 3DF1 and 3CA5 scFv-AP/S recombinant antibodies expressed in *E. coli* as a fusion protein with the alkaline phosphatase enzyme. The sensitivity of the immunoprinting-ELISA method was the same as immunocapture-PCR, but it was more reliable. A kit has been designed and evaluated under nursery conditions. This kit has been successfully used by nurserymen to test more than 600,000 plants over the last five years.

5.3.15 Flash Kits

Flash kits (or Immunostrips) are quick detection kits based on the immunochromatography system. The kits are as reliable as ELISA tests used in laboratories. Flash kits are user-friendly and allow knowing, in just a few minutes, whether a tree is infected or not. Available for CTV, *Phytophthora* spp. and Xac Flashkits® are easy and reliable. In addition, they can be used in the production site rather than having to go to a laboratory. This detection test is simple as it does not require special equipment or technical qualifications. It is composed of a plastic bag and strip. The sample is ground in the bag containing a specific extraction buffer, and a strip is then introduced vertically inside the bag. After 15–30 minutes, the grower can read the results and take the corresponding decisions. If one line appears on the strip, the test is negative; if two lines appear the test is considered positive for the pathogen tested.

5.3.16 PepMV Test Kit

The Pepino mosaic virus test is designed to be used in the field or glasshouse and can be used by operators with little or no training. Pocket Diagnostic test results form an important part of the information mix required for sound decision making in routine screening or disease control situations.

The Pocket Diagnostic Pepino mosaic virus test kit is used widely by growers of protected and field tomato and ornamental crops for detection of the pathogen. Testing is usually carried out on symptomatic plants where disease is suspected. Samples for testing should be taken from leaves or flowers with symptoms or skin from suspect fruit. The tomato flesh should not be used as this inhibits the test. All Pocket Diagnostic kits put disease diagnosis in the hands of growers, consultants, and inspectors. Sampling and testing can be carried out in a few minutes in the field or glasshouse. Pocket Diagnostic kits use a unique 'ball and bottle' extraction method which optimizes separation of the pathogen from the plant tissues in no more than 60 seconds. Results appear in the viewing window of the test device approximately 5 minutes after starting the test. The result is easy to read, allowing the user to make simple 'yes/no' interpretation.

5.3.17 New Kit to Diagnose Pathogens

Agdia Biofords Ltd. has developed tests for citrus diseases such as *Phytophthora, Xanthomonas* (bacterial canker), *Spiroplasma* (stubborn), citrus greening (HLB), *Xylella*, etc. In the Agdia Biofords catalog, more than two hundred diagnostic tests for vegetable pathogens such as viruses, bacteria, fungi, phytoplasmas and viroids are included. All of the products have been tested to assess their sensitivity, specificity, accuracy, and robustness. The performances of products are checked first of all at an internal level and then in cooperation with external laboratories, universities and research institutes in various parts of the world.

Rapid and accurate detection of plant pathogens in the field is crucial to prevent the proliferation of infected crops. Polymerase chain reaction (PCR) process is the most reliable and accepted method for plant pathogen diagnosis. However, current conventional PCR machines are not portable and require additional postprocessing steps to detect the amplified DNA (amplicon) of pathogens (Koo *et al.*, 2013). Real-time PCR can directly quantify the amplicon during the DNA amplification without the need for post-processing, thus more suitable for field operations, however still takes time and require large instruments that are costly and not portable. Microchip PCR systems have emerged in the past decade to miniaturize conventional PCR systems and to reduce operation time and cost. Real-time microchip PCR systems have also emerged, but, unfortunately, all reported portable real-time microchip PCR systems require various auxiliary instruments. A stand-alone real-time microchip PCR system was described, which is composed of a PCR reaction chamber microchip with integrated thin-film heater, a compact fluorescence detector to detect amplified DNA, a microcontroller to control the entire thermocycling operation with data acquisition capability and a battery.

The disposable microchip requires only 8 mL sample volume, and a single PCR run consumes 110 mAh of power. A DNA extraction protocol, notably without the use of liquid nitrogen, chemicals, and other large laboratory equipment, was developed for field operations. The developed real-time microchip PCR system and the DNA extraction protocol were used to successfully detect six different fungal and bacterial plant pathogens with 100% success rate with a detection limit of 5 ng/8 mL sample.

5.3.18 Nanodiagnostic Techniques

The application of nanodiagostic techniques in plant pathology has been reviewed by Mohammad A. Khiyami *et al.* (2014). Traditional molecular diagnostic methods are widely used in laboratories all over the world to identify plant pathogenic organisms with a high degree of sensitivity and specificity. Still, most of these procedures cannot be applied in the field (on-site detection) or in developing countries where incomes are small. Furthermore, the high price and short shelf half-life of some molecular biology reagents, such as enzymes and primers, limit the application of traditional molecular methods in developing countries. Nanotechnology may have actual solutions to many agriculture problems like plant disease control. Nanobased materials will be presented which will increase the efficacy of fungicides and pesticides, allowing only minor doses to be used (Abd-Elsalam, 2013). Moreover, nanodiagnostic and microfluidics offer novel tools to improve the sample preparation step that remains difficult to integrate into a miniaturized platform. The signal amplification approaches could challenge those of target amplification. Quick on-site detection of plant pathogens using, nanobased kits, nanosenser, nanobiosensors, nanobarcodes and other portable diagnostic systems will also help the agricultural and food industry to manage different plant diseases. In this review, we describe the concepts and the current state of the nanotechnology application in plant pathology including nanodiagnosis by portable polymerase chain reaction (PCR) systems, nanopore sequencing tools, nanodiagonastic kit, gold nanoparticles, quantum dots (QDs), nanobarcodes and nanosensors.

5.3.19 Loop-Mediated Isothermal Amplification (LAMP)

Loop-mediated isothermal amplification (Notomi *et al.*, 2000) is a relatively recent and novel technique for the specific amplification of DNA templates at a single temperature, hence requiring less specialized equipment than conventional PCR technologies. This technique has been used to detect several vertebrate and plant pathogens, including bacteria and viruses and shows promise as a rapid and effective method for the detection of important pathogens (Lisa Ward *et al.*, 2010). Loop-mediated isothermal amplification (LAMP) of DNA is a simple, cost-effective, and rapid method for the specific on-site detection method of genomic DNA using a set of six oligonucleotide primers with eight binding sites hybridizing specifically to different regions of a target gene and a thermophilic DNA polymerase from *Geobacillus stearothermophilus* for DNA amplification. There are several detection devices varying from gold nanoparticles tagged with short fragments of DNA to multicolor optical coding for biological tests that have been achieved by embedding different-sized QDs into polymeric microbeads (Jain, 2003). The thiol-modified oligonucleotides (Au-nanoprobes) technique discards the use of electrophoretic analysis of LAMP-PCR products, thus increasing speed, specificity of results and decreasing costs. LAMP has the potential to implement early detection of plant pathogens at a local level (e.g., on a farm) instead of in a laboratory or less well-resourced settings by molecular methods for monitoring inoculum levels in the air and is currently developing models to predict when and where a pathogen will first occur in the agricultural fields.

For instance, LAMP-PCR of DNA is an easy, commercial and quick method for the specific detection of genomic DNA using a set of six oligo primers with eight binding sites hybridizing, specifically to various regions of a target gene [(Fig. 5.6) (Lippolis *et al.*, 2012; Lattanzio *et al.*, 2012)] and a thermophilic DNA polymerase from

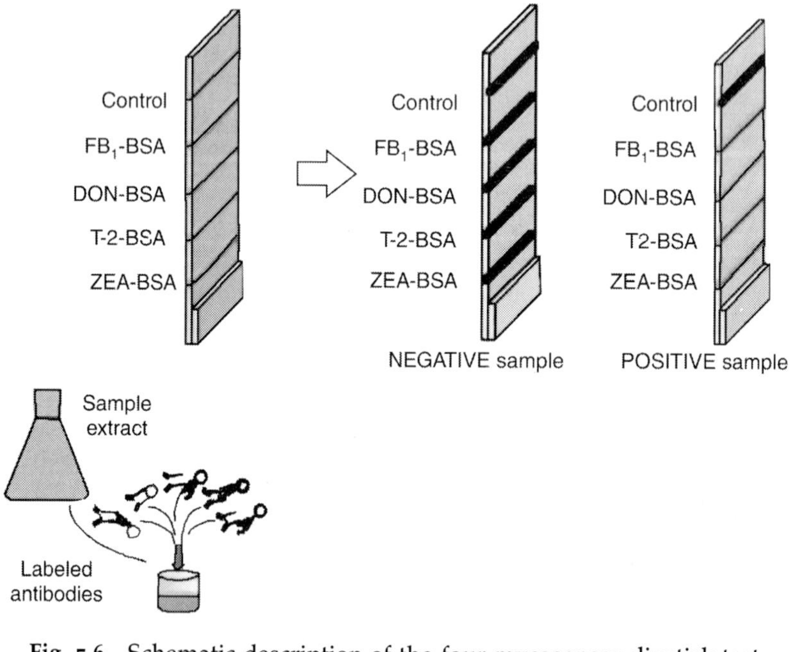

Fig. 5.6 Schematic description of the four mycosensor dipstick test

G. stearothermophilus for DNA amplification of *Fusarium graminearum*. Therefore, in the LAMP technique, diagnostics can be performed without examination of the amplification product during or after the amplification. Predominantly, positive reactions will be expected to show bright green fluorescence, while a negative reaction would remain light orange. This would be the expected result when SYBR Green I dye is added, meaning that the LAMP amplicons will be visualized directly and seen with the naked eye or under ultraviolet transillumination.

Positive results were obtained with *F. graminearum* isolates, while all reactions with other *Fusarium* and fungal species included for specificity remained negative, with a response identical to the negative control. This methodology is used to detect plant pathogenic fungi to rhizospheric microorganisms such as *Phytophthora sojae, P. ramorum* and *P. kernoviae, Aspergillus flavus* and *A. parasiticus* and toxigenic *Fusarium* (Niessen *et al.*, 2012).

5.3.19.1 The LAMP reaction

The basis of the LAMP method is the strand displacement activity of the *Bacillus subtilis* derived Bst DNA polymerase with specific activity between 60–65°C. Many other polymerases, including the bacteriophage Φ29 and Escherichia coli DNA polymerase I, display strand displacement activity although at significantly lower temperature optima which exclude their use in a LAMP reaction. LAMP operates through the priming of heat-denatured positive and negative strands with two template-specific inner primers on which the Bst polymerase binds and releases single-strand DNA. This is then targeted by outer primers from which template extension, or more simply, second strand synthesis can occur. LAMP differs from normal PCR in that the 3′ and 5′ ends of the inner primers bind to two different, though neighboring regions within the template, with a small, generally 4-mer nonspecific oligo 'hinge' between, that when bound to ssDNA template forms a 'dumbbell-like' shape that initiates further extension. An additional two loop primers trigger secondary strand displacement reactions within extended template (Lisa Ward *et al.*, 2010). As the reaction proceeds the DNA template resembles a cauliflower-like shape with multiple copies joined by the loop-forming inner primers, and being displaced by outer and loop primer pairs. The end result of strand-displacement, loop formation, and synthesis is the single-temperature amplification of a highly-specific fragment from a DNA template at far greater titer than an equivalent PCR reaction.

5.3.19.2 Comparison to conventional and real-time PCR

The LAMP reaction differs in several major aspects to conventional PCR amplification and its variants. The use of six primers provides a greater level of target specificity than can be achieved with the two primers normally used in PCR (Notomi *et al.*, 2000). This is both an advantage and a shortcoming of the LAMP process in that polymorphism, particularly at the 3′ end of the F1 and 5′ end of the B1 loop primers within the target organism can reduce the likelihood of a successful amplification. LAMP is, however, less susceptible to interference from nontarget DNA presence in the reaction, comparable to nested PCR (Lisa Ward *et al.*, 2010). The sensitivity of LAMP for detection of low copy number templates, i.e., less than 100 copies/reaction is largely dependent on the target selected, template purity and the presence of inhibitory substances. Unlike normal PCR, LAMP does not require a dedicated thermocycler and reactions can readily be achieved in standard heat-block or water-bath apparatus.

The cost the LAMP assay may be considered. The LAMP assay, with its requirement for specialized enzymes and reagents (especially dNTP usage and the need for HPLC-purified loop primers), costs approximately five times more for consumables than conventional or real-time PCR. However, this does not include the cost of specialized equipment such as real-time thermocyclers. The cost of LAMP may be a limitation for large-scale surveys; yet this method offers several advantages; there would be a huge saving in time if reactions are carried out in the field, it would reduce the movement of infected tissue across the country for laboratory testing, and nonscientific staff could be trained to use the method due to its simplicity.

It may be stated the LAMP rimM assay (the optimized protocol is described in Appendix 1) has a high level of specificity for the detection and diagnosis of the major subspecies of *X. fastidiosa*. Provided that care is taken to avoid contamination by running LAMP as a closed-tube reaction with HNB as a reporter dye, LAMP has the potential to be used in the field (Lisa Ward *et al.*, 2010).

5.3.20 PrepGEM Extraction Kit for *Xylella fastidiosa*

Xylella fastidiosa is a regulated plant pathogen in many parts of the world. To increase the diagnostic capability of *X. fastidiosa* in the field a loop-mediated isothermal amplification (LAMP) was developed to the rimM gene of *Xylella fastidiosa*, and evaluated for specificity, sensitivity, and reliability. The assay was more robust than existing published assays for detection of *X. fastidiosa* when screened against 20 isolates representing the four major subgroups of the bacterium, and a range of host species. No cross-reaction was observed with DNA from healthy host tissue or other bacterial species. The LAMP assay could detect as little as 250 copies of the rimM gene. The LAMP assay was also found to have greater sensitivity than conventional PCR if less than a published real-time PCR assay. A series of endpoint detection methods, including fluorescent and colorimetric dyes, and lateral flow devices were also assessed. It was found that a colorimetric dye, hydroxynapthol blue performed best and could indicate the presence of 500 copies of the template with a purple to blue color change after 60 minutes incubation; hydroxynapthol blue was also found to be the most suitable for field use as the reaction could be completed in a closed-tube format. Techniques suitable for DNA extraction from plant tissue and insect *in situ* were compared to a standard silica-column-based laboratory extraction method, and it was found that a portable PickPen® and magnetic bead system can be used to successfully extract DNA from infected tissue and could be used in conjunction with LAMP in the field, whilst *X. fastidiosa* was successfully amplified from insect tissue extracted using the commercial PrepGEM extraction kit.

5.3.21 QualiPlate™ Kit for Lettuce Mosaic Virus (LMV)

Lettuce Mosaic Virus (LMV) is a potyvirus disease of solanaceous plants, in particular, lettuce, with considerable worldwide economic impact. LMV is seed-borne in all lettuce types; however, aphids can efficiently transmit the virus contributing to the secondary spread of LMV. The primary host of LMV is all lettuce types, escarole, endive, as well as pea, sweet pea, and marigolds. The long list of host plants and the globalization of seed distribution contributed to the spread of this virus to virtually every country where lettuce is grown. Seed-borne infected plants show mosaic and stunting and will never develop marketable heads. Later infections (secondary spread) could result in mosaic, leaf puckering, and deep or accentuated serration of the leaf margins. It may be difficult to diagnosis this disease in the field due to the presence of other viruses on the plants.

Portland, ME– EnviroLogix, a leader in the development and manufacture of diagnostic test kits for GMOs and plant pathogens, has released the QualiPlate for Lettuce Mosaic Virus (LMV) (Anonymous, 2007). This test simplifies the LMV testing process enabling analysis laboratories to deliver test results faster and with greater data consistency and integrity. The 'ready-coated' QualiPlate takes only two hours to perform compared with other LMV tests kits. The increased productivity and throughput allowed by the faster test time of the EnviroLogix QualiPlate for LMV will enable laboratories to test more plates during very time critical periods.

The key behind the EnviroLogix QualiPlate for LMV is that the plates are precoated by EnviroLogix in their production laboratory to precise standards. This ensures uniformity of plate preparation and provides the users the convenience of not having to coat plates in advance of running the test.

5.3.22 INSV ImmunoStrips Kit for Impatiens Necrotic Spot Virus

Impatiens Necrotic Spot Virus (INSV) has been seen in all types of ornamental plants, including annuals and perennials. This virus causes symptoms that vary depending on the plant affected, but include leaf and stem spots and/or discoloration, and stunted growth. The virus is rapidly spread by the insect vector, the Western flower thrips, and there is no cure for the disease. Due to its potential spread to other plants grown in the same area, it is very important to act quickly to remove and destroy affected plants. Therefore, a rapid diagnosis of INSV is crucial so that growers can determine if plants are diseased and act without delay. INSV ImmunoStrips from Agdia can play a key role in testing and diagnosis (Mara Grossman *et al.*, 2014). These test kits allow growers to quickly determine if a plant has INSV on-site, without any special equipment.

Test results appear within 30 minutes, but may be visible as quickly as 5 minutes. When the test is completed, one pink line will appear on the test strip as a control. The presence of this line indicates that the test is working; if no pink line appears, the test is invalid and should be repeated. If the control line is the only line that appears, this indicates a negative result (no INSV detected). The presence of two pink lines (the control line and a second test line) indicates a positive result (INSV is present).

5.3.23 AmplifyRP Acceler8 LChV2 Kit for Little Cherry 2 Virus

Agdia-Biofords, a company located on the Genopole campus and an affiliate of the American company Agdia Inc., has made available a new rapid molecular diagnostic kit for the detection of the virus responsible for little cherry disease, which results in small, badly formed and insipid cherries (Évry, 2014). Little cherry 2 virus (LChV2), the pathogen responsible for the little cherry disease, is an RNA virus that can infect all types of cherry trees, be they for production or decoration. The disease can be transmitted through grafting or by Coccidae insects that feed on the sap of the tree. Trees infected with the virus produce small, pale and bland cherries that are difficult to commercialize. To aid in the battle against this disease, Agdia Inc. has developed a novel, easy to use, rapid detection kit for LChV2. The kit is built upon the company's isothermal amplification-based AmplifyRPTM Acceler8TM technology that enables the amplification of the virus's RNA from a sample of leaves. It works at a single constant temperature (no thermal cycling is needed) and requires no particular technical skills. The result is read from a flow strip (similarly to a pregnancy test) after only 30 minutes. The kit has been validated on numerous strains of LChV2 and has shown nonreaction with other pathogens, such as little cherry virus 1 or western X phytoplasma, which are known to cause similar symptoms.

5.3.24 Crop Doctor Field Diagnostic Kit or Block Kit

An initiative of TOPCROP West, *Crop Doctor*, allows growers to make an on-the-spot diagnosis of any unhealthy plant symptoms visible above ground. In South Australia, TOPCROP has produced the *Block Kit* to assist growers in the practical identification of root diseases in the paddock. Separately or together, the two kits take much of the guesswork out of crop-disease diagnosis and put some rigor into the process (Anonymous, 1999). *Crop Doctor* consists of six diagnostic cards, one each for cereals, lupins, canola, chickpeas, faba beans and field peas. The diagnostic cards are keyed to critical stages for each crop from the seedling stage to maturity. For many crop diseases, it's crucial to be right on top of the situation

The *Block Kit* offers a collection of disease-affected plants, individually preserved and packaged in perspex so that the tell-tale symptoms of root diseases such as take-all and cereal cyst nematode are immediately evident. It fits in well with the field-based monitoring done by TOPCROP groups. Both kits come with a range of tools from a field magnifier and test sample bags to field guides on plant samples.

5.3.25 Fungal RNA and DNA Isolation Kit

Norgen Biotek Corp., an innovative Canadian biotechnology company focusing primarily on advancing powerful tools for RNA, DNA, and protein purifications, has developed four new kits geared towards the isolation of RNA and DNA from plants and fungi for research and diagnostic applications in response to customer demands and emerging needs, which are all available in spin-column formats as well as in 96-well plate formats for high-throughput applications (Yousef Haj-Ahmad, 2008.). This kit allows for the rapid isolation and purification of genomic DNA from the fungi found on the surface of symptomatic plant samples. A major benefit of this kit is that it does not require the use of physical homogenization steps to isolate the fungal DNA, which are often time-consuming and tedious procedures. While a mortar and pestle, a rotor-stator homogenizer or bead mills associated with liquid nitrogen may enhance the DNA extraction yield, they are not suitable methods for use in the analysis of large numbers of samples, and in the area of pathogen diagnosis simple and convenient sample preparation steps are a critical factor in supporting high-throughput analysis. The procedure yields sufficient fungal DNA to be used in a number of downstream applications including Real-Time PCR, sequencing, Southern blotting and SNP analysis and the total procedure can be completed less than 60 minutes.

5.3.26 DNA Capture Kit

Banana wilt caused by the bacterium *Xanthomonas campestris* pv. *musacearum* and banana bunchy top virus disease (BBTD), caused by the banana bunchy top virus (BBTV), are two of the most devastating diseases of banana. The diagnosis of *Xanthomonas* wilt in the field based on disease symptoms is often confused with another disease that causes wilting that is due to the fungus *Fusarium oxysporum* f. sp. *cubense*, so diagnostic tools that are easy to use, cost-effective and reliable are required to help field workers. BBTD symptoms can also be confused with nutrient deficiencies or varietal differences. To this end, three types of portable kits were evaluated for their performance in capturing pathogen DNA from the field for use in precise, molecular-based pathogen diagnostics under controlled laboratory conditions (Ramathani and Beed, 2013). All three prototypes tested – FTA cards, the PhytoPASS kit, and two-minute DNA extraction dipsticks – gave excellent results and methods were optimized for sample collection in the field, DNA extraction and PCR-based diagnostics for *X. campestris* pv. *musacearum* and BBTV.

The benefits of using DNA capture kits include a collection of pathogen DNA in a cheap and practical manner, safe and fast transfer of high integrity pathogen DNA across country borders, rapid and precise diagnosis using state-of-the-art molecular technologies and direct comparison of results from geographically diverse samples. This is possible because the analysis is performed in a uniform manner in terms of method, date, equipment, reagents and technical staff. Technical support was provided to workers in the field to help them differentiate symptoms based on diagnostic results from laboratory analyses; such support could also be provided to regulatory officials at borders to ensure that, for example, planting material is free from disease.

5.3.27 Pocket Diagnostic Kit for Potato Viruses

Express diagnostic system ('pocket diagnostic') was used. Potato Y virus, as well as mixed infection X and Y viruses, were noted on the following potato sorts: 'Jeli,' 'Agria,' 'Rozelinda,' and 'Marfona.' In particular, Y virus was identified on the sorts of 'Jeli,' 'Agria' and 'Marfona.' X and Y viruses were identified at the same time on the 'Rozelinda' sort. It was shown that 'pocket diagnostic' is a modern and one of the perspectives and precise methods for field and laboratory diagnostic and fast identification of viral diseases (Jakeli and Megrelidze, 2014).

Express test system consists of analyzing plate with a container of serological whey. After dropping an analyzing material on the right part of the plate, one control blue zone appears on the mini screen. In the case of positive result of the test, another blue zone shows up. There are plates with specific whey for all strains of viruses. These diagnostic kits with a serological test for pathogens of viral ethology, help especially in-field conditions, to identify pathogens fast and without any extra cost to carry out more fundamental further researches. The diagnostic plate is transportable and valid for a long period, and the so-called 'pocket diagnostic' practically excludes the use of expensive and labor intensive diagnostic, such as a method of indicator plants. Consequently, 'pocket diagnostic' is a modern and one of the perspectives and precise methods for field and laboratory diagnostics for fast identification of viral diseases.

5.3.28 Nematode Diagnostic Kit

Diagnosis of nematode species based on morphology is often difficult as the characters used show considerable intraspecific variation. Moreover, the identification of many morphologically identical nematode races or pathotypes is impossible using physical characteristics alone. Molecular biology offers a number of exciting new approaches that are revolutionizing parasite identification, but with a few notable exceptions these techniques have been largely neglected by plant nematologists (Paul R. Bumows, 1990). For routine identification of plant-parasitic nematodes, specific DNA probes have an important advantage over RFLPs in that they can give a positive or negative assay for nematode species, race or pathotype without the need for restriction digestion of the DNA or time-consuming electrophoresis. It seems that a major impediment to the wider acceptance and use of this technique for routine diagnostics will be the need to detect probes using radioisotopes.

Radiolabeled probes are expensive, hazardous to work with and once labeled the activity decays within a few weeks necessitating replacement or frequent relabeling. The full potential of specific DNA probes for the identification of plant-parasitic nematodes will only be realized when suitable nonradioactive methods for the detection of bound probes are developed or adopted. Specific DNA probes can be visualized by a number of different nonradioactive methods, but they are often considered to be too insensitive. Although this is generally correct, one particular group of techniques based on biotin is very promising. Biotin-1 1-dUTP is a nucleotide analog that may be incorporated into a probe by nick translation in the same way as a radionucleotide. Hybridization of a biotin-labeled probe with nematode DNA bound to a nitrocellulose filter can be detected by conjugating a preformed complex of avidin and alkaline phosphatase to the biotin. Enzyme catalyzed the production of an insoluble blue pigment at the site of hybridization reveals the presence of the probe.

Modifications have been made to a basic biotin technique and its subsequent use to visualize DNA from *Globodera pallida* using a *G. pallida* specific DNA probe in dot blot assays. The sensitivity of the method in this application has been improved, and nonspecific background staining has been almost eliminated even when using crude DNA preparations. This work is an important step towards the production of diagnostic kits for agriculturally important plant parasitic nematodes. Such kits will use small amounts of nematode material, give a reliable result within a few hours and require little operator skill or knowledge of nematode morphology.

5.3.29 DNA Extraction and Purification Kit for Nematode Suspensions and Multiple Cysts

Plant parasitic nematodes are surrounded by a cuticle, a relatively thick, physically strong (outer) layer. The inner structure of this protective layer shows similarity to a radial tire and is refractory to biodegradation. This characteristic hampers the efficient, quantitative isolation of DNA from plant parasitic nematodes or cysts. This has changed with the release of the ClearDetections extraction and purification kit. The hot detergent lysis buffer rapidly and quantitatively extracts DNA from a wide variety of nematodes and cyst making it an indispensable tool for any plant pathogen diagnostic laboratory. Because nematode suspensions from soil are unavoidably contaminated with PCR-inhibiting components, especially in soils with high organic matter content, it is essential to clean the lysate before use. In the purification step of the ClearDetections extraction and purification kit, a simple silica-based column is used to remove organic PCR inhibitors such as humic and fulvic acids. Purification efficiency can be monitored with external control DNA (included in kit). External control primers allow users to accurately monitor losses during the purification process (and – in a next step – compensate for it). This nematode DNA extraction and purification kit is an efficient tool for every diagnostic plant pathogen laboratory analyzing the presence of pathogenic nematodes in nematode suspensions.

ClearDetections, a joint initiative of BLGG and Wageningen University has developed nematodes diagnostic, mainly for DNA extraction and purification, species identification and species quantification. Renske Landeweert and Hans Helder (2012) have mined the world's largest nematode DNA database developed at the Laboratory of Nematology (Wageningen University). Unique DNA motives were identified for major plant parasite, allowing the detection and quantification of individual nematode species in highly complex DNA backgrounds. These principles combined with the expertise of BLGG, the Netherlands' leading agricultural service laboratory, in product development resulted in series of proprietary DNA-based detection tests. ClearDetections delivers these tests as all-inclusive diagnostic kits (Q-PCR based) and dedicated systems for molecular detection of plant diseases, with special emphasis on harmful nematodes. In doing so, ClearDetections offers agricultural (inspection) laboratories a turnkey solution that could completely replace the current routine nematode detection through microscopic examination, enabling large-scale, sensitive and accurate monitoring of soil quality.

The kit includes ClearDetections unique and proprietary extraction buffer (96 samples), proteinase K solution, exogenous external control DNA and matching primers

(96 samples), ClearDetections 2x SYBR Green PCR mix (96 reactions), 96 well silica-filter purification plate and disposables and buffers needed for extraction and purification

5.3.29.1 *Major advantages of this all-inclusive kit*

Extract and purify PCR-ready DNA from nematode suspensions and multiple cysts in less than one hour (excluding incubation time). There is no need for dedicated equipment or bead-beating. It is applicable to a wide variety of nematode species. Method is essentially unbiased with respect to species identity and contains free exogenous external control DNA, matching primers and real-time PCR mix to check purification efficiency.

A diagnostic test has been developed by Xiaohong Wang (2011) from the Agricultural Research Service of the US Department of Agriculture to identify potato cyst nematode (PCN) infestation (Anonymous, 2011). By cloning and sequencing key genes, Xiaohong Wang improved on traditional methods of distinguishing PCN species that are time-consuming and require large samples of nematode cysts. Being more sensitive than other systems, the diagnostic test is expected to be widely used in regulatory and quarantine programs. Since breeders have yet to develop potatoes that resist pale cyst nematodes, it's important to be able to distinguish the pale cyst from the golden nematode. The pale cyst nematode is genetically more diverse than the golden nematode, and that makes it harder to come up with resistant germplasm in potato breeding programs.

Traditional methods used to distinguish between the two PCN species have relied on morphological analyses and PCR (polymerase chain reaction) assays that are time-consuming and require extensive training in nematode morphology and relatively large samples of nematode cysts. Wang has developed a way to distinguish between the two species of PCN that uses a new genetic marker and is more sensitive and reliable. The system is capable of testing much smaller samples of cyst material, making it a thousand times more sensitive than the current PCR technologies.

They have developed their system by cloning the parasitism gene that these nematodes use to produce an enzyme, chorismate mutase that plays an important role in nematode infection. The researchers then sequenced those chorismate mutase genes, compared the sequences, and identified unique regions in each sequence. They then developed a 'TaqMan probe' capable of recognizing the unique regions in the nematode's DNA. It's the first time that a parasitism gene has been used as a diagnostic marker for identifying plant-parasitic nematodes. The system is one of several new technologies developed to distinguish the golden nematode from the pale cyst nematode. But with its high degree of sensitivity and reliability, it should prove to be a valuable tool for nematode regulatory and quarantine programs throughout the United States.

5.3.30 Diagnostic Q-PCR Test Kit

Several agricultural laboratories have invested in the setup of a molecular laboratory in the expectation that DNA-based identification and detection of plant pathogens, including nematodes, will be the golden standard in the near future (Landeweert *et al.*, 2012). However, as molecular methods for the detection of some specific (quarantine) nematodes have the potential to fully replace the current microscopic routine screening of (soil) samples, there is a strong need to check for their accuracy and performance. Whereas a molecular method in itself may be far more precise and accurate than a microscopic method ever can be, its accuracy will largely depend on the biological information that was used to develop it. Its performance will depend on the robustness of the method and the test performed. On the basis of a phylum-wide, small subunit ribosomal DNA (SSU rDNA)-based framework consisting of ~ 2,500 full-length nematode sequences, we have developed over 20 diagnostic Q-PCR tests for plant pathogenic nematodes. Some nematode species have a dubious reputation as parasites of major crops such as potato, sugar beet, and soybean. Among the most notorious ones are cyst (*Globodera rostochiensis* and *G. pallida*), root knot (*Meloidogyne chitwoodi, M. fallax* and *M. minor*), stem and bulb (*Ditylenchus dipsaci* and *D. destructor*), and foliar nematodes (various *Aphelenchoides* species). By using this SSU rDNA framework, we could define unique DNA sequence signatures that enable 'blind' identification of these individual nematode species in highly complex DNA backgrounds.

A PCR-based diagnostic assay was developed for early detection and identification of *Aphelenchoides fragariae* directly in host plant tissues using the species-specific primers AFragFl and AFragRl that amplify a 169-bp fragment in the internal transcribed spacer (ITS1) region of ribosomal DNA (Jamie L. McCuiston *et al.*, 2007). These species-specific primers did not amplify DNA from *Aphelenchoides besseyi* or *A. ritzemabosi*. The PCR assay was sensitive, detecting a single nematode in a background of plant tissue extract. The assay accurately detected *A. fragariae* in more than 100 naturally infected, ornamental plant samples collected in North Carolina nurseries, garden centers, and landscapes, including 50 plant species not previously reported as hosts of *Aphelenchoides* spp. The detection sensitivity of the PCR-based assay was higher for infected yet asymptomatic plants when compared to the traditional, water extraction method for *Aphelenchoides* spp. detection. The utility of using NaOH extraction for rapid preparation of total DNA from plant samples infected with *A. fragariae* was demonstrated.

A. fragariae-specific PCR assay has application in plant disease diagnostic clinics in assisting growers with screening plant material for early detection of foliar nematodes, in regulatory laboratories where accurate identification is required for phytosanitary permits and as a tool to help study *A. fragariae* biology to improve and develop more effective foliar nematode management strategies for control of this pest in ornamental plant nurseries.

5.4 NANODIAGNOSIS AND NANOPHYTOPATHOLOGY

The extension/integration of molecular diagnostics on a nanoscale is a promising technology for identifying pathogens. Nanomolecular diagnostic is the use of nanobiotechnology to diagnose plant diseases, and this can be termed as nanodiagnostics. In particular, several nanodevices and nanosystems are used for sequencing single molecules of DNA (Fig. 5.7). Assays with the use of nanosize devices to investigate DNA sequences and diagnose disease are becoming faster, more flexible and more sensitive.

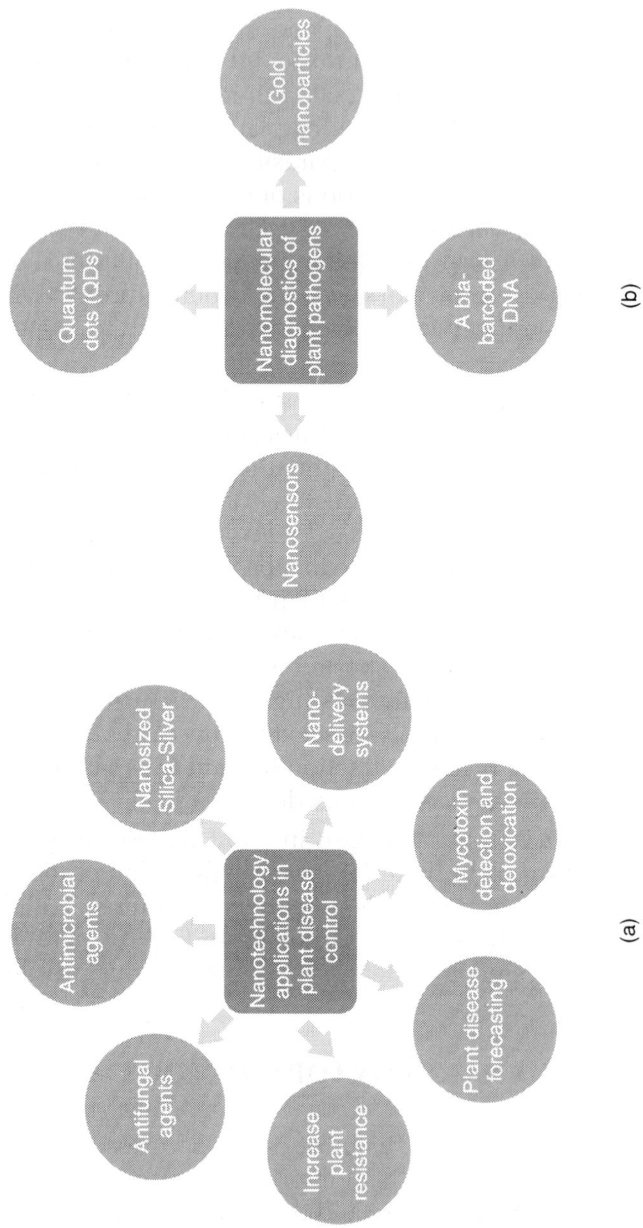

Fig. 5.7 Potential nanotechnology applications in plant pathology; (a) Plant disease control and (b) Detection of plant pathogens

It is noteworthy that new detection techniques involving nanobiosensors for pathogen identification will likely be a cornerstone to this trend. During the 1980s and 1990s, phytopathologists relied on visual assessment to identify plant diseases. Identifying plant pathogens via conventional techniques may take several days, and, therefore, researchers need rapid detection tools that can provide results within a few hours. To develop such detection tools, phytopathologists are constantly working with nanotechnologists (Rai and Ingle, 2012). Nano- and/or phytopathology researchers are attempting to develop an easy assay that is moveable and accurate and does not need any difficult method for the procedure so that farmers can use the mobile laboratory themselves to detect specific diseases (Mohammad A. Khiyami *et al.*, 2014).

Newly developed nanomaterials with special nanoscale characteristics could present a tremendous breakthrough in pathogen and contaminant detection. Nanotechnology is also driving the development of lab-on-a-chip systems for detecting toxicity in waters, observing nutrients in irrigation water and controlling the quality of food production (Gardeniers and van den Berg, 2004).

Nanophytopathology is a cutting-edge science that uses nanotechnology for detecting, diagnosing and controlling plant disease and their pathogens at an early stage, owing to crop protection from epidemic diseases. The modern plant pathologist strives to apply his knowledge in nanophytopathology to enable understanding of controlling factors of plant diseases and to develop and/or evaluate ecofriendly diagnostic measurements. Modern nanomolecular techniques are used for monitoring or understanding of pathogen population genetics, plant-microbe interactions and gene transfer between pathogens and even the host. Furthermore, nanoparticles such as nanosized silica-silver have recently been applied as antimicrobial and antifungal agents. Additionally, nanomaterials can be used for mycotoxin detection and detoxication, increasing plant resistance, plant disease forecasting and nanomolecular diagnostics of plant pathogens. Potential applications of nanophytopathology are shown in Fig. 5.7 (Mohammad A. Khiyami *et al.*, 2014).

A first step towards managing a plant disease under greenhouse and field conditions is to correctly identify the pathogen. Current technologies, such as quantitative polymerase chain reaction (Q-PCR), require a relatively large amount of target tissue and rely on multiple assays to accurately identify distinct plant pathogens. The common disadvantage of the traditional diagnostic methods is that they are time-consuming and lack high sensitivity. Consequently, developing low-cost methods to improve the accuracy and rapidity of plant pathogens diagnosis are needed (Mohammad A. Khiyami *et al.*, 2014).

Nanotechnology, nanoparticles and quantum dots (QDs) have emerged as essential tools for fast detection of a particular biological marker with extreme accuracy. Biosensor, QDs, nanostructured platforms, nanoimaging and nanopore DNA sequencing tools have the potential to raise sensitivity, specificity and speed of the pathogen detection, facilitate high-throughput analysis, and to be used for high-quality monitoring and crop protection. Furthermore, nanodiagnostic kit equipment can easily and quickly detect potential serious plant pathogens, allowing experts to help farmers in the prevention of epidemic diseases. The review by Mohammad A. Khiyami *et al.* (2014) deals with the application of nanotechnology for quicker, more cost-effective and precise diagnostic procedures of plant diseases. Such an accurate technology may help to design a properly integrated disease management system which may modify crop environments to adversely affect crop pathogens.

5.5 PORTABLE DIAGNOSTIC EQUIPMENT

5.5.1 Portable PCR Systems

The capability to produce a fast PCR amplicon in a completely moveable setup will not only allow plant pathologists to achieve more effective and predictable laboratory trials but also considerably simplify other downstream applications such as those requiring in situ genomic tools for detection (Mohammad A. Khiyami *et al.*, 2014).

Palm PCR, developed by Ahram Biosystems Company in Korea, is a manageable thermal cycler designed for easy use, accuracy and high efficiency. Despite its tiny size, this controlling device delivers extremely well-organized and quick amplification for different types of DNA samples including the plant genome. The DNA can be amplified in less than 25 minutes to enough quantity for ideal agarose gel detection. The portable system presents a highly functional and user-friendly way to perform different types of PCR tests for both beginner and experienced researchers.

Compared to other diagnostic testing systems, Twista quantitative and portable real-time fluorometer is a customized device developed for examining recombinase polymerase amplification (RPA) reactions and represents a new revolution in DNA diagnostics, as well as, combining advantage in speed, portability and ease of use with good sensitivity and specificity. Twista RPA fluorometer will supply immediate fast diagnosis, allowing an on-time suitable treatment compared with traditional microbiological assays, which require at least hours and molecular assays that typically need centralized equipment.

5.5.2 Portable Genome Sequencer (Nanopore Sequencing System)

Many companies are exploring the idea of using nanopore technology. There are two main challenges that must be addressed in any nanopore sequencing, viz., how to distinguish the nucleotides as the strand passes through the nanopore and how to control the speed of the DNA strand as it passes through the nanopore. Even with these unique challenges the nanopore sequencing platform is thought to be simple and straightforward because theoretically very long reads can be generated from a low quantity of nucleic acid (Ozsolak, 2012). The protein nanopore and enzyme were designed to control a single strand of DNA, and as the DNA goes through the nanopore, a direct electronic analysis is conducted. Nanotechnology can be applied to plant science research to analyze plant genomics, gene function, and pathogens detection as well as improvement of crop species.

5.5.3 Nanodiagonastic Kit

Plant pathologists plan to study this 'lab in a box' very soon. This refers to packing sophisticated measuring devices, reagents, power supply and other features that now take up laboratory space into a parcel no larger or heavier than a briefcase (Goluch *et al.*, 2006).

A briefcase-sized kit is transferred to a field where crops are growing to search for pathogens that could infect and reduce the yield. This is a quick and precise procedure. Nanodiagnostic kit equipment can easily and quickly detect potential serious plant pathogens, allowing experts to help farmers in the prevention of disease epidemics from breaking out. For example, 4mycosensor is a tetraplex competitive antibody-based assay in a dipstick format for the real-time detection of ZEA, T-2/HT-2, DON and FB1/FB2 mycotoxins on the same single strip in corn, wheat, oat and barley samples at or below their respective European maximum residue limits [(MRLs) (Lattanzio *et al.*, 2012)]. The proposed immunoassay protocol is fast, cheap, easy-to-use and suitable for the purpose of quick screening of mycotoxins in cereals (Mohammad A. Khiyami *et al.*, 2014).

5.5.4 Gold Nanoparticles as Biosensor

Utilization of nanomaterials or nanoparticles in biosensors allows the development and use of some novel signal detection procedures and equipment. Different strategies such as antibody–antigen, adhesion–receptor, antibiotic and complementary DNA sequence recognitions have been discovered for a specific detection of distinction between target phytopathogenic cells and biofunctionalized nanomaterials (Fan *et al.*, 2003).

Gold nanoparticles are excellent markers to be used in biosensors as several optical, or electrochemical procedures can be modified to identify pathogens (Mohammad A. Khiyami *et al.*, 2014). A number of nanoparticle-based experiments have been performed to develop biomolecular detection with DNA- or protein-functionalized gold nanoparticles, which are used as the target-specific probes. Several nanobiosensors for the molecular diagnosis of food-borne pathogens and agroterrorism agents can be found in recent studies. These detection methods include conductive polymer nanowires, carbon nanotubes, nanoporous silicon and gold nanoparticles. Dubertret *et al.* (2001) highlighted the ability of gold nanoparticles to act as fluorescence quenchers and, therefore, it could be used to solve major drawbacks in molecular biology experiments. For example, a DNA oligonucleotide could be synthesized, fluorescently labeled at its 5′ end and conjugated at the 3′ end with gold nanoparticles. These oligonucleotides can be used in diagnostic procedures, particularly in cases where DNA analysis cannot be circumvented, and in the diagnosis of the phytoplasma associated with the flavescence dorée (FD) of grapevine.

Fan *et al.* (2003) reported that the gold nanoparticles efficiently quench the fluorescence of light harvester polymers, such as polyfluorene, and will open new perspectives in the development of the optical performances of nanobiotransducers for diagnostic purposes. Furthermore, a diagnostic probe made of a specific oligonucleotide bearing a fluorescein at its 5′ end and 2-nm gold particles at its 3′ end acts as a nanobiotransducer in DNA hybridization. It produces a stronger fluorescence signal when hybridized to target DNA.

5.5.5 Quantum Dots (QDs)

QDs are semiconductor nanoparticles that fluoresce when stimulated by an excitation light source. Furthermore, QDs are inorganic fluorophores presenting major advantages over traditional organic fluorophores used as markers on nucleic acids or proteins for visual detection (Arya *et al.*, 2005). The mycosynthesis of semiconductor nanomaterials was first reported in unicellular yeast, which was shown to be capable of producing cadmium sulfide (CdS) crystallites in response to a cadmium salt stress. Different microbes have also been used for the biosynthesis of CdS; however, few studies have focused on its luminescent properties. Very luminescent CdSe QDs were created by the fungus, *F. oxysporum* when incubated with a mixture of $CdCl_2$ and $SeCl_4$ at room temperature. A proficient mycomediated synthesis of highly fluorescent CdTe QDs was accomplished by the *F. oxysporum* isolates when reacted with a mixture of $CdCl_2$ and $TeCl_2$ at ambient conditions. Description of these biosynthesized CdTe nanoparticles was performed by transmission electron microscopy (TEM) and selected-area electron diffraction (SAED).

An effective biosynthesis method to prepare easily harvested biocompatible cadmium telluride (CdTe) QDs with tunable fluorescence emission using yeast cells was developed. The confocal images of the yeast cells show obvious green emission from the synthesized CdTe QDs (Mohammad A. Khiyami *et al.*, 2014). The CdTe QD units were spread only in the cytoplasm and nucleus of yeast cells while none was found in the cell membrane.

Biocompatible semiconductor crystals are composed of a nucleus and a shell allowing the binding of ligands and thus the attachment of this fluorescent marker to the target pathogen. A hybridization-based biosensor known as the quartz-crystal microbalance biosensor is a good example (Mohammad A. Khiyami *et al.*, 2014). The quartz-crystal microbalance vibrates under electric stimulus, and a change in mass due to the attachment of any compound to its surface can be detected through a reduction in the frequency of its vibration. When a nucleic acid probe is attached to the surface of a quartz-crystal microbalance biosensor and then exposed to complementary PCR product, a hybrid forms and causes the resonance frequency of the quartz-crystal microbalance biosensor to decrease dramatically. This system can be combined with fast PCR protocols to decrease the time for specific environmental detection of plant pathogens (Sharon *et al.*, 2010).

5.5.6 Nanobarcodes – A Biobarcoded DNA (b-DNA)

The biobarcode assay is an ultrasensitive method for amplification and detection of proteins or nucleic acids. DNA biobarcoded tests employ oligonucleotide-modified magnetic gold nanoparticles (AuMNPs) for signal amplification and simple separation of a target protein from the sample. The large b-DNA-to-recognition agent ratio affords a means of substantial signal amplification. It is also promising by allowing the quick detection of numerous protein targets at low-attomolar concentrations and nucleic acids at high-zeptomolar levels under optimized conditions. The concept of the biobarcode assay is unique and represents a potential alternative to the PCR technique.

5.5.7 Nanostructured Platform for Mycotoxin Detection

Nanostructured platform for mycotoxin detection and detoxification mycotoxins are secondary metabolites produced by pathogenic filamentous fungi causing disease and death in humans and other animals. Mycotoxicoses are diseases caused by the ingestion of mycotoxins by humans and animals, mostly through the consumption of polluted food. They can induce various different biological effects, i.e., carcinogenic, mutagenic, teratogenic, estrogenic, immunotoxic, nephrotoxic, and neurotoxic effects (Kankkunen *et al.*, 2009).

A variety of biosensors have been developed and described in the recent literature for mycotoxin analysis some of which have the potential of multiarray mycotoxins identification (Mohammad A. Khiyami *et al.*, 2014).

Nanosensors could be placed directly on the packaging material, where they would serve as 'electronic noses' which can detect chemicals released during food spoilage. Some nanosensors are based on microfluidic devices, (Baeummer, 2004) and they can be used to identify pathogens proficiently in a short time with high sensitivity. Sol–gel derived nano-ZnO coat can be used for the immobilization of r-IgGs while bovine serum albumin (BSA) can be used for blocking nonspecific binding sites of r-IgGs to identify ochratoxin A (OTA). BSA/r-IgGs/nano-ZnO/ITO platform for detection of mycotoxins such as aflatoxins (APH), ochratoxin B, citrinin, patulin, ergot alkaloids, fumonisins, trichothecenes and zearalenone (ZEA) is still under development. A novel ultrasensitive magnetic nanoparticle immunoassay for mycotoxin detection was developed to provide real-time quantitative results for detecting more than one mycotoxin. Moreover, real-time assays were performed upon the addition of magnetic nanotags onto the spin-valve sensor surface immobilized with capture antibodies for detection of mycotoxins (aflatoxin-B1, zearalenone and HT-2) (Mohammad A. Khiyami *et al.*, 2014). Lattanzio *et al.* (2012) described a new method, signal transduction by ion nanogating sensors which are used as an analytical technique for the ultrasensitive detection of mycotoxins, with a detection limit of up to 100 ng/mL.

The ozonization and adsorption efficiency of modified nanodiamonds to decrease the content of aflatoxin-B1 has been examined by Puzyr *et al.* (2010). Silica and clays are most efficient in combination with smaller sized water molecules and smaller mycotoxins such as aflatoxins and ochratoxins. However, clays are less efficient in binding the larger mycotoxins such as fumonisin and deoxynivalenol (vomitoxin) because the distance among clay layers is not sufficient to accommodate the larger molecules. By using nanosized clay, the space between the layers of clay has been prolonged 10 times. As a result, the nanoclay can bind the whole family of mycotoxins. A rapid enzyme-linked immunosorbent assay (ELISA) method was investigated by reducing the coating, blocking and competition time required in usual ELISA method using superparamagnetic nanoparticles (Radoi *et al.*, 2008). This method was effective for detecting aflatoxin M1 (AFM1) in milk in a linear working range of 4–250 ng/L.

A moveable machine has recently been developed that can concurrently identify various bacterial, fungal toxins and pathogens in stored food. Obviously, biosensors can be an exciting alternative to the conventional techniques for the detection of mycotoxins and pathogens in food.

5.5.8 Nanosensors and Plant Disease Predicting

Plant disease forecasting is a management method used to forecast the chances or severity of plant diseases and to help farmers make cost-effective decisions for controlling diseases. Presently research is being carried out by using nanosensors to improve pathogen detection methods in crop systems. Many electronic companies have been examining the use of electrically conducting polymers such as polyaniline, polythiophene, and polypyrrole. These polymers can also be used to fabricate sensors that can detect molecular signals with very low intensity of spoilage and food-borne pathogens within minutes.

The bionanosensor has the potential of increased sensitivity and, therefore, a significantly reduced response time to discover potential disease problems. Such bioanalytical nanosensors were utilized to detect and quantify minute amounts of contaminants such as viruses, bacteria, fungi, toxins and other biohazardous substances in the agriculture and food systems (Mohammad A. Khiyami *et al.*, 2014). Therefore, these biosensors may have a huge impact on the precision farming methods (Rai and Ingle, 2012). Nanosensors can be linked to a GPS for real-time monitoring of disease and distributed throughout the field to monitor soil conditions and crop health. The union of biotechnological and nanotechnological approaches in biosensors will be used to construct equipment with increased sensitivity, allowing an earlier response to ecological changes and disease prevalence. Nanosensors will allow us to identify plant diseases before visible symptoms appear and thus facilitate their control. Precision farming will allow improved agriculture production by providing precise data, helping growers to make better decisions.

To detect pathogens and toxins in food, the fibreoptic biosensor has commonly used tapered fibreoptic probes coated with antibodies. An approach has been developed for rapid determination of *Escherichia coli* using a flow-injection system. Electrochemical measurement of $K_3Fe\ (CN)_6$, reduced by microbial metabolism, allowed the quantitative determination of bacteria and fungi in 20 minutes. Hashimoto *et al.* (2008) developed a new biosensor system for the rapid diagnosis of soil-borne diseases, consisting of two biosensors. The system was constructed using equal quantities of two different microbes, each individually immobilized on an electrode.

Fluorescent silica nanoparticles (FSNP) combined with antibody molecules successfully detected plant pathogens such as *Xanthomonas axonopodis* pv. *vesicatoria* which causes bacterial spot disease in tomatoes and peppers. Copper oxide (CuO) nanoparticles and nanolayers were synthesized by sol–gel and spray pyrolysis methods, respectively. Both CuO nanoparticles and nanostructural layer biosensors were used for detecting the *A. niger* fungi (Etefagh *et al.*, 2013)

5.5.9 Nanofabrication Imaging

Diagnostic imaging refers to a broad slew of technologies used to look inside or outside plant tissues to diagnose various plant pathogens (Mohammad A. Khiyami *et al.*, 2014). The rapidly expanding use of diagnostic imaging technologies continues to push the boundaries of plant pathologist diagnostic capabilities. Using these techniques, plant doctors are able to diagnose crop diseases earlier and more precisely (Rosen *et al.*, 2011). Nanotechnology offers us the opportunity to precisely tune and control the chemical and physical properties of contrast materials to overcome concerns with toxicity, useful imaging time, tissue specificity and signal strength. In the 'mesoscopic' size range of 5–100 nm diameter, nanoparticles also have large surface areas and functional groups for conjugating to multiple diagnosis tools. Thus, progress in the field of nanoscale contrast agents will play a key role in the continued enhancement of our diagnostic imaging capabilities in the coming years.

Electron beam and photolithography techniques were used to fabricate topographies that mimic leaf surface features as well as the internal plumbing of plants, and then nanoimaging technologies were used to study how bacteria and fungi invade and colonize the leaf. Lithography was used to nanofabricate a pillared surface on silicon wafers. This lawn of miniature pillars was between 1.4 and 20 μm wide and spaced various distances apart. It was used to examine the movement across the surface by the fungus that mimicked some of the characteristics of the host plant. Images of the *Colletotrichum graminicola* crawling across the nanofabricated surface assisted the researchers to determine that the fungus needs to make a minimum contact of at least 4.5 μm before it starts to develop appressoria. To develop disease resistant cultivars, the infection process and behavior of bacterium pathogen causing Pierce's disease inside grapevine xylem were studied using nanofabrication methods (Meng *et al.*, 2005).

The portable diagnostic equipment, nanoparticle-based, biobarcoded DNA sensor, and the QD have potential applications in the multiple detection of plant pathogens and toxigenic fungi (Mohammad A. Khiyami *et al.*, 2014). To date, mobile diagnostic assays have been developed to rapidly detect plant disease and may be used to prevent epidemics. These nanobased diagnostic kits not only increase the speed of pathogen detection but also increase the accuracy of the diagnosis. Additionally, the combination of nanotechnology with microfluidic systems has been effectively applied in molecular plant pathology and can be adapted to detect specific pathogens and toxins. A good example is the microPCR where 40 cycles of PCR can be performed in less than six minutes. In the near future, nanoscale devices with novel properties could be used to make smart agricultural systems. For instance, these nanodevices could be used to identify plant health issues before these become observable to the grower. Such devices may be capable of responding to special situations, identifying the problem and taking an appropriate disease management action. In this way, nanosmart devices will act as both a protective and an early warning system. During the next decade, nanodevices, which can make thousands of measurements speedily and very cheaply, will become available. Future prospects in plant disease diagnostic will continue in miniaturization of biochip technology to the nanoscale range. Nanophytopathology can be applied as a tool to understand plant–pathogen interactions, providing new methods for crop protection. Specific nanodevices and DNA nanodevices could enable accurate tracking, detection and diagnosis of plant pathogens in the early stages of plant disease.

Major application prospects of this technology, according to Mohammad A. Khiyami *et al.* (2014), include quick response within integrated disease management system via external and implanted nanosensor systems, improvement of rapid laboratory biosensors to detect plant pathogens in the field or postharvest, development of retrieval nanosystems for a specific sampling from soil, air and plant samples, provide rapid and reliable nanomethods for detection of mycotoxins and toxigenic fungi and detection of pesticide residues in food and feeds.

The significance of advanced techniques for detecting plant diseases were reviewed by Sindhuja Sankaran *et al.* (2010), who discussed the need for developing a rapid, cost-effective, and reliable health monitoring sensor that would facilitate advancements in agriculture. The review described the currently used technologies that can be used for developing a ground-based sensor system to assist in monitoring health and diseases in plants under field conditions. These technologies include spectroscopic and imaging-based and volatile profiling-based plant disease detection methods. They also compared the benefits and limitations of these potential methods.

Disease detection techniques can be broadly classified into direct and indirect methods. Figure 5.8 summarizes some of these methods of disease detection. An advanced plant disease detection technique can provide rapid, accurate, and reliable detection of plant diseases in early stages for economic, production and agricultural benefits.

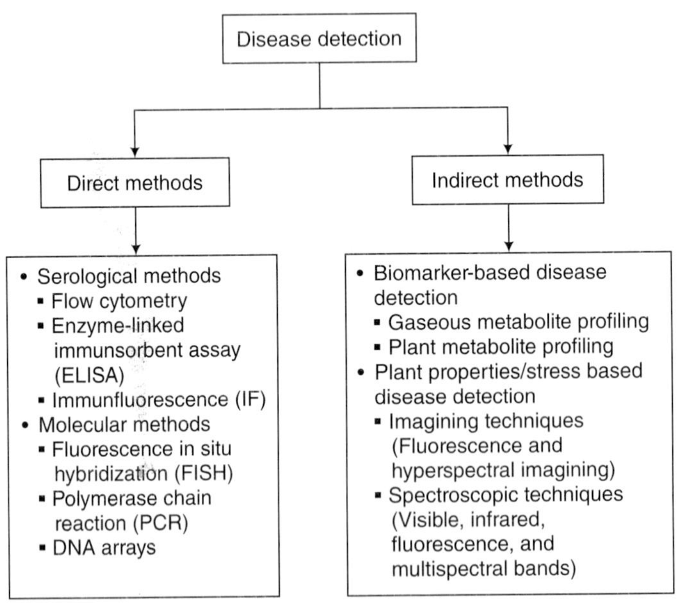

Fig. 5.8 Methods of plant disease detection

The advanced techniques of ground-based disease detection that could be possibly integrated with an automated agricultural vehicle were reviewed (Sindhuja Sankaran *et al.*, 2010). In ground-based disease detection studies, both field-based and laboratory-based experiments were discussed. The field-based studies refer to studies that involve spectral data collection under field conditions, whereas laboratory-based studies refer to data collection under laboratory conditions. The laboratory-based experiments provide strong background knowledge (such as the experimental protocol and statistical algorithm for classification) for the field-based applications.

Two approaches were considered to detect plant diseases. The first approach involves the application of spectroscopic and imaging techniques for disease detection, whereas the second approach describes the application of volatile organic metabolites as possible biomarkers for disease detection. These two approaches were selected as they could be easily integrated with an agricultural vehicle for a fast, reliable, and real-time plant disease monitoring for disease control and management. The early detection of plant diseases (before the onset of disease symptoms) could be a valuable source of information for executing proper pest management strategies and disease control measures to prevent the development and the spread of diseases

Apart from molecular techniques of plant disease detection, spectroscopic and imaging techniques for disease detection (Fluorescence spectroscopy, visible and infrared spectroscopy, fluorescence imaging, hyperspectral imaging, etc.), profiling of plant volatile organic compounds for disease detection, electronic nose system and GC–MS are of much significance.

5.6 SPECTROSCOPIC AND IMAGING TECHNIQUES FOR DISEASE DETECTION

Recent developments in agricultural technology have lead to a demand for a new era of automated nondestructive methods of plant disease detection. It is desirable that the plant disease detection tool should be rapid, specific to a particular disease, and sensitive for detection at the early onset of the symptoms. The spectroscopic and imaging techniques are unique disease monitoring methods that have been used to detect diseases and stress due to various factors, in plants and trees. Current research activities are towards the development of such technologies to create a practical tool for a large-scale real-time disease monitoring under field conditions.

Various spectroscopic and imaging techniques have been studied for the detection of symptomatic and asymptomatic plant diseases. Some of the methods are fluorescence imaging, multispectral or hyperspectral imaging, infrared spectroscopy, fluorescence spectroscopy, visible/multiband spectroscopy and nuclear magnetic resonance (NMR) spectroscopy and multiple methods (sensors and algorithms) for pathogen detection, with special emphasis on postharvest diseases (Sindhuja Sankaran *et al.*, 2010).

The spectroscopic and imaging techniques could be integrated with an autonomous agricultural vehicle that can provide information on disease detection at early stages to control the spread of plant diseases. This technology can also be applied to identify stress levels and nutrient deficiencies in plants. In regard to plant disease detection, significant research is ongoing on the prospective of this technology from last few decades. Spectroscopic technology has been successfully applied for plant stress detection such as water-stress detection and nutrient-stress detection. In addition, there have also been significant applications for monitoring the quality of postharvest fruits and vegetables. Some of the commonly used spectroscopic and imaging techniques are described in the following sections. Tables 5.1 and 5.2 summarize few studies on plant disease detection using spectroscopic and imaging techniques. Some postharvest applications have also been discussed in this paper, as the knowledge of statistical models and methodology for data collection for postharvest applications is transferrable for real-time plant disease detection.

Table 5.1 Examples of Studies on Plant Disease Detection Using Spectroscopic Techniques

Plant	*Disease/Damage*	*Statistical methods*	*Optimum spectral range*
(i) Fluorescence Spectroscopy			
Citrus	Citrus canker	–	452, 685 and 735 nm
(ii) Visible and Infrared Spectroscopy			
Rice	Infested with brown planthopper (*Nilaparvata luge*)	–	737–925 nm
Wheat	Powdery mildew and take-all disease	Analysis of variance, correlation and regression analysis	490_{780}, 510_{780}, 516_{1300} and 540_{1300} nm
Rice	Brown planthopper and leaf folder infestation	Linear regression models	426 nm
Kiwifruit	Gray mold, Sclerotinia rot	Principal component analysis	–
Wheat	Yellow rust	Regression	–
Tomato	Leaf miner damage	–	800 to 1100 nm, 1450 and 1900 nm
Grapevine	Grapevine leafroll disease	Discriminant analysis	752, 684 and 970 nm
(iii) Nuclear Magnetic Resonance (NMR) Spectroscopy			
Catharanthus roseus (ornamental plant)	Phytoplasmas causing yellowing disease	Principal component analysis	–

Table 5.2 Examples of Studies on Plant Disease Detection Using Imaging Techniques

Plant	*Disease/Damage*	*Statistical methods*	*Optimum spectral range*
Wheat	Scab (*Fusarium* head blight)	Step discrimination and discriminant analysis	568, 715 nm (550, 605, 623, 660, 697 and 733 nm)
Tomato	Late blight disease	Minimum noise fraction transformation and spectral angle mapping-based classification	700–750 nm, 750–930 nm, 950–1,030 nm, and 1,040–1,130 nm
Wheat	Yellow rust, nutrient deficiency	Self-organizing map neural network, quadratic discriminant analysis	680, 725 and 750 nm
Wheat	Yellow rust	Regression analysis	–
Grapefruit (fruit)	Citrus canker	Principal component analysis	553, 677, 718 and 858 nm
Vidalia sweet onions	Sour skin disease	Image analysis	1,150–1,280 nm
Sweet orange	Blue mold, Browning rot	Difference in reflectance	540–680 nm

5.6.1 Fluorescence Spectroscopy

Fluorescence spectroscopy refers to a type of spectroscopic method, where the fluorescence from the object of interest is measured after excitation with a beam of light (usually ultraviolet spectra). For the last twenty years, the laser-induced fluorescence has been used for vegetative studies, such as to monitor stress levels and physiological states in plants (Belasque *et al.*, 2008). Two types of fluorescence, viz., blue-green fluorescence in about 400–600 nm range and chlorophyll fluorescence in about 650–800 nm range, are produced by green leaves. The fluorescence spectroscopy can be utilized to monitor nutrient deficiencies, environmental conditions based stress levels and diseases in plants.

Fluorescence spectroscopy was employed to detect stress caused by citrus canker (a bacterial disease caused by *Xanthomonas citri–X. axonopodis* pv. *citri*) and mechanical injury (Belasque *et al.*, 2008). The studies reported the potential of fluorescence spectroscopy for disease detection and discrimination between the mechanical and disease stress. A similar approach was taken to detect water stress and differentiate citrus canker leaves from variegated chlorosis leaves (Marcassa *et al.*, 2006). The above studies could classify healthy from citrus canker-affected leaves, but were unable to identify water stress and distinguish between variegated chlorosis and citrus canker-infected leaves. The above-mentioned scientists did not yet present any statistical analysis to evaluate the ability of the technique to discriminate or classify different plant condition.

Field experiments were conducted to discriminate citrus canker-stressed leaves from chlorosis-infected (caused by *Xylella fastidiosa*) and healthy leaves (Lins *et al.*, 2009). In addition, they conducted leaf detachment experiments to monitor the effect of time (up to 12 hr) on fluorescence of detached leaves using fluorescence spectroscopy. This report recommended the application of a range of spectral data rather than fluorescence alone for a reliable discrimination results, and further work needs to be done to classify different plant conditions using statistical tools. Methods such as principal component analysis (PCA), discriminant analysis and neural network-based classification algorithms can be applied to analyze the results obtained from fluorescence spectroscopy. Methods such as PCA, parallel factor analysis, cluster analysis, partial least square (PLS) regression, and Fischer's linear discriminant analysis (LDA) can be applied for classifying fluorescent spectrometric data having two or more classes.

5.6.2 Visible and Infrared Spectroscopy

Similar to fluorescence spectroscopy, visible and infrared spectroscopy have been used as a rapid, nondestructive and cost-effective method for the detection of plant diseases. Some of these wavebands specific to a disease can be used to detect plant diseases, even before the symptoms are visible. In general, visible spectroscopy is used for disease detection in plants in combination with infrared spectroscopy.

The near infrared (NIR)-based technique was assessed for detecting fire blight disease in the asymptomatic pear plants under greenhouse conditions (Spinelli *et al.*, 2006). The NIR technique did not exhibit potential for classifying infected plants from that of healthy ones while electronic nose system showed a better potential to classify diseased plants. The authors reported that the possible reason for the inability of the NIR based technique to distinguish diseased from healthy plants could be due to a very small leaf scan area (2 mm^2 in this study). The authors recommended that a multidimensional image analysis could provide more information on the diseased plants rather than a smaller field of view. This signifies that the spectroscopy-based imaging techniques could be more robust in disease identification at early stages in plant diseases than spectroscopic methods alone.

The data analysis procedure used for the classification of postharvest food products can be applicable for plant disease detection. Thus, the knowledge from one application (on the statistical algorithm, data processing, experimental protocol, etc.) can be transferred to other possible applications. Sirisomboon *et al.* (2009) used visible spectra along with NIR spectra (600–1100 nm) to identify defective pods during soybean processing. In addition to PCA, soft independent modeling of class analogy (SIMCA) and PLS-discriminant analysis were used to classify the groups. The scientists reported that the SIMCA-based model was able to discriminate different groups of soybean better than the PLS-based model.

Naidu *et al.* (2009) used leaf spectral reflectance to identify viral infection (under field conditions) in grapevines (*Vitis vinifera* L.) that cause grapevine leafroll disease. The application of hyperspectral reflectance to identify cotton canopy infected with *Verticillium* wilt was studied by Chen *et al.* (2008), whose results indicated that the spectral range from 426 to 1450 nm showed the maximum correlation intensity. The changes in spectral properties were low in visible and ultraviolet (UV) range, whereas the infrared region (740–2400 nm) yielded the maximum change in spectral signature. Delalieux *et al.* (2007) used hyperspectral reflectance data (350–2500 nm) to detect apple scab caused by *Venturia inaequalis*. The study involved the identification of infected trees and selection of wavelengths best suited for classifying the infected leaves from those of the healthy leaves.

Various studies have used different methods/models for the classification of diseases/conditions of plants based on spectral data. For an instance, Roggo *et al.* (2003) utilized eight classification models (linear discriminant analysis, *k*-nearest neighbors, soft independent modeling of class analogy, discriminant partial least squares (DPLS), procrustes discriminant analysis (PDA), classification and regression tree, probabilistic neural network, and learning vector quantization-based neural network) for qualitative determination of sugar beet. They found that SIMCA, DPLS, and PDA yielded the highest classification accuracies that those of other models. Wu *et al.* (2008) used the PCA-based back-propagation neural network (BPNN) model and PLS wavelength-based BPNN for detection of *Botrytis cinerea*-affected eggplant leaves prior to the visibility of symptoms under laboratory conditions. The BPNN model yielded a maximum of 85% classification accuracy for predicting fungal infections.

In addition to the statistical models for classification, many spectroscopy-based studies use different vegetative indices for evaluating the change in spectral reflectance at different plant conditions (diseased or healthy plant).

5.6.3 Fluorescence Imaging

The change in blue-green fluorescence and chlorophyll fluorescence of the plants upon ultraviolet excitation could provide the status of the physiological condition of the plant. Fluorescence imaging is an advancement of fluorescence spectroscopy, where fluorescence images (rather than single spectra) are obtained using a camera. A xenon or halogen lamp is used as a UV light source for fluorescence excitation, and the fluorescence at specific wavelengths is recorded using the charge-coupled device (CCD)-based camera system (Lenk *et al.*, 2007).

The chlorophyll fluorescence imaging can be an effective tool in monitoring leaf diseases. Chaerle *et al.* (2007) used blue-green fluorescence to evaluate the effectiveness of this technique in observing the development of tobacco mosaic virus (TMV) infection in tobacco plants. A temporal effect of TMV infection on the fluorescence (blue-green and chlorophyll fluorescence) of infected plants was observed. The reflectance image at 550 and 800 nm were acquired and considered as the reference images. The authors reported an increase in blue, green, and chlorophyll fluorescence after about 40–55 h upon inoculation with TMV. The fluorescence imaging demonstrated a visible difference between the infected and noninfected leaves in a short period of time (50 h) in comparison to the reference images (14 days for visible symptoms of infection).

Bravo *et al.* (2004) used fluorescence imaging for detecting yellow rust in winter wheat. They acquired two fluorescence images: a background image without the xenon lamp source and a fluorescence image with the xenon lamp source during the experiments. The fluorescence image utilized for the analysis was obtained by subtracting fluorescence image from the background image. The fluorescence was measured at 450, 550, 690, and 740 nm. The authors stated that the difference between the fluorescence at 550 and 690 nm were higher in the diseased portion of the leaves while it was very low for healthy regions of the leaves. Quadratic discriminant analysis (QDA) was utilized to differentiate the healthy from the infected plants. Results indicated that though the methods were not effective in differentiating the healthy from mildly infected plants, QDA could classify healthy and diseased plants with an accuracy of 71% and 96%, respectively.

The applicability of hyperspectral reflectance imaging in combination with multispectral fluorescence imaging was investigated through sensor fusion to detect yellow rust (*Puccinia striiformis*) disease of winter wheat (Moshou *et al.*, 2005). The hyperspectral imaging was performed under an ambient condition in winter wheat plots, whereas fluorescence images were procured upon UV-excitation. The authors reported that when the sensor information from the fluorescence and multispectral imaging were combined, the QDA-based classification accuracy of the healthy plants improved from 71–90% to 97%. The classification accuracy of the diseased plants and healthy plants further improved to 98.7% and 99.4%, respectively, when the self-organizing map (SOM)-based neural network was used for the plant classification. The above studies indicate the possibility of using imaging techniques for disease identification. The imaging techniques are an improvement over spectroscopic techniques as these methods acquire spectral information over a larger area and provide three-dimensional spectral information in the form of images.

5.6.4 Hyperspectral Imaging

In recent years, hyperspectral imaging is gaining considerable interest for its application in precision agriculture (Okamoto *et al.*, 2009). In the hyperspectral imaging, the spectral reflectance of each pixel is acquired for a range of wavelengths in the electromagnetic spectra. The wavelengths may include the visible and infrared regions of the electromagnetic spectra. The hyperspectral imaging is similar to multispectral imaging, the difference being a broader range of wavelengths (more number of spectral bands) being scanned for each pixel in the hyperspectral imaging. The resulting information is a set of pixel values (intensity of the reflectance) at each wavelength of the spectra in the form of an image.

Some of the major challenges in hyperspectral imaging-based plant disease detection are the selection of disease-specific spectral band and selection of statistical classification algorithm for a particular application, which depends on the data acquisition set-up under field conditions. Blasco *et al.* (2007) applied multispectral computer vision using nonvisible (ultraviolet, IR, and fluorescence) and visible multiple spectra for citrus sorting. The anthracnose was classified better with NIR images (86%), whereas green mold was more accurately classified with fluorescence imaging (94%). The stem-end injury was classified up to 100% using the ultraviolet spectra in this study. This study showed the utilization of hyperspectral bands for detecting different aspects of a single problem. Similarly, the hyperspectral imaging could be used for detecting different features within a plant to identify diseases. Bravo *et al.* (2003) investigated the application of visible-NIR hyperspectral imaging for the early detection of yellow rust disease (*Puccinia striiformis*) in winter wheat. A discrimination model was developed using quadratic discriminant analysis for the classification of diseases from the healthy plants. The classification model yielded about 92–98% classification accuracy while classifying diseased plants.

Shafri and Hamdan (2009) used airborne hyperspectral imaging for the detection of Ganoderma basal stem rot disease in oil palm plantations. The authors used various vegetative indices and red edge techniques to classify the diseased from the healthy plantations. The reported classification accuracies using different methods ranged from 73 to 84%. The results indicated that an aerial hyperspectral imaging could be used for disease detection and management of plantations in large scale. Qin *et al.* (2009) obtained hyperspectral images in the wavelength range 450–930 nm to detect citrus canker and other damages to Ruby red grapefruit.

5.6.5 Other Imaging Techniques

The other imaging techniques that can used for detecting plant diseases are infrared thermography, terahertz spectroscopy, NMR spectroscopy, and X-ray imaging. Infrared thermography refers to an imaging technique that utilizes the thermal energy of the infrared band and transforms the procured information into a visible image. Infrared thermography similar to other imaging techniques can be used for nondestructive monitoring of the physiological status of plants. Lenthe *et al.* (2007) used infrared thermography to examine the possible relationship between the leaf microclimate and fungal diseases in wheat fields.

Although microclimate can be determined from infrared thermography, direct diseased leaf area identification could not be established. Chaerle *et al.* (2003) reviewed the application of imaging techniques in agronomy and stress detection. They reported that the stomatal changes in the leaves of the plants upon pathogen infection could be monitored by thermography (e.g., hydrogen peroxide produced by *Pseudomonas syringae* induces stomata in the leaves to close). Similarly, local temperature changes due to plant defense mechanisms against diseases can also help in monitoring plant diseases. Tobacco leaves produce salicylic acid (promoting thermal and stomatal change) that can be monitored using thermography. Other studies include presymptomatic detection of cucumber downy mildew-*Pseudoperonospora cubensis*, fungal infection by *Cercospora beticola* in sugar beet, and *Brassicanapus* infection with *Phomalingam* among others.

Nuclear magnetic resonance and X-ray-based imaging techniques can also be used for detecting infections, different types of stress, and other health conditions in trees/fruits. Goodman *et al.* (1992) used NMR microscopic imaging for identification of the fungal pathogen *Botrytis cinerea* in red raspberry. Narvankar *et al.* (2009) applied X-ray imaging to identify fungal infections in wheat. The authors employed statistical discriminant models and ANN to classify the images of the wheat kernels. Statistical classifiers, especially the Mahalanobis discriminant classifier performed better (92–99% accuracy) than the ANN-based classification.

5.7 PROFILING OF PLANT VOLATILE ORGANIC COMPOUNDS FOR DISEASE DETECTION

The volatile organic compounds (VOC) released by plants and trees contribute about two-thirds of the total VOC emissions present in the atmosphere. These plant volatiles in turn influence their relationship between the plants and other organisms including pathogens. Dudareva *et al.* (2006) reviewed a range of plant volatiles released by the plants due to biotic and abiotic interactions. Some of the commonly found secondary plant volatiles are terpenoids, volatile fatty acids (such as *trans*-2-hexenal, *cis*-3-hexenol and methyl jasmonate), phenylpropanoids and benzenoids, and amino acid volatiles (such as aldehydes, alcohols, esters, acids, and nitrogen- and sulphur-containing volatiles derived from amino acids). The abiotic and biotic stresses can result in a change in the volatile profile of the plants that can be utilized for the plant disease detection. Cevallos-Cevallos *et al.* (2009) reported that the compounds (extracted from leaves) such as hesperidin, naringenin, and quercetin present in the leaves could be used as a biomarker to identify huanglongbing diseases in citrus trees.

The VOCs released by the plants change when the plant is infected with a disease due to change in its physiology. These emissions are expected to vary from the VOCs released under normal plant health conditions. This technology could facilitate the detection of plant diseases in real-time, thereby preventing the spread of plant diseases. Such techniques will provide agricultural and financial benefits for the growth and development of our economy. A handful of studies has been performed to detect diseases or infections in postharvest fruits/vegetables. The two common methods used for assessing the profile of volatile metabolites released by plants are electronic nose system-based and using gas chromatography (GC)-based techniques.

5.7.1 Electronic Nose System

An electronic nose system consists of a series of gas sensors that are sensitive to a range of organic compounds. As each sensor has specific sensitivities, the sensitivities of a series of sensors could be used to discriminate different compounds present in the atmosphere. Electronic nose systems have been used for multiple applications. Li *et al.* (2009b) used a Cyranose® 320 (an array of 32 conducting polymer-based sensors) to detect postharvest fungal disease in blueberries in a controlled environment. Blueberries were disinfected with ethanol to eliminate any naturally present fungal spores and bacteria. Once the blueberries were rinsed with distilled water to remove residual ethanol, they were inoculated with spore suspensions of three fungal species: *Botrytis cinerea*, *Colletotrichum gloeosporioides*, and *Alternaria* spp. that cause gray mold, anthracnose, and *Alternaria* fruit rot in postharvest blueberries, respectively. The berries with *C. gloeosporioides* could be distinctively differentiated from the other groups though there was some overlap in the VOC profiles of the berries infected with *B. cinerea* and *Alternaria* spp. The authors suggested that as *B. cinerea* and *Alternaria* spp. mainly infect the same region (stem scar region) of blueberry fruit that might have resulted in a similar VOC profile. When linear Bayesian classifier was used, the study found that the type I error for classifying infections caused by *B. cinerea*, *C. gloeosporioides*, and *Alternaria* spp. were 18%, 11%, and 13%, respectively.

The near infrared and electronic nose system-based techniques were evaluated to detect fire blight (asymptomatic stage) in pear plants (Spinelli *et al.*, 2006). It was reported that the electronic nose system was able to provide a distinct olfactory signature required to identify the disease. The disease could be detected as early as six days after the infection. The study indicated that electronic nose system could be used as an effective tool for the early diagnosis of the plant disease under natural conditions. An electronic nose system was used to detect basal stem rot disease in oil palm plantation during field experiments Markom *et al.* (2009). The scientists reported that two principal components based on PCA were able to account for 99.32% variability in the data, and the MLP-based artificial neural network could classify infected from healthy trees with high accuracy. Studies revealed the prospect of the electronic nose-based VOC monitoring systems in conjunction with advanced statistical techniques for detecting different stress and health conditions in plants. This indicates that the profile of volatile metabolites released by plants could be used as a disease-monitoring tool for early and rapid detection of plant diseases.

5.7.2 Gas Chromatography–Mass Spectrography (GC–MS)

The GC–MS is a commonly used technique for a qualitative as well as quantitative analysis of volatile metabolites released by plants/trees in different environmental and physiological conditions. The GC–MS studies have been performed to evaluate the change in volatiles caused by a bacterial or fungal infection in various food products. Prithiviraj *et al.* (2004) assessed the variability in the volatiles released from onion bulbs infected with bacterial (*Erwinia carotovora* causing soft rot) and fungal species (*Fusarium oxysporum* and *Botrytis allii* causing basal and neck rots) using HAPSITE, commercial portable GC–MS instrument. The study indicated that 25 volatile compounds (among the 59 consistently detected compounds) released from onion can be used to identify the disease based on VOC profiling. Although no statistical analyses were performed to determine the discriminatory ability of an algorithm in classifying the VOC profiles for disease detection, model development and software development were recommended for the purpose.

Similar studies on potato tubers inoculated with *Erwinia carotovora* subsp. *carotovora*, *E. carotovora* subsp. *atroseptica*, *Pythium ultimum*, *Phytophthora infestans*, or *Fusarium sambucinum* using solid phase microextraction (SPME) fiber along with GC-flame ionization detector (FID) indicated the potential of the VOC profiling for disease detection.

Moalemiyan *et al.* (2007) employed VOC profiling to detect fungal diseases (*Lasiodiplodia theobromae* causing stem-end rot and *Colletotrichum gloeosporioides* causing anthracnose) in mangoes. Discriminant analysis models were used to classify the groups using the gaseous metabolite profile and mass ions. Though the methods showed potential for detecting fungal diseases, the classification accuracy of the models needs to be further increased to make it a feasible method for postharvest disease detection.

5.8 FUTURE DIRECTIONS

Recent reports in the literature support the notion that both volatile profiling and changes in spectral reflectance can be used for noninvasive field monitoring of plant diseases. Plants and trees release volatile organic compounds (VOCs) as a result of the metabolic activities taking place in its shoots, leaves, flowers, or fruits. The VOC profile of each plant differs significantly based on its physiological condition and the species. Various factors influence the profile of VOCs from a particular plant or tree, which may include changes in plant metabolism as a result of environmental changes, the age of plant, developmental stage of a plant, effect of stress on plants, and the presence of disease/ herbivore in a plant.

One of the biggest challenges in the utilization of gaseous plant metabolites as an indicator for the presence of plant diseases is the natural variation in the VOC profile within plant species. The variation in VOCs released by plants may mask the changes due to the stress and the presence of diseases. Therefore, there is a need to identify distinct volatile biomarker specific for a particular plant and disease that would be different from the VOCs produced due to an environmental or nutrient stress. For practical applications, development of a robust and reliable system for real-time monitoring of plant diseases is required.

Similar to VOC profile of the plants, the environmental conditions affect the spectral reflectance of the object (Griffin and Burke, 2003). Therefore, there is a need to identify a suitable approach to overcome this problem. The possible way to overcome this problem could be the identification of wavelength range or index that is not only sensitive to a specific plant disease but also is least affected by the changes in the environmental condition.

It is feasible to incorporate the imaging and VOC profiling techniques into an autonomous robot as these techniques are well established for other industrial applications. Once these techniques are well established for a specific disease detection application, these methods can be integrated with an autonomous agricultural vehicle for real-time monitoring for plant diseases. The overall comparisons of three major techniques are summarized in Table 5.3.

Table 5.3 Comparison of Various Types of Plant Disease Detection

Characteristics	*Molecular techniques*	*Imaging and spectroscopic techniques*	*VOCs profiling-based techniques*
(i) Accuracy of the method	Molecular techniques are presently the most accurate method for plant disease detection.	The accuracy of imaging and spectroscopic techniques is plant and disease specific.	Accuracy is currently unknown, as this method is in the developmental stages and has been utilized in recent years.
	Efforts are ongoing to make molecular methods more reliable and simpler as well as develop molecular detection kits for field applications. However, it is difficult to develop kits for all diseases. They are usually focused towards commonly found and harmful diseases.	Higher the visible symptoms better are the accuracy of the technique. Nevertheless, the nonvisible regions of spectra can be utilized for improving the accuracy of the method.	Identification of disease-specific biomarker volatiles (most challenging step) can improve the accuracy significantly.

	Moderately expensive. Quite often: is labor intensive and requires specific instrumentation.	Expensive, especially if techniques as hyperspectral imaging are used. Fewer the wave bands used, cheaper will be the instrument.	The cost of technique depends on the desired accuracy for VOC profiling.
	Trained personnel are also required for careful handling of samples and results.	Require computers/ laptops for data analysis.	The cost can range depending on the detector required for biomarker identification.
(ii) Applicability for rapid detection	Speed depends on the samples required to be analyzed, the number of personnel, and equipment and materials.	The focus on the development of this technique is due to its ability for rapid detection.	The technique shows the potential for rapid plant disease detection.
	The technique is not fast for a huge amount of samples.		
(iii) Applicability for field work/ Ruggedness	Field kits are being developed. However, it is difficult to develop kits for all diseases.	Moderately ruggedness. The ruggedness of the spectrometer or the scanner depends on the base on which the sensor is mounted.	Moderately rugged, depending on the detector used for sensing VOCs.
	The field kits are rugged but require good accuracy for reliable results.		
	May require 24-48 h for reliable results.	Once this technique is established, may require minutes for disease detection.	This method may require significantly less time if proven as an effective method for a particular disease.
	Molecular kits are faster.	The speed depends on the computational speed of the computer as well as the speed of the scanner.	Speed would depend on the detector speed and computational speed.
(iv) Other aspects	One of the limitations of this technique is that it is difficult to automate the process for rapid detection.	This method can be automated, or disease detection can be performed through remote operation easily.	Selection of biomarker is a critical step in determining the applicability of this technique.
	Requires personnel for monitoring plant disease by this method.	This method can be used as a preliminary screening of diseases so that fewer samples can be confirmed by molecular technique.	Once established this method can be automated with a robotic vehicle for plant disease detection.

Contd...

Contd...

		The methods in combination with other methods as molecular detection can be effective in rapid detection of plant diseases	

The probability that a particular pathogen will be used as an agent of bioterrorism is low, but the threat is nonetheless real. It is much more likely that pathogens capable of causing severe economic and social problems will continue to be introduced inadvertently. NPPOs must make every effort to safeguard plants within their country's borders, but it is also crucial to prepare for these pathogens preemptively so that they can be managed effectively when they do arrive. An initiative such as 'Predict and Prevent,' currently directed toward human and animal diseases would be timely and appropriate for plant systems as well. Such an initiative would involve a global effort in early detection of emerging diseases (identifying hot spots), monitoring pathogen movement, and preparing mitigating responses well in advance.

Enhanced networking of diagnostic and research laboratories and expert plant pathologists beyond their own borders and sharing information on a global scale will be necessary to accomplish this goal. Communications tools made possible by the internet are available to enhance interactions among the appropriate partners. The recent NATO project 'Tools for Crop Biosecurity' was a foray into cross networking, developing a broad multinational consortium representing the United States, the European Union, and Israel in the Eastern Hemisphere Plant Diagnostic Network. Such consortiums have value in bringing like-minded scientists together to promote dialogue and information exchange. It is hoped that such dialogue will form the basis for, or at least encourage, interactions between NPPOs to foster the formation of official ties with other countries and regions.

In this age of shrinking support for agricultural programs worldwide, the issue of sustainability of diagnostic systems and networks, in fact of plant health programs, in general, is of utmost concern. All too often, governments and funding agencies do not commit to sustained support of a program, regardless of its success. It is likely that the private sector, at least in the developed world, will need to play a larger role in support of diagnostic networks. This may mean a greater willingness to pay more of the cost for diagnoses and advice, although, in the experience of most public diagnostic laboratories, when the private sector is asked to pay more, the sample volume decreases. It is, however, encouraging that protection of plant-based agriculture from invasive pathogen threats has been recognized and investments made in recent years. It will be necessary to continue to stress the need for public support. Donors to programs in developing countries expect that sponsored programs will become institutionalized under public funding or adopted by the private sector. Both alternatives have merit, but the latter is unlikely in the near term at least. Therefore, governments in these countries will need to recognize the importance of both safeguarding natural and agricultural systems and assisting farmers in developing disease management practices that maximize income generation, and prioritize these areas for support. Centers of excellence will provide a means of increasing regional capacity for disease diagnosis and pathogen detection at a reasonable cost. It is important that these centers establish strong interconnecting linkages that ultimately tie them directly to farmers. This approach will help to direct their efforts toward the most important issues and thereby maximize their impact across the region.

Investments in human resource capacity are desperately needed in both developing and developed countries. The shrinking supply not only of taxonomists but of diagnosticians and plant disease management specialists who are linked directly to farmers erodes capacity to discover threatening pathogens and meet the challenges of managing them. Graduate programs must continue to train students in applied plant pathology; however their training must encompass not only traditional disease diagnosis and management but provide a good foundation in molecular biology to prepare them to develop, adapt, and/or adopt the new technologies that will become available to agricultural systems in the future.

New and exciting diagnostic technology for plant pathogens will continue to develop, as there is sufficient demand for diagnostics in the field of human and veterinary medicine to drive innovation. The number of organizations developing diagnostic technology for plant-based agriculture is small, but nonetheless, selective adaptation of medical technology will take place, the driving factors being applicability and cost (including intellectual property considerations). Many laboratories have adopted ELISA and ordinary PCR for routine use, whereas the highest tier laboratories, including centers of excellence and those responsible for identification of regulated pathogens, are currently using qPCR on a routine basis, usually under appropriate SOPs. Inexpensive field tests such as LFDs should be developed that can precisely diagnose new disease outbreaks to rapidly stimulate decision-making processes for disease management.

The next phase in technology adoption is likely to be a move towards generic platform technology such as microarrays. Microarray technology is currently too costly for wide utilization for plant pathogen diagnosis, but as the price for development of diagnostic chips reduces and demand increases, adoption rates will also increase. Local or regional development and manufacture of diagnostic assays in poorer regions of the world will be necessary to keep costs low enough for the adoption of the technology. Supply chain improvements that increase the ease of access to and cost of basic diagnostic supplies and reagents are critical to technology adoption as well. The developing world is leading the way in adaptation of mobile phone technology for information exchange, and innovations in this area to enhance disease diagnosis, surveillance, and management may find their way into agricultural production systems in developed countries. Whatever the platform, diagnostic technologies will continue to advance, and the extent of their applications will be driven by ease of use, cost, and the implications of the results they produce. Tests that directly inform a decision, whether it is the implementation of phytosanitary measures, change in cultural practices or application of a particular fungicide, for example, are the most likely to be adopted.

6 Plant Quarantine and Pest Risk Analysis

The first legal restrictions to hinder the spread of disease were enacted against human disease. 'Plant quarantine' may be defined as the restriction imposed by duly constituted authorities on the production, movement and existence of plants or plant materials or animals or animal products or any other article or material or normal activity of persons and is brought under regulation so that the introduction or spread of a pest may be prevented or limited or so that the pest already introduced may be controlled or to avoid losses that would otherwise occur through the damage done by the pest or through the continuing cost of their control. When plant pathogens are introduced into an area in which they did not previously exist, they are likely to cause much more catastrophic epidemics than do the existing pathogens. Some of the worst plant disease epidemics that have occurred throughout the world, for example, the downy mildew of grapes in Europe and the bacterial canker of citrus, the chestnut operation of the quarantine regulations. It is extremely difficult to predict accurately whether an exotic organism will become established, and once established, become economically important. Factors which affect entry and establishment of an organism are hitchhiking potential compared with natural dispersal, the ecological range of the pest as compared with the ecological range of its host, weather, and ease of colonization including reproductive potential and agricultural practices including pest management.

The quantum of import and export of plant commodities have been increased during the recent years; there is a distinct possibility of moving insect pests and diseases from their original native habitation to a new location. To prevent the introduction of exotic pests, diseases, and weeds from foreign countries or within a country, legal restrictions are enforced commonly known as Quarantine. Plant Quarantine regulatory measures are taken at the national level (Domestic Quarantine) as well as international level (Foreign Quarantine). The enforcement of the quarantine measures is supported by legal enactments, called quarantine laws. It acts as an important tool in excluding pests from the crop. Effective implementation of quarantine is highly emphasized for the management of pests, which in turn helps in maintaining the productivity of crops (Rai Vijay Laxmi *et al.*, 2014).

Plant quarantine regulations are promulgated by the national and the state governments to prevent the introduction and spread of harmful pests and pathogens.

Protection of the plant and plant products by quarantine however only become the Governments' prerogative at the turn of this century, following a series of catastrophic pest and diseases epidemics in different parts of the world. Exotic organisms when introduced have caused extensive damage, the classical example of introductions of late blight of potato (*Phytopthora infestans*) was one of the famous examples for what an introduced disease can do to change to course of history, which occurred in epidemic form in 1845 and devastated the entire potato crop in Irish fields leading to death of millions of people due to starvation. In view of increases in the quantum of import and export of plant commodities during the recent years, there is a distinct possibility of moving insect pests and diseases from their original native habitation to a new location. One of the methods of crop protection is to exclude the pests from entering into a new area.

The method of exclusion of the pests is enforced through certain legal measures commonly known as quarantine. The term 'quarantine' is a French word that literally means 40-day period. Another definition of the term 'quarantine' is "A legal restriction on the movement of agricultural commodities for the purpose of exclusion, prevention or delay in the spread of plant pests and diseases in uninfected areas" (Khetarpal and Gupta, 2005).

Major responsibilities include the quarantine of imported and exported plants and plant products, onboard quarantine for imported bulk grains and cereals, fumigation/ disinfection or disinfestation service, passenger quarantine of Kaohsiung harbor and airport, on-site inspection and quarantine in the exporting country, isolation quarantine of the designated plants and plant products, and postal parcel quarantine (Anonymous, 1996).

6.1 TYPES OF QUARANTINE

6.1.1 Importation Quarantine

This refers to the quarantine of plants and plant products imported or exported including vessels, harbor area, passenger terminal, container terminal and warehouse to check out any invasion of injurious pests and diseases, especially to detect quarantine pests and diseases which are prohibited or permitted under certain conditions. The materials subjected to import quarantine are fruits, vegetables, crops, flower plants, bulbs, seedlings, seeds, woods, nursery and other plant products. Recently, woods quarantine is the major one (Anonymous, 1996).

6.1.2 Exportation Quarantine

Conforming to the requirements and phytosanitary regulations of the importing country, this type is to provide the service to quarantine the plants and plant products and issues phytosanitary certificate for exportation.

6.1.3 Passenger Quarantine

All plants or plant products carried by immigrated passengers by cruise ship or aircraft must declare when it is arrived in the seaport/ International airport. Failure to declare will be fined. Live plants and plant products must be in accordance with the requirements of Plant Protection and Quarantine Act. These plant or their products must be applied to quarantine, viz., live plants (Including seeds and bulbs), live insects and soil; vegetables (Including fresh or chilled); cut flowers and cut branches; dried plants or its products (Including herbs and spices) and cereals.

6.1.4 Postal Parcel Quarantine

As informed by the post office, our quarantine officer will inspect the mails or parcels including plants or plant products at the post office as soon as possible, so as not to delay the mail delivery. Present day plant quarantine is essentially a legally based defense system against new arrivals of exotic organisms called pests that compete with human beings for food, shelter, and health or otherwise threaten human comfort and welfare. Plant Protection Act No. 35 of 1999 of Sri Lanka considers a pest as a 'biotic agent capable of causing injury or damage or loss to plants or plant products and shall include weeds'. The same Act defines weed as 'a plant, which is competitive, persistent, pernicious and interfering with man and his activities'.

A new pest, after arrival and establishment, can rapidly develop into destructive proportions in the absence of natural enemies in the new environment. Plant pest epidemics could result in the loss of agricultural produce. This may adversely affect the food production or an existing export trade. Therefore, the main aim of plant quarantine is to prevent the entry of exotic pests into the country by enforcing laws without which it is hard to obtain compliance.

6.1.4.1 Services

Various procedures of plants and plant products quarantine are outlined in Fig. 6.1.

6.1.5 Quarantine Pest

Each country may determine the dangerous pests that must be kept out, controlled, or eradicated by enforcing legal provisions. The Plant Protection Act No. 35 of 1999 considers a Quarantine pest as 'a pest of potential economic or environmental importance to any area within Sri Lanka and not yet present there, or present but not widely distributed and being officially controlled. The Asia and Pacific Plant Protection Commission (APPPC) has made two lists (A1 and A2) of dangerous pests to help the member countries in deciding on quarantine pests, viz., the A1 list contains dangerous pests that are not yet recorded in the Asia and Pacific region, and the A2 list contains dangerous pests present but of limited distribution in some countries within the Asia and Pacific region.

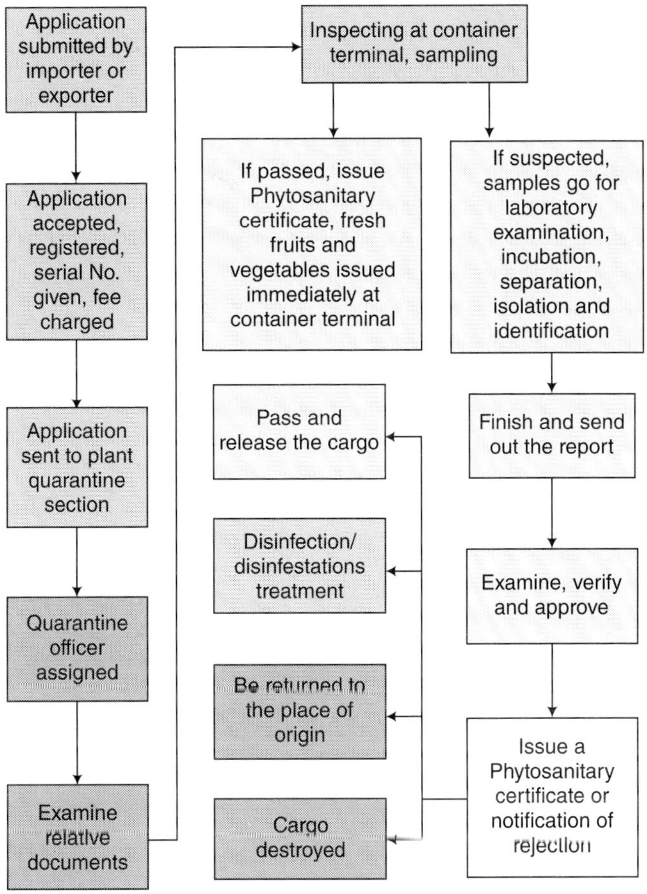

Fig. 6.1 Procedures of plants and plant products quarantine

6.1.6 Regulated Pest

Some pests present in Sri Lanka may cause serious outbreaks and hence continuous control programs must be carried out. They are considered as regulated pests once the minister declares them to be so.

6.1.6.1 Entry of alien pests into the country

Pests can enter a country by several ways. Some are natural means of pest spread, which cannot be prevented by enforcing laws. In contrast, there are manmade pathways of pest spread. These include (but not limited to) aircraft and ships, passengers, baggage, plant products imported for consumption (fruits and vegetables), plants including vegetative planting materials and seeds, products manufactured from plant materials, cargo and containerized cargo, used vehicles and farm implements, dunnage, pallets and packing material, mail, and garbage.

6.1.6.2 *Quarantine strategies*

To prevent the entry of dangerous pests into Sri Lanka, plant quarantine could adopt several strategies, such as embargo, inspection at the port of entry, inspection at the port of dispatch, field inspection during growing season, treatment as a condition of entry and postentry growing and inspection. They could be imported following the procedure given subsequently in this book. However, depending on the perceived risk of introduction of dangerous pests into the country, the importation of some plants and plant products are either prohibited or restricted.

6.1.6.3 *Prohibited materials*

These include coconut, tea, rubber, rice and cacao-planting materials and could only be imported in restricted amounts for research by the respective institutes.

6.1.6.4 *Restricted materials*

These materials could be imported by fulfilling the prescribed requirements.

6.1.7 Import Permit

The permit is a document that contains a statement of conditions that must be observed for the importation of plant materials allowed. The issue of the import permit depends upon the risks involved with respect to the plant materials imported and or the country of origin.

6.1.8 Phytosanitary Certificate

It is a document, which states that the material is free from injurious pests and should confirm to the plant import rules of the importing country. Phytosanitary certificate issued by any country should conform to the model appended to International Plant Protection Convention (IPPC). It is issued only after thorough examination of the materials. If necessary, treatments are given so as to conform to the conditions imposed by the importing country. Issuance must be done within 14 days prior to shipment of the material.

Although the materials have been examined in the country of origin, there is a possibility of contamination by pests during transit. Hence, all plant materials are subjected to quarantine examination at the port of entry. The plant materials may be retained for detailed examination, for treatment, destruction, or postentry quarantine observation at the discretion of the plant quarantine officer.

6.2 POSTENTRY QUARANTINE

The main purpose of postentry quarantine is to observe the latent development of symptoms of pests, particularly diseases, for a period of time in a pest-proof house with close drainage and double entry doors.

Four methods were tested to assess the fire blight disease response on grafted pear plants. The leaves of the plants were inoculated with *Erwinia amylovora* suspensions by pricking with clamps, cutting with scissors, local infiltration, and painting a bacterial suspension onto the leaves with a paintbrush (Ruz *et al.*, 2008). The effects of the inoculation methods were studied in dose-time-response experiments carried out in climate chambers under quarantine conditions. A modified Gompertz model was used to analyze the disease-time relationships and provided information on the rate of infection progression (rg) and time delay to the start of symptoms (t0). The disease-pathogen-dose relationships were analyzed according to a hyperbolic saturation model in which the median effective dose (ED50) of the pathogen and maximum disease level (ymax) were determined. Localized infiltration into the leaf mesophile resulted in the early (short t0) but slow (low rg) development of infection whereas in leaves pricked with clamps disease symptoms developed late (long t0) but rapidly (high rg). Paintbrush inoculation of the plants resulted in an incubation period of medium length, a moderate rate of infection progression, and low ymax values. In leaves inoculated with scissors, fireblight symptoms developed early (short t0) and rapidly (high rg), and with the lowest ED50 and the highest ymax.

6.2.1 Plant Protection and Quarantine (PPQ)

It is a program within the Animal and Plant Health Inspection Service (APHIS), an agency of the United States Department of Agriculture (USDA). The PPQ program attempts to safeguard agriculture and natural resources in the USA against the entry, establishment, and spread of animal and plant pests and noxious weeds. PPQ also supports trade and exports of U.S. agricultural products. PQ responds to many new introductions of plant pests to eradicate, suppress, or contain them through various programs in cooperation with state departments of agriculture and other government agencies within the USA. These may be an emergency or longer-term domestic programs that target a specific pest. Plant Protection and Quarantine safeguards agriculture and natural resources from the risks associated with the entry, establishment, or spread of animal and plant pests and noxious weeds to ensure an abundant, high-quality, and varied food supply.

6.2.1.1 Pest detection

The goal of PPQ's pest detection program is to protect America's agricultural and ecological resources by ensuring the early detection of harmful or economically significant plant pest and weeds. View this page to learn more about this program. The Pest Detection program supports APHIS' goal of safeguarding U.S. agricultural and environmental resources by ensuring that new introductions of harmful plant pests and diseases are detected as soon as possible before they have a chance to cause significant damage. A strong domestic agricultural pest detection system is an essential element in providing a continuum of checks from offshore preclearance programs, domestic port inspections, and surveys in rural and urban sites across the United States.

6.2.1.2 Center for Plant Health Science and Technology (CPHST)

The Center for Plant Health Science and Technology is the scientific support division for Plant Protection and Quarantine. CPHST is responsible for ensuring that PPQ has the information, tools, and technology to make the most scientifically valid regulatory and policy decisions possible. In addition, CPHST ensures PPQ's operations have the most scientifically viable and practical tools for pest exclusion, detection, and management. Center for Plant Health Science and Technology provides scientific support for PPQ regulatory decisions and operations.

6.3 PROJECT AREAS

6.3.1 Trade Issues and Risk Analysis

It means assessing the potential impact of new invasive plant pests to US agriculture and the pest risks associated with imported plant products.

6.3.2 Treatment Technology

This refers to developing new treatment methods for plant products to prevent movement of invasive pests through international trade.

6.3.3 Pest Detection

Refers to developing tools and techniques to improve early detection of exotic pests in surveillance programs.

6.3.4 Identification and Diagnostics

Developing and validating new technologies to identify exotic pests and accrediting external laboratories to perform diagnostics for high consequence pests.

6.3.5 Emergency Response

Providing scientific support during plant health emergencies.

6.3.6 Arthropod Pests

Developing methods to manage invasive arthropods.

6.3.7 Plant Diseases

Developing methods to manage invasive plant diseases.

6.3.8 Noxious and Invasive Plants

Implementing existing methods and developing new technologies for the identification, exclusion, eradication, and management of invasive weeds and regulated plants.

6.3.9 Biological Control

Developing technologies that allow natural enemies to effectively mitigate the impacts of invasive arthropods, weeds, and plant pathogens.

6.4 PEST IDENTIFICATION

The National Identification Services (NIS) coordinates the identification of plant pests in support of USDA's regulatory programs. Accurate and timely identifications provide the foundation for quarantine action decisions and are essential in the effort to safeguard the nation's agricultural and natural resources. NIS collaborates with scientists who specialize in various plant pest groups, including weeds, insects, mites, snails and plant diseases. These scientists are stationed at a variety of institutions around the country, including federal research laboratories, plant inspection stations, land-grant universities, and natural history museums. NIS supports the use of alternative diagnostic methods to enhance the speed and precision of the identification process. The Remote Pest Identification Program utilizes digital imaging technology to capture detailed images of suspected pests which can then be transmitted electronically to qualified specialists for identification. Additionally, the Molecular Diagnostics Laboratory is responsible for providing biochemical testing services in support of the agency's pest monitoring programs.

6.4.1 Programs

Agriculture Quarantine Inspection (AQI), Remote Pest Identification Program (RPIP), procedures, plant pest confirmation notifications to the states, resources, and identification aids and services.

6.4.2 Plant Import and Export

PPQ serves to allow legitimate international and interstate trade while preventing the introduction of foreign plant pests. View this page to learn more about import and export requirements for live plants and plant products, such as fresh fruits and vegetables, and certification information on solid wood packing material. Links to PPQ's electronic manuals are also available on this site. To learn about information related to what roles APHIS facilitates for Import and Export, click here. Plant Protection and Quarantine (PPQ) maintains the export program for the United States exporters of the United States and foreign-origin agricultural commodities. The export program does not require certification of any exports but does provide certification of commodities as a service to United States exporters. See: Phytosanitary Certificates; Phytosanitary Certificate Issuance and Tracking System. The USDA APHIS PPQ Export program follow the IPPC Standards and NAPPO Standards.

6.4.3 IPPC Standards

The International Plant Protection Convention (IPPC) is an international plant health agreement, established in 1952, that aims to protect cultivated and wild plants by preventing the introduction and spread of pests. The IPPC Standards consist of International Standards for Phytosanitary Measures (ISPM).

6.5 NATIONAL PLANT PROTECTION ORGANIZATION (NAPPO) STANDARDS

PPQ is delegated as the National Plant Protection Organization (NPPO) and assumes the responsibilities for ensuring the U.S. export program meets international standards. USDA APHIS PPQ policies and procedures are designed to be consistent with IPPC and NAPPO standards. The NAPPO depends on a wide range of stakeholders including regulators, scientists, academics, producers and national industry associations to achieve its mission. NAPPO Mission and Strategic Goals: "Provide a forum for public and private sectors in Canada, the United States and Mexico to collaborate in the development of science-based standards intended to protect agricultural, forest and other plant resources against regulated plant pests, while facilitating trade. Participate in related international cooperative efforts". The NAPPO Standards consist of Regional Standards for Phytosanitary Measures (RSPM).

6.5.1 Permits

Permits are required under specific USDA Animal and Plant Health Inspection Service Regulatory Authorities to import designated plants, plant products and soil into the USA, transport designated plants and plant products through the USA, import plant pests and biological control organisms into the USA, and move plant pests and biological control organisms between the States.

6.5.1.1 Plant, organism and soil permits

Permits are required for the importation, transit, domestic movement and environmental release of Organisms that impact plants, and the importation and transit of Plants and Plant Products under the authority of the Plant Protection and Honeybee Acts.

6.5.1.2 Organism and soil permits

Organism Permits include Plant Pests such as insects and snails; Plant Pathogens such as fungi, bacteria, and virus'; Biological Control Agents, Bees, Plant Pest Diagnostic

Laboratories, Soil Microbe Isolation Laboratories, Federal Noxious Weeds and Parasitic Plants.

6.5.1.3 *Plants and plant products permits*

Plant and plant product permits include plants for planting such as nursery stock, small lots of seed and postentry; plant products such as fruits and vegetable, timber, cotton and cut flowers; protected plants and plant products such as orchids, and threatened and endangered plant species; transit permits to ship regulated articles into, through, and out of the USA; and departmental permits to import prohibited plant materials for research. Permits are offered through the on-line web-based e-permits system.

6.5.2 Crop Biosecurity and Emergency Response

PPQ, the Federal response agency for plant health emergencies, develops and delivers strategic science-based regulatory programs designed to protect US crops and natural resources. View this page to learn about PPQ's infrastructure, programs, and activities that strive to deliver an effective systems approach to mitigate risks posed by select agents and regulated pests.

6.5.3 Accreditation, Certification and Network Services

The Accreditation, Certification, and Network Services (ACNS) unit manages the National Seed Health System; the US Nursery Certification Program; the US Greenhouse Certification Program; the State National Harmonization Program for Seed Potatoes; Special Foreign Inspection and Certification Programs; Plants in Growing Media; Postentry Quarantine, Audit-based Certification Systems pertaining to section 10201(d)(1) of the Farm Bill; and the National Clean Plant Network pertaining to Section 10202 of the Farm Bill.

6.5.4 Guidelines for Import of Germplasm

To prevent entry of pathogens through germplasm, following are the guidelines to be considered for import of germplasm – it is important to import from a country where the pathogen(s) is absent; import from a country with an efficient plant quarantine service, so that inspection and treatment are done; obtain planting material from the safest known source within the selected country; obtain nontreated seeds so that detection of seed-borne pathogens is facilitated; obtain clean, healthy-looking seeds of type of impurities; obtain an official certificate of freedom from pests and diseases from the exporting country; import the smallest possible amount of planting material; the smaller the amount, the less the chance of its carrying infection. It will also simplify postentry inspection; inspect material carefully on arrival and treat; if other precautions are not adequate, subject the material to intermediate or postentry quarantine and salvage infected seeds (Rai Vijay Laxmi *et al.*, 2014).

6.5.5 Agencies Involved in Plant Quarantine

On a global level, International Plant Protection Convention (IPPC) under FAO was established to prevent the introduction and spread of diseases and pests through legislation and Organization across international boundaries. This convention provided a model phytosanitary certificates to be adopted by member countries within this convention. This ten-region plant protection organization have been refurbished on the basis of biogeographical areas.

6.6 INSPECTION PROCEDURES FOLLOWED IN QUARANTINE STATION

6.6.1 Inspection Method

(i) ***Visual inspection:*** To detect sclerotia, nematode galls, buntgalls, smuts, insect infestations, weed seeds insect eggs, inert mater, etc.

(ii) ***X-ray test:*** Insect infestation (hidden).

(iii) ***Washing test:*** Spores of fungus eggs of insect adhering to seeds, nematode galls,

(iv) ***Sedimentation test:*** Stem and bulb nematode (*Ditylenchus dipsaci*) (Baerman Funnel Test).

(v) ***Incubation test:*** Seed borne fungi/bacteria (Blotter/agar test).

(vi) ***Grow out test:*** Seed borne bacteria/viruses/downy mildews.

(vii) ***Electron microscopy:*** Potentially used for identification and characterization of all plant viruses.

(viii) ***Serological methods:*** (a) ELISA (Enzyme Linked ImmunoSorbant Assay) (b) DIBA (Dot Immuno-Binding Assay) (c) ISEM (ImmunoSorbant Electron Microscopy) (d) Latex agglutination test.

(ix) ***Nucleic acid hybridization***

(x) ***Polymerized chain reaction (PCR)***

6.6.2 Plant Quarantine Measures

The measures suggested for effective plant quarantine are as follows (Khetarpal R.K. and Gupta K., 2002).

6.6.2.1 Import control: Regulations of importing country

(i) ***Embargoes:*** Total prohibition of import of certain kind of infected material from particular region/country. This is the most effective measure to exclude infected plant material (Rai Vijay Laxmi *et al.*, 2014).

(ii) ***Inspection of seed lots:*** The examination of seed samples must be by the most sensitive and reliable method for detection of dangerous pathogens. A sample may subject to more than one method.

(iii) ***Postentry quarantine (PEQ):*** Because it is difficult to detect all types of seed borne pathogens by simple tests, it may be necessary to subject seeds to PEQ. Seeds are subjected to a period of growing in the quarantine station under strict supervision in importing countries.

6.6.2.2 *Export control: Regulations of the exporting country*

(i) ***Field inspection of field crops:*** The seed crop is inspected regularly for diseases, infected plants are routed. The crop should meet requirements of the importing country (Rai Vijay Laxmi *et al.*, 2014).

(ii) ***Inspection of seed lots:*** The seed lot is thoroughly examined for the presence or absence of the microorganisms before export. The sample should meet the importing country.

(iii) ***Seed treatment:*** The seed lot should be treated as per the requirements of the exporting country. However, the treatment should confine to regulations of the importing country.

(iv) ***Phytosanitary certificate:*** Phytosanitary certificates are issued by the exporting country along with the seed as per the international plant protection convention of 1951.

6.6.2.3 *Intermediate quarantine*

This is an international cooperative effort to lower the risk of introducing a pathogen to one country with the germplasm from the other by passing this germplasm through isolation or quarantine in a third country because either the crop is not grown there or the pathogen, even if it escapes, will not become established because of the environment. The third country quarantine locations are Plant quarantine facility, Glenn Dale, US Subtropical Horticultural Research Unit, Miami, Kew Botanical Gardens, the UK, Royal Imperial Institute, Wageningen, The Netherlands, and IRAT at Nogentsur Marne, France. The US serves as the third country for the international exchange of coffee, tea, rubber, and cocoa.

6.7 PEST SURVEILLANCE IN PLANT QUARANTINE

As far as quarantine pests are concerned we generally discuss about those exotic pests, which have not entered into India but have potential and probability of entering into India through international trade or otherwise. We also discuss the exotic pests, which have already entered in our country through the movement of imported agricultural material. There are various illegal ways and means by which certain pests have been introduced in our country and for such pests the plant quarantine regulations are not applicable. Since plant quarantine is a legal strategy to restrict the entry of any exotic pest, other possibilities of the entry of exotic pests such as barrier-free border, smuggling, etc., have largely been ignored. Leaf curl virus of cotton has entered in Rajasthan and Punjab bordering Pakistan during the early nineties. Pest surveillance should be an integral component of plant quarantine services to make them useful and effective. Pest surveillance is essential to detect the occurrence of exotic pest/disease in the early stages of their introduction so as to adopted and enforce domestic quarantine regulation to ward off their further spread or to check the movement of seed and seed material to other areas (Rai Vijay Laxmi *et al.*, 2014).

6.8 PLANTS SUBJECT TO POSTENTRY QUARANTINE

- Bulbs of flowers
- Tubers of potato or tuberous roots of sweet potato
- Seedlings of Alpine strawberry, and seedling, cuttings, and scions of rose tree
- Seedlings, cuttings and scions of cherry trees and fruit trees *'Seeds' of plants imported under import permit or imported for agricultural genetic resources
- Imported seeds which are not cultivated but are exported
- Seeds believed to have low risk of introduction of quarantine pests as a result of pest risk analysis and designated by Director General of QIA
- Seeds agreed by the government of the exporting country not to undergo postentry quarantine
- Seeds of flower bulbs, potato tubers, sweet potato tubers, and Alpine strawberry seedlings which have taken inspection at the growing fields by the government of the exporting country and described in the phytosanitary certificate that they meet the standards notified by Director General of QIA
- Imported bulbs of less than 100 in quantity which are not for sale
- Fruit trees for potting plants that have established tree forms of less than ten weeks old or cherry tree seedlings that are not for sale
- Scions and cuttings of cut roses with leaves
- Seeds permitted to import which are intended for destruction after the experiment and research
- Alpine strawberry seedlings where leaves are kept whole

The Post Entry Quarantine (PEQ) enables State inspectors to perform the tasks like complete appropriate form for the Permit Unit, complete appropriate forms for the Post Entry Quarantine Unit (PEQU), conduct a survey of the growing site to determine whether to approve the site, and to inspect post entered plant material during the growing period for pests, primarily plant pathogens. The Post Entry Quarantine Manual covers procedures for conducting PEQ tasks beginning with the request to approve a growing site and ending with the release or refusal of the plant material.

6.8.1 Postentry Quarantine Sites

Plants subject to postentry quarantine shall be inspected at two national quarantine sites but because all imported volumes cannot be accommodated, only limited quantity are accepted at the national quarantine field sites, and those that exceed the capacity shall be sent to designated field sites equipped with certain requirements for isolated cultivation. Plants subject to national quarantine field sites. Flower bulbs (applicable to those whose import volumes are more than 3,000 (10 kg) and 50 per inspection unit), seedlings and cuttings of Alpine strawberry and rose tree (50 per inspection unit), potato tubers and sweet potato tubers (10 kg per inspection unit), seedlings and cuttings of cherry trees and fruit trees: 50 per inspection unit, Designated field sites of private farms, those that exceed the capacity of national field sites and flower bulbs whose imported volume is less than 3,000 (10 kg). Inspection unit is determined by the variety, or by growing field, grower or production areas of the exporting country.

6.8.2 Requirements for Postentry Quarantine Facilities

The postentry quarantine stations must be equipped with isolation facilities such as a greenhouse, net screen house, and vinyl house, and ventilation windows of the facilities shall be covered with a net screen of mesh size of no more than 0.5 × 0.7 mm or less. Nevertheless, in the case of flower bulbs, potato tubers, sweet potato tuberous roots, Alpine strawberry seedlings, and seedlings, cuttings and scions of rose trees, isolated cultivation can be conducted by inspection unit even if isolation facilities are not installed by dividing lands to ensure that no plants of the same family come into contact. However, if pests are transmitted to nearby plants and result in treatment or destruction of plants due to lack of isolation facilities, the relevant cost will be borne by the owner of the seeds. When temporarily storing seeds before planting seeds subject to postentry quarantine, one must select a storage where no pests can fly in, and obtain approval from a plant quarantine inspector in advance.

6.8.3 Postentry Quarantine Period

In the case of a herby plant such as bulbs, postentry quarantine period shall be less than one generation; rose tree shall be until the first shipment, and seedlings of cherry tree and fruit trees as well as import prohibited plants that obtained import approval shall be within two years. There will be standard operating procedures for certification of facilities for growing imported plants/plant material including transgenic plant material and other regulated articles such as biological control agents/beneficial organisms and living modified organisms under postentry quarantine and inspecting the same to ensure free from quarantine pests and other regulated non-quarantine pests. It facilitates adoption of standard operating procedures by all the inspection authorities notified by the Ministry of Agriculture (Department of Agriculture and Cooperation)/ PQOs (with PEQ responsibilities) for undertaking certification of postentry quarantine inspection facilities as well as conducting postentry quarantine inspection of growing plants and other regulated articles under the supervision of concerned inspection authority/PQOs (with PEQ responsibilities) in accordance with phytosanitary requirements specified under the Plant Quarantine and amendments issued thereunder to prevent the introduction and spread of destructive pests that affects plants and other regulated articles such as biological control agents and beneficial organisms.

6.9 DEFINITIONS AND TERMS

Some common definitions and terms used in plant quarantine are listed in Table 6.1.

Table 6.1 Common Definitions and Terms Used in Plant Quarantine

Additional declaration	A statement that is required by an importing country to be entered in Phytosanitary Certificate and which provides specific additional information pertinent to the phytosanitary condition of a consignment
Biological control agent	A natural enemy, antagonist, competitor or other organism used for pest control
Bulbs and tubers	A commodity class for dormant underground parts of plant intended for planting (includes corms and rhizomes)
Certificate	An official document which attests to the phytosanitary status of any consignment affected by phytosanitary regulations
Commodity	A type of plant, plant product, or other article being moved for trade or other purpose
Compliance procedure	Official procedure used to verify that a consignment complies with stated phytosanitary requirements
Consignment	A quantity of plants, plant products and/or other regulated articles being moved from one country to another and covered by a single phytosanitary certificate (a consignment may be composed of one or more lots)
Country of origin (of a consignment of plants)	Country where the plants were grown
Country of origin (of regulated articles other than plants and plant products)	Country where the regulated articles were first exposed to contamination by pests
Field	A plot of land with defined boundaries within a place of production on which a commodity was grown
Fumigation	Treatment with a chemical agent that reaches the commodity wholly or primarily in gaseous stage
Germplasm	Plants intended for use in breeding or conservation programs
Import permit	Official document authorizing importation of a commodity or a biological control agent in accordance with specified phytosanitary requirements
Inspection	Official visual examination of plants, plant products or other regulated articles to determine if pests are present and/or to determine compliance with phytosanitary regulations
Inspector	A trained technical staff assigned with the responsibility of inspection/ sampling of consignments of plants/plant products and other regulated articles to ensure free from pests

IPPC	International Plant Protection Convention, as deposited in 1951 with FAO in Rome and as subsequently amended
International Standard for Phytosanitary Measures (ISPM)	An international standard adopted by the conference of FAO, the interim commission on phytosanitary measures or the commission on phytosanitary measures established under IPPC
Inspection authority	Any institute/research organization notified by the Ministry of Agriculture Department of Agriculture and Cooperation), for the purpose of certification of postentry quarantine facilities and conducting of inspection of growing plants and other regulated articles under postentry quarantine
Isolated condition	A place where imported material is grown
Lot	A number of units of a single commodity, identifiable by its homogeneity of composition, origin, etc., forming part of a consignment
National Plant Protection Organization (NPPO)	Official service established by a government to discharge the functions specified by the IPPC
Official	Established, authorized or performed by a National Plant Protection Organization
Pest	Any species, strain or biotype of plant, or pathogenic agent, injurious to plants or plant products
Phytosanitary certificate	Certificate patterned after the model certificates of IPPC
Phytosanitary certification	Use of phytosanitary procedures leading to the issue of a Phytosanitary Certificate
Phytosanitary regulation	Official rule to prevent the introduction and/or spread of quarantine pests or to limit the economic impact of regulated nonquarantine pests including the establishment of procedures for phytosanitary certification
Plants	Living plants and parts thereof, including seeds and germplasm
Postentry Quarantine (PEQ)	Growing of plants in isolation for any specified period in a glass-house, a facility, area or nursery and/or holding of biological control agents and beneficial organisms under contained facility
Quarantine pest	A pest of potential economic importance to the area endangered and not yet present there, or present but not widely distributed and being officially controlled
Regulated article	Any plant, plant product, storage place, packaging, conveyance container, soil and any other organism, object or material capable of harboring or spread of pests deemed to require phytosanitary measures, particularly, where international transportation is involved
Regulated nonquarantine pest	A nonquarantine pest whose presence in plants for planting affects the intended use of those plants with an economically unacceptable impact and which is therefore regulated within the territory of the importing contracting party

Contd...

Contd...

Seeds	Seeds for planting or intended for planting and not for consumption or processing
Treatment	Official procedure for the killing, inactivation, or removal of pests or rendering pests infertile or devitalization
Visual examination	The physical examination of plants, plant products, or other regulated articles using the unaided eye, lens, stereoscope or microscope to detect pests or contaminants without testing or processing

6.10 PLANT QUARANTINE REGULATORY MEASURES

Plant quarantine regulatory measures are operative through the 'Destructive Insects and Pests Act, 1914 (Act 2 of 1914) in the country. The purpose and intent of this Act is to prevent the introduction of any insect, fungus or another pest, which is or may be destructive to crops. The import of agricultural commodities is presently regulated through the Plant Quarantine (Regulation of Import into India) Order, 2003 the provisions of New Policy on Seed Development, 1988. Further, the significance of Plant Quarantine has increased in view of Globalization and liberalization in International trade of plants and plant material in the wake of Sanitary and Phytosanitary (SPS) Agreement under WTO. The phytosanitary certification of agricultural commodities being exported is also undertaken as per International Plant Protection Convention (IPPC), 1951.

In line with the New Policy on Seed Development, 1988 and the provisions of PQ Order, 2003, the specified planting material for propagation (viz., cuttings, saplings, bud woods, etc.) require growing under Post Entry Quarantine (PEQ) for a specified period. The import permit for such planting material is granted based on a certificate from Designated Inspection Authorities of the concerned jurisdiction stating that the importer possesses the post entry quarantine facility for the imported planting material. Such consignments are released with the intimation to the concerned Inspection Authorities for conducting further PEQ inspections, and the final clearance is granted based on the PEQ Inspection Report. The pest risk analysis is mandatory for all the plants/plant material prior to its import into India as per PQ Order, 2003. The risk of exotic pests and diseases is minimized by identifying the potential pests which can get into the country with the specified commodity and seeking export certification for their freedom/ pest free area status, etc. from the exporting country. Inspection of agricultural commodities meant for export as per the requirements of importing countries under International Plant Protection Convention (IPPC) 1951 of FAO, which is now replaced by revised text of IPPC as per the model certificates prescribed under IPPC and issue Phytosanitary Certificate. The export inspections are carried out to facilitate certification of exportable plants and plant material as per the requirement of importing country in line with the above Convention. The export inspections involve sampling and detailed laboratory tests in case of seeds and planting material for propagation whereas visual examination with a hand lens and washing tests, etc. are carried out for plant material meant for consumption. The export inspections are conducted at exporters' premises also to facilitate exports for agricultural commodities meant for consumption. As per Plant Quarantine (Regulation of Import into India) Order, 2003 a total of 60 entry points including 34 seaports, 12 airports, and 14 land customs stations are notified points of entry for import of plants and plant material. Besides, 60 Inland Container Depot/ Container Freight Station, 11 Foreign Post Offices have also been notified of the entry of plants/plant material under the PQ Order, 2003.

6.10.1 Objectives

Inspection of imported agricultural commodities for preventing the introduction of exotic pests and diseases inimical to Indian fauna and flora are done through implementation of DIP Act, 1914 and the Plant Quarantine (Regulation of Import into India) Order, 2003 issued thereunder. Inspection of plants and plant material meant for export as per the requirements under International Plant Protection Convention (IPPC) 1951 of FAO to facilitate pest free trade; Detection of exotic pests and diseases for their containment by adopting domestic quarantine regulations, if introduced.

The ongoing activities assigned under the scheme include issue import permits with additional declarations and special conditions to facilitate safe imports of agricultural products, to undertake quarantine inspection and laboratory testing of plants and plant material to ensure freedom from exotic pests, to undertake phytosanitary certification (for the issuance of Phytosanitary Certificates (PSCs). One hundred and fifty officers from entral/state/UT governments have been authorized for this purpose, to undertake fumigation/disinfestations/disinfections of commodities to control infestation/infection, to undertake certification of postentry quarantine facilities and inspection of imported growing plants and plant material. Forty-one inspection authorities have been designated, to support export market access for India's Agriculture products from the phytosanitary point of view, to facilitate safe global trade in agriculture by assisting the producers and exporters by providing a technically competent and reliable phytosanitary certificate system to meet the requirements of trading partners, to provide grants-in-aid to designated inspection authorities to meet the travel expenses and also to state PSC issuing authorities for equipping them with minimal equipment required for export inspection/certification, granting approval/accreditation of treatment providers in line with the requirement of ISPM-15, and to undertake PRAs of different agricultural commodities with respect to their import or export in relation to the countries concerned.

With a view to modernize, upgrade, standardize and enhance the somewhat outdated plant quarantine system, its capacities and the related legal and administrative framework, the Government of India has recently approved the notification of a new Plant Quarantine Order for the country. This new Order is a step forward in harmonizing India's regulatory framework with the International Plant Protection Convention and internationally accepted standards and the tenets of the SPS Agreement of the World Trade Organization. Other supporting and managerial steps are also being taken to improve, to international standards, the entire gamut of the country's quarantine activity and phytosanitary border controls, including import and export inspections, on-field surveillance for pests and vectors, treatment standards and processes, and certification methodology. India is making imports of plants and plant materials subject to pest risk analysis to protect its crops from the risk of introduction of alien pests. Efforts are also under way to improve the export certification process and standards to ensure that such phytosanitary certification gives an assurance of freedom from quarantine and regulated pests and vectors, including alien species for importing countries.

Pest risk analysis is one of the core elements for protecting plant health today, as reflected by the International Plant Protection Convention and its international standards for phytosanitary measures (Bram de Hoop, 2003). The first international pest risk analysis standard became available in February 1996. Seven years later, in April 2003, the international plant protection community adopted a standard (supplementary to another pest risk analysis standard of 2001) to allow for specific focus on environmental risk. This paper discusses the requirements of international agreements for technical justification of phytosanitary measures, the elements of pest risk analysis, the need for more international standards, recent developments in addressing environmental risk and the potential for dispute settlement mechanisms to contribute to protecting the environment.

Hazardous risks of plant pests have increased over the past century with the increasing international trade in plants and plant products. Plant pests can result in unacceptable damage to both cultivated and uncultivated plants. This has led to the establishment of national plant protection organizations throughout the world. The Netherlands Plant Protection Organization, for example, was established in 1899, with the recognition of 'plant protection' as a leading concept for governmental action (Koeman and Zadoks, 1999). However, national plant protection organizations are now faced with substantial changes in international agreements, which will have a direct impact on international trade and plant production systems.

Three key international agreements relate to plant protection: the International Plant Protection Convention, the Agreement on the Application of Sanitary and Phytosanitary Measures of the World Trade Organization and the Convention on Biological Diversity. International standards for phytosanitary measures developed within the framework of the IPPC are recognized within the SPS Agreement as the appropriate standards on which WTO members should base their phytosanitary measures. For furthering the protection of biological diversity worldwide, guidelines adopted under the CBD in 2002 aim to assist countries in dealing with the impacts of invasive alien species, including plant pests. The IPPC framework, including the national plant protection organizations of all its contracting parties, could well provide an institutional framework that serves the objectives of the CBD and WTO as well. A key element of all three international agreements is the application of pest risk analysis.

6.11 PHYTOSANITARY MEASURES

Countries are putting in place new phytosanitary regulations affecting the trade of other countries have two options for complying with international agreements. Technical justification can be attained by completing a pest risk analysis. Alternatively, countries can base measures on relevant international standards if these exist (Bram de Hoop, 2003). The standard setting within the framework of the IPPC started in 1995. To date only one specific commodity standard has been developed, namely, ISPM 15: *Guidelines for regulating wood packaging material in international trade*. Because ISPMs are recognized by the WTO, standard setting is acknowledged as the most important tool for promoting international harmonization, both for safeguarding countries against phytosanitary risks and for preventing unnecessary inhibitions to trade. The more standards, the better, it is often argued. There are several pros and cons to this hypothesis, as outlined below.

6.11.1 More Specific Standards

Whenever there is no trade (or trade history) of a specific commodity between two countries, the importing country often prohibits the initiation of trade. For trade to take place, an exporting country must file a request, provide information and then wait for completion of the pest risk analysis by the importing country, which may take several years. In principle, importing countries are justified in imposing import prohibitions in view of uncertainties about the possible impact of plant pests associated with the commodity. If an internationally agreed standard existed for phytosanitary measures for all possible pests of the commodity concerned, then phytosanitary safeguards, as well as facilitation of trade, could more readily take place. The exporting country would have to demonstrate to the importing country only its capability for complying with the standard.

More specific standards could lead to more phytosanitary regulations, which may result in new obstacles to trade. However, if more international standards were available, importing countries would not have to put in place so many new phytosanitary regulations. Also, exporting countries would face fewer difficulties from having to provide the information necessary for completing pest risk analysis by the importing country. The first specific commodity standard developed within the IPPC framework, ISPM 15, is concerned with regulating packaging material. This is noteworthy from an environmental viewpoint. Wood packaging in international trade may serve as a pathway for plant pests posing a threat mainly to living trees. The standard is of benefit, first, to forest plant health because new regulations for its protection can be implemented more readily worldwide. Plant health regulators and trade also benefit by having clear guidelines for the commodity (Bram de Hoop, 2003). Unfortunately, such commodity standards are difficult to develop. Many ICPM resources will be needed in attempting the development of more specific standards, as already demonstrated by the citrus canker standard, which has been under development for several years.

Adding to the difficulties of developing international standards for specific commodities and pests of environmental consequence, ecosystems always differ from one country to another. Therefore, countries face different types of environmental plant health problems. In the case of different pest risks, countries will usually opt for different phytosanitary requirements or pest risk management options. Harmonization of measures based on international standards may then be less useful. However, specific standards could be especially helpful for harmonizing management options for similar pest risks. For instance, wood packaging material is recognized as an important pathway for forest pests by many countries; this accord was crucial for the adoption of ISPM 15. If the appropriate phytosanitary standard is not available, then countries have the alternative of risk analysis to support their restrictive measures. Pest risk analysis helps countries to focus on the specific pest risks to their plant biodiversity and, for this reason, will always be necessary.

6.11.2 Developments

The contracting parties to the IPPC recognized the need to address environmental concerns relating to plant pests and developed a reference standard specifically to target the analysis of environmental risks posed by plant pests. This was adopted in April 2003 as a supplement (*Analysis of environmental risks*) to ISPM 11 [2001]: *Pest risk analysis for quarantine pests* (FAO, 2003). In particular, plants being traded can be considered as potential plant pests because exotic plants can become weeds for cultivated plants or affect plant biological diversity through competition. These guidelines take into account the CBD's guiding principles for the prevention, introduction, and mitigation of impacts of invasive alien species.

6.12 DISPUTE SETTLEMENT AND JURISPRUDENCE

International harmonization of phytosanitary measures can be successful only if contracting parties implement standards. Control systems are needed to verify whether this is done. A breakthrough of the agreements establishing the WTO was the development of a strengthened dispute settlement mechanism (WTO, 1994). If countries cannot resolve trade disputes bilaterally, a WTO dispute procedure can be initiated. The outcome of such a procedure is binding. Compliance by concerned countries with international standards is an important judgment criterion for decisions by WTO dispute panels. A major disadvantage of WTO dispute settlement procedures is the high costs for concerned parties. Moreover, if the losing party does not comply with the outcome, the other party may 'retaliate' by imposing import restrictions on other commodities with a similar market volume. Such an outcome would never compensate for the losses incurred by the concerned industry of the winning country, nor would it be helpful for protecting plant health and plant biodiversity.

From a plant health perspective, phytosanitary disputes may better be resolved within the framework of the IPPC. In 2001, the ICPM adopted procedures for dispute settlement (FAO, 2001) and the following year established the subsidiary body responsible for administering them. The IPPC procedures rely on the voluntary cooperation of both disputing parties, who are not bound by the findings of the subsidiary body. The use of the IPPC dispute settlement mechanism could provide valuable jurisprudence for consideration in phytosanitary disputes carried out through the WTO mechanism. To date, no dispute settlement has taken place within the IPPC framework. Reference standards concerning the main rights and obligations of contracting parties to the IPPC have largely been developed. The resources of the IPPC may now better be targeted at identifying and resolving phytosanitary issues between members. Again, the outcome of such disputes could add to a compilation of jurisprudence, which, in this instance, could be used as the basis for developing specific standards. In effect, this could truly improve plant health benefiting all concerned parties, including growers, consumers, and the environment.

Over the short history of the use of pest risk analysis as a harmonized approach to decision-making in plant health, several relevant international standards for phytosanitary measures (ISPMs) have been endorsed; more are approaching endorsement (Mary Megan Quinlan *et al.*, 2003). Primary among these standards is the revision of ISPM 11 clarifying the scope of coverage for national plant protection organizations to include uncultivated plant systems. Historically, the emphasis of these organizations has been on protecting agriculture from pests. Now, with this suite of current and coming ISPMs to use, many countries will be wondering when and how to apply the expanded guidance on environmental risks. They discussed approaches for scoping environmental risks before applying the pest risk analysis process. It also considered more complicated issues that may arise in a pest risk analysis, such as indirect or secondary effects from the introduction of the species in question.

Quantitative microbial risk assessment models for estimating the annual risk of enteric virus infection associated with consuming raw vegetables that have been overhead irrigated with non-disinfected secondary treated reclaimed water were constructed (Hamilton *et al.*, 2006). They ran models for several different scenarios of crop type, viral concentration in effluent, and time since last irrigation event. The mean annual risk of infection was always less for cucumber than for broccoli, cabbage, or lettuce. Across the various crops, effluent qualities, and viral decay rates considered, the annual risk of infection ranged from 10^{-3} to 10^{-1} when reclaimed-water irrigation ceased one day before harvest and from 10^{-9} to 10^{-3} when it ceased two weeks before harvest. Two previously published decay coefficients were used to describe the die-off of viruses in the environment. For all combinations of crop type and effluent quality, application of the more aggressive decay coefficient led to annual risks of infection that satisfied the commonly propounded benchmark of $\leq 10^{-4}$, i.e., one infection or less per 10,000 people per year, providing that 14 days had elapsed since irrigation with reclaimed water. Conversely, this benchmark was not attained for any combination of crop and water quality when this withholding period was one day. The lower decay rate conferred markedly less protection, with broccoli and cucumber being the only crops satisfying the 10^{-4} standard for all water qualities after a 14-day withholding period. Sensitivity analyzes on the models revealed that in nearly all cases, variation in the amount of produce consumed had the most significant effect on the total uncertainty surrounding the estimate of annual infection risk. The models presented cover what would generally be considered to be worst-case scenarios: overhead irrigation and consumption of vegetables raw. Practices such as subsurface, furrow, or drip irrigation and postharvest washing/disinfection and food preparation could substantially lower risks and need to be considered in future models, particularly for developed nations where these extra risk reduction measures are more common.

There are many useful data sets and tools that can be employed in pest risk analysis. Although some of the examples given may seem complex, it is important to stress that pest risk analyses do not need to be long and detailed. ISPM 11 does not state that a risk analysis must achieve a specific level of detail or quantification for it to be acceptable. A risk analysis can be very short. For example, if a key component of establishment in the PRA area (e.g., a suitable host plant) is absent no further work may be necessary (Richard Baker and Alan MacLeod, 2001). Further work needs to be carried out on two fronts. First, good pest risk analysis practice needs to be captured and distilled in some form of manual enabling those new to pest risk analysis to learn the discipline as quickly and easily as possible. Increased publication of pest risk analyses would also help to demonstrate how pest risk analyses should be constructed. However, a reliable mechanism for the publication of pest risk analyses is needed, because many journal editors find the PRA format incompatible with their requirements. Some mechanism to foster links between PRA practitioners worldwide would also be invaluable. Second, additional resources need to be put into enhancing the science of pest risk analysis. There are many challenging components of pest risk analysis that would benefit from additional scientific input, such as modeling pest spread, increasing the temporal and spatial resolution of risk maps, scaling up from one outbreak to many, modeling how impacts change over time, quantifying environmental impacts and managing uncertainty. The increasing cooperation between plant health pest risk analysts and those analyzing the risks posed by invasive alien species to biodiversity is already proving to be highly beneficial.

6.13 PLANT QUARANTINE STRUCTURE AND POLICY IN INDIA

Plant quarantine operations in India are carried out by the Directorate of Plant Protection, Quarantine and Storage, which functions under the aegis of the Ministry of Agriculture. The administrative structure of plant quarantine is shown in the organizational chart (Fig. 6.2). The development of the new Plant Quarantine (Regulation of Import into India) Order, 2003 (referred to hereafter as 'the new Order') reflects the primary plant quarantine concerns of the Government of India. These include, to prevent the introduction and spread of exotic pests that are destructive to the country, by regulating the import of plants and plant products through adequate policy and statutory measures, to support India's agricultural exports through credible export certification, and to facilitate safe global trade in agriculture by assisting producers, exporters, and importers and by providing technically comprehensive and credible phytosanitary certification.

Before the new Order Gazette on 18 November 2003, the hitherto existing plant quarantine statute and regulations dated back as early as 1914. Until recently, regulatory measures in India operated on the basis of The Destructive Insects and Pests Act, 1914, which was promulgated to prevent introduction and spread of destructive pests affecting crops, and the Plant, Fruits and Seeds (Regulation of Import into India) Order, 1989.

Some other Rules were promulgated for regulating the import of live insects (1941), fungi (1943) and cotton (1972). The New Seed Policy, 1988 was formulated to provide Indian farmers with access to the best available seeds and planting material, domestic and imported (Anonymous, 2003).

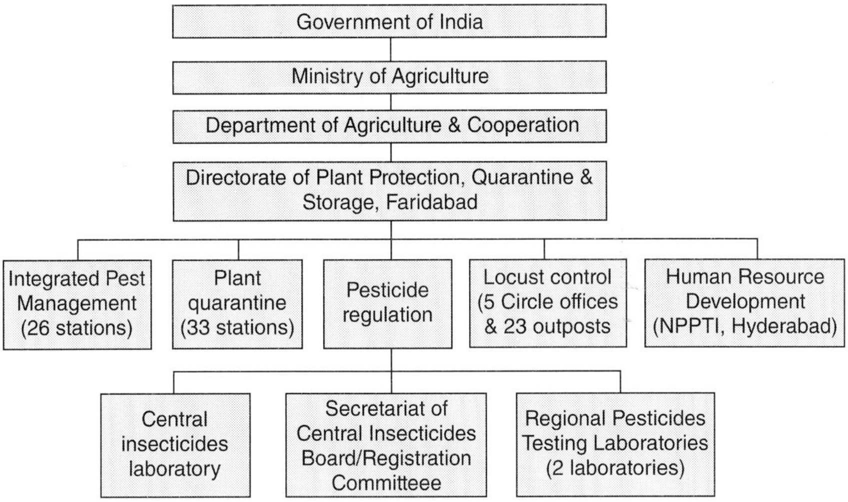

Fig. 6.2 Organizational chart of Indian plant quarantine structure

The Plants, Fruits and Seeds (Regulation of Import into India) Order, 1989, prohibiting and regulating the import into India of plants, plant materials and the like, is based on postentry quarantine checks. It has now been replaced by the new Order of 2003.

6.14 NEW PLANT QUARANTINE REGULATIONS

There were many reasons prompting the development of new legislative provisions for plant quarantine in India. They include the following the liberalized trade in agriculture as a consequence of the WTO Agreements, although offering economic opportunities, also implied fresh challenges of amplified pest risks, a corollary to the increased volumes and array of international trade in agricultural commodities. With diverse agro climatic zones, varied agricultural produce and surplus food production India finds itself in a position to exponentially expand its agricultural trade. However, what were required were accreditable national standards for all critical phytosanitary activities. Relevant entities and other interested parties invariably felt the need for compliance with stringent international phytosanitary regulations. There was an urgent need to fill the gaps in the existing plants, fruits and seeds order of 1989, namely, by regulating the import of germplasm, genetically modified organisms and transgenic plant material, as well as live insects and fungi (including biocontrol agents, etc.). India needed to facilitate the safe conduct of global trade in agriculture and thereby fulfill its legal obligations under the relevant international agreements. It was necessary to protect the interest of the country's farmers by preventing the entry, establishment and spread of destructive pests, vectors and alien species. It was necessary to protect the national plant life and the environment. It was necessary to safeguard the national biodiversity from threats of invasive alien species. The need was felt for incorporation in the regulation of additional special declarations for the freedom of import commodities from quarantine and alien pests. These declarations would be based on standardized pest risk analysis, particularly for seeds and planting materials.

6.14.1 Salient Features of the New Order

The 2003 new Order for plant quarantine in India has widened the scope of plant quarantine activities with the incorporation of additional definitions. It also makes pest risk analysis a precondition for imports (Anonymous, 2003). It places a prohibition on the import of commodities contaminated with weeds and/or alien species. Import of packaging material of plant origin is restricted unless the material has been treated.

The new Order includes provisions for regulating the import of soil, peat and sphagnum moss, germplasm, genetically modified organisms and transgenic material for research, live insects, microbial cultures and biocontrol agents and timber and wooden logs. Agricultural imports are now classified as prohibited plant species, restricted species where import is permitted only by authorized institutions, restricted species permitted only with additional declarations of freedom from quarantine pests and subject to specified treatment certifications and plant material imported for consumption or industrial processing permitted with normal phytosanitary certification.

A permit requirement is now enforced on imports of seeds, including flower seeds, propagating material and mushroom spawn cultures. Additional declarations are specified in the new Order for the import of 144 agricultural commodities, specifically listing as many as 590 quarantine pests and 61 weed species. Notified points of entry have increased dramatically: there are now 130 such entry points, where previously there were 59. The new Order also rationalizes the structure of certification fees and inspection charges.

6.14.2 Harmony with International Plant Protection Convention (IPPC)

India's new order for plant quarantine aligns with the framework of the International Plant Protection Convention in several important ways, viz., the phytosanitary measures are designed to prevent the global spread of noxious pests and are based on justified scientific principles with pest risk analysis as the cornerstone. Provisions have been made applicable to packages, and transportation, inspection and certification provisions accord with Article IV of the IPPC, phytosanitary certificates are in the format of Article V of the IPPC and according to the plant quarantine requirements of the importing country. They are to be issued after careful inspection and the required treatment, there is an emphasis on capacity enhancement and development and training of staff, the new Order has been hosted on the Web site of the Ministry of Agriculture (it is available at http://www.agricoop.nic.in/) and made accessible to everyone. It is transparent and applies uniformly to all exporting countries or parties.

6.15 ISSUE OF IMPORT PERMITS

Features of the new regulation of imports include the following. Separate formats have been devised for applications for the issue of import permits and also for the permit letters issued for consumption purposes as opposed to those for propagative plant materials (Anonymous, 2003). Commercial imports of seeds of coarse cereals, pulses, oilseed, fodder crops and planting materials of fruit plant species require prior clearance. Applications for seeds and planting materials must be accompanied by a registration certificate issued by the National Seeds Corporation or the Director of Agriculture or Director of Horticulture of the state government and a certificate of approval of post entry quarantine facilities issued by the designated inspection authority. Permits are to be issued within a maximum period of three working days of submission of an application. Pest risk analysis has been made a precondition for import of new agricultural commodities. Permits for import of soil or peat and import of live insects, microbial cultures or biocontrol agents are to be issued only by the Plant Protection Adviser, the technical head of the plant quarantine service. Permits for import of germplasm, genetically modified organisms, and transgenic material are to be issued by the director of the National Bureau of Plant Genetic Resources, New Delhi, India. Issued permits are valid for six months. This may be extended a further six months. Permits are not transferable and no permits are issued for landed consignments. Relaxations from the conditions of the new Order, necessitated by an emergency or unforeseen circumstances, are to rest with the Ministry of Agriculture.

6.15.1 Strengthening and Modernization of Plant Quarantine Facilities

The World Trade Organization – Agreement on application of Sanitary and Phytosanitary Measures (WTO-SPS Agreement) facilitates movement of plants, planting materials globally. Increased volume of trade poses greater risks of introduction of exotic pests/ diseases in the country. Plant Quarantine regulatory measures derive legal sanctity from the Destructive Insects and Pests Act, 1914 (Act 2 of 1914). The purpose and intent of this Act is to prevent the introduction of any insect, fungus or pest which is or could be destructive to crops. Import of agricultural commodities into India is regulated by Plant Quarantine (Regulation of Imports into India) Order 2003 notified under Section 3 (1) of Destructive Insects and Pests Act, 1914. This Order is a significant step in harmonizing India's regulatory framework with the International Plant Protection Convention and internationally accepted standards and the tenets of the SPS Agreement of the World Trade Organization. Steps are also being taken to improve the entire gamut of quarantine activities and Phytosanitary border controls, related to import and export inspections, on-field surveillance for pests and vectors, treatment standards and processes, and certification methodologies. India has made imports of plants and plant materials subject to pest risk analysis to help protect its crops from the risks of introduction of alien pests.

6.16 INTERNATIONAL QUARANTINE

Its major objectives include strengthening pest risk analysis (PRA) for import of agricultural commodities, be it plant or animal and other risk factors associated with GMOs, biocontrol agents, and other microorganisms. Upgraded Plant Quarantine Stations, in terms of trained manpower and infrastructure, continue the review of the national quarantine legislation in the light of Agreement on Application of Sanitary and Phytosanitary (SPS) Measures of WTO and the status of our compliance with SPS/WTO, initiate preimport inspection and strengthen the postentry quarantine (PEQ) growing and inspection of imported material, and to establish a National Biosecurity Centre to deal with biosecurity in a comprehensive manner to achieve food safety, protect animal and plant life and health, and also protect the environment.

Plant quarantine, while being national in execution, is international in character. Therefore, international/regional cooperation is very necessary for achieving the objectives since plant genetic resources are a world resource meant for the welfare of the human race as a whole. Cooperation on the following lines would greatly help in safe exchange of germplasm materials.

6.16.1 Consortium of Plant Quarantine Stations

It is an excellent concept to facilitate the exchange of genetic stocks and scientific information at international/regional level. Any material passing through a plant quarantine station will have a very low pest/pathogen risk. The material so generated should be exchanged with other plant quarantine stations promptly before it is distributed locally and gets contaminated with local pests/pathogens. Several quarantine stations working independently may be processing the same material (same crop or even same variety) at each station. Under the proposed consortium concept, different quarantine stations would undertake the processing of different materials (several accessions of the same crop or a group of crops at each station) and then share the material. This would avoid duplication of efforts, reduce costs of processing and more material would be available with adequate quarantine safeguards. In the same spirit, scientific information (detection techniques, treatments, distribution of pests/pathogens) and antisera for serodiagnosis of viruses and bacteria could be shared by quarantine stations of different countries.

6.16.2 Establishment of Central Seed Health Testing Laboratories

At present, thousands of seed samples of a variety of crops are being exchanged by different countries for breeding purposes and for conducting multilocation international trials. This has exposed many countries, particularly the developing ones, to the hazards of serious and new seed borne pests/pathogens. Sometimes, the volume of material may be so large that it is not physically possible for a quarantine service to process it with any degree of surety during the time at its disposal before the planting date. In many countries, seed health testing facilities may not exist, or they may be inadequate. Neergaard (1977) proposed the concept of establishing a few central seed health testing laboratories well equipped with required facilities and trained personnel. These may be coordinated with regional gene banks or international centers, but should be independent of the organization for which they are working. Such an arrangement would lower pest risk to a considerable extent and avoid duplication of efforts, thus reducing costs of processing. Plant quarantine services of different countries will have faith in such laboratories and will accept their certification.

6.16.3 Third Country/Intermediate Quarantine

The concept of third country quarantine is another example where international cooperation could play an important role in the safe transfer of plant genetic resources. This is particularly helpful for transferring high-risk tropical/subtropical plant genera from one country to another. The material could be grown, tested/indexed for hazardous plant pests and pathogens in a temperate country without much risk because either the possible hosts are not present there, or the environment is unfavorable for their establishment. Some centers which have been providing third country quarantine facilities for transferring genetic resources include Subtropical Horticulture Research Station, Miami, the USA for cocoa, coffee, rubber, and tea; the University of Reading, the UK for groundnut; and Institute Nazionale Per Piante de Legno, Torino, Italy for cassava.

6.16.4 Biogeographical Regions

The concept of biogeographical regions is an example of regional cooperation. The eight biogeographical regions proposed for effective quarantine are separated by natural barriers like sea, high mountains, and deserts, making pest/pathogen dissemination extremely difficult so long as the exchange of genetic resources is judiciously regulated. Accordingly, all countries in such a region must have common quarantine regulations since the larger the land mass covered by the same set of regulations, the greater is the protection afforded to the agriculture of the region. Based on this concept, countries of the ASEAN region have come to an official level understanding and have formulated a common set of quarantine regulations to protect the region against alien pests/ pathogens. Understanding on the same lines among SAARC countries along with Afghanistan and Burma would be very useful in protecting the agriculture of the entire region. Such an arrangement will reduce duplication of efforts and operational costs. Countries like Sri Lanka and Maldives or even some islands of India (Andaman group of islands) could even serve as a third country quarantine for temperate, high-risk genera. SAARC countries have already developed cooperative programs in many crops and quarantine could be incorporated into this agreement.

6.17 DOMESTIC QUARANTINE

Major aims of domestic quarantine are to review the status of existing domestic quarantine for establishment of interstate quarantine check-posts to monitor movement of pests of significance, review the national regulatory framework and develop a mechanism for distribution or sale of pest-free seeds/plants/planting material within the country, be it seed distribution for multilocation testing under All India Coordinated Research Projects, inland supply of germplasm by NBPGR or seed distribution by the national/state seed corporations/private organizations, to strengthen state certification mechanism to ensure the supply of pest-free nursery material, developing an early warning system and rapid response team to curtail spread/eradicate the recently introduced pests in an area so as to avoid their further spread and establishment, to amend the Biological Diversity Act, 2002 for making provisions to check international/interstate movement of invasive pests, and to harmonize the Act with the quarantine and export-import (EXIM) regulations for implementation.

6.17.1 Research and HRD

Reprioritize research in areas that need immediate emphasis such as survey and surveillance of diseases/pests, development of digitized keys, development of diagnostic protocols using serological and molecular techniques for detection of exotic pests/transgenes and to develop a strong programme of awareness generation on importance of biosecurity for various stakeholders such as academia, policy makers, custom officials, students, and general public.

6.17.2 Review of Domestic Quarantine Regulations

The legislature measures to prevent the introduction and spread of destructive pests of crops are operative through the 'Destructive Insect and Pest Act, 1914.' The domestic quarantine regulations are operative by the central government through powers vested under Section 4 A, B, C and D, and Section 5 authorizes the state government to enact similar regulations and Section 5A provides for the penalties.

The first domestic quarantine notification was issued by the central government in 1944 against fluted scale (*Icerya purchasi*) and San Jose scale (*Quadraspidiotus perniciosus*) is 1953. To prevent the spread of Banana bunchy top virus disease from states of Assam, Kerala, Orissa, Tamil Nadu Government of India issued a notification in 1951. The bunchy top is believed to have been introduced from Sri Lanka in 1940. A notification was issued by the central government in 1959 against potato wart (*Synchytrium endobioticum*), prohibiting the movement of potato from states of West Bengal. The wart disease was reported first from Darjeeling district of West Bengal in 1953 and believed to have been introduced from the Netherlands. To prevent the spread of cyst nematodes of potato (*Globodera rostochiensis*) from the states of Tamil Nadu, the central government has issued a notification in 1983, though the cyst nematodes have been reported as early as 1961 from Ooty area of Tamil Nadu state and were believed to have been introduced from Europe. To prevent the spread of apple scab (*Venturia inequalis*) from the state of Himachal Pradesh, the central government issued a notification in 1977, prohibiting the export of planting materials of apple from Himachal Pradesh.

6.18 ISSUES CONCERNING PLANT QUARANTINE IN INDIA

There are a number of issues related to quarantine in exchange of plant/planting material. The legislative issues are specific to the material being exchanged in bulk or as small samples meant for research, whereas technical issues on quarantine processing and methodology are applicable to both (Pathak, 2010).

6.18.1 Legislative Issues

The national quarantine legislation needs to be in harmony with the international norms laid down by the International Plant Protection Convention (IPPC), and most of the countries are in the process of trying to gear up their activities to comply with it. India has attempted to do so by bringing out the PQ Order 2003.

6.18.2 Issues Specific to Germplasm Exchange

The various schedules, viz., V (restricted species where import is permitted only by authorized institutions), VI (restricted species whose import is permitted only with additional declarations of freedoms from quarantine/regulated pests and subject to specified treatment certifications), and VII (plant material whose import for consumption/industrial processing is permitted with a normal Phytosanitary Certificate) of the PQ Order give lists of crops for which a generic PRA is given and detailed PRA is not required. In the case of germplasm, a large number of species of cultivated crops (and their wild relatives/land races) with useful traits are imported but may not find mention in any of the schedules. Hence, a PRA becomes obligatory for them, and the modalities of its preparation have been developed by Department of Agriculture and Cooperation (DAC), Ministry of Agriculture. Since, 2007, Director, NBPGR is empowered to issue Import Permit for germplasm after undertaking PRA to facilitate import of planting material. The PRA undertaken for germplasm needs to be considered case-by-case for listing the pests under additional declarations at the time of issue of Import Permit. This would greatly help the indenters in procuring germplasm of their interest from varied sources.

6.18.3 Issues Specific to Bulk Material in Trade

In India, the quarantine officials at ports of entry do not have the authority to search and seize the personal baggage of the incoming travelers, and the powers are vested only with customs officials who are not professionally trained to understand the importance of pests. The customs officials need orientation programs for appreciating the importance of quarantine. The provisions of the DIP Act, 1914 and its amendments for domestic quarantine to restrict the interstate movement of nine invasive pests need to be reviewed for their current distribution and regulatory measures implemented to check their further spread into neighboring states within the country. Establishment of interstate barriers or check-posts on the movement of seeds and planting materials of their host species could be a possible solution. Besides, some of the recently introduced pests like sunflower downy mildew (*Plasmopara halstedii*) in 1984, American serpentine leafminer (*Liriomyza trifolii*) in 1991, spiraling whitefly (*Aleurodicus dispersus*) in 1993 and biotype 'B' of whitefly *Bemisia tabaci* in 1999, etc. also need to be considered for inclusion under domestic quarantine.

6.18.4 Technical Issues

The quarantine laboratories adopt various general and specific detection techniques for different pests based on the nature and the volume of the material imported. The imported material generally includes seeds (true seed), grains, vegetative propagules, seedlings, rootstocks, stem cuttings, whole plants, fruits and *in vitro* cultures. The quarantine procedures integrate various methodologies of Pathology, Entomology, Nematology and Plant Taxonomy for detection of pests and also for salvaging the infected/infested samples. Over the years, a large number of pests have been intercepted in imported bulk consignments and germplasm and other research material during quarantine processing. The intercepted pests can be divided into the following categories.

(i) Pests not known to occur in India
(ii) Pests with different races/biotypes/strains not known to occur in India
(iii) Pests intercepted on a new host or are from a country from where they have never been reported before
(iv) New pest species hitherto unreported in science
(v) Pests reported to be present in India but with a wide host range

These interceptions, especially of pests and their variability not yet reported from India [Category (i) and (ii)] signify the importance of quarantine in preventing the introduction of destructive exotic pests. The categories (iii) and (iv) pests are not expected in the sample as per the PRA, which is literature based and since no records are available on the pest/host, their presence is unexpected and hence, important from quarantine viewpoint. The last category (v) – pests with a wide host range are critical and could become invasive in case they find suitable biotic and abiotic environment. The significant interceptions made, which are yet not reported from India, include insects like *Acanthoscelides obtectus* in *Cajanus cajan* and *Phaseolus vulgaris* from Brazil, Colombia, Italy and Nigeria, *Anthonomus grandis* in *Gossypium* spp. from the USA, *Ephestia elutella* in *Macadamia* nuts and *Vigna* spp. from the USA, *Quadrastichodella eucalyptii* in *Eucalyptus* from Australia, nematodes like *Heterodera schachtii* from Denmark, Germany and Italy,

Ditylenchus dipsaci, D. destructor, Rhadinaphelenchus cocophilus, etc. in soil clods and plant debris, fungi like *Claviceps purpurea* in seeds of wheat, barley, etc. and *Peronospora manshurica* on soybean from several countries, *Fusarium nivale* on wheat barley and *Aegilops* sp. from Germany, Italy, Hungary, Sweden and the UK, *Uromyces betae* on sugar beet from Belgium, Germany, Italy, the UK and the USA, bacteria like *Xanthomonas campestris* pv. *campestris* on *Brassica* spp. from Canada, France, Pakistan, Sweden, Taiwan, the UK and the USA and viruses like barley stripe mosaic virus on wheat from the USA, cowpea mottle virus on cowpea from Philippines, raspberry ringspot virus on soybean from AVRDC (Taiwan), Sri Lanka, Thailand, the USA, etc. Such interceptions signify the success of quarantine. Otherwise, these pests could have entered the country and played havoc with the plant biodiversity and agriculture had they not been detected and stopped during quarantine processing.

The other technical issues pertain to quarantine processing, i.e., methodology for detection and salvaging of pests. In the case of bulk consignments, a proper sampling of the lot is to be done as per the norms for ensuring an effective processing of the material. However, due to the stacking of containers of bulk material or inaccessibility, proper sampling procedures are not followed leading to an oversight during quarantine processing. Despite the fact that the five major plant quarantine stations have been modernized, they remain poorly staffed. There is a need for trained personnel to deal effectively with the quarantine processing of huge amount of samples received annually. Besides, the stations especially those seaports that deal with timber log imports need a complete revamping in terms of space, expertise and facilities to ensure pest-free timber imports.

The material known/suspected to carry viruses needs to be grown in isolation and the seedlings showing symptoms should be tested for viruses by a combination of biological/physical/serological/molecular tests. The inspection authorities authorized under the PQ Order need to follow the SOPs notified in 2007 for postentry quarantine inspection. Although this might delay the release and use of the material, it would ensure its healthy status.

For accurate diagnostics, the reagents such as antisera for viruses/bacteria are often not easily available especially for exotic pests for which a repository of antisera needs to be established. Also, there is a need to establish a repository of seeds of indicator hosts. Expertise is also required to identify new pathogens or strains unknown to a region. The identity of the pest, when intercepted in quarantine, needs to be established speedily to determine the quarantine risk posed by the pest and suitable disinfestations procedure to be applied. For example, identification of quarantine insect pests is difficult because few institutions have an insect reference collection, especially for exotic insect pests and identification keys, are not readily available. Moreover, even if information on the species is available, it is scattered in various journals and quick retrieval is a problem. Therefore, there is a need to consolidate the taxonomic information through the preparation of digitized keys for quick and accurate identification of intercepted pests.

6.18.5 Legislative Frameworks for Phytosanitary Measures Information Digest of Section 3

Section 3 discusses legal regimes for plant health regulation, pointing out aspects relevant to the management of invasive alien species. Chapter 7 of phytosanitary measures offers a plan of action for countries that might be looking for ways to adapt national phytosanitary legislation and systems to better control invasive alien species. Chapters 8 and 9 discuss recent developments in national phytosanitary regulation in particular countries, namely, China and India.

6.18.5.1 Assessing a phytosanitary regime for control of invasive alien species

A methodology that could be used by countries in identifying and assessing their institutional and legislative frameworks for the control of invasive alien species has to be described. Based on work in Ghana, the methodology merits consideration by other jurisdictions (especially developing states that adhere to the common law tradition) for its applicability to their situation. Elements and stages of the methodology are as follows.

6.18.5.2 Identification of the nature and sources of introduction of invasive alien species

Assessment of causes and/or sources of introduction of invasive alien species (intentional or unintentional and through various pathways, including air, land and sea) allows determination of the priority areas for action. Accurate information on plant health status in a country provides an informed basis for the provision of preventive or eradication measures under the legislation.

6.18.5.3 Identification of the applicable global and regional instruments governing invasive alien species

The major global, regional and soft law instruments that address the subject of invasive alien species are documented in this publication and elsewhere. It is the provisions of these instruments that constitute the benchmarks against which national standards may be measured. In this regard, the Convention on Biological Diversity and its guiding principles on invasive alien species and the International Plant Protection Convention and the international standards for phytosanitary measures are of paramount importance.

6.18.5.4 Determination of the ratification status or otherwise of the instruments, and if not ratified, the necessary steps for ratification

In common law jurisdictions, treaties entered into by states are usually not directly enforceable in the courts unless enacted as legislation at the domestic level. This contrasts with the situation in other jurisdictions where treaties form part of the law. The treaty-making or ratification power is governed by the constitutional order. A treaty ratified by a state is binding on that state. Accordingly, the state would be enjoined, where the treaty is not directly applicable at the domestic level, to enact appropriate legislation to give force and effect to the provisions of the treaty.

6.18.5.5 Identification of the existing relevant legislation and institutions

In all jurisdictions, there are sources of law from which to ascertain the relevant legislation on a subject. In the common law tradition, these include the constitution, parliamentary enactments, subsidiary legislation and the common law. In many jurisdictions, legislative enactments provide the main body of law, emanating from a single legislature in unitary states or a two-tier system of legislation in federal states. Subsidiary legislation in the form of regulations is most often employed as a means of providing detailed implementation of the law. There are also codes of practices, directives or standards, which are not legally binding but serve as guides. The components of legislation in the control of invasive alien species are likely to be those listed in Exhibit 6.1.

Exhibit 6.1: General Components of Legislation on Control of Invasive Alien Species

The **scope of the legislation** explains precisely and concisely what the law covers and its stated objectives; for example, 'An Act for the prevention of the introduction and spread of pest of plants, plant products, and related matters'.

The **definition of terms** could be adopted from the relevant instruments, especially the CBD and the IPPC.

The **institutional** or **organizational basis** for the control of invasive alien species could include provision for a plant protection organization and a technical advisory body, for example.

Enforcement provisions could include specific powers or functions for enforcement officers, conditions for the importation and exportation of plant and plant products and quarantine provisions.

Offenses and **penalties** include penalties for convictions for violations on the basis of due process.

Financial provisions include the power to levy fees and financial autonomy for the NPPO.

Special procedures could include immunities for officials for acts done in good faith.

Appeal processes and **procedures** should be described.

Provision for **international cooperation** and **collaboration** (especially with neighboring states) is included because the source of invasive alien species is usually external.

The **power to make regulations** is given to the minister or relevant authority to enable the law to meet changing circumstances.

The next step is sometimes an arduous task: an in-depth examination of each of the various sources of law to ascertain whether or not any provisions address the subject of invasive alien species. In the absence of a specific enactment on invasive alien species, the legislation in each relevant sector (e.g., forestry, wildlife, wetlands, marine environment, and plant protection) could be examined. The large number of relevant institutions in the field of control of invasive alien species (e.g., ministries or agencies for agriculture, health, environment, transport, justice, interior and trade, environmental protection, customs, and immigration) may lead to conflicts, overlaps and gaps in the regime. This may require the strengthening of the lead agency or the creation of a coordinating body within the existing regime. The choice will influence the form of legislation to be adopted.

6.18.5.6 *Assessment of the existing legislation in the light of the international instruments*

At this stage, the inventoried existing legislation must be examined to ascertain whether it conforms to the provisions of the various international instruments. Gaps or inconsistencies would be noted for appropriate corrective measures in any proposed legislation. For example, the IPPC enjoins parties to establish national plant protection organizations with specified functions for pest regulation, phytosanitary certification, importation and exportation regulations, etc.; if there were no national plant protection organization, it would have to be catered for under the proposed legislation.

6.18.5.7 *Determination of the form and content of the legislation*

In providing an enactment on invasive species, care must be taken to ensure that its provisions conform to, or are consistent with, the fundamental law of the land, the constitution. If the proposed legislation is to give force to the international instruments such as the CBD and IPPC, it must reflect the relevant provisions of these instruments. In the preparation of the legislation, one of three identified approaches that could be adopted are a review of existing laws and their consolidation into a single piece of legislation, the enactment of one core framework legislation under which various pieces of existing legislation would be implemented, and the harmonization of all relevant laws to address conflicts, gaps or overlaps.

6.18.5.8 *Assessment of the capacity for implementation*

Laws, when enacted, will not achieve the desired objective unless they are backed by adequate resources, both human and financial, for their implementation.

6.19 DEVELOPMENTS IN PHYTOSANITARY REGULATION IN CHINA

China has analyzed its existing legislation and regulations in relation to the IPPC framework and is updating plant quarantine regulations to harmonize them with international agreements and standards. It has also initiated efforts to prevent and contain invasive alien species, which have proved a significant problem to China (Exhibit 6.2).

Exhibit 6.2: The Problem of Invasive Alien Species in China

Invasive alien species have caused notable economic losses in agricultural and forestry production. Examples include three alien species of leafminer *(Liriomyza sativae, L. huidobrensis, and L. trifolii)*, pinewood nematode (*Bursaphelenchus xylophilus*) and fall webworm (*Hyphantria cunea*). Other alien organisms have had serious consequences for biodiversity, in particular, crofton weed (*Eupatorium adenophorum*), common cordgrass (*Spartina anglica*) and water hyacinth (*Eichhornia crassipes*).

Some of the reasons why invasive alien species have been a major problem in China are:

- increasing international trade of plants and their products
- suitable habitats for invasive alien species within the variety of ecosystems in China's vast territory
- poor integration of the national phytosanitary management system, with separate external and internal quarantine systems and sometimes inconsistent laws, rules and regulations
- a shortage of quarantine facilities and basic instruments for quarantine and inspection, and lack of a reliable early-warning system.

Initiatives to update the legislation and build phytosanitary capacity will help in the country's efforts in risk identification and management of invasive alien species.

The two principal pieces of legislation relating to the phytosanitary system are the law of the People's Republic of China on the entry and exit animal and plant quarantine (implemented in April 1992) which provide specific stipulations for entry and exit quarantine work, including phytosanitary inspections and treatments and other phytosanitary requirements, and the regulation of plant quarantine (implemented in May 1992 after a revision) prescribes the purpose, the domestic organization, plant quarantine scope and the measures for pest risk management.

The latter is being updated to meet the requirements of the IPPC framework and relevant international standards. The proposed new legislation will attempt to align all the definitions with the international standards. Items such as pest free area and pest risk analysis will be included. China is putting a lot of effort into conforming well to the market economy and international law. A nation-wide, general investigation of agricultural plant pests has been conducted to gain information on epidemic situations and to establish a pest information system. It is the basis for a new pest early-warning system. China's program to harmonize national phytosanitary standards with the provisions of the IPPC includes drawing up or revising national phytosanitary standards, application of regulations under the IPPC framework, setting up a reporting system for invasive alien species, increased emphasis on pest risk analysis and the establishment of pest free areas.

In recent years, China has adopted many international standards in its routine work. In general, however, there are still not enough national standards for plant quarantine. To deal with the problem of invasive alien species, China recognizes that it is necessary to cooperate with other countries or areas and to use the measures of the country of origin of an alien species as a reference point. It needs the techniques, methods of management, information and economic assistance to prevent the further spread and negative impacts of invasive alien species.

6.20 DEVELOPMENTS IN PHYTOSANITARY REGULATION IN INDIA

India has approved a new Plant Quarantine (Regulation of Import into India) Order, 2003 ('the new Order') as a step forward in harmonizing India's regulatory framework with the IPPC and internationally accepted standards and the World Trade Organization's Agreement on the Application of Sanitary and Phytosanitary Measures. Supporting improvements cover the entire gamut of the country's quarantine activity and phytosanitary border controls, including import and export inspections, on-field surveillance for pests and vectors, treatment standards and processes, and certification methodology.

Plant quarantine efforts are focused on reducing the risk of invasive alien species by making imports of plants and plant materials subject to pest risk analysis, supporting agricultural exports by credible export certification assuring freedom from quarantine and regulated pests and vectors, including alien species for importing countries and facilitating safe global trade in agriculture by assisting producers, exporters and importers and by providing technically comprehensive and credible phytosanitary certification.

Noteworthy changes resulting from the 2003 new Order for plant quarantine include a wider scope of plant quarantine activities and pest risk analysis as a precondition for certain imports. There has also been a dramatic increase in notified points of entry, which have increased from 59 to 130. The new Order places a prohibition on the import of commodities contaminated with weeds and/or alien species. Import of packaging material of plant origin is restricted unless the material has been treated. Agricultural imports are now classified as one of the following: prohibited plant species; restricted species where import is permitted only by authorized institutions; restricted species permitted only with additional declarations of freedom from quarantine and/or regulated pests and subject to specified treatment certifications; or plant material imported for consumption or industrial processing permitted with normal phytosanitary certification.

India's new order for plant quarantine aligns with the IPPC framework in several important ways, viz., phytosanitary measures are designed to prevent a global spread of plant pests and are based on scientific principles with pest risk analysis as the cornerstone. Provisions have been made applicable to packages and transportation, inspection and certification provisions accord with Article IV of the IPPC, Phytosanitary certificates are in the format of Article V of the IPPC and according to the plant quarantine requirements of the importing country. There is an emphasis on capacity enhancement and development and training of staff, and the new order has been published on the Internet, is transparent and applies uniformly to all exporting countries or parties.

India's action plan for pest risk analysis took effect from December 2003. A major feature is the establishment of a national pest risk analysis unit. The plan also includes organizing PRA training, establishing working groups and holding a workshop attended by national and international experts to prioritize crops and commodities for pest risk analysis. Commodities selected for detailed pest risk analysis over the next few years include fruit (e.g., banana, apple, and citrus fruits), pulses (e.g., black and green gram), cereals (e.g., pearl millet, maize, and sorghum), oilseeds (e.g., sunflower, linseed, and rapeseed) and fibres (e.g., cotton and jute). India's national integrated pest management (IPM) program is considered to be the mechanism with which to prevent and control the threat posed by invasive alien species within the country. State governments, nongovernmental organizations, private sector organizations, research institutions and farmer self-help groups are involved in the surveillance and detection of pests and diseases. They are capable of taking environmentally friendly corrective action within the IPM scheme.

6.20.1 Points to Note

Section 3 provides one generalized scheme and two country-specific accounts of legislative review and change to modernize national phytosanitary frameworks and include measures relevant to invasive alien species that are plant pests. Aspects of general relevance include gathering and analyzing accurate information on plant health status which helps determining priorities for action. The conformity of existing legislation with international instruments must be assessed before introducing change to plant health frameworks, new legislation, and other changes to phytosanitary frameworks must be given adequate resources for implementation. The use of pest risk analysis is critical and increasingly an employed component of any modernized phytosanitary system.

6.21 SANITARY AND PHYTOSANITARY REQUIREMENTS (SPS)

In WTO agreement, protection of human safety or health, protection of animal and plant life or health and also protection of the environment is of prime importance. All these are broadly covered in sanitary and phytosanitary (SPS) requirements. The agreement on the application of sanitary and phytosanitary measures is an integral part of Final Act GATT 1994 and (b) Agreement on Agriculture. The SPS measures are defined as any measure applied in the following ways:

- To protect animal or plant life or health within the territory of the member from risks arising from the entry, establishment or spread or pests, disease and disease-carrying organisms
- To protect human or animal life or health within the territory of the member from risks arising from additives, contaminants, toxins or disease-causing organisms in foods, beverages or feedstuffs
- To protect human life or health within the territory of the member from risks arising from the disease carried by animals, plants or products thereof, or from the entry, establishment or spread of pests or
- To prevent or limit other damages within the territory of the member from the entry, establishment or spread of the pests.

The SPS measures include all laws, decrees, regulation requirements, and procedures including *inter alia*, end product criteria, process and production methods, testing, inspection certification and approval procedures, quarantine treatments including relevant requirements associated with the transport of animals or plants or with the materials necessary for their survival during transportation, provisions on the relevant statistical methods, sampling procedures, and methods of risk assessment and packaging and labeling requirements directly related to food safety.

These measures thus include food additives, contaminants, toxins, drug or pesticide residues in food, certificate of food, animal or plant health safety, processing methods, food labeling, plant or animal quarantine, the requirement for prevention, control or establishment of pest or disease and sanitary requirements for imports. Whereas the sanitary provisions relate to food and animal health, phytosanitary provisions relate to plant health aspects of products.

6.21.1 Phytosanitary Certification (PSC)

The phytosanitary certificate is an official declaration stating that plants and plant material that move across international boundaries are free from pests and diseases so as to prevent introduction and spread of any pests in the newer areas. PSC paves the way for rapid screening and expeditious release of imported plants/plant material at the port of arrival. In WTO agreement, its adoption has assumed greater significance.

6.21.1.1 Facilities for issue of PSC

The requisite arrangements have been made in the country for the inspection of PSC. All Plant Quarantine field stations are authorized to issue PSC in respect of exportable agricultural items, whereas the identified officers of the Argil/Horticulture Departments of States/Union territories are to issue PSC only for perishable commodities comprising fresh fruits and vegetables.

6.22 PHYTOSANITARY CAPACITY EVALUATION (PCE)

Phytosanitary Capacity Evaluation (PCE) is a diagnostic tool enabling a country to assess the weaknesses and strengths of its phytosanitary system in relation to its ability to fully implement the International Plant Protection Convention and other international phytosanitary obligations (Felipe Canale, 2003). The pooled results of the supervised application of the tool in over 35 countries are presented. PCE has also been applied as a cross-disciplinary tool among the sanitary, phytosanitary and food safety areas in the Andean subregion. Since 'quarantine pests,' as defined by the convention, constitute a subset of invasive alien species, PCE results are already useful in relation to invasive alien species. PCE methodology has the potential to be further developed to cover a country's needs in implementing Article 8(h) of the Convention on Biological Diversity.

Article XX of the International Plant Protection Convention establishes the agreement by contracting parties to promote the provision of technical assistance to facilitate implementation of the IPPC. Phytosanitary Capacity Evaluation (PCE) is the first tool approved by the Interim Commission of Phytosanitary Measures to address that commitment, assisting contracting parties to perform a needs assessment and identification of their constraints for achieving full implementation of the IPPC at the national level.

6.22.1 PCE Objectives

PCE is intended primarily as a tool to assist countries in the modernization of their phytosanitary systems; in this context, the modernity of a national phytosanitary system is the capacity of a country to meet international phytosanitary obligations in an efficient and sustainable manner. Other potential uses of the tool are also envisaged; for instance, PCE could be used as a mutual recognition tool or as an instrument to measure IPPC implementation at the global level.

6.22.2 PCE Rationale

The rationale for PCE is based on the considerations that the contemporary needs of a national phytosanitary system can be estimated by a measure of its capacity to meet the international obligations of the country, in an efficient and sustainable manner and this capacity can be measured through an inventory of the functions and resources that must be available to comply with each international standard for phytosanitary measures and with the major categories of activities of a phytosanitary system established under the IPPC (technical justification, surveillance, certification, etc.).

Consequently, PCE has been developed as a set of modules addressing the main categories of phytosanitary activities covered by the ISPMs and the IPPC. At present, those categories include pest risk analysis, pest surveillance, pest diagnostic capabilities, pest free areas, pest eradication, inspection systems, pest reporting, export certification, legislation, institutional issues and country background. However, PCE is not a static tool, and new modules will be added as required when new ISPMs are adopted. Each module comprises several sections: a questionnaire addressing the main functions and resources required for the activity covered in the module; a strengths and weaknesses identification section to be filled as a result of the responses given to the questionnaire; a section for the identification of the activities that must be performed in order to overcome the weaknesses identified in the previous section; and a logical frame matrix to consolidate the results of the previous sections. Thus, PCE goes beyond the needs assessment by including SWOT analysis (i.e., strengths, weaknesses, opportunities, and threats) and some elements of strategic planning in each module.

6.22.3 PCE Strategy

PCE is designed to enable a country to conduct a self-assessment of its phytosanitary system. When the capacity evaluation is performed with international cooperation, there is a shift in the role of international consultants from being experts to acting more as facilitators of the process. Instead, national counterparts now function in the role of experts, conducting an analysis of their own circumstances. Optimum results from performing the PCE have been obtained through provision of international assistance to facilitate the work of a national group of experts, and then the submission of their conclusions to a national team (composed of representatives of all the sectors and institutions related to the national phytosanitary system) for revision and validation within the context of a national plan.

As a technical assistance tool, PCE provides for a self-assessment of national weaknesses and the identification of priorities by a national group of experts, followed by validation at the national level by a larger group, representing all the actors of the national phytosanitary system. This helps in promoting national awareness and consensus, focusing on the attention of the national experts on the important issues and also facilitates analysis of the phytosanitary system in a systematic manner. This helps in reducing the variation among the judgments of experts facing the same situation, so that the assessment becomes more objective.

6.22.4 PCE Outcomes

The pooled results of PCE conducted with international assistance during the period 2000–2003 in 36 developing countries distributed in Africa, Asia, Europe, Latin America (Andean subregion) and the Caribbean Community (CARICOM) countries, were presented. The main limiting factors (priorities 1 to 5) identified in the 36 countries for each module have been grouped into three basic categories according to their characteristics: L-legislation and institutional issues; P-management (including documented procedures) and training; and E-infrastructure and equipment.

6.22.4.1 L1-Legislation

Shortcomings include general obsolescence of legislation in relation to the new revised text of IPPC and other relevant international agreements, country obligations and national plant protection organization functions not addressed in legislation, lack of legal powers to seize and search, no obligation for Customs to report to NPPO, lack of specific mandate to provide technical justification of phytosanitary measures, fragmentation of the authority (phytosanitary functions scattered among several organizations), lack of provisions for an emergency exotic pest response fund, lack of provisions for meaningful participation of stakeholders (private and official sectors), absence of provisions for accreditation mechanisms, overlapping of national and subnational authorities, lack of specification of the procedures for the adoption of phytosanitary requirements, and rather weak penalties and fees.

6.22.4.2 L2-institutional issues

Weaknesses include fragmentation of the authority (e.g., among ministries, inside the Ministry of Agriculture, and between the national and subnational governments), lack of cost recovery policies, salaries not sufficiently competitive, institutional organization of NPPO according to products and not by functions and programmes in relation to the main categories of phytosanitary activity, lack of regular training programmes, lack of a special unit in charge of international liaison, lack of trained managers in charge of national programmes, services understaffed with consequent multitasking (e.g., inspectors conduct pest risk analysis, perform pest diagnostic in laboratories and do inspections at the points of entry) and no differentiation between scientific and management staffing patterns.

6.22.4.3 *P-Documented procedures, management and training*

Deficiencies include lack of documented procedures and managers for pest surveillance, pest listing, pest risk analysis, export certification, inspection of pest diagnostic, exotic pest response, pest free areas, places and sites, lack of operational manuals, lack of internal audit systems, lack of computerized systems, deficiencies in budget allocation related to the lack of annual operational plans and proposed programme and budget for the main categories of phytosanitary activities, and lack of regular training programmes.

6.22.4.4 *E-Infrastructure and equipment*

Typical deficiencies are found to be inspection facilities at points of entry, laboratory facilities and equipment, incinerators and scanning equipment, library, computer systems and networks and glasshouses and screen houses.

6.22.5 PCE Conclusions

It is possible to visualize that the pooled PCE results from 36 countries demonstrate a common pattern where the legislative framework and institutional issues become the main limiting factor in most categories considered. The second limiting factor appears to be the weakness grouped under management, documented procedures, and training. Third, the limitations are those deficiencies related to infrastructure and equipment. This indicates that it would not be profitable to make investments in infrastructure and equipment before the elimination of the main limiting factors categorized as L and P. This conclusion is relevant for technical assistance programs, which traditionally are more concerned with investments in infrastructure and equipment than in legal and institutional issues. The nature of the main limiting factors (categories L and P) explains the intrinsic difficulty faced by many developing countries in fully implementing the IPPC and its international standards. It also highlights why traditional technical assistance approaches (mainly focused on category E) are not the best suited to resolve the problems of those countries.

An additional problem arises because of the lack of specific ISPMs (e.g., pest-specific, product-specific, pathway-specific standards) under the IPPC. (This is in contrast to the situation for standards on food safety under Codex Alimentarius or on animals and animal products under Office International des Épizooties.) Contracting parties to the IPPC nearly always need to justify their phytosanitary measures through pest risk analysis. The combined effect of the intrinsic difficulties in implementing generic ISPMs and the lack of specific ones seems to be provoking a 'domino effect' in certain developing countries. As an example, if is not possible to run a good surveillance programme then it is not possible to justify the list of pests present in the country. Consequently, it will not be possible to categorize the pests into regulated or unregulated categories. Thus, it will not be possible to conduct pest risk analyses and hence provide the technical justification for phytosanitary measures. The end result is a reduction in a country's ability to compete globally.

6.22.6 PCE Results in Relation to Invasive Alien Species

As discussed in previous chapters, under the Convention on Biological Diversity and its guiding principles, invasive alien species are "alien species that threaten ecosystems, habitats or species" (CBD, 2002). An invasive species has also been defined as, "an alien species which becomes established in natural or seminatural ecosystems or habitat, is an agent of change, and threatens native biological diversity" (IUCN, 2000).

Again as discussed previously, under the IPPC, a [plant] pest is "any species, strain or biotype of plant, animal or pathogenic agent injurious to plants or plant products." The scope of the IPPC includes wild flora; it also covers diseases, weeds and other species that may have indirect effects on plants. Inasmuch as an invasive alien species may be considered to be a plant pest, it falls within the scope of the IPPC and the corresponding standards and procedures. Plant quarantine pests are therefore a subset of invasive alien species. The CBD's guiding principles on invasive alien species are consistent with IPPC and its international standards for quarantine pests. The IPPC, therefore, is in a position to make an important contribution to the conservation of plant diversity.

PCE results and conclusions can be extrapolated to invasive alien species, with the reservation that the scientific, administrative and legislative framework for considering invasive alien species is largely more fragmented at both the national and international level than the plant protection framework. This fragmentation of the framework for invasive alien species is likely to magnify the problems already detected by PCE; the impact may be felt by many countries, but could constitute an additional disadvantage for developing countries. Phytosanitary Capacity Evaluation can be further improved to address the issues of invasive alien species in the context of a strategic cross-disciplinary tool. This would constitute not only a contribution to the national needs assessment but also a practical first step in fostering the required cooperation between international bodies with responsibilities in the matter.

It is clear that cell–cell communication, often referred to as quorum sensing (QS), is the norm in the prokaryotic kingdom, and this community-wide genetic regulatory mechanism has been adopted for the regulation of many important biological functions (He and Zhang, 2008). Since the 1980s, several types of QS signals have been identified, which are associated commonly with different types of QS mechanisms. Among them, the diffusible signal factor (DSF)-dependent QS system, originally discovered from bacterial pathogen *Xanthomonas campestris* pv. *Campestris,* is a relatively new regulatory mechanism. The rapid research progress over the last few years has identified the chemical structure of the QS signal DSF, established the DSF regulon, and unveiled the general signaling pathways and mechanisms. Particular noteworthy are that DSF biosynthesis is modulated by a novel posttranslational autoinduction mechanism involving protein–protein interaction between the DSF synthase RpfF and the sensor RpfC and that QS signal sensing is coupled to intracellular regulatory networks through a second messenger cyclic-di-GMP and a global regulator Clp. Genomic and genetic analyses show that the DSF QS-signaling pathway regulates diverse biological functions including virulence, biofilm dispersal, and ecological competence. Moreover, the evidence is emerging that the DSF QS system is conserved in a range of plant and human bacterial pathogens.

6.23 PHYTOSANITARY CERTIFICATE ISSUANCE AND TRACKING SYSTEM (PCIT)

Animal and Plant Health Inspection Service (APHIS) and Plant Protection and Quarantine (PPQ) are responsible for safeguarding agriculture and natural resources from the risks associated with the entry, establishment, or spread of animal and plant pests and noxious weeds. The PCIT system tracks the inspection of agricultural products and certifies compliance with plant health standards of importing countries. This capability provides APHIS/PPQ better security, reporting functions, and monitoring capabilities for exported commodities. The PCIT also provides a link to the PExD. PExD is a repository of phytosanitary import requirements of US origin commodities to foreign countries. PExD enhances Plant Protection and Quarantine's (PPQ's) ability to maintain the export program for United States exporters. PExD will support PPQ's Phytosanitary Issues Management (PIM) Export Services (ES) with the task of entering export summaries for foreign countries. PExD provides reusable text to facilitate uniform entry of export summaries. PExD also provides PPQ staff, State and County cooperators, and exporters' easy access to export summaries via direct user queries. PExD interfaces with the Phytosanitary Certificate Issuance and Tracking System (PCIT) application processing. During application processing, users are provided with pertinent export summaries based on the consignee country, applicable commodities, and other data contained in an application.

6.23.1 Phytosanitary Certificate

Certificate patterned after the model certificates of the IPPC [FAO, 1990]. In the U.S., this is an official document (PPQ Form 577) that attests to the phytosanitary condition of commodities and is issued by an Authorized Certification Official.

6.23.1.1 Phytosanitary certification

Phytosanitary certification is used to attest that consignments meet phytosanitary import requirements and is undertaken by an NPPO (National Plant Protection Organization). A phytosanitary certificate for export or re-export can be issued only by a public officer who is technically qualified and duly authorized by an NPPO (ISPM 12). A phytosanitary certificate for export is usually issued by the NPPO of the country where the plants, plant products or regulated articles were grown or processed. Phytosanitary certificates are issued to indicate that consignments of plants, plant products or other regulated articles meet specified phytosanitary import requirements and are in conformity with the certifying statement of the appropriate model certificate. Phytosanitary certificates should only be issued for this purpose.

6.23.2 USDA APHIS Plant Protection and Quarantine

The US designated NPPO – Plant Protection and Quarantine maintains the export program for the United States exporters of the United States and foreign-origin agricultural commodities. The export program does not require certification of any exports but does provide certification of commodities as a service to United States exporters. APHIS and PPQ are responsible for safeguarding agriculture and natural resources from the risks associated with the entry, establishment, or spread of animal and plant pests and noxious weeds. Phytosanitary certification is provided as a service to U.S. applicants based on the phytosanitary requirements of foreign countries. After assessing the phytosanitary condition of the commodities intended for export, an ACO (Authorized Certification Official) issues these internationally recognized phytosanitary certificates, viz., PPQ Form 577, Phytosanitary Certificate and PPQ Form 579, Phytosanitary Certificate for Re-export.

6.23.3 Authorized Certification Official (ACO)

A public officer who is authorized by the National Plant Protection Organization (NPPO) and accredited for the signing of phytosanitary certificates, who possesses the required education, experience and training and who has written confirmation of having successfully passed an approved examination (RSPM No. 8, Accreditation) [NAPPO, 2004].

6.23.4 Export Certification Specialists

The USDA APHIS PPQ Export Certification Specialists are responsible for maintaining the quality and credibility of Phytosanitary Exports Certification. Additional questions regarding the export program may be directed to the Certification Specialist in your State or the State of export.

6.23.5 PCIT-Phytosanitary Certificate Issuance and Tracking System

The USDA APHIS PPQ issues phytosanitary certificates for export via the PCIT Phytosanitary Certificate Issuance and Tracking System. This is a program that tracks the inspection of agricultural products and certifies compliance with statutory and regulatory plant health standards of importing countries. This capability provides

APHIS/PPQ better security, reporting functions, and monitoring capabilities for exported commodities.

6.24 INTERNATIONAL PLANT PROTECTION CONVENTION (IPPC)

The International Plant Protection Convention is a 1951 multilateral treaty overseen by the FAO that aims to secure coordinated, effective action to prevent and to control the introduction and spread of pests of plants and plant products. The convention extends beyond the protection of cultivated plants to the protection of natural flora and plant products. It also takes into consideration both direct and indirect damage by pests, so it includes weeds. The Convention created a governing body consisting of each party, known as the Commission on Phytosanitary Measures, which oversees the implementation of the convention. As of December 2013, the convention has 181 parties, which includes 178 United Nations member states, the Cook Islands, Niue and the European Union. The convention is recognized by the World Trade Organization's (WTO) Agreement on the Application of Sanitary and Phytosanitary Measures (the SPS Agreement) as the only international standard setting body for plant health. While the IPPC's primary focus is on plants and plant products moving in international trade, the convention also covers research materials, biological control organisms, germplasm banks, containment facilities, food aid, emergency aid and anything else that can act as a vector for the spread of plant pests, for example, containers, packaging materials, soil, vehicles, vessels, and machinery.

The IPPC was created by member countries of the FAO of the United Nations. The IPPC places emphasis on three core areas: international standard setting, information exchange and capacity development for the implementation of the IPPC and associated international phytosanitary standards. The Secretariat of the IPPC is housed at FAO headquarters in Rome, Italy and is responsible for the coordination of core activities under the IPPC work program. In recent years the Commission of Phytosanitary Measures of the IPPC has developed a strategic framework with the objectives of protecting sustainable agriculture and enhancing global food security through the prevention of pest spread, protecting the environment, forests and biodiversity from plant pests, facilitating economic and trade development through the promotion of harmonized scientifically based phytosanitary measures, and developing phytosanitary capacity for members to accomplish the preceding three objectives.

By focusing the convention's efforts on these objectives, the Commission on Phytosanitary Measures of the IPPC intends to protect farmers from economically devastating pest and disease outbreaks, protect the environment from the loss of species diversity, protect ecosystems from the loss of viability and function as a result of pest invasions, protect industries and consumers from the costs of pest control or eradication, facilitate trade through international standards that regulate the safe movements of plants and plant products, and protect livelihoods and food security by preventing the entry and spread of new pests of plants into a country.

6.25 NORTH AMERICAN PLANT PROTECTION ORGANIZATION (NAPPO)

The North American Plant Protection Organization (NAPPO) is the phytosanitary standard setting organization recognized by the North American free Trade Agreement (NAFTA). It was created in 1976 as a regional organization of the International Plant Protection Convention (IPPC) of the FAO of the United States. NAPPO has a global and a regional mandate. The global mandate comes from Article IX of the New Revised Text (1997) of the International Plant Protection Convention (IPPC) of the Food and Agriculture Organization (FAO) of the United Nations. The main activity under this mandate is to cooperate with the IPPC Secretary in achieving the objectives of the Convention and, where appropriate, cooperating with the Secretary and the Commission in developing international standards.

The goal of the IPPC is to protect the world's cultivated and natural plant resources from the spread and introduction of plant pests while minimizing interference with the international movement of goods and people. The regional mandate for NAPPO was formalized by Canada, the United States and Mexico in a Cooperative Agreement signed in 1976 at the Minister/Secretary of Agriculture level. The NAPPO Constitution and Bylaws confirm that NAPPO is accountable to the Minister/Secretary of Agriculture in NAPPO member countries. NAPPO's mission is to provide a forum for public and private sectors in Canada, the United States, and Mexico to collaborate in the development of science-based standards intended to protect agricultural, forest and other plant resources against regulated plant pests while facilitating trade and to participate in related international cooperative efforts.

6.25.1 Strategic Goals

To accomplish this mission, various strategic goals being established include protecting plant resources and the environment, capacity building, communicating results, building partnerships, offering an effective dispute settlement mechanism, working with sound management practices and working with a stable funding base.

6.25.1.1 Responsibilities

NAPPO has responsibilities at the regional, hemispheric and global level.

6.25.1.2 Regional

Developing regional phytosanitary standards which facilitate the safe movement of plants, plant products and other regulated articles into and within the NAPPO region is a key responsibility. Once a subject has been identified, experts representing the member countries meet to develop a draft standard which is submitted to a wider country consultation before final approval. Industry participation in the entire process is encouraged. Examples of standards developed so far include standards for commodities such as potatoes, stone and pome fruit trees, grapevines, citrus and plants for planting; standards for the regulation of biocontrol agents, products of biotechnology and non-*Apis* pollinating insects; standards for other regulated articles such as wood packing materials and other generic standards such as accreditation of officials to issue phytosanitary certificates. In addition, NAPPO provides technical support to the Sanitary and Phytosanitary Committee of the North American Free Trade Agreement.

6.25.1.3 *Hemispheric*

NAPPO is a member of the Inter American Coordinating Group in Plant Protection which brings together the regional plant protection organizations of the Americas to coordinate their efforts in protecting the hemisphere from the entry, establishment, and spread of regulated pests. This group provides the opportunity to discuss emerging pest situations, to coordinate our control efforts and to provide technical assistance. NAPPO also works with other hemispheric organizations such as OIRSA and COSAVE on topics of mutual interest.

6.25.1.4 *Global*

On a global level NAPPO supports the Secretariat of the International Plant Protection Convention by assisting in the development of international standards for phytosanitary measures and monitoring their application in the NAPPO region. It works with other Regional Plant Protection Organizations by exchanging information on current phytosanitary issues at their Annual Technical Consultation and to support the work program of the Commission on Phytosanitary Measures of the IPPC. The National Plant Quarantine Station at Rangpuri, New Delhi and four Regional Plant Quarantine Stations at Amritsar, Chennai, Kolkata and Mumbai in India are the major stations and are located at places having international airport/seaport/land frontier with neighboring countries. National Bureau of Plant Genetic Resources, New Delhi, which is the nodal institution for the exchange of plant genetic resources has been empowered under PQ order to handle quarantine processing of germplasm and transgenic planting material being imported for research purposes in the country. As per the recent amendments made under the PQ order, the Advanced Centre for Plant Virology at JARI, New Delhi, Indian Institute of Horticultural Research, Bangalore and Institute of Himalayan Bioresource Technology, Palampur have been identified for ensuring virus-free status in the imported *in vitro* material. Under the DIP Act, there is a provision of Domestic Quarantine to restrict the interstate movement of nine invasive pests, viz., flute scale, San Jose scale, coffee berry borer, codling moth, banana bunchy top, and mosaic viruses, potato cyst nematode, potato wart and apple scab. Keeping in view the biosafety concerns associated with the growing of imported transgenic material, a containment Facility of Level-4 has been established at NBPGR, New Delhi for their quarantine processing (Anonymous, 1990).

6.26 PLANT QUARANTINE FACILITIES IN INDIA

Plant Quarantine regulatory measures are operative through the 'Destructive Insects and Pests Act, 1914 (Act 2 of 1914)' in the country. The purpose and intent of this Act are to prevent the introduction of any insect, fungus or other pests, which are or may be destructive to crops. The import of agricultural commodities is presently regulated through the Plant Quarantine (Regulation of Import into India) Order, 2003 issued under DIP Act, 1914 incorporating the provisions of New Policy on Seed Development, 1988. Further, the significance of Plant Quarantine has increased in view of globalization and liberalization in international trade of plants and plant material in the wake of Sanitary and Phytosanitary (SPS) Agreement under WTO. The phytosanitary certification of agricultural commodities being exported is also undertaken through the scheme as per International Plant Protection Convention (IPPC), 1951.

The primary objectives are to prevent the introduction and spread of exotic pests that are destructive to crops by regulating/restricting the import of plants/plant products and to facilitate safe global trade in agriculture by assisting the producers and exporters by providing a technically competent and reliable phytosanitary certificate system to meet the requirements of trading partners.

The major activities include inspection of imported agricultural commodities for preventing the introduction of exotic pests and diseases inimical to Indian fauna and flora, inspection of agricultural commodities meant for export as per the requirements of importing countries under International Plant Protection Convention (IPPC), detection of exotic pests and diseases already introduced for containing/controlling them by adopting domestic quarantine regulations, undertaking Postentry Quarantine Inspection in respect of identified planting materials and conducting the Pest Risk Analysis (PRA) to finalize phytosanitary requirements for import of plant/plant material.

There are 35 plant quarantine stations at different airports, aeaports and land frontiers implementing the plant quarantine regulations. The NPQS, New Delhi, and RPQSs at Chennai, Kolkata, Amritsar and Mumbai have been strengthened with modern equipment for plant quarantine testing, etc., to facilitate speedy clearance of imports and exports under the FAO-UNDP Project. The Sanitary and Phytosanitary Agreement of WTO envisages application of Phytosanitary measures based on scientific justifications, therefore, it is imperative to conduct all Plant Quarantine inspections as per the International Standards/guidelines (Anonymous, 1990). Accordingly, the National Standards for Phytosanitary Measures for some of the important activities have already been developed and adopted including the Guidelines for Development of National Standards for Phytosanitary Measures, and six draft National Standards are under the process of approval. The Standards which are critical for our exports have been prioritized. Further, a National Integrated Fruit Fly Surveillance Programme has also been prepared with a view to establish pest-free areas against fruit flies. To streamline the Plant Quarantine activities, efforts are being made to computerize the Plant Quarantine Stations for speedy and transparent functioning.

Substantial efforts have also been made towards the international cooperation on the phytosanitary matters including participation in IPPC Standard Committee meetings for development of International Standards for Phytosanitary Measures, Training of fumigation service providers and regulators in methyl bromide fumigation under the Australian Fumigation Accreditation Scheme, finalization of SPS Protocol for export of mango to China, harmonization of phytosanitary measures with reference to Seed trade through the Workshops organized by Asia pacific Seed Association (APSA), negotiations on free trade agreement with Thailand and Singapore, Memorandum of Understanding and work plan with Chile for cooperation on Phytosanitary matters and Indo-Nepal Joint trade and economic development negotiations.

6.27 BIOSECURITY IMPLICATIONS

Despite these limitations, making the best systematic predictions of impact and risk possible, based on as much data as can be obtained is preferable to the alternative of empirical measurements of actual damage following real introductions of potentially damaging species. Where real introductions of exotic nematodes have occurred and been measured, the impacts have mostly been substantial. Of course, by this stage, eradication is seldom an option (Hodda *et al.*, 2008). More recently nematode species have been used as bioindicators during quarantine inspections. In the UK, *Helicotylenchus dihystera* is often intercepted with planting materials supposedly grown in sterile conditions, indicating that phytosanitary standards were not met (Hockland and Anderson, 2012). Other microscopic plant pathogens such as fungi, bacteria and viruses may also use similar pathways to nematodes (Grousset *et al.*, 2012), so targeting the generally larger nematodes during quarantine inspections has assisted in reducing the risks from other microscopic quarantine organisms generally. Phytosanitary risks are specific to countries or regions. Regulated organisms differ between countries depending on species distributions and regulatory or biosecurity policy in different countries. Formal pest risk analysis processes.

6.28 NEMATODES OF QUARANTINE IMPORTANCE

Phytonematodes have ever been an unseen and often underestimated menace in crops around the Globe. Besides causing damage due to their own feeding, nematodes also incite or aggravate fungal and bacterial diseases and serve as vectors of many plant viruses. The maxim "prevention is better than cure" is particularly true with nematodes. Nematodes that establish once in a given area are almost impossible to eliminate. Substantial crop losses from phytonematodes could be much greater if species currently causing localized damage became more widespread. Many phytonematodes have low impact in their native range, but much greater impact when introduced to new areas. For example, Pine Wood Nematode, *Bursaphelenchus xylophilus* in Japan, China, Portugal and most recently in Spain (Robertson *et al.*, 2011) and the Potato Cyst Nematodes in Europe, the USA, Canada and Australia (Sun *et al.*, 2007). Quarantine measures against known damaging nematodes are effective in preventing their spread, thus effectively and economically preventing crop losses. The soil habitat that elaborates survival mechanisms of nematodes protect them and enable their quick establishment in a new locality. The low environmental resistance in a new locality also allows their quick growth.

Quarantine refers to regulatory actions aimed at preventing or retarding the introduction, establishment and spread of dangerous pests in crop protection (Mass, 1987). In quarantine and certification programs, intensive sampling may be needed to determine if lots or shipments of plants, pots, cuttings, or other units are free of plant-parasitic nematodes and other plant pests. Because it is seldom feasible, or even possible, to examine and test entire lots for such harmful pests, these determinations must nearly always be made on the basis of samples drawn from the lots.

A quarantine nematode refers to "a nematode species that can cause a pest of potential economic importance to the area, which can be one specific country or a whole continent, endangered thereby and not yet present there, or present but not widely distributed and being officially controlled," as given by the International Plant Protection Convention (IPPC). Opportunities for the introduction of new species of nematodes have been numerous, but its introduction does not necessarily mean its establishment as a pest. Often a complicated set of conditions and events must be present before an organism can survive in a new environment. Frequent introductions, particularly of large numbers of the pest organism, increase that organism's chance of becoming established in a new area. Once a nematode is introduced and established, it may be years before numbers sufficient to cause severe crop injury are produced. Ordinary plant quarantine regulatory action is not well adapted to the detection of plant parasitic nematodes.

The potential phytosanitary importance of all named plant-parasitic nematode species is determined by evaluating available information on species characteristics, association with economically-important crop hosts, and ability to act as vectors of viruses or form disease complexes with other pathogens. Most named species of phytonematodes are poorly known, recorded from a single location only, not associated with economically important crops, and not known to be associated with other plant disease organisms.

Phytonematodes during quarantine inspections may also act as a bioindicator for consignments that do not meet the phytosanitary requirements of plants being grown in sterile environments and could be carrying other plant pathogens and microorganisms (Hockland and Anderson, 2012). Quarantine measures can also prevent the spread of nontarget species which are potentially invasive (Schrader and Unger, 2003).

Quarantine and other phytosanitary measures are particularly important for PPN because other management methods such as chemical control or crop resistance can be far more costly and difficult to implement without other adverse effects (Nicol *et al.*, 2011).

The first step in implementing quarantine measures is to determine which species should be regulated under international trade rules. The international plant pest convention defines a quarantine pest as 'a pest of potential economic importance to the area endangered thereby and not yet present there, or present but not widely distributed and being officially controlled. Organisms which meet this definition can be regulated. Countries determine their lists of regulated pests according to guidelines set by the International Plant Protection Convention (IPPC) and International Standards for Phytosanitary Measures (ISPM). Countries may also establish lists of regulated nonquarantine pests. The IPPC defines these as follows: 'A nonquarantine pest whose presence in plants for planting affects the intended use of those plants with an economically unacceptable impact and which is therefore regulated within the territory of the importing contracting party.' All lists of regulated pests are dynamic, with species added or removed from lists as phytosanitary risks change or as new species emerge as pests.

Species other than those on official lists of quarantine pests may also pose a biosecurity risk. Only a small proportion of nematode species have been described (Blaxter, 2011; Hodda, 2011). New species are being described regularly, and species new to science have been intercepted during phytosanitary inspections [e.g., *Radopholus bridgei* and *Meloidogyne thailandica,* both described from material intercepted in quarantine (Handoo *et al.*, 2005). A list of major phytonematodes of quarantine importance in horticultural crops has been mentioned in Tables 6.2 and 6.3.

Table 6.2 Nematodes of Quarantine Importance Specific to Crop and Country

Nematode species	*Crop*	*Country*
Globodera rostochiensis *Ditylenchus destructor* *Meloidogyne* spp. *Pratylenchus* spp.	Potato	India
Radopholus similis *Helicotylenchus multicinctus*	Banana	India
Radinaphelenchus cocophilus	Palms	Japan, East Asian Nations
Heterodera schachtii	Sugar beet	European countries
Heterodera geottingiana	Vegetables	Europe
H. humuli	Hops	Europe
H. cacti	Cacti	Europe
Aphelenchoides fragariae	Narcissus	Europe
Pratylenchus vulnus	Temperate fruits	Europe and the USA

Table 6.3 A General List of Quarantine Nematodes

Nematode species
Meloidogyne chitwoodi, Meloidogyne fallax
Globodera rostochiensis, Globodera pallida, Heterodera glycines
Hirschmanniella spp. (except *H. gracilis* and *H. loofi*), *Nacobbus aberrans, Radopholus similis*
Ditylenchus dipsaci, Ditylenchus destructor

Contd...

Contd...

Aphelenchoides besseyi, Bursaphelenchus xylophilus
Longidorus diadecturus, Xiphinema americanum, Xiphinema bricolens, Xiphinema californicum, Xiphinema rivesi
Heterodera schachtii, Cactodera cacti, Radopholus citrophilus
Aphelenchoides fragariae, A. ritzemabosi, Heterodera humuli, Heterodera goettingiana
Rhadinaphelenchus cocophilus, Aphelenchoides arachidis

Yield loss caused by phytonematodes is often used to determine economic importance, but there are limitations in its calculation. Yield loss calculations from different studies and countries do not necessarily use the same methods, with some reporting damage as percentage yield loss and others reporting as tons per hectare or as percentage yield gained after application of nematicides or as correlations of yield gains with declining nematode abundance. Indirect losses also generally remain unaccounted for in most yield loss figures. Yield loss estimates based on work done long ago and under different nematode management regimes are also likely to require updating. Impacts can also be difficult to estimate or severely underestimated because nematodes may injure plants in many different ways. One way is by direct feeding action, but this may take many different forms including direct damage, root galls, root stunting or withdrawal of resources from other parts of the plant (Norton and Niblack, 1991). Another aspect to consider when assessing phytosanitary status is intraspecific variation. This is often not considered because most phytosanitary regulation is on the taxonomic level of the species. However, species from the economically important genera *Belonolaimus, Ditylenchus, Globodera, Heterodera, Meloidogyne, Nacobbus, Radopholus, Rotylenchulus* and *Tylenchulus,* all have pathotypes or races with distinctive host responses and differences in host range.

6.28.1 Phyonematodes Under Quarantine Act

Most quarantine action utilizes visual inspection-supplemented by a hand lens or low power microscope to detect parasitic organisms themselves or to check symptoms of infection. Symptoms of nematode injury usually are not diagnostic, because they are largely such plant reactions as poor growth, chlorosis and other symptoms equally indicative of root injury from many other causes. Specific symptoms also may vary with the reaction of different plants. Large obvious lesions are formed on the roots of certain plants attacked by root-lesion nematodes while such lesions are not found on the roots of many other kinds of plants attacked by the same nematode, although serious injury may be caused on both. Nematodes also may enter and be carried in roots of plants which are not preferred hosts. These nematodes later may prove to be of importance on other crops in the area or to subsequent crops on the same land. Nursery stock is a common means of disseminating plant parasitic nematodes, either as parasites of nursery plants involved or as contaminants in or on these plants. Regulations established by Department of Horticulture require that locations, where nursery stock is grown or sold, be inspected for phytonematodes.

Fortunately, the active dispersal of nematodes is rather limited but this is more than compensated by the ease of passive dispersal with the movement of infested soil, plant material and water. Man has aided in the indiscriminate spread of nematode while carrying the plant materials and other items infested with nematodes for research, trade, charity or other purposes. With the recent trade liberalization, the bulk movement of commodities has become common and consequently the risk of entry of pests and diseases has increased many folds. Several devastating phytonematodes have entered into new locations and established successfully on major horticultural crops, leading to losses.

Nematodes under the following families have been included under quarantine act. Families, viz., Anguinidae, Aphelenchoididae, Atylenchidae, Belonolaimidae, Caloosiidae, Criconematidae, Dolichodoridae, Ecphyadophoridae, Hemicycliophoridae, Heteroderidae, Hoplolaimidae, Longidoridae, Meloidogynidae, Neotylechidae, Parasitaphelenchidae, Paratylenchidae, Pratylenchidae, Psilenchidae, Rotylenchulidae, Sphaeronematidae, Telotylenchidae, Trichodoridae, Tylenchidae, Tylenchulidae and Tylodoridae. Major phytonematodes include *Dityienchus angustus, Globodera pallida, Globodera rostochiensis, Heterodera goettingiana, Heterodera schachtii, Heterodera zeae, Meloidogyne chitwoodi, Meloidogyne fallax, Nacobbus aberrans, Radopholus citrophilus, Radopholus similis and Xiphinema index.* However, of all the phytonematodes, in most countries, quarantine measures have been strictly followed in potato cyst nematodes.

6.28.1.1 Potato cyst nematodes

Golden cyst nematode of potato (*Globodera rostochiensis*) and Pale cyst nematode (*Globodera pallida*) are the major cyst nematodes infesting potato.

6.28.1.2 Host range

Potatoes are by far the most important host crop. Tomatoes and aubergines are also attacked. Other *Solanum* spp. and their hybrids can also act as hosts. Both species of *Globodera* have several different pathotypes (Kort, 1974). The pathotypes are characterized by their ability to multiply on certain tuberous *Solanum* clones and hybrids used in breeding. Five pathotypes are recognized within *G. rostochiensis* (Ro1-Ro5 international notation) and three in *G. pallida* (Pa1-Pa3) (Kort *et al.*, 1977). Some of these pathotypes are recognized by their almost total inability to multiply on specific cultivars of potato (single-gene resistance); for example, the most commonly grown resistant potato cultivars (based on gene H1 derived from clones of *S. tuberosum* subsp. *andigena*) are resistant to pathotype Ro1 of *G. rostochiensis* only. Other pathotypes show different levels of ability to multiply on different cultivars; the testing of this form of resistance is discussed by Mugniéry *et al.* (1989). The internationally recognized system of classification of pathotypes refers mainly to the pathotypes present in Europe (and spread from there) and may not have relevance to South America (Kort *et al.*, 1977). It is probable that pathotypes exist there which were never transferred from the Andean region (Canto-Saenz and Mayer de Scurrah, 1978).

The golden nematode is one of the world's most damaging potato pests. First detected in the United States in 1941 in Nassau County on Long Island, New York, it was subsequently found in eight other New York counties. For over 60 years, an effective Federal and State quarantine program has confined the pest to nine counties in New York. If golden nematode were to become more widely established in U.S. potato, tomato, and eggplant production areas, annual crop losses could reach $4.8 billion. The golden nematode is primarily a pest of potatoes and is also referred to as a species of potato cyst nematode. In the larval stage, it bores into the roots of host plants and feeds on their juices. Because the above-ground damage is not visible during the early stages of infestation, the pest can remain undetected for years. The first sign of a golden nematode infestation is poor plant growth in one or more areas of a potato field. As an infestation builds, the damaged area increases and eventually the entire field displays poor plant growth. Large numbers of the nematodes cause wilting, stunted growth, poor root development, and early plant death. Golden nematodes can also reproduce on the roots of eggplant, tomatoes, and several solanaceous weeds.

The pale cyst nematode, *Globodera pallida*, is a major pest of potato crops in cool-temperate areas. It primarily affects plants within the potato family including tomatoes, eggplants, and some weeds. If left uncontrolled, pale cyst nematodes can cause up to 80 percent yield loss in potato fields. The pale cyst nematode is widely distributed in many potato-growing regions throughout the world. In North America, besides the current find in Idaho, the nematode is also known to be present on the island of Newfoundland, Canada. Pale cyst nematode infestations may be associated with patches of poor growth in potato fields. At high nematode populations, affected potato plants may exhibit yellowing, wilting or death of foliage – none of which has been observed in Idaho potato fields.

6.28.1.3 *Diagnosis*

When suspected nematode introductions are found in diagnostic samples, their identity should be confirmed, voucher specimen should be deposited in an appropriate reference collection, and quarantine authorities should be notified. In addition to detecting newly introduced nematodes, diagnostic services may become involved in quarantine issues through their role in assessing samples of produce designated for the overseas export market. As international trade increases, there will be an increasing demand from overseas quarantine authorities to ensure that agricultural products (Ex. Tubers, rhizomes, rooted plants) are free of specific nematodes. If protocols for collecting and processing samples are set by the importing country, these must be followed by the diagnostic service. In the absence of specific protocols, the level of risk should be assessed, and appropriate assessment procedures need to be developed.

6.28.1.4 *Detection and inspection methods*

Ordinary plant quarantine regulatory action is not well adapted for the detection of plant parasitic nematodes. Most quarantine action utilizes visual inspection-supplemented by a hand lens or low power microscope to detect parasitic organisms themselves or to check symptoms of infection. Symptoms of nematode injury usually are not diagnostic, because they are largely such plant reactions as poor growth, chlorosis and other symptoms equally indicative of root injury from many other causes. Specific symptoms also may vary with the reaction of different plants. Large, obvious lesions are formed on the roots of certain plants attacked by root-lesion nematodes while such lesions are not found on the roots of many other kinds of plants attacked by the same nematode, although serious injury may be caused on both.

The symptoms described for *Globodera* spp. can have many other causes and cannot be taken as proof of the presence of nematodes. For positive detection, it is necessary to find cysts in soil samples or females or cysts on host roots. Mature females and cysts are just visible to the naked eye and can be seen as minute white or yellow globes on the root surface. Specific identification is just possible by observation of the female color at the appropriate stage of development, either a change from white to yellow in *G. rostochiensis* or prolonged white (slightly cream but not yellow phase) in *G. pallida*. Several methods are available for extracting juveniles or larvae from the soil (Southey, 1986; OEPP/EPPO, 1991) and thereafter specialist microscopic examination of juveniles, females or cysts are necessary for precise identification. The species can also be distinguished by morphological characters and measurements of 2nd-stage juveniles and cysts. The use of biochemical techniques such as DNA probes is being investigated as morphological identification can be difficult.

6.28.1.5 Means of movement and dispersal

These nematodes have no natural means of dispersal, and can only move the short distances traveled by juveniles attracted towards roots in the soil. They are spread into new areas as cysts on, in order of importance, seed potatoes, nursery stock, soil, flower bulbs, and potatoes for consumption or processing. The last named are only important if there is a risk of their being planted or if care is not taken with the disposal of waste soil.

6.28.1.6 Phytosanitary risk

Both species of potato cyst nematode are A2 quarantine pests for EPPO (OEPP/EPPO, 1978; 1981). They are also of quarantine significance for APPPC and NAPPO. In addition, *G. rostochiensis* is a quarantine pest for CPPC and IAPSC. The nematodes are already established in most or all areas in the EPPO region that are important for the cultivation of potatoes for consumption or the production of starch; therefore, regular attention to control is needed in such areas. Where domestic legislative measures are in force, import regulations are justified to ensure comparable standards for imported material. It is essential that areas of seed potato production be kept as free as possible from these nematodes. *G. pallida* is generally less common than *G. rostochiensis* in most of the EPPO region (with the exception of the southern part of the UK) and is absent from some countries; it, therefore, merits greater attention from the phytosanitary standpoint. At some time in the future, it may be worthwhile to consider the individual pathotypes as being the quarantine organisms, rather than the two species themselves. It is obvious that some pathotypes are very widely distributed, and furthermore some are more important economically than others. Unfortunately, information is still limited on the precise distribution of the pathotypes.

6.28.1.7 *Phytosanitary measures*

Measures to prevent the introduction of the nematodes to areas where they are not already established include soil sampling surveys and regulations concerning the movement of *G. rostochiensis* and *G. pallida* seed potatoes, nursery stock, flower bulbs, and soil. These apply nationally as well as internationally (CEC, 1969). Consignments of potato tubers, rooted plants, and bulbs from countries where the nematodes occur may be examined to check amounts of adhering soil, if any, or to take samples of soil for laboratory examination. Additional safeguards during transit of consignments could be the washing of tubers and flower bulbs to remove soil, although it should be noted that cysts can remain embedded in tubers, especially in the eyes. Alternatively, tubers may be dipped in dilute sodium hypochlorite solution (Wood and Foot, 1975). The EPPO specific quarantine requirements (OEPP/EPPO, 1990) for these nematodes require that the field in which seed potatoes or rooted plants being imported were grown was inspected by taking soil samples according to an EPPO-recommended method (OEPP/EPPO, 1991) and found free of viable cysts of both species. The sampling must have been performed after harvest and after removal of the previous potato crop.

6.28.1.8 *Best management schedule*

Once established, these nematodes are nearly impossible to eliminate as cysts can lie dormant in the soil without a host crop for decades. A new generation of PCN is produced each time a host crop is grown. However, there are best management practices to help prevent infection and reduce the nematode population over time.

(a) ***Minimize soil movement:*** Avoid sharing farm machinery, equipment, tools and containers, do not spread tare dirt or debris onto agricultural land or areas where it could be spread to another agricultural land. Tare dirt or debris is the loose soil knocked off either during potato processing operations or storage filling/emptying operations, never use bags, containers, etc. more than once for potato transport unless they are free of soil. Be sure all commercial transport vehicles are free of soil.

(b) ***Keep farm equipment and vehicles clean:*** Ensure equipment is available to conduct proper cleaning and disinfection of farm machinery and vehicles; clean and disinfect all used equipment before using on a farm; clean and disinfect all machinery, vehicles and other equipment before going between fields. This includes those of temporary help, custom applicators, and utility companies.

(c) ***Healthy seed:*** Plant certified seed potatoes produced on land determined not to be infested with the nematode. Avoid continuous planting of potatoes in the same field(s). A long rotation cycle is critical for managing and preventing PCN. During interim years, plant nonhost crops. Contain water and soil during tuber washing to avoid contaminating farmland. Processing facilities should not return soil, water, cull tubers and debris to agricultural land. This includes managing and limiting waste used as livestock feed. Segregate potatoes in the storage - each field should be stored separately.

(d) ***Best land practices:*** Plant cover crops when fields are not in use to prevent wind and water from moving soil. Keep hedgerows, sod barriers or sod strips between fields and along highways to provide a physical barrier to soil movement. Do not use common headlands, farm roads and public roads as turning areas.

(e) ***Long-term management:*** Successful management or eradication of cyst nematodes is complex and takes many years. It is important to develop appropriate long-term strategies to deal with the pest. These strategies will help protect the local, provincial and national potato industries as well as other affected industries while being sensitive to the situation of positive growers. With appropriate management, it will be possible to grow most nonhost crops on affected fields and it may, over several years, become possible to grow potatoes in an appropriate rotation.

(f) ***Regulatory measures:*** In the event of nematode cyst detection, immediate regulatory measures are to be taken to contain potential sources of spread. Surveys and investigations to trace the infestation back to an infested field and to trace the infestation forward are also conducted to prevent any further spread. No potatoes could be grown on land known to be infested through soil sampling. Movement of farm machinery and construction equipment out of the regulated area was permitted only after steam cleaning and fumigation. No soil nor sod could be moved out of the restricted area. Potato grading stations could be operated only under permit, and the tare soil had to be treated. Potatoes grown on Long Island could not be used for seed; potatoes shipped for table stock must be washed. Potatoes must be packaged in non-reusable paper bags.

6.28.2 Stem and Bulb Nematode, *Ditylenchus dipsaci*

D. dipsaci is known to attack over 450 different plant species, including many weeds. However, it occurs in more than ten biological 'races' some of which have a limited host range. The race(s) that breed on rye, oats and onions seem to be polyphagous and can also infest several other crops, whereas those breeding on lucerne, *Trifolium pratense* and strawberries are virtually specific for their named hosts and seem to have relatively few alternative host plants. The tulip race will also infest *Narcissus* whereas another race commonly found in *Narcissus* does not breed on tulip. It is known that some of the races can interbreed and that their progeny has different host preferences (Eriksson, 1974). The principal hosts are faba beans, garlic, *Hyacinthus orientalis*, leeks, lucerne, maize, *Narcissus pseudonarcissus*, oats, onions, peas, *Phlox drummondii*, *P. paniculata*, potatoes, rye, strawberries, *sugar beet*, tobacco, *Trifolium pratense*, *T. repens*, tulips. It has also been reported on carnations, celery, *Hydrangea*, lentils, rape, parsley, sunflowers, and wheat.

6.28.2.1 *Detection and identification*

Major symptoms, in general, include swellings and distortion of aerial plant parts and necrosis or rotting of stem bases, bulbs, tubers, and rhizomes.

6.28.2.1.1 On Allium spp. (onions, garlic, leeks, etc.)

Penetration of onion leaves by *D. dipsaci* causes leaf deformation and leaf swellings or blister-like areas on the surface. The leaves grow in a disorderly fashion, often hang as if wilted and become chlorotic. Young plants can be killed by high infestations. The inner scales of the bulb are usually more severely attacked than the outer scales. As the season advances the bulbs become soft and when cut open show browning of the scales in concentric circles. Conversely, *D. dipsaci* on garlic does not induce deformation or swellings but causes leaf yellowing and death (Netscher and Sikora, 1990).

6.28.2.1.2 On lucerne

The crop declines in patches in the field and damage is more serious in humid climates. The whole plant becomes desiccated and presents symptoms of stunting and swelling at the base of the stem with conspicuously shortened internodes. With heavy infestation, plants can be killed.

6.28.2.1.3 On tobacco

Invasion by the nematode of the lower part of the stem causes stunting and deformation of the plant followed by 'stem break.'

6.28.2.1.4 On faba beans

D. dipsaci causes swelling and deformation of stem tissue or lesions which turn reddish-brown then black, depending on cultivar and environmental factors. Newly formed pods take on a dark-brown appearance. The lesions envelop the stem and increase in length, often advancing to the edge of an internode. Leaf and petiole necrosis is also common under heavy infestations, but can be confused with symptoms induced by fungal leaf pathogens. Infected seeds are darker, distorted, smaller in size and may have speckle-like spots on the surface. Heavy infestations often kill the main shoots, stimulating secondary tiller formation. The more severe symptoms are usually induced by the 'giant race' on faba beans (Sikora and Greco, 1990).

6.28.2.2 *Detection and inspection methods*

D. dipsaci can be isolated from samples of suspected seed material (according to symptoms) by dissection in water at 20 times magnification. Nematodes leave the dissected tissue and swim actively in the water. Microscopic examination at 800 times magnification is necessary for correct identification of the nematode species.

6.28.2.3 *Means of movement and dispersal*

In international trade *D. dipsaci* is liable to be carried on dry seeds and planting material of host plants. In the field, the fourth-stage juvenile can withstand desiccation for many years, and although soil densities seem to decrease rapidly, the nematode can survive for years without a host plant. Nematode survival and damage are greater in heavy soils as compared to sandy soils. It can also survive on a number of weeds. Irrigation water and cultivation by contaminated farm tools and machinery are other sources of inoculum dissemination. Nematode-free (certified) seeds and planting material are most essential to prevent crop damage by *D. dipsaci*. Hot-water treatments with different temperature-time combinations, depending on type and state of seed material, are effective against nematodes.

6.28.2.4 *Phytosanitary risk*

At present, the distribution of the different races throughout the region is patchy, and some countries apply official control measures to limit spread. Other countries regard the pest as being a quality pest which can be effectively controlled by production and use of healthy planting material. It is certainly true that, without control, *D. dipsaci* may cause complete failure of host crops within the EPPO region. EPPO lists it as an A2 quarantine pest, and CPPC, IAPSC, and NAPPO also consider it to be of quarantine significance.

6.28.2.5 *Phytosanitary measures*

The implementation of certification schemes for the production of host plants of *D. dipsaci* can provide planting material free from the pest. Imports of soil and plants for planting and seeds of host plants from countries where this nematode occurs should be restricted.

6.28.3 Potato Rot Nematode, *Ditylenchus destructor*

6.28.3.1 *Symptoms*

There are, in general, no obvious symptoms in the aerial parts of the plant, although heavily infested potato tubers give rise to weak plants which usually die. Early infections can be detected by peeling the tuber which can reveal small, off-white spots in the otherwise healthy flesh. These later enlarge, darken, are wooly in texture and may be slightly hollow at the center.

On badly affected tubers there are typically slightly sunken areas with cracked and wrinkled skin which is detached in places from the underlying flesh. The flesh has a dry and mealy appearance, varying in color from grayish to dark brown or black. This discoloration is largely due to secondary invasion by fungi, bacteria and free-living nematodes (the latter are easily confused with *D. destructor*).

In contrast, the skin of potatoes infested with *D. dipsaci* is not usually cracked, and the rot darkens towards the inside of the tuber. The symptoms are more obvious in the foliage, which is shortened and malformed.

6.28.3.1.1 On Iris and tulips

Infestations usually begin at the base and extend up the fleshy scales, causing gray to black lesions; roots may be blackened, and leaves poorly developed, with yellow tips.

6.28.3.2 *On groundnuts*

Hulls of groundnuts show black discoloration which appears first along the longitudinal veins. The kernels are shrunken. The infected testae are brown to black, and the embryo shows a brown discoloration (Jones and De Waele, 1988).

6.28.4 Detection and Inspection Methods

Prior to planting, soil can be sampled using a standard extraction procedure for nematodes of this size (Hooper, 1986). It is difficult to detect the presence of *D. destructor* from external tuber appearance alone. Sample tubers should be cut or peeled to look for the characteristic whitish pockets in which most of the nematodes are found. Microscopic examination of the nematode is necessary for correct identification of the species.

6.28.5 Means of Movement and Dispersal

The nematodes can move only short distances in the soil and have no natural means of long-range movement. The main means of dispersal is with infested potato tubers or other subterranean organs of host plants, e.g., bulbs and rhizomes (especially of *Iris*). Transport in infested soil is another important means of spread. Irrigation water can also carry the nematodes.

6.28.6 Phytosanitary Risk

D. destructor was considered to be an EPPO A2 quarantine pest (OEPP/EPPO, 1978) but was deleted from the quarantine list in 1984 because of its minor importance and very wide distribution throughout the EPPO region, in particular in those areas where it would be likely to cause crop damage. *D. destructor* is of quarantine significance for the APPPC and COSAVE. The requirement of the nematode for high relative humidity means it would be unlikely to become a problem in areas with warm, dry soils; it may, therefore, be of concern to potato production only in the northern parts of the EPPO region. However, its establishment as a groundnut pathogen in South Africa has shown its potential to adjust to different (and normally unfavorable) climatic conditions (De Waele and Wilken, 1990).

6.28.7 Phytosanitary Measures

Fumigation under vacuum (650 mm Hg) with hydrogen cyanide (initial dose, 4 g/m^3) for 1 h at above 10°C gives good control of the nematode in bulbs, rhizomes, and tubers, and especially asparagus roots and strawberry plants. Infestation in *Iris* bulbs can be controlled by immersion in water containing 0.5% formaldehyde at 43.5°C for 2-3 h, but some varieties may be injured during this treatment. In garlic bulbs, nematodes were controlled by drying at 34–36°C for 12–17 days (Fujimura *et al.*, 1989).

6.28.8 Burrowing Nematode, *Radopholus Similis* on Citrus

6.28.8.1 *Parasitism and ecology of the burrowing nematode on citrus*

The burrowing nematode is a dimorphic species. Females have a well-developed stylet and are endoparasitic. Males have a poorly developed stylet and are not considered to be parasitic. Development and reproduction take place inside root tissues of plant hosts. Nematode feeding and migration in roots cause large cavities and necrosis in the cortical and vascular tissues. Nematode life stages require healthy root tissue for development and reproduction. They migrate out of roots that have been colonized and damaged as a result of nematode feeding activity and the subsequent decay caused by the invasion of bacteria, fungi, mites and other organisms. The migratory habit of the nematodes in soil provides the opportunity for burrowing nematodes to search for healthy roots. Burrowing nematodes can survive in the absence of a host for about six months.

6.28.8.2 *Races of burrowing nematode*

Studies have shown that there are two races with different host preferences (Inserra *et al.*, 2005). One race called the banana race, parasitizes banana, but not citrus. The other race called the citrus race, parasitizes both citrus and banana. Both races have extended, overlapping host ranges that include ornamentals, agronomic crops, and weeds. There are no obvious and specific morphological differences between the two races. The citrus race was eventually elevated to the status of species; *Radopholus citrophilus* had differences between the two races in oocyte maturation, sex pheromones, enzymes, and proteins. However, recent molecular and mating studies by Kaplan and Opperman (2000) show that *R. similis* and *R. citrophilus* are nonspecific, and they have been returned to their previous status as races of *R. similis*.

Several biological factors need to be considered for regulatory management of different phytonematodes, including burrowing nematode [(Table 6.4) (O'Bannon and Esser, 1987)].

Table 6.4 Comparative Biological Factors to be Considered for Regulatory Management of Major Phytonematodes

Factor	*Burrowing nematode*	*Golden cyst nematode*	*Soybean cyst nematode*
Soil depth occurrence	15 cm–5 m	7 cm–30 cm	7 cm–30 cm
Longevity, without food	6 months	8–10 years	4 years
Host range (plant species)	290	3	500
Parasitic habit	Endoparasitic	Semi-endoparasitic	Semi-endoparasitic
Natural protection	None	Cyst	Cyst
Climate	Tropical	Temperate	Temperate

6.28.8.3 *Damage and symptoms*

The burrowing nematode severely damages the fibrous roots of citrus trees. The debilitated root system does not adequately support the above-ground portion of the tree which exhibits sparse foliage, weakened and dead branches. This condition results in severely reduced yields of as much as 80% fruit loss compared to healthy trees, rendering infested orchards economically nonviable.

Soil factors play a major role in the expression of 'Spreading Decline' symptoms, which occur mainly in deep well-drained sands (90–95% sand) poor in organic matter (< 1%). This soil type makes up much of the subsoil in some locations known as the 'Ridge' and favors the burrowing nematode. The combination of low soil moisture conditions (5–7%) in the subsoil and nematode damage result in 'Spreading Decline' symptoms. Soils rich in organic matter, silt and clay components, and high moisture levels inhibit the increase of burrowing nematode populations and are not conducive to expression of the disease (O'Bannon and Tomerlin, 1971).

6.28.8.4 *Quarantine actions and management of the burrowing nematode on citrus*

Long-term research has shown that the burrowing nematodes were spread in the 'ridge' with infected citrus propagative material originating from infested nurseries. The nematode occurs mainly in areas where burrowing nematode-infected plants, such as citrus, banana, and many ornamentals, were introduced by human intervention. Nematodes may also occur on introduced nematode-infected aquatic plants, such as aquatic aroids. It is essential to implement appropriate phytosanitary actions to prevent the spread of the nematode in uncultivated areas and the adoption of effective management practices in citrus orchards.

(a) *Eradication:* A burrowing nematode eradication program, called 'push and treat,' was attempted in Florida by removing infected citrus trees from orchards and applying high doses of fumigant nematicides. This program was discontinued due to environmental concern and chemical contamination of the water table. However, these eradication attempts on more than 6,000 hectares were effective in containing the spread of the nematode from infested orchards into noninfested ones.

Isolation: Infested orchards can be separated from healthy orchards by areas (called barriers). In the past, these barriers were kept weed and root-free by application of large amounts of fumigant nematicides to chemically devitalize the citrus roots in the barrier and prevent nematode migration. Today they are maintained by mechanical plowing and root pruning. Maintenance of the mechanical barriers is costly.

(b) *Certification:* In several locations like Florida, the Citrus Nursery Certification Programme has been implemented to prevent the spread of citrus nematode pests in Florida citrus orchards (Rule Chapter 5B- 44.003, Florida Administrative Code). The Citrus Nursery Certification Programme requires that: Commercial citrus nurseries produce propagative material free of citrus nematode pests. Soil, peat, and rock material for use in citrus orchards and nurseries must be certified free of these nematodes by the Florida Department of Agriculture and Consumer Services.

6.28.8.5 *The citrus nursery certification programme*

The Citrus Nursery Certification Programme consists of three phases, viz., site approval, pit approval and premovement certification of young citrus trees before they are moved from the nurseries and transplanted into orchards.

(a) ***Site approval:*** For economic reasons, Florida commercial citrus nurseries are established directly on the ground. Pasture lands are often used for commercial citrus nursery sites. These potential nursery sites are sampled to check for the presence of nematode pests and certified if found free from regulated citrus nematode pests. *Tylenchulus* spp. populations occur in several pastures and uncultivated lands. At one time, these populations were called 'wild' strains of the citrus nematode and because of their morphological similarity, were confused with the regulated citrus nematode.

(b) ***Pit approval:*** Land destined for citrus nurseries and certified nematode-free must be planted with nematode-free citrus rootstock seedlings, which are grown from seeds or tissue culture in soil mixes containing peat and sand free from regulated nematode pests that parasitize citrus. In Florida, peat and sand for use in citrus nurseries are mined from deposits located in natural fields free from the citrus nematode pests. Construction material, such as gravel, clay, and shell, to be used in citrus orchards for the construction of roads or other purposes, must be extracted from sites free from nematode pests of citrus. These deposits of construction material are sampled as for pit and nursery sites.

(c) ***Premovement:*** Citrus rootstock seedlings are grown in certified media on elevated benches that prevent nematode contamination from the ground and transplanted into nematode-free growing areas. The Citrus Nursery Certification Programme requires that any soil or infill unloaded within 30 meters of the borders of a citrus nursery or orchard must be certified free of any citrus nematode pest. This successful Citrus Nursery Certification Programme resulted in a rapid reduction in the number of burrowing nematode-infested citrus nurseries. Hence, the implementation of the eradication, isolation of Barrier and Citrus Nursery Certification Programmes initiated in the 1950s reduced the area of orchards infested with the burrowing nematode to less than 4,000 hectares in 1984. Present distribution of burrowing nematode in citrus orchards is < 1% of the total area.

6.28.8.6 *Nature of damage*

Since nematodes affect the storage/planting materials which are dormant, their tissues are physiologically different from actively growing plants. Nematodes employ mechanical energy and biochemical components, viz., enzymes and toxins, (Table 6.5) to cause diseases. Nematodes either degrade the planting material (rotting or drying) or the growth and development of the plant. Nematodes come in the way of respiration and photosynthesis. 'True seeds' may be shriveled/distorted, become chaffy resulting in crop loss.

Table 6.5 Enzymes/toxins Released from Nematode-infected Plant Tissues

Crop	*Nematode*	*Enzymes/toxins released*
Potato	*D. destructor*	Polygalacturonase (PG) Pectinase
Onion	*D. dipsaci*	Endo-polygalacturonase (E-PG) Alkaline Phosphatase
Potato	*G. rostochiensis*	Amylase β-galactosidase, β-glucosidase
Tubers/ cuttings	*Meloidogyne* spp.	Acid phosphatase, Alkaline Phosphatase, Amylase, Proteolytic cellulase, Esterase, Proteases, Pectinase
Banana	*R. similis*	Cellulase, Invertase
Sugar beet	*H. schactii*	Amylase, Cellulase
Citrus	*T. semipenetrans*	Alkaline Phosphatase, Amylase

6.28.8.7 *Procedure for quarantine inspection for nematodes*

Following is the practical procedure for quarantine inspection of plant material for nematode infestation/contamination

- Maintain of records relating to the type of plant material, country of origin, destination, date of receipt and date of disposal, phytosanitary certificate, etc.
- Visual examination of underground and aerial parts of plant material for symptoms of nematode injury, if any, for example, surface or deep necrosis of roots, root rot, root galls, lesions, excessive root branching, stubby root, coarse root, curly root tips, dead or devitalized buds, crinkled or distorted stems and foliage, seed galls, leaf lesions or spots, leaf galls, distorted or discolored seed, etc. Each of these could give indications of infestation by nematodes which cause these specific symptoms.
- Use of appropriate for the recovery of nematodes from plant tissue; separation of plant debris, soil clods, free soil and other extraneous inert matter mixed with true seed or adhering to the seed or plants. Examination of such soil and debris for the incidence of nematodes, using appropriate techniques.
- Detailed examination and identification of the nematodes intercepted, under a microscope, followed by risk analysis of the species intercepted.
- Decision on the disposal of the plant sample (clearance or rejection) in view of the regulations, the risk involved and the possibility of salvaging the material.
- Salvaging of infested material by fumigation, hot water treatment, nematicidal dip, manual cleaning, washing, and drying, etc. whenever appropriate methods of denematization are available. It should be mentioned here that the salvaging methods used in quarantine must ensure a hundred percent kill of the nematode without injury to the planting material. Since tested and standardized methods (for

quarantine operations) are available only for a limited number of plants and planting material (and these too pertain to limited number of species of endoparasitic nematodes), the most ideal procedure for vegetatively propagated ones is to grow the material in nematode-free areas, especially when the plant consignments are large and the suspected nematodes have not been recorded from the country.

6.29 MOLECULAR AIDS TO NEMATODE DIAGNOSIS

Traditional morphological methods of identification have formed the basis of nematode taxonomy for more than 100 years and the foreseeable future, they will continue to be important. However, rapid advances in biotechnology have provided taxonomists with a wide range of new tools, which may eventually revolutionize nematode identification. These new technologies are already widely used in areas such as medical and veterinary diagnosis, where they have been adopted because of their discriminatory power, low cost and their ease of use and interpretation. Protein electrophoresis and DNA-based technologies are major methods that may eventually be used for the routine identification of nematodes.

Morphological characters have long been used for nematode taxonomy and will continue to be the mainstay. It is an essential component of any higher level classification of nematodes and in many cases morphology provides a quick and unambiguous diagnosis to species (Ravichandra, 2008). Nevertheless, morphological taxonomy suffers from some limitations. Morphological taxonomy cannot be used for intra-sub-species groups such as races, pathotypes, and strains. Identification by other methods including host range testing is very time-consuming and not full proof. Since the genotype of the nematode is examined directly in case of molecular markers, problems associated with phenotypic variation in taxonomic characters are avoided.

Furthermore, greater discrimination is possible because the entire genome is available for study, including 75–80% of the genome, which is noncoding and contains many highly variable sequences. This DNA sequence variability provides a large pool of information that can provide useful diagnostic characters for separation of taxa and intra-sub-specific categories for the unraveling of evolutionary relationships between nematode groups. Diagnostic kits can be developed for the identification, which can be used for advisory services and intercepting the consignments for domestic as well as international quarantine purposes. Another advantage of molecular markers is that it is independent of the stage of the nematode to be identified, and one or few nematodes are enough. The results are interchangeable among laboratories worldwide and reproducible. DNA analysis also helps is to understand the genetic pool similarity of any nematode species in a given geographical area and this information is highly useful in resistance breeding. There are various molecular markers available at present, which can be used in nematode identification depending on the availability of time, facilities, purpose and quantity of nematodes.

6.29.1 To Avoid Infiltration of Nematodes/Restrict Spread

- Set up a nematode diagnostic unit with basic apparatus for extraction, observation and decontamination and a trained nematologist at each port of entry. Measures like hot water treatment of the planting stock can detect and kill nematodes
- Set up regional SPS laboratories, with equipment and two nematologists (one taxonomist and one applied nematologist)
- Establish a centralized database of indigenous and exotic nematode pests
- Carry out pest risk analyses for exotic and localized indigenous nematode pests
- Introduce mandatory phytosanitary inspection of bulk imports of plant materials at the time of entry
- Introduce post entry periodic monitoring of sites where imported materials are sown/planted
- Introduce mandatory destruction at entry time and postentry decontamination and if necessary, destruction of infested material

Use of treated planting stock is one of the best measures to avoid nematodes, which may be carried out in following manners.

1. **Heat:** Steam (Potato Cyst Nematode, Stemand bulb nematode) and autoclaving small quantities of soil
2. **Hot water:** More common. Time and temperature specific
3. **Irradiation:** *G. rostochiensis, D. myceliophagus*
4. **Washing processes:** Proper and careful washing of material to remove adhering nematodes
5. **Seed cleaning:** Mechanical seed cleaning to remove dormant structures in the seed lot

6.29.2 Some Requirements and Solutions

Basic information on occurrence and distribution of nematode species already present in the country, categorization of nematodes based on the information collected, into groups based on their damage potential-prohibited species, restricted species and unrestricted species, import policy-bulk assignments are difficult to examine. Importing small quantities and multiplying of seed/planting material is suitable, post entry quarantine facilities— postentry isolation growing facilities are essential to all quarantine agencies. Mere lab examinations are difficult to detect the nematodes, operational requirements-socio economic factors of the country, lack of public awareness on the nematodes, programs of educating the public through films, lectures, pamphlets, etc. require immediate attention of the authorities and training and communication on various aspects of nematodes

6.29.3 Future Thrusts

Information on the occurrence, distribution, pathogenicity and host range of phytonematodes of quarantine significance, classification of nematode species into categories according to their damage potential to be incorporated into quarantine regulations under DIP Act, regulation of quantity of hazardous material by the authorities, postentry quarantine facilities to be improved, attempts to create more public awareness, qualified nematologists to be appointed at quarantine centers, minimum usage of chemicals, and safer and eco-friendly approaches like bioagents and use of genetically modified planting material.

6.30 SOME COMMON TERMINOLOGIES USED IN PLANT QUARANTINE

AAPGA .... Australian Apple and Pear Growers' Association

ABARE .... Australian Bureau of Agricultural and Resource Economics

etiology .... The science of the cause of disease

ALOP .... Appropriate level of protection is the level of protection deemed appropriate by the country establishing a sanitary or phytosanitary measure to protect human, animal or plant life or health

AQIS .... Australian Quarantine and Inspection Service

BA .... Biosecurity Australia

DNA .... Deoxyribonucleic Acid, usually consisting of two helical chains of polynucleotides that are responsible for the transfer of genetic characteristics

DNRE .... Department of Natural Resources and Environment

HRDC .... Horticultural Research and Development Corporation

IPC .... International Phytosanitary Certificate

IPPC .... International Plant Protection Convention

Pathogen .... A parasite able to cause disease in a particular host or range of hosts

PCR .... Polymerase Chain Reaction, where conserved nucleotide sequences are replicated to levels that can be detected by gel electrophoresis

PEQ .... Postentry Quarantine

Pest .... Any species, strain or biotype of plant, animal, or pathogenic agent, injurious to plants or plant products

Phytosanitary measure .... Any legislation, regulation or official procedure having the purpose to prevent the introduction and/ or spread of quarantine pests

Quarantine pest	....	A pest of potential economic importance to the area endangered thereby and not yet present there or present but not widely distributed and under official control
RNA	....	Ribonucleic Acid, consisting of a single helical chain of polynucleotides, the main function of which is the translation of the genetic code into protein synthesis
Rootstock	....	Plant into which bud or graft is inserted
Scion	....	Shoots that develop from a bud or graft on a rootstock
SPS Agreement	....	World Trade Organization Agreement on the Application of Sanitary and Phytosanitary Measures
Synonymy	....	Listing of illegitimate name(s) not currently employed for the species or group under the current name
Vector	....	An organism that carries and transmits a pathogen to the host which it attacks
WTO	....	World Trade Organization

6.31 NEW QUARANTINE DISEASES ADDED DURING 2007–2009

Table 6.6 lists the major diseases added during 2007, 2008 and 2009.

Table 6.6 New Quarantine Diseases Added during 2007 – 2009

Sl. No.	*Name of disease*
1.	Coconut cadang–cadang viroid
2.	*Arabis mosaic* virus
3.	*Strawberry latent* ringspot virus
4.	*Asparagus* virus 1
5.	*Asparagus* virus 2
6.	*Acremonium strictum*
7.	Soybean dwarf virus
8.	Ergot (*Claviceps purpurea*)
9.	*Artichoke* Italian latent virus
10.	*Celery* mosaic virus
11.	*Phomopsis longicolla* (*Pod and stem blight*)
12.	*Atelocauda digitata*
13.	*Ceratobasidium cereale*
14.	*Pleospora herbarum* (Leaf blight of onion)
15.	*Anthonomus grandis* (Mexican cotton boll weevil)
16.	*Phytophthora cryptogea* tomato foot rot)

17.	*Phomopsis longicolla*
18.	*Cercospora abelmoschi*
19.	*Colletotrichum higginsianum*
20.	*Pseudomonas syringae* pv. *maculicola* (cabbage leaf spot)
21.	*Xanthomonas campestris* pv. *raphani* (leaf spot)
22.	*Pseudomonas viridiflava* (bacterial leaf blight of tomato)
23.	*Pseudomonas syringae* pv. *atropurpurea*
24.	*Pseudomonas syringae* pv. *corona faciens* (halo blight)
25.	*Rhizobium rhizogenes* (gall)
26.	*Burkholderia plantarii*
27.	*Burkholderia glumae*
28.	*Nectria radicicola*
29.	*Monographella nivalis*
30.	*Fusarium oxysporum* f. sp. *passiflorae*
31.	*Phaeosphaeria avenaria* f. pv. *Avenaria* (leaf spot of oat)
32.	*Peronospora hyoscyami* f. sp. *tabacina*
33.	Pepino mosaic virus
34.	Tomato bushy stunt virus
	Nematode
1.	*Ditylenchus dipsaci*

6.32 PEST RISK ANALYSIS (PRA)

'Pest' refers to any species, strain or biotype of plant, animal or pathogenic agent injurious to plants and plant products (FAO, 1996). For example insects, mites, nematodes, rodents, birds, snails, fungi, bacteria, virus and weeds pests challenge International and interstate borders continuously both by natural movement (climate change) and by human assisted means. Pests are highly mobile with a vast array of sensory organs that enable them to disperse and find hosts, can cause significant problems when introduced into new areas. Pests are classified either as 'native pests' which are originated from that region or as 'introduced pests' are not native to a region, but introduced accidentally or intentionally, and the third one is 'quarantine' pest which means a pest of potential economic importance to the area endangered thereby and not yet present there or present but not widely distributed and being officially controlled.

6.32.1 Process of PRA

It is a process of investigation, evaluation of information and decision making with respect to a certain pest, that starts once it is known or determined that this pest is a quarantine pest (FAO, 1996). It is conducted to protect the country's agriculture from damages that can be caused by harmful (quarantine) pests this can be brought in along with imported commodities. It evaluates the likelihood of the entry, establishment or spread of a pest and the associated potential biological and economic consequences (Fig. 6.3). It requires international cooperation.

6.33 REGIONAL PLANT PROTECTION ORGANIZATIONS (RPPOs)

There are currently 10 RPPOs, viz., Asia and Pacific Plant Protection Commission (APPPC), Comunidad Andina (CA), Comite de Sanidad Vegetal del Cono Sur (COSAVE), Caribbean Plant Protection Commission (CPPC), European and Mediterranean Plant Protection Organization (EPPO), Inter-African Phytosanitary Council (IAPSC), Near East Plant Protection Organization (NEPPO), North American Plant Protection Organization (NAPPO), Organismo Internacional Regional de Sanidad Agropecuaria (OIRSA), and Pacific Plant Protection Organization (PPPO).

PRA evaluates the probability of entry, establishment, and spread of a pest in a given area, magnitude of the impacts and the ability of selected measures to reduce the risk to an acceptable level.

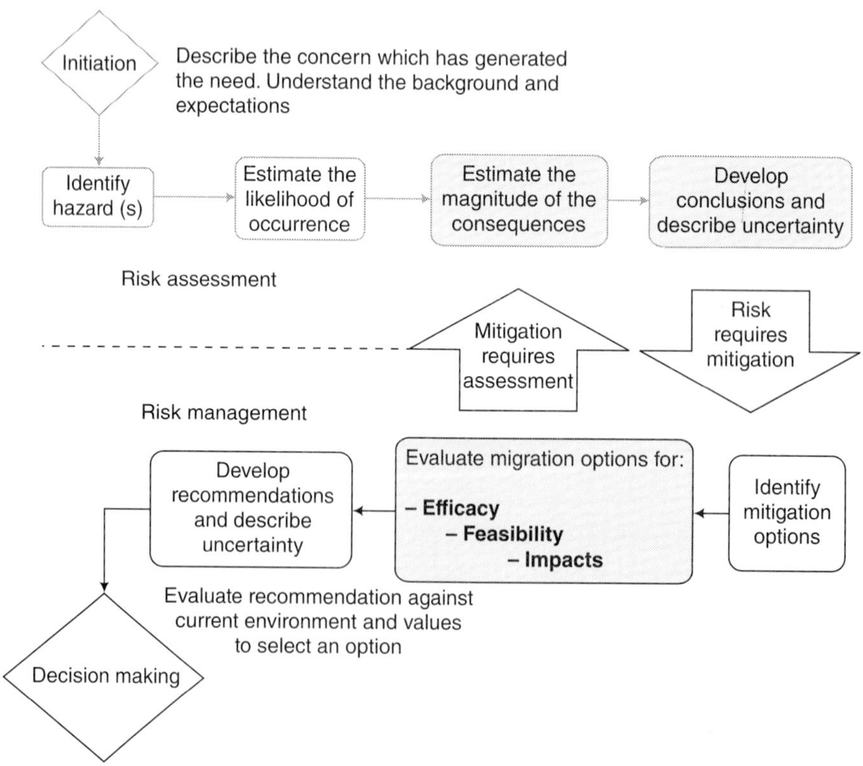

Fig. 6.3 A process overview for pest risk analysis

6.33.1 Pest Risk Assessment and Pest Categorization

Elements of categorization are the identity of the pest, presence or absence in the PRA area, regulatory status, potential for establishment and spread in PRA area and potential for economic consequences in the PRA area. There are generally two initiation points for a PRA, viz., the identification of a pathway, usually an imported commodity that may allow the introduction and/or spread of quarantine pests and the identification of a pest that may qualify as a quarantine pest.

6.33.2 PRA Initiated by a Pathway

A requirement for a new or revised PRA originating from a specific pathway will most frequently arise in the following situations.

1. Quarantine pest– a pest of potential economic importance to the area, endangered thereby and not yet present there, or present but not widely distributed and being officially controlled.
2. International trade is initiated in a new commodity (Plant or Plant Product) or commodity from a new origin. The PRA may be triggered by a request for import, or by the appearance in trade of consignments of a commodity. The pathway may concern a single area of origin or several.
3. New plant species are imported for selection and scientific research purposes.
4. A pathway other than commodity import is identified (natural spread, mail, garbage, passenger's baggage, etc.).
5. A policy decision is taken to establish or revise phytosanitary regulations or requirements concerning specific commodities.
6. A new treatment, system or process, or new information impacts on an earlier decision.

The pests which are likely to follow the pathway (e.g., carried by the commodity) are then listed, and each is then subjected to a next stage in the PRA process. If no potential quarantine pests are identified as likely to follow the pathway, the PRA stops at this point.

6.33.3 PRA Initiated by a Pest

A requirement for a new or revised PRA originating from a specific pest will most frequently arise in the following situations:

1. An emergency arises on discovery of an established infestation or an outbreak of a new pest within a PRA area
2. An emergency arises on interception of a new pest on an imported commodity
3. A new pest risk is identified by scientific research
4. A pest is introduced into a new area other than the PRA area
5. A pest is reported to be more damaging in a new area other than the PRA area itself, than in its area of origin

6. Audits reveal that a particular pest is repeatedly intercepted
7. A request is made to import, such as an organism, for example, by researchers, educators, biological practitioners, businesses (pet store owners), the food industry (e.g., snails for consumption) or hobbyists (aquatic plants for aquaria) 2 PRA area, i.e., area in relation to which a pest risk analysis is conducted.
8. A policy decision is taken to revise phytosanitary regulations or requirements concerning specific pests
9. A proposal is made by another country or by an international organization (RPPO, FAO)
10. A new treatment system, process or new information impacts on an earlier decision
11. Stages of PRA-Initiation risk assessment probability, consequences, risk management transparency, uncertainty initiation stage, pest-initiated detection (interception or incursion), research or biological control pathway commodity evaluation, other regulated article policy review or change, etc.

6.34 NECESSARY INFORMATION ON A PRODUCT WHICH IS TO BE SUBJECTED TO THE PRA PROCESS

1. Scientific name, genus, species, and family of the plant, product or by-product of interest
2. Localization, altitude, and latitude of the areas of production designated to exportation in the country of origin
3. Map of the country showing the areas of production designated to exportation and other areas
4. Climatically conditions in the areas of production: Maximum and minimum temperatures, level of precipitation, predominant winds and relative humidity
5. Phenology of the crop, emphasizing the most important phases of growth according to the use and destiny of the product: leaf development, flowering, and fructification
6. Phytosanitary management of the crop, showing dates and stages of major pest incidences
7. List of quarantine pests by stages of development of the crop, emphasizing the important pests related to the part of the plant which is being imported
8. List of pests of quarantine importance according to the A1 and A2 pest lists, both of the importing and exporting country
9. Biology, and actual situation, distribution, economic damage of the important quarantine pests in the production zone designated for export
10. Pre and post harvest phytosanitary treatments for important quarantine pests
11. Interior phytosanitary regulations of the exporting country related to the crop of interest, or pests identified as quarantine pests if such regulations are present in the country

12. Vigilance and monitoring systems to prevent the outbreak of pests of quarantine importance if such systems are present in the country
13. Infrastructure for the application of recognized quarantine treatments for the pests of quarantine importance
14. Volumes of production and exportation
15. List of natural enemies of the pests of quarantine importance, if they exist in the exporting country related to the plant, product or byproduct of interest.

6.34.1 PRA Initiated by Review or Revision of Policy

A requirement for a new or revised PRA originating from policy concerns will most frequently arise in the following situations, viz., National decision is taken to review phytosanitary regulations, requirements or operations; the proposal made by another country or by an international organization (RPPO, FAO) is reviewed; a new treatment or loss of a treatment system, a new process, or new information impacts on an earlier decision; a dispute that arise on phytosanitary measures; the phytosanitary situation in a country changes, for a new country is created, or political boundaries changed.

6.34.2 Risk Assessment Stage

Presence/absence/distribution, Official control, introduction potential, establishment potential, spread potential and elements of risk assessment. The probability includes, with pathway escapes detection, favorable location overcomes resistance, able to reproduce and spread. Consequences may be direct economic, indirect economic, environmental, political, social, aesthetic pathway for the importation of fresh fruit, etc.

6.35 ASSESSMENT OF THE PROBABILITY OF INTRODUCTION AND SPREAD

6.35.1 Probability of Entry of Pest

These include probability of the pest being associated with the pathway, probability of survival during transport or storage, probability of pest surviving existing pest management procedures, probability of transfer to a suitable host, availability of suitable hosts, alternate hosts and vectors in the PRA area, suitability of environment, cultural practices and control measures and other characteristics of the pest affecting the probability of establishment.

6.35.2 Probability of Spread after Establishment

Probability of the pest being associated with the pathway at the origin and probability of the pest being associated spatially or temporally with the pathway at origin should be estimated. Factors to be considered are prevalence of the pest in the source area, occurrence of the pest in a life-stage that would be associated with commodities, containers or conveyances, volume and frequency of movement along the pathway, seasonal timing and pest management, cultural and commercial procedures applied at the place of origin (application of plant protection products, handling, culling, routing, and grading).

6.35.3 Probability of Survival During Transport or Storage

Factors to be considered include speed and conditions of transport and duration of the life cycle of the pest in relation to time in transport and storage, vulnerability of the life-stages during transport or storage, prevalence of pest likely to be associated with a consignment and commercial procedures (like refrigeration) applied to consignments in the country of origin and country of destination or in transport or storage.

6.35.4 Probability of Pest Surviving Existing Pest Management Procedures

Existing pest management procedures (including phytosanitary procedures) applied to consignments against other pests from origin to end-use should be evaluated for effectiveness against the pest in question. The probability that the pest will go undetected during inspection or survive other existing phytosanitary procedures should be estimated.

6.35.5 Probability of Transfer to Suitable Host

Factors to consider are dispersal mechanisms including vectors to allow movement from the pathway to a suitable host, whether the imported commodity is to be sent to a few or many destination points in the PRA area, proximity of entry, transit and destination points to suitable hosts, time of year at which import takes place, intended use of the commodity (e.g., for planting, processing and consumption) and risks from by-products and waste.

Some uses are associated with a much higher probability of introduction (planting) than others (processing). The probability associated with any growth, processing, or disposal of the commodity in the vicinity of suitable hosts should also be considered. For LMOs, the probability of gene flow and gene transfer should also be considered when there is a trait of phytosanitary concern that may be transferred.

6.35.6 Probability of Establishment

To estimate the probability of establishment of a pest, reliable biological information (life cycle, host range, epidemiology, survival, etc.) should be obtained from the areas where the pest currently occurs. The situation in the PRA area can then be compared with that in the areas where it currently occurs (taking account also of protected environments such as glass- or greenhouses) and expert judgment used to assess the probability of establishment. Case histories concerning comparable pests can be considered. Examples of the factors to consider include the availability, quantity and distribution of hosts in the PRA area, environmental suitability in the PRA area, potential for adaptation of the pest reproductive strategy of the pest, method of pest survival and cultural practices and control measures.

In considering probability of establishment, it should be noted that a transient pest may not be able to establish in the PRA area, because of unsuitable climatic conditions, but could still have unacceptable economic consequences.

6.35.7 Availability of Suitable Hosts, Alternate Hosts and Vectors in PRA Area

Factors to be considered are whether hosts and alternate hosts are present and how abundant or widely distributed they may be, whether hosts and alternate hosts occur within sufficient geographic proximity to allow the pest to complete its life cycle, whether there are other plant species which could prove to be suitable hosts in the absence of the usual host species, whether a vector, if needed for dispersal of the pest, is already present in the PRA area or likely to be introduced, and whether another vector species occurs in the PRA area.

The taxonomic level at which hosts are considered should normally be the 'species.' The use of higher or lower taxonomic levels should be justified by scientifically sound rationale.

6.35.8 Suitability of Environment

It is important to identify the factors in the environment (e.g., suitability of climate, soil, pest and host competition) that are critical to the development of the pest, its host and if applicable its vector and to their ability to survive periods of climatic stress and complete their life cycles. It should be noted that the environment is likely to have different effects on the pest, its host, and its vector. This needs to be recognized in determining whether the interaction between these organisms in the area of origin is maintained in the PRA area to the benefit or detriment of the pest. The probability of establishment in a protected environment, e.g., in glass houses, should also be considered. Climatic modeling systems may be used to compare climatic data on the known distribution of a pest with that in the PRA area.

6.35.9 Cultural Practices and Control Measures

Where applicable, practices employed during the cultivation/production of the host crops should be compared to determine if there are differences in such practices between the PRA area and the origin of the pest that may influence its ability to establish. For plants that are LMOs, it may also be appropriate to consider specific cultural, control or management practices.

Pest control programs or natural enemies already in the PRA area which reduce the probability of establishment may be considered. Pests for which control is not feasible should be considered to present a greater risk than those for which treatment is easily accomplished. The availability (or lack) of suitable methods for eradication should also be considered.

6.35.10 Conclusion on Probability of Introduction and Spread

The overall probability of introduction should be expressed in terms most suitable for the data, the methods used for analysis and the intended audience. This may be quantitative or qualitative since either output is in any case the result of a combination of both quantitative and qualitative information. The probability of introduction may be expressed as a comparison with that obtained from PRAs on other pests.

6.35.11 Assessment of Potential Economic Consequences

Requirements described in this step indicate what information relative to the pest and its potential host plants should be assembled, and suggest levels of economic analysis that may be carried out using that information to assess all the effects of the pest, i.e., the potential economic consequences. Wherever appropriate, quantitative data that will provide monetary values should be obtained. Qualitative data may also be used. Consultation with an economist may be useful. In many instances, detailed analysis of the estimated economic consequences is not necessary if there is sufficient evidence or it is widely agreed that the introduction of a pest will have unacceptable economic consequences (including environmental consequences). In such cases, the risk assessment will primarily focus on the probability of introduction and spread. It will, however, be necessary to examine economic factors in greater detail when the level of economic consequences is in question, or when the level of economic consequences is needed to evaluate the strength of measures used for risk management or in assessing the cost-benefit of exclusion or control.

6.35.12 Pest Effects

To estimate the potential economic importance of the pest, information should be obtained from areas where the pest occurs naturally or has been introduced. This information should be compared with the situation in the PRA area. Case histories concerning comparable pests can usefully be considered. The effects considered may be direct or indirect.

(a) The basic method for estimating the potential economic importance of pests in this section also applies to pests affecting uncultivated/unmanaged plants, weeds and/or invasive plants and pests affecting plants through effects on other organisms.
(b) In the case of direct and indirect environmental effects, specific evidence is needed.
(c) In the case of plants to be imported for planting, the long-term consequences for the intended habitat may be included in the assessment. Planting may affect further use or have a harmful effect on the intended habitat. Environmental effects

and consequences considered should result from effects on plants. Such effects, however, on plants may be less significant than the effects and/or consequences on other organisms or systems. For example, a minor weed may be significantly allergenic for humans or a minor plant pathogen may produce toxins that seriously affect livestock. However, the regulation of plants solely on the basis of their effects on other organisms or systems (e.g., on human or animal health) is outside the scope of this standard. If the PRA process reveals evidence of a potential hazard to other organisms or systems, this should be communicated to the appropriate authorities who have the legal responsibility to deal with the issue.

6.35.13 Direct Pest Effects

For identification and characterization of the direct effects of the pest on each potential host in the PRA area or those effects which are host-specific, the following are examples that could be considered: known or potential host plants (in the field, under protected cultivation, or in the wild); types, amount and frequency of damage; crop losses in yield and quality; biotic factors (e.g., adaptability and virulence of the pest) affecting damage and losses; biotic factors (e.g., climate) affecting damage and losses; rate of spread; rate of reproduction; control measures (including existing measures), their efficacy and cost; and effect on existing production practices and environmental effects.

For each of the potential hosts, the total area of the crop and area potentially endangered should be estimated in relation to the elements given above.

In the case of the analysis of environmental risks, examples of direct pest effects on plants and/or their environmental consequences that could be considered include reduction of keystone plant species, reduction of plant species that are major components of ecosystems (in terms of abundance or size), endangered native plant species (including effects below species level where there is evidence of such effects being significant) and significant reduction, displacement or elimination of other plant species.

6.35.14 Indirect Pest Effects

For identification and characterization of the indirect effects of the pest in the PRA area or those effects that are not host-specific, the following are examples that could be considered, viz., effects on domestic and export markets including in particular effects on export market access. The potential consequences for market access which may result if the pest becomes established should be estimated. This involves considering the extent of any phytosanitary regulations imposed (or likely to be imposed) by trading partners, changes to producer costs or input demands including control costs, changes to domestic or foreign consumer demand for a product resulting from quality changes, environmental and other undesired effects of control measures, feasibility and cost of eradication or containment, capacity to act as a vector for other pests, resources needed for additional research and advice, and social and other effects (e.g., tourism).

Pest risk analysis is not a new concept. During previous centuries, it was recognized that plant pests could be controlled best by regulatory measures developed and executed by a national authority (Bram de Hoop, 2003). Assessments of the impact of plant pests and scientific information can provide the justification for such measures. The first international pest risk analysis standard, ISPM 2: *Guidelines for pest risk analysis* became available in February 1996. Five years later, ISPM 11 [2001]: *Pest risk analysis for quarantine pests* was adopted. Opinions about the feasibility of prevention of the international spread of pests differ (Bos, 1988). Bos noted that pathologists tend to be conservative and to play safe. Breeders take a more liberal view and focus first on the benefits of introducing new varieties, which may include resistance to new pests as well. Despite these seemingly opposing views, it is clear that without the international movement of plant material, especially plant propagation material, agriculture would not be what it is today. Increasingly so, such international movement of plants and plant products is required to keep feeding the ever-growing world population. Therefore, plant quarantine measures should be realistic and aim for "Improvement of the rapid and safe global transfer of (plants and plant products)" (Chiarappa and Karpati, 1981).

To combine speed and safety of global movement implies compromise and acceptance of certain risks. It is here that 'good governance' is essential for individual countries to determine those pests that could cause unacceptable damage to plants or plant products, the means and the likelihood of their introduction and the options (measures) available to prevent such introduction. Together these three steps constitute the process known as pest risk analysis. The end result of the application of the pest risk analysis process is usually an advice to a national authority favoring or disfavoring phytosanitary measures (Hopper, 1993).

Pest risk analysis is one of the core elements for protecting plant health today, as reflected by the International Plant Protection Convention and its international standards for phytosanitary measures. The first international pest risk analysis standard became available in February 1996. Seven years later, in April 2003, the international plant protection community adopted a standard (supplementary to another pest risk analysis standard of 2001) to allow for specific focus on environmental risk.

6.36 OPTIONS OF INTERNATIONAL STANDARDS OR PEST RISK ANALYSIS

Pest risk analysis is a key element of all three major international agreements relating to plant protection, namely, the International Plant Protection Convention, the World Trade Organization's Agreement on the Application of Sanitary and Phytosanitary Measures and the Convention on Biological Diversity. The IPPC framework, including the national plant protection organizations of all its contracting parties, could well provide an institutional framework that serves the objectives of the CBD and WTO as well.

Pest risk analysis is a framework for organizing biological and other scientific and economic information to calculate risk (essentially the probability of an unwanted event or hazard occurring multiplied by the magnitude of the consequences if it does happen). This assessment of risk is used to identify appropriate measures to reduce risk to an acceptable level. To comply with international agreements, countries imposing new phytosanitary regulations must either support them through a pest risk analysis or base them on relevant international standards.

More specific international standards would be useful in harmonizing management options for pest risks. However, the wide variation in pest risks faced by individual countries in relation to introduction and spread of a pest (or pests) associated with a particular commodity usually makes it difficult to develop an agreed international standard. To date, only one specific commodity standard has been accepted under the IPPC as an international standard for phytosanitary measures, namely, ISPM 15: *Guidelines for regulating wood packaging material in international trade.* This standard's acceptance was helped in that many countries were in agreement on the risks associated with wood packaging material as a pathway for pests. If the appropriate phytosanitary standard is not available, countries have the alternative of risk analysis to support restrictive measures.

6.36.1 Dispute Settlement

International harmonization of phytosanitary measures will be successful only if contracting parties implement standards. Control systems are needed to verify whether this is done. The WTO has a dispute settlement mechanism in which the outcome of a dispute settlement is binding on the parties concerned. Phytosanitary disputes may be resolved within the IPPC framework, in a dispute settlement process which relies on the voluntary cooperation of both disputing parties, who are not bound by the findings. The use of the IPPC dispute settlement mechanism could provide valuable jurisprudence for consideration in phytosanitary disputes carried out through the WTO mechanism. To date, no dispute settlement has taken place within the IPPC framework.

6.37 ADDRESSING ENVIRONMENTAL RISKS ASSOCIATED WITH PLANT PESTS

The contracting parties to the IPPC have developed a reference standard specifically to target the analysis of environmental risks posed by plant pests. This was adopted in April 2003 as a supplement (*Analysis of environmental risks*) to ISPM 11 [2001]: *Pest risk analysis for quarantine pests.* The standard has been further amended and supplemented to ISPM 11 [2004]: *Pest risk analysis for quarantine pests,* including analysis of environmental risks and living modified organisms. The process of revising ISPM 11 [2001] took into account the CBD's guiding principles for the prevention, introduction, and mitigation of impacts of invasive alien species. ISPM 11 and other relevant ISPMs provide specific guidance to countries on how to use the information available to make timely decisions about potential threats to plants and plant communities from international trade. The international glossary of phytosanitary terms, ISPM 5, has been supplemented in 2001 with *Guidelines on the interpretation and application of the concept of official control for regulated pests* and in 2003 with *Guidelines on the understanding of the potential economic importance and related terms* including reference to environmental considerations. These supplements have helped clarify what should be covered by pest risk analysis. Several other recent new or revised standards support pest risk analysis.

Systems for the classification of pests of potential economic significance, generally arthropods and pathogens affecting agriculture or forestry, are well established; identifying species or pathways that pose an environmental risk is a relatively new challenge for NPPOs. Following are the three different approaches for developing an overall PRA strategy to the management of invasive alien species.

(i) ***To expand monitoring and regulation to cover taxa and pathways that have not been covered previously:*** This is particularly pertinent in a country that has not had legislation covering weeds. An NPPO could set up a quick screening process or a partial pest risk analysis on all plant species imported for planting. It would reduce the risk of introducing species of plants that 'escape' their intended use and become countryside weeds or that become more invasive than in their native environment. If a species is found to be suspicious, a full pest risk analysis may be required. Although the focus is on plants for planting, any organism intentionally introduced that will establish a breeding population may warrant a pest risk analysis.

(ii) ***To identify the most important natural systems and/or native plants and review all potential threats to these:*** NPPOs work with a broad range of stakeholders, including environmental nongovernmental organizations, to decide which species or ecosystems that are not already the focus for plant protection effort are the most important to start protecting.

(iii) ***To focus efforts on particular geographic areas or areas with specific levels of protected status:*** Limiting an NPPO's efforts to particular geographic areas, such as protected areas or parks, may be a good starting point if resources are limited. Information on such an area will exist, including ecological data; other agencies may have the mandate of protecting the area and add their authority to the outcome of the analysis. The constant threat of reentry of a pest to a protected area from the surrounding area could reduce the cost-effectiveness of this approach. However, it could suit countries with natural barriers that will reduce the likelihood of internal spread.

6.37.1 Pests Causing Indirect Injury to Plants

An annex to the revision of ISPM 11 [2001] states that the scope of the IPPC also extends to organisms which are pests because they "indirectly affect plants through effects on other organisms." In addition to pests that directly affect host plants, there are those, like most weeds/invasive plants, which affect plants primarily by other processes such as competition. Such 'secondary pests' are a difficult category for risk analysis. NPPOs could find themselves asked to regulate new pests and pathways for pests that were outside the experience, knowledge and training of their officials. Some of the organisms in this category of pests may be predicted (e.g., parasites of biological control agents or pollinators, species of exotic ants); for others, ecologists may have to sound the warning of the potential threat before an NPPO will know to act.

6.37.2 Resources for Pest Risk Assessments

The principal tools and resources available for the pest risk assessment component of pest risk analysis include the following.

(i) ***National data sources:*** National data will be needed from the PRA area and to assess entry potential, from the country of origin.

(ii) ***International data sources:*** Examples include CAB International's Crop Protection Compendium, which can be used to select species that might be present in a commodity based on the pest distribution and host range in the exporting country, climatological information; abstracts of scientific literature, books, pest alerts from regional plant protection organizations, and web search engines.

(iii) ***Data sources for assessing entry potential:*** Data on trade pathways and interceptions (detections in consignments) are most important but usually require specific enquiries for unpublished data. It includes data sources for assessing establishment potential, if the area of origin and the host plants are known, climates in the area of origin can be compared with climates in the area under threat and the distribution of host plants determined and data sources for assessing economic, environmental and social impacts.

Assembling comprehensive data on a pest's impacts in its current range, sufficient biological data to predict its spread and population dynamics in the PRA area coupled with financial, economic, environmental and social data for the enterprises, ecosystems and people likely to be affected may prove difficult. However, assessments can still be made even if data are lacking.

6.37.3 Adaptation of a Regional PRA Scheme

EPPO supplies a user-friendly questionnaire to assess the risks of organisms that are harmful to plants and to apply pest risk analysis in the framework of the IPPC. This scheme provides technical justification for the regulation of certain pests for the EPPO region (or parts of it) as the PRA area. It has now been adapted to the revised ISPM 11 and will be applicable to pest plants (invasive plants or weeds) and to indirect plant pests (pests that have negative impacts on other organisms resulting in harm to plants). The structure of the scheme has been altered to accord with the sections of ISPM 11 and a rearrangement of questions to follow it more closely.

The most important changes regarding environmental aspects are the PRA scheme's applicability to plants as pests. Invasive plants have effects on the environment; impacts are generally described in qualitative rather than in economic terms. This will be reflected in the revision of the scheme, for example by the question, how important is environmental harm caused by the pest within its existing geographic range. In instances of intentional introduction of a plant, its potential for establishment in an unintended habitat has to be assessed. Intrinsic characteristics of invasiveness should be identified as far as possible, as covered by the question, "does the plant have intrinsic attributes that indicate that it could cause significant harm to plants or plant communities?"

6.38 PRA - CASE STUDIES

6.38.1 Fungal Plant Pathogens

Assessing risk from a novel pest or pathogen requires knowing which local plant species are susceptible (Gregory S. Gilbert *et al.*, 2012). Empirical data on the local host range of novel pests are usually lacking, but it is known that some pests are more likely to attack closely related plant species than species separated by greater evolutionary distance. The Global Pest and Disease Database was used, an internal database maintained by the United States Department of Agriculture Animal and Plant Health Inspection Service-Plant Protection and Quarantine Division (USDA APHIS-PPQ), to evaluate the strength of the phylogenetic signal in host range for nine major groups of plant pests and pathogens. Eight of nine groups showed significant phylogenetic signal in host range. Additionally, pests and pathogens with more known hosts attacked a phylogenetically broader range of hosts. This suggested that easily obtained data - the number of known hosts and the phylogenetic distance between known hosts and other species of interest - could be used to predict which plant species are likely to be susceptible to a particular pest. This can facilitate rapid assessment of risk from novel pests and pathogens when empirical host range data are not yet available and guide the efficient collection of empirical data for risk evaluation.

Pest risk assessment studies are aimed at evaluating if weather conditions are suitable for the potential entry and establishment of an organism in a new environment (Simone Bregaglioa *et al.*, 2012). For fungal plant pathogens like *Puccinia striiformis, Venturia inaequalis, Bipolaris oryzae* and *Cercospora carotae*, the crucial aspect to be explored is the fulfillment of the infection process that constitutes the first phase of the development of an epidemic as mainly driven by temperature and leaf wetness duration. This is of particular interest for climate change studies, because the modified pattern of temperature and moisture regimes could completely alter the known distribution and severity of plant disease epidemics. Biophysical process-based models could effectively be used in such studies, because they allow, within their applicability range, estimating organisms responses to climatic drivers in environmental conditions not yet experienced. One of the prerequisites for their adoption in operational contexts is a sensitivity analysis assessment aimed at understanding their ability to differentiate the responses according to different parameterizations and to be sensitive to the variability of the input data. In this study, a generic potential fungal infection model simulating four pathogens chosen to provide a wide range in temperature and moisture requirements was analyzed. The model was run under diverse climatic conditions. The sensitivity of the model significantly changed according to the pathogen tested, and the relevance of its parameters in explaining model output resulted strongly linked to the environmental conditions tested, indicating it to be used in pest risk assessment studies.

The EFSA panel on Plant Health has delivered a pest risk assessment on the risk posed by *Monilinia fructicola* to the EU territory and has identified risk management options and evaluated their effectiveness in reducing the risk to plant health posed by this organism (Anonymous, 2011). The panel has also analyzed the effectiveness of the special requirements presently listed in Annex IV, Part A, Section I of Council Directive 2000/29/EC, in reducing the risk of introduction of this pest into the EU territory. The panel concluded that the main pathways for entry into the EU territory are plant material for propagation purposes and fruit of host genera and that, with the exception of dried fruit, the probability of entry is very likely. The probability of establishment is also very likely due to the suitable environmental conditions and to the widespread presence of host species, susceptible for most of the year, on most of the risk assessment area. Cultural practices and control measures currently applied and competition with other *Monilinia* species cannot prevent the establishment of *M. fructicola*. The probability of spread is very likely because of the multiple ways of dispersal of the pest. The overall impact in the endangered area is estimated to be moderate. Neither additional cultural measures nor increased fungicide treatments would be needed to control of brown rot in the orchard after the introduction of *M. fructicola*.

Following a request from the European Commission, the Panel on Plant Health was asked to deliver a scientific opinion on the risk posed by *Monilinia fructicola* (Winter) Honey to the EU territory and to identify risk management options and to evaluate their effectiveness in reducing the risk to plant health posed by this organism. The panel was also requested to provide an opinion on the effectiveness of the special requirements linked to *M. fructicola*, presently listed in Annex IV, Part A, Section I of Council Directive 2000/29/EC4, in reducing the risk of introduction of this pest into the EU territory. Having given due consideration to the evidence, the panel concluded as follows:

(a) Entry of *M. fructicola* by means of plant propagation material, fresh fruits of susceptible genera and by natural means from infested European non-EU countries is very likely. It is very unlikely in the case of dried fruit and natural means from infested non-European countries. In both cases, the level of uncertainty is low.

(b) Establishment of *M. fructicola* in the risk assessment area is very likely with a low level of uncertainty because of the availability of host plants with a long period of susceptibility and of suitable environmental conditions. Competition from other *Monilinia* species (*M. laxa* and *M. fructigena*) and currently applied cultural practices and control measures cannot prevent the establishment of the pest. In addition, the pest has already been detected in several member states in the risk assessment area (France, Germany, Hungary, Italy, Poland, Romania, Slovenia, and Spain).

(c) The spread of *M. fructicola* within the risk assessment area is very likely with a low level of uncertainty because of its multiple ways to spread (natural and human-assisted), to the wide distribution of host species in the risk assessment area and the absence of effective barriers.

(d) The potential for yield reduction and negative effects on fruit production in orchards is estimated as moderate, with a medium level of uncertainty mainly because of the incompleteness of data from the current area of distribution of the pest. Incidence and severity of the disease caused by the brown rot fungi, on flowers and twigs/branches are unlikely to increase compared to the situation in which only *M. fructigena* and *M. laxa* are present.

The panel identified the following risk management options as highly effective in reducing the likelihood of entry of *M. fructicola;* certification systems for plants for planting, control of the movement of fruit or propagation material consignments by legislation from infested non-European countries and management of fruit waste. The likelihood of establishment of *M. fructicola:* certification systems for plants for planting. The likelihood of spread and impact of *M. fructicola:* certification systems for plants for planting and packaging of fruit, sanitation of packaging, storage facilities and means of transport.

As described by the International Plant Protection Convention (IPPC), Pest Risk Analysis (PRA) includes three stages: initiation, pest risk assessment, and pest risk management. The Risk Management Document (RMD) is part of the pest risk management and communication stage. This RMD includes a summary of the findings of a pest risk assessment and records the pest risk management process for the identified issue. Consultation on the RMD, from a range of potentially affected stakeholders, is a key step prior to the approval and implementation of regulatory changes to the status of a pest (Anonymous, 2003).

In April 2002, a PRA for *Fusarium foetens* was finalized after the finding of an outbreak of the fungus in November 2000 in the Netherlands on a breeding/propagation company (Dirk Jan van der Gaag and Marcel Raak, 2010). At that time, the fungus and disease had not been described before. Morphological en phylogenetic studies described the fungus as a new species, distinct from *Fusarium begoniae* and other species within the *F. oxysporum* complex. The fungus causes basal rot, vein yellowing, and wilting symptoms causing rapid death of its primary host plant, *Begonia elatior* hybrids. It was suggested to list the fungus on Annex II A2 of the EU directive 2000/29/EC and to regulate the fungus for propagation material of *Begonia elatior*. Since an outbreak in January 2002 on a breeding/propagation company, the fungus is under official control for propagation material by the Plant Protection Service (NPPO of the Netherlands). In 2005, *Fusarium foetens* was added to the EPPO Alert list (EPPO Reporting Service 2005/111). In 2007, the fungus obtained an EPPO A2 status and at present it is on the EPPO Action List.

Since the first version of the PRA in April 2002, new information has become available about *Fusarium foetens,* especially about its impact under practical conditions and possibilities to control the disease. Since the detection of the disease, breeding, propagation and retail marketing companies of *Begonia elatior* have taken measures to control the pest. In 2004, The Dutch Plant Protection Service detected a second outbreak in propagation material and measures were imposed for eradication. To date, no other additional findings in propagation material are known in the Netherlands but the fungus is still present in companies growing pot plants for retail market. In this update of the first PRA, they described the experiences with the disease and evaluate control measures taken by companies since 2002. PRA area was the Netherlands.

Detected in 2000 for the first time, *F. foetens* was a new pathogen. At the time of first finding, the fungus and disease were new and had not been described before. In April

2002, a PRA was finalized by the NPPO of the Netherlands, and it was suggested to list the fungus on Annex II A2 of the EU directive 2000/29/EC and to regulate the fungus for propagation material of *Begonia elatior* because of its potentially high impact for Begonia pot plant industry. However, the fungus has not been regulated by the EU so far. Since an outbreak in January 2002 on a breeding company, the fungus is under official control for propagation material by the Plant Protection Service (NPPO of the Netherlands). The present PRA is an update of the previous PRA and has included new information that has become available since April 2002.

The conclusions of this updated PRA were infected propagation material was probably the only relevant pathway for *F. foetens*. Without any control measures, the fungus can easily be introduced and spread by import and trade of propagation material. *F. foetens* can potentially cause high yield losses, up to 100% in *Begonia* pot plants. *F. foetens* is still present on *Begonia* pot plant producing companies in the Netherlands despite official measures for propagation material implemented since January 2002. Hygienic measures taken by *Begonia* pot plant nurseries, especially disinfestations of the drain water after irrigation, are effective in controlling *F. foetens* and plant losses at pot plant producing companies due to *F. foetens* are presently low or minimal. It is not possible to assess to which extent the official measures implemented since 2002 in the Netherlands have contributed to the control of the pathogen. Observations suggested that voluntary hygienic measures taken by both pot plant producing companies and propagation companies have lead to effective control of the pathogen. Control/hygienic measures at pot plant producing companies will also be needed in the future even if the emergency measures for propagation material in the Netherlands are being continued because pot plant producing companies have not been able to completely eradicate the pathogen.

6.38.2 Bacterial Plant Pathogens

Following a request from the European Commission, the panel on Plant Health was asked to deliver a scientific opinion on 30 pest risk assessments made by France on organisms, which are considered by France as harmful in four French overseas departments of French Guiana, Guadeloupe, Martinique and Réunion (Anonymous, 2007). In particular, the panel was asked whether these organisms can be considered as harmful organisms for the endangered area of the above departments, in the meaning of the definition mentioned in Article 2.1.(e) of Directive 2000/29/EC and thus potentially eligible for addition to the list of harmful organisms in Directive 2000/29/EC. This document presents the opinion of the Panel on Plant Health on the simplified pest risk assessment conducted by France on '*Ralstonia* sp. bacterium responsible for banana blood disease' with French Guiana, Guadeloupe, Martinique and Réunion considered as endangered area. The panel examined in detail the risk assessment provided, and considered the accuracy and quality of the information and methods applied for pest risk assessment purposes. The review was based on the principles and terminology of the International Standard on Phytosanitary Measures ISPM No. 113: Pest risk analysis for quarantine pests including analysis of environmental risks and living modified organisms (2004) by the International Plant Protection Convention (FAO, 2007).

The panel observed that the pest risk assessment provides a very limited amount of information supported by scientific data to justify the conclusions reached. The probabilities of entry, establishment and spread and the potential impacts of BDB in the PRA area should have been assessed separately for each department, due to the differing importance of banana crops in each of the departments. Furthermore, the new situation in the PRA area, in case the current phytosanitary regulations governing the imports of banana plant material are lifted, should have been taken into account in the document. The panel also noticed the entry pathways should have been considered and assessed in the document in terms of importance and risk. French Guiana, which is part of the PRA area in the document, was not mentioned in the assessor's general conclusion.

The panel concludes that the identity of the pathogen can be defined as '*Ralstonia* sp. agent of banana blood disease,' also called blood disease bacterium (BDB). Although the taxonomical position of the pathogen is not yet assigned by the International Committee on Taxonomy of Plant Pathogenic Bacteria, BDB can be identified by cultural and molecular methods. The entry of BDB into the French overseas departments through natural dispersal is not likely due to their great distance from the areas of the pathogen's present distribution. Under the current regulations, the probability of entry of BDB on Musa (and ornamental *Heliconia* if proven a natural host) plant propagation material (other than vitroplants), banana fruit and soil would be low for French Guiana, Guadeloupe, Martinique and Réunion due to the import restrictions. Should the current regulations governing the imports of banana plant material be lifted, the probability of entry of BDB on Musa (and ornamental *Heliconia* if proven a natural host) plant propagation material (other than vitroplants), banana fruit and soil would be moderate. The probability of BDB entering the PRA area on vitroplants would be very low as such production systems are characterized by high sanitation practices that ensure pathogen-free propagation material, provided that mother plants are carefully screened prior to collection of material for *in vitro* multiplication. The probability of establishment and spread, should BDB gain entry into the PRA area, would be high, given that most of the endangered area has a climate similar to that of the pathogen's present distribution, and that Musa host plants are present in the PRA area. Within commercial plantations of bananas, the disease effects could be mitigated by existing practices of de-budding and bagging of male flowers. However, additional routine sanitation measures would be necessary at extra costs. With the possibility of continuous reinfestation from the environment, the production costs would be structurally increased. The additional control costs may reduce the competitiveness of the banana industry in Guadeloupe and Martinique. As a result, employment may be reduced causing negative social impacts. There is uncertainty regarding the level of the cost increase.

The bacterium *Xanthomonas fragariae* (Xf) causes angular leaf spot disease on strawberry (*Fragaria* spp.). It primarily affects the leaves including the fruit calyx on which it forms water-soaked angular spots (Van der Gaag *et al.*, 2013). Generally, (latently) infected planting material is the primary infection source of Xf in strawberry fruit cultivation, but introduction with contaminated machinery, tools, field workers, etc. is also possible. The bacterium may also survive in contaminated plant debris in the field from which it may be transmitted to new plantings. Natural spread of the disease from affected plants is (usually) limited to a few meters and results from dissemination of the bacteria by splashing of rain or irrigation water. Dispersal of Xf may also occur through manipulations of the plants before planting or in the field. The bacterial exudates can also dry on the leaf surface and form flaky silvery pellets containing viable Xf cells. These cells can survive for extended periods under dry conditions and during frost, and they get dispersed naturally or by human assistance. See the EPPO- and CABI-datasheet for more details on the biology of Xf.

6.38.2.1 Distribution of Xanthomonas fragariae

Xf is known to be present in at least 10 EU-countries. It is possibly more widespread than records indicate. Presence in Hungary and Slovakia, for example, is plausible because the pathogen has been intercepted on plants originating from these countries.

6.38.2.2 Reason for performing the PRA

At present, EU Council directive 2000/29/EC lists Xf as an Annex IIAII statutory organism on *Fragaria* plants intended for planting, other than seeds. *Fragaria* plants must either originate from a pest free country or area (within the EU) or a production place where no symptoms of Xf have been observed since the beginning of the last complete cycle of vegetation. However, the requirement of absence of symptoms does not ensure pest freedom, mainly because of the possibility of symptomless infection of planting material with Xf. Because Xf is already present in many EU-countries and is generally not considered a pest of major economic importance, removal of Xf from the EU quarantine list has been suggested. The current PRA evaluates the risk of Xf for the EU and can be used by risk managers to reconsider the current statutory position of the pathogen

6.38.2.3 Risk reduction options

Risk reduction of Xf dissemination through control of planting material; the current EU-requirement absence of Xf symptoms in *Fragaria* plants is not sufficient to ensure Xf freedom mainly because of the possibility of latent infections. Testing methods are available to detect latent infections, but the probability of detection largely depends on the sampling intensity and detection method used. Five possible (de)regulatory options for production and trade of planting material, ranging from most to least stringent, were discussed in the PRA, viz., pest free area, pest free production place based on visual inspections and testing for latent infections, pest freedom of the crop (field) based on visual inspections and testing for latent infections, pest freedom of the crop (field) based on visual inspections, and no EU regulation (Xf can be implemented in national certification schemes).

6.38.2.4 *Uncertainties*

The main uncertainties in the present PRA include the current distribution of Xf in the EU. Xf may be more widely distributed than present records indicate. Xf may have been spread throughout most of the strawberry producing areas in the PRA-area through trade of latently infected planting material, but only reported in areas where conditions are favorable for disease development and where growers have suffered damage.

6.38.2.5 *Current impact of Xf in the EU*

Only a few publications are available on yield effects. The current risk associated with import and trade of planting material because of possible latent infections. The cultivation methods in the different strawberry producing areas within in the EU, related potential impact of Xf and indication of most endangered areas.

6.38.2.6 *Possibility of seed transmission*

No studies are known on seed transmission; the host plant status of *Potentilla* spp. and *Fragaria* spp. other than *Fragaria xananassa* and the effect of Xf on the susceptibility of strawberry plants to other diseases.

6.38.3 Viral Diseases

Following a request from the European Commission, the EFSA panel on Plant Health was asked to deliver a scientific opinion on 30 pest risk assessments made by France on organisms which are considered by France as harmful in four French overseas departments, i.e., French Guiana, Guadeloupe, Martinique and Réunion (Anonymous, 2008). In particular, the panel was asked whether these organisms can be considered as harmful organisms for the endangered area of the above departments, in the meaning of the definition mentioned in Article 2.1.(e) of Directive 2000/29/EC and thus potentially eligible for addition to the list of harmful organisms in Directive 2000/29/EC. This document presented the opinion of the panel on Plant Health on the simplified pest risk assessment conducted by France on Citrus yellow mosaic virus or Citrus mosaic badnavirus, with French Guiana, Guadeloupe, Martinique and Réunion considered as endangered area. The panel has used the internationally accepted name of Citrus mosaic virus (CMBV) in the text.

CMBV is a member of the family Caulimoviridae, genus *Badnavirus,* which causes a mosaic disease in citrus that has been termed citrus mosaic disease or citrus yellow mosaic disease. This virus is currently restricted to India. The causal agent is transmitted by grafting and experimentally, by mechanical inoculation and by the citrus mealybug, *Planococcus citri* (Risso). CMBV causes mosaic and yellow flecking along the veins of the leaves. Trees affected by the disease not only show significant yield reduction, but also a decrease in the quality of fruits (less juice and ascorbic acid). The virus can infect and cause symptoms in almost all citrus species and cultivars. The panel examined in detail the risk assessment provided, and considered the accuracy and quality of the information provided and methods applied for pest risk assessment purposes. The review was based on the principles and terminology of the International Standard on Phytosanitary Measures ISPM No. 113 [*Pest risk analysis for quarantine pests including analysis of environmental risks and living modified organisms* (2004)] and the terminology of ISPM No. 5 [*Glossary of Phytosanitary Terms* (2007)] by the International Plant Protection Convention (FAO, 2007b).

The French document concludes that the phytosanitary risk associated with CMBV and citrus mosaic disease is low to moderate. The panel notes that the main concern highlighted in the French document relates to the uncertainty in the regulatory status of CMBV, caused by past confusion in taxonomy. An analysis of the French document, the references cited in it and additional literature consulted by the panel, has allowed the panel to reach to following conclusions, i.e., citrus mosaic virus (CMBV) is the causal agent of a mosaic disease in citrus that has been termed citrus mosaic disease or citrus yellow mosaic disease. It is listed (Annex IIAI) in Directive 2000/29/EC as citrus mosaic virus but has been previously known as citrus yellow mosaic virus or citrus mosaic badnavirus. CMBV is a well-characterized badnavirus, and there are reliable molecular methods available to distinguish it from another virus causing mosaic symptoms in citrus. Satsuma dwarf virus (SDV), a Sadwavirus which was previously referred to as citrus mosaic virus (CiMV) and is also listed in Directive 2000/29/EC. CMBV occurs in India, but not in the PRA area nor other citrus-growing regions worldwide. The known host range of the CMBV is limited to *Citrus* spp. The main pathway of entry in the PRA area is through the import from India of citrus plants or bud wood for multiplication. There is no demonstration that CMBV is spread in nature by its experimental vector *P. citri.* Mealy bugs are generally considered to be inefficient vectors and spread of CMBV in the PRA area by *P. citri* is therefore estimated to be of limited potential impact. Substantial uncertainty is however attached to this evaluation: if CMBV is introduced in the PRA area in infected citrus material, it will remain for at least as long as the infected material is cultivated and propagated. The citrus mosaic disease can cause severe losses in sweet orange, mandarins, pummelo and grapefruit orchards and ornamental citrus plants, but not in lime. It, therefore, has the potential to cause significant damage to the citrus growing industry of the various French overseas departments under consideration.

The overall conclusion of the panel was that CMBV could pose a risk to citrus production in the French overseas departments and that this pathogen is listed in Annex IIAI of Directive 2000/29/EC as citrus mosaic virus.

Following a request from the European Commission, the Panel on Plant Health was asked to deliver a scientific opinion on 30 pest risk assessments made by France on organisms which are considered by France as harmful in four French overseas departments, i.e., French Guiana, Guadeloupe, Martinique, and Réunion. In particular, the panel was asked whether these organisms can be considered as harmful organisms for the endangered area of the above departments, in the meaning of the definition mentioned in Article 2.1.(e) of Directive 2000/29/EC and thus potentially eligible for addition to the list of harmful organisms in Directive 2000/29/EC. This document contains the opinion of the Panel on Plant Health on the full pest risk assessment conducted by France on Banana mild mosaic virus (BanMMV) with French Guiana, Guadeloupe, Martinique and Réunion considered as endangered area. Banana mild mosaic virus (BanMMV) is a recently characterized virus of the family Flexiviridae which infects banana and plantains. BanMMV has a very wide geographic distribution and its presence has been reported in Guadeloupe and Martinique.

The panel examined in detail the risk assessment provided, and considered the accuracy and quality of the information provided and methods applied for pest risk assessment purposes. The review was based on the principles of the International Standard on Phytosanitary Measures ISPM No. 113: Pest risk analysis for quarantine pests including analysis of environmental risks and living modified organisms (2004) by the International Plant Protection Convention (FAO, 2007b).

Although the French pest risk assessment could be improved upon, the panel agrees with the conclusion of the French document that the phytosanitary risk of BanMMV as a single virus is extremely low. The panel agreed with the French analysis that BanMMV has a high probability of entry and establishment and that the international movement of large quantities of vitroplants represents the most important entry pathway. The panel further considered that virus indexing of mother plants used for vitroplants production could provide an effective means to reduce the probability of entry of BanMMV in the area under consideration. Although BanMMV is a pathogenic agent for banana, the panel concluded that the economic consequences of its further entry in the French overseas departments could be considered low since BanMMV, which is present in most if not all banana-growing areas of the world, does not appear to be associated with significant yield loss in banana.

However, the panel recognized that there are several areas of BanMMV biology where information is incomplete or lacking. These areas of uncertainty include clarification of the host range of BanMMV, its symptom expression (in particular when in mixed infection with other banana-infecting viruses) and the possibility that the virus may have further potential for spread by a vector. The overall conclusion of the panel was that BanMMV was not appropriate for the analysis of risk management options for French Guiana, Guadeloupe, Martinique and Réunion and was thus not potentially eligible for addition to the list of harmful organisms of Directive 2000/29/EC.

6.38.4 Phytonematodes

Heterodera glycines, the soybean cyst nematode, is currently on the list of pests regulated by Canada. Schedule II of the Plant Protection Regulations restricts the domestic movement of soybean seed for propagation, soybean harvesting equipment, machinery, soil and any other thing suspected of being infested with the pest. To prevent the introduction of *H. glycines,* the Canadian Food Inspection Agency (CFIA) regulates imports of soybean seed, potatoes, root crops, and soil and soil-related matter, alone or in association with plants. Soybean cyst nematode has been considered one of the most destructive pests of soybean. The soil is the primary pathway for the movement of this pest. Plant commodities, farm equipment and natural means such as water, air, and birds, also move infested soil.

The Pest Risk Assessment conducted by the CFIA in 2010 for *H. glycines* determined that the overall risk of the pest was 'low' and supported the possibility of deregulation. In Canada, surveys have detected this nematode in most Ontario soybean producing counties, since the first detection in Kent County in 1987. *Heterodera glycines* infested counties account for about 75% of the soybeans produced in Ontario and one-half of the total Canadian soybean production. The limitation in enforcing domestic regulations and movement restrictions, along with the lack of movement compliance, may have been a contributing factor to the spread of the pest to new areas in Ontario. The biology of *H. glycines* indicates that cryptic spread is ongoing both within the province of Ontario and potentially to other regions. The United States of America has not regulated this pest since 1972, either at the federal or state level because domestic regulations and movement restrictions were unsuccessful in preventing the spread. The nematode has been effectively managed through the use of best management practices, including crop rotation with non-host crops and planting resistant soybean varieties.

Risk management consideration was given to the aspects like the challenges in enforcing domestic movement regulations to prevent the spread of the pest from infested areas, the lack of control over the natural spread of the pest. Due to a discrepancy over strict import requirements versus limitations in enforcing domestic movement requirements, Canada faces challenges in meeting its obligations under the IPPC. Limitation of soil surveys to map the distribution of the pest and highly expensive and resource intensive quarantine measures that need to be enforced if *status quo* is to be maintained (Anonymous, 2003).

After consultation and due consideration of stakeholder feedback, the CFIA is deregulating *H. glycines* in Canada and communicating the decision through this RMD. The deregulation is not expected to cause any significant market access issues, as most of Canada's trading partners, including the USA, have never regulated or have already de-regulated this pest. The purpose was to deregulate *Heterodera glycines* Ichinohe, the soybean cyst nematode, in Canada. This pest risk management document pertained to Canada's policies (plant import and domestic movement) to address the risks from *H. glycines*. *Heterodera glycines* is currently on the *List of Pests Regulated by Canada*. Under Schedule II of the *Plant Protection Regulations*, a movement certificate is to be produced for items like soil or compost material, soybean harvesting equipment, machinery and any other thing suspected of being infested with the pest including soybean (*Glycine max*) seed for propagation that is transported from an area of infestation to any other area of Canada. In Canada, *H. glycines* has been regulated for more than three decades. The following Canadian Food Inspection Agency (CFIA) Plant Health Directives include requirements that regulate imports to prevent the entry of *H. glycines* into Canada or prevent its spread, through infested soil, within Canada:

Heterodera glycines has been considered one of the most destructive pests of soybean. Wrather and coworkers (2001) reported that, both in Canada and the USA, yield loss due to *H. glycines* was higher than that caused by any other pest or pathogen of soybean. Yield loss to this nematode is an ongoing issue in regions where the pest is not effectively managed through the use of crop rotation with non-host crops, and the use of appropriate resistant soybean varieties. Soil is the primary pathway for the movement of this pest.

The spread of *H. glycines* via soil associated with crops, commodities, seed, horticultural plants, machinery, footwear, etc., is primarily human mediated. Cysts can also move via the feet of birds and other wildlife; via water (e.g., flooding) and wind, suggesting that the pest will eventually spread to all soybean production areas in North America.

Since its first detection in Kent County, Ontario in 1987, the pest has been detected in soybean fields of several counties in Ontario including Essex, Elgin, Huron, Haldimand, the Municipality of Chatham-Kent, Lambton, Middlesex, Norfolk, Oxford, the United Counties of Prescott-Russell, Brant, Regional Municipality of Niagara, Northumberland, the City of Ottawa, the Regional Municipality of Peel, Perth, the United Counties of Stormont, Dundas and Glengarry, and City of Kawartha Lakes. *Heterodera glycines*-infested countries account for about 75% of the soybeans produced in Ontario and one-half of the total Canadian soybean production. In 2009, a single cyst was detected in a soil sample from the Rural Municipality of South Norfolk in Manitoba. However, a more intensive follow-up survey of the same field in 2010 did not confirm the presence of the pest (Anonymous, 2003).

Domestic movement restrictions of soil from *H. glycines* infested areas have been difficult to enforce due to the wide geographic area of the infested counties in Ontario, the lack of knowledge of movement activities and compliance of growers to the movement of regulated articles, which includes equipment and agricultural commodities associated with infested soil. The expanding range of *H. glycines* in Ontario attests to the continuous spread from infested counties to other areas. The pest, since its reported first detection in Ontario, has been controlled through effective best management practices, which include development and use of *H. glycines* resistant soybean varieties through industry-government partnerships. The *H. glycines* resistant soybean varieties are grown throughout the soybean production areas of Ontario.

Prior to 1972, United States Department of Agriculture's (USDA) domestic quarantines for *H. glycines* contained provisions that were equivalent to, if not more stringent than, those contained in current Canadian regulations relating to *H. glycines*. Despite this, the USDA abandoned their federal quarantines in 1972 as they had proven ineffective in limiting the spread of *H. glycines*, due in part to limitations in enforcement, lack of compliance with regulatory requirements, and natural means of spread of the pest to new areas. The pest is considered widespread in the soybean production areas of the USA. In the absence of USA regulation, *H. glycines* will continue to spread undetected to new areas of the USA, and repeatedly into Canada via established pathways of introduction. New infestations have almost certainly occurred, but will not be reported until populations build to detectable levels. In North Dakota, it was estimated that *H. glycines* had been present for 5 to 7 years before the symptoms in the field led to detection by soil tests (Bradley *et al.*, 2003).

Given that numerous states of the USA could be infested and remain unregulated, that soil surveys are of limited sensitivity, and that some *H. glycines* positive counties in Ontario are not currently regulated; current Canadian regulations and practices are unable to prevent further spread of this pest. The challenges in enforcing strict domestic movement requirements versus the rigorous enforcement of import control on host material are inconsistent with Canada's obligations under the World Trade Organization-Sanitary and Phytosanitary Measures (WTO-SPS).

6.39 RISK MANAGEMENT CONSIDERATIONS

International Standards for Phytosanitary Measures and Canada's obligations: The IPPC has been identified as the international phytosanitary standard-setting authority by the WTO. The IPPC Standard for Phytosanitary Measures (ISPM) No. 19, Guidelines on Lists of Regulated Pests (2003), requires that pests regulated by National Plant Protection Organizations meet the criteria for either quarantine pests or regulated nonquarantine pests. To be considered a quarantine pest according to the IPPC's definition, an organism must be "a pest of potential economic importance to the area endangered thereby and not yet present there, or present but not widely distributed and being officially controlled". In 2001, the Interim Commission on Phytosanitary Measures approved Guidelines on the Interpretation and Application of the Concept of Official Control for Regulated Pests (see Supplement No. 1 in ISPM No. 5). These guidelines include, among other elements, that measures be mandatory, and that domestic and import requirements should have an equivalent effect. Thus, for Canada to comply with WTO and IPPC guidelines and continue to regulate *H. glycines* as a quarantine pest on imports, it would have to strengthen significantly the enforcement of domestic movement restriction from any infested area in Canada (Anonymous, 2003).

6.40 CONSIDERATIONS FOR PEST RISK MANAGEMENT

6.40.1 Domestic Movement Regulations

The limitation in enforcing domestic regulation along with the lack of movement compliance may have been a contributing factor to the spread of the pest to new areas in Ontario. The recent case of *H. glycines* in Manitoba is a good example which reiterates the challenges in enforcing domestic movement regulations over a wider area. The single cyst detection was in a field well outside the flood zone of the Red River, and on a site where there was no record of previous soybean production. Thus, the likelihood of the pathway of the introduction of *H. glycines* to this site may have been the seed from an infested field or spread of infested soil using agricultural equipment from other fields where soybeans are produced. However, it is noted that further cysts were not detected during a intensive follow-up survey of that particular field in Manitoba.

6.40.2 Spread by Natural Pathways

Natural pathways, such as wind, water, and birds, also contribute to the spread of the nematode to new areas. These natural pathways and the spread cannot be controlled by regulations and enforcements. The continuous spread of the pest in Ontario to new production areas though cryptic could be attributed to these natural pathways. The Manitoba soybean production area is in the Red River Valley, which is contiguous with the production areas of Minnesota and North Dakota, where *H. glycines* has been reported within 300 km of the border. Considering the potential for annual flooding in this area it is very likely that *H. glycines* has spread into the soybean production areas of Manitoba along the Red River. Similarly, *H. glycines* has been reported in Eastern Ontario in 2008 and is likely present in the neighboring soybean production area extending down the St. Lawrence valley.

6.40.3 Soil Surveys

To effectively regulate *H. glycines* in Canada, additional extensive surveys would be required to determine the current distribution of the pest in soybean production areas. Soil surveys, which are highly resource-intensive exercises, would never fully describe the true distribution of this pest, given its biology and pathways of spread. It is reported that the population of the pest has to build to a detectable level before it can be confirmed by soil surveys.

6.40.4 Enforcement Requirements on Infested Fields

To maintain the regulated status of *H. glycines,* containment measures would then have to be implemented by the CFIA that would parallel the current measures in place to contain the two species of potato cyst nematode that are present in Canada, *Globodera pallida* and *G. rostochiensis,* as per the Golden Nematode Order, SOR/80-260 (Saanich, British Columbia) and Golden Nematode Infested Places Order (Quebec) (Anonymous, 2003).

Quarantine measures would include immediate issuance of a notice of quarantine or equivalent regulatory order, categorization of lands in the infested zone as infested, exposed and adjacent. All these categorized lands will be placed under quarantine – restriction on the production of all host and minor hosts of *H. glycines,* prohibition on movement of soil and equipment associated with infested soil, prohibition on movement of plant and plant parts with infested soil, restriction on the movement of equipment and plant material from other category fields in the quarantine area, restriction on movement of grains intended for other end uses, and establishment of various compliance agreements for facilities handling regulated plant and plant parts originating from regulated areas. It would be prohibitively costly and resource intensive to implement such a rigorous regulatory program, and it would severely impact most soybean seed growers and those agricultural sectors, such as equipment operators in Ontario, that move soil, either intentionally or unintentionally.

6.40.5 Impact on Exports and Imports

The de-regulation of *H. glycines* in Canada is not expected to have any impact on market access based on the current requirements of trading partners. Freedom from the soil in exports of seed and grain from Canada addresses the phytosanitary concerns of trading partners for this pest. The costs of implementing stricter quarantine measures, combined with the lack of control over the natural spread of *H. glycines,* make an investment in strengthening a regulatory program questionable in terms of benefit to stakeholders, particularly when market access for soybeans is not an issue, and the pest is managed by growers through current crop management practices. Once deregulated, Canada can remove phytosanitary requirements on the import of soybean seed, as well as potatoes (seed, table, and processing) from some states of the USA, which are currently regulated by the presence of soybean cyst nematode. This is crucial in complying with the existing obligations under the WTO and IPPC.

6.40.6 Risk Management Decision

The initial risk management discussion document, proposing the recommended option of deregulating *H. glycines* all across Canada, was circulated for stakeholder consultation in February 2011, and stakeholder feedback was received (Anonymous, 2003). While stakeholders from other provinces expressed support for deregulation of *H. glycines*, stakeholders from Quebec and Manitoba preferred continued regulation of *H. glycines*. Further, in December 2011, a Question and Answer document was distributed, which addressed specific concerns raised by stakeholders. Further discussions with concerned Quebec and Manitoba stakeholders were conducted in summer of 2012 to better understand their position, address specific concerns and provide a timeline for the implementation of the deregulation of *H. glycines*. There was a concern from Manitoba and Quebec that the proposed deregulation will introduce *H. glycines* in their soybean growing areas, where this pest has not been reported yet. During the consultation, the CFIA emphasized that all imports of seed and grain are required to be free of soil, as the soil is the primary pathway for introduction of *H. glycines*.

The mandate of the CFIA is to prevent the introduction of pests/pathogens and to protect the agricultural resource base of Canada. The current status of *H. glycines*, widespread presence in Ontario, is such that it needs to move from being a quarantine pest to a management pest. It was also noted that despite the presence of *H. glycines*, the soybean acreage in Ontario has been increasing over the last two decades, which is largely attributed to the use of best management practices involving the development and use of *H. glycines* resistant soybean varieties. The discrepancy between Canada's strict import requirements and inconsistent enforcement of domestic regulations for *H. glycines* can be perceived as unjustified discrimination or a disguised restriction on international trade of some plant commodities. This makes Canada vulnerable to WTO challenges from trading partners. The decision was *Heterodera glycines* no longer fits the definition of a quarantine pest.

By way of this risk management document, the CFIA is announcing the deregulation of *H. glycines* in Canada. As of November 25, 2013, the CFIA will not enforce the import regulations related to *H. glycines*. The CFIA will not be enforcing the Schedule II of the *Plant Protection Regulations* (*PPR*) pertaining to the domestic movement restriction of *H. glycines* infested material. The risk-based decision was made considering several factors, including the low risk of the pest, the utilization of *H. glycines* best management practices and discrepancy in domestic and import regulations. The soil is a prohibited substance, and all imports are required to be free of soil. The implementation of the deregulation of *H. glycines* would require that relevant plant health import directives be amended to remove reference to requirements for freedom from *H. glycines* or revoked.

The Automated Import Reference Systems (AIRS) will also be updated to reflect the deregulation of *H. glycines*. As part of the overall regulatory review process, the CFIA will work towards the removal of *H. glycines* from the Schedule II of the *Plant Protection Regulations* and the *List of Pests Regulated by Canada*.

6.41 PEST RISK MANAGEMENT OPTIONS

Table 6.7 discusses three pest risk management options considered, with the advantages and disadvantages listed for each (Anonymous, 2003).

Table 6.7 Pest Risk Management Options

Options	*Advantages*	*Disadvantages*
Status quo: Regulated Pest Status of *Heterodera glycines* Retain *H. glycines* in the *List of Pests Regulated by Canada.* and *Schedule II of Plant Protection Regulations* on Restricted Movement within Canada. and Implement official control measures if *H. glycines* is found in a new area.	Control over the import of seeds, potatoes, and associated soil. Control over Domestic Movement of seed, and soil associated with seed, potato tubers, horticultural root crops, plant material, farm machinery and equipment, etc. Authority to respond to positive finds and implementing official control measures. Authority to contain the pest by enforcing mitigation measures at the point of origin, if from an infested area.	Already present in large areas and numerous soybean production counties of Ontario. No control over natural pathways of pest introduction, such as wind, floods, birds, etc. More resources needed to enforce domestic movement restrictions. Potential additional costs to the growers to follow quarantine measures. Additional costs in establishing and maintaining compliance agreements to handle regulated plant and plant parts from regulated areas Potential additional costs to equipment operators (e.g., cleaning costs). Potential for negative impact on trading partners and trading relationships due to continued oversight of imported commodities. Potential for WTO challenges from trading partners over discrepancy in domestic and import regulations, if domestic regulations are not enforced.
Declare Ontario as 'infested area' and continue regulation of *H. glycines* in rest of Canada. Retain *H. glycines* in the *List of Pests Regulated by Canada.* and	Authority to respond to positive finds at a location outside Ontario and implement official quarantine measures. Quarantine measures would include:	Domestic movement restrictions apply from Ontario to rest of Canada. No control over natural pathways of pest introduction, such as wind, floods, birds, etc. Potential impact on trade from Ontario.

Schedule II of Plant Protection Regulations on Restricted Movement within Canada. and Make changes to *Schedule II of Plant Protection Regulations* on Restricted Movement from Ontario to the rest of Canada. and Implement official control measures if *H. glycines* is found in a new area.	1. Declaring Ontario as a 'soybean cyst nematode-infested area.' 2. Immediate issuance of notice of quarantine or equivalent regulatory order. 3. Categorization of lands in the infested zone, as infested, exposed and adjacent. All these categorized lands will be under quarantine. 4. Restriction on production of all host and minor hosts of *H. glycines*. 5. Restriction on the movement of soil and equipment associated with infested soil. 6. Restriction on the movement of plant and plant parts with infested soil. 7. Restriction on the movement of equipment and plant material from other category fields in the quarantine area. 8. Restriction on the movement of grains intended for other end uses. Control over the import of seeds, potatoes, horticultural root crops and associated soil into Canada except Ontario. Requirement for procurement of seed and seed potatoes intended for planting outside Ontario, from a field or area free of *H. glycines* Control over Domestic Movement of seed, and soil associated with seed, potato tubers, horticultural root crops, plant material, farm machinery and equipment, etc., from Ontario to the rest of Canada. Better preparedness through official surveys, in soybean production areas outside Ontario.	More resources, potential partnerships with provinces and service providers needed to: 1. Apply potential quarantine measures. 2. Enforce domestic movement restrictions. 3. To conduct extensive annual surveys. Potential additional costs to the grower/regulated field owner when quarantine measures implemented: restriction on sale and movement of equipment, cleaning costs, etc. Potential impact on soybean and potato seed producers in Ontario. Additional costs in establishing and maintaining compliance agreements to handle regulated plant and plant parts from regulated areas. Potential additional costs to Ontario equipment operators, like cleaning costs. Potential for negative impact on trading partners and trading relationships due to more stringent oversight of imported commodities. Potential for WTO challenges from trading partners over a discrepancy in domestic and import regulations, if domestic regulations are not enforced.
De-regulate *H. glycines* across Canada	Meet international obligations to IPPC, WTO-SPS, by adopting a harmonized approach to imports and domestic policies.	No authority to regulate *H. glycines* on imports

Contd...

Contd...

Remove *H. glycines* from the *List of Pests Regulated by Canada.* and Remove *H. glycines* from Schedule II of the *Plant Protection Regulations* on Domestic Movement within Canada	No additional resources and costs for enforcement of regulations pertaining to *H. glycines*. No additional requirements and costs, related to *H. glycines* certification, for exporters to Canada. No domestic movement restrictions for commodities and equipment associated with *H. glycines*. Unrestricted movement between *H. glycines* infested counties/ provinces, states in the USA asnd the rest of Canada. Soil – The most significant pathway for movement of *H. glycines* is regulated in imports.	

6.42 PEST RISK ANALYSIS (PRA) IN PLANT QUARANTINE

In view of the manifold increase in the quantum of import and export of the plant commodities during the current years, the existing plant quarantine process seems to be far from being satisfactorily equipped. To cope with these difficulties there hamper free market accessibility to the plant commodities and the quarantine should be based on the PRA (Rajak *et al.*, 1999). Risk can be defined as expected magnitude of loss. Hence, pest risk analysis aimed at identifying, quantifying and reporting of the possible management options on account of probable, introduction of a pest along with an imported plant commodity. Prime goal of PRA is to ensure healthy transaction of agricultural commodities in the global trade.

Analysis of pest risk while introducing seeds planting materials is essential to determine the potential of a post to cause damage. In general, risks are more with the introduction vegetative propagules than true seed. In the case of true seed, risks are more with deep-seated infections than surface contaminated seed. Further, risks are more with virus, viroids, smuts, downy mildew, etc. Risk increases with the size of import. Bulk imports create more quarantine problems. (*Example: Fusarium oxysporum* f. sp. *gladioli* on bulk consignments of gladioli bulbs imported from Holland through Bombay airport in 1989-90.) *Agrobacterium tumefaciens* was intercepted at Madras port in 40000 rooted cuttings imported from Holland during 1988 and in two consignments each of 20000 rose cuttings imported through Bombay airport in 1991 (Agarwal *et al.*, 1990).

6.42.1 Stage 1 - Initiation

The aim of the initiation stage is to identify the pest(s) and pathways which are of quarantine concern and should be considered for risk analysis in relation to the identified PRA area.

6.42.1.1 Initiation points

The PRA process may be initiated as a result of the identification of a pathway that presents a potential pest hazard, the identification of a pest that may require phytosanitary measures and the review or revision of phytosanitary policies and priorities. The initiation points frequently refer to 'pests.' The IPPC defines a pest as "any species, strain or biotype of plant, animal, or pathogenic agent, injurious to plants or plant products." In applying these initiation points to the specific case of plants as pests, it is important to note that the plants concerned should satisfy this definition. Pests directly affecting plants satisfy this definition. In addition, many organisms indirectly affecting plants also satisfy this definition (such as weeds/invasive plants). The fact that they are injurious to plants can be based on evidence obtained in an area where they occur. In the case of organisms where there is insufficient evidence that they affect plants indirectly, it may nevertheless be appropriate to assess on the basis of available pertinent information, whether they are potentially injurious in the PRA area by using a clearly documented, consistently applied and transparent system. This is particularly important for plant species or cultivars that are imported for planting.

6.42.1.2 PRA initiated by the identification of a pathway

The need for a new or revised PRA of a specific pathway may arise in the situations like international trade which is initiated in a commodity not previously imported into the country (usually a plant or plant product, including genetically altered plants), or a commodity from a new area or new country of origin. New plant species are imported for selection and scientific research purposes and a pathway other than commodity import is identified (natural spread, packing material, mail, garbage, passenger baggage, etc.). A list of pests likely to be associated with the pathway (e.g., carried by the commodity) may be generated by any combination of official sources, databases, scientific and other literature, or expert consultation. It is preferable to prioritize the listing, based on expert judgment on pest distribution and types of pests. If no potential quarantine pests are identified as likely to follow the pathway, the PRA may stop at this point.

6.42.1.3 PRA initiated by the identification of a pest

A requirement for a new or revised PRA on a specific pest may arise in the situations some of which are – an emergency arises on discovery of an established infestation or an outbreak of a new pest within a PRA area, an emergency arises on interception of a new pest on an imported commodity, scientific research identifies a new pest risk, a pest is introduced into an area, a pest is reported to be more damaging in an area other than in its area of origin, a pest is repeatedly intercepted, a request is made to import an organism, an organism is identified as a vector for other pests, and an organism is genetically altered in a way which clearly identifies its potential as a plant pest.

6.42.1.4 *PRA initiated by the review or revision of a policy*

The requirement for a new or revised PRA originating from policy concerns will most frequently arise in the situations like - when a national decision is taken to review phytosanitary regulations, requirements or operations, a proposal is made by another country or by an international organization (RPPO, FAO) is reviewed, a new treatment or loss of a treatment system, a new process, or new information impacts on an earlier decision, a dispute arises on phytosanitary measures and the phytosanitary situation in a country changes, and a new country is created or political boundaries have changed.

6.42.1.5 *Identification of PRA area*

The PRA area should be defined as precisely as possible to identify the area for which information is needed.

6.42.1.6 *Information*

Information gathering is an essential element of all stages of PRA. It is important at the initiation stage to clarify the identity of the pest(s), its/their present distribution and association with host plants, commodities, etc. Other information will be gathered as required to reach necessary decisions as the PRA continues. Information for PRA may come from a variety of sources. The provision of official information regarding pest status is an obligation under the IPPC (Art. VIII.1c) facilitated by official contact points (Art. VIII.2). For environmental risks, the variety of sources of information will generally be wider than traditionally used by NPPOs. Broader inputs may be required. These sources may include environmental impact assessments, but it should be recognized that such assessments usually do not have the same purpose as PRA and cannot substitute for PRA.

6.42.1.7 *Previous PRA*

A check should also be made as to whether pathways, pests or policies have already been subjected to the PRA process, either nationally or internationally. If a PRA exists, its validity should be checked as circumstances and information may have changed. The possibility of using a PRA from a similar pathway or pest that may partly or entirely replace the need for a new PRA, should also be investigated.

6.42.1.8 *Conclusion of initiation*

At the end of Stage 1, the initiation point, the pests, and pathways of concern and the PRA area will have been identified. Relevant information has been collected, and pests have been identified as possible candidates for phytosanitary measures, either individually or in association with a pathway.

6.42.2 Stage 2 – Pest Risk Assessment

The process of pest risk assessment can be broadly divided into three interrelated steps, viz., pest categorization, assessment of the probability of introduction and spread and assessment of potential economic consequences (including environmental impacts). In most cases, these steps will be applied sequentially in a PRA but it is not essential to follow a particular sequence. Pest risk assessment needs to be only as complex as is technically justified by the circumstances. This standard allows a specific PRA to be judged against the principles of necessity, minimal impact, transparency, equivalence, risk analysis, managed risk and nondiscrimination set out in ISPM No. 1: *Principles of plant quarantine as related to international trade* (FAO, 1995).

6.42.2.1 *Pest categorization*

At the outset, it may not be clear which pest(s) identified in Stage 1 require a PRA. The categorization process examines for each pest whether the criteria in the definition for a quarantine pest are satisfied. In the evaluation of a pathway associated with a commodity, a number of individual PRAs may be necessary for the various pests potentially associated with the pathway. The opportunity to eliminate an organism or organisms from consideration before the in-depth examination is undertaken is a valuable characteristic of the categorization process. An advantage of pest categorization is that it can be done with relatively little information; however information should be sufficient to adequately carry out the categorization.

6.42.2.2 *Elements of categorization*

The categorization of a pest as a quarantine pest includes the following primary elements, viz., the identity of the pest, presence or absence in the PRA area, regulatory status, potential for establishment and spread in PRA area and potential for economic consequences (including environmental consequences) in the PRA area.

6.42.2.3 *Identity of the pest*

The identity of the pest should be clearly defined to ensure that the assessment is being performed on a distinct organism and that biological and other information used in the assessment is relevant to the organism in question. If this is not possible because the causal agent of particular symptoms has not yet been fully identified, then it should have been shown to produce consistent symptoms and to be transmissible. The taxonomic unit for the pest is generally species. The use of a higher or lower taxonomic level should be supported by scientifically sound rationale. In the case of levels below the species, this should include evidence demonstrating that factors such as differences in virulence, host range or vector relationships are significant enough to affect phytosanitary status. In cases where a vector is involved, the vector may also be considered a pest to the extent that it is associated with the causal organism and is required for transmission of the pest.

6.42.2.4 *Presence or absence in PRA area*

The pest should be absent from all or a defined part of the PRA area.

6.42.2.5 *Regulatory status*

If the pest is present but not widely distributed in the PRA area, it should be under official control or expected to be under official control in the near future. Official control of pests presenting an environmental risk may involve agencies other than the NPPO. However, it is recognized that ISPM No. 5 Glossary of phytosanitary terms, Supplement No. 1 on official control, in particular Section 5.7, applies.

6.42.2.6 *Potential for establishment and spread in PRA area*

Evidence should be available to support the conclusion that the pest could become established or spread in the PRA area. The PRA area should have ecological/climatic conditions including those in protected conditions suitable for the establishment and spread of the pest and where relevant, host species (or near relatives); alternate hosts and vectors should be present in the PRA area.

6.42.2.7 *Potential for economic consequences in PRA area*

There should be clear indications that the pest is likely to have an unacceptable economic impact (including environmental impact) in the PRA area. The unacceptable economic impact is described in ISPM No. 5 *Glossary of phytosanitary terms*, Supplement No. 2: *Guidelines on the understanding of potential economic importance and related terms.*

6.42.2.8 *Conclusion of pest categorization*

If it has been determined that the pest has the potential to be a quarantine pest, the PRA process should continue. If a pest does not fulfill all of the criteria for a quarantine pest, the PRA process for that pest may stop. In the absence of sufficient information, the uncertainties should be identified, and the PRA process should continue.

6.42.2.9 *Assessment of the probability of introduction and spread*

Pest introduction is comprised of both entry and establishment. Assessing the probability of introduction requires an analysis of each of the pathways with which a pest may be associated from its origin to its establishment in the PRA area. In a PRA initiated by a specific pathway (usually an imported commodity), the probability of pest entry is evaluated for the pathway in question. The probabilities for pest entry associated with other pathways need to be investigated as well. For risk analyses that have been initiated for a specific pest, with no particular commodity or pathway under consideration, the potential of all probable pathways should be considered. The assessment of the probability of spread is based primarily on biological considerations similar to those for entry and establishment. With respect to a plant being assessed as a pest with indirect effects, wherever a reference is made to a host or a host range, this should be understood to refer instead to a suitable habitat (that is a place where the plant can grow) in the PRA area. The intended habitat is the place where the plants are intended to grow, and the unintended habitat is the place where the plants are not intended to grow. In the case of plants to be imported, the concepts of entry, establishment and spread have to be considered differently. Plants for planting that are imported will enter and then be maintained in an intended habitat, probably in substantial numbers and for an indeterminate period. Accordingly, Section 2.2.1 on Entry does not apply. The risk arises because of the probability that the plant may spread from the intended habitat to unintended habitats within the PRA area and then establish in those habitats. Accordingly, section 2.2.3 may be considered before Section 2.2.2. Unintended habitats may occur in the vicinity of the intended habitat in the PRA area. Imported plants not intended to be planted may be used for different purposes (e.g., used as bird seed, as fodder, or for processing). The risk arises because of the probability that the plant may escape or be diverted from the intended use to an unintended habitat and establish there.

6.42.2.10 *Probability of entry of a pest*

The probability of entry of a pest depends on the pathways from the exporting country to the destination and the frequency and quantity of pests associated with them. The higher the number of pathways, the greater is the probability of the pest entering the PRA area. Documented pathways for the pest to enter new areas should be noted. Potential pathways, which may not currently exist, should be assessed. Pest interception data may provide evidence of the ability of a pest to be associated with a pathway and to survive in transport or storage. In the case of plants to be imported, the plants will enter, and an assessment of the probability of entry will not be required. Therefore, this section does not apply. However, this section does apply to pests that may be carried by such plants (e.g., weed seeds with seeds imported for planting).

6.42.2.11 *Identification of pathways for a PRA initiated by a pest*

All relevant pathways should be considered. They can be identified principally in relation to the geographical distribution and host range of the pest. Consignments of plants and plant products moving in international trade are the principal pathways of concern, and existing patterns of such trade will, to a substantial extent, determine which pathways are relevant. Other pathways such as other types of commodities, packing materials, persons, baggage, mail, conveyances and the exchange of scientific material should be considered where appropriate. Entry by natural means should also be assessed, as the natural spread is likely to reduce the effectiveness of phytosanitary measures.

6.42.2.12 *Probability of the pest being associated with the pathway at origin*

The probability of the pest being associated, spatially or temporally, with the pathway at origin should be estimated. Factors to consider are prevalence of the pest in the source area, occurrence of the pest in a life-stage that would be associated with commodities, containers, or conveyances, volume and frequency of movement along the pathway, seasonal timing and pest management, and cultural and commercial procedures applied at the place of origin (application of plant protection products, handling, culling, roguing, and grading).

6.42.2.13 *Probability of survival during transport or storage*

Examples of factors to consider include speed and conditions of transport and duration of the life cycle of the pest in relation to time in transport and storage, vulnerability of the life-stages during transport or storage, prevalence of pest likely to be associated with a consignment, and commercial procedures (e.g., refrigeration) applied to consignments in the country of origin, country of destination, or in transport or storage.

6.42.2.14 *Probability of pest surviving existing pest management procedures*

Existing pest management procedures (including phytosanitary procedures) applied to consignments against other pests from origin to end-use, should be evaluated for effectiveness against the pest in question. The probability that the pest will go undetected during inspection or survive other existing phytosanitary procedures should be estimated.

6.42.2.15 *Probability of transfer to a suitable host*

Factors to consider are dispersal mechanisms, including vectors to allow movement from the pathway to a suitable host, whether the imported commodity is to be sent to a few or many destination points in the PRA area, proximity of entry, transit and destination points to suitable hosts, time of year at which import takes place, intended use of the commodity (e.g., for planting, processing and consumption), and risks from by-products and waste.

Some uses are associated with a much higher probability of introduction (e.g., planting) than others (e.g., processing). The probability associated with any growth, processing, or disposal of the commodity in the vicinity of suitable hosts should also be considered.

6.42.2.16 *Probability of establishment*

To estimate the probability of establishment of a pest, reliable biological information (life cycle, host range, epidemiology, survival, etc.) should be obtained from the areas where the pest currently occurs. The situation in the PRA area can then be compared with that in the areas where it currently occurs (taking account also of protected environments such as glass- or greenhouses) and expert judgment used to assess the probability of establishment. Case histories concerning comparable pests can be considered. Examples of the factors to consider are availability, quantity and distribution of hosts in the PRA area, environmental suitability in the PRA area, the potential for adaptation of the pest, reproductive strategy of the pest, method of pest survival, and cultural practices and control measures.

In considering the probability of establishment, it should be noted that a transient pest may not be able to establish in the PRA area (e.g., because of unsuitable climatic conditions) but could still have unacceptable economic consequences. In the case of plants to be imported, the assessment of the probability of establishment concerns the unintended habitats.

6.42.2.17 *Availability of suitable hosts, alternate hosts and vectors in the PRA area*

Factors to consider are whether hosts and alternate hosts are present and how abundant or widely distributed they may be, whether hosts and alternate hosts occur within sufficient geographic proximity to allow the pest to complete its life cycle, whether there are other plant species, which could prove to be suitable hosts in the absence of the usual host species, whether a vector, if needed for dispersal of the pest, is already present in the PRA area or likely to be introduced and whether another vector species occurs in the PRA area.

The taxonomic level at which hosts are considered should normally be the 'species.' The use of higher or lower taxonomic levels should be justified by scientifically sound rationale.

6.42.2.18 *Suitability of environment*

Factors in the environment (e.g., suitability of climate, soil, pest and host competition) that are critical to the development of the pest, its host and if applicable its vector, and to their ability to survive periods of climatic stress and complete their life cycles, should be identified. It should be noted that the environment is likely to have different effects on the pest, its host, and its vector. This needs to be recognized in determining whether the interaction between these organisms in the area of origin is maintained in the PRA area to the benefit or detriment of the pest. The probability of establishment in a protected environment, e.g., in glass houses, should also be considered. Climatic modeling systems may be used to compare climatic data on the known distribution of a pest with that in the PRA area.

6.42.2.19 *Cultural practices and control measures*

Where applicable, practices employed during the cultivation/production of the host crops should be compared to determine if there are differences in such practices between the PRA area and the origin of the pest that may influence its ability to establish. Pest control programs or natural enemies already in the PRA area which reduce the probability of establishment may be considered. Pests for which control is not feasible should be considered to present a greater risk than those for which treatment is easily accomplished. The availability (or lack) of suitable methods for eradication should also be considered.

6.42.2.20 Other characteristics of the pest affecting the probability of establishment

These include reproductive strategy of the pests, and method of pest survival, the characteristics which enable the pest to reproduce effectively in the new environment, such as parthenogenesis/self-crossing, duration of the life cycle, number of generations per year, resting stage, etc., should be identified. Also genetic adaptability; whether the species is polymorphic and the degree to which the pest has demonstrated the ability to adapt to conditions like those in the PRA area should be considered, e.g., host-specific races or races adapted to a wider range of habitats or new hosts. This genotypic (and phenotypic) variability facilitates a pest's ability to withstand environmental fluctuations, to adapt to a wider range of habitats, to develop pesticide resistance and to overcome host resistance. In addition, minimum population needed for the establishment, and, if possible, the threshold population that is required for establishment should be estimated.

6.42.2.21 Probability of spread after establishment

A pest with a high potential for spread may also have a high potential for the establishment, and possibilities for its successful containment and/or eradication are more limited. To estimate the probability of spread of the pest, reliable biological information should be obtained from areas where the pest currently occurs. The situation in the PRA area can then be carefully compared with that in the areas where the pest currently occurs and expert judgment used to assess the probability of spread. Case histories concerning comparable pests can usefully be considered. Examples of the factors to consider include suitability of the natural and/or managed environment for natural spread of the pest, presence of natural barriers, the potential for movement with commodities or conveyances, intended use of the commodity, potential vectors of the pest in the PRA area, and potential natural enemies of the pest in the PRA area.

In the case of plants to be imported, the assessment of spread concerns spread from the intended habitat or the intended use of an unintended habitat, where the pest may establish. Further spread may then occur to other unintended habitats. The information on the probability of spread is used to estimate how rapidly a pest's potential economic importance may be expressed within the PRA area. This also has significance if the pest is liable to enter and establish in an area of low potential economic importance and then spread to an area of high potential economic importance. In addition, it may be important in the risk management stage when considering the feasibility of containment or eradication of an introduced pest. Certain pests may not cause injurious effects on plants immediately after they establish and, in particular, may only spread after a certain time. In assessing the probability of spread, this should be considered, based on evidence of such behavior. The overall probability of introduction should be expressed in terms most suitable for the data, the methods used for analysis, and the intended audience. This may be quantitative or qualitative since either output is in any case the result of a combination of both quantitative and qualitative information. The probability of introduction may be expressed as a comparison with that obtained from PRAs on other pests. Conclusion regarding endangered areas: The part of the PRA area where ecological factors favor the establishment of the pest should be identified to define the endangered area. This may be the whole of the PRA area or a part of the area.

6.43 ANALYSIS OF ECONOMIC CONSEQUENCES

Estimations made in the previous section related to a hypothetical situation where the pest is supposed to have been introduced and to be fully expressing its potential economic consequences (per year) in the PRA area. In practice, however, economic consequences are expressed with time, and may concern one year, several years or an indeterminate period. Various scenarios should be considered. The total economic consequences over more than one year can be expressed as the net present value of annual economic consequences, and an appropriate discount rate selected to calculate net present value.

Other scenarios could concern whether the pest occurs at one, few or many points in the PRA area and the expression of potential economic consequences will depend on the rate and manner of spread in the PRA area. The rate of spread may be envisaged to be slow or rapid; in some cases, it may be supposed that spread can be prevented. Appropriate analysis may be used to estimate potential economic consequences over the period of time when a pest is spreading in the PRA area. In addition, many of the factors or effects considered above could be expected to change over time, with the consequent effects of potential economic consequences. Expert judgment and estimations will be required.

6.43.1 Analysis of Commercial Consequences

As determined above, most of the direct effects of a pest and some of the indirect effects will be of a commercial nature, or have consequences for an identified market. These effects, which may be positive or negative, should be identified and quantified. These may usefully be considered; effect of pest, induced changes to producer profits that result from changes in production costs, yields or prices and effect of pest-induced changes in quantities demanded or prices paid for commodities by domestic and international consumers. This could include quality changes in products and/or quarantine-related trade restrictions resulting from a pest introduction.

6.43.2 Analytical Techniques

There are analytical techniques which can be used in consultation with experts in economics to make a more detailed analysis of the potential economic effects of a quarantine pest. These should incorporate all of the effects that have been identified. These techniques may include partial budgeting (this will be adequate if the economic effects induced by the action of the pest to producer profits are generally limited to producers and are considered to be relatively minor), partial equilibrium. Partial equilibrium analysis is necessary to measure welfare changes, or the net changes arising from the pest impacts on producers and consumers. General equilibrium (if the economic changes are significant to a national economy, and could cause changes to factors such as wages, interest rates or exchange rates, then general equilibrium analysis could be used to establish the full range of economic effects). The use of analytical techniques is often limited by lack of data, by uncertainties in the data, and by the fact that for certain effects only qualitative information can be provided.

6.43.3 Noncommercial and Environmental Consequences

Some of the direct and indirect effects of the introduction of a pest will be of an economic nature, or affect some type of value, but not have an existing market which can be easily identified. As a result, the effects may not be adequately measured in terms of prices in established product or service markets. Examples include in particular environmental effects (such as ecosystem stability, biodiversity, amenity value) and social effects (such as employment, tourism) arising from a pest introduction. These impacts could be approximated with an appropriate nonmarket valuation method. More details on the environment are given below.

If quantitative measurement of such consequences is not feasible, qualitative information about the consequences may be provided. An explanation of how this information has been incorporated into decisions should also be provided. Application of this standard to environmental hazards requires clear categorization of environmental values and how they can be assessed. The environment can be valued using different methodologies, but these methodologies are best used in consultation with experts in economics. Methodologies may include consideration of 'use' and 'nonuse' values. 'Use' values arise from consumption of an element of the environment, such as accessing clean water, or fishing in a lake, and also those that are nonconsumptive, such as use of forests for leisure activities.

'Nonuse' values may be subdivided into option value (value for use at a later date), existence value (knowledge that an element of the environment exists) and bequest value (knowledge that an element of the environment is available for future generations).

Whether the element of the environment is being assessed in terms of use or nonuse values, methods exist for their valuation, such as market-based approaches, surrogate markets, simulated markets, and benefits transfer. Each has advantages, disadvantages, and situations where it is particularly useful. The assessment of consequences may be either quantitative or qualitative and in many cases, qualitative data is sufficient. A quantitative method may not exist to address a situation (e.g., catastrophic effects on a keystone species), or quantitative analysis may not be possible (no methods available).

Useful analyses can be based on nonmonetary valuations (number of species affected, water quality), or expert judgment, if the analyses follow documented, consistent and transparent procedures.

6.43.4 Conclusion of Assessment of Economic Consequences

Wherever appropriate, the output of the assessment of economic consequences described in this step should be in terms of monetary value. The economic consequences can also be expressed qualitatively or using quantitative measures without monetary terms. Sources of information, assumptions, and methods of analysis should be clearly specified.

6.43.5 Endangered Area

The part of the PRA area where the presence of the pest will result in an economically important loss should be identified as appropriate. This is needed to define the endangered area.

6.43.6 Degree of Uncertainty

Estimation of the probability of introduction of a pest and its economic consequences involves many uncertainties. In particular, this estimation is an extrapolation from the situation where the pest occurs to the hypothetical situation in the PRA area. It is important to document the areas of uncertainty and the degree of uncertainty in the assessment and to indicate where expert judgment has been used. This is necessary for transparency and may also be useful for identifying and prioritizing research needs. It should be noted that the assessment of the probability and consequences of environmental hazards of pests of uncultivated and unmanaged plants often involves greater uncertainty than for pests of cultivated or managed plants. This is due to the lack of information, the additional complexity associated with ecosystems, and variability associated with pests, hosts or habitats.

6.43.7 Conclusion of the Pest Risk Assessment Stage

As a result of the pest risk assessment, all or some of the categorized pests may be considered appropriate for pest risk management. For each pest, all or part of the PRA area may be identified as an endangered area. A quantitative or qualitative estimate of the probability of introduction of a pest or pests, and a corresponding quantitative or qualitative estimate of economic consequences (including environmental consequences) have been obtained and documented, or an overall rating could have been assigned. These estimates, with associated uncertainties, are utilized in the pest risk management stage of the PRA.

6.43.8 Stage 3: Pest Risk Management

The conclusions from pest risk assessment are used to decide whether risk management is required and the strength of measures to be used. Since zero-risk is not a reasonable option, the guiding principle for risk management should be to manage risk to achieve the required degree of safety that can be justified and is feasible within the limits of available options and resources. Pest risk management (in the analytical sense) is the process of identifying ways to react to a perceived risk, evaluating the efficacy of these actions, and identifying the most appropriate options. The uncertainty noted in the assessments of economic consequences, and the probability of introduction should also be considered and included in the selection of a pest management option. In considering the management of environmental risks, it should be stressed that phytosanitary measures are intended to account for uncertainty and should be designed in proportion to the risk. Pest risk management options should be identified, taking account of the degree of uncertainty in the assessment of economic consequences, the probability of introduction, and the respective technical justification of those options. In this respect, the management of risks to the environment caused by plant pests does not differ from the management of other plant pest risks. The level of risk includes the principle of 'managed risk' (ISPM No. 1: *Principles of plant quarantine as related to international trade*) states: "Because some risk of introduction of a quarantine pest always exists, countries shall agree to a policy of risk management when formulating phytosanitary measures." In implementing this principle, countries should decide what level of risk is acceptable to them. The acceptable level of risk may be expressed in a number of ways, such as a reference to existing phytosanitary requirements, indexed to estimated economic losses, expressed on a scale of risk tolerance and compared with the level of risk accepted by other countries.

6.43.8.1 Technical information required

The decisions to be made in the pest risk management process will be based on the information collected during the preceding stages of PRA. This information will be composed of reasons for initiating the process, estimation of the probability of introduction to the PRA area and evaluation of potential economic consequences in the PRA area.

6.43.8.2 Acceptability of risk

Overall risk is determined by the examination of the outputs of the assessments of the probability of introduction and the economic impact. If the risk is found to be unacceptable, then the first step in risk management is to identify possible phytosanitary measures that will reduce the risk to, or below an acceptable level. Measures are not justified if the risk is already acceptable or must be accepted because it is not manageable (as may be the case with natural spread). Countries may decide that a low level of monitoring or audit is maintained to ensure that future changes in the pest risk are identified.

6.44 IDENTIFICATION AND SELECTION OF APPROPRIATE RISK MANAGEMENT OPTIONS

Appropriate measures should be chosen based on their effectiveness in reducing the probability of introduction of the pest. The choice should be based on the following considerations, which include several of the *Principles of plant quarantine as related to international trade* (ISPM No. 1).

(i) ***Phytosanitary measures shown to be cost-effective and feasible:*** The benefit from the use of phytosanitary measures is that the pest will not be introduced, and the PRA area will, consequently, not be subjected to the potential economic consequences. The cost-benefit analysis for each of the minimum measures found to provide acceptable security may be estimated. Those measures with an acceptable benefit-to-cost ratio should be considered.

(ii) ***Principle of 'minimal impact':*** Measures should not be more trade restrictive than necessary. Measures should be applied to the minimum area necessary for the effective protection of the endangered area.

(iii) ***Reassessment of previous requirements:*** No additional measures should be imposed if existing measures are effective.

(iv) ***Principle of 'equivalence':*** If different phytosanitary measures with the same effect are identified, they should be accepted as alternatives.

(v) ***Principle of 'nondiscrimination':*** If the pest under consideration is established in the PRA area but of limited distribution and under official control, the phytosanitary measures in relation to import should not be more stringent than those applied within the PRA area. Likewise, phytosanitary measures should not discriminate between exporting countries of the same phytosanitary status.

The principle of nondiscrimination and the concept of official control also apply to pests affecting uncultivated/unmanaged plants, weeds and/or invasive plants and pests affecting plants through effects on other organisms. If any of these become established in the PRA area and if official control is applied, then phytosanitary measures at import should not be more stringent than the official control measures. The major risk of introduction of plant pests is with imported consignments of plants and plant products, but (especially for a PRA performed on a particular pest) it is necessary to consider the risk of introduction with other types of pathways (e.g., packing materials, conveyances, travelers and their luggage, and the natural spread of a pest).

The measures listed below are examples of those that are most commonly applied to traded commodities. They are applied to pathways, usually consignments of a host, from a specific origin. The measures should be as precise as possible as to consignment type (hosts, parts of plants) and origin so as not to act as barriers to trade by limiting the import of products where this is not justified. Combinations of two or more measures may be needed to reduce the risk to an acceptable level. The available measures can be classified into broad categories which relate to the pest status of the pathway in the country of origin. These include measures applied to the consignment, applied to prevent or reduce original infestation in the crop, to ensure the area or place of production is free from the pest concerning the prohibition of commodities and other options may arise in the PRA area (restrictions on the use of a commodity), control measures, the introduction of a biological control agent, eradication, and containment. Such options should also be evaluated and will apply in particular if the pest is already present but not widely distributed in the PRA area.

6.44.1 Options for Consignments

Measures may include any combinations of the inspection or testing for freedom from a pest or to a specified pest tolerance. The sample size should be adequate to give an acceptable probability of detecting the pest, prohibition of parts of the host, a preentry or postentry quarantine system. This system could be considered to be the most intensive form of inspection or testing where suitable facilities and resources are available, and may be the only option for certain pests not detectable on entry, specified conditions of preparation of the consignment (e.g., handling to prevent infestation or reinfestation), and specified treatment of the consignment (such treatments are applied postharvest and could include chemical, thermal, irradiation or other physical methods and restrictions on end use, distribution and periods of entry of the commodity).

Measures may also be applied to restrict the import of consignments of pests. The concept of consignments of pests may be applied to the import of plants considered to be pests. These consignments may be restricted to species or varieties posing less risk.

6.44.2 Options Preventing or Reducing Infestation in Crop

Measures may include treatment of the crop, field, or place of production, restriction of the composition of a consignment so that it is composed of plants belonging to resistant or less susceptible species, growing plants under specially protected conditions (glass house and isolation), harvesting of plants at a certain age or a specified time of year and production in a certification scheme. An officially monitored plant production scheme usually involves a number of carefully controlled generations, beginning with nuclear stock plants of high health status. It may be specified that the plants be derived from plants within a limited number of generations.

6.44.3 Options Ensuring that Area, Place or Site of Production or Crop is Free from Pest

Measures may include pest-free area, pest-free place of production or pest-free production site and inspection of crop to confirm pest freedom.

6.44.4 Options for Other Types of Pathways

For many types of pathways, the measures considered above for plants and plant products to detect the pest in the consignment or to prevent infestation of the consignment may also be used or adapted. For certain types of pathways, the factors that should be considered are as follows.

(i) Natural spread of a pest includes movement of the pest by flight, wind dispersal, transport by vectors such as insects or birds and natural migration. If the pest is entering the PRA area by natural spread or is likely to enter in the immediate future, phytosanitary measures may have little effect. Control measures applied in the area of origin could be considered. Similarly, containment or eradication, supported by suppression and surveillance, in the PRA area after entry of the pest could be considered.

(ii) Measures for human travelers and their baggage could include targeted inspections, publicity, and fines or incentives. In a few cases, treatments may be possible.

(iii) Contaminated machinery or modes of transport (ships, trains, planes, and road transport) could be subjected to cleaning or disinfestation.

6.44.5 Options within Importing Country

Certain measures applied within the importing country may also be used. These could include careful surveillance to try and detect the entry of the pest as early as possible, eradication programs to eliminate any foci of infestation and/or containment action to limit spread. For plants to be imported, where there is a high level of uncertainty regarding pest risk, it may be decided not to take phytosanitary measures at import, but only to apply surveillance or other procedures after entry (e.g., by or under the supervision of the NPPO).

6.44.6 Prohibition of Commodities

If no satisfactory measure to reduce risk to an acceptable level can be found, the final option may be to prohibit the importation of the relevant commodities. This should be viewed as a measure of last resort and should be considered in light of the anticipated efficacy, especially in instances where the incentives for illegal import may be significant.

6.45 PHYTOSANITARY CERTIFICATES AND OTHER COMPLIANCE MEASURES

Risk management includes the consideration of appropriate compliance procedures. The most important of these is export certification (see ISPM No. 7: *Export certification system*). The issuance of phytosanitary certificates (see ISPM No. 12: *Guidelines for Phytosanitary Certificates*) provides official assurance that a consignment is "considered to be free from the quarantine pests specified by the importing contracting party and to conform to the current phytosanitary requirements of the importing contracting party." It thus confirms that the specified risk management options have been followed. An additional declaration may be required to indicate that a particular measure has been carried out. Other compliance measures may be used subject to bilateral or multilateral agreement.

The result of the pest risk management procedure will be either that no measures are identified which are considered appropriate or the selection of one or more management options that have been found to lower the risk associated with the pest(s) to an acceptable level. These management options form the basis of phytosanitary regulations or requirements. The application and maintenance of such regulations are subject to certain obligations, in the case of contracting parties to the IPPC. Phytosanitary measures taken in relation to environmental hazards should, as appropriate, be notified to relevant competent authorities responsible for national biodiversity policies, strategies, and action plans. It is noted that the communication of risks associated with environmental hazards is of particular importance to promote awareness.

6.45.1 Monitoring and Review of Phytosanitary Measures

The principle of 'modification' states: "As conditions change, and as new facts become available, phytosanitary measures shall be modified promptly, either by inclusion of prohibitions, restrictions or requirements necessary for their success, or by removal of those found to be unnecessary" (ISPM No. 1: *Principles of plant quarantine as related to international trade*). Thus, the implementation of particular phytosanitary measures should not be considered to be permanent. After application, the success of the measures in achieving their aim should be determined by monitoring during use. This is often achieved by inspection of the commodity on arrival, noting any interceptions or any entries of the pest to the PRA area. The information supporting the pest risk analysis should be periodically reviewed to ensure that any new information that becomes available does not invalidate the decision taken.

6.45.2 Documentation Requirements for Pest Risk Analysis

The IPPC and the principle of 'transparency' (ISPM No. 1: *Principles of plant quarantine as related to international trade*) require that countries should, on request, make available the rationale for phytosanitary requirements. The whole process from initiation to pest risk management should be sufficiently documented so that when a review or dispute arises, the sources of information and rationale used in reaching the management decision can be clearly demonstrated. The main elements of documentation are a purpose for the PRA, pest, pest list, pathways, PRA area, endangered area, sources of information, categorized pest list, conclusions of risk assessment, probability, consequences, risk management, options identified and options selected.

6.46 DEFRA'S APPROACH TO PEST RISK ANALYSIS

(i) Plant pests include insects, other invertebrates, bacteria, fungi, viruses and other pathogens which affect the health of plants by feeding on them or causing disease.

(ii) Pest Risk Analysis (PRA) is defined under the International Plant Protection Convention as: "The process of evaluating biological or other scientific and economic evidence to determine whether a pest should be regulated and the strength of any phytosanitary [plant health] measures to be taken against it." PRAs may also be conducted on commodity imports to determine whether they provide a pathway for pests to enter the importing country or area.

(iii) There are international standards that set out what to consider when conducting a PRA. PRAs vary in length from simple expert judgments ('does not feed on plants' or 'tropical, no host plants outdoors in Europe') to long documents with assessment of biological factors, climate maps, and estimated costs and benefits of the measures needed to keep a pest out of an area, or to eradicate an outbreak. Different schemes are being developed for PRA. The UK uses both its own short scheme and a more detailed scheme from the European and Mediterranean Plant Protection Organization (EPPO). In most cases, the PRAs published on the Defra website provide summaries of more detailed analyses.

(iv) PRAs are written by plant health scientists to inform decisions by plant health policy makers. They are more technical than the information we prepare for the general public but more accessible than most scientific papers. They only give a snapshot of the analysis at a particular time, and although we do revise them, they may go out of date as new information comes to light.

(v) Specific treatments may be mentioned in a PRA as examples of control measures which are applied to areas of the world where the pest is established. These should not be regarded as recommendations: indeed the treatments may not be approved for use in the UK. All suspect findings of new pests in England or Wales should be reported to the Plant Health and Seeds Inspectorate (HQ tel 01904 455174).

(vi) For some organisms, the conclusion of the PRA is that the risk is low and statutory action by the plant health service to exclude or eradicate the organism is not warranted. However, it is an offence under Section 14 of the Wildlife and Countryside Act 1981 for any person to release or allow to escape into the wild any animal which "is of a kind which is not ordinarily resident in and is not a regular visitor to Great Britain in a wild state, or which is listed on Part I of Schedule 9 to the Act." The release does not have to be deliberate or intentional, as this offense is one of strict liability.

(vii) All risk analyses have to cope with areas of uncertainty where evidence is lacking, inconclusive or contradictory. For organisms that do not occur here, our experts have to use their judgment in extrapolating from available data to assess how a pest might behave under our conditions on our crops and wild plants. If there is a wide range of uncertainty about the risk we may commission research to fill gaps in the risk analysis. Until more is known, we take a precautionary approach.

(viii) Defra's PRAs focus on the risk to the UK and the European Community. Rules on imports and movement of plants are coordinated in the EC's Standing

Committee on Plant Health and set out in the Plant Health Directive. The committee takes account of the risks from the pest, and the cost of risk management measures, across the whole of the EC. When the risk is confined to one area, there is provision in the directive for action to be restricted to that area by recognizing it as a 'protected zone.'

(ix) This is a very brief summary of Defra's approach to Pest Risk Analysis. Further information can be obtained from a number of websites shown below.

6.46.1 Action Plan for Pest Risk Analysis

Pest risk analysis plays a key role in the new order for plant quarantine in India. An action plan for pest risk analysis was drawn up, to take effect from December 2003. A major feature of the plan is the establishment of a national pest risk analysis unit. The action plan includes organizing PRA training, establishing working groups and holding a workshop attended by national and international experts to prioritize crops and commodities for pest risk analysis. India's priority crops for PRA are furnished in Table 6.8.

Table 6.8 India's Priority Crops for Pest Risk Analysis

Strawberry	*Banana*	*Kiwi*
Musk melon	Watermelon	Pears
Mandarin	Cashew nut	Apple
Grape	Citrus fruits	Lentil
Red beans	Chickpea	Jute
Black gram	Green gram	Cotton
Wheat	Rice	Barley
Maize	Baby corn	Pearl millet
Sorghum	Lettuce	Garlic
Broccoli	Potato	Chinese cabbage
Mustard	Sunflower	Safflower
Linseed	Castor	Rapeseed

6.47 POLICY FOR THE CONTROL OF INVASIVE ALIEN SPECIES IN INDIA AND RELATED POLICY INITIATIVES

The national integrated pest management (IPM) program is considered to be the mechanism with which to prevent and control the threat posed by invasive alien species within the country. State governments, nongovernmental organizations, private sector organizations, research institutions and farmer self-help groups are all increasingly involved in the surveillance and detection of pests and diseases. They are capable of taking environmentally friendly corrective action within the IPM scheme. International cooperation has helped in dealing with migratory locust, a pest of great concern for the Asian region. India maintains active coordination with FAO and with neighboring countries for surveillance, early detection and control measures for locust. There was no major incidence reported in the region in 2003.

A peculiar cyclic problem, thankfully confined to a small hilly region of north-eastern India, relates to unexplained but sudden surges in rodent population and activity. The menace reaches peak proportions at the same time as the periodic gregarious flowering of bamboo. The problem has surfaced again and is likely to peak in 2006–2007 when the next mass flowering is predicted, causing crop losses. Research and preventive control measures under way include the study of the rodent characteristics, damage capacity, pathways associated with the pest and an environment-friendly control strategy. The traditional knowledge of the local agrarian community of the region is also utilized. Of particular interest to India is research being conducted to study the impact of climate change on the threat posed by invasive alien species. The topic is of greater importance since a serious white wooly aphid infestation of the sugarcane crop in parts of peninsular India in 2002 caused substantial crop damage and losses. This pest had never before infested sugarcane in India.

The task of research, future prevention and control measures for white wooly aphid is being handled by the Ministry of Agriculture in coordination with other central government departments, concerned state governments, the Indian Council of Agricultural Research, other research institutions and agriculture universities, private sector organizations and sugar factories. The severity of the white wooly aphid infestation, recorded in 2002 in over 200 000 ha of sugarcane, has subsequently reduced substantially. However, almost 75 000 ha of the crop was still infested in 2003, and the matter continues to be of concern (Anonymous, 2003).

Logistic regression models for wheat Fusarium head blight were developed using information collected at 50 location-years, including four states, representing three different US wheat production regions. Nonparametric correlation analysis and stepwise logistic regression analysis identified combinations of temperature, relative humidity, and rainfall or durations of specified weather conditions, for seven days prior to anthesis, and ten days beginning at crop anthesis, as potential predictor variables. Prediction accuracy of developed logistic regression models ranged from 62 to 85%. Models suitable for application as a disease warning system were identified based on model prediction accuracy, sensitivity, specificity, and availability of weather variables at crop anthesis. Four of the identified models correctly classified 84% of the 50 location-years. A fifth model that used only preanthesis weather conditions correctly classified 70% of the location-years. The most useful predictor variables were the duration (h) of precipitation seven days prior to anthesis, duration (h) when the temperature was between 15 and 30°C seven days prior to anthesis, and the duration (h) when the temperature was between 15 and 30°C and relative humidity was greater than or equal to 90%. When the model performance was evaluated with an independent validation set (n = 9), prediction accuracy was only 6% lower than the accuracy for the original data sets. These results indicate that narrow time periods around crop anthesis can be used to predict Fusarium head blight epidemics (De Wolf *et al.*, 2003).

The international standard for phytosanitary measures ISPM 11 [2001]: *Pest risk analysis for quarantine pests* (and its revisions) provides a clear description of the procedures to be followed in conducting pest risk analyses but, apart from statements such as "climatic modeling systems may be used," does not give further guidance on the tools and resources that can support the pest risk analyst. This paper summarizes the principal tools and resources available for the pest risk assessment component of pest risk analysis, giving examples of how they can be used. It also highlights the principal challenges for the future (Anonymous, 2003).

The European and Mediterranean Plant Protection Organization (EPPO) provides a scheme to assess the risk of organisms that are harmful to plants. This scheme is designed as a user-friendly questionnaire and has been available since 1997. It is based on the first international standard for phytosanitary measures (ISPM) on pest risk analysis, ISPM 2, and addresses environmental risks of plant pests only generally. Two factors necessitated an adaptation of the EPPO scheme: the development of an international standard for *Pest risk analysis for quarantine pests* (ISPM 11) and its supplement on the analysis of environmental risks; and a new policy of EPPO focusing in more detail on the risks of plant pests to biodiversity and human use. These changes demanded that the scheme is changed to accord with the revision of ISPM 11. The paper describes the structure of the EPPO scheme and how it is used. It explains the adaptations of the EPPO scheme to ISPM 11, in particular to environmental aspects, and gives an outlook on how the EPPO scheme can be used.

6.47.1 EPPO Pest Risk Assessment Scheme

The IPPC standards on pest risk analysis are useful general bases to the assessment of pest risks, containing all relevant and necessary steps and being recognized by the WTO via the SPS Agreement (WTO, 1994). Nevertheless, their applicability can be enhanced when transformed into a questionnaire. EPPO supplies a user-friendly questionnaire (EPPO, 1997) to assess the risks of organisms that are harmful to plants and to apply pest risk analysis in the framework of the IPPC. This scheme provides technical justification for the regulation of certain pests for the EPPO region (or parts of it) as the PRA area. It is based on ISPM 2, addressing environmental risks of plant pests only generally. The structure, as well as the use of the EPPO scheme, is described below and its adaptation to the revised ISPM 11, in particular to its applicability to plants as pests.

6.47.2 Structure of EPPO PRA Scheme

The (more or less) simple, clearly arranged scheme includes a sequence of questions for deciding whether an organism could present a pest risk and gives detailed instructions for the several stages of pest risk analysis for quarantine pests, viz., initiation, pest categorization, probability of introduction (entry and establishment), economic impact assessment, and final evaluation. In the initiation stage, the reasons for performing a pest risk analysis have to be given, the identity of the organism has to be indicated, and the PRA area (i.e., the region for which the risk posed by the organism is to be assessed) has to be defined. Then follows a qualitative assessment to determine; if the organism could be a pest and present a risk in the PRA area (section A). Geographical and regulatory criteria have to be considered. An estimation is made of whether the organism would be able to establish in the PRA area and whether there is potential for economic, environmental and social damage and if the organism does not fulfill certain criteria, the assessment could be terminated (Anonymous, 2003).

If the organism is deemed to present a possible risk to or in the PRA area, a quantitative assessment is made (section B), with an in-depth analysis of its probability of introduction and spread and a more detailed economic impact assessment. This is evaluated with the help of.a series of questions, for example, how many pathways (from different origin) could the pest be carried on? How likely is the pest to be able to transfer from the pathway to a suitable host? How extensive are the host plants in the PRA area? And how similar are the climatic conditions that would affect pest establishment in the PRA area and the area of origin? Replies to these are expressed as scores (1 to 9). It is recommended that a preliminary consideration of questions be conducted to allow a quick decision on whether there is a very obvious risk or no risk at all. If a quick decision is not possible, a full assessment has to be done. For a pest risk analysis, a collection of adequate information is essential. This information is to be interpreted by the risk assessor. To facilitate the collection of data, EPPO provides a checklist (EPPO, 1998), which lists the necessary information that should be considered before deciding if a particular organism qualifies to be declared a quarantine pest. References should be noted for all items of information.

In the final evaluation, a comment of the assessor is requested to indicate (important) uncertainties and to estimate the level of risk. The theoretical pest risk is the introduction potential of the pest combined with potential economic (including environmental and social) impacts. However, the scheme is not designed to provide estimates of meaningful units, and it does not include a procedure for the calculation of units. Instead, expert judgment to estimate pest risk is essential. To make the pest risk analysis as transparent as possible and also to allow for any future reevaluation of the pest risk analysis, all steps of the procedure have to be fully documented, indicating who performed the evaluation, how each decision was reached, the date on which the information was collected, and on what information it was based.

6.47.3 Adaptations to Revised ISPM 11

The development of the revised standard including assessment of environmental risks, together with the new policy of EPPO focusing in more detail on risks of plant pests to biodiversity and human use, necessitated the revision of the EPPO PRA scheme. This process has been started. The revised scheme will also be applicable to pest plants (invasive plants or weeds) and to indirect plant pests (pests that have negative impacts on other organisms resulting in harm to plants) and will be in line with the IPPC standards (Anonymous, 2003).

The most important changes are an adjustment to accord with the sections of ISPM 11 and a rearrangement of questions to follow it more closely. In Section A (pest categorization), additional questions are under consideration to cover more precisely the cases of plants as pests (see below). The current scheme does not refer to habitats but to host plants (*Example:* How extensive are the host plants in the PRA area?). The revision of the scheme will consider habitats as well. To make the EPPO scheme easier to apply, questions of the scheme are simplified, and questions not relevant to plants are indicated. The scoring system (scores from 1 to 9) will possibly be replaced by words and be reduced to five levels (e.g., very unlikely, unlikely, possible, likely, and very likely). To shorten the PRA process, a pest may immediately be declared a (potential) invasive alien species or a quarantine pest if an answer to a critical question (e.g., How important is environmental damage likely to be in the PRA area?) indicates a high impact. The following questions can then be left out. A computerized system planned to run the scheme will avoid going through questions that are not relevant. The revision is still in process, and further discussions in panels are necessary. An EPPO workshop in October 2004 reviewed work on an adaptation of the scheme, which is now expected to be finalized in mid-2005.

6.47.4 Applicability of EPPO Scheme to Plants as Pests

The most important changes regarding environmental aspects are up to now the PRA scheme's applicability to plants as pests (Anonymous, 2003). Invasive plants have effects on the environment; impacts are generally described in qualitative rather than in economic terms. This will be reflected in the revision of the scheme, for example by the question, 'how important is environmental harm caused by the pest within its existing geographic range?'.

Effects of pests may include reduction of keystone species, reduction of species that are major components of ecosystems and of endangered species, significant reduction, displacement or elimination of other native species, indirect effects on plant communities (species richness, and biodiversity), significant effects on designated environmentally sensitive areas, significant change in ecological processes and the structure, and stability or processes of an ecosystem (including further effects on plant species).

It will be possible to assess not only the unintentional introduction of, for example, seeds or other propagules contaminating imported commodities, but also the intentional import for agriculture or forestry or for horticultural and ornamental purposes. In instances of intentional introduction of plants, the entry of the plant is not part of the assessment. Instead, it is very important to consider the pathway(s) from the garden, field, park, pond or whatever may be the intended habitat for the plant to the unintended habitat where the plant is not wanted. The plant's establishment in the intended habitat is very likely and, indeed, often promoted. Its establishment in an unintended habitat, however, can cause severe problems. Therefore, its potential for establishment in the unintended habitat has to be assessed. To evaluate the potential invasiveness of a plant is a difficult task. Intrinsic characteristics of invasiveness should be identified as far as possible. This will be covered by the question, "does the plant have intrinsic attributes that indicate that it could cause significant harm to plants or plant communities?" It is intended to enclose an annex to the scheme, listing information on intrinsic attributes for invasiveness. For example, a very strong hint for invasiveness is given when a plant is already invasive elsewhere. High propagule pressure, effective reproduction strategies, rapid growth and broad ecological amplitude are also factors for increased invasiveness. Principle 10 of the CBD guiding principles on invasive alien species recommends that no intentional first-time introduction of an alien species should take place without prior authorization from a competent authority of the recipient state(s). Hence, a pest risk analysis could technically justify import prohibitions of invasive alien plants or special requirements for import or use.

7 Biosafety and Biosecurity

Concepts of biosafety and biosecurity deal with related but distinctly different issues. Biosafety is a well-established concept with a widely-accepted meaning and international guidance on how it is put into practice at the national level. Biosecurity is a comparatively new term, with divergent meanings depending upon the setting in which it is used. Bioprospecting is the process of discovery and commercialization of new products based on biological resources. It also includes biopiracy.

The common understanding of biosafety is derived from the practical guidance issued by the World Health Organization on techniques for use in laboratories. The WHO Laboratory Biosafety Manual (LBM) considers biosafety to be "the containment principles, technologies, and practices that are implemented to prevent unintentional exposure to pathogens and toxins or their accidental release". The LBM contains expert guidance on how to implement the relevant principles, technologies, and practices. The WHO encourages all states to consider such concepts when developing and enhancing national regulatory regimes. The international guidance is then tailored to fit precise national requirements (Anonymous, 2004). Such concepts are consistent across public, animal and plant health sectors, and close cooperation among the WHO, FAO and OIE contributes to the development of relevant guidance and understandings. Biosafety correlates to the obligation under the convention to ensure that the necessary safety precautions are taken, when conducting activities not prohibited by the BWC, to protect populations and the environment.

A general definition of 'biosafety' encompasses the practices, procedures, and use of equipment needed to ensure adequate safety conditions in all facilities that work with potentially infectious microorganisms and other biological hazards. These include health care settings, clinical and diagnostic laboratories that handle human clinical samples, veterinary facilities that work with animal tissue samples, biological research laboratories, and teaching laboratories. All of these facilities must seek to reduce the risks associated with handling potential biological hazards by employing a continuous process of hazard recognition, risk assessment, and hazard mitigation.

'Laboratory biosafety' refers to the application of combinations of laboratory practices and procedures, laboratory facilities, safety equipment, and appropriate occupational health programs to mitigate the risks associated with handling potentially infectious microorganisms and other biological hazards (biohazards). The key principles of laboratory biosafety are hazard recognition, risk assessment, and hazard mitigation, including appropriate biocontainment. 'Laboratory biocontainment' refers to the primary and secondary physical containment barriers in a facility such as contained dressing and shower rooms, sealed service penetrations, specialized doors, entry and exit avenues to prevent cross-contamination, specialized air handling systems for contamination control, personal protective equipment, biosafety cabinets, etc. Current biosafety and biocontainment practices and procedures are designed to reduce the exposure of laboratory personnel, the public, agriculture, and the environment to potentially hazardous biological agents.

The term biosecurity is more complex as it can have different meanings in different contexts. According to the relevant WHO guidance, the phrase has evolved simultaneously in various processes and is used differently in each (Anonymous, 2004). In the setting of the BWC, it is most commonly used to refer to mechanisms to establish and maintain the security and oversight of pathogenic microorganisms, toxins, and relevant resources, as discussed during the 2003 meetings of the convention. The term biosecurity does not seem to have a standardized meaning across human, animal and plant health sectors. In veterinary and agricultural fields, the term has come to denote protecting biological resources from foreign or invasive species. The connotations of biosecurity in public health settings, however, relate more closely to the BWC. The Foreword to the Third Edition of the LBM, released in 2004 notes that in public health, biosecurity concerns with the protection of microbiological assets from theft, loss or diversion, which could lead to the inappropriate use of these agents to cause public health harm.

Efforts continue to evolve within WHO to differentiate between this specific meaning and applications to which biosecurity is put in other settings. In 2006, when the WHO released its first dedicated guidance publication on this issue, the phrase had evolved into laboratory biosecurity. Laboratory biosecurity describes "the protection, control and accountability for valuable biological materials within laboratories, to prevent their unauthorized access, loss, theft, misuse, diversion or intentional release." Such concepts are not restricted to laboratories but have also been expanded to a variety of other facilities that work with resources which could be used for purposes prohibited by the Convention. The OECD, for example, has developed best practice guidelines for biosecurity at ancillary facilities (Biological Resource Centres). In this context, biosecurity means "institutional and personal security measures and procedures designed to prevent the loss, theft, misuse, diversion or intentional release of pathogens, or parts of them, and toxin-producing organisms, as well as such toxins that are held, transferred and/ or supplied by Biological Resource Centres."

Differentiating biosafety from biosecurity: Biosafety protects people from germs while biosecurity protects germs from people (Anonymous, 2004). While these objectives are related, and both are linked to the Convention, their purposes remain distinct. As a result, biosecurity concepts differ from biosafety concepts. The approaches used to achieve them are often similar or mutually reinforcing, but in some cases may conflict.

A common example of conflict arises with the transport of dangerous pathogens: in the interests of biosafety, such pathogens should be clearly labeled during transport, but from a biosecurity perspective, labeling the pathogen being shipped may increase the risk of theft or diversion.

7.1 BIOSAFETY

Biohazards (biological hazards) are infectious agents or other hazardous biologic materials that present a risk or potential risk to the health of humans, animals, or the environment. The risk can be direct through infection or indirect through damage to the environment. Biohazards include certain types of recombinant DNA, organisms and viruses infectious to humans, animals or plants (e.g., parasites, viruses, bacteria, fungi, nematodes, prions and rickettsia) and biologically active agents like toxins, allergens, and venoms, that can cause disease in other living organisms or cause significant impact to the environment or community.

'Biosafety levels' (BSLs) are designations of laboratories in ascending order based on the degree of risk associated with the work being conducted. The designations BSL-1, BSL-2, BSL-3, and BSL-4, are for work with human and zoonotic pathogens (disease-causing organisms that arise in animals but can be transmitted to humans) and are based on the utilization of combinations of engineering controls, facility design, safe work practices, and safety equipment. Each combination is specifically appropriate for the operations performed the documented or suspected routes of transmission of the infectious agents, and the laboratory function or activity. The assignment of a BSL to a particular work process or research protocol is made through protocol-driven risk assessment so that potential hazards specific to the work can be identified and mitigated effectively. (The 'BSL' term for laboratory designation does not apply to plant pathogens. However, plant pathogens are typically contained in laboratories and greenhouse facilities with containment features similar to those described for BSL-1, BSL-2 and BSL-3 laboratories).

'Biorisk' is the combination of the likelihood of the occurrence of an adverse event involving exposure to biological agents and toxins and the consequence (in terms of accidental infection, toxicity or allergy or unauthorized access, loss, theft, misuse, diversion or release of biological agents) of such an exposure. Biosafety is used to describe efforts to reduce and eliminate the potential risks resulting from biotechnology and its products. It has similarly been defined as "the avoidance of risk to human health and safety and the conservation of the environment, as a result of the use for research and commerce of infectious or genetically modified organisms." Relevant scientific disciplines that underpin biosafety studies include molecular biology, plant breeding, genetics, plant pathology, agronomy, weed science, entomology and ecology, among others (Zaid, 2001). Therefore a large amount of scientific knowledge and data has a direct impact on biosafety, and it can be a difficult process to collate this information in a credible way to create a balanced view.

7.1.1 Biosafety Principle

Biological hazards include pathogenic microbes (human, animal or plant pathogens), toxins, venoms, human blood, certain body fluids, cells/tissues, recombinant deoxyribonucleic acid, and genetically modified agents. Biological safety (i.e., biosafety) or biohazard control is the management of biological hazards through the proper application of engineered containment and administrative controls. The term containment refers to a series of safe methods for managing infectious agents in the laboratory. The purpose of containment is to reduce or eliminate human and environmental exposure to potentially harmful agents. The primary principle of biosafety is thus, containment.

7.1.1.1 Primary and secondary containment

There are two levels of biological containment: primary and secondary. Primary containment protects people and the immediate laboratory environment from exposure to infectious agents. Good microbiological techniques and safety equipment provide sufficient primary containment. Examples of primary barriers include safety equipment such as biological safety cabinets (BSCs), enclosed containers, and safety centrifuge cups. Occasionally, when it is impractical to work in BSCs, personal protective equipment, such as laboratory coats and gloves may act as the primary barrier between personnel and infectious materials. Secondary containment protects the environment external to the laboratory from exposure to infectious materials. Good facility design and operational practices provide secondary containment. Examples of secondary barriers include work areas that are separate from public areas, decontamination facilities, hand-washing facilities, special ventilation systems, and airlocks.

7.1.1.2 Elements of containment

The three key elements of biological containment include laboratory practices, safety equipment, and facility design. To ensure minimal exposure, employees must assess the hazards associated with their work and determine how to apply the biosafety principle appropriately. Employees working with infectious agents or potentially infectious materials must be aware of the hazards associated with their work. Employees must be trained and proficient in biosafety procedures and techniques.

7.2 GENERAL BIOSAFETY GUIDELINES

General guidelines include the following:

(i) Recognizing the revolutionary economic potential of the new biotechnology in agriculture, health, industry, environment and energy sectors; and appreciating the concerns that mixing of genes from unrelated organisms might create natural imbalance that is not yet adequately understood; and considering the fears that manipulated genes or products thereof if allowed to move around freely in nature, may pose potential hazards; and realizing the apprehensions that certain transgenic organisms may be harmful or become harmful to economic plants, animals, and human being; and discharging the obligations of the Convention of Biological Diversity (CBD) and Cartagena Protocols to which Pakistan is a signatory; the Minister of Environment constituted a National Biosafety Expert Committee to deliberate on these issues to regulate the safe release of Genetically Modified Organisms (GMOs) and products thereof (Anonymous, 2005).

(ii) Consequent upon extensive deliberations/consultation with all the stakeholders, the national committee concluded that there is a national need to develop biosafety guidelines to control laboratory research, field studies and commercial release of GMOs and products thereof, as have been done by all the developed countries and many developing countries.

(iii) Accordingly, a set of National Biosafety Guidelines has been developed through a national forum participated by all the stakeholders including the academic institutions, R&D organization, industry, nongovernment organizations and human right societies.

(iv) These guidelines have been prepared keeping in view the guidelines prepared by UNIDO, FAO, WHO, UNEP, and all the developed and developing countries with modification to suit our unique and specific socio-economic and geographic environment.

(v) The objective of these guidelines is to prevent unintentional negligence leading to misuse and irresponsibility by laboratory workers/researchers as well as the end users.

(vi) To define the bounds guidelines/regulatory jurisdiction of these guidelines, biotechnology has been defined as processes using living organisms or parts thereof to make or modify products; and to improve plants, animals, or microorganisms for specific uses. 'Recombinant DNA' has been defined as molecules developed outside living cells by joining natural or synthetic DNA segments to DNA that can replicate in a living cell and those DNA molecules that result from the replication of such DNA.

(vii) For the purposes of these guidelines, regulated material includes all genetically modified materials (DNA & RNA preparations, viroids, viruses, cells and organisms, modified or constructed through genetic engineering), derivatives thereof and wastes or by-products of genetic engineering practices (containing viable organisms or otherwise).

(viii) The scope of these guidelines embraces all works related to gene manipulation employing recombinant DNA technology for all purposes including the development of transgenic plants, animals and microorganisms; production of vaccines; industrial manufacturing of Genetically Modified Organisms and products thereof, and their release into the environment for field trials as well as for commercial uses.

(ix) The Guidelines consist of two parts; the first part relates to regulated work in laboratory research and field trials, and the second part deals with procedures for approvals which must be obtained to deregulate the regulated materials to allow their free movement and commercial uses.

(x) All regulated laboratory research works are classified into (a) Minimal risk, (b) Low risk, (c) Considerable level of risk and laboratory containment conditions are accordingly prescribed. For regulated fieldwork, comprehensive containment conditions have been prescribed separately for genetically modified microorganisms plants and animals.

(xi) The mechanism of monitoring and implementation of the proposed guidelines is built on three tiers as specified in the Biosafety Rules, 2005, namely, a National Biosafety Committee (NBC); a Technical Advisory Committee (TAC); and Institutional Biosafety Committees (IBCs) at the institutional levels. The NBC headed by the Secretary, Ministry of Environment will be responsible to oversee all laboratory work, field trials and allow commercial releases of GMOs and their products.

(xii) Enforcement of various clauses of the National Biosafety Guidelines will be administered by the three monitoring implementation bodies, as per legal authority under Clause 7(g) of the Pakistan Environment Protection Act 1997.

(xiii) The IBC may recommend to NBC for awarding exempt status for Laboratory work/field work with genetically modified organisms if there is sufficient information/ground available to consider the work as having no risk and NBC may consider for formal approval.

(xiv) Permission for deregulation granted by the NBC can be withdrawn if sufficient technical data/evidence becomes available after the approval, which warrants its deregulation.

(xv) To standardize processing, separate format for application/proposal for obtaining permission to undertake laboratory genetic manipulation work, field trial and commercialization have been designed.

(xvi) A separate application format has also been designed for the movement of regulated materials and/or exempt status.

(xvii) Instructions for the preparation of applications as well as for its assessment have also been prescribed to streamline and standardize the procedures for assessment and evaluation

(xviii) These biosafety guidelines have been developed on the basis of technical information presently available and may change in the future as more know-how becomes available. Therefore, revision of these guidelines will be a continuous process.

(xix) These biosafety guidelines will be supplemented with annexures explaining/ elaborating concepts, procedures, and protocols required in monitoring/ implementation of the guidelines.

7.3 RECOMBINANT DNA SAFETY GUIDELINES

Remarkable developments have ensured in the last few years in the field of genetic manipulation, and the scenario has shifted from the laboratories to the marketplace elsewhere. In India, there is a growing awareness of the commercial potential of Biotechnology and efforts are being made to promote the large-scale use of indigenously relevant biotechnologies (Anonymous, 2003). A large number of research institutions in Government, Universities, and private R&D labs have active biotech programs where research is being done in both in basic and applied fronts utilizing microorganisms plant and animals, tissue culture and cell lines and on the development of vaccines towards communicable diseases of both men and animals. A good deal of effort is being made in the areas of diagnostics, biofertilizers, biocides, fertility control, tissue culture of high-value crops to develop technologies and useful products. The successes in indigenous research efforts would soon be translated into commercially viable technologies through clearing houses with major R&D centers, University shops with academic institutions and by the industry itself.

The Biotechnology Safety Guidelines could never be one-time exercise as knowledge is ever expanding and the Department of Biotechnology which has the mandate in this area, set up the rDNA Committee to prepare a modified draft on the basis of current scientific information and from the experience gained locally and outside the country on the use of the new technique in the area of research, possible manufacture, and applications. The guidelines cover areas of research involving genetically engineered organism. It also deals with the genetic transformation of green plants, rDNA technology in vaccine development and on large scale production and deliberate/accidental release of organisms, plants, animals and products derived from rDNA technology into the environment. The issues relating to Genetic Engineering of human embryos, use of embryos and fetuses in research and human germline gene therapy are excluded from the scope of the guidelines.

While preparing the revised guidelines the Committee and its subgroups have met four times and have taken note of the guidelines currently in use in other countries. The evolution of the guidelines and their update have gone through the process of consultation with experts, academics, agencies and industry and the concerned ministries with a view to gain general acceptance and broad consensus. The guidelines are in respect of safety measures for the research activities, large-scale use and also the environmental impact during field applications of genetically altered material products.

7.4 SCOPE OF THE REVISED GUIDELINES

7.4.1 Research

The levels of the risk and the classification of the organisms at these levels based on pathogenicity and local prevalence of diseases and epidemic causing strains in India are defined in the guidelines. Some of the microorganisms not native to the country have been assigned to a special category requiring the highest degree of safety. These include Lassa virus, Yellow fever virus, etc. Appropriate practices, equipment, and facilities are recommended for necessary safeguards in handling organisms, plants and animals in various risk groups. The guidelines employ the concept of physical and biological containment and also based on the principle of good laboratory practice (GLP). In this context, biosafety practices as recommended in the WHO laboratory safety Manual on genetic engineering techniques involving microorganisms of different risk groups have incorporated in the guidelines.

7.4.2 Large-scale Operations

The concern does not diminish when it comes to the use of recombinant organisms scale fermentation operations on large scale fermentation operations or applications of it in the environment. As such, the guidelines prescribe criteria for good large scale practices (GLSP) for using recombinant organisms. These include measures such as proper engineering for containment, quality control, personnel protection, medical surveillance, etc.

7.4.3 Environmental Risks

Application and release of engineered organisms into the environment could lead to ecological consequences and potential risks unless necessary safeguards are taken into account. The guidelines prescribe the criteria for assessment of the ecological aspects on a case by case basis for planned introduction of rDNA organism into the environment. It also suggests regulatory measures to ensure safety for import of genetically engineered materials, plants and animals. The recommendations also cover the various quality control methods needed to establish the safety, purity and efficacy of rDNA products.

7.4.3.1 Definition of recombinant DNA

Recombinant deoxyribonucleic acid (rDNA) by definition involves *in vitro* introduction of different segments of DNA (one being the vector and the others normally unrelated DNA sequences) that are capable of replication in a host cell either autonomously or as an integral part of host's genome and maintenance of their continued propagation. This will include all types of cell fusion, microinjection of DNA or RNA or parts or all of the chromosomes, genetic engineering including self-cloning and deletion as well as cell hybridization, transformation and other types of virus or pathogen introduction into unnatural hosts. The organisms involved may belong to various categories, viz., intergeneric organisms, well-defined organisms with noncoding regulatory regions or biological agents whose source of DNA is a pathogen, organisms that are generally recognized as nonpathogenic and may imbibe the characteristics of a pathogen on genetic manipulation.

7.4.4 Classification of Pathogenic Microorganisms

The classification of infective microorganisms is drawn up into four risk groups in increasing order of risk based on the parameters like pathogenicity, modes of transmission and host range, availability of effective preventive treatments or curative medicines, the capability to cause diseases to humans/animals/plants and epidemic causing strains. These parameters may be influenced by levels of immunity, density and movement of the host population, the presence of vectors for transmission and standards of environmental hygiene.

The scientific considerations for assessment of potential risks in the handling of pathogenic organisms include the characterization of donor and recipient organisms, characterization of the modified organism and expression and properties of the gene product. Based on the risk assessment information, the probability of risk could be further assigned certain quantitative values for categorization of experiments in terms of the access factor of the organism, expression factor of DNA and damage factor of the biologically active substance.

7.5 CONTAINMENT

Containment facilitates for different Risk Groups as per the recommendations of World Health Organization (WHO). The term 'containment' is used in describing the safe methods for managing infectious agents in the laboratory environment where they are being handled or maintained.

Purpose of containment: To reduce exposure of laboratory workers, other persons, and outside environment to potentially hazardous agents.

7.5.1 Types of Containment

7.5.1.1 Biological Containment (BC)

In consideration of biological containment, the vector (plasmid, organelle, or virus) for the recombinant DNA and the host (bacterial, plant, or animal cell) in which the vector is propagated in the laboratory will be considered together. Any combination of vector and host which is to provide biological containment must be chosen or constructed to limit the infectivity of the vector to specific hosts and control the host-vector survival in the environment. These have been categorized into two levels – one permitting standard biological containment and the other even higher that relates to normal and disabled host-vector systems respectively (Chapter V: A3).

7.5.1.2 Physical Containment (PC)

The objective of physical containment is to confine recombinant organisms thereby preventing the exposure of the researcher and the environment to the harmful agents. Physical containment is achieved through the use of (i) Laboratory Practice, (ii) Containment Equipment, and (iii) Special Laboratory Design. The protection of personnel and the immediate laboratory environment from exposure to infectious agents is provided by good microbiological techniques and the use of appropriate safety equipment, (Primary Containment). The protection of the environment external to the laboratory from exposure to infectious materials is provided by a combination of facility design and operational practices, (Secondary Containment).

7.5.2 Elements of Containment

The three elements of containment include laboratory practice and technique, safety equipment, and facility design.

7.5.2.1 Laboratory practice and technique

Strict adherence to standard microbiological practices and techniques, awareness of potential hazards; providing/arranging for appropriate training of personnel, selection of safety practices in addition to standard laboratory practices if required and developing of adopting a biosafety or operations manual which identifies the hazards.

7.5.2.2 Safety equipment (primary barriers)

Safety equipment includes biological safety cabinets and a variety of enclosed containers (e.g., safety centrifuge cup). The biological safety cabinet (BSC) is the principal device used to provide containment of infectious aerosols generated by many microbiological procedures. Three types of BSCs (Class I, II, and III) are used in microbiological laboratories. Safety equipment also includes items for personal protection such as gloves, coats, gowns, shoe covers, boots, respirators, face shields and safety glasses, etc.

7.5.2.3 Facility design (secondary barriers)

The design of the facility is important in providing a barrier to protect persons working in the facility but outside of the laboratory and those in the community from infectious agents which may be accidentally released from the laboratory. There are three types of facility designs: viz., the Basic Laboratory (for Risk Group I and II), the Containment Laboratory (for Risk Group III) and the Maximum Containment Laboratory (for Risk Group IV).

7.6 BIOSAFETY LEVELS

It consists of a combination of laboratory practices and techniques, safety equipment and laboratory facilities appropriate for the operations performed and the hazard posed by the infectious agents. The guidelines for Microbiological and Biomedical Laboratories suggest four biosafety levels in incremental order depending on the nature of work. Additional flexibility in containment levels can be obtained by a combination of the physical with the biological barriers. The proposed safety levels for work with recombinant DNA technique take into consideration the source of the donor DNA and its disease-producing potential. These four levels correspond to ($P_1 < P_2 < P_3 < P_4$) facilities approximate to four risk groups assigned to etiologic agents. These levels and the appropriate conditions are enumerated as follows.

7.6.1 Biosafety Level 1

These practices, safety equipment, and facilities are appropriate for undergraduate and secondary educational training and teaching laboratories and for other facilities in which work is done with defined and characterized strains of viable microorganisms not known to cause disease in healthy adult human. No special accommodation or equipment is required, but the laboratory personnel is required to have specific training and to be supervised by a scientist with general training in microbiology or a related science.

7.6.2 Biosafety Level 2

These practices, safety equipment, and facilities are applicable in clinical, diagnostic, teaching and other facilities in which work is done with the broad spectrum of indigenous moderate-risk agents present in the community and associated with human disease of varying severity. Laboratory workers are required to have specific training in handling pathogenic agents and to be supervised by competent scientists. Accommodation and facilities including safety cabinets are prescribed, especially for handling large volume are high concentrations of agents when aerosols are likely to be created. Access to the laboratory is controlled.

7.6.3 Biosafety Level 3

These practices, safety equipment, and facilities are applicable to clinical, diagnostic, teaching research or production facilities in which work is done with indigenous or exotic agents where the potential for infection by aerosols is real, and the disease may have serious or lethal consequences. Personnel are required to have specific training in work with these agents and to be supervised by scientists experienced in this kind of microbiology. Specially designed laboratories and precautions including the use of safety cabinets are prescribed, and the access is strictly controlled.

7.6.4 Biosafety Level 4

These practices, safety equipment, and facilities are applicable to work with dangerous and exotic agents which pose a high individual risk of life-threatening disease. Strict training and supervision are required, and the work is done in specially designed laboratories under stringent safety conditions, including the use of safety cabinets and positive pressure personnel suits. Access is strictly limited.

A specially designed suit area may be provided in the facility. Personnel who enter this area wear a one-piece positive pressure suit that is ventilated by a life support system. The life support system is provided with alarms and emergency break-up breathing air tanks. Entry to this area is through an airlock fitted with airtight doors. A chemical shower is provided to decontaminate the surface of the suit before the worker leaves the area. The exhaust air from the suit area is filtered by two sets of HEPA filters installed in the series. A duplicate filtration unit, exhaust fan, and an automatically starting emergency power source are provided. The air pressure within the suit area is lower than that of any adjacent area. Emergency lighting and communication systems are provided. All penetrations into the inner shell of the suit area are sealed. A double door autoclave is provided for decontamination of disposable waste materials from the suit area.

7.7 GUIDELINES FOR RDNA RESEARCH ACTIVITIES

The guidelines stipulate three categories of research activities.

Category I: Which are exempt for the purpose of intimation and approval of competent authority? The experiments involving self-cloning, using strains and also interspecies cloning belonging to organism in the same exchanger group (Vide Chapter-V A4, A5), organelle DNA including those from chloroplasts and mitochondria and host-vector systems consisting of cells in culture and vectors, either nonviral or viral containing defective viral genomes (except from cells known to harbor class III, IV and special category etiologic agents listed under Chapter V: A1.

Category II: Those requiring prior intimation of the competent authority.

(i) Experiments falling under containment levels II, III, and IV.

(ii) Experiment wherein DNA or RNA molecules derived from any source except for eukaryotic viral genome may be transferred to any nonhuman vertebrate or any invertebrate organisms and propagated under conditions of physical containment PC1 and appropriate to the organism under study.

(iii) Experiments involving nonpathogen DNA vector systems and regeneration from single cells.

(iv) Large scale use of recombinants made by self-cloning in systems belonging to exempt category (e.g., *E. coli, Saccharomyces,* and *B. subtilis*).

Category III: Those requiring review and approval of competent authority before commencement.

7.8 TOXIN GENE CLONINGS

A list of toxins classified based on their potential toxicity is listed in Chapter V-A6. The number of plasmid toxin gene clonings at present going on are only three, viz., *B. subtilis* and *B. sphericus* toxin genes are cloned in *B. subtilis* and cholera toxin genes and *B. thuringiensis* crystal protein genes cloned in *E. coli* K12. These toxins gene cloning are being done under PC1 and BC1 Containment conditions. All toxin gene cloning experiments producing LD50 less than 50 μg/kg of body weight of vertebrates (Chapter V-A6) or large scale growing may be referred to Institutional Biosafety Committee (IBSC) for clearance.

7.8.1 Cloning of Genes for Vaccine Production

Rinderpest and leprosy antigens have been classified under Risk Group II in view of the common incidence of the disease in India though they are listed under special category in the Centres for Disease Control & National Institute of Health (CDC-NIH) system. Similarly, leprosy afflicts a large segment of the population which calls for concerted program to control the disease by vaccination and detection at early stages through immunodiagnostic tests. The containment should be decided by Review Committee on Genetic Manipulation (RCGM) on a case by case basis on experiment utilizing DNA from nondefective genomes of organisms recognized as a pathogen. In view of no demonstrated risk from handling free *M. laprae* antigens, inactivated whole cells, as well as antigens, can be assigned to Risk Group I.

(i) Cloning of mosquito and tick DNA experiments should be prescribed on a case by case basis since these are natural vectors for certain endemic viral and parasitic diseases.

(ii) Genes coding for antibiotic resistance into pathogenic organisms which do not naturally possess such resistance.

(iii) Introduction into cultured human cells of recombinant DNA molecules containing complete genes of potentially oncogenic viruses or transformed cellular genes.

(iv) Introduction into animal cells of unidentified DNA molecules derived from cancer cells or *in vitro* transformed cells.

(v) Experiments involving the use of infectious animal and plant viruses in tissue culture systems.

(vi) Experiments involving gene transfer to whole plants and animals.

(vii) Cell fusion experiments of Animal cells containing sequences from viral vectors if the sequence leads to transmissible infection either directly or indirectly as a result of complementation or recombination in the animals. For experiments involving recombinant DNA of higher class organisms using whole animals will be approved on a case by case following IBSC review.

(viii) Transgenosis in animal experiments: Transgenosis method is used to transform animal cells with foreign DNA by using viruses as vectors or by microinjection of DNA into eggs and pre-embryos. The expression of an inserted gene can be influenced both by the regulatory sequences associated with the gene and the sequences present at the site of integration of host genome. At present, there is no way to control where a gene is inserted into the chromosome of either an animal or plant cell. Yet this site of insertion can affect not only the expression of the interested gene but also the regulation of the host cells' DNA, e.g., by nonspecific activation of cellular proton-co-genes.

(ix) All experiments involving the genetic manipulation of plant pathogens and the use of such genetically manipulated plant pathogens would require the approval of competent authority (IBSC).

(x) Transfer of genes with known toxicity to plants using *Agrobacterium tumefaciens* or other vectors. Attempts are under way using Ti-plasmid, *A. tumefaciens,* and other vectors to transfer toxin-encoding genes that enable plants to make their own insecticide, resist infections or tolerate a variety of environmental stresses. Case by case clearance is needed though exemption may be made for the use of well-characterized vectors and nontoxic genes.

(xi) In the case of plant viruses, permission may be obtained only when it is known that there is a chance of nonspecies specific spread of infection to plants that could produce changes in pathogenicity, host range or vector transmissibility. The growth of whole plants, propagation of genetically manipulated organisms in plants, regeneration of plants from cells transformed by manipulated plant pathogen vector would require containment conditions that are elaborated in Chapter V: C2.

(xii) Experiments requiring field testing and release of rDNA-engineered microorganisms and plants (Chapter V: C3).

(xiii) Experiments involving engineered microbes with deletions and certain rearrangements.

(xiv) Diagnostics: No major risk can be foreseen on diagnostics involving *in vitro* tests. But for diagnostics involving *in vivo* tests, specific containment levels have to be prescribed on a case by case basis. For example, tuberculin moiety could be cloned and used for *in vivo* hypersensitivity test as a diagnostic method.

(xv) Gene therapy for hereditary diseases of genetic disorders.

7.8.2 Large Scale Experiments

Large scale production of biomolecules from genetically engineered microorganisms has not just been taken up in the country. However, the use of recombinant organisms in large scale operations is expected in the near future.

(i) In the guidelines, experiments beyond 20 liters capacity for research as well as industrial purposes are included in the category of large scale experimentation/operations.

(ii) For such activities, it is recommended that one should seek approval of the competent authority as described in Chapter-III. To seek approval, it will be necessary to furnish the relevant details in a prescribed format on the lines suggested by GEAC.

(iii) For good large scale practice (GLSP) as well as levels of containment, the following principles of occupational safety and hygiene will be applied.

(iv) To keep workplace and environment exposure to any physical, chemical or biological agent to the lowest practicable level, to exercise engineering control measures at the source and to supplement these with appropriate personal protective clothing and equipment when necessary; to test adequately and maintain control measures and equipment; to test when necessary for the presence of viable process organisms outside the primary physical containment; to provide training of personnel; to formulate and implement a local code of practice for the safety of personnel.

(v) The following safety criteria are to be compiled with for good large scale practice. The host organism should not be a pathogen, should not contain adventitious agents, and should have an extended history of safe use, or have built-in environmental limitations that permit optimum growth in the bioreactor but limited survival with no adverse consequences in the environment. The vector/insert should be well characterized and free from known harmful sequences; the DNA should be limited in size as much as possible to perform the intended function: it should not increase the stability of the recombinant in the environment unless that is a requirement for the intended function; should be poorly mobilizable; and should not transfer any resistance markers to microorganisms not known to acquire them naturally if such acquisition could compromise the use of a drug to control disease agents in human or veterinary medicine or agriculture. The genetically manipulated organism should not be a pathogen and should be assessed as being as safe in the bioreactor as the host organism, and without adverse consequences for the environment. The physical containment conditions need to be ensured for large-scale experiments and production activities.

7.9 RELEASE TO ENVIRONMENT

Depending on the types of organisms handled and assessment of potential risks involved appropriate containment facilities must be provided to ensure the safety of the worker and to prevent unwanted release in the environment. Biowastes are resulting from laboratory experiments, in industrial operations, should be properly treated so that the pathogenicity of genetically engineered organisms are either destroyed or rendered harmless before disposal in the environment. Special facilities should be created for disposal of experimental animals. All refuse and carcasses must be incinerated. Exemption/relaxation of safety measures in specific cases may be considered based on the risk assessment criteria.

For planned release of organisms into the environment, the following points should be taken into consideration.

(i) Geographical location, size, and nature of the site of release and physical and biological proximity to man and other significant biota. In the case of plants, proximity to plants which might be cross pollinated.

(ii) Details of target ecosystem and the predicted effects of a release on that ecosystem.

(iii) Method and amount of release, rate frequency and duration of application.

(iv) Monitoring capabilities and intentions: how many novel organisms are traced, e.g., to measure the effectiveness of the application.

(v) Onsite worker safety procedures and facilities.

(vi) Contingency plans in the event of unanticipated effects of novel organisms.

It is important to evaluate rDNA-modified organism for potential risk prior to application in agriculture and environment. Prior to the introduction of microorganisms, properties of the organism, the possible interaction with other disease-causing agents and the infected wild plant species should be evaluated. An independent review of potential risks should be conducted on a case by case basis prior to application. Details of points to be taken into account for risk assessment of genetically altered organisms while making proposals for release applications. The biohazard evaluation of viral, bacterial, insecticidal agents for field applications are to be provided. Development of organisms for agricultural or environmental applications should be conducted in a stepwise fashion, moving where appropriate, from the laboratory to the growth chamber and greenhouse under containment conditions and good laboratory practice. It should be done under the expert advice of competent authority with regard to the area to be covered taking into account the experimental design and condition of isolation. The release of any strain for field testing should be done with the permission of Genetic Engineering Approval Committee (GEAC) as mentioned in Chapter III.

Though, manipulation of plants under containment would not require regulatory clearance of GEAC, testing of altered plant material in the environment, however, should follow regulatory guidelines seeking experimental field use permit from GEAC even though prima facie, plant material appears safe to test under containment conditions. License for large scale release in case of genetically engineered plants tested pathogens is required.

7.10 IMPORT AND SHIPMENT

1. The import or receipt of etiologic agents and vectors of human and animal disease or their carriers is subject to the quarantine regulations. Permits authorizing the import or receipt of regulated materials for research (e.g., toxin genes, hybridomas, cell cultures, organelle) and specifying conditions under which the agent or vector is shipped, handled and used are issued by the Review Committee on Genetic Manipulation while large-scale imports for industrial use are regulated by Genetic Engineering Approval Committee. Safety testing may be required to ensure that it is far from risk.
2. The interstate shipment of indigenous etiologic agents, diagnostic specimens, and biological products is subject to applicable packaging, labeling and shipping requirements specified for etiologic agents. Packaging and labeling requirements for Interstate shipment of etiologic agents are summarized and illustrated in the rDNA booklet. All such shipments would need the clearance of Institutional Biosafety Committee.

7.10.1 Quality Control of Biologicals Produced by rDNA Technology

The general regulations normally applicable for biologicals are applicable to the recombinant DNA products. The specific relevant aspects of a particular product should be discussed with the appropriate government agency on a case by case basis.

(i) A new license for the product or drug application would be required on products made of recombinant DNA technology even if the product is considered to be chemically and physically similar to the naturally occurring substance or previously approved product produced in the conventional system.

(ii) A recombinant DNA product demonstrated to be identical to normally occurring substance would not require toxicological and pharmacological data if the information is already available at dose levels of intended use, but fresh clinical trials will be necessary on all such products.

(iii) The booklet prescribes the various control methods needed to establish the safety, purity and efficiency of rDNA products.

(iv) Animal feeds: The prevention of food adulteration Act 1954 makes it an offense to sell any material for use as a feeding stuff containing any ingredient which is deleterious to animals.

The use of stilbesterol, vitamin B12, antibiotics, direct or indirect sources of nitrogen such as urea and its derivatives, amino acids as additives in forage and animal feed to enhance nutritive effect are in practice. The possibilities of introduction of products derived by a biotechnological process such as single cell protein, enzymes and also the growing interest in probiotics, i.e., living organisms that are fed to animals to improve performance and use of microorganisms as silage aids may find means to improve the overall health of animals. The control of these products is the same in principle whether they are produced by chemical or biotechnological process provided the purity criteria are met. The products derived from animals for human consumption such as meat and milk should be free from any contaminants or residue effect resultant on the use of feedstuffs containing additives produced by biotechnological processes.

Biohazardous materials require special safety precautions and procedures. Following guidelines are to be borne in mind when working with infectious agents.

7.11 GUIDELINES

7.11.1 Clothing Guidelines

Always wear appropriate Personal Protective Equipment (PPE) such as a laboratory coat, gloves, and a mask, if applicable, when working with infectious agents or infected animals. Add boot/shoe covers and eye protection/face shields when necessary; wear gloves over laboratory coat cuffs; never wear contact lenses when working with infectious agents; do not wear potentially contaminated clothing outside the laboratory area.

7.11.2 Remove Contaminated Clothing

Remove boot/shoe covers; remove gloves by peeling them from the inside out. Do not let the glove make a snapping sound; since that would splash any material on the glove's surface to the air; take- off eye protection/face shield; remove overall clothing protection.

If clothing protection is reused, hang in approved, controlled area. Otherwise discard; remove mask/respirator by untying or lifting the straps up and over from the back of your head and away from face; properly dispose of used personal protective equipment in containers lined with a red biohazard bag and marked with the biohazard symbol. These containers must be closeable and prevent leakage during collection, handling, processing, storage, transport, or shipping.

7.11.3 Infectious Agents Handling Guidelines

Use mechanical pipetting devices and minimize aerosol production; add disinfectant to water baths for infectious substances; use sealed rotors, sealed buckets, or a guard bowl cover complete with gasket as well as safety centrifuge; tubes (tube or bottle carrier with sealable cap or 'O' ring cap) for potentially infectious samples or otherwise hazardous samples. Before use, the tubes should be checked for cracks. Always wait at least 10–30 minutes after the centrifuge has completely stopped to allow any aerosolized leakage to settle, especially if the samples are potentially infectious; always use secondary leak-proof containers when transporting samples, cultures, inoculated Petri dishes, and other containers of biohazardous materials.

7.11.4 Working with Sharps Guidelines

To avoid accidental sticks, place hypodermic needles directly into the sharp containers and do not recap, bend, break, clip, or remove from disposable syringes; use needle-locking or disposable needle units; do not attempt to treat (decontaminate) sharps yourself for any biohazard; do not allow the containers to become overfilled. They should not be more than 3/4 full when picked up ; do not force anything into a sharps container. If it is full, start a new one; n ever put your hands in a used sharps container; do not dispose of these containers with the regular trash or incinerate them.

Biological material is used intensively in biological and biomedical research. In this type of research, the use of modern molecular biological techniques like recombinant-DNA technology is still on the increase. Three types of safety measures can be distinguished: measures to protect the worker, measures to protect the experiment and measures to protect humans and the environment. The safe use of biological material is also required by two types of legislation, viz., the legislation on the protection of workers from activities with biological agents and the legislation on the environment with regard to working with genetically modified organisms.

Biological material and living organisms are neither intrinsically dangerous nor intrinsically safe (René Custers, 2004). Any danger will depend on the characteristics of the material or the organisms. Following are the characteristics that represent a danger.

7.11.4.1 Pathogenicity

The pathogenicity of an organism indicates whether an organism – for instance, a bacterium, a virus, fungus or a parasite – is able to cause a disease in a plant, animal or human. Factors like infectious dose, virulence and the production of toxins by the pathogen play a role in the extent to which the organism is able to cause disease.

7.11.4.2 Toxicity

Toxicity means poisoning. Most substances are not poisonous when they are used under normal circumstances. The toxicity of a substance is mostly given as an LD50 for vertebrates in weight units per kilogram body weight. The LD50 (LD stands for lethal dose) is the amount at which exposure to the substance leads to the death of 50% of the animals exposed. When the toxicity of living organisms (especially bacteria) is considered, toxicity often coincides with pathogenicity.

7.11.4.3 Allergenicity

Allergenicity in an animal body is a nontoxic, immune system mediated, undesired reaction of the body to a substance or agent. Immune globulin E (IgE) and mast cells (immune system cells that, among other things, produce heparin) often play a role in the allergic reaction.

7.11.4.4 Disturbance of ecological balances

The aspect of disturbance of ecological balances is especially relevant for activities involving GMOs. Disturbance of an ecological balance may happen when a GMO possessing a certain characteristic is accidentally spread to the environment or when genetic material originating from that organism spreads to other organisms in the environment.

7.11.4.5 Other harmful effects

Sometimes there are other unwanted effects that urge one to be even more cautious when handling biological material. It is not possible to give an exhaustive list of these effects. What matters is that one stops to think about the characteristics of the biological material, before starting to work with it. One important class of genes that should be looked at carefully is genes that produce proteins with immune modulating properties, although not all immune modulations are harmful. For certainty about the possible level of harm the effects of the immune modulation should be thought through carefully, and quite often consultation with experts will be necessary. One example is the handling of a vaccinia virus in which a gene responsible for immune suppression is cloned. Immune suppression may lead to the body not being able to fight an infection by the virus. In some exceptional cases, infections with vaccinia viruses may lead to fatal encephalitis.

7.12 RISK ASSESSMENT

GMOs, like non-GMOs, are neither intrinsically hazardous nor intrinsically safe. That is why risk assessment is performed on a case-by-case basis. The risk assessment procedure consists of three subsequent steps (René Custers, 2004). Firstly, the characteristics of the host, vector and donor sequences that are potentially hazardous like pathogenicity, toxicity, the possibility of uncontrolled spreading of the organism or its genetic material, are identified. This leads to a preliminary identification of the risk level. Secondly, the circumstances under which the organisms can be handled safely are determined, taking into account the following aspects: The characteristics of the environment that could be exposed to the GMOs, the type and scale of the activity, any nonstandard activities or actions. Finally, a risk class is determined, based on the results of the first two steps.

As for pathogens, four risk classes have been determined for GMOs (Table 7.1).

Table 7.1 Risk Classes for GMOs

Risk class 1	GMO activities holding no or a negligible risk	Activities for which level 1 containment is appropriate to protect human health as well as the environment
Risk class 2	GMO activities holding a low risk	Activities for which level 2 containment is appropriate to protect human health as well as the environment
Risk class 3	GMO activities holding a moderate risk	Activities for which level 3 containment is appropriate to protect human health as well as the environment
Risk class 4	GMO activities holding a high risk	Activities for which level 4 containment is appropriate to protect human health as well as the environment

After determining a first classification, and depending on the type and scale of activity, it should also be determined whether or not specific extra measures will be necessary for addition to the basic requirements of that containment level, or whether some measures can be left out.

Below a comparison is given between the categorization of pathogens and GMOs in their risk classes and the accompanying basic containment levels [(Table 7.2) (René Custers, 2004)].

Table 7.2 Comparative Risk Classes and Levels Among Pathogens and GMOs

Class	*Pathogens*	*GMOs*	*Basic containment level*
Risk class 1	Nonpathogens	No, or negligible risk	Level 1 for GMOs, SMP for nonmodified microorganisms or cells*
Risk class 2	Mild pathogens	Low risk	Level 2
Risk class 3	Moderate pathogens	Moderate risk	Level 3
Risk class 4	Strong pathogens	High risk	Level 4

***Note:** Level 1 gives requirements for both the infrastructure and the working practices, although the requirements for the infrastructure are very limited. SMP (Safe Microbiological Practices) only describes requirements for the working practices.

7.12.1 Undesired Spread of Organisms or Genetic Material

It may have become clear that the spread of hazardous organisms represents a danger both to yourself and to your colleagues. When it is possible for organisms to spread to a colleague, they may spread to the environment as well. This dissemination of organisms or genetic material to the environment is often undesired since it may involve the spread of pathogens or toxins, or lead to the disruption of ecological balances. This is undoubtedly true with regard to organisms belonging to risk classes 2, 3 and 4. However, even the spread of organisms (and their genetic material) belonging to risk class 1, and thus presenting only a minor risk, should be limited.

7.13 TRANSGENIC PLANTS

Transgenic plants are grown *in vitro,* in growth chambers or greenhouses, and the plants are not able to disseminate just like that (Anonymous, 2012). Nevertheless, the undesired spread of transgenic plants deserves special attention (René Custers, 2004). If no proper containment measures are taken, pollen may be disseminated to the environment through the air or aided by insects. Whether or not this presents a genuine risk, depends on how the plant reproduces: by self-pollination or by cross-pollination. The spread of pollen by strict self-pollinators has no effect whatsoever, but when a cross-pollinator is involved, it should be carefully checked whether any of its wild relatives, which it might successfully hybridize with, is growing in the vicinity. In addition to pollen, seeds originating from transgenic plants may sometimes easily be disseminated in the environment. Especially when they are very small or sticky, these seeds are very likely to be accidentally taken along by a researcher leaving the growth chamber or greenhouse. It is not only pollen or seeds that may be responsible for the undesired spread of transgenic plants. Some plant parts may grow and turn into whole new plants themselves. These reproductive parts of plants should not be discarded without destroying them properly. For example, the branch of a willow can grow roots and leaves very easily, and the stem base of a cabbage can also grow roots. This is why laboratory staff handling transgenic plants or plant material should pay special attention to the possible spread of plant parts that are still able to reproduce. If there is a genuine possibility that a transgenic plant will be able to establish itself in the environment, or that it will hybridize with wild relatives, reproductive plant parts should be destroyed before they are discarded as waste.

7.14 RESPONSIBILITIES

7.14.1 Responsibilities of Institutional Biosafety Committee (IBC)

It reviews and approves all procedures, policies, and guidelines regarding the storage, usage, transport and disposal of biohazardous materials (Anonymous, 2012). It provides technical advice to users of biohazardous materials, reviews permit applications and risk assessments involving biological agents and materials and meets quarterly or as needed at the request of any Committee member.

7.14.2 Responsibilities of the Biosafety Officer

Biosafety Officer provides a liaison between Laurier and the Office of Laboratory Security, Public Health Agency of Canada; provides technical support, consultation, and advice on all aspects of biohazardous work; develops effective procedures for the implementation of current standards, and guidelines; oversees a biosafety program that complies with the regulations for use of or exposure to biohazardous materials; provides training for persons using biological hazards, develops procedures for shipping all biohazardous materials; provides guidance for appropriate decontamination procedures for transport and after a spill or an accident involving biohazardous materials; provides guidance on the proper disposal of biohazardous waste and recommends waste disposal procedures; ensures the university's registration and license for the 'Human Pathogens and Toxins Act' is up to date and accurate and completes renewals and annual reporting as necessary and advises the Director, SHERM, on policy matters concerned with the protection of personnel from biohazardous agents.

7.14.3 Responsibilities of the Supervisors/Principal Investigators

Supervisors or Principal Investigators (PIs) are responsible for implementing the Biosafety Program within the scope of their projects (Anonymous, 2012). They must ensure that all biohazardous materials are registered, and that appropriate permits and risk assessment forms are completed prior to purchasing/obtaining the agent(s). In the case of environmental isolates and genetically modified organisms, identification of the organism should be established for isolates and a risk assessment conducted for both isolates and genetically modified organisms (GMOs).

- Ensure that appropriate containment levels are established in consultation with the Biosafety Officer; ensure adherence to these levels.
- Implement engineering controls to the extent feasible including containment equipment (primary barriers) and facility design (secondary barriers) appropriate for the laboratory function, hazard, and risk.
- Ensure strict adherence to biosafety practices and techniques.
- Implement procedures for safe handling of specimens of human or primate origin.
- Ensure that all individuals under their supervision are trained and knowledgeable about the hazards associated with handling biohazardous materials.
- Direct laboratory activities: Ensure appropriate training of personnel about the potential hazards of biohazardous materials and the practices and techniques required for their safe handling.

7.14.4 Responsibilities of User

Individuals are responsible for their own safety and the safety of others they may affect with biohazardous agents/substances (Anonymous, 2012). Users must work with biohazardous agents/substances using the appropriate containment level, as directed by the Lab Supervisor and the Biosafety Officer, review containment level requirements with their Supervisor, work with strict adherence to biosafety practices and techniques, participate in health surveillance programs where appropriate, report unsafe conditions or any reason that duties cannot be completed safely and report any accidents/incidents to the laboratory supervisor and SHERM.

7.14.5 Pathogen Safety Data Sheets

A limited number of Pathogen Safety Data Sheets (PSDSs) are available from the Public Health Agency of Canada website at http://www.phac-aspc.gc.ca/lab-bio/res/psds-ftss/index-eng.php. PSDSs contain health hazard information, recommended precautions, spill clean-up procedures and other information that is relevant specifically to the laboratory setting. They serve as an additional safety resource for laboratory personnel working with biological agents.

7.14.6 Additional Safety Considerations for Plant Pathogens

The Containment Standards for Facilities Handling Plant Pests, published by the Canadian Food Inspection Agency, describes the minimum acceptable physical and operational requirements for facilities that work with plant pests other than weeds, soil, genetically modified plants and arthropod biological control agents (Anonymous, 2012). Any persons who grow, raise, culture or produce anything that is a pest or is infected or infested with a pest must adhere to these containment standards. They provide guidance on the operation of plant pest containment facilities such as laboratories, greenhouses, and screen houses. Compliance with these standards and with documents such as import permits will help to ensure that economically, and environmentally significant plant pests do not inadvertently escape into the environment and become established in Canada.

7.14.6.1 General precautions

All workers should routinely use appropriate barrier precautions to prevent skin and mucous membrane exposure when contact with human blood or other body fluids is anticipated (Anonymous, 2012).

- Gloves must be worn when touching blood and body fluids, mucous membranes, or nonintact skin, for handling items or surfaces soiled with blood or body fluids, and for performing venipuncture and other vascular access procedures. If a glove is torn or damaged during use, it should be removed and a new glove used as promptly as safety permits. Disposable gloves must not be washed or disinfected for reuse. Washing with surfactants may enhance penetration of liquids through undetected holes in the glove. Disinfecting agents may cause deterioration of the glove material.

- Masks and protective eyewear or face shields should be worn during procedures that are likely to generate droplets of blood or other body fluids to prevent exposure of mucous membranes of the mouth, nose, and eyes.
- Gowns or aprons should be worn during procedures that are likely to generate splashes of blood or other body fluids. Protective clothing should be removed before leaving the area.
- Hands and other skin surfaces must be washed immediately and thoroughly if contaminated with blood or other body fluids. Hands should be washed immediately after gloves are removed since no barrier is 100% effective.
- Workers should take precautions to prevent injuries caused by needles, scalpels, and other sharp instruments or devices during procedures, when cleaning used instruments, during disposal of used needles, and when handling sharp instruments after procedures. Needles and syringes should be used only in those situations when there is no alternative. To prevent needle-stick injuries, needles must not be recapped, purposely bent or broken by hand, removed from disposable syringes, or otherwise manipulated by hand. After they are used, disposable syringes and needles, scalpel blades, and other sharp items should be placed in puncture-resistant containers for disposal. The puncture-resistant container should be located as close to the use area as practical.
- Workers who have exudative lesions, weeping dermatitis, cuts, open wounds or other breaks in the skin must either refrain from all direct contact with blood and other body fluids until the condition resolves or utilize protective barriers to reduce the risk of exposure.
- Pregnant workers should be aware that some BBPs may be particularly harmful to a developing fetus.

7.15 BIOSAFETY SIGNS AND LABELS

The Public Health Agency of Canada requires that warning signs and/or symbols be used to inform personnel and visitors of the potential of hazards in the workplace. Specifically, with regard to biohazards, the universal biohazard warning symbol (Fig. 7.1), must be used to "signify the actual or potential presence of a biohazard and to identify equipment, containers, rooms, materials, experimental animals or combinations thereof, which contain and/or are contaminated with, viable hazardous agents" (Anonymous, 2012).

Fig. 7.1 Universal biohazard warning symbol

The universal biohazard symbol must be used to designate the presence of agents/ substances that are believed to be biohazardous. All laboratories and work areas utilizing and/or storing biohazardous substances must have the appropriate biohazardous caution sign posted prominently. If RG2 agents are used, a biohazard sign must be located outside the laboratory door to indicate the nature of the hazard, the biohazard level, special provisions for entry and contact information for the Principal Investigator and/ or another responsible person(s). Principal Investigators/Supervisors are responsible for ensuring that all hazard signs are current and accurate. Notify the Biosafety Officer if changes are necessary for laboratory door signage and/or equipment labeling.

7.15.1 Working in a Biological Safety Cabinet

7.15.1.1 Start-up procedures

When preparing for work in the BSC (Anonymous, 2012) follow the following instructions.

- Turn off UV lights if in use and ensure that the sash is in the appropriate position.
- Turn on fluorescent light and cabinet blower.
- Check the air intake and exhaust grilles for obstructions.
- If the cabinet is equipped with an alarm, test the alarm and switch it to the 'on' position.
- Confirm inward airflow by holding a tissue in the middle of the edge of the viewing panel and ensuring that it is drawn in.
- Disinfect the interior surfaces with a suitable, noncorrosive disinfectant (e.g., 70% alcohol).
- Assemble all materials required for the procedure and load them into the cabinet; do not obstruct the air grilles; the working surface may be lined with absorbent paper with plastic backing; segregate 'clean' items from 'contaminated' items.
- Wait 5 minutes to purge airborne contaminants from the work area.

7.15.1.2 Working in the cabinet

When working in the BSC,

- Put on a protective lab coat and gloves as appropriate.
- Perform operations as far to the rear of the work area as possible.
- Avoid movement of materials or excessive movement of hands and arms through the front access opening during use; when you do enter or exit the cabinet, do so from straight on; allow the cabinet to stabilize before resuming work.
- Keep discarded, contaminated material to the rear of the cabinet; do not discard materials in containers outside of the cabinet.
- Do not work with open flames inside the cabinet.

- If there is a spill during use, surface decontaminate all objects in the cabinet; disinfect the working area of the cabinet while it is still in operation (do not turn the cabinet off).
- Ensure small items such as Kim Wipes don't get sucked into the vents, as they can disrupt motor operations.

7.15.1.3 *Shut-down procedures*

Follow these procedures upon completion of the work.

- Allow the cabinet to run for 5 minutes with no activity.
- Close or cover open containers before removing them from the cabinet.
- Surface disinfect by spraying objects with 70% alcohol in contact with the contaminated material before removal from the cabinet.
- Remove contaminated gloves and dispose of them as appropriate; wash hands.
- Put on clean gloves, and ensure that all contaminated materials are placed into biohazard bags within the cabinet.
- Using a suitable noncorrosive disinfectant (e.g., 70% ethanol), disinfect interior surfaces of the cabinet; periodically remove the work surface and disinfect the area beneath it (including the catch pan) and wipe the surface of the UV light with disinfectant.
- Turn off the fluorescent light and cabinet blower.
- Turn on the UV light if appropriate (do not turn on when people are working close by).

7.16 CERTIFICATION OF BIOLOGICAL SAFETY CABINETS (BSCs)

The certification of the Biological Safety Cabinets is essential to their safe and effective use. All Biological Safety Cabinets must be certified on installation, when the filters are changed, before and after a move or transfer and annually. Cabinets must not be moved without first undergoing a decontamination process. The Biosafety Officer must approve any modifications to any Biological Safety Cabinets. Cabinets must undergo certification following modification. The certification process is the responsibility of SHERM. If problems are encountered in operating a biological safety cabinet, do not continue to use it; contact Physical Resources immediately.

7.16.1 International Efforts in Biosafety

Heightened global environmental awareness and concern over accelerating ecological degradation during the latter quarter of the 20th Century resulted in a desire by the international community to push the protection of the environment higher up the political agenda. These efforts came to fruition in 1992 when the Convention on Biological Diversity (CBD) came into force. Its objectives include "the conservation of biological diversity, the sustainable use of its components and the fair and equitable sharing of the benefits arising out of the utilization of genetic resources." During the elaboration of the Convention, negotiators recognized that biotechnology could contribute to achieving these objectives if developed and used with adequate safety measures for both the environment and human health. Accordingly, procedures were developed to address the safe transfer, to handle and use of any LMO ('living modified organism'; used interchangeably with 'genetically modified organism' [GMO] in these web pages) resulting from biotechnology that may have an adverse effect on the conservation and sustainable use of biological diversity (Article 19.3, CBD). These procedures formed the Cartagena Protocol on Biosafety, which came into force in 2003 and had 138 signatory countries to date (January 2007). Parties lacking a cohesive biosafety policy undertook, or are currently undertaking, a number of initiatives to put a national framework in place to comply with the CPB. This has resulted in a great demand for biosafety-related information, training programs, and capacity-building projects.

7.16.2 Global GM Crop Cultivation

This period of heightened political activity in environmental protection has coincided with a concomitant rise in genetically modified crop cultivation.

In 2010, the accumulated hectarage planted during the 15 years, 1996 to 2010, exceeded for the first time, 1 billion hectares, a record 87-fold increase in hectarage between 1996 and 2010, making biotech crops the fastest adopted crop technology in the history of modern agriculture. (James, C. 2010). The number of countries electing to grow biotech crops has been increasing consistently from 6 in 1996 to 29 in 2010, is worth mentioning that out of the 29 countries growing GMOs in 2010, 19 were developing countries. Among the top 10 GM crop-growing countries by area, the USA, Argentina, Canada, Uruguay, and Australia are currently not parties to the CPB. At the same time, many developing countries that have ratified the CPB are still in the process of elaborating a regulatory framework governing the import or cultivation of GM crops. This has led to the current situation where different strategies and standards have been adopted at the national level, caused by the different infrastructures available in developed and developing countries, and has resulted in much confusion and difficulty in harmonizing environment and trade agreements and regulations.

7.16.3 Topics of Concern

Based upon ICGEB's long-standing activities in biosafety, we have identified the main issues derived from the deliberate introduction of GM crops (and their derived products) into the environment or onto the market of concern today. These have been classified as,

(i) **Risks for animal and human health:** Toxicity & food/feed quality/safety; allergies; pathogen drug resistance (antibiotic resistance), impact of selectable marker

(ii) **Risks for the environment:** Persistency of gene or transgene (volunteers, increased fitness of GM crop, invasiveness) or of transgene products (accumulative effects); susceptibility of nontarget organisms; change in use of chemicals in agriculture; unpredictable gene expression or transgene instability (gene silencing); environmentally-induced (abiotic) changes in transgene expression; ecological fitness; changes to biodiversity (interference of tri-trophic interactions); and impact on soil fertility/soil degradation of organic material

(iii) **Horizontal gene transfer:** Genetic pollution through pollen or seed dispersal and horizontal gene transfer (transgene or promoter dispersion); transfer of foreign gene to microorganisms (DNA uptake) or generation of new live viruses by recombination (transcapsidation, complementation, etc.).

(iv) **Risks for agriculture:** Resistance/tolerance of target organisms; weeds or super weeds; alteration of nutritional value (attractiveness of the organism to pests); change in the cost of agriculture; pest/weed management; unpredictable variation in active product availability; loss of familiarity/changes in agricultural practice.

(v) **General concerns:** Detection and analytical methods; ethical issues (e.g., labeling); substantial equivalence; risk assessment/risk management; general biosafety; public attitudes, perception; legislation (including liability and redress); monitoring; socio-economics (e.g., situation of poor farmers in developing countries); and IPR (Intellectual Property Rights); and GM traceability/commodity segregation.

Biosafety also refers to "the means to regulate, manage or control the risks associated with the use and release of living modified organisms (LMOs) resulting from biotechnology which are likely to have adverse environmental impacts that could affect the conservation and sustainable use of biological diversity, taking also into account the risks to human health." As such, biosafety does not represent an individual biosecurity sector as it is cross-cutting in scope.

7.16.4 Adoption of a Risk-based Approach

It includes rapid proliferation of new genetic technologies, emphasis on rapid establishment of credible and effective controls for LMOs and GMOs so as to maximize the benefits of biotechnology while minimizing associated risks, development of detailed national strategies for conservation and protection of the environment, increasing 'public good' regulation for sustainable use of biological resources, and greater inclusion of indigenous and local communities in decision-making.

LMOs are increasingly being released on a worldwide basis. While they may have potential benefits for human well-being and achieving sustainable economic development, their proliferation could have unintended adverse effects on the environment, including the destruction of native flora and fauna, as well as adverse effects on human health. This could be especially significant in developing countries that do not have the capacity to track releases of these organisms and, therefore, cannot adequately safeguard national interests.

Regulatory requirements covering the safe transfer, handling and use of LMOs resulting from modern biotechnology are a new focus point in biosecurity and are triggering strong cross-sectoral interest in more holistic approaches to their management.

However, controls on transboundary movements currently vary considerably between countries in terms of their development, importation, field testing or release. Food may also be derived from (or traits introduced) by modern biotechnology. Although international guidelines on assessment of the safety of foods derived from GMOs are being developed, the adequacy of current processes is a continuing issue of public concern.

As with plant biotechnology in the early 1990s, animal biotechnology has reached a point where developers are beginning to market products derived in this manner. This may, in the near future, include agri-food applications. As an example, transgenic animals derived from recombinant DNA technology or by cloning (somatic cell nuclear transfer) is a means to generate animals with preferred traits. These animals and/or their products are likely to trigger regulatory requirements in most countries, but guidance on safety assessment is still at the developmental stage.

7.17 ENVIRONMENTAL PROTECTION

Environmental protection in a broad sense is also a biosecurity activity. While not excluding any aspects of the above sectors, specific biosecurity cross-sectoral environmental initiatives may be undertaken by competent authorities, especially in the management of biological resources to ensure sustainable agriculture while maintaining the full biological diversity of genetic resources.

7.17.1 Practical Guide to Containment for Greenhouse Research with Transgenic Plants and Microbes

A practical guide to containment for greenhouse research with transgenic plants and microbes was developed by Patricia *et al.* (2001), which comprises various aspects like regulation and oversight of Genetically Modified Organisms, biosafety levels, elements of containment, management practices, retrofitting for containment, design of new containment facilities, etc.

This guide discusses issues like relevant information on four levels of biosafety containment; physical and biological strategies that provide containment; suggested facility modifications to achieve prescribed containment levels; suggestions for day-to-day greenhouse management; methods for proper handling of GMOs; discussions of selected design issues for new or renovated facilities; descriptions of equipment and supplies; a sample floor plan and; sources for additional information. This guide applies to greenhouses—controlled environment structures having a transparent or translucent covering and used for growing plants—that contain genetically modified plants or plant-associated organisms. The wide ranges of microorganisms that are plant-associated include viruses, bacteria, fungi, protozoa, mycoplasma-like organisms, nematodes, insects, mites, and others. Other contained plant growth facilities, such as growth chambers, biosafety cabinets, incubators, and tissue culture tables or rooms, often are an integral part of the process leading to the preparation of GMO materials for greenhouse studies or field tests.

7.17.2 Criteria to be Considered while Making a Biosafety Level Assignment

The purpose of containment is to prevent the transfer of recombinant DNA from transgenic organisms inside the greenhouse to populations outside the greenhouse. Section III of the NIH Guidelines describes four physical containment levels for experiments involving recombinant DNA molecules. It further categorizes experiments according to specific risk criteria and assigns them to one of the four biosafety levels, BL1-P through BL4-P. Following are the criteria.

(i) **Source and nature of the introduced DNA:** Whether from an exotic infectious agent or pathogenic organism; and whether a fragment of DNA or complete genome

(ii) **Recipient organism:** Mode and ease of dissemination; invasiveness; whether a noxious weed or capable of interbreeding with noxious weeds; potential for outcrossing between recipient organisms and nearby related species; and potential for detrimental impact on natural or managed ecosystems

(iii) **Nature of expressed protein:** Whether a vertebrate toxin or potential or known allergen; and whether toxic to other organisms in local environment

(iv) **Local environment:** Nature and importance of nearby crops; presence of sexually compatible wild or weedy species

(v) **Experimental procedures:** Transfer to or from greenhouse and necessary containment measures.

The guidelines state that the principle purpose of GMO containment is to avoid unintentional transmission of rDNA-containing plant genomes or release of rDNA-derived organisms associated with plants, minimize the possibility of unanticipated deleterious effects on organisms and ecosystems outside of the experimental facility, avoid the inadvertent spread of a serious pathogen from a greenhouse to a local agricultural crop, and avoid the unintentional introduction and establishment of an organism in a new ecosystem.

Environmental protection is the predominant goal; the key to achieving it lies in understanding the biological systems involved and accepted scientific research practices. Containment is accomplished through a combination of management practices, physical barriers, and biological methods intended to prevent GMO transfer or survival. In general, containment requirements are more stringent if plant-associated materials, such as insects and microorganisms, are included in the experiment. If insect quarantine measures are required, regardless of the presence of rDNA material, managers should contact APHIS for guidance.

Research involving transgenic plants at the BL1-P or BL2-P containment levels requires little more than the basic facilities, equipment, and protocols common to most research greenhouses. However, greenhouses that offer high-level BL3-P and BL4-P containment are expensive to build and operate. The cost of greenhouse containment at these levels may be prohibitive for many institutions. Other means of attaining a high level of containment, such as the use of a growth chamber or growth room, may provide a suitable alternative at a fraction of the cost.

7.17.3 Microorganisms

Effective physical containment of bacteria, viruses, and other microbes can be extremely difficult because they cannot be seen and, once dispersed, cannot be recovered. However, many will not survive and persist if they are dispersed. Biological measures often provide better containment options. The following methods may help prevent dissemination of genetically modified microorganisms.

7.17.4 Avoid Creating Aerosols When Inoculating Plants with Transgenic Microbes

Provide adequate distance between an infected plant and another susceptible host, especially if the microorganism can be disseminated through the air or by leaf contact, grow experimental plants and microbes at a time of year when nearby susceptible plants are not growing, eliminate vectors for insect-borne microorganisms, choose microorganisms having an obligate association with the host plant, genetically disable the microorganism to minimize survival, and reproduction and treat or evaporate runoff water.

7.18 FEDERAL GOVERNMENT ENTITIES

Federal Government entities are responsible for oversight of laboratory biosafety and biocontainment. The Federal entities that have primary regulatory oversight responsibility for facilities that possess, use, or transfer biohazards are Department of Labor (DOL), Occupational Safety and Health Administration (OSHA), HHS Centers for Disease Control and Prevention (CDC) and USDA Animal and Plant Health Inspection Service (APHIS). USDA/APHIS works in a variety of ways to oversee, protect, and improve the health, quality, and marketability of animals (including various wildlife species), animal products, and veterinary biologics and USDA/APHIS Plant Protection and Quarantine (USDA/APHIS/PPQ). USDA/APHIS/PPQ is the primary Federal agency charged with overseeing the system for safeguarding plants.

The Federal biosafety and biocontainment regulations most relevant to research and related activities involving potentially hazardous biological materials are as follows.

- OSHA General Duty Clause (15 USC § 654; Duties of employers and employees), Bloodborne Pathogens Standard (29 CFR § 1910.1030), and Personal Protective Equipment Standards)
- HHS and USDA Select Agent Regulations (42 CFR part 73, 9 CFR part 121, 7 CFR part 331; for more information (http://www.selectagents.gov/)
- USDA regulations that require permits for work with high-consequence animal and plant pathogens (http://www.aphis.usda.gov/)
- HHS/CDC regulations that require a permit for the import of any infectious agent known or suspected to cause disease in humans (HHS/CDC Foreign Quarantine Regulations; 42 CFR § 71.54)

- Federal guidelines pertaining to laboratory biosafety and biocontainment include:
- NIH and CDC manual entitled Biosafety in Microbiological and Biochemical Laboratories (BMBL)
- NIH Guidelines for research involving recombinant DNA molecules (NIH Guidelines)

For more than two decades, the BMBL and the NIH Guidelines have been the codes of practice for biosafety and biocontainment in the United States. Federal guidelines pertaining to plants and plant pests containing recombinant DNA-modified pathogens are described in Appendix P of the NIH Guidelines. These documents are amended and revised to reflect advances in science and technology.

7.19 BIOSAFETY POLICY

Biosecurity is generally understood to mean all policies and regulatory framework that manages the biological risks associated with food and agriculture, including relevant environmental risks. Biosafety is the biosecurity measures used to protect human health and the environment from the possible adverse effects of the products of modern biotechnology. At the 1992 Earth Summit in Rio de Janeiro, Brazil, world leaders agreed on a comprehensive strategy for 'sustainable development' – meeting our needs for health, environment and biodiversity while ensuring that we leave a healthy and viable world for future generations. One of the key agreements adopted at Rio was Convention on Biological Diversity (CBD) which establishes three main goals: the conservation of biological diversity, the sustainable use of its components, and the fair and equitable sharing of the benefits arising from the use of genetic resources (Anonymous, 2009).

Article 28 of the CBD provides for the formulation of Protocols to address the implementation of various aspects of the Agreement. The use of modern biotechnology is an issue of growing international concern due to the need to protect human health and the environment from possible adverse effects. However, it is also important to recognize modern biotechnology's potential for addressing the World's critical needs for food and health care. As a result of these concerns the Cartagena Protocol on Biosafety (CPB) was adopted in Montreal on the 29th January 2000 at an extraordinary meeting of the Conference of the Parties to the Convention on Biological Diversity.

The Cartagena Protocol on Biosafety makes provisions to regulate, manage or control risks associated with transfer, handling and use of organisms and derived products resulting from modern biotechnology that may have adverse effects on the conservation and sustainable use of biological diversity focusing on their transboundary movement. Belize signed on to the Convention on Biological Diversity on 13th June 1992 and ratified it on 30th December 1993. With regards to the Cartagena Protocol on Biosafety, Belize ratified this Protocol on 12th February 2004 and is a party by accession since 12th May 2004. As a signatory to the CBD and CPB Belize is obliged to implement the articles of the CPB and develop its own national regulatory framework for the safe transfer, handling, use and release of Genetically Modified Organisms (GMOs) and products resulting from modern biotechnology.

7.19.1 Justification

Biotechnology brings with it a number of known and unknown risks. With our current knowledge, the potential risks associated with the application of modern biotechnology can be categorized into human, plant and animal health risks, risks associated with the conservation of biodiversity and agricultural sustainability, and ethical and socio-economic risks. Genetic modification could be used to produce GMO crops that, for example, may be resistant to a specific disease and so provide a significant advantage to the farmer who is producing a food product destined for human or animal consumption (Anonymous, 2009). Genetic modification can also be used to produce animal food products. However, the use of products from such sources has a potential risk to human and animal health as well as to the health of the plant or animal modified. This is because such genetic modification alters the naturally occurring expression or production levels of proteins at different stages of growth, as well as in different organs. Such changes have the potential of increasing natural or foreign protein levels in food or food derivatives and with this the possibility of creating toxic, pathogenic and/or allergic effects from the use or consumption of the product. Other risks could include unexpected products and effects, and changes in the nutrition, composition or digestibility of the products. The genetic modification requires the insertion of a gene into a chromosome (DNA) of the recipient organism. The gene inserted may impart a specific character to the recipient organism; such as disease or herbicide resistance in crop plants. However, such modified crop plants may breed naturally with nearby wild relatives, and this may lead to the transfer of the inserted gene into natural plant populations. Such an occurrence could have an adverse impact on biodiversity as any wild recipient of such a gene may have a competitive advantage in natural ecosystems. In addition, the unexpected persistence of such a transferred gene and products in the environment, the proliferation of volunteer plants and effects on nontarget organisms may occur. Agricultural production sustainability could be compromised by the development of resistant, unmanageable pests (which include weeds) through such a gene transfer process or through natural selection if such risks are not adequately assessed, and production protocols that use GMOs are not strictly regulated. The Government of Belize shall ensure that such risks are adequately assessed, and production protocols are adequately regulated (Anonymous, 2009).

Ethical and socio-economic concerns include the value of biological diversity to local communities including the preservation of indigenous germplasm and ecosystems. Of particular concern to Belize is the potential loss of export opportunities and access to niche markets that exist for organic and fair trade products. The Government has an obligation to ensure that if the production and use of genetically modified organisms are permitted, it is done in an ethically and socially acceptable manner. Any development, application, and release of GMOs must be conducted in a manner that will ensure the conservation and sustainable use of natural resources. Biosafety is, therefore, not only an international obligation in terms of the Cartagena Protocol but also the desired endeavor that the Government of Belize wants to ensure for its people. There is an urgent need to establish biosafety measures for Belize, not only for the genetically modified organisms (GMOs) or genetically modified products used in food, feed and processing (FFP) that may be produced locally in the future but also for those that may be imported into the country. There remains considerable uncertainty about potential risks associated with modern biotechnology. The possible costs of mitigating or reversing any harm that may occur as a result of the use of modern biotechnology may also prove to be immense and far-reaching, especially to the Government, which is ultimately responsible for assuring the health status and food security of the Belizean population. This policy sets the overall framework in which adequate safety measures will be developed and put into force so that Belize can minimize possible risks to human health and the environment while extracting maximum benefit from any potential that modern biotechnology may offer. This National Policy on Biosafety is also an important tool to ensure that the knowledge, practices and benefits of Belize's traditional techniques are safeguarded.

The establishment of this National Policy on Biosafety is also an important step in meeting Belize's obligations to the Convention on Biological Diversity (CBD) and the Cartagena Protocol on Biosafety (Anonymous, 2009).

7.20 CARTAGENA PROTOCOL ON BIOSAFETY

Although several nations have enacted national biosafety legislations to ensure the safe use of Genetically modified organisms (GMOs) and products thereof, biotechnology being a global industry and GMOs traded across borders, international rules are needed as well. Cartagena protocol on biosafety is an attempt to produce a globally harmonized regime for biosafety under the Convention of Biological Diversity (CBD). Named after the Columbian city where the final round of talks was launched, the Cartagena Protocol sets out a comprehensive regulatory system to ensure safe transfer, handling and use of Living Modified Organisms (LMOs) subject to transboundary movement. In daily usage LMOs are regarded to be similar to GMOs though definitions and interpretations vary (Anonymous, 2007). The protocol deals mainly with LMOs that are to be intentionally introduced into the environment like trees, seeds, etc., and with genetically modified farm commodities like maize/grain used for food. It does not, however, cover pharmaceuticals for humans addressed by other international agreements and organizations or products derived from LMOs like cooking oil from genetically modified maize. The main purposes of CBD (1992) includes the biological diversity conservation, sustainable use of fits components and the fair and equitable sharing of the benefits arising out of the utilization of genetic resources.

The protocol has started its action from September 11, 2003, with objectives of contributing to ensure an adequate level of protection in the field of the safe transfer, handling and use of LMOs leading from modern biotechnology that may possess adverse effects on the conservation and sustainable use of biological diversity, taking into account risks to human health and focusing on transboundary movements.

7.20.1 Risk Assessment

The protocol empowers Governments to make its decisions in accordance with scientifically sound risk assessments (Anonymous, 2007). These assessments aim to identify and evaluate the potential adverse effects that LMO may have on the conservation and sustainable use of biodiversity in the receiving environments. They are to be undertaken in a scientific way using recognized risk assessment techniques. While the nation considering permitting the import of an LMO is responsible for ensuring that a risk assessment is carried out, it has the right to require the exporter to do the work or to bear the cost. This is specifically important for several developing countries.

7.20.2 Risk Management

The protocol needs each nation to manage and control any risks that may be identified by a risk assessment. Key elements of effective risk management include monitoring systems, research programs, technical training and improved domestic coordination among Government agencies and services. The protocol also needs each Government to notify and consult other affected or potentially affected Governments when it becomes aware that LMOs under its jurisdiction may cross international borders due to illegal trade or release into the environment. This will enable them to pursue emergency measures or other appropriate action. Governments must establish official contact points for emergencies as a way of improving international coordination.

7.20.3 Benefits of Becoming Party to Protocol

Various benefits include the influence on the implementation of the protocol and shaping of its further development through participation in the decision-making processes of the Conference of the Parties, Parties to the Protocol (COP-MOP); for developing country parties and parties with economics in transition, eligibility for financial support from the global environment facility for capacity-building as well as other support for implementation of the protocol and participation in its processes; enhanced visibility and credibility of national systems for regulating biosafety within the global community; contribution to harmonized rules, procedures and practices in managing the transboundary movement of LMOs; facilitation of mechanisms and opportunities for governments to collaborate with other governments, the private sector and civil society on strengthening biosafety; improved access to relevant technologies and data and benefitting from a regular exchange of information and expertise; demonstration of commitment to conversation; and sustainable use of biological diversity through the implementation of biosafety measures.

7.21 OTHER INTERNATIONAL AGREEMENTS

There are several other international instruments and standards-setting processes that address several issues related to LMOs/GMOs, although the Cartagena Protocol on

Biosafety is the only international instrument dealing exclusively with LMOs/GMOs (Anonymous, 2007). The protocol states that its provisions are intended neither to override nor to be subordinate these existing international agreements. Following are the few such agreements.

7.21.1 International Plant Protection Convention (IPPC)

This protects plant health by assessing and managing the risks of plant pests. The IPPC is in the process of setting standards to address the plant pest risks associated with GMOs and invasive species. Any GMO that could be considered as a plant pest falls within the scope of this treaty. The IPPC allows governments to take action to prevent the introduction and spread of such pests. It established procedures for analyzing pest risks, including impacts on natural vegetation.

7.21.2 Codex Alimentarius Commission (CAC)

This addresses food safety and consumer health. The commission has established an *ad hoc* Intergovernmental Task Force on Foods derived from Biotechnologies that is responsible for developing standards and guidelines for genetically modified foods. The commission is also considering the issue of labeling biotech foods to allow the consumer to make an informed choice.

7.21.3 The World Organization for Animal Health (OIE)

It develops standards and guidelines designed to prevent the introduction of infectious agents and diseases into the importing country during international trade in animals, animal genetic material and animal products. In 2000, the OIE Standards Commission published the Manual of Standards for Diagnostic Tests and Vaccines. Some of the tests and vaccines described are genetically engineered. Likewise, the OIE working group on biotechnology has produced several technical publications relating to animal production and risk analysis.

7.21.4 World Trade Organization (WTO)

The major WTO agreements that contain provisions relevant to LMOs/GMOs are the Agreement on the Application of Sanitary and Phytosanitary Measures (SPS) and the Technical Barriers to Trade (TBT) Agreement. Though both Cartagena Protocol and WTO advocate the use of science-based risk assessments as a means to justify trade-related measures, there are areas of potential conflicts mainly with respect to the application of precautionary approach in the conditions that scientific evidence is insufficient and the socio-economic considerations in the decision of importing LMOs.

7.21.5 Scope

This policy provides the framework to protect the natural resources of Belize and the health of the people living in the country from the adverse effects that may arise from the development and application of GMOs and its derived products including pharmaceuticals. This will be achieved by regulating, evaluating and monitoring the development and use of GMOs in Belize, establishing criteria for assessing the risks associated with GMO use, developing the capacity in Belize to effectively manage and mitigate such risks, promoting the establishment of collaborative links with regional countries and institutions on biosafety, establishing mechanisms for assessing the benefits to be derived from GMO use and ensuring that public education, participation, and consultation is critical to the implementation of this policy (Anonymous, 2009).

7.21.6 Policy Framework

7.21.6.1 Goal

To ensure an appropriate level of protection of human, animal and plant health and life in the development and application of modern biotechnology, while ensuring the well-being of the country of Belize.

7.21.6.2 Policy objectives

1. Implementation of biosafety measures to ensure that there will be no adverse effects of modern biotechnology on human health, the environment, food security, biodiversity or existing agricultural activities and markets.
2. Ensure effective mechanisms are established to regulate the importation, management and use of GMOs and GMOs used in FFP.
3. Regulate local production of GMOs with respect to research and development.
4. Support and facilitate capacity building in Biosafety, with particular reference to regulatory management, risk assessment, risk management, risk mitigation and risk communication, including the development of a roster of experts in biosafety.
5. Promote dissemination of knowledge in the safe use and probable hazards of modern biotechnology Policy Statement.

The Biosafety Policy reflects the commitment of the Government of Belize in ensuring appropriate levels of protection in the safe use of modern biotechnology based on the precautionary principle and in accordance with other national policies, within the framework of sustainable development of the country for the benefit of present and future generations of Belizeans. All matters related to the handling and/or use of GMOs and its products in Belize shall be in accordance with the goal and objectives as expressed in this policy. Provide an institutional framework for national decision making, networking, monitoring R&D, and international cooperation in all matters relating to biosafety.

7.21.6.3 Policy principles

Major policy principles include recognizing the importance of protecting its people, environment and biodiversity while promoting sustainable social and economic development of Belize, the human health, environmental and socio-economic risks that may be incurred by careless, illegal, unscrupulous or unethical development or use of modern biotechnology and its products, realizing the need for developing Belize's own capabilities in biosafety through research, development and training (Anonymous, 2009), recognizing that niche markets have already been established and are flourishing in Belize for organic and fair-trade products and that these markets expressly forbid the use of or risk of contamination from GMOs or their derivatives, recognizing that the rich natural environment of Belize has one of the highest levels of biodiversity in Central America and that GMOs have the potential to significantly reduce biodiversity levels, recognizing that GMOs have the potential to damage (contaminate, disrupt, destroy) – through gene transfer – indigenous ecosystems (germplasm, landraces, seed sources, agricultural crops, and agricultural production practices), recognizing that there are existing traditional alternative methods to be used for sustainable agricultural practices, recognizing that GMOs may have potential benefits with respect to food security and pharmaceuticals and reaffirming the commitment to the obligations of CBD and CPB.

Belize shall ensure that biosafety regulations are established based on the precautionary principle and the advanced informed agreements, the production, use, import, export, sale, or transboundary movements of modern biotechnology applications; practices and products conform fully to all relevant national legislations and international agreements and obligations to which Belize is signatory. This will include a mechanism to ensure the traceability of GMO products. Adequate regulations and procedures are established to address national food security needs in the event of an emergency. Public awareness, education and participation in the decision-making processes are made essential for ensuring the ethical and judicious use of modern biotechnological applications, practices and products for socio-economic development without jeopardizing the environment, indigenous cultures and practices, biodiversity or human health, risk assessment and management of GMO and GMOs used in FFP shall be carried out according to national biosafety regulations. Decisions shall be based on evaluation of the risks that may result from a biotechnology product, application or procedure, recognition and respect for intellectual property rights with respect to GMOs and their products, which will only be granted if it contributes to the promotion of technological innovation and to the transfer and dissemination of technology, to the mutual advantage of producers and users of technological knowledge and is done in a manner that is conducive to social and economic welfare and to a balance of rights and obligations [(Article 7 of the Agreement on Trade-Related aspects of Intellectual Property Rights (TRIPS) (Anonymous, 2009)].

Obligations under the CBD treaty, Belize shall provide a mechanism that will allow fair and equitable sharing of the benefits arising from the commercial and other utilization of genetic resources. Access to genetic resources and traditional knowledge for such commercial purpose must fulfill the stated requirements and have the prior informed consent of the Government Ministry responsible for biological resources and the consent of any indigenous communities involved, with both government and indigenous community sharing the proceeds deriving from the commercial exploitation of the material or knowledge concerned. Intellectual property rights can only be obtained if the aforementioned requirements are met. Persons involved in the use or sale of modern biotechnology or its derived products shall ensure that consumers are fully informed of the nature of the technology or product. Data, resulting from safety tests/research on the technology or product, (for agricultural biotechnology, human genetic testing, manipulations or applications, for example), shall be fully disclosed and made public. Decisions on biosafety issues shall not favor commercial considerations over public health, environmental and safety interests. Existing and potential markets for organic and fair-trade products are protected. Labeling of genetically modified products using international norms shall be mandatory so that the consumer may make an informed choice. The consumer has a right to choose not to consume foods that are derived from GMO and GMO products (Anonymous, 2009).

National safety guidelines and implementation practices can be adopted by industries using modern biotechnology or their products. The guidelines will cover all related aspects, including material handling, equipment, storage, waste disposal, laboratory safety, etc. Analytical laboratories legally recognized and proficient for GMO detection be identified, certified, and or established and supported. Priorities in Human Resource Development in Biosafety are determined, ranked and implemented. Public awareness of modern biotechnology in relation to the assessment of potential risks/benefits/ alternatives and management techniques shall be enhanced, involving the community at large, including policy makers, legislators, administrators, the private sector and biotechnology industries. Research into the risks to the environment and human health that can be caused by modern biotechnology be supported.

A comprehensive regime for liability as well as for adequate and prompt compensation (redress) for damage resulting from the transfer, handling or use (including illegal trafficking) of a GMO or its products under the national jurisdiction of Belize shall be provided for. This regime shall be consistent with the polluter pays principle and shall cover property damage, economic damage (including socio-economic damage to local and indigenous communities) damage to biodiversity, preventive measures, injury or disease and the cost of reinstatement and reinstatement or remediation of an impaired environment. Given the risk involved this legislation shall have provisions for the implementation of a moratorium on the development or importation of GMOs for planting for commercial purposes into Belize until such time that Belize has this capacity. Persons involved in the production, use or sale of modern biotechnology or its derived products recognize their obligation to ensure that contamination and/or damage to ecosystems, animal, and human health do not occur. This policy is supported by legislation. In the interim period leading up to the establishment of the infrastructure and capacity necessary to enable full implementation and compliance with the National Policy on Biosafety, appropriate regulations will be drafted under the laws in existence (Anonymous, 2009).

7.21.6.4 Strategy

To achieve the goal and objectives and facilitate adherence to the policy principles, Belize will use the following organizational arrangements and mechanisms for implementation recognizing the Ministry of Agriculture and Fisheries as the lead Ministry.

7.22 MANAGEMENT AND COORDINATION

A National Biosafety Council shall be established. The council shall be comprised of technical representatives from the Ministries of Agriculture, Health, Natural Resources and Environment, Regulatory agencies/bodies involved in agricultural health, food safety and standards, International and Regional Organizations, the NGO community, civil society and the private sector. The Council will oversee the management and coordination process for production, importation and use of GMO material in Belize. A multisectoral approach will be used.

1. The council shall be supported by a secretariat and coordinator and shall meet periodically to review the implementation of the policy, legislation and programs and shall report to the Cabinet of Belize, develop and recommend policy improvements, strategic and operational plans, allocate human, technical, financial resources and monitor and evaluate program operations and achievements.
2. The Council will make decisions with respect to production, importation and use of GMO and GMO-derived products in Belize. The Belize Agricultural Health Authority (BAHA) is the National Competent Authority for Biosafety in Belize. The National Biosafety Council shall establish a National Biosafety Clearing House to facilitate the exchange of scientific, technical, environmental and legal information on, and experience with, genetically modified organisms and assist Belize as a party, to implement the Protocol in accordance with Article 20 of the Cartagena Protocol on Biosafety National Focal Point (NFP). In accordance with Article 19 of the Cartagena Protocol, the National Biosafety Council shall designate an NFP responsible for liaising with international institutions, government, and the public. The Focal Point of the Cartagena Protocol on Biosafety will be under the Belize Agricultural Health Authority (BAHA) (Anonymous, 2009).

7.22.1 Capacity Building

The Council will support and facilitate capacity building in biosafety (risk assessment, risk management, risk communication, etc.) to ensure the effective regulatory management of GMOs in Belize.

7.22.2 Financial Implications

The Council shall prepare an annual budget to ensure effective and sustainable implementation of the policy. This will include the use of a cost recovery mechanism along with government and other institutional support.

7.22.3 Public Awareness, Education and Participation

The council will establish a Public Education Committee with the responsibility to ensure public awareness and participation through periodic public consultation and the dissemination of up to date information on GMO issues, using all available media.

7.22.4 National and Global Security

The council will keep abreast of international developments/systems and agreements related to biosafety issues with a view to updating and enhancing Belize's policy and regulatory framework.

7.22.5 Indicators of Performance

The biosafety regulatory framework is in accordance with national, regional and international requirements. All activities with GMOs are conducted in accordance with the goals, objectives, and principles of this policy and the provisions of the biosafety regulations. There is an increased public consultation, awareness, education and participation in the biosafety regulatory framework, there is an increased capacity in the field of GMOs especially with regard to risk assessment, risk management, and risk communication. An updated database of current biosafety issues in other countries, especially countries in the region is established (Anonymous, 2009). An effective monitoring mechanism of GMO use in Belize is established. There is an increased public awareness of emerging biosafety issues. There is the periodic dissemination of information on the state of biosafety in Belize.

7.23 DEFINITIONS OF SOME COMMON TERMINOLOGIES

7.23.1 Accountability

Accountability ensures that valuable biological materials (VBM, see definition below) are controlled and traced as intended, by formally associating the specified materials with the individuals who provide oversight and are held responsible for them.

7.23.2 Bioethics

It is the study of the ethical and moral implications of biological discoveries, biomedical advances, and their applications as in the fields of genetic engineering and drug research. In this document, bioethics is one of the three components that contribute to a successful biorisk management culture.

7.23.3 Biological Laboratory

A facility within which microorganisms, their components or their derivatives are collected handled and/or stored. Biological laboratories include clinical laboratories, diagnostic facilities, regional and/national reference centers, public health laboratories, research centers (academic, pharmaceutical, environmental, etc.) and production facilities (manufacturers of vaccines, pharmaceuticals, large scale GMOs, etc.) for human, veterinary and agricultural purposes.

7.23.4 Biorisk

The probability or chance that a particular adverse event (in the context of this document: accidental infection or unauthorized access, loss, theft, misuse, diversion or intentional release), possibly leading to harm, will occur.

7.23.4.1 Biorisk assessment

The process to identify acceptable and unacceptable risks (embracing biosafety risks (risks of accidental infection) and laboratory biosecurity risks (risks of unauthorized access, loss, theft, misuse, diversion or intentional release) and their potential consequences.

7.23.4.2 Biorisk management

The analysis of ways and development of strategies to minimize the likelihood of the occurrence of biorisks. The management of biorisk places responsibility on the facility and its manager (director) to demonstrate that appropriate and valid biorisk reduction (minimization) procedures have been established and are implemented. A biorisk management committee should be established to assist the facility director in identifying, developing and reaching biorisk management goals.

7.23.5 Biosafety

Laboratory biosafety describes the containment principles, technologies, and practices that are implemented to prevent the unintentional exposure to pathogens and toxins or their accidental release.

7.23.6 Code of Conduct, Code of Ethics, and Code of Practice

Nonlegislated guidelines which one or more organizations and individuals voluntarily agree to abide by that set out the standard of conduct or behavior with respect to a particular activity.

7.23.7 Control

Control is the combination of engineered and procedural measures that ensure valuable biological material (VBM, see definition below) are used only as intended.

7.23.8 Dual-use

Initially used to refer to the aspects of certain materials, information and technologies that are useful in both military and civilian spheres. The expression is increasingly being used to refer not only to military and civilian purposes but also to harmful misuse and peaceful activities.

7.23.9 Genetically Modified Organisms (GMO)

Organisms whose genetic material has been altered using techniques generally known as 'recombinant DNA technology.' Recombinant DNA technology is the ability to combine DNA molecules from different sources into one molecule in a test tube. GMOs are often not reproducible in nature, and the term generally does not cover organisms whose genetic composition has been altered by conventional cross-breeding or by 'mutagenesis' breeding, as these methods predate the discovery (1973) of recombinant DNA techniques.

7.23.10 Hazard

A danger or source of danger; the potential to cause harm.

7.23.11 Laboratory Biosecurity

Laboratory biosecurity describes the protection, control and accountability for valuable biological materials (VBM, see definition below) within laboratories, to prevent their unauthorized access, loss, theft, misuse, diversion or intentional release.

7.23.12 Misuse

The misuse of valuable biological materials (VBM, see definition below) describes their inappropriate or illegitimate use, despite existing and subscribed agreements, treaties and conventions.

7.23.13 Threat

The likelihood of an adverse event to occur, as an expression of intention to inflict evil, injury, disruption or damage.

7.23.14 Transfer of VBM

Legal and/or administrative policies and procedures relating to the oversight and approval process for the transfer of custody and/or ownership of valuable biological materials (VBM, see definition below) between countries, entities (organizations, institutions, facilities, etc.) or individuals.

7.23.15 Transport of VBM

Procedures and practices to correctly categorize, package, document and safely and securely transport valuable biological materials (VBM, see definition below) from one place to another, following applicable national and/or international regulations.

7.23.16 Valuable Biological Materials (VBM)

Biological materials that require (according to their owners, users, custodians, caretakers or regulators) administrative oversight, control, accountability, and specific protective and monitoring measures in laboratories to protect their economic and historical (archival) value, and/or the population from their potential to cause harm. VBM may include pathogens and toxins, as well as nonpathogenic organisms, vaccine strains, foods, genetically modified organisms (GMOs), cell components, genetic elements, and extraterrestrial samples.

On a very practical level, there may be differences between means to prevent an unintended release into the environment (sometimes referred to as 'biosafety') and means to prevent abuse or theft (sometimes referred to as 'biosecurity'). But conceptually, legally, and in terms of organization and implementation, there are broad overlapping areas. A wide and shared conceptual framework that integrates sectoral notions of 'biosafety' and 'biosecurity' is required for each sector (agriculture, human and animal health, disarmament, and environment) that addresses them to be efficient and effective.

7.24 BIOSECURITY

Biosecurity is emerging as one of the most pressing issues facing developed, developing and transition countries. Globalization, the increased movement of people, agricultural and food products across borders, greater attention to biodiversity and the environment, the emergence and spread of transboundary diseases, changes in the way food, plants, and animals are produced, processed and distributed, uncertainties surrounding new technologies, as well as international legal obligations are some of the trends that drive this growing interest, and highlight the importance of adequate biosecurity capacity. Biosecurity is a strategic and integrated approach to analyzing and managing relevant risks to human, animal and plant life and health, and associated risks to the environment. It is based on recognition of the critical linkages between sectors. Biosecurity hazards of various types exist in each sector and have high potential to move between sectors. For that reason, inadequate controls in one sector can have far-reaching consequences for other sectors.

Harmonizing and integrating national biosecurity systems and controls whenever possible provides a means to take advantage of the synergies that exist across sectors. This will considerably enhance the capability of countries to protect human health, agricultural production systems, and the people and industries that depend on them. In addition, there are likely to be other benefits. A harmonized and integrated approach to biosecurity will help to safeguard the environment and protect against the uncertainties associated with new technologies. It will further enhance the capacity of countries to meet obligations under relevant international agreements and to take full advantage of opportunities associated with the global trade in food and other agricultural products.

Plant biosecurity is a strategic socio-economic issue concerning the protection of natural and managed plant systems from alien and emerging indigenous pests. Globalization has opened up state borders which, along with climate change and landscape modification, have facilitated the dissemination and introduction of alien pests and the evolution of new races, biotypes and strains of indigenous pests, leading to emerging diseases. Pathogens represent one of the most important threats affecting plant ecosystems by compromising their productivity, sustainability and biodiversity. There are many examples of pathogens resulting in high-impact epidemics – both historical and more recent, from the *Phytophthora infestans* causing the well-known Irish potato famine of the 1840s, to the recent outbreaks of *Pseudomonas syringae* pv. *actinidiae*, the etiological agent of kiwifruit bacterial canker, which severely affects the main kiwifruit producing areas worldwide (Vincenza Ilardi and Stefania Loreti, 2014).

A plant biosecurity system requires early detection, accurate diagnosis and a rapid response to prevent the establishment and dispersal of pests, thus minimizing the subsequent impact. Many intergovernmental organizations promoting cooperation in plant health have been established within the framework of the International Plant Protection Convention (IPPC, https://www.ippc.int/). In Europe and the Mediterranean basin, the 'European and Mediterranean Plant Protection Organization' (EPPO http://www.eppo.int), develops international strategies against the introduction and spread of dangerous pests, promotes pest control methods, and drafts diagnostic protocols.

The diagnosis of fungi, bacteria, nematodes, phytoplasmas, viruses and viroids can be carried out using conventional (morphological, biological, biochemical, physiological, and nutritional tests), serological and molecular (nucleic acid-based) methods. In the case of typical symptomatology, a presumptive diagnosis can be performed. However, plants can show unspecific symptoms that can be confused with different biotic or abiotic stresses, as in the case of symptoms caused in *Citrus* by 'Candidatus Liberibacter' spp. which very much resembles a zinc deficiency. In these cases, a detection test is necessary to identify the disease's cause. Often multiple assays are needed to confirm or support the diagnosis. To ensure reliable results, a diagnostic method should thus fulfill certain requisites: sensitivity, specificity, reproducibility, repeatability, selectivity (Janse, 2012). The ease of use, speed and cost should also be taken into account, each characteristic depending on the circumstances of use. In the case of routine diagnosis or surveillance, speed and cost may be more critical than sensitivity or specificity. Conversely, where a pathogen is not known to occur, diagnostic tests are preferable with high levels of specificity, reliability and reproducibility.

Tests and procedures need to be validated to prevent a disconnection between 'developers' (researchers) and 'users' (diagnosticians) of diagnostic assays. Thus interlaboratory studies (ring-tests) and proficiency tests aimed at evaluating the protocols and the skills of the laboratory, respectively, are also desirable. The use of appropriate validation guidelines may produce universally acceptable results which are useful for phytosanitary action and regulatory decisions (Vincenza Ilardi and Stefania Loreti, 2014).

To intercept pathogens early, it is important to implement monitoring and sampling. The detection of pathogens in symptomless plant material is crucial for prevention. Statistics and modeling are needed to develop correct advanced sampling strategies. This is particularly true in surveillance programs of global trade products for 'on-site,' at 'point of entry,' and also 'at-source' inspections. The bacterium *Xylella fastidiosa* is a typical example of a disease that requires careful monitoring. Pierce's disease of grapevine caused by the subspecies *fastidiosa* is known to have been present in the USA since 1887, but it has never been reported in Europe with the unconfirmed exception in Kosovo. Great concern has always been placed on the threat of *X. fastidiosa* entry in the EPPO region due to the large import of hosts of the bacterium from the USA.

Recently, a rapidly spreading decline of aged olive trees has taken place in the major Italian olive producing region, Apulia (Southern Italy). In October 2013 *X. fastidiosa* was associated with these diseased plants and also with oleander and almond (Saponari *et al.*, 2013). Considering the high number of plant hosts of this bacterium, it is essential to identify the cause of its introduction and the characterization of its subspecies.

Pathogens quickly evolve to survive and perpetuate in various environmental conditions. This variability may reflect the fitness *in planta* and/or the virulence of the pathogen. One case of the convergent evolution of different pathogen populations in the same plant genus is the occurrence of at least four genetically distinct populations of *P. syringae* pv. *actinidiae* which, to different extents, infect *Actinidia* spp. worldwide. One of these four populations is responsible, as mentioned above, for the severe outbreak of bacterial canker, which is considered as a pandemic disease. In these cases, in addition to methods for the detection of the pathovar, it is important to identify the virulent population to prevent possible new outbreaks (Vincenza Ilardi and Stefania Loreti, 2014). The improvement and harmonization of detection and diagnosis procedures are of primary importance to coordinate a timely response to contain the spread of pathogens and guarantee plant biosecurity worldwide.

Biosecurity is a process of managing biological risks associated with food and agriculture in a holistic manner. Besides enhanced productivity, sustainability and profitability, interest in biosecurity is increasing as national regulatory and export certification systems are being challenged by large increases in the volume of food and agricultural products being traded internationally, by the expanding variety of imported products and by the growing number of countries from which these imports are originating. Increased travel is also creating more pathways to spread pests, diseases, and other hazards that are moving faster and farther than ever before. Improved coordination is being sought by national bodies responsible for enforcing sanitary, phytosanitary and zoosanitary measures to better protect human, animal and plant life and health without creating unnecessary technical barriers to trade.

The strengthening of policy and regulatory frameworks for biosecurity in food and agriculture must be among the highest priorities. These policy and legislative frameworks need to be extended to include biosafety needs within the overall framework of biosecurity. This will provide optimization of scarce human and financial resources, improving the cohesiveness of advice on all aspects of biosecurity, including biosafety, recognition of the special importance of biosafety to food and agriculture as well as the special impacts of food and agriculture on biosafety. Further, this will seek the development of appropriate standards, guidelines, and other recommendations for food safety and the protection of plant, animal and aquatic life and health based on risk assessment and taking into account relevant aspects of biosafety, including environmental health.

Thus, risk analysis and management as a framework for biosecurity become the central binding force across various sectors. It provides an opportunity to harmonize terminology and methodology while respecting the need for individual sectors to tailor risk analysis procedures to the characteristics of the risks involved. It should be recognized that risk analysis procedures should provide an appropriate science-based and transparent basis for biosecurity. Fast expanding trade has intensified the need for effective risk analysis capacities and for bilaterally and multilaterally agreed on standards. There are several commonalities in risk assessment and management across the subsectors and organisms which must ideally be addressed by the National Agricultural Biosecurity System. Since Risk Analysis and Management will constitute the bulk of the animal, plant and fish health management, in time to come, biosecurity would automatically occupy the center stage in this field.

7.24.1 Actions

The various international standards issued through several international conventions and agreements notwithstanding, in the wake of implementation of WTO/SPS Agreement and to meet TRIPS and TBT; the country is required to undertake the following actions.

(i) Designating a single Central Government authority as responsible for implementation of SPS measures.

(ii) Reviewing and updating of legislation and regulations related to SPS to give effect to international agreement and establishing a nodal point for inquiries and information exchange.

(iii) Establishing national standards on SPS measures in line with international standards.

(iv) Establishing a notification procedure.

(v) Undertaking pest risk analysis and identifying and maintaining pest-free areas for plants and animals as per international standards and safety assessment for food.

(vi) Providing scientific justification for high-level protection in the absence of pest risk assessment.

(vii) Recognizing of equivalence of specific measures through bilateral or multilateral agreements.

(viii) Identifying researchable issues and strengthening backup research.

(ix) Capacity building in terms of infrastructure and expertise.

(x) Awareness building and catalyzing attitude change.

(xi) Developing functional public-private-NGO partnerships.

Requirements of several of the other international, regional and bilateral agreements and national regulations on various agricultural commodities and ago-ecological and socio-economic regimes, whether on plants or animals or fisheries or microbes, are analogous. Establishing and operationalizing separate facilities for each of the requirements amounts to not only gross duplication of scarce resources and efforts but also the erosion of effectiveness, efficacy, and synergy which are essential for achieving sustainable, sound and competitive outcomes. A National Agricultural Biosecurity System, which involves the management of biological risks in a comprehensive and synergistic manner, is a strong force of convergence of the various paths of sustainable development, a win-win situation for all partners of development.

The Consultation considered the use of the English term, *Biosecurity*, bearing in mind the need for translation and to harmonize terminology. Delegates noted that the term *Biosecurity* is used widely, and that usage varies among countries. They also noted that the term presents translation challenges, particularly for Spanish and French translation. Following considerable discussion on terminology, delegates agreed that the term *Biosecurity* in food and agriculture best describes the concept as used by FAO, and recommended that for the purposes of the consultation and this report, the English term, *Biosecurity* be used in all languages and that it be italicized and capitalized, and not be translated.

The consultation considered that *Biosecurity* involves the management of biological risks in a comprehensive manner to achieve food safety, protect animal and plant life and health, protect the environment and contribute to its sustainable use. Achieving *Biosecurity* requires an understanding of, and the ability to analyze diverse and complex risks, and determine and apply measures in a coherent manner while respecting differences among sectors and organizations. Risk analysis is the most important unifying concept across different *Biosecurity* sectors. *Biosecurity* frameworks should not create unjustified barriers to international trade.

7.24.2 Recommendations

Committee on Agriculture during its Seventeenth Session on "*Biosecurity* in Food and Agriculture" held at Rome, from 31 March–4 April 2003, gave the following recommendations under Item 9 of the Provisional Agenda.

(i) Countries should determine the potential for synergies and harmonization within their national and subnational regulatory frameworks that would result from a holistic and coordinated approach to *Biosecurity*. Policy-makers should recognize the importance of *Biosecurity* as a key element of sustainable development, and the benefits, including in trade that can be gained from comprehensive approaches to *Biosecurity*.

(ii) Recognizing the efficiencies that may emanate from regional and subregional approaches to risk analysis, particularly in relation to animal and plant life and health, and living modified organisms, countries should also cooperate to address *Biosecurity* issues at regional and subregional levels.

(iii) Risk analysis and management frameworks are essential to achieve *Biosecurity*. In the past, such frameworks have been mostly sectoral or used to address specific technical issues. In future, such frameworks should seek to improve collaboration among diverse interests and institutions (particularly agriculture, public health, environment, trade, and their associated stakeholders) to achieve *Biosecurity* in a mutually supportive manner, thus avoiding duplication and possible inconsistencies.

(iv) General principles for risk analysis for biological risk analysis in food and agriculture are the same, although procedures may differ depending on the hazards addressed. The IPPC, the *Codex* Alimentarius, the OIE, the CBD and its Cartagena Protocol (noting that the Protocol has not yet entered into force), where appropriate, should apply coherent risk analysis methodologies in different sectors by jointly analyzing differences and commonalities in approaches, and use of terms in risk analysis.

(v) Many developing countries and countries with economies in transition have limited infrastructure and limited capacity to undertake risk analysis, and to enforce risk management decisions. International standards should thus be developed with due consideration of their implications and impacts on developing countries and countries with economies in transition, including the effect on their ability to participate in international trade. The participation of countries in the development of such standards should be supported.

(vi) Countries should implement a more coherent and holistic approach to biological risk management in food and agriculture by respective government authorities to strengthen the achievement of common *Biosecurity* objectives.

(vii) FAO, in collaboration with relevant international and regional organizations, should provide guidance and develop guidelines to assist countries to develop and implement national *Biosecurity* frameworks in harmony with their international obligations. The FAO, in collaboration with other relevant international and regional organizations, should consider undertaking further analysis to better understand and advance *Biosecurity*, including analysis of differences, similarities, duplications and gaps, across the various sectors of *Biosecurity*, the implications for developing countries and countries with economies in transition to *Biosecurity* standards, procedures and technical regulations; and measures required to establish coherent and mutually supportive *Biosecurity* approaches in relation to food safety, animal health and life, plant health and life, and the environment (Anonymous, 2003).

7.25 BIOSECURITY TOOLKIT

This Toolkit offers countries guidance to develop and implement national Biosecurity systems in accordance with their international obligations. The benefits of a more holistic and integrated approach to Biosecurity are presented, and the toolkit seeks to increase knowledge on the broader development and implementation of Biosecurity policies and frameworks at the national level. This includes enhancement of Biosecurity capacity through the assessment of needs and the generic application of risk analysis principles as an essential element of Biosecurity. The target audience is Competent Authorities who are responsible for one or more Biosecurity sectors within a national setting (Anonymous, 2003a).

The Toolkit comprises the following three documents.

(i) **Modern Biosecurity: Principles and Components (Part 1)** which provides a contemporary context for development and implementation of a more harmonized and integrated approach to *Biosecurity* across all sectors.

(ii) **Biosecurity Capacity Needs Assessment Tool (Part 2)** which provides a process for assessing all dimensions of *Biosecurity* capacity across all sectors and sector organizations.

(iii) **Biosecurity Risk Analysis Manual (Part 3)** which presents a generic framework to structure and guide the application of risk analysis principles in *Biosecurity*.

The Toolkit is respectful of variation in *Biosecurity* needs across countries, sectors and sector organizations and fully acknowledges that there is no universally acceptable or standard policy or infrastructure that governs national systems. The Toolkit develops the thesis that risk analysis provides a common foundation for *Biosecurity* and examples are presented to illustrate the experiences of Competent Authorities that have moved towards an integrated approach to *Biosecurity* in recent times.

7.26 PERSONS/ORGANIZATIONS INVOLVED

7.26.1 National Stakeholders

Biosecurity involves many different kinds of stakeholders at the national level. Government agencies have a primary interest, but industry, scientific research institutes, specialist interest groups, NGOs and the general public all have a vital role to play. Several branches of government, at both the national and subnational levels, are involved in *Biosecurity*. The Competent Authorities responsible for the sectors usually associated with *Biosecurity*, agriculture, forestry, fisheries, food safety and public health, environment, play the primary role in a contemporary integrated approach to *Biosecurity*. However, other parts of government responsible for sectors such as trade, customs, transport, finance and tourism can also play a role, depending on national circumstances [(Fig. 7.2) (Anonymous, 2003a)]. In addition, 'third party' organizations are often contracted by Competent Authorities to deliver a range of core *Biosecurity* functions including surveillance programs, incursion response activities and laboratory diagnostic services.

7.26.2 International Stakeholders

At the global level, international standard-setting organizations, international bodies, and international legal instruments and agreements play important and complementary roles in *Biosecurity*. Although the standards and related texts developed by the Codex Alimentarius Commission (CAC), the World Organization for Animal Health (OIE), and the International Plant Protection Commission (IPPC) and are not legally binding in and of them, they have become international reference points through the WTO SPS Agreement. International organizations like FAO and WHO contribute to the international *Biosecurity* environment by providing guidance texts, manuals, and web-based

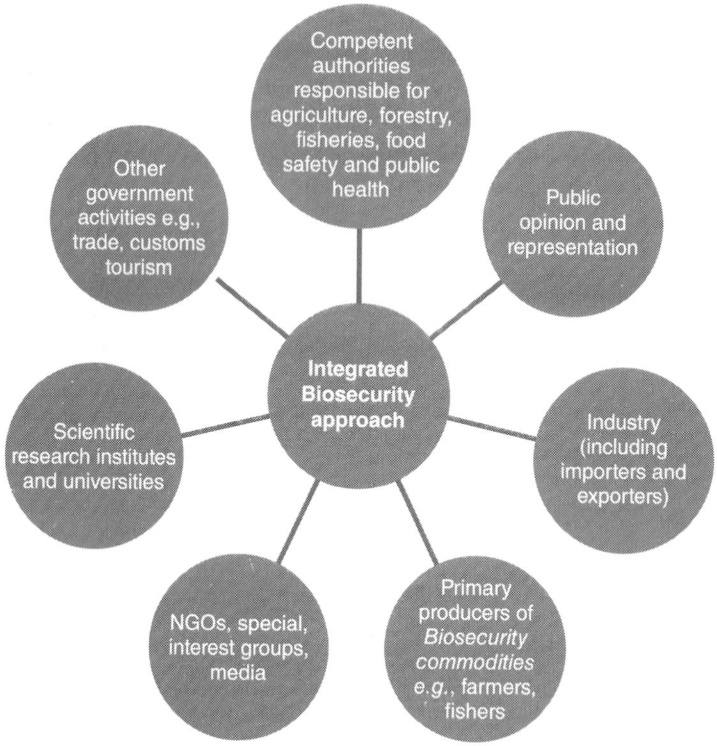

Fig. 7.2 Sector interests that are important to an integrated approach to biosecurity

informational portals that support the application of standards developed by the international bodies. They also provide essential inputs to the development of these standards in terms of good science and risk assessment. Several international legal instruments, agreements, and texts are relevant to *Biosecurity* but generally have a single sector perspective, e.g., food safety, environmental safety, biosafety, biological diversity, plant/animal/human health, nature conservation, wetland protection and marine resources (see Annex. 4). However, they do share certain common characteristics including risk analysis principles, notification procedures and information exchange (Anonymous, 2003a).

A toolkit was developed by FAO that provides practical guidance and support to develop and implement national biosecurity frameworks at the country level. It presents the benefits of a harmonized and integrated approach to biosecurity and illustrates the experiences of countries, including Belize, Norway, and New Zealand, which have adopted such an approach in recent times. By providing a framework to identify cross-cutting biosecurity capacity needs based on an integrated approach, this toolkit addresses the gaps inherent in a purely sectoral approach to biosecurity. The purpose is to support governments to better manage biosecurity as a means to protect public health, agricultural production, and the environment. At the same time, this will enhance the ability of countries to comply with international agreements, regulations and requirements focused on sanitary and phytosanitary measures, contributing to economic development and trade.

The toolkit comprises three separate but linked documents. All three documents are developed on the premise that biosecurity concerns different parts of government, that biosecurity risks are interrelated, and that the best way to manage the risks faced is through coordinated action across the relevant sectors, contributing to improved outcomes and efficiencies.

7.26.2.1 Part 1: Biosecurity principles and components

The first part of the toolkit provides a broad introduction to biosecurity and outlines the contemporary context for development and implementation of a harmonized and integrated biosecurity approach across all sectors. It shows how such an approach can enhance the protection of human, animal and plant life and health and the environment by taking advantage of synergies across sectors, as well as generating a number of other tangible benefits.

7.26.2.2 Part 2: Guide to assess biosecurity capacity

The second part of the toolkit provides guidance on how to assess dimensions of biosecurity capacity across all sectors and sector organizations in accordance with the requirements of an integrated biosecurity approach as presented in Part 1. Use of this guide will enable governments to increase awareness of the synergies and interdependencies that exist across biosecurity sectors. It will further help to generate an understanding of existing biosecurity capacity and performance, a medium-term vision for national biosecurity and a strategy and action plan to enhance biosecurity capacity based on an identification of capacity needs. Part 3: An Overview and Framework Manual for Biosecurity.

7.26.3 Contributions of International Agreements on Crop Biosecurity

During the last three decades, there have been numerous meetings and consultations at the international level for streamlining the availability and utilization of existing biological resources in an equitable manner. This has been necessitated by revolutionary developments in the life sciences and possibilities of generating immense economic benefits. This situation is in contrast to the era of the Green Revolution when all the germplasm developed was easily available throughout the world, and many developing countries greatly benefited from it. At that time, no issues regarding IPRs were raised. However, now that the world situation has changed with the 'free market economy' as the dominant force, agriculture is viewed as an industry or business subject to all the regulatory measures. It is in this context that agriculture figures prominently in all the deliberations of the World Trade Organization (WTO) wherein various aspects have been strongly contested between countries of different blocks with different interests.

7.27 FOOD AND AGRICULTURE ORGANIZATION (FAO) COMMITMENTS FOR CROP BIOSECURITY

To protect the farmers' right to plant variety, a revision of the *International Undertaking on Plant Genetic Resource* is being negotiated within the FAO Commission on Genetic Resources. The proposed revision recognizes the protection of traditional knowledge, the right to participate in sharing the benefit that arises from the use of plant variety, and the right to participate in decision-making concerning their management. The revision also recognizes that no limit should be put on the farmers' rights to sale, use and exchange of seeds including every right over what they have produced. On the issue of protection of bioassays, the FAO undertakes to protect the farmer's right that is contrary to Article 15 of the Convention on Biological Diversity (CBD) which recognizes the sovereign right of the state on biodiversity.

7.27.1 Convention on Biological Diversity (CBD)

The idea of an umbrella convention addressing the whole gamut of biodiversity first emerged at the Third World National Parks Congress in Bali, Indonesia, 1982. Notwithstanding that United Nations Environmental Programme (UNEP) had established an *ad hoc* working group in 1987 to decide upon the necessity of such a convention. Like many other international negotiations, this also turned into a North-South debate, which on this occasion was between the technology-rich North and the resource-rich South. However, since the technology of the North depends on the South's resources, the Northern countries need a conducive international legal regime to assure raw material supply for their biotechnology-oriented agrofertilizer and pharmaceutical industries as well as a new legal regime in harmony with the evolving international trade regime under WTO. Moreover, the developed countries are not ready to change their development paradigm for biodiversity conservation. The South, on the other hand, wanted to exchange their resources with technology from the North for agriculture. To achieve these goals both groups had heated debates on access to genetic resources, transfer of technology, biosafety, biosecurity and financial mechanisms for conservation.

7.27.2 Requirements for Harmonized and Integrated Approach to Biosecurity

The successful implementation of a harmonized and integrated biosecurity approach requires a clear policy and legal framework, an institutional framework that defines the roles and responsibilities of relevant stakeholders, adequate technical and scientific capability (including the use of risk analysis), a well-functioning infrastructure, and a system for communication and information exchange. The Guide to Assess Biosecurity Capacity (Part 2) provides a process for assessing biosecurity capacity needs across all sectors and all sector organizations, which will help to identify requirements to pursue a harmonized and integrated biosecurity approach.

7.27.2.1 *Policy framework*

A biosecurity policy framework sets out a broad course of action to address biosecurity risks in food and agriculture. It is based on appropriate public goals and a set of beliefs about the best way of achieving those goals. It provides a common basis for assessing biosecurity risks and priorities for action and gives direction and guidance to all the parties concerned.

7.27.2.2 *Legal framework*

Sound biosecurity legislation (encompassing laws and regulations) is necessary to create an enabling environment of predictability and certainty through good governance and respect for the rule of law. The law clarifies the roles, responsibilities and rights of different stakeholders, including those parts of government with policy and delivery roles for biosecurity outcomes and programs, to ensure consistency and accountability. It also defines appropriate powers to act, which is essential for enforcement.

7.27.2.3 *Institutional framework*

A clear institutional framework within which to manage biosecurity is an important part of a more harmonized and integrated approach to biosecurity. The institutional framework identifies the competent authority or authorities responsible for establishing biosecurity controls and ensuring their implementation, as well as any other stakeholders involved. It also sets out the rules and procedures governing their roles and defines the mechanisms through which they work towards shared goals. The choice of institutional framework will be determined by factors which are specific to a country and biosecurity context (e.g., historical traditions, political orientation, financial and other resources).

7.27.2.4 *Communication and information exchange*

The complexity inherent in managing biosecurity requires communication and information exchange among a wide range of national stakeholders including government agencies, the private sector (agricultural producers, processors, enterprises, importers/ exporters, etc.), the scientific and research community, and the general public. Transparency obligations under international agreements such as the SPS Agreement require governments to ensure transparency in the adoption of their sanitary and phytosanitary rules. This includes publishing proposed rules in advance and allowing time for comments from the public, as well as the establishment of inquiry points for consultations on rules and inspection and control procedures applicable to imports and exports. They also must open to scrutiny how they apply their food safety and animal and plant health regulations. National, regional and global networks all contribute to meeting the information needs of an integrated biosecurity system.

7.28 RISK ANALYSIS

The third part of the toolkit presents a generic framework to structure and guide the application of risk analysis principles in biosecurity. Risk analysis is at the heart of modern approaches and is rapidly emerging as a unifying discipline across all biosecurity sectors. International standard-setting organizations and bodies have embraced risk assessment as an essential tool to achieve their goals, and competent national authorities are obliged by international agreements to similarly utilize risk assessment wherever possible and practical. Part 3 of the toolkit explores the processes and methods common to cross-sectoral risk analysis and illustrates the role of this discipline in forging better linkages and promoting more efficient use of technical resources. Many aspects of a risk-based approach to biosecurity are shared by the different sectors concerned, and this provides an essential impetus to risk analysis as a unifying discipline in biosecurity. Risk analysis is composed of three distinct but closely connected components - risk assessment, risk management, and risk communication - which are explained in detail in the Overview and Framework Manual for Biosecurity Risk Analysis (Part 3 of this toolkit). International standard-setting organizations and bodies involved with different components of biosecurity have embraced risk assessment as an essential tool to achieve their goals. Biosecurity risk assessment involves a scientific process to estimate risks to life and health that may be associated with a particular food, animal, plant or specific organism.

Prevention, reduction or elimination of those risks can take many forms. Prior to the enactment of the SPS Agreement, biosecurity systems were not necessarily based on robust and transparent scientific inputs to standard-setting processes, especially those for traded agricultural goods. Now, the importance of good science and risk assessment to biosecurity cannot be overemphasized, and this places considerable technical demands on relevant stakeholders. Biosecurity risk management incorporates considerably different processes to risk assessment. Core decisions involve the balancing of scientific findings against questions of life and health expectations, likely economic and social impacts and the technical feasibility and cost-effectiveness of controls. The merging of policies and values with science in biosecurity risk management presents considerable challenges and has different expression in different countries. Both risk assessment and risk management should be wrapped in a 'sea of communication' that includes all stakeholders as appropriate. Successful risk communication is a prerequisite for effective risk assessment and risk management and facilitates the iterative and ongoing nature of risk analysis.

7.28.1 Plant Health

Application of regulatory controls to protect plant health is an important biosecurity domain. This also covers threats to wild plants. Plant health can be adversely affected by different types of pests (i.e., plants themselves, and animals or pathogenic organisms which are injurious to plants or plant products). Management of pathways and vectors is an important aspect of plant health biosecurity. Establishment and spread of a pest often depend directly on biological factors such as availability of suitable plant hosts and vectors, crop cultivation practices, suitability of the environment and natural enemies. As with animal health biosecurity, adverse plant health impacts are usually evaluated in direct economic terms. Approaches to plant health biosecurity are undergoing changes similar to those in other biosecurity sectors. With an increasing interest in environmental issues, competent authorities responsible for plant health must also manage environmental pests that primarily affect other organisms, thereby causing harmful effects on plants and plant ecosystems. Organisms produced by modern biotechnology also may threaten the plant environment such as by out-crossing to create more aggressive weeds or wild relatives that upset the ecological balance and decrease biodiversity.

While competent authorities can be proactive in preventing the import of pests, risk management programs are needed to control pests that have become established within the borders of a country. As with animal health, 'pest-free' geographical compartments can be established within countries or regions so that plants and their products can continue to be traded.

7.29 NATIONAL BIOSECURITY STRATEGY

A national biosecurity strategy can provide an impetus and unifying force to support the achievement of a harmonized and integrated approach to biosecurity. This concept has gained prominence in a number of countries in recent years. A national biosecurity strategy translates high-level policy into objectives to achieve specific outputs and outcomes. It gives direction and guidance to all the parties concerned with the implementation of biosecurity measures. A national biosecurity strategy should be developed in consultation with all stakeholder groups and incorporate a 'whole of government' approach. It should also include reference to the international regulatory environment. A 'national vision' for biosecurity is agreed upon by all stakeholder groups.

7.29.1 Components of National Biosecurity Strategy

These components are availability of sufficient financial and technical resources, mechanisms for establishing national· risk-based priorities; coordination between competent authorities working within and between biosecurity sector; a culture of collaboration between competent authorities, especially in areas where control structures are decentralized, and local and national priorities are different; recognition of international biosecurity obligations, and participation in international standard-setting organizations and bodies and effective representation of national interests.

7.29.2 Better Risk Analysis

There are considerable advantages from a harmonized and integrated approach to risk analysis at the national level. While international risk assessment processes differ in part between sectors, many aspects are common (e.g., recognition of the benefits of probabilistic modeling of hazard pathways to better represent and describe the complexity of real-world situations). Utilization of the expertise and experience gained in all biosecurity situations has the potential to improve risk analysis both within and between sectors, provide for consistency in approaches and outputs and facilitate better uptake and understanding by competent authorities and other stakeholders. A more integrated and holistic approach will help in ensuring public confidence in overarching regulatory frameworks and assist in optimization of scarce biosecurity resources in developing countries. Expanded uptake of risk assessment methodologies by competent authorities and more systematic risk management processes will result in enhanced implementation of integrated national biosecurity goals. If a national biosecurity strategy has been developed, an integrated risk management approach enables the overall use of government resources to be prioritized according to a broad ranking of biosecurity issues.

7.29.3 Improved Biosecurity Capability

National level A harmonized and integrated biosecurity approach considerably improves the ability of competent authorities to achieve their mandates. Taking advantage of the interdependencies of competent authorities is increasingly reflected in shared technical capability. The resulting improvements in biosecurity capacity may be manifest in many ways and include the opportunity to develop a national strategy for biosecurity.

7.29.3.1 Improved National biosecurity capability resulting from increasing interdependence of competent authorities and convergence of biosecurity issues

These include simplification of legislation and condensing of biosecurity jurisdictions, development of a national biosecurity strategy and establishment of cross-sectoral priorities, better use of resources (e.g., sharing of methodologies, and border inspection systems, and training), rationalization of controls (e.g., opportunity to develop a single import health standard for an agricultural product that meets all biosecurity needs), shared certification where appropriate, improved data acquisition and quality, improved emergency preparedness and response (including contingency planning), integrated response to new and emerging diseases (e.g., combining veterinary, public health and food safety aspects of zoonoses), integrated pest management (IPM) programs (e.g., appropriate use of pesticides to achieve pest control goals while ensuring human health, protection of the environment and sustainability of agriculture), integrated surveillance (e.g., systems capable of detecting any unexpected adverse public health or environmental effects that may be associated with LMOs), integrated traceability systems, greater acceptance of privatization of some biosecurity services, restructuring of competent authorities and consolidation of multiple legislative and functional activities that were previously spread over several jurisdictions is progressing in different ways in different countries.

7.29.3.2 International level

The rapidly accelerating volumes and diversity of food and other agricultural commodities in international trade are contributing to the ever-increasing interdependence of competent authorities operating in different countries and illustrate the convergence of sector issues. This is significantly influencing biosecurity strategies and processes to the advantage of the global community. In Canada, a new regulatory initiative is the consolidation and modernization of biosecurity inspection and enforcement activities in the areas of food, agricultural and aquatic commodities, agricultural inputs (e.g., seed, feed, and fertilizer), animals and plants.

7.29.3.3 Laboratory biosecurity

The term laboratory biosecurity refers to the protection of hazardous biological agents, including toxins, from loss, theft, diversion, or intentional misuse (Anonymous, 2014).

7.29.4 Biological Select Agents and Toxins (BSAT)

Biological Select Agents and Toxins are federally regulated microbial pathogens and toxins that have the potential to pose a severe threat to public, animal, or plant health, or to animal or plant products and whose possession, use and transfer are regulated by the Select Agent Regulations (7 CFR Part 331, 9 CFR Part 121, and 42 CFR Part 73). The Department of Health and Human Services, Centers for Disease Control and Prevention (HHS/CDC), regulates the domestic possession, use, and transfer of Select Agents and Toxins that have the potential to pose a severe threat to public health and safety. The United States Department of Agriculture, Animal and Plant Health Inspection Service (USDA/APHIS), regulates the domestic possession, use, and transfer of Select Agents and Toxins that have the potential to pose a severe threat to public health and safety, to animal health, to plant health, or to animal or plant products. Select Agents and Toxins that are regulated by both HHS/CDC and USDA/APHIS are referred to as 'overlap' Select Agents and Toxins. The Department of Commerce, Bureau of Industry and Security (DOC/BIS) regulates the export of Select Agents and Toxins, relevant biological materials, and the technology associated with these pathogens and toxins. The unlawful possession, use, and transfer of Select Agents and Toxins are punishable by criminal and civil penalties.

7.29.5 Select Agent Regulations

The Select Agent Regulations are the U.S. Department of Health and Human Services and the U.S. Department of Agriculture regulations implementing the Public Health and Bioterrorism Preparedness and Response Act of 2002 and the Agricultural Bioterrorism Protection Act of 2002. Responsibility for the promulgation of the Select Agent Regulations and the implementation of Select Agent regulatory oversight resides with the Department of Health and Human Services, Centers for Disease Control and Prevention (HHS/CDC), and the United States Department of Agriculture, Animal and Plant Health Inspection Service (USDA/APHIS).

7.29.6 Select Agent Program

The Public Health Security and Bioterrorism Preparedness and Response Act of 2002 and the Agricultural Bioterrorism Protection Act of 2002 (the Acts) require entities to register with the Department of Health and Human Services, Centers for Disease Control and Prevention (HHS/CDC), or the United States Department of Agriculture, Animal and Plant Health Inspection Service (USDA/APHIS), if they possess, use, or transfer biological agents or toxins (i.e., select agents and toxins) that could pose a severe threat to public health and safety; to animal or plant health; or animal or plant products. The Acts also require increased safeguards and security measures for these agents, including controlling access, screening entities and personnel (i.e., security risk assessments), and establishing a comprehensive and detailed national database of registered entities. The Select Agent Program implements the requirements of these Acts through the Select Agent Regulations. USDA/APHIS and HHS/CDC work closely with Department of Justice's Federal Bureau of Investigation, Criminal Justice Information Service (CJIS) to identify those individuals who are prohibited from access to Select Agents and based on the restrictions identified in the USA PATRIOT Act.

Biosecurity is a strategic and integrated approach that encompasses the policy and regulatory frameworks (including instruments and activities) that analyze and manage risks in the sectors of food safety, animal life and health, and plant life and health, including associated environmental risk. "Biosecurity covers the introduction of plant pests, animal pests, and diseases, and zoonoses, the introduction and release of genetically modified organisms (GMOs) and their products, and the introduction and management of invasive alien species and genotypes" (Anonymous, 2003).

7.29.7 Cartagena Biosafety Protocol, 2000

In accordance with the precautionary approach contained in Principle 15 of the Rio Declaration on Environment and Development, the objective of this protocol is to contribute to ensuring an adequate level of protection in the field of the safe transfer, handling and use of living modified organisms resulting from modern biotechnology that may have adverse effects on the conservation and sustainable use of biological diversity, also taking into account risks to human health, and specifically focusing on transboundary movements.

7.29.8 World Conservation Union (IUCN), 1999, Definition of 'Biosecurity Threats'

They refer to those matters or activities which, individually or collectively, may constitute a biological risk to the ecological welfare or the well-being of humans, animals or plants, of a country.

Recent acute, but unintentional, introductions of non-indigenous plant pathogens (or their vectors) demonstrate the range of damage and consequences associated with newly introduced exotic pathogens (Table 7.3). If such an outbreak were to be caused by the intentional release of a naturally occurring or engineered biological agent, the political, economic and societal impacts would be considerable. In addition to natural outbreaks, crops are vulnerable to deliberate biological attack. Agriculture can be an attractive target for the introduction of pathogens as bioweapons because of its critical role in the infrastructure of most nations. Because regular surveillance of such extensive areas is not feasible, long lag times (months or even years) may pass between the introduction of a threatening pathogen and its detection, even if symptoms and signs are present. For example, an outbreak of citrus canker in Florida was first diagnosed in 1995 (164), but the causal bacterium was probably introduced two years prior to its detection. Similarly, plum pox virus (PPV) was first detected in Pennsylvania in 1999 but is now thought to have been introduced as much as 6 to 8 years earlier. Many of our valuable crop and forest species, bred for consistent quality and high yield, are planted as monocultures. Low genetic diversity increases the potential for pathogen spread and widespread damage. Finally, plant pathogens are generally easy to obtain, increase, transport, and deploy. Most pose little if any threat to the health of their human handlers, and the ethical barriers to deployment of a plant pathogen may be lower than those for human or animal pathogens.

A deliberate pathogen introduction event could result in severe negative impacts on crop yield and quality, cause significant public shock and/or panic due to a loss of confidence in a portion of the food supply and negatively impact the national economy, particularly in rural and agricultural sectors, due to yield decreases, reduced income from commodities sold, and the potential effects of quarantines and/or the loss of international markets. In countries of the third world, particularly those in which a single crop (such as rice or cassava) is the primary food commodity for a large segment of the population, it could lead to human hunger, suffering, and political disruption, as have resulted in the past from natural and accidental plant pathogen introductions. Strict regulations, registrations, restrictions, and security are required for their handling and investigation. Such regulations may be useful in attributing a crime involving a select agent. However, a plant pathogen select agent introduced into the United States with little likelihood of eradication may be delisted to facilitate the research needed for effective postintroduction disease management. The recent removal of soybean rust from the APHIS select agent list after the natural introduction of the causal fungus into the United States in the fall of 2004 and the concomitant delisting of PPV are two examples of this policy.

Table 7.3 Recent Acute, but Unintentional Examples of the of Non-indigenous Plant Pathogens and their Consequences Arrival in the United States

Disease	*Pathogen*
Asiatic citrus canker	*Xanthomonas axonopodis* pv. citri
Bacterial wilt	*Ralstonia solanacearum* race 3, biovar 2
Sudden oak death	*Phytophthora ramorum*
Pierce's disease of grapevine	*Xylella fastidiosa*
Karnal bunt of wheat	*Tilletia indica*
Asian soybean rust	*Phakopsora pachyrhizi*

7.30 PLANT PATHOGENS AS BIOWEAPONS

Biological warfare against agricultural targets is not a new idea. Over many decades, various state-sponsored research programs have been established with the goal of using weaponized microorganisms as part of the country's military arsenal (Fletcher *et al.*, 2006). The United States conducted research involving a number of pathogenic microbes, including the biological agents causing anthrax, foot and mouth disease and rice blast. Germany had programs during both the First and Second World Wars, whereas the former Soviet Union conducted programs from the Second World War through the Cold War, as did Iraq, beginning with the Iran-Iraq War. Other countries are thought to have had biological weapon programs aimed at agriculture and food production. Evidence found in caves in Afghanistan suggested interest by Islamic militants in the weaponization of the fungus that causes wheat rust. Other countries that have explored microbes as potential weapons include Canada, France, Japan, and the United Kingdom. Agricultural bioterrorism may be an outgrowth of such thinking and activities. There have been no documented cases, as yet, of the deliberate use of pathogens to attack crops or other plants. However, a posture of preparedness dictates that reasonable step is taken to ensure that appropriate crop biosecurity capabilities are in place before a devastating event occurs, not afterward. To that end, the examination of natural or accidental plant pathogen or pest introductions can provide insight into the possible impacts of a successful deliberate attack. For example, the potato blight epidemic in Ireland (1845 to 1846) led to extensive famine, resulting in the deaths of 1 million and the migration of an additional 1.5 million Irish. Brown spot of rice contributed to the Great Bengal Famine of 1943. In the United State, a leaf blight in 1970 destroyed about 20% of a corn crop valued at $1 billion.

7.30.1 Role of Microbial Eorensics in Crop Biosecurity

Whether plant pathogens (or their products, e.g., toxins) are used deliberately as weapons to cause social or economic damage or are introduced inadvertently into a new area, it is important to determine the source, the method, and the time of the introduction, as well as those responsible for it [(Fletcher *et al.*, 2006) (Table 7.4)]. Analyses related to such determinations fall into both the forensic sciences and epidemiology. Forensic science is the application of scientific methods in the investigation of possible violations of the law, where scientific knowledge and technology provide evidence in both criminal and civil matters. The discipline provides specific support for investigative and law enforcement efforts in which the ultimate goal is attribution, the determination of the perpetrator of a criminal act. Forensic analyses include the use of tools for sampling, packaging, shipping, storage, microbial identification and discrimination, chemical analyses, epidemiological modeling, bioinformatics, and other considerations. Because the safety and confidence of the public must be preserved, and because the legal ramifications of criminal attribution and prosecution require unusual investigative validation and stringency, a forensic investigation involving a plant pathogen may require use of methodologies that address different questions and/or have higher resolution than those normally used in disease diagnosis and plant pathogen identification. The forensic application requires the confidence necessary for ultimate attribution.

The degree of confidence with which forensic analyses can support the identification of a specific microbe, reconstruction of its method of introduction into a particular location, and identification of the perpetrator depends on many factors. Lag times may occur at several stages of the investigation. The first is the time between the introduction of a pathogen and its detection, which is affected by many factors, including weather conditions before, during, and after the introduction. A second lag is the time required to develop and execute an appropriate sampling protocol. Protocols must include validated techniques that minimize the time between on-site sample collection and arrival at a forensics laboratory. Third, the time required for stringent laboratory assessment and the resolution, reliability, and repeatability of the chosen analytical methods affects the success of a forensic investigation. In many cases, it may be far easier to determine exclusion (assurance that a particular pathogen or person is not involved in the incident) than absolute attribution (evidence that may uniquely associate a particular pathogen/isolate or person to the incident). A critical first question with respect to a plant disease outbreak is whether a crime has occurred. Many diseases are already endemic, and once-exotic diseases that are not eradicated may become endemic relatively quickly after introduction.

An intentional introduction of a plant pathogen as a biocrime or bioterrorist event might not be recognized as such (Fletcher *et al.*, 2006). Thus, tools for better pathogen resolution, more relevant background information, and more robust surveillance mechanisms are needed to better evaluate whether a disease is natural or human incited. On the positive side, crop producers and plant pathologists are already poised to move quickly to apply management strategies to control disease. Therefore, rapid determination of whether criminal activity has occurred is crucial so that responders know if the event should be handled as a crime, with appropriate steps for attribution, or solely as a containment effort. Forensic science can assist in this endeavor.

Table 7.4 Proposed Plant Pathosystems Developed for the Assessment of Current Microbial Forensics Capabilities, Gaps, and Needs

Pathogen	*Disease*	*Vector/transmission*
Soybean mosaic virus	Soybean mosaic	Aphids, seed
Tomato spotted wilt virus	Tomato spotted wilt	Thrips
Plum pox virus	Plum pox (Sharka)	Aphids, cuttings, grafting
Pseudomonas syringae pv. tomato	Bacterial speck of tomato, pepper, and edible brassica	Wind-blown rain; infested soil
Corynebacterium michiganensis subsp. *sepedonicus*		Ring rot of potato Infested tubers
Ustilago maydis	Smut of maize and teosinte	Wind blown

As the nascent discipline of plant pathogen forensics develops, standard crime scene processing and evidence handling protocols must be validated and adapted to plant pathogen forensics applications. It may be appropriate to develop some new technologies specific for crime scenes involving crops, forests, nurseries, orchards, or rangelands. A thorough analysis is required to identify and assess the information, capabilities, tools, and resources already in existence. Once these are brought to bear on the new applications of forensic science, it will be possible to identify remaining gaps and the needed capabilities to fill them, a step that will serve as the basis for the development of a forward strategy. This subdiscipline of forensics specifically targeted toward microbial pathogens, as applied to bioterrorism and biocrimes involving humans and animals, has been developing over the past few years. However, few if any field or laboratory methods, standard operating procedures (SOPs), or protocols have yet been specifically developed and rigorously validated for application to plant pathogens. As plant pathogen forensics becomes established as a separate subdiscipline of forensic science, a major early area of opportunity will be to critically assess, select, and shepherd existing methods, SOPs, and protocols through an appropriate process so that 'sets' of validated 'tools' are available and defensible should a crop bioterrorism event occur (Fletcher *et al.*, 2006). To accomplish these near- and long-term goals, plant pathologists and forensic scientists (especially those working in microbial forensics) need to plan and work together.

7.30.2 Components of Strong Microbial Forensics Capability

The application of forensic methods to a plant disease outbreak requires specific steps in forensic microbiology (Nutter, 2004), including careful documentation of disease characteristics; sampling the potential crime scene; identification of the pathogen to the race, strain, isolate, or isolate mixture; selection of appropriate mitigation response measures; further characterization of the pathogen to identify likely sources; and attribution or exclusion of pathogens. Mistakes made during early stages of this process, i.e., during disease characterization and sampling, cannot be corrected at later stages of the investigation (Fletcher *et al.*, 2006).

7.31 FORENSIC TYPING METHODS AND CRITERIA FOR SELECTING APPROPRIATE METHODS

Critical characteristics of microbial typing for forensic applications include universality, the ability to type all organisms within the taxon using a particular method; sensitivity, the percentage of actual positive samples detected (with no false negatives); specificity, the percentage of actual negative samples identified correctly (with no false positives); efficiency, the total percentage of correct test results; reproducibility, the same result obtained consistently when a particular isolate is tested repeatedly; and resolution, the degree of attribution that can be obtained with a method. For most plant pathogens, multiple methods of microbial identification and typing are available. Having results from multiple tests will increase the level of accuracy and confidence in microbial forensics investigations. Typing methods currently in use for plant pathogens include both nucleic acid-based and nonnucleic acid-based technologies.

7.31.1 Nonnucleic Acid-based Methods

The first and most important assays for both microbial forensics and management are those that determine the species of the pathogenic agent. A number of traditional methods, in use long before the advent of molecular biology, remain effective for some applications and may provide significant clues for pathogen identification in a forensics setting. Symptomatology, the ability to cause either no reaction or a 'hypersensitive' (resistance) reaction on a nonhost plant; plant host range; insect vector specificity and pathogen morphology (of bacterial colonies or cells, fungal colonies or fruiting bodies, or virus particles or inclusion bodies, etc.) are often the first steps in identification (Schaad *et al.*, 2001). The pathogen's host range and the host's specific response to the pathogen also are used for typing of plant pathogens. Species of many plant pathogenic bacteria are further divided into pathovars, based solely on the host range of the bacterium and methods that define physiological processes, or the complement of certain molecules are also used to define taxa. For example, BIOLOG (Hayward, Calif.) and other substrate utilization tests provide profiles of metabolic capabilities, while fatty acid methyl ester (FAME) analysis produces a profile of the microbe's fatty acid composition. In both cases, the profiles of the test strain are compared with those in a database of species and strains for the closest match. The accuracy of such assays for microbial identification is limited by the population of characterized strains in the databases.

Like bacteria, plant virus strains belonging to the same species also may be discriminated by the comparative reactions of a set of plant species or cultivars within a species, known as differentials. Fungi, because of their large genomes and complex life cycles, present particular typing challenges. For example, the fungal mating type, determined by plate mating assays, is the primary mode of identification of the model fungal plant pathogen *U. maydis*. Although the mating assay is reliable and accurate, it is not very definitive for strain attribution because it does not measure other variations in the genome; on average about 1 of 36 of the cells will possess a given mating-type genotype in a random population. Thus, the mating-type distinction is a good, exclusionary tool but will not achieve absolute attribution.

7.31.2 Serological Techniques

ELISA and indirect fluorescent antibody staining are serological assays commonly used for identification of plant pathogens, particularly viruses and bacteria. The sensitivity and specificity of serological assays vary with the titer and specificity of the antibody and whether the antibody is monoclonal or polyclonal. Recent adaptations by diagnostic industries for dipstick convenience and portability have enhanced the usefulness of these immunology-based technologies in the field. Cross-reactivity among closely related strains may be a problem; for example, the seven strains of SMV, identified on differential host cultivars, are chemically and serologically homogeneous at the coat protein level (Fletcher *et al.*, 2006).

7.31.3 Nucleic Acid-based Methods

The popularity of nucleic acid based technologies has grown rapidly. Older methods, such as restriction fragment length polymorphism (RFLP), DNA fingerprinting and phage typing are still valid and useful. However, complete genome sequences are now available for many economically important viruses and bacteria and a few fungi and nematodes, and others are in progress. Thus, many genetic markers are available for analyses. DNA probes, constructed for taxon-specific marker genes, are used widely. Sequencing of particular genome regions known to provide informative data, such as the 16S rRNA, the internal transcribed spacer region between the sequences coding for the16S and the 23S rRNA, or the *groE* or *recA* genes, is often used for bacterial identification. The sensitivity, specificity, and versatility of PCR have made it a method of choice for applications related to sequence analysis and comparison. PCR-based assays have been used widely, for example, in the typing of DNA viruses. The discovery of repeated sequences in many bacterial genomes has given rise to a version called rep-PCR, in which electrophoretic banding patterns reflect different numbers and positions of repeated sequences (Fletcher *et al.*, 2006).

Real-time PCR has been used for genetic characterization of bacteria, viruses and fungi. Although the technique is rapid and can be very specific, it may not be as sensitive as culture-based assays to detect pathogens present in plant extracts when PCR inhibitors are present or with very small sample volumes, both of which can reduce sensitivity. For cultivable bacteria, PCR can be combined with isolation in BIOPCR. In this assay, viable cells of the target bacterium are enriched in medium and thereby detected at extremely low original levels in seeds and other propagation materials. No DNA extraction is needed since the cells lyse during the initial denaturation step. For higher levels of specificity, BIOPCR can be performed on membranes, although a possible disadvantage of membrane use is the chance for cross contamination. Multiplex PCR, in which primers against more than one target are combined in a single reaction mixture, can be employed to detect more than one species of bacterium or virus in the same sample. Assay and detection of multiple sites of a microorganism's genome can increase confidence in the identification.

Such systems are currently in use for certifying vegetative plant propagules as virus free. Another PCR variant, reverse transcription-PCR (RT-PCR) is useful for plant viruses having RNA genomes. Reasonably 'universal' primers have been developed for some virus families, genera, and 'species' targeted to taxon-specific sequences. RTPCR clearly differentiates between some strains of common viruses: for example, between PPV strains D and M and SMV strains G2 and G7, as well as between the common strain of *Potato virus Y*, PVYO, and strain PVYN (Mills *et al.*, 2003). Kim *et al.* (2003) conducted RFLP of RT-PCR products ('restrictotyping') to differentiate five Korean SMV strains. Many other new variations and assay combinations, such as multiplex PCRELISA and immunocapture PCR, have been developed (120). The methods mentioned above are all limited by their reliance on a minute fraction of a taxon's many defining features. For nucleic acid-based methods, the accuracy of the comparison to some degree will be proportional to the length of the fragment (and/or the site) used in the analysis. Direct

DNA-DNA hybridization methods provide information about the degree of similarity of entire genomes, without the need for actual sequence information.

Direct comparisons to rank the sensitivity and specificity of certain detection/ diagnostic methods have been carried out for some plant pathogens. Many research programs shifted from traditional methods (symptomatology, electron microscopy, and host differential reactions) to PCR-based or immunological tests when the latter were demonstrated to be more sensitive or specific. In limited cases, methods have been standardized among laboratories to ensure that comparisons between/among the groups were reliable. However, such test comparisons and standardizations are not frequently done, because the validation of methods at a level necessary for more rigorous challenge adds significant cost and generally is not required for managing a natural disease outbreak. Thus, for many diagnostic systems the relative effectiveness of one technology over another for critical identification is not known, and in fact, the 'best' test will often depend on the 'diagnostic' sites and methods available for a given taxon and the databases of information collected on the species and closely related strains and species. For example, in a recent analysis of multiple plant pathogenic strains of the ubiquitous bacterium *Serratia marcescens*, different 'identifications' were provided by BIOLOG, FAME, 16S rRNA and *groE* sequencing, and DNA-DNA hybridization because each of these tests measures or compares a different genome region, gene product, or phenotype. While exclusion and inclusion interpretations can be made with all of these methods, research is needed to establish the most reliable and informational methods for high-priority plant pathogens and to develop the reagents and databases for them.

The diversity of organisms within a microbial population must also be considered in evaluating typing methods. In a given plant, field or region, some pathogen populations, such as those of *U. maydis* are relatively homogeneous. For others, such as PPV, high population-level species diversity means that an individual sample from a single host will contain many mixed sequence variants (Fletcher *et al.*, 2006). Researchers in the United States and the European Union are investigating the evolutionary tempo and drift of various regions within the PPV genome, and mutation 'hot spots' within the genome could be targets for forensic analysis. Also relevant is the rate at which the pathogens change. For example, the genome of SMV appears to be more stable than that of PPV, although new strains of SMV are reported relatively frequently (Kim *et al.*, 2003). Furthermore, some isolates described as new may actually be strains or recombinants of existing viruses. Tomato spotted wilt virus (TSWV) is quite variable because of reassortment of genomic segments and other mutations. For TSWV and related viruses, three loci (the N and NSm genes and the intergenic region) are used for comparisons. A rule of thumb is that isolates having 90% amino acid sequence identity in their N-protein sequences are distinct strains.

7.31.4 Molecular Markers for Forensic Analysis

As stated above, most plant pathogens lack a standard marker system. For those that do, markers are based on any of a variety of biomolecules and are taxon-dependent. Rarely does a marker satisfy all three of the criteria defined above, inherent, diagnostic and Boolean. Many marker systems were developed for use in systematics, the hierarchical assignment of organisms to a taxonomic classification. Several series of terms describe the hierarchy, strain, subspecies, species, genus, family, etc. There is a well-defined system for classification of viruses into families, genera, and species, but these taxa are being redefined with molecular techniques. Below the species level, further subdivision into subtypes, groups, or strains is common. Taxonomic levels do not reflect consistent evolutionary time periods. Although systematics has provided many useful markers, it does not always follow evolutionary descent patterns (Fletcher *et al.*, 2006) and it is less relevant to microbial forensic considerations. Phylogenetics, in contrast, reconstructs the order of the organism's divergence from a common ancestor, so that under a certain set of assumptions, placement of a crime scene isolate on a phylogenetic tree identifies its closest known relatives or most recent common ancestor. Phylogenetic analyses, using a specific gene or genomic region, have been conducted on many plant pathogens.

Meaningful trees require considerable sequence variation in the targeted portions of the genome, but such variability should not interfere with reliable sequence alignment. Few plant pathogen marker systems have been used widely enough to acquire good estimates of the frequencies of particular markers in populations. For most, the scope of the population is geographically restricted to a plot/field, county, state, or region, sometimes a nation. Seldom are data on worldwide frequencies calculable. Of marker types, the nucleotide sequence is the best developed because more isolates have been examined by this method than by others. However, the diversity of microbes of potential phytopathogenic threat is so large that obtaining complete sequences of enough microbes for a good level of background knowledge is a daunting task. Genome microarrays and subtractive hybridization approaches may be good alternative technologies for resequencing. Indeed 'high-resolution differentiation between closely related' microbes was obtained in a microarray study in which 295 of 300 pairs of bacterial strains were statistically differentiable (Willse *et al.*, 2004).

7.31.5 Building Plant Pathogen Forensic Capability

The discipline of plant pathogen forensics, currently in its infancy, will require intellectual and financial investment to develop an effective level of robustness and utility. The existing physical and human infrastructure in the United States for plant pathogen detection, diagnosis, forensics, and research has become strained as officials, and the scientific community worked to develop and implement new programs to address crop biosecurity issues. There are significant gaps in research, technology, education, and training. Current efforts to establish plant pathogen forensics programs compete for limited resources with more established human and animal biosecurity programs. However, purpose-driven action and targeted financial resources are essential to build the necessary programs, personnel, and infrastructure (Fletcher *et al.*, 2006). The next stage in building capacity in plant pathogen forensics should be the development of a detailed strategic plan, to include specific and prioritized issues and needs as defined and identified by forensic specialists working together with agricultural leaders.

A thorough systems analysis of the existing science is needed to further define current and future threats and risks to the U.S. agricultural enterprise from the deliberate use of plant pathogens. The plan also should include a time line for the research, development, testing, validation, and implementation of methods, technologies, and practices that will establish plant pathogen forensics and incorporate it into the national agricultural biodefense 'toolbox.'

Various initiatives on the part of existing U.S. agencies to redirect efforts and to initiate new programs to enhance the national security of our crops, rangelands and forests have strengthened national agricultural security. However, the establishment of a single federal coordinating body, a National Center for Plant Biosecurity (NCPB), for plant pathogens and crop health could provide focus, coordination, and strategic planning for such efforts. The NCPB, as described in a proposal by the American Phytopathological Society and coauthored by a number of other scientific societies (Fletcher *et al.*, 2006) would be able to build on, enhance, and provide leadership and coordination of national efforts for documenting, monitoring and protecting crops, forests, and rangelands against new or emerging plant diseases.

Because most high-risk pathogens, by definition, are not now present in the United States, much of the scientific research on them is done overseas. Furthermore, because an intentional introduction of such pathogens into the United States could be an international incident, forensic investigation of such events will stretch across national borders. Plant pathogens do not recognize political boundaries; the introduction of a plant pathogen into agricultural areas in a country adjoining the United States often results in the eventual movement of the pathogen into the United States. Thus, no national security plan can be optimally effective on its own. Plans developed, adopted, and monitored at a global level will be most effective.

The National Research Council (NRC) has called for harmonized international oversight of biosecurity efforts (Fletcher *et al.*, 2006). Whether this is sought by a meeting of nations (an International Forum on Biological Security, sponsored by governmental and nongovernmental entities, was proposed by the NRC) or by a series of more focused interactions, the goal would be to strengthen research and operational links for the enhancement of global security.

7.31.6 Plant Pathogen Forensics

A biological attack on crops, rangelands, or forests could reduce yield and quality, erode consumer confidence, affect economic health and the environment, and possibly impact human nutrition and international relations. Preparedness for a crop bioterror event requires a strong national security plan that includes steps for microbial forensics and criminal attribution. However, U.S. crop producers, consultants, and agricultural scientists have traditionally focused primarily on strategies for prevention and management of diseases introduced naturally or unintentionally rather than on responding appropriately to an intentional pathogen introduction. We assess currently available information, technologies, and resources that were developed originally to ensure plant health but also could be utilized for postintroduction plant pathogen forensics. Recommendations for prioritization of efforts and resource expenditures needed to enhance our plant pathogen forensics capabilities were presented (Fletcher *et al.*, 2006).

A biological attack on U.S. crops, rangelands or forests could reduce the yield and quality of agricultural products, erode consumer confidence, affect nutrition, adversely impact the nation's economic health and international markets, and cause harm to the environment. Preparedness for a crop bioterror event must begin with recognition of the threat, which must lead to the development of a plan to thwart, or to respond to and attribute, an attack. However, most USA crop producers, crop consultants, and agricultural scientists have not focused on the possibility of deliberate plant pathogen introduction but instead have directed their efforts toward the prevention of accidental or natural introductions and the development of cost-effective disease management strategies. Should these efforts fail, responses typically address rapid eradication and/ or long-term disease management. A case was built for the integration of the traditional discipline of plant pathology and the specialized field of forensic science, envisioning the birth of the new field of plant pathogen forensics. The potential threat to plant resources was described and then focus on one part of a strong national security plan: the development of a program in microbial forensics and criminal attribution that addresses crop and other plant resource targets. Although they were developed originally for plant health or economic applications, many may also be utilized for plant pathogen forensics. Recommendations were presented for the prioritization of activities, including critical research, development of new technologies and infrastructure, and allocation of the human and financial resources needed to ensure a strong capability in plant pathogen forensics.

Worldwide, losses for the eight major crops that comprise half of the global croplands were estimated at $300 billion from 1988 to 1990. In the United States alone, plants are subject to attack by over 50,000 different pathogens, primarily fungi, viruses, bacteria, and nematodes (Fletcher *et al.*, 2006). For any given region and crop, producers may deal with up to 10 to 15 serious plant diseases that can cause severe economic repercussions. About 65% of US crop losses are due to nonindigenous (introduced) pathogens, amounting to an estimated cost of $137 billion annually (Pimentel *et al.*, 2000). All crop pests (pathogens, arthropods, and weeds) combined cause preharvest losses of 42% and an additional 10% loss after harvest. Of these, 13% are due to plant pathogens, 15% to arthropods, and 13% to weeds.

Recent acute, but unintentional, introductions of nonindigenous plant pathogens/their vectors) demonstrate the range of damage and consequences associated with newly introduced exotic pathogens. If such an outbreak were to be caused by the intentional release of a naturally occurring or engineered biological agent, the political, economic, and societal impacts would be considerable. In addition to natural outbreaks, crops are vulnerable to deliberate biological attack. Agriculture can be an attractive target for the introduction of pathogens as bioweapons because of its critical role in the infrastructure of most nations. In the United States, for example, agriculture and related industries comprise about one-sixth of the gross domestic product or about $1 trillion annually. Moreover, 17% of U.S. employment is related to agriculture. Agricultural products comprise a major component of U.S. exports. Crops, rangelands, and forests occupy vast areas in the United States, covering half a billion hectares. Because regular surveillance of such extensive areas is not feasible, long lag times (months or even years) may pass between the introduction of a threatening pathogen and its detection, even if symptoms and signs are present. In wealthy countries, a deliberate pathogen introduction event could result in severe negative impacts on crop yield and quality, cause significant public shock and/or panic due to a loss of confidence in a portion of the food supply and negatively impact the national economy, particularly in rural and agricultural sectors, due to yield decreases, reduced income from commodities sold, and the potential effects of quarantines and/or the loss of international markets. In countries of the third world, particularly those in which a single crop (such as rice or cassava) is the primary food commodity for a large segment of the population, it could lead to human hunger, suffering, and political disruption, as have resulted in the past from natural and accidental plant pathogen introductions.

Plant pathogens of high risk for the United States that are designated select agents under the Code of Federal Regulations, Title 7, Part 331, by the Agricultural Bioterrorism Protection Act of 2002 and the United States Department of Agriculture (USDA) Animal and Plant Health Inspection Service (APHIS), are listed. This list is parallel to select agent lists for human and zoonotic diseases, except that all plant pathogen select agents are nonindigenous pathogens that are not yet known to occur in the United States (termed 'exotic' in the Plant Pest Act). Strict regulations, registrations, restrictions, and security are required for their handling and investigation.

Such regulations may be useful in attributing a crime involving a select agent. However, a plant pathogen select agent introduced into the United States with little likelihood of eradication may be delisted to facilitate the research needed for effective post introduction disease management. The recent removal of soybean rust from the APHIS select agent list after the natural introduction of the causal fungus into the United States in the fall of 2004 (Stokstad, 2004) and the concomitant delisting of PPV are two examples of this policy. Although no endemic (or indigenous) plant pathogens are on the select agent list, many pose threats to agriculture that are equal in severity to those on the list. Further, many of the endemic pathogens are readily accessible and could be used in an illicit manner against crops. Since criminal penalties may now be imposed for unlawful possession or use of listed (as well as illicit use of unlisted) pathogens, a sound and reliable forensic analysis system is needed for purposes of attribution.

7.31.7 Use of Surrogate Pathosystems as Models

Fletcher *et al.* (2006) explored forensics with a focus on plant pathogens. Since it would be impossible to address all plants and pathogens of interest, they focused on several specific model pathosystems (systems inclusive of the pathogen, the plant host, the insect vector [if applicable] and the environment) that, in future research initiatives, can be exploited easily outside of biocontainment facilities. In some cases, where information and/or examples are lacking among these model pathosystems, examples were cited from other plant pathogen-host systems. Although research focuses on case studies from a U.S. perspective, these principles and concepts apply globally.

Five of the six model pathosystems are damaging pathogen-host-environment relationships that are currently established in the United States, where they cause substantial losses to the nation's agricultural productivity. High-caliber research programs are already in place, and a significant body of knowledge has been accumulated for each pathosystem (Fletcher *et al.*, 2006). These pathosystems were selected to represent unique features of critical importance to forensic science so that research on the models will address real issues and answer real problems. The long-term utility of these model pathosystems (and of others not described herein) will be in the application of general principles, established from rigorous research to any pathogen that might be deliberately introduced. To address the possible challenges of such a transfer of principles, our sixth model, PPV was chosen to represent a USDA-APHIS-listed select agent. Although PPV was recently removed from the select agent list, this model pathosystem highlights the differences, both beneficial and restrictive, of working on an exotic and regulated pathogen. Our sample pathosystem models provide insight for the near- and long-term research investments required to achieve a strong and effective US program in plant pathogen forensics.

7.31.8 Components of Strong Microbial Forensics Capability

The application of forensic methods to a plant disease outbreak requires specific steps in forensic microbiology, including careful documentation of disease characteristics; sampling the potential crime scene; identification of the pathogen to the race, strain, isolate or isolate mixture; selection of appropriate mitigation response measures; and further characterization of the pathogen to identify likely sources and attribution or exclusion of pathogens. Mistakes made during early stages of this process, i.e., during disease characterization and sampling, cannot be corrected at later stages of the investigation.

7.31.9 On-site Disease Assessment

Initial assessment of the disease in the field should be done prior to any disturbance of the site. This analysis should include the pattern of disease occurrence and any other field characteristics deemed relevant or unusual. Certain disease incidence patterns may be typical (or atypical) of a natural event (Fletcher *et al.*, 2006). If possible, aerial and satellite images should be obtained prior to any ground disturbance. On the ground, areas with symptomatic plants should be delimited by application of global positioning system (GPS) technology prior to commencing with detailed sampling in affected areas. SOPs for the collection of microbial forensic field samples must allow for variation among crop species and suspected pathogens. Forensic field samples may include whole plants, selected plant parts, plant surface swabs or exudates, soil with or without root tissue, suspected insect vectors, natural or irrigation water in or near the fields, air samples, and/or biological samples (alternative weed hosts and soil or aquatic organisms, etc.). Containers must be clean and unused, and the samples must be collected directly into the container. Minimum documentation includes an administrative log, a sample log, the complete chain of custody, a collection sitemap(s) sufficient to allow repeat sampling from the same location (within or among fields), and a laboratory submission or transferal document providing detailed information on the crop, field history, and environment. Photographs showing symptoms, field layout, and other relevant details may supplement, but not replace this documentation.

7.31.10 Characteristics of Good Sample

A 'good' sample varies depending on the patterns of disease intensity, the pathogen, and the host. If multiple disease foci are present in the field, samples should be collected from a representative number of these locations, as well as from outside the focal areas (Fletcher *et al.*, 2006). Since pathogen titers often differ in leaves, stems, roots, and flowers and with the distance from the site(s) of initial infection, it is important to sample from different plants and different plant parts. The ease of pathogen collection may vary with the season; for example, tree fruit phytoplasmas overwinter in tree roots and move into above-ground branches in the spring, while PPV is absent from tree samples collected in summer when the temperature rises above 30°C. When pathogen detection is carried out with sensitive assays such as enzyme-linked immunosorbent assay (ELISA) or PCR, it is often practical to combine tissue samples from several plants and analyze them together as a pooled sample. This form of group testing allows a larger proportion of the plant population to be tested with a minimal number of assays, thereby improving the detection limit and reducing laboratory resource demands. A drawback of sample pooling is the loss of spatial and disease incidence information that would result from individual sample processing, however, plants from positive pooled samples could be retested individually in a second round of assays if information on the exact location and disease intensity of each infected plant is needed.

If necrotic lesions are present, it is best to sample from the lesion edges, where living plant tissue better supports active pathogen growth, as the lesion centers may subsequently be invaded by saprophytic microbes. Seeds are a good source of seed-borne pathogens such as soybean mosaic virus (SMV), and underground stems and tubers may serve as a source of pathogens as occurs with the potato ring rot pathogen, *Clavibacter michiganensis* subsp. *sepedonicus*. Certain specialized pathogen structures, such as the galls of the corn smut fungus, *Ustilago maydis* or the tumors produced by the crown gall bacterium, *Agrobacterium tumefaciens*, may be collected directly.

7.31.11 Sample Size and Sampling Pattern

The number of samples collected should be representative of the impacted area and be defensible scientifically and legally. Sampling pattern and sampling size considerations have been described for numerous pathosystems, including those that involve exotic or once-exotic pathogens such as PPV or the citrus canker pathogen, *Xanthomonas axonopodis* pv. *citri* (Mavrodieva *et al.*, 2004). In general, sampling for detection (i.e., presence or absence of a pathogen or disease) in a given field requires a different sampling pattern and sample size than sampling to determine the disease incidence or severity in the field.

7.31.12 Logistics

Collection and documentation of 30 to 40 samples could occupy a two-person team for 8 to 12 h, particularly if travel is required between sample locations. Protocols for forensic field sampling should be designed in consideration of the number of available personnel; similarly, sample numbers should be reasonable in the context of the analytic and volume capabilities of the facility performing the laboratory analyses and/or diagnoses. However, no protocol should be designed solely based on resource limitations. The use of custody seals will alert the recipient to any tampering between the time of collection and receipt of the samples.

7.31.13 Sample Storage

Long-term storage of forensic microbial samples prior to analysis may be necessary; thus, great care must be taken to preserve the integrity and security of the samples. Some plant pathogens may be stored more successfully than others. Storage of viable pathogen cultures is a matter very different from the preservation of desiccated leaf tissue, seeds, or fruits. Documentation of environmental conditions during storage is required, and chain-of-custody records must reflect all aspects of storage conditions and exposure to the environment, including records of individuals who may have access to the samples (Fletcher *et al.*, 2006).

7.31.14 First Detectors and First Responders

'First detectors' on the scene of a deliberate plant pathogen introduction are likely to be growers, crop consultants, master gardeners, extension agents, or other local personnel not affiliated with the government. 'First responders,' individuals authorized to respond and take action after a potential deliberate introduction, generally arrive on the scene later, after being notified by first detectors. Note that the designation 'first responders' differs here from that traditionally used for human targets, where the first responders are police or firefighters, etc. Clearly, timely and effective management of a crime scene will be impossible unless first detectors and first responders are equipped with the knowledge and skills to recognize that a crime has occurred and to react appropriately.

7.31.15 Epidemiological Tools and Models to Support Forensic Analysis

The close relationships between public health and veterinary science with respect to epidemiological and forensic investigations are well recognized, as both seek to determine the origin and source of an outbreak even though subsequent response pathways may differ. Many parallels also exist between these two disciplines and that of plant pathogen forensics. The abilities to trace plant pathogens and to attribute their source(s) require analyses at levels from the whole plant to the landscape. Critical issues include detection and quantification of the organism within the plant, field, or landscape, estimates of the initial infection time, based on anticipated environmental conditions and host growth stage and the likelihood of spread to adjacent crops based on environmental conditions and/or vector activity (Fletcher *et al.*, 2006). Such analyses require the use of epidemiological tools. Two broad epidemiological approaches that can support forensic analyses can be distinguished. Proactive approaches help investigators to understand variations in the intensity or geographical extent of a disease against a natural background while reconstructive methodologies enable understanding of an event and facilitate the establishment of the temporal sequence that comprised, as well as resulted from, it. The modeling tools used with both approaches are largely identical.

7.31.16 Post-translational Protein Modification

The complement of proteins that result from the transcription and translation of a pathogen's genes can provide a fingerprint of that organism that reflects environmental influences not revealed in the genome. Posttranslational protein modifications most commonly reported for plant pathogen proteins include glycosylation and phosphorylation. For *U. maydis*, the phosphorylation of some protein kinases is a critical component in the regulation of genes involved in morphogenesis and pathogenesis (Fletcher *et al.*, 2006). Some viral genomes, particularly in the *Potyviridae*, are translated as a single, large polyprotein that is subsequently cleaved into individual proteins by virus-encoded proteases. Regulation of the process is presumably by the interaction between the plant translation mechanisms and the viral genome. For other viruses, the coat protein is transcribed separately from a unique mRNA.

7.31.17 Building Plant Pathogen Forensic Capability

The discipline of plant pathogen forensics, currently in its infancy, will require intellectual and financial investment to develop an effective level of robustness and utility. The existing physical and human infrastructure in the United States for plant pathogen detection, diagnosis, forensics, and research has become strained as officials, and the scientific community worked to develop and implement new programs to address crop biosecurity issues. There are significant gaps in research, technology, education, and training. Current efforts to establish plant pathogen forensics programs compete for limited resources with more established human and animal biosecurity programs. However, purpose-driven action and targeted financial resources are essential to build the necessary programs, personnel, and infrastructure (Fletcher *et al.*, 2006).

The next stage in building capacity in plant pathogen forensics should be the development of a detailed strategic plan, to include specific and prioritized issues and needs as defined and identified by forensic specialists working together with agricultural leaders. A thorough systems analysis of the existing science is needed to further define current and future threats and risks to the US agricultural enterprise from the deliberate use of plant pathogens. The plan also should include a time line for the research, development, testing, validation, and implementation of methods, technologies, and practices that will establish plant pathogen forensics and incorporate it into the national agricultural biodefense 'toolbox.'

Because most high-risk pathogens (by definition) are not now present in the United States, much of the scientific research on them is done overseas. Furthermore, because an intentional introduction of such pathogens into the United States could be an international incident, forensic investigation of such events will stretch across national borders. Plant pathogens do not recognize political boundaries; the introduction of a plant pathogen into agricultural areas in a country adjoining the United States often results in the eventual movement of the pathogen into the United States. Thus, no national security plan can be optimally effective on its own. Plans developed, adopted, and monitored at a global level will be most effective. The National Research Council (NRC) has called for harmonized international oversight of biosecurity efforts (Fletcher *et al.*, 2006). Whether this is sought by a meeting of nations (an International Forum on Biological Security, sponsored by governmental and nongovernmental entities, was proposed by the NRC) or by a series of more focused interactions, the goal would be to strengthen research and operational links for the enhancement of global security.

Strategies for securing our plant resources (crops, rangelands, and forests) must address both prevention and preparedness. Capabilities needed to respond to an attack, should one occur, must include microbial forensic technology and practice to allow determination of the source of the pathogen and to provide the evidence needed to attribute the act to those responsible. This component should include strategies for assuring high stringency (validation, confidence, statistical significance, and consistency), tracing pathogen origin and movement, discerning the timing and site of the initial introduction, identification of the perpetrators, collection of evidence for criminal attribution, and establishment of links with the law enforcement and security communities (Fletcher *et al.*, 2006). New investments in research, infrastructure, personnel, training and strategic planning will be critical in meeting the needs of our overall national security programs.

The occurrence of ergot bodies (sclerotia of *Claviceps purpurea*) in cereals presents a high toxicity risk for animals and humans due to the alkaloid content. To reduce this risk, the European Commission fixed an ergot concentration limit of 0.1% in all feedstuffs containing ungrounded cereals, and a limit of 0.05% in 'intervention' cereals destined for humans (Ph.Vermeulen *et al.*, 2012). This study sought to develop a procedure based on near infrared hyperspectral imaging and multivariate image analysis to detect and quantify ergot contamination in cereals. Hyperspectral images were collected using an NIR hyperspectral line-scan combined with a conveyor belt. All images consisted of lines of 320 pixels that were acquired at 209 wavelength channels (1100–2400 nm). To test the procedure, several wheat samples with different levels of ergot contamination were prepared. The results showed a correlation higher than 0.99 between the predicted values obtained using chemometric tools such as partial least squares discriminant analysis or support vector machine and the reference values. For a wheat sample with a level of ergot contamination as low as 0.01%, it was possible to identify groups of pixels detected as ergot to conclude that the sample was contaminated. In addition, no false positives were obtained with noncontaminated samples. The limit of detection was found to be 145 mg/kg and the limit of quantification 341 mg/kg. The reproducibility tests of the measurements performed over several weeks showed that the results were always within the limits allowed. Additional studies were done to optimize the parameters in terms of number of samples analyzed per unit of time or conveyor belt speed. It was shown that ergot can be detected using a speed of 1–100 mm/s and that a sample of 250 g can be analyzed in 1 minute.

An investigation was carried out with the idea of developing an Online Pest Management Information System (PMISNET) on major agricultural crops containing information on various aspects of IPM, viz., general information, insect/pests, diseases and weeds on major crops, IPM strategies, etc. (Soubhratra Das *et al.*, 2002). PMISNET had a three layer client-server architecture. The software runs on HTTP server and served the request of the client that may on any computer connected to The Internet and having a graphic web browser. There was provision to browse, insert, update and delete the information through a user-friendly interface. It had inbuilt help and indexing of information to facilitate smooth navigation. For demonstration purpose two crops, viz., sugarcane and cucurbits were used. More crops could be added in PMISNET for which information on IPM is available.

7.31.18 Issues Related to Release of Genetically Modified Crops

The issues relating to the Genetically Modified Foods have generated intense public debate in many parts of the world. Even though the issues under debate include the costs and benefits of the GM crops and the inherent safety concerns, the outcome of the debate differs from country to country, depending on its geographical location, strength and resilience of the farm sector, attitudes of people towards food and so on (Anonymous, 2009).

7.32 GENETICALLY MODIFIED ORGANISMS (GMOs)

Genetically Modified Organisms are the ones in which the genetic material (DNA) has been altered in such a way as to get the required quality. This technology is often called 'gene technology,' or 'recombinant DNA technology' or 'genetic engineering' and the resulting organism is said to be 'genetically modified,' 'genetically engineered' or 'transgenic.' GM products (current or those in development) include medicines and vaccines, foods and food ingredients, feeds and fiber.

7.32.1 Genetic Engineering

All living organisms, from viruses to human beings, are made up of cells, with a nucleus at the center, which contains a unique set of instructions regarding their size, strength, and other qualities. These instructions are found on a long molecule called DNA, which is divided into small sections called genes. It is the sequencing of genes on DNA that determines an organism's characteristics. Very simple organisms such as bacteria may have fewer genes than the more complicated ones. In simple terms, the complete set of genetic material of an organism, i.e., the entire DNA contained in an organism, is called a genome. The process of isolating gene(s) from the genome of one organism and inserting the same into the genome of another organism is known as Genetic Engineering. In nature, exchange of genes happens only between compatible or closely related species. However, the modern technique of genetic engineering facilitates the removal of a group of genes from one species and insertion into another, there being no need for compatibility. The transfer process involves shifting the desired gene from the chromosome of a particular plant or animal or any other organism into a cell. This genetically modified cell is then regenerated to produce a 'genetically modified organism' (GMOs). The modified organism passes the new gene onto its progeny. Such methods are now being used to create GM plants, of desired quality, growth, and strength. The basic idea is to have plant varieties with high yield, disease resistant, or other such qualities mainly for better marketability and durability. This is different from the processes of modifying crops/plants from their wild ancestors through selective breeding or mutation breeding, which have been practiced by farmers as part of their regular farming activity.

7.32.2 Commercial Venture

The first commercially grown GM food crop was tomato (called Flavr Savr), modified to ripen without softening by a Californian company Calgene, which took the initiative to obtain approval for its release in 1994. Currently, a number of food crops such as soybean, corn, cotton, tomatoes, Hawaiian papaya, potatoes, rapeseed (canola), sugarcane, sugar beet, field corn as well as sweet corn, rice and a few others have been genetically modified to enhance either their yield or size or durability, etc. Scientists are also working on crops which they hope will be useful for industry, such as plants that produce oil for the cosmetics industry, crops with altered nutritional value, and even crops that produce pharmaceutical drugs. Major producers of transgenic crops include USA, Argentina, Brazil, India, Canada, China, Paraguay, South Africa, among others.

7.32.3 Issues of Crop Protection

The initial objective of developing GM plants was to improve crop protection. The GM crops currently in the market are mainly aimed at an increased level of crop protection through the use of one of the three basic traits, resistance to insect damage, resistance to viral infections and tolerance towards herbicides. All the genes used to modify crops so far are derived from microorganisms.

(i) Insect resistance is achieved by incorporating the gene for toxin production into the food plant from the bacterium, *Baclus thuringiensis* (Bt). This toxin is used as a conventional insecticide in agriculture and is safe for human consumption. GM crops that permanently produce this toxin have been shown to require lower quantities of insecticides.

(ii) Virus resistance is achieved through the introduction of a gene from certain viruses which cause disease in plants. Virus resistance makes plants less susceptible to diseases caused by such viruses, resulting in higher crop yields

(iii) Herbicide tolerance is achieved through the introduction of a gene from a bacterium conveying resistance to some herbicides. In situations where weed pressure is high, the use of such crops has resulted in a reduction in the quantity of the herbicides used.

7.32.4 Understanding Risks and Benefits

The risk-benefit analysis of the GM crops can be summarized as below.

7.32.4.1 Benefits

Improved resistance to diseases, pests and herbicides, improved tolerance to cold/heat, improved tolerance to drought/salinity, reduced maturation time, increased nutrients, yields, quality and stress tolerance, food with greater shelf life or food with medicinal benefits, such as edible vaccines, (for example, bananas with bacterial or rotavirus antigens), increased food security for growing population.

7.32.4.2 Issues of concern (human health risks and environmental safety concerns)

Potential impact on human health including allergens, transfer of antibiotic resistance markers and 'outcrossing.' The movement of genes from GM plants into conventional crops or related species in the wild (referred to as 'outcrossing'), as well as the mixing of crops derived from conventional seeds with those grown using GM seeds, may have an indirect effect on food safety and food security. It has been found that genes inserted into GM food survive digestive processes and are transferred to the human gut. Potential impact will be on the environment including the transfer of transgenes through cross-pollination, unknown effects on other organisms (e.g., soil microbes) and loss of flora and fauna biodiversity.

7.32.4.3 Some other concerns

Critics of genetically modified food have also pointed out certain other aspects apart from human health risks and environmental safety concerns. They include the following.

7.32.4.4 Access and intellectual property rights

Critics claim that patent laws give developers of the GM crops a dangerous degree of control over the food supply, domination of world food production by a few companies, increasing the dependence of developing countries on industrialized nations, biopiracy or foreign exploitation of natural resources.

7.32.4.5 *Ethical concerns*

Violation of natural organisms' intrinsic values by mixing among species, objections to consuming animal genes in plants.

7.32.5 Safety Assessment

The starting point for the safety assessment of genetically engineered food products is to assess if the food is 'substantially equivalent' to its natural counterpart. In deciding whether a modified product is substantially equivalent, the product is tested by the manufacturer for unexpected changes in a limited set of components such as toxins, nutrients or allergens that are present in the unmodified food. The data is then assessed by an independent regulatory body. If these tests show no significant difference between the modified and the unmodified products, then no further food safety testing is required. However, if the product has no natural equivalent, or shows significant differences from the unmodified food, then further safety testing is carried out. This method has, however, been severely criticized by some scientists since it is not clear what level of similarity makes something 'substantially equivalent.'

7.33 GENETIC ENGINEERING AND ITS APPLICATION IN AGRICULTURE

As the Food and Agriculture Organization (FAO) had rightly pointed out in 2004, "Science cannot declare any technology completely risk free. Genetically engineered crops can reduce some environmental risks associated with conventional agriculture, but will also introduce new challenges that must be addressed. Society will have to decide when and where genetic engineering is safe enough." Arguments both for and against the cultivation and use of the GM crops are varied, and there is a wide consensus that assessment should take place on a case-by-case basis before the genetically modified food is brought to the market. These assessments should be done by government or an independent, credible regulatory authority or private agencies, and these should not be driven by any commercial interests. Moreover, educating public opinion is also very important as food is always a sensitive cultural issue. Merely indicating whether a product is genetically modified or not, without providing any additional vital information, would. For instance, in India, application of biotechnology in agriculture is being dealt with by three different Ministries/Departments, viz., Ministry of Agriculture, Ministry of Environment and Forests and Department of Biotechnology, Ministry of Science and Technology. The legislative framework on agrotechnology rests mainly with the Ministry of Environment and Forests, Government of India. Under the Environment (Protection) Act, 1986, the Ministry of Environment and Forests has notified the Rules for the Manufacture, Use, Import, Export and Storage of Hazardous Micro-Organisms/Genetically Engineered Organisms or Cells, 1989, or in short, the Rules, 1989. These rules and regulations cover the areas of research as well as large scale applications of the GMOs and such products throughout India. These rules also define the competent authorities and composition of such authorities for handling of various aspects of the rules. Presently, there are six competent authorities. Institutional framework and the governing rules especially in the context of India, where the majority of the population depends on agriculture as a mainstay of livelihood, involves too many questions that do not serve any purpose, rather information on its content and possible risks or benefits should be provided.

To sum up in the words of M.S. Swaminathan, "GM foods have the potential to solve many of the world's hunger and malnutrition problems and to help protect and preserve the environment by increasing yield and reducing reliance on chemical pesticides. Yet there are many challenges ahead for governments, especially in the areas of safety testing, regulation, industrial policy and food labeling."

7.33.1 Competent Authorities Dealing with GMOs

(i) Recombinant DNA Advisory Committee (RDAC) under the Department of Biotechnology, Ministry of Science and Technology, to recommend appropriate safety regulations in recombinant research, use, and applications.

(ii) Institutional Biosafety Committees (IBSC) under the Department of Biotechnology, Ministry of Science and Technology, to prepare site-specific plans for use of genetically engineered microorganisms.

(iii) Review Committee on Genetic Manipulation (RCGM) under the Department of Biotechnology, to monitor safety-related aspects in respect of ongoing research projects and activities involving genetically engineered organisms. It lays down procedures/regulations regarding research, production, sale, import and use of genetically engineered organisms with a view to ensure environment safety.

(iv) Genetic Engineering Approval Committee (GEAC) under the Ministry of Environment and Forests, to look into approval for large-scale releases and commercialization of the GMOs.

(v) State Biotechnology Coordination Committee (SBCC) in the States wherever necessary to inspect, investigate and take punitive action in case of violations of safety and control measures in the handling of genetically engineered organisms.

(vi) District Level Committee (DLC) in the districts wherever necessary under the District Collectors to monitor safety regulations in installations engaged in the use of genetically modified organisms and their applications in the environment.

7.34 BIOSAFETY REGULATIONS FOLLOWED IN INDIA

Regulation of genetically engineered crops is extremely important to address the biosafety concerns associated with these products. As more genetically engineered plants are being developed for commercial release, concerns have been expressed about their safety. The concept of food safety assurance has assumed importance as with any method of genetic manipulation; there is a possibility of introducing unintended changes along with intended changes, which in turn have an impact on the health and nutritional status of the consumer. Unintended effects can result from the random insertion of DNA sequences into the plant genome which may cause disruption or silencing of existing genes, activation of silent genes, or modifications in the expression of existing genes. The international food code or the Codex Alimentarius has been used as a point of reference. In India, the Rules 1989, cover the areas of research as well as large scale applications of GMOs.

However, with continuous advancements in the area of research and development of GM crops, fresh guidelines have been framed from time to time. Some of these include the following:

(i) Recombinant DNA Safety Guidelines, 1990 by Department of Biotechnology covering research in biotechnology, field trials and commercial applications.

(ii) Revised Guidelines for Research in Transgenic Plants, 1998 by Department of Biotechnology.

(iii) Protocols for Food and Feed Safety Assessment of GE Crops, 2008 by Department of Biotechnology.

(iv) Guidelines for the Safety Assessment of Foods Derived from Genetically Engineered Plants, 2008 by Indian Council of Medical Research (ICMR).

(v) Guidelines and Standard Operating Procedures (SOPs) for Confined Field Trials of Regulated, Genetically Engineered (GE) Plants, 2008 by Department of Biotechnology and Ministry of Environment and Forests.

The Government of India follows a policy of case-by-case approval of transgenic crops. As per the guidelines framed by the ICMR, the safety assessment is designed to identify whether a hazard, nutritional or another safety concern is present, and if present, to collect and analyze information on its nature and severity following a structured and integrated approach performed on a case-by-case basis. Factors taken into account in the safety assessment include: identity, source, domposition, effects of processing/ cooking, transformation process, the recombinant DNA (e.g., stability of insertion, potential for gene transfer), expression product of the novel DNA, possible secondary effects from gene expression or the disruption of the host DNA or metabolic pathways, including composition of critical macro, micronutrients, antinutrients, endogenous toxicants, allergens and physiologically active substances; and potential intake and dietary impact of the introduction of the GM food.

With a view to streamline the regulatory mechanisms for biosafety evaluation and release system/protocol, the Task Force on Application of Biotechnology in Agriculture suggested as below.

Once an extant/transgene has been declared safe, its derivatives need not always be evaluated for biosafety to the same extent again. Such derivative crop varieties may be considered for biosafety clearance after case verification and need-based trial by RCGM. Studies on gene stability and expression levels will, however, have to be repeated for new varieties. For example, cotton Cry 1Ac gene has been found to be safe. Therefore, the use of this gene for improvement of other varieties in the same crop need not be subjected to the same degree of biosafety assessment (Kavitha Kuruganti and Ramanjaneyulu, 2007). Doubts have been expressed in some quarters that the case-by-case policy of approval followed by the Government is time-consuming and arduous and provides ample scope for spreading misinformation about GM crops. But the opposing view argues that genes cannot be assessed in isolation, but only as part of the genome into which they are introduced, and so testing should be done on a case-by-case basis.

7.34.1 The Case of *Bt* Cotton

The Maharashtra Hybrids Seed Company (Mahyco) jointly with the US seed company Monsanto developed the genetically modified *Bt* Cotton to tackle the bollworm problem that had devastated cotton crops in the past, by introducing into the cotton seed a gene of the common soil microbe, *Bacillus thuringiensis* that encoded an insecticidal protein lethal to the bollworm (hence the name *Bt.* Cotton). In 2002, *Bt* Cotton became the first and only transgenic crop approved by the GEAC for commercial cultivation in six States, namely, Andhra Pradesh, Gujarat, Karnataka, Madhya Pradesh, Maharashtra and Tamil Nadu. It has been further extended to Punjab and Haryana. The *Bt* Cotton seeds were marketed by the Monsanto-Mahyco joint venture.

Though the public opinion has been divided on this issue, the government has indicated the satisfactory performance of the *Bt* Cotton. As per the government figures, the area under *Bt* Cotton has increased from 0.70 lakh acres in *Kharif*–2002 to 2.30 lakh acres in *Kharif*–2003 and further increased to 12.00 lakh acres in *Kharif*–2004 (Latha Jishnu, 2009). As per the latest unofficial report (Lola Nayar, 2009), India has become the fourth largest adopter of biotech crops in 2008 with cotton alone occupying 7.6 million hectares.

On the one hand, it has been claimed as the '*Bt* Cotton Revolution' with transgenic cotton being grown in 90 percent of the cotton growing areas, increasing yields by as much as 50 percent in certain regions. However, the critics, especially various civil society groups have contested this claim. It has been argued that *Bt* Cotton cultivation has resulted in adverse economics for farmers, highly priced seeds, changed pest ecology in cotton fields, increased incidence of diseases (requiring more pesticides to control these), unpredictable crop performance and more resources being used by farmers as part of their risk insurance mechanisms (use of more irrigation, fertilizers, etc.). Stress tolerance of the *Bt* Cotton, such as surviving adverse weather conditions, has been said to be very low. There have been reports of adverse impact on soils, human health (allergic symptoms) as well as toxicity in animals grazing on the *Bt* Cotton fields. There have also been reports of large-scale contamination and rapid proliferation of various illegal varieties.

In India, brinjal is grown all over the country and is one of the most popular vegetables. It is mainly grown in small plots as a cash crop by the farmers. The main growing areas are in the States of Andhra Pradesh, Bihar, Karnataka, Maharashtra,

Orissa, Tamil Nadu, Uttar Pradesh and West Bengal. There are many local varieties in India, in addition to improved varieties and hybrids. It is estimated that the damage caused by the Fruit and Shoot Borer, which has been the major pest for the past two decades or so, ranges from 50 to 70 percent. It is to lend tolerance to this pest primarily that the *Bt* Brinjal has been developed. *Bt* Brinjal is a transgenic brinjal created out of inserting a gene [Cry 1Ac] from the soil bacterium, *Bacillus thuringiensis* into brinjal. The insertion of the gene into the brinjal cell is being done along with other genes like promoters, markers, etc.

This is said to give the brinjal plant resistance against lepidopteron insects like the brinjal fruit and shoot borer. It is reported that upon the ingestion of the *Bt* toxin by the insect, there would be disruption of digestive processes, ultimately resulting in the death of the insect. *Bt* Brinjal is being developed in India by the Maharashtra Hybrid Seeds Company (Mahyco). Recently, the GEAC has cleared *Bt* Brinjal-country's first genetically modified food for commercial use. However, amidst protests and campaigns by the anti-GM groups, the Government of India would take a final call on this matter.

7.34.2 Other GM Crops

As per the Indian GMO Research Information System (IGMORIS), the GM crops, apart from brinjal that are being currently tested are listed in Table 7.5.

Table 7.5 Important GM Crops

Sl. No.	*GM crop*	*Trait*
1.	Cabbage and Cauliflower	Insect resistance
2.	Potato	Transgenic dwarf potato Disease resistance, reduction in cold-induced sweetening and chip color improvement
3.	Cotton	Insect resistance
4.	RRF Cotton	Insect resistance and herbicide tolerance
5.	Corn	Insect resistance and herbicide tolerance
6.	Rice	Insect resistance
7.	Groundnut	Virus resistance
8.	Sorghum	Insect resistance

7.34.3 GM Crops — Opposing Views

There have been claims and counter-claims regarding growing of the GM crops and as of now with limited facts and figures, it is very difficult to analyze the risks and benefits associated with this technological breakthrough in agriculture. A glance at some of the arguments put forth by the opposing viewpoints would shed some light on this aspect (Table 7.6).

Table 7.6 'For' and 'Against' Arguments on GM Crops

For	*Against*
1. No significant difference was noted between *Bt* Brinjal and non-*BT* brinjal as per biosafety tests like acute oral toxicity, subchronic oral toxicity in rats, the allergenicity of protein to rats, germination, weediness and aggressiveness tests, soil studies, etc. According to Prof. M.S. Swaminathan (2009) "*Bt* is a naturally occurring bacterium that produces crystal proteins that are lethal only to insect larvae. *Bt* genes are lethal only in the acidic, insect gut environment and do not get activated in an alkaline environment, prevalent in humans and other animals that feed on these plants."	1. The current safety assessments are inadequate to catch most of the harmful effects of the GM crops. GM technology is unpredictable and imprecise that too when released in an open environment situation. Moreover, the regulatory regime in India with regard to the GM crops has never been assessed thoroughly with regard to the GM risk assessment in Indian conditions.
2. Human health concerns due to pesticide use can be addressed with this transgenic brinjal with its inbuilt tolerance to pests resulting in the lesser use of pesticides.	2. Several studies in *Bt* crops show that there are many potential health hazards. With *Bt* crops, allergies have been reported. Itching skin, eruptions on the body, swollen faces, etc., have been reported, correlated with levels of exposure to *Bt* Cotton. Many groups are also studying the decline in fertility and milk yield of cattle due to GM Cottonseed cattle feed. *Bt* toxin has caused powerful immune responses and abnormal cell growth in mice. Cry proteins in *Bt* crops have amino acid sequence similar to known allergens and are hence potential allergens.
3. With this inbuilt tolerance against pests in *Bt* Brinjal, there would be a substantial increase in marketable yields resulting in higher incomes for farmers. With reduced waste13, farmers would expect to rake in an additional ₹ 4000 crores annually.	3. Apprehension has been expressed that the target pest would grow resistance to the *Bt* toxin with time. Not enough studies on soil ecology have been done to understand the impact of *Bt* toxin. Farmers from various parts of the country have reported a decline in their soil productivity after growing *Bt* Cotton. The contention of the farmers' organizations meeting in October 2009 under the aegis of the Coordination Committee of Indian Farmers' Movements was that the pesticide use would not come down with the introduction of *Bt* Brinjal because Mahyco's technology would take care of only the fruit and shoot borer. There are other pests like aphids, jassids and white fly apart from problems like fungal disease. The farmers have also questioned the need for a GM Brinjal when there is surplus production leading to a glut at times.

Contd...

Contd...

4. Pricing of the seeds would be based on a cost recovery model, making it affordable for all farmers. Moreover, farmers would be able to save and reuse their seeds for the hybrids.	4. With the promotion of GM agriculture in general and with *Bt* Brinjal, in this case, the rights of non-GM farmers to stay GM-free get badly affected. Moreover, the pricing policy has also been questioned. Cost-recovery would be much higher for *Bt* Brinjal seeds due to the research and marketing involved. In addition, patent infringement is a big concern of agribusiness. Litigation on the part of the company with regard to pricing and use of GM seeds is not new.

7.34.4 Advantages of GM Foods

With the world population increasing and the amount of land available for farming decreasing, many have touted genetically modified (GM) food as the answer to combating world hunger (Anonymous, 2014). Plant genes can be altered to make crops resistant to diseases or pests, or to improve their yield. According to a study done by Nuffield Council on Bioethics, "The Use of Genetically Modified Crops in Developing Countries," farmers who plant these crops gain many advantages. "Although GM crops primarily benefit large-scale farmers, many small-scale farmers in China and South Africa have already successfully grown GM cotton. In China, yields were estimated to have increased by 10% compared to non-GM crops, and the amount of pesticide used fell by as much as 80%, leading to an increase in profits. The efficiency of agriculture has a major impact on the standard of living in most developing countries." GM foods have many other benefits as well. Other ecological benefits are that GM crops are not only pest and disease resistant, but can be modified to grow under many different conditions including cold, drought, or salinity. Food can grow in areas where it was not possible before, leading to a greater yield, which in turn can help alleviate world hunger. Another benefit is that plants can be altered to use less herbicide and fertilizer. This addresses one of the major environmental problems connected with farming. As Arcadia Biosciences explains, "Agriculture is the second-leading source of global greenhouse gas and nitrogen fertilizer represents a significant cause of these emissions." Statistics from the International Service for the Acquisition of Agri-biotech Application (ISAAA) show that from 1996 to 2006 pesticide use decreased by 224,000 tons. Thus, using GM plants can significantly reduce the ecological impact of pesticides and fertilizers.

Additional benefits of GM plants include soil, water, and energy conservation. Although initial plantings produced less food than traditional farming, future crops are being modified for greater yield. GM can also make crops tastier and add nutrients. For example, golden grain, a strain of rice enriched with beta-carotene, can help prevent blindness. Rice with iron and other vitamins and minerals can enhance the diets of the poor. Animal species developed through GM will also be healthier and hardier. Livestock can be genetically modified to increase milk or egg production, or to be meatier. Ongoing developments include adding vaccines or other valuable additives to milk.

7.34.5 Claimed Disadvantages of GM Foods

Though GM foods seem to be the ideal solution for increasing the world food supply and helping the poor, they have stimulated a great deal of controversies (Anonymous, 2014). One of the foremost concerns is their long-term effects on health, on both people and the environment.

7.34.5.1 Health problems

Although many sources point to the fifteen or so years GM foods have been used to indicate that they are safe for humans, many groups disagree. They say that sufficient testing has not been conducted. A study published in 2009 in the *International Journal of Biological Sciences* revealed adverse effects of GM corn on rats' livers and kidneys, particularly male rats. The company who markets the corn, however, refuted the results and challenged the validity of the experiment. Scientific researcher Arpad Pusztai examined the studies that had been conducted on GM foods, many of which indicated that mice developed health problems or even died during the experiments, but concluded that none of the studies were done under rigorous enough conditions to be valid. Pusztai expressed concern about the lack of properly conducted research in his article, "Genetically Modified Foods: Are They a Risk to Human/Animal Health?." "We need more and better testing methods before making GM foods available for human consumption. Our present database is woefully inadequate. Moreover, the scientific quality of what has been published is, in most instances, not up to expected standards." Conversely, Wu Yongning, a food safety specialist with the Chinese Center for Disease Control and Prevention, believes studies have not proven that genetically modified food is harmful to human health. He also said, "I am not ruling out all possible risks, but those risks of genetically-modified food are no greater than that of traditional ones, given the heavy use of pesticide in growing traditional food."

Many people, however, feel that GM food should not be put on the market until studies have shown without a doubt that it is safe for human consumption. In addition to the unknown health risks, critics list other negative effects of using GM foods. New crop species could contain allergens or toxins, or they might cause people's bodies to resist antibiotics.

7.34.5.2 Environmental concerns

These include the long-term impact of GM products on the environment is also unknown. A few difficulties have already developed. The transfer of modified genes to other nearby plants can occur through cross-pollination. GM plants that were bred for pest resistance and hardiness have passed those genes on to the weeds growing in the fields, making them more tenacious and resistant to herbicides, so additional amounts of weed-killers must be applied. This same concern could also apply to pests. Insects may soon evolve into stronger, more virulent species that are tolerant to the pesticide embedded in GM plants. Some people also decry the loss of biodiversity. As plants become more specialized through GM technologies, many regular crops will no longer be planted. Those species could eventually become extinct.

7.34.5.3 *Access issues*

Costs for testing each new GM crop costs millions of dollars, so developers patent their inventions to protect their intellectual property, which, although understandable, generally means that the technology remains in the hands of a few companies. Statistics given by OECD reveal that GM foods are dominated by a small segment of the population: "Between 1995 and 1999, 146 firms applied for at least one GM field trial. Ten years later the number declined by almost half to 76 firms that applied for a field trial between 2005 and 2009." Thus, world food production is concentrated in the hands of a few producers, leading to developing countries' increasing economic dependence on developed countries. The high cost of development and the monopoly on intellectual property also result in high prices, making seed costs prohibitive for most independent farmers, particularly those in developing countries. Thus, a technology that originally was invented to prevent world hunger is not accessible to those who need it most.

7.34.6 Ethical Questions

Another criticism leveled at GM is tampering with nature and mixing genes from different species. Some critics wonder about the violation of organisms and object to mixing animal and plant genes. They also raise the issue of what GM does to the animals themselves. Others also point to the effects on the experimental animals that are fed GM foods during testing. Along with these concerns is that of labeling GM products. The United States has no mandatory labeling of GM ingredients. Activists believe that purchasers should be informed of what they are eating, but the costs to develop standardized labeling and inspections would increase the price of food. So, although GM technology looks promising in many ways, it is also fraught with difficulties that need to be addressed, not only by America but by the global community.

7.35 BIOTECHNOLOGY AROUND THE WORLD

In 1999, many countries of the world met in Cartagena, Columbia, to adopt the Cartagena Protocol on Biosafety. The protocol was not finalized and accepted, however, until the following year in Montreal, Canada. By 2009, 157 countries had ratified the document and had begun reaching the objectives the group had identified as paramount: "ensuring the safe transfer, handling, and use of living modified organisms (LMOs) that result from modern biotechnology and that may have adverse effects on biological diversity, also taking into account risks to human health." Signers included forty-five African, forty-one Asia and Pacific, twenty-two Central and Eastern European, twenty-eight Latin American/Caribbean, and twenty-one Western European countries. Noticeably absent from the list are the United States and Canada. One of the protocol's other objectives is "to promote and facilitate public awareness, education, and participation." Meeting this requirement has proved difficult for many of the participants. The Secretariat of the Convention on Biological Diversity discussed this issue in the 2009 edition of *Biosafety Protocol News*: "Most developing countries and countries with economies in transition lack the financial resources and technical capabilities to promote public awareness and education concerning LMOs." In addition, although those countries were responsible for almost half of all GM plantings in 2008, they did not have trained scientists to provide information about possible health risks. Member countries try to rectify this inequity by offering Internet sessions and translating broadcasts into a variety of dialects and languages.

Many countries passed legislation to implement the protocol. The European Union, for example, adopted laws that require labels to indicate the presence of GM foods. Difficulties arise with this, though, especially when only trace ingredients contain GMOs. Nevertheless, the international community has forged ahead of the United States, who in 2010 still had set no regulations. Part of the problem in America is that GM foods fall under different jurisdictions. The EPA, which deals with environmental safety, supervises pesticides so that it would be responsible for the pest-resistant crops. The USDA determines whether plants are safe to grow, whereas the FDA evaluates whether products made from the plant are safe to eat. All of these divisions would need to become involved in setting standards. For now, the FDA has classified GM food as a 'substantial equivalent' of natural foods, so it is not subject to regulation.

In spite of worldwide initiatives, many questions still remain unanswered, including the long-term health effects of GMOs on humans. Although some people insist that the health risks are comparable to nongenetically modified foods, the safety of GMOs has not been assessed over an extended period of time. This testing should be of paramount concern because these products represent an important trend in the future.

7.36 FUTURE OF BIOTECHNOLOGY

As OECD stated in "The Bioeconomy to 2030," their International Futures Programme report, "GM or MAS [marker-assisted selection] varieties of soybeans and maize could be responsible for the vast majority of total plantings by 2012." OECD also predicts that by 2015 about half of all global food and livestock feed will come from plants that have been genetically modified. The OECD report continues, "Demand for food, feed, and fiber is expected to increase substantially in the future due to population and income increases across the globe. To meet increased demand, a diverse range of solutions are going to need to work in concert. Biotechnological solutions will play a major role, but will not provide a silver bullet. They will need to be combined with other strategies to modernize agricultural methods and increase agricultural productivity (e.g., through farmer education, improved water management and conservation, and precision farming)."

7.36.1 A Balanced View

Arguments are strong for both sides of the issue of GM foods as Deborah B. Whitman, senior editor of *Community Supported Agriculture* succinctly sums up in her article, "Genetically Modified Foods: Harmful or Helpful?": "Genetically-modified foods have the potential to solve many of the world's hunger and malnutrition problems, and to help protect and preserve the environment by increasing yield and reducing reliance upon chemical pesticides and herbicides. Yet there are many challenges ahead for governments, especially in the areas of safety testing, regulation, international policy, and food labeling. Many people feel that genetic engineering is the inevitable wave of the future and that we cannot afford to ignore a technology that has such enormous potential benefits. However, we must proceed with caution to avoid causing unintended harm to human health and the environment as a result of our enthusiasm for this powerful technology" (Anonymous, 2014). To sum up the claimed disadvantages, Genetically Modified Foods pose huge health risk to pregnant women and babies; food designed to produce toxin, GMOs provoke immune reactions, animals dying in large numbers; the worst finding of all-GMOs remain inside humans, warnings by government scientists ignored and denied, GM foods have not been 'properly tested' and 'pose a serious health risk.

7.36.2 Protecting Bioassets: An Emerging Challenges to Crop Biosecurity

Worldwide attempts to preserve biodiversity and the information and materials generated by the genomics revolution present a significant new challenge to governments and industry. What biological resources should be preserved? Where should they be preserved? Who should be responsible for their preservation? How can governments cooperate to enhance efficiency? How can *ex situ* collections of plant genetic resources cope with the wealth of biodiversity and the vast quantity of information and products emerging from the genomics revolution?

7.36.3 Conservation of Diversity for Crop Biosecurity

During the 1970s and 1980s, developing nations expressed concern about the free flow of plant genetic resource materials or germplasm, from the South to the North. Why, they asked, were patented seeds of southern origin bringing tremendous profits to multinational seed companies without compensation for the developing world? In the culmination of what was dubbed as the Seed Wars of the Eighties, Third World leaders managed to air their concerns in an international arena via the FAO. In 1983, FAO established the Global System for the Conservation and Utilization of Plant Genetic Resources that includes a legally nonbinding set of guidelines called the "International 'Undertaking' on Plant Genetic Resources" and an intergovernmental Commission on Plant Genetic Resources. The Commission monitors the implementation of the Undertaking and more generally discusses the use, control and conservation of plant genetic resources. It operates on the principle of one country, one vote. The purpose of the 'Undertaking' is to "ensure that plant genetic resources of economic or social interest, particularly for agriculture, will be explored, preserved, evaluated and made available for plant breeding and scientific purposes." The underlying notion is the common heritage principle – that "plant genetic resources are a heritage of mankind and consequently should be available without restriction." The principle is extended to include not only native plant materials but also farmer-developed varieties and new products of biotechnology. This guarantee of access without restriction caused eight industrialized countries to register reservations; the US and Canada still do not adhere to the 'undertaking.'

7.36.4 Challenges of Crop Biosecurity for Developing Countries

The South Asian region is one of the largest gene-rich regions of the world and equally rich in traditional and indigenous knowledge (Kauser Abdulla Malik and Yusuf Zafar, 2005). The rich socio-cultural heritage of the developing countries is evident in that the plant variety has always remained freely accessible to all since times immemorial. Forthcoming legislations concerning the conservation of crop biosecurity should encompass the following points.

(i) Regulate crop biosecurity with reference to access to the biological resources of the country for purposes of securing an equitable share of benefits arising out of the use of biological and associated knowledge relating to bioresources.

(ii) Conserve and sustain the use of biological diversity.

(iii) Respect and protect knowledge of local communities related to biodiversity.

(iv) Secure sharing of benefits with local people as conservers of biological resources and holders of traditional knowledge and information.

(v) Conserve and develop areas from the standpoint of rich bioresources.

(vi) Protect and rehabilitate endangered fauna and flora.

(vii) Ensure active participation of the private sector, NGOs and local people in the broad sense of schemes for policy implementation for crop biosecurity.

(viii) The proposed legal reforms on the subject of crop biosecurity should address issues concerning access to genetic resources associated with knowledge by foreign individuals, institutions and equitable sharing of benefits arising out of these resources and knowledge between the host countries and the local people. In this connection, exemptions should be made in the case of the following:

(ix) Free access to biological resources for intracountry uses by country's own nationals other than commercial uses.

(x) Free access by country's own citizens to use bioresources within the country for research purposes.

(xi) Ensure the plant biosecurity and protection of the farmers' right to save, use, exchange or share the farm produce of a protected variety of crop without any limitation.

(xii) Recognize the contribution of the farming community for the development of a new crop variety with due consideration to financial compensation following the commercial use of the new variety.

(xiii) A total ban on 'terminator technologies' that force the farmers to buy seeds every time they sow the crop.

(xiv) Use diversified genetic materials. To obtain more high yielding varieties concerning the legal access to bioresources capital would without doubt provide a safeguard to the interests of the local people.

Developing countries, however, can check the adverse implications on crop biosecurity by using certain provisions like Articles 6, 8 and 10, etc., which provide a mandate to chalk out strategies to protect their biodiversity. The reasoning is simple; because Third World interest cannot be protected by the Western notion of conservation and biosecurity. Any attempt at conservation and sustainable use of biodiversity would not succeed without changing the development model. The existing production pattern encourages monoculture, homogeneity, overproduction and overexploitation of nature. Therefore, the success of biodiversity conservation with the present development model is bound to be a futile exercise.

7.37 BIOPROSPECTING AND BIOPIRACY

Bioprospecting is the process of discovery and commercialization of new products based on biological resources. Despite being intuitively helpful, bioprospecting has only recently begun to incorporate indigenous knowledge in focusing screening efforts for bioactive compounds. Bioprospecting also includes biopiracy, the exploitative appropriation of indigenous forms of knowledge by commercial actors. As well as the search for previously unknown compounds in organisms that have never been used in traditional medicine before.

The Action Group on Erosion, Technology and Concentration (ETC Group) defines it as "the appropriation of the knowledge and genetic resources of farming and indigenous communities by individuals or institutions seeking exclusive monopoly control (usually patents or plant breeders' rights) over these resources and knowledge" (Kauser Abdulla Malik and Yusuf Zafar, 2005).

Biopiracy, therefore, represents the gratuitous extraction of the environmental heritage and cultural knowledge from regions of the Earth for economic exploitation and industrial monopolization (Gavin Stenton, 2003). The contemporary phenomenon of biopiracy continues to be justified by a variety of legally biased rules founded upon Western concepts and politico-economic objectives, which seek to enforce a worldwide high level and nondiscriminatory patent regime. After the successful production of commercially useful products from those organisms, the company applies for a patent in its own name on those products. In most cases, the inventor not even acknowledges in his patent application that his product was derived from information provided by a local community. Biopiracy, therefore, can be described as the unjustified extraction of the environmental heritage and traditional knowledge from various regions of the earth for economic exploitation and industrial monopolization (Sayan Bhattacharya, 2014).

The following have been described as 'biopiracy':

(i) Granting of 'wrong' patents for inventions that are neither novel nor inventive having regard to traditional knowledge already in the public domain. Such patents may have been granted due either to oversights during a patent examination or simply because the patent examiner did not have access to the knowledge. This anomaly may be because the knowledge is written down but not accessible using the tools available to the examiner, or because it is unwritten knowledge. A WIPO initiative to document and classify traditional knowledge seeks to address some of these problems was made in 2001.

(ii) Granting of 'right' patents in accordance with the national law on inventions derived from a community's traditional knowledge or genetic resources. This 'biopiracy' results since: Patenting standards are too low for inventions that amount to little more than discoveries. Alternatively, the national patent regime (for example, in the US) may not recognize some forms of public disclosure of traditional knowledge as prior art. Even if the patent represents a genuine invention, however, defined, no arrangements may have been made to obtain the prior informed consent (PIC) of the communities providing the knowledge or resource, and for sharing the benefits of commercialization to reward them appropriately in accordance with the principles of the CBD.

7.37.1 Categories of Biopiracy

There are three different categories of biopiracy (Sayan Bhattacharya, 2014).

(i) Patent-based biopiracy: The patenting of (often spurious) inventions based on biological resources and/or traditional knowledge that are extracted without adequate authorization and benefit-sharing from other (usually developing) countries, indigenous or local communities.

(ii) Nonpatent biopiracy: Other intellectual property control (through plant variety protection or deceptive trademarks) based on biological resources and/or traditional knowledge that have been extracted without adequate authorization and benefit-sharing from other (usually developing) countries, indigenous or local communities

(iii) Misappropriation: The unauthorized extraction of biological resources and/or traditional knowledge from other (usually developing) countries, indigenous or local communities, without adequate benefit-sharing.

7.38 CONVENTION ON BIOLOGICAL DIVERSITY (CBD)

The CBD came into force in 1993. It secured rights to control access to genetic resources for the countries in which those resources are located. One objective of the CBD is to enable lesser-developed countries to benefit better from their resources and traditional knowledge. Under the rules of the CBD, bioprospectors are required to obtain informed consent to access such resources, and must share any benefits with the biodiversity-rich country (Cluis Corinne, 2013). However, some critics believe that the CBD has failed to establish appropriate regulations to prevent biopiracy. Others claim that the main problem is the failure of national governments to pass appropriate laws implementing the provisions of the CBD. The Nagoya Protocol to the CBD (negotiated in 2010, expected to come into force in 2014) will provide further regulations. The CBD has been ratified by all countries in the world except for Andorra, Holy See, and the United States. The 1994 Agreement on Trade-Related Aspects of Intellectual Property Rights (TRIPs) and the 2001 International Treaty on Plant Genetic Resources for Food and Agriculture are further relevant international agreements.

Genetic diversity is the key to life on this planet. Biodiversity on earth is necessary for the survival of humanity. The conservation and utilization of plant genetic resources is vital to improving agricultural productivity and sustainability (Anonymous, 1998). Organisms evolve by natural selection, and variation exists within a population. Loss of this variation leads to the inevitable extinction of species unable to evolve adaptations to the environment. This process has occurred a countless number of times on this earth. However, the rate of extinction has increased by our interference in natural processes. Push for productivity kills off world's hardiest breeds of farm animals. The process of gene hunting and their exploitation is referred to as 'bioprospecting': Relatively common or inexpensive raw materials are transformed through detailed scientific work and high technology into highly refined commodities and desperately needed medicines.

Endophytic microorganisms are to be found in virtually every plant on earth. These organisms reside in the living tissues of the host plant and do so in a variety of relationships, ranging from symbiotic to slightly pathogenic. Because of what appears to be their contribution to the host plant, the endophytes may produce a plethora of substances of potential use to modern medicine, agriculture, and industry (Gary Strobel and Bryn daisy, 2003). Novel antibiotics, antimycotics, immunosuppressants and anticancer compounds are only a few examples of what has been found after the isolation, culture, purification, and characterization of some choice endophytes in the recent past. The potential prospects of finding new drugs that may be effective candidates for treating newly developing diseases in humans, plants, and animals are great. Entophytic actinomycetes are now being tested and seriously considered for use in controlling plant diseases. The ecological and physiological explanation for the wide distribution of fungi that make paclitaxel seems to be related to the fact that paclitaxel is a fungicide and the organisms with the most sensitivity to it are plant pathogens such as *Pythium* spp. and *Phytophthora* spp. These pythiaceous organisms are some of the world's most important plant pathogens and are strong competitors with endophytic fungi for niches within plants. In fact, their sensitivity to paclitaxel is based on their interaction with tubulin in a manner identical to that in rapidly dividing human cancer cells. Thus, bonafide endophytes may be producing paclitaxel to protect their respective host plant from degradation and disease caused by these pathogens.

7.38.1 Biopiracy and its Economic Implications

Over the last few decades, new developments in science, biotechnology and intellectual property rights (IPR) regimes have come together to fuel a (re)turn to nature as a site for cosmetic, pharmaceutical, chemical and agricultural discovery work. In the early 1990s, this particular configuration of actors and regulatory regimes fuelled millennial claims about the possibilities inherent in such 'chemical prospecting' and the potential for massive benefits that could be garnered from the genetic wealth of areas with a high concentration of biodiversity (Chris Hamilton, 2006). At the time, one of the most promising strategies employed to identify how this material and the attendant knowledge of its uses could be collected and used was a practice called 'bioprospecting.' The term is now generally used generically for any program that endeavors to collect genetic material and/or the knowledge of its uses, usually from areas with high concentrations of biodiversity.

Some developing countries are unhappy about the activities of biotechnology firms from advanced countries. These companies are searching for plants that give improved crop yields or contain substances of pharmaceutical value in a process called as bioprospecting. The complaint is not against making use of these genetic resources for the benefit of mankind, but the complaint is against such bioprospecting by these firms 'without permission,' 'without payment' and even 'without acknowledgment' of the original owners of these resources (Anonymous, 1998). There are many disputes and accusations. For example in 1997, attempts by an Australian Government Agency to patent chickpea obtained from an international gene bank, at Hyderabad in India. Also, Brazil is fighting to control companies prospecting of useful genetic material in Amazonian rainforests. Biotech patents spawn biopiracy of genetic resource of the third world. The recent explosive growth in biotech patents in drugs, insecticides, and agro-products in advanced countries is based on the genetic resources of the third world. Several examples of biopiracy can be cited. Genebanks in the affluent north are another highway for biopiracy. Exports of developing countries are going to be greatly affected by biopiracy. FAO is well aware of the conflicts. UNEP also has a role to arrive at an agreed protocol on biosafety.

Another example is the case of the then-new, high-yield rice varieties that were protected from grassy stunt virus during the 1970s by breeding with a wild species of rice from India (Plucknett *et al.*, 1987). Because of many similar successes, the collection and protection of crop germplasm for use in crop breeding remains a very high priority worldwide.

7.38.2 Raising of Voice Against Biopiracy

The issue of biopiracy mostly raised by underdeveloped biodiversity rich countries (e.g., Pakistan, India, Brazil, and Malaysia, among others) and by some NGOs (e.g., GRAIN, ETC (previously RAFI) and the Third World Network) is not always acknowledged by corporations and governments. According to some, biopiracy is not only happening at the expense of Third World farmers but also occurs at the expense of Northern farmers. Some American and European farmers consider themselves to be compelled with contracts limiting their independence. Some companies argue that they are losing millions of dollars per year because of lack of respect of patents and that underdeveloped countries are themselves guilty of piracy believing that the Southern countries do not have adequate and efficient intellectual property protection laws. These companies have been applying pressure for the strengthening of IP issues within the WTO. Companies say access to biological resources allows them to develop new products that could help solve food and health essential issues and argue that research, development and commercialization authorizations have a cost that must be balanced by the protection of the resulting product. Patents offer this much-needed revenue and favor innovation.

One of the solutions suggested to solve this North-South disagreement was to institute bilateral bioprospecting contracts between source-country and pharmaceutical or seed companies which define and lay down the rules of benefits sharing, and that can potentially bring substantial royalties to Southern countries (Kauser Abdulla Malik and Yusuf Zafar, 2005). Such agreements can result in high potential benefits for the source-country. However, there are several reasons why this usually does not happen since bilateral contracts are not always respected as they do not propose a fair trade. By admitting that the principle of compensation of the populations is retained, there remains the issue of which and what amount could be an equitable and realistic remuneration? How could the rights be redistributed? Other considerations include lack of awareness of the potential value of the products - very few of the samples collected actually lead to a new profitable product; lost ownership in the case of genetic modification — majority of concerned species to be found in several countries at the same time, thus preventing some of them from taking advantage of the product, or diluting the benefits for all and protection of collective knowledge doesn't fit within the legal systems of IPR protection (e.g., patents, copyrights, and trademarks).

Options considered by Southern countries include the documentation of traditional knowledge, registration and innovation system, easier and less expensive patenting system, development of a sui generis system, development of own research and creation of alliances of source-countries.

The implications for the source-country are as follows:

- The possible necessity for the farmers to use the new patented variety, implying increasing dependence on the company owning the seed, especially in the case of monopoly.
- Prohibition of the use of the seed for any further breeding.
- Increased dependence on developed countries for the product.
- Loss of biodiversity resulting from increased monoculture, monospecies culture and lack of access to seed stocks.
- Possible financial loss when fair agreement on benefits sharing between the source country and the company was not achieved.
- Possible loss of traditional community knowledge, with limited to nil compensation.

The concept of biopiracy assumes that it is a natural right to own plants, animals, and human genes. Some do not follow this principle and consider that users from all over the world need to be free to manufacture drugs, free to cultivate and to raise plants and animals. For many indigenous people, nature and culture are indissociable. Resources belong to the community, private property has no meaning. They argue that what is 'wrong' is not so much the appropriation of somebody else's property, but rather to consider as private, natural resources that should stay public.

A search of the World Intellectual Property Organization's Patentscope database was performed to identify claims on tomato (*Solanum lycopersicum*) filed since 2007. Eight patent applications were selected for further investigation (Edward Hammond, 2013). In seven of these cases, the materials claimed come from a wild relative of the cultivated tomato obtained from a seed bank. Genes from these related species can be introduced into commercial tomato varieties and, in some cases, other organisms. In the eighth case, the patent application contains insufficient data to determine the origin of genes claimed. The expansion of industrial tomato farming has been accompanied by a rise in patent claims over tomato traits and genes which include claims over seedless tomatoes, disease resistance, growth habits, higher yields and harder fruit (a desirable trait in the industry). Other claims cover tomato genes that yield precursor molecules for the pharmaceutical and chemical industries.

A company has claimed tomato genes and plants resistant to *Botrytis,* a mold infecting many fruits and vegetables (Edward Hammond, 2013). The resistance comes from *Solanum habrochaites,* another wild tomato relative. Monsanto's source for the resistance is LYC 4,16 a seed from the IPK Gatersleben seed bank in Germany. Although IPK Gatersleben's seed database lists LYC 4 as being of unknown origin, native populations of *S. habrochaites* are found in Peru and Ecuador.

7.39 THREAT OF BIOTERRORISM TO AGRICULTURAL CROP BIOSECURITY

An agroterrorism attack could result in any of the two forms, namely, agricultural biowarfare and bioterrorism. (Kauser Abdulla Malik and Yusuf Zafar, 2005) Undue agri-technology protection leading to adverse effects on the food security of a country should also be considered as a form of agricultural biowarfare. This may cause a loss to crop biosecurity in the following ways:

- Reduced production of food (although unlikely to be an issue in a rich country, starvation could occur in developing/poor countries with limited resources).
- Dramatic economic loss within the affected agricultural sector or sectors (through direct loss of crops or animals, the cost of containment activities, or reduced domestic demand).
- Export embargoes that would remove agricultural products from the global market. – Destabilization of related economic markets (such as revenue losses to shippers, processors, exporters, and others).
- Creation of social instability causing the local population to lose confidence in the safety of the food supply and by inciting fear and a sense of vulnerability.
- A bioterrorist attack on agricultural targets has been considered by some to be a 'high consequence–high probability event' 23 for the following reasons:
- The technological barriers to obtaining and weaponizing agricultural pathogens are relatively minimal.
- Many crop and animal pathogens can be isolated from the environment or obtained from laboratories without substantial difficulty.
- An attack on crops or livestock could be carried out relatively easily without sophisticated equipment or expertise.

- Only a small quantity of the affecting agent would be needed since many of the agents of concern are highly transmissible between animals or, for plant diseases, via the air.
- Crops are openly exposed and relatively vulnerable to attack.
- Livestock and poultry often are raised under conditions involving high concentrations of animals.

A biological attack on the agricultural sector is likely to be covert. Such attacks will be options for perpetrators only to the extent that they are able to maintain the plausibility that such events are natural events. Increased epidemiological capacity, especially in strain identification from molecular sequence data, makes it increasingly difficult to escape detection, and thus would act as a substantial deterrent.

7.39.1 Policy and Infrastructure Requirements for Crop Biosecurity

Policy and infrastructure requirements are needed to ensure that biotechnology will benefit the communities (Kauser Abdulla Malik Yusuf Zafar, 2005) such as, generate, adapt, and/or negotiate access to biotechnology innovations for crop biosecurity, generate good quality animal and plant germplasm where biotechnology can be used, generate capacity and capability to identify and prioritize critical problems affecting biosecurity concerning biotechnology, development of a technology and information delivery system, development of a science-based transparent biosecurity regulatory system and capability of the public sector and the IARCs to negotiate, promote and deliver private-public partnerships in an environment where biotechnologies for crop biosecurity can be considered public goods.

Public and private sector research institutions need to emphasize biotechnology innovations that will express the following traits.

- Pest resistance that offers benefits to farmers in need of genetic control mechanisms where cultural practices are not effective and where a reduction in pesticides is advantageous leading to the need for more research to assess the sustainability of resistance expressed in transgenic pest-protected plants.
- Improved yield through isolation of dwarfing genes originally used to increase yields of cereals during the green revolution has now been shown to have the same effect on the potential to increase yield in other crops.
- Tolerance to biotic and abiotic stresses such as genetic control of the rice yellow mottle virus is one example of how transgenics can accomplish resistance when transgenics work to extend such benefits to smallholder farming communities.
- Nutritional benefits such as increasing nutritional elements of many plant varieties such as transgenic research that enhances vitamin A content and elevated iron levels for developing countries.
- Reduced environmental impact – producing crops that tolerate stressful conditions, by introducing GM traits that control root diseases and that will help farmers cultivate crops where reduced tillage is essential. Conventional approaches failed to do so illustrating the need for public funded research.

8 Intellectual Property Rights (IPRs) and World Trade Organization (WTO) in Relation to Phytopathology

8.1 INTELLECTUAL PROPERTY RIGHTs (IPRs)

Intellectual Property Rights are the legally recognized exclusive rights to creations of the mind. Under intellectual property law, owners are granted certain exclusive rights to a variety of intangible assets, such as musical, literary and artistic works, discoveries and inventions and words, phrases, symbols, and designs. Common types of intellectual property rights include copyright, trademarks, patents, industrial design rights, trade dress and in some jurisdictions trade secrets. Although many of the legal principles governing intellectual property rights have evolved over centuries, it was not until the 19th century that the term intellectual property began to be used, and not until the late 20th century that it became commonplace in the majority of the world. The British Statute of Anne (1710) and the Statute of Monopolies (1624) are now seen as the origins of copyright and patent law respectively. Intellectual property rights (IPRs) are important internationally, in public, nonprofit and private institutions, both as incentives for innovation and as constraints on freedom to operate in agricultural research and development.

The challenge of securing the world's food supply calls for further research and incentives to develop innovative agricultural solutions. The revenue potential of intellectual property (IP) rights is a key driver of innovation. In the sphere of agriculture, patent law, plant variety protection rights (breeder's rights) and rights over genetic resources are particularly relevant. To ensure global food security, agricultural innovations need to be affordable, and farmers need an incentive to adopt them – in sum, the economic benefit of using these technologies needs to outweigh their cost. While some commentators argue that this does not necessarily translate into higher prices per unit of farm produce, it seems clear that if farmers' incomes do not rise in real terms then they will be unable to pay for the new technologies and new varieties required to boost agricultural productivity. If farm-gate prices stagnate the question of access to these technologies is likely to become the subject of hot public debate (Jelle Bruinsma, 2009). Rising levels of food insecurity are likely to intensify debates about the patenting of seeds and fuel calls for compulsory licensing provisions akin to those established to deal with public health crises. Similar debates may also ensue in relation to products that protect plants against pests and disease.

Any debate on seeds must consider the UPOV system and its flexibilities which reside in exceptions to the breeder's right. The exception to the breeder's right to use protected varieties to breed new varieties without the authorization of the right holder accelerates breeding and innovation. The so-called 'farmers' privilege' which UPOV members have the option to introduce into their national legislation can, however, be a double-edged sword. While it sounds reasonable that a small farmer should be able to use seeds produced on his or her own farm without paying a license fee, excessive use of this exception can have serious implications for plant breeders and their ability to develop locally adapted varieties. Finding the appropriate balance is a thorny but necessary question for agricultural policy-makers. The provisions of the International Treaty on Plant Genetic Resources for Food and Agriculture (ITPGR) introduce an important perspective to this question. They seek to establish a concrete balance between access to biodiversity for incremental innovation and benefit-sharing to reward farmers for on-farm conservation and management of such biodiversity. The principal aim of the ITPGR is to "facilitate the exchange of seeds and other germplasm for research, breeding, and crop development. It essentially creates a multilateral gene pool. Those creating commercial products that incorporate its plant genetic resources must pay a percentage of their profits into a fund used to promote conservation and sustainable use of plant genetic resources except when such a product is available without restriction to others for further research and breeding (e.g., plant varieties protected according to the UPOV system). In such a case, a voluntary payment is encouraged. The financial viability of the system hinges on the ability of private parties to be able to create and commercialize derivative products using the bank's materials. In sum, the treaty seeks to manage the intellectual property associated with a defined set of genetic resources resulting from a combination of collective and individual innovation to conserve a public good.

The importance of offering incentives to develop the innovative new technologies that will enable us to meet the challenge of food security in a context of climate change and rapid population growth cannot be overstated. The intellectual property system will, without a doubt, have a key role to play in providing the incentives to foster the innovation required if we are to meet this challenge.

Modern usage of the term intellectual property goes back at least as far as 1867 with the founding of the North German Confederation whose constitution granted legislative power over the protection of intellectual property (*Schutz des geistigen Eigentums*) to the confederation. When the administrative secretariats established by the Paris Convention (1883) and the Berne Convention (1886) merged in 1893, they located in Berne, and also adopted the term intellectual property in their new combined title, the United International Bureau for the protection of intellectual property. The organization subsequently relocated to Geneva in 1960 and was succeeded in 1967 with the establishment of the World Intellectual Property Organization (WIPO) by treaty as an agency of the United Nations. According to Lemley, it was only at this point that the term really began to be used in the United States (which had not been a party to the Berne Convention) and it did not enter popular usage until passage of the Bayh-Dole Act in 1980.

"The history of patents does not begin with inventions, but rather with royal grants by Queen Elizabeth (1558–1603) for monopoly privileges. Approximately 200 years after the end of Elizabeth's reign, however, a patent represents a legal [right] obtained by an inventor providing for exclusive control over the production and sale of his mechanical or scientific invention... [demonstrating] the evolution of patents from royal prerogative to common-law doctrine."

The term intellectual property can be found used in an October 1845 Massachusetts Circuit Court ruling in the patent case Davoll *et al.* vs. Brown., in which Justice Charles L. Woodbury wrote, "...only in this way can we protect intellectual property, the labors of the mind, productions and interests are as much a man's own...as the wheat he cultivates, or the flocks he rears." The statement that "discoveries are...property" goes back earlier. Section 1 of the French law of 1791 stated, "All new discoveries are the property of the author; to assure the inventor the property and temporary enjoyment of his discovery, there shall be delivered to him a patent for five, ten or fifteen years." In Europe, French author A. Nion mentioned *propriété intellectuelle* in his *Droits civils des auteurs, artistes et inventeurs*, published in 1846.

Until recently, the purpose of intellectual property law was to give as little protection possible to encourage innovation. Historically, therefore, they were granted only when they were necessary to encourage invention, limited in time and scope. The concept's origins can potentially be traced back further. Jewish Law includes several considerations whose effects are similar to those of modern intellectual property laws though the notion of intellectual creations as property does not seem to exist – notably the principle of Hasagat Ge'vul (unfair encroachment) was used to justify limited-term publisher (but not the author) copyright in the 16th century. In 500 BCE, the government of the Greek state of Sybaris offered one year's patent "to all who should discover any new refinement in luxury."

8.1.1 Forms of IPR

There are different forms of IPR that play a key role in crop biosecurity. Common types of intellectual property rights include patents, copyright, industrial design, trademarks, trade dress, and in some jurisdictions trade secrets (Kauser Abdulla Malik and Yusuf Zafar, 2005). There are also more specialized varieties of *sui generis* exclusive rights, such as circuit design rights (called mask work rights in the US law, protected under the Intellectual Circuit Topography Act in Canadian law and in European Union Law by Directive 87/54/EEC of 16 December 1986 on the legal protection of topographies of semiconductor products), plant breeders' rights, plant variety rights, industrial design rights, supplementary protection certificates for pharmaceutical products and database rights (in European law).

8.1.2 Patent

A patent is an exclusive right granted to an inventor. Once issued, a patent gives the inventor the legal right to create a limited monopoly by excluding others from creating, producing, selling or importing the invention. This right is of limited duration, for a minimum period of 20 years from the date of filing the patent application. In exchange for the right of exclusion, the inventor must disclose all details describing the invention, so that when the 20-year patent right expires, the public may have the opportunity to develop and profit from the use of the invention. The patent system has many objectives. It aims to protect inventors; promotes the disclosure of inventions, as against secrecy, through the publication of patent applications; and to stimulate others to 'invent around' patents in devising new solutions to technical problems. Therefore, insofar as a patent gives its holder the exclusive right to benefit from his/her particular solution, others may be induced to find alternative solutions which can be used without infringing the patent in question.

The granting of a patent is subject to the fulfillment of the three conditions, viz., usefulness or industrial application, newness or novelty, in the sense that the invention was not previously known to the public and nonobviousness, or inventive step so that the invention constitutes an acknowledged extension of prior knowledge. The right conferred by the patent grant is, in the language of the statute and of the grant itself, "the right to exclude others from making, using, offering for sale, or selling" the invention in the United States or 'importing' the invention into the United States. What is granted is not the right to make, use, offer for sale, sell or import, but the right to exclude others from making, using, offering for sale, selling or importing the invention (Anonymous, 2014).

There are three types of patents:

1. **Utility patents:** These patents may be granted to anyone who invents or discovers any new and useful process, machine, article of manufacture, or composition of matter, or any new and useful improvement thereof;
2. **Design patents:** They may be granted to anyone who invents a new, original, and ornamental design for an article of manufacture; and
3. **Plant patents:** These patents may be granted to anyone who invents or discovers and asexually reproduces any distinct and new variety of plant.

8.1.3 Forms of Protection

IPRs can be defined as the rights given to people over the creation of their minds. They usually give the creator an exclusive right over the use of his/her creation for a certain period of time. Intellectual property includes patents, copyrights, trademarks, geographical indications, industrial designs, integrated circuits and trade secrets (Chawla and Singh, 2005). The protection of IPRs is binding and legally enforceable (Chawla, 2010). A patent is a government-granted exclusive right to an inventor to prevent others from practicing, i.e., making, using or selling the invention. A patent is personal property, which can be licensed or sold like any other property. The purpose of a patent is to encourage and develop new innovations. The Patent Law recognizes the exclusive right of a patentee to gain commercial advantage out of his invention. There are three criteria to issue a patent for the innovation.

(i) **Novelty:** The inventor must establish that the invention is new or novel. The novelty requirement refers to the prior existence of an invention. If an invention is identical to an already patented invention, the novelty requirement is not met, so a patent cannot be issued.

(ii) **Inventiveness (Non-obviousness):** It is an invention and not merely a discovery. It is nonobvious to one skilled in the field. The non-obvious requirement refers to the level of difficulty required to invent the technology. If an invention is so obvious that anyone having ordinary skill would have thought of it, then it does not meet this requirement.

(iii) **Usefulness (Industrial application):** It has a utility or is useful for the society. The useful requirement refers to the practical use of the invention. If an invention provides a product that is required or needed in some manner, then it meets this requirement. In the patent, adequate disclosure should be made so that others can also work on it. It should have the features like be a written description, enables other persons to follow, adequate and deposit mechanism (Chawla, 2010). The present law, Patents Act 1970, Amendment 2005 is effective from January 1, 2005.

8.1.4 Trademark or Service Mark

A trademark is a word, name, symbol, or device that is used in trade with goods to indicate the source of the goods and to distinguish them from the goods of others. A service mark is the same as a trademark except that it identifies and distinguishes the source of a service rather than a product. The terms 'trademark' and 'mark' are commonly used to refer to both trademarks and service marks.

Trademark rights may be used to prevent others from using a confusingly similar mark, but not to prevent others from making the same goods or from selling the same goods or services under a clearly different mark. Trademarks that are used in interstate or foreign commerce may be registered with the USPTO. The registration procedure for trademarks and general information concerning trademarks can be found in the separate book entitled *Basic Facts about Trademarks.*

8.1.5 Patent Laws

The Constitution of the United States gives Congress the power to enact laws relating to patents, in Article I, Section 8, which reads, "Congress shall have power... to promote the progress of science and useful arts, by securing for limited times to authors and inventors the exclusive right to their respective writings and discoveries." Under this power, Congress has from time to time enacted various laws relating to patents. The first patent law was enacted in 1790. The patent laws underwent a general revision which was enacted July 19, 1952, and which came into effect January 1, 1953. It is codified in Title 35, United States Code. Additionally, on November 29, 1999, Congress enacted the American Inventors Protection Act of 1999 (AIPA), which further revised the patent laws (Anonymous, 2014).

The patent law specifies the subject matter for which a patent may be obtained and the conditions for patentability. The law establishes the United States Patent and Trademark Office to administer the law relating to the granting of patents and contains various other provisions relating to patents.

8.1.6 International Convention for Protection of New Varieties of Plants (UPOV)

The UPOV Convention was signed in 1961; entered into force in 1968; and was then revised several times, in 1972, 1978 and 1991. This Convention allows countries to protect plant variety by patents. Plant Breeders' Rights (PBRs, or Plant Variety Protection (PVP): The PBRs allows for the protection of new plant varieties for a term of 20 years (25 for tree crops). A country can develop its own system of protection, referred to as a sui generis system, i.e., a system of rights designed to fit a particular context and need that is a unique alternative to standard patent protection (Kauser Abdulla Malik and Yusuf Zafar, 2005).

8.1.7 Geographical Indications (GIs)

Geographical indications (GIs) GIs identify the specific geographical origin of a product, and the associated qualities, reputation or other characteristics and usually consist of the name of the place of origin. For example, food products sometimes have qualities that derive from their place of production and local environmental factors. The geographical indication prevents unauthorized parties from using a protected GI for products not from that region or from misleading the public as to the true origin of the product.

8.1.8 Database Protection

The EU has adopted legislation to provide *sui generis* protection in respect of databases, preventing unauthorized use of data compilations even if nonoriginal. Exclusive rights to extract or utilize all or a substantial part of the contents of the protected database are granted. Over the last two decades there has been an unprecedented increase in the level, scope, territorial extent and role of intellectual property (IP) protection in crop biosafety and biosecurity which include the following trends:

(i) The patenting of living things and materials found in nature, as opposed to man-made products and processes more readily recognizable to the layman as inventions.

(ii) The modification of protection regimes to accommodate new technologies (particularly, biotechnology and information technology).

(iii) A new emphasis on the protection of new knowledge and technologies produced in the public sector.

(iv) The focus on the relationship between IP protection and traditional knowledge, folklore and genetic resources.

(v) The geographical extension of minimum standards for IP protection through the TRIPS agreement and of higher standards through bilateral and regional trade and investment agreements; and

(vi) The widening of exclusive rights, extension of the duration of protection, and strengthening of enforcement mechanisms.

8.1.9 Copyright

Copyright is a form of protection provided to the authors of 'original works of authorship' including literary, dramatic, musical, artistic and certain other intellectual works, both published and unpublished. The 1976 Copyright Act generally gives the owner of copyright the exclusive right to reproduce the copyrighted work, to prepare derivative works, to distribute copies or phonorecords of the copyrighted work, to perform the copyrighted work publicly, or to display the copyrighted work publicly (Anonymous, 2014).

The copyright protects the form of expression rather than the subject matter of the writing. For example, a description of a machine could be copyrighted, but this would only prevent others from copying the description; it would not prevent others from writing a description of their own or from making and using the machine. Copyrights are registered by the *Copyright Office of the Library of Congress.*

8.1.10 Industrial Design Rights

An industrial design right protects the visual design of objects that are not purely utilitarian. An industrial design consists of the creation of a shape, configuration or composition of pattern or color, or a combination of pattern and color in a three-dimensional form containing an aesthetic value. An industrial design can be a two- or three-dimensional pattern used to produce a product, industrial commodity or handicraft.

8.1.11 Trade Dress

It is a legal term of art that generally refers to characteristics of the visual appearance of a product or its packaging (or even the design of a building) that signify the source of the product to consumers.

8.1.12 Trade Secrets

A trade secret is a formula, practice, process, design, instrument, pattern or compilation of information which is not generally known or reasonably ascertainable, by which a business can obtain an economic advantage over competitors or customers. In the United States, trade secret law is primarily handled at the state level under the Uniform Trade Secrets Act, which most states have adopted, and federal law, the Economic Espionage Act of 1996, which makes the theft or misappropriation of a trade secret a federal crime. This law contains two provisions criminalizing two sorts of activity. The first, 18 U.S.C.

§ 1831(a), criminalizes the theft of trade secrets to benefit foreign powers. The second, 18 U.S.C. § 1832 criminalizes their theft for commercial or economic purposes. (The statutory penalties are different for the two offenses.) Trade secret law varies from country to country.

These consist of commercially valuable information about production methods, business plans, clientele, etc. They are protected as long as they remain secret by laws which prevent acquisition by commercially unfair means and unauthorized disclosure.

8.1.13 Licenses

Licenses are formal contracts between a technology owner and a licensee extending permission to use technology covered by any form of property right according to the terms and conditions of the contract. Permission may be granted on an exclusive or nonexclusive basis, and the conditions of the contract may define or restrict the scope of use allowed by the patent monopoly. Payment for licenses can take many forms, from an up-front lump sum to a running royalty, dependent on value or volume of production. US courts have held that royalties and patent fees cannot be collected beyond the life of a patent, or once a patent has been declared invalid. Licenses might be restricted to research uses, or permit commercial use. Contrary to popular perceptions, few countries have liberal legal exemptions for use of patented technology in research, even research done by public or nonprofit agencies.

Internationally, patent licenses are often traded in combination with 'know-how' contracts, generating royalties and fees that are major sources of revenue flows for technology providers.

8.1.14 Material Transfer Agreement (MTA)

An MTA is a contract for transfer and use of an input to the research process, either for basic research or commercial use. The transferred material must have some independent means of intellectual property protection (for example, a patent or trade secrecy), or be a defined piece of tangible property, to provide a basis for preventing its appropriation by third parties. MTAs are often means of transferring material under trade secret protection. They may restrict the user's rights to modify, improve, resell or commercialize the biological material. MTAs are being used by the research centers of the Consultative Group on International Agricultural Research (CGIAR) to control the use of plant varieties held 'in trust' on behalf of the countries of origin in their gene banks (Binenbaum *et al.*, 2003). If access to the materials is not otherwise available, this protection may be effective in preserving the provider's rights over the germplasm. MTAs typically prohibit transfer of material to third parties, although recipients do not always pay attention to these restrictions.

8.1.15 Bag Label Contract

In selling seeds in developed countries, a contract is often described on a seed bag label. Users of the product are presumed to have agreed to comply with the contract if they break the label to open the bag. Thus, a bag label contract seems analogous to the 'shrink-wrap' contract common in software transactions. If the use of the seed for breeding is found to violate such a contract, then bag label contracts are an additional means of protection of intellectual property. The strength of such contracts is not yet firmly established in case law.

The patent law specifies the general field of subject matter that can be patented and the conditions under which a patent may be obtained. In the language of the statute, any person who "invents or discovers any new and useful process, machine, manufacture, or composition of matter, or any new and useful improvement thereof, may obtain a patent," subject to the conditions and requirements of the law. The word 'process' is defined by law as a process, act, or method, and primarily includes industrial or technical processes (Anonymous, 2014.). The term 'machine' used in the statute needs no explanation. The term 'manufacture' refers to articles that are made and includes all manufactured articles. The term 'composition of matter' relates to chemical compositions and may include mixtures of ingredients as well as new chemical compounds. These classes of subject matter taken together include practically everything that is made by man and the processes for making the products.

The Atomic Energy Act of 1954 excludes the patenting of inventions useful solely in the utilization of special nuclear material or atomic energy in an atomic weapon. See 42 U.S.C. 2181(a).

The patent law specifies that the subject matter must be 'useful.' The term 'useful' in this connection refers to the condition that the subject matter has a useful purpose and also includes operativeness, that is, a machine which will not operate to perform the intended purpose would not be called useful, and, therefore, would not be granted a patent. Interpretations of the statute by the courts have defined the limits of the field of subject matter that can be patented. Thus it has been held that the laws of nature, physical phenomena, and abstract ideas are not patentable subject matter.

A patent cannot be obtained from a mere idea or suggestion. The patent is granted upon the new machine, manufacture, etc., as has been said, and not upon the idea or suggestion of the new machine. A complete description of the actual machine or another subject matter for which a patent is sought is required.

8.1.16 Conditions for Obtaining Patent

For an invention to be patentable it must be new as defined in the patent law, which provides that an invention cannot be patented if,

(i) The claimed invention was patented, described in a printed publication, or in public use, on sale, or otherwise available to the public before the effective filing date of the claimed invention or

(ii) The claimed invention was described in a patent issued [by the US] or in an application for patent published or deemed published [by the US], in which the patent or application, as the case may be, names another inventor and was effectively filed before the effective filing date of the claimed invention.

There are certain limited patent law exceptions to patent prohibitions (i) and (ii) above. Notably, an exception may apply to a "disclosure made one year or less before the effective filing date of the claimed invention," but only if "the disclosure was made by the inventor or joint inventor or by another who obtained the subject matter disclosed… from the inventor or a joint inventor" (Anonymous, 2014).

In patent prohibition (i), the term 'otherwise available to the public' refers to other types of disclosures of the claimed invention such as, for example, an oral presentation at a scientific meeting, a demonstration at a trade show, a lecture or speech, a statement made on a radio talk show, a YouTube™ video, or a website or other on-line material. The effective filing date of the claimed invention: This term appears in patent prohibitions (i) and (ii). For an US nonprovisional patent application that is the first application containing the claimed subject matter, the term 'effective filing date of the claimed invention' means the actual filing date of the US nonprovisional patent application. For an US nonprovisional application that claims the benefit of a corresponding prior-filed US provisional application, 'effective filing date of the claimed invention' can be the filing date of the prior-filed provisional application provided the provisional application sufficiently describes the claimed invention. Similarly, for an US nonprovisional application that is a continuation or division of a prior-filed US nonprovisional application, 'effective filing date of the claimed invention' can be the filing date of the prior-filed nonprovisional application that sufficiently describes the claimed invention. Finally, 'effective filing date of the claimed invention' may be the filing date of a prior-filed foreign patent application to which foreign priority is claimed provided the foreign patent application sufficiently describes the claimed invention. Even if the subject matter sought to be patented is not exactly shown by the prior art, and involves one or more differences over the most nearly similar thing already known, a patent may still be refused if the differences would be obvious. The subject matter sought to be patented must be sufficiently different from what has been used or described before that it may be said to be nonobvious to a person having ordinary skill in the area of technology related to the invention. For example, the substitution of one color for another or changes in size are ordinarily not patentable.

8.1.16.1 *Who may apply for patent*

According to the law, the inventor or a person to whom the inventor has assigned or is under an obligation to assign the invention, may apply for a patent, with certain exceptions. If the inventor is deceased, the application may be made by legal representatives, that is, the administrator or executor of the estate. If the inventor is legally incapacitated, the application for a patent may be made by a legal representative (e.g., guardian). If an inventor refuses to apply for a patent or cannot be found, a joint inventor may apply on behalf of the nonsigning inventor.

If two or more persons make an invention jointly, they apply for a patent as joint inventors. A person who makes only a financial contribution is not a joint inventor and cannot be joined in the application as an inventor. It is possible to correct an innocent mistake in erroneously omitting an inventor or in erroneously naming a person as an inventor. Officers and employees of the United States Patent and Trademark Office are prohibited by law from applying for a patent or acquiring, directly or indirectly, except by inheritance or bequest, any patent or any right or interest in any patent.

8.1.16.2 *Nature of patent and patent rights*

The patent is issued in the name of the United States under the seal of the United States Patent and Trademark Office, and is either signed by the Director of the USPTO or is electronically written thereon and attested by an Office official. The patent contains a grant to the patentee, and a printed copy of the specification and drawing is annexed to the patent and forms a part of it. The grant confers "the right to exclude others from making, using, offering for sale, or selling the invention throughout the United States or importing the invention into the United States" and its territories and possessions for which the term of the patent shall generally be 20 years from the date on which the application for the patent was filed in the United States or, if the application contains a specific reference to an earlier filed application under 35 U.S.C. 120, 121 or 365(c), from the date of the earliest such application, was filed, and subject to the payment of maintenance fees as provided by law (Anonymous, 2014).

The exact nature of the right conferred must be carefully distinguished, and the key is in the words 'right to exclude' in the phrase just quoted. The patent does not grant the right to make, use, offer for sale or sell or import the invention but only grants the exclusive nature of the right. Any person is ordinarily free to make, use, offer for sale or sell or import anything he or she pleases, and a grant from the government is not necessary. The patent only grants the right to exclude others from making, using, offering for sale or selling or importing the invention. Since the patent does not grant the right to make, use, offer for sale, or sell, or import the invention, the patentee's own right to do so is dependent upon the rights of others and whatever general laws might be applicable. A patentee, merely because he or she has received a patent for an invention, is not thereby authorized to make, use, offer for sale, or sell, or import the invention if doing so would violate any law.

An inventor of a new automobile who has obtained a patent thereon would not be entitled to use the patented automobile in violation of the laws of a state requiring a license, nor may a patentee sell an article, the sale of which may be forbidden by law, merely because a patent has been obtained. Neither may a patentee make, use, offer for sale or sell or import his or her own invention if doing so would infringe the prior rights of others. A patentee may not violate the federal antitrust laws, such as by resale price agreements or entering into combination in restraints of trade, or the pure food and drug laws, by virtue of having a patent. Ordinarily there is nothing that prohibits a patentee from making, using, offering for sale, or selling, or importing his or her own invention, unless he or she thereby infringes another's patent that is still in force. For example, a patent for an improvement of an original device already patented would be subject to the patent on the device.

The term of the patent shall generally be 20 years from the date on which the application for the patent was filed in the United States or if the application contains a specific reference to an earlier filed application under 35 U.S.C. 120, 121 or 365(c), from the date of the earliest such application was filed, and subject to the payment of maintenance fees as provided by law. A maintenance fee is due 3.5, 7.5 and 11.5 years after the original grant for all patents issuing from the applications filed on and after December 12, 1980. The maintenance fee must be paid at the stipulated times to maintain the patent in force. After the patent has expired anyone may make, use, offer for sale, or sell or import the invention without permission of the patentee, provided that matter covered by other unexpired patents is not used. The terms may be extended for certain pharmaceuticals and under certain circumstances as provided by law.

8.1.16.3 *Correction of patents*

Once the patent is granted, it is outside the jurisdiction of the USPTO except in a few respects. The Office may issue without charge a certificate correcting a clerical error it has made in the patent when the printed patent does not correspond to the record in the Office. These are mostly corrections of typographical errors made in printing. Some minor errors of a typographical nature made by the applicant may be corrected by a certificate of correction for which a fee is required. The patentee may disclaim one or more claims of his or her patent by filing in the Office a disclaimer as provided by the statute (35 U.S.C. 253).

When the patent is defective in certain respects, the law provides that the patentee may apply for a reissue patent. Following an examination in which the proposed changes correcting any defects in the original patent are evaluated, a reissue patent would be granted to replace the original and is granted only for the balance of the unexpired term. However, the nature of the changes that can be made by means of the reissue are rather limited; new matter cannot be added. In a different type of proceeding, any person may file a request for reexamination of a patent, along with the required fee, on the basis of prior art consisting of patents or printed publications. At the conclusion of the reexamination proceedings, a certificate setting forth the results of the reexamination proceeding is issued.

8.1.16.4 *Assignments and licenses*

A patent is personal property and may be sold to others or mortgaged; it may be bequeathed by a will, and it may pass to the heirs of a deceased patentee. The patent law provides for the transfer or sale of a patent, or of an application for patent, by an instrument in writing. Such an instrument is referred to as an assignment and may transfer the entire interest in the patent (Anonymous, 2014). The assignee, when the patent is assigned to him or her, becomes the owner of the patent and has the same rights that the original patentee had.

The statute also provides for the assignment of a part interest, that is, a half interest, a fourth interest, etc., in a patent. There may also be a grant that conveys the same character of interest as an assignment but only for a particularly specified part of the United States. A mortgage of patent property passes ownership thereof to the mortgagee or lender until the mortgage has been satisfied and a retransfer from the mortgagee back to the mortgagor, the borrower, is made. A conditional assignment also passes ownership of the patent and is regarded as absolute until canceled by the parties or by the decree of a competent court. An assignment, grant, or conveyance of any patent or application for patent should be acknowledged before a notary public or officer authorized to administer oaths or perform notarial acts. The certificate of such acknowledgment constitutes prima facie evidence of the execution of the assignment, grant, or conveyance.

8.1.16.5 *Joint ownership*

Patents may be owned jointly by two or more persons as in the case of a patent granted to joint inventors or in the case of the assignment of a part interest in a patent. Any joint owner of a patent, no matter how small the part interest, may make, use, offer for sale and sell and import the invention for his or her own profit provided they do not infringe another's patent rights, without regard to the other owners, and may sell the interest or any part of it, or grant licenses to others, without regard to the other joint owner, unless the joint owners have made a contract governing their relation to each other. It is accordingly dangerous to assign a part interest without a definite agreement between the parties as to the extent of their respective rights and their obligations to each other if the above result is to be avoided. The owner of a patent may grant licenses to others. Since the patentee has the right to exclude others from making, using, offering for sale, or selling or importing the invention, no one else may do any of these things without his or her permission.

A patent license agreement is, in essence, nothing more than a promise by the licensor not to sue the licensee. No particular form of license is required; a license is a contract and may include whatever provisions the parties agree upon, including the payment of royalties, etc. The drawing up of a license agreement (as well as assignments) is within the field of an attorney at law. Such an attorney should be familiar with patent matters as well. A few states have prescribed certain formalities to be observed in connection with the sale of patent rights.

8.1.16.6 *Patent marking and patent pending*

A patentee who makes or sells patented articles, or a person who does so for or under the patentee is required to mark the articles with the word 'patent' and the number of the patent. The penalty for failure to mark is that the patentee may not recover damages from an infringer unless the infringer was duly notified of the infringement and continued to infringe after the notice (Anonymous, 2014).

The marking of an article as patented when it is not in fact patented is against the law and subjects the offender to a penalty. Some persons mark articles sold with the terms 'Patent Applied For' or 'Patent Pending.' These phrases have no legal effect, but only give information that an application for patent has been filed in the USPTO. The protection afforded by a patent does not start until the actual grant of the patent. False use of these phrases or their equivalent is prohibited.

8.1.16.7 *Plant patents*

The law also provides for the granting of a patent to anyone who has invented or discovered and asexually reproduced any distinct and new variety of plant, including cultivated sports, mutants, hybrids, and newly found seedlings, other than a tuber-propagated plant or a plant found in an uncultivated state. Asexually propagated plants are those that are reproduced by means other than from seeds, such as by the rooting of cuttings, by layering, budding, grafting, inarching, etc.

With reference to tuber-propagated plants, for which a plant patent cannot be obtained, the term 'tuber' is used in its narrow horticultural sense as meaning a short, thickened portion of an underground branch. Such plants covered by the term 'tuber-propagated' are the Irish potato and the Jerusalem artichoke. An application for a plant patent consists of the same parts as other applications. The term of a plant patent shall be 20 years from the date on which the application for the patent was filed in the United States or if the application contains a specific reference to an earlier filed application under 35 U.S.C. 120, 121 or 365(c), from the date the earliest such application was filed.

The specification should include a complete detailed description of the plant and the characteristics thereof that distinguish the same over related known varieties, and its antecedents, expressed in botanical terms in the general form followed in standard botanical textbooks or publications dealing with the varieties of the kind of plant involved (evergreen tree, dahlia plant, rose plant, apple tree, etc.), rather than a mere broad nonbotanical characterization such as commonly found in nursery or seed catalogs. The specification should also include the origin or parentage of the plant variety sought to be patented and must particularly point out where and in what manner the variety of plant has been asexually reproduced. The Latin name of the genus and species of the plant should be stated. Where color is a distinctive feature of the plant, the color should be positively identified in the specification by reference to a designated color as given by a recognized color dictionary. Where the plant variety originated as a newly found seedling, the specification must fully describe the conditions (cultivation, environment, etc.) under which the seedling was found growing to establish that it was not found in an uncultivated state.

A plant patent is granted on the entire plant. It, therefore, follows that only one claim is necessary, and only one is permitted. The oath or declaration required of the applicant in addition to the statements required for other applications must include the statement that the applicant has asexually reproduced the new plant variety. If the plant is a newly found plant, the oath or declaration must also state that the plant was found in a cultivated area (Anonymous, 2014).

Plant patent drawings are not mechanical drawings and should be artistically and competently executed. The drawing must disclose all the distinctive characteristics of the plant capable of visual representation. When color is a distinguishing characteristic of the new variety, the drawing must be in color. Two duplicate copies of color drawings must be submitted. All color drawings should include a one-inch margin at the top for Office markings when the patent is printed. Specimens of the plant variety, its flower or fruit, should not be submitted unless specifically called for by the examiner. The filing fee on each plant application and the issue fee can be found in the fee schedule. For a qualifying small entity, most fees are reduced by half. 35 U.S.C. 41(h)(1). Plant patent applications may be published pursuant to Title 35, United States Code, Section 122(b), but the publication fee is not reduced for small entities.

A plant patent application is the only type of patent application filed at the USPTO, which is not permitted to be filed via EFS-Web. All inquiries relating to plant patents and pending plant patent applications should be directed to the United States Patent and Trademark Office and not to the Department of Agriculture. The Plant Variety Protection Act (Public Law 91577), approved December 24, 1970, provides for a system of protection for sexually reproduced varieties, for which protection was not previously provided, under the administration of a Plant Variety Protection Office within the Department of Agriculture.

8.2 PROTECTION OF INTELLECTUAL PROPERTY

The second part of the TRIPS agreement looks at different kinds of intellectual property rights and how to protect them. The purpose is to ensure that adequate standards of protection exist in all member countries. Here, the starting point is the obligations of the main international agreements of the World Intellectual Property Organization (WIPO) that already existed before the WTO was created, namely, the Paris Convention for the Protection of Industrial Property (patents, industrial designs etc.) and the Berne Convention for the Protection of Literary and Artistic Works (copyright). Some areas are not covered by these conventions. In some cases, the standards of protection prescribed were thought inadequate. So the TRIPS agreement adds a significant number of new or higher standards. The TRIPS agreement ensures that computer programs will be protected as literary works under the Berne Convention and outlines how databases should be protected.

It also expands international copyright rules to cover rental rights. Authors of computer programs and producers of sound recordings must have the right to prohibit the commercial rental of their works to the public. A similar exclusive right applies to films where commercial rental has led to widespread copying, affecting copyright-owners' potential earnings from their films. The agreement says performers must also have the right to prevent unauthorized recording, reproduction and broadcast of live performances (bootlegging) for no less than 50 years. Producers of sound recordings must have the right to prevent the unauthorized reproduction of recordings for a period of 50 years.

8.2.1 Evolution of International Arrangements for Protecting Intellectual Property

The current international environment for intellectual property relevant to plant biotechnology is a snapshot of the results of continuing complex interactions between many sets of actors representing many different interests, in many different settings

(Sara Boettiger *et al.*, 2004). Before trying to predict how this environment will evolve, it is useful to try to form a balanced appreciation of the current situation. This is in itself a real challenge. Our approach here is to consider in turn four quite distinct, and in some ways conflicting international agreements.

8.2.2 Convention on Biological Diversity (CBD)

The Convention on Biological Diversity is the third major international treaty affecting plant IPRs. The CBD began as an international initiative to conserve flora and fauna in situ, responding to field research that suggested an alarmingly high rate of species extinction. The International Union for the Conservation of Nature (IUCN), an international nongovernmental organization, played a major role in the CBD. In 1987, the United States proposed the Convention to the Governing Council of the United Nations Environment Programme (UNEP), which began drafting an international treaty in 1989 that sought to address the conservation of biodiversity from a sustainable use perspective (Sara Boettiger *et al.*, 2004). The CBD was opened for signature in 1992 and entered into force in 1993. 186 countries are now party to the convention. All developed countries except the United States have ratified the convention. The language of the CBD reaches well beyond the conservation objectives envisaged by the United States and the IUCN.

Developing nations encompassing centers of biodiversity saw the initial conservation agenda as threatening their rights to exploit the economic benefits of their forests and other biodiversity resources and pressed for fair and equitable sharing of the benefits of conservation sought by nations and nongovernmental organizations based in the North. Aware that many pharmaceutical products worth billions of dollars were based on products of nature, some participants envisioned that royalties from biodiversity-derived products could be the basis for sustained in situ conservation of biodiversity in developing countries, and constitute a significant source of wealth for those countries rich in biological diversity. This vision was encouraged by the hugely publicized bioprospecting agreement between Merck and INBIO of Costa Rica, involving an initial commitment of over one million dollars, plus unspecified future royalties.

A notable impact of the CBD has been to inject the interests of developing countries rich in natural resources, and of environmentalists (mainly from developed countries), into the IPR debate. The biennial meetings of the conference of parties on implementation of the CBD provide an alternate forum for debate on trade and IPR policy, involving partners in developed and less developed nations with a more conservation and less commercially oriented agenda than typically represented in WTO and TRIPS.

8.3 WTO AGREEMENT ON TRADE RELATED ASPECTS OF INTELLECTUAL PROPERTY RIGHTS (TRIPS)

The TRIPS Agreement, adopted in 1994, requires that 'patents shall be available for any inventions, whether products or processes, in all fields of technology' (Sara Boettiger *et al.*, 2004). However, it allows countries to exclude from protecting 'plants and animals other than microorganisms.' It does require that countries provide for the protection of plant varieties either by patents or by an effective sui generis system (i.e., PVP) or both. The TRIPS Agreement permits countries some flexibility in the precise form and the extent of protection and promotes the fundamental idea of extending IPRs to agricultural genetic resources. The general objectives of the TRIPS Agreement are the protection and enforcement of IPRs, the promotion of technological innovation, and the transfer and dissemination of technology. A WTO member country must be nondiscriminatory and extend the same treatment to all other members that it affords one member. Most developing countries are opposed to the use of patent systems in agriculture.

8.3.1 IPR and Developing Countries

The task to consider is how and whether IPRs could play a role in helping the world meet the targets of crop biosecurity, in particular by reducing poverty, helping combat crop diseases, enhancing access to crop biodiversity, contributing to sustainable development, and in identifying and removing obstacles to meeting those targets. Some argue rather strongly that IPRs are necessary to stimulate economic growth, which, in turn, contributes to poverty reduction through crop biosecurity (Kauser Abdulla Malik and Yusuf Zafar, 2005.). Stimulating invention and new technologies lead to increases in agricultural or industrial production. Hence, these proponents of IPRs are of the view that there is no reason a system that works for developed countries could not do the same for developing countries. Others argue with equal vehemence that IPRs rights do little to stimulate invention and research in developing countries since the necessary human and technical capacity may be absent; and moreover the poor will not be able to afford the products, even if developed. IPRs limit the option of technological learning through imitation, allow foreign firms to drive out domestic competition by obtaining patent protection, and to service the market through imports rather than through domestic manufacture. Moreover, IPRs increase the costs of essential medicines and agricultural inputs affecting poor people and farmers rather badly.

Developing countries are not a homogeneous group, a self-evident fact that is often forgotten. Not only do their scientific and technical capacities vary, but so also their social and economic structures, and their inequalities of income and wealth. The determinants of poverty, and, therefore, the appropriate policies to address it, will vary accordingly between countries. The same applies to policies on IPRs. Policies required in countries with a relatively advanced technological capability where most poor people happen to live, for instance in Pakistan, India or China, may well differ from those in other countries such as many countries in sub-Saharan Africa with a weak capability. The impact of IP policies on poor people will also vary according to socio-economic circumstances. The ability of IPR owners to charge above marginal cost derives from the exclusivity/'monopoly' which they are granted. Not necessarily a monopoly in an antitrust sense, it does limit competition to some degree. The effect on competition depends on the nature and extent of IPRs granted and the extent to which close substitutes are, or are likely to be, available. For example, patent rights provide exclusive rights over ideas, whereas copyright only provides protection against copying particular expressions of ideas; copyright does not provide protection against independent creation of an identical expression, or different expression of the same ideas. In this regard, patents are likely to have a much greater impact on competition than copyright.

8.3.2 Scope and Objectives of Intellectual Property Rights

All countries need to determine and design IP policies and rules appropriate to their needs and level of development for crop biosecurity. IP is a means, not an end in itself. The one-size-fits-all approach is widely rejected, yet efforts continue by some transnational corporations and developed countries to have one set of global rules enforced by a global system (Kauser Abdulla Malik and Yusuf Zafar, 2005.). The question of IPRs and plant genetic material has thus become linked to 'farmers' rights' in the case of the FAO undertaking, and, in the case of the Biodiversity Convention, to the "equitable sharing of the benefits arising from the utilization of plant genetic resources." Despite the undertaking by the signatories of the TRIPs agreement to introduce IPRs in one form or another to cover plant genetic material, plant varieties, and plant parts, opinions differ widely over the possession and control of genetic resources and the role played by IPR regimes.

The stated objective of most intellectual property law (with the exception of trademarks) is to 'Promote progress.' By exchanging limited exclusive rights for disclosure of inventions and creative works, society and the patentee/copyright owner mutually benefit, and an incentive is created for inventors and authors to create and disclose their work. Some commentators have noted that the objective of intellectual property legislators and those who support its implementation appears to be 'absolute protection.' If some intellectual property is desirable because it encourages innovation, they reason, more is better. The thinking is that creators will not have sufficient incentive to invent unless they are legally entitled to capture the full social value of their inventions." This absolute protection or full value view treats intellectual property as another type of 'real' property, typically adopting its law and rhetoric. Other recent developments in intellectual property law, such as the America Invents Act, stress international harmonization.

8.3.3 Role of Public Sector in IPR

The public sector R&D is well developed in a number of developed countries. With structural adjustment and liberalization, public research institutions are under pressure to become involved in income-generating activities (Kauser Abdulla Malik and Yusuf Zafar, 2005). Revenues from licensing or royalties, or from the provision of services, could, therefore, become important for those institutions, which, until now, have made their 'innovations' freely available to both public and private sectors. The need for public sector research in agriculture will remain despite the private sector assuming a growing role in innovation. The introduction of a national IPR system may even facilitate a more rational public/private division of biosecurity roles. One of the important constraints in elucidating this biosecurity role is a very weak private sector R&D activity related to biotechnology and other crop biosecurity related issues in a majority of the developing countries. Most of the biotech research, at whatever level, is being undertaken in the public sector of the developing countries. For any public/private partnership, the private sector from the developed world has to be involved. This is all the more true in cases where cells, organelles, genes or molecular constructs are under patent with some of the big multinational biotech companies. In many cases, the material under patent has its origin from the germplasm or material located at some other place. Moreover, based on legal interpretations of IPR laws; some advanced countries have been providing protection to biological processes and products which essentially do not fulfill the stringent application of the condition of novelty or discovery. Some such examples have already been stated above. Complications are also known to arise while making available any biotechnology product to the poor farming communities.

8.3.4 Intellectual Property Rights (IPRs) and Crop Biosecurity

IPRs can be defined as a set of laws devised for the purpose of protecting or rewarding inventors or creators of new knowledge. Precisely because knowledge, unlike consumable goods, can be shared by any number of persons without being diminished, creators are dependent on legal protection to prevent direct copying or the utilization of the product or process they have invented without the payment of compensation. IPRs are thus intended to confer exclusive rights for inventors or discoverers, for a fixed period of time (Kauser Abdulla Malik and Yusuf Zafar, 2005.) Biological materials and data have long been preserved in and disseminated by repositories of microbial culture collections, seed banks, etc. and were a source of crop biosecurity. These biological collections face great challenges but also great opportunities owing to the explosive increase in biological materials and data in the field of crop safety and biosecurity.

The fact that the richest nations are home to the smallest pockets of biodiversity while the poor are stewards of the richest resources underlines the interdependence of all nations and the urgency of formulating common strategies for sustaining biodiversity, eliminating biopiracy and genetic drain and ultimately ensuring crop biosecurity. A biosecurity guarantee attempts to ensure that ecologies sustaining either people or animals are maintained. Crop biosecurity is the maintenance and conservation of crop biodiversity, checking the threat of bioterrorism, judicially and wisely using crop genetic diversity for crop improvement, reducing the risk of biopiracy and genetic erosion, and protecting the crops from other hazards such as insect pests and diseases for the welfare of humankind.

8.3.5 Challenges of IPRs in Crop Biosecurity

Countries should increase their efforts to achieve greater international harmonization of the laws, rules and practices governing access. Legal and regulatory differences among countries, particularly with regard to health and safety, use of plant materials and IPR can lead to unnecessary restriction of access to and exchange of plant genetic resource materials. Governments must remain alert to possible negative long-term consequences of restrictive gene licensing practices on crop biosecurity and plant biotechnology and on the rights of research institutions to provide access to genetic biological resources, and must be prepared to address these negative consequences when they can be documented (Kauser Abdulla Malik and Yusuf Zafar, 2005). Governments need to work towards international harmonization of the broad variety of laws, regulations and practices aimed at protecting the health and safety of humans, animals, plants and the environment from potentially hazardous biological materials. This will discourage inappropriate uses of biological resources and will ensure crop biosecurity. Governments need to develop internationally compatible laws, standards and practices regarding plant materials and data, to protect crops, so that the unethical handling of biological materials and data from plants can be guaranteed.

8.3.5.1 Morality

According to Article 27 of the Universal Declaration of Human Rights, "everyone has the right to the protection of the moral and material interests resulting from any scientific, literary or artistic production of which he is the author." Although the relationship between intellectual property and human rights is a complex one, there are moral arguments for intellectual property (Ronald V. Bettig, 1996). The arguments that justify intellectual property fall into three major categories. Personality theorists believe intellectual property is an extension of an individual. Utilitarians believe that intellectual property stimulates social progress and pushes people to further innovation. Lockeans argue that intellectual property is justified based on deservedness and hard work.

Various moral justifications for private property can be used to argue in favor of the morality of intellectual property, such as the following.

1. **Natural Rights/Justice Argument:** This argument is based on Locke's idea that a person has a natural right to the labor and/or products which is produced by his/her body. Appropriating these products is viewed as unjust. Although Locke had never explicitly stated that natural right applied to products of the mind, it is possible to apply his argument to intellectual property rights, in which it would be unjust for people to misuse another's ideas. Locke's argument for intellectual property is based upon the idea that laborers have the right to control that which they create. They argue that we own our bodies which are the laborers; this right of ownership extends to what we create. Thus, intellectual property ensures this right when it comes to production.
2. **Utilitarian-Pragmatic Argument:** According to this rationale, a society that protects private property is more effective and prosperous than societies that do not. Innovation and invention in 19th century America has been said to be attributed to the development of the patent system. By providing innovators with "durable and tangible return on their investment of time, labor, and other resources," intellectual property rights seek to maximize social utility. The presumption is that they promote public welfare by encouraging the "creation, production, and distribution of intellectual works" (Spinello and Richard, 2007).

Utilitarians argue that without intellectual property there would be a lack of incentive to produce new ideas. Systems of protection such as Intellectual property optimize social utility.

3. **Personality Argument:** This argument is based on a quote from Hegel: "Every man has the right to turn his will upon a thing or make the thing an object of his will, that is to say, to set aside the mere thing and recreate it as his own." European intellectual property law is shaped by this notion that ideas are an 'extension of oneself and one's personality.' Personality theorists argue that by being a creator of something one is inherently at risk and vulnerable to having their ideas and designs stolen and/or altered. Intellectual property protects these moral claims that have to do with personality.

8.3.5.2 Expansion in nature and scope of intellectual property laws

Other criticism of intellectual property law concerns the expansion of intellectual property, both in duration and in scope. In addition, as scientific knowledge has expanded and allowed new industries to arise in fields such as biotechnology and nanotechnology, originators of technology have sought IP protection for the new technologies. Patents have been granted for living organisms, (and in the United States, certain living organisms have been patentable for over a century).

The increase in terms of protection is particularly seen in relation to copyright, which has recently been the subject of serial extensions in the United States and Europe (Dennis Wharton, 1992). With no need for registration or copyright notices, this is thought to have led to an increase in orphan works (copyrighted works for which the copyright owner cannot be contacted), a problem that has been noticed and addressed by governmental bodies around the world.

Also with respect to copyright, the American film industry helped to change the social construct of intellectual property via its trade organization, the Motion Picture Association of America. In amicus briefs in important cases, in lobbying before Congress, and in its statements to the public, the MPAA has advocated strong protection of intellectual property rights. In framing its presentations, the association has claimed that people are entitled to the property that is produced by their labor. Additionally Congress's awareness of the position of the United States as the world's largest producer of films has made it convenient to expand the conception of intellectual property. These doctrinal reforms have further strengthened the industry, lending the MPAA even more power and authority.

The growth of the Internet and particularly distributed search engines like Kazaa and Gnutella, have represented a challenge for copyright policy. The Recording Industry Association of America, in particular, has been on the front lines of the fight against copyright infringement, which the industry calls 'piracy.' The industry has had victories against some services, including a highly publicized case against the file-sharing company Napster, and some people have been prosecuted for sharing files in violation of copyright. The electronic age has seen an increase in the attempt to use software-based digital rights management tools to restrict the copying and use of digitally based works. Laws such as the Digital Millennium Copyright Act have been enacted; that use criminal law to prevent any circumvention of software used to enforce digital rights management systems. Equivalent provisions, to prevent circumvention of copyright protection have existed in EU for some time and are being expanded in, for example, Articles 6 and 7 of the Copyright Directive. Other examples are Article 7 of the Software Directive of 1991 (91/250/EEC) and the Conditional Access Directive of 1998 (98/84/EEC). This can hinder legal uses, affecting public domain works, limitations and exceptions to copyright or uses allowed by the copyright holder. Some copyleft licenses, like GNU GPL 3, are designed to counter that (Katherine Beckman and Christa Pletcher, 2009). Laws may permit circumvention under specific conditions like when it is necessary to achieve interoperability with the circumventor's program, or for accessibility reasons; however, distribution of circumvention tools or instructions may be illegal. In the context of trademarks, this expansion has been driven by international efforts to harmonize the definition of 'trademark,' as exemplified by the Agreement on Trade-Related Aspects of Intellectual property rights ratified in 1994, which formalized regulations for IP rights that had been handled by common law, or not at all, in member states. Pursuant to TRIPs, any sign which is 'capable of distinguishing' the products or services of one business from the products or services of another business is capable of constituting a trademark.

Intellectual property rights (IPRs) are being introduced or strengthened in developing countries as a result of the Agreement on Trade-Related Aspects of Intellectual Property (TRIPS) of the World Trade Organization, bilateral trade negotiations and pressure from export-oriented sectors in agriculture. IPRs in agriculture provide a temporary exclusive right on the commercialization of the protected subject matter (the invention) and are primarily meant to stimulate investment in breeding and seed supply (Chawla, 2010).

This is a recent phenomenon. Intellectual property law has recently received attention as a motor for technological innovation and industrial change. It has also been seen as a tool for promoting the conservation of biological diversity, sustainable use of its components, and for ensuring that benefits arising from the utilization of genetic resources are shared in a fair and equitable manner among the relevant stakeholders. Critics argue that intellectual property protection increases the costs of products, promotes genetic monoculture by concentrating industrial and agricultural activities on a few cultivated varieties or species, and, when extended to plants and animals, is in conflict with the morals of many societies (Chawla, 2010). Intellectual property laws vary in nature and scope from one country to another. Intellectual property protected in one country may not be recognized in another country. Despite the existence of various international agreements that attempt to harmonize intellectual property protection, there are still differences among national laws, especially those regarding patents. For example, while the USA and countries in the European Union allow patent protection for genetically engineered organisms which meet the normal requirements for patentability, many other countries are opposed to extending patents to such subject matter.

There are also differences in the duration of patent protection. The period for which an inventor is granted a patent varies from one country to another. In addition, different countries have different conditions for the disclosure of information concerning the invention. While some (for example, the USA and the European Union countries) have strict conditions and mechanisms for enforcing patent application requirements, others (particularly those of the developing world) have weak institutional arrangements for ensuring compliance with disclosure requirements. These differences in national application of intellectual property law are at the center of much of the debate on the intellectual property rights of indigenous and local peoples. The case of traditional knowledge of indigenous and local peoples has opened debate on the adequacy and ethics of intellectual property protection. The debate (particularly the absence of consensus on whether and how to extend intellectual property protection to traditional knowledge) has so far shown these issues are complex and controversial. This is partly because of differences in conceptual treatment and often a lack of clarity on the two concepts of traditional knowledge and intellectual property. It is also because a scant body of information is available to those responsible for policy and law making, at both national and international levels. In addition, these issues are often debated in isolated United Nations, business sector and nongovernmental organizations' conferences, each with its distinct sectoral interest and focus in the subject. For example, dialogue (within the ILO and the United Nations Working Group on Indigenous Populations, amongst others) on the human rights of indigenous peoples has seldom addressed, at least consistently, issues of intellectual property rights in traditional knowledge. The World Trade Organization (WTO) regime has not confronted the implications of its Agreement on Trade Related Aspects of Intellectual Property Rights (the TRIPS Agreement) for the protection and use of traditional knowledge. On the whole, international debate on issues of intellectual property protection in general and rights in traditional knowledge, in particular, are characterized by tension and inconsistency (Chawla, 2010).

WTO administers not only the Multilateral Trade Agreements (MTAs) in goods but also the General Agreement on Trade in Services (GATS) and the Agreement on Trade Related Aspects of Intellectual Property Rights (TRIPS), which came into existence with WTO. All the agreements annexed to the Agreement establishing the WTO were signed as part of a package deal. Member countries did not have the option of choosing some and rejecting others. Another important difference with the erstwhile GATT is that WTO has a stronger compliance mechanism that did GATT— a member's failure to meet the obligations can invoke retaliation across agreements and sectors (Chawla, 2007a). As one of the WTO agreements, TRIPS is binding on all member countries of WTO. TRIPS aims at establishing strong minimum standards for intellectual property rights (IPRs). Apart from patents, intellectual property includes copyrights, trademarks, geographical indications, industrial designs, integrated circuits and trade secrets. The protection of IPRs is binding and legally enforceable (Chawla, 2010).

Intellectual property is a product of the mind. Intellectual property is intangible in contrast to real property (land) or physical property, which one can see, feel and use. With any type of property, there are property rights. When IPs are expressed in a tangible form, they can also be protected. Intellectual property rights (IPRs) have been created to protect the right of individuals to enjoy their creations and discoveries. In fact, IPRs can be traced back to the fourteenth century, when European monarchs granted proprietary rights to writers for their literary works. IPRs have been created to ensure protection against unfair trade practice. Owners of IP are granted protection by a state and/or country under varying conditions and periods of time. This protection includes the right to defend their rights to the property they have created, prevent others from taking advantage of their ingenuity, encourage their continuing innovativeness and creativity and assure the world a flow of useful, informative and intellectual works.

Product patents on all items including food, agrochemical and pharmaceuticals have also been allowed making the Patents Act fully TRIPS compliant. The patent system was developed as a means to reward inventions which would be useful to the society. However, to ensure the interests of society, as per the Indian Patents Act, certain things have been excluded from the purview of patentability.

The sections relevant to plant material and agriculture which are excluded from patentability are as follows:

- Section 3(h): a method of agriculture and horticulture
- Section 3(i): any process for medicinal, surgical, curative, prophylactic (diagnostic therapeutic) or other treatment of human beings or any process for a similar treatment of animals to render them free of disease or to increase their economic value or that of their products
- Section 3(j): plants and animals in whole or any part thereof other than microorganisms but including seeds, varieties and species and essentially biological processes for production or propagation of plants and animals
- Section 3(p): an invention which in effect, is traditional knowledge or which is an aggregation or duplication of known properties of the traditionally known component or components.

Further, the mere discovery of any new property or new use for a known substance or the mere use of a known process, machine or apparatus unless such known process results in a new product or employ at least one new reactant is not patentable. Also, a patent claim for a substance obtained by merely mixing ingredients resulting only in the aggregation of the properties of the components is not a patentable invention (Chawla, 2010).

Microorganisms *per se* can be claimed for protection provided they are not the mere discovery of organisms. It is mandatory to deposit the biological material in International Depositary Authority (IDA). In India, Institute of Microbial Technology (IMTECH), Chandigarh is a recognized international depositary for some category of microorganisms. If an applicant mentions a biological material in the patent specification, then disclosure requirements prescribed for biological materials have been notified in the list of the Central Government or for indicating its source and geographical origin [Section: 10,4(d)]. However, in India, method for rendering plants free of diseases or to increase their economic value or that of their products can be claimed for patent protection. The purpose of a patent is to promote the progress of science and useful arts. The patent law promotes this progress by giving the inventor the right of exclusion. In exchange for this right to exclude others, the inventor must disclose all details describing the invention, so that when the patent period expires, the public may have the opportunity to develop and profit from the use of the invention. A patent is enforced in the country which issues it, meaning thereby territorial in nature. For each country, a separate application is to be filed in that country where protection is sought.

8.3.6 Microorganism Patents

The first patent on living organism was granted to *Pseudomonas*. The first example was the classical judgment in the *Diamond* vs. *Chakrabarty* case in 1980. In the Chakrabarty case, USPTO rejected the patent application on the ground of product of nature but the US Supreme Court decided that a microorganism was not precluded from patentability solely because it was alive. Thus, a *Pseudomonas* bacterium manipulated to contain more than one plasmid (four plasmids were present) controlling the breakdown of hydrocarbons (therefore more useful in dispersing oil slicks than the natural organism containing only one such plasmid) was "a new bacterium with markedly different characteristics from any found in nature" and hence not nature's handiwork but that of the inventor. The 'product of nature' objection therefore failed, and the modified organisms were held patentable. This precedent is being followed even today to define the patentability of microorganisms (Chawla, 2007b).

8.3.7 Plant Patents

Plant patents are obtainable in the US, Europe, and Japan. The US Plant Patent Act of 1930 (PPA) granted property rights for privately developed plant varieties of asexually reproducing plants. These rights were extended to new and distinct asexual varieties for a period of seventeen years. Advances in breeding technology provided the momentum for the 1970 Plant Variety Protection Act (PVPA). The PVPA provided protection for sexually reproducing plants, including seed germination. In 1980 *Diamond* vs. *Chakrabarty* case set in motion the trend towards the legal acceptance of the commodification of germplasm. The court held that a live, man-made bacterium was patentable under the PPA and the 'product of nature' objection therefore failed, and the modified organisms were held patentable. Life forms of plants and animals except microorganisms are not patentable in India. In pursuance of the TRIPS agreement, India has enacted Protection of Plant Varieties and Farmers' Rights (PPV&FR) Act, 2001, a *sui generis* system of plant variety protection which has been described in detail separately (Chawla, 2010).

8.4 PLANT VARIETY PROTECTION IN INDIA

As stated, India is a signatory to WTO agreements, and it has to abide by the TRIPS regulations. As per article 27.3(b) of the TRIPS, which demand that member countries should protect their plant varieties either by patent, or an effective system of *sui generis* protection, or a combination of these two. In this context, India chose a *sui generis* system for protection of plant varieties. An Act named as Protection of Plant Varieties and Farmers' Rights (PPV&FR) Act 2001 has been passed and rules have been framed.

PPV&FR Authority has been constituted with its head office located at Delhi. The PPV&FR Act is TRIPS compliant and compatible with UPOV system of plant variety protection (Anonymous, 2003).

The PPV&FR Act 2001 provides protection to these types of plant varieties (Anonymous, 2003), viz., newly bred varieties, extant varieties (the varieties which were released under Indian Seeds Act, 1966 and have not completed 15 years as on the date of application for their protection), farmer's varieties (the varieties which have been traditionally cultivated, including landraces and their wild relatives which are in common knowledge, as well as those evolved by farmers), essentially derived varieties, and transgenic varieties.

An application for registration can be made by any person claiming to be the breeder of the variety, the successor of the breeder, assignee, any farmer or group of community of farmers, any person authorized for the above-mentioned categories or any university or publicly funded agricultural institution claiming to be the breeder of the variety. It is pertinent to note that the Act recognizes the farmer as a cultivator, conserver, and breeder. This embraces all farmers, landed or landless, male and female. To qualify for registration under the Act, a new variety has to conform to the criteria of novelty (N), distinctiveness (D), uniformity (U) and stability (S). Besides, a denomination has to be given for the registration of variety. Denomination refers to the label or title of the variety. It is the denomination that is registered. For extant and farmers' varieties which are in the public domain, the DUS features will be considered while the novelty feature will not be taken because these varieties are not new and are in public domain. In this Act, a special clause has been put which states that any variety with terminator gene sequences will not be registered. Thus, any transgenic material with genetic use restriction technology (GURT) sequences will not be registered (Chawla, 2010).

All the varieties will be registered with PPV&FR Authority. DUS guidelines for 35 crops have been prepared by ICAR while guidelines for 12 crop species have been notified by PPV&FR Authority in the Gazette. PPV&FR Authority has established testing centers for each and every crop species. In the first phase which has started in May 2007, the registration of varieties will be done for 12 crop species of cereals and legumes. The registration will then extend to 35 crops which include cereals, pulses, oilseeds, vegetable and two flower species. DUS guidelines are also being prepared for medicinal and aromatic plants, spices, ornamentals and forest trees for which task forces have been constituted by the PPV&FR Authority.

Indian PPV&FR Act allows farmers to save, use, sow, resow, exchange, share or sell his farm produce including seed of a variety protected under this Act, but it prohibits that the farmer shall not be entitled to sell branded seed of a variety protected under the Act [Sec. 39, 1(iv)]. The farmers have been given the right to register farmers varieties themselves [Sec. 39,1(i)], the right to claim compensation for underperformance of a protected variety from the promised level [Sec. 39(2)], benefits sharing for use of biodiversity conserved by farming community [Sec. 41]. According to the concept of benefits sharing, whenever a variety submitted for protection is bred with the possible use of a landrace, extant variety or farmer's variety, a claim can be referred either on behalf of the local community or institution for a share of the royalty [Sec. 41(1)] (Anonymous, 2003). In the Act, a provision of a compulsory license has also been put. According to this, after the expiry of three years from the date of issue of certificate of registration of a variety, any person interested can claim in an application to the authority alleging that reasonable requirements of the public for seeds or other propagating material have not been satisfied or that the seed or other propagating material is not available to the public at a reasonable price and pray for the grant of a compulsory license to undertake production, distribution and sale of the seed or other propagating material of that variety [Sec. 47(1)] (Anonymous, 2003).

The Act had laid down the norms for registration of plant varieties, fee structure, provisions of opposition, DUS testing of material, etc. If any farmer or association of farmers is applying for registration of a plant variety, then this category is not required to pay any fee for either registration or DUS testing. Once the variety has been tested for its features, then the Registrar of the Authority will issue the certificate of registration. It shall have the validity of nine years initially in the case of trees and vines with renewal up to a period of 18 years. For other crops certificate of registration will be issued for six years initially with renewal up to 15 years. In the case of extant varieties, the validity period is 15 years from the date of notification of that variety by the Central Government under section 5 of the Seeds Act 1966 (Chawla, 2010).

Research endeavors are being affected by issues involving intellectual property (patents, copyrights, and trademarks). The acquisition of rights in intellectual property by universities can result in the transfer of new innovations to the private sector, with the university recouping a share of the profits for support of further scientific research. Intellectual property rights available for new plant cultivars include plant patents, plant variety protection certificates, plant breeder's rights, and utility patents. Under the patent laws, there is no explicit exemption for research use, so researchers are increasingly being required to execute materials transfer agreements to obtain permission to use patented materials, such as techniques, genes, seeds, and cell lines, in laboratory research and breeding programs. Research scientists must educate themselves on these issues so that they can make informed decisions regarding their research practices and the licensing of their discoveries (Janice A. Kimpel, 1999).

Private intellectual property rights have historically been of little relevance in agricultural research. For more than a century, mechanical inventions, produced mainly by farmers and mechanics, could be protected in many countries via utility patents. But the financial rewards were only modest, at best. Inventions tended to be easy to copy and licensing contracts difficult to enforce. Biological innovations such novel technologies were not patentable and had no protection from duplication. The role of intellectual property rights in international trade, the global economy, and international relation has grown considerably, especially since the 1970s (Kalpana Sastry, 2007). IPR-protected products, technologies, and services are major exports and rights manifesting in the form of licenses to use the patented processes, products, designs, trademarks or copyrights. All these developments necessitated legal protection mechanisms to be in place.

Over the last few decades, several agreements at various international fora have been negotiated and adopted to enhance and better the livelihood and to help the nations achieve the targets set in the millennium development goals. Some of these are of direct relevance to sustainable agriculture, enhanced trade and ensuring a better environment. The agreements include Convention of Biological Diversity (CBD), 1992; Convention on Trade in Endangered Species (CITES), 1973; International Plant Protection Convention (IPPC), 1997; International Union for the Protection of Plant Varieties (UPOV), 1978 and 1991; Cartagena Protocol on Biosafety, 2000; Trade Related Intellectual Property Rights (TRIPS), 1994; and the International Treaty on Plant Genetic Resources for Food and Agriculture (ITPGRFA), 2001. These international conventions/treaties/agreements have comprehensive provisions for conservation and sustainable use of and access to genetic resources and sharing of benefits derived from their use. Concurrently, new emerging regimes in protection mechanisms for innovations at the global levels are impacting the access, transfer, and use of biological/genetic resources for furthering the research and developmental activities in all fields of agriculture.

The developments towards providing a better platform for international trading including agriculture have led to private property rights and have necessitated the need to relook at the present framework for the issues like ownership of these resources. Apprehensions on the marked dichotomy between the use of unrestricted biological resources and property rights regimes are becoming conscientious issues leading to questionable access to control on and ownership of agro-biodiversity, especially in the developing countries. Monopolies on new technologies have resulted in problems of excess ownership rights and mandatory gains.

8.5 LEGAL MECHANISMS IN INDIA

"Protection of Plant Varieties and Farmers Rights Act, 2001" in India, is perhaps the most progressive Act on plant variety protection (PVP) adopted by a developing country. The will to provide equal rights to breeders and farmers is considered farsighted insofar as it indicates a clear understanding that shaping the new regime requires the assignment of property rights to all concerned players where appropriation takes place. With operationalization of the Act with effect from October 2005, the onus now rests with the authorities to implement it in its true element (Kalpana Sastry, 2007). The CBD has its origins in environmental concerns and recognizes the members' sovereign rights over their natural biological resources.

The IP issues form only a part of the concerns of CBD and address directly the origin, value, rights, and benefits associated with the natural resources and the development of traditional knowledge. To address the challenges of India for asserting its rights over natural resources, the Biological Diversity Act 2002 (No.18 of 2003) aims to provide conservation of biological diversity, sustainable use of its components, and fair and equitable sharing of the benefits arising out of the use of biological resources and knowledge. It strives to put limits on access to the biological resources or related knowledge including prior intimation to the National Biodiversity Authority (NBA) and

State Biodiversity Boards (SBB). On issues of IPRs too, prior consent from the concerned authority is now required thus enforcing a certain amount of discipline to the IPR system. It also entitles the authorities to oppose the grant of IPRs outside India on any biological resources obtained from India. Benefits sharing concept has also been integrated into the Act thus addressing rights of holders of local knowledge and helping towards facilitating better living standards to benefit claimers (BD Act, 2002).

The present tools for implementation are material transfer agreements (MTAs), prior informed consent (PIC) and applications for import/export of materials between countries. Since India is a party to the Convention and as a consequence of the enactment of BD Act, 2002, the Government of India notified the Biological Diversity Rules, 2004. The essence is now on guidance and compliance by the stakeholders. This Act fulfills the Indian obligations under CBD-TRIPS regime with the main aim to protect biological diversity for sustainable use and to uphold the sovereignty over the biological resources of the country.

The WTO-TRIPS Agreement of 1995, which is binding on all member countries, provides minimum norms and standards in respect of protection of IPRs in several categories, viz., patents, copyrights, trademarks, geographical indications, industrial designs, layout-designs of integrated circuits, and trade secrets. TRIPS incorporates provisions from many existing international IP agreements like the Paris and Berne Conventions. It also provided a transition period of five years (until 1 January 2000) to give effect to the provisions of the agreement. In the case of product patents in some areas of technology, this period was extended up to 1 January 2005. India has amended its various laws and acts (Copyright Law, Patent Act, Trademark Act, Geographical Indication of Goods Act, Designs Act, and Semiconductor Integrated Circuits Layout Design Act) to be in line with the minimum requirements specified by the Agreement. It may be recalled that prior to this Agreement innovation in living organisms (plants and animals) or the biological processes that produce them were not protectable as IP.

The patent law also excludes from patentability all inventions arising out of the use of traditional knowledge (Patent Act, 1970). The legal position on patents on plant parts, cells, cell lines, genes, and mitochondria, all of which are already patentable in developed countries, is still in development stage, and initial case laws on this would clarify the thinking (Kalpana Sastry, 2007). Based on the available options through the legal mechanisms in the present day innovation systems in India for agriculture, the possible IPs may be patents that can be used to protect and get benefits from their IP (with regard to their traditional practice-based innovations or inventions) or can be used to defend their traditional knowledge (in instances where the existence of traditional knowledge as prior art is revealed during the examination of a patent application, thereby providing grounds for refusal); geographical indications and trademarks, which could be used to protect plant protection products derived from traditional knowledge in rural sector and can offer a strong element of cultural identification; trade secrets, which could protect traditional agri-based knowledge, especially where such knowledge is held exclusively by a particular group within a community towards crop protection; designs, which can protect the delivery mechanisms used in agriculture like pest control, fertilizer applications; agreements and contracts based on principles of prior informed consent (PIC), mutually agreed terms (MAT), which can pave for amicable *modus operandi* in the highly complex and dynamic environment of high-end agricultural research leading to development of technological innovations and products derived from the use of various biological resources.

It is necessary to realize that a great care must be taken to ensure that the rights provided are neither insufficient nor excessive in terms of enhancing social and economic welfare. They can perhaps lead to reducing the rate of innovation, which may be a roadblock in the race to keep pace with global developments. However, the present attempts in India are to balance and adopt a wait and watch approach as the statutory system evolves. Even as the emerging framework gets evolved, it is unclear whether the farming community can exercise its right to oppose the grant of patent or IP. It is also unclear what kind of evidence is needed to prove priority and how these three legislations would ensure access and benefits sharing (ABS). The grants of patent rights on wound-healing properties of turmeric (US Patent 5,40,1,504), fungi-toxic properties of neem extract (EP No. 43,6257) or basmati rice (US Patent 5,66,3,484) 11 and several other cases indicate the dangers that patents or IPs can create in biopiracy rather than bioprospecting. There is a need for digital databases of aforesaid registration, to link local information to global and from one agency to another and need to constantly search Indian and global databases of IPR application claims. This electronic vigilance will help NBA and other agencies to locate any claims infringing on prior local knowledge, so as to oppose the grant of IPR protection.

It is evident that the policy environment and governance affecting agriculture at large and agri-based products, in particular, is undergoing a significant change during the past decade. All stakeholders of agricultural research in India need to be aware of the statutory requirements as they develop new products and processes. The system needs to be able to support knowledge holders in rural communities who often lack the know-how and financial resources to take advantage of the IP system, either in its present or any further evolved form (Kalpana Sastry, 2007). For this, it is essential to develop next generation manpower in IP management as related to agri-based activities.

8.6 FUTURE OF INTERNATIONAL IPR AND FREEDOM TO OPERATE

Internationally the future of freedom to operate and private incentives in plant biotechnology hinges on the evolution of international negotiations of TRIPS and on the levels of enforcement actually exercised in countries worldwide (Sara Boettiger *et al.*, 2004). The developed country stance on IPR has recently been effectively spearheaded by the United States, where utility patents are coming to dominate the IPR system for plant biotechnology and breeding. The United States is pressuring other countries globally to adopt similar IP regimes. Yet, as a template for worldwide adoption, the US patent system suffers from a number of apparent shortcomings. Foremost among these are its overly permissive patenting criteria (Barton, 2003), new and surprising evidence of which has recently come to light. Quillen (2002) estimates that in the mid-1990s the allowance rate (patents granted per patent filed with a two-year examination lag), corrected for continuing applications, was 83%, whereas the allowance rates in Europe and Japan were 68% and 65%, respectively. Critics also claim US patent examiners tend to ignore prior art beyond the scope of their own electronic database. Others point to the cursory examination processes of countries such as Australia and South Africa as being more problematic. To control the abuse of blocking patents, the United States (along with European countries, among others) rely on their antitrust mechanisms to control against market manipulations based on patent positions, mechanisms that are not available in many developing countries.

Some kind of licensing market segmentation is also a key part of any strategy to take account of the needs of those currently not well served by commercial agricultural biotechnology. Effective segmentation could create an IP exchange mechanism whereby the public sector's two-thirds of the global investment in agricultural R&D could develop and deliver biotechnological solutions that may have beneficial social impacts but that are not feasible on commercial terms. Implementation, however, might be quite problematical (Lybert, 2002).

It is important to recognize that aspects of recent initiatives in IPR could be very advantageous for developing countries. Harmonization of administration and the centralization of function (if not control) of the patenting process, as advocated by WIPO in the SPLT proposal, makes very good sense in a system where three-fourths or more of the work and the cost of national patent offices is duplicated elsewhere in the world. Further centralization could be especially useful for countries with scant administrative capacity. Indeed, the absence of substantial investment in administrative infrastructure, it is hard to see how intellectual property regimes in conformity with TRIPS could effectively be implemented in many developing countries (Smith, 2002). Lack of such infrastructure creates a disincentive for local innovators, and countries are unable to benefit from the fruits of any domestic research or from flows of foreign direct investment and the technology transfer they often bring.

An economically optimal outcome for international intellectual property arrangements over plants and plant biotechnologies, from the point of view of global public welfare, could be the harmonization and centralization of functions while allowing some local choice of patentability, as in TRIPS, along with policies to encourage after-market segmentation, recognizing differences between developed and developing countries and between commercial and humanitarian use (Sara Boettiger *et al.*, 2004). With time, certain limitations on the scope and the value of private innovation in agricultural biotechnology will become more widely recognized. To the extent that agricultural biotechnology fulfills its great promise, most of the benefits will likely be realized in feeding a large and growing world population at low cost, and not in producer profit margins that are in any way comparable to those earned in pharmaceuticals. IPRs do have an increasingly important role in encouraging agricultural innovation but continued and increased funding of public and nonprofit research beyond a handful of high-value commercial crops will remain a crucial source of the continued technical process in global agriculture.

8.7 WORLD TRADE ORGANIZATION (WTO)

The World Trade Organization (WTO) is an organization that intends to supervise and liberalize international trade. It is the only international organization dealing with global rules of trade between nations. Its main function is to ensure that trade flows as smoothly, predictably and freely as possible. The result is assurance. Consumers and producers know that they can enjoy secure supplies and greater choice of the finished products, components, raw materials and services that they use. Producers and exporters know that foreign markets will remain open to them. The result is also a more prosperous, peaceful and accountable economic world. Decisions in the WTO are typically taken by consensus among all member countries, and they are ratified by respective Parliaments. Trade friction is channeled into the WTO's dispute settlement process where the focus is on interpreting agreements and commitments, and how to ensure that countries' trade policies conform to them (Anonymous, 2001). That way, the risk of disputes spilling over into political or military conflict is reduced. By lowering trade barriers, the WTO's system also breaks down other barriers between peoples and nations.

At the heart of the system, known as the multilateral trading system, are the WTO's agreements, negotiated and signed by a large majority of the world's trading nations, and ratified in their parliaments. These agreements are the legal ground-rules for international commerce. Essentially, they are contracts, guaranteeing member countries important trade rights. They also bind governments to keep their trade policies within agreed limits to everybody's benefit. The agreements were negotiated and signed by governments. But their purpose is to help producers of goods and services, exporters, and importers conduct their business. The goal is to improve the welfare of the people of the member countries.

The organization officially commenced on 1 January 1995 under the Marrakech Agreement, replacing the General Agreement on Tariffs and Trade (GATT), which commenced in 1948. The organization deals with regulation of trade between participating countries by providing a framework for negotiating and formalizing trade agreements and a dispute resolution process aimed at enforcing participant's adherence to WTO agreements, which are signed by representatives of member governments and ratified by their parliaments. Most of the issues that the WTO focuses on derive from previous trade negotiations, especially from the Uruguay Round (1986–1994) (Malanczuk, 1999).

As of the youngest of the international organizations, the WTO is the successor to the General Agreement on Tariffs and Trade (GATT) established in the wake of the Second World War. So, while the WTO is still young, the multilateral trading system that was originally set up under GATT is well over 50 years old (Anonymous, 2001). The past 50 years have seen exceptional growth in world trade. Merchandise exports grew on average by 6% annually. Total trade in 2000 was 22-times the level of 1950. GATT and the WTO have helped to create a strong and prosperous trading system contributing to unprecedented growth. The system was developed through a series of trade negotiations, or rounds, held under GATT. The first rounds dealt mainly with tariff reductions, but later negotiations included other areas such as antidumping and nontariff measures. The last round – the 1986–94 Uruguay Round – led to WTO's creation.

8.7.1 WTO Agreements

The WTO's rules, the agreements, are the result of negotiations between the members. The current set were the outcome of the 1986–94 Uruguay Round negotiations which included a major revision of the original General Agreement on Tariffs and Trade (GATT). GATT is now the WTO's principal rule-book for trade in goods (Anonymous, 2001). The Uruguay Round also created new rules for dealing with trade in services, relevant aspects of intellectual property, dispute settlement, and trade policy reviews. The complete set runs to some 30,000 pages consisting of about 30 agreements and separate commitments (called schedules) made by individual members in specific areas such as lower customs duty rates and services market-opening. Through these agreements, WTO members operate a nondiscriminatory trading system that spells out their rights and their obligations. Each country receives guarantees that its exports will be treated fairly and consistently in other countries' markets. Each promises to do the same for imports into its own market. The system also gives developing countries some flexibility in implementing their commitments.

8.7.2 Structure and Functions

The WTO has 153 members, accounting for almost 95% of world trade. Around 30 others are negotiating membership. Decisions are made by the entire membership. This is typically by consensus. A majority vote is also possible, but it has never been used in the WTO and was extremely rare under the WTO's predecessor, the General Agreement on Tariffs and Trade (GATT). The WTO's agreements have been ratified in all members' parliaments. The WTO's top-level decision-making body is the Ministerial Conference, which meets at least once every two years. Below this is the General Council (normally ambassadors and heads of delegation in Geneva but sometimes officials sent from members' capitals) which meets several times a year in the Geneva headquarters. The General Council also meets as the Trade Policy Review Body and also as dispute settlement body.

At the next level, the Goods Council, Services Council, and Intellectual Property (TRIPS) Council report to the General Council. Numerous specialized committees, working groups, and working parties deal with the individual agreements and other areas such as the environment, development, membership applications and regional trade agreements (Anonymous, 2001).

The WTO Location	:	Geneva, Switzerland
Established	:	1 January 1995
Created by	:	Uruguay Round negotiations (1986–94)
Membership	:	153 countries (on 23 July 2008)
Budget	:	189 million Swiss francs for 2009
Secretariat staff	:	625
Head	:	Director-General, Pascal Lamy

Major functions include administering WTO trade agreements, forum for trade negotiations, handling trade disputes, monitoring national trade policies, technical assistance and training for developing countries and cooperation with other international organizations.

The organization is attempting to complete negotiations on the Doha Development Round, which was launched in 2001 with an explicit focus on addressing the needs of developing countries. As of June 2012, the future of the Doha Round remained uncertain: the work program lists 21 subjects in which the original deadline of 1 January 2005 was missed, and the round is still incomplete. The conflict between free trade on industrial goods and services but retention of protectionism on farm substudies to domestic agricultural sector (requested by developed countries) and the substantiation of the international liberalization of fair trade on agricultural products (requested by developing countries) remain the major obstacles. These points of contention have hindered any progress to launch new WTO negotiations beyond the Doha Development Round. As a result of this impasse, there has been an increasing number of bilateral free trade agreements signed. As of July 2012, there were various negotiation groups in the WTO system for the current agricultural trade negotiation which is in the condition of stalemate.

8.7.3 Major Roles

Among the various functions of the WTO, the most important are, WTO oversees the implementation, administration, and operation of the covered agreements and provides a forum for negotiations and for settling disputes (Steinberg, Richard, 2002). Additionally, it is the WTO's duty to review and propagate the national trade policies, and to ensure the coherence and transparency of trade policies through surveillance in global economic policy-making. Another priority of the WTO is the assistance of developing, least developed and low-income countries in transition to adjust to WTO rules and disciplines through technical cooperation and training.

- (i) The WTO shall facilitate the implementation, administration, and operation and further the objectives of this Agreement and the Multilateral Trade Agreements, and shall also provide the framework for the implementation, administration, and operation of the multilateral trade agreements.
- (ii) The WTO shall provide the forum for negotiations among its members concerning their multilateral trade relations in matters dealt with under the Agreement in the Annexes to this Agreement.
- (iii) The WTO shall administer the Understanding on Rules and Procedures Governing the Settlement of Disputes.
- (iv) The WTO shall administer Trade Policy Review Mechanism.
- (v) With a view to achieving greater coherence in global economic policy-making, the WTO shall cooperate, as appropriate, with the International Monetary Fund (IMF) and with the International Bank for Reconstruction and Development (IBRD) and its affiliated agencies.

The above five listings are the additional functions of the World Trade Organization. As globalization proceeds in today's society, the necessity of an International Organization to manage the trading systems has been of vital importance. As the trade volume increases, issues such as protectionism, trade barriers, subsidies, violation of intellectual property arise due to the differences in the trading rules of every nation. The World Trade Organization serves as the mediator between the nations when such problems arise. WTO could be referred to as the product of globalization and also as one of the most important organizations in today's globalized society (Malanczuk, 1999).

The WTO is also a center of economic research and analysis: regular assessments of the global trade picture in its annual publications and research reports on specific topics are produced by the organization. Finally, the WTO cooperates closely with the two other components of the Bretton Woods system, the IMF, and the World Bank.

8.7.4 Principles of Trading System

The WTO establishes a framework for trade policies; it does not define or specify outcomes. That is, it is concerned with setting the rules of the trade policy games. Following five principles are of particular importance in understanding both the pre-1994 GATT and the WTO.

1. **Nondiscrimination:** It has two major components: the most favored nation (MFN) rule, and the national treatment policy. Both are embedded in the main WTO rules on goods, services, and intellectual property, but their precise scope and nature differ across these areas. The MFN rule requires that a WTO member must apply the same conditions on all trade with other WTO members, i.e., a WTO member has to grant the most favorable conditions under which it allows trade in a certain product type to all other WTO members."Grant someone a special favor and you have to do the same for all other WTO members." National treatment means that imported goods should be treated no less favorably than domestically produced goods (at least after the foreign goods have entered the market) and was introduced to tackle nontariff barriers to trade (e.g., technical standards, security standards *et al.* discriminating against imported goods).
2. **Reciprocity:** It reflects both a desire to limit the scope of free-riding that may arise because of the MFN rule and a desire to obtain better access to foreign markets. A related point is that for a nation to negotiate, it is necessary that the gain from doing so be greater than the gain available from unilateral liberalization; reciprocal concessions intend to ensure that such gains will materialize.
3. **Binding and enforceable commitments:** The tariff commitments made by WTO members in a multilateral trade negotiation and on accession are enumerated in a schedule (list) of concessions. These schedules establish 'ceiling bindings': a country can change its bindings, but only after negotiating with its trading partners, which could mean compensating them for loss of trade. If satisfaction is not obtained, the complaining country may invoke the WTO dispute settlement procedures.

4. **Transparency:** The WTO members are required to publish their trade regulations, to maintain institutions allowing for the review of administrative decisions affecting trade, to respond to requests for information by other members, and to notify changes in trade policies to the WTO. These internal transparency requirements are supplemented and facilitated by periodic country-specific reports (trade policy reviews) through the Trade Policy Review Mechanism (TPRM). The WTO system tries also to improve predictability and stability, discouraging the use of quotas and other measures used to set limits on quantities of imports.
5. **Safety valves:** In specific circumstances, governments are able to restrict trade. The WTO's agreements permit members to take measures to protect not only the environment but also public health, animal health, and plant health.

The three types of provision in this direction include articles allowing for the use of trade measures to attain noneconomic objectives; articles aimed at ensuring 'fair competition'; members must not use environmental protection measures as a means of disguising protectionist policies and provisions permitting intervention in trade for economic reasons. Exceptions to the MFN principle also allow for preferential treatment of developing countries, regional free trade areas, and customs unions.

8.7.5 Decision Making

The WTO describes itself as "a rules-based, member-driven organization, all decisions are made by the member governments, and the rules are the outcome of negotiations among members." The WTO Agreement foresees votes where consensus cannot be reached, but the practice of consensus dominates the process of decision-making.

8.7.6 Dispute Settlement

In 1994, the WTO members agreed on the Understanding on Rules and Procedures Governing the Settlement of Disputes (DSU) annexed to the 'Final Act' signed in Marrakesh in 1994. Dispute settlement is regarded by the WTO as the central pillar of the multilateral trading system, and as a "unique contribution to the stability of the global economy." WTO members have agreed that, if they believe fellow-members are violating trade rules, they will use the multilateral system of settling disputes instead of taking action unilaterally. The operation of the WTO dispute settlement process involves the DSB panels, the Appellate Body, the WTO Secretariat, arbitrators, independent experts and several specialized institutions. Bodies involved in the dispute settlement process, World Trade Organization.

8.7.7 Agreements

The WTO oversees about 60 different agreements which have the status of international legal texts. Member countries must sign and ratify all WTO agreements on accession. A discussion of some of the most important agreements follows. The Agreement on

Agriculture came into effect with the establishment of the WTO at the beginning of 1995. The AoA has three central concepts, or 'pillars': domestic support, market access, and export subsidies. The General Agreement on Trade in Services was created to extend the multilateral trading system to service sector, in the same way as the General Agreement on Tariffs and Trade (GATT) provided such a system for merchandise trade. The agreement entered into force in January 1995. The Agreement on Trade-Related Aspects of Intellectual Property Rights sets down minimum standards for many forms of intellectual property (IP) regulations. It was negotiated at the end of the Uruguay Round of the General Agreement on Tariffs and Trade (GATT) in 1994.

The Agreement on the Application of Sanitary and Phytosanitary measures, also known as the SPS Agreement, was negotiated during the Uruguay Round of GATT and entered into force with the establishment of the WTO at the beginning of 1995. Under the SPS agreement, the WTO sets constraints on members' policies relating to food safety (bacterial contaminants, pesticides, inspection and labeling) as well as animal and plant health (imported pests and diseases). The Agreement on Technical Barriers to Trade is an international treaty of the World Trade Organization. It was negotiated during the Uruguay Round of the General Agreement on Tariffs and Trade and entered into force with the establishment of the WTO at the end of 1994. The object ensures that technical negotiations and standards, as well as testing and certification procedures, do not create unnecessary obstacles to trade." The Agreement on Customs Valuation, formally known as the Agreement on Implementation of Article VII of GATT, prescribes methods of customs valuation that Members are to follow. Chiefly, it adopts the 'transaction value' approach. In December 2013, the biggest agreement within the WTO was signed and known as the Bali Package.

8.7.7.1 WTO Agreement on sanitary and phytosanitary measures

The Agreement on the Application of Sanitary and Phytosanitary Measures (the 'SPS Agreement') entered into force with the establishment of the World Trade Organization on 1 January 1995. It concerns the application of food safety and animal and plant health regulations. The Agreement on the Application of Sanitary and Phytosanitary Measures sets out the basic rules for food safety and animal and plant health standards.

For the purposes of the SPS Agreement, sanitary and phytosanitary measures are defined as any measures applied to protect human or animal life from risks arising from additives, contaminants, toxins or disease-causing organisms in their food; to protect human life from plant- or animal-carried diseases; to protect animal or plant life from pests, diseases, or disease-causing organisms and to prevent or limit other damage to a country from the entry, establishment or spread of pests.

These include sanitary and phytosanitary measures taken to protect the health of fish and wild fauna, as well as of forests and wild flora. Measures for environmental protection (other than as defined above), to protect consumer interests, or for the welfare of animals are not covered by the SPS Agreement. These concerns, however, are addressed by other WTO agreements. It allows countries to set their own standards. But it also says regulations must be based on science. They should be applied only to the extent necessary to protect human, animal or plant life or health. And they should not arbitrarily or unjustifiably discriminate between countries where identical or similar conditions prevail. Member countries are encouraged to use international standards, guidelines, and recommendations where they exist. However, members may use measures which result in higher standards if there is scientific justification. They can also set higher standards based on appropriate assessment of risks so long as the approach is consistent, not arbitrary. The agreement still allows countries to use different standards and different methods of inspecting products

8.7.7.1.1 Key features

All countries maintain measures to ensure that food is safe for consumers, and to prevent the spread of pests or diseases among animals and plants. These sanitary and phytosanitary measures can take many forms, such as requiring products to come from a disease-free area, inspection of products, specific treatment or processing of products, setting of maximum allowable levels of pesticide residues or permitted use of only certain additives in food. Sanitary (human and animal health) and phytosanitary (plant health) measures apply to domestically produced food or local animal and plant diseases, as well as to products coming from other countries.

Sanitary and phytosanitary measures, by their very nature, may result in restrictions on trade. All governments accept the fact that some trade restrictions may be necessary to ensure food safety and animal and plant health protection. However, governments are sometimes pressured to go beyond what is needed for health protection and to use sanitary and phytosanitary restrictions to shield domestic producers from economic competition. Such pressure is likely to increase as other trade barriers are reduced as a result of the Uruguay Round agreements. A sanitary or phytosanitary restriction which is not actually required for health reasons can be a very effective protectionist device, and because of its technical complexity, a particularly deceptive and difficult barrier to challenge.

The Agreement on Sanitary and Phytosanitary Measures (SPS) builds on previous GATT rules to restrict the use of unjustified sanitary and phytosanitary measures for the purpose of trade protection. The basic aim of the SPS Agreement is to maintain the sovereign right of any government to provide the level of health protection it deems appropriate, but to ensure that these sovereign rights are not misused for protectionist purposes and do not result in unnecessary barriers to international trade.

8.8 SANITARY AND PHYTOSANITARY REQUIREMENTS (SPS)

In WTO agreement, protection of human safety or health, protection of animal and plant life or health and also protection of the environment is of prime importance, all these are broadly covered in Sanitary and Phytosanitary (SPS) requirements. The agreement on the application of Sanitary and Phytosanitary Measures is an integral part of Final Act GATT 1994 (b) and Agreement on Agriculture. SPS measures are defined as any measure applied in the following ways.

(i) To protect animal or plant life or health within the territory of the member from risks arising from the entry, establishment or spread or pests, disease and disease-carrying organisms;

(ii) To protect human or animal life or health within the territory of the member from risks arising from additives, contaminants, toxins or disease-causing organisms in foods, beverages or feedstuffs;

(iii) To protect human life or health within the territory of the member from risks arising from the disease carried by animals, plants or products thereof, or from the entry, establishment or spread of pests; or

(iv) To prevent or limit other damages within the territory of the member from the entry, establishment or spread of the pests.

The SPS measures include all laws, decrees, regulation requirements and procedures, including *inter alia*, end product criteria, process and production methods, testing, inspection certification and approval procedures, quarantine treatments including relevant requirements associated with the transport of animals or plants or with the materials necessary for their survival during transportation, provisions on the relevant statistical methods, sampling procedures and methods of risk assessment and packaging and labeling requirements directly related to food safety.

These measures thus include food additives, contaminants, toxins, drug or pesticide residues in food, certificate of food, animal or plant health safety, processing methods, food labeling, plant or animal quarantine, requirement for prevention, control or establishment of pest or disease and sanitary requirements for imports, whereas the sanitary provisions relate to food and animal health, phytosanitary provisions relate to plant health aspects of products.

8.8.1 Justification of Measures

The SPS Agreement, while permitting governments to maintain appropriate sanitary and phytosanitary protection, reduces possible arbitrariness of decisions and encourages consistent decision-making. It requires that sanitary and phytosanitary measures be applied for no other purpose than that of ensuring food safety and animal and plant health. In particular, the agreement clarifies which factors should be taken into account in the assessment of the risk involved. Measures to ensure food safety and to protect the health of animals and plants should be based as far as possible on the analysis and assessment of objective and accurate scientific data.

8.8.2 International Standards

The SPS Agreement encourages governments to establish national SPS measures consistent with international standards, guidelines, and recommendations. This process is often referred to as 'harmonization.' The WTO itself does not and will not develop such standards. However, most of the WTO's member governments (132 at the date of drafting) participate in the development of these standards in other international bodies.

The standards are developed by leading scientists in the field and governmental experts on health protection and are subject to international scrutiny and review. International standards are often higher than the national requirements of many countries, including developed countries, but the SPS Agreement explicitly permits governments to choose not to use the international standards. However, if the national requirement results in a greater restriction of trade, a country may be asked to provide scientific justification, demonstrating that the relevant international standard would not result in the level of health protection the country considered appropriate.

8.8.3 Adapting to Conditions

Due to differences in climate, existing pests or diseases or food safety conditions, it is not always appropriate to impose the same sanitary and phytosanitary requirements on food, animal or plant products coming from different countries. Therefore, sanitary and phytosanitary measures sometimes vary, depending on the country of origin of the food, animal or plant product concerned. This is taken into account in the SPS Agreement. Governments should also recognize disease-free areas which may not correspond to political boundaries, and appropriately adapt their requirements to products from these areas. The agreement, however, checks unjustified discrimination in the use of sanitary and phytosanitary measures, whether in favor of domestic producers or among foreign suppliers.

8.8.4 Alternative Measures

An acceptable level of risk can often be achieved in alternative ways. Among the alternatives and on the assumption that they are technically and economically feasible and provide the same level of food safety or animal and plant health, Governments should select those which are not more trade-restrictive than required to meet their health objective. Furthermore, if another country can show that the measures it applies provide the same level of health protection, these should be accepted as equivalent. This helps ensure that protection is maintained while providing the greatest quantity and variety of safe foodstuffs for consumers, the best availability of safe inputs for producers, and healthy economic competition.

8.8.5 Risk Assessment

The SPS Agreement increases the transparency of sanitary and phytosanitary measures. Countries must establish SPS measures on the basis of an appropriate assessment of the actual risks involved, and, if requested, make known what factors they took into consideration, the assessment procedures they used and the level of risk they determined to be acceptable. Although many governments already use risk assessment in their management of food safety and animal and plant health, the SPS Agreement encourages the wider use of systematic risk assessment among all WTO member governments and for all relevant products.

8.8.6 Transparency

Governments are required to notify other countries of any new or changed sanitary and phytosanitary requirements which affect trade, and to set up offices (called 'Enquiry Points') to respond to requests for more information on new or existing measures. They also must open to scrutiny how they apply their food safety and animal and plant health regulations. The systematic communication of information and exchange of experiences among the WTO's member governments provides a better basis for national standards. Such increased transparency also protects the interests of consumers, as well as of trading partners, from hidden protectionism through unnecessary technical requirements. A special Committee has been established within the WTO as a forum for the exchange of information among member governments on all aspects related to the implementation of the SPS Agreement. The SPS Committee reviews compliance with the agreement, discusses matters with potential trade impacts and maintains close cooperation with the appropriate technical organizations. In a trade dispute regarding a sanitary or phytosanitary measure, the normal WTO dispute settlement procedures are used, and advice from appropriate scientific experts can be sought.

Since 1948 national food safety, animal and plant health measures which affect trade were subject to GATT rule (Article I of the GATT), the most-favored-nation clause, required nondiscriminatory treatment of imported products from different foreign suppliers, and Article III required that such products be treated no less favorably than domestically produced goods with respect to any laws or requirements affecting their sale. These rules applied, for instance, to pesticide residue and food additive limits, as well as to restrictions for animal or plant health purposes. ATT Article XX on General Exceptions lays out a number of specific instances in which WTO members may be exempted from GATT rules. Two exceptions are of particular relevance to the protection of the environment: paragraphs (b) and (g) of Article XX. Pursuant to these two paragraphs, WTO members may adopt policy measures that are inconsistent with GATT disciplines, but necessary to protect human, animal or plant life or health (paragraph (b)), or relating to the conservation of exhaustible natural resources (paragraph (g)).

GATT Article XX on General Exceptions consists of two cumulative requirements. For a GATT-inconsistent environmental measure to be justified under Article XX, a member must perform a two-tier analysis proving, first, that its measure falls under at least one of the exceptions (e.g., paragraphs (b) to (g), two of the ten exceptions under Article XX) and, then, that the measure satisfies the requirements of the introductory paragraph (the "chapeau" of Article XX), i.e., that it is not applied in a manner which would constitute "a means of arbitrary or unjustifiable discrimination between countries where the same conditions prevail," and is not "a disguised restriction on international trade."

In the Tokyo Round of multilateral trade negotiations (1974–79), an Agreement on Technical Barriers to Trade was negotiated (the 1979 TBT Agreement or 'Standards Code'). Although this agreement was not developed primarily for the purpose of regulating sanitary and phytosanitary measures, it covered technical requirements resulting from food safety and animal and plant health measures, including pesticide residue limits, inspection requirements, and labeling. Governments which were members of the 1979 TBT Agreement agreed to use relevant international standards (such as those for food safety developed by the Codex) except when they considered that these standards would not adequately protect health. They also agreed to notify other governments, through the GATT Secretariat, of any technical regulations which were not based on international standards. The 1979 TBT Agreement included provisions for settling trade disputes arising from the use of food safety and other technical restrictions.

8.8.7 Beneficiaries of Implementation of SPS Agreement

8.8.7.1 Consumers

Consumers in all countries benefit. The SPS Agreement helps ensure, and in many cases enhances, the safety of their food as it encourages the systematic use of scientific information in this regard, thus reducing the scope for arbitrary and unjustified decisions. More information will increasingly become available to consumers as a result of greater transparency in governmental procedures and on the basis of their food safety, animal and plant health decisions. The elimination of unnecessary trade barriers allows consumers to benefit from a greater choice of safe foods and healthy international competition among producers.

8.8.7.2 Developing countries

Specific sanitary and phytosanitary requirements are most frequently applied on a bilateral basis between trading countries. Developing countries benefit from the SPS Agreement as it provides an international framework for sanitary and phytosanitary arrangements among countries, irrespective of their political and economic strength or technological capacity. Without such an agreement, developing countries could be at a disadvantage when challenging unjustified trade restrictions. Furthermore, under the SPS Agreement, governments must accept imported products that meet their safety requirements, whether these products are the result of simpler, less sophisticated methods or the most modern technology. Increased technical assistance to help developing countries in the area of food safety and animal and plant health, whether bilateral or through international organizations, is also an element of the SPS Agreement.

8.8.7.3 Exporters

Exporters of agricultural products in all countries benefit from the elimination of unjustified barriers to their products. The SPS Agreement reduces uncertainty about the conditions for selling to a specific market. Efforts to produce safe food for another market should not be thwarted by regulations imposed for protectionist purposes under the guise of health measures.

8.8.7.4 Importers

Importers of food and other agricultural products also benefit from the greater certainty regarding border measures. The basis for sanitary and phytosanitary measures which restrict trade are made clearer by the SPS Agreement, as well as the basis for challenging requirements which may be unjustified. This also benefits the many processors and commercial users of imported food, animal or plant products.

8.8.8 Phytosanitary Certification (PSC)

The phytosanitary certificate is an official declaration stating that plants and plant material that move across international boundaries are free from pests and diseases so as to prevent introduction and spread of any pests in the newer areas. PSC paves the way for rapid screening and expeditious release of imported plants/plant material at the port of arrival. In WTO agreement, its adoption has assumed greater significance.

8.8.8.1 Facilities for issue of PSC

The requisite arrangements have been made in the country for the inspection of PSC. All Plant Quarantine field stations are authorized to issue PSC in respect of exportable agricultural items, whereas the identified officers of the Argil/Horticulture Departments of States/Union territories are to issue PSC only for perishable commodities comprising fresh fruits and vegetables.

8.8.8.2 Procedure

Various steps include application (exporter has to apply to the PSC Issuing Officer-PIO), intimation from PIO (about date, place, time to produce exportable material for certification), presentation of consignment (exporter will do it at his own expenses), inspection, facility for inspection, sampling, post inspection formalities, validity of PSC, realization of fees (to be paid by exporter), and fumigation of exportable consignment (approval of Pest Control Operator–PCO, is required).

As per the plant, fruits and seeds regulation of import into India, Order 1989 (PFS order), Biosecurity and sanitary/phytosanitary import permit will be issued by the Directorate of Plant Protection Quarantine and Storage, Department of Agriculture and Cooperation, Government of India.

8.8.9 Seed Health Certification and Transboundary Movement of Seeds Under WTO Regime

Seed certification for a crop comprises of legal norms for ensuring the genetic identity, physical purity, germinability and freedom from seed-transmitted pathogens and weeds (Khetarpal, 2004). International Seed Testing Association (ISTA) and Association of

Official Seed Certifying Agencies (AOSCA), among others, have introduced minimum seed certification standards. Essentially, the certification procedures ensure the genetic purity and quality of seed production in the field, during harvest, processing, storage and, finally, inspection in the market. Seeds are distributed to farmers under the guarantee of quality, in terms of genetic and physical purity and germination capacity. Certification for seed-borne pathogens is followed only as and when required depending on the impact of the pathogen on yields. In the context of quality control for seed-transmitted diseases at the seed stage, it is essential to fix seed standards (tolerance limits) for the diseases before testing the seed lots by drawing workable samples as per norms. The detection of seed-borne pathogens is then carried out by the approved or available techniques. The technique has to be adapted to group testing of seeds, and the results are inferred based on the notified seed standards for the diseases.

Seed is an excellent carrier for the dissemination of plant pathogens and insect pests to long geographical distances. Bulk import of seeds without proper phytosanitary measures, indiscriminate exchange of germplasm and the distribution of seed by international agencies have increased the possibility of dissemination of pathogens and pests in areas previously considered disease-free (Khetarpal, 2004). Further, the threat may become severe if more virulent strains or races of the pathogens are introduced. Even a low seed transmission rate of a pathogen, especially viruses, may lead to an epiphytotic proportion of the disease in the field if other conditions of field spread and climate are favorable.

Establishment of the World Trade Organization (WTO) in 1995 was heralded as ushering in a new era of trade liberalization. Although trade in agricultural commodities has expanded as a result of obligations imposed by the WTO Agreement on the Application of Sanitary and Phytosanitary Measures (the SPS Agreement), the goal has eluded many developing countries. A major problem for the countries of South East Asia is that most of them cannot provide an adequate description of the health status of their agricultural industries. This hampers exports because quarantine agencies in prospective markets cannot assess the risk of moving pests in traded commodities and leads to trade disputes. The problem is the small number of specimen-based records contained in plant disease herbaria. The arthropod pest collections are generally much better populated, but most would benefit from assistance to bring the collections up to contemporary international standards. Many of the herbarium and arthropod specimens that do exist are not supported by essential 'passport' data and do not meet the international standard for a pest record (ISPM 8) (Lum Keng Yeang, 2003). Recognizing the need to build specimen-based records of agricultural pests to underpin market access negotiations, several ASEAN countries have turned to external expertise and funding assistance. The Australian Agency for International Development (AusAID), through various bilateral and multilateral programs, provides assistance that addresses the capacity of plant health scientists in recipient countries to diagnose pests of agriculture, to preserve specimens and maintain collections over time. It is also building capacity to manage data held in these collections that can contain tens of thousands of specimens, or more.

Under Article 9 of the WTO SPS Agreement, developed countries agreed to provide technical assistance to developing country members to build capacity in relation to SPS measures. Australia has responded to its obligations in a number of ways, including by introducing an SPS Capacity Building Program (SPSCBP) funded through the Australian Agency for International Development (AusAID). SPS capacity building is also included in the work program for the AFTA-CER Closer Economic Partnership. Objectives are to build regional cooperation in SPS matters, strengthen quarantine capacities and build the technical, scientific and managerial capacities of regional and national organizations responsible for SPS matters within the ASEAN region (Lum Keng Yeang, 2003).

8.8.10 World Trade Organization Regime and its Implications

The World Trade Organization (WTO), established on January 1, 1995, is the legal and institutional foundation of the multilateral trading system. The main purpose of the WTO is to promote free trade, serve as a forum for trade negotiations and settles disputes based upon the principles of nondiscrimination, equivalence, and predictability. The WTO Agreement contains more than 60 agreements covered under some 29 individual legal texts encompassing everything from services to government procurement, rules of origin and intellectual property. Of these, the following four agreements have a direct bearing on agriculture and related activities (Khetarpal, 2004).

(i) Agreement on agriculture is designed to ensure increased fairness in farm trade Agreement on Trade Related Intellectual Property Rights is aimed at improving conditions of competition where ideas and inventions are involved.

(ii) Agreement on technical barriers to trade is to ensure that technical regulations and certification procedures do not create obstacles to trade.

(iii) Agreement on the Application of Sanitary and Phytosanitary (SPS) Measures is, in fact, the one which is going to have major implications on seed/plant health in trade (Khetarpal *et al.*, 2002).

(iv) Agreement on the application of SPS measures.

The SPS Agreement covers measures to protect human health from diseases, human or animal life from food-borne risks and animals and plants from pests and diseases. Therefore, the specific aims of SPS measures are to ensure food safety and to prevent the spread of diseases among animals and plants. To achieve these objectives, international standards need to be developed for which WTO has assigned the responsibilities to three following organizations. These three organizations are often referred to as the 'Three Sisters' who are observers and contributors to the SPS committee meetings. They also serve as experts to give advice to WTO dispute settlement panels.

8.8.10.1 For food safety

The Codex Alimentarius Commission (CAC), Vienna, a subsidiary organ of the Food and Agriculture Organization (FAO) of the United Nations and the World Health Organization (WHO) has been authorized for all matters related to food safety evaluation and harmonization (Khetarpal, 2004).

8.8.10.2 *For animal health and zoonosis*

The International Office of Epizootics (IOE), Paris, develops the standards, guidelines, and recommendations.

8.8.10.3 *For plant health*

The International Plant Protection Convention (IPPC) at FAO, Rome, is the source for International Standards for the Phytosanitary Measures (ISPMs) affecting trade. The IPPC has developed 21 phytosanitary standards so far.

8.8.11 Implications of SPS Measures on Seed Health Management

The implications of SPS Agreement cannot be dealt with separately. Moreover, we need to take into account the issues related to environment and the biodiversity. The Ministry of Environment and Forests is the nodal agency for matters related to biodiversity and deals with CBD issues. It is unfortunate that low priority has been accorded by the Ministry in the National Report submitted to CBD on the implementation of Article 8 (h) dealing with invasive alien species (Khetarpal, 2004). The phytosanitary aspects of SPS Agreement require Ministry of Agriculture to revise and harmonize our regulations with international norms, to develop national standards in tune with ISPMs, to carry out PRA and to identify pest-free areas (PFAs) (Khetarpal and Gupta, 2002 a). To increase production and to have a vibrant seed industry, we are expected to overcome the seed health related impediments in seed production, seed trade and also to market access in the WTO era. It is amply clear that we need to harmonize our national legislation with SPS norms, develop national standards and appropriate methodologies and procedures for testing, and to reorient our research projects. Information access and exchange on seed health certification and quarantine are crucial for effective working. Establishment of a national database such as a Seed Health Information System (SHIS) linking the research laboratories with seed testing and quarantine stations would be the backbone for strengthening the program on seed health certification. The SHIS can be a storehouse of information on the biology of seed-borne diseases, seed health testing procedures and policies, international standards and related issues.

Regional collaboration in quarantine among the South Asian countries including SAARC countries has to be set up to facilitate the seed trade due to commonalities in agro-ecological conditions and crop spectra (Khetarpal and Gupta, 2003). For an effective networking lists of quarantine pests not known to occur in South Asian countries as such, and pests that may be known to occur in the region but not known in a particular country of the region need to be prepared. This would help in safeguarding the interest of the region and the countries that are free from a specific quarantine pest while trading within and outside the region. The international bodies like Asia and Pacific Plant Protection Organization, and Asia and Pacific Seed Association and others are expected to take a lead to develop guidelines for disease-free seed production and safe movement of planting materials. The International Plant Genetic Resources Institute, which has published a series of technical guidelines for safe movement of germplasm, needs to support an effective quarantine networking on plant genetic resources under exchange and conservation in the region (Dhillon *et al.*, 2000). The public-private sector partnership is discussed in many fora, but a concrete action plan is still awaited in the seed sector. Hopefully, the critical appraisal of the subject would contribute significantly to generating awareness among the researchers, policy makers, and other stakeholders in the coming years for producing and trading in healthy seeds the importance of which India first proved to the world in 400 BC (Khetarpal, 2004).

The World Trade Organization Agreement on the Application of Sanitary and Phytosanitary Measures (WTO SPS) provide the rights and obligations for members to take phytosanitary measures to protect animal, plant and human life or health. Using the guidelines of the WTO SPS Agreement Article 7 this study evaluated the current SPS transparency capacity of the South African regulatory system (Theyse, 2009). Based on the outcome of the evaluation a Best Practice Model for WTO SPS notification and information management was developed to improve WTO SPS compliance for South Africa Phytosanitary capacities of regulatory systems are challenged with increased global agricultural trade and a proliferation of international and regional phytosanitary standards. International Standards for Phytosanitary measures (ISPMs) are developed by the International Plant Protection Convention (IPPC). The concept of phytosanitary capacity was analyzed and the International Plant Protection Convention (IPPC) Phytosanitary Capacity Evaluation (PCE) tool evaluated in terms of its scope, purpose, and usefulness. South Africa has attempted to address some of its phytosanitary capacity challenges system and organizational challenges by restructuring and strengthening the capacity of its National Plant Protection Organization (NPPO) to meet the demands of international phytosanitary obligations and commitments. This study evaluates the phytosanitary capacity of South Africa and uses the IPPC Phytosanitary Capacity Evaluation (PCE) tool to identify and assess the current constraints impacting on the capacity. Based on the outcome of the PCE recommendations to address information management and capacity constraints are made. The study used the IPPC Phytosanitary Capacity Evaluation (PCE) tool to evaluate phytosanitary regulatory capacity constraints impacting on the phytosanitary capacity of Malawi. The results from the PCE for Malawi was compared with the results obtained from the PCE for South Africa. The results highlighted the different levels of phytosanitary capacity between a developing country such as South Africa and a Least Developed Country such as Malawi and made recommendations to address the country-specific constraints.

8.8.12 Risk Arising from International Plant Health Protocols

In response to expanding world trade and concern over spread of plant diseases, international protocols were set up in the 1950s via the International Plant Protection Convention (IPPC) of the FAO and World Trade Organization (WTO) rules to regulate the process of trade and to reduce the likelihood of accidental introductions of organisms of phytosanitary concern. Today, protecting a state from invasive plant pathogens is often referred to as plant biosecurity (Brasier, 2008). In most of Europe plant biosecurity protocols are applied via the plant health regulations of the European Union (EU). These broadly follow the Sanitary and Phytosanitary Agreement (SPS) of the World Trade Organization as consolidated in the 1990s. In the UK, EU regulations are usually regulated and operated to a high standard (plant health teams within the Department for Environment, Food and Rural Affairs (Defra) and the UK Forestry Commission (FC) have many skilled officers and scientists). Equally, many involved in the UK plant trade aim to adhere to the protocols and to minimize the risks involved. However, in the light of recent developments in the plant trade itself and of regular breaches of UK plant biosecurity, some tenets underlying the protocols must now be viewed as outdated and seriously flawed.

8.8.13 Problems with Identifying Risk

The SPS Agreement of the World Trade Organization aims to minimize any disruption to trade that plant health regulation might impose (Brasier, 2008). The intention is to ensure that global commercial trade in plants is not unduly hindered by artificial barriers; apparently without question as to whether such international trade is a fundamentally sound or unsound process based on scientific and global environmental grounds. The protocols principally involve the production of lists of named harmful organisms. These tend to concentrate on organisms likely to affect widely grown agricultural commodities and timber. The case for inclusion of each organism must be founded on 'sound science.' By definition, all 'unlisted' organisms remain unregulated. However, the lists principally comprise pathogens that have already escaped from their geographical centers of origin and started to cause overt disease in another part of the globe. Many of these 'newly escaped' organisms were previously unknown to science and were not therefore on any international list before they escaped (Brasier, 2005). Dutch elm disease, sudden oak death, *Phytophthora* disease of alder and box blight in the UK are all examples of major disease episodes caused by previously unknown pathogens. Based on these and similar examples, and on estimates that only 7–10% of all fungal species having so far been identified (Crous and Groenwald, 2005), some 90% of pathogens may be unknown to science. The number of unknown species of *Phytophthora*, for example, arguably the world's most destructive group of plant pathogens, may be between 100 and 500 (Brasier, 2008).

As indicated in the introduction, Darwinian evolution predicts that, being adapted to and coevolved with their hosts, many of these pathogens may be between 100 and 500 (Brasier, 2008). As indicated in the introduction, Darwinian evolution predicts that, being adapted to and coevolved with their hosts, many of these pathogens are unlikely to do noticeable damage to their native ecosystems, and so are less likely to be detected. Thus, a previous survey in the Himalayas led to the discovery of a third species of Dutch elm disease fungus, unknown to science, highly aggressive to European elms, yet apparently benign on Himalayan elm species (Brasier and Mehrotra, 1995). Both practical experience and predictive science, therefore, dictate that current SPS protocols are flawed. First because they tend to concentrate on only the most noticeable escapees and so come into effect only after a problem is identified. Second, because they may cover only a minority of the organisms which pose a threat. Moreover, since they largely ignore the risk from benign, coevolved, unescaped organisms, the protocols may ignore the risk from 90% of potential pathogens. In this sense, therefore, they are non-Darwinian. Rather than focus on already escaped organisms, it is paramount to concentrate on scientific facts and principles which indicate that pathogens need to be contained within their centers of origin, not distributed around the world and subject to regulation only when causing visible damage beyond their natural range.

Under current protocols, once an organism is recognized as a potential threat the scientific case for its regulation or listing is usually based on a 'Pest Risk Analysis' or PRA. A PRA is typically initiated in response to a severe new disease episode occurring in some part of the world (Brasier, 2008). Usually, this will involve a newly escaped or an emerging pathogen. The PRA is developing into an acceptable tool for summarizing current risk information on well-characterized pathogens. It is also proving useful for identifying what extra research is needed to better evaluate a newly emerged threat. Indeed, it often highlights the fact that insufficient information is available on the potential susceptibility of local hosts to make a meaningful assessment of risk (oak wilt, caused by *Ceratocystis fagacearum*; Gibbs, 2003). In addition, a PRA can be a valuable tool for assessing what action might be undertaken to eradicate or reduce the rate of spread of a new invasive. However, PRAs cannot cover the multiplicity of unknown organisms that may pose a threat. There was no PRA in existence for *O. novo-ulmi*, *P. ramorum*, the alder *Phytophthora*, *P. kernoviae*, or *Cylindrocladium buxicola* prior to their arrival, i.e., until it was too late to prevent them escaping their origin. As with much current international plant biosecurity protocol, the PRA process tends to be reactive, slowing the spread of the fire rather than preventing its ignition. In addition, many dangerous pathogens identified as a threat in one part of the world still do not have PRAs or are not on regional quarantine schedules, in countries or areas where they might cause damage.

The 'new' Japanese oak wilt pathogen *Raffaelea quercivora*, implicated in heavy oak mortality in Japan since the 1990s, might be a serious threat to the UK and European oaks but is not on European quarantine schedules. Conversely *P. alni* subsp. *alni* (PAA), killing alders across Europe for over a decade, could be a major threat to the alders of North America but, although subject to a risk assessment by USDA, it has yet to be scheduled by APHIS. There are many such examples. The process of listing such organisms on schedules is often slow, perhaps too slow to be sufficiently effective given the speed of modern international trade and with so many emerging outbreaks of pests and pathogens worldwide, the system may be in danger of being overwhelmed, effectively reduced to dealing with already damaging organisms, frequently too late for effective mediation (Brasier, 2008).

The Office of the United States Trade Representative (USTR) published its fourth annual report on Sanitary and Phytosanitary Measures (SPS Report). This report was created to respond to the concerns of US farmers, ranchers, manufacturers, and workers who confront sanitary and phytosanitary (SPS) trade barriers as they seek to export high-quality American food and agricultural products globally. SPS measures are rules and procedures that governments use to ensure that foods and beverages are safe to consume and to protect animals and plants from pests and diseases (Anonymous, 2013).

The 2013 Report on Sanitary and Phytosanitary Measures (SPS Report) is a specialized report dedicated to describing significant barriers to the US food, farm, and ranch exports arising from measures that foreign governments apply on the grounds that such measures are necessary to protect human, animal, or plant life or health from risks arising from the entry or spread of plant or animal-borne pests or diseases, or from additives, contaminants, toxins, or disease-causing organisms in foods, beverages, or feedstuffs. These measures, known in World Trade Organization (WTO) parlance as 'sanitary and phytosanitary (SPS) measures,' play an increasingly critical role in shaping the flow of global trade. The United States strongly supports the right of governments through robust regulatory frameworks to protect their people, animals, and plants from health risks of this kind. This report focuses on SPS measures that appear to be unscientific, unduly burdensome, discriminatory or otherwise unwarranted and create significant barriers to US exports. Many of these measures can present particular challenges for small and medium-sized enterprises that typically lack the resources to identify and address such barriers. This report was intended to describe and advance US efforts to identify and eliminate these unwarranted measures; report presents an overview of SPS measures, describes the relevant international agreements governing these measures and discusses the US and international mechanisms for addressing them.

In particular, section II covers the topics like the genesis of this report, the growing importance of SPS measures in global trade, rules governing SPS measures under the WTO's Agreement on the Application of Sanitary and Phytosanitary Measures (SPS Agreement), rules and mechanisms regarding SPS measures in the US free trade agreements (FTAs), international standard setting in the SPS area, the role of various US Government agencies in addressing SPS-related trade issues, sources of information about SPS trade barriers and US trade policy mechanisms for considering and addressing SPS measures, including bilateral engagement and WTO dispute settlement.

Section III discusses important unwarranted SPS barriers that impede US exports to multiple foreign markets. Among the most significant of these cross-cutting barriers are restrictions related to export certifications, agricultural biotechnology, bovine spongiform encephalopathy (BSE), avian influenza (AI), and maximum residue limits (MRLs) for pesticides. The focal point of this report is section IV, which identifies and describes significant unwarranted SPS-related trade barriers currently facing US exporters, along with US Government initiatives to eliminate or reduce the impact of these barriers (Anonymous, 2013). The report identified SPS measures in the following countries and groups of countries: Argentina, Australia, Bahrain, Bolivia, Bosnia and Herzegovina, Brazil, Chile, China, Colombia, Croatia, the Dominican Republic, Ecuador, Egypt, El Salvador, Ethiopia, the European Union, India, Indonesia, Israel, Jamaica, Japan, Kazakhstan, Kenya, Korea, Kuwait, Kyrgyzstan, Macedonia, Malaysia, Mexico, Morocco, New Zealand, Norway, Peru, Philippines, Russia, Saudi Arabia, Serbia, Singapore, South Africa, the South African Development Community, Sri Lanka, Switzerland, Taiwan, Thailand, Turkey, Ukraine, Uruguay, and Vietnam. The report also discussed the US Government's efforts to provide technical assistance to developing countries on SPS issues. Such assistance was instrumental in US efforts to ensure that countries adopt and maintain science-based SPS measures, and helped to eliminate impediments to US food and agricultural exports (Anonymous, 2013).

Over the last 60 years, global trade has changed significantly with the formation of the World Trade Organization (WTO) in 1995 (Anonymous, 1995) and the establishment of multiple trade agreements between countries and jurisdictions all aiming at reducing obstacles to international trade. Agreements such as the Sanitary and Phytosanitary (SPS) Agreement have promoted trade liberalization and harmonization of biosecurity measures and agri-food standards but have also facilitated the intracontinental and intercontinental movement of insects and pathogenic microbes that disturb native and exotic ecosystems. The frequency of incursions of unwanted organisms into these globalized pathways is increasing and leads to the emergence of diseases and pests that are costly to control and/or eradicate (Robin MacDiarmid *et al.*, 2013).

At least three parties are involved in international trading transactions, exporters, biosecurity agencies, and importers, but the responsibility for identifying whether new viruses are present in a country or region and associated with a risk to the local flora is not always clear. It is common for the importing country to adopt a conservative stance and limit the importation of plant material based on the risk of a pathogen that may be associated with the imported material. Therefore, a shared burden of proof exists between trading countries for identifying the biosecurity risk of an organism. Human activity is causing new encounters between viruses and plants. Anthropogenic interventions include changing land use, decreasing biodiversity, trade, the introduction of new plant and vector species to native landscapes, and changing atmospheric and climatic conditions. The discovery of thousands of new viruses, especially those associated with healthy-appearing native plants, is shifting the paradigm for their role within the ecosystem from foe to friend. The cost of new plant virus incursions can be high and result in the loss of trade and/or production for short or extended periods. Three recommendations were presented and justified for plant biosecurity to improve communication about plant viruses, assist with the identification of viruses and their impacts and protect the high economic, social, environmental and cultural value of our respective nations' unique flora, viz., as part of the burden of proof, countries and jurisdictions should identify what pests already exist in, and which pests pose a risk to, their native flora, plant virus sequences not associated with a recognized virus infection are designated as 'uncultured virus' and tentatively named using the host plant species of greatest known prevalence, the word 'virus,' a general location identifier, and a serial number and invest in basic research to determine the ecology of known and new viruses with existing and potential new plant hosts and vectors and develop host-virus pathogenicity prediction tools (Robin MacDiarmid *et al.*, 2013). These recommendations have implications for researchers, risk analysts, biosecurity authorities and policy makersat both a national and an international level.

9 Nanotechnology in Plant Disease Management

The term nanotechnology, the buzzword of present day science owes its origin from the Greek word 'nano' literally meaning dwarf. When it is expressed in terms of dimension one nanometer equals to one billionth of a meter (1 nm = 10^{-9} m). The subject nanotechnology deals with manufacturing, study and manipulation of matter at the nanoscale (or atomic scale) in the size range of 1–100 nm which may be called as 'nanoparticles' (Rajan, 2004). Any material, when attenuated at the nanometer scale (less than 100 nm), exhibits new properties that are entirely different from its bulk counterpart due to the small size and high surface to volume ratio. It is shown in terms of higher plasticity at high temperature, higher hardness, breaking strength and toughness at low temperature, higher chemical reactivity and surface energy and high mobility in the body of an organism including cellular entry.

Nanotechnology refers to controlling, building and restructuring materials and devices on the scale of atoms and molecules. A nanometer (nm) is one-billionth of a meter. To get the sense of the nanoscales, the width of the human hair is 80,000 nm, and the smallest things visible to the naked human eye are 10,000 nm across (Table 9.1). At nanoscales, the basic rules of chemistry and physics are not applicable. One example of this technology is the carbon nanotube discovered in 1991, which is only a few nanometers in diameter but can conduct electricity better than copper, 100 times stronger than steel but only one-sixth of its weight. Novel agricultural systems can make excellent use of nanotech-enabled 'smart' devices that can perform a dual role of being a preventive and early warning system. These devices can identify plant related health issues even before they become visible to the farmers and simultaneously provide remedial measures. These nanotech systems can also be used to monitor the delivery of pesticides. User-friendly and eco-friendly nanodelivery systems for pesticides have started to find their place in the market. These can allow the use of pesticides with the absolute minimum risk of environmental damage. Companies have implemented nanoemulsions in commercial pesticide products.

Nanotechnology is the manipulation or self-assembly of individual atoms, molecules or molecular clusters into structures to create materials and devices with new or vastly different properties (Sunkara V Manorama, 2004). Nanotechnology has the potential to revolutionize the agricultural and food industry with new tools for the molecular treatment of diseases, rapid disease detection, enhancing the ability of plants to absorb nutrients, etc. Smart sensors and smart delivery systems will help the agricultural industry combat plant pathogens. In the near future, nanostructured catalysts will be available which will increase the efficiency of pesticides and herbicides, allowing lower doses to be used. Nanotechnology will also protect the environment indirectly through the use of alternative (renewable) energy supplies, and filters or catalysts to reduce pollution and clean-up existing pollutants. Nanotechnology is the understanding and control of matter at dimensions between approximately 1 and 100 nanometers, or nanoscale. Unusual physical, chemical, and biological properties can emerge in materials at the nanoscale. These properties may differ in important ways from the properties of bulk materials and single atoms or molecules. Encompassing nanoscale science, engineering, and technology, nanotechnology involves imaging, measuring, modeling, and manipulating matter at this length scale. The sizes of organisms and different biosubjects are given in Table 9.1.

Table 9.1 Size of different Organisms and Biomolecules on the Micro- and Nanometric Scale

S. No.	*Nature of organisms and different biosubjects*	*Size range (μm)*	*Size on nanometric scale (nm)*
1.	Plant, animal cell	10–100	10 000–10 0000
2.	Bacteria	≤ 1–10	≤ 1000–100
3.	Virus	0.03–0.1	30–100
4.	Simple molecules (proteins, DNA turns)	0.001–0.01	1–10
5.	Atoms (DNA 'base')	0.0001–0.001	0.1–1

For comparison, the wavelength of visible light is between 400 and 700 nm. A living cell has dimensions of microns [(thousands of nanometers) (Table 9.2)].

Table 9.2 Comparisons of Scale — from Macro to Molecular

Size (nm)	*Examples*	*Terminology*
0.1–0.5	Individual chemical bonds	Molecular/atomic
0.5–1.0	Small molecules, pores in zeolites	Molecular
1–1000	Proteins, DNA, mesopores, inorganic nanoparticles	Nano
103–104	Microfluidic channels, MEMS, Devices on a silicon chip, living cells	Micro
> 104	Normal bulk matter	Macro

Nanotechnology-based products and its applications in agriculture include nanofertilizers, nanoherbicides, nanopesticides, recalcitrant contaminants from water, nanoscale carriers, nanosensors, veterinary care, fisheries and aquaculture, detection of nutrient deficiencies, preservation, photocatalysis, nanobarcode, quantum dots, etc. This fast growing technology is already having a significant commercial impact, which will certainly increase in the future. While nanotechnologies offer many opportunities for innovation, the use of nanomaterials in food and agriculture has also raised a number of safety, environmental, ethical, policy and regulatory issues. Nanotechnology is emerging as the sixth revolutionary technology in the current era after the Industrial Revolution of the mid-1700s, Nuclear Energy Revolution of the 1940s, The Green Revolution of 1960s, Information Technology Revolution of 1980s, and Biotechnology Revolution of the 1990s.

A survey by Salamanca–Buentella *et al.* (2005) predicted several nanotechnology applications for agricultural production for developing countries within next ten years. These included nanoforms zeolites for slow release and efficient dosage of water and fertilizers for plants; drugs for livestock, nanocapsules, and herbicide delivery; nanosensors for soil quality and plant health monitoring, nanosensors for pests detection; nanomagnets for removal of soil contaminants and nanoparticles for new pesticides, insecticides, and insect repellents.

Nanotechnology is the design, fabrication and utilization of materials, devices and systems through control of matter on the nanometer length scale and exploitation of novel phenomena and properties (physical, chemical, biological) at that length scale. It is now more properly labeled as 'molecular nanotechnology' (MNT) or 'nanoscale-engineering.' By taking advantage of quantum-level properties, nanotechnology allows for unprecedented control of the material world, at the nanoscale, providing the means by which systems and materials can be built with exact specifications and characteristics, allowing materials to be lighter, stronger, smarter, cheaper, cleaner and more precise. Nanotechnology has the potential to advance agricultural productivity through genetic improvement of plants, delivery of genes and drug molecules to specific sites at cellular levels, and nanoarray-based gene technologies for gene expressions in plants and animals under stress conditions. The potential is increasing with suitable techniques and sensors being identified for precision agriculture, natural resource management, early detection of pathogens and contaminants in food products, smart delivery systems for agrochemicals like fertilizers and pesticides, smart systems integration for food processing, packaging and other areas like monitoring agricultural and food system security (Abhay Kumar, 2008). Further developments in nanotechnology in this sector can be expected to become the main economic driving forces in the long run and benefits consumers, producers, farmers, ecosystems, and the general society at large.

Several metallic nanoparticles such as silver and copper have the property of antimicrobial activity. Polymer-based copper nanocompounds have been investigated with the antifungal activity against plant-infecting fungi. Efficacy of silica-silver nanoparticles was studied against *Rhizoctonia solani, Pythium ultimum, Botrytis cinerea, Magnaporthe grisea* and *Colletotrichum gloeosporioides* for their control. Similarly, nanobased products were used for the control of pumpkin disease, the powdery mildew and it was demonstrated that the infecting pathogens on the leaves causing the disease were vanished within three days after spraying the product (Shafqat Saeed *et al.*, 2014).

'Nanotechnology,' is a brainchild of late Richard Feynman, Nobel Laureate in Physics in the year 1965. Nanotechnology, the frontier science of materials having unique properties than their macroscopic or bulk counterparts, has wider applications in material sciences and biology as well (Amarendra N. Misra *et al.*, 2013). The ability of nanomaterials to work at the molecular level, atom by atom, to create large structures with the fundamentally new molecular organization is the key to its applications. Nanotechnology will leave no field untouched by its groundbreaking scientific innovations. The agriculture and food industry is no exception. Nanotechnology has provided new solutions to problems in plants, plant products, and postharvest technology in enhancing the quality of plant products.

Nanoparticles show properties sharp contrast to their bulk in many respects which are utilized for use in 'nanotechnology', viz., small size (1–100 nm), large surface to volume ratio, chemically alterable physical properties, change in the chemical and physical properties with respect to size and shape. Structural sturdiness in spite of atomic granularity, enhanced or delayed particles aggregation depending on the type of the surface modification, enhanced photoemission, high electrical and heat conductivity and improved surface catalytic activity (Amarendra N. Misra *et al.*, 2013).

Nanotechnology, the process to generate, manipulate, and deploy nanomaterials, represents an area holding significant promise for the agricultural scenario (Vandana Ghormade *et al.*, 2011). Nanotechnology employs nanoparticles (NPs) having one or more dimensions in the order of 100 nm or less. Other authors refer to NPs as colloidal particulate systems with size ranging between 10 and 1000 nm. Nanomaterials hold great promise regarding their application in plant protection and nutrition due to their size-dependent qualities, high surface-to-volume ratio, and unique optical properties. A wide variety of materials is used to make NPs, such as metal oxides, ceramics, silicates, magnetic materials, semiconductor quantum dots (QDs), lipids, polymers, dendrimers, and emulsions. Polymers display controlled release of ingredients, a character useful for developing polymeric NPs as agrochemical carriers. Metal nanoparticles display size-dependent properties such as magnetism (magnetic NPs), fluorescence (QDs) or photocatalytic degradation (metal oxide NPs) that have biotechnological applications in sensor development, agrochemical degradation and soil remediation (Table 9.3).

Table 9.3 Applications of Nanotechnology in Agriculture

Application pesticide delivery	*Nanoparticles*
(a) Chemical	
Avermectin	Porous hollow silica (15 nm)
Ethiprole or phenylpyrazole	Poly-caprolactone (135 nm)
Gamma-cyhalothrin	Solid lipid (300 nm)
Tebuconazole/chlorothalonil	Polyvinylpyridine and polyvinylpyridine-co-styrene (100 nm)
(b) Biopesticides	
Plant origin: nanosilica	Nanosilica (3–5 nm)
essential oil encapsulation	Solid lipid (200–294 nm)

Contd...

Contd...

Microorganisms: *Lagenidium giganteum* cells in emulsion	Silica (7–14 nm)
Microbial product: absorption of Myrothrecium verrucaria enzyme complex	Chitosan/kaolin (250–350 nm)
(c) Controlled delivery	
NPK controlled delivery	Nanocoating of sulfur (100 nm layer) Chitosan (78 nm)
Genetic material delivery DNA	Gold (10–15 nm), Gold (5–25 nm), Starch (50–100 nm)
Double-stranded RNA	Chitosan (100–200 nm)
(d) Pesticide sensor	
Carbofuran/triazophos	Gold (40 nm)
DDT	Gold (30 nm)
Dimethoate	Iron oxide (30 nm), zirconium oxide (31.5 nm)
Organophosphate	Zirconium oxide (50 nm)
Paraoxon	Silica (100–500 nm), Carbon nanotubes
Pyrethroid	Iron oxide (22 nm)
(e) Pesticide degradation	
Lindane	Iron sulfide (200 nm)
Imidacloprid	Titanium oxide (30 nm)

Potential applications of nanotechnology in agriculture are: delivery of nanocides–pesticides encapsulated in nanomaterials for controlled release; stabilization of biopesticides with nanomaterials; slow release of nanomaterial assisted fertilizers, biofertilizers, and micronutrients for efficient use; and field applications of agrochemicals, nanomaterials assisted delivery of genetic material for crop improvement. Nanosensors for plant pathogen and pesticide detection and NPs for soil conservation or remediation are other areas in agriculture that can benefit from nanotechnology. Enzyme immobilization for nanobiosensor using nanomaterials involves the high-value low volume application of enzymes (Kim *et al.*, 2006). Usually costly, large enzyme volumes are required for biocontrol in agricultural fields that would be practical if spray applications combined high volume with low value. Cost-effectiveness of such biocontrol preparations can be achieved by immobilization of enzyme/inhibitors on nanostructures, providing large surface areas, to increase the effective concentration of the preparation.

Nanotechnology for the control of plant diseases is a promising technique in plant pathology either by providing controlled delivery of functional molecules or as a diagnostic tool for disease detection, an important step in plant disease treatment (Sharon *et al.*, 2010). Several studies were conducted using Nanosized particles to control fungal pathogens including such as *Pythium ultimum, Magnaporthe grisea, Colletotrichum gloeosporioides, Botrytis cinerea* and, *Rhyzoctonia solani,* as well as bacterial disease including *Bacillus subtilis, Azotobacter chrococuum, Rhizobium tropici, Pseudomonas syringae* and *Xanthomonas compestris* pv. *vesicatoria* (Park *et al.*, 2006).

9.1 MAJOR CLASSIFICATION OF NANOPARTICLES

Nanoparticles can be broadly classified into two groups, organic (carbon) and inorganic nanoparticles (Noncarbon). Organic nanoparticles include carbon nanoparticles (fullerenes); inorganic nanoparticles include magnetic nanoparticles, noble metal nanoparticles (like gold and silver) and semiconductor nanoparticles (like titanium dioxide and zinc oxide) (Ratna Kumari and Sobha, 2013). Inorganic nanoparticles are of special interest as they provide superior material properties with functional versatility and are therefore potential tools for medical imaging as well as for treating diseases and used in cellular delivery due to their versatile features like wide availability, rich functionality, good biocompatibility, the capability of targeted drug delivery and controlled release of drugs. Inorganic nanoparticles exhibit intrinsic optical properties which may enhance the transparency of polymer-particle composites. Both organic and inorganic nanoparticles are widely employed in environmental remediation.

9.1.1 Carbon-based Nanoparticles

Carbon as a nonmetal element can be found in every living organism and millions of various compounds. Based on the structure, electronic, mechanical, optical and chemical characters, carbon nanoparticles are classified into various dimension dependent categories ranging from zero-dimensional (fullerene, carbon-encapsulated metal nanoparticles, nanodiamond, and onion-like carbons) to one-dimensional (carbon nanofibers and carbon nanotubes) and two-dimensional (graphene and carbon nanowalls) depending on their nanoscale range (< 100 nm) in different spatial directions.

9.1.1.1 Zero-dimensional carbon nanomaterials (0-Dcns)

9.1.1.1.1 Fullerene

Fullerenes are large spherical, ellipsoid or cylindrical carbon caged molecules forming polyhedral structures consisting of pentagons and hexagons (Fig. 9.1). These are inert and indefinitely modifiable. The suffix 'ene' indicates that each C atom is covalently bonded to three others (instead of the maximum of four). The cylindrical fullerenes are known as buckytubes or nanotubes. A fullerene nanotube has about 20 times high tensile strength than that of a high strength steel alloy and a density half that of aluminum (Ratna Kumari and Sobha, 2013). The first synthesized and most commonly used fullerene is C60; the other types of fullerene are C70, C76, and C84 (the subscripts indicates the number of carbon atoms found in the structure).

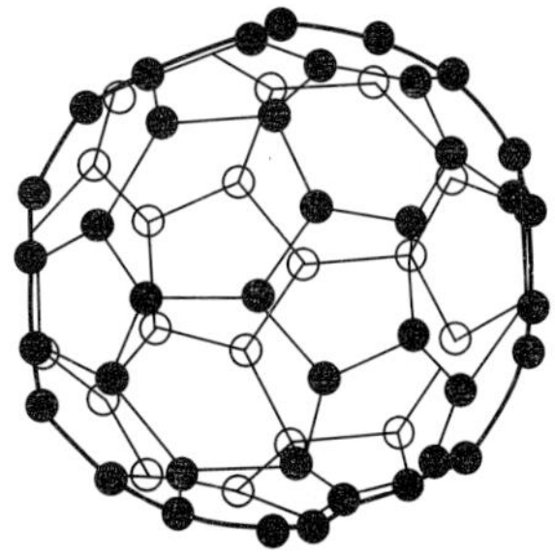

Fig. 9.1 The sixty-member carbon 'buckyball' molecule

9.1.1.1.2 Nanodiamond

Diamond, spherical and truncated octahedron diamond with predominant 3-bonded carbon (Fig. 9.2) is one of the hardest materials known to date and regarded as the king of all gemstones because of its scientific qualities in hardness, chemical corrosiveness, thermal expansion and conductivity, electrical insulation, and biocompatibility. Nanodiamond (ND) is a cubic structural diamond, with structure and properties as that of a diamond and is 5 nm in diameter on average. Nanodiamond contains a variety of diamond based materials at the nanoscale (the length scale of approximately 1–100 nm) including pure-phase diamond films, diamond particles, and their structural assemblies. Nanodiamonds have a wide range of applications in electrochemical coatings, polymer compositions, antifriction coatings, polishing, lubricants, biosensors, imaging probes, implant coatings and drug carriers (Ratna Kumari and Sobha, 2013).

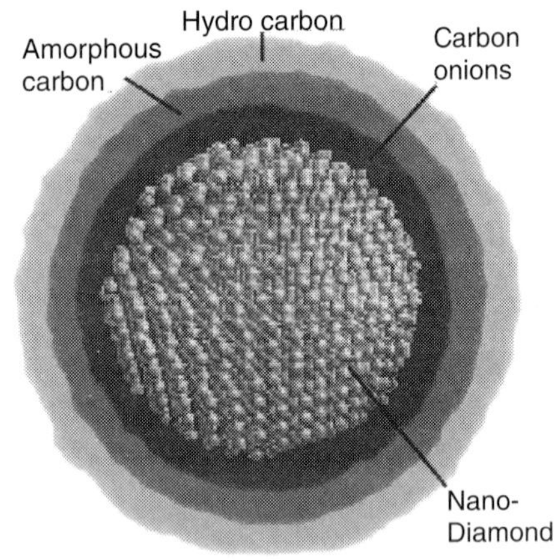

Fig. 9.2 Nanodiamond

9.1.1.1.3 Onion-like carbons (OLCs)

The onion-like carbons (OLCs) are polyhedral nanoparticles with well aligned concentric rings and show a high degree of symmetry. Carbon onions are quasi-spherical nanoparticles, possess a hollow core encased with three to eight closed graphitic shell structures (Fig. 9.3). The outer diameters are in the range of 20–100 nm. They can serve as nanocapsules for drug delivery systems. The external graphite acts as a template for the attachment of desirable functional groups and protects the substances present inside.

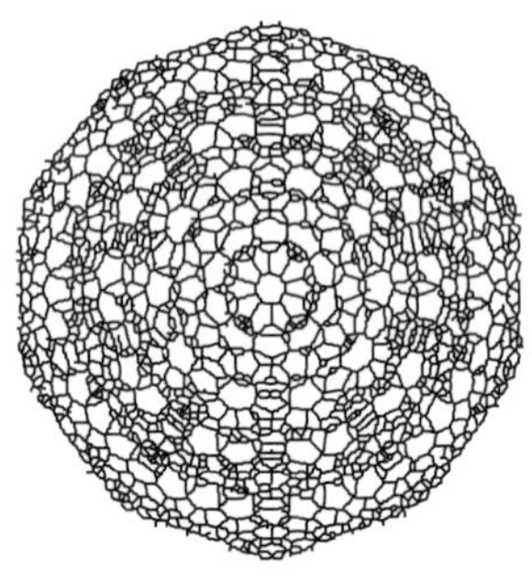

Fig. 9.3 Onion-like carbons

9.1.1.2 One-dimensional carbon nanomaterials (1-Dcns)

9.1.1.2.1 Carbon nanofibers

These are cylindrical or conical, filamentous carbon nanostructures from a quasi-onedimensional (1D) filament (Fig. 9.4). Their diameter ranges from a few to hundred nanometers, whereas lengths range from less than a micrometer to millimeters. The morphological structure is often divided into plate CNFs, ribbon-like CNFs, herringbone

CNFs (Rassaei *et al.*, 2007). These are used for electrode materials, as anodes for rechargeable lithium-ion batteries.

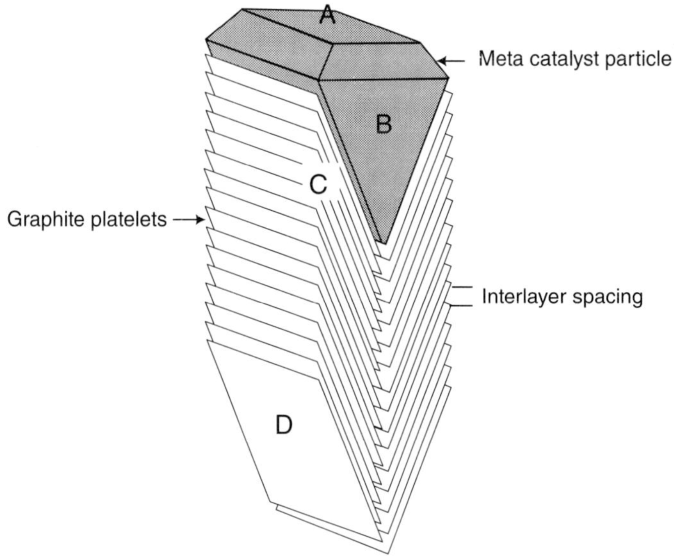

Fig. 9.4 Schematic diagram of a catalytically grown carbon nanofiber

9.1.1.2.2 Carbon nanotubes

These are the strongest material discovered to date. Carbon nanotubes are the most important members of the hollow 1-D nanosystem. They are tubular structures with nanometer diameter and large length ratio. The nanotubes consist of three different types of carbon nanotubes, viz., single wall nanotubes (SWNTs) made of single layers of graphene with a diameter of the order of 1.4 nm, multiwall nanotubes (MWNTs) made of 4–24 concentric cylinders of graphene layers with adjacent shells separation of 0.34 nm and a diameter of 10–20 nm, and fishbone nanotubes in which the graphene layers are not parallel to the growing axis (Fig. 9.5). These are widely used in nanotechnology.

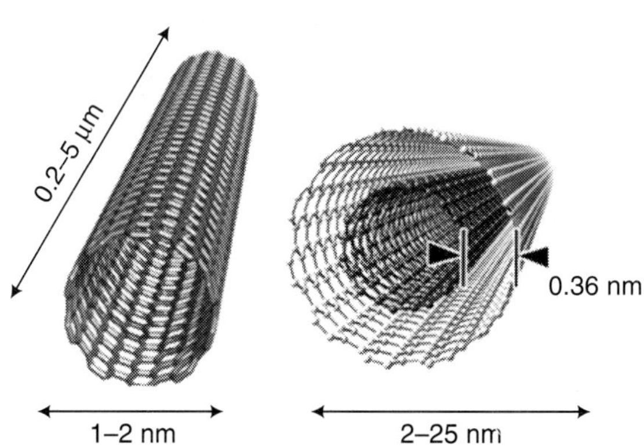

Fig. 9.5 (a) SWNTs (b) MWNTs

9.1.1.3 Two-dimensional carbon nanomaterials (2-Dcns)

9.1.1.3.1 Graphene

It is the basic building block of (0-D) fullerenes, (1-D) carbon nanotubes, and (3-D) graphite. Graphene is arranged densely in a two-dimensional hexagonal honeycomb crystal lattice (Fig. 9.6). The three strong σ bonds provide mechanical stability of the carbon sheet, and the π orbitals are responsible for the electron conduction. It is one of the strongest materials in nature roughly 200 times the strength of steel. It is very thin and flexible in nature. Graphene is a high electron mobility material with low resistivity at room temperature.

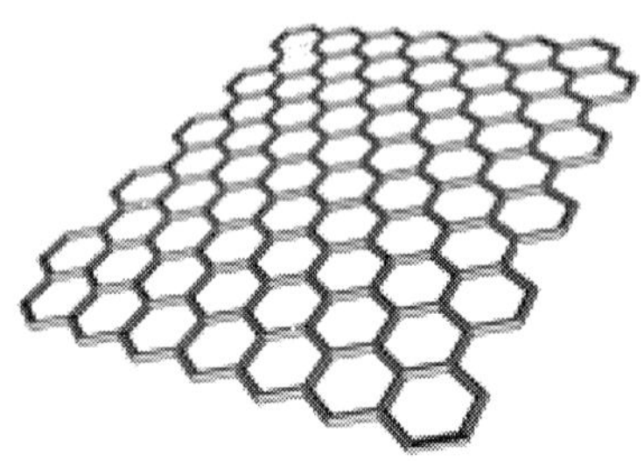

Fig. 9.6 Graphene

9.1.1.3.2 Carbon nanowalls

These are also referred to as 'carbon nanoflakes.' Carbon nanowalls (CNWs) consist of vertically aligned graphene sheets standing on the substrates, form two-dimensional wall structure with large surface areas and sharp edges (Fig. 9.7). The thickness of CNWs ranges from a few nm to a few tens nm. CNWs can also be used as a template for loading other nanomaterials, and the resulting hybrid nanostructures are used as photocatalysts.

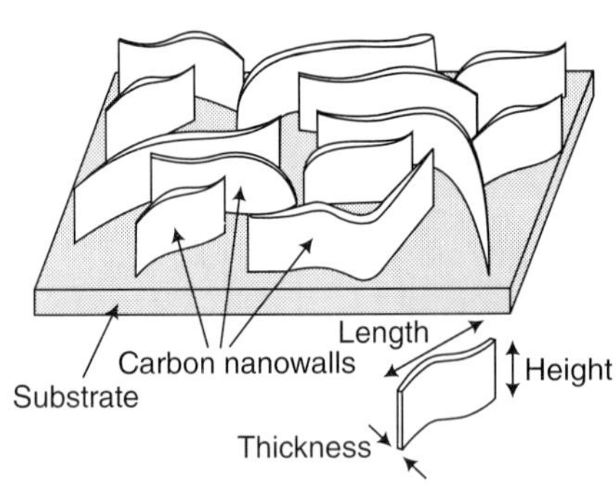

Fig. 9.7 Carbon nanowalls

9.1.2 Non-carbon Based Nanoparticles

9.1.2.1 Magnetic nanoparticles

Nanoparticles that can be manipulated by the application of magnetic field are known as magnetic nanoparticles. The magnetic elements such as iron, nickel and cobalt and their chemical compounds are commonly used. Magnetic nanoparticles are available in the past four decades in different compositions, sizes, and stabilizers. The magnetic nanoparticles attain magnetic property due to central iron oxides typically Fe_2O_3 and Fe_2O_4 (Tartaj *et al.*, 2003). The central core is surrounded by a silicon coat or a layer of polymer such as dextran or polyacrylamide (Fig. 9.8). As the particle size is reduced to nanorange, it increases the magnetic behavior. These nanoparticles lose magnetic property when removed from the magnetic field. Magnetic nanoparticles can be made to interact with the biological entity. These particles are used *in vitro* biosensing agents to detect cellular biomarkers (Shao *et al.*, 2012).

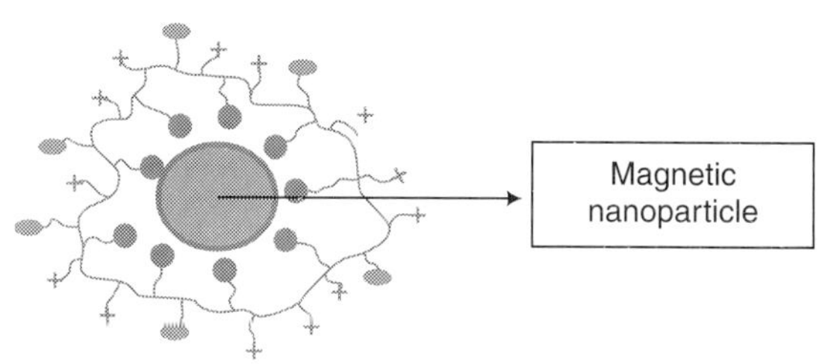

Fig. 9.8 Polymer coated magnetic nanoparticle

9.1.2.2 Nanocapsule

A 'Nanocapsule' is a nanoparticle that consists of a shell and a 'space,' in which desired substances may be placed (Fig. 9.9). For example, polyelectrolyte (PE) capsules prepared by means of the layer-by-layer technique with tailored properties, fulfill the requirements for nanoorganized systems in a satisfactory manner. With these shells, material can be delivered to predefined organs. Nanocapsules are prepared using molecules called phospholipids, which are hydrophobic (water-repellant) on one end and hydrophilic (water-loving) on the other. When such molecules are placed in an aqueous environment, they can spontaneously form capsules in which the hydrophobic portions are inside, protecting them from contact with water. In particular, these nanocapsules are used for the delivery of fragile biological substances.

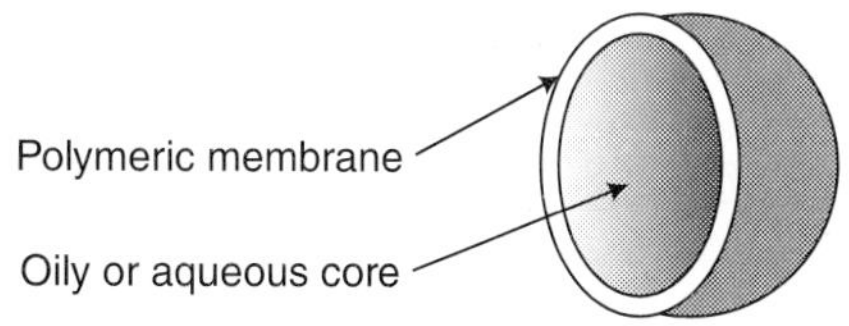

Fig. 9.9 Nanocapsule

9.1.2.3 *Liposomes*

Liposomes are artificially constructed spherical cells of phospholipid bilayers. The bilayers are stabilized with charged nanoparticles (carboxyl-modified polystyrene) which can act as vehicles for drug delivery and are also biodegradable (Fig. 9.10). These nanoparticle-stabilized liposomes are interesting from several functional perspectives. Firstly, they comprise a novel colloidal-sized sensor. These nanoparticles are used in medicine for drug and small molecule delivery. These are effectively used for encapsulation of cosmetics and environmental remediation.

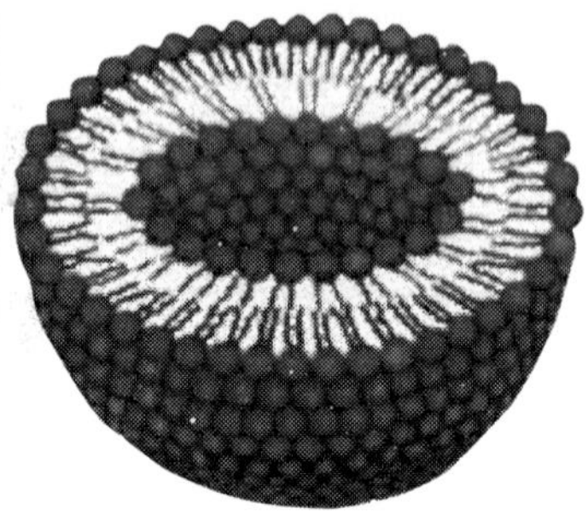

Fig. 9.10 Cutaway drawing showing the bilayer structure of phospholipid liposomes

9.1.2.4 *Dendrimers*

Dendrimers are a relatively new class of biodegradable macromolecules which are synthetic, hyperbranched, symmetric, nanospherical (Fig. 9.11). The dendritic molecules can be divided into two classes, viz., one is low-molecular weight species which includes dendrimers and dendrons and the other is high molecular weight species which includes dendronized polymers, hyperbranched polymers, and the polymer brush. The

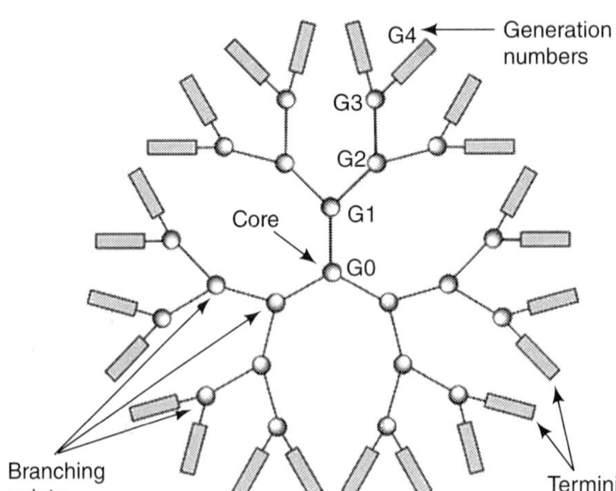

Fig. 9.11 Dendrimer

dendronized polymers are cylindrical structures that possess a linear polymeric backbone and dendritic side chains. Dendrimers can also act as biosensors, for example, multifunctional Au nanoparticles. Dendrimers can act as vehicles for drug delivery (Sontjens *et al.*, 2006), DNA (transfection), imaging agents and tissue engineering scaffolding. Cadmium-sulfide/polypropylenimine tetrahexacontaamine dendrimer composites are used to detect fluorescence signal quenching, and poly propylenamine first and second generation dendrimers are used for metal cation photodetection (Campos Bruno *et al.*, 2010).

9.1.2.5 *Yolk/Shell nanoparticles (YSNs)*

Yolk/shell nanoparticles (YSNs) consist of a nanoparticle core surrounded by a hollow shell and are named after an egg with a shell, interstitial space, and the core – the yolk (Fig. 9.12). Yolk-shell structures have been prepared with various types of movable cores, such as gold, SiO_2, and magnetic Fe_3O_4. Yolk/shell nanoparticles are one of the important nanomaterials for a wide variety of applications such as catalysis, delivery, lithium-ion batteries and biosensors due to their tailorability and functionality in both the cores and hollow shells (Jian Liu *et al.*, 2011). Due to its tiny size, they provide a large surface area. These nanostructures reduce drug risks, by breathable pores the drug enters the cell and do not flow.

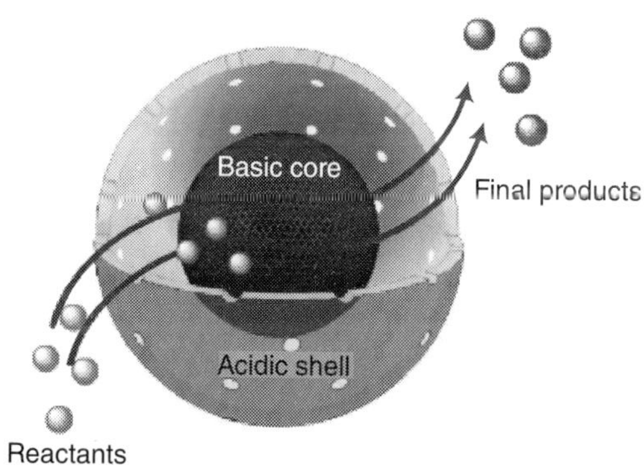

Fig. 9.12 Yolk/shell nanoparticles

9.1.2.6 *Bionanoparticles*

The biogenic method of nanoparticles production is simple, eco-friendly and is of great interest due to the possible application in catalysis, medicine, and nanooptoelectronics. Biologically fabricated nanostructures offer substantially different properties like good adhesion, tribologically good properties (like lubrication – due to this it is able to interact with surfaces), optical and electrical properties. It is possible to get the nanoparticles from simple microbes and bacteria up to more evolved plants. For the synthesis of metal and metal sulfide nanoparticles, fungi are the most efficient biological agents. *Verticillium* sp. and *Fusarium oxysporum* are used for the synthesis of metal nanoparticles, as they can reduce the metal ions extracellularly. An extracellular enzyme, hydrogenase in the

F. oxysporum shows excellent redox properties, and it can act as an electron shuttle in metal reduction. The electron shuttle or other reducing agents (e.g., hydrochinones) released by microorganisms are capable of reducing ions to nanoparticles. When challenged with silver nitrate in an aqueous medium the fungal members *Aspergillus flavus, Aspergillus fumigatus* and Phanerochaete *Chrysosporium* produce stable silver nanoparticles (Ratna Kumari and Sobha, 2013).

9.1.2.7 *Nanogold*

A strong absorption band in the visible region is shown with gold nanoparticles when the frequency of the electromagnetic field is resonant with the coherent electron motion, which is called surface plasmon resonance absorption (Fig. 9.13) (Sara Biglari *et al.*, 2014). Interaction with the electric field results in a polarization of the electrons with respect to and relative to the ionic core of a nanoparticle. The dipole oscillations of the free electrons with respect to the ionic core of a spherical nanoparticle result in the so-called surface plasmon absorption. When the frequency of the electromagnetic field becomes resonant with the coherent electron motion, a strong absorption band around 520 nm in the spectrum is observed, which is the origin of the observed brilliant color of the nanoparticles in solution. The extracellular synthesis of nanogold particles (NGP) by *Fusarium oxysporum* and actinomycete *Thermomonospora* sp., respectively has been reported. The same group has reported the intracellular synthesis of gold nanoparticles by fungus *Verticillium* sp. as well. The reaction of $AuCl^{4-}$ ions with the extract of geranium leaves and an endophytic fungus, *Colletotrichum* sp., present in the leaves, leads to the formation of AuNPs. Gold nanoparticles modified electrodes are used in the assembly of electrochemical DNA biosensors. They constitute useful analytical tools for sequence-specific DNA diagnosis and detection due to their inherent advantages of low cost, sensitivity and rapidity of response. Pathogens, bacteria, and viruses, can be detected via their unique, respective nucleic acid sequences.

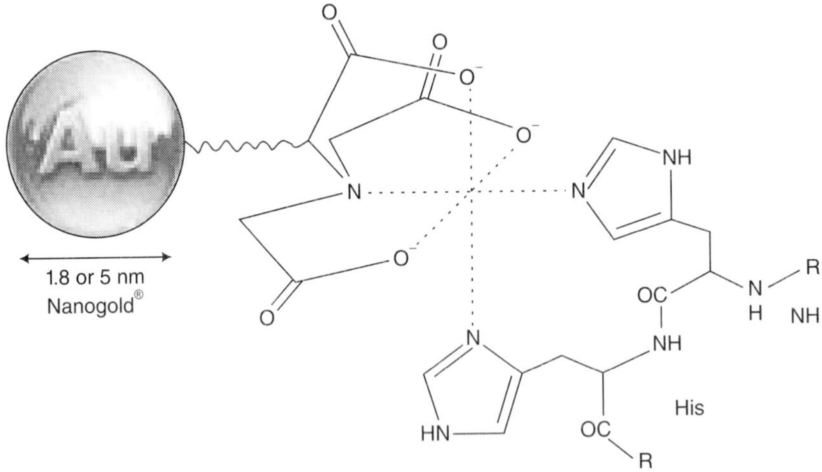

Fig. 9.13 Nanogold structure

Exploring a green chemistry approach, this study brings to the fore, the anthelmintic efficacy of gold nanoparticles, highlighting the plausible usage of myconanotechnology. Gold nanoparticles of ~6 to ~18 nm diameter were synthesized by treating the mycelia-free culture filtrate of the phytopathogenic fungus with gold chloride. Their size and morphology were confirmed by UV-Vis spectroscopy, DLS data, AFM and TEM images. The XRD studies reveal a crystalline nature of the nanoparticles, which are in cubic phase. The FTIR spectroscopic studies before and after the formation of nanoparticles show the presence of possible functional groups responsible for the bioreduction and capping of the synthesized gold nanoparticles. The latter were tested as vermifungal agents against a model cestode *Raillietina* sp., an intestinal parasite of domestic fowl. Further, ultrastructural and biochemical parameters were used to corroborate the efficacy study (Kar *et al.*, 2014). Treating chloroauric acid solutions with this bacterium resulted in a reduction of gold ions and formation of stable NGPs. TEM and SEM electron micrographs of NGPs indicated size range from 2–25 nm with an average of 9.09 nm produced intracellularly by the bacterium. SEM electron micrographs revealed the morphology of spores and mycelia. The amplified PCR fragment of 16S rRNA gene was cloned and sequenced from both sides; it consisted of 741 nucleotides. According to NCBI GenBank, the bacterium had 97.1% homology with *Streptomyces djakartensis* strain RT-49. The GenBank accession number for partial 16S rRNA gene was recorded as JX162550. Xiaoning Li *et al.* (2014) presented the use of functionalized gold nanoparticles (AuNPs) to combat multi-drug-resistant pathogenic bacteria. Tuning of the functional groups on the nanoparticle surface provided gold nanoparticles that were effective against both Gram-negative and Gram-positive uropathogens, including multi-drug-resistant pathogens. These AuNPs exhibited low toxicity to mammalian cells, and bacterial resistance was not observed after 20 generations. A strong structure–activity relationship was observed as a function of AuNP functionality, providing guidance to activity prediction and rational design of effective antimicrobial nanoparticles.

Karnal bunt (KB) of wheat incited by *Tilletia indica* is an economically important quarantined fungal pathogen. The similarity in teliospore configuration makes it difficult to differentiate KB teliospores from the teliospores of other bunt fungi using conventional approaches. To determine the correct identity of KB teliospore, an infectious entity, it is essential to develop specific diagnostic probes and high quality of immunological reagents against infectious entities. The immuno-dipstick assay was developed by Sadhna Singh *et al.* (2010), which was quite sensitive to detect the antigens of even five teliospores. For on-site testing of KB, a nanogold based lateral flow immuno-dipstick assay (LFID) was also developed using antiteliospore antibodies. For the development of LFID, colloidal gold was prepared and conjugated with antiteliospore antibodies which were further characterized. However, species-specific detection is yet to be achieved by generation of monoclonal antibodies against diagnostic antigens (34 and 66kD teliospore's proteins). Based on the experience of nanogold based LFID, it was proposed to develop surface plasmon resonance (SPR) based label-free detection system. The same may be employed not only for the determination of the quality of immunological reagents in terms of sensitivity, specificity, and precision but also the development of suitable design of immunosensor for differential diagnosis of *T. indica*. The approach involved the use of a mouse monoclonal antibody against diagnostic antigens and a suitable design of SPR sensor by the preparation of immuno-affinity layer over self-assembled monolayer surfaces. The signal to noise ratio could further be enhanced using nanogold particles. This was 'the first' attempt for real-time monitoring of KB teliospores in wheat lots by SPR sensor and in a fully automated manner for establishing seed certification and plant quarantine standards.

9.1.2.8 *Nanosilica silver*

Silicon (Si) is known to be absorbed into plants to increase disease resistance and stress resistance. Silica itself has no direct effect on pathogenic microorganisms and has no effect on diseases. So a new composition of nanosized silica-silver for control of various plant diseases has been developed. Aluminosilicate nanotubes sprayed on plant surfaces are also developed and are easily picked up in insect hairs. Insects actively groom and consume pesticide-filled nanotubes. These are biologically more active and relatively more environmentally safe pesticides. Mesoporous silica nanoparticles can deliver DNA and chemicals into Plants thus, creating a powerful new tool for targeted delivery into plant cells (Amarendra N. Misra *et al.*, 2013). It has been successfully used to introduce DNA and chemicals into Arabidopsis, tobacco and corn plants. The other advantage is that with the mesoporous nanoparticles, one can deliver two biogenic species at the same time.

9.1.2.9 *Nanosized silver*

Silver (Ag) is known to have antimicrobial activity both in ionic or nanoparticle forms. The powerful antimicrobial effect of silver, especially in unicellular microorganisms, is believed to be brought about by enzyme inactivation [(Kim *et al.*, 1998) (Fig. 9.15)]. Nanosilver whose antimicrobial effect has been tested against many diseases causing pathogens of animals and plants is the most studied and utilized nanoparticle. Silver is also an excellent plant growth stimulator (Sharon *et al.*, 2010). Antifungal effect of nanosilver colloids (average diameter of 1.5 nm) was studied against the powdery mildew pathogen of rose caused by *Sphaerotheca pannosa* var. *rosae*. Silver is now an accepted agrochemical replacement' and maximum number of patents are filed for 'nanosilver' for preservation and treatment of diseases in agriculture field (Sharon *et al.*, 2010). Application of silver in the management of plant diseases has been tested by Jo *et al.* (2009) with reference to two fungal pathogens of cereals, viz., *Bipolaris sorokiniana* (spot blotch of wheat) and *Magnaporthe grisea* (rice blast). *In vitro* assays indicated that silver both in ionic and nanoparticle forms inhibited colony growth of both the pathogens but *M. grisea* was comparatively more sensitive to the silver application. When tested *in vivo* with perennial ryegrass (*Lolium perenne*) silver ions and nanoparticles brought significant reduction in disease severity when applied 3 hours prior to pathogen inoculation. In another study, silver nanoparticles synthesized extracellularly by *Alternaria alternata* were found to cause significant enhancement of the antifungal action of the triazole fungicide fluconazole against *Candida albicans, Phoma glomerata* and *Trichoderma* spp. (Gajbhiye *et al.*, 2009). However, no significant enhancement was observed with respect to the fungi *Phoma herbarum* and *Fusarium semitectum*.

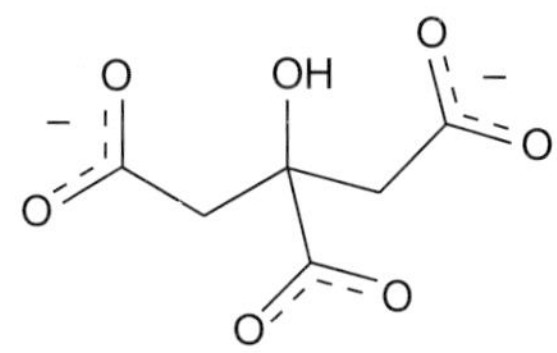

Fig. 9.14 Nanosilver

Silver nanoparticles have shown promise against Gram-positive *S. aureus*. As metal nanoparticle embedded paints have been synthesized using vegetable oils and have been found to have good ant microbiology activity (Abdul Hameed and Al-Samarrai, 2012). The use of nanosized silver particles as antimicrobial agents has become more common as technological advances make their production more economical. One of the potential applications in which silver can be utilized is in the management of plant diseases. A significant reduction in mycelial growth was observed for spores incubated with silver nanoparticles. The sporulation test showed that, relative to control samples, the number of spores formed by mycelia increased in the culture after contact with silver nanoparticles, especially on the nutrient-poor PDA medium. The 24 h incubation of FC spores with a 2.5 ppm solution of silver nanoparticles greatly reduced the number of germinating fragments and sprout length relative to the control. Silver nanoparticles may be synthesized using plant extracts or microbes, viz., fungi, bacteria, yeast and actinomycetes [(Table 9.4) (Sahayaraj and Rajesh, 2011)].

Table 9.4 Antimicrobial Activities of Bio-nanoparticles Synthesized Using Silver with Botanicals and Microbes

Biological entity	*Test microorganisms*
(a) Using silver with botanicals	
Argemone mexicana	*Escherichia coli and Pseudomonas aeruginosa, Aspergillus flavus*
Aloe Vera	*E. coli*
Caulerpa scalpelliformis	*Xanthomonas campestris* pv. *malvacearum*
Caulerpa veravalensis	*X. campestris* pv. *malvacearum*
Desmodium triflorom	*Staphylococcus* sp., *E. coli, Bacillus subtilis*
Cinnamon zeylanicum	*E. coli* BL-21
Gliricidia sepium	*S. aureus, E. coli, P. aeruginosa, Klebsiella pneumoniae*
Ocimum sanctum & Vitex negundo	*S. aureus, Vibrio cholerae, Proteus vulgaris, P. aeruginosa*
Padina pavonica	*X. campestris* pv. *malvacearum*
Sagassum wightii	*S. aureus, B. rhizoids, E. coli, P. aeruginosa*
Vernonia cinerea	*X. campestris* pv. *malvacearum*

Contd...

Contd...

Trianthema decandra	*E. coli, P. aeruginosa*
Solanus torvum	*P. aeruginosa, S. aureus, Aspergillus flavus, A. niger*
Svensonia hyderabadensis	*A. niger, Fusarium oxysporum, Curvularia lunata, Rhizopus arrhizus*
(b) Silver with various microbes	
Amylomyces rouxii KSU-09	*Shigella dysenteriae* type I, *Staphylococcus aureus, Citrobacter* sp., *Escherichia coli, Pseudomonas aeruginosa, Bacillus subtilis, Candida albicans and Fusarium oxysporum*
Alternaria alternata	*Phoma glomerata, P. herbarum, F. semitectum, Trichoderma* spp. and *C. albicans*
Aspergillus sp.	*Staphylococcus* sp.
Euphorbia hirta	*S. aureus, E. coli, K. pneumoniae, B. cereus and P. aeruginosa*
Klebsiella pneumoniae	*S. aureus, E. coli*
Fusarium acuminatum	*S. aureus, Salmonella typhi, Staphylococcus epidermidis, E. coli*
Phytophthora infestans	*E. coli, S. typhi, K. pneumoniae, P. vulgaris, B. subtilis, S. aureus*
Trichoderma viride	*S. typhi, E. coli, S. aureus, Micrococcus luteus*

Pawan Kaur *et al.* (2012) reported the fungicidal properties of nanosize silver/chitosan nanoformulations (NFs) used as an agent for antifungal treatment of various seed borne plant pathogens. Fungal phytopathogens, especially seed borne disease-causing species, *Rhizoctonia solani, Aspergillus flavus, Alternaria alterneta* were isolated from chickpea seeds. Differences were observed in the antifungal activity of the silver nanoparticle, chitosan nanoparticles, and silver/chitosan nanocomposite, upon the mycelial growth and zone of inhibition of the fungi. The extent of fungal inhibition was derived using the above information, to evaluate the antifungal efficacy of these nanoformulations against pathogens. Tests for the fungal growth revealed that the NFs showed significant inhibition effectiveness. This study suggested the possibility to use silver/chitosan NFs as an alternative to fungicides for controlling seed-borne phytopathogens. The effect of silver nanoparticles was evaluated against pepper anthracnose (*Colletotrichum* species) under different culture conditions. Silver nanoparticles (WA-PR-WB13R) were applied at various concentrations to determine antifungal activities *in vitro* and the field (Kabir Lamsal *et al.*, 2011). The application of 100 ppm concentration of silver nanoparticles produced maximum inhibition of the growth of fungal hyphae as well as conidial germination in comparison to the control *in vitro*. In field trials, the inhibition of fungi was significantly high when silver nanoparticles were applied before disease outbreak on the plants. Scanning electron microscopy results indicated that the silver nanoparticles caused a detrimental effect on mycelial growth of *Colletotrichum* species.

9.1.2.9.1 Mechanism of silver nanoparticles on bacteria

The mechanisms behind the activity of nanosilver on bacteria are not yet fully elucidated. The three most common mechanisms of toxicity proposed up to now are uptake of free silver ions followed by disruption of ATP production and DNA replication, formation of Reactive Oxygen Species (ROS), and direct damage to cell membranes (Sahayaraj and Rajesh, 2011).

9.1.2.9.2 Uptake of free silver ions

Antimicrobial properties of ionic silver are well known; thus, it is likely that eluted ions from silver nanoparticles are accountable for at least a part of their antibacterial properties. At submicromolar concentrations, Ag+ interacts with enzymes of the respiratory chain such as NADH dehydrogenase ensuing in the uncoupling of respiration from ATP synthesis. Ionic silver also binds with transport protein leading to proton leakage, inducing collapse of the proton motive force. Ionic silver hampers the uptake of phosphate and thus causes the efflux of intracellular phosphate. The interaction with respiratory and transport proteins is due to the high affinity of Ag+ with thiol groups present in the cysteine residues of those proteins. In addition, it has been reported that Ag+ increases DNA mutation frequencies during polymerase chain reactions (Yang *et al.*, 2009). Bacterial cells exposed to millimolar Ag+ Doses suffer morphological changes such as DNA condensation and localization in an electron-light region in the center of the cell, cytoplasm contraction and disintegration of cell wall membrane, and cell membrane degradation allowing leakage of intracellular contents. Physiological changes take place together with the morphological changes. Bacterial cells enter a lively, but nonculturable state in which physiological levels can be measured but bacteria are not able to growth and replicate.

9.1.2.10 Nanosilica

Silicon (Si) increases disease resistance and stress resistance in plants. It also stimulates the physiological activity and growth of plants (Chowdappa and Shivakumar Gowda, 2013 (Fig. 9.15). Honeycomb mesoporous silica nanoparticle (MSN) system with 3 nm pores has been used to deliver DNA and chemic also into plant cells and intact leaves. They loaded the gene and its chemical inducer into the MSN system and capped the ends with gold nanoparticles and studied the release pattern of chemicals and induction of gene expression in the plants under controlled-release conditions. This study showed an application of silica nanoparticles in the target-specific delivery of proteins, nucleotides, and chemicals in plant biotechnology.

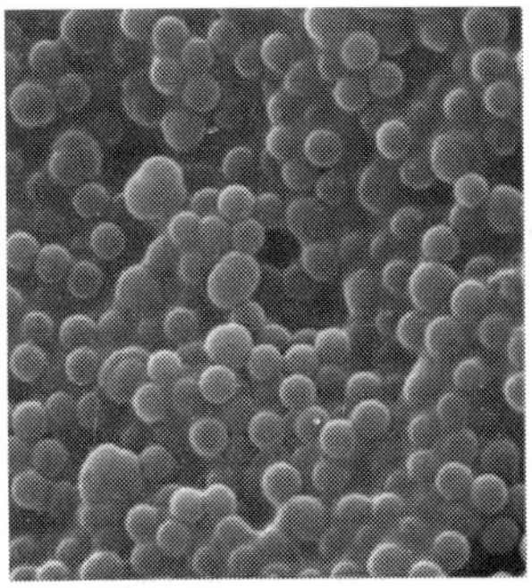

Fig. 9.15 SEM images of silica nanoparticles

9.1.2.11 *Copper nanoparticle*

Copper-based fungicides produce highly reactive hydroxyl radicals which can damage lipids, proteins, DNA, and other biomolecules. It plays an important role in disease prevention and treatment of a large variety of plants (Fig. 9.16). Complexation of copper with chitosan nanogels was shown to have strong synergistic effect between chitosan and copper in inhibiting the growth of *Fusarium graminearum.* Because of its biocompatibility, these nanohydrogels are included as a new generation of copper-based biopesticides, and it could also be developed into an efficient delivery system for copper based fungicides for plant protection (Chowdappa and Shivakumar Gowda, 2013). Low melting point soda-lime glass powder containing copper nanoparticles showed efficient antimicrobial activity against Gram-positive, Gram-negative bacteria, yeast, and fungi, a key reason for the increased antimicrobial activity is because of the inhibitory synergistic effect of the Ca^{2+} lixiviated from the glass.

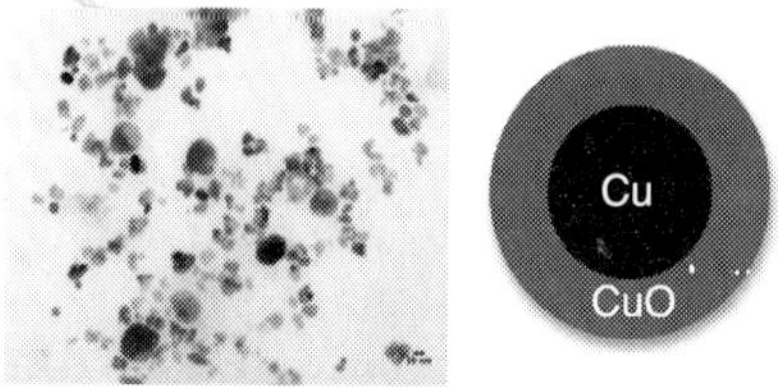

Fig. 9.16 Copper nanoparticles (Image courtesy: Imali A. Mudunkotuwa, Grassian research group, University of Iowa)

Figure 9.16 shows a transmission electron microscopy image of aged copper nanoparticles. These nanoparticles are oxidized giving a core-shell morphology with a metallic copper (Cu) core and a fully oxidized copper oxide (CuO) shell (see cartoon representation). Additionally, there appear to be larger particles in the image from particle coalescence. These aged nanoparticles show different chemical behavior compared to new Cu nanoparticles and fully oxidized CuO nanoparticles.

9.1.2.12 *Nanocomposites*

Silver has been studied as an antibacterial agent and less as an antifungal agent. Preparation and antifungal activity of composite films of pullulan and Ag nanoparticles (NP) against *Aspergillus niger* as a model system have been reported. They found that these composite films showed strong inhibitory action on fungal sporulation, which was confirmed by disruption of the spores cells when observed under SEM. Silver in an ionic state exhibits high antimicrobial activity. A new nanosized Silica-Silver composite has been developed for control of various plant diseases (Chowdappa and Shivakumar Gowda, 2013). Composite showed good antifungal activity where pathogens disappeared from the infected leaves within three days of spraying, and the plants remained healthy thereafter. An attempt was done to determine the effective concentration of composites and also used it effectively for suppression of growth of many pathogens. Nanocomposites showed 100% growth inhibition of *Pythium ultimum, Magnaporthe grisea, Colletotrichum gloeosporioides, Botrytis cinerea* and *Rhyzoctonia solani* at 10 ppm concentration. Whereas, *Bacillus subtilis, Azotobacter chrococuum, Rhizobium tropici, Pseudomonas syringae* and *Xanthomonas compestris* pv. *vesicatoria* showed 100% growth inhibition at 100 ppm concentration.

Chowdappa and Shivakumar Gowda (2013) used chitosansilver nanoparticle (chitosan-Ag NP) composite (size distribution from 10 nm to 15 nm) for inhibition of conidial germination in Colletotrichum gloeosporioides. They observed complete inhibition of spore at 100 μg/ml concentration of the composite, but chitosan alone did not show significant inhibition at the same concentration. They observed the shrinking of spores and other morphological changes in the spores treated with CS-Ag nanocomposite.

9.2 BIONANOTECHNOLOGY

Bionanotechnology has probably been the most developing field in the last decade as an integration of biotechnology on the molecular level and nanotechnology. It provided new avenues of research and development to produce novel and new nanomaterials of medicinal applications which paved the way to the vibrant field of nanomedicine (Adel M. Mahasneh, 2013). This is dependent on combining biological norms with unique characteristics of nanosized particles to carry out very specific functions. This will hopefully lead to the elucidation of our understanding of disease therapy and the biology of different life forms. The prospects of using nanoparticles in such fields would include drug and gene therapy, tumor control, detection of pathogens and proteins, tissue and cell engineering, DNA probing, pest control in agricultural fields and many other applications. Knowing the expected implications of novel nanoparticles applications in nanoparticle-based medicine and research, it became important to compile literature relevant to this subject in a single publication.

Plant diseases are major factors limiting crop yields. The problem with the disease management lies with the detection of the exact stage of prevention. Most of the time appropriate plant protection chemicals are applied to the crop as a precautionary measure leading to avoidable environmental hazards, or else applications are made after the appearance of the disease symptoms, thereby causing some amount of crop losses. Among the different diseases, the viral diseases are the most difficult to control, as one has to stop the spread of the disease by the vectors. But once it starts showing its symptoms, pesticide application would not be of much use. Therefore, detection of the exact stage such as stage of viral DNA replication or the production of initial viral protein is the key to the success of control of viral diseases (Devakumar and Chhonkar, 2013). Nanobased viral diagnostics, including multiplexed diagnostics kits development, have taken momentum to detect the exact strain of virus and the stage of application of some therapeutic to stop the disease. Detection and utilization of biomarkers, that accurately indicate disease stages, is also an emerging area of research in bionanotechnology. Measuring differential protein production in both healthy and diseased states leads to the identification of the development of several proteins during the infection cycle.

Adverse effects: The ultrasmall sizes that make the nanoparticles of immense usefulness, unfortunately, the same characteristic also causes several adverse effects and may represent significant hazards to the environment, animals, human beings and plants when used nonjudiciously. These are the possible hazards, viz., e-nanoparticles as pesticides, fertilizers or in other formulations, when airborne, may deposit on above-ground parts of plants. They may plug stomata and create a fine physical and toxic barrier layer on stigma preventing pollen tube penetration. They may also enter the vascular tissue and impair translocation of water, minerals and photosynthate. Nanopesticides may reduce environmental contamination through the reduction in pesticide application rates, but they may also create new kinds of contamination of soils and waterways due to enhanced transport, longer persistence, and higher toxicity.

9.3 NANOPESTICIDES

Presence of active ingredient of a formulation in the root zone of a plant at the initial stage of crop growth helps in the protection of the plant from pathogen invasion and bringing down its population below economic threshold level (Mujeebur Rahman Khan and Tanveer Fatima Rizvi, 2014). Since, all the propagules/spores of a pathogen do not approach/invade the host at one time, rather they attack intermittently. Hence persistence or slow/gradual release of active ingredient in the root zone shall enhance the effectiveness of the formulation. Moreover, timely and slow release of the chemical shall reduce the net amount of pesticide required for the disease control. This would subsequently minimize the adverse effects of the pesticide on the man and environment. The controlled release of the active ingredient can be achieved through a nanotechnological approach by using nanomaterials as a carrier for the chemicals. Such formulations may greatly decrease the amount of pesticide input and associated environmental hazards. Nanopesticides shall reduce the rate of application because the quantity of the chemical actually being effective is at least 10–15 times smaller than that applied to classical formulations, hence a much smaller than the normal amount could be required to achieve satisfactory control of the disease. Due to better kinetic stability, smaller size, low viscosity and optical transparency, nanoemulsions can serve as better pesticide delivery systems. The micro or nanoemulsion, as a carrier for pesticide delivery, can improve the solubility and bioavailability of the active ingredients of the chemical (Mujeebur Rahman Khan and Tanveer Fatima Rizvi, 2014).

Nanopesticides contain very small particles of pesticide's active ingredients or other small engineered structures with useful pesticidal properties. Nanopesticides can increase the dispersion and wettability of agricultural formulations (reduction in organic solvent runoff) and unwanted pesticide movement. Nanomaterials and biocomposites exhibit useful properties such as stiffness, permeability, crystallinity, thermal stability, solubility and biodegradability needed to formulate the nanopesticides. The nanopesticide formulations also offer large specific surface area and hence, increased affinity to the target (Mujeebur Rahman Khan and Tanveer Fatima Rizvi, 2014). Nanoemulsions, nanoencapsulates, nanocontainers, and nanocages are some of the nanopesticide delivery techniques that may prove effective in plant protection programs; the possibility of utilizing chitosan nanoparticles, a highly degradable antibacterial material was explored for slow release of NPK fertilizer. Kaolin clay-based nanolayers were developed to be used as a cementing and coating material for control release of fertilizers. Primarily, nanoclay materials offer interactive surfaces with high aspect ratio for encapsulating 'agrochemicals such as fertilizers, plant growth promoters, and pesticides.' Generally, there are three kinds of Controlled Release Systems (CRS), viz., zero-order release, first-order release and square root time release which may be selected in view of the environmental condition and pest/pathogen biology.

Biodegradable microbial polymers like polyhydroxyl alkanates have proved effective for controlled release of pesticides (Mujeebur Rahman Khan and Tanveer Fatima Rizvi, 2014). Temperature sensitive polymers, e.g., intelimer can regulate the release of pesticide with the progress of season depending on the temperature fluctuations, thereby protecting the active ingredient from unwanted leaching and degradation). Some important controlled release devices are microcapsules, microspheres, coated granules and granular matrices that may be used in formulating the nanopesticides. A microcapsule is a reservoir system where active ingredient of the pesticide is contained within the core which is surrounded by a membrane. Microencapsulation has a potential scope in formulating nanopesticides. A microsphere is a monolithic system consisting of spherical or irregular shape particles (20 nm-2000 μm size) in which the active agent is dissolved or dispersed in a polymer matrix). Various naturally occurring polymers (polysaccharides and proteins), synthetic polymers (polystyrene, polyacrylamide, polyamides, polyesters, etc., and inorganic materials (silica, zeolites, inorganic oxides, glass beads, ceramic, etc.) can be explored to test their suitability in formulating nanopesticides helpful in crop protection. Basically, the nanoformulations should degrade faster in the soil and slowly in plants with residue levels below the regulatory criteria in foodstuffs.

The development of Sodium Dodecyl Sulphate (SDS) modified photocatalytic TiO_2/Ag nanomaterial conjugated with dimethomorph (DMM) has been reported which met the above criterion. Khot *et al.* (2012) used common pesticides to prepare the nanopesticide. Modified formulation with 96 nm average granularity, increased dispersivity and decomposition of the pesticide in soil have much greater effectiveness in the seedling of cabbage and cucumber. The modification of the nanomaterials using SDS greatly increased the absorption of DMM. Encapsulated nanoimidacloprid was formulated with above properties that could be used for pests control during vegetable production. The SDS-modified Ag/TiO_2 imidacloprid nanoformulation was developed using a microencapsulation technique that used chitosan and alginate. It was tested on soybean plants that were transplanted to soil pH of 6.2. The formulation residues in soil and the plants degraded faster during the first eight days and were minimal to undetectable after 20 days. The SDS in the above applications was used to increase the photo-degradation of the nanoparticles in soil (Mujeebur Rahman Khan and Tanveer Fatima Rizvi, 2014). Alternatively, Mohamed and Khairou (2011) developed highly photo-degradable Ag/TiO_2 particles (57 nm) which were synthesized using polyoxyethylenelauryl ether (POL). They tested Ag/TiO_2 nanoparticles synthesized using both POL and SDS for the degradation of 2, 4-D herbicide under visible and UV radiation. Results confirmed that the POL synthesized nanoparticles photo-degraded faster during the same exposure period. All the above results cannot be considered as safeguarded due to lack of any national/international nanoregularity institutional framework/ standards related to the use of nanomaterials in agriculture.

9.4 DETECTION AND OTHER USES OF NANOTECHNOLOGY IN PLANT PATHOLOGY

Various nanoparticles employed in agriculture and plant disease management are listed in Table 9.3. Biopolymer nanoparticles: Development of nanoformulation for field application of agrochemicals requires the use of readily biodegradable, nontoxic, environment-friendly, safe and low-cost materials. So, use of biopolymers produced by natural sources with good physical and chemical properties is a fascinating approach to prevent the use of petrochemical and toxic chemical substances in the production of nanomaterial.

9.4.1 Chitosan

Chitosan nanoparticles have got various applications in biology due to its biodegradable and nontoxic properties. In acidic condition, the free amino groups of chitosan protonates and contributes to its positive charge (Phaechamud and Ritthidej, 2008). The inhibition mode of chitosan against fungi is defined by three mechanisms as described here. The positive charge of chitosan interacts with negatively charged phospholipid components of fungi membrane, which in turn alter cell permeability of plasma membrane and causes the leakage of cellular contents, which consequently leads to the death of the cell (GarcÌaRincÛn *et al.*, 2010). Chitosan chelates with metal ions, which has been implicated as a possible mode of antimicrobial action (Rabea *et al.*, 2003). On binding to trace elements, it interrupts normal growth of fungi by making the essential nutrients unavailable for its development (Roller and Covill, 1999) and it is suggested that chitosan could penetrate fungal cell wall and bind to its DNA and inhibit the synthesis of mRNA and, in turn, affect the production of essential proteins and enzymes (Sudarshan *et al.*, 1992; Kong *et al.*, 2010).

In search of natural antimicrobials to avoid harmful synthetic chemicals, chitosan, and chitosan nanoparticles are found to be more effective against plant pathogens like *Fusarium solani*. The inhibitory effect was also influenced by particle size and zeta potential of chitosan nanoparticles. The chitosan, therefore, could be formulated and applied as a natural antifungal agent in nanoparticles form to enhance its antifungal activity (Ing *et al.*, 2012). Chen *et al.* (2010) studied antibacterial activities of low molecular weight chitosan products (molecular weight ranges from 6.9 kDa to 22.4 kDa) with average nanoparticle sizes of 117 nm to 965 nm. They emphasized that antimicrobial activity of chitosan nanoparticles depends on its zeta potential, which plays a significant role in binding with the negatively charged microbial membrane. Antimicrobial activity of chitosan, chitosan derivatives bound metal ions, and nanoparticles are well studied (Sanpui *et al.*, 2008; Jagadish *et al.*, 2012; Kaur *et al.*, 2012).

9.4.2 Nanosized Metals as Diagnostic Probes

As the techniques have not been perfected yet, and things are in a budding stage in terms of plant pathogen detection researchers are looking forward to harness the benefits of nanomaterials by addressing all possible drawbacks as found in currently available diagnostic tools. Nanoparticles are different from their bulk counterparts, which, when reduced to nanosize (1–100 nm) achieve certain properties which make them suitable for development as diagnostic probes (Sharon *et al.*, 2010). These properties are large aspect ratio (surface to volume ratio), chemically alterable physical properties, strong affinity to target (particularly of gold nanoparticles to proteins), structural sturdiness in spite of atomic granularity and enhanced or delayed particles aggregation depending on the type of the surface modification, enhanced photoemission, high electrical and heat conductivity, and improved surface catalytic activity (Liu, 2006; Garg *et al.*, 2008; McNeil, 2005; Shrestha *et al.*, 2007). Fluorescent silica nanoprobes have potential for rapid diagnosis of plant diseases (Fig. 9.17). Fluorescent silica nanoprobes conjugated with the secondary antibody of goat antirabbit IgG (Yao *et al.*, 2009) was used for detection of a bacterial plant pathogen *Xanthomonas axonopodis* pv. *vesicatoria* (bacterial spot on solanaceous plants). An organic dye tris-2,2′-bipyridyldichlororuthenium (II) hexahydrate (Rubpy) was incorporated into the core of circular silica nanoparticles with average diameter of 50 ± 4.2 nm. Thus, silica nanoparticles became fluorescent which was photostable. Metallic nanoparticles possess unique chemical and physical properties, small size, huge surface to volume ratio, structural stability and strong affinity to their targets (Kumar *et al.*, 2010). Metal nanoparticles can be used as new antimicrobial agents and an alternative to synthetic fungicide to delay or inhibit the growth of many pathogens species because of its multiple modes of inhibition.

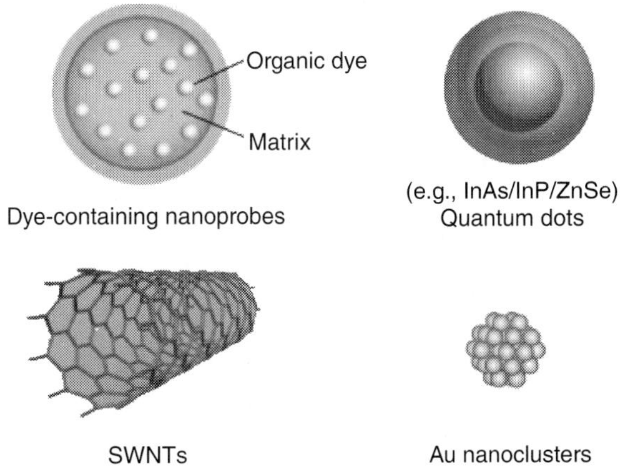

Fig. 9.17 Representative nanomaterial-based NIRF nanoprobes

Nanoparticles can be used in the preparation of new formulations like pesticides. Researchers have reported nanosilver and titanium dioxide nanoparticle applications in the management of plant diseases. The fungicidal efficiency of sulfur nanoparticles against two phytopathogens, *Fusarium solani* (isolated from an infected tomato leaf, responsible for early blight and *Fusarium* wilt diseases) and *Venturia inaequalis* (responsible for the apple scab disease) has been reported (Bhupinder Singh Sekhon, 2014).

9.5 NANOSCALE BIOSENSOR/NANOSENSORS

Sensors are a class of devices that have found widespread use, ranging from the detection of gas molecules to the real-time tracking of chemical signals in biological cells (Fig. 9.18). In general, a sensor comprises an active sensing element and a signal transducer and produces an electrical, optical, thermal or magnetic output signal. While the sensing element is responsible for the selective detection of the analyte, the transducer converts a chemical event into an appropriate signal that can be used with or without amplification to determine the analyte concentration in a given test sample. Biosensors differ from classical chemical sensors in that the sensing element consists of a biological material such as proteins (e.g., cell receptors, enzymes, antibodies) oligo- or polynucleotides, microorganisms, or even whole biological tissues and the sensor is used to monitor biological processes or for the recognition of biomolecules. Smart sensors, which are obtained by nanotechnology, are the powerful tools for track detect and control with animal and plant pathogen. Detection of very small amounts of a chemical contaminant, virus, or bacteria in agricultural and food systems is envisioned from the integration of chemical, physical and biological devices working together as an integrated sensor at the nanoscale. The bioanalytical nanosensors either use biology as a part of the sensor or are used for biological samples (Scott and Chen, 2003). One of the major roles for nanotechnology-enabled devices will be the increased use of autonomous sensors linked into a GPS system for real-time monitoring. These nanosensors could be distributed throughout the field where they can monitor soil conditions and crop growth. Ultimately, precision farming, with the help of smart sensors, will allow enhanced productivity in agriculture by providing accurate information, thus helping farmers to make better decisions (Sayed Roholla Mousavi and Maryam Rezae, 2011).

Fig. 9.18 Different nanosensor configurations

Nanosensors dispersed in the field can also detect the presence of plant viruses and the level of soil nutrients. The nanosensors which would be small and portable would provide rapid response and real-time processing with accurate, quantitative, reliable, reproducible, robust, specific and stable results. Detection of infection in nonsymptomatic plant followed by targeted delivery of treatment would be an essential component for precision farming. Use of micromechanical cantilever arrays for detection of fungal spore (*Aspergillus niger* and *Saccharomyces cerevisiae*) was demonstrated by Nugaeva *et al.* (2005). Proteins like concanavalin A, fibronectin or immunoglobulin G were surface grafted on microfabricated uncoated as well as gold-coated silicon cantilevers. These proteins were found to have different affinities to bind to the molecular structures present on the fungal cell surface. Spore immobilization and germination of the test fungi led to shift in resonance frequency which was measured by dynamically operated cantilever arrays. This took a few hours in contrast to several days in conventional techniques. The finding that shift was proportional to the mass of single fungal spore can be used for quantitative estimation. The biosensors detected the target fungi in the range of 103–106 cfu ml^{-1} in the investigation made by Nugaeva *et al.* (2005).

Sensors are sophisticated instruments which respond to physicochemical and biological aspects and transfer that response into a signal or output that can be used by humans. They allow the detection of contaminants such as microbes, pests, nutrient content and plant stress due to drought, temperature, insect or pathogen pressure, or lack of nutrients. Nanosensors have the potential to allow farmers to utilize inputs more efficiently by indicating the nutrient or water status of crop plants over fine spatial and temporal scales. This makes the farmers apply nutrients, water, or crop protection (insecticide, fungicide or herbicide) only where necessary (Ram Prasad *et al.*, 2014). One of the major roles of nanotechnology-enabled devices is to increase the use of autonomous sensors linked to a global positioning system (GPS) system for real-time monitoring. These nanosensors could be distributed throughout the field where they can monitor soil conditions and crop growth. Nanoparticles or nanosurfaces can be engineered to trigger an electrical or chemical signal in the presence of a contaminant such as a bacterium. Ultimately, precision farming, with the help of smart sensors leads to enhanced productivity in agriculture by providing accurate information, thus helping farmers to make better decisions.

Currently, several methods of microbial identification and typing are available for most plant pathogens. Traditional culture-based methods are time-consuming (Vandana Ghormade *et al.*, 2011). Techniques based on biochemical profiles for identification are limited by the population of characterized strains in the databases. Sensitive and specific serological techniques such as ELISA and indirect fluorescent antibody staining depend on the titer and specificity of the antibody either monoclonal or polyclonal. Cross-reactivity among closely related strains prevents their clear identification. Costly, nucleic acid based polymerase chain reaction methods such as restriction fragment length polymorphism, DNA fingerprinting and amplification of the internal transcribed spacer region from rRNA gene increase specificity of identification. (A novel microbial detection technology based on NPs is being developed. Silica-based NPs (60 nm) were filled with a fluorescent dye and conjugated to an antibody specific to a surface antigen of the microbe of interest. Detection of a single bacterial cell was possible using this technique. This method has potential for sensitive detection of plant pathogens.

Increased understanding of plant–pathogen interactions was mainly driven by the development of new techniques that facilitated analyses on a subcellular and molecular level. The development of labeling technologies, which allowed the visualization and localization of cellular structures and proteins in live cell imaging, promoted the use of fluorescence and laser scanning microscopy in the field of plant–pathogen interactions (Dorothea Elinger and Christian A. Voigt, 2014). Recent advances in new microscopic technologies opened their application in plant science and the investigation of plant diseases. In this regard, *in planta* Förster/Fluorescence resonance energy transfer has demonstrated to facilitate the measurement of protein–protein interactions within the living tissue, supporting the analysis of regulatory pathways involved in plant immunity and putative host–pathogen interactions on a nanoscale level. It was felt that localization microscopy, an emerging, noninvasive microscopic technology, will allow investigations with a nanoscale resolution leading to new possibilities in the understanding of molecular processes.

The plant disease such as Cucumber Mosaic Virus (CMV) and Papaya Ring Spot Virus (PRSV) is a most dangerous disease that can decrease productivity and quality of the vegetable and fruit. Besides that, it also can destroy and kill those plants in the long term when infected and to tackle this problem at early stages, the nanowire-based biosensor application is a most reliable sensor nowadays because of advantages towards detecting biological molecule especially plant diseases (Shahrul A.B. Ariffin *et al.*, 2013). They opined that dealing with a tiny form of molecules such as virus is very difficult and due to the nanostructure uniqueness such as nanowire, it can be done by undergoing formation of nanowire process. The result will be elaborated about how nanowires work in environment in order to detecting those virus. Nanosensors may be used to diagnose oil disease (caused by infecting soil microorganisms, such as viruses, bacteria, and fungi) via the quantitative measurement of differential oxygen consumption in the respiration (relative activity) of good microbes and bad microbes in the soil.

Efficient detection systems which monitor and detect pathogen invasion, infection, nutrition requirement and uptake and contamination have been developed. As the nanosensors enable us to detect and eradicate infectious diseases of plants and animals before visible symptoms appear, the heavy economic losses that occur otherwise are reduced by many folds. Nanosensor devices that use carbon nanotubes or nanocantilevers are small enough to trap and measure individual proteins or small molecules. A contaminant can be detected by specially engineered nanoparticles or nanosurfaces which trigger an electrical or chemical signal. Some nanosensors work by initiating enzymatic reactions or by using nanoengineered branching molecules called dendrimers as probes to bind to target chemicals and proteins (Madhuri Sharon, 2010). Pathogen and contaminant detection is possible with increased sensitivity and decreased response time due to nanosensors.

Nanosensors are expected to impact agricultural, food, and environmental sectors. The nanotechnology signature initiative *Nanotechnology for Sensors and Sensors for*

Nanotechnology: Improving and Protecting Health, Safety, and the Environment is the fifth to be launched by agencies of the National Nanotechnology Initiative. Portable nanodevices can rapidly detect insects, diseases, pathogens, chemicals, and contaminants and can result in faster treatments (Bhupinder Singh Sekhon, 2014).

9.5.1 Nanobiosensors for Plant Pathogen Detection

Currently, several methods of microbial identification and typing are available for most plant pathogens. Traditional culture-based methods are time-consuming (Vandana Ghormade *et al.*, 2011). Techniques based on biochemical profiles for identification are limited by the population of characterized strains in the databases. Sensitive and specific serological techniques such as ELISA and indirect fluorescent antibody staining depend on the titer and specificity of the antibody either monoclonal or polyclonal. Cross-reactivity among closely related strains prevents their clear identification. Costly, nucleic acid based polymerase chain reaction methods such as restriction fragment length polymorphism, DNA fingerprinting and amplification of the internal transcribed spacer region from rRNA gene increase specificity of identification. A novel microbial detection technology based on NPs is being developed. Silica-based NPs (60 nm) were filled with a fluorescent dye and conjugated to an antibody specific to a surface antigen of the microbe of interest. Detection of a single bacterial cell was possible using this technique (Zhao *et al.*, 2004). This method has potential for sensitive detection of plant pathogens.

9.5.2 Nanobiosensors for Pesticide Detection

Several studies were conducted for the development of analytical tools for pesticide residue estimation. In parallel with typical chromatography, immunochemical assays based on biomolecules were employed as an alternative to pesticide measurement by virtue of its high selectivity, sensitivity, and reliability as well as its rapidity. Traditional immunoassay is performed as discrete tests, that is, one assay for one analyte and several detecting runs for all of the components in a complex system. Application of nanomaterials adds further advantages to such systems, such as, further miniaturization, measurement of more variables, greater sensitivity, less sample material required, faster detection rates, read-outs in real time and application of novel detection methodologies [(e.g., electronic, colorimetric, fluorometric, and mass changes) (Vandana Ghormade *et al.*, 2011)].

Currently unimolecular and array type of nanomaterial-based biosensors are being developed for detection of pesticides. However, the format of biosensors varies, from free biomolecules to those conjugated to a substrate such as NPs, nanowires, nanotubes and thin films. Interaction of the target with the biosensor can be measured either directly or indirectly by recording the changes in color, fluorescence or electrical potential. In array technologies, multiple biomolecules are fixed to a substrate allowing multiple analytes to be measured simultaneously. Among the unimolecular array systems a gold NPs (30 nm) based dipstick competitive immunoassay with a sensitivity of 27 ng ml^{-1} was developed to detect organochlorine pesticide such as DDT. Gold NPs have the property of agglomeration associated with color production that was used for pesticide detection. Development of a color signal aided easy visual detected when gold NPs labeled antibodies bound to the pesticide residues. The gold NP based dipstick technique was suitable for the detection of several toxins in food and environmental samples and can be applied for rapid on-site testing of pesticides. Wang *et al.* (2009) developed a zirconium oxide NPs (~50 nm) based immunoassay for sensitive detection (0.02 nm) of organophosphate pesticides using phosphorylated enzyme acetylcholinesterase as a potential biomarker. Guo *et al.* (2009) developed a gold NPs (40 nm) based portable one-step strip assay allowing simultaneous screening for two pesticides carbofuran and triazophos within a short time (8–10 min) without any equipment. A bispecific monoclonal antibody (McAb), against both carbofuran and triazophos, was labeled with nanogold in comparison to the nanogold labeled anticarbofuran McAb and antitriazaphos McAb as detector reagents. The scope for application of nanotechnology for the development of sensors that are specific, fast, handy and inexpensive is vast, and the key challenge would be the commercialization of biosensors.

Furthermore, combining nanotechnology with microfluidic devices, that allow the flow of microvolumes of liquid through channels, will promote further miniaturization and rapid sampling. Cantilever nanomechanical devices, with springboard like-mechanism, are interesting as they detect the presence of specific target molecules when the target binds a reporter molecule attached to the cantilever. Cantilever arrays can give rise to a sensitive nanosensor electronic 'nose' or 'tongue.' The future would see the development of multianalyte array sensors, microfluidic devices, and cantilever arrays to make environmental monitoring easy.

9.6 PHYTOTOXICITY OF NANOPARTICLES

Plants are an essential base component of all ecosystems and play a critical role in the fate and transport of engineered nanoparticles (ENPs) in the environment through plant uptake and bioaccumulation (Shweta Agrawal and Pragya Rathore, 2014). It is also important to mention that the bioaccumulation, biomagnifications and biotransformation of engineered nanoparticles in food crops are still not well understood. Very few nanoparticles and plant species have been studied with respect to the accumulation and subsequent availability of nanoparticles in food crops. Most commonly encountered ENPs in the environment fall into one of the five following categories: carbonaceous nanoparticles, metal oxides, quantum dots, zero valent metals, and nanopolymers. These ENPs closely interact with their surrounding environment and, as a result, ENPs will inevitably interact with plants and these interactions such as uptake and accumulation in plant biomass will greatly affect their fate and transport in the environment. ENPs could also adhere to plant roots and exert physical or chemical toxicity on plants.

For interacting with plants, ENPs have to penetrate cell walls and plasma membranes of epidermal layers in roots to enter vascular tissues (xylem) to be taken up, and translocated through stems to leaves. Cell walls, through which water molecules and other solutes must pass to enter into roots, are a porous network of polysaccharide fiber matrices. ENP aggregates with a size smaller than the largest pore are expected to pass through and reach the plasma membrane, and the larger particle aggregates will not enter into plant cells. But the authors also admitted that ENPs may induce the formation of new and large size pores which allow the internalization of large ENPs through cell walls. Once micro- and macromolecules enter plant cell walls, the molecules can be transported through plasmadesmata, the intercellular organelles of 20–50 nm in diameter.

Selective and nonselective pathways through plasmadesmata are found to transport regulatory proteins and RNAs in short distances (Shweta Agrawal and Pragya Rathore, 2014). In one of the study, that reports phytotoxicity of five types of nanoparticles (multiwalled carbon nanotube, aluminum-alumina, zinc, and zinc oxide) on seed germination and root growth of six higher plant species (radish, rape, rye-grass, lettuce, corn, and cucumber) were investigated. Seed germination was not affected except for the inhibition of nanoscale zinc (nanoZn) on ryegrass and zinc oxide (nanoZnO) on corn at 2000 mg/L. Inhibition of root growth varied greatly among nanoparticles and plants. Suspensions of 2000 mg/L nanoZn or nanoZnO practically terminated root elongation of the tested plant species. Fifty percent inhibitory concentrations (IC50) of nanoZn and nanoZnO were estimated to be near 50 mg/L for radish, and about 20 mg/L for rape and ryegrass. The inhibition occurred during the seed incubation process rather than seed soaking stage. These results are significant in terms of use and disposal of engineered nanoparticles. On the contrary, the impacts of nanocerium oxide were different: even at low levels of exposure to this material, plant growth and the size of soybean pods were reduced. Furthermore, it appeared that the nanocerium oxide entered the roots and root nodules, the latter of which are important to the nitrogen fixing process that soybean crops perform. With higher amounts of nanocerium oxide in the soil, soybeans were nearly incapable of fixing nitrogen. Further, the manufactured nanomaterial may pose a potential risk to humans and animals if it enters the food chain in an unregulated way.

However, it was also observed that a very high concentration of nanosilica silver produced some chemical injuries on the tested plants (cucumber leaves and pansy flowers). Several plants grown hydroponically are affected by myriad Manufactured nanomaterials (MNMs), raising concerns regarding the long-term effects of these materials on the food supply (Shweta Agrawal and Pragya Rathore, 2014) However, MNMs may not be bioavailable (i.e., accessible to organisms) in the soil.

Protection of the soil health and the environment requires the rapid, sensitive detection of pollutants and pathogens with molecular precision. Soil fertility evaluation is being carried out for the past sixty years with the same set of protocols which may be obsolete for the current production systems and in the context of precision farming approaches. Accurate sensors are needed for *in situ* detection, as miniaturized portable devices, and as remote sensors, for the real-time monitoring of large areas in the field. These instruments are able to reduce the time required for lengthy microbial testing and immunoassays. Biosensors provide high-performance capabilities for use in detecting contaminants in food or environmental media (NAAS, 2013). They offer high specificity and sensitivity, rapid response, user-friendly operation, and compact size at a low cost

(Amine *et al.*, 2006). While the direct enzyme inhibition sensors currently lack the analytical ability to discriminate between multiple toxic substances in a sample (such as the simultaneous presence of heavy metal and pesticide), they may prove useful as a screening tool to determine when a sample contains one or more contaminants. These methods are amenable to deployment in single-use test strips (making them useful to those in the field). Detection of multiple residues of organophosphorus pesticides has been accomplished using a nanomagnetic particle in an enzyme-linked immunosorbent assay (ELISA) test.

Phytotoxicity of five types of nanoparticles was evaluated by Thul *et al.* (2013). Only Zn and ZnO particles were observed to have significant inhibition of seed germination and root growth of the six plant species. Inhibition was predominant in the seed incubation process rather than the seed soaking process. In a similar study wherein, the toxicity effect of ZnO and TiO_2 nanoparticles was evaluated in rice seed germination. No significant reduction in the percent seed germination from both nanoparticles was observed. However, nanoZnO shows detrimental effects on rice roots at early seedling stage. NanoZnO is found responsible to stunt roots length and reduce the number of roots, whereas nanoTiO_2 found to have no effect on root length. Effect of alumina nanoparticles loaded with and without phenanthrene was investigated by means of root elongation in hydroponic experiments. The nonloaded particles showed decreased inhibition of root elongations in this study. Similarly, in hydroponic experiment, Ma *et al.* demonstrated that plant seedlings of two common plant species: cattail (*Typha latifolia*) and hybrid poplars (*Populous deltoids* × *Populous nigra*) at different concentrations of nZVI (0–1000 mg/L) were found to exhibit strong toxic effect on *Typha* at higher concentrations (>200 mg/L) but enhanced plant growth at lower concentrations. It has been also noted that nZVI significantly reduced the transpiration and growth of hybrid poplars at higher concentrations. Microscopic studies indicated that large amount of nZVI coated on plant root surface as irregular aggregates and some nZVI penetrated into several layers of epidermal cells of poplar root cells.

Among the studied nanoparticles, most of these studies are focused on the potential toxicity of nanoparticles to plants and both positive and negative or inconsequential effects have been reported (Menard *et al.*, 2011). Among the positive effect reports on plants, nanoTiO_2 was observed to promote the growth of Spinach through an increase in photosynthetic rate and nitrogen metabolism. In one of the studies both positive and negative effects were observed by Lopez-Moreno *et al.* In this study, seeds of alfalfa (*Medicago sativa*), corn (*Zea mays*), cucumber (*Cucumis sativus*), and tomato (*Lycopersicon esculentum*) treated with nanoceria at 0–4000 mg L^{-1}. Uptake of nanoceria were significantly correlated with reduced corn germination (about 30% at 2000 mg L^{-1}, $p < 0.05$), and at 2000 mg L^{-1} germination of tomato and cucumber was reduced by 30% and 20%, respectively ($p < 0.05$). On the other hand, root growth was observed to be significantly promoted ($p < 0.05$) in cucumber and corn except in alfalfa and tomato where it was reduced ($p < 0.05$). Almost at all concentrations nanoceria promoted shoot elongation in these four plant species.

In line with above studies, ZnO and TiO_2 NPs reduced the wheat's biomass and thus were reported to be harmful to the plant. The TiO_2 NPs, considered having low solubility, remained in the soil for long periods and stuck to the plants' cell walls, which might create potential environmental risks for deeper soil layers. Additionally, a few individual small-sized TiO_2 NPs (around 20 nm) were able to penetrate the cell wall. The ZnO NPs, known to have a higher solubility than TiO_2, dissolved in the soil and increased the wheat's uptake of toxic zinc. Copper oxide is used in agriculture as a fungicide to protect coffee, cocoa, tea, banana, citrus, and other plants from major fungal leaf and fruit diseases such as blight, downy mildew, and rust. Copper is toxic to life at levels that vary depending on the organism [(Table 9.5) (Thul *et al.*, 2013)].

Table 9.5 Nanophytotoxicity on Some Food and Agricultural Crops

Crops	*Toxicity*	*Nanoparticles*	*Particle size (nm)*
1. Onion	Cell wall breaks, deformed chromosomes, indistinct disturbed metaphase and mitosis	Ag	< 100
2. Flax, ryegrass, barley	Reduced germination and shoot length	Ag	0.6 – 2
3. Flax, barley	Reduced germination and shoot length	Ag	5
4. Ryegrass, barley	Reduced germination and shoot length	Ag	20
5. Corn, lettuce	Decline in root length	Al	
6. Ryegrass	Reduced germination and decreased root length	Al	
7. Corn	Reduced root length	Al_2O_3	
8. Carrots, cabbage, cucumber, maize	Decline in root growth	Al_2O_3	13
9. Alfalfa, cucumber, maize, soybean, tomato	Reduced germination, biomass, shoot and root growth	CeO_2	7
10. Mungbean	Reduced seedling and shoot growth	Cu	
11. Wheat	Decline in root and seedling growth	Cu	
12. Onion	DNA damage, lipid peroxidation	TiO_2	10
13. Ryegrass, radish, rape, lettuce, cucumber, corn	Decline in root growth	Zn	
14. Lettuce, radish, rape, ryegrass, corn, cucumber	Reduced germination and root growth	ZnO	
15. Ryegrass	Seedlings failed to develop root hairs, reduced biomass, had highly vacuolated and collapsed cortical cells, broken epidermis, and root cap	ZnO	9–37

Contd...

Contd...

16. Soybean	Decline in root growth	ZnO	8
17. *Typha latifolia*, hybrid poplars *(Populous deltoids × Populous nigra)*	Strong toxic effect, reduced the transpiration and growth, penetrated into several layers of epidermal cells	nZVI	

9.7 QUANTUM DOTS

These are minute semiconductor structures on which the next generation of lasers is based upon. Quantum dots are typically only 10 to 50 nm across and contain just a few hundred or thousand atoms (Fig. 9.19). They are designed to confine the movements of the electrons responsible for semiconducting activity within the dot. Figure 9.20 illustrates a semiconductor nanoparticle (Quantum dot) of lead sulfide with complete passivation by oleic acid, oleyl, and hydroxyl (size ~5 nm). The result is an object with electrons trapped in discrete quantized energy levels, as in an atom. These electrons can be excited into higher energy states, at the same time leaving behind a positively charged 'hole.' Recombination of the electron-hole pairs results in the emission of light whose wavelength depends on the size, shape and configuration of the dots. Quantum dots form when a very thin film of one semiconductor is fabricated on the surface of another with a slightly different crystal-lattice size, for example, indium arsenide grown on a substrate of gallium arsenide (InAs/GaAs). The mismatch creates stresses which cause the film to buckle and form tiny three-dimensional blobs. Further layers of the semiconductors can be grown on top to build up a lasing device comprising several layers of millions of dots.

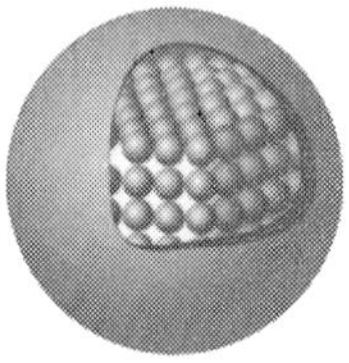

Fig. 9.19 Quantum dot

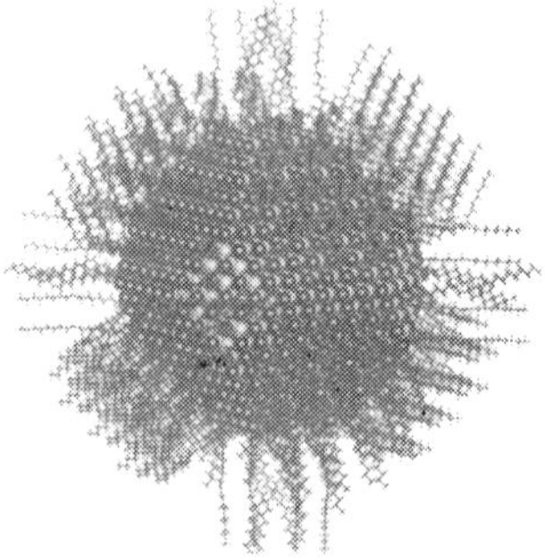

Fig. 9.20 Semiconductor nanoparticle

Quantum dots often described as 'artificial atoms,' exhibit discrete energy levels, and their band gap can be precisely modulated by varying the size. QDs are inorganic semiconductor nanocrystals, typically composed of a cadmium selenide (CdSe) core and a zinc sulfide (ZnS) shell and whose excitons (excited electron-hole-pairs) are confined in all three dimensions, giving rise to characteristic fluorescent properties. QDs are extremely photostable, bright and are characterized by broad absorption profiles, high extinction coefficients and narrow and spectrally tunable emission profiles (Ana R. Santos *et al.*, 2010).

Methods have been developed for introducing a linear nucleic acid molecule of interest into a cell comprising a cell wall, include the use of nanoparticles coated with polyethylene glycol. In some embodiments, the cell comprising a cell wall is a plant cell. Methods include genetically or otherwise modifying plants and for treating or preventing disease in plant cells comprising a cell wall. Transgenic plants include a nucleic acid molecule of interest produced by regeneration of whole plants from plant cells transformed with linear nucleic acid molecules. The plasmodiophoromycete, *Polymyxa betae*, and *P. graminis* are eukaryotic biotrophic parasites residing in the roots of Chenopodiacae and Gramineae plants. They are natural transmitting agents of several important plant viruses such as are beet necrotic yellow vein virus (BNYVV), beet soilborne mosaic virus (BSBMV), wheat soil-borne mosaic virus (WSBMV). Developing advanced high-throughput diagnostic methods capable of accurate detection of these pathogens could assist with the screening programs and consequently with the development of disease-resistant germplasms. In the present study, a previously developed quantum dots (QDs) FRET-based nanobiosensor was upgraded to a high-throughput version. QDs were functionalized with a specific antibody against the *P. betae's* specific glutathione-S-transferase (GST) protein. On the other hand, GST was conjugated to Rhodamine dye. Ninety-six well polystyrene plates were used as the detection platform. The mutual affinity for the antigen and the antibody brought the CdTe QDs and rhodamine together close enough to allow the resonance dipole-dipole coupling required for fluorescence resonance energy transfer (FRET) to occur. The immunosensor constructed showed a high sensitivity and specificity of 100%, and was successfully used for high-throughput screening of 96 real samples with consistent results within the course of less than 30 minutes (Hossein Safarpour *et al.*, 2012).

Nanotechnology, nanoparticles and quantum dots (QD) have emerged as a pivotal tool for detection of a particular biological marker with extreme accuracy. Diagnosis of a disease in its very early stage can play an important role in treatment. Due to phenomenal advancement in nanotechnology, QDs have emerged as a pivotal tool for detection of a particular biological marker with extreme accuracy. QDS being very photo-stable and optically sensitive can be used as labeling and can be easily traced with ordinary equipment. Early detection of tumor markers using quantum dots is proved to be a boon for cancer diagnosis. Use of QDs has also helped in unlocking complex neurological phenomenon, such as molecular activities at synapse during neurotransmission. QDs also give important information about receptor movement if tagged with suitable antibodies. In short, optical stability and easy to handle properties have made QDs to remain at the apex of medical diagnostics. QDs are few nm in diameter, roughly spherical (some QDs have rod-like structures), fluorescent, crystalline particles of semiconductors whose excitons are confined in all the three spatial dimensions. QDs have emerged as an important tool for detection of a specific biological marker in the medical field with extreme accuracy. They have been used in cell labeling, cell tracking, *in vivo* imaging and DNA detection (Sharon *et al.*, 2010).

Diagnosis of a disease in its very early stage can play an important role in treatment. Due to phenomenal advancement in nanotechnology, QDs have emerged as a pivotal tool for detection of a particular biological marker with extreme accuracy. QDs being very photo-stable and optically sensitive can be used as labeling and can be easily traced with ordinary equipment. Early detection of diseases using quantum dots is proved to be a boon.

The witches' broom disease of lime (WBDL) supposedly caused by *Candidatus* Phytoplasma aurantifolia (Ca. *P. aurantifolia*) is the most devastating disease of lime in Oman and southern parts of Iran. The present study describes developing a quantum dot (QD)-based nanobiosensor for the highly sensitive detection of phytoplasma in infected trees. The immunodominant membrane protein (IMP) expressed on the surface of phytoplasma was selected as a target protein for the construction of a specific binding antibody. The antibody (antiIMP) was effectively conjugated to thioglycolic acid-modified cadmium telluride quantum dots (CdTe-QDs) synthesized in an aqueous solution via electrostatic interaction. Dye (rhodamine) molecules were attached to the IMP, and then, the donor-acceptor complexes (QDs-Ab-IMP-Rhodamine) were formed based on the antigen-antibody interaction. The mutual affinity for the antigen and the antibody brought the CdTe QDs and rhodamine together close enough to allow the fluorescence resonance energy transfer (FRET) to occur. The immunosensor constructed showed a high sensitivity and specificity of 100%, a detection limit of 5 ca. *P. aurantifolia*/µl, and acceptable stability so that it could be used for detection with consistent results.

Glutathione-S-transferase (GST), a specific immunogenic protein, is a critical enzyme expressed in zoospores, sporangia and resting spores of *P. betae* that could be regarded as a good candidate for the development of the biobase of a nanobiosensor. In fact, the pathogen expresses GST at high levels to over some host defense mechanisms (Mutasa-Gottgens *et al.*, 2000).

Successful vector-mediated plant virus transmission entails an intricate but poorly understood interplay of interactions among virus, vector, and plant. The complexity of interactions requires continually improving/evaluating tools and methods for investigating the determinants that are central to mediating virus transmission. A recent study using an organic fluorophore (Alexa Fluor)-based immunofluorescent localization assay demonstrated that specific retention of Lettuce infectious yellows virus (LIYV) virions in the anterior foregut or cibarium of whitefly vector is required for virus transmission (Ng, 2013). Continuous exposure of organic fluorophore to high excitation light intensity could result in diminished or loss of signals, potentially confounding the identification of important interactions associated with virus transmission. This limitation could be circumvented by incorporation of photostable fluorescent nanocrystals, such as quantum dots (QDs), into the assay. A QD-immunofluorescent labeling method was developed and evaluated for the *in vitro* and *in situ* localization of LIYV virions based on the recognition specificity of streptavidin-conjugated QD_{605}(S-QD_{605}) for biotin-conjugated anti-LIYV IgG (B-αIgG). IgG biotinylation was verified in a blot overlay assay by probing SDS-PAGE separated B-αIgG with S-QD_{605}. Immunoblot analyses of LIYV using B-αIgG and S-QD_{605} resulted in a virus detection limit comparable to that of DAS-ELISA. In membrane feeding experiments, QD signals were observed in the anterior foregut or cibarium of virion-fed whitefly vectors but absent in those of virion-fed whitefly nonvectors. Specific virion retention in whitefly vectors corresponded with successful virus transmission. A fluorescence photobleaching assay of viruliferous whiteflies fed B-αIgG and S-QD_{605} vs. those fed antiLIYV IgG, and Alexa Fluor 488-conjugated IgG revealed that QD signal was stable and deteriorated approx. seven- to eightfold slower than that of Alexa Fluor.

Molecular beacons have become an important fluorescent probe for sequence-specific DNA detection. To improve the sensitivity and robustness of molecular beacon assays, fluorescent semiconductor quantum dots (QDs) are now being used as the fluorescent moiety for molecular beacon synthesis. Multiple linkage strategies can be used for attaching molecular beacon DNA to QDs, and multiple quenchers, including gold particles, can be used for fluorescence quenching (Nathaniel C. Cady, 2009). Covalent attachment of QDs to DNA can be achieved through an amide linkage, and affinity-based attachment can be achieved with streptavidin-biotin linkage. We have shown that these linkage strategies can be used to successfully create quantum dot molecular beacons that can be used in DNA detection assays with high specificity.

To detect bacteria, dye staining method is used. To stain bacteria, the most commonly used biolabels are organic dyes, but these are expensive, and their fluorescence degrades with time. Fluorescent labeling by quantum dots (QDs) with biorecognition molecules has been discovered through the recent developments in the field of luminescent nanocrystals (NAAS, 2013). QDs are better than conventional organic fluorophores (dyes) due to their more efficient luminescence compared to the organic dyes, narrow emission spectra, excellent photostability, symmetry and tenability according to the particles sizes and material composition. By a single excitation light source, they can be excited to all colors of the QDs due to their broad absorption spectra. Biolabeled *Bacillus* bacteria with nanoparticle consisting of ZnS and Mn_2^+ capped with biocompatible 'chitosan' gave an organ glow when viewed under a fluorescence microscope.

9.8 CARBON NANOMATERIAL AS A SENSOR

Carbon nanomaterials have been developed to act as an electrode for electrochemical analysis (Sharon and Sharon, 2008). They have the potential to be developed as an electrochemical sensor to detect pesticide residue in plants. Though there has been no patent filed so far exclusively for diagnosis for plant diseases through nanotechnology method, the methods developed for the diagnosis of animal diseases can be applicable to plants as well (Kalpana Sastry *et al.*, 2010). Carbon nanotubes (CNTs) exhibit a unique combination of excellent mechanical, electrical and electrochemical properties, which has stimulated increasing interest in the application of CNTs as components in (bio) sensors (Fig. 9.21). Their review highlights various design methodologies for CNT-based biosensors and their employment for the detection of a number of biomolecules. The discovery of fullerenes and carbon nanotubes has opened a new chapter in carbon chemistry. Superconducting and magnetic fullerides, atoms trapped inside the fullerene cage, chemically bonded fullerene complexes, and nanometer-scale helical carbon nanotubes are some of the leading areas that have generated much excitement. The creation of the hollow carbon buckminsterfullerene molecule as well as methods to produce and purify bulk quantities of it has triggered an explosive growth of research in the field. Carbon nanotubes, in particular, hold tremendous potential for applications because of their unique properties, such as high thermal and electrical conductivities, high strength, high stiffness, and special adsorption properties (Mansoori *et al.*, 2008).

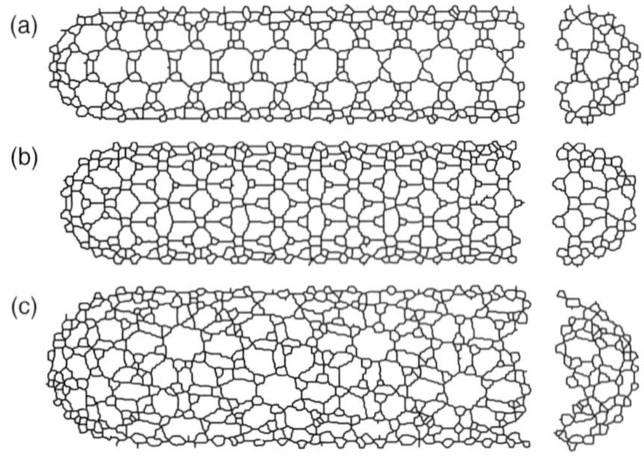

Fig. 9.21 Some SWNTs (single-walled carbon nanotubes) with different chiralities. The difference in structure is easily shown at the open end of the tubes. (a) armchair structure, (b) zigzag structure, and (c) chiral structure

Carbon nanotubes have cylindrical pores and adsorbent molecules interact with their carbon atoms on the surrounding walls (Figs. 9.22 and 9.23). This interaction between molecules and solid surface depends on the pore size and geometry of pores. When a molecule is placed in between two flat surfaces, i.e., in a slit-shaped pore, it interacts with both surfaces and the potentials on the two surfaces overlap. The extent of the overlap depends on the pore size. However, for cylindrical and spherical pores, the potentials are greater because more surface atoms interact with the adsorbed molecule. In addition, carbon nanotubes are highly graphitic (much more than the activated carbons). Hence, the carbon nanotubes can adsorb molecules much stronger than activated carbons, which have slit-shaped or wedge-shaped pores.

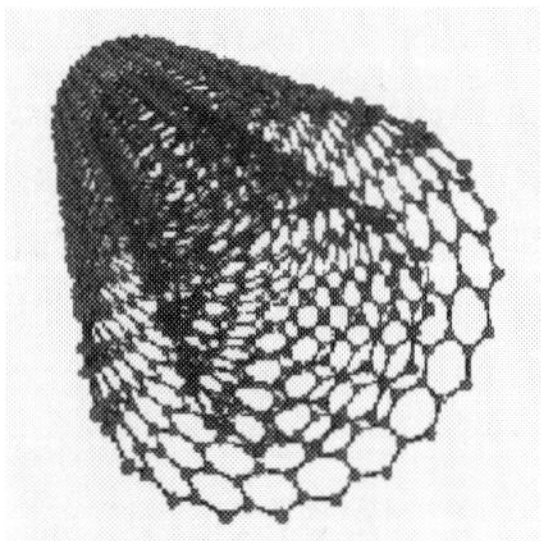

Fig. 9.22 Structure model of multiwall carbon nanotube

The cyclic voltammetric behavior of three common pesticides such as isoproturon (ISO), voltage (VOL) and dicofol (DCF) was investigated at glassy carbon electrode (GCE), multiwalled carbon nanotubes modified GCE (MWCNTs/GCE), polyaniline (PANI) and polypyrrole (PPY) deposited MWCNT/GCE (Manishankar *et al.*, 2008). The modified electrode film was characterized by scanning electron microscopy (SEM) and X-ray diffraction analysis (XRD). The electroactive behavior of the pesticides was realized from the cyclic voltammetric studies. The differential pulse voltammetric principle was used to analyze the above-mentioned pesticides using MWCNT/GCE, PANI/MWCNT/GCE, and PPY/MWCNT/GCE. Effects of accumulation potential, accumulation time, Initial scan potential, amplitude and pulse width were examined for the optimization of stripping conditions. The PANI/MWCNT/GCE performed well among the three electrode systems and the determination range obtained was 0.01–100 mg L^{-1} for ISO, VOL and DCF respectively. The limit of detection (LOD) was 0.1 μg L^{-1} for ISO, 0.01 μg L^{-1} for VOL and 0.05 μg L^{-1} for DCF on PANI/MWCNT/GCE modified system. It is significant to note that the PANI/MWCNT/GCE modified system results in the lowest LOD in comparison with the earlier reports. Suitability of this method for the trace determination of pesticide in spiked samples was also realized.

9.9 NANOFABRICATION

Nanofabrication techniques have been used in creating artificial plant parts such as stomata and xylem vessel which are then used to study the infection process and behavior of pathogens inside host plant, for example, *Uromyces appendiculatus* (fungus causing rust disease of bean), *Colletotrichum graminicola* (fungus causing anthracnose in corn) and *Xylella fastidiosa* [(xylem limited bacterium causing Pierce's disease of grapevine) (Meng *et al.*, 2005)]. Such study would help breeder to look for specific stomatal characters to prevent entry inside host through stomata or leaf surface characters to prevent appressorium formation before penetration of fungi or vascular characters to obstruct the movement of vascular pathogens like bacteria, fungi, etc. In other words, it would help the formation of a proper breeding strategy to screen for or to develop disease resistant crop plants.

The advancement in nanofabrication and characterization tools have enabled studies of physical, chemical and biological interactions between plant cell organelles and various disease causing pathogens, i.e., plant pathology. A better understanding of plant pathogenic mechanisms such as flagella motility and biofilm formation will lead to improved treatment strategies to control the diseases and protect production (Cursino *et al.*, 2009). For example, spatial and temporal studies of plant pathogenic xylem inhabiting bacteria have traditionally been conducted by monitoring changes in bacterial populations through destructive sampling techniques of tissues at various distances from inoculation sites. This approach seriously limits the information that can be obtained regarding colonization, biofilm development, and subsequent movement and recolonization at new areas, primarily because the same region or sample site cannot be followed temporally. Microfabricated xylem vessels with nanosize features have been shown very useful in gaining an appreciation of the mechanisms and kinetics of bacterial colonization in xylem vessels such that novel disease control strategies may be developed (Hongda Chen and Rickey Yada, 2011).

Surface engineering techniques for cellular micropatterning are emerging as important tools to clarify the effects of the microenvironment on cellular behavior, as cells usually integrate and respond the microscale environment, such as chemical and mechanical properties of the surrounding fluid and extracellular matrix, soluble protein factors, small signal molecules, and contacts with neighboring cells. Furthermore, recent progress in cellular micropatterning has contributed to the development of cell-based biosensors for the functional characterization and detection of drugs, pathogens, toxicants, and odorants. In this regard, the ability to control the shape and spreading of attached cells and cell-cell contacts through the form and dimension of the cell-adhesive patches with high precision is important. The commitment of stem cells to different specific lineages depends strongly on cell shape, implying that controlled microenvironments through engineered surfaces may not only be a valuable approach towards fundamental cellulaar biological studies but also of great importance for the design of cell culture substrates for tissue engineering (Otsuka, 2010). To develop this kind of cellular microarray composed of a cell-resistant surface and cell attachment region, micropatterning a protein-repellent surface is important because cellular adhesion and proliferation are regulated by protein adsorption. The review was focused on the surface engineering aspects of biologically motivated micropatterning of two-dimensional surfaces with the aim to provide an introductory overview described in the literature. It also discussed, in particular, the importance of nonfouling surface chemistries.

9.10 'SMART' DELIVERY

An interesting and fascinating area of nanoparticles is 'smart' or targeted drug delivery in the biological system. This is being vigorously pursued in cancer treatment (Kukowska-Latallo *et al.*, 2005). Use of nanoparticles in 'smart' delivery was postulated as 'magic bullets' more than a century ago by P. Ehrlich (Himmelweit, 1960). To develop a smart treatment delivery system in plant, Gonza´lez-Melendi *et al.* (2008) worked with Cucurbita pepo, which were treated with carbon-coated Fe nanoparticles *in vitro*. The magnetic core consisting of Fe nanoparticles allow themselves to be guided to a place of interest in the body (affected part) of an organism using small magnets that create a magnetic field. The carbon coating provides biocompatibility and acts as a surface for adsorption where various types of molecules of interest (drug/DNA/chemical/enzyme) can be adsorbed. Gonza´lez-Melendi *et al.* (2008) were the first to report the penetration and transport of nanoparticles inside the whole plant. These results indicate the possibility and potential of nanoparticles in the delivery of substances inhibitory to various plant pathogens.

Technologies such as encapsulation and controlled release methods have revolutionized the use of pesticides and herbicides. Many companies make formulations which contain nanoparticles within the 100–250 nm size range that are able to dissolve in water more effectively than existing ones (thus increasing their activity). Other companies employ suspensions of nanoscale particles (nanoemulsions), which can be either water or oil-based and contain uniform suspensions of pesticidal or herbicidal nanoparticles in the range of 200–400 nm. These can be easily incorporated in various media such as gels, creams, liquids, etc., and have multiple applications for preventative measures, treatment or preservation of the harvested product (Tiju Joseph and Mark Morrison, 2006).

Nanotechnology enables companies to manipulate the properties of the outer shell of a capsule to control the release of the substance to be delivered. 'Controlled release' strategy is highly prized in medicine since it allows drugs absorbed more slowly, at a specific location in the body or the say-so of an external trigger. Pesticides containing nanoscale active ingredients are already in the market, and many of the world's leading agrochemical firms are conducting research on the development of new nanoscale formulations of pesticides. Nanoparticles comprising a crop protection agent involve an active ingredient whose ideal particle size is between 10 and 150 nm (Chinnamuthu and Murugesa Boopathi, 2009). The advantage of nanoformulation is that the pesticide dissolves more easily in water; it is more stable, and the killing capacity of the chemical (herbicide, insecticide or fungicide) is optimized.

Nanopesticides cover a wide variety of products, some of which are already on the market. They cannot be considered as a single entity; rather such nanoformulations combine several surfactants, polymers (organic), and metal nanoparticles (inorganic) in the nanometer size range. The lack of water solubility is one of the limiting factors in the development of crop-protecting agents. Microencapsulation has been used as a versatile tool for hydrophobic pesticides, enhancing their dispersion in aqueous media and allowing a controlled release of the active compound. Controlled-release (CR) formulations of carbofuran are in the market. Nanomaterials including polymeric nanoparticles, iron oxide nanoparticles, gold nanoparticles, and silver ions have been exploited as pesticides. Researchers have reported various aspects of nanoparticle formulation, characterization, the effect of their characteristics, and their applications in the management of plant diseases (Bhupinder Singh Sekhon, 2014). Nanoencapsulation helps the slow release of a chemical to the particular host for insect pest control through release mechanisms that include dissolution, biodegradation, diffusion, and osmotic pressure with specific pH. Particles suspended in water have been used since at least 1931, in a product known as Bouisol as a fungicide in the growing of grapes and fruit trees.

In the research and development stage, nanosized agrochemicals or nanoagrochemicals are mostly nanoreformulations of existing pesticides and fungicides. Nanoformulations are generally expected to increase the apparent solubility of poorly soluble active ingredients, to release the active ingredient in a slow/targeted manner, and/or to protect against premature degradation. Nanopesticides offer a way to both control delivery of pesticide and achieve greater effects with lower chemical dose. Many companies make formulations that contain nanoparticles within the 1 00–250 nm size range that are able to dissolve in water more effectively than existing ones, thus increasing their activity (Kumar *et al.*, 2010). Nanoemulsions, which can be either water or oil-based and contain uniform suspensions of pesticidal or herbicidal nanoparticles, have many potential applications as disease and pest prevention measures. Other companies employ suspensions of nanoscale particles/nanoemulsions, which can be either water-based or oil-based and contain uniform suspensions of pesticidal or herbicidal nanoparticles in the range of 200–400 nm. Viral capsids can be altered by mutagenesis to achieve different configurations and deliver specific nucleic acids, enzymes, or antimicrobial peptides acting against the parasites.

Silver nanoparticles at 100 mg/kg inhibited mycelial growth and conidial germination on cucurbits and pumpkins against powdery mildew disease (Jo *et al.*, 2009). Silver nanoparticles have received significant attention as a pesticide for agricultural applications. Antifungal activities of polymer-based copper nanocomposites against pathogenic fungi and silica–silver nanoparticles against *Botrytis cinerea, Rhizoctonia solani, Calllectotrichum gloeosporioides, Bipolaris sorokiniana,* and *Magnaporthe grisea* have been reported. Copper nanoparticles in soda lime glass powder showed efficient antimicrobial activity against Gram-positive and Gram-negative bacteria and fungi. Nanoparticles can be used in the preparation of new formulations like pesticides, insecticides, and insect repellents. Researchers have reported nanosilver and titanium dioxide nanoparticle applications in the management of plant diseases. The fungicidal efficiency of sulfur nanoparticles against two phytopathogens, *Fusarium solani* (isolated from an infected tomato leaf, responsible for early blight and *Fusarium* wilt diseases) and *Venturia inaequalis* (responsible for the apple scab disease) has been reported (Rao and Paria, 2013).

As a potential 'green' nanocarrier of pesticides, novel amphiphilic self-assembled octahydrogenated retinoic acid-glycol chitosan (OR-GC) nanoparticles have been synthesized and characterized, which can effectively load the poorly water-soluble pesticide AZA (about 74%). Compared with free AZA in organic solvents, the AZA-OR-GC nanoparticles showed favorable inhibition of cell proliferation and sustained drug release performance in Sl-1 cells *in vitro* (Wei Lu *et al.*, 2013).

9.11 NANOTECHNOLOGY-ENABLED DELIVERY OF PESTICIDES

Nanoscale carriers can be utilized for the efficient delivery of fertilizers, pesticides, herbicides, plant growth regulators, etc. The mechanisms involved in the efficient delivery, better storage and controlled release include encapsulation and entrapment, polymers and dendrimers, surface ionic and weak bond attachments among others (Allah Ditta, 2012). These mechanisms help improve stability against degradation in the environment and ultimately reduce the amount to be applied, which reduces chemical runoff and alleviates environmental problems. These carriers can be designed in such a way that they can anchor the plant roots to the surrounding soil structure and organic matter. This can only be possible through the understanding of molecular and conformational mechanisms between the delivery nanoscale structure and targeted structures and matter in the soil. These advances will help in slowing the uptake of active ingredients, thereby reducing the amount of inputs to be used and also the waste produced.

Many nanoscale carriers, including encapsulation and entrapment, polymers and dendrimers, surface ionic and weak bond attachments and other mechanisms may be used to store, protect, deliver, and release by control of intended payloads in crop production processes. One of the advantages of nanoscale delivery vehicles in agronomic applications is its improved stability of the payloads against degradation in the environment, thereby increasing its effectiveness while reducing the amount applied. This reduction helps address agricultural chemicals runoff and alleviate the environmental consequence. The nanoscale delivery vehicles may be designed to 'anchor' to plant roots or the surrounding soil structures and organic matter if molecular or conformational affinity between the delivery nanoscale structure and targeted structures and matters in soil could be utilized. Controlled release mechanisms allow the active ingredients to be slowly taken up, hence, avoiding temporal overdose, reducing the amount of agricultural chemicals used, and minimizing the input and waste. Environmental consideration including precision farming can reduce pollution to a minimum. The advancement in nanotechnology-enabled gene sequencing is expected to introduce rapid and cost-effective capability within a decade (Hongda Chen and Rickey Yada, 2011).

9.11.1 Nanocides-Pesticides *via* Encapsulation

A more sophisticated approach to formulating nanoscale pesticides involves encapsulation-packaging the nanoscale active ingredient within a kind of tiny 'envelope' or 'shell.' Both food ingredients and agrochemicals in microencapsulated form have been on the market for several decades. According to industry, reformulation of pesticides in microcapsules has triggered 'revolutionary changes,' including the ability to control under what conditions the active ingredient is released. According to the agrochemical industry, reformulating pesticides in microcapsules can also extend patent protection, increase solubility, reduce the contact of active ingredients with agricultural workers and may have environmental advantages such as reducing runoff rates.

Nanotechnology operates on the same scale as a virus or disease-infecting particle, and thus holds the potential for very early detection and eradication. Smart treatment delivery systems are envisioned for biology and bioactive systems such as drugs, pesticides, nutrients, probiotics, nutraceuticals and implantable cell bioreactors (Chinnamuthu and Murugesa Boopathi, 2009). Experts in agriculture and nanoscience identified some of the important areas in the field of agriculture that have enormous potential for the application of nanotechnology. Some of the products like nanopesticides have arrived already in the market while many others are under developing stage, and it may take many years before they are commercialized. These applications are largely intended to address some of the limitations and challenges like problem weed management, slow release fertilizers, conditional release of pesticides and herbicides, precise micromanagement of soils, the more efficient and targeted use of inputs and new toxin formulations for pest control.

Nanobased viral diagnostics, including multiplexed diagnostic kit development, have taken momentum to detect the exact strain of virus and stage of application. Detection and utilization of biomarkers that accurately indicate disease stages with differential protein production in both healthy and diseased states lead to the identification of the development of several proteins during the infection cycle. These nanobased diagnostic kits not only increase the speed of detection but also increase the power of the detection. These nano and microformulations are being developed and patented by agribusiness and food corporations such as Monsanto, Syngenta and Kraft (Chinnamuthu and Murugesa Boopathi, 2009). Pesticides containing nanoscale active ingredients are already in the market, and many of the world's leading agrochemical firms are conducting research on the development of new nanoscale formulations of pesticides. For example, BASF of Germany, the world's fourth-ranking agrochemical corporation, recognizes nanotech's potential usefulness in the formulation of pesticides. BASF is conducting basic research and has applied for a patent on a pesticide formulation, "Nanoparticles Comprising a Crop Protection Agent," that involves an active ingredient whose ideal particle size is between 10 and 150 nm. The advantage of nanoformulation is that the pesticide dissolves more easily in water; it is more stable, and the killing capacity of the chemical (herbicide, insecticide or fungicide) is optimized. Bayer Crop Science of Germany has also applied for a patent on agrochemicals in the form of an emulsion in which the active ingredient is made up of nanoscale droplets in the range of 10-400 nm. The company refers to the invention as a 'microemulsion concentrate' with advantages such as reduced application rate, 'a more rapid and reliable activity' and 'extended long-term activity.'

Like Bayer Crop Science, Syngenta claims that its Primo MAXX Plant Growth Regulator and Banner MAXX fungicide are extremely small particle size of about 100 nm (or 0.1 micron) prevents spray tank filters from clogging, and the chemicals mix so completely in water that they neither settle in the spray tank not separate from water for up to one year. Additionally the fungicides will be absorbed into the plant's system and cannot be washed off by rain or irrigation.

9.12 CLAY NANOTUBES

Clay nanotubes (Halloysite) have been developed as carriers of pesticides for low cost, extended release and better contact with plants, and they will reduce the amount of pesticides by 70–80%, hence reducing the cost of pesticide and also the impact on water streams (Murphy, 2008). Halloysite nanotubes are ultra tiny hollow tubes with diameters typically smaller than 100 nanometers (100 billionths of a meter), with lengths typically ranging from about 500 nanometers to over 1.2 microns (millionths of a meter) (Fig. 9.23). The functional characteristics desired for specific applications can be controlled through selection of nanotube diameter and length. Diameters typically range from about 40 nm to 200 nm and in a variety of lengths, allowing for a wide range of applications. Halloysite nanotubes can be coated with metallic and other substances to achieve a wide variety of electrical, chemical, and physical properties. They can be filled with such things as active ingredients including many that are used for pesticides and, pest repellents (Anonymous, 2014).

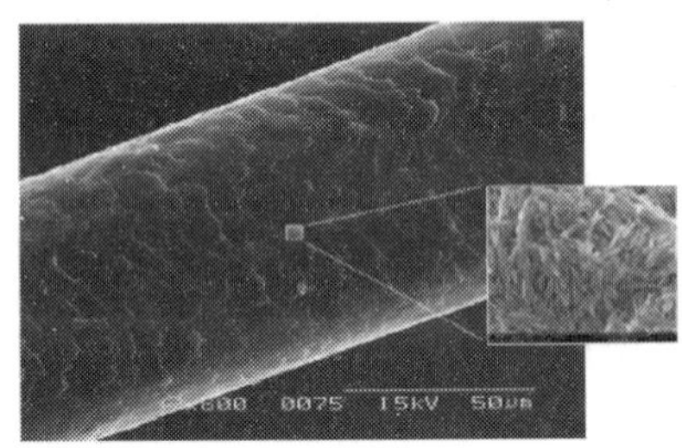

Fig. 9.23 A bundle of natural nanohalloysite nanotubes compared to the width of a human hair

Halloysite nanotubes are not equivalent to carbon nanotubes. However, both are examples of hollow tubes with nanometer scale diameters. The nanotube is made up of lignin from plant material obtained from biofuel pilot facility. Lignin is an integral part of the secondary cell walls of plants and enables water movement from the roots to the leaves, but it is not used to make biofuels and would otherwise be burned to generate heat or electricity at the biofuel plant. The lignin nanotubes are about 500 times smaller than a human eyelash. The lignin nanotubes can be made from a variety of plant residues, including sorghum, poplar, loblolly pine, and sugar cane. The Ag nanoparticles, when biosynthesized by fungi, showed toxic effects against fungal species like *Aspergillus niger* (Jaidev and Narasimha, 2010). Another improvement in this sense might be the availability of nanostructured catalysts which will increase the efficiency of pesticides and insecticides and also reduce the dose level required for plants (Muhammad Azam Ali *et al.*, 2014).

9.13 BIOREMEDIATION OF RESISTANT PESTICIDES

Nanoparticles can be used for the bioremediation of resistant or slowly degradable compounds like pesticides. These harmful compounds tend to join the positive holes, are degraded and converted into nontoxic compounds. Otherwise, these harmful compounds enter the food chain and result in serious problems for the body. Nanoparticles, hence, can be used for environmental safety (Lhomme *et al.*, 2008).

Bioremediation is the application of biological systems to the clean-up of organic and inorganic pollution, with bacteria and fungi being the most important organisms for reclamation, immobilization or detoxification of metallic and radionuclide pollutants (Geoffrey Michel Gadd, 2010). Some biominerals or metallic elements deposited by microbes have catalytic, and other properties in the nanoparticle, crystalline or colloidal forms, and these are relevant to the development of novel biomaterials for technological and antimicrobial purposes. On the negative side, metal and mineral transformations by microbes may result in spoilage and destruction of natural and synthetic materials, rock and mineral-based building materials (e.g., concrete), acid mine drainage and associated metal pollution, biocorrosion of metals, alloys and related substances, and adverse effects on radionuclide speciation, mobility and containment, all with immense social and economic consequences. The ubiquity and importance of microbes in biosphere processes make geomicrobiology one of the most important concepts in microbiology, and one requiring an interdisciplinary approach to define environmental and applied significance and underpin exploitation in biotechnology.

Enzymes are versatile catalysts in laboratories and on an industrial scale; improving their immobilization would be beneficial to broadening their applicability and ensuring their (re)use. Lipid-coated nanomagnets produced by magnetotactic bacteria are suitable for a universally applicable single-step method of enzyme immobilization. By genetically functionalizing the membrane surrounding these magnetite particles with a phosphohydrolase, Ginet *et al.* (2011) engineered an easy-to-purify, robust and recyclable biocatalyst to degrade ethyl-paraoxon, a commonly used pesticide. For this, we genetically fused the *opd* gene from *Flavobacterium* sp. ATCC 27551 encoding a paraoxonase to *mamC*, an abundant protein of the magnetosome membrane in *Magnetospirillum magneticum* AMB-1. The MamC protein acts as an anchor for the paraoxonase to the magnetosome surface, thus producing magnetic nanoparticles displaying phosphohydrolase activity.

Magnetosomes functionalized with Opd were easily recovered from genetically modified AMB-1 cells: after cellular disruption with a French press, the magnetic nanoparticles are purified using a commercially available magnetic separation system. The catalytic properties of the immobilized Opd were measured on ethyl-paraoxon hydrolysis: they are comparable with the purified enzyme, with K_m (and k_{cat}) values of 58 μM (and 178 s^{-1}) and 43 μM (and 314 s^{-1}) for the immobilized and purified enzyme respectively. The Opd, a metalloenzyme requiring a zinc cofactor, is thus properly matured in AMB-1. The recycling of the functionalized magnetosomes was investigated, and their catalytic activity proved to be stable over repeated use for pesticide degradation. This study demonstrated the easy production of functionalized magnetic nanoparticles with suitably genetically modified magnetotactic bacteria that are efficient as a reusable nanobiocatalyst for pesticides bioremediation in contaminated effluents.

Annually thousands of people die or suffer from organophosphate (pesticide) poisoning. To remove these toxic compounds from the environment, the use of enzymes as bioremediators has been proposed. Lena J. Daumann *et al.* (2014) reported a Ser127Ala mutant based on the enzyme glycerophosphodiesterase (GpdQ) from *Enterobacter aerogenes*. The mutant, with improved metal binding abilities, has been immobilized using glutaraldehyde on PAMAM dendrimer-modified magnetite nanoparticles. The immobilized system was characterized using elemental analysis as well as infrared, transmission electron, and X-ray photoelectron spectroscopies. The amount of GpdQ that was immobilized with the optimized procedure was 1.488 nmol per g MNP. A kinetic assay has been designed to evaluate the activity of the system towards organophosphoester substrates. The specific activity towards BPNPP directly after immobilization was 3.55 μmol mg^{-1} min^{-1}, after one week, 3.39 μmol mg^{-1} min^{-1} and after 120 days, 3.36 μ mol mg^{-1} min^{-1}, demonstrating that the immobilized enzyme was active for multiple cycles and could be stored on the nanoparticles for a prolonged period.

Glyphosate is a member of the amino acid herbicide family, and its mode of action is through inhibition of 5-enolpyruvylshikimate-3-phosphate synthase (EPSP), an enzyme of the shikimic acid pathway. This enzyme is important in the biosynthesis of the aromatic amino acids phenylalanine, tyrosine, and tryptophan. A blockage of the shikimic acid pathway leads to a depletion of the free pool of aromatic amino acids in higher plants consumption of it results in symptoms such as intestinal pain, vomiting, excess fluid in the lungs, pneumonia, clouding of consciousness, and destruction of red blood cells. Short-term exposure to glyphosate can cause breathing difficulties, loss of muscle control and convulsions. Isoprothiolane (di-isopropyl 1, 3-dithiolan-2-ylidenemalonate), a systemic fungicide, is effective against *Pyricularia oryzae*, which causes blast disease in rice. It is applied at 3.2–4.8 kg/ha. Because of a high rate of application, there is an appreciable risk of groundwater and surface water contamination by isoprothiolane, particularly in the agro-climatic zone having light soil and high rainfall. With continuous usage, these pesticides enter the human body via the food chain. Residues of these pesticides have been detected in soil, water, and air. A microbial consortium capable of degrading glyphosate and isoprothiolane was isolated in the laboratory by long-term enrichment. The consortium was made of ten bacterial isolates of which seven belonged to *Pseudomonas* sp. and one each of *Flavobacterium, Vibrio,* and *Burkholderia* sp. Conditions were optimized by response surface methodology. The maximum predicted percentage degradation of 100, 100, 100, 82 and 73 was obtained respectively for, 20, 40, 50, 75 and 100 ppm initial glyphosate concentrations at pH 6.60 with an incubation temperature of 30°C and inoculum concentration of g protein/ml. Similarly, the maximum predicted percentage degradation of μ 1146 100, 100, 100, 100 and 95.5 were obtained respectively for 5, 10, 20, 30 and 50 ppm initial isoprothiolane concentrations at pH levels 7.7, 6.8, 6.2, 4.7 and 4.6 gm with an incubation temperature of 30°C and inoculum concentration of 50 protein/mL. The degradation of these pesticides by the microbial consortium increased with time. The degradation reached 100% by 120 h of incubation for these pesticides. Residues of glyphosate and isoprothiolane are generally analyzed by thin layer chromatography, gas chromatography, and high-performance liquid chromatography which are expensive, labor-intensive, require extensive clean up and are not suited for on field assay. Immunoassay is highly sensitive, specific, on-field method, does not need sample clean up and has a high throughput of samples. In our laboratory, we attempted to develop an immunoassay for the detection of glyphosate and isoprothiolane by using avian antibodies. Glyphosate and isoprothiolane levels up to 2 ng could be detected by Well-Dot. Signal amplification improved the detection signal with the same sensitivity. Immunosensor based on DNA coated CdS nanoparticles and antibody coated CdS nanoparticles were developed for the detection of glyphosate and isoprothiolane. The detection limit by these nanoparticles based methods was 0.2 ng/mL (Arul Selvi, 2011).

A Ser127Ala mutant based on the enzyme glycerophosphodiesterase (GpdQ) from *Enterobacter aerogenes* (Daumann *et al.*, 2014). The mutant, with improved metal binding, was immobilized using glutaraldehyde on PAMAM dendrimer modified magnetite nanoparticles. The immobilized system was characterized using elemental analysis as well as infrared, transmission electron, and X-ray photoelectron spectroscopies. The amount of GpdQ that was immobilized with the optimized procedure was 1.4888 nmol per g MNP. A kinetic assay has been designed to evaluate the activity of the system towards organophosphoester substrates. The specific activity towards BPNPP directly after immobilization was 3.55 micromol mg^{-1} min^{-1}, after one week 3.39 micromol mg^{-1} min^{-1} and after 120 days 3.36 micromol mg^{-1} min^{-1}, demonstrating that the immobilized enzyme was active for multiple cycles and could be stored on the nanoparticles for a prolonged period.

Currently spraying of pesticides involves either knapsacks that deliver large droplets (9–266 μm) associated with splash loss or ultralight volume sprayers for controlled droplet application with smaller droplets (3–28 μm) causing spray drift (Vandana Ghormade *et al.*, 2011). Constraints due to droplet size may be overcome by using NP-encapsulated or nanosized pesticides that will contribute to efficient spraying and reduction of spray drift and splash losses. Another practical problem faced during pesticide application in the field is a settlement of formulation components in the spray tank and clogging of spray nozzles. The recent nanosized fungicide (~100 nm, BannerMAXX, Syngenta) prevented spray tank filters from clogging, did not required mixing and did not settle out in the spray tank due to the smaller sized particles (Vandana

Ghormade *et al.*, 2011). Furthermore, this fungicide did not separate from water for up to one year due to nanosize, whereas fungicides that contained larger particle size ingredients typically required agitation every 2 h to prevent clogging in the tank.

9.14 DISINFECTANTS

Disinfection is one of the most important directions of the complex of measures for controlling infectious diseases and pathogenic flora. Nevertheless, conventional disinfecting agents no more meet strict requirements for such products due to objective reasons. According to the opinion of scientists of many countries, disinfectants based on nanoscale structures with a wide range of biocidal properties will constitute the next generation. The electron-hole pair, especially the negative electrons resulting from the excitation of nanoparticles can also be used as a disinfectant of bacteria, as when bacteria make contact with nanoparticles, the excited electrons are injected into their bodies, which results in the bacterial removal from the object concerned, as in fruit packaging and food engineering (Melemeni *et al.*, 2009).

Engineered nanomaterials display significant advantages due to their unique nanostructure, along with their tuneable properties for the designed application. Silver nanoparticles (Ag-NPs) have drawn attention due to their use as potent bactericidal agents and were characterized in this research to provide an understanding of the interaction between nanomaterials and bacteria (Chamakura *et al.*, 2011). This work presented the bactericidal performance of Ag-NPs using *Escherichia coli* as a model microorganism. Several state-of-the-art techniques, such as high-angle annular dark-field detector in scanning transmission electron microscopy, and energy filtered imaging in electron energy loss spectroscopy, were employed to obtain nanostructural and elemental information. The bactericidal activities of Ag-NPs were then compared with two commonly used disinfectants, sodium hypochlorite (NaClO) and phenol (C(6)H(5)OH). These two chemical disinfectants exhibited rapid bactericidal activity, showing a minimal bactericidal concentration (MBC) of 16 parts-per-million (ppm) and 16 part-per-thousand (ppth), respectively for NaClO and C(6)H(5)OH within about 10 min. In contrast, Ag-NPs exhibit slow but long-term bactericidal effect demonstrating MBCs of 0.6 parts-per-million (ppm) within 6h when used as disinfectant. An advantage using Ag-NPs to inactivate *E. coli* at low dosages is a negligible environmental waste or hazardous by-products. The results showed that Ag-NPs caused bacterial inactivation by a mechanism involving several steps, including cell wall and cytoplasmic membrane damage.

9.15 ZigBee – MESH NETWORKING STANDARD

ZigBee is a wireless mesh networking standard with low-cost and utilizes low-power. It has given the concept of 'Smart Fields' and 'SoilNet.' It consists of one or more sensors for environmental data (temperature, humidity, etc.), a signal conditioning block, a microprocessor/microcontroller with an external memory chip and a radio module for wireless communication between the sensor nodes and/or a base station. It can be used for the identification and monitoring of pests and diseases, drought or increased moisture levels to counterbalance their adverse effects on crop production (Kalra A Chechi and Khanna, 2010)]. Through this wireless sensor technology with nanoscale sensitivity, we can control plant viruses and level of soil nutrients as the plant surfaces can be changed at the nanoscale with specific proteins. This technology is important in realizing the vision of smart fields in particular. Wireless network sensor technology can also be used for monitoring the optimal conditions for mobile plants biotechnology. Nanotechnology fits in well to 'Precision farming,' which is also known as site-specific management, describes a bundle of new information technologies applied to the management of large-scale, commercial agriculture. Precision farming technologies include, for example personal computers, satellite-positioning systems, geographic information systems, automated machine guidance, remote sensing devices and telecommunications. Precision farming has been a long-desired goal to maximize output (i.e., crop yields) while minimizing input (i.e., fertilizers, pesticides, herbicides, etc.) through monitoring environmental variables and applying targeted action. Benefits include very less chance of pathogens attack, enhanced productivity in agriculture and help in reducing agricultural waste, thus keep environmental pollution minimum.

The role of networks of wireless nanosensors in precision farming: If they function as designed, ubiquitous wireless sensors will become an essential tool for bringing this vision of precision farming to maturity. When scattered on fields, networked sensors are expected to provide detailed data on crop and soil conditions and relay that information in real time to a remote location so that crop scouting will no longer require the farmer (or agribusiness executive) to get their boots dirty. Since many of the conditions that a farmer may want to monitor (e.g., the presence of plant viruses or the level of soil nutrients) which operate at the nanoscale, and because surfaces can be altered at the nanoscale to bind selectively with particular biological proteins, sensors with nanoscale sensitivity will be particularly important in realizing this vision.

9.16 NANOMATERIALS FOR MANAGEMENT OF PLANT DISEASES

For the past few decades, there has been considerable in research interest in the area of natural product delivery using particulate for controlling plant pathology. The secondary metabolites in plants have been used in the formulation of nanoparticles through increase the effectiveness of therapeutic compounds used to reduce the spread of plant diseases while minimizing side effects of being a rich source of bioactive chemicals, biodegradable in nature and nonpolluting [(ecofriendly) (Abdul Hameed and Al-Samarrai, 2012)]. Particulate systems like nanoparticles have been used as a physical approach to alter and improve the effect of properties of some types of synthetic chemical pesticides or in the production of biopesticides directly.

Some of the nanoparticles that have entered into the arena of controlling plant diseases are nanoforms of carbon, silver, silica and aluminosilicates. Many scientists have come forward to explain carbon's uniqueness in a unique way. But the fact is that there are many concepts yet to be understood. At such a situation, nanotechnology has astonished scientific community because, at the nanolevel, the material shows different properties. Thus, exposed to a huge spectrum of nanosciences, wherein there are totally new materials, new technologies and new hope for existing problems related to agrochemicals, pesticides, herbicides-regulation and smart utilization. Brazil is trying to improve the value of its exports by developing agricultural nanotechnologies. Brazilian agriculture research corporation's (Embrapa) areas of focus are to include research for producing carbon nanofibers to strengthen natural fibers, for example, those from coconuts and sisal and making nanoparticles that contain pesticides and control their release. Scientists are mostly concentrating on carbon nanotubes (CNT). Carbon nanotubes are allotropes of carbon whose nanostructure is cylindrical in shape. These nanotubes have many applications, especially in the fields of nanotechnology, electronics, and architecture. Often used as thermal conductors, these nanotubes also host unique electrical properties and are surprisingly strong. Recently scientists (Khodakovsky *et al.*, 2010) have reported that when they planted tomato seeds in a soil that contained carbon nanotubes; these CNTs could not only penetrate into the hard coat of germinating tomato seeds but also exerted a growth enhancing effect. They envisaged that the enhanced growth was due to increased water uptake caused by penetration of CNT. This could be a boon for using CNT as a vehicle to deliver desired molecules into the seeds during germination that can protect them from the diseases. Since it is growth promoting, it will not have any toxic or inhibiting or adverse effect on the plant.

Use of nanoparticles has been considered an alternative and effective approach which is eco-friendly and cost effective for the control of pathogenic microbes (Ram Prasad *et al.*, 2014). These nanoparticles have a great potential in the management of plant diseases compared to synthetic fungicides. Zinc oxide (ZnO) and magnesium oxide (MgO) nanoparticles are effective antibacterial and antiodor agents (Shah and Towkeer, 2010). The increased ease in dispensability, optical transparency, and smoothness make ZnO and MgO nanostructures an attractive antibacterial ingredient in many products. Both have also been proposed as an antimicrobial preservative for wood or food products. Properly functionalized nanocapsules provide better penetration through the cuticle and allow slow and controlled release of active ingredients on reaching the target weed. The use of such nanobiopesticide is more acceptable since they are safe for plants and cause less environmental pollution in comparison to conventional chemical pesticides.

Nanosilver is the most studied and utilized nanoparticle for biosystem. It has long been known to have strong inhibitory and bactericidal effects as well as a broad spectrum of antimicrobial activities. Silver nanoparticles, which have high surface area and a high fraction of surface atoms, have a high antimicrobial effect compared to the bulk. Fungicidal properties of the nanosize silver colloidal solution are used as an agent for antifungal treatment of various plant pathogens; the most significant inhibition of plant pathogenic fungi was observed on potato dextrose agar (PDA) and 100 ppm of Ag NPs (Kim *et al.*, 2012).

Plant pathologists made a late start in harnessing the benefits of nanomaterials for the management of plant pathogens. There have been some exciting results obtained especially in plant disease management aspects concerning fungi, bacteria, and flowering plant parasite by developing nanoparticles of different metals, pesticides and growth promoters.

The fungicidal properties of the nanosize silver colloidal solution as an agent for antifungal treatment of various plant pathogens was studied (Sang Woo Kim *et al.*, 2012). WA-CV-WA13B, WA-AT-WB13R, and WA-PR-WB13R silver nanoparticles (Ag NPs) were used at concentrations of 10, 25, 50, and 100 ppm. Eighteen different plant pathogenic fungi were treated with these Ag NPs on potato dextrose agar (PDA), malt extract agar, and corn meal agar plates. Fungal inhibition was calculated to evaluate the antifungal efficacy of silver nanoparticles against pathogens. The results indicated that Ag NPs possess antifungal properties against these plant pathogens at various levels. Treatment with WA-CV-WB13R Ag NPs resulted in maximum inhibition of most fungi. Results also showed that the most significant inhibition of plant pathogenic fungi was observed on PDA and 100 ppm of Ag NPs.

Copper nanoparticles are capable of killing fungi that infect important agricultural crops. These nanoparticles could be used as a next-generation agricultural pesticide. Copper compounds have long been used in agricultural fungicides. However, fungi have grown resistant to existing pesticides. Furthermore, existing synthetic pesticides are harmful to humans. There is thus a need to develop new pesticides that can be used to realize sustainable agriculture without having harmful effects on humans. To produce a new eco-friendly pesticide, the researchers synthesized copper nanoparticles by reacting copper nitrate and isopropyl alcohol in the presence of a surfactant, cetyltrimethylammonium bromide (Prachi Kanhed *et al.*, 2014). They then tested the antifungal efficacy of the nanoparticles against four pathogenic fungi: *Phoma destructiva, Curvularia lunata, Alternaria alternata* and *Fusarium oxysporum*. They also compared the antifungal efficacy of the nanoparticles with that of a commercial antifungal agent, bavistin. The nanoparticles showed remarkable antifungal activity against the four pathogenic fungi. Two of the pathogenic fungi (*C. lunata* and *A. alternate*) exhibited resistance to bavistin but were sensitive to the nanoparticles. The nanoparticles showed maximum antifungal activity against *C. lunata* followed by *A. alternate* and minimum activity against *P. destructiva*. As copper is a micronutrient of crops, plants will take up any excess nanoparticles. In addition, excess nanoparticles can be detoxified naturally by the benign fungi that are present in the roots of plants. "The nanoparticles can be used as a novel antifungal agent in agriculture to control plant pathogenic fungi as well as a potent disinfectant in poultry and animal husbandry."

Principally, plants grown in soilless culture may be attacked by the same pests and diseases as cultivated traditionally in the soil. The most destructive phytopathogens are fungi, such as *Phythium, Phytophthora,* and *Fusarium,* followed by viruses, bacteria, and nematodes. Amooaghaie (2011) investigated the effect of carbon nanotube filters on disease management of soilless culture. Tomato seedlings were transplanted in plastic pots filled with a soilless media of vermiculite. The crop was irrigated and fertilized using a hydroponic nutrient solution. Carbon nanotube filters were used for nutrient solution disinfection. The results showed that carbon nanotube filtration significantly reduced pathogens on tomato plants. Fungal elimination (*Fusarium oxysporum* and

Pythium spp.) was usually successful at about 96 to 99.9% all over the cultural season. In tomato soilless culture, nanofiltration constituted a reliable method that allows control of the development of diseases caused by pathogenic fungi. The need for natural antimicrobials is paramount to avoid harmful synthetic chemicals.

A study by Ling Yien Ing *et al.* (2012) was aimed to determine the antifungal activity of natural compound chitosan and its nanoparticles forms against *Candida albicans, Fusarium solani,* and *Aspergillus niger.* Chitosan nanoparticles were prepared from low (LMW), high molecular weight (HMW) chitosan and its derivative, trimethyl chitosan (TMC). Particle size was increased when chitosan/TMC concentration was increased from 1 to 3 mg/mL. Their zeta potential ranged from +22 to +55 mV. Chitosan nanoparticles prepared from different concentrations of LMW and HMW were also found to serve a better inhibitory activity against *C. albicans* (MIC(LMW) = 0.25–0.86 mg/mL and MIC(HMW) = 0.6–1.0 mg/mL) and *F. solani* (MIC(LMW) = 0.86–1.2 mg/mL and MIC(HMW) = 0.5-1.2 mg/mL) compared to the solution form (MIC = 3 mg/mL for both MWs and species). This inhibitory effect was also influenced by particle size and zeta potential of chitosan nanoparticles. Besides, *Aspergillus niger* was found to be resistant to chitosan nanoparticles except for nanoparticles prepared from higher concentrations of HMW. Antifungal activity of nanoparticles prepared from TMC was negligible. The parent compound, therefore, could be formulated and applied as a natural antifungal agent into nanoparticles form to enhance its antifungal activity.

Metal oxide nanomaterials like ZnO and CuO have been used industrially for several purposes, including cosmetics, paints, plastics, and textiles. A common feature that these nanoparticles exhibit is their antimicrobial behavior against pathogenic bacteria. In this report, we demonstrate the antimicrobial activity of ZnO, CuO, and Fe_2O_3 nanoparticles against Gram-positive and Gram-negative bacteria (Ameer Azam *et al.*, 2012). Among the three metal oxide nanomaterials, ZnO showed greatest antimicrobial activity against both Gram-positive and Gram-negative bacteria used in this study. It was observed that ZnO nanoparticles have excellent bactericidal potential while Fe_2O_3 nanoparticles exhibited the least bactericidal activity. The order of antibacterial activity was demonstrated as ZnO > CuO > Fe_2O_3.

Antifungal activities of zinc oxide nanoparticles (ZnO NPs) and their mode of action against two postharvest pathogenic fungi (*Botrytis cinerea* and *Penicillium expansum*) were investigated (Lili He *et al.*, 2011). ZnO NPs with sizes of 70 ± 15 nm and concentrations of 0, 3, 6 and 12 m mol l^{-1} were used. Traditional microbiological plating, scanning electron microscopy (SEM), and Raman spectroscopy were used to study antifungal activities of ZnO NPs and to characterize the changes in morphology and cellular compositions of fungal hyphae treated with ZnO NPs. Results show that ZnO NPs at concentrations greater than 3 m mol l^{-1} can significantly inhibit the growth of *B. cinerea* and *P. expansum. P. expansum* was more sensitive to the treatment with ZnO NPs than *B. cinerea.* SEM images and Raman spectra indicate two different antifungal activities of ZnO NPs against *B. cinerea* and *P. expansum.* ZnO NPs inhibited the growth of *B. cinerea* by affecting cellular functions, which caused deformation in fungal hyphae. In comparison, ZnO NPs prevented the development of conidiophores and conidia of

P. expansum, which eventually led to the death of fungal hyphae. These results suggest that ZnO NPs could be used as an effective fungicide in agricultural and food safety applications.

Nanoparticles containing a fungicide, tebuconazole or chlorothalonil, were prepared as aqueous dispersions to treat sapwood of southern yellow pine (SYP) and birch using conventional pressure treatment methods (Liu *et al.*, 2003). The maximum amount of active ingredient (a.i.) delivered into wood was ca. 2 kg a.i./m^3 wood with tebuconazole-containing and 1.6 kg a.i/m^3 wood with chlorothalonil-containing nanoparticles. Nanoparticles were prepared using polyvinylpyridine (PVPy) and poly(vinylpyridine-co-styrene) with 10% styrene (PVPy-10%-Sty) and with 30% styrene (PVPy-30%-Sty) as matrices. Soil jar tests were performed where nanoparticle-treated SYP was exposed to *Gloeophyllum trabeum* for 16 weeks and nanoparticle-treated birch was exposed to *T. versicolor* for 12 weeks. Following exposure to *G. trabeum*, SYP lost ca. 5% of its mass when treated with tebuconazole-containing nanoparticles (PVPy, PVPy-10%-Sty, PVPy-30%-Sty) with a tebuconazole content of ca. 0.5 kg a.i./m^3 SYP. Similar results were obtained with chlorothalonil-treated SYP (0.5 kg a.i./m^3) with PVPy-10%-Sty and PVPy-30%-Sty. Birch exposed to *T. versicolor* lost ca. 5% mass when tebuconazole (in PVPy and PVPy-30%-Sty) was ca. 0.5 kg a.i./m^3, while similar results were found with chlorothalonil (PVPy, PVPy-10%-Sty, PVPy-30%-Sty) only at a biocide content of ca. 1.6 kg/m^3 wood.

9.17 FORMATION OF REACTIVE OXYGEN SPECIES (ROS)

Reactive oxygen species are likely natural by-products of the metabolism of respiring organisms. While small levels can be guarded by the antioxidant defenses of the cells such glutathione/glutathione disulfide (GSH/GSSG) ratio, excess ROS production may produce oxidative stress. The additional generation of free radicals can assault membrane lipids and lead to a breakdown of the membrane and mitochondrial function or cause DNA damage. Metals can act as catalysts and generate ROS in the presence of dissolved oxygen. In this context, silver nanoparticles may catalyze reactions with oxygen directing to excess free radical generation. Studies performed in eukaryotic cells suggest that silver nanoparticles inhibit the antioxidant defense by interacting directly with GSH, binding GSH reductase or other GSH maintenance enzymes (Carlson *et al.*, 2008). This could minimize the GSH/GSSG ratio and, subsequently, accumulate ROS in the cell. In bacterial cells, silver ions would likely induce the generation of ROS by abrogating the respiratory chain enzymes through direct interactions with thiol groups in these enzymes or the superoxide radical scavenging enzymes such as superoxide dismutases (Park *et al.*, 2009). Kim *et al.* determined the presence of free radicals from silver nanoparticles by means of spin resonance measurements. They observed that silver nanoparticles and silver nitrate toxicity was eliminated in the presence of an antioxidant, these results led them to confer that the antimicrobial mechanisms of silver nanoparticles against *S. aureus* and *E. coli* was related to the formation of free radicals from the surface of silver nanoparticles and subsequent free radical induced membrane damage.

9.18 DIRECT DAMAGE TO CELL MEMBRANES

Nanosilver interacts with the bacterial membrane and is capable of penetrating the cell. TEM studies show that silver nanoparticles adhere to and penetrate into *E. coli* cells and also are able to induce the development of small pits in the cell membrane. Silver nanoparticles have been observed within *E. coli* cells at sizes much smaller than the original particles; moreover, silver nanoparticles with oxidized surfaces induce the formation of huge holes in *E. coli* surfaces after the interaction, and large portion of the cellular content seemed to be 'eaten away.' The detailed mechanism by which silver nanoparticles interact with cytoplasmic membranes and are able to penetrate inside cells is not fully determined. One hypothesis explains that the interaction between nanoparticles and bacterial cells is due to electrostatic attraction between negatively charged cell membranes and positively charged nanoparticles. However, this mechanism does not likely explain the adhesion and uptake of negatively charged silver nanoparticles. It has been also proposed that the preferential sites of interaction between silver nanoparticles and membrane cells might be sulfur containing proteins, in a similar way as silver interacts with thiol groups of respiratory chain proteins and transport proteins, interfering with their proper function. Despite the mechanism of interaction involved, it is evident that silver nanoparticles attached to bacterial cell membranes increase permeability and disturb respiration. Proteomic data show the accumulation of envelope protein precursors in *E. coli* cells after exposure to silver nanoparticles. Energy from ATP and proton motive force is required to newly synthesize envelope proteins to be translocated to the membrane. Therefore cytoplasmic accumulation of protein precursors suggests dissipation of proton motive force and depletion of intracellular levels of ATP. One of research subject focused on the measurement of a biological binding signal between antigen and antibody using the triangular Ag-nanoparticles. For biological applications, quantum dots and nanoparticles are conjugated with biospecific molecules such as antibodies, DNA, or enzymes. Most of these applications are based on the specific optical properties of gold or silver (Sahayaraj and Rajesh, 2011).

NanoSilica-Silver composite Silicon (Si) is known to be absorbed into plants to increase disease resistance and stress resistance (Susanta Kumar Biswal *et al.*, 2012). Aqueous silicate solution, used to treat plants, is reported to exhibit excellent preventive effects on pathogenic microorganisms causing powdery mildew or downy mildew in plants. Moreover, it promotes the physiological activity and growth of plants and induces disease and stress resistance in plants. But, since silica has no direct disinfection effects on pathogenic microorganisms in plants, it does not exhibit any effect on established diseases. Further, the effects of silica significantly vary with the physiological environment and thus, they are not registered as an agricultural chemical. As mentioned above Silver is known as a powerful disinfecting agent. It kills unicellular microorganisms by inactivating enzymes having metabolic functions in the microorganisms by oligodynamic action and is known to exhibit superb inhibitory effects on algal growth also. Silver in an ionic state exhibits high antimicrobial activity. However, ionic silver is unstable due to its high reactivity and thus gets easily oxidized or reduced into a metal depending on the surrounding media and it does not continuously exert antimicrobial activity. Silver in the form of a metal or oxide is stable in the environment, but because of its low antimicrobial activity it is used in relatively increased amount, which is not very desirable.

A new composition of nanosized silica-silver for control of various plant diseases has been developed, which consisted of nanosilver combined with silica molecules and water soluble polymer, prepared by exposing a solution including silver salt, silicate, and water soluble polymer to radioactive rays. It showed antifungal activity and controlled powdery mildews of pumpkin at 0.3 ppm in both field and greenhouse tests. The pathogens disappeared from the infected leaves three days after spray and the plants remained healthy thereafter. Also studies on the 'effective concentration' of nanosized silica-silver on suppression of growth of many fungi resulted that *Pythium ultimum, Magnaporthe grisea, Colletotrichum gloeosporioides, Botrytis cinerea* and *Rhyzoctonia solani,* showed 100% growth inhibition at 10 ppm of the nanosized silica-silver, whereas, *Bacillus subtilis, Azotobacter chrococum, Rhizobium tropici, Pseudomonas syringae* and *Xanthomonas compestris* pv. *vesicatoria* showed 100% growth inhibition at 100 ppm (Susanta Kumar Biswal *et al.,* 2012). They have also reported chemical injuries caused by a higher concentration of nanosized silica-silver on cucumber and pansy plant when they were sprayed with a high concentration of 3200 ppm.

The extracellular synthesis of silver nanoparticles (Ag NPs) was studied using culture supernatant of an agriculturally important bacterium, *Serratia* sp. BHU-S4 and demonstrated its effective application for the management of spot blotch disease in wheat (Mishra *et al.,* 2014). The biosynthesis of Ag NPs by *Serratia* sp. BHU-S4 (denoted as bsAg NPs) was monitored by UV–visible spectrum that showed the surface plasmon resonance (SPR) peak at 410 nm, an important characteristic of Ag NPs. Furthermore, the structural, morphological, elemental, functional and thermal characterization of bsAg NPs was carried out using the X-ray diffraction (XRD), electron and atomic microscopy, energy dispersive X-ray (EDAX) spectrometer, FTIR spectroscopy and thermogravimetric analyzer (TGA), respectively. The bsAg NPs were spherical in shape with a size range of ~10 to 20 nm. The XRD and EDAX analysis confirmed successful biosynthesis and crystalline nature of Ag NPs. The bsAg NPs exhibited a strong antifungal activity against *Bipolaris sorokiniana,* the spot blotch pathogen of wheat. Interestingly, 2, 4 and 10 µg/mL concentrations of bsAg NPs accounted for complete inhibition of conidial germination, whereas, in the absence of bsAg NPs, conidial germination was 100%. A detached leaf bioassay revealed prominent conidial germination on wheat leaves infected with *B. sorokiniana* conidial suspension alone while the germination of conidia was totally inhibited when the leaves were treated with bsAg NPs. The results were further authenticated under greenhouse conditions, where application of bsAg NPs significantly reduced *B. sorokiniana* infection in wheat plants. Histochemical staining revealed a significant role of bsAg NPs treatment in inducing lignin deposition in vascular bundles. The findings represented the efficient application of bsAg NPs in plant disease management, indicating the exciting possibilities of nanofungicide employing agriculturally important bacteria.

The effect of silver nanoparticles was evaluated against pepper anthracnose disease (*Colleotrichum* spp.) under different culture conditions (Kabir Lamsal *et al.*, 2011). Silver nanoparticles (WA-PR-WB13R) were applied at various concentrations to determine antifungal activities *in vitro* and the field. The application of 100 ppm concentration of silver nanoparticles produced maximum inhibition of the growth of fungal hyphae as well as conidial germination in comparison to the control *in vitro*. In field trials, the inhibition of fungi was significantly high when silver nanoparticles were applied before disease outbreak on the plants. Scanning electron microscopy results indicated that the silver nanoparticles caused a detrimental effect on mycelial growth of *Colletotrichum* species.

Silver nanoparticles have been synthesized biologically using *Aspergillus foetidus* MTCC8876 and then characterized by various biophysical and biochemical techniques (Swarup Roy and Tapan Kumar, 2013). It was observed that this synthesized nanoparticle had very potential antifungal activity against some pathogenic fungi. Some synthetic and natural fungicides tebuconazole, trifloxystrobin and hexaconazole, as well as plant extract of ginger and teak, were used. The effect of silver nanoparticles in combination with systematic as well as a natural fungicide against some *Aspergillus* sp. and *Fusarium* sp. was observed. It was noticed that the diameter of the inhibitory zone of fungicide + silver nanoparticles was higher compared to only fungicide or only silver nanoparticles against pathogenic fungi. It was assumed that the silver nanoparticles are very useful antifungal substance in combination with systematic and natural fungicide to control the pathogenic fungi.

Silver ions and nanoparticles were evaluated to determine the antifungal action on *Bipolaris sorokiniana* and *Magnaporthe grisea*. The *in vitro* and *in vivo* evaluations of both silver ions and nanoparticles decrease disease development of phytopathogenic fungi. Min *et al.* (2009) evaluated the antifungal effects of silver nanoparticles, especially on sclerotia forming phytopathogenic fungi. The antifungal activity of silver nanoparticles assessed against filamentous ambrosia fungi in South Korea. The effect of silver nanoparticles on plant pathogenic spores of *Fusarium culmorum* was studied. The silver nanoparticles also found to exhibit antifungal activity against *Fusarium oxysporum*. Silver nanoparticles strikingly decreased the number of germinating fragments and sprout length, relative to the control. Zinc oxide nanoparticles inhibited the fungal growth of *Botrytis cinerea* by influencing cellular functions, which caused deformation in mycelial mats. In addition, zinc oxide NPs inhibited the growth of conidiophores and conidia of *Penicillium expansum*, which finally led to the death of fungal mats. The mycosynthesized silver nanoparticles may be nontoxic to people and animals, than synthetic fungicides. Moreover, in addition to the toxicity that nanoparticles may cause on algae, plants, and fungi, they may also have some positive effects (*Abd-Elsalam, 2012)*).

The antifungal activity of the silver nanoparticles was evaluated on *Colletotrichum gloeosporioides*, which is responsible for anthracnose in a wide range of fruits. Silver nanoparticles significantly reduced the mycelia growth of *Colletotrichum gloeosporioides* in a dose-dependent manner. Antifungal properties of silver nanoparticles, silver ions, acrylate paint and cotton fabric impregnated with silver nanoparticles were assessed against *A. niger, Aureobasidium pullulans,* and *Penicillium phoeniceum*. Bioassay of elemental and nanosulfur against *Aspergillus niger* showed that nanosulfur was more efficient than its elemental structure (Choudhury *et al.*, 2010).

The synthesis of metal nanoparticles using biological systems is an expanding research area due to the potential applications in nanomedicines. Nanoparticles synthesized by the chemical method are not eco-friendly. The biological synthesis of silver nanoparticles is a convenient and extracellular method, which is environmentally safe. In the study, the silver nanoparticles synthesized rapidly by using the stem barks of endemic medicinal plants, *Boswellia ovalifoliolata* and *Shorea tumbuggaia* (Savithramma *et al.*, 2011). After assessing the formation of silver nanoparticles with the help of UV-Visible spectroscopy and were characterized by using EDAX and SEM. Diversity was observed in size and shape of the silver nanoparticles synthesized in two plants. Phytosynthesized silver nanoparticles were tested for the antifungal activity. The test cultures of *Aspergillus, Fusarium, Curvularia,* and *Rhizopus* species were used. The fungal property of silver nanoparticles was analyzed by measuring the inhibitory zone. The silver nanoparticles were synthesized from bark extract of *Boswellia ovalifoliolata* and *Shorea tumbuggaia*. *Boswellia ovalifoliolata* showed moderately toxic to the *Aspergillus, Curvularia* and *Rhizopus* species and highly toxic to *Fusarium* species. Biologically synthesized silver nanoparticles of *Shorea tumbuggaia* bark extract were moderately toxic to *Aspergillus, Fusarium* and *Rhizopus* species and highly toxic to *Curvularia* species. The important outcome of the study was the development of value-added products from medicinal plants of India for biomedical and nanotechnology-based industries.

The synthesis of silver nanoparticles of varying sizes using an aqueous extract of 15 medicinal plants at room temperature was carried out by Prashanth *et al.* (2011). The synthesized nanoparticles were evaluated for their antimicrobial activity against human pathogens as well as fungal phytopathogens. Synthesized nanoparticles were characterized using a UV-visible spectrophotometer and two nanoparticles which showed highest inhibitory activity against microbes were also characterized with Scanning Electron Microscope (SEM) and X-ray diffractometer (XRD). The plant extract was very cost effective and eco-friendly and thus can be an economic and effective alternative for the large-scale synthesis of silver nanoparticles.

Different concentrations of nanosized silica-silver were evaluated for growth inhibition of phytopathogenic bacteria and fungi, and it was found that 100% growth inhibition of *Pseudomonas syringae* and *Xanthomonas compestris pv. vesicatoria* occurred at 100 ppm. *Magnaporthe grisea, Botrytis cinerea, Colletotrichum gloeosporioides, Pythium ultimum,* and *Rhizoctonia solani* showed 100% growth inhibition at 10 ppm of the nanosized silica-silver. Antimycotic activity of some nanoparticles of silver has also been reported on some fungi like wood rotting fungi, *Fusarium* species, and other phytopathogenic fungi. Nanodispersed formulations can be prepared in a simple, cost-effective manner, and are suited for developing new forms of fungicidal materials. There are some records of the nanosized or nanoformulation for agrochemicals development of existing pesticides, fungicides, plant, soil and seed treatments. Syngenta's Banner MAXX™ is a systemic fungicide offering broad-spectrum disease control in turf and ornamental plants. It is commercialized as a microemulsion concentrate formulation, providing excellent tank mix compatibility and stability. Resistance in plants would help in the management of agents mentioned above to overcome the problem of economic loss. Nanoparticles mediated plant transformation has the potential for genetic modification of plants for further improvement. In particular, application of nanoparticles technology in agriculture targets specific problems in plant protection, pathogen detection, plant–pathogen interactions and offers new methods of plant disease management (Mahendra *et al.*, 2012). For example, the introduction of resistance genes in plant cells using nanotechnological approaches may lead to the development of resistant varieties, which will minimize expenses on agrochemicals required for disease control. Presently, research is being carried out by using nanosensors to improve pathogen detection methods and plant disease forecasting in crop systems. Nanosensors can be linked to a GPS system for real-time monitoring of disease, and distributed throughout the field to monitor soil conditions and crop health. NanoPhytopathology can be applied as a tool to understand plant-pathogen interactions, which will provide new methods for crop protection.

Antimicrobial activity of silver was different depending on microbial species (Sahar M. Ouda, 2014). Silver nanoparticles can significantly delay mycelia growth in a dose-dependent manner *in vitro*. Silver nanoparticles may directly attach to and penetrate the cell membrane to kill spores, although penetration of silver nanoparticles into microbial cell membranes is not completely understood. The highest antifungal properties were observed in the case of treatment with 15 mg L^{-1} silver, silver/copper, and copper nanoparticles, respectively. Therefore, the microscopic observation revealed that the three types of nanoparticles clearly damaged the hyphae of *Alternaria alternata* and *Botrytis cinerea* compared with control. Microscopic observation revealed that silver nanoparticles caused detrimental effects not only on fungal hyphae but also on conidial germination. There were also, disrupting the structure of the cell membrane and inhibiting the normal budding process of *Candida albicans* due to the destruction of the membrane integrity. From this study, it was concluded that silver nanoparticles were more effective than copper/silver or copper nanoparticles for the control of both plant pathogens *Alternaria alternata* and *Botrytis cinerea.* Thus, silver nanoparticles prepared by a cost effective method has great promise as an antimicrobial agent and antifungal activity, and it has a great potential for use in controlling spore-producing fungal plant pathogens.

A silica-based nanoformulation and method was used to treat citrus canker, inhibit the growth of mold and mildew, and add nutrients to soil used for agricultural purposes (Swadeshmukul Santra, 2012). The nanotechnology-enabled copper-loaded, silica nanoformulation (CuSiNP/NG) design is a 'revolutionary reinvention' of Cu for safe application because it provides a formulation with maximum abundance of ionic Cu, provides sustained and optimal Cu ion release for long-term disease protection, better adherence to plant surfaces and structural surfaces due to gel-based nanostructure of CuSiNG, thus avoiding multiple spray applications and reducing the amount of Cu used in comparison to existing Cu compounds without compromising antibacterial activity. Thus, the silica-based nanoformulation releases copper in nontoxic quantities to the environment, and the silica matrix provides an environmentally safe host material with a flexible design that is optimized to provide specific antifungal and antibacterial remediation using infrequent applications.

The effect of silver nanoparticles (20–30 nm) against rice leaf blast fungus, *Magnaporthe grisea,* was evaluated under different cultivation conditions both *in vitro* and *in vivo.* (Elamawi *et al.,* 2013). Under lab conditions, the application of four concentrations of silver nanoparticles to the culture of *M. grisea* showed significant inhibition of both hyphal growth and number of colonies formed in a dose-dependent manner (Table 9.6). The number of spores/mL decreased within all treatment. Under greenhouse conditions, Silver nanoparticles were sprayed in concentrations 0, 25, 50, 100 and 200 ppm in rice seedling leaves at three times (three hours before inoculation, one and five days after artificial inoculation with spore suspension). Damaged Leaf Area Percentage (DLA %) indicated that the application of 100 ppm silver nanoparticles was highly efficient before and after inoculation (26.7, 15.3 and 20%), respectively compared to the untreated plants of 80%. The chemical fungicides isoprothiolane (Fuji one) and azoxystrobin (Amistar) at a concentration of 100 ppm each showed the lowest DLA (19.6% and 14.7%, respectively). Scanning electron microscopy results revealed that the silver nanoparticles caused a detrimental effect on mycelial growth.

Table 9.6 Effect of Silver Nanoparticles on Hyphal Growth and Colony Formation Compared to Fungicides

Concentration (ug ml^{-1})	*Growth diameter (cm)*	**No. of spores/mL*	*Number of formed colonies (cm^2)*
Control (water)	8.23	91.67	17.33
25	3.87	13.19	16.33
50	5.03	40.97	15.67
100	5.27	10.42	10.67
Fuji-one (100)	1.33	4.167	3.33
Amistar (100)	1.17	4.861	3.33
LSD 5%	0.473	25.7	3.984
F	**	**	**
*No. of spores × 104			

A study by Mishra *et al.* (2014) was focused on the extracellular synthesis of silver nanoparticles (Ag NPs) using culture supernatant of an agriculturally important bacterium, *Serratia* sp. BHU-S4 and demonstrates its effective application for the management of spot blotch disease in wheat. The biosynthesis of Ag NPs by *Serratia* sp. BHU-S4 (denoted as bsAg NPs) was monitored by UV–visible spectrum that showed the surface plasmon resonance (SPR) peak at 410 nm, an important characteristic of Ag NPs. The structural, morphological, elemental, functional and thermal characterization of bsAg NPs was carried out using the X-ray diffraction (XRD), electron and atomic microscopy, energy dispersive X-ray (EDAX) spectrometer, FTIR spectroscopy and thermogravimetric analyzer (TGA), respectively. The bsAg NPs were spherical in shape with a size range of ~10 to 20 nm. The XRD and EDAX analysis confirmed successful biosynthesis and crystalline nature of Ag NPs. The bsAg NPs exhibited a strong antifungal activity against *Bipolaris sorokiniana*, the spot blotch pathogen of wheat. Interestingly, 2, 4 and 10 μg/mL concentrations of bsAg NPs accounted for complete inhibition of conidial germination, whereas, in the absence of bsAg NPs, conidial germination was 100%. A detached leaf bioassay revealed prominent conidial germination on wheat leaves infected with *B. sorokiniana* conidial suspension alone while the germination of conidia was totally inhibited when the leaves were treated with bsAg NPs. The results were further authenticated under greenhouse conditions, where the application of bsAg NPs significantly reduced *B. sorokiniana* infection in wheat plants. Histochemical staining revealed a significant role of bsAg NPs treatment in inducing lignin deposition in vascular bundles.

The development of green processes for the synthesis of nanoparticles is evolving into an important branch of nanotechnology because biological methods are considered safe and ecologically sound for the nanomaterial fabrication as an alternative to conventional physical and chemical methods. The green synthesis techniques are generally synthetic routes that utilize relatively nontoxic chemicals to synthesize nanomaterials and include the use of nontoxic solvents such as water, biological extracts, biological systems and microwave assisted synthesis.

Silica is well known to enhance stress resistance to plants including plant diseases (Brecht *et al.*, 2003) through the promotion of plant physiological activity and growth (Kanto *et al.*, 2004) but it has no direct antimicrobial effect. On the other hand, silver is known to have excellent antimicrobial effect as discussed earlier. Thus, a new composition of nanosilica-silver was developed (Park *et al.*, 2006) to combat plant diseases. Nanosized silica-silver (Si-Ag) particles were produced and tested by Park *et al.* (2006) against a number of fungal and bacterial pathogens. *In vitro* test showed higher effectiveness of silica-silver nanoparticles towards fungi at the dose of 10 ppm causing 100% inhibition of vegetative growth. It was found that smaller size of silver nanoparticles was more effective against fungi. Most of the bacteria tested were inhibited completely with only 100 ppm of silica-silver nanoparticles. When nanosized silica-silver particles were applied in field condition to control powdery mildew diseases of cucurbits, 100% control was achieved after three weeks (Park *et al.*, 2006). These nanoparticles were found to be phytotoxic only at a very high dose of 3200 ppm when tested in cucumber and pansy plants. Nanosized silica silver inhibited the growth and development of both Gram-positive and Gram-negative bacteria.

Salome Egger *et al.* (2009) evaluated the properties of a novel silver-silica nanocomposite material (HeiQ AGS-20; HeiQ Materials, Bad Zurzach, Switzerland) used as an antimicrobial additive and compared its efficacy to the efficacies of the conventional silver additives silver nitrate ($AgNO_3$; 63.5% Ag) and silver zeolite (38% Ag bound to type A zeolite; Sigma-Aldrich, Buchs, Switzerland). The novel silver-silica nanocomposite material was produced using an industrial flame spray pyrolysis process. This process involved combustion of a flammable solvent containing homogeneously dissolved compounds as the source of components for the synthesis of the material. It was concluded that, upon contact with moisture, the pure silver particles act as a source that releases silver ions, which represent the active antimicrobial principle (Radeshkumar and Münstedt, 2006). Some key advantages of the novel nanocomposite are the dispersion of the discrete silver particles throughout the silica (which prevents agglomeration of the silver particles), the small diameter of the silver particles (which results in a large surface area and release of a large amount of Ag^+, which results in high antimicrobial efficiency), and the small size of the silver-silica composite (ca. 1 µm) (which allows the material to be uniformly dispersed and readily incorporated into a variety of substrates). A further advantage is that the immobilization of silver nanoparticles within the silica structure limits the potential for release and disposal of the nanoparticles themselves. This property may be highly desirable because of the possible abilities of nanoparticles to cross biological membranes and other barriers. The growth of *Candida albicans* was inhibited by 125 µg/mL silver nanocomposite, and the minimal fungicidal concentration was 2 mg/mL. Development of visible colonies of *A. niger* on agar plates was also completely inhibited by 2 mg silver nanocomposite per mL agar. In general, Gram-positive bacteria are more tolerant to silver than Gram-negative cells except for *Enterococcus faecalis*, for which the MICs and MBCs were similar to those for Gram-negative bacteria. It has previously been reported that Gram-positive bacteria are less susceptible to the antimicrobial activity of silver. It was speculated that this may be due to differences in the cell wall structure. The cell wall of Gram-positive bacteria contains multiple layers of peptidoglycan compared to the cell wall of Gram-negative bacteria. Peptidoglycan is a complex structure and often contains teichoic acids or lipoteichoic acids which have a strong negative charge, which may contribute to sequestration of free Ag^+ ions. Thus, Gram-positive bacteria may allow less Ag^+ to reach the cytoplasmic membrane than Gram-negative bacteria allow and may, therefore, be less susceptible.

Spherical silver nanoparticles (nanoAg) were synthesized, and their antifungal effects on fungal pathogens of the skin were investigated (Kim *et al.*, 2008). NanoAg showed potent activity against clinical isolates and ATCC strains of Trichophyton mentagrophytes and Candida species (IC80, 1-7 µg/mL). The activity of nanoAg was comparable to that of amphotericin B, but superior to that of fluconazole (amphotericin B IC80, 1–5 µg/mL; fluconazole IC80, 10–30 µg/mL). Additionally, we investigated their effects on the dimorphism of *Candida albicans*. The results showed nanoAg exerted activity on the mycelia. It was indicated that nanoAg may have considerable antifungal activity, deserving further investigation for clinical applications. Due to high thermal and chemical stability, high surface area and good biocompatibility, the interest on silica nanoparticles as a system to deliver drugs such as antibiotics is increasing. Colloidal silica is a very versatile material as it can be prepared in a wide variety of forms and sizes, its surface is easy to modify, and it can be obtained from relatively cheap precursors (Camporotondi *et al.*, 2013).

Several strategies have been developed for delivering NO to bacteria, including NO-releasing silica nanoparticles. Nitric oxide releasing silica nanoparticles have been used to successfully control the formation of biofilms by microorganisms such as *Pseudomonas aeruginosa* and *Escherichia coli*. Two different types of precursor were used for nanoparticle synthesis, N-Methylaminopropyltrimethoxysilane (MAP3) and

N-(6-inohexyl) aminopropyltrimethoxysilane (AHAP3). The authors found that the MAP3 silica nanoparticles had 1000-fold greater efficacy against *P. aeruginosa* biofilms than the AHAP3 nanoparticles. The greater amount of NO released by MAP3 nanoparticles, the more rapid delivery leading to a greater instantaneous concentration of NO in solution and the smaller size of MAP3 nanoparticles that allow them to penetrate biofilm matrix more effectively could explain these results.

Mesoporous silica nanoparticles are silica (SiO_2) nanoparticles with regularly arranged pores which increase the surface area of the nanoparticles. Targeted delivery of chemicals and DNA can be made by mesoporous silica nanoparticles (Wang *et al.*, 2002). It offers the possibility of genetic manipulation of plants, delivery of chemicals at the targeted site in the plant, improve the efficiency of used chemical and reduce the chemical residue problem to the minimum. Lin's research group has developed porous, silica nanoparticles systems that are spherical in shape and the particles have arrays of independent porous channels (Madhuri Sharon *et al.*, 2010). The channels form a honeycomb-like structure that can be filled with chemicals or molecules. These nanoparticles have a unique 'capping' strategy that seals the chemical inside. They have also demonstrated that the caps can be chemically activated to pop open and release the cargo into the cells where it is delivered. This unique feature provides total control for timing the delivery. Plant cells have a rigid cell wall. Hence, to penetrate it they had to modify the surface of the particle with a chemical coating. It has been successfully used to introduce DNA and chemicals into Arabidopsis, tobacco and corn plants. The other advantage is that with the mesoporous nanoparticles, one can deliver two biogenic species at the same time.

Nanocopper was reported to be highly effective in controlling bacterial diseases, viz., bacterial blight of rice (*Xanthomonas oryzae* pv. *oryzae*) and leaf spot of mung [(*X. campestris* pv. *phaseoli*) (Gogoi *et al.*, 2009)]. Ionic NanoCopper + Zinc is a unique mineral-based foliar spray that offers superior performance at low doses to control and prevent fungal and bacterial plant diseases in greenhouses, nurseries and fields. It utilizes the natural antimicrobial properties of mineral ions and presents minerals in their most biologically active form. Its liquid concentrate provides dust-free mixing and application apart from superior biological efficacy of pathogen control with low application rates. It is effective against a wide range of bacterial and fungal plant diseases and also controls nuisance algae growth in turfgrass. Above all, it is environmentally responsible, safe, and cost effective. Unlike copper sulfate, ionic nanocopper's copper does not sink to the bottom. It naturally seeks a balanced disbursement leading to even performance results. In contrast, copper sulfate is commonly plagued by 'hot spots' and inconsistent effectiveness. Ionic nanocopper is also longer lasting. It remains bioavailable through a wide range of conditions until assimilated by a target microorganism allowing for on-going water protection. And unlike copper sulfate and chlorine-based antimicrobial applications, it is not affected by sunlight and does not evaporate nor precipitate. It is also capable of remaining biologically active through the rigors of high pH, alkalinity, and organic content.

Bacterial spot caused by *Xanthomonas perforans* is a major disease of tomatoes, leading to a reduction in production by 10–50%. While copper (Cu)-based bactericides have been used for disease management, most of the *X. perforans* strains isolated from tomatoes in Florida and other locations worldwide are Cu-resistant. Ismail Oscoy *et al.* (2013) developed DNA-directed silver (Ag) nanoparticles (NPs) grown on graphene oxide (GO). These Ag@dsDNA@GO composites effectively decreased *X. perforans* cell viability in culture and on plants (Fig. 9.24). At the very low concentration of 16 ppm of Ag@dsDNA@GO, composites showed excellent antibacterial capability in culture with significant advantages in improved stability, enhanced antibacterial activity, and stronger adsorption properties. Application of Ag@dsDNA@GO at 100 ppm on tomato transplants in a greenhouse experiment significantly reduced the severity of bacterial spot disease compared to untreated plants, giving results similar to those of the current grower standard treatment, with no phytotoxicity.

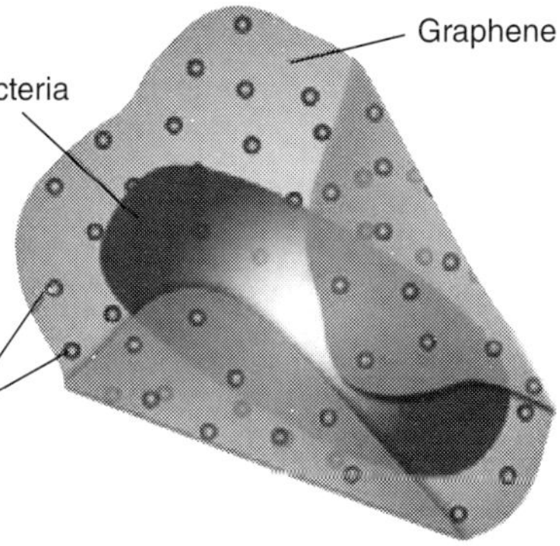

Fig. 9.24 DNA-directed silver nanoparticles (Ag NPs) grown on graphene oxide

The biological effects of TiO_2 sol derived from peroxotitanic acid solution (PAS TiO_2 sol) has been investigated on the control of plant diseases (caused by *Pseudoperonospora cubens, Pseudomonas syringae pv. lachrymans* and *Xanthomonas campestris* pv. *vesicatoria*. These diseases significantly reduce cucumber plant) and growth regulation were studied (Haixin Cui *et al.*, 2013). The PAS TiO_2 sol is a neutral, viscous aqueous colloid of TiO_2 containing 1.6% TiO_2 with a size of between 10–50 nm. Experiments were conducted under normal laboratory and greenhouse growing conditions. For the bactericidal effects of the PAS TiO_2 experiments, TiO_2 coated glass slides were inoculated with pathogenic bacteria in the laboratory. For the Greenhouse experiments on disease control and photosynthesis, all test plants (treatment block and control block) were inoculated with pathogenic bacteria. While plants in the treatment blocks were sprayed with PAS TiO_2, plants in the control blocks were sprayed with water. X-ray diffraction (XRD) results demonstrated that the crystal form of TiO_2 nanoparticles was anatase. Results from greenhouse experiments using cucumbers showed that PAS TiO_2 sol formed a well-spread, transparent, and sticky layer of TiO_2 photocatalyst on the plants' surface. The PAS TiO_2 photocatalyst layer prevented pathogenic infection and improved photosynthetic performance. These results suggested that PAS TiO_2 sol could be used as a potentially environmentally friendly plant germicide and growth regulator.

The antifungal activity of silver (Ag NPs), copper (CuNPs and silver/copper (Ag/CuNPs) nanoparticles were investigated against two plant pathogenic fungi *Alternaria alternata* and *Botrytis cinerea*., which have been responsible for the damage to a large number of plants (Sahar M. Ouda, 2014). Metals nanoparticles were applied at various concentrations to determine antifungal activities *in vitro*. The application of 15 mg L^{-1} concentration of silver nanoparticles produced maximum inhibition of the growth of fungal hyphae. The effectiveness of combining the silver and copper nanoparticles was also assessed. Microscopic observation revealed that nanoparticles caused damage effects on fungal hyphae and conidia. In addition to Ag-nanoparticles had a detrimental effect on sugar, protein, n-acetyl glucosamine and lipid of culture filtrate and cell wall components of both plant pathogens. The efficacy of nanocopper in the suppression of growth as well as in the water-soaked lesions induced by *Xanthomonas axonopodis* pv. *punicae*, (Xap) the causal agent of bacterial blight of pomegranate (*Punica granatum*), was studied (Kalyan K Mondal and Chander Mani, 2012). The nanocopper suppressed Xap growth at 0.2 ppm, i.e., >10,000 times lower than that usually recommended for Cu-oxychloride. Scanning electron microscopy revealed cell wall degradation in nanocopper-treated bacterial cells that failed to colonize plant tissues as well as to produce the characteristic intense water soaking. This study suggested that nanocopper could be employed as an environmentally friendly strategy for the management of pomegranate bacterial blight.

The release of heavy metal-containing nanoparticles (NP) into the environment may be harmful to the efficacy of beneficial microbes that function in element cycling, pollutant degradation, and plant growth. Nanoparticles of Ag, CuO and ZnO, are of interest as antimicrobials against pathogenic bacteria. The antimicrobial activity against the beneficial soil microbe, *Pseudomonas putida* KT2440 has been demonstrated (Gajjar *et al.*, 2009). Toxicity was detected in a KT2440 construct possessing a plasmid bearing the luxAB reporter genes. 'As manufactured' preparations of nanoAg, -CuO and -ZnO caused rapid dose-dependent loss of light output in the biosensor. Cell death accompanied loss in Lux activity with treatments by nanoAg and -CuO, but with -ZnO the treatments were bacteriostatic rather than bactericidal. Bulk equivalents of these products showed no inhibitory activity, indicating that particle size was determinant in activity. Flow Field-Flow Fractionation (FlFFF) of an aqueous suspension of the nanoCuO and ZnO revealed a small proportion of 5 nm NP and aggregated particulates with sizes ranging between 70 nm and 300 nm; the majority portion of the material was aggregated into particles larger than 300 nm in size. Thus, within the commercial preparation there may be microbially active and inactive forms. The 'as-made' NP of Ag, CuO and ZnO had toxic effects on a beneficial soil microbe, leading to bactericidal or bacteriostatic effects depending on the NP employed. The lack of toxicity from bulk materials suggested that aggregation of the NP into larger particles, possibly by factors present in the environment might reduce their nontarget antimicrobial activity.

Growing resistance of microorganisms to potent antibiotics has renewed a great interest towards investigating bactericidal properties of nanoparticles and their nanocomposites as an alternative. A study was carried out to investigate and compare the antibacterial properties of copper (Cu) and zerovalent iron (Fe^{O}) nanoparticles (Selvarani and Prema, 2013). Nanoparticles were synthesized by reducing the aqueous solution of the respective salt solution with sodium borohydride. The synthesized particles were further characterized by X-Ray Diffractogram and Scanning Electron Microscopy techniques to analyze size and morphology respectively. Antibacterial efficacy of metal nanoparticles was evaluated by agar well diffusion method. The average size of the particles was found to be 17.25 nm (Cu) and 44.87 nm. The energy-dispersive spectrum of the nanoparticles dispersion confirmed the presence of elemental metals. The mechanism of microorganism inactivation is considered as species-dependent. *Pseudomonas aeruginosa* exhibited highest antibacterial sensitivity (26.00 ± 0.41 mm) to copper nanoparticles whereas *Bacillus cereus* offered a maximum zone of inhibition (23.33 ± 0.89 mm) to Fe^O nanoparticles. These findings signified that the Cu and Fe^O nanoparticles potentiate the antibacterial action of both Gram positive and Gram negative bacteria.

Prolonged use of Cu biocides would lead to undesirable Cu levels in the agricultural soil. In the absence of a suitable alternative, prudent use of Cu biocides is required. (Mikaeel Young and Swadeshmukul Santra, 2011) reported for the first time a composite material where sol-gel silica host matrix loaded with mixed-valence Cu as an alternative to conventional biocides. In this composite material, Cu was present in different oxidation states. The hydrophilic silica matrix served as a water-dispersible delivery vehicle for antimicrobial Cu. It was hypothesized that mixed-valence Cu system, specifically enriched with Cu (O) and Cu(I) exhibited enhanced antimicrobial efficacy over traditional Cu (II) compounds. Materials were characterized by High-Resolution Transmission Electron Microscopy and X-ray Photoelectron Spectroscopy for the determination of particle size, morphology, crystallinity and Cu oxidation states. Antimicrobial studies against *Xanthomonas alfalfae* and *Escherichia coli* (Minimum Inhibitory Concentration) showed improved efficacy in MV-CuSiNG treatment compared to CuSiNG and other controls. Phytotoxicity studies performed (in *Vinca* sp. and *Hamlin orange)* under greenhouse conditions showed that the newly prepared nanocomposite was safe to plants, demonstrating the potential usefulness of the material in agricultural biocides.

9.19 NANOALUMINOSILICATE

Leading chemical companies are now formulating efficient pesticides at the nanoscale. One of such effort is the use of aluminosilicate nanotubes with active ingredients. The advantage is that aluminosilicate nanotubes sprayed on plant surfaces are easily picked up in insect hairs. Insects actively groom and consume pesticide-filled nanotubes. They are biologically more active and relatively more environmentally-safe pesticides (Madhuri Sharon *et al.*, 2010).

9.19.1 Imogolite, Aluminosilicate Nanotube

Structure of Imogolite was discovered as early as in 1962, and detail investigation using electron diffraction analysis by Cradwick *et al.* in 1972 confirmed its composition [$(A_{l2}O_3 \cdot SiO_2 \cdot 2H_2O)$ (Wei Ma *et al.*, 2012) (Fig. 9.25)]. The schematic representation of imogolite is shown in Fig. 9.26. The gas adsorption data of N_2, CO_2, and CH_4 concluded that imogolite possesses an inner pore diameter of 1 nm. The wall structure of imogolite comprises a layer of gibbsite on the outer wall and a layer of silicate on the inner wall.

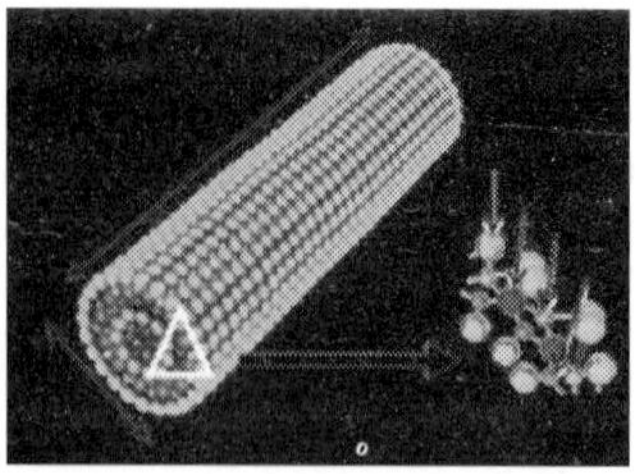

Fig. 9.25 Schematic illustration of imogolite nanotube structure

The formation mechanism of imogolite is not well understood, but an early study suggested that the evolution of the tubular morphology is started by the binding of isolated silicate groups to the gibbsite sheet, in which the tetravalent Si atoms pull the oxygen atoms of the gibbsite sheet into the curvature cylinder. Attempts to tune the imogolite dimensions appear to be futile, as the tubular structure does not change significantly throughout the synthesis process and the formation of nanotubes occurs at an early stage. Individually dispersing imogolite nanotubes on a transmission electron microscopy (TEM) carbon grid, by means of the droplet evaporation technique, has made a semiquantitative analysis of imogolite dimensions possible. Semiquantitative analysis on the TEM images supports the kinetic growth mechanism, in which protoimogolites and short imogolites are observed in the initial stage of synthesis and the average length of the nanotubes increases rapidly with reaction time (Wei Ma *et al.*, 2012).

9.20 CALCIUM CARBONATE

Improvements in nanotechnology in the past decade have created various opportunities for the evaluation of biologic effects such as antibacterial effects of nanoparticles. Ataee *et al.* (2011) evaluated the antibacterial activity of calcium carbonate nanoparticles on two different bacteria including *Agrobacterium tumefaciens* and *Staphylococcus aureus*. The antibacterial effect of calcium carbonate nanoparticles against mentioned bacteria was evaluated by dilution in agar containing medium and dilution in Broth medium. Each of prepared Broth media (10 mL) was inoculated with 1 mL of bacterial suspension (106 CFU/mL) and incubated. A sampling of culture media was performed at specific intervals and diluted as 10-1 to 10-6. Then 100 μl of each sample was transferred to agar plates and was spread carefully and then incubated. Grown colonies were counted, and MIC and MBC were determined. Calcium carbonate nanoparticles showed very good antibacterial effect and after 16 hours the bacteria were totally diminished. The lowest and highest MIC concentration of these nanoparticles in the solid medium was 31.2 and 125 μg/mL respectively. The MIC of calcium carbonate nanoparticles in Broth medium was two times more than the MIC concentration in the solid medium while different concentrations of ordinary calcium carbonate not only revealed antibacterial effects but also supported the bacterial growth.

9.21 NANOIRON

Movement and behavior of nanoparticles and their curative effect are being studied more extensively involving humans. A similar study to deliver the nanoparticles to the targeted site of a diseased plant has been done by Corredor *et al.* (2009). They applied iron nanoparticles coated with carbon to pumpkin plants for treating specific plant part that is infected. Antimycotic activity of nanoparticles of magnesium, iron, and zinc, were investigated under *in vitro* conditions and was observed that all the nanoparticles at different concentrations brought about significant inhibition in the germination of spores of *Penicillium notatum, Aspergillus niger,* and *Nigrospora oryzae* (Wani *et al.*, 2012). However, the highest inhibition of the germination of all the test fungi was observed at higher concentrations followed by lower concentrations of nanoparticles. The nanoMgO at highest concentration was found most effective in reducing the spore germination followed by nanoFeO and nanoZnO at the same concentration respectively.

9.22 ZINC OXIDE

The antibacterial potentiality of zinc oxide (ZnO) nanoparticles (NPs), compared with conventional ZnO powder, against nine bacterial strains, mostly foodborne including pathogens, was evaluated using qualitative and quantitative assays. ZnO NP was more efficient as an antibacterial agent than powder. Gram-positive bacteria were generally more sensitive to ZnO than Gram-negatives. The exposure of *Salmonella typhimurium* and *Staphylococcus aureus* to their relevant minimal inhibitory concentrations from ZnO NP reduced the cell number to zero within 8 and 4 h, respectively. Scanning electron micrographs of the treated bacteria with NPs exhibited that the disruptive effect of ZnO on *S. aureus* was vigorous as all treated cells were completely exploded or lysed after only 4 h from exposure. Promising results of ZnO NP antibacterial activity suggest its usage in food systems as a preservative agent after further required investigations and risk assessments (Ahamed *et al.*, 2011).

The release of heavy metal-containing nanoparticles (NP) into the environment may be harmful to the efficacy of beneficial microbes that function in element cycling, pollutant degradation, and plant growth. Nanoparticles of Ag, CuO and ZnO, are of interest as antimicrobials against pathogenic bacteria. The antimicrobial activity against the beneficial soil microbe, *Pseudomonas putida* KT2440 was demonstrated (Priyanka Gajjar *et al.*, 2009). Toxicity was detected in a KT2440 construct possessing a plasmid bearing the *luxAB* reporter genes. 'As manufactured' preparations of nanoAg, -CuO and -ZnO caused rapid dose-dependent loss of light output in the biosensor. Cell death accompanied loss in Lux activity with treatments by nanoAg and -CuO, but with -ZnO the treatments were bacteriostatic rather than bactericidal. Bulk equivalents of these products showed no inhibitory activity, indicating that particle size was determinant in activity. Flow Field-Flow Fractionation (FlFFF) of an aqueous suspension of the nanoCuO and ZnO revealed a small proportion of 5 nm NP and aggregated particulates with sizes ranging between 70 nm and 300 nm; the majority portion of the material was aggregated into particles larger than 300 nm in size. Thus within the commercial preparation there may be microbially active and inactive forms. The 'as-made' NP of Ag, CuO and ZnO had toxic effects on a beneficial soil microbe, leading to bactericidal or bacteriostatic effects depending on the NP employed. The lack of toxicity of bulk materials suggested that aggregation of the NP into larger particles, possibly by factors present in the environment might reduce their nontarget antimicrobial activity. Carbon nanotubes have shown growth enhancing the effect on tomato when grown in soil containing carbon nanotubes (Khodakovsky *et al.*, 2000). It is believed that carbon nanotubes entered the germinating tomato seeds thus facilitating water uptake and plant growth.

Silicon (Si) is known to be absorbed into plants to increase disease resistance and stress resistance (Madhuri Sharon *et al.*, 2010). Aqueous silicate solution, used to treat plants, is reported to exhibit excellent preventive effects on pathogenic microorganisms causing powdery mildew or downy mildew in plants. Moreover, it promotes the physiological activity and growth of plants and induces disease and stress resistance in plants. But, since silica has no direct disinfection effects on pathogenic microorganisms in plants, it does not exhibit any effect on established diseases. Further the effects of silica significantly vary with the physiological environment and thus, they are not registered as an agricultural chemical. As mentioned above silver is known as a powerful disinfecting agent. It kills unicellular microorganisms by inactivating enzymes having metabolic functions in the microorganisms by oligodynamic action and is known to exhibit superb inhibitory effects on algal growth also.

Silver in an ionic state exhibits high antimicrobial activity. However, ionic silver is unstable due to its high reactivity and thus gets easily oxidized or reduced into a metal depending on the surrounding media and it does not continuously exert antimicrobial activity. Silver in the form of a metal or oxide is stable in the environment, but because of its low antimicrobial activity it is used in relatively increased amount, which is not very desirable. A new composition of nanosized silica-silver for control of various plant diseases has been developed by Park *et al.* (2006), which consisted of nanosilver combined with silica molecules and water soluble polymer, prepared by exposing a solution including silver salt, silicate, and water soluble polymer to radioactive rays. It showed antifungal activity and controlled powdery mildews of pumpkin at 0.3 ppm in both field and greenhouse tests. The pathogens disappeared from the infected leaves three days after the spray, and the plants remained healthy thereafter. Park *et al.* (2006) also studied the 'effective concentration' of nanosized silica-silver on the suppression of growth of many fungi; and found that *Pythium ultimum, Magnaporthe grisea, Colletotrichum gloeosporioides, Botrytis cinere* and *Rhyzoctonia solani* showed 100% growth inhibition at 10 ppm of the nanosized silica-silver. Whereas, *Bacillus subtilis, Azotobacter chrococuum, Rhizobium tropici, Pseudomonas syringae* and *Xanthomonas compestris* pv. *vesicatoria* showed 100% growth inhibition at 100 ppm. They have also reported chemical injuries caused by a higher concentration of nanosized silica-silver on cucumber and pansy plant when they were sprayed with a high concentration of 3200 ppm.

9.23 SULFUR NANOFUNGICIDES

Sulfur can be considered a high-efficiency pesticide that used in agriculture where it has good effect against a wide range of powdery mildew diseases as well as black spot (Mohammed Suleiman *et al.*, 2013). Sulfur nanoparticles can be prepared by different successful methods such as acid hydrolysis of sodium thiosulphate, oxidation H_2S gas using Fe-chelate, ultrasonic treatment of sulfur-cystine solution, aqueous surfactant solutions, microemulsion technique and electrochemical method (Fig. 9.26). Microemulsion method is one of the most important methods due to the easy control of the particle size using this method. Different shapes and size of Sulfur NPs were obtained depending on the used method and parameters. The sulfur NPs shapes were found to be spherical nanoparticles, monoclinic, orthorhombic, rhombic and nanowires. Different shapes and size of sulfur nanoparticles were characterized by using the transmission electron microscopy (TEM), scanning electron microscope (SEM) and X-ray diffraction (XRD).

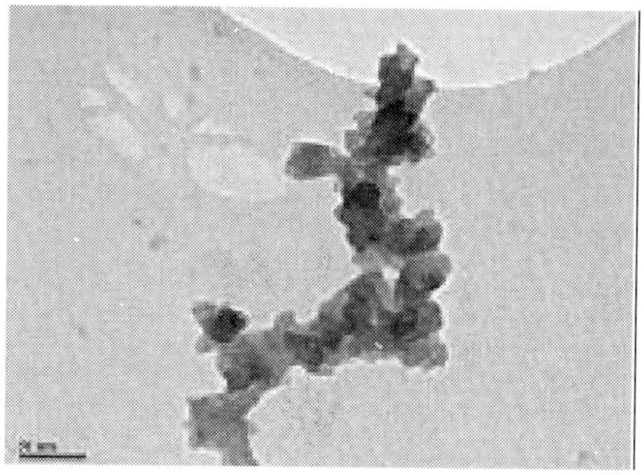

Fig. 9.26 TEM image of sulfur nanoparticles

An electrochemical method is presented for the preparation of sulfur nanoparticles (S-NPs) from thiosulfate ion. The particle size of the S-NPs can be adjusted between 35 and 65 nm by varying parameters such as the initial concentration of thiosulfate. As Orthorhombic (spherical ~10 nm) and monoclinic (cylindrical ~50 nm) (SNPs) were synthesized and examined for their effects on the total lipid content and desaturase enzymes of *Aspergillus niger* (Choudhury *et al.*, 2012).

Despite the antifungal efficacy of sulfur, its use is now restricted to farmers and agrochemical industries mainly for two reasons, firstly sulfur is required in bulk quantities for application in agricultural fields and secondly it is also likely to induce resistance in the target species (Samrat Roy Choudhury *et al.*, 2010). Sulfur nanoparticles (NPs) were synthesized bottom-up via a liquid synthesis method with average particle size in the range of 50–80 nm, and the shapes of the NPs were spherical. It is desirable to considerably reduce the amount so that the cost is cut down and at the same time, the build up of resistance in target pathogens is minimized. Intelligent bottom up nanoscience application is the most appropriate tool to resolve the problems above. Nanosulfur proved to be efficacious as a fungi-toxic agent over the commercially available elemental sulfur. The synthesized nanoformulation retarded the growth of *Aspergillus niger* but also inhibited sporulation. In comparison, elemental sulfur retarded fungal growth in a dose-dependent manner but failed to inhibit spore formation.

Investigations of antifungal activity of micron and nanoscale forms of sulfur on two types of pathogenic Schutte fungi were studied (Massalimov *et al.*, 2013), viz., as micronized sulfur used crushed in a roller mill and as nanosized sulfur precipitated from a solution of sodium polysulfide. The size and shape of the particles of sulfur were characterized using a laser analyzer and probe microscope, the structure by X-ray diffractometer. Antifungal effects of sulfur particles in Sabouraud medium and field experience were studied. It was found that in all cases of laboratory and field experience, the antifungal activity of nanosized sulfur with an average particle size of 25 nm, was 5–10 times higher than the sulfur micron with an average particle size of 8 microns. The results could be used to create more effective than conventional sulfur plant protection products.

Application of fungicides interferes with plant-mycorrhiza symbiosis and inhibits root system development, absorption, and growth. An experiment was conducted, to evaluate the effect of different fungicides on mycorrhizal symbiosis and quality of sunflower (Maryam Abbasian *et al.*, 2012). The study was conducted in factorial in the form of a randomized complete block design with four replications and two factors. The first factor was fungicide (without, F0; chemical, F1; nano, F2; biologic, F3) and the second factor was mycorrhiza species (without, M0; *Glomus mosseae*, M1; *G. etunicatum*, M2; *G. intraradices*, M3). Results indicated that fungicide had only a significant effect on grain protein content, and mycorrhiza had a significant effect on grain oil content and root colonization. F2M1 was the best treatment with the highest effect which increased grain yield, protein yield and oil yield by 60.77, 60.93 and 69.90%, respectively, compared with the control (F0M0).

New nanosulfur synthesized at Indian Agricultural Research Institute, New Delhi, India and three other commercial products, namely, commercial sulfur (Merck), commercial nanosulfur (MK Impex, Canada) and sulfur 80 WP (Corel Insecticide) were evaluated *in vitro* for fungicidal efficacy at 1000 ppm against Erysiphe cichoracearum of okra (Robin Gogoi *et al.*, 2013). All the sulfur fungicides significantly reduced the germination of conidia of *E. cichoracearum* as compared to control. Least conidial germination was recorded in IARI nanosulfur (4.56%) followed by Canadian nanosulfur (14.17%), Merck sulfur (15.53%), sulfur 80 WP (15.97%) and control (23.09%). Nongerminated conidia count was also high in case of IARI nanosulfur followed by Canadian nanosulfur, Merck sulfur, and sulfur 80 WP. Apart from inhibition of conidial germination, cleistothecial appendages were also disrupted in contact with nanosulfur and the cleistothecia became sterile. The study proved that IARI nanosulfur was more effective than the commercial formulations and could be applied in lower amount for controlling powdery mildew disease for its better efficacy.

Applications of nanoparticle-based fungicides have been increasing gradually in recent years because of many advantages over conventional chemical fungicides. A study by (Jagajjanani Rao and Santanu Paria, 2013) presented the fungicidal efficiency of sulfur nanoparticles (SNPs) against two phytopathogens, *Fusarium solani* (isolated from an infected tomato leaf, responsible for early blight and *Fusarium* wilt diseases) and *Venturia inaequalis* (responsible for the apple scab disease). Different parameters, such as particle size, particle concentration, surfactant media, etc., which affect the efficiency of the fungicidal effect were also studied. It has been found that small sized particles (~35 nm) were very effective in preventing the fungal growth. The microscopic study confirmed that the fungicidal effect was mainly because of the deposition of particles on the cell wall and subsequent damage to the cell wall. A plausible imbalance in the cell wall structure was also supported by a Biuret assay test. SNP-based fungicide could be useful for the protection of important crops such as tomato, potato, apple, grape, etc., from different diseases, mainly for 'organic' farming.

9.24 PESTICIDE NANOFORMULATIONS

Conventional methods to control the pathogens and pests have affected both the environment and economy of farmers as 90% of the applied pesticides are lost to the air during application and as runoff, affecting both the environment and application costs to the farmer. Additionally, indiscriminate usage of pesticide increases pathogen and pest resistance, reduces soil biodiversity, diminishes nitrogen fixation; contributes to bioaccumulation of pesticides, pollinator decline and destroys habitat for birds (Susanta Banik and Pratibha Sharma, 2011). Manufacturers are developing nanoformulations of existing fungicidal compounds by reducing the size of active ingredients to nanoscale and also by nanoencapsulating them. Syngenta has developed fungicide formulation containing nanoparticles, for example, Banner MAXX Fungicide (active ingredient propiconazole), Apron MAXX (active ingredient fludioxonil) RFC for seed treatments.

Similarly, cyclopropyl derivative of cyclohexenone (Primo MAXX) has been developed as plant growth regulator but it helps the plant in withstanding abiotic as well as biotic stresses including plant pathogens (Gogoi *et al.*, 2009). 'Nano5' is a marketed product and is projected as natural mucilage organic solution to control several plant pathogens and pests besides improving crop yield. A few examples of plant diseases against which 'Nano5' was found effective (at a dilution of 1:500). A product of nanotechnology research in agriculture with the name of 'NanoGro' has been launched (Agro Nanotechnology Corp., Florida, http://www.agronano.com). Plants treated with 'NanoGro' show an average yield increase of 20% with a maximum of 50% in case of grain yield of sunflower; increase in protein and sugar content by about 10% and plants can fight various diseases. The product is certified to be an organic one and harmless to plants and soil. 'NanoGreen' a product prepared by mixing several biobased chemicals was reported to eliminate blast disease (*Magnaporthe grisea*) from infected rice plant. The test was conducted at the University of Georgia, and the product was found to outperform any other pesticide or fungicides currently in use in agriculture (Gogoi *et al.*, 2009).

The use of Ag nanoparticles as an alternative to pesticides for the control of *Sclerotium* forming phytopathogenic fungi was investigated. Exposure of fungal hyphae to Ag NPs caused severe damage by the separation of layers of hyphal wall and collapse of hyphae. Detrimental effects of Ag NPs on unidentified fungal species of the genus *Raffaelea* causing mortality of oak trees was also investigated and studies showed harmful effects of Ag NPs on conidial germination (Nair *et al.*, 2010). Bacteria, viruses, and fungi can function as biological control agents against insect pests. Bacterial and viral formulations are susceptible to desiccation, heat, and UV inactivation and need to be ingested by an insect for action. Fungal biocontrol agents or mycopesticides are promising as they act by contact and do not need ingestion can be easily mass produced, and are relatively specific. The use of nanoformulations may offer new ways to enhance the stability of these biological agents (Shweta Agrawal and Pragya Rathore, 2014).

Manufacturers are developing nanoformulations of existing fungicidal compounds by reducing the size of active ingredients to nanoscale and also by nanoencapsulating them. Syngenta has developed fungicide formulation containing nanoparticles, for example, Banner MAXX Fungicide (active ingredient propiconazole), Apron MAXX (active ingredient fludioxonil) RFC for seed treatments. Similarly, cyclopropyl derivative of cyclohexenone (Primo MAXX) has been developed as plant growth regulator but it helps the plant in withstanding abiotic as well as biotic stresses including plant pathogens (Gogoi *et al.*, 2009). 'Nano5' is a marketed product and is projected as natural mucilage organic solution to control several plant pathogens and pests besides improving crop yield.

The nanotechnology, through encapsulation of active ingredients, has showed an important way to avoid problems with quick degradation of the pesticide molecules. Thus, neem (*Azadirachta indica*) oil nanoformulations containing β-ciclodextrin and poli-ε-caprolactone (PCL) were tested as to their control efficiency against eggs and nymphs of *Bemisia tabaci* (Genn.) biotype B reared in soybean. The Lethal Concentration (LC50) was estimated using commercial neem oil (Organic Neem®) on first-instar nymphs to establish the adequate volume of the nanoformulations per treatment. After that, they were sprayed on eggs and first-instar nymphs in the laboratory and greenhouse and on third-instar nymphs in the greenhouse (Sheila Salles Carvalho1 *et al.*, 2011). The commercial neem oil and distilled water were used as controls. Egg viability was not affected by any treatment. Among six nanoformulations, only one was efficient against the first-instar nymphs in laboratory conditions. However, its effective period was not increased as expected. In the greenhouse, first-instar nymphs were more affected by two nanoformulations which were significantly different from the commercial neem oil, the most effective one. No mortality differences among the formulations in the third-instar test were observed. The nanoformulations were less efficient to control the *B. tabaci* biotype B nymphs than the commercial neem oil.

Currently, the most common pesticide formulations for poorly water-soluble a.i. are emulsifiable concentrates (ECs) and oil-in-water (O/W) emulsions. ECs still represented about 28% of the total number of formulations listed in the Pesticide Manual in 2007. These concentrates consist of an a.i. dissolved in an organic solvent (originally xylene, but safer alternatives are now used) and a blend of surfactant emulsifiers to ensure spontaneous emulsification into the water in the spray tank. The main disadvantages of ECs relate to the relatively poor stability after dilution (droplets of about 10 μm) and the use of organic solvents, leading to increases in the cost and flammability, as well as in dermal toxicity for the handlers. O/W emulsions have been proposed as an alternative to ECs. O/W emulsions generally consist of a mixture of a nonionic surfactant, block polymers, and polymeric surfactants. The drawback to O/W emulsion is that emulsification requires a high energy input, which is provided by high shear mixers (typically producing droplets of 2 μm diameter) or high-pressure valve homogenizers (droplets down to 500 nm; Knowles, 2005). Most nanoformulations aiming to increase the solubility of a.i. derive directly from the two previously mentioned approaches.

9.24.1 Microemulsions

Microemulsions are thermodynamically stable water-based formulations consisting of dissolved a.i. in oil, surfactant solubilizers (blend) along with a cosurfactant (often medium chain aliphatic alcohol), and water. Once formulation design is established, microemulsions form spontaneously upon addition of water and gentle stirring (unlike classical emulsions where preparation requires large energy input. The particle size in microemulsions may be about 250 times smaller than typical pesticide particles and several reports have suggested diameters of less than 100 nm. Measuring the size distribution of the micelles in microemulsions is difficult and information on how they may evolve with time is relatively scarce. Scattering techniques (light, neutron, and X-ray) and pulsed field gradient nuclear magnetic resonance can be used to determine the microstructure of microemulsions (Kah *et al.*, 2013). However, these techniques rely on the concentration of disperse phases to be low enough to avoid particle-particle interactions. This becomes a serious limitation for microemulsions, which can undergo changes upon dilution, due to cosurfactants located between the oil and water phases. Droplet sizes reported in the literature range from 6 nm in a microemulsion of novaluron (determined by small angle X-ray scattering and confirmed by derivation from diffusion coefficient measured by self-diffusion NMR), up to 50 nm for a microemulsion of Chlorpyrifos. Products by Syngenta also marketed as microemulsion concentrate formulations include Banner MAXX (systemic fungicide for broad-spectrum disease control in turf and ornamentals), Subdue MAXX (systemic fungicide), and ApronMAXX (disease protection for soybean. The fact that the previously mentioned products are marketed as microemulsion concentrates has been strongly criticized in some reports, pointing out that using the prefix *micro* when talking about a formulation containing entities in the nm size range is misleading. Disadvantages of microemulsions include a low a.i. content (<30%), a high concentration of surfactants (usually in the region of 20%) and the limited number of suitable surfactant systems. Finally, provided an enhanced uptake is confirmed microemulsion formulations may also present phytotoxicity and handler toxicity issues.

9.24.2 Nanoemulsions

Nanoemulsions, also referred to as miniemulsions, ultrafine emulsions, and submicron emulsions, are emulsions with a droplet size that can overlap with those of microemulsions (Kah *et al.*, 2013). Although there still seems to be some disagreement between authors regarding the appropriate terminology, the main difference between microemulsions and nanoemulsions is generally based on equilibrium status. While microemulsions are thermodynamically stable; nanoemulsions have the tendency to separate into the constituent phases (mainly by Oswald ripening). Nanoemulsions may nevertheless possess a relatively high kinetic (meta) stability and are often said to be metastable nanoemulsions contain a lower concentration of surfactants than microemulsions (typically 5–10% of surfactant, as opposed to about 20% in microemulsion), and many preparation methods include a step that consists of diluting a microemulsion. The range of droplet sizes typically quoted is 20–200 nm.

9.24.3 Nanodispersion

Dispersion of nanocrystals (crystalline or amorphous particles consisting of 100% a.i.) in liquid media leads to the formation of nanodispersions having similar properties to solutions. They are also called nanosuspensions (Müller and Junghanns, 2006). The approach aims to maximize the surface area to increase the dissolution velocity and solubility saturation of poorly water-soluble a.i. The greatest increase in solubility is expected for crystals <50 nm. A method was described that rapidly converted a microemulsion of novaluron into powders consisting of a.i. and surfactants. The use of a spray-drier (or lyophilizer) makes the method easy to upscale, as it does not require high shear equipment. The nanoparticles were determined by dynamic light scattering and consisted of aggregates of smaller nanoparticles (30–100 nm, by cryotransmission electronic microscopy).

Nanoformulations aiming to increase the solubility of an a.i. are likely to affect the fate of the a.i. More experiments performed under realistic conditions are required to evaluate whether these effects will have a significant impact on the distribution, transport, and degradation processes of a given a.i. A key question relates to the stability of nanoformulations following application. The stability of some nanoformulations is limited, and aggregation/agglomeration is likely to occur soon after they come into contact with the soil solution. In other cases, dilution may occur sufficiently rapidly for the fate of the different ingredients to be assessed separately. It is worth mentioning that these questions may also apply to more classical pesticide formulations (Kah *et al.*, 2013).

9.25 NANOTECHNOLOGY FOR DETECTING PLANT DISEASES

The problem with the disease management lies with the detection of the exact stage of prevention. Most of the times pesticides are applied as a precautionary manner leading to the residual toxicity and environmental hazards and, on the other hand, application of pesticides after the appearance of the disease leads to some amount of crop losses (Sayed Roholla Mousavi and Maryam Rezaei, 2011). Among the different diseases, the viral diseases are the most difficult to control, as one has to stop the spread of the disease by the vectors. But, once it starts showing its symptoms, pesticide application would not be of much use. Therefore, detection of the exact stage such as stage of viral DNA replication or the production of initial viral protein is the key to the success of control of diseases particularly viral diseases. Nanobased viral diagnostics, including multiplexed diagnostic kit development, have taken momentum to detect the exact strain of virus and stage of application of some therapeutic to stop the disease. Detection and utilization of biomarkers that accurately indicate disease stages is also a new area of research.

Measuring differential protein production in both healthy and diseased states leads to the identification of the development of several proteins during the infection cycle. These nanobased diagnostic kits not only increase the speed of detection but also increase the power of the detection (Prasanna, 2007).

In the future, nanoscale devices with novel properties could be used to make agricultural systems 'smart.' For example, devices could be used to identify plant health issues before these become visible to the farmer. Such devices may be capable of responding to different situations by taking appropriate remedial action. If not, they will alert the farmer to the problem. In this way, smart devices will act as both a preventive and an early warning system. Such devices could be used to deliver chemicals in a controlled and targeted manner in the same way as nanomedicine has implications for drug delivery in humans.

9.25.1 Nanofibers

Nanotechnology with the use of biological, chemical and physical processes plays a role in recycling the residual materials of agricultural products to energy and industrial chemicals. For example when cotton is processed into fabric or garment, some of the cellulose or the fibers are discarded as waste or used for low-value products such as cotton balls, yarns and cotton batting (Sayed Roholla Mousavi and Maryam Rezaei, 2011). With the use of newly-developed solvents and a technique called electrospinning, scientists produce 100 nanometer diameter fibers that can be used as a fertilizer or pesticide absorbent. These high-performance absorbents allow targeted application at desired time and location (Lang, 2003). Nanofibers are also used for encapsulating chemical pesticides, for prevention of scattering of chemical pesticides in the environment and water and soil pollution. This technology increases the chemical pesticides durability and security applications. When the fibers are degraded through biological, chemical materials are released slowly into the soil.

When hydrophobic organic pollutants are enters to the soil through water, easily absorbed by the water insoluble solids. Porous nanopolymers have a very similar to the pollutants molecules and considered the most suitable means for separating organic pollutants of soil and water. Similar nanofiber-based fabrics are being used as a detection technology platform to capture and isolate pathogens. The nanofibers in this fabric are embedded with antibodies against specific pathogens. The fabric can be wiped across a surface and tested to determine whether the pathogens are present, perhaps indicating their presence by a change in color.

9.25.2 Nanotech Sensor

Smart sensors, which are obtained by nanotechnology, are the powerful tools for track detect and control with animal and plant pathogen. Detection of very small amounts of a chemical contaminant, virus, or bacteria in agricultural and food systems is envisioned from the integration of chemical, physical and biological devices working together as an integrated sensor at the nanoscale. The bioanalytical nanosensors either use biology as a part of the sensor or are used for biological samples (Scott and Chen, 2003). At the University of Manitoba in Winnipeg, microelectronics and nanotechnology have been combined to create a tiny sensor that can help farmers in the early detection of grain spoilage during storage. The sensor was developed by Suresh Neethirajan, research, and development engineer in the Department of Biosystems and Engineering, along with colleagues from electrical engineering, entomology, and chemistry (Sayed Roholla Mousavi and Maryam Rezaei, 2011). The stand-alone sensor is the size of a dime and can detect parts per billion levels of carbon dioxide and odor-causing chemicals to determine the level and cause of spoilage. If the grain is being infested with fungus, it produces different chemicals. This particular sensor has seven chips in it and identifies which fungus is causing the spoilage. The sensor also measures changes in carbon dioxide to detect incipient and ongoing deterioration of stored grains. Once the cause of spoilage is identified, specific treatment can be used to correct the problem. "We are also looking at building the sensor to be wireless," says Neethirajan. He envisions that multiple sensors would be distributed throughout the grain to pinpoint problem areas and would communicate with a central hub. The central hub would automatically update a household computer, website, or personal mobile device such as a cell phone so that a farmer could monitor the grain on a daily basis without having to visit the grain bin. By catching and treating spoilage before it becomes severe, additional benefits might include reduced chemical usage and better grain quality.

One of the major roles for nanotechnology-enabled devices will be the increased use of autonomous sensors linked into a GPS system for real-time monitoring. These nanosensors could be distributed throughout the field where they can monitor soil conditions and crop growth. Ultimately, precision farming, with the help of smart sensors, will allow enhanced productivity in agriculture by providing accurate information, thus helping farmers to make better decisions. Nanosmart dust and gas sensors are used in determining the amount of pollutants and dust in the air. It is possible evaluation the presence of pollutants in the environment by these sensors (that are made by nanotechnology) in the few minutes (Scott and Chen, 2003).

Fluorescent silica nanoprobes have the potential for rapid diagnosis of plant diseases. Fluorescent silica nanoprobes conjugated with the secondary antibody of goat antirabbit IgG (Yao *et al.*, 2009) was used for detection of a bacterial plant pathogen *Xanthomonas axonopodis* pv. *vesicatoria* (bacterial spot on solanaceous plants). An organic dye tris-2,2′-bipyridyldichlororuthenium (II) hexahydrate (Rubpy) was incorporated into the core of circular silica nanoparticles with average diameter of 50 ± 4.2 nm. Thus, silica nanoparticles became fluorescent which was photostable.

A need for detecting plant disease at an early stage so that tons of food can be protected from the possible outbreak; has tempted nanotechnologists to look for a nanosolution for protecting the food and agriculture from bacteria, fungus and viral agents (Susanta Kumar Biswal *et al.*, 2012). A detection technique that takes less time and that can give results within a few hours, that are simple, portable and accurate and does not require any complicated technique for operation so that even a simple farmer can use the portable system. If an autonomous nanosensors linked into a GPS system for real-time monitoring can be distributed throughout the field to monitor soil conditions and crop, it would be of great help. The union of biotechnology and nanotechnology in sensors will create equipment of increased sensitivity, allowing an earlier response to environmental changes and diseases.

9.26 MICROORGANISMS IN THE SYNTHESIS OF NANOPARTICLES

The research on nanoscience and nanotechnology essentially involves preparation and use of nanoparticles of various elements and compounds. Among various uses, nanoparticles are also being used as antimicrobial agents for plant disease management. Formation of nanoparticles can be achieved via several processes which may be either physical or chemical. The safe method of nanoparticle production is the biological systems especially microorganisms (Mansoori, 2005). Microorganisms offer several advantages like maneuverability for desired result using biotechnology, ease of handling especially fungi (Vigneshwaran *et al.*, 2006), the cheapness of production, easy scaling up of the process high efficiency (Goodsell, 2004), simplicity and nature of green chemistry or ecofriendliness. Microorganisms have been regarded as 'biofactories' for the production of metallic nanoparticles.

9.26.1 Fungi

Fungi are relatively recent in their use in the synthesis of nanoparticles. There has been a shift from bacteria to fungi to be used as natural 'nanofactories' owing to easy downstream processing, easy handling (Mandal *et al.*, 2006) and their ability to secrete a large amount of enzymes. However, fungi being eukaryotes are less amenable to genetic manipulation compared to prokaryotes. Therefore, any alteration of fungi at genetic level for the synthesis of more nanoparticles would not be so easy. A good number of fungi have been tried to synthesize metallic nanoparticles till date (Tables 9.7 and 9.8). It is important to know the mechanism of synthesis of nanoparticles (NPs) in microbial systems to get better control over shape, size and other desired properties of the synthesized nanomaterials (Susanta Banik and Pratibha Sharma, 2011).

Table 9.7 *In vitro* Antimicrobial Activity Test of Nanosized Silica-silver

Microorganisms	*0.3 ppm*	*3.0 ppm*	*10 ppm*	*100 ppm*
Pythium ultimum	15.5	66.7	100	100
Magnaporthe grisea	1.4	27.0	100	100
Colletotrichum gloeosporioides	11.8	21.6	100	100
Botrytis cinerea	2.6	82.7	100	100
Rhizoctonia solani	54.8	94.8	100	100
Bacillus subtilis	0.0	0.0	50	100

Contd...

Contd...

Azotobacter chroococcum	0.0	0.0	0.0	100
Rhizobium tropici	0.0	0.0	0.0	100
Pseudomonas syringae	0.0	0.0	0.0	100
Xanthomonas campestris pv. *vesicatoria*	0.0	0.0	0.0	100

Since physicochemical properties of nanoforms from its macroform vary greatly, it becomes important to examine the effect of nanoparticles on microorganisms to harness the benefit of this technology in the plant protection especially against phytopathogens (Mujeebur Rahman Khan and Tanveer Fatima Rizvi, 2014). Nanoparticles because of ultrasmall size, even smaller than a virus particle and high reactivity, may affect the activity of microorganisms. The silver has been generally found noninjurious to microorganisms. However, silver NPs inhibited the colonization of *Staphylococcus aureus, Pseudomonas aeruginosa, Escherichia coli* and *Klebsiella pneumonia*. The highest antimicrobial activity of silver nanoparticles (30 mm) synthesized by *Solanum tricobatum* and *Ocimum tenuiflorum* leaf extracts was found against *S. aureus* and *E. coli*, respectively. The information so far available on this aspect has shown that the nanoparticles have a definite effect on the colonization of bacteria and fungi. However, these effects are suppressive as well as stimulatory and hence cannot be generalized.

The nanoparticles have also been found suppressive to fungi. Singh *et al.* (2013) reported that among nanoforms of 15 micronutrients, $CuSo_4$ and $Na_2B_4O_7$ were found most effective in controlling rust disease of field peas. Microelements such as manganese and zinc also suppressed the damping off and charcoal rot diseases in sunflower. The Ag NPs/PVP were tested for fungicidal activity against different yeasts and molds such as *Candida albicans, C. krusei, C. tropicalis, C. glabrata* and *Aspergillus brasiliensis*. The hybrid materials showed strong antifungal effects against the tested microbes. Fungicidal effect of zinc oxide nanoparticles (ZnO NPs) against two postharvest pathogenic fungi, *Botrytis cinerea,* and *Penicillium expansum* has been reported. Traditional microbiological plating, Scanning Electron Microscopy (SEM) and Raman spectroscopy were used to study antifungal activities of ZnO NPs and to characterize the changes in morphology and cellular compositions of fungal hyphae. The ZnO NPs (70±15 nm) at the concentrations greater than 3 mmol L^{-1} significantly inhibited the growth of *B. cinerea* and *P. expansum,* the later fungus was found more sensitive to the treatments. The NP treatments caused deformation in the hyphae of *B. cinerea* and prevented the development of conidiophores and conidia in *P. expansum* which eventually led to the death of fungal hyphae. Krishnaraj *et al.* (2012) examined the effect of silver nanoparticles on plant pathogenic fungi, *Alternaria alternata, Sclerotinia sclerotiorum, Macrophomina phaseolina, Rhizoctonia solani, B. cinerea* and *Curvularia lunata* and found that 15 mg L^{-1} concentration of NPs greatly inhibited the activity of all the tested pathogens. The zinc nanoparticles (25 mg mL^{-1}) suppressed the colonization of *A. flavus.*

At present, nanotechnology, which has already made some imprints in many areas of plant pathology is the second craze after biotechnology for innovative research. There are a plethora of examples of animal/human pathogen detection by the use of nanosensors, but in plant pathogen nanotechnological application is still in its infancy. As far as the use of microbes and plant pathogens in the biosynthesis of nanoparticles is concerned, more and more reports are coming up. More bionanotechnological studies on the physiology of host and pathogen, infection process, disease diagnosis will help in formulating novel disease management strategies. If 'smart' delivery method is perfected in plant system and complete mechanisms of penetration, transport and accumulation of nanoparticles at targeted site are elucidated precision crop protection will get a heavy boost. Nanoformulations are viewed to be safer and environment-friendly option for plant disease management, but the high toxicity of nanoparticles inadvertently released in the environment may pose a greater threat to man and other organisms. Therefore, nanotechnological progress is to be viewed with caution and dealt accordingly.

The antifungal activity of zinc oxide (ZnO) and magnesium oxide (MgO) nanoparticles prepared by biosafe method was evaluated for *Alternaria alternate, Fusarium oxysporum, Rhizopus stolonifer* and *Mucor plumbeus* (Wani and Shah, 2012). It was observed from the study that all the nanoparticles at different concentrations brought about significant inhibition in the germination of spores of *Alternaria alternata, Fusarium oxysporum, Rhizopus stolonifer* and *Mucor plumbeus*. However, the highest inhibition of the germination of all the test fungi was observed at higher concentrations followed by lower concentrations of nanoparticles. Nanoparticles of the highest concentration of MgO were found most effective in reducing the spore germination followed by nanoparticles of ZnO at the same concentration.

9.26.2 Bacteria

Among microbes, prokaryotes have received the most attention for biosynthesis of nanoparticles (Mandal *et al.*, 2006) some of which are presented (Table 9.8). Bacteria have been used to biosynthesize mostly silver, gold, FeS, and magnetite nanoparticles and quantum dots of cadmium sulfide (CdS), zinc sulfide (ZnS) and lead sulfide (PbS).

Table 9.8 Examples of Fungi and Bacteria Used for Biosynthesis of Nanoparticles

Sl. No.	*Nanoparticle*	*Fungi*
1.	Silver nanoparticles	*Verticillium* sp, *Phoma* sp., *Fusarium oxysporum, Phaenerochaete chrysosporium, Aspergillus flavus, Aspergillus niger, Fusarium semitectum, Fusarium solani, Cladosporium cladosporioides, Pleurotus sajor caju, Alternaria alternate, Penicillium brevicompactum, Bipolaris nodulosa, Trichoderma viride, Aspergillus clavatus*
2.	Nanocrystalline silver	*Trichoderma asperellum*
3.	Gold nanoparticles	*Colletotrichum* sp., *Verticillium* sp.

Contd...

Contd...

4.	Gold and gold-silver alloy nanoparticles	*Fusarium semitectum*
5.	Bimetallic gold-silver alloy nanoparticle	*F. oxysporum*
6.	Gold, silver (5–50 nm) and gold-silver alloy nanoparticle (8–14 nm)	*F. oxysporum*
7.	Cadmium sulphide	*Coriolus versicolor*
8.	Cadmium sulphide nanoparticles	*Fusarium* sp.
9.	Zirconia nanoparticles	*Fusarium oxysporum*
10.	Nanoparticulate magnetite	*Fusarium oxysporum and Verticillium* sp.
		Bacteria
1.	Silver nanoparticle	*Clostridium versicolor, Bacillus subtilis, Brevibacterium casei, Escherichia coli, Staphylococcus aureus*
2.	Silver, silver sulphide	*Pseudomonas stutzeri*
3.	Silver, gold, and alloy of silver and gold	*Lactobacillus*
4.	Triangular gold nanoprisms	*Actinomycete*
5.	Gold nanoparticle	*Rhodococcus* sp., *Bacillus subtilis* 168

Bacteria have been used to biosynthesize mostly silver, gold, FeS, and magnetite nanoparticles and quantum dots of cadmium sulfide (CdS), zinc sulfide (ZnS) and lead sulfide (PbS).

While copper (Cu)-based bactericides have been used for disease management, most of the *Xanthomonas perforans* (pathogen of Bacterial spot of tomato) strains isolated from tomatoes in Florida and other locations worldwide are Cu-resistant. Ocsoy *et al.* (2013) developed DNA-directed silver (Ag) nanoparticles (NPs) grown on graphene oxide (GO). These Ag@dsDNA@GO composites effectively decrease *X. perforans* cell viability in culture and on plants. At the very low concentration of 16 ppm of Ag@dsDNA@GO, composites showed excellent antibacterial capability in culture with significant advantages in improved stability, enhanced antibacterial activity, and stronger adsorption properties.

A DNA-directed, facile and one-step approach to synthesize Ag@dsDNA@GO composites was reported that display high antibacterial activity toward *X. perforans*, a model plant pathogenic bacterium Ocsoy *et al.* (2013). It was proposed that the use of dsDNA as a template for growing Ag NPs on GO enhances the synergistic effect between Ag NPs and GO by controlling size, aggregation, and distribution of Ag NPs and by increasing the adhesive force between Ag@GO composites and bacterial cell membranes.

Antibacterial activity of zinc nanoparticles against *P. aeruginosa* has been reported by Jayaseelan *et al.* (2012). The maximum zone of inhibition in the colonization of the bacteria (22 ± 1.8 mm) was recorded at 25 ng mL^{-1} ZnO NPs. The study showed that the ZnO NPs proved to be a novel antimicrobial material. The antibacterial activity of the synthesized Ag NPs/PVP (hybrid materials based on polyvinylpyrrolidone with silver nanoparticles) against three different groups of *Staphylococcus aureus* (Gram-positive bacteria), *E. coli* (Gram-negative bacteria), *P. aeruginosa* (nonfermenting Gram-negative bacteria), as well as against spores of *Bacillus subtilis* has been studied. The antibacterial activities of CuO NPs have also been reported against *S. aureus, Bacillus subtilis, P. aeruginosa* and *E. coli.* Silver nanoparticles showed high antimicrobial and bactericidal activity against Gram-positive bacteria such as *E. coli, P. aureginosa* and *S. aureus* which are highly methicillin-resistant strains. The antibacterial activity of nanoparticles was found to be dependent on NPs concentration, physiology, metabolism, the intracellular selective permeability of membranes and the type of microbial cell (Mujeebur Rahman Khan and Tanveer Fatima Rizvi, 2014).

9.26.3 Plant Viruses

Plant viruses, especially spherical/icosahedral viruses represent the examples of naturally occurring nanomaterials or nanoparticles. The smallest plant virus known till date is satellite Tobacco necrosis virus measuring only 18 nm in diameter (Hoglund, 1968). By nature's design, plant viruses are ought to be used for the advancement of nanoscience and nanotechnology. Plant viruses are made up of single or double stranded RNA/DNA as genome which is encapsidated by a protein coat. The protein coat/shell structurally and functionally appears like a container carrying the nucleic acid molecule as cargo from one host to another. Their ability to infect, deliver nucleic acid genome to a specific site in a host cell, replicate, package nucleic acid and come out of host cell precisely in an orderly manner have necessitated them to be used in nanotechnology. Plant viruses have been used as template for synthesis of various types of nanomaterials (Table 9.9). A complete review on the use of plant viruses as biotemplates for nanomaterials and their application has been done recently by Young *et al.* (2008). Plant virus especially spherical/icosahedral viruses represent the examples of naturally occurring nanomaterials or nanoparticles. The smallest plant virus known till date is satellite Tobacco necrosis virus measuring only 18 nm in diameter (Hoglund, 1968).

By nature's design, plant viruses are ought to be used for the advancement of nanoscience and nanotechnology. Plant viruses are made up of single or double stranded RNA/DNA as genome which is encapsidated by a protein coat. The protein coat/shell structurally and functionally appears like a container carrying the nucleic acid molecule as cargo from one host to another. Their ability to infect, deliver nucleic acid genome to a specific site in a host cell, replicate, package nucleic acid and come out of host cell precisely in an orderly manner have necessitated them to be used in nanotechnology. Plant viruses have been used as a template for synthesis of various types of nanomaterials. A complete review on the use of plant viruses as biotemplates for nanomaterials and their application has been done recently by Young *et al.* (2008).

Table 9.9 Some Examples of Plant Virus Capsids Used as Biotemplates for Nanomaterial Production

Sl. No.	*Plant virus*	*Application*
1.	Cowpea mosaic virus (CMV), an engineered CMV	Iron-platinum nanoparticle (30 nm diameter) synthesis
2.	Cowpea chlorotic mottle virus (CCMV)	Gold nanoparticle synthesis

Contd...

Contd...

3.	Cowpea chlorotic mottle virus (CCMV)	As reaction vessel for nanomaterial synthesis
4.	Tobacco mosaic virus (TMV) synthesis	Ag and Ni nanoparticle
5.	Tobacco mosaic virus (TMV)	Synthesis of nanowire of nickel and cobalt
6.	Tobacco mosaic virus (TMV)	Synthesis of bimetallic alloys of CoPt, $CoPt_3$ and $FePt_3$ nanowires
7.	Brome mosaic virus	Gold nanoparticle synthesis, Iron oxide synthesis
8.	Red clover necrotic mosaic virus	Au, $CoFe_2O_4$, and CdSe nanoparticles synthesis

To protect crops and animals from the effects of pathogens, polymeric and lipid-based edible film layers have been developed. Several metal ions, such as Zn2þ, Cu2þ, and Agþ, have been used to inactivate bacterial growth, although the mechanisms are not fully understood. Because of the rather low antibacterial activity of metal ions, many researchers have focused on the use of nanomaterials to increase the effective antibacterial impact. This has led to the development of several types of nanomaterials, including the metallic nanoparticles Ag, Cu, CuO, ZnO and TiO_2 with carbon-based nanomaterials, such as carbon nanotubes (CNTs), as well as graphene oxide (GO) and metallic nanoparticles formed on silica nanoparticles.

Among the different types of nanoparticles, Ag NPs have been used as effective biocides against a variety of pathogens, fungi, and viruses. Researchers have debated the antibacterial activity of Agþ ions and Ag NPs and the mechanism underlying the bactericidal process. For example, one plausible theory holds that Agþ ions interact with thiol, carboxyl, hydroxyl, amino, phosphate, and imidazole groups in proteins and enzymes on bacterial membranes, leading to serious structural deformation of the cell membrane. This is followed by the penetration of Agþ ions into cells through the vulnerable membranes to inactivate enzymes, leading to suffocation, inhibition of cell replication, and eventual cell death.

9.26.4 Nematodes

Researchers at North Carolina State University have developed a plant-virus-based, small molecule delivery system for the controlled delivery of nematicides. This delivery system comprises nematicide-loaded plant viral nanoparticles combined with tailored seed coating formulations. Encapsulation of nematicides by the plant viral nanoparticles will eliminate nontarget adverse effects, and incorporation into seed coatings is a more economical and environment-friendly delivery mechanism than granular nematicide applications. In addition, the coating formulation can be tailored for controlled release based on biodegradation rate in the soil, making the system adaptable to enable delivery of active ingredients to nematodes during the critical early growth stage of plants. This plant-virus-based, small molecule delivery system can also be used for the controlled delivery of a wide range of other small molecules critical for plant development, including other pest repellants, nutrients, and hormones. This nanotechnology delivery system has the ideal attributes for agricultural application – it is robust and viable in a wide range of environments; it is stable in buffers, the soil, and physiological fluids; it has low leakage rates for small molecule cargo; it presents targeting molecules conveniently, and it can be triggered to release cargo specifically inside a targeted cell (Steven A. Lommel *et al.*, 2011).

9.26.5 Advantages

Major advantages include elimination of nontarget adverse effects through encapsulation of nematicides by plant viral nanoparticles; loading of active agents into plant viral nanoparticles is repeatable and scalable; economical and environmentally friendly relative to current nematicide control technologies; suitability of being tailored for controlled time release of nematicide during necessary treatment windows; enhanced cargo capacity and colloidal stability in solution compared to other nanoscale delivery vehicle strategies; and less leakage than other nanoscale delivery vehicle strategies.

Toxicity of nanoparticulate and bulk ZnO, Al_2O_3 and TiO_2 were examined to the nematode *Caenorhabditis elegans* with *Escherichia coli* as a food source. Parallel experiments with dissolved metal ions from NPs were also conducted (Huanhua Wang and Robert Wick, 2009). The 24-h median lethal concentration (LC_{50}) and sublethal endpoints were assessed. Both NPs and their bulk counterparts were toxic, inhibiting growth and especially the reproductive capability of the nematode. The 24-h LC_{50} for ZnO NPs (2.3 mg L^{-1}) and bulk ZnO was not significantly different, but significantly different between Al_2O_3 NPs (82 mg L^{-1}) and bulk Al_2O_3 (153 mg L^{-1}), and between TiO_2 NPs (80 mg L^{-1}) and bulk TiO_2 (136 mg L^{-1}). Oxide solubility influenced the toxicity of ZnO and Al_2O_3 NPs, but nanoparticle-dependent toxicity was indeed observed for the investigated NPs.

The ecotoxicity of silver nanoparticles (Ag NPs) was investigated in *Caenorhabditis elegans* using survival, growth, and reproduction, as the ecotoxicological endpoints, as well as stress response gene expression. Whole genome microarray was used to screen global changes in *C. elegans* transcription profiles after Ag NPs exposure, followed by quantitative analysis of selected genes (Ji-yeon Roh *et al.*, 2009). The integration of gene expression with organism and population level endpoints was investigated using *C. elegans* functional genomics tools, to test the ecotoxicological relevance of Ag NPs-induced gene expression. Ag NPs exerted considerable toxicity in *C. elegans,* most clearly as dramatically decreased reproduction potential. Increased expression of the superoxide dismutases-3 (*sod-3)* and abnormal dauer formation protein (*daf-12*) genes with 0.1 and 0.5 mg/L of Ag NPs exposures occurred concurrently with significant decreases in reproduction ability. Overall results of functional genomic studies using mutant analyses suggested that the *sod-3* and *daf-12* gene expressions may have been related to the Ag NPs-induced reproductive failure in *C. elegans,* and that oxidative stress may have been an important mechanism in Ag NPs' toxicity.

Invertebrate animal models such as the nematode *Caenorhabditis elegans* (*C. elegans*) are increasingly used in nanotechnological applications (Andrea Scharf *et al.*, 2013).

Research in this area covers a wide range of remote control of worm behavior by nanoparticles (NPs) to the evaluation of organismal nanomaterial safety. Despite the broad spectrum of investigated NP–bio-interactions, little is known about the role of nanomaterials with respect to aging processes in *C. elegans.* NPs in single cells of adult *C. elegans* were traced and correlated particle distribution with the worm's metabolism and organ function. By confocal microscopy analysis of fluorescently labeled NPs in living worms, two entry portals for the uptake of nanomaterials were identified *via* the pharynx to the intestinal system and *via* the *vulva* to the reproductive system. NPs were localized throughout the cytoplasm and the cell nucleus in single intestinal, and vulval B and D cells. Silica NPs induced an untimely accumulation of insoluble ubiquitinated proteins, nuclear amyloid and reduction of pharyngeal pumping that taken together constituted a premature aging phenotype of *C. elegans* on the molecular and behavioral level, respectively. Screening of different nanomaterials for their effects on protein solubility showed that polystyrene or silver NPs did not induce accumulation of ubiquitinated proteins suggesting that alteration of protein homeostasis is a unique property of silica NPs. The nematode *C. elegans* represents an excellent model to investigate the effect of different types of nanomaterials on aging at the molecule, cell, and whole organism level.

The new nanotechnology with materials having unique properties than their macroscopic or bulk counterparts has promised applications in various fields. The essence of Nanotechnology is the ability to work at the molecular level, atom by atom, to create large structures with the fundamentally new molecular organization. The aim is to exploit these properties by gaining control of structures and devices at atomic, molecular, and supramolecular levels and to learn to efficiently manufacture and use these devices (Madhuri Sharon *et al.*, 2010). Nanotechnology has provided new solutions to problems in plants and food science (postharvest products) and offers new approaches to the rational selection of raw materials, or the processing of such materials to enhance the quality of plant products. The heart of nanotechnology lies in the ability to compress the tools and devices to the nanometer range, and to accumulate atoms and molecules into bulkier structures while the size remains very small. However, it must be said here that phytopathologists took comparatively longer time in utilizing the uniqueness of nanomaterials for their benefit.

9.27 UTILIZATION OF CARBON NANOMATERIAL AS A SENSOR

Carbon is already used as an electrode for electrochemical analysis. Work on Selection of the Carbon Nanomaterial is in progress for the electrochemical sensor by measuring the electrochemical properties of various synthesized CNM from different raw materials in response to different pesticides. Residual pesticides in the plants are not only harmful to human consumers but are also not very healthy for the plants. This involves preparation of electrochemical cell having CNM as an electrode in this electrochemical sensor, to assign appropriate oxidation/reduction potential values of a particular analyte. Nanosensor devices that use carbon nanotubes or nanocantilevers are small enough to trap and measure individual proteins or small molecules. A contaminant can be detected by specially engineered nanoparticles or nanosurfaces which trigger an electrical or chemical signal. Some nanosensors work by initiating enzymatic reactions or by using Nanoengineered branching molecules called dendrimers as probes to bind to target chemicals and proteins. Pathogen and contaminant detection is possible with increased sensitivity and decreased response time due to nanosensors.

9.27.1 Detection of Phytopathogens

A prior and accurate detection of pathogens would help in the timely application of pesticides/fertilizers in the field to protect the crops from diseases (Mujeebur Rahman Khan and Tanveer Fatima Rizvi, 2014). Nanoparticles can be used as biomarkers or as a rapid diagnostic tool for detection of phytobacteria, viruses, and fungi. Nanoparticles based sensors might offer improved detection limits for detecting viral pathogens in plants. Nanoparticles can either be directly modified for use in pathogen detection or for a diagnostic tool to detect compounds indicative of a diseased condition. Nanochips are types of microarrays that contain fluorescent oligo capture probes through which the hybridization can be detected. The nanochips are known for their high sensitivity and specificity for detecting single nucleotide change in bacteria and viruses. Fluorescence silica nanoparticles were utilized in combination with antibodies to detect *Xanthomonas axonopodis* pv. *vesicatoria* that causes bacterial spot disease in plants. Nanogold based immunosensors were used that could detect Karnal bunt disease in wheat (*Tilletia indica*) using Surface Plasmon Resonance (SPR). Application of SPR sensor in the seed certification and plant quarantines may prove highly effective and accurate in the detection of infection in plants. Research on pathogen detecting nanosensors for their field application would be highly valuable for rapid diagnosis and disease management.

Plants respond to different stress conditions through physiological changes. One such response is the induction of systemic defense which is thought to be regulated by jasmonic acid, methyl jasmonate and salicylic acid (Khan and Haque, 2013). Harnessing this indirect stimulus a sensitive electrochemical sensor was developed, using modified gold electrode with copper nanoparticles, to monitor the levels of salicylic acid in oil seeds to detect the pathogenic fungus, *Sclerotinia sclerotiorum*. They successfully and accurately measured salicylic acid using this sensor. Research on similar sensors and sensing techniques needs to be expanded for detecting pathogens, their by-products or monitor physiological changes in plants.

9.28 NANOTECHNOLOGY IN POSTHARVEST DISEASE MANAGEMENT

There is a need to increase production efficiency and decrease postharvest wastage with the application of emerging technologies like biotechnology and nanotechnology in postharvest products. Nanotechnology has been effectively applied in agricultural and horticultural products by increasing shelf life, controlling the growth of microorganisms by nanofilms and coatings, controlling the influence of gasses and the harmful rays (UV), using Nanobiosensors for detection of quality and spoilage (Chowdappa and

Shivakumar Gowda, 2013). Nanotechnology can be applied in postharvest operations such as drying, storage and preservation of agricultural products. Chitosan, a deacetylated derivative of chitin, is found to be very effective in reducing postharvest decay of fruit and vegetables. Chitosan at a concentration of 1 g/L has found to be very effective in reducing the growth of several phytopathogenic fungi causing postharvest spoilage of fruit and vegetables.

The effect of 1% chitosan film with 0.04% nanosilicon dioxide was studied on the qualitative properties of harvested jujube after 32 days of storage under ambient temperature, they studied the related defense enzymes in fruits and found that coated sample showed lower red indices, decay incidence, respiration rate, and weight loss. Shi *et al.* (2013) studied the novel chitosan/nanosilica hybrid film and its effect on preservation quality of longan fruits under ambient temperature. Coating extended shelf life, reduced browning index, retarded weight loss and inhibited the increase of malondialdehyde amount and polyphenoloxidase activity in fresh longan fruit. An investigation was carried out on the effects of nanosilver (with 2ñ5 nm diameters) on the postharvest shelf life of cut gerbera flowers. They observed that, pulsing for 24 h with 5 mg/L nanosolution extended vase life and inhibited the bacteria growth in vase solution for initial two days when observed *in vitro* under a microscope. A postharvest treatment of nanosilver was extensively studied. Nanosilver was found to be very significant in prolonging the vase life by inhibiting the growth of bacteria on several cut flowers with different cultivar variety, including rose, gerbera and *Acacia holosericea.*

In a study, Chowdappa and Shivakumar Gowda (2013) evaluated the applicability of the CS-Ag NP composite as a fruit coating material to inhibit the growth of *C. gloeosporioides* associated with mango anthracnose. We found that nanocomposites showed a significant effect in reducing the percentage of rotting fruit tissue (71.28% at 1% concentration). The addition of 0.1% nonionic surfactant Tween 80 enhanced the wettability and adhesion property of coating solution and exhibited significant disease reduction compared to control (84.55% at 1% concentration). Thus, these nanocomposites can be utilized as a coating material in preventing quiescent infections of *Colletotrichum gloeosporioides* on mango to prevent postharvest losses. Nanomaterial has important implication for the management of postharvest diseases. Research findings showed the better applicability and advantages of nanopacking materials over conventional normal packing material on the physicochemical and physiological quality of stored fruits, vegetables, and other horticultural crops.

Antimicrobial packaging of edible food films made with cinnamon or oregano oil, or nanoparticles of zinc, calcium other materials that kill bacteria is being tried. Green packaging using nanofibers made from lobster shells or organic corn (both are antimicrobial and biodegradable) is also a food safety effort. Improved food packaging needs packaging materials having strength, barrier properties, and stability to heat and cold. These are being achieved using nanocomposite materials. Bayer Polymers have produced a nanocomposite film 'Durethan,' It is a film enriched with silicate nanoparticles which reduce the entrance of oxygen and other gasses, and preserves moisture, thus preventing food from spoiling. In future, incorporation of silver, magnesium oxide or zinc oxide nanoparticles (which can kill harmful microorganisms) in food or beer packages will save the contamination. Antimicrobial activity can also be imparted by addition of nanosensors to food packages is also anticipated in the future (Madhuri Sharon *et al.*, 2010). McHugh has suggested that these nanosensors could be used to detect chemicals, pathogens, and toxins in foods. Radio Frequency Identification (RFID) tags could be incorporated into food packages in the future. These do not require line-of-sight for reading like bar-codes and enable registration of hundreds of tags in a second. Use of nanowheels, nanofibers and nanotubes are being tried to improve the qualities of food packages.

Nanotechnology can be used for combating the plant diseases either by controlled delivery of functional molecules or as a diagnostic tool for disease detection. Nanotechnology, nanoparticles and quantum dots have emerged as a pivotal tool for detection of a particular biological marker with extreme accuracy. Diagnosis of a disease in its very early stage can play an important role in treatment. Due to phenomenal advancement in nanotechnology, QDs have emerged as a pivotal tool for detection of a particular biological marker with extreme accuracy. QDS being very photo-stable and optically sensitive can be used as labeling and can be easily traced with ordinary equipment (Madhuri Sharon *et al.*, 2010). Early detection of tumor markers using quantum dots is proved to be a boon for cancer diagnosis. Use of QDs has also helped in the unlocking complex neurological phenomenon, such as molecular activities at synapse during neurotransmission. QDs also give important information about receptor movement if tagged with suitable antibodies. In short, optical stability and easy to handle properties have made QDs remain at the apex of medical diagnostics.

Plant pathogens and pests are of the major factors limiting crop productivity. In crop sciences, nanotechnology can be used for the production of nanocapsules for delivery of pesticides, fertilizers, and other agrochemicals (Jha *et al.*, 2011). Nanotechnology for the control of plant diseases is a promising technique in plant pathology either by providing controlled delivery of functional molecules or as a diagnostic tool for disease detection, an important step in plant disease treatment (Sharon *et al.*, 2010). Encapsulation of herbicides could provide an improvement in their application. For example, Sulfonylurea herbicides are applied to the soil to control *Orobanche* spp., but several applications are needed to achieve effective control (Joel *et al.*, 2007). Several studies were conducted using Nanosized particles to control fungal pathogens including such as *Pythium ultimum, Magnaporthe grisea, Colletotrichum gloeosporioides, Botrytis cinerea* and, *Rhyzoctonia solani,* as well as bacterial disease including *Bacillus subtilis, Azotobacter chrococuum, Rhizobium tropici, Pseudomonas syringae* and *Xanthomonas compestris* pv. *vesicatoria* (Park *et al.*, 2006). In Palestine, nanotechnology might be used for the control of several plant pathogens such as powdery mildews on grapevine and olive leaf spot on olive trees.

Nanotechnology is the manipulation or self-assembly of individual atoms, molecules, or molecular clusters into structures to create materials and devices with new or vastly different properties. Nanotechnology has been described as the new industrial revolution, and both developed and developing countries are investing in this technology to secure a market share. At present the USA leads with a four year, 3.7 billion USD investments through its National Nanotechnology Initiative (NNI) (Neha Bhalla, 2012). The USA is followed by Japan and the European Union, which have both committed substantial funds (750 million and 1.2 billion, including individual country contributions, respectively per year). The level of funding in developing countries may be comparatively lower. However, this has not lessened the impact of some countries on the global stage. A need for detecting plant disease at an early stage so that tons of food can be protected from the possible outbreak; has tempted nanotechnologists to look for a nanosolution for protecting food and agriculture from bacteria, fungi, and viral agents. A detection technique that takes less time and that can give results within a few hours, that is simple, portable and accurate and does not require any complicated technique for operation so that even a simple farmer can use the portable system.

In recent years, nanotechnology has been increasingly applied to the development of novel antimicrobials for the management of pathogenic bacteria affecting agricultural crops, humans, and animals. In particular, significant development in nanomaterial synthesis, such as polymeric, carbon-based, and metallic, has attracted researchers' attention toward applications in managing plant diseases caused by bacteria. Toxicity considerations, including negative environmental effects, have also led to the redesign of nanomaterial by tuning the size and shape and by surface modification, leading to increased antimicrobial activity and decreased ecological toxicity.

Most recently, an extensive study of the photocatalytic activity of TiO_2/Zn, TiO_2/Ag, and TiO_2 NPs on *X. perforans* has been carried out, indicating the potential of using nanoparticles in managing bacterial spot in tomato (Paret *et al.*, 2010). Because of the rather low antibacterial activity of metal ions, many researchers have focused on the use of nanomaterials to increase the effective antibacterial impact. This has led to the development of several types of nanomaterials, including the metallic nanoparticles Ag, Cu, CuO, ZnO, and TiO_2 10 with carbon-based nanomaterials, such as carbon nanotubes (CNTs), as well as graphene oxide (GO)14 and metallic nanoparticles formed on silica nanoparticles.

Among the different types of nanoparticles, Ag NPs have been used as effective biocides against a variety of pathogens, fungi, and viruses (Zodrow *et al.*, 2009). Researchers have debated the antibacterial activity of Agþ ions and Ag NPs and the mechanism underlying the bactericidal process. For example, one plausible theory holds that Agþ ions interact with thiol, carboxyl, hydroxyl, amino, phosphate, and imidazole groups in proteins and enzymes on bacterial membranes, leading to serious structural deformation of the cell membrane. This is followed by the penetration of Agþ ions into cells through the vulnerable membranes to inactivate enzymes, leading to suffocation, inhibition of cell replication, and eventual cell death. Several experiments have been performed on bacteria to evaluate the bactericidal effect of single Ag NPs. It is assumed that Ag NPs interact with the bacterial membrane by way of Agþ ions and that bacteria are killed by a combination of the bactericidal effects of Ag NPs and released Agþ ions, as noted above. To enhance the antibacterial activity of Ag NPs and to increase the amount of released Agþ ions, Ag NPs have been prepared with different sizes, shapes, types of surface coatings, and surface charges for use in both aerobic and anaerobic environmental conditions. However, the dependence of toxicity on these parameters is currently a major drawback in the use of bare Ag NPs. In addition, agglomeration of bare Ag NPs in contact with bacteria results in the loss of active surface area and weakening of antibacterial activity.

To overcome these issues, graphene oxide, composed of a single layer of sp2-bonded carbon atoms with active surface hydroxyl, epoxy, and carboxyl groups can be used as a support material to grow Ag NPs or to load presynthesized Ag NPs. As a consequence of the synergistic effects of Ag NPs and GO, these conjugates display extraordinary antibacterial activity compared to bare Ag NPs (Das *et al.*, 2011).

Copper nanoparticles were synthesized by chemical reduction of Cu^{2+} in the presence of Cetyl Trimethyl Ammonium Bromide and isopropyl alcohol. Characterizations of copper nanoparticles were carried out by UV–visible spectroscopy, nanoparticles tracking analysis (NTA), Fourier transform infrared spectroscopy (FTIR) and transmission electron microscopy (TEM) which revealed that synthesized nanoparticles were coated by cetyltrimethylammonium bromide (CTAB) having particle size of 3–10 nm (Prachi Kanhed *et al.*, 2014). Copper nanoparticles demonstrated significant antifungal activity against plant pathogenic fungi, *Phoma destructiva* (DBT-66), *Curvularia lunata* (MTCC no. 2030), *Alternaria alternata* (MTCC No. 6572) and *Fusarium oxysporum* (MTCC No. 1755). Since for the synthesis of copper nanoparticles the present chemical method by using C-TAB-IPA was found to be simple, economical and fast, the synthesized copper nanoparticles could be used as a novel antifungal agent in agriculture to control the plant pathogenic fungi as well as a potent disinfectant in poultry and animal husbandry.

Use of nanoparticles has been considered an alternative and effective approach which is eco-friendly and cost effective for the control of pathogenic microbes. These nanoparticles have a great potential in the management of plant diseases compared to synthetic fungicides. Zinc oxide (ZnO) and magnesium oxide (Mg O) nanoparticles are effective antibacterial and antiodor agents (Ram Prasad *et al.*, 2014). The increased ease in dispensability, optical transparency, and smoothness make ZnO and MgO nanostructures an attractive antibacterial ingredient in many products. Both have also been proposed as an antimicrobial preservative for wood or food products. Properly functionalized nanocapsules provide better penetration through the cuticle and allow slow and controlled release of active ingredients on reaching the target weed. The use of such nanobiopesticide is more acceptable since they are safe for plants and cause less environmental pollution in comparison to conventional chemical pesticides.

Nanotechnology holds the promise of controlled delivery of agrochemicals to improve disease resistance, plant growth enhancement, and nutrient utilization. Nanoencapsulation shows the benefit of more efficient and targeted use of pesticides, herbicides, and insecticides in environment-friendly greener way. Research and development in postharvest nanotechnology can help in the preservation of the freshness and quality and prevent diseases in a relatively safer way (Chowdappa and Shivakumar Gowda, 2013). With the advancement of nanotechnology, application of green chemistry in the synthesis of nanomaterials by using plant extracts and living cells has reduced the use of toxic solvents and guarantees ecoprotection. Nanotechnology in conjunction with biotechnology has significantly extended the applicability of nanomaterials in crop protection and production. Even though the toxicity of nanomaterials has not yet clearly understood, it plays a significant role in crop protection because of its unique physical and chemical properties. The application of nanomaterials is relatively new in the field of agriculture, and it needs further research investigations. Barring the minuscule limitations, nanomaterials have a tremendous potential for making crop protection methodologies cost effective and environmental friendly.

9.29 LIMITATIONS OF NANOTECHNOLOGY

Nanotechnology may pose negative effects on the environment, ecosystem, and humans. The potential risks associated with releasing nanomaterials into the environment (soil and water organisms) are still unclear. Recent findings showed the potentially harmful effects of nanomaterials on the digestive systems of a beneficial soil organism, earthworm. The increased safety over application of nanomaterials in food and agriculture is of concern. Studies have been conducted on main exposure routes and determinants of nanotoxicity involving particle size, surface, structure, chemical composition and dosage. The toxic effect of metallic nanomaterials has been reported on aquatic organisms, viz., zebrafish, daphnids, and an algal species, which were exposed them to silver, copper, aluminum, nickel and cobalt in both soluble salts and nanoparticle form. Nanosilver and nanocopper caused toxicity in all tested organisms at median lethal concentrations. With the technological advancement, a large amount of engineered nanomaterials is reaching environment through consumer and commercial products (Chowdappa and Shivakumar Gowda, 2013). The adverse impact of nanosilver has also been studied in plants and microorganisms. Nanosilver treatment led to changes in microbial community composition, biomass, extracellular enzyme activity and affected some ground plant species. All nanostructures may not cause the same toxicity risks. Biopolymers, liposomes and natural organic compounds based nanocapsules such as lipids and chitin are potentially less hazardous than heavy metals nanoparticles. Current important limitation in the application of nanocarriers to agriculture is the production scale and cost. Large scale manufacturing of nanomaterials and its effective application to agriculture will bring down the cost to a drastic range. There are many gaps and unresolved problems and new challenges concerning the biological effects of nanoparticles.

Both positive and negative effects of nanoparticles on higher plants have been reported (Monica and Cremonini, 2009). With the rapid growth in nanotechnology, there is an urgent need for right regulation over the indiscriminate use of nanomaterials and their effective disposal. With the increasing nanomaterial production and their subsequent release into the environment, a concern has been increased by their effect on ecosystem health. Many commercial nanomaterials pose greater toxicity risks than the same materials in their larger particle form if a substance has been approved in bulk form; it remains legal to sell it in nanoform. The potential application of nanomaterials in different agricultural applications needs further research investigation with respect to synthesis, toxicology and its effective application at field level. Nanotechnology has got applicability in various fields, but research and development are still at bench-top scale. Great efforts are required in the commercialization of nanomaterials for agricultural applications, which requires proper protection needs, testing priorities, risk assessment and regulatory guidance at the global level.

10 Innovations in Teaching, Research and Extension in Plant Pathology

10.1 TEACHING INNOVATIONS IN PLANT PATHOLOGY

The teaching environment for plant pathology is changing in both positive and negative ways. Teaching expectations are increasing, and resources are decreasing, however, recent educational research and instructional technology offers new approaches to meet these challenges. Plant pathologists are teaching courses that may attract new students to the discipline or at least improve agricultural awareness. The internet offers rapid access to information and images for both students and instructors. Instructional technology provides new tools for classroom presentations, communication with students, reaching new audiences, and distance learning, but using these new tools to enhance learning requires skilled and creative instructors. In the past, many plant pathology instructors worked in relative isolation but new communication and technologies and publishing opportunities for teaching scholarship should improve the sharing of instructional resources and methods. Because computer technology is evolving at an incredible pace, the instructional technologies discussed in 1978 seem primitive by today's standards but the problems associated with them - cost, time for development, technical support and the recognition that technology is not a substitute for teaching, remain the same. After 25 years, it is certainly appropriate to revisit this subject and examine some recent innovations and trends. Plant pathologists teach in many forms and venues, from graduate student advising to extension/outreach programs to traditional classrooms. Many creative plant pathology instructors have not published their innovations, either for lack of time and/or opportunity.

Innovation refers to 'any idea, practice or object that is perceived as new by an individual or another unit of adoption'. Innovative teaching is a proactive approach to integrating new teaching strategies and methods into a classroom. Technology plays a key role in innovative teaching. Innovative teachers use new technology to enhance or expand upon the student experience. The transition from the traditional blackboard and overhead projector instruction into computer-aided presentations were innovative. The teacher serves as a guide or consultant as students participate. A primary motive for innovative teaching is encouraging students to engage more in the learning process.

Teaching expectations in phytopathology are increasing, and recent educational research and instructional technology offer new approaches to meet these challenges. New communication technologies and publishing opportunities have improved the sharing of instructional resources and methods. Any success of teaching reflects on the relationship between students and teachers (Schumann, 2003). These interactions are governed by traits of both student and teacher. The various roles played by the Plant Pathological Societies contribute to pathology teaching.

There is no one right way to teach, just as there is no one right way to learn. Highly motivated students will learn under nearly any circumstances. Students who may not be academically oriented but require an understanding of plant diseases to make their living from plants are also looking for a useful education. An alert teacher continually assesses the level of learning in a classroom by eliciting questions from students or by observing a look of confusion.

The challenge lies in the delivery of quality education to a good student who do not yet appreciate the new discipline. Teaching programs are committed to provide best-unbiased educational information on disease management and responsive diagnosis to facilitate high implementation of management strategies. The goal is achieved by effective interdepartmental and interdisciplinary programming across the functions of research, extension and instruction. Sufficient assistantships, fellowships training/refreshing courses and more pathology positions would probably promote greater interest.

This is an exciting time for education in plant pathology. Students are being taught with different commodity interests and preparing courses for new audiences where students may not even be science majors. We also have a far wider choice of tools that may enhance learning. Our students now have rapid access to images, animations, simulations, and vast amounts of information. Plant pathology instructors have new ways to communicate with each other, to share instructional materials, and, perhaps, to gain some recognition for their creative scholarly contributions in teaching. Any discussion of teaching usually ends with successful teachers reflecting on the relationship between students and instructors and how these interactions are governed by traits of both the student and the instructor.

No computer program can replace these qualities in an instructor. They are the people who understand the human value of teaching and learning. Anthropologists have speculated that human evolution was strongly influenced by speech because this means of communication could be used to quickly spread new information and ideas. There is a very strong urge to share new ideas and insights that we find exciting with those around us. Instructors with that same urge to share the excitement will inspire the next generation of plant pathologists and, of equal importance, people who appreciate plant pathologists.

There is a need for a better understanding of advances in learning theory, how to incorporate new instructional technologies to improve learning and the academic rewards for teaching relative to research and extension responsibilities. Because computer technology is evolving at an incredible pace, the instructional technologies discussed earlier seem primitive by today's standards, but the problems associated with them cost time for development, technical support, and the recognition that technology is not a substitute for teaching remain the same.

Plant pathologists teach in many forms and venues, from graduate student advising to extension/outreach programs to traditional classrooms. It does seem that many factors that influence the teaching of plant pathology are changing unusually rapidly. There are both positive and negative influences. Plant pathology remains a graduate major primarily; other programs offer undergraduate majors with titles that include the words 'plant protection,' 'crop protection' or 'plant health management.'

10.1.1 Challenges

10.1.1.1 Student recruitment

It is essential to cope with the fact that currently students are not very much interested in agriculture; thus, it is very difficult to attract to our field the best and brightest students. There is much to do in this respect, trying to make our discipline more attractive to young people. Brightest students should help in attracting new students, providing them with real-world examples. Topics such as forensics, food safety, food security, saving forests and wild plants, environment protection are very relevant and suitable for attracting young people to this profession.

10.1.1.2 Student education

There is a strong disconnection between what University teaches and what employers need. To remain relevant to the agro-food industry, teaching programs need to put more emphasis on management, interpersonal and technical skills. In this respect, the industry could help Universities offering internships for students. Field and lab work could help attracting students. It will also be important to invest in summer outreach programs for students. On-farm experience for students should be reintroduced. Students should be encouraged to attend local professional meetings to interact with potential employers. Moreover, more incentives are needed for Ph.D. students and post-docs in applied areas. An increased movement of international students (North-South and South-South) should help increasing enrollments agricultural studies are still relevant in developing countries.

10.1.1.3 Student support

There is a need for funding for supporting experiential training as well as of grant programs in applied sciences. Again, internships with industry and governmental agencies could help. Moreover, stronger ties with industry should influence and help colleges to maintain basic agricultural disciplines.

10.1.1.4 Public relations and communication

There is a need for a new brand image: the public should understand that farmers are not the enemies of the earth, but, on the contrary, that life on earth depends on plants.

There is a need to collect and share data related to the percent of graduates employed, salaries, placement, etc. Such data should reach people who have influence.

10.1.1.5 Teaching

New teaching methods and tools should be adopted more extensively. The Internet permits rapid access to information and images for both students and instructors. The internet is also very useful for a life-long program used for extension services. Instructional technology provides new tools for classroom presentations, communications with students, reaching new audiences, and distance learning (Schuman, 2003). To improve teaching, an exchange of faculty and industry and extension service professionals would be very useful. Introducing new skills and abilities, such as communication, critical thinking, writing, international experience are very useful. Students need to be adaptable to future changes. It is important to avoid clonal propagation of ourselves.

10.1.2 Current Teaching Environment

Most authors who accept the challenge of writing about an impossibly broad subject usually describe the situation as a time of great change or transition. Of course, life is never static, but it does seem that many factors that influence the teaching of plant pathology are changing unusually rapidly. There are both positive and negative influences. It is impossible to predict how some aspects of the teaching environment will evolve, but general directions are becoming evident. There are no simple solutions for some of the most pressing constraints and limitations, but one can feel near euphoria about the progress in other areas.

10.1.2.1 Existing programs

Plant pathology remains primarily a graduate major, with some institutions offering a B.S. degree, although other programs offer undergraduate majors with titles that include the words 'plant protection,' 'crop protection,' or 'plant health management.' In the past ten years, many independent plant pathology programs have been merged to create multidisciplinary departments, or the faculties were dispersed to several departments. In several institutions, additional departments that were already multidisciplinary were merged into new units with titles that no longer include the words plant pathology.

10.1.2.2 Distance learning

Faculty members reported conflicting responses to the questions of whether their institution encouraged them to participate in distance education (most did) and if they did participate (most did not). They cited the potential benefits of reaching new audiences with commodity-specific instruction in a time-independent manner. Commonly listed difficulties in implementing distance learning included time constraints, poor technical support, and lack of new resources. Problems with an assessment of the quality of learning, the loss of personal interaction, and difficulties regarding a lab component were also noted (Schumann, 2003). Plant pathology is about problems in plants caused by physiological damage or irritation at a cellular level. It is concerned with problems that affect parts of the plant cell, leading to malfunctions in the normal processes which occur. There are potential benefits of reaching new audiences with commodity-specific instruction in a time-independent manner. Commonly listed difficulties in implementing distance learning included time constraints, poor technical support, and lack of new resources. Problems with an assessment of the quality of learning, the loss of personal interaction, and difficulties regarding a lab component were also noted.

10.1.2.3 Instructional technology

Many faculty members are now using some modes of instructional technology including PowerPoint presentations, web pages, simulations or other decision-making software and APSnet Education Center materials. Only a few programs offer online plant pathology courses at this time. Major constraints to using instructional technology include technical support, classroom equipment, and time.

10.1.2.4 Teaching expectations

Three-quarters of the programs surveyed reported that teaching expectations for their departments had increased in the absence of increased resources, fulfilling these expectations detracts from research and extension programs. Plant pathology faculty are participating in teaching biology and microbiology courses for majors, biology courses for general education requirements and 'science and society'-type courses.

Industry, policy makers and scientists all recognize that plant pathology is essential in addressing the issue of food security and in mitigating the effects of climate change. One must retain the ability to teach and train plant pathologists able to tackle these important new challenges. This audit, carried out on behalf of the British Society for Plant Pathology in the UK, surveyed the current levels of plant pathology teaching and training in the UK (Anonymous, 2012). Overall, there is less plant pathology teaching in UK Higher Education Institutes (HEI) than there used to be. Less than half of the 103 HEI offering biology, agriculture, horticulture or forestry courses at B.Sc., level provide plant pathology teaching, and this can be as little as 1-2 lectures. Only half of these HEI offer practical classes in plant pathology. The modular nature of HEI courses means that students can, and do, opt out of plant-based modules. Students should have the opportunity to encounter plant pathology and other plant sciences in the first year, making them more likely to choose more advanced plant pathology modules in subsequent years. A wide range of teaching and training links exist between HEI and both UK research institutes and industry. These bring practical plant pathology knowledge directly into the lecture theater. Employers in this survey and elsewhere have highlighted the need for graduates to see beyond the gene and understand plant pathology at the whole organism level.

The age profile of HEI plant pathologists is of major concern, and the strategic provision of plant pathology teaching needs to be addressed. Across HEI in general, academics are retiring and not being replaced. Plant pathology is no longer taught at five HEI due to staff retirement, and staffing levels have been reduced at a further six HEI. This implies that plant pathology is not considered a priority area across a range of HEI and it is not known how many will retain the capacity to teach plant pathology in 5–10 years time. New departmental appointments and RAE/REF assessments are driven in part by the Impact Factor (IF) of scientific publications. The highly specialized nature of much plant pathology research means that many publications are of low IF. Hence, in terms of this particular measure of esteem, plant pathologists may struggle to compete with scientists from other disciplines. There are many ways in which outreach and public engagement work can promote plant pathology to a wider audience. The British Society for Plant Pathology has appointed an Outreach Officer.

Today's students are tomorrow's teachers. A reduction in HEI teaching of plant pathology and other highly specialized subjects crucial to the UK well being and the economy will result in fewer teachers able to teach these subjects in future. A vicious circle is developing which must be challenged. A clear and constructive case must be made to policy makers of the importance of plant pathology and the need to teach and train the next generation of plant pathologists.

10.1.3 Future of Plant Pathology as Discipline

Although graduate students in plant pathology work on dissertation subjects that span the breadth of biology, one cannot be an expert in all areas. It asked if there are core competencies that one should expect from a Ph.D. candidate in plant pathology, nearly everyone might agree that there were, but defining them was much more difficult (Schumann, 2003). Many respondents in surveys noted that their programs were currently trying to determine the answer to this very question. There was near unanimity that plant pathology should continue as a discipline, but widespread concern about the future of the discipline. Within this climate of tight fiscal resources, a changing student population, and a rapidly changing science that is becoming ever broader while individual scientists are becoming increasingly specialized, instructors are trying to provide an education for students. Some significant innovations have developed since 1978 even though the problems to be solved have not really changed. Instructors are still seeking ways to improve learning and provide information in more interesting and useful ways. In this review, several important areas of innovation are discussed: changing teaching roles for plant pathologists, including teaching methods and teaching to new audiences; instructional technology for traditional classrooms, distance learning, and outreach to new audiences; and instructor communication and scholarship.

Plant Pathologists will be an increasingly international and interdisciplinary group of scholars, practitioners and educators dedicated to the study of plant-microbe interactions, plant disease management and information delivery on a global and collaborative scale (Anonymous, 2005). Through new technologies, more efficient communication tools and broad educational backgrounds, plant pathology practitioners will be more effective in devising and disseminating practical solutions for management of plant disease.

A challenge overarching that of capturing student interest in plant pathology is the growing challenge of finding students to attract in the first place (James MacDonald, 2009). In reality, plant pathology can only hope to attract a subset of those college students who generally are interested in the plant, molecular, and microbial sciences, and, therefore, are seeking relevant work experiences and classes. However, in the past two decades, there has been a decline in the numbers of students enrolled in plant science related majors in the United States, which has had the concomitant effect of reducing the total number of undergraduate students who might be exposed to plant pathology through work or classroom experiences. Decreased student interest in the plant sciences may grow as an area of concern for graduate programs; however, the greatest concern currently, as expressed by the program heads, is financial. In response to a question regarding the constraints that need to be overcome for their programs to have greater success in student recruitment, over 80% of the program heads indicated "Greater capability of student support." A heavy reliance on grant funds to support students and when this fact is coupled with the statements of graduate program heads that extramural funds are increasingly difficult to obtain, it explains why the heads feel constrained with regard to student support.

The reliance on grant funds to support students and the competitive nature of those funds may be a factor in the perception of employers that many graduates lack sufficient breadth of training. One obvious hypothesis to explain this is that faculty mentors and supported students feel pressured to accomplish their specific research project objectives to improve their chances of continuation funding since it often takes more than a single grant cycle for students to complete advanced degrees. This focused effort may be a factor contributing to the high level of confidence among employers that they will be able to find future employees with an in-depth knowledge of a particular subject, and their lesser confidence about finding future employees with a broad knowledge of many subjects.

Human-driven changes in the earth's atmospheric composition are likely to alter plant diseases of the future. Researchers are studying the impact of elevated carbon dioxide, elevated ozone and higher atmospheric temperatures on plant diseases that could challenge crops in these changing conditions (Anonymous, 2010). Elevated carbon dioxide levels are more likely to have a direct effect on plant diseases through changes to the plant hosts rather than the plant pathogens. Plants growing in a high carbon dioxide environment tend to grow faster and larger, and they have relatively denser canopies. These dense plant canopies favor the development of some diseases because the low light levels and reduced air circulation allow higher relative humidity levels to develop, and this promotes the growth and sporulation of many plant pathogens. At the same time, plants grown in high carbon dioxide environments also close their stomata, pores in the leaves that allow the plant to take in carbon dioxide and release oxygen, more often. Because plant pathogens often enter the plant through the stomata, the more frequent closing of the stomata may help prevent some pathogens from getting into the plant. In elevated ozone, plant growth is inhibited and results in shorter plants with less dense canopies. This can slow the growth and reproduction of certain pathogens. However, ozone also damages plant tissues that can help pathogens infect the plant more easily. Elevated levels of carbon dioxide and ozone can make a plant more susceptible to some diseases, but less susceptible to others.

Plant diseases can drastically abate the crop yields as the degree of disease outbreak is getting severe around the world. Therefore, plant disease management has always been one of the main objectives of any crop improvement program. Plant disease resistance (R) genes have the ability to detect a pathogen attack and facilitate a counterattack against the pathogen. Numerous plant R-genes have been used with varying degree of success in crop improvement programs in the past, and many of them are being continuously exploited (Mayank Anand Gururani *et al.*, 2012). With the onset of recent genomic, bioinformatics and molecular biology techniques, it is quite possible to tame the R-genes for efficiently controlling the plant diseases caused by pathogens.

With the advent of high-throughput techniques and efficient genomic approaches, researchers have managed to produce a large amount of experimental data in the form of ESTs, whole genome sequences, gene expression data, etc. Still, the progress in understanding the functional mechanism of resistance genes has been moderate. For instance, little is known about the structural basis of pathogen recognition. Furthermore, there is still an inadequacy of a reference set of sequences to be used as a model for resistance genes that usually cluster in genomic regions with a high number of homologs and pseudogenes. The difficulties in performing the plant-pathogen interaction studies pose another obstacle. Nevertheless, efficacious applications are being continuously developed based on our rather a finite knowledge base. For example, recently PRGdb, a web accessible open source database providing a comprehensive overview of resistance genes has been developed, which definitely is going to help filling some gaps in the models of the plant defense signal transduction network.

Efficient application of functional genomics tools for disease resistance could not only help us better understand the plant defense signaling; but it could also reveal novel insights into the interactions between these signaling pathways and other plant processes (Ramonell and Somerville, 2002). Even though, the progress of the overall plant defense mechanism studies is going on at a considerable pace; it would still be imprudent to expect a great breakthrough in impervious broad-spectrum resistance. However, it is judicious to anticipate an array of highly useful tools aided by other control measures providing adequate protection in certain contexts.

10.1.4 Changing Teaching Roles for Plant Pathologists

Within this climate of tight fiscal resources, a changing student population and a rapidly changing science that is becoming ever broader while individual scientists are becoming increasingly specialized, teachers are trying to provide an education for students. Some significant innovations have developed of late even though the problems to be solved have not really changed. Instructors are still seeking ways to improve learning and provide information in more interesting and useful ways. Several important areas of innovation include changing teaching roles for plant pathologists, including teaching methods and teaching to new audiences; instructional technology for traditional classrooms, distance learning, and outreach to new audiences; and instructor communication and scholarship.

10.1.4.1 *Teaching methods and innovative multimedia*

Pedagogy itself is the subject of extensive research. Some plant pathology instructors are turning to advances in scholarship on how people learn to modify how and what they teach (Lopez and Whittington, 2001). Colleges and universities also are expanding the definition of skills required of students in individual courses. Skills in critical thinking and oral and written communication go beyond simply learning about a discipline. The use of multimedia in industries has been extensive, as it has been effective in increasing productivity and retention rates – where research has shown that people remember 20% of what they see, 40% of what they see and hear, but about 75% of what they see, hear and do simultaneously (Lindstrom, 1994). With multimedia, the communication of the information can be done in a more effective manner, and it can be an effective instructional medium for delivering information. A multisensory experience can be created for the audience, which, in turn, elicits positive attitudes toward the application. Multimedia has also been shown to elicit the highest rate of information retention and result in shorter learning time (Ng and Komiya, 2000).

The availability of interactive multimedia authoring software programs promises to revolutionize the teaching. These programs integrate text, hypertext, graphics, animations, video, and sound. The user interacts with the information on demand in a nonlinear fashion. Beginning students can limit themselves to the general outlines of the subject, and advanced students can explore the information to the limits of their ability. Use of interactive multimedia does not eliminate the need for effective, enthusiastic teachers but provides a mechanism for the efficient transfer of information. An interactive multimedia presentation that supplements lectures in an introductory course is presented as an example of the application of this technology in teaching.

10.1.4.2 *Active learning*

It is important to point out that there is some debate about whether active learning strategies really result in better learning when compared to traditional teaching techniques where instruction is fully explained by the instructor (Emily G. Cantonwine, 2014). That said, some generalities have emerged from the educational literature regarding when and where active learning may best be employed. To summarize, active learning appears to be more appropriate for higher order skill development (i.e., critical thinking, analysis, and application) and the reinforcement of lower-order content (i.e., facts and concepts) than it is for the introduction and transfer of lower-order content. This is, in part, because students require enough background knowledge and skills to support engagement with the activity. In fact, minimally guided activities, such as discovery and experiential learning models, akin to teaching someone to swim by throwing them in a pool, have repeatedly been shown to be less efficient and effective at teaching content and concepts than instruction that is fully explained by the instructor. However, instructor guided approaches, such as cooperative learning and guided inquiry, which integrate active learning with traditional instruction, have been shown to be quite effective. This appears to be especially true for the purpose of reinforcing recently learned content and skills and building higher order thinking skills. The inquiry-based strategy for teaching disease cycles, published in Plant Health Instructor, is a good example of a guided active learning approach designed for the classroom. Prepared slides are excellent educational tools. They ensure that students have access to important observations, provide opportunities to build microscopy and observational skills, and require little time to prepare. However, traditional lab activities where students are given a set of slides and asked to draw and label predefined structures are often completed without meaningful engagement.

10.1.4.3 Managing the active learning environment

Transitioning the lab environment from a traditional setting to an active learning setting is not difficult as long as the instructor has a plan of action and is consistent with the plan and the students know what is expected of them. A strategy to facilitate active learning in the lab and for developing new active learning activities is presented under 10.1.4.3.1 (Emily G. Cantonwine, 2014).

10.1.4.3.1 Choosing an active learning strategy

Of the many guided active learning strategies that are used in the classroom, some, such as problem-based learning and cooperative learning, are very structured with specific guidelines for the instructor and students to follow. Others, often referred to as active learning, guided inquiry, or student-centered learning, are more open-ended. The strategy presented here is a modified version of a cooperative learning model that was used at Florida International University and includes the following components:

1. The activities are designed to be knowledge or skill challenges (i.e., puzzles).
2. The activities are presented as tasks rather than instructions.
3. Students are encouraged to work together to complete the tasks or solve problems.
4. Students are expected to make sure that everyone they are working with understands the material.
5. The instructor's role is that of a facilitator of learning rather than a teacher.

These five guidelines help to support student engagement. First, because the lab activities are designed to be challenging, but fun, they create an atmosphere of engagement that, in my opinion, inspires students to do their best. The small modification of presenting the activities as tasks instead of a list of instructions requires students to think about how they are going to complete each activity, which is an obvious first step of student engagement. Encouraging students to talk to each other about their observations, and expecting them to help each other to understand the material, leads to students using terminology in conversation. This gives the stronger students an opportunity to take their understanding to a higher level as they help their peers with the material. Finally, because the instructor's focus is to facilitate learning, students must find the answers to their content questions instead of relying on the instructor to give them the answers. Instructors are looking for more forms of active learning to engage students' interest and to encourage class participation (Grey and Mathre, 2001) through discussions and the application of classroom materials to real problems and situations. A lab on fungal infection processes is enhanced by a discussion about the stages of infection relative to application timing and efficacy of a particular fungicide. An extension diagnostic problem that resulted in expert legal testimony by a plant pathologist can be turned constructively into a case study. Active learning and applications to real-world problems not only give students satisfaction in their ability to use what they have learned but may also increase the likelihood of their retaining the information.

10.1.4.3.2 Group learning

Inquiry-based learning, group learning, and collaborative learning are strategies frequently discussed by instructors seeking to engage students more fully (Schumann, 2003).The underlying philosophy is that students will learn more when they are encouraged to resolve problems for themselves and explain their findings to each other. Such activities must be carefully planned, however, inquiry-based learning fosters in students the excitement of discovery, but they also need enough background knowledge to make progress towards a solution feasible. Theoretically, students should be able to educate themselves with available university resources, with the instructor effectively guiding their learning. For some students, group learning has some advantages by teaching them to work collaboratively through lab exercises or other projects by discussing what is to be done and why. However, some of the more able students may feel restricted when forced into a group whose members do not share an equal commitment to study.

A clear message of recent research on learning theory is that people learn in different ways, a fact that instructors should try to accommodate. In certain labs, certain experiments are done in groups, although individual students are responsible for their own lab reports. Students are encouraged to work in a group, but are given the option to work alone – there are always some who choose that option. In the first session, care is taken to help students who appear hesitant to join a group. Once established, the groups tend to maintain themselves voluntarily throughout the semester, and the students interact more in other lab activities because of the group work. Group learning and inquiry-based learning function well only when students are not in competition with each other for grades. This requires that students be graded on a straight scale rather than a curve, a practice that also encourages them to participate in study groups.

10.1.4.3.3 Games

Some instructors have adapted television game shows to review sessions in plant pathology, and students actually look forward to them. Without taking a lot of class time, students enjoy the competition for small prizes, and instructors find it easy to develop review materials for most aspects of plant pathology.

10.1.4.3.4 Educational research

Because the science of learning is an academic discipline and because educational research is outside the purview of most plant pathologists, evaluation of the success of different teaching methods is often difficult. Instructors choose new methods from research on science teaching in higher education and communications with colleagues with demonstrated success in the implementation of new methods. Most instructors introduce new methods slowly, so courses evolve over time rather than through a radical change in a single semester or term. However, once some instructors feel that a new approach is beneficial, they want to make it available to the entire class as soon as possible. Most instructors have neither the inclination nor the resources to conduct formal educational research with their limited student numbers, as their responsibilities in research and/ or extension already constrain the time they can commit to teaching. Ideally, when more formal educational research is needed, plant pathologists should be able to avail themselves of services provided by learning specialists within the institution.

10.1.4.3.5 Graduate education

There is currently considerable debate about how best to prepare graduate students for their future roles in academic positions (In addition to research training, more attention is now given to pedagogical preparation (Schumann, 2003). In the past, an apprentice-type system of working as a teaching assistant with a faculty instructor was the informal practice. More institutions are looking for a more formal approach.

10.1.4.3.6 Course content

Most instructors adapt their courses continually to the changing needs of their students. The vast majority of students in introductory plant pathology courses will take one, and only one, course in plant pathology and that course is likely to be commodity oriented. Therefore, it is important to equip students with the concepts and vocabulary necessary to access our science in any future independent inquiry. Because so many students in our introductory courses are now interested in turf and ornamentals, many of the 'classic' diseases are of little interest despite their historical and economic significance. These same courses also contain students who will be the next generation of plant pathologists. Many of them are not aware that this field will attract them until they take a course in plant pathology. This is where the role of the individual instructor becomes critical in inspiring them to continue their studies. For these students, instructors must provide the traditional principles and concepts of plant pathology at the same time as introducing them to the incredible advances in molecular biology.

In every program, there are debates ranging from the content of the introductory course to what graduate students should know if they are granted an advanced degree in plant pathology. The discipline of plant pathology demands of undergraduates completing a general course and graduate students in plant pathology alike an understanding of the fundamental concepts of diagnosis, the basic biology of the major pathogen groups, host-parasite interactions, epidemiology and principles of disease management. The importance of each area must be recognized, as well as having an appreciation for what we know and what we do not know for each area. Shouldn't all students who complete the introductory course be exposed to all aspects of the discipline so that they are well-prepared to continue their studies once their interest is sparked (Schumann, 2003)? It is important to equip students with the concepts and vocabulary necessary to access our science in any future independent inquiry. The courses contain students who will be the next generation of plant pathologists. Many of them are not aware that this field will attract them until they take a course in plant pathology. This is where the role of the individual instructor becomes critical in inspiring them to continue their studies. For these students, instructors must provide the traditional principles and concepts of plant pathology at the same time as introducing them to the incredible advances in molecular biology. Future plant pathologists will be better equipped to prioritize our limited resources if graduate education requires a set of core courses and attendance at a wide variety of seminars such that they can appreciate the value of all aspects of plant pathology.

10.1.4.3.7 New audiences

Although institutional demands for increased teaching by plant pathologists have affected research and extension programs, they have also provided new opportunities for the science of plant pathology and plant biology in general.

10.1.5 Biology and Microbiology Courses

Participation in biology and microbiology courses for science majors can enrich the education of these students by introducing completely new and fascinating aspects of plant-parasite interactions. Plant pathology offers exciting ways to increase the study of plant biology in biology classes perhaps even fostering an interest that leads to graduate studies in the discipline.

10.1.5.1 Courses for nonscience students

Although various instructors have offered courses for nonscience students over the years, the new teaching demands have greatly increased the number of institutions offering this type of general interest course related to plant pathology. Some such courses fulfill defined graduation requirements, mostly in the area of biological science and others fulfill writing or other nontraditional requirements. Courses in plant pathology need little justification for their contribution to the education of students. With only a small percentage (2–3%) of people participating directly in agriculture today, the need for some understanding of food and fiber production by students is rarely mentioned in higher education. Thus, courses in agricultural awareness can help students understand the role of agriculture in the nation's economy and, at the same time, learn about the interesting interactions of organisms that we study in plant pathology. The practical relevance, historical importance, and ties to culture provided in the study of biology using plant pathology are ideal topics for inclusion, in general, education requirements for students.

Many scientists question why separate courses need to be offered for students in nonscience majors. The goals and approaches of such courses are very different. Science majors are beginning a lifetime of detailed learning about their subject, whereas nonscience students need more general concepts. Nonscience students often lack confidence in their ability to do well in a science course, and many have minimal interest, limitations that must be overcome before any learning can occur. There is often a negative view of scientists as people with limited cultural interests and without a moral vision to their work. Stories of plant pathology and plant pathologists and their ties to culture and history help break down these stereotypes. Interest developed in these courses can sometimes encourage students to become science majors and, possibly, consider plant pathology as a career. At the very least, students will learn enough about science and its ethical dimensions to become better-educated citizens in an increasingly technical world. Many controversial scientific issues require input from both scientists and nonscientists who bring a variety of viewpoints to these complex problems. Most significant is introducing students to the issues of agriculture.

Urban and suburban students require some biological background to participate as educated citizens in decisions related to their own health, the appropriate use of genetic resources, land use, rational judgment of genetic engineering, and the risks and benefits of agricultural chemicals. Most college graduates are not science majors, but they will be voters, legislators, judges, and taxpayers. As stated in (Schumann and D'Arcy, 1999), "Never in history has the knowledge of science been more important in society. Never in history have most people been so removed from the production of their food." We are still struggling with the appropriate response to recent terrorist attacks and the potential risks to human and plant health in our country. A better-educated electorate will be able more appropriately to assess the risk involved in 'homeland security' if they have enough background in biology to make reasonable judgments. When people feel ignorant about a subject, they tend to take conservative stances out of fear of the unknown and the inability to judge risk.

10.1.5.2 Lifelong learning

One can no longer claim that the education obtained in a typical undergraduate or even graduate program is sufficient in our rapidly changing world. As career changes over a longer working life become routine, education, both formal and informal, becomes a lifelong preoccupation. Plant pathologists have traditionally provided educational services through extension education programs for growers, county agents, master gardeners, and others. With diminishing resources and the demand for easier access to information, the Internet is becoming a resource for these same audiences.

10.1.5.3 Images

In the past, probably every plant pathology instructor has felt limited by the supply of images of diseases and pathogens, especially for the introductory courses. Now extensive collections of digitized images are available for instructional purposes. This rich supply of images can be used in classroom presentations or online study guides to enhance learning in significant ways. As stated above, students learn in various ways, so the more ways that we can offer them to learn, the more likely it is that an individual student's needs will be met. Study guides of computerized materials help prepare students for what they will see and do in laboratories.

An image resource also is helpful for review of materials from past classes and use when a class is missed. An image collection furnishes information about diseases according to their commodity interests and about disease problems from distant regions and foreign climates. Most students study plant pathology during the winter months when living materials are in short supply in many regions. Electron micrographs can enhance the understanding of host-parasite interactions when combined with standard microscopic views. An exciting new set of short videos combines photography of diseases in the field and microscopic images with realistic animations of sexual reproduction, spore dispersal, and the early stages of infection.

10.1.5.4 Image collections

Access to image collections is rapidly improving. A collection of nearly 10,000 images with an accompanying database for image identification is available on a video disk published by APS Press (Tainter *et al.*, 1995). Introductions to the major pathogen groups and some lessons on example diseases are available at no charge in English, French, and Spanish, together with image numbers. Advances in video disk technology allow a large collection of images to be viewed on a single disc, with repeated uses of images for various applications with minimal equipment (video disk player and television). A student study station can be created with the video disk and a TV, but multiple users at various sites require digitized images in computerized programs. APS Press now offers CD-ROM collections of commodity-based images and a collection of images from previously published slide sets on the major fungal pathogen groups.

10.1.5.5 Image uses

Images can now easily accompany glossaries or hypertext links to new vocabulary in published materials (D'Arcy *et al.*, 2001). Many important terms in plant pathology are essentially meaningless to students without images. Keys used by professionals to identify diseases and pathogens can be modified to include diagrams and photographs along with explanations of new vocabulary to enhance students' understanding of relationships. Without these enhancements, many keys are unusable except by experienced pathologists, who are already familiar with the organisms and their features. Various software packages using digitized images are available for lecturing of which the most popular program probably is PowerPoint. This program makes the preparation of relatively sophisticated presentations fairly simple. However, many classrooms, although equipped with slide projectors, still do not have computer (liquid crystal display/LCD) projectors; instructors are understandably reluctant to carry a computer and LCD projector to each class. Even in equipped classrooms, until recently, the instructor still needed to bring a computer, which might not be compatible with the projector. Computer projectors are now much smaller, less expensive, and easier to transport. In addition, some 'data' computer projectors have a PowerPoint slide viewer feature. Used with a USB CompactFlash card reader/writer that hooks up to the computer's USB port, PowerPoint files can be downloaded to a CompactFlash card, which can then be inserted into the projector without the presence of the computer.

10.1.6 Computer-based Teaching Tools

With diminishing resources and the demand for easier access to information, the Internet is becoming a resource for these same audiences. Electron micrographs can enhance the understanding of host-parasite interactions when combined with standard microscopic views. An exciting new set of short videos combines photography of diseases in the field and microscopic images with realistic animations of sexual reproduction, spore dispersal, and the early stages of infection. CD-ROM collections of commodity-based images and a collection of images from previously published slide sets on the major fungal pathogen groups. Images can now easily accompany glossaries or hypertext links to new vocabulary in published materials. Many important terms in plant pathology are essentially meaningless to students without images.

Various software packages using digitized images are available for lecturing, of which the most popular program probably is PowerPoint. This program makes the preparation of relatively sophisticated presentations fairly simple. However, many classrooms, although equipped with slide projectors, still do not have computer (liquid crystal display/LCD) projectors; instructors are understandably reluctant to carry a computer and LCD projector to each class.

Simulations and other programs can help students make decisions using what they are learning about plant pathology. Diagnosis TM is an excellent program to help students use their powers of observation and simulated laboratory results to diagnose disease problems. Some sample scenarios have been prepared as well as a program 'shell' with which instructors can prepare their own scenarios. Several computer programs including AppleScab, LateBlight, Resistan, and TurfBlight help students apply what they have learned when attempting to manage the various diseases and avoid fungicide resistance. Textbooks and computer resources can be combined in creative ways to enhance teaching in many areas of plant pathology. For example, *Exercises in Plant Disease Epidemiology* includes computer discs that contain public domain software, example data sets, and annotated computer code for statistical analysis.

The Internet allows rapid access to unlimited current information related to plant pathology in presentations that are image-rich, colorful, and animated, all more attractive than standard textbooks. Students can access this information from a computer at any location, a factor that may be more conducive to research on topics of interest than traditional modes of research. Digital search mechanisms are, of course, of unparalleled significance in providing materials of interest to the searcher. Course web pages are now commonly used for routine communications between instructors and students, offering access to materials such as the syllabus, practice quizzes and old exams, and misplaced class handouts. A course website also can give students links to enrichment materials at other sites preselected by the instructor. Instructors can post PowerPoint presentations from classes or post all of their lecture notes relatively quickly. Other instructors are creating elaborate presentations to supplement their class time, but this generally requires more time and resources than are typically available.

Specific courses in Plant Protection are in place in most Agricultural Universities around the world. They are generally given at the MS level, after the concepts of plant protection have been introduced to students in Plant Pathology and Agricultural Entomology courses offered at the BS level. Plant protection, with very few exceptions, is thought worldwide in integrated courses, which combine the expertise of plant pathologists, entomologists and, less frequently, weed scientists. Such fragmentation represents one of the problems encountered, mostly at the student level, because a split view of the different topics, that generally need a holistic approach, is often given. Plant Protection is also the topic of MS courses and Ph.D. programs. The changing role and importance of courses in plant protection in agricultural curricula, the need for such course also in food technology and agricultural biotechnology curricula are critically discussed. The importance of a good balance between theory and practical activities as well as the need for life-long learning in such field is discussed, together with the possible innovations in teaching methodologies (Gullino, 2009). Teaching plant pathology and plant disease management today is confronting with both negative and positive challenges. Despite an urgent need to feed an increasing population, there is a general decreasing role of agriculture, which is strictly connected with an inadequate understanding of agriculture and science amongst the general public. Most public think that food comes from the grocery store, and there is a lack of literacy of the public and, even worse, of many policy-makers. Urban and suburban populations do not have the scientific background needed to participate, as citizens, in decisions related to their own health, land use, appropriate use of genetic resources, rational judgment of genetic engineering, risks linked to the use of pesticides. Today, in our field, even in the most competitive systems, it is difficult to recruit and to retain the best teachers, who are attracted by other fields of research.

Plant pathology is a discipline lost visibility, also due to the disappearance of many plant pathology departments. The discipline is now included in larger multidisciplinary departments, often trying to find its own space and role. Such invisibility impacts profession in many ways, students are not attracted, and public funds are not directed to plant pathology (Gadoury *et al.*, 2009). Due to shrinking students enrollments, we are bound to a loss in courses, degrees and faculty members. Actually, when a group like plant pathology becomes a minority in a larger department, it is difficult to significantly influence votes on questions related to policy, curricula development, or faculty positions.

Simulations and other programs can help students make decisions using what they are learning about plant pathology. DiagnosisTM is an excellent program to help students use their powers of observation and simulated laboratory results to diagnose disease problems. Some sample scenarios have been prepared as well as a program 'shell' with which instructors can prepare their own scenarios. Development of additional scenarios that could be shared would contribute greatly to the value of this program for students, but few instructors have time available. Recently retired plant pathologists might be excellent authors of new diagnostic modules for plant pathology students, based on their years of experience.

Plant diagnosis is both an art and science, and it requires considerable experience to become skilled at the task (Stewart, 2004). How then, can students be best educated in this process? In the absence of the 'the real thing,' one-way experience can be gained to give students the opportunity to explore and analyze diagnostic 'cases' or 'scenarios.' The use of case studies as learning tools in plant pathology has been reported by other authors. In educational terms, this facilitates 'goal-based' or 'problem-based' learning, an effective learning paradigm that has gained wide acceptance among educators. The ability to diagnose an abnormal plant condition is an important skill for the practicing plant pathologist. Those working in the area of plant pathology extension are often called upon to undertake this task in the course of their daily activities. Even lab-based plant pathologists should have an appreciation of the diagnostic process as they may also engage in it at some stage in their careers.

The use and development of DIAGNOSIS training system for Crop Problems (DIAGNOSIS), a computer-based authoring tool and scenario presenter which facilitates training in plant disease diagnosis was developed. The DIAGNOSIS software presents students with a plant diagnostic problem to solve; it captures their diagnosis, justification and solution; and it provides feedback. The exercises are designed to be firmly embedded in a series of lectures and laboratory classes on the diagnostic process, within a plant pathology course. In this way, the software is designed to complement, not replace, hands-on laboratory work.

Contrary to what the name might suggest, 'DIAGNOSIS is not an expert system, but the reverse of an expert system. It assumes students are 'experts' and provides them with simulated observations of a plant problem. The students need to come up with the diagnoses themselves. Therefore the students, not the program, carry out the interpretation and gathering of information. The software simply provides a virtual reality environment in which they can work. It uses the metaphor of the computer 'adventure game,' where players inspect objects and clues and draw inferences, which then enable them to progress to a conclusion.

The core of a DIAGNOSIS exercise is the instructor-constructed scenario. The scenario presents students with an unknown plant health problem, perhaps wilting plants in a potato field, poorly growing trees in an apple orchard or yellow leaves on wheat. In a typical scenario, the students are 'dropped into' a field or environment containing this problem, usually accompanied by a grower who can supply information on management and the history of the crop. Then, by way of navigational links and icons, the student gathers observations about the plant and environment, checks equipment and asks questions of the grower. Students can collect objects such as plant parts, weeds, and soil and take them to a laboratory for testing. Here they can make further observations, look up further information (such as weather data), try to extract the causal organisms and/or conduct a number of tests.

Once students feel they have enough information to make a diagnosis, they enter this into the computer along with a justification and recommendation. Students can then be told the answer and be given immediate feedback or, more commonly, the student input is later extracted by their instructor, printed out and graded. In the case of the latter, feedback is appended to their input text. Feedback takes the form of the diagnosis, a correct interpretation of the significant observations, the common investigative pathway an experienced diagnostician would have taken and a model recommendation to the grower. Some feedback is customized depending on whether or not students carried out particular tasks. Many of the tasks in DIAGNOSIS, such as laboratory tests, have a price tag associated with them. If instructors wish it, students can be given a budget before the exercise and asked not to exceed it. This gives some incentive for the students to limit themselves to tasks and tests appropriate to the likely problem rather than trying all the menu options. It mirrors the real world, where such procedures do indeed cost real money. Scenarios, then, are the 'data' used by the DIAGNOSIS program. The software provides a 'Builder' tool for instructors so that scenarios can be constructed or altered as desired. A scenario 'Player' program then presents the problems to students.

Teaching the diagnostic process is an important but often overlooked aspect of the plant pathology curriculum. For this component, DIAGNOSIS has aided the move from an information-based style of teaching to a problem-based one. However, like all teaching technologies, it is how it is used that matters. The value of DIAGNOSIS as an educational experience is not the Builder and the Player software, useful though these are. The real value of DIAGNOSIS is in the quality of the scenarios and how the exercises are used within a course. The software is a useful tool, but it will always require good teachers and good scenarios to realize its full potential (Schumann, 2003).

10.1.6.1 Computer games

Several computer games, including AppleScab, LateBlight, Resistan, and TurfBlight (Schumann, 1998), help students apply what they have learned when attempting to manage the various diseases and avoid fungicide resistance. Dr. Arneson has recently been involved in modifying three of these for web-based formats. Although the currently available version of AppleScab is an old DOS-based program, it is being used in introductory courses. Student reports clearly demonstrate that they were challenged by keeping track of weather, fungicide applications, inoculum levels, and tree phonological development. Even the most 'organically' oriented students felt a sense of panic when the scab graph continued to rise uncontrollably. Many students expressed sympathy for growers who have to battle this disease.

10.1.6.2 Decision-making tools

Any situation has a value that allows students to apply what they are learning during a course to solve a problem. Entertaining computer-based games can be enjoyable, but there are not many available. Even nongame sites can be interesting and useful, such as the peanut/tomato spotted wilt virus website. In the best applications, students can try many different approaches to solving a problem in a relatively short period of time and at no financial risk.

10.1.6.3 Computer-based instructional materials

Textbooks and computer resources can be combined in creative ways to enhance teaching in many areas of plant pathology. CD-ROM technology can be used to modify existing resources to make them more accessible to a variety of users. For example, the Compendium of Turfgrass Diseases was reformatted into a CD-ROM version to make the information more useful to students and more accessible to working professionals who are not plant pathologists. The core of information on each specific disease was modified so that images and text were placed together, whereas images were on separate plate pages in the printed version of the Compendium. Disease information was organized using the same categories each time so that users could easily find what they needed. The more applied information was separated from the more technical details to reduce confusion by less educated users. Disease cycles were highlighted with pop-up images. Pop-up squares highlight various sections of the disease cycles to guide readers when colored words are clicked on in the accompanying text. Each disease section has links to diseases with similar symptoms. Disease management recommendations were expanded and, in some cases, separated into different lists for lawns and golf courses.

Access to this core of information is possible from a number of entry points, viz., the index, an illustrated key to diseases, the image browser, and the host-specific disease calendars. This helps users find the appropriate disease section depending on information available to them. For all users, the information was 'layered' to first provide a general overview followed by increasingly detailed information. There are many existing plant pathology resources that could be made much more accessible through modification into a CD-ROM format. Extension personnel can bring their expertise 'e' to students and other nonspecialists with the use of CD-ROM technology and extension web pages, and many are already doing so.

10.1.6.4 Course web pages

Course web pages are now commonly used for routine communications between instructors and students, offering access to materials such as the syllabus, practice quizzes and old exams, and misplaced class handouts. A course website also can give students links to enrichment materials at other sites preselected by the instructor. Instructors can post PowerPoint presentations from classes or post all of their lecture notes relatively quickly. Other instructors are creating elaborate presentations to supplement their class time, but this generally requires more time and resources than are typically available (Partridge and Osborne, 1999). The APSnet Education Center was created to prevent redundancies in online study materials for students. A goal of the free site is to provide a central resource where many standard materials, such as disease lessons and an illustrated glossary, can be posted. Experts with high-quality images can contribute to their area of specialty, and students at many institutions can benefit. Instructors of individual courses can then link to these standard materials from their personal websites. The technical support provided to authors by the staff at APS allows instructors to focus on content rather than on the technical basis for instructional technology. Improved technical support will allow all instructors to enhance their teaching with these new tools. As noted above, most instructors are already balancing teaching obligations against research, which is the primary source of monetary reward and career advancement. Many instructors have very creative ideas about how to better present materials to students but need the technical backup that is sadly lacking at many institutions to bring these ideas to fruition.

10.1.7 Distance Learning

Distance learning, which has been available for more than a century through correspondence courses and the more recent television and videotape courses, continues to be offered in various formats. Advances in instructional technology have caused many universities to encourage their faculty to participate in distance learning. Online courses that can be offered asynchronously to anyone with Internet access are becoming popular (Curtis, 2002), with instructor-student and student-students interactions through email and bulletin boards. Students and instructors can even communicate in real-time from multiple locations. Traditional on-campus plant pathology courses with a standard laboratory component allow the instructor to work with students on a weekly basis in a hands-on, informal environment. Laboratory work is an essential complement of formal lectures and is probably most effective when required weekly over a period of time. Offering plant pathology courses with a laboratory component through distance is problematic, although creative solutions have been devised. At the University of Georgia, students come to campus for some all-day or half-day sessions to fulfill the laboratory component. The many forms of instructional technology are still at such early stages that there are few standards and little evaluation of the quality of learning that is derived from it (Meachen, 2002). Educational research aimed at such evaluations is under way, and some plant pathologists have published critiques on the use of instructional technology in their courses (Anonymous, 2001).

10.1.7.1 WebCT program

Agriculturists, such as consultants, managers, and extension agents, often find that they need advanced degrees to keep pace with the rapid changes in their area of specialization. However, they generally cannot attend regular campus classes. The Master of Plant Protection and Pest Management (MPPPM), a nonthesis degree program at the University of Georgia (UGA), is a logical choice for many who are seeking comprehensive, multidisciplinary training in crop production and crop protection (Elizabeth L. Litte, 2001). Therefore, the MPPPM program was targeted for conversion to a distance education format. Two Plant Pathology courses were the first to be converted to a distance education format, viz., Introductory Plant Pathology and Diagnosis and Management of Plant Diseases. Diagnosis and management are the core curricula courses for the MPPPM degree. Introductory Plant Pathology is a prerequisite for the Diagnosis and Management course. In addition, both courses would be attractive to students in the private sector and other colleges in the Southeast who desire to learn the basic principles and applications of plant pathology. Introductory Plant Pathology was converted to a distance education format and taught in the spring of 2000. Diagnosis and Management followed in the Fall of 2000. Students consisted mainly of county extension agents and research personnel at the experiment stations.

The distance education version of these courses was the same in credit and content as the courses taught on campus. The WebCT program, a limited-access instructional resource application, was the primary method to deliver the content for both courses. WebCT features included path editor for delivering course material in modules, a bulletin board for discussions, private mail for student/instructor communication, a glossary with links to the path editor, image gallery, on-line quiz, chat and area for posting grades. An interactive video network (GSAMS), with connections to each county in the state, was used in the Introductory Plant Pathology course to supplement the course material and to provide face-to-face interaction with the students. Hands-on laboratory sessions were an essential part of both courses. Up to five mandatory half-day lab sessions were conducted during the semester. In addition, some exercises were performed independently by the students and kits were provided to students for pathogen isolations and transfers. The distance diagnostics system, administered by the Cooperative Extension Service allowed students to send images of symptoms and signs to the instructor for consultation or the bulletin board for sharing with the other students. Proctored exams were either administered during lab periods or at another arranged time.

10.1.7.2 Bioimaging informatics

Innovative technologies have enhanced the quality of data recording as a digital photograph of pathogen-infested plant parts using mobile/handy digital camera, video photomicrography and movie including time-lapse cinematography. Images are transmitted for diagnosis. Imaging bioinformatics coupled with computational biology is a novel means of acquisition, image processing, data visualization, maintenance and dissemination of biological information. The advancement in Digital Imaging Technology (DIT), huge stored microscopic image can be reprographed into bioimage database as the foundation of bioimaging informatics for imaging bioinformatics of a pathogen for research and teaching. Digital and electronic image analysis and digital photography are being used for detection of pathogens prior to symptom expression, including chemical and sero- and electrodiagnosis.

The following observations have been made based on several course evaluations and student comments.

(i) Students were able to master the technology needed to participate in the on-line course material, although students who had limited computer skills had more difficulties initially. The greatest obstacle to the on-line delivery of information was the low capacity of the connections for transmission of information. Most students were still using analog modems to connect to the internet. Large or numerous images, as well as multiple links, do not work well under these circumstances.

(ii) Students varied in the ease with which they adapted to the distance education style of learning. Distance education involves more independent learning and students must have enough self-discipline to find the time in their schedules for the course. In general, older students adapted more easily and actually preferred the flexibility. Many of the students had been exposed to agronomic problems as part of their jobs, and these students were especially enthusiastic about learning the practical aspects of plant pathology and were eager to share their field experiences in bulletin board discussions. A few students never adapted well to distance education and were more comfortable with traditional classroom lectures. Active learning through mandatory discussion sessions, on-line quizzes, and a problem-solving approach was important in keeping the students engaged and learning. While these techniques are useful in all learning situations, for on-line learning they are essential.

(iii) The hands-on laboratory sessions were the greatest challenge in teaching distance plant pathology courses. For the Introductory Plant Pathology course, students found that there was too much material condensed into each lab session. The Diagnosis and Management of Plant Diseases course was particularly challenging due to the extensive hands-on time students need to learn the plant disease diagnosing techniques. University extension personnel were particularly helpful in working with the students individually.

10.1.7.3 Future directions

The initial years of teaching distance education classes are as much of a learning experience for the instructor as for the students. The goal is to incorporate the following changes in future offerings of these courses:

(i) The course content, as well as images of diseases and pathogen structures, would be more accessible and useful to the students if incorporated into a CD that the students could use anytime without connecting to the internet.

(ii) Mandatory weekly discussion sessions, question sets incorporating a problem-solving approach, and quizzes would still be presented over the WebCT but would become a more frequent and dominant part of the course.

(iii) Independent laboratory modules for the Introductory Plant Pathology course would be developed although group lab sessions would not be entirely omitted. Satellite labs could be established with the cooperation of other institutions around the state as locations where students could use the equipment needed to perform the laboratory modules.

(iv) An extended and intensive field trip at the beginning of the semester would be incorporated into the Diagnosis and Management course. Students would spend several days traveling through the state collecting disease specimens, observing cropping systems, and learning the techniques for diagnosis. The students would then have the skills necessary to compile their independent disease collections as part of the course requirements.

(v) Devoted telephone lines and video-conferencing between scientists with national and international experts specialized in taxonomic groups and other disciplines including diagnostic and advisory services need to be developed.

(vi) Adoption of biosystematics research and imaging in teaching is essential. Pathogen species information is now available in e-form in the websites on-line, including biological data, geographical data, 3-D images, video clips and molecular sequences sans conventional morphological and morphometrical data. Having enough expertise and skill, access to published keys of specific groups, taxonomists may make the best use of web resources and software tools for identification for faster work. The rapid and easy access to newly published documents as well as unpublished documents from superspeciality sources in phyloinformatics and bioinformatics has opened up an opportunity for the systematists to deal with several pathogens. Further, bioinformatics assists systematics to understand phylogeny of plant pathogens by sequencing informative regions of ribosomal DNA (rDNA) and even comparing entire genome sequence to understand major phylogenetic events. However, more primary data are required about genomic sequences about major pathogens. Morphological and molecular archival of pathogen information are useful for easy and accurate diagnosis of pathogens.

10.1.8 Virtual Plant Pathology Lab CD-ROM

The development of diagnostics skills in students of plant pathology requires significant practice and presentation of a wide range of cases to ensure that sufficient exposure is gained (Stewart and Galea, 2006). This allows students to begin to see and understand the patterns of symptom expression and the variations in disease development which may exist among and between cases. More often than not, students fail to appreciate that plant disease diagnosis is a process which follows a series of logical paths determined by critical decision points that are supported by a range of diagnostic observations or tests. Because of the limited possibilities for classroom contact, distance mode students are greatly challenged in their ability to grasp such concepts. The need to keep accurate records of each case, to apply a systematic approach to the investigation and to follow through with the diagnostics task to reach a conclusion are also essential. Additional emphasis on this learning experience can be gained by using realistic case studies, providing instruction on basic diagnostic skills, illustrating how to access to supporting information resources and showing examples of the development of client recommendations.

The development and use of a software package were described, which support the learning and understanding of the plant disease diagnostic process (Galea, 2006). This approach to plant disease diagnosis centers on the use of a logic flowchart and is supported by a range of diagnostic resources and procedures which assist the student to more fully appreciate what is required to manage a real diagnostic case for a client. This approach, although initially developed for use by distance mode students, has proven also to effectively support learning for on-campus students. This tool was designed to prepare distance mode students for the experience of dealing with real plant disease problems in a laboratory class at a residential school. Collaborative work with educational designers and software developers and designers resulted in a highly functional and professional package.

10.1.8.1 Case form

A blank case form, which can be printed for student use from the CD-ROM, provides an organizational structure to the diagnostic procedure, as well as a means for recording observations and the recommendations developed for the client. In the first problem case (silverbeet) provided in this software package, the case form is progressively completed 'on screen' to illustrate its correct application to the procedure. In subsequent cases, the form is not prominent as the scenario progresses; however, a completed case form can be printed upon case conclusion.

Diagnosis of each problem was achieved by systematically working through the procedure outlined in the logic flowchart during which unlikely possibilities are removed from the possible causes list. By this process of systematic elimination, the student begins to appreciate the way in which the range of investigative tools used; the understanding of symptom descriptions and microscopic features associated with each pathogen, and the information inherent in the field case data all come together to resolve the identity of the causal pathogen. Completion of each case was achieved by the development of a management solution for the identified problem. Each solution was developed with integrated disease management in mind, incorporating cultural, physical, chemical and biological options.

10.1.8.2 Web study pages

The World Wide Web can be an effective pedagogical tool to improve student learning. Carefully designed Web Study Pages are easily delivered to students via the World Wide Web. Designing a useful Web Study Page requires careful planning. Students are pragmatic users and exercise a 'print and go' attitude while instructors may focus on innovative and artful delivery anticipating the student will process information at the computer (Partridge and. Osborne, 2001). Web Study Pages are easily updated to meet changing class needs. Web Study Pages afford teachers the unique opportunity to be intrusive into the student's preclass preparation process during the time when they are acquiring factual information. Students who used the Web Study Pages demonstrated better preparation for class and evidenced by improved factual understanding and ability to synthesize information to solve problems. Over the past three years, an extensive set of documents for publication to the World Wide Web (Web) has been developed, modified, and revised (Partridge 1997). These Web Study Pages are designed to support a junior-level university course titled Introductory Plant Pathology. The experience we gained in developing this system may be beneficial to others considering the use of the Web to support their courses.

This mode of teaching requires students to participate in an active intellectual discussion after having assimilated a factual information base through their own per-class preparation, was a new experience for many. Accordingly study guides (handouts) served to direct students in their preparations for class became an integral part of the course materials. These study guides reflected a 'constructivist' view of learning that students must individually discover and process information to integrate it into their own mind. The guides worked well but had some drawbacks, primarily their production and the repetitive, noninteractive nature of multiple handouts. We initially used the World Wide Web to replace handouts. The Web was an unqualified success in that regard. The Web Study Pages have expanded beyond the original handout replacement, but this discussion will be confined to the Web Study Pages.

The World Wide Web allows increased interaction between student and faculty. The active use of faculty-drafted Web Study Pages enables faculty to become an active part of the students' preclass study process by providing intrusive comments and asking questions. These questions are used to stimulate questioning, reasoning, and discussion outside of the classroom period, in preparation for in-class discussion. Students respond positively to this type of guided study and come to class better prepared to engage in discussion. Because textbooks are dated soon after publication and consequently as new research results are published it becomes necessary to update information. The Web Study Page is a ready means to do all of these. Because the Web Study Pages can be easily edited and modified, there was an unexpected response the first time the Web Study Pages were used. Students became exceedingly uneasy when a Web Study Page was edited after it was loaded up to the Web server. Therefore, we adopted a policy of not loading up pages more than three to five days in advance of student needs and not changing them during the semester.

A student accesses the Plant Pathology Web Study Pages through Internet connections at home or at a University-provided computer. On entering the site, a student is presented with the course Home Page. From this starting point, a student can view study materials, read study test questions, and review the course syllabus with hyperlinks to the Web Study Pages for each class period. The student also has the option of contacting the instructor via e-mail from any page. To be successful, an instructor must carefully and specifically, explain to the students the intended use of Web-based materials.

10.1.8.3 Instructor communication and scholarship

Plant pathology instructors often work in relative isolation. Teaching committees of professional societies offer sessions and discussions at annual meetings, but otherwise there is often little communication among instructors. Depending on the university, faculty members who are interested in teaching meet each other, but these meetings need not be necessarily relevant to plant pathology. The Internet again provides new means of access for instructors who wish to interact throughout the year. Teaching committee websites are obvious potential support. Attendance at professional meetings may be limited by financial exigencies or teaching obligations. Instructors from countries with limited resources find it especially difficult to attend international meetings. The past two International Congresses of Plant Pathology held workshops on information and instructional technology. Participants demonstrated their programs but were limited in their interactions with the other participants. Some programs required working time substantially beyond that of the workshop to be fully comprehensible.

10.1.9 Teaching Art and Science of Plant Disease Diagnosis

The ability to diagnose an abnormal plant condition is an important skill for the practicing plant pathologist. Those working in the area of plant pathology extension are often called upon to undertake this task in the course of their daily activities. Even lab-based plant pathologists should have an appreciation of the diagnostic process as they may also engage in it at some stage in their careers (Stewart, 2004). Plant diagnosis is both an art and science, and it requires considerable experience to become skilled at the task. How then, can students be best educated in this process? In the absence of the 'the real thing,' one-way experience can be gained is to give students the opportunity to explore and analyze diagnostic 'cases' or 'scenarios.' The use of case studies as learning tools in plant pathology has been reported by other authors. In educational terms, this facilitates 'goal-based' or 'problem-based' learning, an effective learning paradigm that has gained wide acceptance among educators. Teaching the diagnostic process is an important but often overlooked aspect of the plant pathology curriculum. For this component, DIAGNOSIS has aided the move from an information-based style of teaching to a problem-based one. However, like all teaching technologies, it is how it is used that matters. The value of DIAGNOSIS as an educational experience is not the Builder and the Player software, useful though these are. The real value of DIAGNOSIS is in the quality of the scenarios and how the exercises are used within a course. The software is a useful tool, but it will always require good teachers and good scenarios to realize its full potential.

Teaching should be closely connected to the research activities. Students should be introduced to the most modern technologies utilized for management of plant diseases while providing a firm grounding in fundamental principles of plant pathology. Various courses need to emphasize the importance of ecological processes in achieving sustainable disease management. Web-based courses are to be included in advanced stages of development.

Critical issues affecting the quality of plant pathology teaching and learning in recent times have been a reduction in student contact hours in many academic programs (from 35 to 20 hours per week), the move towards generalist degree programs relying on honours or postgraduate programs for development of specialist disciplinary knowledge, the gradual decline in number of plant pathologists teaching in universities, the challenge to engage the current generation of university students to become excited about plant pathology, and the teaching vs. research conflict.

10.1.10 Virtual Plant Pathology Lab (VPPL) – A Problem-based Learning Primer

The virtual plant pathology lab is designed to introduce logic concepts, investigative skills and case management approaches for plant disease diagnostics. This is created to improve student engagement and learning outcomes in plant pathology.

10.1.10.1 Key features

It uses real case studies in an authentic learning environment, teaches the principles of plant disease case management, uses a logical framework (scaffolding) to reinforce appropriate process, reinforces notion of critical decision-making points, uses high quality images and an engaging interface, demonstrates appropriate clinical laboratory skills and management outcomes, and uses on-line teaching to develop critical understanding of the diagnostic process. Industry, policy makers and scientists all recognize that plant pathology is essential in addressing the issue of food security and in mitigating the effects of climate change. Collaborative learning as a mechanism to reinforce understanding of diagnostic procedures and client relationships is important. The use of the Live Classroom to hold forums which allow online verbal discussion with the aid of PowerPoint presentations, eBoard and a CD-ROM to assist with solving the case studies, is an impressive collection of resources and knowledge that provide students with efficient and effective learning outcomes (Galea, 2006). This enables students to have their questions answered and have immediate gratification. This has been the most interesting style of learning at university. The interaction through the typed messages in Live Classroom is useful in putting forward an idea or just if speaking aloud is not comfortable. In the UK, the ability to teach and train plant pathologists able to tackle these important new challenges was retained. This audit, carried out on behalf of the British Society for Plant Pathology, surveyed the current levels of plant pathology teaching and training in the UK. Overall, there is less plant pathology teaching in UK Higher Education Institutes (HEI) than there used to be. Less than half of the 103 HEI offering biology, agriculture, horticulture or forestry courses at B.Sc., level provide plant pathology teaching, and this can be as little as 1-2 lectures. Only half of these HEI offer practical classes in plant pathology. The modular nature of HEI courses means that students can, and, do opt out plant-based modules.

Students should have the opportunity to encounter plant pathology and other plant sciences in the first year, making them more likely to choose more advanced plant pathology modules in subsequent years. A wide range of teaching and training links exist between HEI and both UK research institutes and industry. These bring practical plant pathology knowledge directly into the lecture theater. Employers in this survey and elsewhere have highlighted the need for graduates to see beyond the gene and understand plant pathology at the whole organism level.

REF assessments are driven in part by the Impact Factor (IF) of scientific publications. The highly specialized nature of much plant pathology research means that many publications are of low IF. So in terms of this particular measure of esteem, plant pathologists may struggle to compete with scientists from other disciplines. There are many ways in which outreach and public engagement work can promote plant pathology to a wider audience. The British Society for Plant Pathology has appointed an Outreach Officer. Today's students are tomorrow's teachers. A reduction in HEI teaching of plant pathology and other highly specialized subjects crucial to the UK well-being and the economy will result in fewer teachers able to teach these subjects in future. A vicious circle is developing which must be challenged. A clear and constructive case must be made to policy makers of the importance of plant pathology and the need to teach and train the next generation of plant pathologists.

Based on a report on the findings of through a series of surveys of key groups, oral reports of the survey results have been presented earlier at APS meetings and in a 'webinar' format on APSnet the education committee as determined. Several internet surveys have been carried out. One targeted graduate student and postdoctoral members of APS to learn what attracted them to plant pathology, to ascertain their career aspirations, and to determine how well prepared they feel for their careers of choice. A second survey targeted the heads of US-based plant pathology graduate programs to learn what the different.

10.1.11 Implementing Case Studies in Plant Pathology

Students in science, including plant pathology, need real-world problem-solving skills. A case teaching method is a pedagogical approach that engages students and develops higher-order thinking skills (Aman Yadav and Janna L. Beckerman, 2009). The case teaching method allows students to become actively involved in their own learning as opposed to merely memorizing facts. Case studies have been used in undergraduate science education as a way to develop students' critical thinking and problem-solving skills. However, little empirical evidence exists on whether this teaching method is having the desired impact on students in plant pathology courses. One educational innovation that has the potential to improve student satisfaction and outcomes in science instruction is the case teaching method, which uses real-life case studies. Case teaching method has been used in medicine, law, and business for over a century to prepare students for real-world activity and allow them to diagnose, detect and act. Cases have the potential to represent the unpredictable world of practice, allow students to develop higher-order thinking skills, and integrate material from many fields (Herreid, 2007). The case teaching method can be used to have students think and act like scientists, as it is an "interactive process in which students become engaged with scientific phenomena and debate with both peers and instructors to develop a full understanding of related phenomena and underlying concepts." The goal of the case teaching method is not so much to teach the content of science but to capture the process of scientific investigation and the way that scientists think. Thus, case teaching method involves a hands-on approach to learning by actively involving students in authentic science inquiry to develop their analytical and decision-making skills.

Aman Yadav and Janna L. Beckerman (2009) evaluated the influence of the case teaching method on students' problem-solving and critical thinking skills in an undergraduate plant pathology course. Specifically, the course utilized both traditional lecture-based teaching as well as case study methodology. Students completed a problem-solving test that gave them four problem scenarios similar to two that were taught by lecture and two that were taught by case studies and examined students' ability to critically think and solve problems. The results of this study indicated that case studies help students develop critical thinking skills required to diagnose plant health problems while actively engaging them in the course content.

Students who learn via cases tend to outperform their peers who learn from a traditional lecture approach. In addition, students who learn from the case teaching method also report higher satisfaction and engagement with the course material. Although the case method is widely used in science courses, research is still lacking on the impact of this teaching technique on students' learning and problem-solving skills within undergraduate science teaching. Most of the support for the use of case studies in science has been based on faculty's personal observations rather than empirical research (Aman Yadav and Janna L. Beckerman, 2009).

Plant pathology is a complex and multidisciplinary field of study that requires an understanding of the biology of plant pathogens (including, but not limited to, mycology, virology, bacteriology, and nematology), in addition to agricultural and horticultural practices (soil science, weed science, etc.), in order to develop management strategies to improve plant health. Thus, the use of case studies in plant pathology lends itself well to "bridge the gap between classroom learning and real-world decision making." Tan and colleagues stated that the use of case teaching method would allow students to analyze complex real-life scenarios and develop solutions and management strategies to address issues presented in the case studies. However, there is little empirical evidence to support the use of case studies and their influence on students in undergraduate plant pathology courses. The purpose of this study examined the influence of case teaching method on students' learning as well as student perceptions of case teaching method on their critical thinking, learning, and engagement with the case study material.

The primary purpose of using case teaching method in science is to actively engage students in their own learning and improve their problem-solving skills (National Research Council, 1997). Results from this study indicate that the case teaching method helped attain this goal. Students in the plant pathology course learned more from the case teaching method and also had positive feelings toward this approach. Specifically, the problem-solving test showed that when students were taught using case studies, they accurately diagnosed the problem and offered multiple plant health management strategies. On the other hand, when taught using the traditional lecture approach, the students either offered only the diagnosis or the management strategies or none. The course is designed for students who are interested in horticulture and turfgrass management to be able to accurately diagnose and offer plant health management strategies. Therefore, the case teaching method was more effective in allowing students to develop problem-solving skills as demonstrated by their ability to transfer their learning from case studies discussed in class to problems presented in the problem-solving test. Students preferred the use of case studies compared with traditional lecture method. Overall, students felt that they were more engaged and active in the class when case studies were used because they added a lot of realism to the class. Students also felt that case studies allowed them to learn more, analyze the basic elements (i.e., plant health problems) of the course, and retain more from the class. The survey results suggested that students believed that the use of case studies increased their critical thinking skills, allowing them to form a deeper understanding of the course concepts, view issues from multiple perspectives, and also allowed for more discussion of course ideas in the class. These positive attitudes toward the use of case studies are consistent with the opinions of science faculty about the influence of case studies on students (Yadav *et al.*, 2007). The case teaching method utilized via the discussion format allowed students to go beyond merely memorizing facts to being actively involved in their own learning. This finding is consistent with previous research on case discussion, which has emphasized the role of social interactions and has found that such an approach to using cases is beneficial to students.

An ideal class size to implement the discussion format of case teaching method is 20 to 60 to allow for a diversity of opinions, which was true in the current research study. However, future research should examine whether learning from case studies leads students to apply their learning to solve real problems in their professional fields. Furthermore, research needs to examine how this teaching approach can be used in large lecture classes and the impact of using the case method in large lecture classes on student learning. To enhance student learning and create more engaging classroom atmospheres, interactive group case studies, reading groups, critical thinking exercises, etc., are just a few of the techniques. One particularly innovative approach involves the use of elaborate, science-based role-playing games, formally known as 'Reacting to the Past' (RTTP) pedagogy. RTTP was originally developed by Professor Mark Carnes at Barnard College with the intents of immersing students in pivotal historical debates, exposing them to classic works of literature and challenging them with interactive case studies that require a mastery of underlying concepts and ideas.

10.1.12 Basic Philosophy Behind Teaching Plant Pathology

Not dissimilar from the way in which clinical understanding is developed in medical or veterinary science students, plant pathology relies on students developing a range of competencies and a particular logical approach when presented with a plant disease problem to solve (Galea, 2008). The diagnostic process is the first (key) step in developing a full understanding of a disease problem and the consequent development of a management solution (Stewart and Galea, 2006). In the 1980s, it was generally believed that computers would automate teaching. It was even feared that somehow, technology may even replace the function of academics. Rather than replacing the role of academics, these technologies have become effective allies in the challenge to engage the current generation of university students. However, without appropriate pedagogy, communications technologies, like any tool will not improve teaching and learning outcomes for students. Students will not engage in course material unless good educational design provides the framework. Using relevant technology is only part of the equation (Anonymous, 2012). Providing relevant learning opportunities is also critical to maintaining student interest.

10.1.12.1 Industry based problems

This lets students lose in the lab working for real clients and to develop skills in diagnosis of plant diseases, develop logical thinking ability, reinforce knowledge of plant pathology, improve communication skills and to develop extension skills. The multimedia format encourages the development of new skills and competencies.

10.1.12.2 Teaching and learning in plant pathology for a new century — what has changed?

Students learn differently from the way we did, and we need to keep pace with their needs if we are to remain relevant to them (Galea, 2006). New communication technologies allow us to communicate more effectively and flexibly with our students bridging many gaps which were previously impassable. Universities now acknowledge and support the development of good teaching and learning practices. Research and further study in education are also valid pursuits for academics.

10.1.12.2.1 Education of plant pathologists

- Graduate programs will provide a spectrum of educational experiences to train the increasingly diverse plant pathologists of the future.
- Graduate education programs will provide abundant active learning opportunities that develop excellent problem-solving skills and a deep conceptual understanding of science.
- International educational opportunities will be essential to prepare students for the increasingly global nature of agriculture and scientific inquiry.
- As the role of plant pathology in building basic knowledge of biological systems increases, concepts of plant pathology will become essential components of the core courses of undergraduate science education.
- Excellence in teaching will be recognized and rewarded.
- Plant pathology will be visible and attractive to undergraduate students, resulting in a large pool of applicants to graduate programs.
- Valuable opportunities for life-long learning will be available to professional plant pathologists.

10.1.12.2.2 Outreach, coordination, and communication

- Plant pathology will be recognized by the public for its contributions to society, food and fiber production, and environmental quality.
- Plant pathologists will contribute to a greater public understanding of science.
- Scientific information will become more readily accessible to the scientific and lay communities.

- Plant pathologists will routinely collaborate internationally.
- Plant pathologists will be an increasingly diverse group with respect to ethnicity, gender, and underrepresented groups.
- The increased need and expectation for multidisciplinary approaches to problems will create stronger connections between plant pathology and allied disciplines.
- Unique benefits will result from coordinated data acquisition and storage systems and databases for meta-analysis internationally and across scientific disciplines.
- Intellectual property will be recognized but will not restrict cooperation or research progress.
- Regulatory policy and regulations will be sufficient and transparent, and compliance will be accomplished without impeding education, research, and disease management.

A sustainable mechanism for future stewardship of valuable scientific resources, especially microbial cultures, will promote long-term studies essential to independent replication and ensure the credibility of findings.

10.1.12.3 What hasn't changed?

Computers have not taken our jobs; they have just become an integral part of our lives. There is no substitute for good educational design and teaching practice with a human touch. Teaching plant pathology remains as rewarding and exciting as it ever was. The lineage of dedicated and passionate plant pathology academics continues across our universities. The generation gap between students and academics continues to exist.

10.1.13 Future Strategic Approaches

Plant pathology is connected with relevant social and economic issues, viz., environmental protection and conservation, food safety and security, climate change. It is continuously challenged to maintain its identity and reaffirm an existence based on the needs of those who grow food and fiber. As the ever-increasing world population demands more to consume: we must respond with improved methods of disease control that are less destructive to the environment. It is important that we keep and help our students keeping "one foot in the furrow" (Mathre, 2002). It will be important to get plant protection and social role of diseases into the early curriculum, also bringing plant pathology teaching into Sciences and Environment Colleges.

Reaching new audiences should be a priority. Bringing plant pathology into biology and microbiology courses can enrich the education of these students, by introducing new and fascinating topics, which are closely related to food quality and safety and environment protection. The undergraduate course "Plants, Pathogens and People" taught at the University of Illinois represents a very good example. This course is followed by students coming from six to eight different colleges or schools, with very different backgrounds in biology. It will be very useful to bring more industry perspective into teaching, as well as contributions from research institutes, government, industry, and botanical gardens. There is a need to integrate, in teaching plant pathology, the knowledge of molecular biology, genetics, biochemistry, physiology, epidemiology of the most important pathogens with applied aspects derived from agronomic questions. Investing in lifelong courses should be another priority, together with investments in critical infrastructures (farms, labs). Strong teams of excellence around crops or problems and more attention to those scientists, capable of working across disciplines should be developed.

Plant pathology remains a critical topic and teaching plant pathology, because of its complexity, remains a major challenge for both teachers and students. The student population is now very different from that of the past, with an increasing number of students with a little biology background. Sometimes we are teaching students with different commodity interest or completely new audiences. New technologies are now available for teaching. Students have now easy access to images, simulation and all type of information. Teachers also have now new possibilities of communicating with them, sharing teaching material and exchanging experiences. More links with Scientific Societies are needed. The International Society for Plant Pathology has a Subject Matter Committee on teaching, and the topic of teaching has been well covered during its International Congresses. The American Phytopathological Society has in place very successful programs, with its APSnet Education Center and established an *ad hoc* Committee dealing with the education of plant pathologists, which prepared a complete analysis of the present situation in the United States (MacDonald *et al.*, 2009).

A new generation of students adaptable to future changes, capable of critical thinking, flexible and able to communicate and work as a team, must be trained. And, finally, we should put in teaching the same passion and enthusiasm that we put in research (Gullino, 2009). Eventually, the world will probably (re) discover that there is a need for agriculture, and the general picture will become more positive. In the surveys conducted by The American Phytopathological Society, (James MacDonald, 2009) some respondents felt to do more to increase student interest in plant pathology. However, the student/postdoctoral survey indicated that two of the three most important factors that attract undergraduate students toward graduate studies in plant pathology are largely under the control of plant pathology faculty, giving students a work experience in plant pathology and exposing them to a class in plant pathology, and argue that the ability to interest students in plant pathology largely resides in academic departments. Thus, departments need to pay a great deal of attention to the quality of these experiences so that they are effective recruitment vehicles. If this could be coordinated on a national level, departments could significantly increase undergraduate exposure to plant pathology and likely increase the overall pool of students interested in plant pathology graduate studies.

A challenge overarching that of capturing student interest in plant pathology is the growing challenge of finding students to attract in the first place. In reality, plant pathology can only hope to attract a subset of those college students who generally are interested in the plant, molecular, and microbial sciences, and, therefore, are seeking relevant work experiences and classes. Decreased student interest in the plant sciences may grow as an area of concern for graduate programs; however, the greatest concern currently, as expressed by the program heads, is financial. In response to a question regarding the constraints that need to be overcome for their programs to have greater success in student recruitment, over 80% of the program heads indicated "Greater capability of student support." Our survey data showed heavy reliance on the grant of funds to support students, and when this fact is coupled with the statements of graduate program heads that extramural funds are increasingly difficult to obtain, it explains why the heads feel constrained with regard to student support.

The reliance on the grant of funds to support students and the competitive nature of those funds may be a factor in the perception of employers that many graduates lack sufficient breadth of training. One obvious hypothesis to explain this is that faculty mentors and supported students feel pressured to accomplish their specific research project objectives to improve their chances of continuation funding since it often takes more than a single grant cycle for students to complete advanced degrees. This focused effort may be a factor contributing to the high level of confidence among employers that they will be able to find future employees with an in-depth knowledge of a particular subject and their lesser confidence about finding future employees with a broad knowledge across many subjects (James MacDonald, 2009). It is not possible to carefully compare the coursework expectations of today with those of 30 years ago-even discounting the significant differences in specific course content as a result of advances in knowledge and technology over the past 30 years. It was puzzling, however, that nearly 80% of graduate program heads considered "Field experience with plant diseases" an important feature of graduate education, while ≤ 50% of the programs included an 'in-field' experience with crop diseases among the courses required or recommended for students.

Because 20% of the responding programs indicated they did not offer such a course, the implication is that approximately 30% of the programs that do offer such a course do not include it among coursework expectations for their students. It is interesting to note that among the skill sets that many employers feel will be in demand are skills (e.g., "Field experience with plant diseases," "Knowledge of disease control," and "Broad knowledge of plant pathology") that fewer graduate programs feel they will be prepared to offer in the future. This may explain the pessimism expressed by employers with regard to finding applicants with these attributes, and while it could be argued that the pessimism among academic employers represents a feedback loop caused by concern for these aspects of their own curricula, there is no self-reinforcement loop to explain the pessimism of government and private sector employers. Clearly, there has been some erosion of confidence in the ability of graduate programs to expose students to certain educational elements valued by employers.

This erosion of confidence also appears to be reflected in the responses of graduate program heads, many of whom felt that there will be fewer free-standing departments in the future, and some of whom expressed a view that plant pathology programs do not thrive in mixed departments. It also may explain the numbers of graduate programs that lacked confidence with regard to maintaining certain elements of their curriculum into the future (James MacDonald, 2009). While there may be an element of bias in some responses from the graduate program heads due to their roles as defenders and promoters of their programs, the gap between the projected needs of employers and what graduate programs feel confident they can provide in the future probably is real because it already is a growing concern in the minds of employers. This developing gap should be a concern for the US agricultural and natural resource interests because the ability of university, government, and private sector entities to successfully research and manage endemic or invasive plant diseases depends upon the continuing ability of agricultural colleges and universities across the United States to prepare students with a strong knowledge of plant pathology and ability to apply that knowledge to the challenge of disease management. However, the historic strength of this vital feeder system is at risk – such that plant pathology seemingly is standing at a very important tipping point in its history, and the challenge is very different from the growing 'Tower of Babel' noted 53 years ago by J.C. Walker (1963), which was a reference made to the emerging subdisciplines within plant pathology, each having unique terminologies and technologies, and which raised fears of disciplinary fragmentation.

Today's challenge also differs from the professional dichotomy noted 57 years ago by J.G. Horsfall (1959), who subsequently argued for two types of terminal degrees in plant pathology, one as a scientist (the Ph.D.) and one as a practitioner (analogous to a medical doctor). The challenge now approaching seems to be one of shrinking educational capacity in the field-related aspects of plant pathology. Thus, the concern is not about differing jargons or whether there should be two types of degrees, but rather it is about the continuing ability of plant pathology programs to prepare students for the mission of disease management, the *raison d'être* for the discipline of plant pathology. This concern becomes particularly acute when considered in light of the fact that plant pathology is rapidly approaching a period that will see many talented and experienced faculty members retire from universities over a relatively brief span of time (Gadoury *et al.*, 2009). In many cases, the people lost from universities will be the very people best able to educate and mentor the broadly trained plant pathologists who employers perceive to be in shrinking supply. This is an issue deserving serious attention. It seems important to keep in mind that the demographic and educational capacity challenges reported here are not unique to plant pathology -- they are shared by a number of other fields of research and education in the nation's system of Land Grant Colleges of Agriculture.

Although there clearly are challenges facing plant pathology and many other agriculturally related disciplines, it seems important to keep these issues in some perspective. One has only to read the many historical articles and perspectives related to plant pathology to realize that almost as soon as the American Phytopathological Society was founded in 1908, people speculated about the demise of plant pathology as a distinct discipline. Indeed, plant pathology has weathered earlier episodes of disciplinary erosion and inadequate research support (Kelman, 1985), and while the challenges we face (in terms of loss of faculty expertise, lack of undergraduate students interested in the plant sciences, etc.) seem daunting, we should remember the words of W.C. Snyder (1971) when he said in 1971, "Today the sciences, including plant pathology, and even the universities are in trouble. Troubled times are continuous and current troubles always seem the most severe." Actions that can be taken now at the level of individual departments include maximizing the opportunities for undergraduate students to have positive exposures to plant pathology through work experience and/ or introductory classes. The survey also indicates the importance of recruiting faculty members who can carry out competitive research programs while also enriching a department's ability to provide field experiences for students.

10.1.14 Conclusions

This is an exciting time for education in plant pathology. Many of us are teaching to a student population very different from that of our own instructors. We are teaching students with different commodity interests and preparing courses for new audiences where students may not even be science majors. We also have a far wider choice of tools that may enhance learning. Our students now have rapid access to images, animations, simulations, and vast amounts of information. Plant pathology instructors have new ways to communicate with each other, to share instructional materials, and, perhaps, to gain some recognition for their creative scholarly contributions in teaching. Any discussion of teaching usually ends with successful teacher's reflecting on the relationship between students and instructors and how these interactions are governed by traits of both the student and the instructor. There is no one right way to teach, just as there is no one right way to learn. Highly motivated students will learn under nearly any circumstances. The challenge lies in the delivery of quality education to good students who do not yet appreciate the new discipline. Students who may not be academically oriented but require an understanding of plant diseases to make their living from plants are also looking for a useful education. An alert teacher continually assesses the level of learning in a classroom by eliciting questions from students or by observing a look of confusion.

It is to be remembered that no computer program can replace these qualities in an instructor. Luckily, despite the resounding negative institutional messages, there will always be teachers who will stop to answer a question, offer advice, or simply make a student feel that someone has a personal interest in him or her. These teachers will be successful with or without computers, digitized images, and multimedia extravaganzas. They are the people who understand the human value of teaching and learning. Anthropologists have speculated that human evolution was strongly influenced by speech because this means of communication could be used to quickly spread new information and ideas. There is a very strong urge to share new ideas and insights that we find exciting with those around us. Instructors with that same urge to share the excitement will inspire the next generation of plant pathologists and equal importance, people who appreciate plant pathologists.

With the advancement of the computer, information, and remote sensing technologies, innovative approaches have been undertaken for documentation of pathogen information. Information and Communication Technology has greatly impacted phytopathology through information generation, supply, and dissemination. Interactive multimedia method and digital imaging of pathogens have made the teaching of phytopathology easy, attractive and convincing to the students within a short time.

10.1.15 Vision for Future

In light of the above, more emphasis on future needs to be concentrated on several aspects including the following practical orientation of the students on plant protection subjects to improve their competence with reference to plant diseases, facelift the existing laboratories with modern facilities for imparting better learning atmosphere; interactive teaching facilities for effective learning process, modernization of staff rooms to create ambiance for carrying out research by the PG students and their major research advisors; emphasis on organic farming and sustainable agriculture for the management of plant diseases, inter-institutional collaboration both at national and international levels in agricultural education; revision of course syllabi based on needs of the stakeholders, introduction of the degree program in plant protection; establishment of centers of excellence in plant pathology; the introduction of interactive teaching methodologies, and development of incubation centers across the campuses of universities and global approaches across the institutions for strengthening teaching and research by the Postgraduate students.

10.2 RESEARCH INNOVATIONS IN PLANT PATHOLOGY

Identification and understanding of plant pathogens and management of plant diseases they cause help to develop sustainable strategies reducing disease impact. Understanding variations in the genetic structure of plant and microbial populations, including biochemical, classical and molecular approaches and the study of pathogen evolution is important. Soil and rhizosphere microbiology includes the study of root diseases, the beneficial interactions between microbes and plant roots, and the impacts of farming on soil biology.

Research on the molecular mechanisms of plant-microbe interactions and applications to sustainable agriculture include the following the study of plant and microbial resistance and genetic modification to improve biological control agents and crop cultivars (Anonymous, 2006) ; improve crop quality, which reduces the need for harmful chemicals and helps ensure economic viability for grapevines, tree fruit and nut and nursery industries; introduce disease forecasting/risk assessment models for grapevine diseases, which allows growers to more accurately schedule sprays and reduce fungicide use, prevent the spread of sudden oak death in natural ecosystems and pitch canker in pines through the implementation of effective regulatory responses and management practices; develop disease-resistant cultivars of cereals and other crops, eliminate seed-borne pathogens in many vegetable crops ; develop alternative disease management strategies that prevent crop failures, improve disease detection and management in rice and other crops by training and certifying industry scouts and regulatory staff; raise awareness and understanding of biotechnology in the plant sciences with educators across the US and internationally; putting in place the necessary scientific infrastructure and networks to support surveillance, regulation, and innovation and ensuring that the scope of

Horizon 2020 takes account of the detailed recommendations for the fundamental and applied research agenda compiled by the scientific community to address emerging risks within the broader context of understanding plant health.

The recommended scope includes pest and disease diagnosis; biology, ecology and epidemiology of plant pests and pathogens and their relationships with hosts and vectors; plant pest and disease resistance; biological and cultural strategies for sustainable pest and disease management; and evaluation of how healthy plants live in association with microbes that provide direct or indirect benefit (Anonymous, 2006). Supporting this multidisciplinary research strategy is important by reducing fragmentation in research capacity and priority-setting across the Member States to sustain critical mass. For example, the ERA-NET initiative EUPHRESCO should be continued and extended apart from increasing use of research and surveillance data in modeling, prediction and extrapolation, including application in coupled crop disease–weather interaction models. This also includes attending to current and impending skill shortages in critical disciplines, including plant pathology and taxonomy, and creating better networking between disciplines and sectors, ensuring the research issues for plant health receive appropriate attention within current European Commission initiatives. For instance, the Joint Programming Initiative on Agriculture, Food Security and Climate Change and the European Innovation Partnership on Agricultural Productivity and Sustainability Innovation, which assign higher priority to better use of research advances in support of innovation and the translation of knowledge from research centers to practical applications in support of plant health (Anonymous, 2006). This involves developing new, durable control approaches to overcome current limitations of pesticides and responding to challenges introduced by EU pesticide product legislation that reduces the number of approved chemical control options, making most of the scientific opportunities for breeding improved plants and durably resistant to biotic stresses. Genetic improvement can be accomplished by more precise breeding techniques (for example, marker-assisted selection), by genetic modification to introduce desirable traits and by other, newer, crop genetic improvement technologies. To deliver these innovations, it is vital that the EU regulatory framework for approving crops developed using genetic improvement technologies is reformed to be proportionate, and to focus objectively on the scientific evidence for benefit–risk for the plant trait, implementing coordinated policy for regulation and innovation, encompassing broader thinking on healthy plants and considering options for building strategic linkages across plant–animal–human health ('One Health').

Scientific advance is leading to rapid developments in diagnostic technologies, surveillance and communication methodologies, and to increased understanding of the current and emerging threats to plant health, and of the means to counter those threats (Anonymous, 2006). The issues to be faced in protecting and promoting plant health are scientific, technological and regulatory, but they cannot be tackled successfully without also raising political and public awareness of the importance of the issues and the need to prepare for future challenges.

10.2.1 Major Themes and the Future Research Niches in Plant Pathology

10.2.1.1 Application for 'omics' technologies in crop protection

Detailed information on 'omics' technologies has been furnished in chapter 1. 'Omics' research programmes should be funded to develop new safe and effective crop protection products and practices, which will have an easier passage through the regulatory process, on the basis of greater knowledge about their mechanisms of action, viz., better understand plant-pathogen/pest interactions (e.g., matching genetic information and population biology), better understand the influence/impact of beneficial microbial interactions with plants, e.g., rhizosphere, phyllosphere and endophyte organisms, on plant health/sickness to enable commercial applications, develop more sensitive diagnostic methods (both DNA and metabolite) for use in the field and 'plant clinics,' better match agrochemistry and crop genetics (cf: personalized medicine), better understand and help counteract resistance mechanisms, develop reliable resistance elicitor-based crop protectants matched to crop genetics, better exploiting the plant's innate resistance mechanisms, and demonstrate safe mode of actions and benefits that might help new crop protection products (e.g., semiochemicals) through the subsequent regulatory process.

10.2.1.2 Research to build knowledge of the biology of plant-microbe interactions

Key discoveries in the study of plant-microbe interactions will continue to lead to fundamental biological principles with applications beyond the realm of plant pathology (Anonymous, 2005). New experimental tools and powerful computation capabilities will facilitate the incorporation of vast data sets into experimental models that accurately predict complex interactions among the pathogen, vector, host, and environment. Knowledge of the structure, biology, physiology, and biochemistry of plant and pathogen cells will be expanded to the point where perturbations on plant cells can be predicted. Through a combination of genomic, molecular, and ecological approaches, plant pathologists will develop a comprehensive view of how microorganisms interact in natural habitats, which will form the conceptual basis for natural processes of plant disease suppression. Genomics will become a cornerstone of plant pathology research. There will be a clear and detailed understanding of plant and pathogen coevolution. With enhanced opportunities to efficiently assess patterns of genetic variation in pathogen and plant populations, novel insights into the epidemiology of plant disease will result. Commonalities of innate defenses in animals and plants will become increasingly apparent, and knowledge of both plant and animal infectious diseases will benefit as a result. The molecular basis of pathogen virulence and quantitative and nonhost disease resistance in plants will be understood.

10.2.1.3 Research to develop knowledge-based disease management strategies

Accurate assessments of plant disease losses and models for prediction of epidemics will become common. A more comprehensive understanding of the interactions between plants, pathogens, vectors, and their environment will allow for the development of more sustainable and profitable disease management practices with minimal impact on the environment. Detection and diagnostic capacity for plant pathogens, with increased accuracy and speed, will approach that for human pathogens. Novel cultural practices, developed from enhanced knowledge of pathogen biology and disease epidemiology, will become increasingly important components of integrated plant health management programs. Disease resistant plants, derived through enhanced plant breeding and/or genetic engineering, will be fundamental to cost-effective and environmentally-sound plant disease management in the future. New chemicals will be discovered and developed that are effective in managing disease at low application rates, pose minimal environmental risk, and have a low potential for the development of pathogen resistance. Natural processes of disease suppression will be better understood and shown to underpin many cultural practices that effectively manage plant disease. Novel aquatic and terrestrial plant production systems will result in new disease problems.

10.2.1.4 The application and practice of plant pathology

The connections between plant pathologists in the private and public sectors will become a crucial component linking research to application in the plant health continuum. The numbers of applied plant pathologists in the private sector will grow as will the need for a mechanism to recognize their professionalism and qualifications. Advances in genomics, robotics, and nanotechnology will dramatically increase the capability of diagnostic laboratories. Worldwide communication among plant disease clinics will be commonplace as international trade and travel necessitate the sharing and leveraging of expertise to combat exotic pathogens. Sensitive, real-time microbial sensing technologies, coupled with low-cost wide-area surveillance technologies, will allow detection of pathogens before the onset of disease. Regulatory policies will be based on sound, peer-reviewed science. New approaches and partnerships will emerge between the public and private sectors to assure that stakeholder needs are addressed through research and that plant pathology is viewed as relevant to society.

10.2.1.5 Novel management agents

Managing diseases is not a trivial issue and never has been. Whether dealing with blight in the 17th, 18th and 19th centuries, dealing with diseases of vines for the last 3000 years, the issues remain the same. It is important to question ourselves on how to get consistent long term control of pests and diseases and is it possible to integrate solutions to avoid the spread and build-up of resistance to any treatments that innovations might bring to the market. Other than innovative farming practices there are three types of intervention.

(i) 'Traditionally,' farmers have been used to a succession of new chemicals to deal with their crop protection problems. More recently, a combination of new legislation, and increased costs associated with dealing with that legislation have eliminated many pesticides and slowed down the introduction of new ones to a trickle (Anonymous, 2013).

(ii) There has been much interest in the use of advanced breeding techniques, including agricultural biotechnology, to introduce insect and disease resistance into major crop plants. This has been very successful in many parts of the world, but commercial activity in Europe is limited essentially to the use of insect-resistant maize in the Iberian Peninsula.

(iii) The use of novel control agents and biopesticides has seen many dawns, some of them false. More recently there has been a flurry of interest in this area, highlighted by high visibility acquisitions of biopesticide companies.

Research needs to be prioritized to understand how interactions with BCAs work, their modes of action and active metabolites; understand how BCAs increase the efficacy of conventional crop protection, and/or help delay/prevent the development of resistance, when used as part of an Integrated Pest Management (IPM) programme (increasing yields in some cases); develop new niche applications for active molecules that might have been discarded from companies' R&D programmes (when looking for more global applications); and apply 'omics' technologies (theme 1) to forecast if resistance can/will develop to these novel control agents apart from establishing public-private partnerships to contribute to the growth agenda. A future proofing of crop protection with more tools available to deal with problems as they arise, including a wider and more systematic use of novel control agents in UK agriculture, capitalizing on past investments in plant and microbial research. More competitive farmers and a more resilient food supply chain that is less reliant on food imports and has less environmental impact.

10.2.1.6 *Agronomic approaches*

Some of the biggest threats to sustainable crop production are policies and resistance issues. Better informed policy, based on scientific evidence and risk rather than hazard, would improve the maintenance and development of safe and effective crop protection products. The number of situations where resistance to pesticides is occurring is increasing. Against a background of stagnating cereal yields, this is a great cause for concern. Resistance across a wider number of targets is occurring at a faster pace than ever seen before. There are diseases resistant to key active ingredients. Any restriction on the number of agrochemical options available to growers could easily exacerbate the situation. At present, there are few alternative solutions to minimize the impact of resistance apart from crop rotation, the use of fallow and to some extent varietal choice/delayed drilling. Looking into the future, there are a lot of exciting technologies that need much more investment to bring about applied developments. These include synthetic biology, use of semiochemicals, allelopathy, and biocontrol agents.

10.2.1.7 *Monitoring resistance to diseases*

Loss of current and new active ingredients in registered pesticides for existing and new pathogens is increasing the risk of crop resistance. The lack of information on the occurrence and spread of resistance among disease organisms hinders forward planning in ensuring effective contingencies for epidemics. There is a need to develop a centrally collated, independent monitoring system for resistance to crop protection agents in key diseases of economic importance, with centralized centers for sample testing and application of diagnostic technologies using consistent sets of molecular markers, and delivering open source information (following the model of EU potato blight initiative). More region-specific information on resistance should be directly available to farmers to allow better decisions in planning and timely application of appropriate control measures, depending on the profile of disease and resistance of varieties sown. Knowledge of pest and disease profile/incidence supports the development of disease resistant varieties, aided by using genomics.

10.2.1.8 Modeling of crop pest epidemics and disease outbreaks

It is more dependable modeling of crop pest epidemics. Development of disease models using 'what if?' scenarios should be considered to manage disease and yield outcome and to capitalize on lessons from models known to work well. A better understanding of disease development and spread; ability to forecast likely disease incidence with weather forecasts; and providing guidance to farmers on prevention and control measures.

10.2.1.9 Weather forecasts

Lack of reliability in short and medium-term weather forecasts makes it difficult for farmers to plan crop management practices. Exploration with Met Office of opportunities is needed to improve the reliability of medium-range weather forecasting, and access to information for farmers. More accuracy in weather forecasts supports farmers in decision making on crop management.

10.2.1.10 Precision farming and diagnostics to improve decision support systems

Developments in precision farming and diagnostics will be influential on crop protection and applicable to several global markets. Technologies include mechanized precision agriculture, advanced molecular diagnostics linked to smartphones, and the ability to bring genomics information to field applications. It also includes the development of technologies and best practice in precision agriculture, including machinery and real-time diagnostics/use of mobile phone apps to support farmers in improving crop protection systems. The gap between potential and actual yields is reduced.

10.2.1.11 Improved management of crop rotations

General focus of farmers in maximizing production yields through traditional applications of fertilizers and pesticides, with limited incentives and knowledge on use and value of crop rotations. Increase research on crop rotations to consider the inclusion of new major crops, including industrial crops, companion crops, and crops which are likely to be displaced from other regions due to climate change. This would entail increasing plant yield in 'underimproved' crops, applying technologies that researchers employ overseas and taking account of improving crop plants not previously economically viable. Improved use of crop rotations, utilizing integrated pest management principles, and development of novel agronomic solutions to reduce risk of attack by diseases

10.2.1.12 Plant listing criteria

Consider changing national and recommended listing criteria for new plant development to include reducing inputs as well as measuring yield. This would require a larger input for 'public good' in the plant breeding process, but could realize major benefits in driving environmental benefits such as reduced nitrogen inputs. Extend plant breeding testing to include reduced inputs as well/in place of increased yield alone. A pipeline leading to new crop varieties better suited to a changing environment, and reduced availability of inputs.

10.2.1.13 Incentives for scientists in translation research

Focus of scientists primarily on peer-reviewed publications for their career development, the absence of recognition by Research Councils of scientists involved with good high quality applied research and lack of incentives for engagement in translating research into use (Anonymous, 2013), encourage a change of culture to incentivize scientists to engage in research to solve practical problems relevant to industry, e.g., by acknowledging the contribution made to applied industrial R&D and associated dissemination to business; encourage inclusion of end-users on project steering groups; change the rule for industry engagement in BBSRC Industrial Partnership Awards to allow in-kind as well as cash contributions; and encourage more input from venture capitalists.

10.2.1.14 National knowledge transfer strategy

Need for better translation and subsequent transfer of crop production/protection science onto the farm to meet practical needs of farmers and their consultants, leading by example, and spread of best practice through the influence of innovative farmers on others. It includes the development of a national strategy for knowledge transfer and translational research to fix the broken pipeline and improve coordination across disparate and fragmented activities by different organizations (leadership from AHDB and input from Landex Colleges), with the engagement of 'nonagricultural' Universities where an enormous potential skill base resides. Stakeholders are signed up to a national strategy and implementation, and working to the same objectives. National Translation Centre established with national and regional roundtable events with researchers and users, and a network of regional experts (using universities, institutes, and existing knowledge transfer organizations). Researchers are actively engaging directly with local farmers and crop associations.

10.2.1.15 Communication/engagement with users

Preventing lack of coordination and communication between stakeholders engaged in crop protection science and putting translation into use; establishment of networks, whilst embracing current mechanisms, to promote better coordination and communication between stakeholders through existing demonstration farms, agronomy/farmers clubs; engagement with users during development and implementation of research to tailor outputs to user needs and establishment of register of expertise and projects; and better links and more effective knowledge exchange between researchers and farmers/other users of research are some aspects to be looked into.

10.2.1.16 Delivery of a national crop protection strategy

Aspects to be looked into are: lack of joined-up delivery of crop protection tools; development and implementation of a national crop protection strategy founded on the principles of integrated management (i.e., using all the tools and developing more) to provide durable control of those pests, diseases and weeds causing the largest losses in UK major arable and horticultural crops; and sufficiently robust and far-sighted to enable adaptation to climate change; continuing reduction in yield gap and improved crop quality.

10.2.1.17 Better regulation

Some of the aspects are: slow speed of product approvals and reregistration; cost, time taken and uncertainty to meet regulatory requirements and uncertainty for the industry regarding future EU pesticide legislation, and lack of new pesticide actives offering solutions; joined up action between government and its regulators with adequate resource, to ensure new active ingredients are approved in timely manner, biological controls are fast tracked, and more effective approaches to regulation of new technologies are developed; with continued engagement in EU to try to ensure EU regulation is proportionate; and encouragement of continued investment in new control measures.

10.2.1.18 Skills, capacity and expertise

Insufficient skills in some areas, particularly bioinformatics, the chemistry of plants, weed science, entomology, plant pathology, knowledge transfer and translational research; investment in training provision among researchers and users to improve the use of new technologies and innovation; vibrant well-skilled sector.

10.2.1.19 Public debate

Some aspects are: improve public awareness and understanding of new technologies to avoid the public perceptions that have impeded the application of GM in agriculture; development of a communication strategy with carefully thought-out narrative and nomenclature for new technologies (e.g., 'omics'); new technologies are available for uptake by users; develop ways to stimulate greater innovation in basic and applied research to support farmers with innovative crop protection practices, so that they start to bridge the yield gap. This will be best supported by a national crop protection strategy, founded on the principles of integrated management (i.e., using all the available tools and developing more) to provide durable control of those pests, diseases, and weeds causing the largest losses in the UK's major arable and horticultural crops. The approach needs to be sufficiently robust and far-sighted to enable adaptation to climate change (Anonymous, 2013). The strategy, owned and delivered by a single organization, should priorities needs at both national and international levels, and be supported by a route for two-way dialogue between the developers and users of agricultural technology. This will enable effective influence of policies controlling agrochemical use and breeding which currently inhibit innovation in crop protection.

Key initiatives supported through the strategy will create a Virtual National 'Omics' Centre, to draw on recent developments in 'omics' technologies to characterize major pests, weeds and diseases and the potential for new control strategies and diagnostics apart from establishing public-private partnerships to develop new niche applications for biological control agents. Also to establish National Translation Centre with national and regional roundtable events with researchers and users, and networks to promote coordination and communication between stakeholders through existing demonstration farmers, farmer clubs etc. It also invovles developing a centrally collated, independent monitoring system with centralized centers for sample testing and application of diagnostic technologies to comprehensively monitor and address pesticide and herbicide resistance at a national level. In addition it also invovles increase research on integrated approaches to crop protection from the molecular to agri-system level including improved management of crop rotations, precision agriculture, and diagnostics to improve decision support systems. Also to make more effective use of existing agricultural data systems and develop more dependable predictive modeling, encourage change of culture to incentivize scientists to engage in research to solve practical problems relevant to industry. Apart from these, to facilitate joined up action between government and its regulators with adequate resource to ensure that new active ingredients are approved in a timely manner. Biological controls are fast tracked and more effective approaches to regulate new technologies are developed and finally to develop a communication strategy that with carefully thought-out narrative and nomenclature for new technologies (e.g., 'omics'). Success for these changes overall is a measurable closing of the yield gap.

10.2.1.20 Research on microclimate and plant diseases

The continuous recording of microclimatic variables is done by Automatic Weather Stations, which are installed in the cropped fields to monitor the microclimate of crops (Kushwaha, 2010). Campbell Scientific Automatic Weather Station has been designed and developed to a very high standard for reliable measurement and recording of a wide range of important micrometeorological variables in and above the crops. The station is soundly engineered and based Campbell's proven 21X Micrologger whose comprehensive specification enables the user to undertake virtually any monitoring task. The automatic weather station is equipped with various micrometeorological instruments for monitoring of micrometeorological weather variables such as air temperature (°C), relative humidity (%), wind speed (m s^{-1}), wind direction (degrees from North), leaf temperature (°C), leaf wetness (% of total wet), solar radiation (W m^{-2}), net radiation (W m^{-2}), rainfall (mm), and soil temperature (°C).

For studying plant disease development, accurate and reliable weather data of nearest meteorological observatory be collected. Quantification studies between weather variables and crop disease can be worked out and in the long run, a suitable disease forecasting system can be development for sustainable management of plant diseases under field conditions. However, for more precise study of the relationship between microclimatic variables and plant diseases for a particular location automatic weather station can be purchased and installed in the field itself in that location.

10.2.1.21 Research on biomolecules

For any of these biological control agents to work, two simple rules must be followed. First, all of these products must be used in conjunction with standard disease cultural controls. Cultural controls include growing plants in a well-drained media, not over watering, keeping the greenhouse relative humidity below 85 percent, practicing strict sanitation and making sure that the nutrient and pH conditions of the host plant are within the ideal range for proper growth and development. This will help assure that the environment is favorable for the growth and development of the beneficial organism (Tripathi, 2010). Second, all of these biocontrol products must be applied at preplant or prior to the onset of disease. In most cases, they will not rescue plants that are already infected.

10.2.1.22 Research on genetically modified organisms (GMOs) or transgenics

Transgenics are produced using genetic engineering methods. Transgenics for resistance against plant pathogens such as viruses, fungi, bacteria and nematodes can be developed. Worldwide, research is being intensified in this aspect. Genetically modified organisms or transgenics are being developed to transfer of useful genes from GP4 to GP1 species (Roy, 2010).

Transgenic plants refer to plants, the DNA of which has been modified using genetic engineering techniques. In other words, these are plants into which one or more genes from other species have been introduced into the genome, using genetic engineering processes. Techniques include the biolistic method, in which a heavy metal is coated with plasmid DNA is shot into cells and *Agrobacterium tumefaciens*-mediated transformation. In most cases, the aim is to introduce a new trait to the plant which does not occur naturally in the species. Examples in food crops include resistance to certain pests, diseases or environmental conditions, reduction of spoilage, or resistance to chemical treatments or improving the nutrient profile of the crop.

In recent years, with the development of genetics molecular biology and plant biotechnology, the vaccination (e.g., genetic engineering subunit vaccine, living vector vaccine, nucleic acid vaccine) programs are taking on a prosperous evolvement. In particular, the technology of the use of transgenic plants to produce human or animal therapeutic vaccines receives increasing attention (Han *et al.*, 2006). Expressing vaccine candidates in vegetables and fruits open up a new avenue for producing oral/edible vaccines. Transgenic plant vaccine disquisitions exhibit a tempting latent exploiting foreground. There are a lot of advantages for transgenic plant vaccines, such as low cost, easiness of storage, and convenient immune-inoculation. Some productions converged in edible tissues so that they can be consumed directly without isolation and purification. Up to now, many transgenic plant vaccine productions have been investigated and developed. Recent advances in plant-derived recombinant protein expression systems, infectious targets, and delivery systems, issues of high concern such as biosafety and public health are to be advanced. The prospects and limitations of transgenic plants are to be considered.

10.2.1.22.1 Methods for developing fungal resistant transgenics

Several types of genes can be transferred to cultivated varieties to provide resistance against fungal pathogens, viz., use of gene for cell wall degrading enzymes such as chitinase and β-1,3-glucanase, ribosome inactivating proteins (RIPs) from different plant species, pathogenesis-related proteins (PRPs), small antifungal peptides and antifungal peptides or proteins from other organisms, use of a foreign phytoalexin, and expression of AOS (active oxygen species).

10.2.1.22.2 Steps involved in the production of transgenics

The steps involved in the development of transgenics include identification of target gene for transfer, creation of plasmid DNA which will contain this target gene and insertion into a vector, say bacteria (*Agrobacterium tumefaciens*), infection of cell culture with bacteria to transfer gene and growing of cell culture into adult plants, and using their seeds to generate GM showing resistance against either a fungus, a virus, a bacterium or a nematode.

10.2.1.22.3 Problems with transgenics

Major problems with transgenics include recombination between transgene and nontarget organism, gene flow of pollen from transgenics to weeds leading to development of herbicide resistant weeds, transmission of unrelated viruses by transencapsidation or enhanced seed or pollen transmission, synergy with unrelated viruses, exposure to new allergens or toxic proteins and increases susceptibility to other pathogens.

Plant biotechnology offers tremendous promise for not only feeding the world's growing population but for also improving the diets of people the world over. Although concerns over ecological and human health safety have led to mistrust over the application of this technology, many of these fears seem unsubstantiated or based on misinformation. A concerted effort must be made to identify valid concerns and risks, and to provide reliable and useful information to the public. Previously, genetically enhanced foods have focused on increased yield and other agronomic properties, which primarily benefits agribusiness corporations and farmers (Neal Stewart, Jr. *et al.*, 2011). The second generation of genetically modified foods will emphasize consumer health benefits. It is only with these new crops that the public will come to accept the uses of genetic modification of foods, and it is here that the overlap of nutrition science, ecology, and plant biotechnology will become most evident. What is needed now is more collaboration between the nutrition and plant science fields to adequately evaluate the functionality of genetically modified foods and to develop new products that could significantly benefit the general population. In addition, biotechnologists need the objective participation of ecological researchers in helping to better determine ecological biosafety of transgenic plants.

In general, scientists must become more proactive in the public debate if agricultural biotechnology is to make a long-lasting and sustainable impact on improving food and fiber production and human health. Currently, the most vocal group in the debate of GM crops has been environmental groups. Their arguments have primarily been based on fear of the unknown and technology plus misinformation based upon the misrepresentation of scientific data. The agricultural biotechnology industry is placed in a precarious position concerning their opinion in the public debate because of their obvious financial stake in the outcome. Academic and public scientists have tended to make two mistakes with regards to the current controversy. The first is their reticence to speak directly to the public and the media for fear of being misunderstood. The second is to assume that the public and the media will take their scientific data at face value. However, media sensationalism of single experiments can lead public opinion to make misinformed decisions. These mistakes have effectively removed the most objective and dispassionate parties (scientists) from the debate, which is unfortunate for public policy. For a scientific and technical society such as ours to properly function, public policy must be informed and shaped by the ideas of those who understand the science and technology best. Scientists need to reexamine their role and professions to include more public and media outreach as a part of their everyday work. Failure to do so will exacerbate the current vacuum of misunderstanding and fear surrounding the current controversy and those to come.

Further, the advancements made with transgenic plants have and will continue to have a great impact on the lives of many. Transgenic plants offer a new approach to producing and administering human antibodies. The recent research with transgenic plants has played a captivating role in providing edible vaccines, which are cheap and easy to administer. The progression of transgenic plant technology now has allowed for the progression of human life and other medicinal advancements. Hiatt was the first to demonstrate that plants could produce human antibodies and since that time, researchers have continued to build upon his findings (Elisa Ferrante and David Simpson, 2001). Since 1983, progress in the field of antibody production in plants has drastically increased, and it is projected that in the near future, many of the necessary human antibodies will have an origin as a plantibody.

10.2.2 Research on Impact of Climate Change on Diseases

Climate changes affect the disease incidence, further spread, and development of the pathogen in several ways (Murty, 2010). The direct effects of climate change on individual plants and plant communities may occur in the absence of pathogens, but may also bring about changes in plants that will affect their interactions with pathogens. Changes in plant architecture may affect microclimate and thus risks of infection. In general, increased plant density will tend to increase leaf surface wetness and leaf surface wetness duration, and so make infection by foliar pathogens more likely. Elevated CO_2 levels tend to result in changed plant structure. At multiple scales, plant organs may increase in size: Increased leaf area, increased leaf thickness, higher numbers of leaves, larger total leaf area per plant, and stems and branches with greater diameter have been observed under elevated CO_2. Enhanced photosynthesis, increased water use efficiency, and reduced damage from ozone are also reported under elevated CO_2. Since many foliar pathogens benefit from denser plant growth and the resulting humid microclimate, there is the potential for these changes in plant architecture to increase infection rates. Also, different populations of the same species may differ in both their genetic structure and the extent to which climate change will push the species to its physiological limits. As a result of climate change, the abundance of particular species may change rapidly as species may lose their ability to recover from other perturbations such as diseases, insect herbivores, and climatic extremes within a background of climate changes. Novel plant communities may result with the increased potential for new patterns of host-sharing by pathogens.

The range of many pathogens is limited by climatic requirements for overwintering or oversummering of the pathogen or vector. For example, higher winter temperatures of −6°C versus −10°C increase survivorship of overwintering rust fungi (*Puccinia graminis*) and increase subsequent disease on *Festuca* and *Lolium*. In the case of *Phytophthora infestans*, the introduction of multiple mating types, allowing sexual reproduction, increases the ability of the pathogen to overwinter. For pathogens subject to an Allee effect, or destabilizing density-dependent reproduction at low population levels, release from overwintering restrictions may have a much stronger effect than expected. Temperature requirements for infection differ among pathogen species. For example, wheat rust fungi differ in their requirements from 2°–15°C for stripe rust, 10°–30°C for leaf rust, and 15°–35°C for stem rust. In a review of the effect of climate change on insect herbivory, Bale *et al.*, make many points relevant to plant pathogens, whether insect-vectored or not. They concluded that temperature was the dominant climate factor in terms of direct effects through effects on overwintering and the potentially important combination of photoperiod and temperature. In many cases, temperature increases are predicted to lead to the geographic expansion of pathogen and vector distributions, bringing pathogens into contact with more potential hosts and providing new opportunities for pathogen hybridization. Increased transportation and human movement may act synergistically with temperature changes.

Temperature governs the rate of reproduction for many pathogens; for example, spore germination of the rust fungus *Puccinia substriata* increases with increasing temperature over a range of temperatures, and the root rot pathogen *Monosporascus cannonballus* reproduces more quickly at higher temperatures. Under climate change, pathogens, like plants, may potentially be unable to migrate or adapt as rapidly as environmental conditions change. But most pathogens will have the advantage over plants because of their shorter generation times and, in many cases, the ability to move readily through wind dispersal. Disease management strategies may require adjustment to climate change. Strategies such as delaying planting to avoid a pathogen may become less reliable. And one of the major problems with applications of biological control for plant disease management in the field has been the vulnerability of biocontrol agent populations to environmental variation. Simulation models are based on theoretical relationships and can be used to predict outcomes under a range of scenarios. Because climate change occurs slowly and variably, it is difficult to study its effects directly (Murty, 2010). Temporal variability in climate can be used to draw inference about the potential effects of climate change through the argument that temporary effects of a year with unusual climatic features are likely to represent the effects of longer-term changes environmental extremes. Models of plant disease have now been developed to incorporate more sophisticated climate predictions from General Circulation Models. In the population level, the adaptive potential of plant and pathogen populations may prove to be one of the most important predictors of the magnitude of climate change effects on plant disease, since, for many species, populations will not be able to migrate quickly enough to keep pace with climate change.

Due to the complex interaction of climate impacts, combined with varying irrigation techniques, regional factors, and differences in crops, the detailed impacts of these factors need to be investigated further. Specific recommendations for further research include precision in climate change prediction with higher resolution on spatial and temporal scales; linking of predictions with agricultural production systems to suggest suitable options for sustaining agricultural production, preparation of a database on climate change impacts on agriculture and development of models for pest/diseases population dynamics. Seed treatments should be given utmost importance for it saves time, energy and money to farmers to combat several major diseases.

10.2.3 Significance of Seed Treatment

Seed will continue to increase and become the delivery vehicle for new traits and functional attributes for improving germination and seedling development. Seed treatment may improve stand quality, protect seed from seed and soil borne pathogens, improve seed shape for planting, increase yields, increase return on investment, improve seed storability or performance and help in fixing nitrogen and enhance uptake of nutrients (Karuna Vishunavat, 2010). Seed treatment is needed when the seed is infected or infested, the field is for seed production, low test weight or older seed is used, when planting in unfavorable germination conditions (dry soil or cold soil), when planting into fields with a history of stand establishment problems, when planting to precise populations, and when replanting will not be feasible if first planting fails. Seed treatments may involve chemical seed treatment, physical, biological, with natural products and biodynamic preparations for seed treatment.

New seed treatments are required in a lower dose which reduces environmental exposure to the user. New seed treatment fungicides contain several active ingredients with different modes of action and higher efficacies and a broader range of disease control. Most of the new seed treatments are water-based and are in ready to apply or RTA formulations. Required in lower dose and thus reduces environmental exposure to the user. New seed treatment fungicides contain several active ingredients with different modes of action and higher efficacies and a broader range of disease control. Coating technologies, pelleting, encrusting and film-coating, are used increasingly to ease seed sowing by altering seed shape, surface properties, and weight. Polymer seed coating, colorants, and inoculants are more widely used. A range of beneficial additives including micronutrients and plant protection agents (high dose systemic pesticides) are used to improve the value addition. Seed vigor or physiological status can be manipulated by hydration treatments such as by priming and steeping. Eradication of fungal and bacterial seed infections can be achieved through heat or biocide treatments although there are challenges in operating these processes on an industrial scale without jeopardizing viability, vigor and shelf life.

It is evident that the relationship between climate change and agriculture is still very much a matter of concern with many uncertainties. Predicted changes in average values of global climate variables (increased temperatures, altered precipitation patterns, increased concentrations of atmospheric CO_2) and changes in the frequency, duration, and degree of extremes (such as frost, heat, drought, hail, storms, and floods) will affect agricultural crops, agroecosystems and agricultural productivity. Forecasts of regional climate changes are still not precise. Overall, shortage of water will be the predominant factor affecting plant growth. As higher temperatures are known to enhance plant development and especially the grain-filling duration of cereals, grain yield losses are possible in a warmer climate. On the other hand, elevated atmospheric CO_2 concentrations are known to stimulate photosynthesis and enhance growth and yield ('CO_2 fertilization'); concomitantly, leaf transpiration is reduced, resulting in improved water use efficiency. Elucidating the interactions between positive and negative effects of climate change is of crucial importance for any prediction of future crop yields. The prediction of the response of crops to climate on both seasonal and decadal timescales shows promise (Murty, 2010). The potential benefit of increasingly accurate prediction is clear: for the season, the mobilization of resources; for the adaptive measures to minimize the adverse impacts of climate change.

10.2.4 Disease Diagnosis in 'Omics' Era

The revolution of 'omic' sciences has introduced integrated high-throughput approaches to address the understanding of the biochemical systems and their dynamic evolution. 'Omics' are depicting a comprehensive view which largely overcomes the merely descriptive approaches of the early proteomic and metabolomic era. The diagnosis of the causal agent of plant diseases is very important for its management and control, so misdiagnosis affects the control actions of the primary causal agent. At present, there are some plant diseases where the causal agent in unknown or the disease symptoms are associated with different pathogens or abiotic agents (F. Victoriano Lara, 2011). In this sense, metagenomics provides a new rationale and effective methodology to identifying the primary causative agents. The study of uncultured microbes may carry out by metagenomics, which gives a broad sight for the investigation of microorganism origins and function in the environment. Metagenomics is based on studies of ecological diversity of uncultured and cultural microorganisms using molecular biology. For the metagenomic analysis of microbial populations in phytopathology, the total content of nucleic acids involved with the disease is used. Since we live in an 'omic' era, remarkable advances in both analytical methodology and computer algorithms pave the way for the discovery of the molecular basis of biology. From the study of the genome to the proteome to the metabolome, our ability to correlate the function of genes as components of a complex interrelated network of organism and environment promises to unlock the secrets of organisms. That disease is a complex interaction of genetic makeup, gene expression, protein action and the environment mandates a holistic study of the complex organism at the level of the common language, chemistry.

10.2.5 'Omic' Technologies

New advances resulting in 'omic' technologies have exploded the areas of genomics, transcriptomics, proteomics, and metabolomics and revealed many fundamental processes driven by both pathogens and commensals. These advances in genome sequencing coupled with other omic technologies such as proteomics and transcriptomics have revolutionized plant pathology.

10.2.5.1 Genomics

It is the new science that deals with the discovery and noting of all the sequences in the entire genome of a particular organism and this has been discussed in chapter 1. The Comprehensive Phytopathogen Genomics Resource (CPGR) provides a web-based portal for plant pathologists and diagnosticians to view the genome and transcriptome sequence status of 806 bacterial, fungal, oomycete, nematode, viral and viroid plant pathogens (John P. Hamilton *et al.*, 2011).

Tools are available to search and analyze annotated genome sequences of 74 bacterial, fungal and oomycete pathogens. Oomycete and fungal genomes are obtained directly from GenBank, whereas bacterial genome sequences are downloaded from the A Systematic Annotation Package (ASAP) database that provides curation of genomes using comparative approaches. Curated lists of bacterial genes relevant to pathogenicity and avirulence are also provided. The Plant Pathogen Transcript Assemblies Database provides annotated assemblies of the transcribed regions of 82 eukaryotic genomes from publicly available single pass Expressed Sequence Tags. Data-mining tools are provided along with tools to create candidate diagnostic markers, an emerging use of genomic sequence data in plant pathology. The Plant Pathogen Ribosomal DNA (rDNA) database is a resource for pathogens that lack genome or transcriptome data sets and contains 131 755 rDNA sequences from GenBank for 17 613 species identified as plant pathogens and related genera.

10.2.5.1.1 Flow cytometers

Flow cytometers are probably the most multipurpose laboratory devices available. They can analyze a vast and very diverse range of cell parameters. This technique has left its mark on cancer, human immunodeficiency virus, and immunology research, and is indispensable in routine clinical diagnostics (D'Hondt *et al.*, 2011). Flow cytometry (FCM) is also a well-known tool for the detection and physiological status assessment of microorganisms in drinking water, marine environments, food and fermentation processes. However, flow cytometers are seldom used in plant pathology, despite FCM's major advantages as both a detection method and a research tool. Potential uses of FCM include the characterization of genome sizes of fungal and oomycete populations, multiplexed pathogen detection and the monitoring of the viability, culturability and gene expression of plant pathogens, and many others. This review provides an overview of the history, advantages and disadvantages of FCM, and focuses on the current applications and future possibilities of FCM in plant pathology.

10.2.5.2 Transcriptome

Transcriptome refers to the set of all RNA molecules, including mRNA, rRNA, tRNA and other noncoding RNA produced in one or a population of cells and has been detailed in chapter 1.

10.2.5.2.1 Mining herbaria for plant pathogen genomes

Nowadays, our understanding of plant pathogens and the diseases they cause great benefits from molecular genetics and genomics. All aspects of plant pathology, from population biology and epidemiology to mechanistic research, are impacted. The polymerase chain reaction (PCR) first enabled access to plant pathogen DNA sequences from historical specimens deposited in herbaria (Yoshida *et al.*, 2014). Historical records in combination with herbarium specimens have turned out to provide powerful tools for understanding the course of past plant epidemics. Recently, thanks to new developments in DNA sequencing technology, it has become possible to reconstruct the genomes of plant pathogens in herbaria. The authors have summarized how whole genome analysis of ancient DNA has been recently used to reconstruct the 19th-century potato-blight epidemic that rapidly spread throughout Europe and triggered the Irish potato famine.

10.2.5.2.2 Potato late blight – From the Irish potato famine to modern epidemics

The oomycete plant pathogen *Phytophthora infestans* causes late blight, the most destructive disease of potatoes and a major disease of tomatoes. Ever since the Irish potato famine in the 1840s, *P. infestans* has remained one of the most serious threats to food production, resulting in losses that add up to enough crop to feed hundreds of millions of people (Fisher *et al.*, 2012). Similar to other oomycetes, *P. infestans* has a complex life cycle that includes both asexual and sexual phases. The ability of *P. infestans* to generate large amounts of asexual spores is a major determinant of its success as a destructive pathogen.

In agricultural systems, asexual reproduction is often dominant, with individual genotypes taking over large portions of the population and occasionally resulting in explosive epidemics. Sexual reproduction is integral to the life cycle of *P. infestans* at its center of diversity in Mexico. In some agricultural systems, notably in the Netherlands and the Nordic countries, sexual reproduction is also common, generating an abundance of new pathogen genotypes. *P. infestans* genotypes with favorable phenotypes, such as increased virulence and spore production, can arise from pockets of sexual reproduction to become invasive when they spread to vulnerable potato production areas. One example is clonal lineage 13_A2, an aggressive genotype that emerged in Europe in the early 21st century, rapidly displaced other genotypes and more recently became pandemic with severe outbreaks in India and China (Chowdappa *et al.*, 2013).

10.2.5.3 Perspectives in archaeogenomics of plant pathogens

Genome analyses of *P. infestans* aDNA have demonstrated that the potential of archaeogenomics to solve important questions in history and evolution does not apply to only humans and their pathogens but also to plant pathology. The budding field of archaeogenomics holds particular promise for plant pathogens because of the excellent preservation of genomic DNA in dried plant samples stored in herbaria. Thus, herbaria are hidden treasures, serving as unexploited archives of the plant pathogen and plant genomes. Beyond the potato late blight, there are numerous interesting targets for investigation of plant pathogens. European herbaria hold over 800,000 specimens of rust fungi, a group of plant pathogens that are a recurring threat to world agriculture (Singh *et al.*, 2011). Another, well-sampled plant pathogen is the oomycete *Plasmopara viticola*, which caused epidemics of downy mildew on the grapevine. Mining these and other historical herbarium samples for plant pathogen genomes should inform past population and evolutionary dynamics of important plant pathogens, ultimately helping us to better prepare for future plant epidemics.

The root transcriptome of the resistant (*Mi*+) and susceptible (*Mi*⁻) cultivars 'Motelle' and 'Moneymaker,' respectively, during a time-course infection by the *Mi*-susceptible root-knot nematode species *Meloidogyne incognita* and the *Mi*-resistant species *Meloidogyne hapla* (Jennifer E. Schaff *et al.*, 2007). In the absence of nematode infection, only a single significantly regulated gene, encoding a glycosyltransferase, was detected. However, nematode infection influenced the expression of broad suites of genes; more than half of the probes on the array identified differential gene regulation between infected and uninfected root tissue at some stage of RKN infection. Totally, 217 genes were discovered, which regulated during the time of nematode infection corresponding to the establishment of feeding sites, and 58 genes that exhibited differential regulation in resistant roots compared to uninfected roots, including the glycosyltransferase. Using virus-induced gene silencing to silence the expression of this gene restored susceptibility to *M. incognita* in 'Motelle,' indicating that this gene is necessary for resistance to nematode. Collectively, this data provided a picture of global gene expression changes in roots during compatible and incompatible associations with nematode and point to candidates for further investigation.

Induced resistance to chemicals such as acibenzolar-S-methyl -ASM (commercialized as Actigard by Syngenta, Inc.) mimics the biological activation of systemic acquired resistance (SAR). ASM takes the place of salicylic acid (SA) in the SAR signal pathway inducing the same molecular markers and range of resistance (Paola Zuluaga *et al.*, 2013). The goal of this work was to understand the downstream molecular events by which ASM confers resistance to *Phytophthora infestans* in tomatoes. To accomplish this goal, gene expression was assayed in ASM-treated plants using a microarray with more than 12,000 tomato ESTs. As many as 300 genes were responsive to ASM. Of these, 117 were detected in most of the biological replications. Basal defense-associated genes, as well as SAR and disease resistance genes (R-like) involved in induced resistance and effector-triggered immunity, were highly expressed. It was attempted to determine the phenotype of 13 of these genes by virus-induced gene silencing (VIGS). These 13 genes were selected on the basis of previous implication in plant defense response and by the reliability of induction by ASM. VIGS was partially successful for three of the 13 genes, but this partial silencing did not lead to a significant reduction in the effect of ASM. The ethylene pathway was also activated in response to ASM, but a tomato mutant not responsive to ethylene remained responsive to ASM. The ASM effect was complex and polygenic, depending on the effect of several genes.

10.2.5.4 Proteomics in phytopathology

Proteomics, transcriptomics and more recently, metabolomics are three of these techniques that have impacted on fungal plant pathology. Used individually, each of these techniques can generate a plethora of data that could occupy a laboratory for years. When used in combination, they have the potential to comprehensively dissect a system at the transcriptional and translational level. Transcriptomics or quantitative gene expression profiling is arguably the most familiar to researchers in the field of fungal plant pathology. Microarrays have been the primary technique for the last decade, but others are now emerging (Tan *et al.*, 2009). Proteomics has also been exploited by the fungal phytopathogen community, but perhaps not to its potential. A lack of genome sequence information has frustrated proteomics researchers and has largely contributed to this technique not fulfilling its potential. The coming of the genome sequencing era has partially alleviated this problem. Metabolomics is the most recent of these techniques to emerge and is concerned with the nontargeted profiling of all metabolites in a given system. Metabolomics studies on fungal plant pathogens are only just beginning to appear, although its potential to dissect many facets of the pathogen and disease will see its popularity increase quickly.

Within the '-omics' techniques, Proteomics constitute, nowadays, priority research for any organism and configures a fundamental discipline in the postgenomic era. It is true that, at 2010, the reality is below the expectations originally generated and that the results gained over the last 15 years have shown that the dynamism, variability, and behavior of proteins are more complex than had ever been imagined (González Fernández and Jorrín Novo, 2010). This especially concerns a number of protein species per gene as a result of alternative splicing, reading frame, and post-translational modifications, trafficking and interactions and considering that protein complexes, rather than individual proteins are the functional units of the biological machines.

The term PROTEOME refers to the 'PROTein complement of a genOME.' Fifteen years later proteomics has become more than just an appendix of genomics or an experimental approach but a complex scientific discipline is dealing with the study of the cell proteome. In the broadest sense, the proteome can be defined as being the total set of protein species present in a biological unit (organule, cell, tissue, organ, individual, species, and ecosystem) at any developmental stage and under specific environmental conditions. By using proteomics, we aim to find out how, where, when and for what are the several hundred thousands of individual protein species produced in a living organism, how they interact with one another and with other molecules to construct the cell building, and how they work with each other to fit in with programmed growth and development and interact with their biotic and abiotic environment. In the last ten years, some excellent reviews and monographs on the fundamentals, concepts, applications, power and limitations of proteomics have appeared.

In Proteomics, several areas can be defined: (i) Descriptive Proteomics, including Intracellular and Subcellular Proteomics; (ii) Differential Expression Proteomics, (iii) Post-translational Modifications; (iv) Interactomics; and (v) Proteinomics (targeted or hypothesis-driven Proteomics). In the case of fungi, a new area can also be defined as Secretomics (the secretome is defined as being the combination of native proteins and cell machinery involved in their secretion), since many fungi secrete a vast number of proteins to accommodate their saprotrophic lifestyle. This would be the case of proteins implicated in the adhesion to the plant surface, host-tissue penetration and invasion effectors, and other virulence factors.

Proteomics, in combination with other techniques, constitutes a powerful tool for providing important information about pathogenicity and virulence factors, thus opening up new possibilities for crop disease diagnosis and crop protection. González Fernández and Jorrín Novo (2010) reviewed how proteomics can contribute to such studies and objectives, with examples of our current research with *Botrytis cinerea*, and reference to the technical platform used, based on electrophoresis-MS-based proteomics, and specific protocols. The study of plant disease and immunity is benefiting tremendously from proteomics. Parallel streams of research from model systems, from pathogens *in vitro* and the relevant pathogen-crop interactions themselves, have begun to reveal a model of how plants succumb to invading pathogens and how they defend themselves without the benefit of a circulating immune system. Rampitsch and Bykova (2012) discussed the contribution of proteomics to these advances – drawing mainly on examples from crop-fungus interactions, from Arabidopsis-bacteria interactions, from elicitor-based model systems and pathogen studies, to highlight the important contribution of noncrop systems to advancing crop protection.

In the current scientific scenario, Proteomics should be considered as being part of a multidisciplinary, Systems Biology, approach. A combination of high-throughput 'Omics' (Genomics, Transcriptomics, Proteomics, and Metabolomics), and classical biochemistry and cell biology techniques should be used for data validation and to deepen the knowledge of living organisms, and, in this respect, fungal pathogens are not the exception. Such studies can be performed with *Botrytis cinerea,* which has been adopted as an important model necrotrophic system in molecular phytopathology.

Although the genomics-based investigation of host-pathogen interactions can provide valuable information on the changes in gene expression, the investigation into changes in protein abundance is also important, to obtain a closer view of the phenotype. Proteomics analysis is an excellent tool that can give us a great deal of information about fungal pathogenicity. This approach has allowed the identification of new fungal virulence factors, characterizing signal transduction or biochemical pathways, studying the fungal life cycle and its lifestyle. We can use this information to provide new targets for crop disease diagnosis focused on fungicide design. Otherwise, the secretome analysis is especially important as it contains an arsenal of extracellular enzymes to break down the plant cell wall for pathogen penetration and nutrient consumption. In this sense, Proteomics allows us to identify numerous differential proteins involved in multiple-player cross-talk normally occurring in nature between plants and pathogens, the so-called 'interaction proteomes.' Therefore, we still have a long way to make a profound the analysis of the phytopathogenic fungi if compared to studies made in humans, bacteria, yeast or plants.

10.2.5.4.1 Development of proteomics-based fungicides

This is a new strategy for environmentally friendly management of fungal plant diseases. Several approaches have been used for studying, for example, tumor development, biomarker discovery, or microbiology. In this 'postgenomic' era, the relevance of these studies has been highlighted as the phenotypes determined by the proteins and not by the genotypes encoding them that is responsible for the final phenotypes. One of the most interesting outcomes of these technologies is the design of new drugs, due to the discovery of new disease factors that may be candidates for new therapeutic targets. To our knowledge, no commercial fungicides have been developed from targeted molecular research, this review will shed some light on future prospects. Francisco Javier Fernández Acero *et al.* (2011) have summarized previous research efforts and discuss future innovations, focused on the fight against one of the main agents causing a devastating crops disease, fungal phytopathogens.

Historically, drugs have been obtained from plant and animals products, from derivates of human endogenous ligands or chemicals or semisynthetic chemicals. Classical methods to control fungal plant diseases are based on the use of chemical compounds. In spite of the success achieved, new criteria for the indiscriminate use of toxic compounds in nature avoid using this technology. Control strategies based on classical fungicides produce serious collateral effects, mainly related to environmental pollution and the development of multidrug resistance. Several changes in the design of chemical fungicides are being tackled by the research community by summarizing the genomic and proteomic information available. Biosynthetic fungicide design has been established as a new focus in fungicide development. Based on an in-depth study of fungal biology, the use of alien or modified natural compounds provides a potential species-specific method of controlling plant pathogens by specific inhibition of those proteins involved in the infection cycle. The use of these compounds minimizes the environmental impact as they are biodegradable, possess high specificity, and poorly integrate into the food chain. In the postgenomic era, new terms related with chemical '-omics' have appeared. The term 'genetic chemical' describes the use of small molecules to selectively perturb gene function. When this concept is applied on a genome-wide scale, it is named 'chemogenomics.' The application of chemogenomics to protein targets is named 'chemoproteomics'; although a more explicit definition is TRAP (targeted related affinity profiling) defined as the use of biology to inform chemistry.

The accumulation of proteomic information of fungal plant pathogens may be an incentive to the development of new and environmentally friendly fungicides. In spite of significant results these technologies have produced in several molecular biology areas, their use against fungal infection constitutes, in some cases, a science fiction exercise. Other technologies described have been used with varying degrees of success, such as the design of new chemical fungicides. However, the wide use of proteomics in crop fields remains unclear. Such approaches could be used in combination with other environmentally friendly initiatives to control fungal plant diseases to minimize the likeliness of resistance in fungal species. The time when these techniques will be common for farmers is near because environmental protection rules already prohibit the free use of fungicides.

10.2.5.5 Metabolomics

Metabolomics refers to the complete set of low molecular weight compounds in a sample and has been discussed in chapter 1. Metabolomics is the comprehensive, nonbiased, high throughput analysis of complex metabolite mixtures allowing the identification and quantification of every individual metabolite ideally. Metabolomics has emerged as a functional genomics methodology that contributes to our understanding of the complex molecular interactions in biological systems (Diego *et al.*, 2013). As such, metabolomics represents the logical progression from large-scale analysis of RNA and proteins at the systems level but unlike transcripts and proteins, the highly diverse molecular identity of metabolites relies on sophisticated instrumentation such as mass spectrometry (MS) and nuclear magnetic resonance spectroscopy (NMR) to be determined. Metabolites are the end products of cellular regulatory processes, and their levels can be regarded as the ultimate response of biological systems to genetic or environmental changes. Metabolites are also highly dynamic in time and space and show an immense range of structures raising the challenges for analytical procedures in their measurement.

Metabolic processes in plants are key components of physiological and biochemical disease resistance. Metabolomics, the analysis of a broad range of small molecule compounds in a biological system, has been used to provide a systems-wide overview of plant metabolism associated with defense responses. Plant immunity has been examined using multiple metabolomics workflows that vary in methods of detection, annotation, and interpretation, and the choice of the workflow can significantly impact the conclusions inferred from a metabolomics investigation (Adam L. Heuberger *et al.*,

2014). The broad range of metabolites involved in plant defense often requires multiple chemical detection platforms and implementation of a nontargeted approach. A review of the current literature reveals a wide range of workflows that are currently used in plant metabolomics and new methods for analyzing and reporting mass spectrometry (MS) data can improve the ability to translate investigative findings among different plant-pathogen systems. In a standard metabolomics workflow, metabolites are assayed by extracting compounds from plant tissue in a solvent, and subsequently detecting and quantifying compounds using various chemical detection platforms (e.g., MS, nuclear magnetic resonance, ultraviolet absorbance). An MS-based metabolite detection is a powerful tool for investigations of plant metabolism due to its sensitivity for low-abundant molecules and flexibility for the detection of multiple chemical molecular classes. Therefore, MS is well-suited to investigate the signaling, physiological, and other chemical events associated with plant defense. Metabolomics studies are also relevant for plant systems as a means to elucidate the mechanisms underlying genetic sources of resistance. Sunflower (*Helianthus annuus*) resistance to the necrotrophic fungal pathogen *Sclerotinia sclerotiorum* is associated with at least 44 quantitative trait loci.

While still considered a relatively new field, the annotation bottlenecks in nontargeted metabolite profiling are predicted to significantly improve as the MS metabolite spectral databases mature. In addition, sharing of unknown spectra as supplemental data upon publication or to a public resource such as Massbank can improve the general understanding of which chemicals, albeit unknown, are consistently associated with plant immunity. It is therefore recommended that mass spectra, which are the information collected in an MS-experiment, should be included in publications of nontargeted metabolomics experiments.

Metabolomics is perhaps the ultimate level of postgenomic analysis as it can reveal changes in metabolite fluxes that are controlled by only minor changes in gene expression measured using transcriptomics and/or by analyzing the proteome that elucidates posttranslational control over enzyme activity. Metabolic change is a major feature of plant genetic modification and plant interactions with pathogens, pests, and their environment. In the assessment of genetically modified plant tissues, metabolomics has been used extensively to explore by-products resulting from transgene expression and scenarios of substantial equivalence (William Allwood *et al.*, 2008). Many studies have concentrated on the physiological development of plant tissues as well as on the stress responses involved in heat shock or treatment with stress-eliciting molecules such as methyl jasmonic acid, yeast elicitor or bacterial lipopolysaccharide. Plant–host interactions represent one of the most biochemically complex and challenging scenarios that are currently being assessed by metabolomic approaches. For example, the mixtures of pathogen-colonized and nonchallenged plant cells represent an extremely heterogeneous and biochemically rich sample; there is also the further complication of identifying which metabolites are derived from the plant host and which are from the interacting pathogen.

An overview of the analytical instrumentation currently applied to plant metabolomic analysis was reviewed (William Allwood *et al.*, 2008), literature within the field was reviewed with particular regard to studies based on plant–host interactions and finally the future prospects for the metabolomic analysis of plants and plant–host interactions were discussed. The current lack of metabolite libraries perhaps impacts most greatly upon plant metabolomics, given that plants possess a huge range of secondary metabolites, many of which we do not currently have the ability to identify. MS libraries and databases are not common because of difficulties in the comparison of data between instruments. This stems from technical difficulties that influence analytical reproducibility. Ultimately, databases are required that are based on a range of organisms from across different kingdoms, which are composed of data from their transcripts, proteins, metabolites, metabolic pathways and the enzymes involved within them (Hall, 2006). Species-specific databases such as AraCyc (www.arabidopsis.org) would benefit greatly from the addition of information on rate constants between reactions within metabolic pathways and between metabolic networks. This would be very useful for systems biology where the relative accumulation of particular metabolites and possibly flux through a pathway could be modeled. Although this is an attractive prospect, one must bear in mind that rate constants are usually established using *in vitro* systems that may not reflect the *in vivo* situation.

Other metabolomic problems stem from sample heterogeneity. Tissues from whole plants are generally pooled, potentially mixing leaves, petioles, shoots, flowers and roots. It should, therefore, be remembered that information regarding the compartmentalization of metabolites to particular organelles is often lost. In the case of plant pathology, there are additional problems such as the sample containing plant, pathogen and infected and noninfected tissues. Differentiation of the metabolome from different tissue and cell types to the different cellular organelles will provide a more complex and yet surely improved appreciation of the plant metabolome. Some groups have already taken on the challenge of compartmentalized metabolomics. Farre *et al.* (2001) studied the compartmentalisation of glycolysis intermediates, sugars, nucleotides, sugar alcohols, amino acids and organic acids, in the potato tubers amyloplast, cytosol, and vacuole. Recently, significant advances have been made for the compartmentalized analysis of metabolites by using stable isotope labeling with for example $15NO_3$, 13C or 15K NO_3.

Multidimensional heteronuclear-NMR-based analysis was performed on the metabolic movement of carbon and nitrogen nuclei in *A. thaliana*, in which distinct ethanol stress responses were investigated, as were nitrogen fluxes in 15N-labeled seeds during germination (Kikuchi *et al.*, 2004). To overcome the current limitations on metabolomic analysis discussed above successfully will require an interdisciplinary-based approach where biologists, chemists, statisticians and instrument manufacturers all have input. Such an approach will maintain that the developments in both instrumentation and computational programs will benefit all research areas within metabolomics. This is especially the case within plant metabolomics with projects such as METAPHOR leading the way. With such coordinated efforts, further developments are inevitable, and so the future of metabolomics within the study of plant and plant–host based systems appears to be rosy and will inevitably lead to a more comprehensive understanding of plant systems than the information provided by transcriptomics and proteomics alone.

10.2.6 Impact of 'Omic' Technologies

Impact of these technologies on the detection of plant pathogens, enhancing resistance with plant genes, pathogen-derived resistance, antimicrobial proteins, plantibodies and high-throughput sequencing of 16S rRNA gene and metagenomics has been discussed in chapter 1.

10.2.7 Prediction of Disease Effects on Ecosystems

Several factors affect ecosystem, viz., pathogen effect on host survival, physiology, behavior and/or reproduction, life stages of a host vulnerable to a pathogen, proportion of individuals/biomass infected at a site, spatial extent and distribution of infection, rate of pathogen effects on hosts in relation to rate of response/recovery by hosts of individuals replacing hosts, frequency and duration of pathogen impact, and functional similarity of infected individuals versus replacements.

Major evidences on the application of 'omic' technologies in plant pathology have been briefed hereunder. Molecular factors enabling microbial pathogens to cause plant diseases have been sought with increasing efficacy over three research eras that successively introduced the tools of disease physiology, single-gene molecular genetics, and genomics. Jeffrey A Kimbrel *et al.* (2010) developed a unified model of the interactions of biotrophic and hemibiotrophic pathogens, which posits that successful pathogens typically defeat two levels of plant defense by translocating cytoplasmic effectors that suppress the first defense while evading the second defense. As is predicted by this model and confirmed by sequence pattern-driven discovery of large repertoires of cytoplasmic effectors in the genomes of many pathogens, the coevolution of hemibiotrophic pathogens and their hosts has generated pathosystems featuring extreme complexity and apparent robustness. These findings highlight the need for a fourth research era of systems biology in which virulence factors are studied as pathosystem components and pathosystems are studied for their emergent properties.

During the past decade, however, genomic technologies have revealed numerous additional bacterial genes that more subtly influence transformation. Members of the genus *Agrobacterium* cause the neoplastic diseases crown gall, hairy root, and cane gall on numerous plant species. Extensive genetic analyses conducted in the 1980s identified key bacterial genes involved in virulence. The results of these genomic analyses allowed scientists to develop a more integrated view of how *Agrobacterium* interacts with host plants. In a similar manner, genomic technologies have identified numerous plant genes important for *Agrobacterium*-mediated genetic transformation. Knowledge of these genes and their roles in transformation has revealed how *Agrobacterium* manipulates its hosts to increase the probability of a successful transformation outcome. The first wave of *Pseudomonas syringae* next-generation genomic studies has revealed insights into host-specific virulence and immunity, genome dynamics and evolution, and genetic and metabolic specialization (Heath E O'Brien *et al.*, 2011). These studies have further enhanced our understanding of type III effector diversity, identified an atypical type III secretion system (T3SS) in a new clade of nonpathogenic *P. syringae*, identified metabolic pathways common to pathogens of woody hosts and revealed extensive genomic diversity among strains that infect common hosts. In general, these discoveries have illustrated the utility of draft genome sequencing for quickly and economically identifying candidate loci for more refined genetic and functional analyses.

Xanthomonas, a major genus of Gram-negative bacteria that causes diseases in hundreds of plant hosts, including many economically important crops. Pathogenic species and pathovars within species show a high degree of host plant specificity and many exhibit tissue specificity, invading either the vascular system or the mesophyll tissue of the host. Robert P. Ryon *et al.* (2011) discussed the insights that functional and comparative genomic studies are providing into the adaptation of this group of bacteria to exploit the extraordinary diversity of plant hosts and different host tissues.

Cultivation-independent terminal restriction fragment length polymorphism analysis of 16S ribosomal DNA directly amplified from plant tissue DNA was compared with molecular characterization of isolates to examine the influence of plant stress, achieved by infection with *Erwinia carotovora* subsp. *atroseptica*, on the endophytic population in two different potato varieties (Birgit Reiter *et al.*, 2002). Community analysis clearly demonstrated increased bacterial diversity in infected plants compared to that in control plants. The pathogen stress had a greater impact on the bacteria population than the plant genotype had. Partial sequencing of the 16S rRNA genes of isolated endophytes revealed a broad phylogenetic spectrum of bacteria, including members of the alpha, beta, and gamma subgroups of the Proteobacteria, high- and low-G+C-content gram-positive organisms, and microbes belonging to the Flexibacter-Cytophaga-Bacteroides group. Screening of the isolates for antagonistic activity against *E. carotovora* subsp. *atroseptica* revealed that 38% of the endophytes protected tissue culture plants from blackleg disease.

Major thrusts for future investigations may include a better understanding of integrated 'omic' studies of host and pathogen responses, as well as communities of soil and plant-associated microbes, gene expression in plants and pathogens in response to climatic factors, multifactor studies of climate change effects, better models of adaptation rates, better data and models related to dispersal, current levels of intraspecific diversity, strength of selection under different climate change scenarios and heritability of traits, long-term large-scale records of pathogen and host distributions, models of regional processes that incorporate disease data and models describing dispersal of propagules and vectors, and integrated multidisciplinary international networks for data collection and synthesis.

The recently born 'omic' sciences, intended as the systematic definition of subsets of specific biomolecules, are finding applications to plant pathogens. The dawn of a new era in plant pathology and plant protection is upon us. 'Omic' technology has rewritten the scope of scientific investigation, broadened the avenues to resistant plants and challenged us to take safe and careful steps. Plant disease has a major impact on agricultural and natural systems. Systems may change more rapidly than in the past, requiring more research and policy attention. Plant pathologists have a long history of bringing together different scales of biology and are noted for having a balanced focus on molecular biology while never losing sight of more complex systems. This need for a careful balance between a systems biology and reductionist approach is becoming increasingly important in the study of plant-microbial interactions. The costs for metagenomic studies are rapidly increasing.

Like any other new technology, much still needs to be done before the full potential of 'Omic' technology is realized. As more and more technology sources become available, studies to evaluate the risks associated with 'Omic' technology must be intensified. Findings from such studies must be easily accessible to all. 'Omic' technology's vast perspective is being confronted, and this astounding view has expanded the very foundation of our understanding of plant pathogens.

The availability of interactive multimedia authoring software programs promises to revolutionize the teaching and research of nematology. These programs integrate text, hypertext, graphics, animations, video, and sound. Documentation is primarily done to record, preserve and maintain documents about the work. Documentation serves as the important asset of the previous work done in the form of audio-visual recording or digital or print form of photographs, published book, journal, monograph and these are put on a website, computer, laptop, CDs/DVDs, etc. Innovative approaches like digitations, offline and online database storage, retrieval, dissemination, up and downloading have been undertaken for documentation of nematode information.

In cyber-taxonomy or e-taxonomy, a taxonomist may use computer-aided identification with suitable software as a more efficient alternative to handle descriptive pathogen data and identify them through interactive keys. Innovative technologies in modern biosystematics are indeed useful. These should be used as a supplementary tool rather than a replacement of conventional ones. Bioinformatics identifies the progress in geographical distribution ecology, life cycle, pathogenicity, and management. Information Technology and widespread computer use have further given a strong leverage for faster dissemination, processing, terabyte level storage, handling, and shipping of pathogen information. Remote Sensing techniques are unique means of recording data of visible as well as invisible part of electromagnetic spectrum. Such data is further processed and analyzed through Geographic Information System. Both GPS and GIS are useful for creating a geographical and obtaining a descriptive biogeographical database of targeted population.

10.2.8 Vision for Future

Future vision should be more focused on areas like exploitation of bioagents and bioproducts and their application in plant disease management, exploitation of nanotechnology and defense-inducing molecules in plant disease management, demonstration and revalidation of Indigenous Technologies Knowledge (ITKs) apart from the effect of climate change on the emergence and reemergence of plant pathogens, vectors and strategies to combat them, development and establishment of disease monitoring and forecasting models/units. It is also important to intensify the basic and applied research on emerging diseases, development of molecular tools for understanding racial pattern in phytopathogens and insect vectors, innovative forecasting and modeling systems for important diseases in agriculture and horticulture crops. Marker Assisted Selections for breeding of crop varieties resistant to diseases is of prime importance along with the documentation of plant pathogens and beneficial microbes in different agro- climatic zones to understand their biodiversity. It is essential to develop farm plans for individual marginal farmers, intensifying research on molecular plant pathology to understand plant microbial interaction and also to develop novel approaches for plant disease management. Most important is to develop technologies for the management of crop diseases under dry land and organic farming, development and commercialization of disease diagnostic kits and new molecules for disease management in addition to the designing of efficient and cost effective plant protection equipments and development and commercialization of bioproducts/bioagents through public-private partnership.

10.3 EXTENSION INNOVATIONS IN PLANT PATHOLOGY

One of the major constraints in the management of phytopathogens is their proper diagnosis under field level. Without a precise field diagnosis, it is difficult to design a suitable management schedule. At this stage, in particular, the role of extension personnel is realized and very much essential to create awareness among the farming community.

The job of an extension specialist has changed over this time, but the need to be knowledgeable about all facets of our science and to have the ability to apply and effectively teach this knowledge to a diverse clientele has remained constant. Extension's mission was and is to be the 'translator' between the university and their clientele, originally the farmer and now farmers, consultants, agribusiness, urban landscape professionals, nurserymen, homeowners, and many others (Jacobsen *et al.*, 2008). Extension work in plant pathology was first undertaken by state land grant university or college faculty in the context of county fairs, farmer institutes, short courses, field demonstrations, farm trains, and other efforts to deliver information to clientele. The concept of a local, state and federal partnership "to provide for cooperative agricultural extension work between the agricultural colleges in the several states ... and the United States Department of Agriculture" was first proposed by Congressman Asbury F. Lever of South Carolina and Senator Hoke Smith of Georgia in 1913. The Smith-Lever Act that created the Cooperative Extension Service was signed into law by President Woodrow Wilson on May 8, 1914. The stated purpose of this act was "to aid in diffusing among the people of the United States useful and practical information on subjects relating to agriculture and home economics and to encourage the application of the same." Further, the act stated that agricultural extension work was to "consist of the giving of instruction and practical demonstrations in agriculture and home economics to persons not attending, or resident in said colleges in several communities and imparting to such person information on said subjects through field demonstrations, publications and otherwise." The Act initially provided that each state receives $10,000 annually with additional amounts to be determined on the basis of rural population. These additional amounts had to be matched by the state, county, local authority, or individual contributions from within a given state. An excellent review of extension activities through its first 75 years can be found in Rasmussen (1989).

Since 1953, the number of plant pathologists with extension appointments and the number of full-time equivalents (FTE) increased dramatically, while the percentage of 100% appointments decreased from approximately 50% to 30%. Currently, 71% of extension specialists have split appointments with most having research responsibilities as part of their appointment, although three-way appointments including research and teaching are more common. The inclusion of teaching appointments has occurred because often only extension plant pathologists have the broad training and experience needed to teach courses on disease control, diagnosis, or diseases of field crops, turf, ornamentals, vegetables, fruits, or other plant groups. Although precise data are not available, we suspect that most of the recent increase in extension FTE is a result of funding from the National Plant Diagnostic Network (NPDN) which started in 2002. In 1989, 36 states reported personnel with plant diagnostic responsibilities. Today, every state and the United States territory has faculty with specific diagnostic responsibilities. Before the establishment of the NPDN, diagnostic responsibilities were commonly the responsibility of extension specialists, who also had many other responsibilities.

The increase in split appointments reflects the increasing specialization of extension specialist assignments. Initially extension plant pathologists were generalists. Today, very few extension specialists have to handle all crop and ornamental disease problems. Most specialists now have assignments limited only to agronomic or horticultural crops and many specialists with < 50% appointments have responsibilities for only a small subset of agronomic or horticultural crops. This specialization reflects the increased size and specialization of modern farm enterprises, the growth of responsibilities to urban and suburban clientele, increased environmental awareness of Americans, the sophistication of clientele and scientific advances in our profession, and other additional factors. The number of split appointments has markedly increased for states in the Pacific and the Southern divisions while decreasing 50% for states in the Potomac Division. The increased number of split appointments in the Pacific and Southern divisions follows the trend that started earlier in the Northeast and North Central divisions. The average FTE per state has increased in all divisions except the Pacific and Potomac.

Extension plant pathologists deliver science-based information that protects the economic value of agricultural and horticultural crops in the United States by educating growers and the general public about plant diseases (Everts *et al.*, 2012). Extension plant pathologists diagnose plant diseases and disorders, provide advice, and conduct applied research on local and regional plant disease problems. During the last century, extension plant pathology programs have adjusted to demographic shifts in the U.S. population and changes in program funding. Extension programs are now more collaborative and more specialized in response to a highly educated clientele. Changes in federal and state budgets and policies have also reduced funding and shifted the source of funding of extension plant pathologists from formula funds towards specialized competitive grants. These competitive grants often favor national over local and regional plant disease issues and typically require a long lead time to secure funding. These changes coupled with a reduction in personnel pose a threat to extension plant pathology programs. Increasing demand for high-quality, unbiased information and the continued reduction in local, state, and federal funds is unsustainable and, if not abated, will lead to a delay in response to emerging diseases, reduce crop yields, increase economic losses, and place U.S. agriculture at a global competitive disadvantage. Four recommendations were outlined to strengthen the role and resources of extension plant pathologists as they guide our nation's food, feed, fuel, fiber, and ornamental producers into an era of increasing technological complexity and global competitiveness.

10.3.1 Changing Clientele

Until the late 1950s, the majority of extension plant pathologists focused on disease problems influencing food production. The educational level of their clientele varied from very few college graduates to many with little formal exposure to science and science-based farming practices. Suspicion of 'book farming' was a major problem for early extension specialists and made skill in educational psychology second only to the need for sound knowledge of the biology of plant pathogens and their control. Today the clientele often includes PhD-holding farmers and consultants in addition to clientele with little formal education in plant production or plant protection. County agents and regional extension personnel remain key for both extension programming and as a focus for educational programs.

In 2007, only 9% of extension plant pathologists listed their responsibilities as general or all plants and 17% had a nonfood crop (e.g., turf, ornamentals, shade trees, floriculture, or others) as their primary area of responsibility. These appointments reflect the increasing urbanization of our clientele and the corresponding increase in the importance of the turf, ornamental and nursery industries that supply products and services to homeowners in urban and suburban landscapes. The remaining ~75% of extension specialists have listed their one or more responsibilities as agronomic crops, cereal crops, forage crops, tree fruit, small fruit, vegetable crops, potato, citrus, mycotoxins, mushrooms, nematology, or others, reflecting the importance and specialization of these industries or specific problems of local importance. Other areas of specialization indicated by extension plant pathology specialists include IPM, pesticide applicator training, diagnostics, urban gardening, water quality, and others. This specialization is reflective of specific governmental funding initiatives for integrative programs (Jacobsen *et al.*, 2008).

The means by which extension clientele receive their information has dramatically changed. Reliance on field demonstrations, local meetings, short courses and farmer institutes, local county extension agents, and printed fact sheets and bulletins has given way to 7-day a week internet access to information. More sophisticated clientele are becoming reliant on the computer for information that is increasingly dynamic in nature. Examples include weather-based disease prediction programs and interactive diagnostic keys. The challenge to extension specialists to meet demands of the increasingly diverse clientele is intimidating, and mechanisms used by specialists to meet these demands must also constantly evolve. Educational psychology remains a useful skill, but today is more commonly known as 'adult learning styles.'

10.3.2 Changing Delivery Techniques

As extension methodology and clientele have evolved, so have the means by which we deliver information to our clientele. One major change is the internet, which has made extension bulletins, newsletters, and diagnostic keys available to the general public and other interested parties on a '24/7' basis. This technology now allows fact sheets and bulletins to be updated in 'real time,' as well as the use of color illustrations at low cost. Often, the only hard copy is printed by the user at home or in the county extension office. In an effort to provide credible, research-based information on the web, extension was launched by the Extension Committee on Organization and Policy in 2006. Currently, there are more than 70 university partners for extension, plus many industry and private sponsors. However, adoption of extension has been slow, and many universities still rely on state-focused fact sheets and information sources.

An example includes accessible and easy-to-use image databases such as IPM Images. IPM Images is one of several accessible and easy-to-use image databases available on the web. These databases also facilitate wiki-based fact sheets and other media that can be updated by users. Other examples include a diagnostic image database for the NPDN and fact sheets for a number of different websites. All of these websites are updated on a regular basis. These are just a few examples of websites that are replacing books and CDs and are often used by authors seeking to regularly update information such as resistant variety or pesticide recommendations. The web sites that incorporate current weather data and dynamic disease prediction models are now widely available to growers and agro-industry. The internet has also facilitated diagnostic efforts via digital photography, allowing county agents, growers, consultants, homeowners, and other clientele to rapidly communicate with extension specialists regarding plant diagnosis and problem solving. The distribution of newsletters has changed significantly, from periodic mailings via the postal service to electronic newsletters, blogs, wikis, and podcasts. An integral feature of the NPDN's Plant Diagnostic Information System (PDIS) is the AgAlert system, which uses e-mail to quickly alert stakeholders about pest threats. This information often reaches additional producers when news media, county extension agents, and others extend the AgAlert via more traditional venues such as radio spots, newsletters, and newspaper articles (Jacobsen *et al.*, 2008).

The internet also encourages direct communication between clientele and the extension specialist through email, IM (instant messaging) and video messaging. More sophisticated clients now seek particular specialists and contact them personally. Of personal note, it is interesting that on a weekly basis we will have growers and others ask us about research articles published in professional scientific journals that are now available on the web. The extension plant pathologists may now find themselves helping people from beyond the state of their assignment. These technologies also encourage communication among coworkers that facilitate the collaboration that expands and maximizes educational efforts.

Other technology coming into broader use includes podcasts and informational videos which are available on the web. Included in this class of technology are online meeting venues such as that offered by the Adobe Connect software. Participants have the opportunity to log onto a website with features such as a chat box, PowerPoint presentation space, and quiz modules to facilitate participant interaction. Members can either call into a conference number on their telephone line or have the audio interaction solely via the web interface. Adobe Connect sessions are replacing conference calls because the software allows experts to share their knowledge with an audience, and because the sessions can be recorded and preserved indefinitely. Therefore, the Adobe Connect session, or portions of that session, can be replayed for the training of additional audiences. There are several other software packages that offer similar technologies for computer-based education delivery, such as Centra Symposium.

Major objectives of extension plant pathology are to provide research-based educational programs and materials on diseases of crops, landscape plants and other plants of interest to Kentucky residents; to provide plant disease diagnostic support for the educational programs of County Extension Agents; to develop effective, environmentally responsible options for managing plant diseases of importance.

Extension specialists can no longer travel to a location, deliver a standard seminar-style presentation, and expect the audience to retain and implement that information. The amount of information a grower must process and implement to make their operation profitable is much greater today compared to 1914 when the extension system was first formed. Today's extension specialist must be creative with hands-on activities, the use of videos for information delivery, and other interactive activities. Considerable research on adult learning styles has been done that we must take into account when designing an effective extension program. In addition, to meet the demands of the administration and granting agencies, the impact of our presentations and implementation of the information must be evaluated by pre- and post-tests or other mechanisms that document audience comprehension (Jacobsen *et al.*, 2008). Therefore, in addition to being experts in our field, we also must make others aware that we are making progress in educating clientele and positively impacting the community both economically and environmentally. Marketing, strategy, and advertising have become central to our responsibilities and are necessary for future funding. While there are many new tools for extension specialists to use in educational programs, it is still critical that our clientele has face-to-face access on a regular basis, if they are to have confidence in our ability to understand their problems and make sound recommendations.

Extension programs are designed to reduce crop losses in the context of sustainable environment by educating and assisting extension agencies and producers through the country delivery system. These programs are committed to provide best-unbiased educational information on nematode disease management and responsive diagnosis. The goal is achieved by effective interdepartmental and interdisciplinary programming across the functionaries of research, extension and instruction.

Education in disease management is a business to maintain desirable environmental quality linked with the results in economic viability. Since most farmers, agronomists, and crop management consultants have only a basic knowledge of pathogens, they may seek professional help when making decisions about their management. Specifically, they will have soil or plant samples assayed for various pathogens, seek advice on whether the pathogens present are likely to cause economic damage and require a recommendation on control. Such services are provided by the government and commercial laboratories. However, in the absence of an accreditation system for pathological laboratories, the quality of the information available to clients cannot be guaranteed. The advisory service has to follow general guidelines on how an advisory service for nematodes should operate. These guidelines will help improve the quality of advice on nematode pests available to farmers.

Disease management measures are specific to certain pathogen species; growers will need accurate species identification if they have to use them successfully. In situations where chemical control remains an option, economic and environmental considerations will mean that growers will be obliged to determine whether pathogen population densities are high enough to warrant treatment. The difficulties involved in diagnosing disease problems and the growing complexity of integrated management programs mean that there will be an increasing demand from farmers for professional help in making decisions on disease management. This help is likely to come initially from agricultural advisers and pest management consultants who work at the farm level and have an intimate knowledge of the cropping system involved. By sending samples to a diagnostic laboratory and liaising with a pathologist, they should be able to make appropriate recommendations.

There are a number of Public and Private Organizations involved in teaching, research, and extension which include State Agricultural Universities (SAU's), ICAR, UGC, CSIR, DST, DBT, Krishi Vignan Kendras, Commodity Boards, Pesticide Industries, Seed Companies, NGOs and several others. Farm and home visit method, group discussion meetings, method demonstrations, result demonstrations, campaigns, exhibitions, field visits, field days, farmers training, extension literature, print and electronic media are various channels through which technology could be effectively disseminated to farmers.

Recently, Kisan Call Centers (KCC) have been set up in the State Agricultural Universities to answer to the queries by the farmers over the telephone (on-line) by the concerned Subject Matter Specialists of the University. Raitha Samparka Kendras (RSK) are located at Hobli Level, looked after by the Agricultural Officers under whom Agricultural Assistants of Gram Panchayats are working. They are actively involved in the transfer of technology at the village level. Agricultural Technology Information Centre (ATIC) has been set up in all the state agricultural universities, which are trying to solve the farmers' problems under one roof. Plant health clinics are envisaged to provide expert services and advice farmers on different aspects of crops and also clinical services. They can play a major role in creating awareness of diseases and their management of a particular locality.

Extension services play an essential role in communicating results of research into the field. They have to identify problems affiliating the farming and agricultural industry and determine the appropriate course of action to obtain remission of the effects of diseases. Above all, the extension agencies can communicate their ideas to the research community, be it at the institutes or at the universities for refinement of technologies to suit the farming community. It is the two-way flow of intelligence between the field and the laboratory bench, i.e., the vital and unique province of the Extension Pathology.

The difficulties involved in diagnosing diseases and the growing complexity of integrated management programs for pathogens mean that there will be an increasing demand from the farmers for greater help in making decisions on disease management in future. This is possible only through extension personnel/consultants who work at the farm level and have an intimate knowledge of the cropping system involved.

10.3.3 Role of Extension Personnel at Various Stages

Sample collection

Soil/plant samples are collected from a field.

↓

Isolation of pathogen/s

Quality control procedures are employed to ensure repeatable results.

↓

Specific pathogen identification

Pathogens are identified to at least genus level under a microscope in the laboratory. An appropriate molecular test may also be used to identify and quantify various pathogens.

↓

Relationship between pathogen population and damage threshold

Estimates of pathogen population density are compared with the damage threshold for the particular pathogen and crop.

↓

Management decision

A decision is made on whether pathogens should be managed. No action may be required, and monitoring may continue, or one or more of the following procedures may be employed, viz., various cultural controls including crop rotation, resistant/tolerant cultivars, organic amendments, bioagents, chemicals and Integrated disease management.

10.3.4 Plant Diagnostic Consultants and Consulting Firms

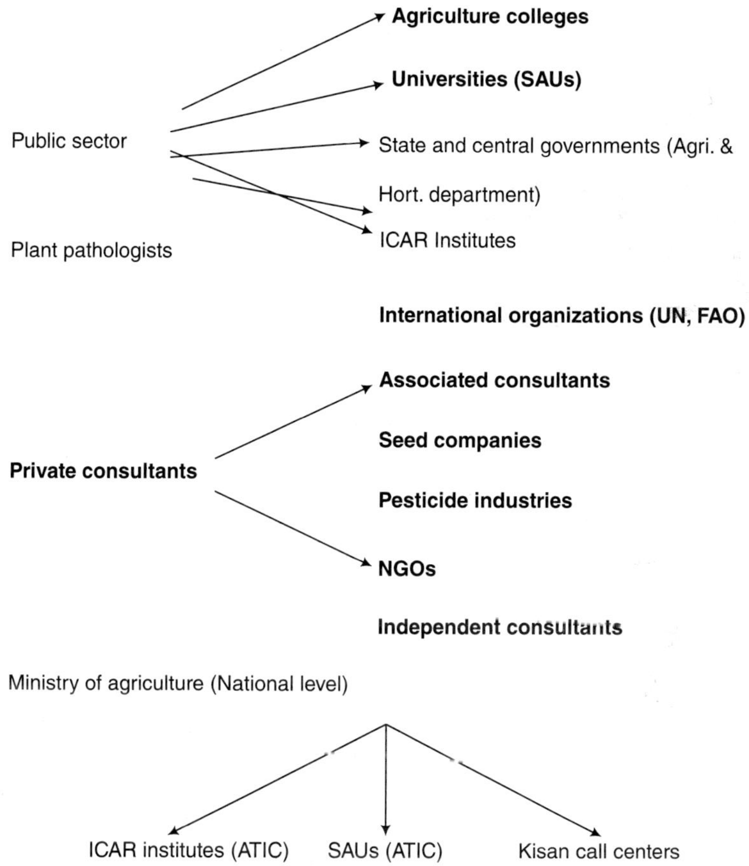

In India, at the national level, Agriculture Technology Information Centres (ATIC) and Kisan Call Centers (KCC) have been established at State Agricultural Universities (SAU) and Indian Council of Agricultural Research Institutes (ICAR).

Plant disease diagnostic clinics generally exist as either a part of the plant pathology department of a state agriculture university, as a part of the state department of agriculture/horticulture or as a private or commercial service. While extension-university-supported facilities, the state department of agriculture laboratories and private laboratories may offer similar services, their motivation and responsibilities may be quite different. Extension-university clinics generally offer disease diagnosis as a service, commonly in conjunction with various aspects of the educational programs of the cooperative extension service. State departments of agriculture diagnostic laboratories operate primarily in a regulatory function with the expressed purpose of identifying pathogens so as to limit their inter- or intrastate dissemination. Commercial or private laboratories also offer diagnostic services, frequently on a more customized basis and commonly in conjunction with customized treatment services.

The concept of plant clinic originally refers to a diagnostic laboratory handling plant protection problems sent to it by the extension service or by farmers. The plant clinic could be part of an advisory center for plant protection or be attached to more research-oriented Plant Pathology laboratories. Plant clinic services play an essential role in communicating results of research into practical advice. Equally, they have to identify problems affiliating the farming and agriculture industry and determine appropriate courses of action to obtain remission from the effects of the disease.

10.3.4.1 Role of extension services

Major role includes the communicating results of research into the field, identify problems affiliating the farming and agricultural industry, determine appropriate course of action, two-way flow of intelligence between the field and the laboratory bench and the vital and unique province of the 'Extension Plant Pathology.'

10.3.4.2 Present challenges faced by extension system

Several extension workers are operating in the tough socio-economic environment. The extension worker has to understand the technical, social, economic, educational, cultural environment he has to operate in (Singh, 2010). For this, it is very important that he appreciates the various problems faced by Indian farmers, viz., small and fragmented land holdings, inferior per hectare yields as compared to international standards, inferior quality of produce, substandard market facilities, poor postharvest and seasonal dependence, multiple produce in small quantities (lack of specialization) leading to wastages, poor storage facilities, problem in availability of adequate and timely credit, distress sale of produce by farmers, poor bargaining power of farmers, inefficient market intelligence, and exploitation of farmer by commission agents.

Research and extension systems need to address the changed paradigm like complex and changing consumer demand, increasing impact of international market, inadequate information with which farmers connect with the market. Research systems should have greater connect with the market. Critical need to build competitiveness to face imports and to increase share in exports will be heightened by the liberalized trade regime. Need to optimize the use of water resources, and logistics and transport cost are becoming extremely important. Factors required are research which acts as source of innovations, discoveries, inventions and continuous improvements, extension system which are capable of disseminating useful information to farmers as well as training and educating them on the utilization of technologies. These should be done for farmers who are willing to improve their productivity and make use of opportunities and efficient market channels for ensuring farmer's benefits. Research in this regard is being conducted by the highly trained researcher in the highly controlled environment, with high inputs and with plenty of labor.

10.3.5 Technology Transfer

In spite of several efforts, the recommendations invariably may not be adopted by the farmers as technology generated is not economically viable, not operationally feasible, not stable, not matching with the farmer's needs and not compatible with the farmer's system. Research and extension should address various queries like 'will it increase the productivity and by how much?,' 'will it decrease the coast and by how much?,' 'will it improve the quality and to what extent?,' 'will it spare, farmer's time and resources and by how much?,' and 'whether the farmer needs this piece of time and resources to allocate them to another activity?

Major steps in technology transfer comprise generation, testing, adaptation, integration, dissemination, adoption and diffusion. However, several times, missing links in technology transfer may include education, training, participation and motivation.

10.3.5.1 *Bottom-up model of technology transfer*

It includes more emphasis on farmer's knowledge, farming situations and farmer's involvement at every step of technology.

10.3.5.1.1 Steps in technology generation through farmer's participation

Diagnosis of the situation, identification of the problem, development of an alternative solution, experimentation, evaluation and finally, diffusion of technology. Therefore, strategy for improved way of knowledge transfer should be by involving farmer as coresearcher/a human resource in whole business from the very beginning.

10.3.6 Role Plant Disease/Health Clinics (PHC/Agriclinic) as Advisory Service

Plant clinics are all about plant health, and the major role of plant clinic lies in diagnosis and advisory. Plant clinics are operating in many nations, but organized clinics are lacking in most of the Asian countries, except India (Srivastava, 2014). Under Centre for Agricultural Biosciences International (CABI)'s Global Plant Clinic Programme (now Plantwise initiative since 2010), walk-in-clinics or community-based plant clinics, lacking physical identity and infrastructure are operating in public/marketplaces, and farmers normally are not aware when and where the next clinic will be held until announced. The situation, therefore, warrants setting up well-organized plant clinics to provide a timely diagnosis of a wide range of ailments and their remedies to reduce losses, increase crop productivity and boost food security of the ever-growing population. Farmers could get right diagnostic and mitigatory advice in time if the Asian nations create well-developed, multispecialty, farmer-centric clinics on the lines of human clinics in rural areas like primary health centers. The clinics should have better physical infrastructure, world-class diagnostic facility with well qualified and experienced pathologists, edaphologist, entomologist and agronomist under one roof along with other facilities like agro-input counter, library, cafe, and toilet. The clinics need to provide services free of cost to plant health and ailments and innovations to achieve higher productivity. The role of a diagnostician is not only diagnosing the problem through visual/microscopic examination or serodiagnostic technique but to provide a need-based mitigatory prescription, with greater reliance on an integrated, holistic approach which is environment-friendly, with least impact on biodiversity and is also easily affordable. It is also important to organize plant health camps, monitor pest scenario, issue pest alert and maintain vigil on pest incursion, and empower farmers with knowledge to tackle pests. During epiphytotic outbreak 'clinic on wheels' must help the farmers in rendering on-the-spot advice to arrest flaring-up of the problem.

Communication with growers needs to be invigorated through print and electronic devices, toll-free telephone, e-mails, and SMSs. Even on-line diagnosis and advisory can be provided as is given in India by XSGrowth Plant Health Clinic— a nonprofit organization (http://www.xsgrowth.com). Plantwise too provides advice on demand to smallholders. Knowledge is crucial to manage plant diseases and ensure sustainability. Therefore, farmers need to be empowered with the latest know-how of managing plant diseases through IPM and 4G fungicides having least impact on biodiversity, using traditional tools of extension and information technology. Besides, maintaining the database, networking, impact analysis and interactions with research, government and pesticide dealers for optimizing techno-resources are also necessary. The important role of classical innovative publications – *Plant Disease Warning and Plant Pathology Courier* in keeping pests at bay also cannot be ignored. To provide a timely diagnosis for a wide range of ailments and providing remedial measures to prevent huge losses due to the wide range of pests, Asian nations must set up organized plant clinics, thus empowering farmers to improve food security.

These agriclinics are basically meant for providing timely advisory services to farmers in respect of both crop production, crop protection and marketing aspects on payment basis. Since these agriclinics operate at gross root level in the system, the situation specific farmer's problems can be easily tackled with the farmers as the key stockholders in the process. Agriclinics are envisaged to provide expert services and advice to farmers on cropping practices, technology dissemination, crop protection from pests and diseases of various crops in the markets and also clinical services for animal health, etc., which would enhance the productivity of crops/animals. For efficient functioning of agriclinics, they should be provided with good infrastructural facilities (laboratory facilities). In the absence of these facilities, they should be assisted by the nearby research stations, State department of agriculture and allied NGOs, colleges and universities working in their vicinity. These clinics should also maintain a complete record of plant disorders and specimens for easier comparison with the field problems. If the specimens are new, they are to be reported to the scientists in research stations/Agricultural Colleges for remedial measures.

10.3.6.1 Global plant clinic

The Global Plant Clinic (GPC) is managed by CABI in alliance with Rothamsted Research and the Central Science Laboratory. The GPC delivers plant health services around the world, working with extension, research, the private sector and governments to make technical support and advice available through plant health clinics. We train plant doctors and scientists, link extension to research and promote new ways to give poor farmers access to the best technologies. Training courses strengthen capacity and foster innovation needed to run regular clinics. Each year the GPC receives queries from over 80 countries, publishes new disease records and extension material and supports more than 60 clinics in the poorest countries of Africa, Asia, and Latin America.

The 1970s and 1980s saw a gradual rise in agricultural projects in developing countries. This was a time when many gained their independence and the previous colonial agriculturists, a proud and notable scientific heritage paid for by the United Kingdom, were becoming 'technical cooperation officers.' The colonial service provided many of the early records of plant diseases in the tropics, as noted by Ainsworth. There was also a notable tradition of research by scientists based in the UK, traveling overseas. The new cadre of postindependence agriculturists made valuable contributions to important diseases such as lethal yellowing of palm in Jamaica, Cocoa swollen shoot virus in West Africa, and the less well known but equally damaging Sumatra disease of cloves in Indonesia (Boa, 2007). Plant clinics are nothing new, but they don't always perform the same functions. The GPC grew out of a 'diagnostic and advisory service,' a unique facility that became a distinct part of the Commonwealth Mycological Institute (CMI) in the 1980s under the fine leadership of Dr. Jim Waller. The GPC name and identity is more recent and has only become well established since around 2000. The GPC is managed by CABI. CMI became the International Mycological Institute before being absorbed into CABI Bioscience. The early history of IMI is described by Aitchison and Hawksworth. The Bioscience label has been dropped and now scientific and publishing activities are grouped under the single identity of CABI.

A more common model is to have a 'clinic' staffed by experts in different types of pathogens (rarely all groups) often with little or no knowledge of growing crops. It would be more accurate to call them diagnostic laboratories. It highlights a lack of clarity and distinction between identification of pest organisms, diagnosis, and advice. (Not so in human health, where pathologists analyze and offer a diagnosis but cannot, at least in the UK, treat patients, even then it is confusing, since pathology labs include clinicians who are also scientists and can treat patients.) When the diagnostic and advisory service was part of IMI the institute was best known for its taxonomic expertise in mycology. There was an uneasy relationship with plant pathology which only improved when CABI Bioscience was created. As the development focus became stronger, it became easier to develop a more integrated US-style clinic, as we now have with the GPC.

Advisory work is usually associated more with the extension than research. Extension and research coexist well in the US but in many countries and among many scientists it would be fair to say that advisory work has an image problem. It appears less glamorous compared to research, with a notion that it requires less effort.

10.3.6.2 Changing the GPC – Plant health service initiatives in US

The pressures to change the diagnostic and advisory service were gradual but persistent. The major impetus eventually came because funders, the UK Department for International Development (DFID), changed. The customer, in this case, DFID, knows best. The nearest equivalent to DFID in the United States is USAID. Both pursue similar aims. The change that ultimately shaped the GPC of today began with the election of the Labour Party in 1997, led by Tony Blair. The new government announced an increased importance and seriousness about international development. Previous governments had been enthusiastic about international development but never, very serious about it. Then DFID published a white paper with a stark title 'Eliminating World Poverty.' It announced a radical change of policy, one matched by increased funds to support a much wider range of activities. This created some unease amongst scientists who read the white paper and found little overt mention of science. Up until 1997 there had been a sizeable chunk of UK government money for agriculture research and hence plant pathology. This shrank as health and education, for example, took on greater importance. Further impetus for change came in 2000 when the millennium development goals were published and strongly supported by the UK government. The millennium development goals set targets for improvements to be met after 15 years.

The main effect of learning more about farmers and agriculture was to work closer with extension agents, giving rise to a new extension method called Going Public. Other innovations include a method for writing fact sheets on plant diseases which are peer-reviewed by farmers and a new method for self-evaluation of technology projects. The GPC still does research on specific plant diseases, though it is limited to major problems supported by additional projects. Examples include Napier grass stunt at Rothamsted Research, and banana *Xanthomonas* wilt at CABI and more recently CSL.

The GPC works in developing countries including Africa, Latin America, and Asia. The GPC continues to provide an expert laboratory diagnostic service for all diseases on all crops, but access to the service is difficult for many in developing countries, even though plant health clinics have improved the sending of samples from growers, particularly poor ones. The cost of sending samples, plus getting permits and other bureaucratic barriers adds to persistent difficulties in collecting suitable material (paying for fuel is a frequent problem). Field diagnosis is the most common method for determining the cause of a plant health problem. It works surprisingly well in many cases though there is little if any formal training on how to do a 'field diagnosis.'

The first training module in "how to become a plant doctor" uses group exercises to compare what people see on sick plants before offering an 'expert' opinion on a sample. This is a standard teaching method we use: people first do a simple exercise, then compare results between themselves before the teacher offers 'expert' advice. Courses are attended by extensionists, scientists and farmers and confidence is often shaky at first, hence the needs to begin gently and avoid emphasizing what people don't know or get wrong. The interpretation (diagnosis) of symptoms begins with a simple distinction between biotic and abiotic before proceeding to particular causes. Pilot clinics have been run in most countries.

10.3.7 Project Based Proposals by Public Sector under National Horticultural Mission, India

Project should be 'development' oriented and not 'research' oriented. The project should be implemented on commercial lines and should be preferably dovetailed with Biological Control Lab and Disease Forecasting Unit.

10.3.7.1 Basic information and implementation guide

(i) Department/Organization seeking assistance should designate experts, lab assistant, and office staff exclusively for the project.

(ii) The project should have a distinct identity with clear signboards and should not become part of other facilities of the organization.

(iii) The organization should identify the building/lab for establishing the plant health clinic.

(iv) The clinic should be located in an area, which is easily accessible to farmers.

(v) The clinic should be self-sustaining after the initial grant, and the revolving fund needs to be created out of profit generated from the income of clinic. Accordingly profitability of the unit has to be worked out by the organization.

(vi) Approximate cost for sample testing that will be available to the farmer has to be worked out along with a clear cut schedule for advising the farmer for remedial measures.

(vii) SHM must ensure quality control through regular checks/inspection.

(viii) Implementation of the project should be monitored through regular checks/ inspection by SHM and Central team. Regular reports are to be submitted to NHM (New Delhi).

(ix) Funds to be recalled if the project is not being implemented as per guidelines or schedule of implementation.

(x) After due diligence and taking a holistic view of the actual requirements, SHM will recommend the proposal to the Central Government with its observations. Mere forwarding of the proposal does not qualify it for sanction of subsidy under NHM.

(xi) Implementing agency to follow proper procedure as per the rules of the organization for procurement of material/equipment.

(xii) Implementing agency will strictly adhere to the implementation schedule and submit a regular progress report to SHM and UC in prescribed format at the end of the year or stipulated time period whichever is earlier.

(xiii) SHM to ensure the release of funds/grant in one installment to implementing agency. Funds are to be released through the finance department/comptroller of the institute/university for following the financial rules, records and auditing purpose.

(xiv) SHM to obtain regular progress and UC from the implementing agency and submit a consolidated report for each component to the Central Government.

(xv) The organization has to maintain a proper record in the stock register for auditing purpose.

(xvi) The organization has to submit a quarterly progress report to the Mission Director.

10.3.7.2 Facilities to be created under the plant health clinic

Major facilities to be created include computer, pH meter, conductivity bridge, microscope, seed tester, BOD incubator, shaker, centrifuge, glassware, Eliza testing kits, lab furniture and racks, water distillation units, provision for skilled manpower on contract basis, construction of lab consumables, etc. The clinics are to be established under supervision and technical support from ICAR or institutes having the capability. The estimated cost is ₹ 20.00 lakh, available as one-time grant-in-aid under NHM.

10.3.7.3 Major channels for technology dissemination to farmers

Important channels include farm and home visits, group discussions/meetings, method demonstrations, result demonstrations, campaigns, exhibitions, field visits, field days, farmers' training, extension literature, and print and electronic media including E-SAP.

10.3.8 e-Pest Surveillance

The objective of e-pest surveillance is scientific control of pests and diseases through meticulous field observations by uploading in time for experts' scientific advice and timely implementation of management interventions based on these advisories so as to ensure ecofriendliness, economical, timely pest control measures with minimum pesticide residue levels. This goal of timely application of pesticides was achieved through three-pronged approaches, viz., recording of pest or disease/disease incidence data at weekly intervals in structured sheet through pest scouts at field level in villages, collation of such pest data into one central server for analysis and interpretation of such data to provide guidance, and issue pest advisory on the basis of economic threshold levels.

It provides field-specific information on pest pressure and crop injury leading to take an appropriate decision on selection and application of pest management procedures by constant monitoring and scouting. It is basically a web-based Decision Support System (DSS) for timely collection of data with regard to pest population and preparation of reports which greatly facilitates the execution of pest management activity. Besides it helps in the creation of database on pests that may also be used to ascertain the impact of climate change on the incidence of pest/ diseases in the mandated crops.

10.3.8.1 e-Pest surveillance device for plant pests and diseases

This is a portable, handheld device, a type of protected Personal Digital Assistant for capturing data on pests and beneficial insects and sending it to institutes where it is made part of databases (Anonymous, 2008). The device can withstand rough field conditions (dust and water, high temperature, poor visibility, etc.) and has high battery life for long field use. This enables application of ICT for farmers' benefit. Such survey and database are prerequisites for reliable pest forecasting. The surveillance apparatus in the field is weak, and part of this inadequate surveillance is a lack of an easy system for capturing field observations in a standard format and immediately transmitting the data upwards. This device will fill this gap.

The introduction of this device is a very important step towards achieving a high level of pest surveillance and issue of farm advisories for effective pest management data would also be maintained beneficial insects. In the long run, the database would be used for understanding long-term implications of climate change and such other phenomenon on crop-pest relationships. The field observation official fills up forms provided by a very simple and user-friendly software in the device. The software makes a summary of the observations, and the official sends the summary to the National Centre for IPM through the internet. The data forms part of the database maintained by the center. The database is geo-referenced using a GPS (geo-positioning system) so as to pinpoint pest data to different geographical areas. This device has been developed by Plant Protection Division of Department of Agriculture and Cooperation with the help of FAO and the National Centre for Integrated Pest Management, ICAR.

10.3.8.2 SXagro

SXagro is a SaaS model agro-informatics platform for real-time agri-data and analytics. It uses remote sensing based satellite data and ground survey to build actionable intelligence for Agri-community (Anonymous, 2008). SXagro has inbuilt specific software algorithms for use in following scenarios. Geographic Information System (GIS), Remote Sensing Image Analysis (RSIA), Global Positioning System (GPS) are employed in various projects to generate spatial information while Modeling and Simulation is the basic tool used to understand and predict biophysical and socioeconomic processes together with outputs of the systems being investigated.

It is the approved e-system by National Centre for Integrated Pest Management (NCIPM), New Delhi, India. GPS-based rugged handhelds SXtreo is used to collect field level data on the pest infections on the crops. Data collect includes the type of disease/ pest, the severity of infection and location (GPS based). This data would be transferred to the central database immediately, and the administrators would have a centralized geospatial pest infection map on the second day itself which would include the weather forecasts. The administrators would have a complete analysis of the whole state's geospatial view of the agricultural situation to take immediate action and send urgent advice to the individual blocks and farmers through SMSes on pest control. With the SXagro from Stesalit, the 21-day cycle thus becomes just a two-day cycle. Pest control through ePest Surveillance system module of our SXagro solution is the most scientific, advanced and cost effective solution, keeping the challenges of agriculture in India in mind.

SXagro is a significant tool to help the nation manage its land and soil resource sustainably, restoring their productivity for future generations. Information includes land use, drainage, weather forecasts, soil type, soil depth, and soil chemical properties and recommendations for each type of land use. Using GIS in the visualization of the results and developing user-friendly graphical interfaces by incorporating analytical models our multidisciplinary research unit has developed SXagro, an SDSS (spatial decision support system) specific to agricultural needs for the Indian subcontinent

10.3.8.3 SXagro modules

10.3.8.3.1 Data capture module

Data Capture is the interface where all historical data entries are available. All geo-reference data inputs including latitude, longitude, pest intensity, weather data entry (temperature, humidity, etc.) is captured.

10.3.8.3.2 Reporting module

Different reports of captured data can be downloaded or viewed using different reporting format such as tabular, graphical and GIS maps. Tabular data report can be sorted or customized for download. All transaction reports, daily, weekly, monthly, quarterly, annually and with given date range. Reports can be downloaded as Spreadsheet, Graphical (PDF). GIS map reports can be searched according to variables like date range, crop, pest, center, block, village, etc. GIS Map will be displayed using Google Map API & Ajax technology for faster processing.

10.3.8.3.3 Data analysis module

This Module will be helpful for Analysis of all captured data inputs. There will be a different statistical method for analysis. Using different GIS tools, all captured data will be analyzed based on required statistical methods, which will be helpful for users to get advisories for pest/disease management.

10.3.9 e-National Pest Reporting and Alert System

Pest surveillance or monitoring is a cornerstone of pest management rather than calendar-based treatments. The purpose of pest surveillance is to know whether pest in present in the fields and hence to apply the appropriate pest management practice. By continuous and systematic pest surveillance, epidemic situations of any pests can be avoided by detecting damage prior to the firm establishment of a higher pest population (Sharma *et al.*, 2013). In fact, pest surveillance provides field-specific information on pest pressure and crop injury leading to take an appropriate decision about selection and application of pest management procedures. Use of information and communication technology (ICT- the internet and mobile) in pest surveillance constitutes e-pest surveillance, i.e., basically a Web-based Decision Support System (DSS) for timely collection of data with regard to pest population and preparation of reports which greatly facilitates the execution of pest management activity.

e-National Pest Reporting and Alert System in pulse crops is a unique ICT based decision support system, which is very effective and easy to operate through a centralized server system at National Centre for Integrated Pest Management, New Delhi, connected with the internet and mobile phones. This system has developed to cater to the needs of rural farmers of India, who grow pulse crops. The useful information is collected, stored, processed, and interpreted, and the appropriate advisories are sent to the registered farmers through the centralized server system. They apply suitable corrective measures as per advisories at the right time, and thereby, heavy loss caused by various pests can be checked/minimized below an economic threshold level. Based on the past experiences and larger response of the stakeholders, Department of Agriculture and Cooperation, Govt. of India, suggested that this program should be extended and implemented in all pulse growing states. This system is quite useful bottom to top level officials/policy makers, involved in the ePest Surveillance program.

In view of the futuristic requirements as well as strategies to meet unforeseen circumstances, Department of Agriculture and Cooperation (DAC) and National Centre for Integrated Pest Management (NCIPM), New Delhi, took the initiative under the ambit of National Food Security Mission (NFSM) to increase the yield by reducing pest infestation, with the help of 'e-Pest Surveillance' and implementation of Area-wide Integrated Pest Management System by involving State Agricultural Departments (SADs), KVK (Krishi Vigyan Kendra) officials and farmers by conveying the clear-cut message to the farmers about appropriate and timely decisions to be taken for pest management, and the response of farmers are highly significant and encouraging. In this program, IPM nuclear model villages have been established.

10.3.10 Information and Communications Technology (ICT)

Information and communications technology are an umbrella term that includes any communication device or application, encompassing: radio, television, cellular phones, computer and network hardware and software, satellite systems and so on, as well as the various services and applications. The term ICT is also used to refer to the convergence of audio-visual and telephone networks with computer networks through a single cabling or link system. There are large economic incentives (huge cost savings due to the elimination of the telephone network) to merge the telephone network with the computer network system using a single unified system of cabling, signal distribution, and management.

The agriculture needs to be developed, and e-Agriculture is an emerging field of providing farmers with relevant agricultural information for their development and business, and, therefore, it need the conceptualization, design, development and application of innovative ways to utilize existing or emerging Information and Communications Technologies (ICTs). In agricultural extension and education, from universities to farmers' fields, ICTs facilitate learning. Scientists communicate to brainstorm ideas and be creative, formulate research questions, solve experimental or theoretical problems, disseminate results, and get feedback.

Information and Communication Technologies (ICTs) are used to produce, store, process, disseminate and exchange information.

ICT → Information Technology (IT) + Communications (C)

The term encompasses a wide range of technologies including the older ICTs – radio (FM), press (digital online), television (phone-in and interactive), film (amateur), walkie-talkies and telephones (call centers, ivrs) and also the newer ICTs – email, CD-ROMs, websites, computers (kiosks), mobile phones, palm pilots, digital video cameras and smartphones, GPS systems, community radio, portals, video conferencing, Skype, voice web, interactive TV, gateways, and social media sites. In extension, ICT is considered as a tool which enhances the flow of information in the application of agricultural extension services. The advancements in ICT can be utilized for providing accurate, timely, relevant information and services to the farmers, thereby facilitating an environment for more remunerative agriculture.

Use of ICT is key to India achieving its real potential in Agricultural Productivity and raising Farm Production to Feed an additional 310 million people expected by 2030. It has multiple roles viz., deliver or provide access to localized, customized and farmer-centric information, to organize the knowledge base into an easy, shareable, searchable and understandable format, to connect people and network them locally, regionally and nationally (including globally) and empower the rural community with livelihood development. ICT technologies provide elements of focus any time, anywhere, any language; it is transparent, equal, equitable, quick, correct, clear, cost-effective, and efficient with legal.

10.3.10.1 ICT-based services

Web or mobile or kiosk or community radio or phone-in programs services include pest and diseases information portals, seed and planting materials information systems, fertilizer information systems, soil/seed testing services and soil health card, agro-advisory services through ICTs, marketing information systems, agricultural mechanization systems and agriculture portals: AGRISNET and Kisan Portal.

10.3.10.2 Innovative usage of ICTs in extension

- e-mail has facilitated the unprecedented collection and exchange of information.
- Website development has resulted in an explosion of online publishing/exchange of information by many organizations.
- Satellite technology has expanded the potential of Radio/TV (particularly, Phone-in programs) broadcasting.
- Open Source software contributed towards knowledge sharing within Organizations and Communities of practice promotes shared use of data, and capture, synthesis and dissemination of (success) lessons learned.
- Telephone hotlines become an important tool to provide information and advice to individuals.
- Print newsletters are published on the web and disseminated by email.

The revolution in web-based technologies has led to great strides in the development and employment of decision support systems for growers and pest management specialists. José Maurício Cunha Fernandes *et al.* (2011) illustrated an approach towards that direction by the use of novel programming languages and technology for the development of a web-based system for model implementation and delivery. SISALERT is a multimodel platform that unleashes the power of hourly weather station data and hour-by-hour weather forecast information using sophisticated disease risk assessment models. These models interpret weather data giving information on the past or recent disease behavior as well as predicted disease risk. The uniqueness of SISALERT format is the modularity that allows coupling crop and disease models depending on which model is under run.

In the apple production area, the current advisory has enabled growers to manage fungicide use, time, and resources much more efficiently and profitably. Improvements to this system could only increase profit and efficiency, and reduce unnecessary dependency on fungicides. Besides extending risk information for a large geographical region, the use of intuitive images representing epidemic risks may facilitate dissemination and understanding of risks to guide decision-making on plant disease management. In addition, maps may be useful for the fine tuning of crop zoning and the identification of quality aspects in postharvest areas. A major obstacle in disease forecasting is the validity of the disease forecasting models. Although accuracy of weather information could be improved by using on-site weather data associated with weather forecast data, the model itself may cause inaccurate forecasting due to its limitation of validity.

There is a need for continuous and systematic effort to develop better models. It is, therefore, imperative that attempts should be made to establish a network for testing and model comparison (José Maurício Cunha Fernandes *et al.*, 2011). The information delivery system in this study can be applied to forecast other diseases and insects pests just by changing forecasting models in the program server. Also, if mesh climatic data are available at a fine scale in the future, the system can produce the disease forecasting information on the same scale without automated weather stations at every site. In conjunction with geographical information systems, this system can be used to generate maps to illustrate disease development over the country.

There has been a sea change in the mindset and the effort put down by the agricultural industry by adapting to the current trends and technologies. One such example is the use of Information and Communication Technology in agriculture which eventually contributes to Precision Agriculture. Presently, plant pathologists follow a tedious technique that mainly relies on naked eye prediction and a disease scoring scale to grade the disease. Tejal Deshpande *et al.* (2014) attempted to automatically grade the bacterial blight disease on the pomegranate plant leaves. This innovative technique would be a boon to many and would have a lot of advantages over the traditional method of grading. Manual grading is not only time consuming but also does not give precise results. They proposed an image processing methodology to deal with disease grading. The results are proved to be accurate and satisfactory in contrast to manual grading and hopefully take a strong leap forward in establishing itself in the market as one of the most efficient and effective processes. The proposed system is also an efficient module that identifies the bacterial blight disease on pomegranate plant.

At first, the captured images are processed for enhancement. Then image segmentation is carried out to get target regions (disease spots) on the leaves and fruits. Later, if the diseased spot on the leaf is bordered by a yellow margin, then it is said that leaf is infected by bacterial blight otherwise not. Similarly when black spots are targeted on fruits, they are checked for whether a crack is passing through these spots. If cracks are passing through the spots, then the disease identified would be bacterial blight. Based on these two characteristics bacterial blight on pomegranate can be appropriately identified. This new technique improved the efficiency and productivity through a robust system which can overcome the shortcomings of the manual process. Looking at the current scenario an approach to automatically grade the disease on plant leaves is very much essential. Grading System built by Machine Vision is very useful for grading the disease. The disadvantages faced through the manual grading would be overcome once this system is adopted and will help the pathologists in terms of complexity and time. Through research and experiments, it has been observed that the results found are precise, accurate and acceptable. Also it is observed that the bigger the bacterial blight spots, the higher grade is obtained.

Due to the manifestation and developments in the fields of sensor networks, robotics, GPS technology, communication systems, etc., precision agriculture started emerging [10]. The objectives of precision agriculture are profit maximization, agricultural input rationalization, and environmental damage reduction by adjusting the agricultural practices to the site demands. In the area of disease management, grade of the disease is determined to provide an accurate and precision treatment advisory (Sanjeev S Sannakki *et al.*, 2011). They introduced an innovative approach to automatically grade the disease on plant leaves. The system effectively inculcates Information and Communication Technology (ICT) in agriculture and hence contributes to Precision Agriculture. Presently, plant pathologists mainly rely on naked eye prediction and a disease scoring scale to grade the disease. This manual grading is not only time consuming but also not feasible. They proposed an image processing based approach to automatically grade the disease spread on plant leaves by employing Fuzzy Logic. The results are proved to be accurate and satisfactory in contrast with manual grading. The Grading System built by Machine Vision and Fuzzy Logic is very useful for grading the disease. This kind of expert system is going to help pathologists as it overcomes almost all the disadvantages of manual grading in terms of complexity and time.

A survey of plant disease detection was taken up by using image processing and Wireless Sensor Network/Wireless Multimedia Sensor Network (WSN/WMSN) techniques in precision agriculture (PA) field (Jinpeng Wang *et al.*, 2014). The machine vision based approach uses size, shape, texture, color and location-based features to detect disease affected leaf or plant. As the development of Internet of Things (IoT) technology, the scalar sensing data and multimedia video/audio data would improve the detections and provide efficient long-term prediction in the PA application. This survey also presented a study about the current existing issues and promoted an emulation based architecture that is able to combine these different emerging technologies practically.

Current plant disease advisory systems tend to be generic and are not tailored to specific plots or farms. This work comprises an agriculture advisory call center similar to a modern call center to provide an agriculture disease mitigation system. The information regarding an individual farm is collected using mobile phones (Sion Hannuna *et al.*, 2011). The image of the diseased/infected crop is also captured using mobile phones and is made available to the expert to provide the advisory. To scale the advisory, an attempt is also made to automate the disease recognition process using image processing. Unfortunately, the photos taken will be sensitive to a number of factors including camera type and lighting incident on the scene. Ideally, the images would be processed in such a way as to provide the expert with a visual representation of the affected crops that reflects the true nature of the scene. They described a framework for standardizing the color of plant images taken using both mobile phones and compact cameras within the context of the advisory system. Also, they presented an effort of using ICT to overcome the information gap by creating a sophisticated call center akin to what is used today in the developed world to provide support services for goods and merchandise sold to individual customers and attempting to automate as far as possible disease recognition process using image processing. This program includes several steps.

10.3.11 Modern Call Centers

The modern call center uses sophisticated technologies. First of all, an extensive data is captured for each customer including specifics of the purchase, the past interaction that he/she would have had with the call center and the complete service history (Sion Hannuna *et al.*, 2011). As the customer calls, the call is routed to an agent and the customer page pops up on the agent's screen. Within seconds, the agent gets a picture of the profile of the customer, the service history making the interaction personalized. Further, the interaction is continuously captured in the database, and a supervisor is added to the conference call whenever the customer is not satisfied with the present interaction.

10.3.11.1 Agriculture advisory call center

The agriculture advisory call center is similar to that used in a modern call center and has the features, viz., uses state-of-the-art technology, has a database of the farmer and farm. The database is constantly updated, provides all the previous interactions with the farmer, integrates relevant information (like recent weather history, at the farmer's land, prices at nearby markets) and presents them on a dashboard, makes images of diseased crop available to assist in providing the advisory, gives the extension worker at the call center a virtual feel of being on the farm and needed basis scientific, and market experts are added to the conference call to provide specific advisory (Sion Hannuna *et al.*, 2011). The agricultural call center has an information repository of the individual farms collected using a mobile phone for data and image capture. Continuous updating of data is carried out, and the interactions with call center advisory are captured. Farmers can upload additional information and pictures of the crop using their mobile phones. The extension worker handles the calls from farmers and conferences in experts as and when necessary.

10.3.12 Computer-based Disease Diagnosis

When a farmer calls the call center for advice, the relevant information of the farm is presented on a dashboard. When needed, the farmer is requested to upload the pictures of diseased crop. The extension worker uses the crop information along with its images to provide the advisory (Sion Hannuna *et al.*, 2011). When the extension personnel fails to identify the disease or the pest or do not have the solution, an expert or a scientist is added to the conference call. The experience is that the solution is found quickly most of the times. While the system does work, it would be difficult to scale the system as more and more farmers start using it. Trained extension personnel, as well as agricultural experts/scientists, are limited in number. Using their services to handle each call would not be scalable. The next step would be to automate the disease/pest detection and advisory system. As the advisory is based on the image, the quality of the image becomes a critical factor. The automatic disease recognition process incorporates color correction, standardizing for lighting and camera variation, deducing a transformation matrix, leaf segmentation, and disease recognition.

The limited field trials show the importance of disease mitigation as a service for farmers. While there are direct and immediate benefits for farmers, this is an important step towards reducing crop losses in India. The work needs to be expanded and tested for an extended period of time to understand the intricacies of the disease mitigation practices and fine-tune it for maximum benefit (Sion Hannuna *et al.*, 2011). At the same time, the early results make us more determined to work on a scalable solution. While the call center would be able to carry out disease mitigation, costs of extension workers would be a matter of concern. Automatic disease recognition will go a long way in reducing costs and provide the ability to scale. The information provided by farmers is used with the computer vision algorithms to assist the expert disseminating advice. An approach to standardizing plant images for pathology diagnosis has been detailed. The efficacy of our custom color chart and the use of a weighted least squares formulation have been demonstrated. It is anticipated that this method will be applicable to other applications utilizing color images that cannot be captured in a repeatable way, for quality control and assessment purposes.

10.3.12.1 Distance diagnostic and identification System (DDIS) for extension

The DDIS for extension is developed jointly by the extension agents, specialists, and faculty of UF/IFAS Information Technology. The system is designed specifically for university agricultural specialists and diagnostic laboratories or clinics (Anonymous, 2014). DDIS provides a collaboration and communication platform for first detectors, extension specialists, and diagnosticians to share information on plant insects and diseases. The system uses field data and digital media as tools for enhancement of diagnosis of plant diseases, insects, weeds, invasive species, plant management, physiology, and nutrient problems. Through interactions on the Internet between extension agents and specialists, problems can be communicated immediately and assessed. Specialists around the state can perform diagnosis and identification and provide the best management practice recommendations to the users. The archived DDIS database becomes a resource for research, educational programs, and classroom teaching. The threat of pests and plant diseases has the potential to seriously damage our agriculture and food supply. The DDIS can be used as a tool to enhance the capacity for screening, early detection, monitoring, pest mapping, and rapid communication to protect agriculture.

Enhanced images have high quality and clarity than original captured images. Computer vision image enhancement (Color conversion and Histogram equalization) is used in different real-time applications such as remote sensing, medical image analysis and plant leaves disease detection. Original captured images are RGB images (Thangadurai and Padmavathi, 2014). RGB images are a combination of primary colors (Red, Green, and Blue). It is difficult to implement the applications because of the range of this color is 0 to 255. Grayscale images have only the range between 0 and 1. So it is easy to implement many applications. Histogram equalization was used to increase the image's clarity. Grayscale conversion and histogram equalization were used in plant leaves disease detection.

10.3.12.2 *Machine learning technology*

Current trends in the prediction of crop pests using Machine Learning Technology were described by Yun Hwan Kim *et al.* (2014). With the advent of data mining, the field of agriculture is also focused on it. Currently, various studies, domestic and overseas, are under progress using machine learning technology, and cases of its utilization are increasing. They classified and introduced SVM (Support Vector Machine), Multiple Linear Regression, Neural Network, and Bayesian Network-based techniques and described some cases of their utilization.

10.3.12.3 *Fractal dimension and chaos technique*

A novel method for detecting rice leaf disease using image processing technique called 'fractal dimension and chaos theory' was proposed by Surendrababu *et al.* (2014). The analysis of a diseased leaf was carried out according to its image pattern and fractal dimension and especially box-counting ratio calculation, and chaos were applied to be able to identify the disease pattern's self-similarity and to recreate the fractal. The image's self-similarity was the disease infected one which was same as when it was fully infected. This method was proposed as preliminary information for the development of an early detection system or for developing knowledge based expert system or decision support system.

The study surveyed the most effective techniques to extract new knowledge and information from existing image processed and extracted data confined with numerous other parameters to return a dynamic solution. More than a few data and image mining coupled with chaos and fractal dimension technique discussed here can be used to detect the rice leaf disease. It has sound theoretical foundations, involves only a dozen examples for training, and is unaffected by the number of dimensions. It can be said that efficient training methods can be developed using the above-said technique. The authors have surveyed a few important research challenges in chaos and fractal dimension in image processing for the detection of rice leaf diseases. The fractal for the leaf pattern of the diseased leaf at the final stage will have the same pattern during the initial stage for every kind of rice leaf disease because of self-similarity. Therefore, the recreation of the fractal would enable the early identification of the disease. The processes are done through fractal dimension and chaos theory to recreate the fractal from the blast disease affected leaf patterns to be done the same way to other types of leaf diseases. In this way, there is identification of every type of disease that has been studied. The recommended methods and findings can also be of much interest and gives scope for future research in the field of chaos and fractal dimension in image processing for other applications and fields.

10.3.12.4 Fuzzy Inference System (FIS)

Plant disease diagnosis is the determination of the cause of a disease in a plant and detection refers to the identification of microorganisms or their products, such as toxins in plant tissues, soil, and water. Failures in pathogen detection and disease diagnosis lead directly to inadequate disease control and reductions in crop production and quality, and hence trade (Sunita I. Naik *et al.*, 2014). When plants are affectionate to diseases, they display a range of visual symptoms such as colored spots, or streaks that can occur in the leaves, stems, and seeds of the plant. These visual symptoms continuously change their color, shape and size as the disease progresses. In most of the cases, identifying diseases relying only on symptoms with bear eye observation is prone to error and misleads crop growers. Therefore, automatic foliar disease identification by visual inspection can be of great benefit to the farming community. The proposed approach aimed to develop an efficient plant disease diagnosis system that successfully recognizes the foliar disease spots, if any and classified it into appropriate category considering pomegranate fungal leaf spots, namely, *Alternaria* leaf spot and *Cercospora* leaf spot as case studies. The system also aimed to develop a treatment advisory wherein disease control measures are provided to farmers, in this case, based on the classification results.

It is very well known that identification of the symptoms of plant diseases by means of image processing techniques is of prime concern in the area of research. There is a need for a plant disease diagnosis system that may support farmers during their daily struggle against disease outbreaks. The proposed work by Sunita I. Naik *et al.* (2014) is a development of an efficient diagnosis system that focuses on plant disease identification by processing acquired digital images of leaves of the plant. These images are made to undergo a set of preprocessing methods for image enhancement. The enhanced image is segmented using *k*-means and thresholding based segmentation approaches to extract the region of interest, i.e., diseased portion. Later, a satisfying set of visual features from the region of interest are extracted by applying first-level Haar Wavelet Transform method for detecting diseases accurately. Based on the extracted features set, a Fuzzy Logic based classifier identifies the type of the disease being occurred. Finally, treatment measures are provided to control the disease. The advisory helps farming community in effective decision making to protect their crop from diseases and increase its productivity. Thereby, the proposed approach improves crop yield and uplifts the economy of the farming community.

10.3.12.5 MATLAB

The most common features of blades which are affected by plant diseases and insect pests have been introduced, the internal relationship between the change of color of the blade surface and the sharp change of grayscale values of the pixels of digital images has been connected, the measure of identifying the edge information of blade surface and locating the affected areas have been analyzed relying on the toolbox of the mathematic software MATLAB (Yi-ch Zhang, 2010). Different operators' edge detection abilities have been compared under the condition of noise. Discussions have been made in aspects such as the principle of edge detection of blade surface, the features of different edge detection operators and so on. A conclusion has been made that the best edge detection algorithm has a quite satisfying performance, as well as the excellent feature against noise such as 'Gaussian' and 'salt and pepper' in plant diseases and insect pests detection.

10.3.13 Electronic Solutions against Agricultural Pests (e-SAP)

In the recent years, interactive relations between various branches of science and technology have improved interdisciplinary fields of science. In fact, most of the research activities take place somewhere among these branches. Therefore, a specialist from one branch usually can propose novel methods, whenever enters a new field, based on his previous knowledge. Taking a look at the extensive problems in the field of agriculture, an expert in the field of Electromagnetic waves can easily suggest some innovative solutions to solve them (Hadi Aliakbarian *et al.*, 2010). The major suffering problems with which a farmer faces are the damages caused by the harmful pests as well as the product freezing in unexpected cold weather. The promising available biological methods of treatment have decreased the need for new treatment methods effectively. However, some advantages of electromagnetic treatment are still without a competitor. The environment-friendly methods we introduce in this chapter are to use electromagnetic waves to kill pest insects without killing the taste or texture of the food they infest.

Electronic Solutions Against Agricultural Pests (e-SAP), an innovative tablet device, promises to be of much help in the field of agriculture. It helps people get instant information about the kind of pest that's ruining the plants, with suggestions on how to fight it. The hand-held device is easy to use. Even a layman can navigate to find the problem in his crop.

The device, Electronic Solutions Against Agricultural Pests (e-SAP), has a web-based application system which facilitates the flow of information from the farmer to the farm scientist and back. The idea behind creating an e-SAP was to help the extension worker, who is not an expert, to collect specimen's right from the farmer's field and send it to the scientists and experts in real time. e-SAP will make the work of the extension service worker easy, enhance their efficiency and at the same time provide the farmers with solutions right in his field in real time. This technology targets one of the critical requirements of a crop cycle, pest management. e-Sap has features that can bring the farmer, extension worker, scientist and policy maker on the same plane, thereby, helping to find solutions that are more practical and in lesser time. e-SAP devices provide continuous feed to the handheld devices with latest farm technologies, modifications, and recommendations.

Scientists across the globe at various universities/institutions/organizations have developed and using this system. The University of Agricultural Sciences, Raichur, Karnataka, India developed a unique IT-enabled handheld device that is providing information to farmers in real time on pest-related problems (Prabhuraj, 2012). e-SAP is unique in the sense that it has a voice-based application system, which guides the farmer and the extension worker in the local language about how to collect the data and the specimens, he added. That apart, it also allows the extension worker and the farmer to do a survey of the pest attack or related problems right in the field, which is then automatically synthesized in the form of graphs and tables along with the decision support intelligence. The University of Agricultural Sciences, Bengaluru, Karnataka, India also has developed this system and is functioning successfully.

Another highlight of e-SAP is the image-based model that captures high-quality images of pests and their symptoms and then guides the user in identifying the pest. This coupled with audio assistance makes it very easy to handle. There is an intuitively built pest-specific survey to quantify the damage caused which is based on images and also utilizes the images captured by the extension worker from the affected field. This survey helps the expert is ascertaining the extent of damage and suggest remedies.

10.3.14 Electronic-nose (e-nose) Instruments

Electronic-nose (e-nose) instruments, derived from numerous types of aroma-sensor technologies, have been developed for a diversity of applications in the broad fields of agriculture and forestry. Recent advances in e-nose technologies within the plant sciences, including improvements in gas-sensor designs, innovations in data analysis and pattern recognition algorithms, and progress in material science and systems integration methods, have led to significant benefits to both industries. Electronic noses have been used in a variety of commercial agricultural-related industries, including the agricultural sectors of agronomy, biochemical processing, botany, cell culture, plant cultivar selections, environmental monitoring, horticulture, pesticide detection, plant physiology and pathology. Applications in forestry include uses in chemotaxonomy, log tracking, wood and paper processing, forest management, forest health protection, and waste management (Wilson, 2013). These aroma-detection applications have improved plant-based product attributes, quality, uniformity, and consistency in ways that have increased the efficiency and effectiveness of production and manufacturing processes. This paper provides a comprehensive review and summary of a broad range of electronic-nose technologies and applications, developed specifically for the agriculture and forestry industries over the past thirty years, which have offered solutions that have greatly improved worldwide agricultural and agroforestry production systems.

Electronic-nose devices are utilized in a wide range of agricultural industries to perform a multitude of functions ranging from quality-control monitoring of agricultural and forestry products, monitoring industrial process controls, food production and storage systems, indoor air-quality control, detection of environmental hazards, gaseous and liquid effluents and other factory waste releases. The most common applications of electronic noses in agriculture are to monitor food quality and production processes, detect crop diseases, and identify insect infestations (Li *et al.*, 2010). Plants utilized in agriculture and forestry release VOCs as a byproduct of normal physiological processes. The specific VOCs produced and the quantities released are indicative of both crop and field conditions. Many factors including humidity, available moisture, light, temperature, soil condition, fertilization, insects, and plant diseases may affect the release of VOCs from agricultural plants. Thus, monitoring VOCs released from plants provide indications of plant health, growing conditions, the presence of environmental stresses, and the presence of adverse factors that may affect plant growth, production and crop yields. Electronic-nose devices offer numerous potential applications in agriculture including such diverse uses as the detection of pesticide residue levels on crops or in the environment, industrial applications including detection of gas leaks and toxic gas emissions, and for homeland security as an early warning system for bioterrorism.

Agronomic uses of e-nose devices have included crop-protection applications in the field to detect hazardous chemicals and microbes (e.g., chemical or biological agents of bioterrorism) as well as pesticides on plant foliage (Wilson, 2012), making selections of plant cultivars of individual crop types for cultivation and to monitor plant cell cultures for growth and behavior. Electronic noses have been utilized for several botanical applications involving the detection and monitoring of volatile biogenic gas emissions and floral odors to determine season variations in plant emissions, for identification of plant host-defense mechanisms, and for plant identifications based on nonfloral volatiles.

Advances in metabolic engineering techniques have provided a better understanding of the biochemical pathways involved in the biosynthesis of volatile secondary metabolite compounds, facilitating the identification of the plant genes and enzymes involved as well as the chemical structures of a large number of new plant volatiles. Plants produce a large diversity of low molecular weight VOCs known as secondary or specialized metabolites. At least 1% of these plant secondary metabolites (PSMs) are lipophilic molecules (consisting primarily of terpenoids, phenylpropanoids/benzenoids, fatty acid and amino acid derivatives) with low boiling points and high vapor pressures at ambient temperatures. Plant secondary metabolites are released from all parts of the plant (e.g., roots, stems, leaves, flowers, and fruits) into the atmosphere. The primary functions of PSMs are to defend plants against insect herbivores and microbial pathogens, attract pollinators, facilitate seed dispersers, promote the growth of beneficial animals and microorganisms, and serve as chemical signals involved in plant-plant and plant-herbivore interactions (Wilson, 2013). Thus, PSMs are important volatiles that contribute to plant defenses as well as survival and reproductive success in natural ecosystems.

Production of PSMs by crop plants also has a significant impact on agronomic and commercial plant characteristics, crop yield and food quality. Consequently, the modification of PSM-volatile production via genetic engineering has the potential to make crop plants less attractive to herbivore enemies and improve the traits of cultivated plant species

Wilson *et al.* (2004) first applied e-nose technologies to plant pathology for the diagnosis of tree diseases, particularly those caused by phytopathogenic microbes, such as vascular wilts and bacterial wet wood, and for the detection and identification of wood decay fungi, causal agents of wood rots in living trees. Subsequent studies have demonstrated the capabilities of several e-nose instruments to detect specific types of wood decays, i.e., those caused by particular wood decay fungi, in different host wood species. The early detection of incipient wood decays in trees with e-noses is particularly important in forested urban environments where tree failures, e.g., breakages of major limbs or the main truck, can cause significant damage to property or result in human fatalities.

A wide range of e-nose instrument types is utilized in the agricultural and forestry industries to perform many diverse functions and applications to facilitate the multitude of steps and processes involved in the production of plant-based products (Table 10.1). The majority of these applications have involved the use of MOS and CP-type sensors, but other e-nose sensor types (CBC, CO_2, ECS, MOSFET, QMB, SAW, and SnO_2 sensors) have been used to detect certain specialized types of gas analytes. The major application sectors to which e-nose gas detections have been applied within the agricultural and forestry industries are in such key areas as crop and food production, chemotaxonomy, environmental protection and monitoring, manufacturing process controls, plant pathology, quality control and quality assurance (QA/QC), waste management, and wood identifications (Wilson, 2013).

Table 10.1 Electronic-noses used for Specific Agricultural and Forestry Applications

Applications	*Electronic-nose*	*Sensors/types*[†]	*Chemicals detected or uses*
Plant pathology	Aromascan A32S	32 CP	Disease detection, pathogen ID, wood decay fungi ID
	LibraNose 2.1	8 QMB	Wood decay and fungi ID
	PEN 3	10 MOS	Wood decay and fungi ID
	Cyranose 320	32 CBC	Postharvest disease detection

[†]Number of sensors and sensor type abbreviations: Carbon black composite (CBC), Carbon dioxide sensor (CO_2), Conducting polymer (CP), electrochemical (EC), Metal oxide semiconductor (MOS), Metal oxide semiconductor field effect transistor (MOSFET), Quartz crystal microbalance (QMB), surface acoustic wave (SAW), and Tin dioxide (SnO_2), a type of MOS sensor.

10.3.14.1 e-nose uses in combination with other sensing technologies

The potential to utilize electronic nose devices in concert with other electronic sensing instruments and new analytical detection methods for additive or synergistic benefits are considerable. The following discussion provides some recent examples of feasible applications, showing how other detection methods might be used in cooperation with e-noses to yield better, more detailed information so critical to effective decision-making required in all phases and types of agricultural and forestry production systems, viz., DNA microarrays, biosensors, chemical aptasensors, electronic tongues and electroconductive hydrogels (Wilson, 2013).

Many other examples show the potential for using e-nose instruments in combination with other electronic sensing devices to help confirm gas-detection determinations for specific application areas. Spinelli *et al.* (2006) evaluated the use of a near infrared (NIR) instrument in combination with an electronic nose system for the early detection of fire blight (disease) in pears. The e-nose system detected the disease prior to symptom development by the distinctive olfactory signature of volatiles released as early as six days after infection. Sankaran *et al.* (2010) reviewed other advanced techniques and instruments for detecting plant diseases which might be used in combination with electronic noses for disease diagnoses.

10.3.15 Major Sources of Advice and Future

Electronic media, print media, radio, television, telephone, Web-enabled internet services, satellite, computer terminal, Agricultural Technology Information Centre (ATIC), Raitha Smapatka Kendras (RSK), Krishi Vigyan Kendra (KVK), Interactive video conferencing, voice call, the internet, mobile call, etc.

10.3.15.1 Future vision

Extension specialists in plant pathology will certainly become more specialized in the future, and most will likely have research appointments. It also is likely that as plant pathology faculty become more basic research oriented that extension personnel will be increasingly asked to teach courses such as plant disease control, plant problem diagnosis and crop disease-related courses in our graduate curricula. Risks associated with this trend are that a fixed teaching schedule will reduce the ability to travel and may cause a teachable moment to be missed in extension programs. Also, fewer students will be trained as generalists because extension specialists capable of providing this training are becoming rarer. Perhaps of equal importance is who will train future extension plant pathologists? Although broad training will be available from institutions that offer Doctor of Plant Medicine degrees, the trend towards split extension and research positions with increasingly specific assignments suggests that the Ph.D. degree will still be needed for future extension plant pathologists, according to Jacobsen *et al.* (2008).

Extension programming will become more specialized and more integrated with research and other disciplines in the future. Examples of such programming include IPM, organic farming, sustainable agriculture, water quality, and other areas brought into focus by government programs. Organic agriculture has become mainstream with 6 to 8% of produce being marketed as organic, and there is increasing demand from urban populations for organic or locally raised produce. Growers responding to that demand are increasingly relying on extension specialists for science-based and organically acceptable solutions to pest problems. We must balance the needs of the organic community with the conventional farming community, and promote integrated pest management to both as cost-effective and sustainable pest management.

The government has recognized the value of marketing the results of federal funding to the general public. For example, the Farm Bill, 2008 requires integrative programs where funded projects must include both research and extension objectives, and money allocated to support both objectives. Federal funding is increasingly dependent on evidence of multistate outcomes. An area where extension specialists excel is bringing people with specialized skills together to complete a task and communicate it to the public, as well as bringing public feedback back to the University researchers. Extension specialists are the key to effective communication about how taxpayer's dollars are being spent to support the community and, therefore, are the key to continued federal funding (Jacobsen *et al.*, 2008). Recognition of extension's role in the community is often still underappreciated, particularly by our urban clientele. As extension's role in the community continues to evolve, those communication skills at which we excel will be tested. Although we will find more novel and effective ways for communicating with our audience, nothing can replace the personal interactions that led to the founding of the extension service in 1914.

In general, the extension programs in Plant Pathology are designed to reduce plant disease losses in the context of a sustainable environment by educating and assisting county faculty and producers through the county delivery system. These programs are committed to providing the best unbiased educational information on disease management, and responsive diagnosis to facilitate timely implementation of those management recommendations. This goal is achieved by effective interdepartmental and interdisciplinary programming across our functions of research, extension, and instruction. Disease management education should be the main business and environmental quality linked with economic viability is the result. Faculty offer comprehensive training opportunities and diagnostic support for county faculty on the vast variety of food, feed, fiber and ornamental crops. Some of these experiences are as plant disease diagnostic clinics where live plant material is used to teach agents; others are multidisciplinary in nature to provide an overview of all aspects of production of a crop. In addition, newsletters, slide sets, videos, web-based materials and other publications are provided on a timely basis as a means of on-going information exchange.

Future vision should be to strengthen extension phytopathology by addressing major frontiers needs, viz., e-training of farmers on plant protection through Village Knowledge Centers (VKC), establishment of quality testing and residue analysis labs for pesticides, bioproducts, seeds and planting materials, adoption of Information and Technology Tools (ICT) for quicker and timely dissemination of information, establishment of Plant Protection Advisory (PPA) services at village level through rural youth, motivation of farmers for intensive use of bioagents/bioproducts through farmer participatory approach, demonstration of organic farming methods of integrated disease management, and creation of knowledge hubs in KVKs, RSKs and Village Panchayats for easy transfer of technology and establishment of contractual services for plant health diagnosis and plant protection assistance.

Bibliography

Abd-Elsalam, K.A., 2013. Nanoplatforms for plant pathogenic fungi management. *Fungal Genome Biol.,* **2:** 1–2.

Abd-Elsalam, K.A., 2012. Nanoplatforms for plant pathogenic fungi management. *Fungal Genome Biol.,* **2:** 107. doi

Abdul Hameed, M. and Al-Samarrai, 2012. Nanoparticles as alternative to pesticides in management plant diseases – A review. *Inter. J. Sci. Res. Publ.,* **2,** Issue 4, April.

Abeles, F.B., Morgan, P.W. and Saltveit, M.E., 1992. *Ethylene in Plant Biology,* Academic Press, New York, 414 pp.

Abeles, F.B., Dunn, L.J., Morgens, P., Callahan, A., Dinterman, R.E. and Schmidt, J., 1988. Induction of 33kD and 60-kD peroxidases during ethylene-induced senescence of cucumber cotyledons. *Plant Physiol.* **87:** 609–615.

Abhay Kumar, 2008. Nanotechnology and its Applications in Crop Improvement, Nano C, Inc.

Aboa, M.E. and Ali Fadhilab, H., 2003. Epidemiology and management of rice viruses and virus diseases. Plant virology in sub-Saharan Africa, National Cereals Research Institute (NCRI) Badeggi, PMB 8, Bida, Nigeria.

Adam, L., Heuberger, Faith, M., Robison, Sarah Marie, A., Lyons Corey, D., Broeckling and Jessica E. Prenni, 2014. Evaluating plant immunity using mass spectrometry-based metabolomics workflows. *Front Plant Sci.,* **5:** 291.

Adel M. Mahasneh, 2013. Bionanotechnology: The novel nanoparticles based approach for disease therapy. *Jordan J. Biol. Sci.,* **6:** 246-251.

Adipala, E., Lipps, P.E. and Madden, L.V., 1995. Use of disease assessment methods in predicting yield loss due to northern leaf blight of maize. *European J. Pl. Pathol.,* **101:** 25-33.

Afgan Farooq, Iqbal Choudhary, M., Atta-ur-Rahmana and, Satoshi Tahara, 2002. Detoxification of terpinolene by plant pathogenic fungus *Botrytis cinerea. Z. Naturforsch,* **57:** 863-866.

Afifi, M.A. and Zayan, S.A.M., 2009. Implementation of EGY-BLIGHTCAST the first computer simulation model for potato late blight in Egypt. *Aspects of Appl. Biol.,* **96:**

103-110.

Aglika Edreva, 2005. Pathogenesis-related proteins: Research progress in the last 15 years. *Gen. Appl. Pl. Physiol.*, **31:** 105-124.

Agrios, G.N., 2005. *Plant Pathology.* 5th ed., Elsevier Academic Press, London, UK.

Agrios, G.N., 1997. Plant diseases caused by fungi. pp. 245–406. *In: Plant Pathology.* San Diego, California, Academic Press.

Agrios, G.N., 2004 (copyright 2005). *Plant Pathology,* 5th ed., Academic Press, San Diego, California, Academic Press.

Ahmed, A.T., Wael, F.T., Shaaban, M.A., and Mohammed, F.S., 2001. Antibacterial action of zinc oxide nanoparticle agents' foodborne pathogens. *J. Food Safety:* **31:** 211–218.

Ahn, J.H. and Walton, J.D. (1997). A fatty acid synthase gene in *Cochliobolus carbonum* required for production of HC-toxin, cyclo (D-prolyl-L-alanyl-Dalanyl- L-2-Amino-9,10-epoxi-8-oxodecanoyl). *Mol. Plant-Microbe Interact.*, **10:** 207–214.

Ahuja, I., Kissen, R. and Bones, A.M., 2012. Phytoalexins in defense against pathogens. *Trends Pl. Sci.*, **17:** 73-90.

Ai Imazaki, Aiko Tanaka, Yoshiaki Harimoto, Mikihiro Yamamoto, Kazuya Akimitsu, Pyoyun Park and Takashi Tsuge, 2010. Contributions of peroxisomes to secondary metabolism and pathogenicity in the fungal plant pathogen *Alternaria alternata. Eukaryot. Cell.*, **9:** 682-694.

Aires, A., Mota, V.R., Saavedra, M.J., Monterio, A.A., Simoes, M., Rosa, E.A.S. and Bennett, R.N., 2009. Initial in vitro evaluations of the antibacterial activities of glucosinolate enzymatic hydrolysis products against plant pathogenic bacteria. *J. App. Microbiol.*, **106:** 2096-2105.

Akella Vani, 2012. Diagnostic kit to detect Citrus tristeza virus in citrus Dip stick format and ELISA format based on Monoclonal antibodies. Indian Institute of Horticultural Research, Hessarghatta Lake, Bangalore-560 089, Karnataka, India.

Akihiko T. and Wei, W., 2009. Endomembrane system of aspen root cells plays a key role in defense against a common fungal root endophyte, *Cryptosporiopsis radicicola. Mycologia,* **101:** 182-189.

Akira, S., Uematsu, S. and Takeuchi, O., 2006. Pathogen recognition and innate immunity. *Cell,* **124:** 783-801.

Alan M. Jones, 2001. Programmed cell death in development and defense. *Plant Physiol.*, **125:** 94-97.

Ales Lebeda, Dagmar Jancova and Lenka Luhovä, 1999. Enzymes in fungal plant pathogenesis. *Pl. Physiol.*, **39:** 51-56.

Alexander, D., Lawton, K., Uknes, S., Ward, E., and Ryals, J., 1994. Defense-related gene induction in plants. *In*: *Genetic Engineering,* Setlow, J. K., (Ed.) Plenum Press, New York, 195-212.

Allah Ditta, 2012. How helpful is nanotechnology in agriculture?. *Adv. Nat. Sci.: Nanosci. Nanotechnol.,* **3:** 1-10.

Almagro, L., Gómez Ros, L.V., Belchi-Navarro, S., Bru, R., Ros Barceló, A. and Pedreño, M.A., 2009. Class III peroxidases in plant defence reactions. *J. Exp. Bot.,* **60:** 377-390.

Altpeter, F., Varshney, A., Abderhalden, O., Douchkov, D., Sautter, C., Kumlehn, J., Dudler, R. Schweizer, P., 2005. Stable expression of a defense-related gene in wheat epidermis under transcriptional control of a novel promoter confers pathogen resistance. *Pl. Mol. Biol.,* **57:** 271-283.

Aman Yadav and Janna L. Beckerman., 2009. Implementing case studies in a plant pathology course: Impact on student learning and engagement *J. Nat. Resour. Life Sci. Educ.,* **38:** 1-6.

Amarendra N. Misra, Meena Misra and Ranjeet Singh, 2013. Nanotechnology in agriculture and food industry. *Int. J. Pure Appl. Sci. Technol.,* **16:** 1-9.

Ameer Azam, Arham S. Ahmed, Mohammad Oves, Mohammad S. Khan, Sami S. Habib and Adnan Memic, 2012. Antimicrobial activity of metal oxide nanoparticles against Gram-positive and Gram-negative bacteria: a comparative study, *Inter. J. Nanomedicine,* **7:** 6003-6009.

Amiri, A. and Bompeix, G., 2005. Diversity and population dynamics of *Penicillium* spp. on apples in pre- and postharvest environments: consequences for decay development. *Pl. Pathol.,* **54:** 74-81.

Amooaghaie, R., 2011. Fungal disinfection by nanofiltration in tomato soilless culture. World Academy of Science. *Engg. & Techn.,* **5:** 8-27.

Amusa, N.A., 2006. Microbially produced phytotoxins and plant disease management. *African J. Biotechnol.,* **5:** 405-414.

Amusa, N.A., Ikotun, Tand Asiedu, R., 1993. Extraction of a phytotoxic substance from *Colletotrichum gloeosporicides* infected yam leaves. *Int. J. Trop. Pl. Dis.,* **11:** 207-211.

Amusa, N.A., 1991. Extraction, characterization and bioassay of metabolities of some plant pathogenic species of *Colletotrichum.* Ph.D. thesis, University of Ibadan, Nigeria, pp. 210.

Ana R Santos, Ana S. Miguel, Leonor Tomaz, Rui Malhó, Christopher Maycock, Maria C. Vaz Patto, Pedro Fevereiro, and Abel Oliva, 2010. The impact of CdSe/ZnS quantum dots in cells of *Medicago sativa* in suspension culture. *J. Nanobiotechnol.,* **8:** 24.

Ana Segura, Manuel Moreno, Antonio Molina, Francisco García-Olmedo, 1998. Novel defensin subfamily from spinach (*Spinacia olerácea*). Laboratorio de Bioquímica y Biología Molecular, Departamento de Biotecnología-UPM, ETS Ingenieros Agrónomos, E-28040 Madrid, Spain.

Anaya-Lopez, J.L., Torres-Pacheco, I., Gonzalez-Chavira, M., Garzon-Tiznado, J.A., Pons-Hernandez, J.L., Guevara-Gonzalez, R.G., Muñoz-Sánchez, C.J., Guevara-Olvera, L., Rivera-Bustamante, R.F. and Hernandez-Verdugo, S., 2003. Resistance to geminivirus mixed infections in mexican wild peppers. *Hort Sci.,* **38:** 251-255.

Anderson Cal Baier, 2009. Regulating nanosilver as a pesticide. Environmental Defense Fund, February 12.

Andrade-Piedra, J.L., Forbes, G.A., Shtienberg, D., Grünwald, N. J., Chacón, M.G., Taipe, M.V., Hijmans, R.J. and Fry, W.E., 2005. Qualification of a plant disease simulation model: Performance of the LATE BLIGHT model across a broad range of environments. *Phytopathology,* **95:** 1412-1422.

Andrea Scharf, Annette Piechulek and Anna von Mikecz, 2013. Effect of Nanoparticles on the Biochemical and Behavioral Aging Phenotype of the Nematode, *Caenorhabditis elegans.*, *ACS Nano.*, **7:** 10695–10703.

Andrés Pinzón, Emiliano Barreto, Adriana Bernal, Luke Achenie, Andres F. González Barrios, Raúl Isea and Silvia Restrepo, 2009. Computational models in plant-pathogen interactions: the case of *Phytophthora infestans. Theoretical Biol. Medical Modelling,* **6:** 24.

Andrew J. Hamilton, Frank Stagnitti, Robert Premier, Anne-Maree Bol and Glenn Hale, 2006. Quantitative microbial risk assessment models for consumption of raw vegetables irrigated with reclaimed water, *Appl. Environ. Microbiol.,* **72:** 3284-3290.

Anita Grover, 2012. Plant chitinases: Genetic diversity and physiological roles. *Critical Rev. in Pl. Sci.,* **31:** 57-73.

Anita, B. and Samiyappan, R., 2012. Induction of systemic resistance in rice by *Pseudomonas fluorescens* against rice root knot nematode, *Meloidogyne graminicola. J. Biopest.,* **5** (Supplementary): 53-59.

Anke, T., Oberwinkler, F., Steglich, W. and Schramm, G., 1977. The strobilurins-new antifungal antibiotics from the basidiomycetes *Strobilurus tenacelus. J. Antibiot.,* **30:** 806.

Anne E. Desjardins and Thomas M. Hohn, 1997. Mycotoxins in plant pathogenesis. *MPML,* **10:** 147-152.

Anne Osbourn, 1996. Saponins and plant defence a soap story. *Trends Pl. Sci.,* **1:** 4-9.

Anne, E.D., Gayland, F.S. and Ronald D.P., 1989. Tolerance and metabolism of furanocoumarins by the phytopathogenic fungus *Gibberella pulicaris* (*Fusarium sambucinum*). *Phytochemistry,* **28:** 2963-2969.

Annett Milling, Lavanya Babujee and Caitilyn Allen. 2011. *Ralstonia solanacearum* Extracellular polysaccharide is a specific elicitor of defense responses in wilt-resistant tomato plants. *PLOS ONE,* **10:** 1371

Annis, S.L. and Goodwin, P.H., 1997. Recent advances in the molecular genetics of plant cell wall-degrading enzymes produced by plant fungi. *European J. Pl. Pathol.,* **103:** 1–14

Anonymous, 1969. CEC, Council Directive 69/465/EEC of 8 December 1969 on control of potato golden nematode. *Official Journal of the European Communities* **L323**, 3-4.

Anonymous, 1978. Data sheets on quarantine organisms No. 124, *Globodera pallida. OEPP/EPPO Bull.,* **8:** 2.

Anonymous, 1981. Data sheets on quarantine organisms No. 125, *Globodera rostochiensis. OEPP/EPPO Bull.,* **11:** 1.

Anonymous, 1990. TNAU Agritech Portal: Exports & Imports, Rashtriya Krishi Vikas Yojana (RKVY), India.

Anonymous, 1991. Quarantine procedure No. 30, *Globodera pallida* & *G. rostochiensis,* soil sampling methods. *OEPP/EPPO Bull.,* **21:** 233-240.

Anonymous, 1994. *Understanding on rules and procedures governing the settlement of disputes.* In: Agreement establishing the World Trade Organization: Annex 2: Dispute settlement understanding. Geneva, Switzerland.

Anonymous, 1995. World Trade Organization (1995) Agreement on the Application of Sanitary and Phytosanitary Measures. In: Results of the Uruguay Round of Multilateral Trade Negotiations: The Legal Texts. Geneva, Switzerland: World Trade Organization. 69–83.

Anonymous, 1996. Bureau of Animal and Plant Health Inspection and Quarantine, Council of Agriculture, Executive Yuan 9F., No. 100, Sec. 2, Heping W. Rd., Zhongzheng Dist., Taipei City 10070, Taiwan, R.O.C.

Anonymous, 1998. Sanitary and Phytosanitary Measures: Introduction Understanding the WTO Agreement on Sanitary and Phytosanitary Measures. World Trade Organization.

Anonymous, 1999. Crop Doctor Field Diagnostic Kit. Grains Research and Development Corpration, Australia, Issue 26.

Anonymous, 2000. *IUCN guidelines for the prevention of biodiversity loss caused by alien invasive species.* Approved by the 51st Meeting of the IUCN Council, Gland, Switzerland.

Anonymous, 2001. Disease Forecasting Systems. Plant Disease Central Specialists Team, Loren Giesler, FLASH, 448, Lincoln NE 68583-0722.

Anonymous, 2001. FAO Report of the Third Interim Commission on Phytosanitary Measures, Rome, Italy, 2-6 April 2001, Appendix XI: *Dispute settlement procedures.*

Anonymous, 2001. Int. Soc. Plant Pathol. (ISPP) Teach. Comm. Online Instr. Technol. Symp. June. Archive: http://www.ispp-itsymposium.org.nz/.

Anonymous, 2001. Plant Sciences Institute Update. Iowa State University. Volume 2, No. 1.

Anonymous, 2001. Training Programme on WTO and Its Implications on Indian Agriculture, National Institute of Agricultural Extension Management (MANAGE) Rajendranagar, Hyderabad – 500 030.

Anonymous, (CBD) 2002. Sixth Conference of the Parties, the Hague, the Netherlands, 7-19 April 2002: Decision VI/23: Alien species that threaten ecosystems, habitats or species to which is annexed Guiding principles for the prevention, introduction and mitigation of impacts of alien species that threaten ecosystems, habitats or species.

Anonymous, 2003. FAO Report of the Fifth Interim Commission on Phytosanitary Measures, Rome, Italy, 7-11 April 2003.

Anonymous, 2003. Biosafety, Biosecurity and Bioweapons. The Sunshine Project, Backgrounder #11, October 2003, German Peace Research Foundation (DSF), Scheplerstr. 78, 22767 Hamburg · Germany.

Anonymous, 2003. Committee on Agriculture, Seventeenth Session on *Biosecurity* in Food and Agriculture, Rome, 31 March-4 April 2003 Item 9 of the Provisional Agenda.

Anonymous, 2003. FAO Report of the Fifth Interim Commission on Phytosanitary Measures, Rome, Italy, 7-11 April 2003.

Anonymous, 2003. Identification of risks and management of invasive alien species using the IPPC framework, *Proceedings of a workshop in Braunschweig, Germany 22-26 September 2003, Secretariat of the International Plant Protection Convention*, Food and Agriculture Organization of the United Nations.

Anonymous, 2003. Microbial Genomic Sequencing Perspectives of the American Phytopathological Society (Revised 2003). The American Phytopathological Society, 3340 Pilot Knob Road St. Paul, MN 55121-2097

Anonymous, 2003. Recombinant DNA Safety Guidelines, 1990, Department of Biotechnology, Ministry of Science and Technology, Govt. of India.

Anonymous, 2003. RMD-11-02: Pest Risk Management Document for Deregulation of *Heterodera glycines* Ichinohe (Soybean Cyst Nematode). Canadian Food Inspeaction Agency. Canada.

Anonymous, 2003. The Protection of Plant Varieties and Farmers' Rights Act, 2001 and Rules, Universal Law Publishing Co., Delhi, 2003.

Anonymous, 2004. Grower 101: Rapid Diagnostic Kit – *Using rapid diagnostic kits to detect viral diseases in the greenhouse is a good way to prevent wide-scale infestation and crop loss., Greenhouse Product News,* **12:** 24.

Anonymous, 2004. WHO, Laboratory Biosafety Manual–Third Edition, http://www.who.int/csr/resources/publications/biosafety/WHO_CDS_CSR_LYO_2004_11/en.

Anonymous, 2005. National Biosafety Guidelines, Government of Pakistan, Pakistan Environmental Protection Agency (Ministry of Environment), Notification No. F.2(7)95-Bio.

Anonymous, 2005. Remote Imagery Aids in Detecting and Managing Plant Disease. *Science Daily. American Phytopathological Society,* 9 May 2005.

Anonymous, 2005. A Vision of Plant Pathology in the 21st Century, *Tech. Bull.,* The American Phytopathological Society (APS), 4.

Anonymous, 2006, Plum Pox Virus RT-PCR Detection Kit Norgen Biotek Corp., 3430 Schmon Parkway, Thorold Ontario, Canada, L2V 4Y6.

Anonymous, 2006. ELISA Test for Banana Bunchy Top Virus (BBTV), University of Hawaii and the Hawaii Department of Agriculture.

Anonymous, 2006. Risks to plant health: European Union priorities for tackling emerging plant pests and diseases, EASAC– the European Academies Science Advisory Council, EASAC Secretariat Deutsche Akademie der Naturforscher Leopoldina German National Academy of Sciences Jägerberg 1, D-06108 Halle (Saale), Germany.

Anonymous, 2007. Cartagena Protocol on Biosafety. Booklet by Project Coordinating and Monitoring Unit (PCMU), Ministry of Environment & Forests, in association with Biotech Consortium India Limited, Government of India, New Delhi-110 003.

Anonymous, 2007. EnviroLogix launches QualiPlate™ Kit for LMV: Greater productivity and data quality for plant testing labs, 00 Riverside Industrial Parkway, Portland, Maine.

Anonymous, 2007. Scientific Opinion of the Panel on Plant Heath on a request from the European Commission on Pest Risk Assessment made by France on *Ralstonia solanacearum* race 2 considered by France as harmful in French overseas departments of French Guiana, Guadeloupe, Martinique and Réunion. *The EFSA J.,* **673:** 1-27.

Anonymous, 2008, QuickStix™ Kit for Tomato Apex Necrosis Virus, SEedQuest, News Section, Portland, Maine.

Anonymous, 2008. SXagro, Pest Information Bureau, Ministry of Agriculture, 26-February, 2008, **16:** 33 IS, New Delhi, India.

Anonymous, 2008. Scientific Opinion of the Panel on Plant Health on a request from the European Commission on Pest Risk Assessment made by France on Citrus yellow mosaic virus or Citrus mosaic badnavirus considered by France as harmful in the French overseas departments of French Guiana, Guadeloupe, Martinique and Réunion. *The EFSA J.,* **686:** 1-16.

Anonymous, 2009. National Biosafety Policy. BELIZE, The National Biosafety Policy for the Government of Belize (Approved March 2009).

Anonymous, 2009. Geneticall genetically modified crops issues and challenges in the context of India. Research Unit (Larrdis) Research Unit (Larrdis), Rajya Sabha Secretariat, New Delhi, India Anymous, 2010. Anonymous, 2011. Diagnostic Kit for Potato Cyst Nematode Detection, *Global Knowledge Center on Crop Biotechnology, International Service for the Aquisition of Agri-Biotech Applications SE Asia Center (ISAAA), Crop Biotech Update.*

Anonymous, 2011. EFSA Panel on Plant Health (PLH); Pest risk assessment of *Monilinia fructicola* for the EU territory and identification and evaluation of risk management options. *EFSA J.,* **9:** 2119.

Anonymous, 2012. Plant Pathology education and training in the UK: An Audit British Society for Plant Pathology http://www.bspp.org.uk.

Anonymous, 2012. Biosafety Manual-Safety, Health, Environment & Risk Management, Safety, Health, Environment & Risk Management (SHERM), Laurier. 75, University Avenue W, Waterloo, Ontario, Canada N2L 3C5.

Anonymous, 2012. Plant pathology education and training in the UK: An Audit- The British Society for Plant Pathology, Marlborough House Basingstoke Road, Spencers Wood Reading RG7 1AG, UK.

Anonymous, 2013. Report on sanitary and phytosanitary measures. Ambassador Demetrios Marantis, Office of the United States Trade Representative, Executive Office of the President Washington, D.C. 20508. 97 p.

Anonymous, 2013. Nanotechnology General News, *Nano Werk, American Chemical Society,* Sep 18,

Anonymous, 2013. Nanotechnology in Agriculture: Scope and Current Relevance. Policy Paper No. 63, *National Academy of Agricultural Sciences,* New Delhi. 20.

Anonymous, 2013. New Innovative Approaches to Crop Protection: Report from Food Research Partnership Stakeholder Workshop. Government Office for Science, 1 Victoria Street, London SW1H 0ET.

Anonymous, 2014. General Information Concerning Patents, United States Patent and Trademark Office, Alexandria, Virginia.

Anonymous, 2014. Genetically Modified Food. *Current Issues: Macmillan Social Science Library*. Detroit: Gale, 2010. *Opposing Viewpoints in Context*. Web. 22 Nov.

Anonymous, 2014. Disease Progress, The American Phytopathological Society (APS), 3340 Pilot Knob Road, St. Paul, MN 55121 USA.

Anonymous, 2014. Institute of Food and Agricultural Sciences Extension, University of Florida.

Anonymous, 2014. Public Health Emergency, US Department of Health and Human Services, 200 Independence Ave, SW, Wahington, DC 20201.

Anonymous, 2015. Science and Advice for Scottish Agriculture (SASA), Monoclonal Antibody Unit, 1 Roddinglaw Road, Edinburgh EH12 9FJ.

Anonymous, 2015. NPDN National Plant Diagnostic Network, Michigan State University.

Anonymous, AGDIA- BIOFORDS, 5, rue Henri Desbruères - Genavenir 8, 91030 Evry Cedex - France.

Anonymous, British Mycological Society, 1948. The measurement of potato blight. Trans. *British Mycol. Soc.,* **31:** 140-141.

Anonymous, 1983. Protection against Trichothecene Mycotoxins. National Research Council, National Academy Press, Washington, D.C.

Anonynous, 2003a. FAO/WHO Technical Consultation on Biological Risk Management in Food and Agriculture.

Antimo Di Maro, Irma Terracciano, Lucia Sticco, Luisa Fiandra, Michelina Ruocco, Giandomenico Corrado, Augusto Parente and Rosa Rao, 2010. Purification and characterization of a viral chitinase active against plant pathogens and herbivores from transgenic tobacco. *J. Biotechnol.,* **147:** 1–6.

Antonio Leon Reyes, T.P. Prins, J.P. van Empel and J.M. van Tuyl. 2005. Differences in Epicuticular Wax Layer in Tulip Can Influence Resistance to *Botrytis tulipae*. Proc. IXth Intl. Symp. on Flower Bulbs, H. Okubo, W.B. Miller and G.A. Chastagner (Eds.), *Acta Hort,* **673**.

Antonio Leon-Reyes, Steven H. Spoel, Elvira S. De Lange, Hiroshi Abe, Masatomo Kobayashi, Shinya Tsuda, Frank F. Millenaar, Rob A.M. Welschen, Tita Ritsema and Corné M.J. Pieterse, 2009. Ethylene modulates the role of nonexpressor of pathogenesis-related genes in cross talk between salicylate and jasmonate signaling. *Pl. Physiol.,* **149:** 1797-1809.

Antonio Leon-Reyes, Steven H. Spoel, evira S. De Lange, Hiroshi Abe, Masatomo Kobayashi, Shinya Tsuda, Frank F. Millenaar, Rob, A.M., Welschen, Tita Ritsema and Corne, M.J. Pieterse, 2009. Aparna Kalawate and Pandey, C.N., 2012. An antimycotic study of Propiconazole for mould inhibition on rubber wood. *Int. J. Fundamental Applied Sci.*, **1:** 24-26.

Arbona, V., Argamasilla, R. and Gómez-Cadenas, A., 2010. Common and divergent physiological, hormonal and metabolic responses of *Arabidopsis thaliana* and *Thellungiella halophila* to water and salt stress. *J. Pl. Physiol.*, **167:** 1342–1350.

Ariane F. Lacerda, Érico A.R. Vasconcelos, and Maria F. Grossi de Sa, 2014. Antifungal defensins and their role in plant defense. *Front Microbiol.*, **5:** 2-50.

Armah, C.N., Mackie, A.R., Roy, C., Price, K., Osbourn, A.E., Bowyer, P. and Ladha, S., 1999. The Membrane-Permeabilizing Effect of Avenacin A-1 Involves the Reorganization of Bilayer Cholesterol. *Biophysical J.*, **76:** 281–290.

Arneson, P.A. and Durbin, R.D., 1968. Studies on the mode of action of tomatine as a fungitoxic agent, *Plant Physiol.*, **43:** 683-686.

Arotupin Daniel Juwon, Akinyosoye, F.A., and Onifade Anthony Kayode, 2012. Purification, characterization and application of polygalacturonase from *Aspergillus niger* CSTRF. *Malaysian J. Microbiol.*, **8:** 175-183.

Arul Selvi, A., 2011. Studies on bioremediation of isoprothiolane, fungicide/insecticide and glyphosate, a herbicide and monitoring of biodegradation using biosensors. Ph.D. thesis, University of Mysore, Karnataka, India.

Arulpandi and Sangeetha, R., 2012. Antibacterial Activity of Fistulin: A Protease Inhibitor Purified from the Leaves of *Cassia fistula*. ISRN Pharmaceutics, Article ID 584073, pp 4.

Arunava Goswami 2013. Suitability of nano-sulphur for biorational management of powdery mildew of Okra (*Abelmoschusss esculentus* Moench) caused by *Erysiphe cichoracearum, Plant Pathol. & Microbiol.*, **4:** 1000171.

Arup K. Mukherjeea, Marie-Jeanne Carpa, Rina Zuchmanc, Tamar Zivc, Benjamin A. Horwitza and Shimon Gepsteina, 2010. Proteomics of the response of *Arabidopsis thaliana* to infection with *Alternaria brassicicola. J. Proteomics,* **73:** 709-720.

Arya, H., Kaul, Z., Wadhwa, R., Taira, K., Hirano, T and Kaul, S.C., 2005. Quantum dots in bio-imaging: revolution by the small. *Biochem Biophys Res. Commun.*, **329:** 1173–1177.

Ataee, R.A., Derakhshanpour, J., Mehrabi Tavana, A. and Eydi, A., 2011. Antibacterial effect of calcium carbonate nanoparticles on *Agrobacterium tumefaciens. Iranian J. Military Medicine,* **13:** 65-70.

Atif Sarwar, Avelina Espinosa and James R. Alfano, 2004. Disabling surveillance: Bacterial type III secretion system effectors that suppress innate immunity. *Cellular Microbiol.,* **6:** 1027-1040.

Ayodele Adesina Adegbite, 2011. Assessment of yield loss of cowpea (*Vigna unguiculata* L.) due to root knot nematode, *Meloidogyne incognita* under field conditions, *American J.f Exptl. Agri.,* **1:** 79-85.

Bacon Charles, Snook Maurice and Hinton Dorothy, 2006. Corn seedling disease, fusaric acid as the wilt toxin and the need for biocontrol of *Fusarium verticillioides* and other *Fusarium species. Aflatoxin/Fumonisin Elimination and Fungal Genomics Workshop* held during October 16-18, 2006 at Fort Worth, TX.

Bacon, C.W., Porter, J.K., Norred, W.P. and Leslie, J. F., 1996, Production of fusaric acid by *Fusarium* species. *Applied Environ. Microbiol.,* **62:** 4039-4043.

Baeummer, A., 2002. Nanosensors identify pathogens in food. *Food Technol.,* **58:** 51–55.

Bagnato V.S., 2006. Fluorescence spectroscopy applied to orange trees. *Laser Phys.,* **16:** 884–888.

Bajar, A., Podila, G.K. and Kolattukudy, P.E., 1991. Identification of a fungal cutinase promoter that is inducible by a plant signal via a phosphorylated trans-acting factor. *Proc. Natl. Acad. Sci., USA*, **88:** 8208–8212.

Baker, B., Zambryski, P., Staskawicz, B. and Dinesh Kumar, S.P., 1997, Signaling in Plant-Microbe Interactions. *Sci.,* **276:** 726-733.

Baker, R.H.A., Sansford., C.E., Jarvis, C.H., Cannon, R.J.C., MacLeod, A. and Walters, K.F.A., 2000. The role of climatic mapping in predicting the potential geographical distribution of non-indigenous pests under current and future climates. *Agric. Ecosyst. Environ.,* **82:** 57–71.

Balasubramanian, V., Vashisht, D., Cletus, J. and Sakthivel, N., 2012. Plant β-1,3-glucanases: their biological functions and transgenic expression against phytopathogenic fungi. *Biotechnol. Lett.,* **34:** 1983-1990.

Ball, A. M., Ashby, A. M., Daniels, M. J., Ingram, D. S., and Johnstone, K., 1991. Evidence for the requirement of extracellular protease in the pathogenic interaction of pyrenopeziza brassicae with oilseed rape. *Physiol. Mol. Plant Pathol.,* **38:** 147-161.

Ballhorn, D.J., Pietrowski, A. and Lieberei, R., 2010. Direct trade-off between cyanogenesis and resistance to a fungal pathogen in lima bean (*Phaseolus lunatus* L.). *J. Ecol.,* **98:** 226-236.

Barbara A. Wilson, Katherine Zdunic, Janine Kinloch and Graeme Behn, 2012. Use of remote sensing to map occurrence and spread of *Phytophthora cinnamomi* in *Banksia* woodlands on the Gnangara groundwater system, Western Australia. *Australian J. Bot.,* **60:** 495-505.

Barras, F., Van Gijsegem, F. and Chatterjee, A.K., 1994. Extracelllar enzymes and pathogenesis of soft-rot Erwinia. *Ann. Rev Phytopathol.,* **32:** 201–234.

Bart, P.H.J., Thomma, Bruno, P.A., Cammue and Karin Thevissen. 2002. Plant defensins. *Planta,* **216:** 193-202.

Barta, T.M., Kinscherf, T.G. and Willis, D.K., 1992. Regulation of tabtoxin production by the lemA gene in *Pseudomonas syringae. J. Bacteriol.,* **174:** 3021–3029.

Bartels-Keith, J.R., 1960. Alternaric acid. Part III. structure. *J. Chem. Soc.,* **47:** 1662-1665.

Battaglia, E., 2011. Regulation and diversity of plant polysaccharide utilization in fungi. *Biologie Proefschriften,* ISBN 978-90-8891-287-0

Beardall, J.M. and Miller, J.D., 1994. Diseases in humans with mycotoxins as possible causes. Pages 487-540. *In: Mycotoxins in Grain: Compounds Other Than Aflatoxin.* J. D. Miller and H. L. Trenholm, (Eds.). Eagan Press, St. Paul, MN.

Beattie, G.A., and Lindow, S.E., 1995. The secret life of foliar bacterial pathogens on leaves. *Ann Rev Phytopathol,* **33:** 145-172.

Beffa, R.S., Hofer, R.M., Thomas, M. and Meins, Jr F., 1996. Decreased susceptibility to viral disease of [beta]-1,3-glucanase-deficient plants generated by antisense transformation. *Plant Cell.,* **8:** 1001–1011.

Belasque, L., Gasparoto, M.C.G. and Marcassa L.G., 2008. Detection of mechanical and disease stresses in citrus plants by fluorescence spectroscopy. *Appl. Optics,* **47:** 1922-1926.

Bender, C.L., Alarcón-Chaidez, F. and Gross, D.C., 1999. Pseudomonas syringae phytotoxins: Mode of action, regulation, and biosynthesis by peptide and polyketide synthetases. *Microbiol. Mol. Biol. Rev.,* **63:** 266-292.

Berger, P.H. and Ferriss, R.S., 1989. Mechanisms of arthropod transmission of plant viruses: Implications for the spread of disease. *In: Spatial Components of Plant Disease Epidemics,* M.J. Jeger (ed.). Prentice Hall, Englewood Cliffs, U.S.A., 40-84.

Bessire, M., Chassot, C., Jacquat, A.C., Humphry, M., Borel, S., Petetot, J.M., Metraux, J.P. and Nawrath, C., 2007. A permeable cuticle in Arabidopsis leads to a strong resistance to *Botrytis cinerea. EMBO J.,* **26:** 2158-2168.

Bevan, M., Shufflebottom, D., Edwards, K., Jefferson, R. and Schuch, W., 1989. Tissue- and cell-specific activity of a phenylalanine ammonia-lyase promoter in transgenic plants. *Embo J.,* **8:** 1899-1906.

Bhupinder Singh Sekhon, 2014. Nanotechnology in agri-food production: An overview. *Nanotechnol. Sci. Appl.,* **7:** 31-53.

Bidochka, M.J., St Leder, R.J., Stuart, A., and Gowanlock, K., 1999. Nuclear rDNA phylogeny in the fungal genus *Verticillium* and the relationship to insect and plant virulence, extracellular proteases and carbohydrases. *Microbiol.,* **145:** 955-963.

Biere, A., Marak, H.B. and Damme, J.M., 2004. Plant chemical defense against herbivores and pathogens: generalized defense or trade-offs? *Oecologia,* **140:** 430-441.

Bilgram, K.S. and Dube, H.C., 1976. Modern Plant Pathology, Vikas Publishing House, New Dehil. pp. 344.

Binutu, O.A. and Cordell, G.A., 2000. Gallic acid derivative from mezoneuron benthamianun leaves. *Pharmacol Biol.,* **38:** 284-286.

Bird, S.M. and Gray, J.E., 2003. Signals from the cuticle affect epidermal cell differentiation. *New Phytol.* **157:** 9-23.

Birgit Reiter, Ulrike Pfeifer, Helmut Schwab and Angela Sessitsch, 2002. Response of endophytic bacterial communities in potato plants to infection with *Erwinia carotovora* subsp. *atroseptica. Applied and Environ. Microbiol.,* **68:** 2261-2268.

Bishop, J.G., Dean, A.M. and Mitchell-Olds, T., 2000. Rapid evolution in plant chitinases: molecular targets of selection in plant-pathogen co-evolution. *Proc. Natl. Acad. Sci.,* USA, **97:** 5322-5327.

Blackwell, B.A., Edwards, O.E., Fruchier, A., ApSimon, J.W., and Miller, J.D., 1996. NMR structural studies of fumonisin B1 and related compounds from *Fusarium moniliforme.* P. 75-92. *In: Fumonisins in Food.* L. S. Jackson, J. W. DeVries and L. L. Bullerman (eds.), Plenum Press, New York.

Blasco, J., Alexios, N., Gómez, J. and Moltó, E., 2007. Citrus sorting by identification of the most common defects using multispectral computer vision. *J. Food Engg.,* **83:** 384-393.

Boa, E., 2007. Plant Healthcare for Poor Farmers: An Introduction to the Work of the Global Plant Clinic. Online. APS*net* Features. doi: 10.1094 /APSnetFeatures-2007-1007.

Bob Asselbergh, Andrew enow achuo, Monica Höfte and Frédérique Van Gijsegem, 2008. Abscisic acid deficiency leads to rapid activation of tomato defence responses upon infection with *Erwinia chrysanthemi. Mol. Pl. Pathol.,* **9:** 11-24.

Boerjan, W., Ralph, J. and Baucher, M., 2003. Lignin biosynthesis. *Ann. Rev. Plant Biol.,* **54:** 519-546.

Bohm, A., 1998. *Introduction to Flavonoids,* Gordon & Breach, Amsterdam, the Netherlands.

Bolkan, H.A., and Reinert, W. R. 1994. Developing and implementing IPM strategies to assist farmers: an industry approach. *Plant Dis.,* **78:** 545-550.

Boller and Plattner, P.A., 1963. Ueber induzierte Abwehrstoffe bei Orchideen. *Helv. Chern. Acta,* **46:** 60.

Boller, T. and Felix, G., 2009. A renaissance of elicitors: perception of microbe-associated molecular patterns and danger signals by pattern-recognition receptors. *Ann. Rev. Pl. Biol.,* **60:** 379-406.

Borad, V. and Sriram Asian, S., 2008. Pathogenesis-related proteins for the plant protection. *J. Exp. Sci.,* **22:** 189-196.

Bos, L., 1988. Germplasm health and international crop improvement, with special reference to viruses. *In* A.W. Jalil, P.L. Manalo, S.S. Sastroutomo, A. Ganapathi, K.C. Chan, L.G. Lim, M.A.A. Rahim, P.S.S. Durai & M.C. Stephensons, eds., *Introduction of Germplasm and Plant Quarantine Procedures,* 28-29. Malaysia, Asean Plant Quarantine Centre and Training Institute.

Boudart, G., Charpentier, M., Lafitte, C., Martinez, Y., Jauneau, A., Gaulin, E., Esquerre-Tugaye, M.T. and Dumas, B., 2003. Elicitor activity of a fungal endopolygalacturonase in tobacco requires a functional catalytic site and cell wall localization. *Pl. Physiol.,* **131:** 93-101.

Boue, S.M., Carter, C.H. and Ehrlich, K.C. and Cleveland, T.E. 2000. Induction of the soybean phytoalexins coumestrol and glyceollin by *Aspergillus. J. Agric. Food Chem.* **48:** 2167.

Bouwmeester, H., Abele, S., Manyong, V.M., Legg, C., Mwangi, M., Nakato, V., Coyne, D. and Sonder, K., 2010. The potential benefits of GIS techniques in disease and pest control: an example based on a regional project in central Africa. *Proc. IC on Banana & Plantain in Africa.* T. Dubois *et al.*, (Eds.), *Acta Hort.*, **879**, ISHS: 333-340

Bowyer, P., Clarke, B.R., Lunness, P., Daniels, M.J. and Osbourn, A.E., 1995. Host range of a plant pathogenic fungus determined by a saponin detoxifying enzyme. *Science*, **267:** 371-374.

Bradley, C, Nelson, B. and Helms, T., 2003. Soybean cyst nematode reported in North Dakota. North Dakota State University. August 28, 2003.

Bram de Hoop, 2003. Pest risk analysis: global harmonization benefits biodiversity, International Standards Adviser, Netherlands Plant Protection Service, P.O. Box 9102, 6700 HC Wageningen, the Netherlands.

Brasier, C.M., 2008. The biosecurity threat to the UK and global environment from international trade in plants. *Plant Pathology*, **57:** 792–808.

Brasier, C.M., 2005. Preventing invasive pathogens: Deficiencies in the system. *The Plantsman*, **4:** 54–57.

Brasier, C.M. and Mehrotra, M.D., 1995. *Ophiostoma himal-ulmi* sp. nov. a new species of dutch elm disease fungus endemic to the Himalayas. *Mycol. Res.*, **99:** 205–215.

Bravo, C., Moshou, D., West, J., McCartney, A. and Ramon, H., 2003. Early disease detection in wheat fields using spectral reflectance. *Biosys. Engg.*, **84:** 137–145.

Bravo, C., Moshou, D., Oberti, R., West, J., McCartney, A., Bodria, L., Ramon, H., 2004. Foliar disease detection in the field using optical sensor fusion. *Agril. Engg. Intern., CIGR J. Scientific Res.& Devt.*, Vol. **VI.**

Brencic, A. and S. C. Winans, 2005. Detection of and response to signals involved in host-microbe interactions by plant-associated bacteria. *Microbiol. Mol. Biol. Rev.*, **69:** 155-194.

Brian C. King., Katrina D. Waxman., Nicholas V. Nenni., Larry P. Walker., Gary C. Bergstrom and Donna M Gibson., 2014. Arsenal of plant cell wall degrading enzymes reflects host preference among plant pathogenic fungi. *Biotech. for Biofuels*, **4:** 4.

Brian J. Staskawicz, 2001. Genetics of plant-pathogen interactions specifying plant disease resistance. *Plant Physiol.*, **125:** 173-176.

Brian, P. W., Elson, G. W., Hemming, H. G. and Joyce M. Wright, 2008. The phytotoxic properties of alternaric acid in relation to the etiology of plant diseases caused by *Alternaria solani* (ELL. & MART.) Jones & Grout. *Annals Appl. Biol.*, **39:** 308–321.

Briggs, S.P. and Johal, G.S.,1994. Genetic patterns of plant host–parasite interactions. *Trends Genet.* **10:** 12–16.

Brigitte Mauch-Mani and Felix Mauch, 2005. The role of abscisic acid in plant–pathogen interactions. *Curr. Opin. Pl. Biol.*, **8:** 409-414.

Brobst, S.W., and Townsend, C.A. 1994. The potential role of fatty acid initiation in the biosynthesis of the fungal aromatic polyketide aflatoxin-B1. *Can. J. Chem.*, **72:** 200-207.

Brogue, K., I., Chet, M., Holliday, R., Cressman and Biddle, P., 1991. Transgenic plants with enhanced resistance to the fungal pathogen *Rhizoctonia solani*. *Science*, **254:** 1194-1197.

Bruno Campus, manuel Algaara and Joaquim C.G. Esteves Da Silva, 2009. Fluorescent properties of a hybrid cadmium sulfide-dendrimer nanocomposite and its quenching with nitromethane. *J. Fluorescence*, **20:** 143–151.

Burdon, J.J., 2012. Evolution of plant-pathogen associations. Tech. Bull., Centre for Plant Biodiversity Research, CSIRD Plant Industry, GPO Box 1600, Canberra ACT 2601, Australia.

Callewaert, R., holo, H., Devreese, B., Van B.J., NesI. and De V.L., 1999. Characterization and production of amylovorin L471, a bacteriocin purified from *Lactobacillus amylovorus* DCE 471 by a novel three-step method. *Microbiol.*, **145:** 2559-68.

Calonnec, A., Cartolaro, P., Naulin, J. M., Bailey, D. and Langlais, M., 2008. A host-pathogen simulation model: powdery mildew of grapevine. *Plant Pathol.*, **57:** 493-508.

Calviño, L., Faulkner, C. and Maule, A., 2011. Plasmodesmata as active conduits for virus cell-to-cell movement. *In: Advances in Plant Virology*, 470. Caranta C., Aranda M. A., Tepfer M., Lopez-Moya J. J.. Norwich, UK: Caister (Eds.), Academic Press.

Cambra, M., Gorris, M. T., Román, M. P., Terrada, E., Garnsey, S. M., Camarasa, E., Olmos, A. and Colomer, M., 2000. Routine Detection of Citrus Tristeza Virus by Direct Immunoprinting-ELISA Method Using Specific Monoclonal and Recombinant Antibodies. Fourteenth IOCV Conference, 2000—Citrus *Tristeza* Virus.

Campbell, C.L. and Madden, L.V., 1990. *Introduction to Plant Disease Epidemiology*. John Wiley & Sons, New York, U.S.A.

Camporotondia, D.E., Fogliaa, M.L., Alvareza, G.S., Meberta, A.M., Diaza, L.E., Coradinb, T. and Desimonea, M.F., 2013. Antimicrobial properties of silica modified nanoparticles: Microbial pathogens and strategies for combating them. *Science, Technology and Education* (A. Méndez-Vilas, Ed.), 283-290.

Campos Bruno, B., Algarra Manuel, Esteves da Silva and Joaquim, C.G., 2010. Fluorescent Properties of a Hybrid Cadmium Sulfide-Dendrimer Nanocomposite and its Quenching with nitromethane. *J. Fluoresc.*, **20:** 143-151.

Cano-Delgado, A., Penfield S., Smith, C., Catley, M. and Bevan, M., 2003. Reduced cellulose synthesis invokes lignification and defense responses in *Arabidopsis thaliana*. *Plant J.*, **34:** 351-362.

Cantonwine Emily, G., 2014. Creating an Active Learning Environment in the Laboratory with Prepared Slides. *The Plant Health Instructor*. DOI: 10.1094/PHI-T-2014-1222-0.

Canto-Saenz, M. and Mayer de Scurrah, M., 1978. Races of potato cyst nematode in the andean region and a new system of classification. *Nematologica*, **23:** 340-349.

Carisse, O., Morisette-Thomas, V. and Van der Heyden, H., 2013. Lagged association between powdery mildew leaf severity, airborne inoculum, weather and crop losses in strawberry. *Phytopathology*, **103:** 811-821.

Carlson, C., Hussain, S.M., Schrand, A.M., Braydich-Stolle, L.K., Hess, K.L., Jones, R.L. and Schlager, J.J., 2008. Unique cellular interaction of silver nanoparticles: Size-dependent generation of reactive oxygen species. *J. Phys. Chem. B.*, **112:** 13608–13619.

Carol L. Friedrich, Dianne Moyles, Terry J. Beveridge and Robert E. W. Hancock, 2000. Antibacterial action of structurally diverse cationic peptides on Gram-positive bacteria. *Antimicrob. Agents Chemother.*, **44:** 82086-2092.

Carolina Escobar, Sigal Brown and Melissa G. Mitchum J. Jones, 2011. Transcriptomic and Proteomic Analysis of the Plant Response to Nematode Infection. Genomics and Molecular Genetics of Plant-Nematode Interactions, Facultad de Ciencias del Medio Ambiente, Universidad de Castilla-La Mancha, Avenida de, Carlos III s/n, 45071 Toledo, Spain, 157-173.

Carroll P. V. and Robert T. S., 1976. Regulation of lignin formation in reed canary grass in relation to disease resistance. *Plant Physiol.*, **57:** 915-919.

Carvalho, A.O. and Gomes, V.M., 2011. Plant defensins and defensin-like peptides – biological activities and biotechnological applications. *Curr. Pharm. Design.*, **17:** 4270–4293.

Casadebaig, P., Quesnel, G., Langlais, M. and Faivre, R,2012. A generic model to simulate airborne diseases as a function of crop architecture. *PLoS ONE,* **7 (11):** e49406.

Cassab, G.I., 1998. Plant cell wall proteins. *Ann. Rev. Physiol. Plant Mol. Biol.*, **49:** 281-309.

Castle, M., Nazarian, A., Yi, S.S. and Tempst, P., 1999. Lethal effects of apidaecin on *Escherichia coli* involve sequential molecular interactions with diverse targets. *J. Biol. Chem.*, **274:** 32555–32564.

Cernadas, R.A., Camillo, L.R. and Benedetti, C.E., 2008. Transcriptional analysis of the sweet orange interaction with the citrus canker pathogens *Xanthomonas axonopodis* pv. *citri* and *Xanthomonas axonopodis pv. Aurantifolii. Mol. Pl. Pathol.*, **9:** 609-631.

Cevallos-Cevallos, J.M., Rouseff, R. and Reyes-De-Corcuera, J.I., 2009. Untargeted metabolite analysis of healthy and Huanglongbing-infected orange leaves by CE-DAD. *Electrophoresis*, **30:** 1-8.

Chaerle, L., Hulsen, K., Hermans, C., Strasser, R.J., Valcke, R., Höfte, M. and Van Der Straeten, D., 2003. Robotized time-lapse imaging to assess in-plant uptake of phenylurea herbicides and their microbial degradation. *Physiologia Plantarium,* **118:** 613-61.

Chaerle, L., Lenk, S., Hagenbeek, D., Buschmann, C. and Van Der Straeten, D., 2007. Multicolor fluorescence imaging for early detection of the hypersensitive reaction to tobacco mosaic virus. *J. Pl. Physiol.*, **164:** 253-262.

Chamakura, K, Perez-Ballestero, R., Luo, Z., Bashir, S. and Liu, J., 2011. Comparison of bactericidal activities of silver nanoparticles with common chemical disinfectants. *Colloids Surf B Biointerfaces,* **84:** 88-96.

Chang, J.H., Desveaux, D. and Creason, A.L., 2014. The ABCs and 123s of bacterial secretion systems in plant pathogenesis. *Ann. Rev. Phytopathol.*, **52:** 317-45.

Chang, P.K., Cary, J.W., Yu, J.J., Bhatnagar, D., and Cleveland, T.E. 1995. The *Aspergillus parasiticus* polyketide synthase gene *pksA*, a homolog of *Aspergillus nidulans wA*, is required for aflatoxin B-1 biosynthesis. *Mol. Gen. Genet.*, **248:** 270-277.

Chapman, E. J., Prokhnevsky, A. I., Gopinath, K., Dolja, V. V. and Carrington J. C., 2004. Viral RNA silencing suppressors inhibit the microRNA pathway at an intermediate step. *Genes Dev.*, **18:** 1179-1186.

Charych, D., Cheng, Q., Reichert, A., Uziemko, G., Stroh, N., Nagy, J., Spevak, W. and Stevens, R., 1996. Principle of formation of clay monolayer containing nanocomposites *Chem. Biol.*, **3:** 113.

Chassot, C., Nawrath, C. and Metraux, J.P., 2007. Cuticular defects lead to full immunity to a major plant pathogen. *Plant J.*, **49:** 972-980.

Chavan, J.K., and Hejgaard, J., 1981. Detection and partial characterisation of subtilisin inhibitors in legume seeds by isoelectric focusing. *J. Sci. Food Agric.*, **32:** 857-862.

Chawla, H.S., 2007b. Intellectual property rights, J. *Eco-friendly Agri.*, **2:** 103-112.

Chawla, H.S., 2010. IPR and WTO in Relation to Plant Protection, Training manual, "Plant Pathology in Practice". G.B. Pant University of Agriculture & Technology, Pantnagar- 263 145, India, 116-121.

Chen, B., Wang, K., Li, S., Wang, J., Bai, J., Xiao, C. and Lai, J., 2008. Spectrum characteristics of cotton canopy infected with *Verticillium* wilt and inversion of severity level. 1169–1180. *In: IFIP- International Federation for Information Processing,* Volume 259; Computer and Computing Technologies in Agriculture, Vol. 2, Daoliang Li, Springer, Boston,

Chen, F., Li, Q., Sun, L. and He, Z., 2006. The rice 14-3-3 gene family and its involvement in responses to biotic and abiotic stress. *DNA Res.*, **13:** 53-63.

Chernin, L.S., De-la-Fuente, L., Sobolev, V., Haran, S., Vorgias, C.E., Oppenheim, A.B. and Chet, I., 1997. Molecular cloning, structural analysis and expression in *Escherichia coli* of a chitinase gene from *Enterobacter agglomerans. Appl. Environ. Microbiol.*, **63:** 834-839.

Chesnokova, O., Coutinho, J.B., Khan, I.H., Mikhail, M.S. and Kado, C.I., 1997. Characterization of flagella genes of *Agrobacterium tumefaciens*, and the effect of a bald strain on virulence. *Mol. Microbiol.*, **23:** 579-590.

Chester, K.S., 1950. Plant disease losses: Their appraisal and interpretation. *Plant Disease Reporter Supplement*, **193:** 189–362.

Chiarappa, L. and Karpati, J.F., 1981. *Safe and rapid transfer of plant genetic resources: A proposal for a global system.* Report of a FAO/UNDP/IBPGR meeting on crop genetic resources. Rome, Italy, Food and Agriculture Organization of the United Nations. pp. 10.

Chinnamuthu, C.R. and Murugesa Boopathi, P., 2009. Nanotechnology and agroecosystem. *Madras Agric. J.*, **96:** 17-31.

Chitra Srivastava, Choudhury, S., Ghosh, M. and Goswami, A., 2012. Sulfur nanoparticles: Synthesis, characterizations and their applications. *Curr Microbio.*, **65:** 91.

Choudhury, S.R., Nair, K.K., Kumar, R., Gogoi, R. and Srivastava, C., 2010. Nanosulfur: A potent fungicide against food pathogen, *Aspergillus niger. AIP Conf. Proc.,* **1276:** 154-157.

Chowdappa, P. and Shivakumar Gowda, 2013. Nanotechnology in crop protection: Status and scope. *Pest Mnt. Hort. Ecosyst.,* **19:** 131-151.

Chowdappa, P., Kumar, N.B.J., Madhura, S., Kumar, M.S.P. and Myers, K.L., 2013. Emergence of 13_A2 blue lineage of *Phytophthora infestans* was responsible for severe outbreaks of late blight on tomato in south-west India. *J. Phytopathol.,* **161:** 49–58.

Chris Hamilton, 2006. Biodiversity, Biopiracy and Benefits: What Allegations of Biopiracy tell us about Intellectual Property. Blackwell Publishing Ltd., 9600 Garsington Road, Oxford OX4 2DQ, UK and 350 Main Street, Malden, MA 02148, USA.

Chris Khadgi Sørensen, 2012. Infection biology and aggressiveness of *Puccinia striiformis* on resistant and susceptible wheat. Ph.D. thesis submitted to the University of Aarhus, Faculty of Science and Technology , Department of Agroecology, Research Center Flakkebjerg.

Christian Voigt, 2013. Early callose confers complete mildew resistance. *Plant Physiology Preview*. January 18, 2013, as DOI:10.1104/pp. 112.211011

Christopher A Gilligan, 2008. Sustainable agriculture and plant diseases: an epidemiological perspective. *Phil. Trans. R. Soc.,* B 27, **363:** 741-759.

Christopher, C., Mundt Kathryn E., Sackett LaRae, D., Wallace, Christina Cowger and Joseph P. Dudley, 2010. Aerial Dispersal and Multiple-Scale Spread of Epidemic Disease. *EcoHealth,* DOI: 10.1007/s10393-009-0251?

Chung, K. R. 2003. Involvement of calcium/calmodulin signaling in cercosporin toxin biosynthesis by *Cercospora nicotianae. Appl. Environ. Microbiol.,* **69:** 1187-1196.

Chunquan, C., Richard, R.B., Nicole, B. and Timothy, C.P., 2000. Defense enzymes induced in cucumber roots by treatment with plant growth-promoting rhizobacteria (PGPR) and *Pythium aphanidermatum. Physiol. Mol. Pl. Pathol.,* **56:** 13–23

Clark, S.J., Templeton, M.D. and Sullivan, P.A., 1997. A secreted aspartic proteinase from glomerella cingulata: Purification of the enzyme and molecular cloning of the cDNA *Microbiology,* **143:** 1395-1403.

Cláudia V. Godoy, Lucimara J. Koga and Marcelo G. Canteri, 2006. Diagrammatic scale for assessment of soybean rust severity. *Fitopatol. Bras.,* **31:** 63-68.

Cluis Corinne, 2013. Bioprospecting: A New Western Blockbuster, After the Gold Rush, the Gene Rush. *The Science Creative Quarterly* (8) — (University of British Columbia). Retrieved 2013-11-04.

Coakley, S.M., Scherm, H. and Chakraborty, S., 1999. Climate change and disease management. *Ann. Rev. Phyto.,* **37:** 399-426.

Coiro, V.M., Segre, A.L., Di Nola, A., Paci, M., Grottesi, A., Veglia, G. and Ballio, A., 1998. Solution conformation of the *Pseudomonas syringae* MSU 16H phytotoxic lipodepsipeptide pseudomycin a determined by computer simulations using distance geometry and molecular dynamics from NMR data. *Eur. J. Biochem.,* **257:** 449-56.

Colagiero, M. and Ciancio, A., 2011. Climate changes and nematodes: Expected effects and perspectives for plant protection. *REDIA,* **XCIV,** 113-118.

Comstock, J.C. and Scheffer, R.P., 1973. Role of host-selective toxin in colonization of corn leaves by *Helminthosporium carbonum. Phytopathology,* **63:** 24–29.

Consonni, C., Bednarek, P., Humphry, M., Francocci, F., Ferrari, S., Harzen, A., Ver Loren van, Themaat, E. and Panstruga, R., 2010. Tryptophan-derived metabolites are required for antifungal defense in the Arabidopsis mlo2 mutant. *Plant Physiol,* **152:** 1544-1561.

Contreras-Medina, L. M., Torres-Pacheco, I., Guevara-González, R. G.,. Romero-Troncoso, R. J., Terol-Villalobos, I. R. and Osornio-Rios, R. A., 2009. Mathematical modeling tendencies in plant pathology. *African J. Biotechnol.,* **8:** 7399-7408.

Cook, C.G., Escobar, D.E., Everitt, J.H., Cavazos, I., Robinson, A.F. and Davis, M.R., 1999. Utilizing airborne video imagery in kenaf management and production. *Industrial Crops and Products,* **9:** 205-210.

Cooke, B.M., 1998. Disease assessment and yield loss. *The Epidemiology of Plant Diseases,* 42-72, Springer Publishers, the Netherlands.

Corley, C.D., Pullum, L.L., Hartley, D.M., Benedum, C. and Noonan, C., 2014. Disease prediction models and operational readiness. *PLoS ONE,* **9(3):** e91989.

Cotty, P. J., 1989. Virulence and cultural characteristics of two *Aspergillus flavus* strains pathogenic on cotton. *Phytopathology,* **79:** 808-814.

Coventry, H. and Dubery, I., 2001. Lipopolysaccharides from *Burkholderia cepacia* contribute to an enhanced defensive capacity and the induction of pathogenesis-related proteins in *Nicotianae tabacum. Physiol. Mol. Pl. Pathol.,* **58:** 149–158.

Crous, P.W. and Groenewald, J.Z., 2005. Hosts, species and genotypes: Opinions versus data. *Australasian Pl. Pathol.,* **34:** 463-470.

Cui, H., Zhang, S., Yang, H., Ji H. and Wang, X., 2001. Gene expression profile analysis of tobacco leaf trichomes. *BMC Plant Biol.,* **11:** 76-86.

Culver, J.N. and Padmanabhan, M.S., 2007. Virus-induced disease: Altering host physiology one interaction at a time. *Ann. Rev. Phytopathol.,* **45:** 221–243.

Cursino, L., Li, Y., Zaini, P.A., De La Fuente, L., Hoch, H.C. and Burr, T. J., 2009. Witching motility and biofilm formation are associated with tonB1 in *Xylella fastidiosa. FEMS Microbiol. Letters,* **299:** 193e199.

Curtis, C.R., 2002. Is asynchronous teaching and learning worth it for you? Top ten take home lessons on starting an online course. *Plant Health Instr.,* DOI:10.1094/ PHI-T-2002-0820-01.

Curtis, M.J. and Wolpert, T.J., 2002. The oat mitochondrial permeability transition and its implication in victorin binding and induced cell death. *Pl. J.,* **29:** 295–312.

Curtis, R.H.C., 2008. Plant-nematode interactions: environmental signals detected by the nematode's chemosensory organs control changes in the surface cuticle and behaviour. *Parasite*, **15:** 310-316.

D'Arcy, C.J., Eastburn, D.M., Schumann, G.L., 2001. Illustrated glossary of plant pathology. *Plant Health Instr.* DOI: 10.1094/PHI I-2001-0219-01.

D'ovido, R., Mattei, B., Roberti, S. and Bellincampi, D., 2004. Polygalacturonases, polygalacturonase-inhibiting proteins and pectic oligomers in plant-pathogen interactions. *Biochem. Biophys. Acta*, **1696:** 237-244.

Daayf, F., El Hadrami, A., El-Bebany, A. F., Henriquez, M. A., Yao, Z., Derksen, H., El-Hadrami, I. and Adam, L. R., 2012. Phenolic compounds in Plant Defense and Pathogen Counter-Defense Mechanisms. *In: Recent Advances in Polyphenol Research*, Vol. **3**, V. Cheynier, P. Sarni-Manchado and S. Quideau (Eds.), Wiley-Blackwell, Oxford, UK. doi: 10.1002/9781118299753.ch8.

Dae Sung Kim and Byung Kook Hwang, 2014. An important role of the pepper phenylalanine ammonia-lyase gene (*PAL1*) in salicylic acid-dependent signalling of the defence response to microbial pathogens. *J. Exp. Bot.*, **10:** 1093.

Dagmar Jancovä and Lenka LuHovä, 1999. Enzymes in fungal plant pathogenesis by ales Lebeda0, Phyton (Austria) special issue: *Plant Physiology*, **39:** 51-56.

Dahm, H. and Strzelczyk, E., 1987. Cellulolytic and pectolytic activity of *Cylindrocarpon destructans* (Zins.) scholt. isolates pathogenic and non-pathogenic to fir (*Abies alba* Mill.) and pine (*Pinus sylvestris* L.). *J. Phytopathol.*, **118:** 76-83.

Dahse, I., Willmer, C.M. and Meidner, H., 1990. Tentoxin suppresses stomatal opening by inhibiting photophosphorylation. *J. Exp. Bot.*, **41:** 1109-1113.

Dangl, J.L. and Jones, J.D.G., 2001. Plantpathogens and integrated defense responses to infection. *Nature*, **411:** 826-833.

Das, M. R., Sarma, R. K., Saikia, R., Kale, V. S., Shelke, M. V. and Sengupta, P., 2011. Synthesis of silver nanoparticles in an aqueous suspension of graphene oxide sheets and its antimicrobial activity. *Colloids Surf.*, **83:** 16-22.

Dasgupta, M.K., 1988. *Principles of Plant Pathology*, Allied Publishers, New Delhi, 1035.

Daub, M.E. and Chung, K-R. 2007. Cercosporin: A Phytoactivated Toxin in Plant Disease. Online. APS*net* Features. doi: 10.1094/APSnetFeature/2007-0207.

Daumann Lena J., Larrabee, James, A., Ollis, David Schenk, Gerhard Gahan, Lawrence R., 2014. Immobilization of the enzyme GpdQ on magnetite nanoparticles for organophosphate pesticide bioremediation. *J. Inorg. Biochem.*, **131:** 1-7.

David C. Sigee, 2009. *Bacterial Plant Pathology – Cell and Molecular Aspects*. Cambridge University Press Publication.

David Guest and John Brown, 1997. *Infection Processes*, 245-262. *In: Plant Pathogens and Plant Diseases*. J.F. Brown and H.J. Ogle (Edrs.). Australasian Plant Pathology Society.

David J. Bonfil, Bella Dolgin, Israel Mufradi and Silvia Asido, 2004. Bioassay to forecast cereal cyst nematode damage to wheat in fields. *Precision Agri.,* **5:** 329-344.

David J. Schneider and Alan Collmer, 2010. Studying plant-pathogen interactions in the genomics Era: Beyond molecular koch's postulates to systems biology. *Ann. Rev. Phytopathol.,* **48:** 457-479.

David O. Niño-liu, Pamela C. Ronald and A. J. Bogdanove, 2006. *Xanthomonas oryzae* pathovars: Model pathogens of a model crop. *Mol. Pl. Pathol.,* **7:** 303-324.

David T. Kaplan and Noel T. Keen, 1980. Mechanisms conferring plant incompatibility to nematodes. *Revue Nématol.,* **3:** 123-134.

Day, R.C., Grossniklaus, U. and Macknight, R.C., 2005. Be more specific! Laser-assisted microdissection of plant cells. *Tre. Pl. Sci.,* **10**, 397-406

De Leo, F., Volpicella, M., Licciulli, F., Liuni, S., Gallerani, R. and Ceci, L.R., 2002. PLANT-PIs: A database for plant protease inhibitors and their genes. *Nucl. Acids Res.,* **30:** 347-348.

De Lorenzo, G., D'Ovidio, R., Ferrari, S., Raiola, A., Di Matteo A. and Mattei, B. 2003. Insights into plant defence strategies by studying PGIPs, polygalacturonases and oligogalacturonides. pp. 233–236. *In: Biology of Plant-Microbe Interactions,* Vol. **4.** Tikhonovic, B. Lugtenberg, and N. Provorov (Eds.). International Society for Molecular Plant-Microbe Interaction, St. Paul, Minn.

De Waard, M.A., Andrade, A.C., Hayashi, K., Schoonbeek, H.J., Stergiopoulos, I. and Zwiers, L.H., 2006. Impact of fungal drug transporters on fungicide sensitivy, multidrug resistance and virulence. *Pest Manag. Sci.,* **62:** 195-207.

De Waele, D. and Wilken, R., 1990. Effect of temperature on the *in vitro* reproduction of *Ditylenchus destructor* isolated from groundnut. *Revue de Nématologie,* **13:** 171-174.

De Wit, C.T. de and Penning de Vries, F. W. T., 1982. L'analyse des systèmes de production primaire. pp 20-27 In: La Prodctivité de Pâturages Sahéliens. F. W. T. Penning de Vries and M. A. Djiteye, eds. Agricultural Research Reports 918. Pudoc, Wageningen.

De Wolf, E. D., Madden, L. V. and Lipps, P. E., 2003. Risk assessment models for wheat fusarium head blight epidemics based on within-season weather data. *Phytopathology,* **93:** 428-435.

Dehne, H.W., 2011. Phytomedicine Detection, identification, and quantification of fungal diseases of sugar beet leaves using imaging and non-imaging hyperspectral techniques, *Anne-Katrin Mahleinaus Ansbach Tech Bull.,* Institute of Crop Science and Rescource Conservation.

Deising, H. and Mendgen, K., 1992. Developmental control of enzyme produdion and cell wall modification in rust fungi and defence reactions of the host plant. *In: Molecular Biology of Filamentovs Fungi,* U. Stahl and P. Tudzynski, (Eds.). pp. 27-44. VCH Publishers: Weinheim, Germany.

Del Pozo,O. and Lam, E., 1998. Caspases and programmed cell death in the hypersensitive response of plants to pathogens. *Curr. Biol.,* **8:** 1129-1132.

Delalieux, S., van Aardt, J., Keulemans, W., Schrevens, E. and Coppin, P., 2007. Detection of biotic stress (*Venturia inaequalis*) in apple trees using hyperspectral data: Non-parametric statistical approaches and physiological implications. *European J. Agron.,* **27:** 130–134.

Dennis Wharton, MPAA's Rebel With Cause Fights for Copyright Coin, *Variety* (August 3, 1992), Vol. **348(2):** p. 18.

Depa James and Glena Mary, C.F., 2014. Secondary metabolites of microbials as potential source of agrochemicals. *Inter.J. Curr. Res.,* **6:** 535-5363

Desvaux, M., N. J., Parham, A., Scott-Tucker and. Henderson, I. R., 2004. The general secretory pathway: A general misnomer? *Trends Microbiol.,* **12:** 306-309.

Devakumar, C. and Chhonkar, P.K., (Eds.), 2013. Nanotechnology in Agriculture: Scope and Current Relevance. National Academy of Agricultural Sciences, New Delhi, p. 26.

Dhillon, B.S., Khetarpal, R.K. and Saxena, S., 2000. Strengthening PGR collaboration in South Asian region: Germplasm exchange, quarantine and conservation. *In: Fifth Meeting of South Asia Network on Plant Genetic Resources (SANPGR),* October 9-11, 2000, New Delhi.

D'Hondt, L., Höfte, M., Van Bockstaele, E. and Leus, L., 2011. Applications of flow cytometry in plant pathology for genome size determination, detection and physiological status. *Mol. Plant Pathol.,* **12:** 815-28.

Diego, F., Gomez-Casati, Maria, I., Zanor and María V. Busi., 2013. Metabolomics in Plants and Humans: Applications in the Prevention and Diagnosis of Diseases. *BioMed Res. Intern.,* Article ID 792527, 11 pp. doi: 10.1155/2013/792527.

Diekman, M.A. and Green, M.L., 1992. Mycotoxins and reproduction in domestic livestock. *J. Anim. Sci.,* **70:** 1615-1627.

DirkJan van der Gaag and Marcel Raak, 2010. Pest Risk Assessment Fusarium foetens, Plant Protection Service Ministry of Agriculture, Nature and Food Quality P.O. BOX 9102, 6700 EH Wageningen, the Netherlands.

Dixon, D.C., Cutt, J.R. and Klessig, D.F., 1991. Differential targeting of the tobacco PR-1 pathogenesis related proteins to the extracellular space and vacuoles of crystal idioblasts. *EMBO J.,* **10:** 1317-1324.

Dixon, R.A. and Steele, C.L., 1999. Flavonoids and isoflavonoids – a gold mine for metabolic engineering. *Trends Plant Sci.,* **4:** 394.

Dixon, R.A., Howles, P.A., Lamb, C., He, X.Z. and Reddy, J.T., 1998. Prospects for the metabolic engineering of bioactive flavonoids and related phenylpropanoid compounds. *Adv. Exp. Med. Biol.,* **439:** 55–66.

Donald, P. A., Pierson, P. E., St. Martin, S. K., Sellers, P. R., Noel, G. R., MacGuidwin, A. E., Faghihi, J., Ferris, V. R., Grau, C. R., Jardine, D. J., Melakeberhan, H., iblack, T. L., Stienstra, W. C., Tylka, G. L., Wheeler, T. A. and Wysong, D. S., 2006. Assessing *Heterodera glycines*-resistant and susceptible cultivar yield response. *J. Nematol.,* **38:** 76-82.

Dorothea ellinger and Christian A. Voigt, 2014. The use of nanoscale fluorescence microscopic to decipher cell wall modifications during fungal penetration. *Front Plant Sci.,* **5:** 270.

D'Ovidio, R., Mattei, B., Roberti, S. and Bellincamp, D., 2004. Polygalacturonases, polygalacturonase-inhibiting proteins and pectic oligomers in plant-pathogen interactions. *Biochim. Biophys. Acta Proteins Proteomics,* **1696:** 237-244.

Dow, J.M. and Callow, J.A., 1978. A possible role for α-tomatine in the varietal-specific resistance of tomato to *Cladosporium fulvum.Phytopathology,* **92:** 211-216.

Duan, C.G., Fang, Y.Y., Zhou, B.J., Zhao, J.H., Hou, W.N., Zhu, H., Ding, S.W. and Guo, H.S., 2012. Suppression of Arabidopsis ARGONAUTE1-mediated slicing, transgene-induced RNA silencing, and DNA methylation by distinct domains of the cucumber mosaic virus 2b protein. *Plant Cell* , **24:** 259-274.

Dubertret, B., Calame, M. and Libchaber, A.J., 2001. Single-mismatch detection using gold-quenched fluorescent oligonucleotides. *Nat Biotechnol., Lattanzio VMT,* Nivarlet N.

Dudareva, N., Negre, F., Nagegowda D.A. and Orlova, I., 2006. Plant volatiles: Recent advances and future perspectives. *Critical Rev. Pl. Sci.,* **25:** 417-440.

Duke, S.O. and Vaughn, K.C., 1982. Lack of involvement of polyphenol oxidase in ortho-hydroxylation of phenolic compounds in mung bean seedlings. *Physiol. Plant.,* **54:** 381-385.

Duroy A. Navarre and Thomas J. Wolpert, 1999. Victorin induction of an apoptotic/ senescence–like response in oats. *The Pl. Cell.,* **11:** 237-249.

Durrant, W.E. and Dong, X., 2004. Systemic acquired resistance. *Ann. Rev. Phytopathol.,* **42:** 185-209.

Dwayne, R.R., David, R.S., Alan, J.C. and Antonet, M.S., 2013. Host exopolysaccharide quantity and composition impact *Erwinia amylovora* bacteriophage pathogenesis. *Appl. Environ. Microbiol.,* **79:** 3249-3256.

Eadie, M.J., 2003. Convulsive ergotism: Epidemics of the serotonin syndrome? *Lancet Neurol.,* **2:** 429-434.

Edward Hammond, 2013. Biopiracy Continues— A compilation of some recent cases — Third World Network, 131 Jalan Macalister, 10400 Penang, Malaysia Anonymous, 1998. BIOPIRACY, Appeared in Chemical Weekly (15th Sept. 1998), Biotech Support Services, S. N. Jogdand, 408, Archana Building, Sector 17, Vashi, Navi Mumbai.

Elaine Roddy, 2012, Evaluation of Easy-to-Use Pocket Diagnostics Test Kit for Confirming Plant Disease, On Vegetabls. Ontario Ministry of Agriuclture, Food and Rural Affairs.

Elamawi Rabab, M. A. and El-Shafey, R. A. S., 2013. Inhibition effects of silver nanoparticles against rice blast disease caused by *Magnaporthe grisea*. *Egypt. J. Agric. Res.,* **91:** 1271.

Elisa Ferrante and David Simpson, 2001. Biological & biomedical sciences, a review of the progression of transgenic plants used to produce plantibodies for human usage, *J. Young Investigators,* **4:** 14-17.

Elizabeth L. Litte, 2001. Distance Education Plant Pathology Courses at the University of Georgia, Department of Plant Pathology, University of Georgia Athens, GA 30602, U.S.A.

Ellen M. Bauske, Suzanne M. Bissonnette and Adrianna D. Hewings, 1997. Yield loss assessment of barley yellow dwarf disease on spring oat in illinois. *Plant Disease,* **81:** 485-488.

Emma Wanjiru Gachomo, Olusola Olusoji Shonukan and Simeon Oloni Kotchoni, 2003. The molecular initiation and subsequent acquisition of disease resistance in plants. *African J. Biotech.,* **2:** 26-32.

Enhanced activity of abscission enzymes predisposes oranges to invasion by Diplodia natalensis during ethylene degreening G. Eldon Brown and Jacqueline K. Burns, 1998. *Postharvest Biol. Technol.,* **14:** 217-227.

Eric A. Schmelza, Fatma Kaplana, Alisa Huffakera, Nicole J. Dafoea, Martha M. Vaughana, Xinzhi Nib, James R. Roccac, Hans T. Alborna and Peter E. Teala, 2011. Identity, regulation, and activity of inducible diterpenoid phytoalexins in maize. *PNAS,* **108:** 5455–5460.

Eric Ceballos-Alcantarill, Antonio Abad-Fuentes, Vincenzo Aloisio, Consuelo Agullo, Antonio Abad-Somovillab and Josep V. Mercader, 2014. Haptens, bioconjugates and antibodies for penthiopyrad immunosensing. *Analyst,* **139:** 5358-5361.

Erickson, F.L., Holzberg, S., Calderon-Urrea, A., Handley, V., Axtell, M., Corr, C.and Baker, B., 1999 The helicase domain of the TMV replicase proteins induces the N-mediated defence response in tobacco. *The Plant J.: Cell and Mol. Biol.,* **18:** 67-75.

Eriksson, K. and Bermek, H., 2009. Applied Microbiology: Industrial Lignin, Lignocellulose, Ligninase. Encyclopedia of Microbiology, Terceraedición. 2009, 373–384

Eriksson, K.B., 1974. Intraspecific variation in *Ditylenchus dipsaci*. I. Compatibility tests with races. *Nematologica,* **20:** 147-162.

Etefagh, R., Azhir, E. and Shahtahmasebi, N., 2013. Synthesis of CuO nanoparticles and fabrication of nanostructural layer biosensors for detecting *Aspergillus niger* fungi. *Sci. Iranian.,* **20:** 1055-1058.

Everts, K.L., Osborne, L., Gevens, A.J., Vasquez, S.J., Gugino, B.K., Ivors, K. and Harmon, C., 2012. Extension plant pathology: strengthening resources to continue serving the public interest. *Phytopathology,* **102:** 652-655.

Évry, 2014. Agdia-Biofords Commercializes is first Rapid Molecular Diagnostic Test to detect the pathogen responsible for little cherry disease, June 16th, 2014, Press release from AGDIA-BIOFORDS.

Fabi, A. and Varvaro, L., 2009. Remote sensing in monitoring the dieback of hazelnut on the 'Monti Cimini' district (Central Italy). *Acta Hort., (ISHS)* **845:** 521-526.

Faharani, G., 2008. Uji Aktifitas Antibakteri Daun Belimbing Wuluh Terhadap Bakteri *Streptocoocus aureus* dan *Eschericia coli* Secara Bioautografi. Jakarta: Fmipa UI.

Fan, C., Wang, S., Hong, J.W., Bazan, G.C., Plaxco, K.W. and Heeger, A.J., 2003. Beyond superquenching: Hyper-efficient energy transfer from conjugated polymers to gold nanoparticles. *Proc. Natl. Acad. Sci.,* **100:** 6297-6301.

Farkas, Á. and Zs. Orosz-Kovács, 2006. Nectary structure of pear cultivars and its relation to fire blight susceptibility. *Acta Hort.* **704:** 1-8.

Farre, E.M., Tiessen, A., Roessner, U., Geigenberger, P., Tretheway, R.N. and Willmitzer, L., 2001. Analysis of compartmentation of glycolytic intermediates, nucleotides, sugars, organic acids, amino acids, and sugar alcohols in potato tubers using a nonaqueous fractionation method. *Plant Physiol,* **127:** 685-700.

Fawe, A., Abou-Zaid, M., Menzies, J. G. and Bélanger, R. R., 1998. Silicon-mediated accumulation of flavonoid phytoalexins in cucumber. *Phytopathol.,* **88:** 396.

Fawzi, A.R. and Mark, A.B., 2003. Reactive oxygen species production in association with suberization: evidence for an NADPH-dependent oxidase. *J. Exp. Bot.,* **54:** 935-941.

Fei Wang, Peng Zhang, Sheng Qiang and and Lang-Lai Xu, 2006. Interaction of plant epicuticular waxes and extracellular esterases of *Curvularia eragrostidis* during infection of *Digitaria sanguinalis* and *Festuca arundinacea* by the fungus. *Int. J. Mol. Sci.,* **7:** 346-357.

Felipe Canale, 2003, Phytosanitary capacity evaluation: the tool, its results and its relation to invasive alien species, Ministerio de Ganadería, Agricultura y Pesca, Melitón González, 1169 - p. 5, Montevideo, Uruguay.

Fengping Chen, Ping Han, Pengfei Liu, Naiguo Si, Junli Liu and Xili Liu, 2014. Activity of the novel fungicide SYP-Z048 against plant pathogens. Scientific Rep., 4, Article number: 6473.

Fengping Chen, Ping Han, Pengfei Liu, Naiguo Si, Junli Liu, & Xili Liu, 2014. Activity of the novel fungicide SYP-Z048 against plant pathogens. *Scientific Reports.*

Ferber, D., 1999. GM crops in the cross hairs. *Science,* **286:** 1662-1665.

Ferrandino, F.J., 2008. Effect of crop growth and canopy filtration on the dynamics of plant disease epidemics spread by aerially dispersed spores. *Phytopathology,* **98:** 492-503.

Ferris, H., 1976. Development of a computer-simulation model for a plant-nematode system. *J. Nematol.,* **8:** 255–263.

Ferris, H., 1978. Modification of a computer simulation model for a plant-nematode system. *J. Nematol.,* **10:** 198–201.

Ferris, H., Duvernay, H. S. and Small, R. H. 1978. Development of a data base on the effects of soil temperature on *Meloidogyne arenaria* eggs for a simulation model. *J. Nematol.,* **10:** 39-42.

Field Kabir Lamsal, Sang Woo Kim, Jin Hee Jung, Yun Seok Kim, Kyong Su Kim and Youn Su Lee. 2011 Application of silver nanoparticles for the control of *Colletotrichum* species *in vitro* and pepper anthracnose disease. *Mycobiology,* **39:** 194–199.

Figen Mert-Turks, 2006.....1995. Saponins versus plant fungal pathogens. *J. Cell Mol. Biol.,* **5:** 13-17.

Fink-Gremmels, J., Blom, M. J., Woutersen van Nijnanten, F. M. A., Jahn, A., deGroene, E. M. and Horbach, G. J. M. J., 1995. Biotransformation processes in the etiology of ochratoxicosis, pp 107-121. *In: Molecular Approaches to Food Safety.* M. Eklund, J. L. Richard, and K. Mise (Eds.), Alaken, Fort Collins, CO.

Fish, W.W. and Madihally, S.V., 2004. Modeling the inhibitor activity and relative binding affinities in enzyme-inhibitor-protein systems: Application to developmental regulation in a PG-PGIP system. *Biotechnol. Prog.,* **20:** 721–727.

Fisher, M.C., Henk, D.A., Briggs, C.J., Brownstein, J.S. and Madoff, L.C., 2012. Emerging fungal threats to animal, plant and ecosystem health. *Nature,* **484:** 186–194.

Fletcher, J., Bender, C., Budowle, B., Cobb, W. T., Gold, S. E., Ishimaru, C. A., Luster, D., Melcher, U., Murch, R., Scherm, H.. Seem, R. C., Sherwood, J. L., Sobral, B. W. and Tolin, S. A., 2006. Plant pathogen forensics: Capabilities, needs and recommendations. *Microbiol. Mol. Biol. Rev.,* **70:** 450-471.

Flors, V., Miralles, M.C., Gonzalez-Bosch, C., Carda, M. and Gracia-Agustin, P., 2003. Induction of protection against the necrotrophic pathogens *Phytophthora citrophthora* and *Alternaria solani* in *Lycopersicon esculentum* Mill by a novel synthetic glycoside combined with amines. *Planta,* **216:** 929-38.

Frank, A., Hoeberichts and Ernst J. Woltering, 2002. Multiple mediators of plant programmed cell death: Interplay of conserved cell death mechanisms and plant-specific regulators. *BioEssays,* **25:** 47-57.

Frederick, M. and Fishel, 2012. Pesticide toxicity profile: Strobilurin pesticides. The Pesticide Information Office, Florida Cooperative Extension Service, Institute of Food and Agricultural Sciences, University of Florida.

Freeman, B.C. and Beattie, G.A., 2008. An Overview of Plant Defenses against Pathogens and Herbivores. *The Plant Health Instructor.* DOI: 10.1094/PHI-I-2008-0226-01.

Fritig, B., Heitz, T. and Legrand, M.,1998. Antimicrobial proteins in induced plant defense. *Curr. Opin. Immunol.,* **10:** 16-22.

Fry, W. E., Goodwin, S. B., Matuszak, J. M., Spielman, L. J. and Drenth, A., 1992. Population genetics and intercontinental migrations of *Phytophthora infestans. Ann. Rev. Phytopathol.,* **30:** 107-129.

Fujimura, T., Ichita, T. and Kimura, T., 1989. Occurrence of potato-rot nematode, *Ditylenchus destructor* Thorne, in garlic and control. 1. Evaluation of treatments applied before planting and after harvest for control. *Japanese J. Nematol.,* **18:** 22-29.

Gadd, G.M. and White, C., 1989. Heavy metal and radionuclide accumulation and toxicity in fungi and yeasts. *In: Metal-Microbe Interaction,* Poole, R.K., Gadd, G.M., (Eds.). IRL Press, Oxford, UK, 19-38.

Gadoury, D. M., Andrews, J., Baumgartner, K., Burr, T. J., Kennelly, M. M., Lichens-Park, A., MacDonald, J., Savary, S., Scherm, H., Tally, A. and Wang, G.L. 2009. Disciplinary, institutional, funding and demographic trends in plant pathology: What does the future hold for the profession? *Plant Dis.* **93:** 1228-1237.

Gadoury, D.M., Andrews, J., Baumgartner, K., Burr, T.J., Kennelly, M.M., Lichens-Park, A., Galagan, J.E., 2003. The genome sequence of the filamentous fungus, *Neurospora crassa. Nature,* **422:** 859-868.

Gajjar, P., Pettee, B., Britt, D.W., Huang, W., Johnson, W.P. and Anderson, A.J., 2009. Antimicrobial activities of commercial nanoparticles against an environmental soil microbe, *Pseudomonas putida KT2440. J. Biol. Eng.,* **3:** 9.

Galea, V.J., 2006. Virtual Plant Pathology Lab CD-ROM: A Tool to Support Plant Pathology Diagnostics Learning for Distance Education, School of Agronomy and Horticulture, University of Queensland, Gatton, Australia.

Galea, V.J., 2006. Virtual Plant Pathology Lab CD-ROM. Teaching & Education Development Institute. The University of Queensland, Australia.

Galea, V.J., 2008. Using Wimba Live Classroom to engage students in Clinical Case Studies. Blended Learning Conference June 19th, University of Queensland, Australia.

Galston, A.W. and Sawhney, R.K., 1990. Polyamines in plant physiology. *Pl. Physiol.,* 94: 406-410.

García-Guzmán, G. and Wennström, A., 2001. Interactions between two rust fungi and their host plant *Anemone nemorosa. Ecography,* **24:** 25–32.

Gardeniers, J.G. and van den Berg, A.A., 2004. Lab-on-a-chip systems for biomedical and environmental monitoring. *Anal Bioanal Chem.,* **378:** 1700–1703.

Garrett, K. A,. Dendy, S. P., Frank, E. E., Rouse, M. N. and Travers, S. E., 2006. Climate change effects on plant disease: Genomes to ecosystems. *Ann. Rev. Phytopathol.,* **44:** 489–509.

Garrett, K.A., Esker, P.D. and Sparks, A.H., 2007. Introduction to the R programming environment. *The Plant Health Instructor.* On-line: DOI: 10.1094/ PHI-A-2007-1226-02.

Gary Lawrence, Roger King, Amber Kelley, John Vickery, 2012. Method for detecting and managing nematode population. US 20060006335 A1(Patent).

Gary Strobel and Bryn Daisy, 2003. Bioprospecting for microbial endophytes and their natural products. *Microbiol. Mol. Biol. Rev.,* **67:** 491-502.

Gary W. Lawrence, Rushabh A. Doshi, Roger L. King, Kathy S. Lawrence and Jacobo Caceres, 2003. Nematode Management using Remote Sensing Technology, Self-Organized Maps and Variable Rate Nematicide Applications. *Tech. Bull.,* Mississippi State University, Mississippi State University, MS.

Gary, G. C., 2009. Diversity and distribution of idioblasts producing calcium oxalate crystals in *Dieffenbachia seguine* (*Araceae*). *American J. Bot.,* **96:** 1245-1254.

Gaumann, E., 1950. *Principle of Plant Infection* pp. 176-242. W.B., Brierly (Ed.) 543 p. Hafner Publish. Co. N.Y.

Gaunt, R.E., 1995. The relationship between plant disease severity and yield. *Ann. Rev. Phytopathol.,* **33:** 119-44.

Gavin Stenton, 2003. Biopiracy within the pharmaceutical industry: A stark illustration of just how abusive, manipulative and perverse the patenting process can be towards countries of the south. *Hertfordshire Law J.,* **1:** 30-47.

Gehan I.Kh. Marei, Mona A. Abdel Rasoul and Samir A.M. Abdelgaleil, 2012. Comparative antifungal activities and biochemical effects of monoterpenes on plant pathogenic fungi. *Pesticide Biochem. Physiol.,* **103:** 56–61.

Geoffrey Michael Gadd, 2010. Metals, minerals and microbes: geomicrobiology and bioremediation. *Microbiol.,* **156:** 609-643.

Geoffroy, M., Legrand, M. and Fritig, B., 1990. Isolation and characterization of a proteinaceous inhibitor of microbial proteinases induced during the hypersensitive reaction of tobacco to Tobacco Mosaic Virus. *Mol.Pl. Microbe Inter.,* **3:** 327-333.

George Mondoh, 2007. Development of an on-farm rapid diagnostic kit for Maize Lethal Necrosis Disease (MLND) causative viruses. Kenya Agricultural and Live stock Research. Organization.

Gerber, I. B., Laukens, K., De Vijlder, T., Witters, E. and Dubery, I. A., 2008. Proteomic profiling of cellular targets of lipopolysaccharide-induced signalling in *Nicotiana tabacum* BY-2 cells. *Biochimica et Biophysica Acta,* **1784:** 1750-1762.

Ghormade, V., Deshpande, M.V and Paknikar, K.M., 2011. Perspectives for nanobiotechnology enabled protection and nutrition of plants. *Biotechnol. Adv.,* **29:** 792-803.

Ghosh, P. P., Mandal, D, Laha, S. and Dasgupta, M. K., 2009. Dynamics and severity model in managing fungal diseases. *J. Pl. Prot. Sci.,* **1:** 55-59.

Gibbs, J.N., 1978a. Intercontinental epidemiology of dutch elm disease. *Ann. Rev. Phytopathol.,* **16:** 287-307.

Giese, A.C., 1980. Hypericism. *Photochem. Photobiol. Rev.,* **5:** 229-255.

Giese, A.C., 1981. The photobiology of *Blepharisma. Photochem. Photobiol. Rev.,* **6:** 139-180.

Gilbert, G.S., 2002. Evolutionary ecology of plant diseases in natural ecosystems. *Ann. Rev Phytopathol.,* **40:** 13-43.

Gilligan, C.A. and Kleczkowski, A., 1997. Population dynamics of botanical epidemics involving primary and secondary infection. *Phil. Trans. R. Soc.,* B. **352:** 591-608.

Gilligan, C.A., 2002. An epidemiological framework for disease management. *Adv. Bot. Res.,* **38:** 1-64.

Ginet, N., Pardoux, R., Adryanczyk, G., Garcia, D. and Brutesco, C., 2011. Single-Step Production of a Recyclable Nanobiocatalyst for Organophosphate Pesticides Biodegradation Using Functionalized Bacterial Magnetosomes. *PLoS ONE,* **6:** e21442.

Gitaitis, R. and Walcott, R., 2007. The epidemiology and management of seedborne bacterial diseases. *Ann. Rev. Phytopathol.,* **45:** 371-97.

Gitte Erbs and Mari-Anne Newman, 2009. MAMPs/PAMPs-elicitors of innate immunity in plants. *In: Plant Pathogenic Bacteria: Genomics and Molecular Biology*, Robert W. Jackson (Ed.), Caister Academic Press, Norfolk, 227-240

Giuseppina Rea, Ouissal Metoui and Riccardo Angelini, 2002. Copper amine oxidase expression in defense responses to wounding and *Ascochyta rabiei* invasion. *Pl. Physiol.*, **128:** 865-875.

Glazebrook, Ausbel; and Ausubel, F.M., 1994. Isolation of phytoalexins-deficient mutants of *Arabidopsis thaliana* and characterization of their interactions with bacterial pathogens. *PNAS,* **91:** 8955-8959.

Gleadow, R.M. and Moller, B.L., 2014. Cyanogenic glycosides: synthesis, physiology, and phenotypic plasticity. *Ann. Rev. Pl. Biol.,* **65:** 155-185

Gogoi, R., Dureja, P. and Singh, P.K., 2009. Nanoformulations – a safer and effective option for agrochemicals. *Indian Farming,* **59:** 7-12

Goluch, E.D., Nam, J.M., Georganopoulou, D.G., Chiesl, T.N., Shaikh, K.A., Ryu, K.S., Barron, A.E., Mirkin, C.A. and Liu, C., 2006. A bio-barcode assay for on-chip attomolar-sensitivity protein detection. *Lab. Chip.,* **6:** 1293-1299.

Gomes, L., Duarte, K., Andrino, F., Leal Jr., G., Garcia, L. , Figueira, A. and de Lira, S. 2014. Alpha-Tomatin against Witches' Broom Disease. *American J. Pl. Sci.,* **5:** 596-604.

Gómez, G., Martínez, G. and Pallás, V., 2009. Interplay between viroid-induced pathogenesis and RNA silencing pathways. *Trends Plant Sci.,* **14:** 264-269.

González Fernández, R. and Jorrín Novo, J.V., 2010. Proteomics of fungal plant pathogens: the case of *Botrytis cinerea. In: Current Research, Technology and Education Topics in Applied Microbiology and Microbial Biotechnology.* A. Mendez-Vilas (Ed.).

González-Lamothe, R., El Oirdi, M., Brisson, N. and Bouarab, K., 2012. The conjugated auxin indole-3-acetic acid-aspartic acid promotes plant disease development. *Pl. Cell.,* **24:** 762-777.

Goodman, B.A., Williamson, B. and Chudek, A., 1992. Non-invasive observation of the development of fungal infection in fruit. *Protoplasma,* **166:** 107-109.

Gorbalenya, A. E. and Snijder, E. J., 1996. Viral cysteine proteases. *Perspect. Drug. Discov. Design.,* **6:** 64-86.

Gosme, M and Lucas, P., 2009. Cascade: An epidemiological model to simulate disease spread and aggregation across multiple scales in a spatial hierarchy. *Phytopathology,* **99:** 823-32.

Goto, K., 1965. Estimating Losses from Rice Blast in Japan. *In: "The Rice Blast Disease".* The Johns Hopkins Press, Baltimore, Maryland, 195-202.

Goud, J.C. and Termorshuizen A.J., 2003. Quality of methods to quantify microsclerotia of *Verticillium dahliae* in soil. *European J. Pl. Pathol.,* **109:** 523-534.

Grant, J. J., Yun, B. W. and Loake, G. J., 2000. Oxidative burst and cognate redox signalling reported by luciferase imaging: Identification of a signal network that functions independently of ethylene, SA and Me-JA but is dependent on MAPKK activity. *Pl. J.,* **24:** 569-582.

Grant, S.R., Fisher, E.J., Chang, J.H., Mole, B.M. and Dangl, J.L. 2006. Subterfuge and manipulation: type III effector proteins of phytopathogenic bacteria. *Ann. Rev. Microbiol.,* **60:** 425–449.

Grassman, J., 2005. Terpenoids as plant antioxidants. *Vitam Horm.,* **72:** 505-535.

Grayer, R.J. and Kokubun, T., 2001. Plant-fungal interactions: the search for phytoalexins and other antifungal compounds from higher plants. *Phytochemistry,* **56:** 253.

Gregory, S., Gilbert, Roger Magarey, Karl Suiter and Campbell O. Webb, 2012. Evolutionary tools for phytosanitary risk analysis: phylogenetic signal as a predictor of host range of plant pests and pathogens, *Evolutionary Applications,* ISSN 1752-4571.

Grey, W.E. and Mathre, D.E., 2001. Plant dis eases in the news. *Plant Health Instr.* DOI 10.1094/PHI-T-2001-1025-02.

Griffin, M.K. and Burke, H.K. 2003. Compensation of hyperspectral data for atmospheric effects. *Lincoln Lab. J.,* **14:** 29–54.

Gritta Schrader, 2003. Adaptation of regional pest risk assessment to the revised ISPM 11, Technical University and Plant Health Department of the Federal Biological Research Centre, Messeweg 11/12, 38104, Braunschweig, Germany.

Grubb, C.D. and Abel, S., 2006. Glucosinolate metabolism and its control. *Trends Plant Sci,* **11:** 89-100.

Gu, C., Kolodziejek, I., Misas-Villamil, J., Shindo, T., Colby, T., Verdoes, M., Richau, K.H., Guest, D. I. and Brown, J.F., 1997. Plant defences against pathogens. pp. 263-286. *In: Plant Pathogens and Plant Diseases.* Brown, J.F. and Ogle, J.H. (Eds.). Rockvale Publications, Armidale, New South Wales.

Guillame, H., Philippe, T. and Marc, O., 2012. PAMPs, DAMPs and others: An update on the diversity of plant immunity elicitors. *Base,* **16:** 257-268.

Gullino, M.L., 2009. Teaching plant pathology and disease and pest management for university students: Some considerations after thirty years of experience. *Arab J. Pl. Protectn.,* **27:** 233-235.

Guo, Y.R., Liu, S.H., Gui, W.J. and Zhu, G.N., 2009. Gold immunochromatographic assay for simultaneous detection of carbofuran and triazophos in water samples. *Anal Biochem.,* **389:** 32-39.

Gutierrez, A.P., DeVay, J.E. and Frieverthauser, G.E., 1983. A model of *Verticillium* wilt in relation to cotton growth and development. *Phytopathology,* **73:** 89-95.

Gutierrez-Campos, R., Torres-Acosta, J.A., Saucedo-Arias, L.J. and Gomez-Lim, M.A., 1999. The use of cysteine proteinase inhibitors to engineer resistance against potyviruses in transgenic tobacco plants. *Nat. Biotechnol.,* **17:** 1223-1236.

Guy, R.K. and Louise- Marie, C.D., 1992. Model for dispersal and epiphytic survival of bacteria applied to crop foliage. Department of Plant, Soil and Entomological Sciences, Univrersity of Idaho, Moscow, ID.

Guy, R. Knudsen and George W. Hudler, 1987. Use of a computer simulation model to evaluate a plant disease biocontrol agent. *Ecol. Modelling,* **35:** 45-62.

Hadi Aliakbarian, Amin Enayati, Maryam Ashayer Soltani, Hossein Ameri Mahabadi and Mahmoud Moghavvemi, 2010. Electromagnetic Solutions for the Agricultural Problems, Advanced Microwave Circuits and Systems, Vitaliy Zhurbenko (Ed.).

Hafez, E.E., Hashem, M., Balbaa, M.M., El-Saadani, M.A. and Ahmed, S.A., 2013. Induction of new defensin genes in tomato plants via pathogens-biocontrol agent interaction. *J. Plant Pathol. Microb.,* **4:** 167.

Hafiz Farhad Ali, Musharaf Ahmad, Muhammad Junaid, Ayesha Bibi, Asad Ali, Muhammad Sharif, Barkat Ali, Khalid Nawab and Amna Sadozai, 2012. Inoculum sources, disease incidence and severity of bacterial blackleg and soft rot of potato. *Pak. J. Bot.,* **44:** 825-830.

Hafiza Tahira Gul, Shafqat Saeed, Fawad Zafar Ahmad Khan and Syed Amir Manzoor, 2014. Potential of nanotechnology in agriculture and crop protection: A review. *Appl. Sci. Busi. Econ.,* **1:** 23-28.

Hails, R.S., 2000. Genetically modified plants–the debate continues. *Tree,* **15:** 14-18.

Haixin Cui, Guochen Yang, Jianfang Jiang, Ping Zhang and WeiGu, 2013. Biological effects of PAS TiO_2 sol on disease control and photosynthesis in cucumber (*Cucumis sativus* L.). *Australian J. Crop Sci.,* **7:** 99-103.

Haliza Katas, 2012. Antifungal activity of chitosan nanoparticles and correlation with their physical properties. *Inter. J. Biomaterials,* 632698.

Hall, R., 2006. Plant metabolomics: from holistic hope, to hype, to hot topic. *New Phytol.,* **169:** 453-468.

Hamed K. Abbas, Mirocha, C. J., Vesonder, R. E. and. Gunther, R., 1990. Acute toxic effects of an isolate of moniliformin-producing *Fusarium oxysporum* and purified moniliformin on rats. *Arch. Environ. Contain. Toxicol.* **19:** 433-436.

Hamid, K. and Strange, R.N., 2000. Phytotoxicity of solanapyrones A and B produced by the chickpea pathogen *Ascochyta rabiei* (Pass.) Labr. and the apparent metabolism of solanapyrone A by chickpea tissues. *Physiol. Mol. Plant Pathol.,* **56:** 235-244.

Hammerschmidt, R., 1999. Phytoalexins: What have we learned after 60 years? *Ann. Rev. Phytopathol.,* **37:** 285.

Hammerschmidt, R., 2003. Phytoalexin accumulation: Response or defense. *Physiol. Mol. Plant Pathol.,* **62:** 125-126.

Hammond-Kosack, K.E. and Jones, J.D.G., 1997. Plant disease resistance genes. *Ann. Rev. Plant. Physiol. Plant. Mol. Biol.,* **48:** 575–607.

Han, M, Su, T., Zu, Y.G. and An, Z.G., 2006. Research advances on transgenic plant vaccines. *Yi Chuan Xue Bao,* **33:** 285-93.

Hans, D., VanEtten, Robert, W., Sandrock, Catherine, C., Wasmann, Scott, D., Soby, Kevin McCluskey and Ping Wang, 1995. Detoxification of phytoanticipins and phytoalexins by phytopathogenic fungi. *Canadian J. Bot.,* **73:** 518-525.

Hara-Nishimura, I. and Hatsugai, N., 2011. The role of vacuole in plant cell death. *Cell Death & Differentiation,* **18:** 1298-1304.

Harborne, J.B., 2003. *Introduction to Ecological Biochemistry*. London: Academic Press.

Harman, G.E., 2000. Myths and dogmas of biocontrol: Changes in perceptions derived from research on *Trichoderma harzianum* T-22. *Plant Dis.*, **84:** 377-93.

Harp, T. L., Godwin, J. R., Scalliet, G ., Walter, H., Stalker, A. D., Bartlett, D. W. and Ranner, D. J., 2011. Isopyrazam, a new generation cereal fungicide, *Aspects of Appl. Biol.*, **106:** 113-120.

Hashimoto, Y., Nakamura, H., Koichi, A.K. and Karube, I., 2008. A new diagnostic method for soil-borne disease using a microbial biosensor. *Microbes Environ.*, **23:** 35–9.

Hassane Ait-Lahsen, Andrés Soler, Manuel Rey, Jesús de la Cruz, Enrique Monte, and Antonio Llobell, 2001. An Antifungal Exo-α-1,3-glucanase (AGN13.1) from the biocontrol fungus *Trichoderma harzianum. Appl. Environ. Microbiol.*, **67:** 5833-5839.

Hatfield, J. L., 1990. Remote detection of crop stress: Application to plant pathology. *Phytopathology*, **80:** 37-39.

Hawes, M. C., and L. Y. Smith. 1989. Requirement for chemotaxis in pathogenicity of *Agrobacterium tumefaciens* on roots of soil-grown pea plants. *J. Bacteriol.*, **171:** 5668-5671.

He, X.Z. and Dixon, R.A., 2000. Genetic manipulation of isoflavone 7-O-methyltransferase enhances biosynthesis of 4′-O-methylated isoflavonoid phytoalexins and disease resistance in alfalfa. *Plant Cell*, **12:** 1689.

He, Y.W. and Zhang, L.-H., 2008. Quorum sensing and virulence regulation in *Xanthomonas campestris. FEMS Microbiol. Rev.*, **32:** 842-857.

Heath E O'Brien, Darrell Desveaux, David S Guttman, 2011. Next-generation genomics of *Pseudomonas syringae. Curr. Opinions in Microbiol.*, **14:** 24-30.

Heidrun Gundlach, Martin, J., Muller, Toni, M. Kutchan, and Meinhart, H. Zenk., 1992. Jasmonic acid is a signal transducer in elicitor-induced plant cell cultures. *Pl. Biol.*, **89:** 2389-2393.

Heiler, S., Mendgen, K. and Deising, H., 1993. Cellulolytic enzymes of the obligately biotrophic rust fungus, *Uromyces viciae-fabae* are regulated differentiation-specifically. *Mycol. Res.*, **97:** 77-85.

Heitz, T., Geoffroy, P., Stintzi, A., Fritig, B. and Legrand, M., 1993. cDNA cloning and gene expression analysis of the microbial proteinase inhibitor of tobacco. *J. Biol. Chem.*, **268:** 16987-16992.

Helen O. Whelan, 1992. Tiie Effeci' of Crop Yield Potential on Disease: Yield Loss Relationships in Barley *(Hordeum vulgare* L.). Ph.D. Thesis submitted to Lincoln University, Canterbury, New Zealand.

Hemetsberger, C., Herrberger, C., Zechmann, B., Hillmer, M. and Doehlemann, G. 2012. The *Ustilago maydis* effector Pep1 suppresses plant immunity by inhibition of host peroxidase activity. *PLoS Pathog*, **8:** e1002684.

Henry, G., Thonart, P. and Ongena, M., 2012. PAMPs, MAMPs, DAMPs and others; An update on the diversity of plant immune elicitors. *Biotechnologie, Agronomie, Société et Environnement*, **16:** 257–268.

Herreid, C.F. 2007. Start with a story: The case study teaching method of teaching college science. National Science Teachers Association Press, Arlington, VA.

Hess, G.R., Randolph, S.E., Arneberg, P., Chemini, C., Furlanello, C., Harwood, J., Roberts, M.G. and Swinton, J., 2002. Spatial aspects of disease dynamics, pp. 102–118. *In: The Ecology of Wildlife Diseases.* Hudson, P.J., Rizzoli, A., Grenfell, B.T., Heesterbeek, H., Dobson, A.P., (Eds.), Oxford University Press; Oxford, UK.

Hijmansa, R. J., Forbesb, G. A. and Walker, T. S., 2000. Estimating the global severity of potato late blight with GIS-linked disease forecast models. *Plant Pathology,* **49:** 697-705.

Himeno Misako, Kitazawa Yugo, Yoshida Tetsuya, Maejima Kensaku, Yamaji Yasuyuki, Oshima Kenro and Namba, Shigetou, 2013. Purple top symptoms are associated with reduction of leaf cell death in phytoplasma-infected plants. *Scientific Reports,* Vol. 4, Article number: 4111, Macmillan Publishers Limited.

Hipskind, J., Wood, K. and Nicholson, R.L., 1996. Localized stimulation of anthocyanin accumulation and delineation of pathogen ingress in maize genetically resistant to *Bipolaris maydis* race O. *Physiol. Mol. Pl. Pathol.,* **49:** 247-256.

Hol, W.H.G., Vrieling, K. and Veen, J.A., 2003. Nutrients decrease pyrrolizidine alkaloid concentrations in *Senecio jacobaea. New Phytol.,* **158:** 175-181.

Holb, I.J., Heijne, B. and Jeger, M.J., 2003. Summer epidemic of apple scab: The relationship between measurements and their implications for the development of predictive models and threshold levels under different disease control levels. *J. Phytopathol.,* **151:** 335-343.

Holland, N., Evron, Y., Jansen, M.A.K., Edelman, M. and Pick, U. 1997. Involvement of thylakoid overenergization in tentoxin-induced chlorosis in *Nicotiana* spp. *Pl. physiol.,* **114:** 21-28.

Hongda Chen and Rickey Yada, 2011. Nanotechnologies in agriculture: New tools for sustainable development. *Trends in Food Sci. & Technol.,* **22:** 123-129.

Hongda Chen and Rickey Yada, 2011. Nanotechnologies in agriculture: New tools for sustainable development. *Trends in Food Sci., & Tech.,* **22:** 585-594.

Hooper, D.J., 1986. Extraction of free-living stages from soil. *In: Laboratory Methods for Work with Plant and Soil Nematodes* Southey, J.F., (Eds.) pp 5-30. *Ministry of Agriculture, Fisheries and Food Reference Book* No. 402. HMSO, London, UK.

Hoover, D.M., Rajashankar, K.R., Blumenthal, R., Puri, A., Oppenheim, J.J., Chertov, O. and Lubkowski, J., 2000. The structure of human β-defensin-2 shows evidence of higher order oligomerization. *J. Biol. Chem.,* **275:** 32911–32918.

Hopper, B.E., 1993. Pest risk analysis: defining the concept. In: *International Approaches to Plant Pest Risk Analysis,* p. 20. Proceedings of the APHIS/NAPPO international workshop on the identification, assessment, and management of risks due to exotic agricultural pests, Alexandria, Virginia, USA, 23-25 October 1991. NAPPO Bulletin no. 11. Ottawa, Canada, North American Plant Protection Organization.

Horgan, R.P. and Kenny, L.C., 2011. Omic' technologies: genomics, transcriptomics, proteomics and metabolomics. *The Obstetrician & Gynaecologist,* **13:** 189-195.

Horsfall, J. G., 1959. A look to the future. The status of plant pathology in biology and agriculture, pp. 63-70. *In: Plant Pathology: Problems and Progress,* 1908-1958. C. S. Holton, G. W. Fischer, R. W. Fulton, H. Hart, S. E. A. Mccallan, (Eds.) University of Wisconsin, Madison.

Hossein Safarpour, Mohammad Reza Safarnejad, Meisam Tabatabaie and Afshin Mohsenifar, 2012. Development of high-throughput quantum dot biosensor against *Polymyxa* species. *Comm. Appl. Biol. Sci.,* **77:** 7-14.

Huang, C.J. and Chen, C.Y., 2008. Synergistic interactions between chitinase ChiCW and fungicides against plant fungal pathogens. *J. Microbiol. Biotechnol.,* **18:** 784-7.

Huang, G., Dong, R., Allen, R., Davis, E.L., Baum, T.J. and Hussey, R.S., 2005. Two chorismate mutase genes from the root-knot nematode, *Meloidogyne incognita. Mol. Pl. Pathol.,* **6:** 23-30.

Huang, J.S., 1986. Ultrastructure of bacterial penetration in plants. *Ann. Rev Phytopathol.,* **24:** 141-157.

Huang, L., Sun, G., Cobessi, D., Wang, A.C., Shen, J.T., Tung, E.Y., Anderson, V.E., Berry, E. A., 2006. 3-Nitropropionic acid is a suicide inhibitor of mitochondrial respiration that, upon oxidation by complex II, forms a covalent adduct with a catalytic base arginine in the active site of the enzyme. *The J. Biol. Chem.,* **281:** 5965-5972.

Huanhua Wang and Robert L. Wick, 2009. Toxicity of nanoparticulate and bulk ZnO, Al_2O_3 and TiO_2 to the nematode *Caenorhabditis elegans.* Baoshan Xing Environ. Pollution, **157:** 1171-1177.

Hüberli, D., Hayden, K.J., Calver, M. and Garbelotto, M., 2012. Intraspecific variation in host susceptibility and climatic factors mediate epidemics of sudden oak death in Western US forests. *Plant Pathology,* **61:** 579-592.

Hudson, J.B. and Towers, G.H.N., 1991. Therapeutic potential of plant photosensitizers. *Pharmacol Ther.,* **49:** 181-222

Huma Habib and Khalid Majid Fazili , 2007. Plant protease inhibitors: a defense strategy in plants. *Biotech. Mol. Biol. Rev.,* **2:** 068-085.

Imteyaz Alam, 2000. Incorporating vector aggregation. *Plant Pathology,* **49:** 435-444.

Inés Ponce de León and Marcos Montesano, 2013. Activation of defense mechanisms against pathogens in mosses and flowering plants. *Int. J. Mol. Sci.,* **14:** 3178-3200.

Inserra, R.N., Stanley, J.D., O'Bannon, J.H. and Esser, R.P., 2005. Nematode quarantine and certification programmes implemented in Florida. *Nematol. Medit.,* **33:** 113-123.

Irfaq, M., Ajab, M., Ma, H. and Khattak, G., 2009. Assessment of genes controlling area under disease progress curve (AUDPC) for stripe rust *(P. striiformis* f. sp. *tritici*) in two wheat (*Triticum aestivum* L.) crosses. *Tsitol Genet,* **43:** 25-38.

Isaaks, E. H. and Srivastava, R. M., 1989. *An Introduction to Applied Geostatistics*. Oxford.

Ishihara A, Miyagawa H, Kuwahara Y, Ueno T, Mayama S. 1996. Involvement of Ca2C ion in phytoalexin induction in oats. *Pl. Sci.*, **115:** 9-16.

Ismail Ocsoy, Mathews L. Paret, Muserref Arslan Ocsoy, Sanju Kunwar, Tao Chen, Mingxu You and Weihong Tan, 2013. Nanotechnology in plant disease management: DNA-directed silver nanoparticles on graphene oxide as an antibacterial against *Xanthomonas perforans, ACS Nano*, **7:** 8972-8980.

Itamar Glazer, Akiva Apelbaum and Daniel Orion, 1984. The role of ethylene in the pathogenic symptoms displayed by *Meloidogyne javanica* nematode infected tomato plants. *Adv. Agril. Biotach.*, **9:** 219-220.

Jackson, A. O. and Tailor, C. B., 1996. Plant-microbe interactions: Life and death at the interface. *Pl. Cell*, **8:** 1651-1668.

Jacobi, J., Peter Krohmann, Kühbauch, W. and Gunter Menz, 2011. Mapping of spatial spreading and infestation level of plant diseases and/or animal pests on a practical scale using cereal and sugar beet fields. A Project by USWCL Remote Sensing Research Program, DFG Research Training Group (Graduiertenkolleg, University of Bonn, Faculty of Agriculture, Germany

Jacobs, A.K., Lipka, V., Burton, R,A., Panstruga, R., Strizhov, N., Schulze-Lefert, P. and Fincher, G.B., 2003. An Arabidopsis callose synthase, GSL5, is required for wound and papillary callose formation. *Plant Cell*, **15:** 2503-2513.

Jacobsen, B.J., Burrows, M.E. and Ong, K.L., 2008. Extension Plant Pathology: Bringing Our Service to the Public. A Century of Service. Online. APS*net* Feature. doi: 10.1094/APSnetFeature-2008-0508.

Jagajjanani Rao, K. and Santanu Paria, 2013. Use of sulfur nanoparticles as a green pesticide on *Fusarium solani* and *Venturia inaequalis* phytopathogens. *RSC Adv.*, **3:** 10471-10478.

Jaidev, L. and Narasimha, G., 2010. Fungal mediated biosynthesis of silver nanoparticles, characterization and antimicrobial activity. *Colloids and surfaces B.: Biointerfaces*, **81:** 430-433.

Jaillais, Y. and Chory, J., 2010. Review-unraveling the paradoxes of plant hormone signaling integration. *Nat. Struct. Mol. Biol.*, **17:** 642-5.

Jaime, G, R., Trinidad, C, R., Felix, G, R., Nelson, M. R. and Orum, T. V., 1996. Spatial and temporal pattern of *Phytophthora infestans* genotypes in the Del Fuerte Valley, Sinaloa, Mexico. (Abstr.) *Phytopathology*, **86:** S84.

Jain, K., 2003. Nanodiagnostics: application of Nanotechnology (NT) in molecular diagnostics. *Expert. Rev. Mol. Diagn.*, **4:** 153–161.

Jakeli, E. and Megrelidze, E., 2014. The modern express diagnostic of viral diseases of potato. *Sci. Agri.*, **5(1):** 22-24.

Jakob, K., Kniskern, J.M. and Bergelson, J., 2007. The role of pectate lyase and the jasmonic acid defense response in *Pseudomonas viridiflava* virulence. *Mol. Pl. Microbe Interact.*, **20:** 146-158.

James MacDonald, 2009. Education in plant pathology: Present status and future challenges. *Plant Disease*, **93:** 1238-1251.

James, J.T. and Drubery, L.A., 2001. Inhibition of PG from *Verticillum dahliae* by a PGIP from cotton. *Phytochemistry*, **57:** 149-156.

James, W.C., 1974. Assessment of plant diseases and losses. *Ann. Rev. Phytopathol.*, **12:** 27-48.

Jamie, L., McCuiston, Laura C. Hudson, Sergei A. Subbotin, Eric L. Davis and Colleen Y. Warfield., 2007. Conventional and PCR detection of *Aphelenchoides fragariae* in diverse ornamental host plant species. *J. Nematol.*, **39:** 343-355.

Janice A. Kimpel, 1999. Freedom to Operate: Intellectual property protection in plant biology and its implications for the conduct of research, *Ann. Rev. Phytopathol.*, **37:** 29-51.

Janki, N., Thakker, Samiksha Patel and Pinakin, C. and Dhandhukia, 2013. Induction of defense-related enzymes in banana plants: effect of live and dead pathogenic strain of *Fusarium oxysporum* f. sp. *cubense*. *ISRN Biotechnology*, Vol. 2013, Article ID 601303, p. 6.

Janse, J.D., 2012. Bacterial diseases that may or do emerge, with (possible) economic damage for europe and the mediterranean basin: Notes on epidemiology, risks, prevention and management on first occurrence. *J. Pl. Pathol.*, **94:** S4.5-S4.29.

Jason, T. H. and Davis, M.R., 2005. Plant defense mechanisms against fungal pathogens: Polygalacturonase inhibitor proteins. *Canadian J. Pl. Pathol.*, **27:** 5-15.

Jay, F., Wang, Y., Yu, A., Taconnat, L., Pelletier, S., Colot, V., Renou, J. P. and Voinnet, O., 2011. Misregulation of AUXIN RESPONSE FACTOR 8 underlies the developmental abnormalities caused by three distinct viral silencing suppressors in *Arabidopsis. PLoS Pathog.*, **7,** e1002035.

Jayaseelan, C., Rahuman, A.A., Kirthi, A.V., Marimuthu, S. and Santhoshkumar, T., 2012. Novel microbial route to synthesize ZnO nanoparticles using *Aeromonas hydrophila* and their activity against pathogenic bacteria and fungi. *Spectrochimica Acta A: Mol. Biomol. Spectrosc.*, **90:** 78-84.

Jeandet, P., Douillet-Breuil, A.-C., Bessis, R., Debord, S., Sbaghi, M. and Adrian, M., 2002. Phytoalexins from the vitaceae: Biosynthesis, phytoalexin gene expression in transgenic plants, antifungal activity, and metabolism. *J. Agric. Food Chem.*, **50:** 2731.

Jeffrey A. Kimbrel, Di, Y., Cumbie, J.S. and, Chang, J.S., 2010. Studying plant-pathogen interaction in the genomics era: Beyond molecular koch's postlates to systems biology. *Ann. Rev. Phytopathol.*, **48:** 457-477.

Jeger, M.J., 1986. The potential of analytic compared with simulation approaches to modelling in plant disease epidemiology, pp. 255–281. *In: Plant Disease Epidemiology, Population Dynamics and Management*, Vol 1, Leonard, K. and Fry, W., (Eds.). Macmillan, New York, USA.

Jeger, M.J., 2004. Analysis of disease progress as a basis for evaluating disease management practices. *Ann. Rev. Phytopatol.*, **42:** 61-82.

Jeger, M.J., Holt, J., Van Den Bosch, F. and Laurence V Maden, 2015. Epidemiology of insect-transmitted plant viruses: Modelling disease dynamics and control interventions. *Research Gate.net.*

Jelle Bruinsma, 2009. How to feed the world in 2050? From the Expert Meeting on FAO, Rome, June 24-26.

Jennifer Parker, Jin Koh, Mi-Jeong Yoo, Ning Zhu, Michelle Feole, Sarah Yi and Sixue Chen, 2013. Quantitative proteomics of tomato defense against *Pseudomonas syringae* infection. *Pl. Mol. Celluar Biol.*, 03/2013; DOI:10.1002/pmic.201200402.

Jennifer, E., Schaff, Dahlia M. Nielsen, Chris P. Smith, Elizabeth H. Scholl and David McK Bird, 2007. Comprehensive transcriptome profiling in tomato reveals a role for glycosyltransferase in *Mi*-mediated nematode resistance, *Pl. Physiol.*, **144:** 1079-1092.

Jespers, A.B.K., 1994. Mode of action of the phenylpyrrole fungicide fenpiclonil in *Fusarium sulphureum. Landbouwuniversiteit Wageningen*, pp. 129.

Jesús, A., Sánchez-Navarro, Mark P. Zwart and Santiago F. Elena, 2013. Effects of the number of genome segments on primary and systemic infections with a multipartite plant RNA virus. *J. Virol.*, **87:** 10805-10815.

Ji Zhou, Thomas Spallek, Christine Faulkner and Silke Robatzek, 2012. Callose measurer: A novel software solution to measure callose deposition and recognise spreading callose patterns. *Plant Methods,* **8:** 49.

Jian Liu, Shi Zhang Qiao, Jun Song Chen, Xiong Wen (David) Lou, Xianran Xing, Gao Qing (Max) Lu, Yolk/shell, 2011. Nanoparticles: new platforms for nanoreactors, drug delivery and lithiumion batteries. *Chem. Commun.*, **47:** 12578-12591.

Jiirgen Siegrist, Wolfgang Jeblick and Heinrich Kauss, 1994. Defense responses in infected and elicited cucumber (*Cucumis sativus*) hypocotyl segments exhibiting acquired resistance. *Plant Physiol.*, **105:** 1365-1374.

Jin, J., 2010. "Functional characterization of Arabidopsis genes changing expression during cyst nematode *Heterodera schachtii* infection". Graduate Theses and Dissertations, Paper 12353, Ph.D. thesis, Iowa State University Ames, Iowa.

Jinpeng Wang, Yibo Chen and Jean-Pierre Chane, 2014. An integrated survey in plant disease detection for precision agriculture using image processing and wireless multimedia sensor network. *Int. J. Infn. Tech. & Comp. Sci., (IJITCS),* **16:** 19-32.

Ji-yeon Roh, Sang Jun Sim, Jongheop Yi, Kwangsik Park, Kyu Hyuck Chung, Dong-young Ryu and Jinhee Choi, 2009. Ecotoxicity of silver nanoparticles on the soil nematode *Caenorhabditis elegans* using functional ecotoxicogenomics. *Environ. Sci. Technol.*, **43:** 3933-3940.

Jo, Y.K., Kim, B.H. and Jung, G., 2009. Antifungal activity of silver ions and nanoparticles on phytopathogenic fungi. *Plant Dis.*, **93:** 1037-1043.

Johal, G. and Briggs, S., 1992. Reductase activity encoded by the *HM1* disease ressitance gene in maize. *Science,* **258:** 985-987.

John A. Steele, Thomas F. Uchytil, Richard D. Durbin, Pradip Bhatnagar and Daniel H.Rich., 1976. Chloroplast coupling factor 1: A species specific receptor for tentoxin. *Proc. Natl. Acad. Sci.,* **73:** 2245- 2248 John Brown and Philip Keane, 2012. Assessment of disease and effects on yield. http://www.appsnet.org/Publications/Brown_Ogle/20%20Disease%20assessment%20%28JFB &PJK%29.

John McDonald, 2010. NGIQ Development Manager. Nursery Papers Technical, March Issue No. 2.

John Mugabe, 1996, Intellectual Property Protection and Traditional Knowledge—An Exploration in International Policy Discourse. In: Greaves, "Tribal Rights" in Brush and Stabinsky (Eds.) *Valuing Local Knowledge: Indigenous Peoples and Intellectual Property Rights.*

John P. Hamilton, Eric C. Neeno-Eckwall, Bishwo N. Adhikari, Nicole T. Perna, Ned Tisserat, Jan E. Leach, C. André Lévesque and C. Robin Buell, 2011. The Comprehensive Phytopathogen Genomics Resource: a web-based resource for data-mining plant pathogen genomes. Database, **10.** 1093/database/bar053.

John, N., Kabashima, James, D., Macdonald, Steve, H., Dreistadt and Diane E. Ullman, 1997. Easy On-Site Tests for Fungi and Viruses in Nurseries and Greenhouses, Division of Agriculture and Natural Resources, University of California.

Johnson, R.D., Johnson, L., Itoh, Y., Kodama, M., Otani, H. and Kohmoto, K., 2000. Cloning and characterization of a cyclic peptide synthetase gene from *Alternaria alternata* apple pathotype whose product is involved in AM-toxin synthesis and pathogenicity. *Mol. Pl. Micr. Interact.,* **13:** 742-753.

Jonathan Blake, 2014. Fungicide management: strobilurin fungicides, Module for Fungicide management in cereals course, ADAS UK Ltd., Wilson Story Venancio, Marco Antônio Tavares Rodrigues, Edson Begliomini, Nilton Luiz de Souza, 2003. Physiological effects of Strobilurin fungicides on plants, *Ponta Grossa,* **9:** 59-68.

Jonathan E. Markham and Jacques Hille, 2001. Host-selective toxins as agents of cell death in plant–fungus interactions. *Mol. Pl. Pathol.,* **2:** 229-239.

Jones J. D. G. and Dangl J. L., 2006. The plant immune system. *Nature,* **444:** 323-329.

Jones, A.M., Thomas, V., Truman, B., Lilley, K., Mansfield, J. and Grant, M., 2004. Specific changes in the Arabidopsis proteome in response to bacterial challenge: differentiating basal and R-gene mediated resistance. *Phytochemistry,* **65:** 1805–1816.

Jones, F.G.W. and Kempton, R.A., 1980, Forecasting Crop damage by nematodes: Nematode population dynamics. *EPPO Bull.,* **10:** 169-180.

Jones, J.D.G. and Dangl, J.L., 2006. The plant immune system. *Nature* , **444:** 323-329.

Joris, J.G., Bernardus, C.J., Schimmel, Juan, M.A., Rocio, E.B., Robert, C.S, and Merjin, R.K., 2012. Plant glandular trichomes as targets for breeding or engineering of resisitance to herbivores. *Int J Mol Sci.,* **13:** 17077-17103.

José A. Del Rio, Ana Gonzalez, María D. Fuster, Juana M. Botia, Pedro Gomez, Víctor Frias, and Ana Ortuño, 2003. Tyloses formation and changes in phenolic compounds of grape roots infected with *Phaeomoniella chlamydospora* and *Phaeoacremonium species. Tech. Bull.*, Plant Biology Department Plant Physiology, Faculty of Biology, University of Murcia, 30100 Murcia, Spain.

Jose M. Belles, Juan Carbonell and Vicente Conejero, 1991. Polyamines in plants infected by citrus exocortis viroid or treated with silver ions and ethephon. *Pl. Physiol.*, **96:** 1053-1059.

José Maurício Cunha Fernandes, Willingthon Pavan and Rosa Maria Sanhueza, 2011. SISALERT – A generic web-based plant disease forecasting system. *In: M. Salampasis, A. Matopoulos (eds.): Proceedings of the International Conference on Information and Communication Technologies for Sustainable Agri-production and Environment (HAICTA 2011)*, Skiathos, 8-11 September, 2011.

José, A., Del Rio, ANa Gonzalez, María D. Fuster, Juana M. Botia, Pedro Gomez, Víctor Frias and Ana Ortuño, 2001. Tylose formation and changes in phenolic compounds of grape roots infected with *Phaeomoniella chlamydospora* and *Phaeoacremonium* species. *Phytopathol. Mediterr.*, **40:** 394-399.

Joseph, L.M., Tan, T.K. and Wong, S.M., 1998. Antifungal effect of hydrogen peroxide and peroxidase of spore germination and mycelia growth of *Pseudocercospora* species. *Canadian J. Bot.*, **76:** 2119-2124.

Josselin Montarry, Philippe Cartolaro, François Delmotte, Jérôme Jolivet and Laetitia Willocquet, 2008. Genetic structure and aggressiveness of *Erysiphe necator* populations during grapevine powdery mildew epidemics. *Appl. Environ. Microbiol.*, **74:** 6327-6332.

Jun Cao, 2012. The Pectin Lyases in *Arabidopsis thaliana*: Evolution, Selection and Expression Profiles. *PLOs ONE*, Oct. 2012 issue

Jun Fan, Lionel Hill, Casey Crooks, Peter Doerner and Chris Lamb., 2009. Abscisic acid has a key role in modulating diverse plant-pathogen interactions. *Pl. Physiol.*, **150:** 1750-1761.

Jun Wu, Lin-tao Wu, Zhi-bin Liu, Lei Qian, Mao-hua Wang, Li-rong Zhou, Yi Yang and Xu-feng Li. 2009. A plant defensin gene from *Orychophragmus violaceus* can improve *Brassica napus'* resistance to *Sclerotinia sclerotiorum. African J. Biotech.*, **8:** 6101-6109.

Justin, B.R., Mark, C.M. and Consuelo, M De M., 2010. Plant defense against parasitic plants. *Pl. Signal Behav.*, **5:** 929-931.

Jyoti Shah, 2005. Lipids, lipases, and lipid-modifying enzymes in plant disease resistance. *Ann. Rev. Phytopathol.*, **43:** 229-260.

Kaemmerer, D.L., Poschenrieder, G., Zellner, M. and Munzert, M., 2007. Epidemiology of bacterial ring rot of potato in plant and soil-consequences for disease management. *J. Pl. Dis & Protn.*, **114:** 159-166.

Ka, M., Beulke, S., Tiede, K. and Hofmann, T., 2013. Nanopesticides: State of knowledge, environmental fate and exposure modeling. *Crit. Rev.* in *Environ. Sci. & Technol.*, **43:** 1823-1867.

Kaisa Marjamaa, Eija M. Kukkola and Kurt V. Fagerstedt, 2009. The role of xylem class III peroxidases in lignification. *J. Exptl. Bot.,* **60:** 367-376.

Kalpana Sastry, R., 2007. Mechanisms for protection of agricultural innovations in India. *DESIDOC Bulletin of Information Technology,* **27:** 3-11.

Kalra, A., Chechi, R. and Khanna, R., 2010. Role of zigbee technology in agriculture sector *National* Conf. on Computational Instrumentation NCCI 2010 CSIO (Chandigarh, India, 19–20 March 2010). 151.

Kalyan, K. and Mondal, Chander Mani, 2012. Investigation of the antibacterial properties of nanocopper against *Xanthomonas axonopodis* pv. *punicae,* the incitant of pomegranate bacterial blight. *Annals of Microbiol.,* **62:** 889-893.

Kameda, Y., Asano, N., Yamaguchi, T. and Matsui, K. 1987. Validoxylamines as trehalase inhibitor. *J. Antibiot.,* **40:** 563-565.

Kamei, K. and. Watanabe, A., 2005. *Aspergillus* mycotoxins and their effect on the host. *Medical Mycol. Suppl.,* **43:** 595-599.

Kankkunen, P., Rintahaka, J., Aalto, A., Leino, M., Majuri, M.L., Alenius, H., Wolff, H. and Matikainen, S., 2009. Trichothecene mycotoxins activate inflammatory response in human macrophages. *J. Immunol.,* **182:** 6418-6425.

Kannangara, R., Branigan, C., Liu, Y., Penfield, T., Rao, V., Mouille, G., Hofte, H., Pauly, M., Riechmann, J.L. and Broun, P., 2007. The transcription factor WIN1/SHN1 regulates Cutin biosynthesis in *Arabidopsis thaliana. Plant Cell.,* **19:** 1278-1294.

Kanto, T., Miyoshi, A., Ogawa, T., Maekawa, K. and Aino, M., 2004. Suppressive effect of potassium silicate on powdery mildew of strawberry in hydroponics. *J. Gen. Plant Pathol.,* **70:** 207-211.

Kaplan, D.T. and Opperman, C.H., 2000. Reproductive strategies and karyotype of the burrowing nematode, *Radopholus similis. J. Nematol.,* **23:** 126-133.

Kar, P.K., Murmu, S., Saha, S., Tandon, V. and Acharya, K., 2014. Anthelmintic efficacy of gold nanoparticles derived from a phytopathogenic fungus, *Nigrospora oryzae. PLoS ONE,* **9(1):** e84693.

Karamat, F., Olry, A., Munakata, R., Koeduka, T., Sugiyama, A., Paris, C., Hehn, A., Bougaud, F. and Yazaki, K., 2014. A coumarin-specific prenyltransferase catalyzes the crucial biosynthetic reaction for furanocoumarin formation in parsley. *Plant J.,* **77:** 627-638.

Karamura, E., Kayobyo, G., Tushemereirwe, W., Benin, S., Blomme, G., Eden Green, S. and Markham, R., 2010. Assessing the Impacts of Banana Bacterial Wilt Disease on Banana (*Musa* spp.) Productivity and Livelihoods of Ugandan Farm Households, Proc. IC on Banana & Plantain in Africa, T. Dubois et al. (Eds.), *Acta Hort.,* **879**, ISHS.

Karaoglanidis, G.S. and Karadimos, D.A., 2006. Efficacy of strobilurins and mixtures with DMI fungicides in controlling powdery mildew in field-grown sugar beet. *Crop Protn.,* **25:** 977-983.

Kar-Chun Tan, Simon V. S. Ipcho, Robert D. Trengove, Richard P. Oliver and Peter S. Solomon, 2009. Assessing the impact of transcriptomics, proteomics and metabolomics on fungal phytopathology. *Mol. Pl. Pathol.,* **10:** 703-715.

Karine Picard, Yves Tirilly and Nicole Benhamous, 2000. Cytological effects of cellulases in the parasitism of *Phytophthora parasitica* by *Pythium oligandrum*. *Appl. Environ. Microbiol.,* **66:** 104305-104314.

Karuna Vishunavat, 2010. Seed Treatment as Innovative Crop Protection Strategies, pp. 57-62. *In: Plant Pathology in Practice,* pp. 101-107. Training manual, G.B. Pant University of Agriculture & Technology, Pantnagar- 263 145, India, pp. 265.

Kasschau, K. D., Xie, Z. X., Allen, E., Llave, C., Chapman, E. J., Krizan, K. A. and Carrington, J. C., 2003. P1/HC-Pro, a viral suppressor of RNA silencing, interferes with Arabidopsis development and miRNA function. *Dev. Cell.,* **4:** 205–217.

Katagiri, Y., Hashidoko, Y. and Tahara, S., 2002. Localization of flavonoids in the yellow lupin seedlings and their UV-B-absorbing potential. *Z. Naturforsch.,* **57c:** 811-816.

Katherine Beckman and Christa Pletcher, 2009. Expanding global trademark regulation wake forest, *Intellectual Property Law J.,* **10:** 215-239.

Katsube, T. and Koshimizu, Y., 1970. Influence of blast disease on harvests in rice plant. 1. Effects of panicle infection on yield components and quality. *Bull. Tohoku Nat. Agric. Exp. Stn.,* **39:** 55-96.

Kauser Abdulla Malik and Yusuf Zafar, 2005. Intellectual property rights in plant biotechnology: A Contribution to Crop Biosecurity. *Asian Biotech. Dev. Rev.,* **8:** 7-43.

Kav Nat, N.V., Srivastava, S., Yajima, W. and Sharma, N., 2007. Application of proteomics to investigate plant microbe interactions. *Curr Proteomics,* **4:** 28-43.

Kavita Sharma, Ravi Sharma, Simanta Shit and Shraddha Gupta, 2012. Nanotechnological Application on Diagnosis of a Plant Disease. *International Conference on Advances in Biological and Medical Sciences* (ICABMS'2012) July 15-16, Singapore.

Kavitha Kuruganti and G.V. Ramanjaneyulu, 2007. Genetic Engineering in Indian Agriculture – An Introductory Handbook, Centre for Sustainable Agriculture, Secunderabad, Andhra Pradesh, India.

Kawai, M., Rich, D.H. and Walton, J.D., 1983. The structure and conformation of HC-toxin. *Biochem. Biophys. Res. Commun.,* **111:** 398-403.

Kell, D.B., 2007. The virtual human: towards a global systems biology of multiscale, distributed biochemical network models. *Lubmb Life,* **59:** 689-95.

Kelman, A., 1985. Plant pathology at the crossroads. *Ann. Rev. Phytopathol.,* **23:** 1-11.

Kemp, G., Stanton, L., Bergmann, C.W., Clay, R.P., Albersheim, P. and Darvill, A. 2004. Polygalacturonases-inhibiting proteins can function as activators of polygalacturonase. *Mol. Pl. Micr. Interact.,* **17:** 888-894.

Kenji Tsubata, Kiyoshi Kuroda, Yoshinobu Yamamoto. And Noriaki Yasokawa, 2006. Development of a novel plant activator for rice diseases, tiadinil. *J. Pesticide Sci.,* **31:** 161-162.

Kessmann, H., Staub, T., Hofmann, C., Maetzke, T., Herzog, J., Ward, E., Uknes, S. and Ryals, J., 1994. Induction of systemic acquired disease resistance in plants by chemicals. *Ann Rev Phytopathol.,* **32:** 439-459.

Kevin C. Vaughn and Stephen O. Duke. 1981. Tentoxin-induced loss of plastidic polyphenol oxidase. *Physiologia Plantarum,* **53:** 421-428.

Khan, M.R. and Z. Haque, 2013. Morphological and biochemical responses of five tobacco cultivars to simultaneous infection with *Pythium aphanidermatum* and *Meloidogyne incognita. Phytopathol. Mediterr.,* **52:** 98-109.

Khatun, S., Bandyopadhyay, P.K. and Chatterjee, N.C., 2009. Phenols with their oxidizing enzymes in defence against black spot of rose (*Rosa centifolia*). *Asian J. Exp. Sci.,* **23:** 249-252.

Khetarpal, R.K. and Gupta, K., 2002. Implication of sanitary and phytosanitary Agreement of seeds under WTO on Plant Protection in India. *Ann. Rev. Pl. Pathol.,* **1:** 1-26.

Khetarpal, R.K. and Gupta, K., 2003. Plant quarantine set-up in India and scope for collaboration in SAARC region. In SAARC Workshop on 'Setting up a Regional Network on Sanitary and Phytosanitary Measures and Quality Control', March 18-19, 2003, Islamabad, Pakistan.

Khetarpal, R.K. and Gupta, K., 2005. Status of plant protection in India in the wake of international agreements. *Indian J. Pl. Protn.,* **33:** 153-163.

Khetarpal, R.K., 2004. A critical appraisal of seed health certification and transboundary movement of seeds under WTO regime. Dr. K.C. Mehta and Manoranjan Mitra Memorial Lecture Award (2004). *Indian Phytopath.,* **57:** 408-421.

Khetarpal, R.K., Gupta, K. and Pandey, B.M., 2002. Sanitary and phytosanitary agreement of WTO and its implications on India. *Indian Phytopathology,* **55:** 356.

Khodakovsky, A., P., Schr¨oder and Sweldens, W., 2000. Progressive geometry compression, in Siggraph 2000, Computer Graphics Proceedings, pp. 271–278.

Khot, L.R., Sankaran, S., Maja, J.M., Ehsani, R. and. Schuster, E.W., 2012. Applications of nanomaterials in agricultural production and crop protection: A review. *Crop Prot.,* **35:** 64-70.

Kikuchi, J., Shinozaki, K. and Hirayama, T., 2004. Stable isotope labeling of *Arabidopsis thaliana* for an NMR-based metabolomics approach. *Pl. Cell Physiol.,* **45:** 1099–1104.

Kilakkathi and Sindhu, P., 2010. A Structural, Pathophysiological, and Ethical Analysis of Phenylketonuria. Molecular Anatomy Project. Doolittle Hall, 123 Bevier Road, Piscataway, NJ 08854.

Kim, D., Lee, D.G., Kim, K.L. and Lee, Y., 2001. Internalization of tenecin 3 by a fungal cellular process is essential for its fungicidal effect on *Candida albicans. Eur. J. Biochem.,* **268:** 4449-4458.

Kim, J., Grate, J.W. and Wang, P., 2006. Nanostructures for enzyme stabilization. *Chem. Engineer. Sci.,* **61:** 1017-26.

Kim, K.S. and Beresford, R.M., 2009. Use of geographic information systems and satellite data for assessing climatic risk of establishment of plant pathogens. *New Zealand Pl. Protn.,* **62:** 109-113.

Kim, Keuk-Jun, Woo Sang Sung, Seok-Ki Moon, Jong-Soo Choi, Jong Guk Kim and Dong Gun Lee, 2008. Antifungal effect of silver nanoparticles on dermatophytes. *J. Microbiol. Biotechnol.,* **18:** 1482-1484.

Kim, M. J., Ciani, S. and Schachtman, D. P., 2010. A peroxidase contributes to ros production during Arabidopsis root response to potassium deficiency. *Mol. Pl.*, **3:** 420–427.

Kim, S.W., Jung, J.H., Lamsal, K., Kim, Y.S., Min, J.S. and Lee, Y.S., 2012. Antifungal effects of silver nanoparticles (AgNPs) against various plant pathogenic fungi. *Mycobiol.*, **40:** 53-58.

Kim, T. N., Feng, Q. L., Kim, J. O., Wu, J., Wang, H., Chen, G. C. and Cui, F. Z., 1998. Antimicrobial effects of metal ions (Ag^+, $Cu2^+$, $Zn2^+$) in hydroxyapatite. *J. Mater. Sci. Mater. Med.*, **9:** 129-134.

Kim, Y. H., Kim, O. S., Lee, B. C., Moon, J. K., . Lee, S. C and Lee, J. Y., 2003. G7H, a new *Soybean mosaic virus* strain: its virulence and nucleotide sequence of the CI gene. *Plant Dis.*, **87:** 1372–1375.

Kimura, M., Anzai, H. and Yamaguchi, I., 2001. Microbial toxins in plant pathogen interactions: Biosynthesis, resistance mechanisms and significance. *J. Gen. Appl. Microbiol.*, **47:** 149-160.

Kinjiro Tamari, Nagahiro Ogasawara, Jun Kaji and Kunihiko TogashI, 1966. On the effect of a piricularin-detoxyfying substance ferulic acid on the tissue resisitance of rice plants to the blast blast-fungal infection. *Japanese J. Phytopathol.*, **32:** 186-193

Kirubakaran, S.I. and Sakthivel, N., 2006. Cloning and overexpression of antifungal barley chitinase gene in *Escherichia coli, Pro. Express. Purific.*, **52:** 159-166.

Kishor K. Bhattarai, Qi Guang Xie, Sophie Mantelin, Usha Bishnoi, Thomas Girke, Duroy A. Navarre and Isgouhi Kaloshian, 2008. Tomato susceptibility to root-knot nematodes requires an intact jasmonic acid signaling pathway. *Mol. Pl. Microbe Inter.*, **21:** 1205-1214.

Kishore Kumar Nair, Knight, T. J., Durbin, R. D., Langston-Unkefer, P. J., 1986. Role of glutamine synthetase adenylation in the self-protection of *Pseudomonas syringae* subsp. "*tabaci*" from its toxin, tabtoxinine-β-lactam. *J. Bacteriol.*, **166:** 224-229.

Kobi Buxdorf, Gilad Rubinsky, Omer Barda, Saul Burdman, Asaph Aharoni and Maggie Levy, 2014. The transcription factor SlSHINE3 modulates defense responses in tomato plants. *Plant Mol. Biol.*, **84:** 37-47.

Koeman, J.H. and Zadoks, J.C., 1999. History and future of plant protection policy, from ancient times to WTO-SPS. *In:* G. Meester, R.D. Woittiez & A. De Zeeuw, eds., *Plants and Politics*, p. 22. Wageningen, the Netherlands, Wageningen Pers. pp. 255.

Koeman, J.H. and Zadoks, J.C., 1999. History and future of plant protection policy, from ancient times to WTO-SPS. *In*: *Plants and Politics*, p. 22. G. Meester, R.D. Woittiez & A. De Zeeuw, (Eds.), Wageningen, the Netherlands, Wageningen Pers. pp. 255.

Kohmoto, K., Itoh, Y., Shimomura, N., Kondoh, Y., Otani, H., Kodama, M., Nishimura, S. and Nakatsuka, S., 1993. Isolation and biological activities of two host specific toxins from the Tangerine pathotype of Alternaria alternata: *Phytopathology*, **83:** 495-502.

Kolattukudy, P.E. and Espelie, K.E., 1989. Chemistry, biochemistry and functions of suberin associated waxes, pp. 235–287. *In Natural Products of Woody Plants* I.J.W. Rowe (Ed.), Springer-Verlag, New York.

Kolattukudy, P.E., 1980. Biopolyester membranes of plants: Cutin and suberin. *Science,* **208:** 990-1000.

Koo, C., Malapi-Wight, M., Kim, H.S., Cifci, O.S. and Vaughn-Diaz, V.L., 2013. Development of a real-time microchip PCR system for portable plant disease diagnosis. *PLoS ONE,* **8:** e82704.

Koornneef, M. and van der Veen, J.H., 1980. Induction and analysis of gibberellin sensitive mutants in *Arabidopsis thaliana* (L.) Heynh. *Theor. Appl. Genet.,* **58:** 257-263.

Koricheva, J., Nykänen, H. and Gianoli, E., 2004. Meta-analysis of trade-offs among plant antiherbivore defenses: Are plants jacks-of-all-trades, masters of all. *Am. Nat.,* **163:** E64-E75.

Kort, J., Ross, H., Rumpenhorst, H.J. and Stone, A.R., 1977. An international scheme for the identification of pathotypes of potato cyst nematodes, *Globodera rostochiensis* and *G. pallida. Nematologica,* **23:** 333-339.

Kort, J., 1974. Identification of pathotypes of the potato cyst nematode. *OEPP/EPPO Bull.,* **4:** 511-518.

Kotchoni, O.S. and Shonukan, O.O., 2002. Regulatory mutations affecting the synthesis of cellulase in *Bacillus pumilus. World J. Microbiol. Biotechnol.,* **18:** 487-491.

Philos Trans R., 1999. Interactions between tobacco mosaic virus and the tobacco N gene. *Soc. Lond. B. Biol. Sci.,* **354:** 653-8.

Krishnaraj, C., Ramachandran, R., Mohan, K. and Kalaichelvan, P.T., 2012. Optimization for rapid synthesis of silver nanoparticles and its effect on phytopathogenic fungi. *Spectrochim. Acta Part A: Mol. Biomol. Spectrosc.,* **93:** 95-99.

Kumar, R., Sharon, M. and Choudhary, A.K., 2010. Nanotechnology in agricultural diseases and food safety. *J. Phytol.,* **2:** 83–92.

Kuo, T. M., Pearce, G., and Ryan, C. A., 1984. Isolation and characterization of proteinase inhibitor I from etiolated tobacco leaves. *Arch. Biochem. Biophys.,* **230:** 504-510.

Kushalappa, A.C. and Ludwig, A., 1982. Calculation of apparent infection rate in plant disease: development of a method to correct for host growth. *Phytopathology,* **72:** 1373-1377.

Kushwaha, H. S., 2010. Micro-climate and plant diseases. *In: Plant Pathology in Practice,* P.101-107. Training manual, G.B. Pant University of Agriculture & Technology, Pantnagar-263 145, India, p. 265.

L'Haridon, F., Besson-Bard, A., Binda, M., Serrano, M., Abou-Mansour E., Balet, F., Schoonbeek, H.J., Hess, S., Mir, R., Leon, J., Lamotte, O. and Metraux, J.P., 2011. A permeable cuticle is associated with the release of reactive oxygen species and induction of innate immunity. *PLoS Pathogy,* **7:** e1002148.

Laborie, M. P. G., 2009. Bacterial cellulose and its polymeric nanocomposites. *The Nanoscience and Technology of Renewable Biomaterials,* L. A. Lucia and O. J. Rojas (Eds.), (Chichester, UK: Wiley).

Lakshman Prasad and Karuna Vishunavat, 2006. Assessment of yield loss in cauliflower seed crop due to *Alternaria* blight. *Indian Phytopath.,* **59:** 185-189.

Lalancette, N. and Hickey, K.D., 1986. Disease progression as a function of plant growth. *Phytopathology,* **76:** 1171-1175.

Lamprecht, S. C., Marasas, W. F. O., Alberts, J. F., Cawood, M. E., Gelderblom, W. C. A., Shephard, G. S., Thiel, P. G., and Calitz, F. J. 1994. Phytotoxicity of fumonisins and TA-toxin to corn and tomato. *Phytopathology,* **84:** 383-391.

Landa, B.B., Cachinero-Diaz, J.M., Lemanceau, P., Jimenez-Diaz, R.M. and Alabouvette, C., 2002. Effect of fusaric acid and phytoanticipins on growth of rhizobacteria and *Fusarium oxysporum. Can. J. Microbiol.,* **48:** 971-985.

Landeweert, R., Mulder, W., Elsen, S.J.J. van den, Mooijman, P.J.W., Helder, J., 2012. Looking for a needle in a hay-stack: diagnostic kits for the detection of plant parasitic nematodes in complex DNA backgrounds, p. 61. The 31st International Symposium of the European Society of Nematologists, Adana, Turkey, 23-27 September 2012. Adana, Turkey.

Lane, C.R., Hobden, E., Laurenson, L., Barton, V.C., Hughes, K.J.D., Swan, H., Boonham, N.and Inman. A.J., 2007. Evaluation of a Rapid Diagnostic Field Test Kit for Identification of *Phytophthora ramorum, P. kernoviae* and other *Phytophthora* species at the Point of Inspection. Proc., the Sudden Oak Death Third Science Symposium, March 5–9, Santa Rosa, California.

Lara F. Victoriano, Castillo Reyes, F., Flores Gallegos, C., Aguilar, C.N. and Rodriguez Herrera, 2011. Metagenomics in plant pathology. *In: Phytopathology in the Omics Era,* Raúl Rodríguez Herrera, Cristobal N. Aguilar, June Kilpatrick Simpson-Williamson, (Edrs.), Research Signpost publishers, p. 308.

Large, E.C., 1966. Measuring plant disease. *Ann. Rev. Phytopathol.,* **4:** 9-28.

Larsen, N.J., Sutherland, P.W., Hallett, I.C., Jones, M.K., Pushparajah, I.P.S. and Everett, K.R., 2012. Infection of ZESPRI® GOLD kiwifruit (*Actinidia chinensis* 'Hort16A') lenticels by *Pseudomonas syringae* pv. *actinidiae.* Poster abstract from *New Zealand Plant Protection Society.*

Latha Jishnu, 2009. Farmers' resounding no to Bt brinjal, *The Business Standard,* 22 October

Lattanzio, V., Di Venere, D., Linsalata, V., Bertolini, P., Ippolito, A. and Salerno, M., 2001. Low temperature metabolism of apple phenolics and quiescence of *Phlyctaena vagabunda. J. Agric. Food Chem.,* **49:** 5817-5821.

Lattanzio, V.M.T., Nivarlet, N., Lippolis, V., Gatta, S.D., Huet, A-C., Delahaut, P., Granier, B. and Visconti, A., 2012. Multiplex dipstick immunoassay for semi-quantitative determination of *Fusarium* mycotoxins in cereals. *Anal. Chim. Acta.,* **718:** 99-108.

Leclerc, M., Doré, T., Gilligan, C.A., Lucas, P. and Filipe, J.A.N., 2014. Estimating the delay between host infection and disease (Incubation Period) and assessing its significance to the epidemiology of plant diseases. *PLoS ONE,* **9(1):** e86568.

Lee, S., Rojas, C.M., Ishiga, Y., Pandey, S. and Mysore, K.S., 2013. Arabidopsis heterotrimeric G-Proteins play a critical role in host and nonhost resistance against *Pseudomonas syringae* pathogens. *PLoS ONE* , **8:** e82445.

Leendert C. van Loon, Bart, P.J., Geraats. and Huub J.M. Linthorst, 2006. Ethylene as a modulator of disease resistance in plants. *Trends Pl. Sci.,* **11:** 184-191.

Leiter, E., Szappanos, H., Oberparleiter, C., Kaiserer, L., Sernoch, L., Pusztahelyi, T., Emri, T., Pócsi I., Salvenmoser, W. and Marx, F., 2005. Antifungal protein PAF severely affects the integrity of the plasma membrane of *Aspergillus nidulans* and Induces an Apoptosis-like phenotype: *Antimicrob. Agents Chemother.*, **49:** 2445–2453.

Lena, J., Daumann, James, A., Larrabee, David Ollis, Gerhard Schenk and Lawrence R. Gahan, 2014. Immobilization of the enzyme GpdQ on magnetite nanoparticles for organophosphate pesticide bioremediation. *J. Inorg. Biochem.*, **131:** 1-7.

Lenk, S., Chaerle, L., Pfündel, E.E., Langsdorf, G., Hagenbeek, D., Lichtenthaler, H.K., Van Der Straeten, D. and Buschmann, C., 2007. Multispectral fluorescence and reflectance imaging at the leaf level and its possible applications. *J. Exp. Bot.*, **58:** 807-814.

Lenthe, J.H., Oerke, E.C. and Dehne, H.W., 2007. Digital infrared thermography for monitoring canopy health of wheat. *Precision Agri.*, **8:** 15-26.

Levi, C. and Durbin, R. D., 1986. The isolation and properties of a tabtoxin-hydrolyzing aminopeptidase from the periplasm of *Pseudomonas syringae* pv. *tabaci*. *Physiol. Mol. Plant Pathol.*, **28:** 345-352.

Lhomme, L., Brosillon, S. and Wolbert, D., 2008. Photocatalytic degradation of pesticides in pure water and a commercial agricultural solution on TiO_2 coated media. *Chemosphere*, **70:** 381.

Li, S., Simonian A. and Chin, B.A., 2010. Sensors for agriculture and the food industry. *Electrochem. Soc. Interfac.*, **19:** 41-46.

Li, X., Longfu, Z., Lili T., Linlin, L., Daojun Y., Li, J., Lu, L. and Xianlong, Z., 2011. Lignin metabolism has a central role in the resistance of cotton to the wilt fungus *Verticillium dahliae* as revealed by RNA-Seq-dependent transcriptional analysis and histochemistry. *J. Exp. Bot.*, doi: 10.1093/jxb/err 245.

Li Ying-Zhang, Zheng Xiao-Hua, Tang Hai-Lin, Zhu Jian-Wei, Yang Jing-Ming, 2003. Increase of β-1, 3-glucanase and chitinase activities in cotton callus cells treated by salicylic acid and toxin of *Verticillium dahlia. Acta Botanica Sinica,* **45:** 802-808.

Li, C., Krewer, G. and Kays, S. J., 2009b. Blueberry postharvest disease detection using an electronic nose. ASABE Paper No. 096783, ASABE Annual International Meeting, Reno, NV, June 21–June 24, 2009.

Li, F., Huang, C., Li, Z. and Zhou, X., 2014. Suppression of RNA silencing by a plant DNA virus satellite requires a host calmodulin-like protein to repress *RDR6* expression. *PLoS Pathog*, **10:** e1003921.

Li, F., Pignatta, D., Bendix, C., Brunkard, J.O., Cohn, M.M., Tung, J., Sun, H., Kumar, P. and Baker, B., 2012. Micro RNA regulation of plant innate immune receptors. *Proc. Natl. Acad. Sci.*, U.S.A., **109:** 1790-1705.

Li, L. and Steffens, J.C., 2002. Overexpression of polyphenol oxidase in transgenic tomato plants results in enhanced bacterial disease resistance. *Planta*, **215:** 239-247.

Li, Y., Beisson, F., Koo, A.J., Molina, I., Pollard, M., and Ohlrogge, J., 2007. Dentification of acyltransferases required for cutin biosynthesis and production of cutin with suberin-like monomers. *Proc. Natl. Acad. Sci.*, U.S.A., **104:** 18339-18344.

Li, Y., Fester, T. and Taylor, C.G., 2009. Transcriptomic analysis of nematode infestation. *In: Cell Biology of Plant Nematode Parasitism*. Howard, R. and Christopher, G.T., (Eds.) Springer-Verlag, Berlin.

Lietzke, S. E., Keen, N.T., Yoder, M. D. and Jurnak, F., 1994. The three-dimensional structure of pectate lyase E, a plant virulence factor from *Erwinia chrysanthemi*. *Pl. Physiol.*, **106:** 849-862.

Lili He, Yang Liu[1], Azlin Mustapha, Mengshi Lin, 2011. Antifungal activity of zinc oxide nanoparticles against *Botrytis cinerea* and *Penicillium expansum*. *Microbiol. Res.*, **166:** 207-215.

Lindgren PB, 1997. The role of hrp genes during plant-bacterial interactions. *Ann. Rev. Phytopathol.* **35:** 129-52.

Lindow, S.E., and Brandl, M.T., 2003. Microbiology of the phyllosphere. *Appl Environ Microbiol*, **69:** 1875-1883.

Lindstrom, R., 1994. The Business Week Guide to Multimedia Presentations: Create

Line, R.F., 2002. Stripe rust of wheat and barley in North America: a retrospective historical review. *Ann. Rev Phytopathol.* **40:** 75-118.

Ling Yien Ing, Ling-Ling XU, Yi-Ling LAIb, Lin WANGb, and Xing-Zhong LIU, 2010. Effects of abscisic acid and nitric oxide on trap formation and trapping of nematodes by the fungus, *Drechslerella stenobrocha* AS6.1. *Fungal Biol.*, **30:** 1-5.

Lins, E.C. , Belasque Junior, J., Marcassa, L.G., 2009. Detection of citrus canker in citrus plants using laser induced fluorescence spectroscopy. *Precision Agri.* **10:** 319-330.

Lippolis, V., Gatta, S.D., Huet, A.C., Delahaut, P., Granier, B. and Visconti, A., 2012. Multiplex dipstick immunoassay for semi-quantitative determination of *Fusarium mycotoxins* in cereals. *Anal. Chim. Acta.*, **718:** 99-108.

Lisa Ward, Scott Harper and Gerard Clover, 2010. Development of a LAMP assay for *Xylella fastidiosa*). *MAF Biosecurity New Zealand Tech.* Paper No: 2010/14.

Liu, Y., Laks, P. and Heiden, P., 2003. Nanoparticles for the Controlled Release of Fungicides in Wood: Soil Jar Studies Using G.Trabeum and T. Versicolor Wood Decay Fungi. *Holzforschung,* **57:** 135-139.

Liu, H., Wang, J.Z., Lu, S., Wang, J., Jiang, W., Ma, Z. and Zhou, L., 2009. Isoquinoline alkaloids from macleaya cordata active against plant microbial pathogens. *Nat. Prod. Commun.*, **4:** 1557-1560.

Loegering, W.Q. and Ellingboe, A.H., 1987. H.H. Flor: Poineer in phytopathology. *Ann. Rev. Phytopathol.*, **25:** 59-66.

Lohmeier, U., Farahani-Kofoet, R. D., Kofoet, A. and Grosch, R., 2013. Factors affecting incidence and severity of leaf spot disease on lettuce caused by *Septoria birgitae*. *Annals Applied Biol.*, **162:** 221.

Lola Nayar, 2009. Ubergene Cometh, *Outlook,* 10 August.

Lolle, S.J. and Pruitt, R.E., 1999. Epidermal cell interactions: a case for local talk. *Trends Plant Sci.*, **4:** 14-20.

Lolle, S.J., Berlyn, G.P., Engstrom, E.M., Krolikowski, K.A., Reiter, W.D., and Pruitt, R.E., 1997. Developmental regulation of cell interactions in the Arabidopsis fiddlehead-1 mutant: A role for the epidermal cell wall and cuticle. *Dev. Biol.,* **189:** 311-321.

Lopez, J. and Whittington, M.S., 2001. Higher order thinking in a college course: A case study. *NACTA J.,* **45:** 22-28.

Lotte, J. and Johannes A. van Veen, 2011. Defensive properties of pyrrolizidine alkaloids against microorganisms. *Phytochem Rev.,* **10:** 127-136.

Lttning, H.U. and SchlSsser, E., 1975. Role of saponins in antifungal resistance V. Enzymatic activation of avenacosides, *Z. Pflanzenkrankh. Pflanzenschutz.,* **82:** 699-703.

Lucas W. J., 2006. Plant viral movement proteins: Agents for cell-to-cell trafficking of viral genomes. *Virology,* **344:** 169-184.

Lulai, E.C., 2005. Non-wound induced suberization of tuber parenchyma cells: A physiological response to the wilt disease pathogen, *Verticillium dahliae. American J. Potato Res.,* **82:** 433-440.

Lulai, E.C. and Corsini, D.L., 1998. Differential deposition of suberin phenolic and aliphatic domains and their roles in resistance to infection during potato tuber (*Solanum tuberosum* L.) wound healing. *Physiol. Mol. Plant Biol.,* **53:** 209–222.

Lum Keng Yeang, 2003. Standards and priorities for developing pest lists in countries of South East Asia, Workshop outcome summary presentd by ASEANET Chairman.

Luna, E., Pastor, V., Robeert, J., Flors, V., Mauch-Mani, B. and Ton, J., 2011. Callose deposition: a multifaceted plant defense response. *Mol. Pl. Microbe Interact.,* **24:** 183-193

Lybert, T.J., 2002. Technology transfer for humanitarian use: economic issues and market segmentation approaches. *IP Strategy Today* No. 5-2002.

Lytic Enzymes of *Trichoderma*: Their Role in Plant Defense Saba Hasan, Garima Gupta, Shreya Anand and Harpreet Kaur, 2014. *Inter. J. Appl. Res. Stud.,* **3:** 1-5.

Ma, J. F., Goto, S., Tami, K. and Ichii, M., 2001. Role of root hairs and lateral roots in silicon uptake by rice. *Plant Physiol.,* **127:** 1773-1780.

Maas, P.W.T., 1987. Physical methods and quarantine. *In: Principles and Practice of Nematode Control in Crops.* Brown, R.H. and Kerry, B.R., (Eds.), Academic Press, London, Australia, 265–291.

MacDonald J., Savary, S., Scherm, H., Tally, A. and Wang, G.L., 2009. Disciplinary, institutional, funding and demographic trends in plant pathology. What does the future hold for the profession? *Plant Disease,* **93:** 1228-1237.

MacDonald, J., Allen, C., Gadoury, D., Jacobi, W., Kelemu, S., Moyer, J., Murray, T., Ong, K., Pearson, C., Sherwood, J. and Vidaver, A., 2009. Education in plant pathology: Present status and future challenges. *Plant Disease,* **93:** 1238-1251.

Madden, L.V., 2006. Botanical epidemiology: some key advances and its continuing role in disease management. *European J. Pl. Pathol.,* **115:** 3-23.

Madden, L.V., Hughes, G, and van den Bosch, F., 2007. The Study of Plant Disease Epidemics. The American Phytopathological Society, APS Press St. Paul, Minnesota.

Madhuban Gopal, Madhuri Sharon, Ajoy Kr. Choudhary and Rohit Kumar, 2010. Nanotechnology in agricultural diseases and food safety. *J. Phytology.,* **2:** 83-92.

Maeli Melotto, William Underwood, Jessica Koczan, Kinya Nomura, Sheng Yang He, 2006. Plant Stomata Function in Innate Immunity against bacterial invasion. *Cell, Volume,* **126:** 969-980.

Mahaletchumy Arujanan, 2009. Plant Disease Diagnostics. The Malaysian Biotechnology Information Centre (MABIC).

Mahendra, R., Shivaji, D., Aniket, G. and Kamel-Abd-Elsalam, 2012. Strategic nano-particle-mediated gene transfer in plants and animals - a novel approach. *Curr. Nanosci.,* **8:** 170-179.

Maheshwari, R. and Dubey, R. S., 2009. Nickel-induced oxidative stress and the role of antioxidant defence in rice seedlings. *Pl. Growth Regulation,* **59:** 37-49.

Mahlein, A.K., Oerke, E.CH., Dehne, H.W. and Steiner, U., 2011. Effects of Iprovalicarb on the Development and Morphology of Various *Phytophthora* and *Pythium* Species, pp. 279-284 *In: Modern Fungicides and Antifungal Compounds VI,* H.W. Dehne, H.B. Deising, U. Gisi, K. H. Kuck, P.E. Russell, H. Lyr (Eds.), Deutsche Phytomedizinische Gesellschaft, Braunschweig, Germany, p. 438.

Mai Neo and Ken T. K. Neo, 2001. Innovative teaching: Using multimedia in a problem-based learning environment. Neo (Eds.). *Educational Tech. & Soc.,* **4:** 1436-4522.

Makiichi Takagaki, Masami Ozaki, Shuichi Fujimoto and Shunichiro Fukumoto, 2014. Development of a novel fungicide, pyribencarb. *J. Pestic. Sci.,* **39:** 177-178.

Malanczuk, P., 1999. International organisations and space law: World trade organization. *Encyclopaedia Britannica,* **442:** 305.

Maleck, K., *et al.,* 2000. The transcriptome of *Arabidopsis thaliana* during systemic acquired resistance. *Nat. Genet.,* **26:** 403–410.

Mandal, D., Bolander, M.E., Mukhopadhyay, D., Sarkar, G. and Mukherjee, P., 2006. The use of microorganisms for the formation of metal nanoparticles and their application. *Appl. Microbiol. Biotechnol.,* **69:** 485-492.

Mang, H.G., Laluk, K.A., Parsons, E.P., Kosma, D.K., Cooper, B.R., Park, H.C., AbuQamar, S., Boccongelli, C., Miyazaki, S., Consiglio, F., Chilosi, G., Bohnert, H.J., Bressan, R.A., Mengiste, T. and Jenks, M.A., 2009. The Arabidopsis RESURRECTION1 gene regulates a novel antagonistic interaction in plant defense to biotrophs and necrotrophs. *Plant Physiol.,* **151:** 290-305.

Mang, H.G., Qian, W., Zhu, Y., Qian, J., Kang, H.G., Klessig, D.F. and Hua, J., 2012. Abscisic acid deficiency antagonizes high-temperature inhibition of disease resistance through enhancing nuclear accumulation of resistance proteins SNC1 and RPS4 in Arabidopsis. *Pl. Cell,* **24:** 1271-1284.

Manisankar, P., Abirama Sundari, P.L., Sasikumar, R., Palaniappan, S.P., 2008. Electroanalysis of some common pesticides using conducting polymer/multiwalled carbon nanotubes modified glassy carbon electrode. *Talanta,* **76:** 1022-1028.

Mansi Kanungo and Juhie Joshi, 2014. Impact of pyraclostrobin (F-500) on crop plant. *Pl. Sci. Today,* **1:** 174-178.

Mansoori, G.A., Rohani, T., Bastami, A., Ahmadpour, Z. and Eshaghi, 2008. Environmental application of nanotechnology. *Ann. Rev. Nano Res.,* **2:** 1-73.

Mansour, A., Mohammadian, A.B., Robert, S., Hill, B.C. and Jennifer R. Watling, 2009. Stomatal plugs and their impact on fungal invasion in *Agathis robusta. Australian J. Bot.,* **57:** 389-395.

Manuel Plantegenest, Christophe Le May and Frédéric Fabre, 2007. Landscape epidemiology of plant diseases. *J. R. Soc. Interface.,* **4:** 963-972.

Manuel Ruiz-Rubio, Alonso PÉrez-Espinosa, Khalid Lairini, Teresa RoldÁn-Arjona, Antonio Dipietro, Nuria Anaya, 2014. Metabolism of the tomato saponin α-tomatine by phytopathogenic fungi. Studies in Natural products Chemistry, Departamento de Genética, Facultad de Cienias , Universidad de Córdoba. 14071 Córdoba, Spain.

Mara Grossman, Erika Lyons, Margaret Aiken, Mary Ann Hansen and Joyce Latimer, 2014. Virginia tech tips for diagnosing impatiens necrotic spot virus (INSV), *e-Grow Alert,* **3:** (53).

Marasas, W.F.O., and Nelson, P.E. 1987. *Mycotoxicology*. The Pennsylvania State University Press, University Park, PA.

Marasas, W.F.O., Nelson, P.E. and Toussoun, T.A. 1984. *Toxigenic Fusarium* Species: *Identity and Mycotoxicology*. The Pennsylvania State University Press, University Park, PA.

Marcia J. Holden and Heven Sze, 1984. *Helminthosporium maydis* T toxin increased membrane permeability to Ca2 in susceptible corn mitochondria. *Pl. Physiol.,* **75:** 235-237.

Marco Antonio L. Cruz, Valdirene M. Gomes, Kátia V.S. Fernandes, Olga L.T. Machado and José Xavier-Filho, 2002. Identification and partial characterization of a chitinase and a β -1,3-glucanase from *Copernicia cerifera* wax. *Plant Physiol. Biochem.,* **40:** 11-16.

Marco Pautasso and Mike J. Jeger, 2014. Network epidemiology and plant trade networks. *AoB PLANTS,* **6:** plu007.

Marco, S. and Levy, D., 1979. Involvement of ethylene in the development of cucumber mosaic virus-induced chlorotic lesions in cucumber cotyledons. *Physiol. Pl. Pathol.* **14:** 235-244.

Marcus, R. and Raccah, B., 1986. Model for the spread of non-persistent virus diseases. *J. Appl. Stat.,* **13:** 167-175.

Margaret E. Daub and Kuang-Ren Chung, 2009. Photoactivated perylenequinone toxins in plant pathogenesis, pp. 201-219. *In: The Mycota-Plant Relationships,* 2nd Edition, Deising V.H (Ed.)

Maria J Soto, Juan Sanjuan and Jose Olivares, 2006. Rhizobia and plant-pathogenic bacteria: Common infection weapons. *Microbiol.,* **152:** 3167-3174.

Marianne, C., Verberne, Rob Verpoorte, John F. Bo, Jesus Mercado-Blanco and Huub, J.M. Linthorst., 2000. Overproduction of salicylic acid in plants by bacterial transgenes enhances pathogen resistance. *Nature Biotech.,* **18:** 779-783.

Mariela Odjakova and Christina Hadjiivanova, 2001. The complexity of pathogen defense in plants. *Bulg. J. Plant Physiol.,* **27:** 101-109.

Mariya V. Khodakovskaya, Kanishka de Silva, Dmitry A. Nedosekin, Enkeleda Dervishi, Alexandru S. Biris, Evgeny V. Shashkov, Ekaterina I. Galanzha and Vladimir P. Zharov, 2011. Complex genetic, photothermal, and photoacoustic analysis of nanoparticle-plant interactions. *Appl. Biol. Sci., Appl. Phys. Sci.,* **108:** 1028-1033.

Mark S. Sisterson, 2009. Transmission of insect-vectored pathogens: Effects of vector fitness as a function of infectivity status. *Environ. Entomol.,* **38:** 345-55.

Markom, M.A., Md Shakaff, A.Y., Adom, A.H., Ahmad, M.N. , Wahyu Hidayat, Abdullah, A.H. and Ahmad Fikri, N., 2009. Intelligent electronic nose system for basal stem rot disease detection. *Comp. Electron. Agri.,* **66:** 140-146.

Markus Fauth, Patrick Schweizer, Antony Buchala, Claus Markstädter, Markus Riederer, Tadahiro Kato and Heinrich Kauss, 1998. Cutin monomers and surface wax constituents elicit H_2O_2 in conditioned cucumber hypocotyl segments and enhance the activity of other H_2O_2 elicitors. *Pl. Physiol.,* **117:** 1373-1380.

Marra, R., Ambrosino, P., Carbone, V., Vinale, F.and Woo, S.L., 2006. Study of the three-way interaction between *Trichoderma atroviride*, plant and fungal pathogens by using a proteomic approach. *Curr. Genet.,* **50:** 307-321

Martin Schaefera, H., Michael Rentzschb and Michael Breuerc, 2008. Anthocyanins reduce fungal growth in fruits. *Natural Product Communications,* **3:** 1267-1272

Martin, B., Dickman and Robert Fluhr, 2013. Centrality of host cell death in plant-microbe interactions. *Ann. Rev. of Phytopathol.,* **51:** 543-570.

Martin, G.B., Brommonschenkel, S.H., Chunwongse, J., Frary, A., Ganal, M.W., Spivey, R., Wu, T., Earle, E.D. and Tanksley, S.D., 1993. Map-based cloning of a protein kinase gene conferring disease resistance in tomato. *Science.,* **262:** 1432-1436.

Martin-Hernandez, A.M., Dufresne, M., Hugouvieux, V., Melton, R. and Osbourn, A., 2000. Effects of targeted replacement of the tomatinase gene on the interaction of *Septoria lycopersici* with tomato plants. *Mol Plant Microbe Interact.,* **13:** 1301-1311.

Marvin, E. and Bauer, 1971. Remote Sensing as a Means of Detecting Crop Disease. *Tech. Bull.,* Purdue University, Purdue e-Pubs, LARS Technical Reports.

Mary Megan Quinlan, Nick M. Pasiecznik and Soetikno S. Sastroutomo, 2003. Assessing environmental risks of invasive species using ISPM 11: where to start, CAB International, SE Asia Regional Centre, P.O. Box 210, 43400 UPM Serdang, Selangor, Malaysia.

Maryam Abbasian, Ali Kashani, Mohammad Reza Ardakani, Farhad Rejali, Mahdieh Timajchi, Seyedeh Maryam Seifi and Saeed Mafakheri, 2012. The effects of chemical, biological and nano fungicides on mycorrhizal colonization and quality of sunflower. *Annals Biol. Res.,* **3:** 4239-4245.

Massalimov, I.A., Zaynitdinova, R.M., Shaynurova, A.R. and Mustafin, A.G., 2013. The efficacy of micron and nanoscale sulfur the schutte fungi. *Inter. J. Sci.,* **2:** 25-30.

Masunaka, A., Tanaka, A., Tsuge, T., Peever, T.L., Timmer, L.W., Yamamoto, M., Yamamoto, H. and Akimitsu, K., 2000. Distribution and characterization of AKT homologs in the tangerine pathotype of *Alternaria alternata. Phytopathology,* **90:** 762–768.

Mathre, D.E., 2002. One foot in the furrow: Implications to one's career in plant pathology. *Ann. Rev. Phytopathol.,* **40:** 1-11.

Matteo Lorito, Sheridan L. Woo, Gary E. Harman and Enrique Monte., 2010. Translational research on trichoderma: From 'omics to the field. *Ann. Rev. Phytopathol.,* **48:** 19.1–19.23.

Matthysse, A. G., 1983. Role of bacterial cellulose fibrils in *Agrobacterium tumefaciens* infection. *J. Bacteriol.,* **154:** 906-915.

Matthysse, A. G., and J. W. Kijne. 1998. Attachment of Rhizobiaceae to plant cells. P 235–249. *In: The Rhizobiaceae,* H. P. Spaink, A. Kondorosi, P. J. J. Hooykaas (Eds.), Kluwer Academic Publishers, Dordrecht, the Netherlands.

Mauch, F., Mauch-Mani, B. and Boller, T., 1988. Antifungal hydrolases in pea tissue II. Inhibition of fungal growth by combinations of chitinase and β-1,3-glucanase. *Pl. Physiol.,* **88:** 936-942.

Maule A., Leh V. and Lederer C., 2002. The dialogue between viruses and hosts in compatible interactions. *Curr. Opin. Plant Biol.,* **5:** 279-284.

Mavrodieva, V., Levy, L. and Gabriel, D. W., 2004. Improved sampling methods for real-time polymerase chain reaction diagnosis of citrus canker from field samples. *Phytopathology,* **94:** 61-68.

Maximina H., Yun, Pablo, S., Torres and Adrián A. Vojnov, 2006. Xanthan induces plant susceptibility by suppressing callose deposition. *Pl. Physiol.,* **141:** 178-187.

Maximina, H.Y., Pablo, S.T., Mohamed, E.O., Luciano, A.R., Rocio, G.L., Marıa, R.M., Atilio, P., Castagnaro, Marcelo, A.D., Kamal, B. and Adrian, A.V., 2006. Xanthan induces plant susceptibility by suppressing callose deposition. *Plant Physiol.,* **141:** 178-187.

Mayank Anand Gururani., Jelli Venkatesh, Chandrama Prakash Upadhyaya, Akula Nookaraju, Shashank Kumar Pandey, Se Won Park, 2012. Plant disease resistance genes: Current status and future directions, *Physiol. & Mol. Pl. Pathol.,* **78:** 51-65

Mazid, M., Khan, T.A. and Mohammad., 2011. Role of secondary metabolites in defense mechanisms of plants. Review Article, *Biology and Medicine,* **3:** 232-249.

McKinley, E. R. and Carlton, W. W., 1991. Patulin, p. 191-236. *In: Mycotoxins and Phytoalexins.* R. P. Sharma and D. K. Salunkhe (Eds.). CRC Press, Boca Raton, FL.

Md. Tofazzal Islam, Toshiaki Ito, Mitsuyoshi Sakasai and Satoshi Tahara, 2002. Zoosporicidal activity of polyflavonoid tannin identified in *Lannea coromandelica* stem bark against phytopathogenic oomycete *Aphanomyces cochlioides. J. Agric. Food Chem.,* **50:** 6697-6703.

Meachen, E.W., 2002. The arrival of standards for web-based instruction and the re-thinking of distributed learning. *Phytopathology*, **92:** S112.

Mehanathan Muthamilarasan and Manoj Prasad, 2013. Plant innate immunity: An updated insight into defense mechanism. *J. Biosci.*, **38:** 433–449.

Mehrothra, R.S., 1985. *Plant Pathology*. Tata-McGraw-Hill Publishing Co. Ltd., New Delhi, p. 770.

Meigy Nelce Mailoa, Meta Mahendradatta, Amran Laga and Natsir Djide, 2014. Antimicrobial activities of tannins extract from guava leaves (*Psidium Guajava L*) on pathogens microbial. *Intl. J. Scientific & Tech. Res.*, **3:** 2277-8616.

Melcher, U., Murch, R., Scherm, H., Seem, R. C., Sherwood, J. L., Sobral, B. W. and Tolin, S. A., 2006. Plant pathogen forensics: Capabilities, needs, and recommendations. *Microbiol.& Mol. Biol. Rev.*, **70:** 450-471.

Melemeni, M., Stamatakis, D., Xekoukoulotakis, N. P., Mantzavinos, D. and Kalogerakis, N., 2009. Disinfection of municipal wastewater by TiO_2 photocatalysis with UV-A, visible and solar irradiation and BDD electrolysis. *Global Nest J.*, **11:** 357.

Melen Leclerc Mail, Thierry Doré, Christopher A. Gilligan, Philippe Lucas and João A. N. Filipe, 2014. Estimating the delay between host infection and disease (Incubation Period) and assessing its significance to the epidemiology of plant diseases. *e-PLOS ONE*, January 22, 2014.

Melotto, M., Underwood, W., Koczan, J., Nomura, K. and He, S.Y., 2006. Plant stomata function in innate immunity against bacterial invasion. *Cell* , **126:** 969-980.

Menard, A., Drobne, D. and Jemec, A., 2011. Ecotoxicity of nanosized TiO_2. Review of *in vivo* data. *Environ Pollut.*, **159:** 677-684.

Mendgen, K. and Hahn, M., 2002. Plant infection and the establishment of fungal biotrophy. *Trends Plant Sci.*, **7:** 352-356.

Meng, X. and Li, Z., 2010. The dynamics of plant disease models with continuous and impulsive cultural control strategies. *J. Theor. Biol.*, **266:** 29-40.

Meng, Y., Li, Y., Galvani, C.D., Hao, G., Turner, J.N., Burr, T.J. and Hoch, H.C., 2005. Upstream migration of *Xylella fastidiosa* via pilus-driven twitching motility. *J. Bacteriol.*, **187:** 5560-5567.

Merritt R. Nelson, Thomas V. Orum, Ramon Jaime-Garcia and Athar Nadeem, 1999. Applications of geographic information systems and geostatistics in plant disease epidemiology and management. *Plant Disease*, **83:** 308-319.

Meyers, B.C., Kozik, A., Griego, A, Kuang, H. and Michel more, R.W., 2003. Genome-wide analysis of NBS-LRR encoding genes in *Arabidopsis*. *Plant Cell*, **15:** 809–834.

Meyers, B., Dickerman, A., Michelmore, R., Sivaramakrishnan, S., Sobral, B. and Young, N., 1999. Plant disease resistance genes encode members of an ancient and diverse protein family within the nucleotide-binding superfamily. *Plant J.*, **20:** 317-332.

Miaoying Tian, Brett Benedetti, and Sophien Kamoun, 2005. A second kazal-like protease inhibitor from *Phytophthora infestans* inhibits and interacts with the apoplastic pathogenesis-related protease P69B of tomato. *Pl. Physiol.*, **138:** 1785-1793.

Michael K. Grimmer, John Foulkes, M. and Neil D. Paveley, 2012. Foliar pathogenesis and plant water relations: A review. *J. Exptl. Bot.*, **63:** 4321-4331.

Mikaeel Young, Y, U. and Swadeshmukul Santra, 2011. Copper (cu)-silica nanocomposite containing valence-engineered Cu: A new strategy for improving antimicrobial efficacy of cu biocides. Publication by *Burnett School of Biomedical Sciences, NanoScience Technology Center, Department of Chemistry and* University of Central Florida, 12424 Research Parkway, Suite 400, Orlando, FL 32826.

Miller, S. A., 2012. IPM Symposium, Memphis, TN, The OHIO Univeristy.

Miller, S.A., Kinyua, Z.M., Beed, F., Harmon, C.L., Xin, J., Vergot, P., Momol, T., Gilbertson, R. and Garcia, L., 2010. The International Plant Diagnostic Network (IPDN) in Africa: Improving Capacity for Diagnosing Diseases of Banana (*Musa* spp.) and Other African Crops. Proc. IC on Banana & Plantain in Africa. T. Dubois *et al.*, (Eds.). *Acta Hort.*, **879**, ISHS 2010.

Mills, D. and Russell, B.W., 2003. Parameters for specific detection of *Clavibacter michiganensis* subsp. *sepedonicus* in potato stems and tubers by multiplexed PCR-ELISA. *Am. J. Potato Res.* **80:** 223-233.

Mills, W. D. and LaPlante, A. A., 1951. Diseases and insects in the orchard. *Cornell Extn. Bull.*, **711:** 21-27.

Min, B. R., Pinchak, W. E., Merkel, R., Walker, S., Tomita, G. and Anderson, R. C., 2008. Comparative antimicrobial activity of tannin extracts from perennial plants on mastitis pathogens. *Scientific Res. and Essay*, **3:** 66-73.

Min, J.S., Kim, K.S., Kim, S.W., Jung, J.H. and Lamsal, K., 2009. Effects of colloidal silver nanoparticles on sclerotium-forming phytopathogenic fungi. *Plant Pathol. J.*, **25:** 376-380.

Minghua Zhang, Zhihao and QinXue Liu, 2005. Remote sensed spectral imagery to detect late blight in field tomatoes. *Precision Agriculture*, **6:** 489-508.

Ming-wang Shi, 2009. MASS '09, International Conference on Management and Service Science, 20-22 Sept., Henan Instititue of Science and technology, Xinxiang, China: Qing-lian Wang; Xi-ling Chen; Xi-ling Chen; Ju-huai.

Miranda, M., Ralph, S.G., Mellway, R., White, R., Heath, M.C., Bohlmann, J. and Constabel, C.P., 2007. The transcriptional response of hybrid poplar (*Populus trichocarpa* x *P. deltoides*) to infection by *Melampsora medusae* leaf rust involves induction of flavonoid pathway genes leading to the accumulation of proanthocyanidins. *Mol. Pl. Microbe Interact.*, **20:** 816-831.

Mirela Paraschivu, Otilia Cotuna and Paraschivu, M., 2013. The use of the area under the disease progress curve (Audpc) to assess the epidemics of *Septoria tritici* in winter wheat. *Res. J. Agril. Sci.*, **45:** 193.

Mirik, M., Ansley, R., Price, J., Workneh, F. and Rush, C., 2013. Remote monitoring of wheat streak mosaic progression using sub-pixel classification of landsat 5 TM imagery for site specific disease management in winter wheat. *Adv. Remote Sensing*, **2:** 16-28.

Mishra, S., Singh, B.R., Singh, A., Keswani, C. and Naqvi, A.H., 2014. Biofabricated silver nanoparticles act as a strong fungicide against *Bipolaris sorokiniana* causing spot blotch disease in wheat. *PLoS ONE* **9(5):** e97881.

Mitchell, R.E., 1991. Implications of toxins in the ecology and evolution of plant pathogenic microorganisms: bacteria. *Experientia,* **47:** 791-803.

Mittler, R., Herr, E. H. and Orvar, B. L., 1999. Transgenic tobacco plants with reduced capability to detoxify reactive oxygen intermediates are hyperresponsive to pathogen infection. *Proc., National Academy of Sciences of the United States of America,* **96:** 14165-14170.

Mittler, R., Shulaev, V. and Lam, E., 1995. Coordinated activation of programmed cell death and defense mechanisms in transgenic tobacco plants expressing a bacterial proton pump. *Plant Cell,* **7:** 29-42.

Miyai, S., 1991. Modelling the effects of aphid vector (*Aulacorthum solani*) movement on the spread of soybean dwarf virus. *In: Proc. of International Seminar on Migration and Dispersal of Agricultural Insects.* National Institute of Agro-Environmental Sciences, Tsukuba, Japan, pp. 145-154.

Miyashita, M., Nakamori, T., Murai, T., Yonemoto, T., Miyagawa, H., Akamatsu, M. and Ueno, T., 2001. Structure-activity relationship study of host-specific phytotoxins (AM-toxin analogs) using a new assay method with leaves from apple meristem culture. *Z. Naturoforsch C.,* **56:** 1029-1037.

Moalemiyan, M., Vikram, A. and Kushalappa A.C., 2007. Detection and discrimination of two fungal diseases of mango (cv. Keitt) fruits based on volatile metabolite profiles using GC/MS. *Postharvest Biol. & Technol.,* **45:** 117-125.

Mohamed, M.M. and Khairou, K.S., 2011. Preparation and characterization of nano-silver/mesoporous titania photocatalysts for herbicide degradation. *Microporous Mesoporous Mater.,* **142:** 130-138.

Mohammad A. Khiyami, Hassan Almoammar, Yasser M. Awad, Mousa A. Alghuthaymi and Kamel A. Abd-Elsalam, 2014. Plant pathogen nanodiagnostic techniques: Forthcoming changes? *Biotech. & Biotechnological Equip.,* **28,** Issue 5

Mohammad Ali Aghajani, Naser Safaie and Azizollah Alizadeh, 2013. Yield loss assessment of *Sclerotinia* stem rot of canola in Iran. *J. Crop Prot.,* **2:** 229-240.

Mohammed Suleiman, Anas Al Ali, Ayman Hussein, Belkheir Hammouti Taibi B. Hadda and Ismail Warad, 2013. Sulfur nanoparticles: Synthesis, characterizations and their applications. *J. Mater. Environ. Sci.,* **4:** 1029-1033.

Mohammod Hossain, 2010. The role of gibberellin in the response of rice to *Meloidogyne graminicola* and *Hirschmanniella oryzae* infection. Thesis submitted in fulfilment of the requirement for the degree of European Master of Science in Nematology, Faculty of Science, Department of Biology, UGent, Belgium.

Mohapatra, N. K., Mukherjee, A. K., Suriya Rao, A. V. and Nayak, P., 2008. Disease progress curves in the rice blast pathosystem compared with the logistic and gompertz models. *ARPN J. Agril. and Biol. Sci.,* **3**(1).

Molineros, J., De Wolf, E. and Haran, M., 2004. Application of hotspot detection analysis to the prediction of *Fusarium* Head Blight Epidemics. Session 3: Etiology, Epidemiology and Disease Forecasting.

Monica, R. C. and Cremonini, R. 2009. Nanoparticles and higher plants. *Caryologia,* **62:** 161.

Morris, R.J., Lewis, O.T. and Godfray, H.C., 2004. Experimental evidence for apparent competition in a tropical forest food web. *Nature,* **428:** 310-313.

Morris, R.O., Blevins, D.G., Dietrich, J.T., Durley, R.C., Gelvin, S.B., Gray, J., Hommes, N.G., Kaminek, M., Mathews, L.J., Meilan, R., Reinbott, T.M. and Sayavedra-Soto, L., 1993. Cytokinins in plant pathogenic bacteria and developing cereal grains. *Australian J. Pl. Physiol.,* **20:** 621-637.

Morrissey, J.P and Osbourn, A.E. 1999. Fungal resistance to plant antibiotics as a mechanism of pathogenesis. *Microbiol. Mol. Biol. Revs.,* **63:** 708-724.

Moshou, D., Bravo, C., Oberti, R., West, J., Bodria, L., McCartney, A. and Ramon, H., 2005. Plant disease detection based on data fusion of hyper-spectral and multi-spectral fluorescence imaging using Kohonen maps. *Real-Time Imaging,* **11:** 75-83.

Mosolov, V. V., Loginova, M. D., Fedurkina, N. V., and Benken, I. I., 1976. The biological significance of proteinase inhibitors in plants. *Plant Sci. Lett:,* **7:** 77-80.

Moura, M.L., Jacques, M.A., Brito, L.M., Mourão, I.M. and Duclos, J. 2005. Tomato pith necrosis (tpn) caused by *P. corrugata* and *P. mediterranea*: severity of damages and crop loss assessment. *Acta Hort.,* (ISHS) **695:** 365-372.

Mousanejad, S., Alizadeh, A. and Safaie, N., 2010. Assessment of yield loss due to rice blast disease in Iran. *J. Agr. Sci. Tech.,* **12:** 357-364.

Movahedy, S. and Heale, J., 1990. The action of proteinase inhibitors from plants upon the enzymes. *Physiol. Mol. Plant Pathol.,* **36:** 303-324.

Movahedy, S., and Heale, J., 1990. The roles of aspartic proteinase and endo-pectin lyase nzymes in the primary stages of infection and pathogensis of various host tissues by different isolates of *Botrytis cinerea* pers ex pers. Physiol. Mol. Plant Pathol., **36:** 303-324.

Mugniéry, D., Phillips, M.S., Rumpenhorst, H.J., Stone, A.R., Treur, A. and Trudgill, D.L., 1989. Assessment of partial resistance of potato, and pathotype and virulence differences in potato cyst nematodes. *OEPP/EPPO Bull.,* **19:** 7-25.

Muhammad Azam Ali, Iqra Rehman, Adnan Iqbal, Salah ud Din, Abdul Qayyum Rao, Ayesha Latif, Tahir Rehman Samiullah, Saira Azam and Tayyab Husnain, 2014. Nanotechnology: A new frontier in agriculture. *Adv. Life Sci.,* **1:** 129-138.

Muhammad Hamayun, Sumera Afzal khan, Abdul Latif Khan, Gauhar Rehman, Youn Ha-Kim, Ilyas Iqbal, Javed Hussain, Eun Young Sohn and In-Jung Lee, 2010. Gibberellin production and plant growth promotion from pure cultures of *Cladosporium* sp. MH-6 isolated from cucumber (*Cucumis sativus* L.). *Mycologia,* **102:** 989-995.

Mujeebur Rahman Khan and Tanveer Fatima Rizvi, 2014. Nanotechnology: Scope and application in plant disease management. *Pl. Pathol. J.,* **13:** 214-231.

Muller, C. and Riederer, M., 2005. Plant surface properties in chemical ecology. *J. Chem. Ecol.*, **31:** 2621-2651.

Muller, K.O., Borger, H., 1940. Experimentelle untersuchugen uber die *Phytophthora* resistenz der kartoffel. *Arbeit. Biol. Reichsant Land Forstwirtsch*, **23:** 189-231.

Murphy, J. M. and Walton, J. D., 1996. Three extracellular proteases from *Cochliobolus carbonum*: cloning and targeted disruption of ALP1. *Mol. Plant Microbe Interact.*, **9:** 290-297.

Murphy, K., 2008. *Nanotechnology: Agriculture's Next "Industrial" Revolution* (Williston, VT: Financial Partner, Yankee Farm Credit, ACA) (ed.), pp. 3–5 Spring.

Murty, N.S., 2010. Climate Change and Plant Diseases. *In: Plant Pathology in Practice*, pp. 45-49. Training manual, G.B. Pant University of Agriculture & Technology, Pantnagar-263 145, India, pp. 265.

Mutasa-Gottgens, E.S., Chwarszczynska, D.M., Halsey, K. and Asher, M.J.C., 2000. Specific polyclonal antibodies for the obligate plant parasite *Polymyxa*– A targeted recombinant DNA approach. *Pl. Pathol.*, **49:** 276-287.

Mysore, K.S. and Ryu, C.M., 2004. Nonhost resistance: How much do we know? *Trends Plant Sci.*, **9:** 97-104.

Myung, J., 2003. Tutorial on maximum likelihood estimation. *J. Mathematical Psychol.*, **47:** 90-100.

Nahar, K., Kyndt, T., Nzogela, Y.b., Gheysen, G., 2012. Abscisic acid interacts antagonistically with classical defense pathways in rice-migratory nematode interaction. *New Phytol.*, **196:** 901-913.

Nahid Sarlak, Asghar Taherifar and Fatemeh Salehi, 2014. Synthesis of nanopesticides by encapsulating pesticide nanoparticles using functionalized carbon nanotubes and application of new nanocomposite for plant disease treatment. *J. Agric. Food Chem.*, **62:** 4833-4838.

Naidu, R.A., Perry, E.M., Pierce, F.J. and Mekuria, T., 2009. The potential of spectral reflectance technique for the detection of grapevine leafroll-associated virus-3 in two red-berried wine grape cultivars. *Computers & Electronics in Agri.*, **66:** 38-45.

Nair, R., Varghese, S.H., Nair, B.G., Maekawa, T., Yoshida, Y. and Kumar, D.S., 2010.

Nakayashiki, H., 2005. RNA silencing in fungi: Mechanisms and applications. *FEBS Lett.*, **579:** 5950–5957.

Namai, T., Kato, T., Yamaguchi, Y. and Hirukawa, T., 1993. Anti-rice blast activity and resistance induction of C-18 oxygenated fatty acids. *Biosci. Biotech. Biochem.*, **57:** 611–613.

Nandini P. Shetty, Jens D. Jensen, Anne Knudsen, Christine Finnie, Naomi Geshi, Andreas Blennow, David B. Collinge and Hans J. Lyngs Jørgensen, 2009. Effects of β-1,3-glucan from septoria tritici on structural defence responses in wheat. *J. Exptl. Bot.*, **60:** 4287-4300.

Naresh Mehta, 2014. Epidemiology and forecasting for the managemetn of rapeseed-mustard diseases. *J. Mycol. Pl. Pathol.*, **44:** 131-147.

Narvankar, D.S., Singh, C.B., Jayas, D.S. and White, N.D.G., 2009. Assessment of soft X-ray imaging for detection of fungal infection in wheat. *Biosys. Engg.,* **103:** 49-56.

NASC, Dev Prakash Shastry Marg, New Delhi-110 012, India.

Nathaniel C. Cady, 2009. Quantum Dot Molecular Beacons for DNA Detection, Methods in Molecular Biology. Volume: 544, Pub. date: Jun-01- Micro and Nano Technologies in Bioanalysis: Methods and Protocols.

Navarre DA, Wolpert TJ. 1995. Inhibition of the glycine decarboxylase mul tienzyme complex by the host-selective toxin, victorin. *Pl. Cell,* **7:** 463–71.

Nawrath, C., 2006. Unraveling the complex network of cuticular structure and function. *Curr. Opin. Plant Biol.,* **9:** 281–287.

Nazmul, H.B., Gopalan, S., Yangdou, W. and John King, 2009. Role of lignification in plant defense. *Pl. Signalling & Behaviour*, **4:** 158-159.

Neal Stewart, C., Jr., 2011. Harold A. Richards, IV, Matthew D. Halfhill, Transgenic Plants and Biosafety: Science, Misconceptions and Public Perceptions. *Bio-techniques,* **29:** 832-843.

Neergaard, P., 1977. Quarantine policy for seed in transfer of genetic resources, pp 309-314. *In: Plant Health and Quarantine in International Transfer of Genetic Resources* W.B. Hewitt and L. Chiarappa, (Eds.), CRC Press, Cleveland, USA.

Neha Bhalla, 2012. Application of Nanotechnology in early detection of plant diseases to ensure food security. A case study for utilizing cloud as a future for Government, eINDIA Posted on December 27, 2012 by admin.

Nelson; M. R., Felix-Gastelum, R., Orum, T. V., Stowell, L. J. and Myers, D. E., 1994. Geographic information systems and geostatistics in the design and validation of regional plant virus management programs. *Phytopathology,* **84:** 898-905.

Nethi Somasekhar and Jonnalagadda Satyanarayana Prasad, 2012. Plant-Nematode interactions: Consequences of climate change, pp. 547-564. *In: Crop Stress and Its Management: Perspectives and Strategies.* Venkareshwarlu, B., Arun K. Shanker and Chitra Shankar (Eds.), pp 612.

Netscher, C. and Sikora, A., 1990. Nematode parasites of vegetables. *In: Plant Parasitic Nematodes in Subtropical and Tropical Agriculture.* Luc, M.; Sikora, R.A.; Bridge, J (Edrs.), pp. 237-283. CAB International, Wallingford, UK.

Newman, M.A., SUndelin, T., Nielsen, J.T. and Erbs, G., 2013. MAMP (microbe-associated molecular pattern) triggered immunity in plants. *Front Pl. Sci.,* **16:** 4:139.

Neysens, P and De V.L., 2011. Carbon dioxide stimulates the production of amylovorin L by Lactobacillus amylovorus DCE 471, while enhanced aeration causes biphasic kinetics of growth and bacteriocin production. Intl. J. Food Microbiol., **105:** 191-202.

Ng, J.C.K., 2013. A quantum dot-immunofluorescent labeling method to investigate the interactions between a crinivirus and its whitefly vector. *Front. Microbiol.,* **4:** 77.

Ng, K. H., and Komiya, R., 2000. Introduction of Intelligent Interface to Virtual Learning Environment. Paper presented at the Multimedia University International Symposium on Information and Communication Technologies 2000 (M2USIC'2000), October 5-6,

2000, Petaling Jaya, Malaysia. *In: Innovative Teaching:* Using *Multimedia in a Problem-based Learning Environment.* Mai Neo and Ken T. K. Neo (Eds.). *Educational Tech. & Soc.,* **4:** 1436-4522.

Nicholson, R.L. and Hammerschmidt, R. 1992. Phenolic compounds and their role in disease resistance. *Ann. Rev. Phytopathol.,* **30:** 369-389.

Nicholson, R.L. and Wood, K.V., 2001. Phytoalexins and secondary products, where are they and how can we measure them? *Physiol. Mol. Plant Pathol.,* **59:** 63-60.

Nicolas Denance, Andrea sanchez-Vallet, Deborah Goffner and Antonio Molina, 2013. Disease resistance or growth: The role of plant hormones in balancing immune responses and fitness costs. *Front Pl. Sci.,* **4:** 155.

Niessen, L., Gräfenhan, T., Vogel and R.F., 2012. ATP citrate lyase 1 (acl1) gene-based loop-mediated amplification assay for the detection of the *Fusarium tricinctum* species complex in pure cultures and in cereal samples. *Int. J. Food Microbiol.,* **158:** 171-185.

Nieves Capote, Ana María Pastrana, Ana Aguado and Paloma Sánchez-Torres, 2012. Molecular Tools for Detection of Plant Pathogenic Fungi and Fungicide Resistance, Plant Pathology, Dr. Christian Joseph Cumagun (Ed.), ISBN: 978-953-51-0489-6.

Nik J. Cunniffe, Britt Koskella, C. Jessica E. Metcalf, Stephen Parnell, Tim R. Gottwald and Christopher A. Gilligan, 2015. Thirteen challenges in modelling plant diseases. *Epidemics,* **10:** 6-10

Nik J. Cunniffe and Christopher A. Gilligan, 2010. Invasion, persistence and control in epidemic models for plant pathogens: the effect of host demography. Interfaaace, DOI: 10.1098/rsif.2009.0226.

Nishimura, S. and Kohmoto, K., 1983. Host-specific toxins and chemical structures from *Alternaria* species. *Ann. Rev. Phytopathol.,* **21:** 87-116.

Noctor, G. and Foyer, C. H., 1998. Ascorbate and glutathione: Keeping active oxygen under control. *Ann. Rev. Plant Biol.,* **49:** 249-279.

Noctor, G., Veljovic-Jovanovic, S., Driscoll, S., Novitskaya, L. and Foyer, C. H., 2002. Drought and oxidative load in the leaves of C_3 plants: A predominant role for photorespiration? *Annals Bot.,* **89:** 841-850.

Nora Olsen, 2012. Using Pathogen Detection Test Kits for Rapid Potato Disease Diagnosis, University of Idaho.

Noraziah Mohamad Zin, Norton, G.A., Holt, J. and Mumford, J.D., 1993. Introduction to pest models. *In: Decision Tools for Pest Management,* pp. 89–99. Norton, G.A. and Mumford, J.D., (Eds.), CAB International, Cambridge, UK.

Notomi, T., Okayama, H., Masubuchi, H., Yonekawa, T., Watanabe, K., Amino, N and Hase, T., 2000. Loop-mediated isothermal amplification of DNA. *Nucleic Acids Res.,* **28:** e63.

Nunes, C.C. and Dean, R.A., 2012. Host-induced gene silencing: a tool for understanding fungal host interaction and for developing novel disease control strategies. *Mol. Plant Pathol.,* **13:** 519-29.

Nürnberger, T. and Lipka, V., 2005. Non-host resistance in plants: New insights into an old phenomenon. *Mol. Plant Pathol.,* **6:** 335-345.

Nutter, F. W. Jr., 2004. Am. Phytopathol. Soc. Annu. Meet., *Phytopathology* , **94** (Suppl.): S77.

Nutter, Jr., F. W., Tylka, G. L., Guan, J., Moreira, A. J. D., Marett, C. C., Rosburg, T. R., Basart, J. P. and Chong, C. S., 2002. Use of remote sensing to detect soybean cyst nematode-induced plant stress. *J. Nematol.,* **34:** 222-231.

Nyadanu, D., Akromah, R., Adomako, B., kwoseh, C., Dzahini-Obiatey, H., Lowor, S.T., Akrofi, A.Y. and Assuah, M.K., 2012. Host plant resistance to *Phytophthora* pod rot in Cacao (*Theobroma cacao* L.): The role of epicuticular wax on pod and leaf surfaces. *Intl. J. Bot.,* **8:** 13-21.

Nzioki, H. S. and Chirchir, A. K., 2001. Yield loss assessment due to nematodes associated with sugarcane in Kenya, Kenya Agricultural Research Institute, P. O. Box 340, Machakos.

O'Bannon, J.H. and Esser, R.P., 1987. Regulatory perspective in nematology, pp. 38-46. *In: Vistas on Nematology: A Commemoration of the Twenty-fifth Anniversary of the Society of Nematologists.* Joseph A. Veech and Donald W. Dickson (Edrs.), Soc. of Nematologists, Hyattsville, Maryland, 509.

O'Bannon, J.H. and Tomerlin, A.T., 1971. Response of citrus seedlings to radopholus similis in two soils. *J. Nematol.,* **3:** 255-260.

Ocsoy, I., Paret, M.L., Ocsoy, M.A., Kunwar, S., Chen, T., You, M. and Tan, W., 2013. Nanotechnology in plant disease management: DNA-directed silver nanoparticles on graphene oxide as an antibacterial against *Xanthomonas perforans. ACS Nano.,* **7:** 8972-8980.

Ohm, R.A., Feau, N., Henrissat, B., Schoch, C.L. and Horwitz, B.A., 2012. Diverse lifestyles and strategies of plant pathogenesis encoded in the genomes of eighteen *Dothideomycetes* Fungi. *PLoS Pathog.,* **8(12):** e1003037.

Okada, K., Saito, T., Nakagawa, T., Kawamukai, M. and Kamiya, Y., 2000. Five geranylgeranyl diphosphate synthases expressed in different organs are localized into three subcellular compartments in *Arabidopsis. Plant Physiol.,* **122:** 1045-1056

Okada, M., Miyagwa, H., Nakagawa,Y. and Ueno, 2000. Binding affinity of the methyl ester of AK-toxin I to membrane fractions from Japanese pear leaves. *BioSci. Biotechnol. Biochem.,* **64:** 2517-2521.

Okamoto, H., Suzuki, Y., Kataoka, T. and Sakai, K., 2009. Unified hyperspectral imaging methodology for agricultural sensing using software framework. *Acta Hort.,* **824:** 49-56.

Oldach, K.H., Becker, D. and Lorz, H., 2001. Heterologous expression of the genes mediating enhanced fungal resistance in transgenic wheat. *Mol. Plant Microbe Interact,* **14:** 832-838.

Olga del Pozo and Eric Lam, 1998. Caspases and programmed cell death in the hypersensitive response of plants to pathogens. *Curr. Biol.,* **8:** 1129-1132.

Olsen, R.A., 1973. Triterpeneglycosides as inhibitors of fungal growth and metabolism. The effect of aescin on fungi with reduced sterol contents, *Physiol. Plant.,* **29:** 145-149.

Onyilagha, J.C. and Grotewold, E., 2004. Recent Research Development in Plant Science, ISBN: 81-7736-239-9, Research Signpost, Trivandrum, India, 1.

Orum, T. V., Bigelow, D. M., Nelson, M. R., Howell, D. R. and Cotty, P. J. 1997. Spatial and temporal patterns of *Aspergillus flavus* strain composition and propagule density in yuma county, arizona, soils. *Plant Disease*, **81:** 911-916.

Osborn, R. W., Desamblanx, G. W., Thevissen, K., Goderis, I., Torrenkens, S. and Vanleuven, F., 1995. Isolation and characterization of plant defensins from seeds of asteraceae, fabaceae, hippocastanaceae and saxifragaceae. *FEBS Lett.*, **368:** 257-262.

Osborne, B.I. and Baker, B., 2012. Movers and shakers: maize transposons as tools for analyzing other plant genomes. *Curr. Opin. Cell. Biol.*, **7:** 406-413.

Osbourn, A., 1996. Saponins and plant defence - a soap story. *Trends Plant Sci.*, **1:** 4-9.

Osbourn, A.E., 1996. Preformed antimicrobial compounds and plant defense againts fungal attack. *Plant Cell.* **8:** 1821-1831.

Osbourn, A.E., 2001. Tox-boxes, fungal secondary metabolites and plant disease. *PNAS*, **98:** 14187-14188.

Ostfeld, R.S., Glass, G.E. and Keesing, F., 2005. Spatial epidemiology: An emerging (or re-emerging) discipline. *Trends Ecol. Evol.*, **20:** 328-36.

Otani, H., Kohmoto, K. and Kodama, M., 1995. *Alternaria* toxins and their effects on host plants. *Can. J. Bot.*, **73:** 453-458.

Otani, H., Kohmoto, K. and Kodama, M., 1995. *Alternaria* toxins and their effects on host plants. *Can. J. Bot.*, **73:** 453–458.

Otsuka, H, 2010. Nanofabrication of nonfouling surfaces for micropatterning of cell and microtissue. *Molecules*, **15:** 5525-5546.

Ozsolak, F., 2012. Third-generation sequencing techniques and applications to *drug discovery. Expert. Opin. Drug Discover.*, **7:** 231-243.

Pablo Velasco, Margarita Lema, Marta Francisco, Pilar Soengas and María Elena Cartea, 2013. *In vivo* and *in vitro* effects of secondary metabolites against *Xanthomonas campestris pv. campestris. Molecules*, **18:** 11131-11143.

Padmanabhan, S. Y., 1965. Estimating Losses from Rice Blast in India. *In: "The Rice Blast Disease"*. The Johns Hopkins Press, Baltimore, Maryland, 203-221.

Pallavi Sharma, Ambuj Bhushan Jha, Rama Shanker Dubey and Mohammad Pessarakli, 2012. Reactive oxygen species, oxidative damage, and antioxidative defense mechanism in plants under stressful conditions. *J. Bot.*, Article ID 217037, 26 pp.

Panaccione, D. 1996. Multiple families of peptide synthetase genes from ergopeptine-producing fungi. *Mycol. Res.*, **100:** 429-436.

Panina, Ya. S., Gerasimova, N. G., Chalenko, G. I., Vasyukova, N. I. and Ozeretskovskaya. O. L., 2005. Salicylic acid and phenylalanine ammonia-lyase in potato plants infected with the causal agent of late blight. *Russian J. Pl. Physiol.*, **52:** 511-515.

Paola Zuluaga, A., Julio Cesar Vega-Arreguín and William E. Fry, 2013.Transcriptome profile of acibenzolar-S-methyl-induced genes in tomato suggests a complex polygenic effect on resistance to *Phytophthora infestans. Physiol. & Mol. Pl. Pathol.*, **81:** 97-106.

Papa M. Diedhiou, E.-C. Oerke and H.-W. Dehne, 2004. Effects of the strobilurin fungicides azoxystrobin and kresoximmethyl on arbuscular mycorrhiza. *J. Pl. Dis & Protn.*, **111:** 545-556

Papaïx, J., Burdon, J.J., Lannou, C. and Thrall, P.H., 2014. Evolution of pathogen specialisation in a host metapopulation: Joint effects of host and pathogen dispersal. *PLoS Comput Biol.*, **10:** e1003633.

Pariaud, B., Ravigné, V., Halkett, F., Goyeau, H., Carlier, J. and Lannou, C., 2009. Aggressiveness and its role in the adaptation of plant pathogens. *Plant Pathology*, **58:** 409–424.

Paris, R. and Lamattina, L., 1999. *Phytophthora infestans* secretes extracellular proteases with necrosis inducing activity on potato. *Eur. J. Plant Pathol.*, **105:** 753-760.

Parissa, T. and Saeed, T., 2012. The role of pathogenesis-related proteins in the tomato-*Rhizoctonia solani* interaction. *J. Bot.*, **20:** 1-6.

Park Hae-Jun, Sung Ho Kim, Hwa Jung Kim and Seong-Ho Choi, 2006. A new composition of nanosized silica-silver for control of various plant diseases. *Plant Pathol. J.*, **22:** 295-302.

Partridge, J.E. and Osborne, L., 2001. Designing and using world wide web study pages to support student learning outside of the classroom, *NACTA J.*, **43:** 43-46.

Passardi, F., Cosio, C., Penel, C. and Dunand, C., 2005. Peroxidases have more functions than a Swiss army knife. *Pl. Cell Rep.*, **24:** 255-265.

Pathak, H.C., 2010. Executive Secretary, National Academy of Agricultural Sciences.

Patricia L. Traynor, Dann Adair and Ruth Irwin. 2001. A practical guide to containment-Greenhouse Research with Transgenic Plants and Microbes. Information Systems for Biotechnology, Virginia Tech., 207, Engel Hall, Blacksburg VA 2406174, p. 74

Patrick Fuerst, E., James, V., Anderson, Ann C. Kennedy and Robert S. Gallagher, 2011. Induction of polyphenol oxidase activity in dormant wild oat (*Avena fatua*) seeds and caryopses: A defense response to seed decay fungi. *Weed Sci.*, **59:** 137-144.

Patrick Siebera, Martine Schordereta, Ulrich Rysera, Antony Buchalaa, Pappachan Kolattukudy, Jean-Pierre Métrauxa and Christiane Nawratha, 2000. Transgenic arabidopsis plants expressing a fungal cutinase show alterations in the structure and properties of the cuticle and postgenital organ fusions. *The Pl. Cell.*, **12:** 5721-5737.

Paul R. Bumows, 1990. The rapid and sensitive nematode detection of the plant parasitic *Globodera pallida* using a non-radioactive biotinylated DNA probe, *Revue Nématol.* **13:** 185-190.

Paul, J., Pinter, Jr., Jerry, L., Hatfield, James S. Schepers, Edward M. Barnes, Susan Moran., Craig, S.T., Daughtry and Dan R. Upchurch, 2003. Remote Sensing for Crop Management. *Photogrammetric Engg. & Remote Sensing*, **69:**. 647-664.

Paulraj, K., Lawrence and Kripa Ram Koundal, 2001. Plant protease inhibitors in control of phytophagous insects. *Pl. Biotech.*, **5:** 21-32.

Pawan Kaur, Rajesh Thakur and Ashok Choudhary, 2012. An *in vitro* study of the antifungal activity of Silver/Chitosan nanoformulations against important seed borne pathogens. *Inter. J. Sci. Techn. Res.,* **1:** 2277-8616.

Paxton, J.D., 1980. A new working definition of the term "phytoalexin. *Plant Dis.,* **64:** 734.

Payne, G. 1983. Nature of field infection of corn by *Aspergillus flavus,* pp. 16-22. *In: Aflatoxins and Aspergillus flavus in Corn.* U. L. Diener, R. L. Asquith and J. W. Dickens, (Eds.), Auburn University, Auburn, AL.

Peeters, N., Guidot, A., Vailleau, F. and Valls, M., 2013. *Ralstonia solanacearum,* a widespread bacterial plant pathogen in the post-genomic era. *Mol. Pl. Pathol.,* **14:** 651–662.

Pel, M. and Pieterse, C.M., 2013. Microbial recognition and evasion of host immunity. *J. Exp. Bot,.* **64:** 1237-48.

Peragine, A., Yoshikawa, M., Wu, G., Albrecht, H.L. and Poethig, R.S., 2004. SGS3 and SGS2/SDE1/RDR6 are required for juvenile development and the production of transacting siRNAs in Arabidopsis. *Genes Dev.,* **18:** 2368-2379.

Peter Constabel, C. and Raymond Barbehenn., 2008. Defensive Roles of Polyphenol Oxidase in Plants. *In: Induced Plant Resistance to Herbivory,* p. 253. A. Schaller (Ed.). C Springer Science+Business Media, B.V. Centre for Forest Biology and Department of Biology, University of Victoria, Victoria, BC, Canada V8W 3N5.

Peter H. Dernoeden, 1998**.** The new generation of fungicides of microbial origin**,** *TURFAX,* **6:** 1-5.

Petter, F., 2008. Introduction to the EPPO Workshop on quality assurance for plant pest laboratories: EPPO's involvement in quality assurance. *EPPO Bull.,* **38:** 167-68.

Ph. Vermeulen, J.A., Fernández Pierna, H.P., van Egmond, Dardenne, P. and Baeten, V., 2012. Online detection and quantification of ergot bodies in cereals using near infrared hyperspectral imaging. Food Additives & Contaminants: Part A , Volume 29, Issue 2.

Phillippe Jeandet, Christophe Clement, Eric Courot and Sylvain Cordelier, 2013. Modulation of phytoalexins biosynthesis in engineeref plants for disease resisitance. *Int. J. Mol. Sci.,* **14:** 14136-14170.

Pimentel, D., Lach, L., Zuniga, R. and Morrison, D., 2000. Environmental and economic costs of nonindigenous species in the united states. *Bioscience,* **50:** 53-65.

Ping He, Libo Shan and Jen Sheen, 2007. Elicitation and suppression of microbe-associated molecular pattern-triggered immunity in plant–microbe interactions. *Cell Microbiol.,* **9:** 1385-1396.

Pinter, Jr., P.J., Stanghellini, M.E., Reginato, R.J., Idso, S.B., Jenkins, A.D. and Jackson, R.D., 1979. Remote detection of biological stresses in plants with infrared thermometry, *Science,* **205:** 585–586.

Piotr Sobiczewski, Zemdirbyste-Agriculture, 2008. Bacterial diseases of plants: Epidemiology, diagnostics and control. *UDK,* **95:** 151–157.

Pitkin, J.W., Nikolskaya, A., Ahn, J.H. and Walton, J.D., 2000. Reduced virulence caused by meiotic instability of the TOX2 chromosome of the maize pathogen *Cochliobolus carbonum*. *Mol. Plant-Microbe Interact.*, **13:** 80–87.

Plucknett, D.L., Smith, N.J.H., Williams, J.T. and Anishetty, N.M.,1987. A case study in rice germplasm: IR36. Gene Banks and the World's Food. Princeton University Press, Princeton, NJ, 171–185.

Podile, A. R. and Laxmi, V. D. V., 2008. Seed Bacterization with Bacillus subtilis AF 1 Increases Phenylalanine Ammonia-lyase and Reduces the Incidence of Fusarial Wilt in Pigeonpea. *J. Phytopathol.*, **146:** 255-259.

Popplestone, C. R. and Unrau, A. M., 1973. Biosynthesis of lycomarasmin. *Canadian J. Chem.*, **51:** 117-216.

Power, A.G., Mitchell, C.E., 2004. Pathogen spillover in disease epidemics. *Am. Nat.*, **164** Suppl **5:** S79-89.

Prabhu, A. S. and DeFaria, J. C., 1982. Relacionamentos quantitatives entre brusone nas folhas e paniculas e seus efeitos sobre enchimento e peso dos graos em arroz de sequeiro. pesqui. agropecu. *Bras. Braz.*, **17:** 219-223.

Prabhuraj Aralimarad, 2012. UAS Raichur builds e-SAP devices to check pest attacks PTI | Dec 25, 2012, 01.07PM IST.

Prachi Kanhed, Sonal Birla, Swapnil Gaikwad, Aniket Gade, Amedea B. Seabra, Olga Rubilar, Nelson Duran and Mahendra Rai., 2014. *In vitro* antifungal efficacy of copper nanoparticles against selected crop pathogenic fungi. *Materials Letters*, **115**: 13-17.

Prasad, D., Schaff, J.E., Laskey, J.G., Kieber, J.J., Bilyeu, K.D. and Bird, D. 2004. Cytokinins play opposite roles in lateral root formation and nematode and rhizobial symbioses. *Pl. J.*, **38**: 203.

Prasanna, B.M., 2007. Nanotechnology in agriculture. ICAR National Fellow, Division of Genetics, I.A.R.I., New Delhi–110012.

Prashanth, S., Menaka, I.R., Muthezhilan and Navin Kumar Sharma, 2011. Synthesis of plant-mediated silver nano particles using medicinal plant extract and evaluation of its anti microbial activities. *Inter. J. Engg.Sci.& Techn.*, **3:** 6235-6250.

Prathibha Sharma, 2014. ELISA-Kit for detection of papaya ring spot virus and other related poyviruses, Indian Agricultural Research Institute, New Delhi, India.

Prelusky, D. B., Rotter, B. A. and Rotter, R. G., 1994. Toxicology of mycotoxins, pp. 359-405. *In: Mycotoxins in Grain: Compounds Other Than Aflatoxin.* J. D. Miller and H. L. Trenholm (Eds.). Eagan Press, St. Paul, MN.

Pringle, R.B. and Scheffer, R.P., 1964. Host-specific plant toxins. *Ann. Rev. Phytopathol.* **2:** 133–156.

Prithiviraj, B., Vikram, A., Kushalappa, A.C. and Yaylayam V., 2004. Volatile metabolite profiling for the discrimination of onion bulbs infected by *Erwinia carotovora* ssp. *carotovora, Fusarium oxysporum* and *Botrytis allii. European J. Pl. Physiol.*, **110:** 371–377.

Priyanka Gajjar, Brian Pettee, David W. Britt, Wenjie Huang, Willium P. Johnson and Anne J. Anderson, 2009. Antimicrobial activities of commercial nanoparticles against an environmental soil microbe, *Pseudomonas putida* KT 2440. *J. Biol. Engg.*, **3:** 9.

Puzyr, A.P., Burov, A.E., Bondar, V.S. and Trusov, Y.N., 2010. Neutralization of aflatoxin b1 by ozone treatment and adsorption by nanodiamonds. *Nanotechnol Russian.*, **5:** 137-141.

Pyoyun Park, Mitsuya Tsuda, Yoshiharu hayasi and Tamio Ueno, 2011. Effect of a host-specific toxin (AM-toxinI) produced by *Alternaria mali*, an apple pathogen, on the ultrastructure of plasma membrane of cells in apple and Japanese pear leaves. *Canadian J. Bot.*, **55:** 2383-2393.

Qin, J., Burks, T.F., Ritenour, M.A. and Bonn, W.G., 2009. Detection of citrus canker using hyperspectral reflectance imaging with spectral information divergence. *J. Food Engg.*, **93:** 183-191.

Qudsia Kanwal, Ishtiaq Hussain, Hamid Latif Siddiqui and Arshad Javaid, 2009. Flavonoids from mango leaves with antibacterial activity. *J. Serb. Chem. Soc.*, **74:** 1389-1399.

Quillen, C.D., 2002. Continuing patent applications and performance of the US patent office-extended. *Federal Circuit Bar. J.*, **12:** 35–55.

Quincke, M. C., Peterson, C. J. and Mundt, C. C., 2012. Relationship between incidence of *Cephalosporium stripe* and yield loss in winter wheat. *Inter. J. Agron.*, Vol. 2012, Article ID 635219, 9.

Rabie, C.J., Marasas, W.F., Thiel, P.G., Lubben, A. and Vleggaar, R., 1982. Moniliformin production and toxicity of different fusarium species from southern *Africa. Appl. Environ. Microbiol.*, **43:** 517-521.

Rabovich, C., Paull, R. and Sipes, B., 2009. Protease inhibitors and reproduction of reniform nematode in pineapple. *Annals of Applied Biol.*, **154:** 127-132.

Rad, F., Mohsenifar, A., Tabatabaei, M., Safarnejad, M.R., Shahryari, F., Safarpour, H., Foroutan, A., Mardi, M., Davoudi, D. and Fotokian, M., 2012. Detection of *Candidatus Phytoplasma Aurantifolia* with a quantum dots fret-based biosensor. *J. Pl. Path.*, 94: 126-132

Radeshkumar, C. and Münstedt, H. 2006. Antimicrobial polymers from polypropylene/silver composites- Ag^+ release measured by anode stripping voltammetry. *React. Funct. Polym.*, **66:** 780-788.

Radoi, A., Targa, M., Prieto-Simon, B. and Marty, J.L., 2008. Enzyme-linked nanoparticles for aflatoxin M1 detection. *Talanta*, **77:** 138–143.

Rai Vijay Laxmi, Geetanjaly and Sharma Preeti, 2014. Plant quarantine: An effective strategy of pest management in India. *Res. J. Agri. For. Sci.*, **2:** 11-16.

Rai, M., Ingle and A., 2012. Role of nanotechnology in agriculture with special reference to management of insect pests. *Appl. Microbiol. Biotechnol.*, **94:** 287–293.

Rajam, B. and Rajam, M.V., 1996. Inhibition of polyamine biosynthesis and growth in plant pathogenic fungi *in vitro. Mycopathologia*, **133:** 95-103.

Rajesh Kumar, Rakesh Kaundal, 2015. Institute for Integrative Genome Biology, University of California, Riverside.

Ralph A. Dean and William E. Timberlake, 1989. Production of cell wall-degrading enzymes by *Aspergillus nidulans*: A model system for fungal pathogenesis of plants. *The Pl. Cell*, **1:** 265-273.

Ram Prasad, Vivek Kumar and Kumar Suranjit Prasad, 2014. Nanotechnology in sustainable agriculture: Present concerns and future aspects. *African J. Biotech.,* **13:** 705-713.

Ramathani, I. and Beed, F., 2013. Use of DNA capture kits to collect *Xanthomonas campestris* pv. *musacearum* and banana bunchy top virus pathogen DNA for molecular diagnostics, p. 109. *In: Banana Systems in the Humid Highlands of Sub-Saharan Africa: Enhancing Resilience and Productivity.* G. Blomme, Bioversity, Uganda, P. van Asten, IITA, Uganda, B. Vanlauwe, (Eds.), p. 268.

Ramonell, K.M. and Somerville, S., 2002. The genomics parade of defense responses: To infinity and beyond. *Curr. Opin. Plant. Biol.,* **5:** 291-4.

Rampitsch, C. and Bykova, N.V., 2012. Proteomics and plant disease: Advances in combating a major threat to the global food supply. *Proteomics,* **12:** 673-90.

Rao, K.J. and Paria, S., 2013. Use of sulfur nanoparticles as a green pesticide on *Fusarium solani* and *Venturia inaequalis* phytopathogens. *RSC Adv.,* **3:** 10471-10478.

Rassaei, L., Sillanpaa, M., Bonn, M.J. and Marken, F., 2007. Carbon nanofiber – polystyrene composite electrode for electroanalytical processes. *Electroanalysis,* **19:** 1461-1466.

Ratna Kumari, A. and Sobha, K., 2013. Nanobiotechnology as a prospective approach For safe environmental remediation – a review. *Int. J. Pharm. Bio Sci.,* **4:** 69-95.

Ravichandra, N.G., 2008. Plant Nematology. I.K. International Publishing House Pvt. Ltd., New Delhi, India p. 695.

Ravichandra, N.G., 2013. Fundamentals of Plant Pahtology. PHI Publishing House, New Delhi, India, p. 645.

Ravichandra, N.G., 2014. Horticultural Nematology, Springers Publications, Netherlands, p. 565.

Ray Hammerschmidth, 1999. PHYTOALEXINS: What Have We Learned After 60 Years? *Ann. Rev. Phytopathol.,* **37:** 285-306.

Rayees A. Ahanger, Hilal A. Bhat, Tauseef A. Bhat, Ganie, S. A., Abid A. Lone, Imtiyaz A. Wani, Shaiq A. Ganai, Shamsul Haq, Owais A. Khan, John Mohd Junaid and Tariq A. Bhat, 2013. Impact of climate change on plant diseases. *Inter. J. Modern Plant & Animal Sci.,* **1:** 105-115.

Raymond Thomas, Xingxiao Fang, Kosala Ranathunge, Terry R. Anderson, Carol A. Peterson and Mark A. Bernards, 2012. Soybean root suberin: Anatomical distribution, chemical composition, and relationship to partial resistance to *Phytophthora sojae. Pl. Physiol.,* **144:** 299-311.

Reagan Kelly, 2008. What are "Omics" Technologies? reagank. com Powered by Movable Type 4.2, University of Michigan.

Reddy, A. P. K., MacKenzie, D. R., Rouse, D. I. and Rao, A. V., 1979. Relationship of bacterial leaf blight severity to grain yield of rice. *Phytopathology,* **69:** 967-969.

Reddy, S.V. and Waliyar, 2012. Proeprties of aflatoxin and its producing fungi. Aflatoxins, http:www.Icrisat.org/aflatoxin.asp.

Reina-Pinto, J.J. and Yephremov, A., 2009. Surface lipids and plant defenses. *Plant Physiol. Biochem.,* **47:** 540-549.

Reinink, K., 1986. Experimental verification and development of EPIPRE, a supervised disease and pest management system for wheat. *European J. Pl. Pathol.,* **92:** 3-14.

Reis Erlei M., Santos, Jones, A.P. and Blum Marta Maria, C., 2007. Critical-point yield model to estimate yield damage caused by *Cercospora zea-maydis* in corn. *Fitopatol. Bras.,* **32:** 110-113.

Reitz, M., Rudolph, K., Schroder, I., Hoffmann-Hergarten, Hallmann, J. and Sikora, R.A., 2000. Lipopolysaccharides of *Rhizobium etli* strain G12 act in potato roots as an inducing agent of systemic resistance to infection by the cyst nematode, *Globodera pallida. Appl. Environ. Microbiol.,* **66:** 3515-3518.

René Custers, 2004. *Biosafetly in the Laboratory.* 3rd revised edition, VIB Publication, Flanders Interuniversity Institute for Biotechnology, p. 71.

Rene Hofer, Isabel Briesen, Martina Beck, Franck Pinot, Lukae Schreiber and Rochus Franke, 2008. The *Arabidopsis* cytochrome P450 *CYP86A1*encodes a fatty acid ω-hydroxylase involved in suberin monomer biosynthesis. *J. Expt. Bot.,* **59:** 2347-2360.

Renske Landeweert & Hans Helder, 2012. Diagnostic kits for identification & detection of nematodes, Clear Detections revolutionary diagnostics— Joint Dutch initiative of Wageningen University and Dutch Sprouts.

Reverchon, S., Nasser, W. and Robert-Baudouy, J., 1991. Characterization of kdgR, a gene of *Erwinia chrysanthemi* that regulates pectin degradation. *Mol. Microbiol.,* **5:** 2203-2216.

Rex B. Dyer, 2015. Plant Pathogen Diagnostics, USDA- ARS - NCAUR Technologies for Transfer National Center for Agricultural Utilization Research.

Richard Baker and Alan Mac Leod, 2001. Pest risk assessments: tools, resources and key challenges, Pest Risk Analysis Sub-Team, Plant Health Group, Central Science Laboratory, York, YO41 1LZ, United Kingdom.

Richard J. Howard and Mehdi Kabbage, 2012. Lifestyle transitions in plant pathogenic *Colletotrichum* fungi deciphered by genome and transcriptome analyses. *Nature Genetics,* **44:** 1060-1065.

Ries, S. M. and Albersheim, P., 1973. Purification of a protease secreted by *Colletotrichum lindemuthianum. Phytopathol.,* **63:** 625-629.

Riley, M.B., Williamson, M.R. and Maloy, O., 2002. Plant disease diagnosis. The Plant Health Instructor. DOI: 10.1094/PHI-I-2002-1021-01.

Robab Davari, Reza Darvishzadeh, Ahmad Majd, Ardashir Kharabian Masoouleh and Yobert Ghosta, 2012. The Infection Processes of *Sclerotinia sclerotiorum* in Basal Stem Tissue of a Susceptible Genotype of *Helianthus annuus* L. *Not Bot Horti Agrobo,* **40:** 143-149.

Robert Faggian, Caroline Donald, Vijay Kaul and Ian J. Porter, 2007. Vegetables Victoria, Victorian State Government and Horticulture Australia Limited.

Robert P. Ryan, Frank Jorg Vorholter, Neha Potnis, Jeffrey B. Jones, Marie Anne Van Sluys, Adam J. Bogdanove and Maxwell Dow, J., 2011. Pathogenicity of *Xanthomonas*: understanding bacterium-plant interactions. *Nature Rev. Microbiol.,* **9:** 344-345.

Roberto Buonaurio, 2008. Infection and plant defense responses during plant-bacterial interaction. *Pl. Microbe Inter.,* **8:** 169-197.

Robertobert, R. M. Paterson, Sariahri Moen and Nelsonelso Lima, 2009. The feasibility of producing oil palm with altered lignin content to control *Ganoderma* disease. *J. Phytopathol.,* **157:** 649-656.

Robin Gogoi, Pradeep Kumar Singh, Robin Gogoi, Pradeep Kumar Singh, Rajesh Kumar, Kishore Kumar Nair, Imteyaz Alam, Chitra Srivastava, Saurabh Yadav, Madhuban Gopal, Samrat Roy Choudhury and Arunava Goswami, 2013. Suitability of nano-sulphur for biorational management of powdery mildew of okra (*Abelmoschus esculentus* Moench) caused by *Erysiphe cichoracearum. J. Pl. Pathol. Microb.,* **4:** 4.

Robin MacDiarmid, Brendan Rodoni, Ulrich Melcher, Francisco Ochoa-Corona and Marilyn Roossinck, 2013. Biosecurity implications of new technology and discovery in plant virus research glenn F. Rall, Editor, *PLoS Pathog.,* **9:** e1003337.

Rocío González-Lamothe, Gabriel Mitchel and Kamal Bouarab, 2009. Plant antimicrobial agents and their effects on plant and human pathogens. *Int. J. Mol. Sci.,* **8:** 3400-3419.

Roddick, J.G., 1974. The steroidal glycoalkaloid α- tomatine. *Phytochemistry,* **13:** 9-25.

Rodriguez-Landa, J.F. and Contreras, C.A., 2003. A review of clinical and experimental observations about antidepressant actions and side effects produced by *Hypericum perforatum* extracts. *Phytomedicine,* **10:** 688-699

Roger A.C. Jones, Moin U. Salam, Timothy J. Maling, Arthur J. Diggle and Deborah J. Thackray., 2010. Principles of predicting plant virus disease epidemics. *Ann. Rev. Phytopathol.,* **48:** 179-203.

Roggo, Y., Duponchel, L. and Huvenne, J.P., 2003. Comparison of supervised pattern recognition methods with McNemar's statistical test: application to qualitative analysis of sugar beet by near-infrared spectroscopy. *Analytica Chimica Acta,* **477:** 187-200.

Ronald P.de Vries and Jaap Visser, 2001. *Aspergillus* Enzymes involved in degradation of plant cell wall polysaccharides. *Microbiol. Mol. Biol. Rev.,* **65:** 4497-522.

Ronald V. Bettig., 1996. Critical Perspectives on the History and Philosophy of Copyright in Copyrighting Culture: The Political Economy of Intellectual Property, by Ronald V. Bettig. (Boulder, CO: Westview Press, 1996), 19–20.

Rong, T. and Ting, Z., 2000. Antifungal activity of monoterpenoids against postharvest pathogens *Botrytis cinerea* and *Monilinia fructicola. J. Esset.Oil.Res.,* **12:** 113-121.

Rosa M. Pérez-Clemente, Vicente Vives, Sara I. Zandalinas, María F. López-Climent, Valeria Muñoz, and Aurelio Gómez-Cadenas, 2013. Biotechnological Approaches to Study Plant Responses to Stress. BioMed Research International, Vol. 2013, Article ID 654120, p. 10.

Rosen, J.E., Yoffe, S., Meerasa, A., Verma, M. and Gu, F.X., 2011. Nanotechnology and diagnostic imaging: New advances in contrast agent technology. *J. Nanomed. Nanotechnol.*, **2:** 1-12.

Rossi, R. E., Borth, P. W. and Tollefson, J. J., 1993. Stochastic simulations for characterizing ecological spatial patterns and appraising risk. *Ecol. Appl.*, **3:** 719-735.

Rossi, V., Racca, P., Giosue, S., Pancaldi, D. and Alberti, I., 1997. A simulation model for the development of brown rust epidemics in winter wheat. *European J. Plant Pathol.*, **103:** 453-465.

Rotem, J., Eidt, J., Wendt, U. and Kranz, J., 1988. Relative effects of *Alternaria alternate* and *A. macrospora* on cotton. *Plant Pathol.*, **37:** 16-19.

Roy, D., 2010. GM as a new tool in resistance tool box. pp. 28-221. *In: Plant Pathology in Practice,* pp. 263-265. Training manual, G.B. Pant University of Agriculture & Technology, Pantnagar-263 145, India, p. 265.

Ruz, L., Moragrega, C. and Montesinos, E., 2008. Evaluation of four whole-plant inoculation methods to analyze the pathogenicity of *Erwinia amylovora* under quarantine conditions. *Int. Microbiol.*, **11:** 111-9.

Ryan, C. A., 1990. Protease inhibitors in plants: genes for improving defenses against insects and pathogens. *Ann. Rev. Phytopathol.*, **28:** 425-449.

Saboki Ebrahim, Usha, K. and Bhupinder Singh, 2012. Plant architectural traits and theirrole in defense mechanism against malformation in mango (*Mangifera Indica* L.). *Scientia Hort.*, **139:** 25-31.

Sadhna Singh, Manoj Singh, Ved Varun Agrawal and Anil Kumar, 2010. An attempt to develop surface plasmon resonance based immunosensor for Karnal bunt (*Tilletia indica*) diagnosis based on the experience of nano-gold based lateral flow immuno-dipstick test.

Sagar, A., Daemmrich, A. and Ashiya, M., 2000. The tragedy of the commoners: biotechnology and its publics. *Nature Biotech.*, **18:** 2-5.

Sahar M. Ouda, 2014. Antifungal activity of silver and copper nanoparticles on two plant pathogens, *Alternaria alternata* and *Botrytis cinerea*. *Res. J. Microbiol.*, **9:** 34-42.

Sahayaraj, K. and Rajesh, S., 2011. Bionanoparticles: synthesis and antimicrobial applications Science against microbial pathogens: communicating current research and technological advances, FORMATEX 2011, Crop Protection Research Centre, Department of Advanced Zoology and Biotechnology, St. Xavier's College (Autonomous), Palayamkottai-627 002, Tamil Nadu, India.

Sahu, D. K., Khare, C. P., Singh, H. K. and Thakur, M. P., 2013. Evaluation of newer fungicide for management of early blight of tomato in Chhattisgarh. *The Bioscan.*, **8:** 1255-1259.

Saladié, M., Matas, A.J., Isaacson, T., Jenks, M.A., Goodwin, S.M., Niklas, K.J., Xiaolin, R., Labavitch, J.M., Shackel, K.A. and Fernie, A.R., 2007. A reevaluation of the key factors that influence tomato fruit softening and integrity. *Plant Physiol.*, **144:** 1012-1028.

Salamanca-Buentello F, Persad DL, Court EB, Martin DK, Daar AS, Singer PA (2005) Nanotechnology and the developing world. *PLoS Med* 2(5): e97.

Salles, I.I., Blount, J.W., Dixon, R.A. and Schubert, K., 2002. Phytoalexin induction and β-1,3-glucanase activities in *Colletotrichum trifolii* infected leaves of alfalfa *(Medicago sativa L). Physiol. Mol. Plant Pathol.,* **61:** 89.

Sally A. Miller, Fen D. Beed and Carrie Lapaire Harmon, 2009. Plant disease diagnostic capabilities and networks, *Annu. Rev. Phytopathol.,* **47:** 15–38.

Salome Egger, Rainer, P., Lehmann, Murray, J., Height, Martin, J., Loessner and Markus Schuppler, 2009. Antimicrobial Properties of a Novel Silver-Silica Nanocomposite Material.

Samrat Roy Choudhury and Samrat Roy Choudhurya, Kishore K. Nairb, Rajesh Kumarb, Robin Gogoic, Chitra Srivastavad, Madhuban Gopalb, Subhramanyamd, B. S., Devakumarb, C. and Arunava Goswamia. 2010. Nanosulfur: A Potent Fungicide against Food Pathogen, *Aspergillus niger. American Inst. Physics, Proc., of International Conference on Adnaced Nanoparticles and Nanotechnology.*

Sánchez-Vallet, A., López, G., Ramos, B., Delgado-Cerezo, M., Riviere, M.P., Llorente, F., Fernández, P.V., Miedes, E., Estevez, J.M., Grant, M. and Molina, A., 2012 Disruption of abscisic acid signaling constitutively activates *Arabidopsis* resistance to the necrotrophic fungus *Plectosphaerella cucumerina. Plant Physiol.,* **160:** 2109-24.

Sang Woo Kim, Jin Hee Jung, Kabir Lamsal, Yun Seok Kim, Ji Seon Min and Youn Su Lee, 2012. Antifungal effects of silver nanoparticles (AgNPs) against various plant pathogenic fungi. *Mycobiology,* **40:** 53–58.

Sanjeev S Sannakki, Vijay S Rajpurohit, Nargund, V. B., Arun Kumar, R., Prema S. Yallur. 2011. Leaf disease grading by machine vision and fuzzy logic. *Int. J. Comp. Tech. Appl.,* **2:** 1709-1716.

Sankaran, S., Mishra, A., Ehsani, R. and Davis, C., 2010. A review of advanced techniques for detecting plant diseases. *Comput. Electron. Agric.,* **72:** 1-13.

Saponari, M., Boscia, D., Nigro, F. and Martelli, G.P., 2013. Identification of DNA sequences related to *Xylella fastidiosa* in oleander, almond and olive trees exhibiting leaf scorch symptoms in Apulia (Southern Italy). *J. Pl. Pathol.,* **95:** 668.

Sara Biglari, Gholam Hosein Sharifi and Gholam Reza Sharifi, 2014. Production of gold nanoparticles by *Streptomyces djakartensis* isolate *B-5, Article 3,* **1:** 229-237.

Sara Boettiger, Gregory D. Graff, Philip G. Pardey, Eric Van Dusen and Brian D. Wright, 2004. Intellectual Property Rights for Plant Biotechnology: International Aspects. *In: Handbook of Plant Biotechnology,* Paul Christou and Harry Klee (Eds.), John Wiley & Sons, Ltd., Chichester.

Sasaki, K., Iwai, T. and Hiraga, S., 2004. Ten rice peroxidases redundantly respond to multiple stresses including infection with rice blast fungus. *Pl. Cell Physiol.,* **45:** 1442-1452.

Sattler, Scott, E., Funnell-Harris and Deanna, L., 2013. Modifying lignin to improve bioenergy feedstocks: Strengthening the barrier against pathogens? *Frontiers in Pl. Sci.,* **4:** 1.

Saurabh Yadav, Savithramma, N., Rao, M. Linga, Basha, S. K. M., 2011. Antifungal efficacy of silver nanoparticles synthesized from the medicinal plants. *Der Pharma Chemica,* **3:** 364.

Savithri Nambeesan, Synan AbuQamar, Kristin Laluk, Autar K. Mattoo, Michael V. Mickelbart, Mario G. Ferruzzi, Tesfaye Mengiste and Avtar K. Handa, 2012. Polyamines attenuate ethylene mediated defense responses to abrogate resistance to *Botrytis cinerea* in tomato.

Sayan Bhattacharya, 2014. Bioprospecting, biopiracy and food security in India: The emerging sides of neoliberalism. *Inter. Letters of Social and Humanistic Sci.,* **12:** 49-56.

Sayed Roholla Mousavi and Maryam Rezae, 2011. Nanotechnology in agriculture and food production. *J. Appl. Environ. Biol. Sci.,* **1:** 414-419.

Schaad, N. W., Jones, J. B. and Chun, W. 2001. Laboratory guide for identification of plant pathogenic bacteria, 3rd ed. American Phytopathological Society, St. Paul, Minn.

Scheffer, R.P., 1976. Host-specific toxins in relation to pathogenesis and disease resistance. *In: Physiological Plant Pathology,* pp. 247-269. R. Heithefus and P.H. Williams (Eds.), Springer-Verlag, Berlin.

Scherm, H., Ngugi, H.K. and Ojiambo, P.S., 2006. Trends in theoretical plant epidemiology. *Eur. J. Plant Pathol.,* **115:** 61-73.

Schmidt, J., Overkleeft, H.S. and Van Der Hoorn, R.A.L., 2010. Proteasome activity profiling: A simple, robust and versatile method revealing subunit-selective inhibitors and cytoplasmic, defense-induced proteasome activities. *Plant J.,* **62:** 160-170.

Schmidt, K., Sikora, R. A. and Richter, O., 1993. Modeling the population dynamics of the sugar beet cyst nematode, *Heterodera schachtii. Crop Protn.,* **12:** 490-496.

Schmidt, S.M. and Panstruga, R., 2011. Pathogenomics of fungal plant parasites: What have we learnt about pathogenesis? *Curr. Opi. Plant Biol.,* **14:** 392-399.

Schmitz, A., Kiewnick, S., Schlang, J., Schmidt, K. and Sikora, R.A., 2010. Use of remote sensing to identify the spatial distribution of the sugar beet cyst nematode, *Heterodera schachtii.* Institute for Plant Diseases, Soil Ecosystem Phytopathology and Nematology, Nussallee 9, 53115 Bonn, Germany.

Schmülling, T., 2004. Cytokinin. *In: Encyclopedia of Biological Chemistry* (Eds. Lennarz, W., Lane, M.D.), Academic Press/Elsevier Science.

Schoeny, A., Jumel, S., Rouault, F., May Le, C. and Tivoli, B., 2007. Assessment of airborne primary inoculum availability and modelling of disease onset of ascochyta blight in field peas. *Eur. J. Plant Pathol.,* **119:** 87-97.

Schreiber, K.J., Ye, D., Fich, E., Jian, A., Lo, T. and Desveaux, D., 2012. A high-throughput forward genetic screen identifies genes required for virulence of *Pseudomonas syringae* pv. *maculicola* ES4326 on *Arabidopsis. PLoS ONE.,* **7(8):** e41461.

Schsnbeck, F. and Schbsser, E., 1976. Preformed substances as potential protectants. *In: Physiological Plant Pathology,* pp. 653-678. Heitefuss, R. and Williams, P.H. (Eds.), Springer-Verlag.

Schumann, G.L., 1998. Electronic teaching aids for students and practitioners. *Tech. Bull.,* 347–57.

Schumann, G.L., 2003, Innovations in teaching plant pathology, *Ann. Rev. Phytopathol.,* **41:** 377-98.

Schumann, G.L. and D'Arcy, C.J., 1999. Plant pathology courses for agricultural aware ness. *Plant Dis.,* **83:** 492-501.

Scott, N. and Chen, H., 2003. Nanoscale science and engineering for agriculture and food systems. A report submitted to cooperative state research, education and extension service, USDA, National Planning Workshop, Washington.

Scott, N.R., 2005. Nanotechnology and animal health. Revue scientifique et technique. *Inter. Off. Epizootics,* **24:** 425-432.

Scurtleff, M. C. and Averre, C. W. 1997. *Glossary of Plant-Pathological Terms,* p. 361. APS Press, St. Paul, Minnesota.

Selvarani, M. and Prema, P., 2013. Evaluation of antibacterial efficacy of chemically synthesized copper and zerovalent iron nanoparticles, *Asian J. Pharm. Clin. Res.,* **6:** 223-227.

Serge Savary and Paul S. Teng, 1998. Assessing the representativeness of data on yield losses, due to rice diseases in tropical Asia. *Plant Disease,* **82:** 593-712.

Shafri, H.Z.M. and Hamdan, N., 2009. Hyperspectral imagery for mapping disease infection in oil palm plantation using vegetation indices and red edge techniques. *American J. Appl., Sci.,* **6:** 1031.1035.

Shah, J., Tsui, F. and Klessig, D., 1997. Characterization of a salicylic acid-insensitive mutant (sai1) of *Arabidopsis thaliana,* identified in a selective screen utilizing the SA-inducible expression of the *TMS2* gene. *Mol. Pl. Microbe Interact.,* **10:** 69-78.

Shahid Siddique, Krzysztof Wieczorek and Holger Bohlmann, 2011. The promoter of a plant defensin gene directs specific expression in nematode-induced syncytia in Arabidopsis roots. *Pl. Physiol. Biochem.,* **49:** 1100-1107.

Shahrul, A.B. Ariffin, Tijjani Adam, Hashim, U., S. Faridah Sfaridah, Ishak Zamri and Muhammad Nur Aiman Uda, 2013, Plant diseases detection using nanowire as biosensor transducer, *Advanced Materials Res.,* **832:** 113-117.

Shalini Purwar, Shanthy Sundaram, Sukrat Sinha, Ankit Gupta, Neha Dobriyall and Anil Kumar, 2013. Expression and *in silico* characterization of phenylalanine ammonium lyase against karnal bunt (*Tilletia indica*) in wheat (*Triticum aestivum*). *Bioinformation.,* **9:** 1013-1018.

Shao, H., Min, C., Issadore, D., Liong, M., Yoon, T.J. and Weisseleder, R., 2012. Magnetic nanoparticles and micro NMR for diagnostic applications. *Theranostics,* **2:** 55-65.

Shaowu Meng, Trudy Torto-Alalibo, Marcus C. Chibucos, Brett M. Tyler and Ralph A. Dean, 2009. Common processes in pathogenesis by fungal and oomycete plant pathogens, described with Gene Ontology terms. *BMC Microbiol.,* **9** (Suppl 1): S7

Sharma, N., Sharma, K.P., Gaur, R.K. and Gupta, V.K., 2011. Role of chitinase in plant defense. *Asian J. Biochem.,* **6:** 29-37.

Sharma, O. P., India Niranjan Singh , Archana Bhardwaj , Vennila, S., Someshwar Bhagat, Neelam Mehta and Kamlesh Kumar, 2013. ICT Based Pest Management System for Sustainable Pulse Production: A Case Study. IGI Global

Sharma, P. N., Sharma, O. P., Padder, B. A. and Kapil, R., 2008. Yield loss assessment in common bean due to anthracnose (*Colletotrichum lindemuthianum*) under sub temperate conditions of north-western Himalayas. *Indian Phytopathol.,* **61:** 323-330.

Sharon, M., Choudhary, A.K. and Kumar, R.J., 2010. Nanotechnology in agricultural diseases and food safety. *J. Phytol.,* **2:** 83-92.

Sharon, M., Kr. Choudhary, A. and Rohit, K., 2010. Review article nanotechnology in agricultural diseases and food safety. *J. Phytology,* **2:** 83-92.

Shehab and Gaber Mohamed Gomaa, 2002. Molecular aspects of resistance to late blight disease in potato (*Solanum tuberosum* L.). Doctoral thesis, Durham University.

Sheila Salles Carvalho, José Djair Vendramim, Rafael Major Pitta and Moacir Rossi Forim, 2011. Efficiency of neem oil nanoformulations to *Bemisia tabaci* (GENN.) Biotype B (Hemiptera: Aleyrodidae**).** *Recebido para publicação,* **33:** 193-291.

Shevchik, V. E., Boccara, M., Vedel, R., and Hugouvieux-Cotte-Pattat, N., 1998. Processing of the pectate lyase pelI by extracellular proteases of *Erwinia chrysanthemi* 3937. *Mol. Microbiol.,* **29:** 1459-1469.

Shi, S., Wang, W., Liu, L., Wu, S., Wei, Y. and Li, W., 2013. Effect of chitosan/nano-silica coating on the physicochemical characteristics of longan fruit under ambient temperature. *J. Food Engg.,* **118:** 125-131.

Shigeru Mitani, Satoshi Araki, Yasuko Takii, Takeshi Ohshima, Norifusa Matsuo and Hideto Miyoshi, 2001. The biochemical mode of action of the novel selective fungicide cyazofamid: Specific inhibition of mitochondrial complex III in *Phythium spinosum. Pesticide Biochem.Physiol.,* **71:** 107–115.

Shim, W. B., and Dunkle, L. D. 2003. *CZK3*, a MAP kinase kinase kinase homolog in *Cercospora zeae-maydis*, regulates cercosporin biosynthesis, fungal development, and pathogenesis. *Mol. Plant-Microbe Interact.,* **16:** 760-768.

Shimon Meir, Sonia Philosoph-Hadas and Amnon Lers, 2011. Identification of defense-related genes newly associated with tomato flower abscission. *Plant Signal Behave.,* **6:** 590-593.

Shinozaki, K., Yamaguchi-Shinozaki, K. and Seki M., 2003. Regulatory network of gene expression in the drought and cold stress responses. *Curr. Opin. Pl. Biol.,* **6:** 410-417.

Shultz, T.R. and Line, R.F., 1992. High-temperature, adult-plant resistance to wheat stripe rust and effects on yield components. *Agronomy J.,* **84:** 170-175.

Shun'ichi Miyai, 2000. Simulation approach to the control of Insect-borne virus diseases. Laboratory of Insect Pest Management Tohoku National Agricultural Experiment Station, Akahira, Shimokuriyagawa, Morioka, Iwate 020-01, Japan.

Shunyi Zhu and Bin gao, 2014. Nematode-derived drosomycin-type antifungal peptides provide evidence for plant-to-ecdysozoan horizontal transfer of a disease resistance gene, *Nature Communications* 5, Article number: 3154.

Shweta Agrawal and Pragya Rathore, 2014. Nanotechnology pros and cons to agriculture: A review. *Int. J. Curr. Microbiol. App. Sci.,* **3:** 43-55.

Siepe, I., Speakman, J.B., Klappach, K., Scherer, M., Manker, D. and Schlehuber, S., 2011. Biological Activity and Mode of Action of Serenade®. pp. 339-346. *In: Modern Fungicides and Antifungal Compounds VI* , H.W. Dehne, H.B. Deising, U. Gisi, K. H. Kuck, P.E. Russell, H. Lyr (Eds.), Deutsche Phytomedizinische Gesellschaft, Braunschweig, Germany, p. 438.

Sikora, R.A. and Greco, N., 1990. Nematode parasites of food legumes. In: *Plant Parasitic Nematodes in Subtropical and Tropical Agriculture.* Luc, M.; Sikora, R.A.; Bridge, J (Eds.), pp. 181-235. CAB International, Wallingford, UK.

Simone Bregaglioa, Giovanni Cappelli and Marcello Donatelli, 2012. Evaluating the suitability of a generic fungal infection model for pest risk assessment studies. *Ecol. Modelling,* **247:** 58-63.

Sinclair, W.A. and Campana, R.J., 1978. Dutch elm disease: Perspectives after 60 years. *Pl. Pathol.,* **1:** 5-52.

Sindhuja Sanakaran, Ashish Mishra, Reza Ehsani and Cristina Davis, 2010. A review of advanced techniques for detecting plant diseases. *Comp. Electron. Agri.,* **72:** 1-13.

Singer, P. A., 2005. Nanotechnology and the developing world. *PLoS Medicine,* **2:** 383-386.

Singh Pushpinder, P., 2011. Epidemiology: A logical science to manage plant diseases. *Pl. Dis. Res.,Year,* **26:** 159.

Singh, D., Kumar, A., Singh, A.K. and Tripathi, H.S., 2013. Induction of resistance in field pea against rust disease through various chemicals/micronutrients and their impact on growth and yield. *Pl. Pathol. J.,* **12:** 36-49.

Singh, K.P., 2010. Knowledge Transfer through Farmer Participatory Training and Research. pp. 28-221. *In: Plant Pathology in Practice,* pp. 101-107. Training manual, G.B. Pant University of Agriculture & Technology, Pantnagar- 263 145, India, pp 265.

Singh, R.P., Hodson, D.P., Huerta-Espino, J., Jin, Y. and Bhavani, S., 2011. The emergence of Ug99 races of the stem rust fungus is a threat to world wheat production. *Ann. Rev. Phytopathol.,* **49:** 465-481.

Singh, R.S., 1984. Introduction to principles of plant pathology. Oxford & IBH Publishing Co., New Delhi, Mumbai, Kolkata, 54.

Sion Hannuna, N. Anantrasirichai, Swarna Subramanian, Suma Prashant, Ashok Jhunjhunwala and Nishan Canagarajah, C., 2011. Agriculture disease mitigation system. Ictact journal on communication technology: special issue on next generation wireless networks and applications, *Tech. Bull.,* **2**: Issue–2

Sirisomboon, P., Hashimoto, Y and Tanaka, M., 2009. Study on non-destructive evaluation methods for defect pods for green soybean processing by near-infrared spectroscopy. *J. Food Engg.,* **93:** 502-512.

Slavov, S., 2005. Phytotoxins and *in vitro* Screening for Improved Disease Resistant Plants. *Biotechnol. & Biotechnol.* 20th Anniversary Agro Bio Institute- R&D, AgroBioInstitute, Sofia, Bulgaria.

Slocombe, S.P., Schauvinhold, I., McQuinn, R.P., Besser, K., Welsby, N.A., Harper, A., Aziz, N., Li Y., Larson, T.R. and Giovannoni, J., 2008. Transcriptomic and reverse genetic analyses of branched-chain fatty acid and acyl sugar production in *Solanum pennellii* and *Nicotiana benthamiana. Plant Physiol.*, **148:** 1830-1846.

Smith, J.J., Waage, J., Woodhall, J.W., Bishop, S.J. and Spence, N.J., 2008. The challenge of providing plant pest diagnostic services for Africa. *European J. Plant Pathol.,* **121:** 365-375.

Smith, P., 2002. Patent rights and trade: Analysis of biological products, medicinals and botanicals, and pharmaceuticals. *Am. J. Agric. Econ.,* **84:** 495-512.

Smith, D.A., 1982. Toxicity of phytoalexins, pp. 218-52. *In: Phytoalexins, Bailey,* J.A., Mansfield, J.W., (Eds.), Blackie: Glasgow and London, UK,

Snyder, W. C., 1971. Plant pathology today. *Ann. Rev. Phytopathol.* **9:** 1-7.

Song, W.Y., 1995. "A receptor kinase-like protein encoded by the rice disease resistance gene, XA21". *Science,* **270:** 1804–1806.

Sontjens, S.H.M., Nettles, D.L., Carnahan, M.A., Setton, L.A. and Grinstaff, M.W., 2006. Biodendrimer-based hydrogel scaffolds for cartilage tissue repair. *Biomacromolecules,* **7:** 310-316.

Soo-Young Park, Rumi Kim, Choong-Min Ryu, Soo-Keun Choi, Choong-Hwan Lee, Jong-Guk Kim and Seung-Hwan Park, 2008. Citrinin, a mycotoxin from *Penicillium citrinum,* plays a role in inducing motility of *Paenibacillus polymyxa. FEMS Microbiol. Ecol.,* **65:** 229-237.

Sophien Kamoun, 2010. The Ohio State University, Ohio Agricultural Research & Development Center Talbot, NJ. 1999. Sci., **285:** 1860.

Soubhratra Das, Basant Kumar and Malhotra, P.K., 2002. Online Pest Management Information System, *J.. Ind. Soc. Ag. Statistics.,* **55:** 184-188.

Southey, J.F., 1986. Laboratory methods for work with plant and soil nematodes. Ministry of Agriculture, Fisheries and Food Reference Book No. 402, 202 pp. HMSO, London, UK.

Sparks, A.H., Esker, P.D., Bates, M,. Dall' Acqua, W., Guo, Z., Segovia, V., Silwal, S.D., Tolos, S. and Garrett, K.A., 2008. Ecology and Epidemiology. In *R: Disease Progress over Time. The Plant Health Instructor.* DOI:10.1094/PHI-A-2008-0129-02.

Spectrin-like Protein V. Toquin, F. Barja, C. Sirven, S. Gamet, M.-P. Latorse, J.-L. Zundel, F. Schmitt and R. Beffa. 2007. A New Mode of Action for Fluopicolide: Modification of the Cellular Localisation of a P.27-38. *In: Modern Fungicides and Antifungal Compounds. Proc., 15th International Reinhardsbrunn Symp.,* H.W. Dehne, H.B. Deising, U. Gisi, K. H. Kuck, P.E. Russell, H. Lyr (Eds.), May 06–10, 2007, Friedrichroda, Germany, p. 360.

Spiegel, Y. and Mcclure, M. A., 1995. The surface coat of plant-parasitic nematodes: chemical composition, origin, and biological rolc a review. *J. Nematol.*, **27:** 127-134.

Spikes, J.D., 1989. Photosensitization. *In: The science of photobiology,* K.C. Smith (Ed.) 2nd edition. Plenum, New York, pp. 79–110.

Spinelli, F., Noferini, M. and Costa, G., 2006. Near infrared spectroscopy (NIRs): Perspective of fire blight detection in asymptomatic plant material. *Acta Hort.,* **704:** 87-90.

Spinello, Richard A. (January 2007). Intellectual property rights. *Library Hi Tech.,* **25:** 12-22.

Srivastava, M. P., 2014. Organized plant health clinics in Asia, *Curr. Sci.,* **106:** 10.

Staskawicz, R.J., Mudgett, M.B., Dangl, J.L. and Galan, J.E., 2001. Common and contrasting themes of plant and animal diseases. *Sci.,* **292:** 2285-2289.

Steinberg and Richard, H., 2002. In the Shadow of Law or Power? Consensus-based Bargaining and Outcomes in the GATT/WTO. *International Organization.* Spring 2002. pp. 339–374.

Steven A. Lommel, Julie Willoughby and Richard H. Guenther, 2014. Plant virus and non-woven fiber nanotechnology delivery system for agricultural applications. *Licensing Associate,* **919:** 515-7199.

Steven H. Spoel and Xinnian Dong, 2012. How do plants achieve immunity? Defence without specialized immune cells. *Immunology,* **12:** 89-100.

Steven R. Herron, Jacques A.E. Benen, Robert D. Scavettat, Jaap Viser and Frances Jurnak, 2000. Structure and function of pectic enzymes: Virulence factors of plant pathogens. *PNAS,* **97:** 8762 8769.

Steven, T., Lunda, Robert E. Stallb and Harry J. Kleea, 1998. Ethylene regulates the susceptible response to pathogen infection in tomato. *The Plant Cell,* **10:** 371-382.

Stewart, T.M. and Galea, V.J., 2006. Approaches to training practitioners in the art and science of plant disease diagnosis. *Plant Disease,* **90:** 539-547.

Stewart, T.M., 2004. Teaching the art and science of plant disease diagnosis: Training students with DIAGNOSIS for crop problems. *The Plant Health Instructor.* DOI:10.1094/PHI-T-2004-0426-01.

Stierle, A, Cardellina, J.H. and Strobel, G.A., 1988. Maculosin, a host-specific phytotoxin for spotted knapweed from *Alternaria alternata. Proc. Natl. Acad. Sci.* USA, **85:** 8008-8013.

Stintzi, A., Heitz, T., Prasad, V., Wiedemann Merdinogu, S., Kauffmann, S. Geoffroy, P., Legrand, M. and Fritig, B., 1993. Plant 'pathogenesis-related' proteins and their role in defense against pathogens. *Biochem.,* **75:** 687-706.

Stokstad, E., 2004. Plant pathologists gear up for battle with dread fungus. *Science,* **306:** 1672-1673.

Stone, B.A., Evans, N.A., Bonig, I., and Clarke, A.E., 1985. The application of sirofluor, a chemically defined fluorochrome from aniline blue for the histochemical detection of callose. *Protoplasma*, **122:** 191–195.

Subramanian, D., 1969. Enzymes in pathogenesis. *Proc. Indian Acad. Sci.,* **69:** 133-141.

Sudisha, J., Amruthesh, K.N., Depak, S.A., Shety, N.P., Sarosh, B.R. and Shekar Shety, H. 2005. Comparative eficacy of strobilurin fungicides against downy mildew of pearl milet. *Pestic. Biochem. Physiol.*, **81:** 18-197.

Sunita I., Naik, P., Vivekanandreddy, Sannakki, S. S., 2014. Plant Disease Diagnosis System for Improved Crop Yield, *Int. J. Innov. Engg. Tech.*, **4:** Issue 1 June.

Sunkara V Manorama, 2004. Application of nanotechnology in agriculture, Inorganic and Physical Chemistry Division Indian Institute of Chemical Technology, Hyderabad-500 607.

Supradip Saha, Suresh Walia, Jitendra Kumar and Balraj S Parmar, 2010. Structure–biological activity relationships in triterpenic saponins: The relative activity of protobassic acid and its derivatives against plant pathogenic fungi. *Pest Mnt. Sci.*, **66:** 825-831.

Surendrababu, V., Sumathi , C.P and Umapathy, E., 2014. Detection of rice leaf diseases using chaos and fractal dimension in image processing, *Inter. J. Comp. Sci. Engg.,* **6:** 69-74

Susanta Banik and Pratibha Sharma, 2011. Plant pathology in the era of nanotechnology, *Indian Phytopath.*, **64:** 120-127

Susanta Kumar Biswal, Ashok Kumar Nayak, Umesh Kumar Parida and P. L. Nayak, 2012. Applications of nanotechnology in agriculture and food sciences**.** *Inter. J. Sci. Innovations and Discoveries,* Volume 2, Issue 1, January-February 2012.

Susi, A. Perini, and E. Olivetti, 2002. Plant disease models. *In: Critical Issues in Development and Use.* A. E. Rizzoli and A. J. Jakeman (Eds.), International Environmental Modelling and Software Society (iEMSs), 426–431, Lugano, Switzerland.

Susumu Hiraga, Katsutomo Sasaki, Hiroyuki Ito, Yuko Ohashi and Hirokazu Matsui, 2001. A large family of class III plant peroxidases. *Pl. Cell Physiol.,* **42:** 462-468.

Sutherst, R.W., 1993. Role of modelling in sustainable pest management. *In: Pest Control and Sustainable Agriculture*, pp. 66–71. Corey, S., Dall, D. and Milne, W., (Eds.), CSIRO, Australia.

Švàbova L. and Lebeda A., 2005. *In vitro* Selection for improved plant resistance to toxin-producing pathogens. *J. Phytopathology,* **153:** 52-64.

Swadeshmukul Santra, 2012. Silica-based antibacterial and antifungal nanoformulation, Patnets, Publi, No., S 8221791 B1.

Swaminathan, M.S., 2009. GM: Food for thought, *The Asian Age,* 26 August.

Swarup Roy Swarup Roy and Tapan Kumar Das, 2013. Activity of biosynthesized silver nanoparticles in combination with synthetic and natural fungicide against some pathogenic fungi. *Asian J. Chem.*, **25:** 315-317.

Tahara, S., 2007. A journey of twenty-five years through the ecological biochemistry of flavonoids. *Biosci. Biotechnol. Biochem.*, **71:** 1387-1404.

Tainter, F.H., Schumann, G.L., Evans, T.A., 1995. Development of the plant pathology laser video. disc. *J. Nat. Resour. Life Sci. Educ.*, **24:** 48-52.

Taku Takahashi and Jun-Ichi Kakehi, 2010. Polyamines: ubiquitous polycations with unique roles in growth and stress responses, *Ann. Bot.*, **105:** 1-6.

Tamaki Uehara-Ichiki, Takuya Shiba, Keiichiro Matsukura, Takanori Ueno, Masahiro Hirae and Takahide Sasaya, 2013. Detection and diagnosis of rice-infecting viruses. Mini Review Article, *Front. Microbiol.*, 09 October 2013.

Tamara Zimaro, Natalia Gottig, Betiana S. Garavaglia, Chris Gehring and Jorgelina Ottado, 2011. Unraveling plant responses to bacterial pathogens through proteomics. *J. Biomedicine and Biotechnol.*, Vol. 2011, Article ID 354801, 12 pages

Tamari, K. and Kaji. J., 1959. The accumulation of coumarin in stunted rice plant caused by the ill-effects of piricularin. *J. Agri. Chem. Soc., Japan*, **33:** 181-183.

Tan, K.C., Ipcho, S.V., Trengove, R.D., Oliver, R.P. and Solomon, P.S., 2009. Assessing the impact of transcriptomics, proteomics and metabolomics on fungal phytopathology. *Mol, Pl. Pathol.*, **10:** 703-715.

Tanaka, A., Shiotani, H., Yamamoto, M. and Tsuge, T., 1999. Insertional mutagenesis and cloning of the genes required for biosynthesis of the host-specific AK-toxin in the Japanese pear pathotype of *Alternaria alternata. Mol. Pl.Micr. Interact.*, **12:** 691–702.

Tarek Hewezi, Peter J. Howe, Tom R. Maier, Richard S. Hussey, Melissa G. Mitchum, Eric L. Davis and Thomas J. Baum, 2010. Arabidopsis spermidine synthase is targeted by an effector protein of the cyst nematode, *Heterodera schachtii. Pl. Physiol.*, **152:** 968-984

Tartaj, P., Del Puerto Morales, M., Veintemillas-Verdaguer, S., Gonzalez-Carreno, T. and Serna, C.J., 2003. The preparation of magnetic nanoparticles for applications in biomedicine. *J. Phys. D: Appl. Phys*, **36:** 13R182-R197.

Tejal Deshpande, Sharmila Sengupta and Raghuvanshi, K.S., 2014. Grading & identification of disease in pomegranate leaf and fruit. *Int. J. Computer Sci., Inf. Techn.*, **5:** 4638-4645.

Tenllado, F. and Díaz-Ruíz, J. R., 2001. Double-stranded RNA-mediated interference with plant virus infection. *J. Virol.*, **75:** 12288-12297.

Thangadurai, K and Padmavathi, K., 2014. Computer Visionimage Enhancement for Plant Leaves Disease Detection. World Congress on Computing and Communication Technologies.

Theorell, H., 1951. The Enzymes. Chemistry and Mechanism of Action. P 397–427 Sumner, J.B. and Myrbäck, K. (Eds.), Academic Press, NY.

Theyse, M., 2009, Development of an effective phytosanitary regulatory information management system framework for WTO SPS compliance, MInstAgrar dissertation, University of Pretoria, Pretoria.

Thipyapong, P., Hunt, M.D. and Steffens, J.C., 2004. Antisense downregulation of polyphenol oxidase results in enhanced disease susceptibility. *Planta,* **220:** 105-117.

Thomas J. Wolpert, Larry D. Dunkle and, Lynda M. Ciuffetti, 2002. Host-elective toxins and avirulence determinants: What's in a name? *Ann. Rev. Phytopathol.,* **40:** 251–85.

Thomas, S. and McCubin, p. 2003. A comparison of the antimicrobial effects of four silvercontaining dressings on three organisms. *J. Wound Care.,* **12:** 101-107.

Thomma, B.P.H.J., 2003. *Alternaria* spp.: from general saprophyte to specific parasite. *Mol. Plant Path.,* **4:** 225-236.

Thompson, C.E., Fernandes C.L., De Souza O.N., Salzano F.M., Bonatto S.L. and Freitas, L.B., 2007. Molecular modeling of pathogenesis-related proteins of family 5. *Cell Biol. Biophys.,* **44:** 385-394.

Thrall, P.H., Burdon, J.J. and Bever, J.D., 2002. Local adaptation in the linum marginale-melampsora lini host-pathogen interaction. *Evolution,* **56:** 1340-51.

Thul, S.T., Sarangi, B.K. andd Pandey, R.A., 2013. Nanotechnology in agroecosystem: implications on plant productivity and its soil environment. *Expert Opin. Environ. Biol.,* **2:** 1.

Tiju Joseph and Mark Morrison, 2006. Nanotechnology in Agriculture and Food. A Nanoforum report, Institute of Nanotechnology, 2-13.

Tim Cushnie, T.P. and Andrew J. Lamb, 2005. Antimicrobial activity of flavonoids. *Intl. J. Antimicrob. Agents,* **26:** 343-356.

Timmons, L., Court, D.L. and Fire, A., 2001. Ingestion of bacterially expressed dsRNAs can produce specific and potent genetic interference in *Caenorhabditis elegans. Gene,* **263:** 103-12.

Timothy, P., Devarenne and Gregory B Martin., 2007. Manipulation of plant programmed cell death pathways during plant-pathogen interactions, *Plant Signal Behav.,* **2:** 188-189.

Ting, J.P. and Davis, B.K., 2005. CATERPILLER: A novel gene family important in immunity, cell death, and diseases. *Ann. Rev Immunol.,* **23:** 387-414.

Tomoya Niki, Ichiro Mitsuhara, Shigemi Seo, Norihiro Ohtsubo and Yuko Ohashi, 1998. Antagonistic effect of salicylic acid and jasmonic acid on the expression of pathogenesis-related (PR) protein genes in wounded mature tobacco leaves. *Pl. Cell Physiol.,* **39:** 500-507.

Tonukari, N.J., Scott-Craig, J.S. and Walton, J.D., 2000. The *Cochliobolus carbonum* SNF1 gene is required for cell wall-degrading enzyme expression and virulence on maize. *Plant Cell,* **12:** 237-248.

Tourte, C. and Manceau, C., 1995. A strain of *Pseudomonas syringae* which does not belong to pathovar *phaseolicola* produces phaseolotoxin. *European J. Pl. Pathol.,* **101:** 483-490.

Trapman, M., 1994. Development and evaluation of a simulation model for ascospore infections of *Venturia inaequalis. Norwegian J. Agril.Sci.,* **17:** 55-67

Treatise. Vol. 2, *How Disease Develops in Populations.* J. G. Horsfall and E. A. Cowling, (Eds.), Academic Press, New York.

Trewavas A., 1999. Gene flow and GM questions. *Trends in Pl. Sci.,* **4:** 339.

Tripathi, H.S., 2010. Bio-Fungicides: Their Role in Plant Disease Management, pp. 108-111. *In: Plant Pathology in Practice,* pp 101-107. Training manual, G.B. Pant University of Agriculture & Technology, Pantnagar- 263 145, India, p. 265.

Trishita Issar, 2009. Omics Technology. I M.Sc., Foods, Nutrition and Dietetics, Roll No. 26, Andhra University, India.

Truscott, J.E. and Gilligan, C.A., 2003. Response of a deterministic epidemiological system to a stochastically varying environment. *Proc. Natl Acad. Sci.,* USA. **100:** 9067–9072.

Tsai, H., Wang, H., Gebler, J. C., Poulter, C. D., and Schardl, C. L. 1995. The *Claviceps purpurea* gene encoding dimethylallytryptophan synthase, the committed step for ergot alkaloid biosynthesis. *Biochem. Biophys. Res. Commun.,* **216:** 119-125.

Tsukada, K., Takahashi, K. and Nabeta, K., 2010. Biosynthesis of jasmonic acid in a plant pathogenic fungus, *Lasiodiplodia theobromae. Phytochem.,* **71:** 2019-2023.

Turkinson, T.K. and Morrall, R.A.A., 1993. Use of petal infestation to forecast sclerotinia stem rot of canola: The influence of inoculums variation over the flowering period and canopy density. *Phytopathology,* **83:** 682–689.

Turner, J.G. and Taha, R.R., 1984. Contribution of tabtoxin to the pathogenicity of *Pseudomonas syringae* pv. *tabaci. Physiol. Plant Pathol.,* **25:** 55-69.

Turner, J.G., 1984. Role of toxins in plant disease. *In: Plant Diseases: Infection, Damage and Loss.* P 3-12. Wood, R.K.S. and Jellis, G.J. (Eds.), Blackwell Scientific Publications, Oxford.

Turner, M.G., 2005. Landscape ecology: What is the state of the science? *Ann. Rev. Ecol. Evol. Systemat.,* **36:** 319–344.

Twengstrom, E., Sigvald, R., Svensson, C. and Yuen, J., 1998. Forecasting *Sclerotinia* stem rot in spring sown oilseed rape. *Crop Protection,* **17:** 405–411.

Tzeng, D.D. and DeVay, J.E., 1985. Physiological responses of *Gossypium hirsutum* L. to infection by defoliating and non-defoliating pathotypes of *Verticillium dahliae* Kleb. *Physiol. Pl. Pathol.,* **26:** 57–72.

Tzfira, T., and V. Citovsky. 2006. *Agrobacterium*-mediated genetic transformation of plants: biology and biotechnology. *Curr. Opin. Biotechnol.,* **17:** 147-154.

Uma Shankar Sagaram, Raghoottama Pandurangi, Jagdeep Kaur, Thomas, J. and Smith, Dilip M. Shah, 2011. Structure-Activity Determinants in Antifungal Plant Defensins MsDef1 and MtDef4 with Different Modes of Action against *Fusarium graminearum. PLOS ONE,* DOI: 10.1371/journal.pone.0018550.

Urdea, M., Penny, LA., Olmsted, S.S,. Giovanni, M.Y. and Kaspar, P., 2006. Requirements for high impact disease diagnostics in the developing world. *Nature,* **S1:** 73–79.

Usha Mina and Parimal Sinha, 2008. Effects of climate change on plant pathogens. *Enviro News, Newsl. ISEB India.* Volume 14, No. 4.

Ute Wittstock and Jonathan Gershenzon, 2002. Constitutive plant toxins and their role in defense against herbivores and pathogens. *Curr. Opinion Pl. Biol.,* **5:** 1-8.

Utpal Dey, Harlapur, S. I., Dhutraj, D. N., Suryawanshi, A. P., Badgujar, S. L., Jagtap, G. P. and Kuldhar, D. P., 2012. Spatiotemporal yield loss assessment in corn due to common rust caused by *Puccinia sorghi* Schw. *African J. Agril. Res.,* **7:** 5265-5269.

Uwe Conrath, 2006. Systemic acquired resistance. *Plant Signal Behav.,* **1:** 179-184.

Valenzeno, D. P., and Pooler, J. P. 1987. Photodynamic action. *BioSci.,* **37:** 270-275.

Valueva, T. A. and Mosolov, V. V., 2004. Role of inhibitors of proteolytic enzymes in plant defense against phytopathogenic microorganisms. *Biochem. (Moscow),* **69:** 1305-1309.

Van Bel, A.J.E. and Van der Schoot, C., 1988. Primary function of the protective layer in contact cells: Buffer against oscillations in hydrostatic pressure in the vessels? *IAWA Bull.,* **9:** 285-288.

Van de Molen, G.E., Beckman, C.H. and Rodehorsssst, E., 1977. Vascular gelation: a general response phenomenon following infection. *Physiol. Plant Pathol.,* **11:** 95-100.

Van Den Bulcke, M., Bauw, G., Castresana, C., Van Montagu, M. and Vandekerckhove, J., 1989. Characterization of vacuolar and extracellular β(1,3)-glucanases of tobacco: Evidence for a strictly compartmentalized plant defense system. *PNAS,* **86:** 2673-2677.

Van der Gaag, D.J., Bergsma-Vlami, M., Van Vaerenbergh, J., Vandroemme, J., Maes, M., 2013. Pest risk analysis for Xanthomonas fragariae. Netherlands Food and Consumer Product Safety Authority, Utrecht, the Netherlands – Institute for Agricultural and Fisheries Research, Merelbeke, Belgium.

Van der Plank, J.E, 1960. Analysis of epidemics. In: *Plant Pathology*. Horsfall J.G. and Cowling E.B. (Eds.). Academic Press, New York, USA.

Van der Plank, J.E., 1963. *Plant Diseases: Epidemics and Control*. Academic Press, New York, London.

Van der Plank, J.E., 1965. Dynamics of epidemics of plant disease: Population bursts of fungi, bacteria, or viruses in field and forest make an interesting dynamical study. *Science,* **8:** 120-124.

Van der plank, J.E., 1982. Host-pathogen interaction in plant disease. Academic Press, New York, USA.

Van Hecke, Dieter Vanderschaeghe, Herman Höfte, Nico Callewaert, Frank Van Breusegem and Monica Höfte, 2010. Abscisic acid deficiency causes changes in cuticle permeability and pectin composition that influence tomato resistance to *Botrytis cinerea. Plant Physiol.,* **154:** 847-860.

Van Ittersum, M. K. and Rabbinge, R., 1997. Ecology for analysis and quantification of agricultural input-output combinations. *Field Crops Res.,* **52:** 197-208.

Van Loon and Van Strien, E.A., 1999. The families of pathogenesis-related proteins, their activities and comparative analysis of PR-1 type proteins. *Physiol. Mol. Pl. Pathol.,* **55:** 85-97.

Van Loon, L.C., Pierpoint, W.S., Boller, Th. and Conejero, V., 1994. Recommendations for naming plant pathgenesis-related proteins. *Pl. Mol. Biol. Rep.,* **12:** 245-264.

Van Loon, L.C., Rep, M. and Pieterse, C.M., 2006. Significance of inducible defense-related proteins in infected plants. *Ann. Rev. Phytopathol.,* **44:** 135-162.

Van Loon, L.C., 1997. Induced resistance in plants and the role of pathogenesis-related proteins. *Eur. J. Plant Pathol.* **103:** 753–765.

Van Maanen, A. and Xu, X.M., 2003. Modelling plant disease epidemics. *European J. Plant Pathol.,* **109:** 669-682.

Van Maanen, A. and Xu, X.M., 2003. Modelling plant disease epidemics. *European J. Pl. Pathol.,* **109:** 669-682.

Vandana Ghormade, Mukund V. Deshpande, Kishore M. Paknikar, 2011. Perspectives for nano-biotechnology enabled protection and nutrition of plants. *Biotech. Adv.,* **29:** 792–803.

Verberne, M.C., Hoekstra, J., Bol, J.F. and Linthorst, H.J.M., 2003. Signaling of systemic acquired resistance in tobacco depends on ethylene perception. *Plant J.,* **35:** 27–32

Vernooji, B., Uknes, S., Ward, E. and Ryals, J., 1994. Salicylic acid as a signal molecule in plant-pathogen interactions. *Curr. Opin. Cell Biol.,* **6:** 275-279.

Viente Pallas and Juan Antonio Garcia, 2011. How do plant viruses induce disease? Interactions and interference with host components. *J. Gen Virol.,* **92:** 2691-2705.

Vincelli, P., 2012. Q_oI (Strobilurin) Fungicides: Benefits and Risks. *The Plant Health Instructor.* DOI: 10.1094/PHI-I-2002-0809-02.

Vincenza Ilardi and Stefania Loreti, 2014. Plant biosecurity: Diagnosis in plant pathology, Ilardi, *Biosafety*, **3:** 1.

Vincenzo Lattanzio, Veronica, M. T., Lattanzio and Angela Cardinali, 2006. Role of phenolics in the resistance mechanisms of plants against fungal pathogens and insects. *Phytochemistry: Advances in Research*, Filippo Imperato (Ed.), Università degli Studi di Foggia, 71100-Foggia, Italy. pp. 23-67.

Vishnudasan, D., Tripathi, M.N., Rao, U. and Khurana, P., 2005. Assessment of nematode resistance in wheat transgenic plants expressing potato proteinase inhibitor (PIN2) gene. *Transgenic Res.,* **14:** 665-75.

Volpi, C., Janni, M., Lionetti, V., Bellincampi, D., Favaron, F. and D'Ovidio, R., 2011. The ectopic expression of a pectin methyl esterase inhibitor increases pectin methyl esterification and limits fungal diseases in wheat. *Mol. Plant Microbe Interact.,* **24:** 1012-1019.

Wagner, G.J., Wang, E. and Shepherd, R.W. 2004. New approaches for studying and exploiting an old protuberance, the plant trichome. *Ann. Bot.,* **93:** 3–11.

Waigmann, E., Ueki, S., Trutnyeva, K. and Citovsky , V., 2004. The ins and outs of nondestructive cell-to-cell and systemic movement of plant viruses. *Crit. Rev. Plant Sci.,* **23:** 195-250.

Walker J. C., 1963. The future of plant pathology. *Ann. Rev. Phytopathol.,* **1:** 1-4.

Wallen, V.R. and Galway, D.A., 1977. Bacterial blight of field bean: disease progress, yield loss, and crop canopy development in principal cultivars in Ontario' . *Canadian Pl. Dis. Sur.*, **57:** 61.

Wallin, J.R. and Waggoner, P.E., 1950. The influence of climate on the development and spread of *Phytophthora infestans* in artificially inoculated potato plots. *Plant Dis. Reptr. Suppl.*, **190:** 19-33.

Walton, J.D., 2006. Molecules of interest: HC-toxin. *Phytochem.*, **67:** 1406-1413.

Wambugu, F., 1999. Why Africa needs agricultural biotech. *Nature,* **400:** 15-16.

Wang, H., Li, J., Bostock, R.M. and Gilchrist, D.G., 1996. Apoptosis: A functional paradigm for programmed plant cell death induced by a host selective phytotoxin and invoked during development. *Pl. Cell,* **8:** 375-91.

Wang, H., Wang, J., Choi, D., Tang, Z., Wu, H. and Lin, Y., 2009. EQCM immunoassay for phosphorylated acetylcholinesterase as a biomarker for organophosphate exposures based on selective zirconia adsorption and enzyme-catalytic precipitation. *Biosense Bioelectron,* **24:** 2377-83.

Wang, H.L., Sudarshana, M.R., Gilbertson, R.L. and Lucas, W.J., 1999. Analysis of cell-to-cell and long-distance movement of a bean dwarf mosaic geminivirus-green fluorescent protein reporter in host and nonhost species: Identification of sites of resistance. *Mol. Plant Microbe Interact.,* **12:** 345-355.

Wang, I.K., Reeves, C. and Gaucher, C. M., 1991. Isolation and sequencing of a genomic DNA clone containing the 3 terminus of the 6-methylsalicylic acid polyketide synthetase gene of *Penicillium urticae. Can. J. Microbiol.,* **37:** 86-95.

Wang, M.B., Masuta, C., Smith, N.A. and Shimura, H., 2012. RNA silencing and plant viral diseases. *Mol. Plant Microbe Interact.,* **25:** 1275-85.

Wang, Y.A., Li, J.J., Chen, H.Y. and Peng, X.G., 2009. Stabilization of inorganic nanocrystals by organic dendrons. *J. Am. Chem. Soc.,* **124:** 2293-2298.

Wang, Z., Guhling, O., Yao, R., Li ,F., Yeats, T.H., Rose, J.K, and Jetter, R., 2011. Two oxidosqualene cyclases responsible for biosynthesis of tomato fruit cuticular triterpenoids. *Plant Physiol.,* **155:** 540–552.

Wani, A. H., Amin, M., Shahnaz, M. and Shah, M. A., 2012. Antimycotic activity of nanoparticles of MgO, FeO and ZnO on some pathogenic fungi. *Inter.J. Manufacturing, Materials & Mech. Engg.,* **2:** 59-70.

Wani, A. H. and Shah, M. A., 2012. A unique and profound effect of MgO and ZnO nanoparticles on some plant pathogenic fungi. *J. Appl. Pharmaceutical Sci.,* **02:** 40-44.

Wan-zhong TAN, Cheng-wen LI, Chao-wei BI, Xian-chao SUN, 2010. A computer software – Epitimulator® for simulating temporal dynamics of plant disease epidemic progress *Agric. Sci. China.,* **9:** 242-248.

Ward, E.W.B., 1986. Biochemical mechanisms involved in resistance of plants of fungi. *In: Biology and molecular biology of plant pathogen interactions.* J.A. Baily (Ed.), Spinger-Verlag KG,Berlin, Germany 107-131.

Webster, R. and Boag, B., 1992. Geostatistical analysis of cyst nematodes in soil. *J. Soil Sci.,* **43:** 583-595.

Wei Lu, Meng-Ling Lu, Qing-Peng Zhang, Yong-Qing Tian, Zhi-Xiang Zhang and Han-Hong Xu, 2013. Octahydrogenated retinoic acid-conjugated glycol chitosan nanoparticles as a novel carrier of azadirachtin: Synthesis, characterization, and *in vitro* evaluation. Part A: Polymer Chemistry, *J. Polymer Sci.,* **51:** 3932-3940

Wei Ma., Weng, On., Yah, Hideyuki Otsuka and Atsushi Takahara, 2012. Surface functionalization of aluminosilicate nanotubes with organic molecules. *Beilstein J. Nanotechnol.,* **3:** 82-100.

Wei Wei, Robert E. Davis, Ing-Ming Lee and Yan Zhao, 2007. Computer-simulated RFLP analysis of 16S rRNA genes: identification of ten new phytoplasma groups. *Inter. J. Systematic and Evolutionary Microbiol.,* **57:** 1855–1867.

Weidenbörner, M., Hindorf. H., Jha, H.C. and Tsotsonos, P., 1990. Antifungal activity of isoflavones in different reduced stages on *Rhizoctonia solani* and *Sclerotium rolfsii. Phytochemistry,* **29:** 801–803.

Weiqing Zeng, Maeli Melotto and Sheng Yang He. 2011. Plant stomata: A checkpoint of host immunity and pathogen virulence. *Curr. Opin. Biotechnol.,* **21:** 599-603.

Welinder, K.G., 1991. Bacterial catalase-peroxidases are gene duplicated members of the plant peroxidase superfamily. *Biochim Biophys Acta,* **1080:** 215-20.

Weltzien, H. C., 1978. Geophytopathology, pp. 339-360. *In: Plant Disease: An Advance.*

Wennströmk, A. and Eriksson, B., 1997. Dispersal patterns of the rust *Gymnosporangium cornutum* between the hosts *Juniperus communis* and *Sorbus aucuparia. Ecoscience,* **4:** 117-119.

Werner, C., Petrini, O. and Hesse, M., 1997. Degradation of the polyamine alkaloid aphelandrine by endophytic fungi isolated from *Aphelandra tetragona. FEMS Microbiol Lett.,* **155:** 147-153.

Wharton, P.S. and Nicholson, R.L., 2000. Temporal synthesis and radio labeling of the sorghum 3-deoxyanthocyanidin phytoalexins and the anthoctanin cyaniding 3-diamalony glucoside. *New Phytol.,* **145:** 457-469.

Wildman, W. E., Bowers, K. W and Bettiga, L. J., 1992. Aerial photography in vineyard pest, soil and water management. *In: Grape Pest Management.* Oakland: University of California Division of Agriculture and Natural Resources Publication, D. L. Flaherty *et al.,* (Eds.), **3343:** 32-38.

Wilkins, M.R., Sanchez, J.C., Gooley. A.A., Appel, R.D., Humphery-Smith, I., Hochstrasser, D.F. and Williams, K.L., 1995. Progress with proteome projects: why all proteins expressed by a genome should be identified and how to do it. *Biotechnol. Genet. Eng. Rev.,* **13:** 19-50.

Willem, F. B., Franky, R. G. T., Bruno, P. A. C. and Rupert, W. O., 1995. Plant defensins: Nove 1 antimicrobial peptides as components of the host defense system'. *Plant Physiol.,* **108:** 1353-1 358.

William Allwood, J., David I. Ellis and Royston Goodacre, 2008. Metabolomic technologies and their application to the study of plants and plant–host interactions. *Physiologia Plantarum*, **132:** 117-135.

William Underwood, Maeli Melotto and Sheng Yang He., 2007. Role of plant stomata in bacterial invasion. *Cellular Microbiol.*, **9:** 1621-1629.

Williamson, V.M. and Hussey, R.S., 1996. Nematode pathogenesis and resistance in plants. *Plant cell*, **8:** 35-1745.

Willocquet, L., Savary, S., Fernandez, L., Elazegui, F. A. and Teng, P. S., 1998. Simulation of yield losses caused by rice diseases, insects, and weeds in tropical Asia. IRRI Discussion Paper Series no 34. IRRI, Los Baños, Philippines.

Willse, A., Straub, T. M., Wunschel, S. C., Small, J. A., Call, D. R., Daly, D. S. and Chandler, D. P., 2004. Quantitative oligonucleotide microarray fingerprinting of *Salmonella enterica* isolates. *Nucleic Acids Res.* **32:** 1848-1856.

Wilson, A.D., 2012. Development of an electronic-nose technology for the rapid detection of agricultural pesticide residues. *Phytopathology*, **102:** 136.

Wilson, A.D., 2013. Diverse applications of electronic-nose technologies in agriculture and forestry. *Sensors (Basel, Switzerland)*, **13:** 2295-2348.

Wilson, A.D., Lester, D.G. and Oberle, C.S., 2004. Development of conductive polymer analysis for the rapid detection and identification of phytopathogenic microbes. *Phytopathology*, **94:** 419-431.

Wilson, C.L. and Wisniewski, M.E., 1989. Biological control of postharvest diseases of fruits and vegetables: An emerging technology. *Ann. Rev. Phytopathol.*, **27:** 425-441.

Wise, R.P., Bronson, C.R., Schnable, P.S. and Horner, H.T., 1999. The genetics, pathology, and molecular biology of T-cytoplasm male sterility in maize. *Adv. Agron.* **65:** 79-130.

Wittstock, U. and Gershenzon, J., 2002. Constitutive plant toxins and their role in defence against herbivores and pathogens. *Ethology*, **5:** 264-269.

Wolpert, T.J., Navarre, D.A., Lorang, J.M. and Moore, D.L., 1995. Molecular-interactions of victorin and oats. *Can. J. Bot.*, **73:** 475-482.

Wong, P.T.W., Mead, J.A. and Croft, M.C., 2002. Effect of temperature, moisture, soil type and *Trichoderma* species on the survival of *Fusarium pseudograminearumin* wheat straw. *Australian Plant Pathol.*, **31:** 253–57.

Woolley, D.W., 1948. Derivatives of aspartic acid Lycomarasmin activity of some Synthesis and determination. *J. Biol. Chem.*, **176:** 1299-1308.

Wu, B. M., van Bruggen, A. H. C., Subbarao, K. V. and Pennings, G. G. H., 2001. Spatial Analysis of Lettuce Downy Mildew Using Geostatistics and Geographic Information Systems.

Wu, D., Feng, L., Zhang, C. and He, Y., 2008. Early detection of *Botrytis cinerea* on eggplant leaves based on visible and near-infrared spectroscopy. *Trans. ASABE*, **51:** 1133-1139.

Wuyts, N., Elsen, A., van Damme, E., de Waele, D., Swennen, R. and Sági, L., 2001. Lectin binding to the banana-parasitic nematode, *Radopholus similis*. Laboratory of Tropical Crop Improvement and Laboratory of Phytopathology and Plant Protection Catholic University of Leuven Kasteelpark Arenberg 13 B-3001 Leuven, Belgium.

Xia, X.J., Wang, Y.J., Zhou, Y.H., Tao, Y., Mao, W.H., Shi, K., Asami, T., Chen, Z. and, Yu, J.Q., 2009. Reactive oxygen species are involved in brassinosteroid-induced stress tolerance in cucumber. *Plant Physiol.*, **150:** 801-814.

Xiao, S., Ellwood, S., Calis, O., Patrick, E., Li, T., Coleman, M. and Turmer, J.G ., 2001. Broad spectrum powdery mildew resistance in *Arabidopsis thaliana* mediated by RPW8. *Science*, **291:** 118-120.

Xiaoning Li, Sandra, M., Robinson , Akash Gupta, Krishnendu Saha, Ziwen Jiang, Daniel, F., Moyano, Ali Sahar, Margaret A. Riley and Vincent M. Rotello, 2014. Functional gold nanoparticles as potent antimicrobial agents against multi-drug-resistant bacteria. *ACS Nano*, **8:** 10682-10686.

Xu, X., 2006. Modelling and interpreting disease progress in time. *In: The epidemiology of plant disease.* Cooke, B.M., Gareth Jones, D. and Kaye, B. (Eds.), Springer, Dordrecht, the Netherlands, 215-238.

Xuming Liu Christie E. Williams, Jill A. Nemacheck, Haiyan Wang, Subhashree Subramayam, Cheng Zheng and Ming-Shun Chen, 2010. Reactive oxygen species are invovled in plant defense against a gall midge. *Pl. Physiol.*, **152:** 985-999.

Yabuta, T., Kambe, K. and Hayashi, T. 1937. Biochemistry of the bakanaefungus. Fusarinic acid, a new product of the bakanae fungus. *J. Agric. Chem. Soc.*, **10:** 1059 1068

Yadav, A., Lundeberg M.A., DeSchryver, M., Dirkin, K H., Schiller, N., Maier, K.S., 2007. Teaching science with case studies: A national survey of faculty perceptions of the benefits and challenges of using cases. *J. College Sci. Teach.*, **37:** 34–38.

Yadav, M.S. and Brar, K.S., 2010. Assessment of yield losses due to mungbean yellow mosaic india virus and evaluation of mungbean genotypes for resistance in South-West Punjab. *Indian Phytopath.*, **63:** 318-320.

Yadeta, K.A. and Thomma, B.P.H.J., 2013. The xylem as battleground for plant hosts and vascular wilt pathogens. *Front. Plant Sci.*, **4:** 97.

Yalpani, N., Silverman, P., Wilson, T. M., Kleier, D. A. and Raskin, I., 1991. Salicylic acid is a systemic signal and an inducer of pathogenesis-related proteins in virus-infected tobacco. *The Pl. Cell.*, **3:** 809-818.

Yang, Q. and Gong, Z., 2002. Purification and Characterization of an Ethylene-Induced Antifungal Protein from Leaves of Guilder Rose (*Hydrangea macrophylla*). *Pro. Express. Purific.*, **24:** 76-82.

Yang, W., Shen, C., Ji, Q., An, H., Wang, J., Liu, Q. and Zhang, Z., 2009. Food storage material silver nanoparticles interfere with DNA replication fidelity and bind with DNA. *Nanotechnology*, **20(8):** 085102.

Yasuomi Tada, Koh Kusaka, Shigeyuki Betsuyaku, Takeshi Shinogi, Masaru Sakamoto, Yasako Ohura, Shingo Hata, Tomoyo Mori, Yukio Tosa and Shigeyuki Mayama, 2005. Victorin triggers programmed cell death and the defense response via interaction with a cell surface mediator. *Pl. Cell Physiol.*, **46:** 1787-1798.

Ya-Wen He and Lian-Hui Zhang, 2008. Quorum sensing and virulence regulation in *Xanthomonas campestris. FEMS Microbiol. Rev.*, **32:** 842-857.

Ye Xia, Qing-Ming Gao, Keshun Yu, Ludmila Lapchyk, DuRoy Navarre, David Hildebrand, Aardra Kachroo and Pradeep Kachroo., 2009. An intact cuticle in distal tissues is essential for the induction of systemic acquired resistance in plants. *Cell Host & Microbe,* **5:** 151-165.

Yedidia, I., Shoresh, M., Kerem, Z., Benhamou, N., Kapulnik, Y. and Chet, I., 2003. Concomitant induction of systemic resistance to *Pseudomonas syringae* pv. *lachrymans* in cucumber by *Trichoderma asperellum* (T-203) and accumulation of phytoalexins. *Appl. Environ. Microb.*, **69:** 7343-7353.

Yi-ch Zhang, 2010. Plant Diseases and Insect Pests Detection Based on MATLAB. 12th Annual Meeting of China Association for Science and Technology on Information and Communication Technology and Smart Grid (AMCST 2010 E-BOOK), 597-601.

Yoder, O.C., 1998. A mechanistic view of the fungal/plant interaction based on host-specific toxin studies. *In: Molecular Genetics of Host-Specific Toxins in Plant Disease,* Kohmoto, K. and Yoder, O.C., (Eds). Dordrecht, Netherlands: Kluwer Academic Publishers, 3–15.

Yoder, M. D., Keen, N. T. and Jurnak, F., 1993. New domain motif: The structure of pectate lyase C, a screted plant virulence factor. *Sci.,* **260:** 1503-1507.

Yoder, M. D., Lietzke, S. E. and Jurnak, F., 1993. Unusual structural features of the parallel β helix of the pectate lyases. *Structure,* **1:** 241–251

Yolanda Pareja-Jaime, M. Isabel G. Roncero, and M. CarmenRuiz-Roldán, 2008. Tomatinase from *Fusarium oxysporum* f. sp. *lycopersici* is required for full virulence on tomato plants. *Mol.Pl.Microbe Inter.*, **21:** 728-736.

Yoshida, K., Burbano, H.A., Krause, J., Thines M. and Weigel, D., 2014. Mining herbaria for plant pathogen genomes: Back to the future. *PLoS Pathog,* **10:** e1004028. doi:10.1371/journal.ppat.1004028.

Young, M., Willits, D., Uchida, M. and Douglas, T., 2008. Plant viruses as biotemplates for materials and their use in nanotechnology. *Ann. Rev. Phytopathol.,* **46:** 361-384.

Yousef Haj-Ahmad, 2008. Norgen Biotek Corporation Launches Four New Kits for the Purification of RNA and DNA from Plants and Fungi, 2008. Industry Newsletter, July 18, 2008.

Yuen, J., 2003. Bayesian approaches to plant disease forecasting. Online. Plant Health Progress, doi:10.1094/PHP-2003-1113-06-RV.

Yun Hwan Kim, Seong Joon Yoo, Yeong Hyeon Gu, Jin Hee Lim, Dongil Han and Sung Wook Baik, 2014. Crop pests prediction method using regression and machine learning technology: Survey. *IERI Procedia*, **6:** 52-56.

Yuri L. Dorokhov, Tatiana V. Komarova, Igor V. Petrunia, Olga Y. Frolova, Denis V. Pozdyshev and Yuri Y. Gleba. 2012. Airborne signals from a wounded leaf facilitate viral spreading and induce antibacterial resistance in neighboring plants. *PLoS Pathog.*, **8(4):** e1002640.

Zadoks, J. C., 1985. On the conceptual basis of crop loss assessment: The threshold theory. *Ann. Rev. Phytopathol.*, **23:** 455-473.

Zamir, K., Punja and Ye-Yan Zhang, 1993. Plant chitinases and their roles in resistance to fungal diseases. *J. Nematol.*, **25:** 526–540.

Zhang X. S., Holt, J. and Colvin, J., 2000. A general model of plant-virus disease infection incorporating vector aggregation. *Plant Pathology*, **49:** 435-444.

Zhang, H., Gao, Z., Zheng, X. and Zhang, Z., 2012. The role of G-proteins in plant immunity. *Plant Signal Behav.*, **7:** 1284-1288.

Zhang, J.H., Wang, C.Z., and Qin, J. D., 2000. The interactions between soybean trypsin inhibitor and delta-endotoxin of *Bacillus thuringiensis* in *Helicoverpa annigera* larva. *J. Invertebrate Pathol.*, **75:** 259-266.

Zhang, X.S. and Holt, J., 2001. Mathematical models of cross protection in the epidemiology of plant-virus diseases. *Phytopathology*, **91:** 924-934.

Zhang, X.S., Holt, J. and Colvin, J., 2000. A general model of plant-virus disease infection incorporating vector aggregation. *Plant Pathology*, **49:** 435-444.

Zhang, X.S., Holt, J. and Colvin, J., 2001. Synergism between plant virus: a mathematical analysis of the epidemiological implications. *Plant Pathol.*, **50:** 732-746.

Zhao, X., Hilliard, L.R., Mechery, S.J., Wang, Y., Bagwe, R.P. and Jin, S.A., 2004. Rapid bioassay for single bacterial cell quantitation using bioconjugated nanoparticles. *Proc. Natl. Acad. Sci.*, **101:** 15027-15032.

Zhonghua Wang, Ortwin Guhling, Ruonan Yao, Fengling Li, Trevor H. Yeats, Jocelyn K.C. Rose and Reinhard Jetter. 2011. Two oxidosqualene cyclases responsible for biosynthesis of tomato fruit cuticular triterpenoids. *Pl. Physiol.*, **155:** 540-552.

Zipfel, C., Kunze, G., Chinchilla, D., Caniard, A., Boller, T. and Felix, G., 2006. Perception of the bacterial PAMP EF-Tu by the receptor EFR restricts *Agrobacterium*-mediated transformation. *Cell*, **125:** 749–760.

Zipfel, C., 2006. Perception of the bacterial PAMP EF-Tu by the receptor EFR restricts *Agrobacterium*-mediated transformation. *Cell*, **125:** 749-760.

Zodrow, K., Brunet, L., Mahendra, S., Li, D., Zhang, A., Li, Q. and Alvarez, P. J. J., 2009. Polysulfone ultrafiltration membranes impregnated with silver nanoparticles show improved biofouling resistance and virus removal. *Water Res.*, **43:** 715-723.

Index

Q

R

S

T